Factor	Prefix	Symbol
10^{24}	Yotta	Y
10^{21}	Zetta	Z
10^{18}	Exa	E
10^{15}	Peta	P
10^{12}	Tera	T
10^{9}	Giga	G
10^{6}	Mega	M
10^{3}	kilo	k
10^{2}	hecto	h
10^{1}	deka	da
10^{-1}	deci	d
10^{-2}	centi	c
10^{-3}	milli	m
10^{-6}	micro	μ
10^{-9}	nano	n
10^{-12}	pico	p
10^{-15}	femto	f
10^{-18}	atto	a
10^{-21}	zepto	z
10^{-24}	yocto	y

Photonics

Springer
Berlin
Heidelberg
New York
Barcelona
Hong Kong
London
Milan
Paris
Singapore
Tokyo

Physics and Astronomy ONLINE LIBRARY

Advanced Texts in Physics

This program of advanced texts covers a broad spectrum of topics which are of current and emerging interest in physics. Each book provides a comprehensive and yet accessible introduction to a field at the forefront of modern research. As such, these texts are intended for senior undergraduate and graduate students at the MS and PhD level; however, research scientists seeking an introduction to particular areas of physics will also benefit from the titles in this collection.

Ralf Menzel

Photonics

Linear and Nonlinear Interactions
of Laser Light and Matter

With 395 Figures

 Springer

Professor Dr. Ralf Menzel
Universität Potsdam
Institut für Physik
Am Neuen Palais 10
14469 Potsdam, Germany
e-mail: photonics_menzel@springer.de

Library of Congress Cataloging-in-Publication Data. Menzel, Ralf, 1949–. Photonics: linear and non-linear interactions of laser light and matter/ Ralf Menzel. p. cm. – (Advanced texts in physics) Includes bibliographical references and index. ISBN 3540670742 (alk. paper) 1. Photonics. I. Title. II. Series. TA 1520.M46 2000 621.36–dc21 00-038820

ISSN 1439-2674

ISBN 3-540-67074-2 Springer-Verlag Berlin Heidelberg New York

Springer-Verlag Berlin Heidelberg New York
a member of BertelsmannSpringer Science+Business Media GmbH

© Springer-Verlag Berlin Heidelberg 2001
Printed in Germany

The use of general descriptive names, registered names, trademarks, etc. in this publication does not imply, even in the absence of a specific statement, that such names are exempt from the relevant protective laws and regulations and therefore free for general use.

Typesetting: Data conversion by Frank Herweg, Leutershausen
Cover design: *design & production* GmbH, Heidelberg

Printed on acid-free paper   SPIN 10633601   56/3141/ba   5 4 3 2 1 0

Preface

Since the invention of the laser in 1960 there has been an enormous increase in the number of applications of this newly available light and its spectacular properties, and there is no end to this development in sight. In many fields of science, technology and medicine laser photons are the driving force of progress. In the near future we will probably experience a further rapid development in this field as a result of the widespread industrial production of semiconductor diode lasers and new nonlinear optical materials. Light from the new lasers may become even cheaper than that from light bulbs. Thus, laser optic devices will influence all sectors of private and public life.

The high power, high brightness, narrow bandwidth, good coherence, special polarization and/or short pulses of laser light beams enable new applications. Many of these processes will be based on nonlinear optical interactions of the laser light with suitable optical material. In these interactions the material is modified by the incident light. The light is then in turn modified by the modified matter. Finally, the nonlinear modification of light as a function of other light becomes possible. Light is modified by light.

To use laser light in this sense in science, technology and medicine, knowledge from different fields of physics, chemistry and engineering is necessary. Besides conventional optics, which is essential in all laser light applications, a large field of new physical phenomena has to be considered. This book assembles the necessary knowledge ranging from the basic principles of quantum physics to the methods describing light and its linear and nonlinear interactions with matter, to practical hints on how the different types of lasers and spectroscopic and other measuring techniques can be applied. So that the book remains handy and readable, the description focuses on newer concepts in a compressed form. Nevertheless, many examples, tables and figures allow direct access for answering practical questions.

In this book, nonlinear physical processes in which laser photons are used as a tool will be summarized under the term *photonics*. This term was introduced by engineers at the Bell Laboratories to describe the optical analogy of electronic devices in electronic communication technologies; here, photons are the information-carrying particles. But the word is used today to cover nonlinear optics and quantum optics, too.

Thus, photonics will become more and more fundamental in the key technologies of the future. Communication and data processing, transportation and traffic, medicine and biotechnologies, new materials and material processing, environmental pollution detection and conservation and power production will be promoted by photonics. As a consequence of this rapid development, scientists and engineers in many fields of research and technology need some basic knowledge in photonics.

Therefore, fundamental laws from the different fields of the large area of photonics are described in this book in a more or less phenomenological way. As far as possible the basic equations are given and the principles of their derivation are mentioned. Exemplary material constants and calculated results are collected in tables to aid direct use of the information. Examples illustrate the physical relations. Thus, this book may be used as a guide to the basics of photonics on the one hand, and as a laboratory manual for designing new experiments and estimating wanted or unwanted laser light effects on the other hand.

The different topics of photonics are described at graduate level. Thus, the book should be useful for students and graduates of physics, electrical engineering, chemistry and biology for learning purposes and as a reference. The articles and textbooks cited should enable extended studies of related topics to be undertaken. Interested non-specialists from other fields may learn at least the basic of photonics by skipping details of the description.

Therefore, the subject is described in combination with practical questions such as: How can I measure this? How do I have to set up this apparatus? What are the physical limits of this application? The representation is based on more than 20 years' experience in laser research and nonlinear spectroscopy as taught in many lectures for physicists and chemists. Of course the description is not complete, and rapid further progress is expected in this area. Nevertheless, it will serve as an introduction to this field.

Photonics uses knowledge from conventional optics, electromagnetism and quantum mechanics. Essential information from these fields is described with respect to their importance. In the first chapter different topics of photonics are described in an overview. The subsequent analysis of the properties and the description of light in the second chapter are essential for the understanding of nonlinear phenomena. Although photonics deals mostly with nonlinear optics, in Chap. 3 some linear interactions of light with matter are treated first. Then the description of nonlinear interactions of light with matter follows in Chap. 4 for transparent matter, and in Chap. 5 for absorbing matter. These two chapters provide basic knowledge for all kinds of photonic applications. Because the laser as a light source is the fundamental tool for almost all photonics, a brief description of the main principles and their consequences is given in Chap. 6. This includes a short description of the main parameters of common laser systems and the principles of generating light with special properties such as short pulses or high brightness. As applications of those subjects, on the one hand, and as a precondition for examining applications, on the other hand, some fundamentals of nonlinear spectroscopy are described in Chap. 7.

A large number of references allows direct access to the detailed scientific research results in the field. The selected articles are cited with all authors, the full title and the number of pages, and are arranged in descending year order per topic. Considering this information and the title of the journal may help to select the most useful articles from the list for the reader's purpose. In addition, the related section is cited as {Sect. ...} and thus the references of a section can be read almost separately. In these references also additional effects and their applications are described. The descriptions in this book allow a general understanding of these specialized articles. It may be worth searching for a special reference in the chapters describing the basics as well as in the applications part of the book because the references are cited usually only once. These references represent mostly current research topics. The pioneering work, if not explicitly given, can be traced back from these articles. Many of the measured material parameters have slightly different values. In the sense used in this book the most probable or averaged values are given without a detailed discussion. For details the references with their cited literature shall be used.

For further general reading some selected textbooks are given (cited as monographs [M...]). The titles and publications years may be used for guidance.

Questions, comments and corrections are welcome and can be sent to the author via the e-mail address: photonics_menzel@springer.de.

Acknowledgments

The list of people I would like to express my thanks to is much too long for the space available. Therefore, I would first like to thank all those people who have contributed to the text in a more or less indirect way and are not quoted here. There is no harm intended if someone is not explicitly mentioned or referred to. I am aware that the overwhelming part of this text has been developed by the common activities of the scientific community in this field and has been published in other textbooks or articles. In some cases it may even be difficult to identify the originator of the ideas and descriptions directly. Therefore, I would like to thank all those open-minded colleagues in science and industry for the possibility of being involved in discussions about these topics over the years.

Special thanks for helping me in the production of this book go to my coworkers. They helped me to collect the data and to work out the figures. In particular, humanitarian support from the coworkers I had the pleasure to work with over the years is acknowledged. Nevertheless, a lot of detailed practical support was necessary to get all the facts collected. Dr. Guido Klemz is especially acknowledged. He carried out many of the calculations for the tables and figures and cross-checked many of the formulas. In addition, I would like to thank, in alphabetical order, Dr. Axel Heuer, Dr. Dieter Lorenz, Dr. Horst Lueck, Dr. Martin Ostermeyer, Dr. Rolf Sander and Dr. Peter Witte who provided me with results from their Ph.D. work. Further gratitude goes to the students Ingo Brandenburg and Lars Ellenberg who produced most of the figures. All of them supplied me with additional information and gave critical comments on the text.

The calculations of the rate equations were made with numerical programs based on partially collaborative developments with Dr. J. Ehlert and Dr. S. Oberländer, supported by Dr. D. Leupold and Prof. J. Hertz since 1975; to them warm thanks are sent.

Further, I would like to thank colleagues from the Physics Department of the University of Potsdam, from the Optical Institute of the Technical University Berlin, from the Chemistry and Physics Departments of the Washington State University, especially Prof. M. Windsor, all colleagues from the Laser Medicine Technique Center Berlin, especially Prof. H. Weber, and our collaborators from the Max Born Institute Berlin. Furthermore, I would like to thank colleagues from TRW, LLNL, HRL and Prof. C. Braeuchle, Prof. H.-J. Eichler, Prof. G. Huber, Prof. A. Müller, Prof. H. Paul, Prof. M. Wilkens, Prof. Welling and again many others including our industrial collaborators, for interesting and constructive discussions.

The staffs of the Verein Deutscher Ingenieure (VDI), the Bundesministerium für Bildung und Forschung (BMBF) and the Deutsche Forschungsgemeinschaft (DFG) are acknowledged for non-bureaucratic financial support of our research activities. I would like to thank all the technical staff in the machinery shops, and the electricians and the secretaries and administration people who helped us.

I thank the editors, especially Dr. H.J. Kölsch and the production team, for supporting me so nicely.

Last but not least I thank all my friends and my family for being patient about my absence for such a long time while I was just writing a book.

Potsdam, December 2000   *Ralf Menzel*

Contents

1. **Topics in Photonics** 1
   1.1 What Does Photonics Mean? 1
   1.2 Photonics and Light Technology 2
   1.3 Scientific Topics 3
   1.4 Technical Topics 5
   1.5 Applications 6
   1.6 Costs of Photons 8

2. **Properties and Description of Light** 11
   2.1 Properties of Photons 11
       2.1.1 Energy, Frequency, Wavelength, Moments,
             Mass, Timing 11
       2.1.2 Uncertainty Principle for Photons 14
             Uncertainty of Position and Momentum 14
             Uncertainty of Energy and Time 15
       2.1.3 Properties of a Light Beam 16
   2.2 Plane Waves Monochromatic Light 18
       2.2.1 Space- and Time Dependent Wave Equation 18
       2.2.2 Complex Representation 21
       2.2.3 Intensity and Energy Density as a Function
             of the Electric Field 22
       2.2.4 Uncertainty of Field Strengths 22
   2.3 Geometrical Optics 23
       2.3.1 Preconditions: Fresnel Number 23
       2.3.2 Theoretical Description 23
       2.3.3 Ray Characteristics 25
       2.3.4 Ray Propagation with Ray Matrices 26
   2.4 Gaussian Beams 27
       2.4.1 Preconditions 27
       2.4.2 Definition and Theoretical Description 27
       2.4.3 Beam Characteristics and Parameter 28
             Rayleigh Length z_R 29
             Beam Radius $w(z)$ 29
             Wave Front Radius $R(z)$ 30

     Divergence Angle θ 31
     Complex Beam Parameter $q(z)$ 31
  2.4.4 Beam Propagation with Ray-Matrices 32
  2.4.5 Determination of w_0 and z_0 33
  2.4.6 How to Use the Formalism 33
 2.5 Ray Matrices 34
  2.5.1 Deriving Ray Matrices 34
  2.5.2 Ray Matrices of Some Optical Elements 35
  2.5.3 Light Passing Through Many Optical Elements 39
  2.5.4 Examples 40
     Focusing with a Lens in Ray Optics 40
     Focusing a Gaussian Beam with a Lens 41
     Imaging with Two Lenses 42
     Focal Length of Thin Spherical Lenses 42
 2.6 Describing Light Polarization 43
  2.6.1 Jones Vectors Characterizing Polarized Light 43
  2.6.2 Jones Matrices of Some Optical Components 46
  2.6.3 Stokes Vectors Characterizing Partially Polarized Light 48
  2.6.4 Mueller Matrices of Some Optical Components 50
  2.6.5 Using the Formalism 51
 2.7 Light Characteristics 51
  2.7.1 Power, Energy and Number of Photons 52
  2.7.2 Average and Peak Power of a Light Pulse 52
  2.7.3 Intensity and Beam Radius 54
  2.7.4 Divergence 57
  2.7.5 Beam Parameter Product Beam Quality 59
  2.7.6 Brightness 61
  2.7.7 Brilliance 61
  2.7.8 Radiation Pressure and Optical Levitation 61
 2.8 Statistical Properties of Photon Fields 62
  2.8.1 Uncertainty of Photon Number and Phase 63
  2.8.2 Description by Elementary Beams 63
  2.8.3 Fluctuations of the Electric Field 65
  2.8.4 Squeezed Light 65
  2.8.5 Zero Point Energy and Vacuum Polarization 66
 2.9 Interference and Coherence of Light 67
  2.9.1 General Aspects 67
  2.9.2 Coherence of Light 68
     Coherence Length 68
     Coherence Time 69
     Lateral Coherence 71
  2.9.3 Two-Beam Interference 72
  2.9.4 Superposition of Two Vertical Polarized Light Beams .. 74
  2.9.5 One-Dimensional Multibeam Interference 76

2.9.6   Fabry–Perot Interferometer 77
2.9.7   Light Beats: Heterodyne Technique.................. 82
2.9.8   Frequency Spectrum of Light Pulses 83

3. Linear Interactions Between Light and Matter 85
3.1   General Description 85
3.2   Refraction and Dispersion 89
3.3   Absorption and Emission................................. 92
   3.3.1   Theoretical Description of Absorption and Emission ... 92
   3.3.2   Properties of Stimulated Emission.................. 98
   3.3.3   Spontaneous Emission 98
   3.3.4   Radiationless Transitions 100
3.4   Measurement of Absorption............................. 101
   3.4.1   Lambert–Beer Law 101
   3.4.2   Cross-Section and Extinction Coefficient 102
   3.4.3   Absorption Spectra of Some Optical Materials
          and Filters 103
3.5   Polarization in Refraction and Reflection (Fresnel's Formula) . 104
   3.5.1   Fresnel's Formula 105
          General Formula.................................. 105
          Transition into Optically Denser Medium 107
          Transition into Optical Thinner Medium 109
   3.5.2   Brewster's Law 111
   3.5.3   Total Reflection................................... 111
3.6   Relation Between Reflection, Absorption and Refraction...... 113
3.7   Birefringence 114
3.8   Optical Activity (Polarization Rotation).................... 118
3.9   Diffraction.. 119
   3.9.1   General Description: Fresnel's Diffraction Integral 120
   3.9.2   Far-Field Pattern: Fraunhofer Diffraction Integral 121
   3.9.3   Diffraction in First-Order Systems: Collins Integral 121
   3.9.4   Diffraction at a One-Dimensional Slit 122
   3.9.5   Diffraction at a Two-Dimensional Slit 124
   3.9.6   Diffraction at a Circular Aperture.................. 126
   3.9.7   Diffraction at Small Objects (Babinet's Theorem) 128
   3.9.8   Spot Size of Foci and Resolution of Optical Images 128
   3.9.9   Modulation Transfer Function (MTF) 129
   3.9.10 Diffraction at a Double-Slit 130
   3.9.11 Diffraction at One-Dimensional Slit Gratings 131
   3.9.12 Diffraction at a Chain of Small Objects 132
   3.9.13 Diffraction at Two-Dimensional Gratings............. 133
   3.9.14 Diffraction at Three-Dimensional Gratings 135
   3.9.15 Bragg Reflection 135
   3.9.16 Amplitude and Phase Gratings 136
   3.9.17 Diffraction at Optically Thin and Thick Gratings 138

3.10 Light Scattering Processes 142
    3.10.1 Rayleigh and Rayleigh Wing Scattering 142
    3.10.2 Mie Scattering 144
    3.10.3 Brillouin Scattering 145
    3.10.4 Raman Scattering 147
3.11 Optical Materials 149

**4. Nonlinear Interactions of Light and Matter
Without Absorption** 151
4.1 General Classification 152
4.2 Nonresonant Interactions 153
4.3 Nonlinear Polarization of the Medium 156
4.4 Second-Order Effects 158
    4.4.1   Generation of the Second Harmonic 158
    4.4.2   Phase Matching 160
        Phase Matching for Second Harmonic Generation 161
        Dispersion of Crystals: Sellmeier Coefficients 164
        Walk-Off Angle 165
        Focusing and Crystal Length 166
        Type I and Type II Phase Matching 166
        Quasi-Phase Matching (qpm) 167
    4.4.3   Frequency Mixing of Two Monochromatic Fields 168
    4.4.4   Parametric Amplifiers and Oscillators 169
    4.4.5   Pockels' Effect 172
    4.4.6   Electro-Optical Beam Deflection 176
    4.4.7   Optical Rectification 177
4.5 Third-Order Effects 179
    4.5.1   Generation of the Third Harmonic 180
    4.5.2   Kerr Effect 181
    4.5.3   Self-Focusing 184
    4.5.4   Spatial Solitons 186
    4.5.5   Self-Diffraction 188
    4.5.6   Self-Focusing in Weakly Absorbing Samples 189
    4.5.7   Self-Phase Modulation 189
    4.5.8   Generation of Temporal Solitons: Soliton Pulses 192
    4.5.9   Stimulated Brillouin Scattering (SBS) 194
    4.5.10 Stimulated Thermal Brillouin Scattering (STBS) 204
    4.5.11 Stimulated Rayleigh (SRLS)
        and Thermal Rayleigh (STRS) Scattering 206
    4.5.12 Stimulated Rayleigh Wing (SRWS) Scattering 207
    4.5.13 Stimulated Raman Techniques 210
        Stimulated Raman Scattering (SRS) 210
        Inverse Raman Spectroscopy (IRS) 215
        Stimulated Raman Gain Spectroscopy (SRGS) 216

       Coherent Anti-Stokes Raman Scattering (CARS) 216
       BOX CARS 218
    4.5.14 Optical Phase Conjugation via Stimulated Scattering .. 219
  4.6 Higher-Order Nonlinear Effects 227
  4.7 Materials for Nonresonant Nonlinear Interactions 228
    4.7.1 Inorganic Crystals 228
    4.7.2 Organic Materials 229
    4.7.3 Liquids 229
    4.7.4 Liquid Crystals 230
    4.7.5 Gases 230

5. Nonlinear Interactions of Light
and Matter with Absorption 231
  5.1 General Remarks 231
  5.2 Homogeneous and Inhomogeneous Broadening 232
  5.3 Incoherent Interaction 235
    5.3.1 Bleaching 236
    5.3.2 Transient Absorption:
           Excited State Absorption (ESA) 238
    5.3.3 Nonlinear Transmission 239
    5.3.4 Stimulated Emission: Superradiance: Laser Action 241
    5.3.5 Spectral Hole Burning 242
    5.3.6 Description with Rate Equations 243
           Basic Equations 243
           Stationary Solutions of Rate Equations 246
           Stationary Two-Level Model 247
           Stationary Four-Level Model 249
           Stationary Model with Two Absorptions 251
           General Stationary Models 253
           Numerical Solution 254
           Considering Spectral Hole Burning with Rate Equations .. 257
    5.3.7 Coherent Light Fields 259
    5.3.8 Induced Transmission
           and Excited State Absorption Gratings 261
    5.3.9 Induced Inversion Gratings 262
    5.3.10 Spatial Hole Burning 263
    5.3.11 Induced Grating Spectroscopy 264
  5.4 Coherent Resonant Interaction 265
    5.4.1 Dephasing Time T_2 265
    5.4.2 Density Matrix Formalism 266
    5.4.3 Modeling Two-Level Scheme 270
    5.4.4 Feynman Diagrams for Nonlinear Optics 275
    5.4.5 Damped Rabi Oscillation and Optical Nutation 278

    5.4.6   Quantum Beat Spectroscopy 280
    5.4.7   Photon Echoes 281
    5.4.8   Self-Induced Transparency: Π Pulses 284
    5.4.9   Superradiance (Superfluorescence) 285
    5.4.10 Amplification Without Inversion 286
  5.5 Two-Photon and Multiphoton Absorption 287
  5.6 Photoionization and Optical Breakdown (OBD) 291
  5.7 Optical Damage....................................... 293
  5.8 Laser Material Processing 295
  5.9 Combined Interactions
       with Diffraction and Absorption Changes................. 297
    5.9.1   Induced Amplitude and Phase Gratings.............. 297
    5.9.2   Four-Wave Mixing (FWM) 300
    5.9.3   Optical Bistability 306
  5.10 Materials in Resonant Nonlinear Optics 309
    5.10.1 Organic Molecules 309
            Structure and Optical Properties 310
            Preparation of the Samples 318
    5.10.2 Anorganic Crystals.............................. 319
    5.10.3 Photorefractive Materials 319
    5.10.4 Semiconductors 321
    5.10.5 Nanometer Structures 321

6. **Lasers**.. 325
  6.1 Principle 325
  6.2 Active Materials: Three- and Four-Level Schemes – Gain 327
  6.3 Pump Mechanism: Quantum Defect and Efficiency 329
    6.3.1   Pumping by Other Lasers 331
    6.3.2   Electrical Pumping in Diode Lasers 337
    6.3.3   Electrical Discharge Pumping...................... 339
    6.3.4   Lamp Pumping 341
    6.3.5   Chemical Pumping 342
    6.3.6   Efficiencies 343
  6.4 Side-Effects from the Pumped Active Material 345
    6.4.1   Thermal Lensing................................. 345
    6.4.2   Thermally Induced Birefringence.................... 348
    6.4.3   Thermal Stress Fracture Limit 351
  6.5 Laser Resonators 352
    6.5.1   Stable Resonators: Resonator Modes 352
    6.5.2   Unstable Resonators 353
  6.6 Transversal Modes of Laser Resonators.................... 354
    6.6.1   Fundamental Mode............................... 354
    6.6.2   Empty Resonator 355
    6.6.3 g Parameter and g Diagram 357
    6.6.4   Selected Stable Empty Resonators 358

    6.6.5   Higher Transversal Modes............................ 362
            Circular Eigenmodes or Gauss–Laguerre Modes 363
            Rectangular or Gauss–Hermite Modes 364
            Hybrid or Donut Modes.............................. 370
    6.6.6   Beam Radii of Higher Transversal Modes
            and Power Content.................................... 370
    6.6.7   Beam Divergence of Higher Transversal Modes......... 373
    6.6.8   Beam Quality of Higher Transversal Modes............ 374
    6.6.9   Propagating Higher Transversal Modes 374
    6.6.10 Fundamental Mode Operation: Mode Apertures......... 375
    6.6.11 Large Mode Volumes: Lenses in the Resonator 377
    6.6.12 Transversal Modes of Lasers
            with a Phase Conjugating Mirror 378
    6.6.13 Misalignment Sensitivity: Stability Ranges 381
    6.6.14 Dynamically Stable Resonators 384
    6.6.15 Measurement
            of the Thermally Induced Refractive Power........... 386
6.7 Longitudinal Modes 387
    6.7.1   Mode Spacing 387
    6.7.2   Bandwidth of Single Longitudinal Modes............. 390
    6.7.3   Spectral Broadening from the Active Material 391
    6.7.4   Methods for Decreasing the Spectral Bandwidth
            of the Laser 392
    6.7.5   Single Mode Laser 394
    6.7.6   Longitudinal Modes of Resonators with an SBS Mirror 396
6.8 Threshold, Gain and Power of Laser Beams.................. 397
    6.8.1   Gain from the Active Material: Parameters........... 398
    6.8.2   Laser Threshold 399
    6.8.3   Laser Intensity and Power.......................... 401
6.9 Spectral Linewidth and Position of Laser Emission 405
    6.9.1   Minimal Spectral Bandwidth 406
    6.9.2   Frequency Pulling.................................. 406
    6.9.3   Broad Band Laser Emission 407
            *Broad-Band Emission
            from Inhomogeneously Broadening*.................... 407
            Broad-Band Emission from Short Pulse Generation ... 408
            Broad-Band Emission from Gain Switching........... 409
6.10 Intensity Modulation and Short Pulse Generation 412
    6.10.1 Spiking Operation: Intensity Fluctuations 412
    6.10.2 Q Switching (Generation of ns Pulses) 415
            Active Q Switching and Cavity Dumping 415
            Passive Q Switching............................... 417
            Theoretical Description of Q Switching 418
    6.10.3 Mode Locking and Generation of ps and fs Pulses 420

Theoretical Description: Bandwidth-Limited Pulses 421
Passive Mode Locking with Nonlinear Absorber 423
Colliding Pulse Mode Locking (CPM Laser) 425
Kerr Lens Mode Locking 426
Additive Pulse Mode Locking 428
Soliton Laser 429
Active Mode Locking with AOM 430
Active Mode Locking by Gain Modulation 431
6.10.4 Other Methods of Short Pulse Generation 432
Distributed Feedback (DFB) Laser 432
Short Resonators 433
Traveling Wave Excitation 434
6.10.5 Chaotic Behavior 434
6.11 Laser Amplifier 436
6.11.1 Gain and Saturation 436
6.11.2 Energy or Power Content: Efficiencies 439
6.11.3 Amplifier Schemes 440
Single Pass Amplifier 440
Double Pass Amplifier 441
Multi Pass Amplifier 442
Regenerative Amplifier 443
Double Pass Amplifier with Phase Conjugating Mirror . 444
6.11.4 Quality Problems 444
Noise 445
Beam Quality 445
Pulse Duration 446
6.12 Laser Classification 446
6.12.1 Classification Parameters 446
6.12.2 Laser Wavelengths 447
6.12.3 Laser Data Checklist 449
Output Data 449
Installation and Connection to Other Devices 450
Operation and Maintenance 450
Prices and Safety 450
6.13 Common Laser Parameters 451
6.13.1 Semiconductor Lasers 451
Single-Diode Lasers 453
Diode Laser Bars, Arrays and Stacks 454
Vertical Cavity Surface-Emitting Lasers (VCSEL) 455
6.13.2 Solid-State Lasers 456
Nd:YAG Lasers 457
Nd:YVO Lasers 458
Nd Glass Laser 459
Yb:YAG Laser 460

Ti:Sapphire Laser 461
Cr:LiCAF and Cr:LiSAF Lasers 462
Alexandrite Laser 463
Erbium (Er), Holmium (Ho), Thulium (Tm) Laser 464
Ruby Laser 465
Fiber Lasers 466
6.13.3 Gas Lasers 467
XeCl, KrF and ArF Excimer Lasers 467
N_2 Laser 468
Home Made N_2 Laser 468
He-Ne Laser 470
He-Cd Laser 471
Ar and Kr Ion Lasers 472
Cu Vapor Lasers 473
CO_2 Lasers 474
6.13.4 Dye Lasers 475
cw and Quasi-cw (Mode-Locked) Dye Lasers 476
Pulsed Dye Lasers 477
6.13.5 Other Lasers 477
6.14 Modification of Pulse Structure 478
6.14.1 Single Pulse Selection 479
6.14.2 Pulse Compression and Optical Gates 479
Pulse Compression of fs Pulses 480
Pulse Compression of ns Pulses 480
Pulse Shortening by Nonlinear Effects 481
Pulse Shortening with Gates 481
Optical Gating with Up-Conversion 481
6.15 Frequency Transformation 482
6.15.1 Harmonic Generation (SHG, THG, FHG, XHG) 482
6.15.2 OPOs and OPAs 483
6.15.3 Raman Shifter 485
6.16 Laser Safety 485

7. Nonlinear Optical Spectroscopy 489
7.1 General Procedure 490
7.1.1 Steps of Analysis 490
7.1.2 Choice of Excitation Light Intensities 492
7.1.3 Choice of Probe Light Intensities 494
7.1.4 Pump and Probe Light Overlap 494
Spatial Overlap 494
Temporal Overlap 496
7.1.5 Light Beam Parameters 496
Polarization and Magic Angle 496
Pulse Width, Delay and Jitter 498
Spectral Width 499

		Focus Size and Rayleigh Length	499
		Coherence Lengths	499
	7.1.6	Sample Parameters...............................	499
		Preparation, Host, Solvent	500
		Concentration, Aggregation	500
		Temperature......................................	501
		Pressure	501
	7.1.7	Possible Measuring Errors........................	501
7.2	Conventional Absorption Measurements		502
	7.2.1	Determination of the Cross-Section.................	502
	7.2.2	Reference Beam Method	503
	7.2.3	Cross-Section of Anisotropic Particles	504
	7.2.4	Further Evaluation of Absorption Spectra	505
		Estimation of Excited State Absorptions (ESA)	505
		Band Shape Analysis	506
	7.2.5	Using Polarized Light	508
7.3	Conventional Emission Measurements		508
	7.3.1	Geometry	508
	7.3.2	Emission Spectra	509
		Fluorescence Spectrum	509
		Phosphorescence Spectrum: Triplet Quenching	510
	7.3.3	Excitation Spectrum: Kasha's Rule..................	511
	7.3.4	Emission Decay Times, Quantum Yield, Cross-Section .	511
		Fluorescence Decay Time	511
		Natural Lifetime	512
		Quantum Yield	512
		Phosphorescence Decay Time	513
		Determination of the Emission Cross Section	513
	7.3.5	Calibration of Spectral Sensitivity of Detection	513
7.4	Nonlinear Transmission Measurements (Bleaching Curves)		514
	7.4.1	Experimental Method	515
	7.4.2	Evaluation of the Nonlinear Absorption Measurement .	517
		Modeling...	517
		Bleaching or Darkening............................	517
		Start of Nonlinearity: Ground State Recovery Time....	518
		Slope, Plateaus, Minima and Maxima................	521
	7.4.3	Variation of Excitation Wavelength..................	522
	7.4.4	Variation of Excitation Pulse Width.................	523
	7.4.5	Variation of Spectral Width of Excitation Pulse.......	524
7.5	z-Scan Measurements		525
	7.5.1	Experimental Method	525
	7.5.2	Theoretical Description	528
	7.5.3	z-Scan with Absorbing Samples	529
7.6	Nonlinear Emission Measurements		530

7.6.1   Excitation Intensity Variation 530
7.6.2   Time-Resolved Measurements 531
7.6.3   Detection of Two-Photon Absorption via Fluorescence . 532
7.6.4   "Blue" Fluorescence 533
7.7 Pump and Probe Measurements 534
7.7.1   Experimental Method 534
7.7.2   Measurements of Transient Spectra 535
7.7.3   Coherence Effects in Pump and Probe Measurements .. 536
7.7.4   Choice of the Excitation Light 537
7.7.5   Probe Light Sources and Detection 538
        Probe Light Pulse Energy 538
        Synchronized Lasers and Frequency Transformations... 539
        White Light Generation with fs Duration 539
        White Light Generation with ps Duration 540
        Fluorescence as Probe Light in the ns Range 541
        Flash Lamps 542
        Spectral Calibration of Detection Systems 542
7.7.6   Steady-State Measurement 543
7.7.7   Polarization Conditions 544
7.7.8   Excited State Absorption (ESA) Measurements 545
        Method 545
        Estimate of the Population Densities 547
        Differentiation of Singlet and Triplet Spectra 549
7.7.9   Decay Time Measurements 550
7.8 Special Pump and Probe Techniques 550
7.8.1   Fractional Bleaching (FB) and Difference Spectra 551
7.8.2   Hole Burning (HB) Measurements 554
        Method 554
        Low Temperature Hole Burning Measurements 556
        Hole Burning Measurements at Room Temperature 557
7.8.3   Measurement with Induced Gratings:
        Four-Wave Mixing 557
7.8.4   Nonlinear Polarization (NLP) Spectroscopy 559
7.8.5   Measurements with Multiple Excitation 562
7.8.6   Detection of Two-Photon Absorption via ESA 563
7.9 Determination of Population Density and Material Parameters 564
7.9.1   Model Calculations 564
7.9.2   Determination of Time Constants for Modeling 566
        Fluorescence Lifetime 566
        Triplet Life Time 566
        Ground State Absorption Recovery Time 567
7.9.3   Fluorescence Intensity Scaling
        for Determining Population 567
7.10 Practical Hints for Determination of Experimental Parameters 569

7.10.1 Excitation Light Intensities.......................... 569
7.10.2 Delay Time 571
7.11 Examples for Spectroscopic Setups 571
7.11.1 ns Regime 572
7.11.2 ps and fs Regime.................................. 573
7.12 Special Sample Conditions............................... 574
7.12.1 Low Temperatures 574
7.12.2 High Pressures 576
7.13 Quantum Chemical Calculations 577
7.13.1 Orbitals and Energy States of Molecules 577
7.13.2 Scheme of Common Approximations................. 578
7.13.3 Ab Initio and Semi-Empirical Calculations 580

Bibliography... 583
Further Reading 583
References.. 585
1.   Topics in Photonics 585
2.   Properties and Description of Light 605
3.   Linear Interactions Between Light and Matter 612
4.   Nonlinear Interactions
      of Light and Matter Without Absorption............. 619
5.   Nonlinear Interactions
      of Light and Matter with Absorption 657
6.   Lasers 709
7.   Nonlinear Optical Spectroscopy 820

List of Tables 853

Subject Index 857

1. Topics in Photonics

In this introductory chapter the term photonics and the topics of this field are explained. In particular the difference between conventional light technologies and the nonlinear optical techniques and their relations are set out. Scientific and practical aspects of photonics are mentioned.

For further reading the monographs [M3, M34, M53] can be particularly recommended.

1.1 What Does Photonics Mean?

Photons as the quantum units of light similar to electrons that build the electrical current carry a certain amount of energy which can be used for a wide variety of applications. This energy can be. e.g. $2.5\,\text{eV} = 4 \cdot 10^{-19}\,\text{J}$ which corresponds to green light with a wavelength of 500 nm. Electrons of this energy are available from a power supply of 2.5 V.

Electrons have been used, e.g. for long-distance communication since the first telegraphs were invented by Morse and Wheatstone in 1837 and even more so after the invention of the electromagnetic telephone by A.G. Bell in 1877. With the invention of electronic devices such as tubes, e.g. the triode invented 1906 by L.D. Forest, radio communication became possible and the first transatlantic connections were available in 1927. Later semiconductor devices such as transistors, invented by W.H. Brattain, J. Bardeen and W.B. Shockley at the Bell Laboratories in 1948, and computer chips were used for this purpose. These electronic devices allow telecommunication with about $10^9\,\text{bit}\,\text{s}^{-1}$.

The electromagnetic field of photons oscillates much faster than is possible for electrons. Thus engineers of the Bell Laboratories invented methods using light for communication purposes once the laser became available after its first realization in 1960 by T.H. Maiman [1.1]. The commonly known fibers for transmitting light over long distances of many hundred kilometers demand several devices for generating, switching and amplifying the light. Thus the engineers at Bell Laboratories created the word photonics to describe the combination of light technologies and electronics in telecommunication.

But laser light photons are useful for many purposes other than communication. Thus the term *photonics* has been extended and now covers almost all processes using laser light in science, medicine and technology, but it does not include illumination and simple conventional optical techniques. Thus photonics includes mainly the nonlinear interactions of light with matter. In this case the characteristic effects are a nonlinear function of the intensity of the applied light. Thus the term has close relations to nonlinear optics and quantum optics. The nonlinear optical processes demand, with a few exceptions, such as e.g. photosynthesis, lasers for providing a sufficiently large number of photons per area and time. Typically more than 10^{18} photons cm^{-2} s^{-1} are needed to reach nonlinearity in materials with fast reaction times. With lasers these intensities can be easily realized and thus photons can be used for many applications as will be shown below.

1.2 Photonics and Light Technology

Light is used for illumination and we are familiar with optical devices such as magnifiers, telescopes, microscopes and mirrors. In these common applications intensities of less than 10^{10} photons cm^{-2} s^{-1} are used and the light is spread over wide spectral ranges. Thus the brilliance of these conventional light sources is more than 10^{10} smaller than in most applications with laser light sources.

Nevertheless conventional light technologies play an important role in linear and nonlinear optics and are needed in many fields of photonics. Thus the relations between photonics and electronics, light technologies and electric technologies, in connection with future key technologies can be represented as in Fig. 1.1.

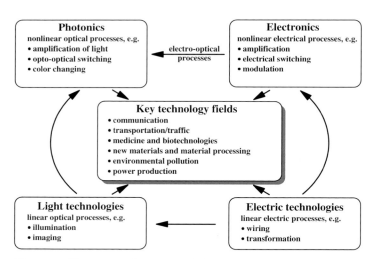

Fig. 1.1. Photonics, electronics, light technologies and electric technologies in connection with key technologies

Typical nonlinear processes in photonic applications are the generation and amplification of light in lasers, the changing of light color by frequency conversion processes or optical switching as used for ultra-short pulse generation. Electronic devices, conventional light technologies and optics are used in these photonic applications as power supplies, in pump sources or in light guiding systems. Electronically driven electro-optic devices such as modulators are used to control the photonic devices and processes. All these elements are necessary for the present key technology fields.

1.3 Scientific Topics

Photonics is based on the physics and devices of conventional optics, on quantum physics and on electromagnetism. But its main topic is the physics of nonlinear optical processes. Thus it is necessary to analyze possible nonlinear processes and investigate suitable materials with methods of nonlinear spectroscopy. Parallel progress in laser physics has to be made (see Fig. 1.2).

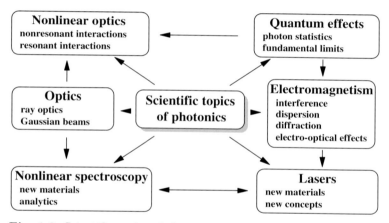

Fig. 1.2. Scientific topics of photonics

Light is built out of photons which are quantum mechanical and relativistic particles. Thus light shows particle and wave properties in the sense of our macroscopic understanding. It moves with the maximum possible speed. Light from lasers shows new statistical properties.

In nonlinear optics all these properties are much more important than in conventional optics. Light-induced changes in the materials responsible for nonlinear effects are mostly functions of these light properties and the superposition of different light beams leads to complicated effects.

Thus it is necessary to learn all the properties of photons and light beams as they are used in photonics. Even quantum statistics may be important

in some cases such as in communications or spectroscopy. Useful parameters
for characterization have to be identified. Then theoretical methods for the
propagation of light beams with regard to diffraction are needed. Finally,
interfering light beams are used for inducing transient or permanent gratings
in nonlinear materials.

Linear interactions of light with matter such as diffraction, refraction, absorption and birefringence produced with light of sufficiently small intensities
are the basis for nonlinear optics. In the nonlinear case processes similar to
those in linear optics are observable but they can be induced by the high
intensities of the light itself. Thus, e.g. materials which are commonly transparent at a certain wavelength can be become highly absorbing under laser
light illumination.

Nonlinear effects can be differentiated in interactions with absorption,
called resonant interactions, and interactions with nonabsorbing, transparent
materials, called nonresonant interactions.

Applications of these nonlinear effects demand sufficient knowledge of the
nonlinear properties of possible materials. Thus the spectroscopic technologies required to investigate these properties and to determine all necessary
material coefficients have to be known in some detail. This seems even more
important as long as most of these measurements are not really standardized.
Thus the published coefficients may sometimes be dependent on experimental
parameters, which are not given in the reference. Based on a comprehensive
knowledge of nonlinear optical effects and the parameters of the laser radiation used the reader should be able to identify experimental differences in
these measurements.

One of the main topics is the invention of new useful materials with high
nonlinear coefficients. Most of today's known materials demand intensities
of more than 10^{20} photons $cm^{-2} s^{-1}$. Some materials such as photorefractives
are applicable with much lower intensities but they show very long reaction
times in the µs-, ms- and s-range. In comparison photosynthesis works with
sunlight and time constants down to femtoseconds. This excellent photonic
"machine" is based on molecular structured organic material. Progress in the
field of new synthetic structures with new nonlinear optical properties can
be expected in the future.

The following scientific topics will not be treated in detail in this book and
therefore some references maybe useful. An overview about quantum effects is
given in, for example, [1.2]. Bose–Einstein condensation [1.3, 1.4] is now investigated in detail [e.g. 1.5–1.17]. Realizing a large number of atoms in the same
quantum state may allow completely new applications such as, for example,
the atom laser [1.18–1.27]. In this context the problems of atom optics may
be investigated [e.g. 1.28–1.30]. The use of specially prepared quantum states
of matter and/or photons may allow new concepts in communication and
data processing technologies. Therefore, entangled states of photons [1.31]
are investigated [1.32–1.54] and the possibilities of quantum cryptography

[1.31] are checked [1.55–1.69]. Concepts of quantum computing [1.70, 1.71] are being developed [1.72–1.84]. Basic research is in progress on quantum non-destructive measurements [1.85–1.89] and the Einstein–Podolski–Rosen (EPR) paradox and Bell's inequalities [1.90–1.94]. Some general aspects are discussed in [1.95–1.99].

1.4 Technical Topics

While planning photonic applications the physical limits as determined by diffraction and other uncertainty rules of the light have to be considered and photons rates should be estimated for designing the beam cross-sections and the measurement devices.

Therefore a detailed knowledge of the technical and economic specifications of commercially available or possibly home-made photonic devices is essential for efficient work (see Fig. 1.3). This may be difficult because of the rapid development of new components on the market. The scientific literature allows new devices, which may become products in the near future, to be evaluated.

In particular, laser light sources and detection/measuring devices are being rapidly developed. The analysis of the basic principles of these devices may serve as a helpful basis to incorporate future developments.

Different lasers allow the generation of light with almost all imaginable properties but usually not in all desirable combinations. Thus there is still a need for new lasers with new combinations of light properties such as e.g.

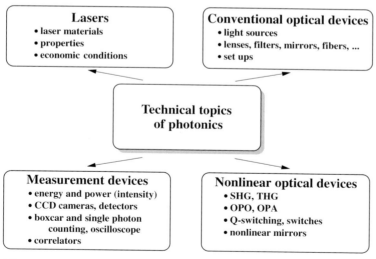

Fig. 1.3. Technical topics of photonics

large spectral tuning ranges or bandwidths with high average output power, good beam qualtity and variable pulse width. In addition increasing efficiency and reliability, while reducing complexity and maintenance, are also common demands for new lasers.

Although many nonlinear optical effects which are useful for photonic applications have been known for a long time, further inventions and materials may be necessary for application in new devices. As an example, optical phase conjugation was first observed in 1972 [1.100]. Since about 1987 extensive research took place and in 1994 the first commercial laser with phase conjugating mirrors with high output power and increased frequency conversion efficiency was brought to the market [1.101]. Other examples are solitons in optical fibers, new frequency conversion technologies such as optical parametric amplifiers, and Kerr lens mode locking in fs lasers.

In all topics of photonics a close connection of basic knowledge about the physical principles on one hand and technical possibilities on the other side is typical.

1.5 Applications

As described in Sect. 1.2 photonics has applications in all key technologies. Thus some examples of great importance in the near future are mentioned in Fig. 1.4.

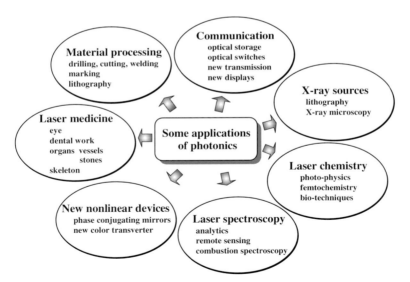

Fig. 1.4. Some applications of photonics

Opto-optical switches may increase the speed of communication by orders of magnitude. Spectral coding may enlarge the capacity of fibers in communication and of optical storage by 10 to 1 000, resulting in optical disks with 1 TByte capacity. Fast optical image processing of huge quantities of graphical information may be developed. New laser displays with high powers of several Watts on large screens of several m are expected to replace other TV displays. In the future quantum computers may be developed supplementing high-power computers in special sectors.

Laser machining will have an increasing impact on almost all technologies. Thus e.g. the efficiency of turbines and airplanes can be increased by better laminar flows as a consequence of laser drilled holes in the material surfaces. Laser cutting and welding may change processes in car production. Laser surface structuring and cleaning may allow the application of new materials which otherwise are not usable for these purposes. Micromachining will enable completely new technologies in biology and medicine to be developed.

Laser-induced X-ray generation in the "water-window" of 2–4 nm wavelength can provide new insights into the mechanism of living organism cells via soft X-ray microscopes. In computer chip production the component density can be increased using these X-ray sources for lithography.

New materials may become available by femtochemistry using short light pulses for well-timed ignition of molecular chemical reactions. In particular, compounds which are needed in small amounts, such as drugs, may than be produced economically. Nonlinear laser spectroscopy may become more important for characterizing and analyzing all kinds of biochemical products.

Detection of environmental pollution is based in many cases on laser spectroscopic measurements. Lidar and other wide-ranging methods are used for the analysis of air pollution. Shorter pulses will allow a further increase of spatial resolution.

New materials are needed for the invention of new solar energy technologies. Nonlinear optical investigations may be necessary for their exploration. Promising first results have been obtained using organic molecules as sensitizers for inorganic solar energy collectors.

New nonlinear devices as, e.g. phase conjugating mirrors, frequency converters and laser materials may result in new photonic applications not commercially available, today.

One of the fastes growing areas in photonics is laser medicine. Almost all parts of the body are treatable. This includes, in particular, eye lens and retina, skin, surgery, vessels and stone demolition. The possibility of guiding light with very high power through thin fibers into the body enables new techniques with minimal invasion to be developed.

An overview of modern photonic applications can be found in [1.102]. In addition to the new quantum information technologies mentioned in Sect. 1.2, the rapid development of other optical switches [1.103–1.105], fibers [1.106–1.111] including wavelength division multiplexing (WDM) [1.112, 1.113] and storage principles [1.114–1.130], is in progress. Laser display techniques are based on new red–green–blue laser sources [1.131–1.136]. Molecular computers [1.137, 1.138] may demand optical interfaces. Progress in laser chemistry [1.139–1.161] is based to a large extent on femtochemistry. Aspects of material processing [1.162, 1.163] are described in Sect. 5.8. A newer topic in this field is laser cleaning [1.164–1.167]. Laser spectroscopy is used for pollution measurements and other environmental detection [1.168–1.198]. New concepts of solar energy converters are investigated [1.199–1.212]. Laser medicine has many applications [1.213–1.247] in, for example, photodynamic therapy [1.222–1.224]. New laser measurement methods [1.225–1.236], and especially optical coherence tomography (OCT) [1.237–1.247], enable new diagnostics in medicine. In a similar way, new bio-technologies were enabled by new measurement methods [1.248–1.262]. Single molecules can be observed with several laser techniques [1.263–1.270]. Gravitational waves can be detected with laser interferometers, as described in Sect. 2.9.2 [1.271–1.299]. Many other new optical measurement techniques have been developed in recent years [e.g. 1.300–1.327]. One of the most spectacular results was achieved by optical sequencing of DNA [e.g. 1.328–1.335]. The generation of light in a water window [1.336–1.338] will allow X-ray microscopy [1.339–1.342] with resolution in the nm range. New x-ray sources [1.343–1.347], also mentioned in Sect. 6.13.5, will increase resolutions in lithography [1.348, 1.349]. Laser ignited fusion, as in the National Ignition Facility (LLNL, USA), demands new concepts for lasers with very high powers [e.g. 1.350–1.353], and particle acceleration seems to be possible with laser pulses [1.354–1.357].

1.6 Costs of Photons

All photonic applications are based on laser light sources and thus the costs of this light are the essential quantity for the invention and commercial exploitation of these new technologies.

The cost of photons with different properties is given in Table 1.1 which includes the operational expense and the purchase of the laser divided by the number of photons produced. The energy of the photons, the price of the laser and its operational costs are also given. Thus CO_2 lasers have 10 times larger wavelength than a Nd:YAG laser and thus they show a ten times larger diameter in the focus for a given focal length. It should further be noted that the beam quality of excimer lasers, diode lasers and light bulbs is much worse than e. g. for solid-state lasers.

Photons of lasers with good beam quality are still more than 300 times as expensive as electrons of the same energy. Nevertheless, semiconductor lasers offer the chance to decrease this price by a factor of 10 or more in the near future by reduced production costs. Photons from flash lamps and bulbs are in any case more expensive than laser photons if beam quality is required. If, in addition, brilliance is necessary then lasers are far ahead. Thus in the

Table 1.1. Roughly estimated costs of some lasers and their operational cost during their lifetime in relation to the photon energy and average output power. The cost of a photon from these lasers and for comparison from bulbs as well as the cost of an electron of different energies is also given

Source	Energy (eV)	Power (W)	Price ($)	Lifetime (h)	Opera- tion ($)	Price/Wh ($)	Photon price ($)
He-Ne laser	2.0	0.005	800	20 000	9	8	7.0×10^{-22}
Argon laser	2.5	5	50 000	20 000	10 000	0.6	6.5×10^{-23}
Excimer laser	4	10	50 000	15 000	1 000	0.34	6.2×10^{-23}
CO_2 laser	0.12	250	50 000	10 000	3 500	0.06	3.3×10^{-25}
Nd:YAG laser	1.2	25	50 000	20 000	10 000	0.12	6.3×10^{-24}
Nd + SHG	2.4	5	40 000	10 000	180	0.8	8.6×10^{-23}
Diode laser	1.6	50	10 000	10 000	200	0.02	1.5×10^{-24}
Light bulb	2	5	0.5	5 000	50	0.004	3.5×10^{-25}
Electron	2					0.0002	(1.8×10^{-26})

case of a flash lamp or diode-pumped solid-state lasers the laser material acts as a converter of beam quality, coherence and brilliance with an opto-optical energy efficiency of typically less then 10% for flash lamp pumping or up to 50% for diode pumping, respectively. Even values of more than 70% can be reached with optimized configurations.

2. Properties and Description of Light

Light is commonly used in photonic applications such as laser beams with a complex distribution of the intensity as a function of wavelength, space, time and polarization. In addition the coherence properties have to be recognized and sometimes just single photons are used. In this chapter different classifications of these physical properties are described. In nonlinear optical processes these properties have to be recognized carefully, because the nonlinearity may depend on these in a complicated manner.

Properties of the single photons determine the beam characteristics of the laser light. In some cases the superposition of light fields and/or their interaction with matter has to be described. In this case plane waves are mostly assumed. They can be realized in the focal range of, e.g. a Gaussian beam. These Gaussian beams can be generated from lasers and their propagation can be calculated in an easy manner. Geometrical optics may be helpful for a first approach in optical systems. Useful additional information may be found especially in the monographs [M5, M16, M18, M21, M27, M31, M34, M38, M42].

2.1 Properties of Photons

Light can be described as an electromagnetic wave or a collection of single photons propagating with speed c, which is a maximum in vacuum and smaller in materials:

$$\textbf{speed} \quad c_{\text{vacuum}} = 2.998 \cdot 10^8 \, \text{m s}^{-1} \qquad (2.1)$$

$$c_{\text{material}} = c_{\text{vacuum}}/n_{\text{material}}. \qquad (2.2)$$

n_{material} is the conventional refractive index of the matter which is 1 for the vacuum (see Chap. 3). Photons are quantum particles and fulfill at least four uncertainty conditions which are described in Sect. 2.1.2. As a consequence we observe diffraction-limited focus sizes and bandwidth-limited pulse widths.

2.1.1 Energy, Frequency, Wavelength, Moments, Mass, Timing

The single photon represents an electromagnetic wave oscillating with frequency ν which determines its energy E:

frequency ν in units s^{-1} or Hz $\hspace{3cm}$ (2.3)

energy $E = h\nu,$ $\hspace{5cm}$ (2.4)

$$h = 6.626 \cdot 10^{-34}\,\text{Js} \hspace{3cm} (2.5)$$

where h is Planck's constant. The wavelength λ of this electromagnetic wave results from

wavelength $\lambda = \dfrac{c}{\nu}.$ $\hspace{4cm}$ (2.6)

If the photon is moving in direction e a wave vector k is defined as:

wave vector $k = \dfrac{2\pi}{\lambda}e$ $\hspace{4cm}$ (2.7)

not to be confused with the wave number $\tilde{\nu}$:

wave number $\tilde{\nu} = \dfrac{1}{\lambda}$ $\hspace{4cm}$ (2.8)

which is proportional to the photon frequency and energy as $\nu = c\tilde{\nu}$ and $E = hc\tilde{\nu}$. This value is used in spectroscopic applications, resulting in handy numbers.

A photon carries the momentum p_{ph}:

momentum $p_{\text{ph}} = \dfrac{h}{2\pi}k$ $\hspace{4cm}$ (2.9)

with the momentum value:

momentum value $p_{\text{ph}} = \dfrac{h}{\lambda} = \dfrac{E}{c}.$ $\hspace{3cm}$ (2.10)

This momentum is obtained if the photon is absorbed, and twice this value will push a 100% reflecting mirror. The photon also has a spin momentum j which is called right or left polarization:

spin momentum (polarization) $j = \pm\hbar = \pm\dfrac{h}{2\pi}$ $\hspace{1.5cm}$ (2.11)

This spin will be received by the matter if the photon is absorbed. If this absorption is combined with an excitation of the atomic or molecular quantum system, selection rules have to be fulfilled to satisfy momentum conservation. If a light beam is built from an equal number of right and left polarized photons it will be linearly polarized. The light beam can also be nonpolarized.

The mass of a photon m_{ph} can be calculated formally from these values as:

mass $m_{\text{ph}} = \dfrac{p_{\text{ph}}}{c} = \dfrac{h\nu}{c^2}$ $\hspace{3cm}$ (2.12)

which is a function of its energy. In the visible range it is about one million less than the mass of an electron. Photons do not exist without moving. In relation to other photons the electromagnetic wave has a phase

phase $\varphi.$ $\hspace{5cm}$ (2.13)

As a particle the single photon can be detected with a diffraction-limited space uncertainty and with a certain energy uncertainty (see Sect. 2.1.2) at a time

time t (2.14)

Photons can in principle interact with each other if the light intensity is extremely high. The cross-section σ of a photon in the visible range is of the order of:

cross section $\sigma \approx 10^{-72}\,\mathrm{cm}^2$ (2.15)

which is more than 50 orders of magnitude smaller than, for example, for atoms. For the observation of a photon photon scattering with light the intensity should exceed 10^{86} photons cm^{-2} s^{-1}. For an experiment at a linear accelerator light by light scattering was reported using Compton-backscattered photons with an energy of about 29 GeV resulting in positron production [2.1]. However, in common laser experiments this high intensity can not be realized, yet. Thus the linear superposition of light is fulfilled in all practical cases.

For illustration in Table 2.1 a collection of relevant values for photons is given. In the visible range at 500 nm the energy of one photon is about $4 \cdot 10^{-19}$ J. Thus in a laser beam of this wavelength with 1 Watt average output power about $2.5 \cdot 10^{18}$ photons per second occur.

Table 2.1. Characteristic values of a photon of different color. $|\mathbf{k}|$ is the value of the wave vector and $|\mathbf{p}|$ the value of the momentum

wavelength	color	energy (J)	frequency (s^{-1})	$\lvert\mathbf{k}\rvert$ (cm^{-1})	$\lvert\mathbf{p}\rvert$ (kg m s^{-1})
10 nm	X-UV	$1.99 \cdot 10^{-17}$	$3.00 \cdot 10^{16}$	$6.28 \cdot 10^{6}$	$6.63 \cdot 10^{-26}$
200 nm	UV	$9.93 \cdot 10^{-19}$	$1.50 \cdot 10^{15}$	$3.14 \cdot 10^{5}$	$3.31 \cdot 10^{-27}$
500 nm	green	$3.97 \cdot 10^{-19}$	$6.00 \cdot 10^{14}$	$1.25 \cdot 10^{5}$	$1.33 \cdot 10^{-27}$
1 μm	IR	$1.99 \cdot 10^{-19}$	$3.00 \cdot 10^{14}$	$6.28 \cdot 10^{4}$	$6.63 \cdot 10^{-28}$
10 μm	far IR	$1.99 \cdot 10^{-20}$	$3.00 \cdot 10^{13}$	$6.28 \cdot 10^{3}$	$6.63 \cdot 10^{-29}$

Sometimes the photon energy is described by other values which are useful for direct comparison with matter parameters with which the photons may interact. Such values are the energy measured in eV relevant for collision excitation by electrons, the inverse wavelength or wave number, which should not be confused with the wave vector, and the temperature T_{emp} of a blackbody whose emission maximum produces photons of the desired wavelength. These values are given in Table 2.2.

Table 2.2. Energy of photons in different measuring units for comparison

wavelength	energy (J)	energy (eV)	$\tilde{\nu} = 1/\lambda$ (cm^{-1})	T (K)
10 nm	$1.99 \cdot 10^{-17}$	124	1 000 000	28 977
200 nm	$9.93 \cdot 10^{-19}$	6.20	50 000	14 488
500 nm	$3.97 \cdot 10^{-19}$	2.48	20 000	5 795
1 μm	$1.99 \cdot 10^{-19}$	1.24	10 000	2 898
10 μm	$1.99 \cdot 10^{-20}$	0.124	1 000	290

2.1.2 Uncertainty Principle for Photons

Photons as quantum particles show wave properties and thus fulfill the uncertainty conditions of their noncommuting physical properties such as position and momentum, photon number and phase. Thus the values of these pairs of physical values cannot be perfectly determined simultaneously. These quantum uncertainties are responsible for practical limits in many photonic applications. As a result in the best case light may be diffraction limited as a consequence of the position and momentum uncertainty, and bandwidth limited as a result of the quantum energy and time relation.

Uncertainty of Position and Momentum

The position–momentum uncertainty occurs independently for both orthogonal coordinates x and y perpendicular to the propagation direction z of the light:

$$1/\text{e}^2 \text{ uncertainty:} \quad \Delta x \Delta p_x \geq \frac{h}{\pi} \text{ and } \Delta y \Delta p_y \geq \frac{h}{\pi} \tag{2.16}$$

where the uncertainty Δx is defined as the value $1/\text{e}^2$ of the transversal distribution function of the photons in an ensemble of measurements or of the transversal intensity distribution of a light beam. This definition is commonly used in photonics e.g. for characterizing the beam diameter of Gaussian beams (see Sect. 2.4). The momentum uncertainty is measured via the position uncertainty in a certain distance resulting in an uncertainty angle, the divergence angle, based on a definition of the position uncertainty. h stands for Planck's constant. The relations in the y coordinate have to be treated in the same way.

As a consequence the propagation direction θ of a photon will be less certain as more precise it is located in the related coordinate e.g. inside the value $2w_0$. For a large number of photons in a light beam this results in an increase of the divergence angle θ (see Sect. 2.7.4) of the beam if its diameter $2w_0$ is decreased in this coordinate, e.g. by use of an aperture (see Fig. 2.1).

With the relations $\Delta x = w_0$, $\Delta \boldsymbol{p} = \Delta \boldsymbol{k} h / 2\pi$ and $\Delta \boldsymbol{k} = \boldsymbol{k}\theta$ follows from (2.16) with the wavelength λ that:

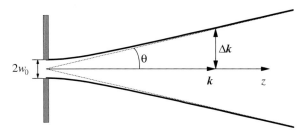

Fig. 2.1. Divergence angle θ of a light beam with the diameter $2w_0$ and the wave vector $\boldsymbol{k}$ with the uncertainty Δk

$$\textbf{position--angle uncertainty} \quad w_0\theta \geq \frac{\lambda}{\pi}. \tag{2.17}$$

If this aperture has sharp edges, as is common, the resulting uncertainty is larger than the minimum of (2.16) by a factor of 1.4 as a consequence of the diffraction at the diaphragm edges. In the case of a Gaussian transmission profile of the diaphragm the lowest possible value of h/π can be reached. A light beam passing such an aperture would be diffraction limited. The minimum divergence angle which can be observed for light with a wavelength of $1\,\mu$m after passing through such a Gaussian aperture (or in a Gaussian beam) with $1\,$mm diameter is $0.32\,$mrad.

Sometimes other definitions for Δx and Δp_x are used. If the $1/e$ value of the distribution is used the resulting uncertainty is only half the value of (2.16) and if the $(1/\sqrt{e})$ value is applied, as in most quantum mechanical textbooks, the resulting uncertainty is $\frac{1}{4}$ times smaller, resulting in $\hbar/2$.

Uncertainty of Energy and Time

Although the uncertainty relation for energy and time cannot be extracted from quantum mechanics in the same way as the position–momentum uncertainty because time is not a quantum mechanical operator, the relations:

$$\textbf{energy--time uncertainty} \quad \Delta E \Delta t \geq \frac{h}{2\pi} \tag{2.18}$$

can be derived from quantum electrodynamics. This equation correlates the energy uncertainty ΔE to a characteristic time interval Δt in which the changes of a given system occur. If the photon energy E is measured for a very short period Δt it will be uncertain in the range of ΔE. The shorter the measurement the larger the absolute energy range ΔE. In the case of absorption of a photon the uncertainty ΔE of the energy transferred to a particle occurs in the time interval Δt. In the case of the emission of a photon follows directly from this equation that the uncertainty relation between the light frequency and the decay time of an emitter is:

$$\textbf{frequency--time uncertainty} \quad \Delta\nu_{\text{FWHM}}\Delta t_{1/e} \geq \frac{1}{2\pi} \tag{2.19}$$

where the frequency uncertainty $\Delta\nu_{FWHM}$ is measured as the full width half maximum of the Lorentzian line shape function and the temporal uncertainty is measured as the $1/e$ value of the decay function.

As a consequence the energy distribution or the spectral bandwidth of a short pulse has a minimum limit and thus the spectral resolution of measurements with very short pulses is limited. In Table 2.3 the spectral uncertainty is given as the distribution of the wave number $\Delta\tilde{\nu}$ and as the wavelength distribution $\Delta\lambda$ as a function of the time window Δt and the latter as a function of the mid-wavelength of the light.

Table 2.3. Spectral uncertainty as a function of the time window Δt and the mid-wavelength

Δt	$\Delta\tilde{\nu}$ (1/cm)	$\Delta\lambda$ (nm) 200 nm	$\Delta\lambda$ (nm) 500 nm	$\Delta\lambda$ (nm) 1 µm	$\Delta\lambda$ (nm) 10 µm
1 µs	$5.31 \cdot 10^{-6}$	$2.1 \cdot 10^{-8}$	$1.33 \cdot 10^{-7}$	$5.31 \cdot 10^{-7}$	$5.31 \cdot 10^{-5}$
1 ns	0.00531	$2.1 \cdot 10^{-5}$	$1.33 \cdot 10^{-4}$	$5.31 \cdot 10^{-4}$	0.0531
1 ps	5.31	0.021	0.133	0.531	53.1
10 fs	531	2.1	13.3	53.1	5310

The uncertainty relation results for short pulses in a minimal bandwidth which can also be developed by describing the pulse as a wave packet via Fourier analysis as given in Sect. 2.9.8. The uncertainty values depend on the pulse shape. The values given are valid for spectral Lorentzian line shapes and temporal exponential decays.

2.1.3 Properties of a Light Beam

In photonic applications light is mostly applied in beams which represent a collection of photons with different properties depending on their generation. These beams have mostly small divergence angles of smaller $30°$ and thus fulfill the *paraxial approximation*. This simplifies optical imaging with lenses and mirrors and their theoretical calculation with e.g. ray matrices (see Sect. 2.5).

Lasers can operate continuously (cw) or pulsed and thus the light beams are cw or can be built from pulses. The pulses can be as short as femtoseconds and can have repetition rates as high as several hundred MHz. Smaller values are possible. High repetition rates, usually above some kHz, are sometimes called quasi-cw.

As described in more detail in Sect. 2.7 these light beams are characterized by their intensity I, their power P or energy E_{pulse} in the case of pulsed light which represents a certain number of photons in the beam. Because of these properties of photons the measured intensity, power or energy are functions of:

space $I, P, E_{\mathrm{pulse}} = f(\boldsymbol{r}) \approx f(w_{\mathrm{waist}}, z_{\mathrm{waist}}, \theta)$ (2.20)

wavelength $I, P, E_{\mathrm{pulse}} = f(\lambda) \approx f(\lambda_{\mathrm{max}}, \Delta\lambda)$ (2.21)

time $I, P, E_{\mathrm{pulse}} = f(t) \approx f(t_{\mathrm{max}}, \Delta t)$ (2.22)

polarization $I, P, E_{\mathrm{pulse}} = f(\varphi) \approx f(\varphi_{\mathrm{max}}, p)$ (2.23)

and thus the detection will be spatially, spectrally, temporal and polarization sensitive.

All these dependencies can be complicated functions with several maxima and minima. In the case of laser beams as mostly used in photonic applications they are often describable by simple Gaussian-like distributions with one maximum and a width. But this approximation has to be checked carefully and the definitions of the width as full width half maximum (FWHM), half width half maximum (HWHM), standard deviation, $1/e^2$ width, $1/e$ width, and so on, has to be given explicitly, as described in Sect. 2.7 in detail.

Using these simple approximations the functions of (2.20)–(2.23) can be simplified as given in the right-hand expressions of these equations. The spatial propagation can be described as a paraxial beam with its waist position z_{waist} and size w_{waist} and the divergence angle θ or the beam quality given by M^2 as described in more detail in Sect. 2.7.5. The beam spectrum can then be characterized by the wavelength of the peak λ_{max} and the spectral width $\Delta\lambda$. Similarly the pulse characteristics are reduced to the time of the maximum t_{max} and the pulse duration Δt. The polarization of the beam is described by the degree of polarization p and the angle of the electric field maximum φ_{max} in relation to laboratory coordinates.

In addition to these four functions the coherence of the beam has to be recognized. This can be done in an approximate and simple way using the coherence length l_{coh} or coherence time τ_{coh} in the axial direction and using the transversal coherence length $l_{\mathrm{coh,transversal}}$ in the transversal direction as described in Sect. 2.9.2. Furthermore the phase of the light field vectors in relation to other fields can be important. Usually it can be determined only in the application itself.

Thus the light intensity I follows in this approximation from:

intensity $I = \dfrac{P}{A} = \dfrac{E}{A \Delta t} = \dfrac{n_{\mathrm{ph}} h\nu}{A \Delta t}$ in $\mathrm{W\,cm}^{-2}$ (2.24)

or as the photon flux density which is often also called intensity:

photon flux intensity $I = \dfrac{I}{h\nu} = \dfrac{E}{A \Delta t h\nu}$ in $\mathrm{photons\,cm}^{-2}\,\mathrm{s}^{-1}$. (2.25)

The two values can easily be distinguished by the measuring unit and by the numerical values which are different by about 19 orders of magnitude. The cross-section A of the light beam usually contains 86.5% of the beam energy or power and Δt is the full width half maximum pulse duration. In the case of more complicated distributions averages are usually used by integrating over a certain area. For non-Gaussian distributions the momentum method has to be used as described in Sect. 2.7.

Describing the power, energy or intensity based on the number of photons is obviously convenient if interactions with countable particles is to be modeled. But in this case the energy, wavelength or frequency of the photons have to be considered, separately. This needs additional attention e. g. if the beam contains a broad spectrum.

2.2 Plane Waves Monochromatic Light

Plane waves of monochromatic light are theoretically the simplest kind of light. They can be realized with good accuracy inside the Rayleigh range of Gaussian laser beams (see Sect. 2.4.3) with narrow bandwidth. They are useful for complicated theoretical descriptions of nonlinear interactions of light with matter.

2.2.1 Space- and Time Dependent Wave Equation

Maxwell's equations have a periodic solution for the electric field vector $\boldsymbol{E}$ and magnetic field vector $\boldsymbol{H}$ in space and time which represents light if the frequency is in the range 10^{13}–10^{15} Hz. From the vector equations for the electric field E:

$$\operatorname{curl} \boldsymbol{E} = -\mu_0 \frac{\partial \boldsymbol{H}}{\partial t} \tag{2.26}$$

and the magnetic field H:

$$\operatorname{curl} \boldsymbol{H} = \varepsilon_0 \frac{\partial \boldsymbol{E}}{\partial t} + \frac{\partial \boldsymbol{P}}{\partial t} \tag{2.27}$$

with material polarization P it follows the differential equation for the electric field vector, measured in $\mathrm{V\,m^{-1}}$:

$$\textbf{wave equation} \quad \Delta \boldsymbol{E} - \frac{1}{c_0^2} \frac{\partial^2 \boldsymbol{E}}{\partial t^2} - \operatorname{grad} \operatorname{div} \boldsymbol{E} = \mu_0 \frac{\partial^2 \boldsymbol{P}(\boldsymbol{E})}{\partial t^2} \tag{2.28}$$

with the Laplace operator

$$\Delta = \left\{ \frac{\partial^2}{\partial x^2} + \frac{\partial^2}{\partial y^2} + \frac{\partial^2}{\partial z^2} \right\}.$$

All light–matter interactions are considered in this equation as linear and/or nonlinear functions of the polarization $P = f(E)$.

These interactions of light with matter will be described in Chaps. 3, 4 and 5 in detail but here as the simplest case, we assume a vacuum or a material which is:

- homogeneous
- isotropic
- nonconductive

- uncharged
- nonmagnetic
- linear (see Chaps. 4 and 5).

Then all material properties can be summarized by the refractive index n which in this case is not dependent on the orientation of the material or on the polarization, wavelength or intensity of the light.

The wave equation can be simplified to:

$$\Delta \boldsymbol{E} - \frac{1}{c^2}\ddot{\boldsymbol{E}} - \operatorname{grad} \operatorname{div} \boldsymbol{E} = 0 \qquad (2.29)$$

with the speed of light c in this material:

$$c^2 = \frac{c_0^2}{\varepsilon_r \mu_r} = \frac{c_0^2}{n^2} = \frac{1}{\mu_0 \varepsilon_0 \mu_r \varepsilon_r} \qquad (2.30)$$

and the values:

vacuum permittivity $\varepsilon_0 = 8.854 \cdot 10^{-12} \dfrac{\mathrm{A\,s}}{\mathrm{V\,m}} = \dfrac{1}{\mu_0 c_0^2}$ $\qquad (2.31)$

vacuum permeability $\mu_0 = 4\pi \cdot 10^{-7} \dfrac{\mathrm{V\,s}}{\mathrm{A\,m}}.$ $\qquad (2.32)$

The specific electric permittivity ε_r and magnetic permeability μ_r are material parameters. μ_r is typically 1 for optical matter and ε_r is 1 for the vacuum. This equation can be solved by a propagating wave with:

$$\boldsymbol{E} = \boldsymbol{E}_0 \, \cos(2\pi\nu t - \boldsymbol{k} \cdot \boldsymbol{r}) \qquad (2.33)$$

where $|\boldsymbol{E}_0|$ is the maximal value of the electric field and the vector describes the polarization. The electric field vector may be measured in three dimensions, as described in [2.2]. The frequency ν, the wave vector $\boldsymbol{k}$ and the phase φ are known from Sect. 2.1. Under the above assumptions the analogous equation for the magnetic field $\boldsymbol{H}$ can be solved to give:

$$\boldsymbol{H} = \boldsymbol{H}_0 \, \cos(2\pi\nu t - \boldsymbol{k} \cdot \boldsymbol{r}). \qquad (2.34)$$

In the simplest case it will be assumed that this monochromatic wave is planar and propagates in the z direction. This wave then has an indefinite dimension in the x and y directions. As a good approximation a dimension of more than 100 times the wavelength may be sufficient. The magnetic field vector $\boldsymbol{H}$ is, under these assumptions, perpendicular to the electric field vector $\boldsymbol{E}$ (see Fig. 2.2). Both fields are in phase because the distance from the emitter is large compared to the wavelength of the light. At any time both fields are present as shown in the figure and they are moving forward in space. It is a continuous (cw) beam of photons with the same energy. In common optical applications the interaction between the magnetic field and matter can be neglected.

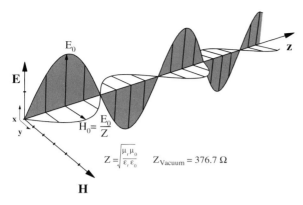

Fig. 2.2. Monochromatic planar wave propagating in the z direction in an isotropic, homogeneous, nonelectric and nonmagnetic, uncharged, linear material

The values of the electric and magnetic field are related by:

$$\frac{E_0}{H_0} = \sqrt{\frac{\mu_0 \mu_\mathrm{r}}{\varepsilon_0 \varepsilon_\mathrm{r}}} \tag{2.35}$$

and the impedance Z of this wave propagating in the vacuum with c_0 results from:

$$Z_\mathrm{vacuum} = \sqrt{\frac{\mu_0}{\varepsilon_0}} = 376.7\,\Omega. \tag{2.36}$$

These waves can be superimposed in any way as long as the intensity is small enough to guarante linear interactions (see Chaps. 4 and 5). Thus in the time domain light beats and pulses are possible by mixing different light frequencies. The polarization of light can be linear, circular or elliptic by mixing fields with different $\boldsymbol{E}_0$ and φ.

In an optical resonator the light wave is propagating back and forth between the two mirrors and thus a standing wave occurs. The electric field for the monochromatic planar wave can be written as:

$$\boldsymbol{E} = \boldsymbol{E}_0 \, \cos(2\pi\nu t) \, \sin(kz) \tag{2.37}$$

with knots at the mirrors. The magnetic field is then given by:

$$\boldsymbol{H} = \boldsymbol{H}_0 \, \sin(2\pi\nu t) \, \cos(kz) \tag{2.38}$$

and thus a phase difference of $\pi/2$ can be observed in this case (see Fig. 2.3).

The energy of the light oscillates between the electric field and the magnetic field. At certain periodic times the electric field is zero along the z axis.

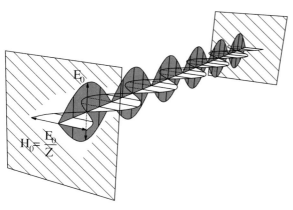

Fig. 2.3. Standing monochromatic planar light wave in an optical resonator filled with an isotropic, homogeneous, nonelectric and nonmagnetic, uncharged, linear material (e.g. air)

2.2.2 Complex Representation

More elegantly the field vectors of these light waves can be written in complex form as:

$$\boldsymbol{E} = \frac{\boldsymbol{E}_0}{2}\, \mathrm{e}^{\mathrm{i}(2\pi\nu t \pm \boldsymbol{k}\cdot\boldsymbol{r}+\varphi)} + \frac{\boldsymbol{E}_0}{2}\, \mathrm{e}^{-\mathrm{i}(2\pi\nu t \pm \boldsymbol{k}\cdot\boldsymbol{r}+\varphi)} \qquad (2.39)$$

which is mathematically identical to (2.33) but is more handy especially if several fields have to be superimposed. The complex representation avoids complicated algebra of trigonometric functions.

For simplicity the phase φ can be assigned to the field amplitude:

$$\boldsymbol{E}_0\, \mathrm{e}^{\mathrm{i}\varphi} \rightarrow \boldsymbol{E}_0(\varphi) \qquad (2.40)$$

This phase of the wave front of the electric field can be a function in space in the case of nonplanar waves and thus the field amplitude will overtake this dependency:

$$\textbf{nonplanar wave} \quad \boldsymbol{E}_0(\varphi) \xrightarrow{\varphi(\boldsymbol{r})} \boldsymbol{E}_0(\boldsymbol{r}) \qquad (2.41)$$

A simple example is the spherical field distribution in the case of a point light source. In this case the field vector describes spheres but in addition the amplitude decreases by $1/r^2$.

For even simpler formulas the electric field vector can be written in the complex form:

$$\boldsymbol{E}_\mathrm{c} = \boldsymbol{E}_0\, \mathrm{e}^{\mathrm{i}(2\pi\nu t - \boldsymbol{k}\cdot\boldsymbol{r})} \qquad (2.42)$$

In this case the real field amplitude has to be recalculated from the complex form by:

$$E = \frac{1}{2}(E_c + E_c^*) \tag{2.43}$$

All these formulas are valid for the magnetic field vector analog but usually the magnetic field does not need to be calculated explicitly. It can be derived from the electric field by:

$$H = c_0 \varepsilon_0 n \left[\frac{k}{k} \times E \right] \tag{2.44}$$

where k/k is a unit wave vector pointing towards the propagation direction of the wave.

2.2.3 Intensity and Energy Density of the Electric Light Field

The intensity I of this light wave follows from the size of the electric field E_0 by:

$$\textbf{intensity} \quad I = \frac{1}{2} c_0 \varepsilon_0 n |E_0|^2 = |\overline{E \times H}| \tag{2.45}$$

which has the important consequence of a quadratic increase of intensities if light beams are superimposed, e.g. in interference experiments.

The energy density of the light field can be calculated from:

$$\textbf{energy density} \quad \rho_E = \frac{1}{2} \varepsilon_0 n |E_0|^2 \tag{2.46}$$

and the light power P is related by:

$$\textbf{power} \quad P = \frac{1}{2} c_0 \varepsilon_0 n \int_A |E_0|^2 \mathrm{d}A. \tag{2.47}$$

As an example a laser beam may have an intensity of:

$$I = 1\,\mathrm{MW\,cm^{-2}} : \Rightarrow E = 30\,\mathrm{kV\,cm^{-1}}$$
$$H = 70\,A\,V^{-1}$$

The sun light has an intensity of several $100\,\mathrm{W\,m^{-2}}$ in Europe.

2.2.4 Uncertainty of Field Strengths

As a quantum mechanical consequence the strength of the electric and the magnetic field cannot be determined exactly at the same point. The uncertainties ΔE_x and ΔH_y for the related electric and magnetic field components measured at a distance L are:

$$\Delta E_x \Delta H_y \geq \frac{hc^2}{4\pi L^4}. \tag{2.48}$$

This formula is in agreement with the uncertainties of position and momentum or energy and time described above.

2.3 Geometrical Optics

Geometrical optics or ray optics is useful for analyzing complex optical imaging in a first overview. It neglects all diffraction phenomena and thus in most photonic applications it is not sufficient and an analysis using at least Gaussian beams is necessary.

2.3.1 Preconditions: Fresnel Number

The main assumption is neglect of diffraction. Thus it has to be proven that the Fresnel number F which is defined as:

$$\textbf{Fresnel number} \quad F = \frac{D^2_{\text{aperture}}}{\lambda L} \tag{2.49}$$

is large compared to 1:

$$\textbf{geometrical optics} \quad F \gg 1 \tag{2.50}$$

where D_{aperture} is the diameter of the last aperture and L the distance between the point to describe and the aperture. With this relation it is considered that both the diaphragm diameter is large compared to wavelength and the observation distance is small to suppress diffraction effects. The influence of intermediate Fresnel numbers will be discussed in Sect. 3.10.3 in detail.

Furthermore it has to be proven that the light beams are paraxial and the medium is isotropic and only slightly inhomogeneous. Transitions into matter with different refractive index will be included by simple refractive laws as described in more detail in Chap. 3.

2.3.2 Theoretical Description

From (2.28) it follows, with the assumptions of Sect. 2.2.1 except that now the material can be slightly inhomogeneous and using a monochromatic wave with wave vector $\textbf{k}_0$, that

$$-\Delta \textbf{E} + n(\textbf{r})^2 k_0^2 \textbf{E} = \text{grad div } \textbf{E} \tag{2.51}$$

with refractive index n. div E can be different from zero and thus this equation can be solved by the complex ansatz:

$$\textbf{E} = \textbf{E}_0 \exp\{ik_0 L_{\text{eikonal}}(\textbf{r})\} \tag{2.52}$$

with the eikonal L_{eikonal}. The eikonal fulfills the condition:

$$(\text{grad } L_{\text{eikonal}})^2 = n^2 \tag{2.53}$$

characterizing the shape and propagation of the wave fronts of the electric (and magnetic) field by:

$$k_0 L_{\text{eikonal}} - 2\pi\nu t = \text{const.} \tag{2.54}$$

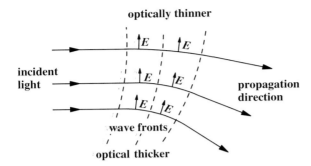

Fig. 2.4. Light propagation, electric field vector and eikonal in slightly inhomogeneous matter

From (2.53) and (2.54) the propagation direction of the light wave can be determined to be:

$$\boldsymbol{e} = \frac{\boldsymbol{k}_0}{k_0} = \frac{1}{n} \operatorname{grad} L_{\text{eikonal}} \tag{2.55}$$

and the phase of the wave will propagate with speed:

$$c_{\text{ph}} = \frac{dL_{\text{eikonal}}}{dt} = \frac{2\pi\nu}{nk_0} = \frac{c}{n}. \tag{2.56}$$

The optical path length L_{path} between the start point a and the final point b along the geometrical path s follows from:

$$\text{optical path length} \quad L_{\text{path}} = \int_a^b dL_{\text{eikonal}} = \int_a^b n \, ds \tag{2.57}$$

With Fermats principle of fastest optical paths the propagation of a beam described by its local wave vector $\boldsymbol{k}(\boldsymbol{r})$ can be determined in this approximation from:

$$\frac{d}{ds}\left[n\frac{dr(s)}{ds}\right] = \operatorname{grad} n(r) \tag{2.58}$$

for a homogeneous medium to give:

$$\boldsymbol{r} = \boldsymbol{r}_0 + \boldsymbol{e}s \tag{2.59}$$

which is the description of a straight line. Thus the parts of a light wave front propagate in a homogeneous material in straight lines, and in slightly inhomogeneous material as given by (2.58).

Thus in geometrical optics light beams are obtained as mathematical straight lines propagating in homogeneous matter. These lines are not completly coincident with light beams produced by lasers. Geometrical optics may be used for analyzing the imaging of light from large incoherent sources. In this sense super-radiation of e.g. nitrogen or excimer lasers may be handled in this rough approximation, too.

2.3.3 Ray Characteristics

Geometrical optics is useful for paraxial rays. The optical axis is commonly defined as the z direction (see Fig. 2.5). Often the analysis of rays which are in the same plane as the z axis is sufficient. The distance of such a ray from this axis is w and its slope is w'.

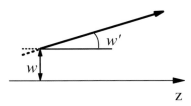

Fig. 2.5. Beam characteristics of optical rays in geometrical optics

Both parameters are functions of z and can be combined into the ray vector:

$$\textbf{ray vector} \quad \begin{pmatrix} w(z) \\ w'(z) \end{pmatrix} \tag{2.60}$$

which is useful for applying ray matrices for the calculation of ray propagation.

In the case of rays which are not in a plane with the z axis, a ray vector with four parameters is necessary (see Fig. 2.6).

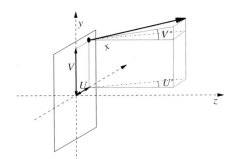

Fig. 2.6. Definition of parameters for rays not in planes with the z axis

These four parameters, two distances u, v and two slopes u' and v', are analogous to the two-parameter case collected in a ray vector:

$$\textbf{ray vector (off plane)} \quad \begin{pmatrix} u(z) \\ v(z) \\ u'(z) \\ v'(z) \end{pmatrix} . \tag{2.61}$$

This ray vector can be calculated with ray matrices, too, but these 4×4 matrices have a maximum of ten independent elements. With these matrices tilted and rotated optical elements can be considered.

2.3.4 Ray Propagation with Ray Matrices

Imaging and illumination with incoherent light can be calculated to a good approximation by determining the beam propagation of the optical rays. This can be done with ray tracing or for paraxial rays with ray matrices. In ray tracing for a large number of geometrical optical rays the propagation is calculated and then superimposed for determining the intensity distributions.

In particular, if many optical elements are in the path the method of ray matrices is very handy. In this formalism the optical path including all optical elements is described with a ray matrix M_{total} (see Fig. 2.7).

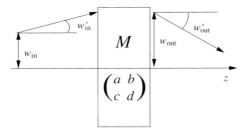

Fig. 2.7. Optical ray passes optical elements with the total matrix M_{total}

Details of the ray matrices will be given below in Sect. 2.5. The ray vector behind a system of optical elements, including the optical paths in vacuum (or air), can be calculated from the incident ray vector and the total ray matrix M_{total} as a simple multiplication:

$$(w_{\text{out}} \; w'_{\text{out}}) = M_{\text{total}} \cdot \begin{pmatrix} w_{\text{in}} \\ w'_{\text{in}} \end{pmatrix} \tag{2.62}$$

or

$$(u_{\text{out}} \; v_{\text{out}} \; u'_{\text{out}} \; v'_{\text{out}}) = M_{\text{total}} \cdot \begin{pmatrix} u_{\text{in}} \\ v_{\text{in}} \\ u'_{\text{in}} \\ v'_{\text{in}} \end{pmatrix} \tag{2.63}$$

respectively.

For calculating the image position and size, one or two rays can be used as in the common image construction. The single ray has to start at the z axis at the bottom of the object with a slope $w'_{\text{in},1} \neq 0$. The position where it crosses the z axis again is the image position z_{image} and the magnification follows from $w'_{\text{in}}/w'_{\text{image}}$ for this ray.

Using two beams, one can be parallel to the optical axis $w'_{in,1} = 0$ and the second should have a slope, e.g. $w'_{in,2} = -w_{object}/a$ with a as the distance from the first lens. The lateral distances $w_{in,i}$ are both set equal to w_{object}. Then both output vectors are calculated as functions of z. From $w_{out,1}(z_{image}) = w_{out,2}(z_{image})$ follows z_{image} and with the known position of the image the size w_{image} can be calculated by $w_{out,1 \text{ or } 2}(z_{image})$.

Further, it can be shown that the image occurs for such distances for which the total ray matrix has the element $b_{total} = 0$. From this condition the distance z_{image} and then the image size w_{image} can easily be calculated, too.

For $a_{total} = 0$ parallel incident light will be focused and for $d_{total} = 0$ an incident point source will result in parallel light after the optical elements represented by the total matrix M_{total}.

2.4 Gaussian Beams

Gaussian beams are three-dimensional solutions of the wave equation derived from Maxwell's equations in free space, or under the same conditions as given in Sect. 2.2.1 as planar waves were calculated [2.3–2.6]. They are diffraction limited as will be described in Sect. 3.9.3 and thus they show the highest possible beam quality. They incorporate the photon position–momentum uncertainty limit. Gaussian beams can be produced by apertures or lasers. They are solutions of transversal laser mode equations. In photonics Gaussian beams are the "work horse" of calculating and applying beams.

2.4.1 Preconditions

Similar to geometrical (or ray) optics the light beams should be paraxial with sufficiently low divergence. Besides the theoretical limits for deriving the models described below, practical limits from lens errors may be even more restrictive.

The medium has to be isotropic and only slightly inhomogeneous. The possibly slight variation of refractive index $n(r)$ will not be recognized explicitly in the formulas. Again, transitions into matter with different refractive index will be included by simple refractive laws, as will be described in more detail in Chap. 3. They are considered in the beam propagation using the ray-matrix formalism as described in Sect. 2.4.4.

2.4.2 Definition and Theoretical Description

Gaussian beams are characterized by the Gaussian shape of the transversal profile of the beam. The electric field is given in the transversal x or y directions which are replaced by r and the propagation direction z as:

$$|\boldsymbol{E}(z,r)| = \mathrm{Re}\{E_{\mathrm{A}}(z,r)\}\mathrm{Re}\left\{e^{i(2\pi\nu t - k_0 z)}\right\} \tag{2.64}$$

with the amplitude $E_{\mathrm{A}}(z,r)$ as:

$$E_{\mathrm{A}}(z,r) = \frac{|\boldsymbol{E}_0|}{1 - i\dfrac{z\lambda}{w_0^2 n\pi}}\, e^{-\dfrac{r^2/w_0^2}{1 - iz\lambda/w_0^2 n\pi}} \tag{2.65}$$

with a maximum $|\boldsymbol{E}_0|$ at $z = 0$ and the important consequence:

> A Gaussian beam is completely determined by the position and size of the waist for a given wavelength and refractive index.

In these formulas it was assumed that the waist position z_0 is at $z = 0$. In other cases z has to be replaced by $z \to z - z_0$. The direction of the electric field vector can point in any direction in the xy plane. It does not have any component in the z direction.

The negative quadratic exponent produces the typical bell shape of the electric field and intensity distribution in the xy plane (see Fig. 2.8):

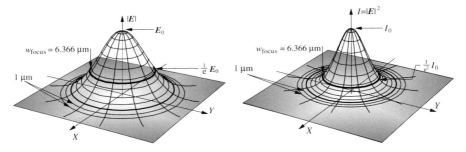

Fig. 2.8. Electric field (left) and intensity (right) distribution of a Gaussian beam transverse to the propagation direction z (The graph represents the focus of a parallel incident beam with a 5-mm diameter and 1.000-nm wavelength behind a lens with a 5-cm focal length)

2.4.3 Beam Characteristics and Parameter

Although the beam radius $w(z)$ or diameter $2w(z)$ of Gaussian beams are completely determined by the position z_{w0} and the size of the waist w_0 plus wavelength and material refractive index, some further parameters are helpful for practical purposes.

Rayleigh Length z_R

The Rayleigh length z_R of a Gaussian beam is defined by:

Rayleigh length $$z_{\mathrm{R}} = \frac{n\pi}{\lambda_0} w_0^2 = \frac{|\boldsymbol{k}|}{2} w_0^2 \qquad (2.66)$$

with wave vector $\boldsymbol{k}$ and refractive index of the material n. It can be used for the simplification of (2.65) to the following form:

$$E_{\mathrm{A}}(z,r) = \frac{|\boldsymbol{E}_0|}{1 - \mathrm{i}\dfrac{z}{z_{\mathrm{R}}}} \, \mathrm{e}^{-\frac{r^2/r_0^2}{1 - \mathrm{i}(z/z_{\mathrm{R}})}} \qquad (2.67)$$

and for the deviation of following formulas. At the Rayleigh length the beam radius (and diameter) are increased by a factor of $\sqrt{2}$:

$$w(z_{\mathrm{R}}) = \sqrt{2}w_0 \qquad (2.68)$$

compared to the waist value and thus the intensity is reduced by a factor of 2 at z_{R}. Some examples for z_{R} are given in Table 2.5.

Beam Radius $w(z)$

The beam radius $w(z)$ is defined as the radius where the electric field amplitude is decreased to its $1/e$ value which is identical with the $1/e^2$ value for the intensity of the beam. Here 86.5% of the whole power of the Gaussian beam is contained inside the area $A(z)$ with diameter $2w(z)$.

The dependency of $E_{\mathrm{A}}(z)$ leads to the following function of the beam radius $w(z)$:

beam radius $$w(z) = w_0 \sqrt{1 + \left(\frac{z\lambda}{w_0^2 n\pi}\right)^2} \qquad (2.69)$$

which can be written by using the Rayleigh length as:

$$w(z) = \sqrt{\frac{\lambda}{n\pi}\left(z_{\mathrm{R}} + \frac{z^2}{z_{\mathrm{R}}}\right)}. \qquad (2.70)$$

This is a hyperbolic function around $z = 0$ with the minimum w_0 at this position and an almost linear increase of the beam radius at large distances (see Fig. 2.9).

This linear increase at large z can be described by the divergence angle θ which will be discussed below. The Gaussian beam size for different beam waist diameters and different wavelengths is given in Table 2.4 and 2.5.

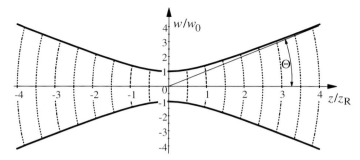

Fig. 2.9. Radius $w(z)$ relative to the radius w_0 at $z = 0$ of the electric field amplitude (or intensity) of a Gaussian beam with a beam waist at $z_w = 0$ as a function of z/z_R

Wave Front Radius $R(z)$

From (2.65) the shape of the phase fronts of the Gaussian beam can also be derived. They have a spherical shape and their radius $R(z)$ is given by:

$$\textbf{wave front radius} \quad R(z) = z + \frac{1}{z}\left(\frac{w_0^2 n\pi}{\lambda}\right)^2 \qquad (2.71)$$

or again by using the Rayleigh length z_R:

$$R(z) = z_R\left[\frac{z}{z_R} + \frac{z_R}{z}\right]. \qquad (2.72)$$

This curvature reaches infinity at two positions: we observe exactly planar wave fronts at the waist position $z = 0$ and for very large distances z to a good approximation. For example, using an aperture of 3 mm diameter at 1 m distance from the waist for a beam with a Rayleigh length of 2 mm would lead to a relative bend error of the wave front of $1.2 \cdot 10^{-6}$. Planar waves may be needed in photonic applications, e.g. for the production of holographic gratings.

The curvature radius is minimal at the Rayleigh length z_R position and has a value of:

$$R_{min} = R(z_R) = 2z_R \qquad (2.73)$$

Some further examples for the beam radius and the wave front curvature of Gaussian beams as a function of the distance from the waist are given in Table 2.4.

The curvature radii of the wave fronts of the Gaussian beams determine the curvature of the interaction zones especially in experiments with interfering beams. Values for Gaussian beams with different wavelengths are given in Table 2.5.

The curvature of the wave front can be determined even for complicated shapes using Shack-Hartmann wavefront sensors or similar measurement schemes as described, for example, in [2.7–2.12].

Table 2.4. Beam radius $w(z)$, beam curvature radius $R(z)$ and local divergence $\theta_{\mathrm{loc}}(z)$ of a Gaussian beam for different distances z from waist at $z = 0$ measured in Rayleigh lengths z_{R}

z/z_{R}	w/w_0	R/z_{R}	$\theta_{\mathrm{loc}}/(w_0/z_{\mathrm{R}})$
0	1	∞	0
0.25	1.045	3.3	0.316
0.5	1.118	2.5	0.447
0.75	1.202	2.16	0.555
1	$\sqrt{2}$	2	$1/\sqrt{2}$
2	$\sqrt{5}$	2.5	0.894
5	5.099	5.2	0.980
10	10.050	10.1	0.995
100	100.005	100.01	0.99995
∞	∞	∞	1

Divergence Angle θ

For large distances l from the waist, positioned at $z = 0$ in our case, compared to the Rayleigh length the Gaussian beam is expanding linearly. This expansion can be determined by the divergence angle θ:

$$\textbf{divergence angle} \quad \theta = \frac{\lambda_0}{n\pi w_0} = \frac{w_0}{z_{\mathrm{R}}} \tag{2.74}$$

From this equation the product $w_0\theta$ can be calculated from the wavelength λ_0 and the refractive index n of the material. This product is called the beam parameter product and describes the beam quality (see Sect. 2.7.5 for more details). It is minimal for Gaussian beams in comparison to all other beams which show a larger divergence for the same waist radius. Thus the quality of a beam has to be measured by both the diameter and the divergence.

Some values of the divergence of Gaussian beams in air with different wavelengths and different beam waists are given in Table 2.5.

The local divergence θ_{loc} of the Gaussian beam changes during propagation along z. It is zero at the waist and maximum at the far-field:

$$\theta_{\mathrm{loc}}(z) = \frac{\mathrm{d}w(z)}{\mathrm{d}z} = \frac{(\lambda/\pi n)^2 z}{w_0\sqrt{(\lambda/\pi n)^2 z^2 + w_0^4}}. \tag{2.75}$$

Complex Beam Parameter $q(z)$

By defining a complex beam parameter $q(z)$ a very elegant method for calculating the propagation of Gaussian beams was established [see e.g. M38]:

$$\textbf{complex beam parameter} \quad \frac{1}{q(z)} = \frac{1}{R(z)} - \frac{i\lambda}{\pi n w(z)^2} \tag{2.76}$$

where the wave front curvature radius $R(z)$ and the beam radius $w(z)$ are combined in a complex vector analogous to the ray vector in Sect. 2.3.3.

Table 2.5. Rayleigh length z_R, divergence θ, beam diameter w $(z = 0.1\,\text{m})$ and curvature radius R $(z = 0.1\,\text{m})$ for Gaussian beams with different wavelength λ and waist radius w_0

λ	w_0 (μm)	z_R (mm)	θ (mrad)	w $(z = 0.1\,\text{m})$ (mm)	R $(z = 0.1\,\text{m})$ (m)
200 nm	10	1.57	6.37	0.636	0.100
	100	157	0.637	0.119	0.347
	1000	15 707	0.0637	1.000	2 468
500 nm	10	0.628	15.9	1.592	0.100
	100	62.8	1.59	0.188	0.140
	1000	6 283	0.159	1.000	395
1000 nm	10	0.314	31.8	3.184	0.100
	100	31.4	3.18	0.333	0.101
	1000	3 142	0.318	1.001	98.80
3 μm	10	0.10	95.5	95.493	0.100
	100	10.5	9.55	0.960	0.101
	1000	1 047	0.955	1.005	11.07
10 μm	10	0.031	318	31.831	0.100
	100	3.14	31.8	3.185	0.100
	1000	314	3.18	1.087	1.087

2.4.4 Beam Propagation with Ray-Matrices

With this complex beam parameter $q(z)$ the propagation of Gaussian beams can be calculated using the same matrices as those in Sect. 2.3.4 and described in more detail in the next chapter [2.13–2.17]. The propagation formalism is based on the final matrix for the whole propagation range including all optical elements and all paths between them. The four matrix elements will again be called a, b, c, d and the incident beam will be indexed by i and the outgoing beam by o as shown in Fig. 2.10.

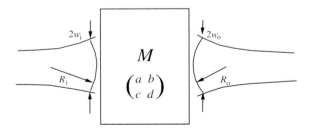

Fig. 2.10. Definition of incoming (left side) and outgoing (right side) Gaussian beam parameters together with the total matrix M with elements a, b, c, d

Based on this definitions the beam parameter of the out coming beam behind the optical system follows from:

beam propagation $\qquad q_o(w_o, r_o) = \dfrac{q_i(w_i, R_i) \cdot a + b}{q_i(w_i, R_i) \cdot c + d}$ $\qquad$ (2.77)

From this complex beam parameter the physically relevant real values of the beam radius w_o and wave front radius R_o can be calculated by:

$$\frac{1}{w_0^2} = -\frac{\pi n}{\lambda} \, \mathrm{Im} \left\{ \frac{1}{q_o} \right\} \tag{2.78}$$

$$\frac{1}{R_o} = \mathrm{Re} \left\{ \frac{1}{q_o} \right\}. \tag{2.79}$$

Using computer programs for analytical calculations these solutions can be derived easily. Thus the complicated calculation of propagation of the diffraction-limited Gaussian beams through a large system of optical elements becomes very easy. Non-Gaussian beams can be propagated in a similar way as will be described in Sect. 6.6.9.

2.4.5 Determination of w_0 and z_0

The beam parameters w_i and R_i are not always known for an existing Gaussian beam, as e.g. for a laser beam. While the beam radius w_i can be determined easily (see Sect. 2.7.3) the curvature radius is usually not directly available. It changes for Gaussian beam with the local radius $w(z)$ as a function of z by:

$$R(z) = \frac{1}{2} \frac{w^2 n^2 \pi^2}{(z - z_0)\lambda^2} \left\{ w^2 + \sqrt{\left| w^4 - 4\frac{\lambda^2}{n^2 \pi^2}(z - z_0)^2 \right|} \right\} \tag{2.80}$$

with the waist position at z_0 and wavelength λ. This formula could be used for the modeling of measured propagation, but a simpler way results from several measurements of $w_i(z_i)$. These can be fitted numerically as $w = f(z)$ using:

$$w(z) = w_0 \sqrt{1 + \left(\frac{(z - z_0)\lambda}{w_0^2 n \pi} \right)^2} \tag{2.81}$$

and thus w_0 and z_0 can be determined. With these values $R_i(z_i)$ can be calculated from:

$$R_i(z_i) = (z_i - z_0) + \frac{1}{(z_i + z_0)} \left(\frac{w_0^2 n \pi}{\lambda} \right)^2. \tag{2.82}$$

Using these values of w_i and R_i as input the further propagation of the Gaussian beam can be calculated as described.

2.4.6 How to Use the Formalism

This calculation can even be simplified to straightforward computation by using the following two formulas. With the substitution:

$$\kappa = \frac{\lambda}{\pi n} \tag{2.83}$$

the general solution of the propagation of a Gaussian beam through an optical system with the elements a, b, c, d of the total matrix is given by:

$$w_{\text{out}} = \sqrt{\frac{a^2 R_{\text{in}}^2 w_{\text{in}}^4 + 2ab R_{\text{in}} w_{\text{in}}^4 + b^2(\kappa^2 R_{\text{in}}^2 + w_{\text{in}}^4)}{R_{\text{in}}^2 w_{\text{in}}^2 (ad - bc)}} \tag{2.84}$$

and

$$R_{\text{out}} = \frac{a^2 R_{\text{in}}^2 w_{\text{in}}^4 + 2ab R_{\text{in}} w_{\text{in}}^4 + b^2(\kappa^2 R_{\text{in}}^2 + w_{\text{in}}^4)}{a R_{\text{in}} w_{\text{in}}^4 (c R_{\text{in}} + d) + b(c R_{\text{in}} w_{\text{in}}^4 + d(\kappa^2 R_{\text{in}}^2 + w_{\text{in}}^4))} \tag{2.85}$$

Using these equations the beam size and the wave font curvature can be calculated directly without solving the complex beam parameter equations. Only the total matrix has to be calculated. This can be done with a spreadsheet computer program. Thus the beam propagation can be drawn as $w = f(z)$ as will be shown in the examples in Sect. 2.5.4 and the foci and divergence can be obtained from these graphs.

In the case when the Gaussian beam has a different diameter and divergence in two orthogonal directions transversal to the propagation direction, these two parameter sets can be calculated separately. Thus the propagation in the xz-plane and in the yz-plane can be computed by a separate set of beam parameters and matrices. The general case of astigmatic beams cannot be solved analytically but special cases can be described (see comments in Sect. 2.7.4).

2.5 Ray Matrices

The following ray matrices can be used for theoretical propagation of simple rays in the sense of geometrical optics or for the propagation of diffraction-limited Gaussian beams described by their complex beam parameter as described in Sects. 2.3.4 and 2.4.4 (see also [M38, M42, 2.13–2.17]).

2.5.1 Deriving Ray Matrices

Ray matrices can be derived by calculating the ray or beam parameters behind the optical element using Maxwell's equations or derived formulas and comparing the coefficients of these equations with the matrix elements. As the simplest example, free space propagation may serve.

In the simplest case of ray propagation in free space, as shown in Fig. 2.11, the ray equations would be:

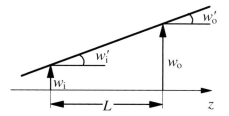

Fig. 2.11. Ray propagation over length L in free space

$$w_o = w_i + w'_i \cdot L$$
$$w'_o = w'_i \tag{2.86}$$

and matrix multiplication:

$$\begin{pmatrix} w_o \\ w'_o \end{pmatrix} = \begin{pmatrix} a & b \\ c & d \end{pmatrix} \cdot \begin{pmatrix} w_i \\ w'_i \end{pmatrix} \tag{2.87}$$

will lead to:

$$\begin{array}{l} w_o = aw_i + bw' \\ w'_o = cw_i + dw' \end{array} \Rightarrow \begin{bmatrix} a = 1 & b = L \\ c = 0 & d = 1 \end{bmatrix}. \tag{2.88}$$

Thus the matrix for any optical element can be developed as long as the light path through these elements is reversible. Beam-cutting apertures cannot be described by matrices with real elements.

The main advantage of using matrices for calculating the light propagation is the easy recognition of many different optical elements. Therefore this formalism can e.g. be used for the calculation of the transversal fundamental mode shape in laser resonators (see Sect. 6.6).

2.5.2 Ray Matrices of Some Optical Elements

Matrices of frequently used optical elements are given in Table 2.6. These matrices can be combined for complex optical elements such as e.g. thick lenses, multiple lens setups or laser resonators.

Table 2.6. Matrices of frequently used optical elements

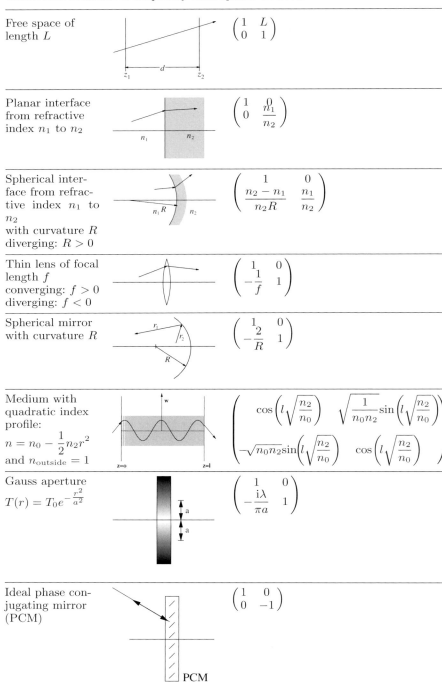

Free space of length L		$\begin{pmatrix} 1 & L \\ 0 & 1 \end{pmatrix}$
Planar interface from refractive index n_1 to n_2		$\begin{pmatrix} 1 & 0 \\ 0 & \dfrac{n_1}{n_2} \end{pmatrix}$
Spherical interface from refractive index n_1 to n_2 with curvature R diverging: $R > 0$		$\begin{pmatrix} 1 & 0 \\ \dfrac{n_2 - n_1}{n_2 R} & \dfrac{n_1}{n_2} \end{pmatrix}$
Thin lens of focal length f converging: $f > 0$ diverging: $f < 0$		$\begin{pmatrix} 1 & 0 \\ -\dfrac{1}{f} & 1 \end{pmatrix}$
Spherical mirror with curvature R		$\begin{pmatrix} 1 & 0 \\ -\dfrac{2}{R} & 1 \end{pmatrix}$
Medium with quadratic index profile: $n = n_0 - \dfrac{1}{2} n_2 r^2$ and $n_{\text{outside}} = 1$		$\begin{pmatrix} \cos\left(l\sqrt{\dfrac{n_2}{n_0}}\right) & \sqrt{\dfrac{1}{n_0 n_2}}\sin\left(l\sqrt{\dfrac{n_2}{n_0}}\right) \\ -\sqrt{n_0 n_2}\sin\left(l\sqrt{\dfrac{n_2}{n_0}}\right) & \cos\left(l\sqrt{\dfrac{n_2}{n_0}}\right) \end{pmatrix}$
Gauss aperture $T(r) = T_0 e^{-\frac{r^2}{a^2}}$		$\begin{pmatrix} 1 & 0 \\ -\dfrac{\mathrm{i}\lambda}{\pi a} & 1 \end{pmatrix}$
Ideal phase conjugating mirror (PCM)		$\begin{pmatrix} 1 & 0 \\ 0 & -1 \end{pmatrix}$

The most useful 4×4 ray matrices are:

- free space length L:

$$\begin{pmatrix} 1 & 0 & L & 0 \\ 0 & 1 & 0 & L \\ 0 & 0 & 1 & 0 \\ 0 & 0 & 0 & 1 \end{pmatrix} \tag{2.89}$$

- spherical interface from refractive index n_1 to n_2 with curvature R ($R > 0$: diverging):

$$\begin{pmatrix} 1 & 0 & 0 & 0 \\ 0 & 1 & 0 & 0 \\ \dfrac{n_1 - n_2}{n_2 R} & 0 & \dfrac{n_1}{n_2} & 0 \\ 0 & \dfrac{n_1 - n_2}{n_2 R} & 0 & \dfrac{n_1}{n_2} \end{pmatrix} \tag{2.90}$$

- lens with focal length f ($f > 0$: converging):

$$\begin{pmatrix} 1 & 0 & 0 & 0 \\ 0 & 1 & 0 & 0 \\ \dfrac{1}{f} & 0 & 1 & 0 \\ 0 & -\dfrac{1}{f} & 0 & 1 \end{pmatrix} \tag{2.91}$$

- cylindrical lens with focal length f_x ($f_x > 0$: converging) in the x direction (see Fig. 2.6):

$$\begin{pmatrix} 1 & 0 & 0 & 0 \\ 0 & 1 & 0 & 0 \\ -\dfrac{1}{f_x} & 0 & 1 & 0 \\ 0 & 0 & 0 & 1 \end{pmatrix} \tag{2.92}$$

- cylindrical lens with focal length f_y ($f_y > 0$: converging) in the y direction (see Fig. 2.6):

$$\begin{pmatrix} 1 & 0 & 0 & 0 \\ 0 & 1 & 0 & 0 \\ 0 & 0 & 1 & 0 \\ 0 & -\dfrac{1}{f_y} & 0 & 1 \end{pmatrix} \tag{2.93}$$

- optical element with matrix M rotated in the xy plane clockwise towards z by angle ϕ:

$$\begin{pmatrix} \sin\phi & -\sin\phi & 0 & 0 \\ \sin\phi & \cos\phi & 0 & 0 \\ 0 & 0 & \cos\phi & -\sin\phi \\ 0 & 0 & \sin\phi & \cos\phi \end{pmatrix} \cdot M \cdot \begin{pmatrix} \cos\phi & \sin\phi & 0 & 0 \\ -\sin\phi & \cos\phi & 0 & 0 \\ 0 & 0 & \cos\phi & \sin\phi \\ 0 & 0 & -\sin\phi & \cos\phi \end{pmatrix} \tag{2.94}$$

The small *tilt of lenses* can be considered by using an effective focal length. The tilt angle may be θ as defined in Fig. 2.12.

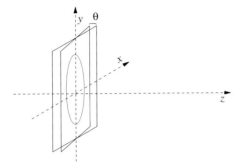

Fig. 2.12. Tilted thin lenses can be calculated with effective focal lengths

Tilting of a thin lens with focal length f around the y axis as in the figure decreases the effective focal length f_x for rays and beam dimension in the xz plane (sagittal) and increases the effective focal length f_y for rays and beam dimension in yz plane (tangential) as:

$$f_x = f \cos \theta \quad \text{effective focal length } f_x \tag{2.95}$$

and

$$f_y = \frac{c}{\cos \theta} \quad \text{effective focal length } f_y \tag{2.96}$$

With these effective focal lengths the elements of the matrices of thin lenses can be modified considering the otherwise complicated calculation. The matrices are then given as:

- tilted spherical interface (analogous to Fig. 2.12) from refractive index n_1 to n_2 with curvature R ($R > 0$: diverging):

$$\begin{pmatrix} 1 & 0 & 0 & 0 \\ 0 & \dfrac{\cos \xi}{\cos \theta} & 0 & 0 \\ \dfrac{n_1 \cos \theta - n_2 \cos \xi}{n_2 R} & 0 & \dfrac{n_1}{n_2} & 0 \\ 0 & \dfrac{n_1 \xi - n_2 \cos \theta}{n_2 R \cos \xi \cos \theta} & 0 & \dfrac{n_1 \cos \theta}{n_2 \cos \xi} \end{pmatrix} \tag{2.97}$$

with

$$\xi = \arcsin \left(\frac{n_1}{n_2} \sin \theta \right) \tag{2.98}$$

- tilted thin lens (see Fig. 2.12) with focal length f ($f > 0$: converging):

$$\begin{pmatrix} 1 & 0 & 0 & 0 \\ 0 & 1 & 0 & 0 \\ -\dfrac{1}{f\cos\theta} & 0 & 1 & 0 \\ 0 & -\dfrac{\cos\theta}{f} & 0 & 1 \end{pmatrix} \tag{2.99}$$

2.5.3 Light Passing Through Many Optical Elements

If light passes n optical elements as, e.g. a sequence of lenses, all lenses and the distances between them have to be recognized by one matrix M_i each (see Fig. 2.13).

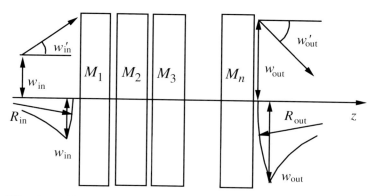

Fig. 2.13. Light passing through a sequence of optical elements described by their matrices M_i

The total matrix is then simply given by the product of all these matrices in the right order:

$$M_{\text{total}} = M_n \cdot M_{n-1} \cdot \ \cdots \ \cdot M_2 \cdot M_1 \tag{2.100}$$

It should be noted that the passed optical element first, with the matrix M_1, is the last one to be multiplied as given in this formula.

Reflecting light with planar or spherical mirrors will change the direction of propagation. Thus the direction of the z axis has to be flipped for correct use of the signs of all further components.

2.5.4 Examples

This handy method for calculating beam propagation through optical systems will be illustrated with a few examples.

Focusing with a Lens in Ray Optics

In geometrical optics all incident rays parallel to the optical axis will be focused perfectly by a lens to the focal point at the optical axis (Fig. 2.14).

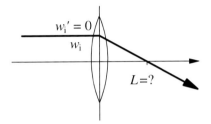

Fig. 2.14. Focusing parallel rays with a lens

Using the ray matrices for the lens with the focal length f and the path L the resulting beam is given by:

$$\begin{pmatrix} w_o \\ w_o' \end{pmatrix} = \begin{pmatrix} 1 & L \\ 0 & 1 \end{pmatrix} \cdot \begin{pmatrix} 1 & 0 \\ -1/f & 1 \end{pmatrix} \cdot \begin{pmatrix} w_i \\ w_i' \end{pmatrix}$$

$$= \begin{pmatrix} 1 - \dfrac{L}{f} & L \\ -\dfrac{1}{f} & 1 \end{pmatrix} \cdot \begin{pmatrix} w_i \\ w_i' \end{pmatrix} \tag{2.101}$$

with $w_i' = 0$

$$w_o = 1 - \frac{L}{f} \quad \text{and} \quad w_o' = -\frac{1}{f}, \tag{2.102}$$

$w_o = 0$ leads to $L = f$ as expected.

It should be noted that with more explicit ray trace calculations based on geometrical optics the focusing of a lens can also be calculated for nonparaxial rays. An example is given in Fig. 2.15.

From this example it can be seen that plano-convex lenses should be used as shown at the left side of the figure with the curved side to the planar wave front for smaller focusing errors.

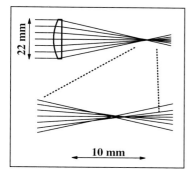 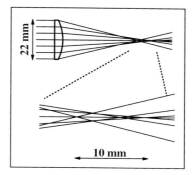

Fig. 2.15. Focusing a beam calculated with a ray tracing computer program based on geometrical optics showing the quality of focusing for nonparaxial beams and using a plano-convex lens in two possible ways. The diameter of the beam was 22.5 mm and the focal length of the lens 50 mm. The left arrangement should be applied for better focusing

Focusing a Gaussian Beam with a Lens

If focusing is calculated for a diffraction-limited (Gaussian) beam, the beam parameter behind the lens has to be calculated from:

$$q_{\text{out}} = \frac{(1 - L/f)q_{\text{lens}} + L}{(-1/f)q_{\text{lens}} + 1} \tag{2.103}$$

with the definition for q_i from (2.76). The solution of this equation can be simplified by using a planar wave front for the incident beam at the lens with $1/R_{\text{lens}} = 0$. The beam will then show its waist at a distance of the focal length. The q parameter behind the lens is, with this assumption, given by:

$$\frac{1}{q_0} = \frac{\kappa^2 f^2 L - w_{\text{lens}}^4 (f - L)}{\kappa^2 f^2 L^2 + w_{\text{lens}}^4 (f - L)^2} - i \frac{\kappa f^2 w_{\text{lens}}^2}{\kappa^2 f^2 L^2 + w_{\text{lens}}^4 (f - L)^2} \tag{2.104}$$

with $\kappa = \lambda/\pi n$ as defined in (2.83). The beam radius at the waist w_{waist} follows to:

$$w_{\text{waist}} = \frac{f\lambda}{w_{\text{lens}}\pi n} \tag{2.105}$$

which shows a reciprocal dependency of the waist diameter on the size of the incident beam in agreement with the above-mentioned properties of Gaussian beams.

This solution can be derived more easily from (2.84) by applying the matrix elements and the assumptions of $1/R_{\text{lens}} = 0$ and $L = f$.

For additional effects in focusing beams, especially of very short pulses, see [2.18–2.20]. In this case the dispersion of the lens may cause additional wave front distortions. The lens diameter can be smaller than the beam diameter and thus diffraction may occur. The resulting effects are described in Sect. 3.10.2 and the references therein.

Imaging with Two Lenses

For imaging an object Gaussian beam with two lenses as in a telescope, five matrices are necessary as shown in Fig. 2.16.

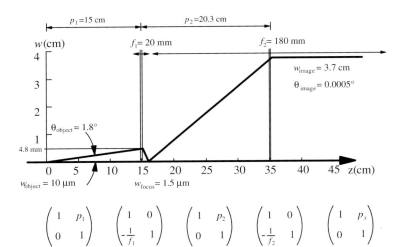

$$
\begin{pmatrix} 1 & p_1 \\ 0 & 1 \end{pmatrix}
\begin{pmatrix} 1 & 0 \\ -\frac{1}{f_1} & 1 \end{pmatrix}
\begin{pmatrix} 1 & p_2 \\ 0 & 1 \end{pmatrix}
\begin{pmatrix} 1 & 0 \\ -\frac{1}{f_2} & 1 \end{pmatrix}
\begin{pmatrix} 1 & p_x \\ 0 & 1 \end{pmatrix}
$$

Fig. 2.16. Imaging an object beam with two lenses – calculation with ray matrices

The resulting total matrix for the beam propagation of this example is given by:

$$
M = \frac{1}{f_1 f_2}\left(
\begin{bmatrix} f_1(f_2 - p_1 - p_2) - \\ -f_2 p_1 + p_1 p_2 \\ f_1 + f_2 - p_2 \end{bmatrix}
\begin{bmatrix} f_1(f_2(p_1 + p_2 + p_x) - p_x(p_1 + p_2)) - \\ -p_1(f_2(p_2 + p_x) - p_2 p_x)] \\ f_1(f_2 - p_x) - f_2(p_2 + p_x) + p_2 p_x \end{bmatrix}
\right)
$$

$$(2.106)$$

with the values as defined in Fig. 2.16. As can be seen from this figure the beam parameter product is constant and the divergence of the out going beam is reduced as the diameter is increased.

Focal Length of Thin Spherical Lenses

Using the matrices of spherical interfaces between air with the refractive index of approximately one and glass with the refractive index n the focal length f of spherical lenses can be calculated as a function of the curvature R of the glass.

For a biconvex lens with curvature radius R at both sides:

biconvex lens $f = \dfrac{1}{2(n-1)}R$ $$(2.107)$$

which leads for BK7 glass with a refractive index of $n = 1.5067$ at a wavelength of $1064\,\text{nm}$ to a relation of $R = (1.0134 \cdot f)$ for this type of lens.

For a plano-convex lens:

plano-convex lens $f = \dfrac{1}{n-1} R$ (2.108)

which results again for BK7 glass, in $R = (0.5067 \cdot f)$.

2.6 Describing Light Polarization

The quantum eigenstates of a single photon are left or right circularly polarized. Nevertheless, it can be prepared with polarizers in mixed states as linearly or elliptically polarized, too. The superposition of many photons can lead to linear, circular, elliptical or nonpolarized light beams. The polarization of the applied light can essentially determine the properties of the nonlinear interaction in nonlinear spectroscopy and photonic devices and the light polarization can be changed by conventional optical elements as, e.g. which beam splitters via Fresnel reflection. Thus the polarization has to be analyzed carefully in nonlinear optics.

The polarization properties can be determined by considering each optical component using Fresnels formula (see Sect. 3.5.1) and all other material influences such as permanent and induced optical anisotropy, birefringence and optical activity.

In more complex cases the use of the following matrix formalism may be helpful. It allows the global calculation of the polarization of the light beam as a function of the polarization of the incident light described by Jones or Stokes matrices and the polarization properties of all optical components described by Jones or Mueller matrices.

The two-element Jones vectors and 2×2 Jones matrices are sufficient for polarized light. With the four-element Stokes vectors and 4×4 Mueller matrices the nonpolarized component of the light can be considered. For more details see [2.21–2.34].

2.6.1 Jones Vectors Characterizing Polarized Light

For the description of linear, circular or elliptical polarized light with Jones vectors, Cartesian coordinates are assumed with z axis pointing in the beam propagation direction (see Fig. 2.17).

The components of the electrical field vector $\boldsymbol{E}(z,t)$ at a certain position and a certain time are $E_x(z,t)$ and $E_y(z,t)$ as shown in the figure. The further temporal and spatial development of these components is a function of the polarization of the light beam.

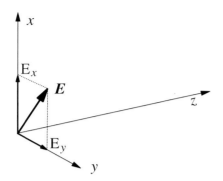

Fig. 2.17. Components of the electric light wave field at a certain moment

In the case of, e.g. linear polarized light, these components can be described by:

$$E_x(z,t) = E_{0,x}\, \mathrm{e}^{\mathrm{i}(2\pi t - kz + \varphi_x)} \tag{2.109}$$

$$E_y(z,t) = E_{0,y}\, \mathrm{e}^{\mathrm{i}(2\pi t - kz + \varphi_y)} \tag{2.110}$$

and by using the total amplitude of the electric field E_0:

$$E_0 = \sqrt{E_{0,x}^2 + E_{0,y}^2} \tag{2.111}$$

the linear polarization of this light beam can be described by the Jones vector $\boldsymbol{J}$ of linear polarized light:

$$\boldsymbol{J} = \frac{1}{E_0} \begin{pmatrix} E_{0,x}\, \mathrm{e}^{\mathrm{i}\varphi_x} \\ E_{0,y}\, \mathrm{e}^{\mathrm{i}\varphi_y} \end{pmatrix} \tag{2.112}$$

or with:

$$\delta = \varphi_y - \varphi_x \tag{2.113}$$

and $\varphi_x = 0$ follows:

$$\boldsymbol{J} = \frac{1}{E_0} \begin{pmatrix} E_{0,x} \\ E_{0,y}\, \mathrm{e}^{\mathrm{i}\delta} \end{pmatrix} \tag{2.114}$$

For some common polarization of light beams the Jones vectors are collected in Table 2.7.

For obtaining the polarization of superimposed light beams these vectors can be added after multiplying by the amplitude of the electric field.

Thus, e.g. the sum of right circular and left circular polarized light of the same intensity results in:

$$\boldsymbol{E}_{0,\mathrm{sum}} = \frac{E_0}{\sqrt{2}} \begin{pmatrix} 1 \\ +\mathrm{i} \end{pmatrix} + \frac{E_0}{\sqrt{2}} \begin{pmatrix} 1 \\ -\mathrm{i} \end{pmatrix} = \frac{2E_0}{\sqrt{2}} \begin{pmatrix} 1 \\ 0 \end{pmatrix} \tag{2.115}$$

which represents linearly polarized light with twice the intensity of each single beam.

Table 2.7. Jones vectors for some common light beam polarizations

Linear polarized: x direction		$\begin{pmatrix} 1 \\ 0 \end{pmatrix}$
Linear polarized: y direction		$\begin{pmatrix} 0 \\ 1 \end{pmatrix}$
Linear polarized: θ direction		$\begin{pmatrix} \cos\theta \\ \sin\theta \end{pmatrix}$
Left circular polarized (viewing into beam)		$\dfrac{1}{\sqrt{2}} \begin{pmatrix} 1 \\ -i \end{pmatrix}$
Right circular polarized (viewing into beam)		$\dfrac{1}{\sqrt{2}} \begin{pmatrix} 1 \\ i \end{pmatrix}$
Left elliptical polarized (axis parallel x and y axis)		$\dfrac{1}{E_0} \begin{pmatrix} E_{0,x} \\ -i\,E_{0,y} \end{pmatrix}$
Right elliptical polarized (axis parallel x and y axis)		$\dfrac{1}{E_0} \begin{pmatrix} E_{0,x} \\ +i\,E_{0,y} \end{pmatrix}$

Table 2.7. Continued

Elliptical polarized with: $E_{0,yr} = E_{0,y} \cos\delta$ $E_{0,yi} = E_{0,y} \sin\delta$	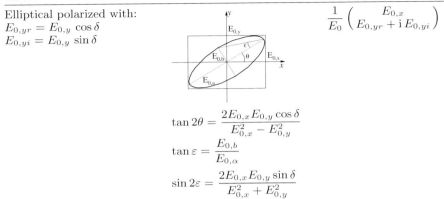	$\dfrac{1}{E_0}\left(\begin{array}{c} E_{0,x} \\ E_{0,yr} + \mathrm{i}\,E_{0,yi} \end{array}\right)$

$$\tan 2\theta = \frac{2E_{0,x}E_{0,y}\cos\delta}{E_{0,x}^2 - E_{0,y}^2}$$

$$\tan\varepsilon = \frac{E_{0,b}}{E_{0,a}}$$

$$\sin 2\varepsilon = \frac{2E_{0,x}E_{0,y}\sin\delta}{E_{0,x}^2 + E_{0,y}^2}$$

2.6.2 Jones Matrices of Some Optical Components

Using the Jones matrices the change of polarization properties for polarized light passing optical elements can be calculated. Such a matrix for any optical element can be determined from the comparison of their matrix elements with the result of a separate calculation analog as it was shown for ray matrices in Sect. 2.5.1. The Jones matrices of some common optical elements are given in Table 2.8:

Table 2.8. Jones matrices for some common optical elements

Polarizer in x direction	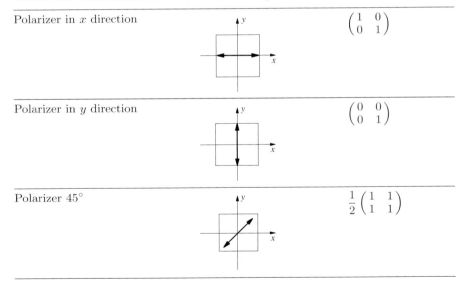	$\begin{pmatrix} 1 & 0 \\ 0 & 1 \end{pmatrix}$
Polarizer in y direction		$\begin{pmatrix} 0 & 0 \\ 0 & 1 \end{pmatrix}$
Polarizer $45°$		$\dfrac{1}{2}\begin{pmatrix} 1 & 1 \\ 1 & 1 \end{pmatrix}$

Table 2.8. Continued

Polarizer $-45°$		$\dfrac{1}{2}\begin{pmatrix} 1 & -1 \\ -1 & 1 \end{pmatrix}$
Phase delay		$\begin{pmatrix} e^{i\varphi_x} & 0 \\ 0 & e^{i\varphi_y} \end{pmatrix}$
Quarter-wave plate, fast axis in x direction		$e^{i\pi/4}\begin{pmatrix} 1 & 0 \\ 0 & -i \end{pmatrix}$
Quarter-wave plate, fast axis in y direction		$e^{-i\pi/4}\begin{pmatrix} 1 & 0 \\ 0 & i \end{pmatrix}$
Half-wave plate, fast axis in x direction		$\underbrace{e^{i\pi/2}}_{=i}\begin{pmatrix} 1 & 0 \\ 0 & -1 \end{pmatrix}$
Half-wave plate, fast axis in y direction		$\underbrace{e^{i\pi/2}}_{=-i}\begin{pmatrix} 1 & 0 \\ 0 & -1 \end{pmatrix}$
Rotator: angle β		$\begin{pmatrix} \cos\beta & +\sin\beta \\ -\sin\beta & \cos\beta \end{pmatrix}$

For a Faraday rotator the rotating angle β can be given as a function of the magnetic field strength H in the direction of the light wave vector $\boldsymbol{k}/k$ and the length of the material used L_{rotator}:

Faraday rotator $\quad \beta = C_{\text{verdet}} H L_{\text{rotator}}$ $\qquad$ (2.116)

using the Verdet constant C_{verdet}. This material constant is about $2.2°$/Tesla cm for water, $2.7°$/Tesla cm for phosphate glass, $2.8°$/Tesla cm for quartz, $5.3°$/Tesla cm for flint glass, $7.1°$/Tesla cm for CS_2 and $40°$/Tesla cm or $77°$/Tesla cm for the Terbium-doped glass or GGG, respectively. Thus $45°$ rotation can be realized with lengths of a few centimeters and strong permanent magnets or electrical coils.

2.6.3 Stokes Vectors Characterizing Partially Polarized Light

Light which is only partially polarized can be described by the Stokes vector $\boldsymbol{S}$ with the four elements $S_0 = 1$, S_1, S_2, and S_3.

$$\text{Stokes vector} \quad \boldsymbol{S} = \begin{pmatrix} 1 \\ S_1 \\ S_2 \\ S_3 \end{pmatrix}.$$ $\qquad$ (2.117)

The components of the Stokes vector have the following meaning:

$$S_1 = \frac{E_{0,x}^2 - E_{0,y}^2}{E_{0,x}^2 + E_{0,y}^2}$$ $\qquad$ (2.118)

represents the reduced difference of the observable intensities linearly polarized in the x and y direction and

$$S_2 = \frac{2 E_{0,x} E_{0,y} \cos \delta}{E_{0,x}^2 E_{0,y}^2}$$ $\qquad$ (2.119)

describes the reduced difference of the observable intensities linearly polarized in $45°$ and $-45°$ direction, whereas

$$S_3 = \frac{2 E_{0,x} E_{0,y} \sin \delta}{E_{0,x}^2 + E_{0,y}^2}$$ $\qquad$ (2.120)

is the reduced difference of right or left circularly polarized light. The degree of polarization p of partially polarized light with the nonpolarized component I_{nonpol} and the polarized component I_{pol} is observed from:

degree of polarization $\quad p = \dfrac{I_{\text{pol}}}{I_{\text{pol}} + I_{\text{nonpol}}}$ $\qquad$ (2.121)

which can be calculated from the Stokes vector by:

$$p = \sqrt{S_1^2 + S_2^2 + S_3^2}.$$ $\qquad$ (2.122)

The Stokes vectors for some polarizations of a light beams are given in Table 2.9.

In combination with the Mueller matrices the degree of polarization can be calculated for light beams propagating through optical elements.

Table 2.9. Stokes vectors for some typical light polarizations

Polarized in x direction		$\begin{pmatrix} 1 \\ 1 \\ 0 \\ 0 \end{pmatrix}$
Polarized in y direction		$\begin{pmatrix} 1 \\ -1 \\ 0 \\ 0 \end{pmatrix}$
Polarized 45°		$\begin{pmatrix} 1 \\ 0 \\ 1 \\ 0 \end{pmatrix}$
Polarized −45°		$\begin{pmatrix} 1 \\ 0 \\ -1 \\ 0 \end{pmatrix}$
Left circular polarized (viewing into beam)		$\begin{pmatrix} 1 \\ 0 \\ 0 \\ -1 \end{pmatrix}$
Right circular polarized (viewing into beam)		$\begin{pmatrix} 1 \\ 0 \\ 0 \\ 1 \end{pmatrix}$
Left elliptical polarized $\tan \varepsilon = \dfrac{b}{a}$		$\begin{pmatrix} 1 \\ \cos 2\varepsilon \, \cos 2\theta \\ \cos 2\varepsilon \, \sin 2\theta \\ \sin 2\varepsilon \end{pmatrix}$
Non polarized		$\begin{pmatrix} 1 \\ 0 \\ 0 \\ 0 \end{pmatrix}$

2.6.4 Mueller Matrices of Some Optical Components

Mueller matrices are useful for the calculation of the polarization properties for partial polarized light passing optical elements [2.26–2.34]. A collection of Mueller matrices for some common optical elements is given in Table 2.10:

Table 2.10. Mueller matrices for some common optical elements

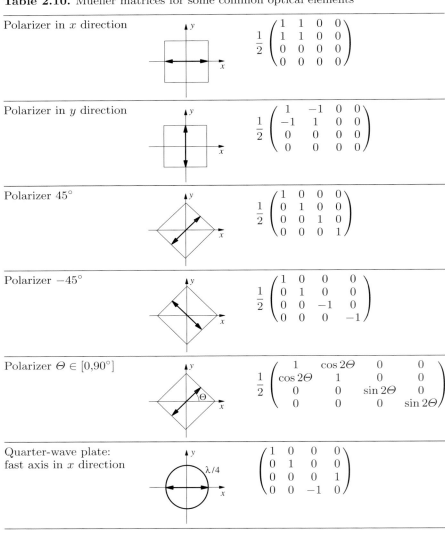

Polarizer in x direction

$$\frac{1}{2}\begin{pmatrix} 1 & 1 & 0 & 0 \\ 1 & 1 & 0 & 0 \\ 0 & 0 & 0 & 0 \\ 0 & 0 & 0 & 0 \end{pmatrix}$$

Polarizer in y direction

$$\frac{1}{2}\begin{pmatrix} 1 & -1 & 0 & 0 \\ -1 & 1 & 0 & 0 \\ 0 & 0 & 0 & 0 \\ 0 & 0 & 0 & 0 \end{pmatrix}$$

Polarizer $45°$

$$\frac{1}{2}\begin{pmatrix} 1 & 0 & 0 & 0 \\ 0 & 1 & 0 & 0 \\ 0 & 0 & 1 & 0 \\ 0 & 0 & 0 & 1 \end{pmatrix}$$

Polarizer $-45°$

$$\frac{1}{2}\begin{pmatrix} 1 & 0 & 0 & 0 \\ 0 & 1 & 0 & 0 \\ 0 & 0 & -1 & 0 \\ 0 & 0 & 0 & -1 \end{pmatrix}$$

Polarizer $\Theta \in [0,90°]$

$$\frac{1}{2}\begin{pmatrix} 1 & \cos 2\Theta & 0 & 0 \\ \cos 2\Theta & 1 & 0 & 0 \\ 0 & 0 & \sin 2\Theta & 0 \\ 0 & 0 & 0 & \sin 2\Theta \end{pmatrix}$$

Quarter-wave plate: fast axis in x direction

$$\begin{pmatrix} 1 & 0 & 0 & 0 \\ 0 & 1 & 0 & 0 \\ 0 & 0 & 0 & 1 \\ 0 & 0 & -1 & 0 \end{pmatrix}$$

Table 2.10. Continued

Quarter wave plate: fast axis in y direction		$\begin{pmatrix} 1 & 0 & 0 & 0 \\ 0 & 1 & 0 & 0 \\ 0 & 0 & 0 & -1 \\ 0 & 0 & 1 & 0 \end{pmatrix}$
Half-wave plate: fast axis in x or y direction		$\begin{pmatrix} 1 & 0 & 0 & 0 \\ 0 & 1 & 0 & 0 \\ 0 & 0 & -1 & 0 \\ 0 & 0 & 0 & -1 \end{pmatrix}$

2.6.5 Using the Formalism

If a light beam passes n optical elements $1, 2, \ldots, n$ with the Jones or Mueller matrices $M_1, M_2, \ldots, M_n$ the polarization of the outgoing beam can be calculated from these matrices and the polarization of the incident beam by:

$$\boldsymbol{J}_{\text{out}} = M_n \cdot \cdots \cdot M_2 \cdot M_1 \cdot \boldsymbol{J}_{\text{in}} \tag{2.123}$$

or

$$\boldsymbol{S}_{\text{out}} = M_n \cdot \cdots \cdot M_2 \cdot M_1 \cdot \boldsymbol{S}_{\text{in}} \tag{2.124}$$

respectively. Please note the order of multiplication of the matrices analogous to the rule for ray matrices. Of course no matrices have to be used for free space propagation in contrast to the case of ray matrices.

The resulting polarization vectors determine the polarization of the beam behind these elements with respect to the kind of polarization and, in case of Stokes vectors, also with respect to the degree of polarization.

2.7 Light Characteristics

Light beams have to be described by their spatial, spectral, temporal and polarization distributions in addition to coherence properties. As fundamental limits the beams can be diffraction and bandwidth limited, linearly polarized and coherent. Usually these limits are not reached or even required in applications. Thus for practical purposes these distributions have to be covered by a suitable number of parameters. In nonlinear optics usually all parameters have to be checked carefully for their influence on the application. Thus a detailed discussion about these parameters and their measurement seems helpful.

2.7.1 Power, Energy and Number of Photons

As shown in Sect. 2.1.3 the light intensity I is a function of the space vector r, wavelength λ, time t and polarization angle φ. It can be determined by measuring the power or energy of the light. With nonlinear spectroscopic methods the direct measurement of the number of photons is possible in principle, e.g. via a photochemical reaction or nonlinear absorption [2.35, 2.36], but usually commercially calibrated devices are used as power meters for continuously operating light sources and energy meters for pulsed light [2.27, 2.38]. The light power P is related to the intensity by:

$$P = \int\int\int I_{\text{pulse}}(r, \lambda, \varphi) \, d\varphi \, d\lambda \, dr \tag{2.125}$$

and the energy of a light pulse E_{pulse} is the integral over all these distributions:

$$E_{\text{pulse}} = \int\int\int\int I_{\text{pulse}}(r, \lambda, \varphi, t) \, dt \, d\varphi \, d\lambda \, dr. \tag{2.126}$$

With the average photon energy $\overline{E}_{\text{photon}}$, which can be used in the case of narrow spectral distributions and calculated from the average wavelength $\overline{\lambda}_{\text{photon}}$, the number of photons in the light pulse can be calculated:

number of photons $n_{\text{photons}} = \dfrac{E_{\text{pulse}}}{\overline{E}_{\text{photon}}} = \dfrac{\overline{\lambda}_{\text{photon}}}{hc} E_{\text{pulse}}.$ $\tag{2.127}$

For example, a light pulse of $1\,\text{mJ}$ energy and an average photon energy corresponding to a wavelength of $500\,\text{nm}$ which represents a photon energy of $4 \cdot 10^{-19}\,\text{J}$ contains $2.5 \cdot 10^{15}$ photons. A continuous stream of $2.5 \cdot 10^{15}$ photons s^{-1} at this wavelength would have a power of $1\,\text{mW}$.

2.7.2 Average and Peak Power of a Light Pulse

The average power $\overline{P}_{\text{pulse}}$ of a light pulse results from its energy E_{pulse} and a characteristic pulse width Δt_{pulse}:

average power $\overline{P}_{\text{pulse}} = \dfrac{E_{\text{pulse}}}{\Delta t_{\text{pulse}}}.$ $\tag{2.128}$

The most useful definition of the pulse width Δt_{pulse} depends on the exponent of the nonlinearity of the problem which should be described. If the pulse power is measured as a function of time $P(t)$ the pulse width Δt_{pulse} can be numerically determined by the second moment of this temporal profile for any pulse shape:

pulse width $\Delta t_{\text{pulse}}^2 = 8 \ln 2 \dfrac{\int\limits_{\text{pulse}} (t - t_0)^2 P(t) \, dt}{\int\limits_{\text{pulse}} P(t) \, dt}.$ $\tag{2.129}$

with the factor $8 \cdot \ln 2 \simeq 5.55$ and the total pulse energy E_{pulse} as the integral over the whole temporal structure:

pulse energy $\displaystyle E_{\text{pulse}} = \int_{\text{pulse}} P(t)\,dt$ (2.130)

and the time of the pulse center t_0:

$$t_0 = \frac{\int\limits_{\text{pulse}} t P(t)\,dt}{\int\limits_{\text{pulse}} P(t)\,dt}.$$ (2.131)

These definitions are useful for any pulse shape and are sufficient for linear optical problems.

Assuming a Gaussian temporal profile (see Fig. 2.18) the pulse width can be measured as the full width at half maximum power (FWHM) which is twice the half width half maximum (HWHM).

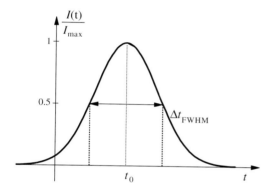

Fig. 2.18. Temporal Gaussian pulse with characteristic times

If the pulse shape is described by:

Gauss pulse $\displaystyle P(t) = P_{\text{max}} \cdot \exp\left(-\frac{(t-t_0)^2}{\sigma_t}\right)$

$$= P_{\text{max}} \cdot \exp\left(-\frac{(t-t_0)^2 4 \ln 2}{\Delta t_{\text{FWHM}}^2}\right)$$ (2.132)

then the pulse width Δt_{pulse} is equal to Δt_{FWHM}:

$$\Delta t_{\text{pulse}} = \Delta t_{\text{FWHM,Gauss}}.$$ (2.133)

The peak power P_{max} of the pulse with Gaussian shape follows from its energy E_{pulse} on:

peak power $\displaystyle P_{\text{max}} = 2\sqrt{\frac{\ln 2}{\pi}}\,\frac{E_{\text{pulse}}}{\Delta t_{\text{FWHM}}} \simeq 0.939\frac{E_{\text{pulse}}}{\Delta t_{\text{FWHM}}}.$ (2.134)

As an example for a pulse with an energy of $1\,\mathrm{mJ}$ and a width of $\Delta t_{\mathrm{FWHM}} = 10\,\mathrm{ns}$ which corresponds to an optical path length of $3\,\mathrm{m}$ the average power would be $42.5\,\mathrm{kW}$ and the peak power $93.9\,\mathrm{kW}$.

In nonlinear optics the pulse width $\Delta t_{\mathrm{pulse}}$ and thus the average power can be adapted to the nonlinearity of the problem. During Δt_{FWHM} centered around t_0, 76.1% of the whole energy of the pulse is transported. The power of a flat-top profile pulse P_{FT} with the original energy and this length would be 1.0645 times P_{max}. Other values are given in Table 2.11.

Table 2.11. Relations of power $P(t_0 - \Delta t_{\mathrm{pulse}}/2)$, power P_{FT} during $\Delta t_{\mathrm{pulse}}$ and energy $E(\Delta t_{\mathrm{pulse}})$ during $\Delta t_{\mathrm{pulse}}$ relative to the peak power P_{max} and the total energy $E_{\mathrm{pulse,tot}}$ for Gaussian pulses. The NLP-exponent describes the nonlinear process which is correctly described by P_{average}

$\Delta t_{\mathrm{pulse}}/\Delta t_{\mathrm{FWHM}}$	$P(t_0 - \Delta t/2)/P_{\mathrm{max}}$	$P_{\mathrm{FT}}(\Delta t)/P_{\mathrm{max}}$	$E(\Delta t)/E_{\mathrm{pulse,tot}}$	NLP-exp
1.505	0.2079	$1/\sqrt{0.7071}$	0.8514	2
1.401	0.2566	0.7598	0.8366	3
$1.201(= \sigma_t)$	$1/e \simeq 0.3679$	0.8862	0.8031	11
1.065	0.4559	1	0.7755	∞
1	0.5	1.0645	0.7610	–
0.5	0.8409	2.1289	0.5949	–

As listed in the table the choice of $\Delta t_{\mathrm{pulse}}/\Delta t_{\mathrm{FWHM}}$ equal to 1.505 and 1.401 for a substituting flat-top pulse is useful for quadratic or cubic nonlinear optical effects, respectively (see Chap. 4 and 5).

For quasi-continuous radiation consisting of a series of pulses with a certain on/off relation the average power during the single pulse and the average power over the whole series have to be differentiated.

Measuring the pulse width of very short laser pulses can be very complicated. Some examples are described in Sects. 5.5 and 7.1.5. Down to pulse widths of about 100 ps electronic devices such as oscilloscopes and boxcar-integrators can be used. Streak cameras can be applied for pulses of a few picoseconds. Below a few picoseconds nonlinear measurements are the only possibility. They usually do not allow the determination of the pulse shape.

2.7.3 Intensity and Beam Radius

In most nonlinear optical applications the intensity of the light (see (2.24) and (2.25)) is the most characteristic parameter. The intensity is usually the "driving force" of the nonlinear process via the pump rates or the field strength. The number of photons per area and time produces excitations which decay with the time rates of the material relaxation processes.

The average intensity $\overline{I}$ for a light beam can be determined from its average power $\overline{P}$ and the characteristic area $\overline{A}$ by:

average intensity $\quad \overline{I} = \dfrac{\overline{P}}{\overline{A}}.$ $\hfill (2.135)$

The characteristic beam cross-section $\overline{A}$ can be determined from measuring the spatial power distribution $P(x, y)$ e.g. with a CCD camera, by:

cross-section $\quad \overline{A} = \dfrac{\iint\limits_{A} (x - x_0)^2 (y - y_0)^2 I_{\text{uncal}}(x, y) \, dx \, dy}{\iint\limits_{A} I_{\text{uncal}}(x, y) \, Dx \, dy}$ $\hfill (2.136)$

with the first momentum of x_0 and y_0 as:

$$x_0 = \dfrac{\iint\limits_{A} x I_{\text{uncal}}(x, y) \, dx \, dy}{\iint\limits_{A} I_{\text{uncal}}(x, y) \, dx \, dy} \quad \text{and} \quad y_0 = \dfrac{\iint\limits_{A} y I_{\text{uncal}}(x, y) \, dx \, dy}{\iint\limits_{A} I_{\text{uncal}}(x, y) \, dx \, dy} \hfill (2.137)$$

describing the average position of the beam.

In the case of pulsed light the power $P(x, y)$ has to be temporally averaged as described in the previous chapter. For rotation-symmetric light beams a characteristic beam radius w_{beam} can be defined as:

beam radius $\quad w_{\text{beam}}^2 = \dfrac{4\pi \int r^3 I_{\text{uncal}}(r) \, dr}{\int I_{\text{uncal}}(r) \, dr}$ $\hfill (2.138)$

assuming the beam is centered at $r = 0$.

In the case of Gaussian beam shapes as described in Sect. 2.4.2 the beam radius is given at the $1/e^2$ value of the intensity or the $1/e$ value for the electric field amplitude. The transversal intensity distribution is given by:

$$I_{\text{Gauss}}(r, z) = I_0 \frac{w_0^2}{w_2(z)} e^{-2\left(\frac{r}{w(z)}\right)^2} \hfill (2.139)$$

with I_0 as the peak intensity at $r = 0$ and $z = 0$ and $w(z)$ as the beam diameter changing with propagation in the z direction as described in Sect. 2.4.4.

The full width half maximum value of the beam radius (see Fig. 2.8) follows from the radius $w(z)$ of (2.139) as:

$$r_{\text{FWHM}}(z) = \sqrt{\ln 2} \, w(z) \approx 0.833 \cdot w(z). \hfill (2.140)$$

Inside the beam radius $w(z)$, 86.5% of the whole beam power is obtained whereas, inside r_{FWHM}, 75% occur for Gaussian shaped beams. The two-dimensional Gaussian curve of the plane cut through the maximum of the three-dimensional intensity distribution contains 95.4% of the whole area inside the radius $w(z)$.

The transversal intensity distribution of a super-Gaussian beam is given by:

super-Gaussian beam

$$I_{\text{SuperGauss}}(r, z) = I_{0,\text{SG}}(z) \, e^{-2\left(\frac{r}{w(z)}\right)^m} \hfill (2.141)$$

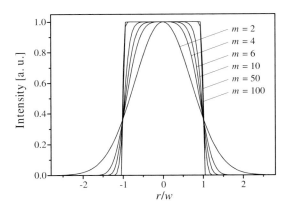

Fig. 2.19. Spatial intensity profile of a super-Gaussian beam shape for exponents m of 2, 4, 6, 10, 50 and 100. For m values of more than 50 an almost flat-top profile is found

with the even super-Gauss exponent $m \geq 2$ realizing a Gaussian profile for $m = 2$. For values above $m = 100$ a flat-top profile is found as shown in Fig. 2.19.

The beam radius can be determined from (2.138) or by analogy to the Gaussian-shaped distribution at the $1/e^2$ value of the intensity. The power contents inside these radii will be different.

For theoretical modeling of nonlinear processes the transversal Gaussian beam shape may be approximated by a beam with a flat-top intensity profile with intensity I_{FT} and beam radius w_{FT}. In this case the flat top should represent the same energy content as the original beam. The radius of this cylindrical beam shape can be selected for different values w_{FT} and the intensity I_{FT} will vary, accordingly. If the radius is equal w_0 of the Gaussian beam the flat-top intensity will than be $I_0/2$. If the nonlinear process is quadratically dependent on the intensity this flat top would produce the same effect as the Gaussian beam. If the nonlinear process depends on the third power the flat-top intensity should be $I_0/\sqrt{3}$ resulting in a radius of $w_0 \cdot \sqrt{\sqrt{3}/2} = w_0 \cdot 0.93$. For a flat-top intensity equal to I_0 the corresponding diameter is equal to $w_0/\sqrt{2}$. These results are summarized in Table 2.12.

The beam profile and beam radius can be measured via different methods [see, for example, 2.39–2.61]. One way for determining the beam radius is the knife edge method [2.44, 2.45]. A sharp edge, such as, e.g. a razor blade, is moved across the beam cross-section and the transmitted relative power P_{kn} or energy E_{kn} is measured as a function of the coordinates x and y. The signal $E_{kn}(x/y)$ will change from 0 to E_{max} as shown in Fig. 2.20.

The coordinate difference between the 16% and 84% value of the measured power P_{kn} or energy E_{kn} gives the beam radius in this direction for 86.5% power content. For Gaussian beams this value is identical with the $1/e^2$ intensity radius w_0.

Table 2.12. Relations of the intensity I_{FT} of a flat-top beam with the same energy as a Gaussian beam as a function of the radius of this beam w_{FT} in comparison of the intensity $I(w_{FT})$ and the energy E inside w_{FT} for the Gaussian beam relative to the peak intensity I_{max} and the total energy E_{tot} of the Gaussian beam. The NLP-exponent describes the nonlinear process which is correctly described by I_{FT}

w_{FT}/w_0	I_{FT}/I_{max}	$I(w_{FT})/I_{max}$	$E(w_{FT})/E_{tot}$	NLP-exp
0.5	2	0.606	0.394	
0.589	0.694	0.5	0.5	
0.707	1 123	0.368	0.632	∞
0.93	0.577	0.177	0.823	3
1	0.5	0.135	0.865	2

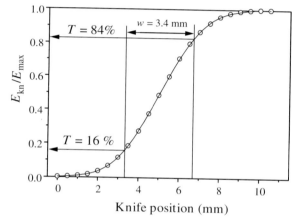

Fig. 2.20. Pulse energy signal P_{kn} of a Nd:YAG laser measured with the knife edge method as a function of the coordinates x (or y) perpendicular to the propagation of a Gaussian beam for determining the beam diameter

2.7.4 Divergence

The divergence of the light beam is measured as the angle θ describing the approximately linear increase of the beam radius w at large distances (far-field) from the beam waist (see Fig. 2.21).

This divergence angle θ can be determined experimentally by measuring the beam radius w_1 and w_2 at two different distances z_1 and z_2 from the waist as shown in Fig. 2.21:

$$\theta = \frac{w_2 - w_1}{z_2 - z_1}. \tag{2.142}$$

For an accuracy of better 98% a minimal distance of five times the Rayleigh length (see Sect. 2.4.3) is necessary and 10 times this distance would lead to an accuracy of 99.5% for Gaussian beams.

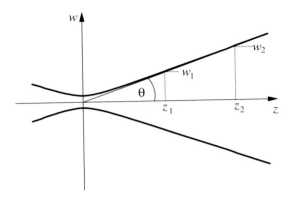

Fig. 2.21. Divergence angle θ of light beam

Another possibility of determining θ is to measure the waist diameter behind a focusing lens with focal length f_{lens} as shown in Fig. 2.22.

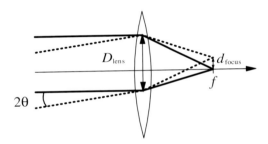

Fig. 2.22. Divergence determined by measuring the waist size behind a focusing lens

The divergence angle can be determined from the focal length f and the diameter d_{focus} in the focal distance of the lens by:

$$\theta = \frac{w_{\mathrm{focus}}}{f} = \frac{d_{\mathrm{focus}}}{2f}. \tag{2.143}$$

Sometimes the full angle 2θ is challed the divergence, which can cause confusion. In any case the divergence is minimal for Gaussian beams and the angle θ can be calculated from the waist radius w_{waist} and vice versa by:

$$\theta = M^2 \frac{\lambda}{\pi n w_{\mathrm{waist}}} \tag{2.144}$$

with the beam propagation factor M^2 which is $M^2_{\mathrm{gauss}} = 1$ for diffraction-limited Gaussian beams.

Laser beams described by higher transversal modes as given in Sect. 6.6.5 have larger divergence angles. A detailed discussion is given in Sects. 6.6.5–6.6.9.

Some light sources emit from areas (e.g. tungsten band light bulbs) or volumes (e.g. some excimer lasers, the luminescence of dyes or synchrotron

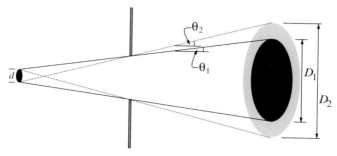

Fig. 2.23. Divergence angles of a conventional light source of diameter d with aperture

radiation) without coherent coupling. The divergence of this light is usually a function of the geometrical dimensions of the source and the apertures in the beam. In some cases even two divergence angles may be useful for the description, as shown in Fig. 2.23.

The possibly very complicated conditions can be analyzed using a combination of geometrical and Gaussian beam propagation formalism. The latter is necessary for the z positions with Fresnel numbers that are not too large compared to 1 (see Sect. 2.3.1).

A general and nice formalism for the experimental characterization and theoretical description of the propagation of any more or less coherent light beam was given in [2.46–2.50]. This formalism is based on a Wigner function for the beam. If this function is determined, further propagation can be calculated based on the ray matrices in a simple way. Unfortunately the experimental determination of this function for a given beam that is only partially coherent in two dimensions is difficult.

2.7.5 Beam Parameter Product Beam Quality

For applications of light beams the characterization of their transversal mode structure is necessary. Both the beam diameter at the waist $2w_{\text{waist}}$ and the beam divergence θ have to be determined for this purpose [2.56, 2.57], although for commercial lasers often only the divergence is given. The quality of light beams can be described by the beam parameter product BP:

beam parameter product $\text{BP} = \theta \cdot w_{\text{waist}}$ (2.145)

which is minimal for diffraction-limited beams.

The values w_{waist} and θ have to be determined experimentally, e.g. from the caustic of the focused beam (see Fig. 2.24).

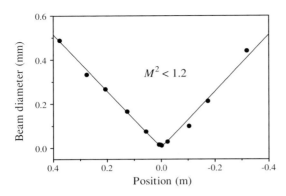

Fig. 2.24. Measurement of the beam diameter using the knife edge method as a function of the z position behind a focusing lens for the determination of the beam propagation factor M^2

The curve is fitted with:

$$w(z) = M w_{\text{waist}} \sqrt{1 + \left(\frac{(z - z_0)\lambda}{w_{\text{waist}}^2 n \pi} \right)^2} \tag{2.146}$$

with position z and wavelength λ. The square of M is the beam propagation factor describing the beam quality as described below. Special care has to be taken to determine the size of the waist diameter. Careful error estimations should be made. From this curve both the beam waist radius w_{waist} and the divergence angle θ can be determined.

Another simpler method is the measurement of the spot size diameter d_{focus} of the beam at the focal length of a lens as shown in Fig. 2.22. The beam parameter product can be calculated from

$$\text{BP}_{\text{beam}} = \theta \cdot w_{\text{waist}} = \frac{D_{\text{lens}} d_{\text{focus}}}{4f} \tag{2.147}$$

with the beam diameter D_{lens} at the position of the lens and focal length f. This method is more crucial for possible measuring errors than the knife edge method.

Further examples are given in references [2.51–2.61]. Commercial beam profilers are available now. They are able to measure the M^2 factor directly in some cases, in a way similar to the knife edge method. For M^2 values larger than 10, and very close to 1, special care has to be taken in the measurements.

The quality of a measured beam is described by the ratio of the beam parameter products of this beam BP_{beam} relative to the best possible value BP_{Gauss} of a Gaussian beam of the same wavelength λ in the same material with refractive index n ($n_{\text{air}} \approx 1$). This ratio is defined as the *beam quality* BQ or the *beam propagation factor* M^2 [2.56–2.58]:

$$\textbf{beam quality} \quad \text{BQ} = M^2 = \frac{\text{BP}_{\text{beam}}}{\text{BP}_{\text{Gauss}}} = (\theta \cdot w_{\text{waist}}) \frac{\pi n}{\lambda}. \tag{2.148}$$

For known transversal laser modes these values are given by the M^2 in Sect. 6.6.6 and 6.6.8. BQ = 1.5 means the beam is "1.5 times diffraction limited" or "1.5 · DL". Thus BQ = 1.5 means the beam parameter product is

1.5 times worse than the best possible value for this wavelength and thus the focus for a given lens would show 1.5 times larger diameter as for a perfect beam.

2.7.6 Brightness

The brightness L of a laser beam describes its potential for realizing high intensities in combination with large Rayleigh lengths or small focusing angles (e.g. a large working distance in material processing). Thus it is calculated from the beam quality BQ and the beam power P by:

$$\textbf{brightness} \quad L = \frac{\pi^2}{\lambda^2} \frac{P}{\text{BQ}^2} = \frac{P}{\text{BP}^2_{\text{beam}}}. \tag{2.149}$$

Thus high brightness demands high power, good beam quality and short wavelengths λ. The brightness can be given for average power or peak power values with respect to different applications. For quasi-continuous radiation the average for the single pulse or the series has to be distinguished. High values are in the range of $100\,\text{W}\,\text{mm}^{-2}\,\text{mrad}^{-2}$ for average powers and 10^{12} times more for fs pulses.

2.7.7 Brilliance

The brilliance BL includes the characterization of the spectral distribution of the radiation. It is defined as the brightness L per 0.1% bandwidth and is thus a function of the wavelength λ or frequency ν of the radiation. It is calculated from:

$$\textbf{brilliance} \quad \text{BL}(\lambda) = L\frac{\lambda/1000}{\Delta\lambda} \quad \text{BL}(\nu) = L\frac{\nu/1000}{\Delta\nu} \tag{2.150}$$

for small spectral ranges.

Thus brilliance is, e.g. useful for comparing synchrotron radiation with laser beams in the different spectral ranges. It is as higher as spectral narrower the emission spectrum. For lasers the brilliance can be more than a million times larger than the brightness. For quasi-continuous radiation, again, the average brilliance and the brilliance during single pulses have to be differentiated, similarly to power or brightness characterization.

2.7.8 Radiation Pressure and Optical Levitation

The moments of all photons in the light beam result in a radiation pressure if light is reflected or absorbed [2.62–2.66]. It has sometimes to be explicitly considered in designing high-power optical setups or in high-resolution spectroscopic measurements [e.g. 2.60]. For a light pulse with energy E_{pulse} the total momentum p_{pulse} for absorption is given by:

$$\textbf{radiation momentum} \quad p_{\text{pulse}} = \frac{E_{\text{pulse}}}{c} \tag{2.151}$$

and for total reflection twice this value is observed.

As an example an iodine molecule with a mass of $4.2 \cdot 10^{-22}$ g will have an additional speed of $0.3\,\mathrm{m\,s}^{-1}$ after absorbing a photon at 530 nm. If a pulse with duration $\Delta t_{\mathrm{pulse}}$ is reflected at a 100% mirror the resulting average force F_{pulse} during this time is:

$$\textbf{force on reflector} \quad F_{\mathrm{pulse}} = \frac{\mathrm{d}p_{\mathrm{sum}}}{\mathrm{d}t} \approx \frac{p_{\max}}{\Delta t_{\mathrm{pulse}}} = \frac{2E_{\mathrm{pulse}}}{\Delta t_{\mathrm{pulse}}c} \qquad (2.152)$$

which can be several kp. A continuous light beam with intensity I produces a radiation pressure P_{light} of:

$$\textbf{radiation pressure} \quad P_{\mathrm{light}} = \frac{I}{c} \qquad (2.153)$$

which can reach several bar for high-power beams with very good beam quality under strong focusing. An excimer laser pulse with an energy of 10 mJ and a duration of 10 ns at a wavelength of 308 nm produces at a high-reflecting mirror a force of 0.067 N which is about the weight of 0.7 g. If this light beam is focused to a diameter of $1\,\mathrm{mm}^2$ the resulting pressure is 6700 Pa.

This effect can be used for *optical levitation* of small particles by laser beams [2.67–2.71]. A particle with transmission T and reflection R will experience a force F_{lev}:

$$\textbf{levitation force} \quad F_{\mathrm{lev}} = (R + 1 - T)P/c \qquad (2.154)$$

Thus in the waist region of a suitable beam with a power of less than a watt the gravitation of particles with diameters in the µm range can be compensated.

Another possibility to fix particles in the waist region of a strong laser beam is based on the interaction of the electric light field with the induced or permanent dipole moment of the particle [2.72–2.82]. Thus it is possible to attract particles into the beam waist and stabilize them there even against the radiation pressure in an optical trap.

Using this light force an optical tweezer can be realized and single molecules can be manipulated, for example they can be stretched, with laser beams [2.83–2.86].

2.8 Statistical Properties of Photon Fields

As described in Sect. 2.1.2 photons as quantum particles fulfill the uncertainty conditions for position and momentum as well as for time and energy. Thus these pairs of values are determined to the limit of h/π and $h/2\pi$, only. Because photons with spin 1 are bosons, they are not distinguishable and can be in the same quantum state. But the quantum mechanical uncertainty results in certain photon statistics [2.87–2.91]. These statistics can be observed if the photon number is small or the measurements are very precise or the experiments are phase sensitive.

2.8.1 Uncertainty of Photon Number and Phase

As a consequence of the energy-time uncertainty (Sect. 2.1.2) the statistical appearance of photons shows an uncertainty, too. The number of photon fluctuations Δn_{ph} and the phase $\Delta\varphi$ of the photons are related to the uncertainty ranges of the energy ΔE and time Δt by:

$$\Delta E = \Delta n_{\mathrm{ph}} h\nu \text{ and } \Delta\varphi = \Delta t 2\pi\nu \tag{2.155}$$

and thus the resulting uncertainty follows from (2.18):

$$\Delta n_{\mathrm{ph}}\Delta\varphi \geq 1 \tag{2.156}$$

which can be important in single photon counting experiments.

2.8.2 Description by Elementary Beams

The best possible light beam is diffraction and bandwidth limited. This beam has the lowest possible products of beam waists w_x and w_y with divergence angles θ_x and θ_y as well as bandwidth $\Delta\nu$ with pulse duration $\Delta t_{\mathrm{pulse}}$. This beam will be called an elementary beam:

$$\textbf{elementary beam} \quad (w_x\theta_x)(w_y\theta_y)(\Delta\nu\Delta t_{\mathrm{pulse}}) = \frac{c^2}{2\pi^3\nu_0^2} \tag{2.157}$$

where ν_0 is the mid-frequency and c the speed of light. All photons in this elementary beam are in principle indistinguishable. These photons can interfere with each other and thus they are coherent, as will be described in Sect. 2.9.2.

The average number of photons per time $\bar{n}$ in this elementary beam is a function of the temperature T and the mid-frequency ν_0 for thermal light sources (blackbody radiation):

$$\textbf{blackbody radiation} \quad \bar{n} = C_{\mathrm{source}}\frac{1}{e^{h\nu_0/k_{\mathrm{B}}T} - 1} \tag{2.158}$$

with the Boltzmann constant k_{B} and a technical constant C_{source} describing the parameters of the source. In laser radiation this average photon number is given by the laser parameter but usually is not a function of the temperature.

If the photons are detected one after the other, e.g. with a photomultiplier, the statistical appearance of the two sources is drastically different. The thermal source will show a Bose–Einstein distribution:

$$\textbf{thermal light} \tag{2.159}$$

$$P_{\mathrm{thermal}}(n_{\mathrm{ph}}) = \frac{1}{(1 + \bar{n}_{\mathrm{ph}})\left(1 + \dfrac{1}{\bar{n}_{\mathrm{ph}}}\right)^{n_{\mathrm{ph}}}} \overset{n_{\mathrm{ph}}\to\infty}{\approx} \frac{1}{\bar{n}_{\mathrm{ph}}}\exp(-n_{\mathrm{ph}}/\bar{n}_{\mathrm{ph}})$$

for the probability $P(n_{\mathrm{ph}})$ of finding n photons per unit time in the elementary beam. Laser light shows a Poisson distribution:

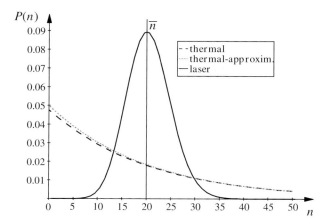

Fig. 2.25. Probability $P(n)$ of finding n photons per time unit for an average photon number of 20 in this time interval

$$\textbf{laser light} \quad P_{\text{laser}}(n_{\text{ph}}) = \frac{1}{n_{\text{ph}}!}(\overline{n}_{\text{ph}})^{n_{\text{ph}}}\,\text{e}^{-n_{\text{ph}}} \tag{2.160}$$

The two distributions are shown in Fig. 2.25.

Note that the thermal distribution shows its probability maximum at zero photons per unit time and a monotone decrease up to high values. Thus the photon number per time is lumped (see Fig. 2.26).

Laser

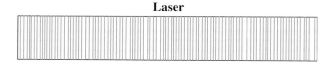

Thermal

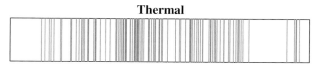

Fig. 2.26. Statistical distribution of photons for a thermal source and a laser as measurable with a single photon multiplier over the time for an average photon number of 20 per time interval as in Fig. 2.25

The width of the distributions is also different. Thermal sources show a width of

$$\textbf{thermal light} \quad \overline{\Delta n^2_{\text{ph,term}}} = \overline{n}_{\text{ph}} + \overline{n}^2_{\text{ph}} \tag{2.161}$$

and laser sources:

laser light $\overline{\Delta n_{\mathrm{ph,laser}}^2} = \overline{n}_{\mathrm{ph}}.$ (2.162)

For large numbers of photons the laser distribution is much narrower compared to thermal sources as is obvious from Fig. 2.25. Therefore the signal-to-noise ratio (SNR) of light behind a beam splitter with transmission T_{BS} is proportional to T_{BS} for laser radiation and $T_{\mathrm{BS}}/(T_{\mathrm{BS}}+1)$ for thermal sources. Thus the SNR can be decreased with increasing power using lasers as planned, e.g. in gravitational wave detectors. In thermal light the SNR is constant for large photon numbers. Changes in the photon statistics can be observed if nonlinear elements are placed in the beam [see, for example, 2.89].

2.8.3 Fluctuations of the Electric Field

Fluctuations in the photon number per unit time result in fluctuations of the amplitude and the phase of the resulting electric field $\boldsymbol{E}$. Thus the electric field and its phase can be determined only with an uncertainty of:

$$\Delta |\boldsymbol{E}|^2 \Delta \varphi \geq \frac{2h\nu}{c_0 \varepsilon_0 n \overline{A} \Delta t}$$ (2.163)

with light frequency ν, speed of light c_0, refractive index n, average cross-section of the beam $\overline{A}$ and duration of the light pulse Δt; see (2.24) and (2.45). In the case of continuous light, set $\Delta t = 1\,\mathrm{s}$. Thus in high-precision interference experiments the size of the electric field and its phase can be determined only to this uncertainty. As a solution for very accurate measurements the light power has to be high, but in this case other problems caused by the radiation pressure, absorption and heating may occur.

As a further uncertainty the components of the electric $\boldsymbol{E}$ and magnetic $\boldsymbol{H}$ fields of light cannot be measured at the same place exactly. The resulting uncertainty relation is given by:

$$\Delta |\boldsymbol{E}|_x \Delta |\boldsymbol{H}|_y \geq \frac{hc_0^2}{2d_{E-H}^4}$$ (2.164)

where d_{E-H} denotes the distance of the measurement of the two vertical components (see Fig. 2.2).

2.8.4 Squeezed Light

Although light has to fulfill all uncertainty relations, as mentioned above, for special purposes the uncertainty can be assigned mostly to one value of the pair. Thus the other value can be determined with very high accuracy. One simple example is to increase the beam diameter which decreases the divergence and thus the momentum uncertainty is decreased at the expense of the position uncertainty.

In other cases it may be important to decrease the uncertainty of the electric field value and thus the noise of the light beam. In this case the phase fluctuations will increase. These light beams are called *squeezed* [2.92–2.116]. For illustration the light beams can be characterized in an E-φ-diagram in polar coordinates (see Fig. 2.27).

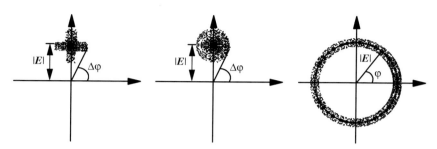

Fig. 2.27. E-φ-diagram in polar coordinates for characterizing amplitude and phase fluctuations of different light sources as squeezed light (left), coherent state (middle) and laser light (right). The monochromatic classical wave is represented by one point in the diagram

If the electric field vector is written as a complex value the diagram axes represent the real (horizontal) and the imaginary (vertical) part of the field.

2.8.5 Zero Point Energy and Vacuum Polarization

Using quantum electrodynamics the minimum energy E_0 of any system limited in space even at zero temperature is given by:

$$\textbf{zero point energy}\quad E_0 = \frac{1}{2}h\nu_{\text{box}} = \frac{c_0 h}{4d_{\text{box}}} \tag{2.165}$$

with the radiation frequency ν_{box} given by the linear box dimension d_{box} as $\nu_{\text{box}} = c_0/d_{\text{box}}$. This radiation will occur as minimum noise in any experiment.

Furthermore the vacuum will change its properties if fields with very high values are applied. If the electric field exceeds

$$\textbf{field for vacuum polarization}\quad |\boldsymbol{E}| \geq 10^{18}\,\frac{\text{V}}{\text{cm}} \tag{2.166}$$

the vacuum will show dispersion and birefringence as in matter. Fortunately this value is so high that the necessary intensities of more than $10^{33}\,\text{W}\,\text{cm}^{-2}$ usually do not occur in photonic applications.

2.9 Interference and Coherence of Light

Superposition of light has to be described by the sum of the local electric field vectors. Because of the temporal and spatial structure of the light and its wavelength, polarization and coherence properties complicated intensity structures can occur.

These intensity modulations can interact nonlinearly with matter as will be discussed in Chaps. 4 and 5. Well-designed interference patterns may result in light-induced *refractive index or absorption gratings* which are useful for nonlinear optical devices. Phase conjugating mirrors, distributed feedback lasers and very sensitive measuring techniques may serve as examples.

These processes have to be analyzed in three steps: the generation of interference patterns, the interaction of the intensity pattern with matter and the scattering of the light by the induced gratings. The first step will be described in this chapter.

2.9.1 General Aspects

The interference pattern $I(\boldsymbol{r}, t)$ can be calculated from the addition of all electric field vectors $\boldsymbol{E}_i(\boldsymbol{r}, t)$ of the superimposed light beams given by:

$$I(\boldsymbol{r}, t) = \left\{ \sum_i \boldsymbol{E}_i(\boldsymbol{r}, t) \right\}^2 \tag{2.167}$$

but only under certain preconditions is the resulting intensity pattern significantly different from the sum of the light beam intensities $I_i = \boldsymbol{E}_i^2(\boldsymbol{r}, t)$. There exists an indefinite number of possibilities for combining light beams with different spatial, temporal, spectral, coherence and polarization structures, but only a few cases are commonly discussed in more detail. Usually only two, three or four different beams are considered.

Most important is a constant phase between the different beams. This demands coherence of the single beam which has to be compared with the conditions of experiment:

coherence length > interaction length

coherence time > observation time $\tag{2.168}$

lateral coherence > transversal interaction range,

These conditions, which will be described in the following chapters in more detail, demand, e.g. sufficiently narrow spectral band widths of the beams. If the beams are orthogonally polarized no intensity pattern will occur. Nevertheless a polarization grating can be induced in polarization-sensitive materials such as liquid crystals and scattering can occur from this. The general case of any polarization can be described by the superposition of the effects from parallel and perpendicular polarization in a simple way.

Interference experiments are used for characterizing the incident light beams or a for investigating optical samples with a very high accuracy. Therefore, several types of interferometers have been developed, such as the Michelson interferometer shown in Fig. 2.28, the Mach–Zehnder interferometer with two separate arms building a rectangle and beam combined with a second beam splitter [M31], the Jamin interferometer [M31], and the Fabry–Perot interferometer, as described in Sect. 2.9.6. Different measuring techniques have been developed [see, for example, 2.117–2.130, and the references of Sect. 1.5].

2.9.2 Coherence of Light

Sufficient light coherence is necessary for interference effects, but even poor coherence can cause induced grating effects as, e.g. very short interaction lengths are applied or very short net observation times are relevant. This net observation time can be much shorter than the laser pulse duration if, e.g. the matter has very short decay times. Thus the coherence conditions and possible interference effects have to be checked carefully.

The light beam coherence [2.131–2.134] can be determined with conventional interference experiments. In complex photonic applications based on induced gratings the coherence should be investigated under the conditions of the application itself.

Coherence Length

The temporally limited coherence of the light can be described by the coherence length and the coherence time of the beam. It can be measured via the observation of the modulation depth in an interference experiment as a function of the optical delay between the beams. A setup for this purpose is shown in Fig. 2.28. The incident beam is split into two equal parts via the beam splitter BS, which can be, e.g. a 50% mirror. The single beams were reflected at the mirrors M_1 and M_2 and superimposed again at BS. By moving the mirror M_1 along the x axis of this Michelson interferometer the light beams can be delayed with respect to each other.

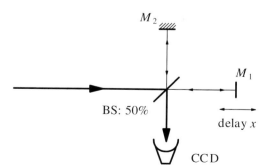

Fig. 2.28. Interferometer for measuring the coherence length l_c of a light beam

The resulting intensity modulation is measured as fringe visibility $V(x)$ via the intensity maxima $I_{max}(x)$ and minima $I_{min}(x)$ across the beam diameter measured, e.g. with a CCD camera as a function of the delay x:

fringe visibility $V(x) = \dfrac{I_{max}(x) - I_{min}(x)}{I_{max}(x) + I_{min}(x)}$ (2.169)

The coherence length l_c is defined as the delay for which the fringe visibility is decreased by $\sqrt{2}$ compared to no delay [e.g. M25]:

coherence length l_c $V(l_c) = \dfrac{1}{\sqrt{2}} V(x = 0)$ (2.170)

Sometimes $1/2$ or $1/e^2$ instead of $1/\sqrt{2}$ is used as factor and thus care has to be taken in comparing different data. In any case the coherence length is a measure of the "length" of the coherent light waves. Dividing by the wavelength of the light gives the number of oscillations in the wave which are synchronized.

Coherence length is related to the spectral bandwidth of the light source. From the full width half maximum wavelength bandwidth $\Delta\lambda_{FWHM}$ at the mid-wavelength λ_{peak} it follows that the coherence length l_c is using the above definition of the $1/\sqrt{2}$ visibility:

coherence length l_c

$$l_c = \frac{2\sqrt{2}}{\pi} \ln 2 \frac{\lambda_{peak}^2}{|\Delta\lambda_{FWHM}|} \approx 0.624 \frac{\lambda_{peak}^2}{|\Delta\lambda_{FWHM}|}$$ (2.171)

The coherence length is directly related to the coherence time τ_c and it can be calculated from (2.172) if τ_c is known.

Coherence Time

The coherence time τ_c of the light wave or pulse is defined in the same way as the coherence length (see above). It is the time the light needs to propagate over the coherence length and so it is calculated from the coherence length l_c (see (2.170)) and the velocity of light in the matter $c_{material}$ in which l_c was measured as:

coherence time $\tau_c = \dfrac{l_c}{c_{material}}$.   (2.172)

Again, multiplying by the light frequency results in the number of synchronized oscillations in the wave. Thus it is a measure of the possible accuracy reachable in interference experiments.

The coherence time is limited by the spectral bandwidth of the light emitters as given above. It can be calculated from the FWHM width $\Delta\nu$ of the spectrum or from the width $\Delta\lambda$ of the related wavelength distribution by:

$$\tau_{\mathrm{c}} = \frac{2\sqrt{2}}{\pi} \ln 2 \frac{1}{|\Delta\nu|} \approx 0.624 \frac{1}{|\Delta\nu|} \tag{2.173}$$

using the relation:

$$\Delta\nu = \nu_{\mathrm{peak}} \frac{\Delta\lambda}{\lambda_{\mathrm{peak}}} = c_{\mathrm{material}} \frac{\Delta\lambda}{\lambda_{\mathrm{peak}}^2} \tag{2.174}$$

where ν_{peak} and λ_{peak} describe the frequency and wavelength of the maximum of the spectrum and it is assumed that the bandwidths are small compared to these values. It should be noted that the coherence time so defined is different from the spontaneous life time of the emitter (about four times larger) or the pulse duration.

In addition to different emitted wavelengths in the light, phase fluctuations can disturb the coherence. If the phases of the emitters are not coupled, as in thermal light sources, phase fluctuations can decrease the coherence length and thus increase the bandwidth of the light (see Fig. 2.29).

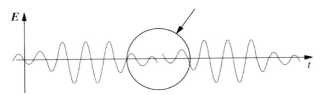

Fig. 2.29. Phase fluctuations of noncoupled emitters decrease the coherence of the light

In nonlinear experiments the coherence time has to be related to the relevant time constants of the experiment (net observation times). They can be shorter than the pulse duration of the excitation pulses. Thus, e.g. the material can have absorption recovery times of several ps. Inducing absorption gratings in such a material would demand coherence times of some ps, only, although the laser pulse may be several ns long. This also means that short coherence does not necessarily exclude induced grating effects or in other words, the appearance of induced gratings has to be checked.

In Table 2.13 the coherence length and the coherence time of several typical light sources are given. In addition the spectral bandwidth $\Delta\nu$ and the value $\nu_0/\Delta\nu$ are depicted. The mid-light-frequency ν_{peak} of these sources is assumed to be in the visible range and thus the mid-frequency is $\nu_{\mathrm{peak}} = 6 \cdot 10^{14}$ Hz.

It can be seen that the coherence properties of laser light cannot be realized with conventional light sources. The coherence of lasers can still be much better than above given values. But commercial lasers usually show

Table 2.13. Coherence length and time, bandwidth $\Delta\nu$ and $\nu_0/\Delta\nu$ of light sources given for a wavelength of 500 nm

Light source	Coherence length [mm]	Coherence time [s]	Bandwidth [Hz]	$\nu_0/\Delta\nu$
Sunlight	$3 \cdot 10^{-4}$	10^{-15}	$6 \cdot 10^{14}$	1
Spectral filter (1 nm)	0.2	$5.2 \cdot 10^{-13}$	$1.2 \cdot 10^{12}$	$5 \cdot 10^2$
Spectral lamp	190	$6.2 \cdot 10^{-10}$	10^9	$6 \cdot 10^5$
Interferometer	1900	$6.2 \cdot 10^{-9}$	10^8	$6 \cdot 10^6$
Laser	10^6	$6.2 \cdot 10^{-6}$	10^5	$6 \cdot 10^9$

bandwidths in the sub-nm range and therefore coherence lengths in the mm to m range, only.

Lateral Coherence

Light from different parts of the cross-section of a beam can be incoherent. Thus this coherence has to be checked with an interference experiment using different shares of the cross-section for a double slit or double "point source" interference experiment as shown in Fig. 2.30.

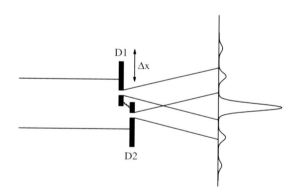

Fig. 2.30. Measuring lateral coherence: two spots out of the beam are selected by apertures D1 and D2 and interfere at the screen. The interference contrast is measured

These apertures can be shifted perpendicularly to the beam propagation direction. The lateral coherence length $l_{c,lat}$ is defined from the visibility in this experiment analogous to the longitudinal coherence length l_c as given in (2.170).

But even in the case of poor lateral coherence of the light source, as, e.g. from thermal emitters, interference experiments can be carried out if the lateral dimension of the source is considered. If different shares of the cross-section were superimposed at a planar screen the phase difference resulting from the different paths has to be smaller than $\lambda/2$. For this negligible influence of the lateral dimension of the incoherent source their distance $z_{\text{screen–source}}$ from the screen should be chosen bigger than:

$$z_{\text{screen–source}} > \frac{D_{\text{screen}} D_{\text{source}}}{\lambda} \tag{2.175}$$

with the diameter D_{screen} of the screen and D_{source} of the light source. The wavelength is described by λ. It should be noted that under this condition Fresnel's number F is not $\gg 1$ and thus the approximation of geometrical optics is not applicable.

In most photonic applications based on interference of different light beams the analysis of path lengths and phases has to be made in three dimensions. Sufficient lateral coherence of the beams is usually necessary for efficient operation.

2.9.3 Two-Beam Interference

Two light beams 1 and 2 produce an intensity modulation in the range of their superposition if the electric field vectors have a parallel component.

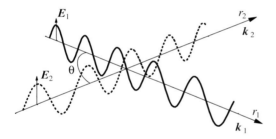

Fig. 2.31. Interference of two light fields

The perpendicular component of the electric fields will show a modulation of the direction of the resulting electric field vector but the intensity will be constant as the sum of the intensities of the two beams, as described in the next chapter.

The intensity pattern from (2.167) can be analyzed using the parallel components of the electric field vector $\boldsymbol{E}_1$ and $\boldsymbol{E}_2$ written in complex form (2.39) but considering the transversal structure of, e.g. a Gaussian beam in the amplitude $\boldsymbol{E}_{0/1,2}(\boldsymbol{r}_{1,2})$:

$$\boldsymbol{E}_{1/2} = \frac{E_{0,1/2}(\boldsymbol{r})}{2} \, \mathrm{e}^{\mathrm{i}(2\pi\nu_{1/2}t \pm \boldsymbol{k}_{1/2}\cdot\boldsymbol{r} + \varphi_{1/2})} + c.c. \tag{2.176}$$

or not complex:

$$\boldsymbol{E}_{1/2} = \boldsymbol{E}_{0,1/2}(\boldsymbol{r}) \cos(2\pi\nu_{1/2}t - \boldsymbol{k}_{1/2}\cdot\boldsymbol{r} + \varphi_{1/2}) \tag{2.177}$$

with wave vectors $\boldsymbol{k}_1$ and $\boldsymbol{k}_2$, frequencies ν_1 and ν_2 and the phases φ_1 and φ_2.

The general description can often be simplified by assuming spectral degeneration of the two beams $\nu_1 = \nu_2$ and fixed phases. With respect to photonic applications the slowly varying part of the modulation may be important. Thus the analysis may be averaged over the time period $1/2\pi\nu$. The intensity I_{sum} is then calculated using (2.45) from:

$$I = \frac{c_0\varepsilon_0 n}{2}\overline{(\boldsymbol{E}_1 + \boldsymbol{E}_2)^2} \tag{2.178}$$

to:

$$I_{\text{ges}} = \frac{c_0\varepsilon_0 n}{2}\left[E_{0,1}^2 + E_{0,2}^2 + 2E_{0,1}E_{0,2}\cos\{\Delta\boldsymbol{k}\cdot\boldsymbol{r} + \Delta\varphi\}\right] \tag{2.179}$$

or

$$I_{\text{ges}} = I_1 + I_2 + 2\sqrt{I_1 I_2}\cos\left\{\frac{2\pi}{\lambda}(\boldsymbol{r}_2 - \boldsymbol{r}_1) + \Delta\varphi\right\}. \tag{2.180}$$

The spatial cosine modulation $I_{\text{max}}/I_{\text{min}}$ is maximum if the amplitudes $E_{0,1}$ and $E_{0,2}$ are equal. The maximum intensity in the interference structure is than 4 times $I = I_1 = I_2$ and the minimum is 0. The modulation wavelength Λ of the intensity maxima in the direction $\boldsymbol{r}_1 - \boldsymbol{r}_2$, which is transversal to the average propagation direction $\boldsymbol{r}_1 + \boldsymbol{r}_2$ of the two beams, is a function of the angle θ between the two beams and their wavelength λ:

$$\Lambda = \lambda\sin\left(\frac{\theta}{2}\right). \tag{2.181}$$

It is zero for parallel beams and maximal as λ for antiparallel beams. This effect is used, e.g. for tuning the emission wavelength of distributed feedback dye lasers by changing the grating constant via the angle of excitation (see Sect. 6.10.4).

As an example the interference pattern resulting from two Gaussian beams crossing in their waist region is shown in Fig. 2.32.

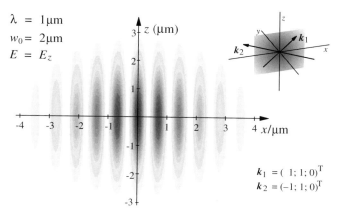

$\lambda = 1\,\mu m$
$w_0 = 2\,\mu m$
$E = E_z$

$\boldsymbol{k}_1 = (\ 1; 1; 0)^T$
$\boldsymbol{k}_2 = (-1; 1; 0)^T$

Fig. 2.32. Interference intensity pattern from two spectrally degenerate Gaussian beams with different propagation directions superimposed at waist

A planar cut through this pattern perpendicular to the plane paper of the results in interference stripes as shown in Fig. 2.33.

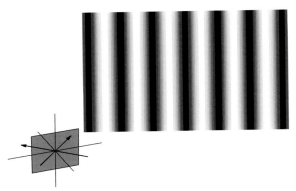

Fig. 2.33. Interference intensity pattern from two spectrally degenerate Gaussian beams superimposed at the waist in the xy plane of Fig. 2.32 which is perpendicular to the paper plane

Two beams with different frequencies result in temporal intensity modulations which will be described in Sect. 2.9.6.

2.9.4 Superposition of Two Vertical Polarized Light Beams

The superposition of two beams with vertical linear polarization does not produce an intensity modulation independent of the direction and phase of the light. But as a function of these parameters the direction of the vector of the resulting electric field will usually vary in space and time in a complicated manner. This vector has to be calculated from the vector addition of the two electric field vectors of the beams. Its direction will always be in the plane of the two polarizations.

If two spectrally degenerate light beams with linear polarization are applied the resulting field will show a grating structure in space:

$$\boldsymbol{E}_{\text{total}} = \boldsymbol{E}_{0,1}\,\text{e}^{\text{i}(2\pi\nu t - \boldsymbol{k}_1\cdot\boldsymbol{r})} + \boldsymbol{E}_{0,2}\,\text{e}^{\text{i}(2\pi\nu t - \boldsymbol{k}_2\cdot\boldsymbol{r} + \Delta\varphi)} \tag{2.182}$$

with the wave vectors $\boldsymbol{k}_1$ and $\boldsymbol{k}_2$ of the two beams, the light frequency $\nu_{1/2} = \frac{c}{2\pi}|\boldsymbol{k}_{1/2}|$ and the phase difference $\Delta\varphi$ between the beams. The resulting nonmodulated intensity can be calculated using (2.45) from:

$$I_{\text{total}} = \frac{c_0\varepsilon_0 n}{2}\left[E_{0,1}^2 + E_{0,2}^2 + \underbrace{2\boldsymbol{E}_{0,1}^2\cdot\boldsymbol{E}_{0,2}^2\,\cos\{\boldsymbol{k}\cdot(\boldsymbol{r}_1 - \boldsymbol{r}_2) + \Delta\varphi\}}_{=0}\right]$$

$$= I_1 + I_2 \tag{2.183}$$

The direction of the field vector can be calculated from:

$$\boldsymbol{E}_{\text{total}} = E_0 \Big[\cos(2\pi\nu t)(\boldsymbol{e}_1 \cos(\boldsymbol{k}_1 \cdot \boldsymbol{r}) + \boldsymbol{e}_2 \cos(\boldsymbol{k}_2 \cdot \boldsymbol{r}))$$
$$- \sin(2\pi\nu t)(\boldsymbol{e}_1 \sin(\boldsymbol{k}_1 \cdot \boldsymbol{r}) + \boldsymbol{e}_2 \sin(\boldsymbol{k}_2 \cdot \boldsymbol{r})) \Big]. \tag{2.184}$$

In case one beam is polarized perpendicular and the other parallel to the plane of the two wave vectors the electric field vector will rotate perpendicular to the wave vector of the latter beam (see Fig. 2.34).

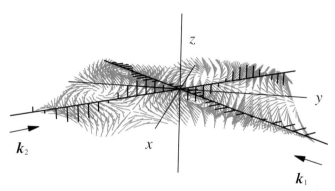

Fig. 2.34. Direction of resulting electric field vector for superposition of two spectrally degenerate beams with different propagation directions and perpendicular polarization to each other, one in the plane of the paper and the other vertical

In the case of 45° polarization of both beams to the plane of their propagation directions the resulting electric field will rotate in a plane of 45° to the propagation plane (see Fig. 2.35).

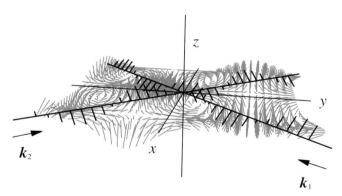

Fig. 2.35. Direction of resulting electric field vector for superposition of two spectrally degenerate beams with different propagation directions and perpendicular polarizations to each other and 45° to the plane of the paper

If these two beams propagate in parallel the resulting beam will be elliptically polarized which can degenerate to circular or linear polarized light as a function of the phases and amplitudes of the two beams.

2.9.5 One-Dimensional Multibeam Interference

Multibeam interference occurs in devices with two reflecting surfaces as, e.g. Fabry–Perot filters. In these cases the light beams have the same wave vector direction. The analysis shall be focused on beams with the same wavelength. For the superposition of p equal light beams with parallel electric field vectors but with constant phase difference Φ:

$$\left.\begin{array}{l} E_1 = \mathrm{Re}\{E_0\,\mathrm{e}^{\mathrm{i}(2\pi\nu t - kz)}\} \\ E_2 = \mathrm{Re}\{E_0\,\mathrm{e}^{\mathrm{i}(2\pi\nu t - kz + \Phi)}\} \\ \vdots \\ E_p = \mathrm{Re}\{E_0\,\mathrm{e}^{\mathrm{i}(2\pi\nu t - kz + (p-1)\Phi)}\} \end{array}\right\} \tag{2.185}$$

the superposition leads to:

$$\begin{aligned} E_{\mathrm{ges}} &= E_1 + E_2 + \cdots E_p \\ &= \mathrm{Re}\{E_0\,\mathrm{e}^{\mathrm{i}(2\pi\nu t - kz)}[1 + \mathrm{e}^{\mathrm{i}\Phi} + \cdots + \mathrm{e}^{\mathrm{i}(p-1)\Phi}]\} \end{aligned} \tag{2.186}$$

which results in:

$$\boldsymbol{E} = \mathrm{Re}\left\{\boldsymbol{E}_0\,\mathrm{e}^{\mathrm{i}(2\pi\nu t - kz)}\frac{1 - \mathrm{e}^{\mathrm{i}p\Phi}}{1 - \mathrm{e}^{\mathrm{i}\Phi}}\right\} \tag{2.187}$$

and the intensity in the slowly varying amplitude approximation follows from this with (2.45) as:

$$I = \frac{c_0\varepsilon_0 n}{2}\,\frac{E_0^2}{2}\,\frac{\sin^2(p\Phi/2)}{\sin^2(\Phi/2)} \tag{2.188}$$

which is illustrated for different numbers p of interfering beams in Fig. 2.36.

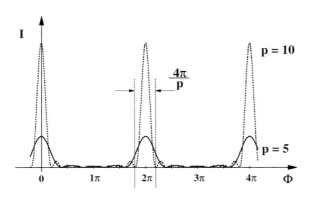

Fig. 2.36. Intensity of p interfering collinear spectrally degenerate beams as a function of their phase shift Φ

The width of these intensity peaks decreases linearly with the number p of interfering beams and the maximum intensity increases with p^2. Thus the higher p the more sensitive is the device to the phase shift. If the same device is used for beams with different wavelengths they are better distinguishable in case of larger p.

If the amplitudes of the interfering single beams decrease with p, as occurs in reflection at a mirror with reflectivity $R < 100\%$,

$$\boldsymbol{E}_{0,p} = \boldsymbol{E}_0 R^p \text{ with } R < 1 \tag{2.189}$$

and infinite reflections are assumed, this formula modifies with (2.45) to:

$$I = \frac{c_0 \varepsilon_0 n E_0^2}{4(1 + R^2 - 2R \cos\varPhi)} \tag{2.190}$$

and Fig. 2.37 shows the influence of R.

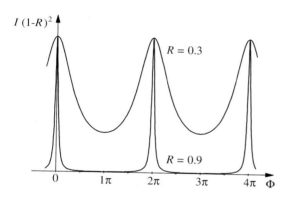

Fig. 2.37. Intensity of infinite interfering collinear spectrally degenerate beams with decreasing amplitude by R^p as a function of their phase shift $\varPhi$

This interference is applied for the investigation of the spectral structure of light or for spectral filtering using interference filters and Fabry–Perot interferometers.

2.9.6 Fabry–Perot Interferometer

The Fabry–Perot interferometer (also called the *Fabry–Perot etalon*) consists of two high-quality planar reflectors in a parallel arrangement as shown in Fig. 2.38.

At the mirror surfaces the R_eth share of the electric light field will be reflected and the T_eth share will be transmitted. The index e indicates the reflectivity and transmission related with the electric field of the light. The reflectivity R and the transmission T related to the intensity will be:

$$R = R_e^2 \text{ and } T = T_e^2. \tag{2.191}$$

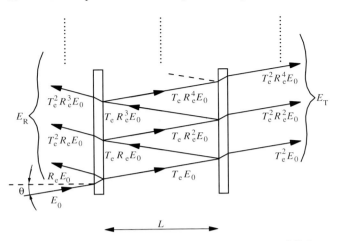

Fig. 2.38. Interference of reflected and transmitted light at a Fabry–Perot interferometer of length L and with two mirrors with reflectivity R and the transmission T

The transmitted and reflected light waves will interfere with their phase relations as a function of the incident wavelength and the thickness and reflectivity of the interferometer. The path length difference between one transmitted beam and the next transmitted beam, which is twice more reflected, is equal to $\Delta x = l_1 + l_2$, as shown in Fig. 2.39.

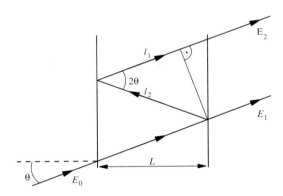

Fig. 2.39. Path length difference between two neighboring transmitted beams in a Fabry–Perot interferometer

This path length difference Δz follows from:

$$\Delta z = l_1 + l_2 = \frac{L}{\cos\theta} + \frac{L\cos 2\theta}{\cos\theta} = 2L\cos\theta \tag{2.192}$$

resulting in a phase difference Φ:

$$\Phi = |\boldsymbol{k}|\Delta z = \frac{4\pi}{\lambda}L\cos\theta \tag{2.193}$$

with light wavelength λ. The total transmitted field $E_{\text{transmitted}}$ follows from:

$$E_{\text{transmitted}} = E_0 T_e^2 \sum_{m=0}^{\infty} R_e^{2m} \, e^{im\Phi}$$

$$= E_0 T_e^2 \frac{1}{1 - R_e^2 \, e^{i\Phi}} \tag{2.194}$$

and thus the total transmitted intensity $I_{\text{transmitted}}$ is given by:

$$I_{\text{transmitted}} = I_0 \frac{T^2}{(1 - R \, e^{i\Phi})^2} \tag{2.195}$$

with transmission T and reflectivity R related to the intensity. The phase shift Φ can also contain possible additional phase shifts from the reflection at the mirrors.

The formula can be written in real form as:

$$I_{\text{transmitted}} = I_0 \frac{T^2}{(1 - R)^2} \frac{1}{1 + \dfrac{4R}{(1 - R)^2} \sin^2\left(\dfrac{\Phi}{2}\right)} \tag{2.196}$$

including the Airy function:

Airy function $\quad f(\Phi) = \dfrac{1}{1 + \dfrac{4R}{(1 - R)^2} \sin^2\left(\dfrac{\Phi}{2}\right)} \tag{2.197}$

which describes the total transmission of a Fabry–Perot interferometer with no absorption losses ($T + R = 1$) as shown in Fig. 2.40.

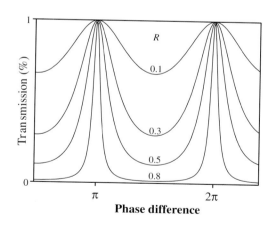

Fig. 2.40. Total transmission (Airy function) of a loss-free Fabry–Perot interferometer as a function of the phase difference $\Phi = k\Delta z$ measured in π for different reflectivity values R

It has to be noticed that even Fabry–Perot interferometers with very high reflectivity values $R > 0.999$ will show 100% total transmission if the wavelength is tuned to the path length L. The mth transmission maximum occurs at $\Phi_{\text{max}} = 2m\pi$ and thus it follows that position of transmission maxima $\quad \dfrac{L}{\lambda} \cos\theta = \dfrac{m}{2}$ $\qquad$ (2.198)

and the transmission minima occur in the middle between these values. The minima will show transmissions of:

transmission minima $\quad T_{\text{minimum}} = \dfrac{T^2}{(1+R)^2} \approx \dfrac{T^2}{4}$ $\qquad$ (2.199)

with the approximation for large reflectivities $R \approx 1$.

The full width half maximum spectral width $\Delta\lambda_{\text{FWHM}}$ of the device follows from:

spectral width $\quad \Delta\lambda_{\text{FWHM}} = \dfrac{4\pi L \cos\theta}{\arcsin\left(\dfrac{1-R}{2\sqrt{R}}\right)} \approx 8L \dfrac{\pi\sqrt{R}}{1-R}$ $\qquad$ (2.200)

again with the approximation for large R and in addition for perpendicular incidence.

The ratio of the free spectral range which is the distance of the wavelength maxima $\Delta\lambda_{\text{freespec}}$ divided by the spectral width $\Delta\lambda_{\text{FWHM}}$ is given by the finesse F:

finesse $\quad F = \dfrac{\Delta\lambda_{\text{freespec}}}{\Delta\lambda_{\text{FWHM}}} = \dfrac{\pi\sqrt{R}}{1-R}.$ $\qquad$ (2.201)

For an interferometer with two mirrors with different reflectivities R_1 and R_2 the finesse is given by:

$$F = \dfrac{\pi(R_1 R_2)^{1/4}}{1-(R_1 R_2)^{1/2}}.$$ $\qquad$ (2.202)

For perpendicular incidence, $\theta = 0$, it follows that the total transmission of the etalon with equal reflectivity R for both mirrors T_{total}:

$$T_{\text{total}} = \dfrac{I_{\text{transmitted}}}{I_0} = \dfrac{T^2}{(1-R)^2 + 4R \sin^2(2\pi L/\lambda)}$$ $\qquad$ (2.203)

and the spectral resolution is then:

spectral resolution $\quad \dfrac{\lambda}{\Delta\lambda_{\text{FWHM}}} = \dfrac{2L}{\lambda}\dfrac{\pi\sqrt{R}}{(1-R)} = \dfrac{2L}{\lambda}F$ $\qquad$ (2.204)

If the Fabry–Perot etalon is realized as a thick material plate, e.g. from glass, with the refractive index n and, for reflection, with two polished and coated surfaces, the wavelength has to be replaced in all formulas by λ/n.

The Fabry–Perot etalon can be used for measuring spectral features with high spectral resolution. An etalon of 10 mm thickness with a reflectivity of 95% of both mirrors will have at a wavelength of 1 μm a spectral bandwidth of 0.83 pm which is equal to a frequency difference of 250 MHz. The finesse of this etalon is $F = 60$.

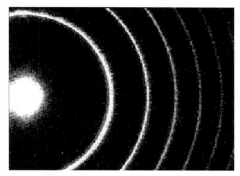

increasing frequency ——————➤

Fig. 2.41. Interference pattern behind a Fabry–Perot interferometer produced by a single mode laser beam. The parameters were $L = 40\,\text{mm}$, $R = 95\%$, $\Delta\nu_{\text{freespec}} = 3.75\,\text{GHz}$, $F = 61$, $\Delta\nu = 62\,\text{MHz}$ and $\lambda = 1064\,\text{nm}$

If the etalon is illuminated with a slightly diverging light beam the interference will produce a ring structure as shown in Fig. 2.41. In these interference rings the larger wavelengths are observable at the inner rings.

Besides their use as a measuring device, Fabry–Perot etalons are used in lasers for decreasing the spectral width of the radiation and the laser resonator itself can be treated as a Fabry–Perot interferometer.

In practical applications the finesse F of a Fabry–Perot etalon is decreased by the roughness of the optical surfaces of the mirrors. The surface quality is measured as the roughness $\Delta x_{\text{surface}}$ which is the deviation from planarity and is compared to the wavelength of the light λ_{light}:

$$\text{roughness} \quad \Delta x_{\text{surface}} = \frac{1}{m}\lambda_{\text{light}} \tag{2.205}$$

which results in the determination of the deviator m which is typically in the range of 2–100. The surface finesse is then given as a function of m by:

$$\textbf{roughness finesse} \quad F_{\text{roughness}} = \frac{m}{2} \tag{2.206}$$

and thus the final finesse of the Fabry–Perot etalon F_{total} as a device is given by:

$$\frac{1}{F_{\text{total}}} = \frac{1}{F_{\text{R}}} + \frac{1}{F_{\text{roughness}}} \tag{2.207}$$

with the finesse F_{R} resulting from the reflection at the mirrors as given in (2.201) and (2.202).

Therefore the quality of the mirror substrates has to be in the range $m > F_{\text{R}}$ to take advantage of the high reflectivity of the mirrors. This can result in very expensive devices. In addition it should be noted that the planarity of plane substrates is often specified at a very long wavelength as, e.g. 1 or even $10\,\mu\text{m}$ whereas the applied wavelength has to be used for the calculations.

2.9.7 Light Beats: Heterodyne Technique

If two light beams with the same amplitude but different frequencies with the electric field amplitudes E_1 and E_2

$$E_{1/2} = \text{Re}\left\{E_0\,e^{i2\pi\nu_{1/2}\left(t-\frac{z}{c}\right)}\right\} \tag{2.208}$$

are superimposed collinearly, the resulting field is:

$$E = E_1 + E_2 = \text{Re}\left\{E_0\,e^{i2\pi\nu_0\left(t-\frac{z}{c}\right)}\cdot\left[e^{-i\pi\Delta\nu\left(t-\frac{z}{c}\right)} + e^{i\pi\Delta\nu\left(t-\frac{z}{c}\right)}\right]\right\} \tag{2.209}$$

with

$$\nu_0 = \frac{\nu_1 + \nu_2}{2} \quad \text{and} \quad \Delta\nu = |\nu_1 - \nu_2|. \tag{2.210}$$

In practical cases the two frequencies can be similar and then the resulting intensity using (2.45) shows a slowly varying modulation with frequency $\Delta\nu$:

$$I = c_0\varepsilon_0 n E_0^2 \cos^2\left(\pi\Delta\nu\left[t - \frac{z}{c}\right]\right). \tag{2.211}$$

This can be used for electronically based detection of small differences in light frequencies up to values of several GHz which is of the order of 10^{-6} of the light frequency as shown in Fig. 2.42.

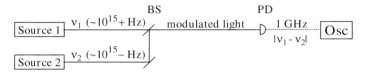

Fig. 2.42. Heterodyne technique for detection of small frequency differences of two light beams with photodiode PD and oscilloscope Osc

In this heterodyne technique the photodiode in combination with an oscilloscope acts as a low-frequency pass filter and detects the slowly varying part of the intensity and not the light frequency itself. Thus as a function of the detection limit of the photodetector and the electronic measuring device, beat frequencies up to 10^{9-10} Hz can be detected.

As an example the superposition of two pulses with slightly different wavelengths is shown in Fig. 2.43. The first pulse was generated by a Q-switched Nd laser with $\lambda_{\text{laser}} = 1064\,\text{nm}$ and the second and delayed pulse is the reflected signal from stimulated Brillouin scattering (see Sect. 4.5.8) in CO_2 at 56 bar. The Fourier analysis of this pulse shows a frequency component of 430 MHz which belongs to the pulse distance of 2.3 ns as shown in the figure. This is the frequency of the hyper-sound wave of the Brillouin process in the nonlinear material.

The method will be more precise for small differences below 1 GHz which are otherwise very difficult to measure. Many schemes have been developed

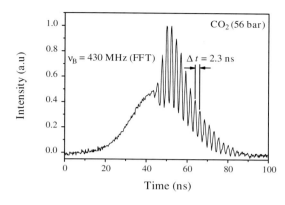

Fig. 2.43. Heterodyne measurement of two light pulses with slightly different frequencies. The second pulse was shifted by stimulated Brillouin scattering in CO_2. The beat frequency of 430 MHz is the frequency of the hyper-sound wave in CO_2

using this method for high-precision measurements [see, for example, 135–149].

2.9.8 Frequency Spectrum of Light Pulses

Light pulses with temporal intensity profile $I(t)$ contain a mixture of different frequencies. The intensity distribution over the frequencies can be calculated by a Fourier analysis:

$$\breve{I}(\nu, \nu_0) = \frac{1}{\sqrt{2\pi}} \int_{-\infty}^{\infty} I(t)\, e^{\pm i 2\pi \nu t}\, dt \qquad (2.212)$$

with the back transformation:

$$I(t) = \frac{1}{\sqrt{2\pi}} \int_{-\infty}^{\infty} \breve{I}(\nu, \nu_0)\, e^{\pm i 2\pi \nu t}\, d\nu \qquad (2.213)$$

For a bandwidth limited *Gaussian temporal shaped pulse* as described in Sect. 2.7.2

$$I(t) = I_0 \exp\left[-4\ln 2 \left(\frac{t - t_0}{\Delta t_{\text{FWHM}}}\right)^2\right] \qquad (2.214)$$

the spectrum follows from

$$\breve{I}(\nu, \nu_0) = I_0 \frac{\Delta t_{\text{FWHM}}^2}{4\ln 2} \exp\left[-\frac{4\pi^2(\nu - \nu_0)^2 \Delta t_{\text{FWHM}}^2}{4\ln 2}\right] \qquad (2.215)$$

which is a Gaussian shaped frequency spectrum. The full width half maximum bandwidth of this spectrum follows from the duration of the pulse width by:

Gaussian bandwidth $\Delta \nu_{\text{FWHM}} = \dfrac{2\ln 2}{\pi} \Delta t_{\text{FWHM}}.$ (2.216)

This is illustrated in Fig. 2.44 for a light pulse with an average wavelength of 500 nm and a pulse length of 1 ps resulting in a FWHM width of $\Delta \nu = 441$ GHz or $\Delta \lambda = 0.368$ nm.

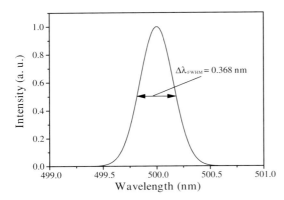

Fig. 2.44. Fourier spectrum of a light pulse with an average wavelength of 500 nm and a pulse length of 1 ps (FWHM) showing a spectral FWHM width of 441 GHz

The same band width limited light beam but with a pulse width of 10 ns would have a 10^4 times narrower bandwidth of 0.037 pm. Thus shorter pulse widths scale linearly with larger frequency spectra which have to be provided by the light emitter. Thus by using shorter pulses spectral resolution becomes worse in the same manner.

Lorentzian line shapes result in a temporal pulse structure constructed from two exponential curves with a peak at the center. The half-width of the line is related to the $1/e$ value of the pulse as described in Sect. 2.1.2.

3. Linear Interactions Between Light and Matter

Linear or conventional optics is the basis of all photonic applications. In these linear interactions of light with matter the relative change of the intensity is not a function of the intensity. Thus in these conventional optical experiments the applied incident intensity is not important and often not even measured. This is in contrast to nonlinear interactions which are crucially dependent on the incident intensity, as will be described in the next chapter.

Dispersion, refraction, reflection, absorption, luminescence, birefringence, optical activity and diffraction are observable in the linear range. But all these linear interactions may become nonlinear if the intensity is high enough.

In this chapter the linear effects will be described briefly to provide the basics for the nonlinear interactions which are discussed in the following chapters. For a more detailed description, especially of the usual linear applications, common optics textbooks should also be used [e.g. M3, M5, M16, M21, M27, M31, M32, M38, M42]. Some aspects of linear interactions were already discussed in the previous chapter, e.g. in Sect. 2.5 for ray matrices.

3.1 General Description

As mentioned in Sects. 2.2 and 2.3 the interaction of light with matter can be described by Maxwell's equations [M21]:

$$\text{curl } \boldsymbol{E} = -\frac{\partial \boldsymbol{B}}{\partial t} \tag{3.1}$$

$$\text{curl } \boldsymbol{H} = -\frac{\partial \boldsymbol{D}}{\partial t} \tag{3.2}$$

$$\text{div } \boldsymbol{D} = \rho \tag{3.3}$$

$$\text{div } \boldsymbol{B} = 0 \tag{3.4}$$

with

electric field	$\boldsymbol{E}$	$\mathrm{V\,m}$
magnetic field	$\boldsymbol{H}$	$\mathrm{A\,m}$
electric displacement	$\boldsymbol{D}$	$\mathrm{A\,s\,m^{-2}}$
magnetic induction	$\boldsymbol{B}$	$\mathrm{V\,m^{-2}}$
current density	$\boldsymbol{j}$	$\mathrm{A\,m^{-2}}$
electrical charge density	ρ	$\mathrm{A\,s\,m^{-3}}$

In these equations the interaction with the material is described by $\boldsymbol{D}$, $\boldsymbol{B}$, $\boldsymbol{j}$ and ρ. This leads to the additional equations:

$$\boldsymbol{D} = \varepsilon_0 \boldsymbol{E} + \boldsymbol{P}(\boldsymbol{E}) \tag{3.5}$$

$$\boldsymbol{B} = \mu_0 \boldsymbol{H} + \boldsymbol{J}(\boldsymbol{H}) \tag{3.6}$$

$$\boldsymbol{j} = \xi \boldsymbol{E} \tag{3.7}$$

with the values:

electrical polarization P $\mathrm{A\,s\,m^{-2}}$
magnetic polarization J $\mathrm{V\,s\,m^{-2}}$
electrical conductivity ξ $\mathrm{A\,V\,m}$

In linear interactions the changes in the matter are linearly dependent on the incident fields and thus:

$$\boldsymbol{P}(\boldsymbol{r}, t) = \varepsilon_0 \, \chi(\boldsymbol{r}) \, \boldsymbol{E}(\boldsymbol{r}, t) \tag{3.8}$$

$$\boldsymbol{J}(\boldsymbol{r}, t) = \mu_0 \, \chi_{\mathrm{m}}(\boldsymbol{r}) \, \boldsymbol{H}(\boldsymbol{r}, t) \tag{3.9}$$

using

electric susceptibility χ
magnetic susceptibility χ_{m}.

With these values the material equations result in:

$$\boldsymbol{D}(\boldsymbol{r}, t) = \varepsilon_0 \boldsymbol{E}(\boldsymbol{r}, t) + \varepsilon_0 \chi(\boldsymbol{r}) \boldsymbol{E}(\boldsymbol{r}, t) = \varepsilon_0 \varepsilon_r(\boldsymbol{r}) \boldsymbol{E}(\boldsymbol{r}, t) \tag{3.10}$$

$$\boldsymbol{B}(\boldsymbol{r}, t) = \mu_0 \boldsymbol{H}(\boldsymbol{r}, t) + \mu_0 \chi_{\mathrm{m}}(\boldsymbol{r}) \boldsymbol{H}(\boldsymbol{r}, t) = \mu_0 \mu_r(\boldsymbol{r}) \boldsymbol{H}(\boldsymbol{r}, t) \tag{3.11}$$

with the:

electric permittivity $\varepsilon_{\mathrm{r}} = 1 + \chi$ (3.12)

magnetic permeability $\mu_{\mathrm{r}} = 1 + \chi_{\mathrm{m}}$.   (3.13)

The material parameters ξ, χ, χ_{m}, ε_{r} and μ_{r} are tensors with nine components in general. In the case of isotropic materials they can degenerate to scalar numbers.

Because of the linearity of the interaction the theoretical analysis can be based on the superposition of fields with different spatial, temporal, spectral and polarization components resulting, e.g. in pulsed light beams with certain spectral distributions and polarization. Thus the single light matter interaction can be described by the wave equation for the electric field vector of monochromatic light as given in the previous chapter. Assuming isotropic matter this results in:

light matter interaction $\Delta \boldsymbol{E} - \dfrac{1}{c_0^2} \dfrac{\partial^2 \boldsymbol{E}}{\partial t^2} = \mu_0 \dfrac{\partial^2 \boldsymbol{P}(\boldsymbol{E})}{\partial t^2}$ (3.14)

As mentioned above in photonics the influence of the magnetic component of the light can usually be neglected. Thus μ_{r} can mostly be set equal to 1.

The tensor of the electrical susceptibility χ is generally complex counting for absorption with the imaginary part $\chi_{\text{absorption}}$ and for phase changes with the real part χ_{phase}. The refractive index then follows from:

$$\text{refractive index} \quad n = \sqrt{1 + \chi_{\text{phase}} - i\chi_{\text{absorption}}}. \tag{3.15}$$

which is also a complex tensor. The detailed description of linear interactions can require the use of these equations directly. The material parameters ξ, χ, χ_m, ε_r and μ_r can be derived from a detailed quantum mechanical description of the interaction. But often phenomenological equations, as given in the following chapters, are sufficient.

The interaction of light with matter results from the forces of Coulomb and Lorentz to the charged particles. This force F acting on an electron with charge e is given by:

$$\boldsymbol{F} = e\boldsymbol{E} + e\boldsymbol{v} \times \boldsymbol{B}. \tag{3.16}$$

If this electron with the mass m_e is elastically bound in a parabolic potential as in the linear approximation and the magnetic force is neglected the differential equation for the linear motion of this electron in the x direction is given by:

$$m_e\ddot{x} + \frac{1}{\tau}m_e\dot{x} + 2\pi\nu_0 m_e x = e|\boldsymbol{E}(\nu)| = E_0\, e^{i2\pi\nu t} \tag{3.17}$$

which can be solved to give:

$$x = \frac{1}{(2\pi\nu_0)^2 - (2\pi\nu)^2 + i2\pi\nu/\tau} \frac{eE_0\, e^{i2\pi\nu t}}{m_e}. \tag{3.18}$$

If no interaction takes place between these electrons (in the linear approximation) the polarization $P(t)$ of the matter with an electron density N_0 is given by:

$$P(t) = eN_0 x(t) \tag{3.19}$$

Using (3.8) the complex susceptibility χ can be determined:

$$\begin{aligned} \chi &= \chi_{\text{phase}} + i\chi_{\text{absorbtion}} \\ &= \left[\frac{1}{(2\pi\nu_0)^2 - (2\pi\nu)^2 + i2\pi\nu/\tau}\right] \frac{e^2 N_0}{\varepsilon_0 m_e} \end{aligned} \tag{3.20}$$

with components:

$$\chi_{\text{phase}} = \left[\frac{\nu_0^2 - \nu^2}{(2\pi)^2(\nu_0^2 - \nu^2) + (\nu/\tau)^2}\right] \frac{e^2 N_0}{\varepsilon_0 m_e} \tag{3.21}$$

$$\chi_{\text{absorbtion}} = \left[\frac{2\pi\nu/\tau}{(2\pi)^4(\nu_0^2 - \nu^2) + (2\pi\nu/\pi)^2}\right] \frac{e^2 N_0}{\varepsilon_0 m_e}. \tag{3.22}$$

The real and imaginary parts of this value show a resonance structure at frequencies close to ν_0 as shown in Fig. 3.1.

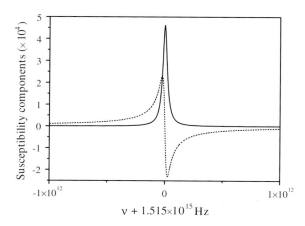

Fig. 3.1. Components of the complex susceptibility of matter with electrons in parabolic potential as a function of the light frequency ν around resonance ν_0. The solid line shows the absorptive and the dashed line the phase component

The imaginary part of this function describes a Lorentzian "absorption" function with a width (FWHM) of $\Delta\nu = 1/2\pi\tau$. At the positions $\nu_0 \pm \Delta\nu/2$ the real part of χ, the phase function, shows a maximum and a minimum, respectively. At large distances from resonance the absorption is negligible and the phase change dominates the interaction, resulting in the dispersion of the material.

In the case of small electron density N_0, meaning also small absorption, the real refractive index is:

$$\text{real refractive index} \quad n_{\text{real}} = \sqrt{1 + \chi_{\text{phase}}} = \sqrt{\varepsilon_r} \tag{3.23}$$

whereas the absorption coefficient a is given by the expression:

$$\text{absorption coefficient} \quad a = \frac{2\pi\nu_0}{c} \frac{\chi_{\text{absorbtion}}}{\sqrt{1 + \chi_{\text{phase}}}}. \tag{3.24}$$

This simple model describes the basics of amplitude and phase changes of the electric field by absorption and dispersion of light from a single absorption transition. This phenomenological description may be helpful in understanding the fundamental principles of linear optics. In nonlinear optics the parameters of the interaction will be functions of the intensity and thus the description is more complicated. Usually quantum effects have to be explicitly considered in nonlinear optics.

From the above equations it follows that in linear optics the real part and the imaginary part of χ can be determined from each other if the spectrum of one part is known for all frequencies. For a measured discrete absorption line spectrum $a(m \cdot \Delta\nu)$ at frequency $m \cdot \Delta\nu$ the discrete values of the dispersion $n(p \cdot \Delta\nu)$ at the frequencies $p \cdot \Delta\nu$ follows from the Kramers–Kronig relation [M5, M34]:

$$n(p \cdot \Delta\nu) - n(\nu = \infty) = \frac{1}{2\pi^2 \Delta\nu} \sum_{m=0, \neq p}^{\infty} \frac{1 - (-1)^{m+p}}{m^2 - p^2} a(m \cdot \Delta\nu) \qquad (3.25)$$

if the spectra are given with the step width $\Delta\nu$.

3.2 Refraction and Dispersion

Conventional optical elements such as lenses, prisms and fibers are based on the refraction and dispersion of light as a consequence of a refractive index larger than one in the material. Thus the speed of light is different in matter compared to the vacuum and is a function of its wavelength or frequency.

If light frequencies are much smaller or larger than the resonance or absorption frequencies ν_0 of the matter, this nonresonant interaction – without absorption – is dominated by phase changes of the light wave. This interaction is based on the forced oscillation of electric dipoles in the matter with the light frequency. The speed of the phase of the light wave c_p is reduced from the vacuum light speed c_{vacuum} to:

$$\text{phase light velocity} \quad c_p = \frac{c_{vacuum}}{n_{matter}} = \frac{1}{\sqrt{\varepsilon_0 \mu_0 \varepsilon_r \mu_r}}$$
$$= \nu_{light} \lambda_{in\text{-}matter} \qquad (3.26)$$

where n_{matter} describes the usual (real) refractive index of the material. In optical materials $\mu_r \simeq 1$ applies as mentioned above. The light frequency in matter is unchanged but the wavelength $\lambda_{in\text{-}matter}$ is shortened to:

$$\lambda_{in\text{-}matter} = \frac{\lambda_{vacuum}}{n_{matter}}. \qquad (3.27)$$

The refractive index is close to 1 for gases but reaches values of more than 2 for special crystals (see Table 3.1). The refractive index can be determined using different methods as given, for example, in [3.1–3.4], or via the measurement of Brewster's angle as described in Sect. 3.5.2 and the references therein.

Table 3.1. Refractive indices of some gases liquids and solids

Material	n	λ (nm)	Material	n	λ (nm)
Air	1.00029	546	Plexiglass	1.49	546
CO_2	1.00045	546	Diamond	2.42	546
Water	1.33	546	Nd:YLF	1.45	1060
Ethanol	1.36	546	Ruby	1.76	694.3
Benzene	1.51	546	Ti:Al$_2$O$_3$	1.76	735
CS_2	1.63	546	Nd:YAG	1.82	1064
Quartz	1.46	546	Nd:YALO	1.90	1078

If the light is a mixture of frequencies the speed of these components will be different as a consequence of a varying refractive index. The refractive index variation as a function of the light frequency $n(\nu)$ or wavelength $n(\lambda)$ is called dispersion:

dispersion $n(\lambda)$ or $n(\nu)$ $\hspace{5cm}$ (3.28)

For spectral ranges without absorption we observe normal dispersion. The conventional refractive index, which is the real part of the complex refractive index, will decrease with increasing wavelength in these ranges:

normal dispersion $\dfrac{\mathrm{d}n}{\mathrm{d}\lambda} < 0$ $\hspace{4cm}$ (3.29)

Thus, e.g. refraction at an air–glass interface leads to a spreading of the colors of a white light beam as sketched in Fig. 3.2.

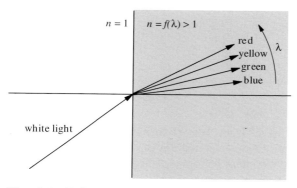

Fig. 3.2. Refraction at an air–glass interface leads to a spreading of the colors given for normal dispersion

For normal dispersion the refraction of the shorter wavelengths is larger than for the longer ones, as it can be demonstrated by Huygen's principle.

In the range of absorption the conventional refractive index increases with the wavelength of the light (see Fig. 3.1) and this is called anomalous dispersion:

anomalous dispersion $\dfrac{\mathrm{d}n}{\mathrm{d}\lambda} > 0$ $\hspace{4cm}$ (3.30)

This anomalous dispersion may sometimes be difficult to observe because the imaginary part of the complex refractive index may be dominant in this spectral range of the absorption. In Sect. 7.5, the z scan method is described which allows the determination of both parts of the complex refractive index in nonlinear spectroscopy.

In any case the velocity of a light mixture is given by the group velocity:

group velocity $c_{\mathrm{g}} = c_{\mathrm{p}} - \lambda \dfrac{\mathrm{d}c_{\mathrm{p}}}{\mathrm{d}\lambda}$ $\hspace{3cm}$ (3.31) and so a group refractive index can be defined as:

$$n_{\text{group}} = n(\lambda) - \lambda \frac{\mathrm{d}n(\lambda)}{\mathrm{d}\lambda}. \tag{3.32}$$

In the case of spectral broad light with several nm bandwidth, as, e.g. in fs pulses, the dispersion will cause significant different delays between the spectrally different components of the light. This is called as chirp (see Section 6.10.3, 6.11.3 and 6.14.2). With the combination of two gratings or prisms or with special mirrors it is possible to compensate this effect.

The conventional refractive index can be determined from the refraction. The angles between the beams and the perpendicular on the mentioned surface are related by the Snellius law (see Fig. 3.3):

refraction law $n_1 \sin \varphi_1 = n_2 \sin \varphi_2 = f(\lambda)$. $\tag{3.33}$

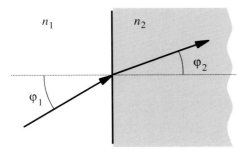

Fig. 3.3. Refraction at an optical surface between two materials from air to quartz with different refractive indices of 1.0 and 1.5

If light beams are not perpendicular to the material surface the refraction will change the intensity of the beam by changing the beam shape in the dimension which is not parallel to the surface (see Sect. 3.5.1).

The group velocity may be determined from the measurement of light pulse propagation through the material. Often the difference can be neglected.

The refractive indices as a function of the wavelength for CS_2 and some technical glasses are shown in Fig. 3.4. The wavelength dependence of the refractive indices is different for different materials and thus the compensation of dispersion is possible, e.g. by combining suitable glasses. If the total refraction is the same for two wavelengths the system will be called *achromatic* and if the compensation works for three wavelengths the system is called *apochromatic*.

For the phenomenalogical description of dispersion as a function of wavelength several suggestions have been made [3.5]. A simple approach was given by Hartmann as:

$$n(\lambda) = n_0 + \frac{C_{\text{disp}}}{(\lambda - \lambda_0)^\alpha} \qquad 0.5 < \alpha < 2 \tag{3.34}$$

with the constants C_{disp} and λ_0. Another approach which can result in useful values from the UV to the IR spectral range was given by Sellmeier as:

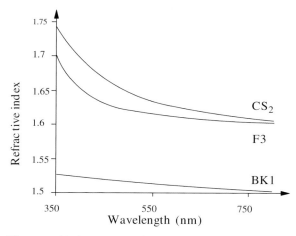

Fig. 3.4. Refractive indices of some materials as a function of wavelength (dispersion)

$$n(\lambda) = 1 + \sum_m \frac{\lambda^2}{\lambda^2 - \lambda_m^2} A_{\mathrm{disp},m} \tag{3.35}$$

with the coefficients $A_{\mathrm{disp},m}$ and the resonance wavelengths λ_{m}. If three terms are used then two of them should have wavelengths in the UV and one in the IR. These coefficients are available especially for crystalline materials as used in nonlinear optics for frequency transformation from the suppliers.

3.3 Absorption and Emission

Absorption and emission of light are complicated quantum processes which can be described correctly only by using quantum electrodynamics. For the purpose of this book some important results of the theoretical analysis will be discussed briefly. Further descriptions will be given in Sect. 5.3 considering nonlinear interactions.

Today we are only able to calculate the absorption and emission spectra for a few interesting materials with good accuracy. Thus the experimental investigation of the absorption and emission properties of matter in the linear, and even more in the nonlinear range are important fields of research. Better understanding of these spectra and the decay mechanisms will help us to design more suitable materials for photonic applications in the future.

3.3.1 Theoretical Description of Absorption and Emission

For analyzing the interaction of the electric field of the photons with the electric charges in the material, i.e. the electron and ion potentials, the time-dependent Schroedinger equation has to be used:

$$\mathsf{H}(\boldsymbol{r},t)\Psi(\boldsymbol{r},t) = \mathrm{i}\frac{h}{2\pi}\frac{\partial\Psi(\boldsymbol{r},t)}{\partial t} \tag{3.36}$$

with the Hamilton operator H representing the total energy of the matter–light system and the wave function ψ representing the quantum state of this system with all detailed spatial and temporal information of all particles in it. Because the energy of the system and the wave function are not known, this eigenvalue equation cannot usually be solved in general. Thus a large number of approximations have to be used (for more details see also Sect. 7.13.2).

First, the stationary Schroedinger equation for matter without any external interaction is usually applied:

$$\mathsf{H}_{\mathrm{matter}}(\boldsymbol{r})\cdot\varphi_m(\boldsymbol{r}) = E_m\cdot\varphi_m(\boldsymbol{r}). \tag{3.37}$$

As a solution of this time-independent equation the eigenstates numbered by m and characterized by the wave functions φ_m as well as by the energy E_m of these states of the investigated material occur. With respect to the interaction with photons it is assumed that the eigenstates of the material will not change under the influence of the light field.

The solution of this equation is analytically still only possible in the simplest cases such as the H atom and numerical solutions need a large number of further approximations even for small molecules. Nevertheless, several principal results about the energy levels and the selection rules can be evaluated from this.

Under these assumptions the wave function can be used with the ansatz:

$$\Psi(\boldsymbol{r},t) = \Psi_1(\boldsymbol{r})\cdot\Psi_2(t) \tag{3.38}$$

and with the solution of the stationary undisturbed system the total wave function Ψ can be described as a superposition of the wave functions of the system with time-dependent coefficients $c_m(t)$:

$$\Psi(\boldsymbol{r},t) = \sum_{m=1}^{\infty} c_m(t)\psi_m(\boldsymbol{r},t) \tag{3.39}$$

with the wavefunctions ψ_{m}:

$$\psi_m(\boldsymbol{r},t) = \mathrm{e}^{-\mathrm{i}(2\pi/h)E_m t}\varphi_m(\boldsymbol{r}). \tag{3.40}$$

The interaction with the light field can be described by first-order perturbation theory. The Hamiltonian of (3.36) is split into the material steady-state Hamiltonian of (3.37) and the Hamiltonian of the interaction as a small disturbance:

$$\mathsf{H}(\boldsymbol{r},t) = \mathsf{H}_{\mathrm{matter}}(\boldsymbol{r}) + \mathsf{H}_{\mathrm{interaction}}(t). \tag{3.41}$$

With this equation the temporal change of the coefficients describing the transitions of the particle under the influence of the light can be calculated from:

$$\frac{\partial}{\partial t}c_{\mathrm{p}}(t) = -\mathrm{i}\frac{2\pi}{h}\sum_{m=1}^{\infty}\left[c_m(t)\int_V \psi_p^*\mathsf{H}_{\mathrm{interaction}}\psi_m\,\mathrm{d}V\right] \tag{3.42}$$

with the integration over the whole volume V of the wavefunctions. Usually the system is in one eigenstate before a transition takes place and than the sum is reduced to one element. The probability of the population of state p is given by the square of c_p and the transition probability $w_{p \leftarrow m}$ for the transition from state m to state p is given by:

$$w_{p \leftarrow m} = \frac{\partial}{\partial t} |c_p(t)|^2 \propto \mu_{p \leftarrow m}^2 \tag{3.43}$$

which is proportional to the square of the transition dipole moment $\mu_{p \leftarrow m}$:

transition dipole moment $\mu_{p \leftarrow m} = \int_V \varphi_p^* \mathsf{H}_{\text{interaction}} \varphi_m \, dV$ (3.44)

The interaction operator is given for a one-electron system in the dipole approximation, assuming a radiation wavelength large compared to the dimension of the particle, by:

$$\mathsf{H}_{\text{interaction}}(t) = -e\boldsymbol{r}\boldsymbol{E}(\boldsymbol{r}_{\text{particle}}) \tag{3.45}$$

with the electrical charge e, the position of the particle center at $\boldsymbol{r}_{\text{particle}}$ and $\boldsymbol{r}$ as the relative position of the charge from the particle center and the electric field vector $\boldsymbol{E}$.

For the more general case, including large molecules the electrical field can be better expressed with the vector potential $\boldsymbol{A}(\boldsymbol{r}, t)$ which is source free:

$$\text{vector potential}\quad \text{div}\, \boldsymbol{A}(\boldsymbol{r}, t) = 0 \tag{3.46}$$

and the electrical field follows from this potential by:

$$\boldsymbol{E}(\boldsymbol{r}, t) = -\frac{1}{c}\frac{\partial}{\partial t}\boldsymbol{A}(\boldsymbol{r}, t) \tag{3.47}$$

and the magnetic field by:

$$\boldsymbol{H}(\boldsymbol{r}, t) = \text{rot}\, \boldsymbol{A}(\boldsymbol{r}, t). \tag{3.48}$$

With respect to the quantum description the vector potential can be written as:

$$\mathsf{A}(\boldsymbol{r}, t) = \sum_m \boldsymbol{e}_m \sqrt{\frac{h\lambda_m}{8\pi^2 V \varepsilon_0 c_0}} \left[\mathsf{b}_m\, e^{i\boldsymbol{k}_m \boldsymbol{r}} + \mathsf{b}_m^+\, e^{-i\boldsymbol{k}_m \boldsymbol{r}} \right] \tag{3.49}$$

with the counter m for the different waves of light and thus of the electric field, $\boldsymbol{e}_m$ as the direction of the field vector, λ_m as the wavelength of the light wave, V as the volume the waves are generated in and $\boldsymbol{k}_m$ as the wave vector of the mth wave. The b_m and b_m^+ are photon absorption and emission operators which would be light amplitudes in the classical case. These operators fulfill the following relations:

$$\begin{aligned} \mathsf{b}_m\mathsf{b}_p^+ - \mathsf{b}_p^+\mathsf{b}_m &= \delta_{mp} \\ \mathsf{b}_m\mathsf{b}_p - \mathsf{b}_p\mathsf{b}_m &= 0 \\ \mathsf{b}_m^+\mathsf{b}_p^+ - \mathsf{b}_m^+\mathsf{b}_m^+ &= 0 \end{aligned} \tag{3.50}$$

which result in the description of the energy of the electrical field by a sum over harmonic oscillators as:

$$\mathsf{H}_{\text{field}} = \sum_m \mathsf{b}_m^+ \mathsf{b}_m h\nu_m \tag{3.51}$$

and the Hamilton operator for a single electron in the potential of the cores V and the electric field $\boldsymbol{A}$ is given by:

$$\mathsf{H}_{\text{electron}} = \frac{1}{2m_{\text{electron}}} \left[\mathsf{p} - e_e\mathsf{A}(r,t)\right]^2 + \mathsf{V}(r,t) \tag{3.52}$$

with the mass m_{electron} and charge e_e of the electron and the pulse operator:

$$\mathsf{p} = -\mathrm{i}\frac{h}{2\pi}\nabla \tag{3.53}$$

With these definitions the interaction operator for a one-electron system follows from:

$$\mathsf{H}_{\text{interaction}}(\boldsymbol{r},t) = -\frac{e_e}{m_{\text{electron}}}\mathsf{A}(\boldsymbol{r},t)\cdot\mathsf{p} + \frac{e_e^2}{2m_{\text{electron}}}\mathsf{A}^2(\boldsymbol{r},t). \tag{3.54}$$

For linear interactions the second term can be neglected. But the interaction has to be considered for all charges in the particle which are, e.g. in molecules for all electrons and core charges. The resulting interaction operator is given by:

$$\mathsf{H}_{\text{interaction}}(\boldsymbol{r},t) = \sum_p \left[-\frac{e_e}{m_{\text{electron}}}\mathsf{A}(\boldsymbol{r}_p,t)\cdot\mathsf{p}_p\right]$$

$$= +\sum_q \left[-\frac{Z_{\text{core},q}e_e}{M_{\text{core}}}\mathsf{A}(\boldsymbol{R}_q,t)\cdot\mathsf{P}_q\right] \tag{3.55}$$

with the charge $Z_{\text{core},q}$ of the qth core, the coordinate R_q of this core and its momentum P_q. In the dipole approximation the interaction operator for such a system can be written as:

$$\mathsf{H}_{\text{interaction}}(\boldsymbol{r},t) = \sum_m \left[-\frac{Z_{\text{charge},m}e_e}{m_{\text{charge},m}}\boldsymbol{E}(\boldsymbol{r}_p,t)\right] \tag{3.56}$$

and thus the transition dipole moment in the dipole approximation follows as:

transition dipole moment

$$\mu_{p\leftarrow m} = e_e \int_V \varphi_p^* \left(\sum_m \left[-\frac{Z_{\text{charge},m}}{m_{\text{charge},m}}\boldsymbol{E}(\boldsymbol{r}_p,t)\right]\right)\varphi_m \,\mathrm{d}V. \tag{3.57}$$

It can be shown that *for absorption or emission of photons* the material has to perform a transition between two eigenstates E_m and E_p of the material and thus the photon energy E_{photon} has to fulfill the resonance condition:

resonance condition $E_{\text{photon}} = h\nu_{\text{photon}} = |E_p - E_m|.$ \hfill (3.58)

Because of the uncertainty relation a certain spectral width for this resonance condition has to be added as a function of the transition time. This bandwidth results from the natural life time of the involved states. The line shape function is described below in Sect. 3.3.3.

In addition to this condition, the resonance condition, as a second condition a certain *overlap of the wave functions* of the initial state φ_m and the final state φ_p of the material is necessary for an absorption or emission process as given in (3.44) and (3.57). The larger this integral and thus the larger the transition dipole moment $\mu_{p \leftarrow m}$ (meaning the larger the overlap of the two wave functions) the larger the probability for the transition.

Other useful values for describing this transition probability are the oscillator strength f, Einstein's coefficients B and the cross-section σ [3.6].

The oscillator strength f [e.g. M42, 3.7] is related to the transition dipole moment for a two-level system as:

$$\textbf{oscillator strength} \quad f_{p \leftarrow m} = \frac{g_p}{g_m} \frac{8\pi^2 m_{\text{electron}} \nu_{p \leftarrow m}}{3 h e_e^2} \mu_{p \leftarrow m}^2 \qquad (3.59)$$

with the electron mass m_{electron}, the multiplicities $g_{j/i}$ of the two states and the frequency of the transition $\nu_{j \leftarrow i}$. The oscillator strength of, e.g. molecules is about one but is larger for strong transitions and much smaller for forbidden transitions:

$f \geq 1$ allowed transitions

$f \ll 1$ forbidden transitions.

The Einstein coefficient for absorption and stimulated emission of this transition is related to the oscillator strength by:

Einstein's coefficients

$$B_{p \leftarrow m} = B_{m \leftarrow p} = \frac{\pi e_e^2}{2\varepsilon_0 m_{\text{electron}} h \nu_{p \leftarrow m}} f_{p \leftarrow m}. \qquad (3.60)$$

Molecules show absorption and emission bands around the electronic transition frequency which result from broadening effects. Thus the oscillator strength of the electronic transition is "distributed" over this band. The experimental absorption band measured, e.g. as the cross-section σ (see Sect. 3.4) as a function of the light frequency or wavelength λ can be related to the oscillator strength by integrating over the cross-section of the band:

$$f_{p \leftarrow m} = \frac{4 m_{\text{electron}} c_0 \varepsilon_0}{e_e^2} \int_{\text{band}} \sigma(\nu) \, \mathrm{d}V$$

$$\simeq 3.76788 \cdot 10^5 \, \frac{\mathrm{s}}{\mathrm{m}^2} \int_{\text{band}} \sigma(\nu) \, \mathrm{d}\nu \qquad (3.61)$$

or in the wavelength scale

$$f_{p \leftarrow m} = \frac{4 m_{\text{electron}} c_0^2 \varepsilon_0}{e_e^2} \int_{\text{band}} \frac{1}{\lambda^2} \sigma(\lambda) \, \mathrm{d}\lambda$$

$$\simeq 1.12958 \cdot 10^{14} \, \frac{1}{\mathrm{m}} \int_{\mathrm{band}} \frac{1}{\lambda^2} \sigma(\lambda) \, \mathrm{d}\lambda \tag{3.62}$$

with the mass m_{electron} and charge e_{e} of the electron. For measured spectra a band shape analysis may be necessary to isolate the investigated transition as described in Sect. 7.2.4. The emission cross-section can be determined from the fluorescence measurements as described in Sect. 7.3.4.

This integral formula can be used for comparing the results of quantum chemical calculations with experimental spectra. Sometimes the extinction coefficient is used to characterize the absorption spectra. The conversion formula is given in Sect. 3.4. Changes in the shape of the absorption spectrum should not change the resulting oscillator strength.

From the analysis a few general selection rules for transitions between two levels can be given. Transitions between vibrational energy levels of molecular systems are allowed for neighboring levels only, changing the vibration quantum number by $\Delta\nu = \pm 1$ and are otherwise strictly forbidden. Transitions between molecular rotation states are allowed for changes of the rotational quantum number by $\Delta J = \pm 1$ and are otherwise theoretically not possible. Electronic transitions are much less probable between electronic states of the same symmetry as even–even or odd–odd. Further transition between electronic systems with different spin such as singlet–triplet transitions are usually forbidden.

Table 3.2. Selection rules for light induced transitions in matter

Type of transition	Allowed	Forbidden
Electronic	Even–odd	Even–even
	Odd–even	Odd–odd
	Singlet–singlet	Singlet–triplet
	Triplet–triplet	Triplet–singlet
Vibronic	$\Delta\nu_{j\leftarrow i} = \pm 1$	Else
Rotational	$\Delta J_{j\leftarrow i} = \pm 1$	Else

In practical cases in almost all systems except atoms in gases commonly not just two energy levels are involved. For example, the energy level density of molecules with 10 atoms reaches values of more than 10^4 states per cm^{-1} energy range for molecular energies of $200\,\mathrm{cm}^{-1}$. This is the molecular energy which is provided at room temperature T_{room} by the Boltzmann energy $kT_{room} \simeq 204\,\mathrm{cm}^{-1}$. Interactions of the matter, e.g. a electronic states of molecules, with their environment, e.g. solvent, or with other internal states, e.g. vibrations, may shift these states by several 10–$100\,\mathrm{cm}^{-1}$. Further the motion of the particles changes the resonance frequency by the Doppler shift. Thus in spectroscopic measurements usually a large number of quantum eigenstates is active. Therefore the measured cross-section commonly shows a continuous spectrum. This spectrum can be homogeneously or inhomogeneously broadened (see Sect. 5.2) which does not usually matter in linear spectroscopy but will be important in many photonic applications.

3.3.2 Properties of Stimulated Emission

As mentioned in the previous chapter the Einstein coefficients and the cross-sections are the same for absorption as for stimulated emission in the case of a two-level system. Stimulated emission is a more complicated process as it includes at least two photons, one which stimulates and the emitted one (see Fig. 3.5).

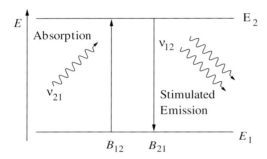

Fig. 3.5. Stimulated emission of a photon in a two-energy-level E_1 and E_2 system

As a most important fact of this process and as the basis of all lasers it should be noted that:

The stimulated photon is perfectly identical with the incident photon.

Thus two identical photons occur after the stimulated emission takes place and therefore by repeated stimulated emission a very large number of identical photons can be generated. In natural systems the probability of stimulated emission is smaller than for absorption because of the lack of high populations in the upper energy state (see Sect. 5.3.4 and 6.1).

3.3.3 Spontaneous Emission

Besides stimulated emission a spontaneous emission of light can happen while the matter undergoes a transition from the upper energy level to the lower one (see also Sect. 7.3 and references there). Of course the resonance condition (3.58) will be fulfilled by the spontaneous emitted photon, too.

This emission is a statistical process which has to be analyzed with quantum electrodynamics including the vacuum fluctuations, and thus the time of emission is not determined. The emission probability is given by the Einstein coefficient $A_{m \leftarrow p}$ which is related to the B coefficients of absorption and stimulated emission under the assumption of constant bandwidth by:

spontaneous emission $A_{m \leftarrow p} = \dfrac{8\pi h}{c^3} \nu^3 B_{m \leftarrow p} = \dfrac{8\pi h}{c^3} \nu^3 B_{p \leftarrow m}$ (3.63)

where ν is the resonance frequency of the transition. From this coefficient it follows that there is a relative increase of the probability of the spontaneous compared to the stimulated emission towards shorter wavelengths. This can, e.g. disturb the emission of UV laser light.

Further the emission probability is constant over time and thus an ensemble of excited particles will show an exponentially decaying spontaneous emission intensity $I_{\text{spont.emission}}$:

$$I_{\text{spont.emission}}(t) = I_0(t = 0)e^{-t/\tau_{m \leftarrow p}} \tag{3.64}$$

with the decay time of this emission:

decay time $\tau_{m \leftarrow p} = \dfrac{1}{A_{m \leftarrow p}}$ (3.65)

which is in the case of a two-level system also the lifetime of the upper energy state with the energy E_p.

From the uncertainty relation it follows that this transition must have a minimum homogeneous spectral width (FWHM) of:

bandwidth $\Delta\nu_{m \leftarrow p} = \dfrac{1}{2\pi\tau_p} + \dfrac{1}{2\pi\tau_m}$ (3.66)

with lifetimes τ_p and τ_m of the upper and the lower energy state. The second term is zero if the lower state is the ground state of the system. This spectral width represents a Lorentzian line shape function:

Lorentzian profile $f(\nu - \nu_0) = \left\{ \dfrac{(\Delta\nu_{\text{FWHM}}/2)^2}{(\nu - \nu_0)^2 + (\Delta\nu_{\text{FWHM}}/2)^2} \right\}$ (3.67)

with the integral:

$$\int_{-\infty}^{\infty} f(\nu - \nu_0) \, d\nu = \pi \Delta\nu_{\text{FWHM}} \tag{3.68}$$

which is taken for practical cases for $\nu > 0$, only. This Lorentzian line shape function is depicted in Fig. 3.6.

This minimal spectral width can be broadened by further parallel decay processes as, e.g. by radiationless transitions, as described in the next chapter, which shorten the lifetimes of the states involved. In practical cases the superposition of homogeneously broadened transitions are frequently observed. In these cases the spectral shape of the resulting inhomogeneously broadened spectrum can mostly be described by Gaussian or Voigt spectra as a result of statistically superimposed narrow Lorentzian spectra (see Sect. 7.2.4).

Spontaneous emission is observed after excitation of the system, for example via the pump mechanisms described in Sect. 6.3. In addition, sonoluminescence was observed and is still being investigated [3.8–3.10]. Several mechanisms are known to reduce the lifetime of the upper state via radiationless transitions as discussed in the following Section.

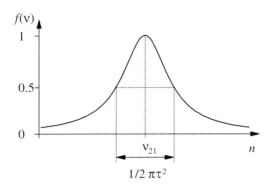

Fig. 3.6. Lorentzian line shape of spontaneous emitted light

3.3.4 Radiationless Transitions

The optically excited energy states of matter can also decay without emitting photons. Further, a part of the absorbed energy can undergo radiationless conversion, which is, e.g. the difference between the energy of the absorbed photons and the emitted photons. This energy is sometimes called the excess energy. For more details see also Sect. 7.3.4.

The radiationless transitions usually result from the coupling of the excited state with states of only slightly smaller or equal energy but of a different kind. For example, the electronically excited but vibrationally nonexcited state of a molecule can transfer to a vibrationally highly excited but electronic ground state. This energy transfer can be followed by a fast transfer of the vibrational energy to the environment of the molecule and thus the dissipation of the energy.

The calculation of the radiationless transitions demands precise knowledge of all energy states which have to be considered and their quantum coupling mechanisms. This is possible under favorable circumstances and only in a phenomenological way.

A further common energy transfer process can be activated by the dipole–dipole interaction, e.g. in molecular systems. Coherent and incoherent coupling between the partners is possible and excitons can occur [3.11–3.14]. In this case new narrow absorption and emission bands may be obtained. For strong coupling the transfer rate is proportional to the third power of the distance between the partners and reaches values of 10^{11}–$10^{14}\,\mathrm{s}^{-1}$. For weak coupling the transfer rate is proportional to the sixth power of the distance and shows values of 10^6–$10^{11}\,\mathrm{s}^{-1}$ [3.15–3.17]. These processes are also known as the Foerster mechanism. Finally transfers over distances up to 10 nm can be observed.

Regardless of the detailed energy dissipation mechanism the resulting energy storage should be considered in the matter. In linear interactions the deposited thermal energy can usually be neglected because of the small excitation light powers.

But the radiationless transitions often shorten the lifetime of the excited states, significantly. If the lifetime of the radiationless transition alone is $\tau_{i,\mathrm{radless}}$ the resulting lifetime of the upper level p will decrease to:

$$\tau_{p,\mathrm{final}} = \left[\frac{1}{\tau_{p,\mathrm{rad}}} + \frac{1}{\tau_{p,\mathrm{radless}}} \right]^{-1} . \tag{3.69}$$

The mechanism of parallel decays is comparable to parallel resistance; the resulting lifetime has to be calculated as the sum of the inverse lifetimes which is the sum of the transition rates (see Sect. 5.3.6).

Thus the radiationless transitions will influence the homogeneous linewidth of optical transitions. Further details are given in Chap. 7.

3.4 Measurement of Absorption

Absorption is observed as a decrease of the number of photons in the beam while transmitting in the material. In linear optics the incremental decrease of the intensity I, e.g. in the z direction, is proportional to the intensity itself:

linear optics $dI(z) = -aI(z)$ $\tag{3.70}$

with the absorption coefficient a measured in cm^{-1}.

3.4.1 Lambert–Beer Law

Integration of this equation leads to the Lambert–Beers law:

Lambert–Beer law $I = I_0\,\mathrm{e}^{-aL}$ $\tag{3.71}$

with the definitions of Fig. 3.7.

The absorption coefficient a is related to the imaginary part of the refractive index $n_{\mathrm{complex}} = n_{\mathrm{real}} + i n_{\mathrm{imag}}$ by:

$$a = \frac{4\pi}{\lambda} n_{\mathrm{imag}} \tag{3.72}$$

as can be compared with (3.24).

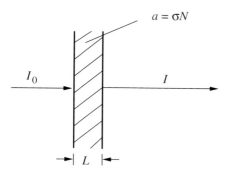

Fig. 3.7. Linear absorption of light in a sample of thickness L

For determining the absorption the incident intensity I_0 and the transmitted intensity I have to be measured. The quotient of these is the transmission or transmittance T:

$$\textbf{transmission} \quad T = \frac{I}{I_0}. \tag{3.73}$$

Besides the well-defined transmission a lot of further definitions are used to characterize the absorption of matter. Quite common is the optical density OD especially for optical filters which is defined as:

$$\textbf{optical density} \quad \text{OD} = -\lg_{10}(T) \text{ or } T = 10^{-\text{OD}}. \tag{3.74}$$

Sometimes "absorption grades", "extinction", "absorption" or other values are used as measures. The definitions of these values can differ from author to author as, e.g. $1 - T$ or $1/T$ and thus it is strongly recommended to relate these values to the clearly defined T.

As the intensity is a five-dimensional function of the photon parameters it is obvious that the transmission can also depend on these parameters. In the linear optics the light has almost no effect on the material and thus the absorption can be characterized by the material parameters alone. It is not a function of the intensity.

3.4.2 Cross-Section and Extinction Coefficient

The absorption coefficient in linear optics will usually be a function of the wavelength λ and sometimes of the polarization ϕ with respect to the sample structure and orientation. In special cases it can be a function of space and time. It will depend on the material parameters such as concentration, temperature and pressure. Thus four further values are commonly used in optics for sample characterization:

$$\textbf{cross-section} \quad \sigma(\lambda) = \frac{a}{N_{\text{part}}} = -\frac{1}{N_{\text{part}}L} \ln(T) \tag{3.75}$$

with

$$[\sigma] = \text{cm}^2 \text{ and } [N_{\text{part}}] = \text{cm}^{-3} \tag{3.76}$$

where the particle density N_{part} is measured in particles per cm^3. In chemistry the extinction coefficient is popular:

$$\textbf{extinction coefficient} \quad \varepsilon_a(\lambda) = \frac{a}{C_{\text{conc}}} = -\frac{1}{C_{\text{conc}}L} \lg_{10}(T) \tag{3.77}$$

with:

$$[\varepsilon_a] = \frac{l}{\text{mol cm}} \text{ and } [C_{\text{conc}}] = \frac{\text{mol}}{l} \tag{3.78}$$

The extinction coefficient and the cross-section are related by:

$$\varepsilon_a(\lambda) = \sigma(\lambda)\frac{N_L}{\ln 10} \simeq \sigma(\lambda) \cdot 2.6154 \cdot 10^{20} \frac{1}{\text{cm}^2} \frac{l}{\text{mol} \cdot \text{cm}} \tag{3.79}$$

with Loschmidt's number $N_L = 6.0221367 \cdot 10^{23} \text{ mol}^{-1}$.

3.4.3 Absorption Spectra of Some Optical Materials and Filters

For photonic applications optically transparent materials for the required wavelengths are needed for conventional optical elements such as lenses, windows and beam splitters [3.18, 3.19]. Thus especially in the infrared (IR) and in the ultra violet (UV) region special materials are needed. In Fig. 3.8 and 3.9 the transmission spectra of some common materials for applications in the UV and IR are shown.

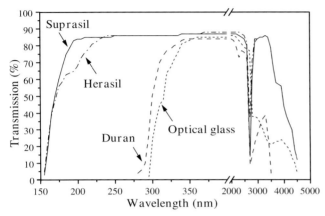

Fig. 3.8. Transmission spectra of some optical glasses for a geometrical thickness of 10 mm

Thus if materials have to be investigated in the far IR spectral range around 10 μm glasses are no longer useful. The most common materials for optical windows in this spectral region are shown in Fig. 3.9. Polymers may be useful in some cases [see, for example, 3.4].

Further Kaliumbromide (KBr) is also used in the IR, e.g. for diluting molecules if their vibronic or rotational transition is to be measured. Thus the mixture of fine KBr with the sample molecule can be pressurized with a special crammer to get transparent samples several 10 mm in diameter.

Special care has to be taken working with UV or with high intensities. Materials may degrade or show color-center formation or other photochemical reactions [3.20, 3.21].

In Fig. 3.10 the spectra of some common filters used in optical setups are shown. They are used in photonic applications in linear and nonlinear measurements to vary the light intensities to illuminate the samples or the detectors in the right way.

As can be seen the absorption of the neutral density (NG) filters is only flat in narrow spectral ranges. Thus the spectral transmission curve has to be considered in the evaluation of spectral measurements. Care has to be taken for nonlinear effects in these filters if laser light is applied as described in Sect. 7.1.7.

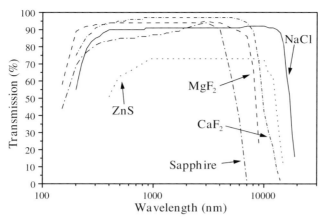

Fig. 3.9. Transmission spectra of materials usable in the IR for a geometrical thickness of 3 mm for MgF_2, CaF_2 and sapphire, of 6 mm for ZnS and of 10 mm for NaCl

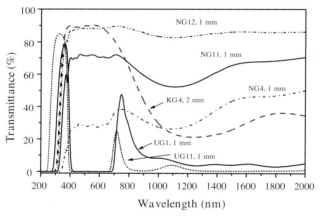

Fig. 3.10. Spectral transmission curves of some common filters as used in photonic applications: neutral density filters NG4, NG11, NG12 and color glass filters UG1, UG11, KG4 (Schott-Glas)

3.5 Polarization in Refraction and Reflection (Fresnel's Formula)

If light waves meet the border of two materials with different refractive indices some light will be reflected and the rest will be refracted and absorbed. If the wave propagation is not perpendicular to the border surface the polarization of the reflected and the refracted light will be changed compared to the polarization of the incident light. The analysis of these processes can be based on Maxwell's equations. Some results will be given in this chapter. They have to be carefully considered in all linear and nonlinear measurements, because neglecting polarization effects can cause serious measuring errors.

3.5.1 Fresnel's Formula

The evaluation of reflection and refraction is carried out first by neglecting absorption. The incident beam with total power P_{inc} meets the surface between the refractive indices n_1 and n_2 at the angle φ relative to the perpendicular to the surface (see Fig. 3.11).

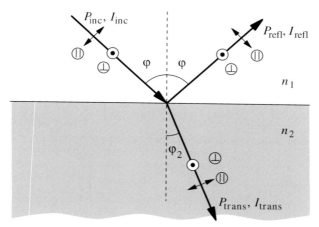

Fig. 3.11. Incident, reflected and refracted light beams with parallel and noparallel polarization components relative to the refractive index boundary

The reflected beam shows the same angle φ towards the perpendicular but the refracted light will show the angle φ_2. One polarization component is parallel to the plane of all beams and is marked by $\parallel$. It has a component perpendicular to the surface. The other component will be perpendicular to this one and will be called perpendicular $\perp$. It is parallel to the border surface polarized.

General Formula

Neglecting absorption the sum of all components of the reflected and transmitted powers have to be equal to the total incident power. The relation of the refractive indices of the homogeneous and isotropic material will be described by:

$$\text{relative index change} \quad n_{rel} = \frac{n_2}{n_1}. \tag{3.80}$$

With this definitions the reflected and transmitted shares are given by:

- reflectivity R of perpendicular polarized intensity components:

$$R_\perp = \frac{P_{\mathrm{refl},\perp}}{P_{\mathrm{inc,bot}}} = \frac{I_{\mathrm{refl},\perp}}{I_{\mathrm{inc},\perp}} = \left\{ \frac{\left(\sqrt{n_{\mathrm{rel}}^2 - \sin^2\varphi} - \cos\varphi\right)^2}{n_{\mathrm{rel}}^2 - 1} \right\}^2 \qquad (3.81)$$

- transmission T of perpendicular polarized intensity components:

$$T_\perp = \frac{P_{\mathrm{trans},\perp}}{P_{\mathrm{inc},\perp}}$$

$$= F(\varphi, n_{\mathrm{rel}}) \left\{ \frac{2\cos\varphi\sqrt{n_{\mathrm{rel}}^2 - \sin^2\varphi} - 2\cos^2\varphi}{n_{\mathrm{rel}}^2 - 1} \right\}^2 \qquad (3.82)$$

- reflectivity R of parallel polarized intensity components:

$$R_\| = \frac{P_{\mathrm{refl},\|}}{P_{\mathrm{inc},\|}} = \frac{I_{\mathrm{refl},\|}}{I_{\mathrm{inc},\|}} = \left\{ \frac{n_{\mathrm{rel}}^2\cos\varphi - \sqrt{n_{\mathrm{rel}}^2 - \sin^2\varphi}}{n_{\mathrm{rel}}^2\cos\varphi + \sqrt{n_{\mathrm{rel}}^2 - \sin^2\varphi}} \right\}^2 \qquad (3.83)$$

- transmission T of parallel polarized intensity components:

$$T_\| = \frac{P_{\mathrm{trans},\|}}{P_{\mathrm{inc},\|}} = F(\varphi, n_{\mathrm{rel}}) \left\{ \frac{2n_{\mathrm{rel}}\cos\varphi}{n_{\mathrm{rel}}^2\cos\varphi + \sqrt{n_{\mathrm{rel}}^2 - \sin^2\varphi}} \right\}^2 . \qquad (3.84)$$

The factor $F(\varphi, n_{\mathrm{rel}})$ considers the change of beam area in the direction not parallel to the surface.

It is given by:

$$F(\varphi, n_{\mathrm{rel}}) = n_{\mathrm{rel}}\sqrt{1 + (1 - 1/n_{\mathrm{rel}}^2)\tan^2\varphi} \qquad (3.85)$$

and the intensities of the transmitted light can be calculated by:

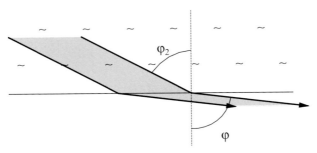

Fig. 3.12. Change of the beam dimension passing an optical surface between different materials via refraction

$$I_{\text{trans},\perp} = T_\perp I_{\text{inc},\perp} F \tag{3.86}$$

and

$$I_{\text{trans},\parallel} = T_\parallel I_{\text{inc},\parallel} F. \tag{3.87}$$

In addition to these formulas the angle φ_2 is given by the Snellius law (3.33). It is important to notice that only for perpendicular incidence no change of polarization occurs. But in any case reflection takes place.

For perpendicular incidence in an isotropic material the reflectivity is obviously, for symmetry reasons, not polarization dependent. The reflectivity for the light power or intensity is given by:

$$R_{0°} = \frac{P_{\text{refl}}}{P_{\text{inc}}} = \frac{I_{\text{refl}}}{I_{\text{inc}}} = \left(\frac{1 - n_{\text{rel}}}{1 + n_{\text{rel}}} \right)^2 \tag{3.88}$$

and the transmitted share $T_{0°}$ is:

$$T_{0°} = \frac{P_{\text{trans}}}{P_{\text{inc}}} = \frac{I_{\text{trans}}}{I_{\text{inc}}} = \frac{4 n_{\text{rel}}}{(1 + n_{\text{rel}})^2}. \tag{3.89}$$

It is worth noting that for the air–glass transition the minimum power or intensity reflection is 4% per surface. All optical elements with two optical surfaces as lenses and glass plates would produce 7.84% losses. High refracting materials such as laser crystals with $n = 1.8$ reflect even more than 8% per surface. Thus the coating of surfaces with thin layers of suitable refractive indices is an important technique in photonics, and the reflection losses of optical surfaces can be decreased to 0.1% or less [e.g. M3].

For illustrating these formulas the special case of an air–glass surface will be shown in graphs below. The results are different if the light passes the border from the optically thinner to the optically denser material ($n_1 < n_2$) or vice versa.

Transition into Optically Denser Medium

If the light passes the material surface from optically thinner to optically denser material, meaning $n_1 < n_2$, the angle φ can be 0–90° and over the whole range reflection and transmission is observed. But at a certain angle φ_B the reflection of the perpendicular polarized component will become zero and thus the reflected light will be perfectly linearly polarized parallel to the surface (see next chapter).

It shall be noticed that:

> The electric field vector of light reflected at an optically denser material shows a phase jump of π.

In Figs. 3.13 and 3.14 the reflectivity and the transmission of the two perpendicular polarized components of the light beams are given as a function of the angle between the surface perpendicular and the propagation direction.

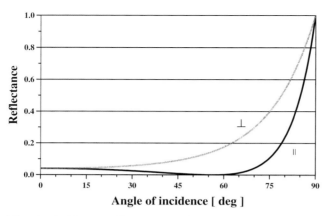

Fig. 3.13. Power reflectivity R of the two perpendicular polarized components of the light beam as a function of the angle of incidence for the transition from the optically thinner air to an optically denser glass with $n = 1.5$

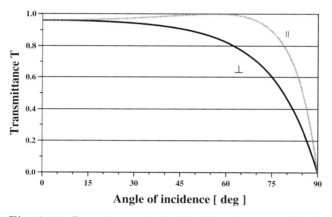

Fig. 3.14. Power transmission T of the two perpendicular polarized components of the light beam as a function of the angle of incidence for the transition from the optically thinner air to an optically denser glass with $n = 1.5$

For nonpolarized light the transmission T and the reflectivity R are given in Fig. 3.15 for the transition of light from air to an optically denser material with $n = 1.5$.

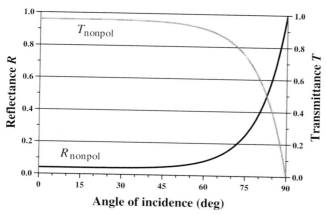

Fig. 3.15. Power reflectivity R and power transmission T of a nonpolarized light beam as a function of the angle of incidence for the transition from the optically thinner air to an optically denser glass with $n = 1.5$

Transition into Optical Thinner Medium

If light passes the optical surface from the higher refracting to the lower refracting material ($n_1 > n_2$) total reflection will occur (see Sect. 3.5.3).

There is no phase jump in the electric field vector, but the magnetic field of the reflected light experiences a shift of π, which is usually not important for photonic applications.

The graphs showing the reflectivity and the transmission for a transition from glass with $n = 1.5$ to air are given in Figs. 3.16 and 3.17.

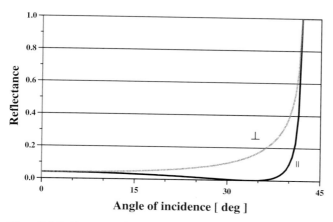

Fig. 3.16. Power reflectivity of the two perpendicular polarized components of the light beam as a function of the angle of incidence for the transition from the optically denser glass with $n = 1.5$ to air

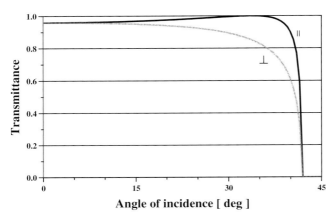

Fig. 3.17. Power transmission of the two perpendicular polarized components of the light beam as a function of the angle of incidence for the transition from the optically denser glass with $n = 1.5$ to air

Again the parallel to the surface polarized light shows zero reflection at an angle φ_B which is smaller than the angle for total reflection. This fact can be used to polarize light with simple glass plates which are applicable for high powers. Again, the transmission T and the reflectivity R are given for nonpolarized light in Fig. 3.18 for the transition of the light from an optically denser material with $n = 1.5$ into air.

Of course all formulas become much more difficult if the material is anisotropic. In photonics anisotropic crystals are widely used for frequency transformation, switching, deflection, etc. If necessary detailed analysis may

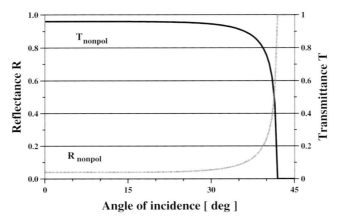

Fig. 3.18. Power reflectivity R and power transmission T of a nonpolarized light beam as a function of the angle of incidence for the transition from the optical denser glass with $n = 1.5$ into air

be needed. Usually these crystals are well transparent for the required light. The antireflection coatings can be designed for the average refractive index and for the applied polarization and entrance angle of the applied light.

3.5.2 Brewster's Law

As mentioned above at a certain angle of incidence φ_B, the Brewster angle, the reflected light is perfectly polarized with a polarization direction parallel to the surface. Under these conditions the reflected and the transmitted beam or the wave vectors of these two waves are perpendicular to each other (see Fig. 3.19).

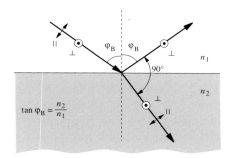

Fig. 3.19. Brewster angle for perfect polarization of the reflected light

The Brewster angle can be determined by:

$$\textbf{Brewster angle} \quad \varphi_B = \arctan\left(\frac{n_2}{n_1}\right) \tag{3.90}$$

with values of $55.6°$–$60.3°$ for the transition from air to different glasses with refractive indices of 1.46 (quartz) to 1.75 (flint). Stacks of such thin glass plates can be used to build technical polarizers for high-power beams in a simple way. Because the Brewster angle can be very precisely measured the refractive indices of optical materials can be determined this way [see, for example, 3.22].

3.5.3 Total Reflection

Total reflection of the incident light beam can be obtained if the light reaches the optical surface from the higher refracting side with n_1 towards the lower refracting material with n_2 ($n_1 > n_2$) (see Fig. 3.20).

Total reflection will occur for all incident angles greater than or equal to φ_{tot}:

$$\textbf{total reflection} \quad \varphi_{tot} = \arcsin\left(\frac{n_2}{n_1}\right). \tag{3.91}$$

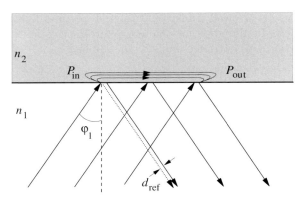

Fig. 3.20. Total reflection of incident light beam at the index transition plane towards a smaller refractive index with beam displacement d and evanescent wave

For transitions from glasses with refractive indices of 1.46 (quartz) to 1.75 (flint) the minimum angle for total reflection will be 43.2°–34.9° which is smaller than 45°. Thus these materials can be used as 45° prisms for 90° reflection of light, independent of its dispersion for different wavelengths.

For angles $\varphi_1 > \varphi_{\text{tot}}$ the reflected light is not linearly but elliptically polarized. The phase shift between the parallel and the perpendicular polarized reflected light δ_{tot} can be calculated from:

$$\tan\left(\frac{\delta_{\text{tot}}}{2}\right) = \frac{1}{\sin^2 \varphi} \sqrt{(1 - \sin^2 \varphi)(\sin^2 \varphi - n_{\text{rel}}^2)} \tag{3.92}$$

with its maximum value:

$$\tan\left(\frac{\delta_{\text{tot,max}}}{2}\right) = \frac{1 - n_{\text{rel}}^2}{2n_{\text{rel}}} \tag{3.93}$$

at the matching angle of incidence $\varphi_{\text{tot,max}}$:

$$\sin \varphi_{\text{tot,max}} = \sqrt{\frac{2n_{\text{rel}}^2}{1 + n_{\text{rel}}^2}} \tag{3.94}$$

For very small n_{rel} close to 1 the phase shift will be a maximum and close to π. Circularly polarized light would occur. Phase differences of $\pi/4$ can easily be achieved with usual glasses and then two reflections would produce circular polarization.

For example, of conservation of total spin there must exist a transmitted light wave which is also depolarized. The phase shift of this light is half of the value of the reflected light. This wave is a maximum after a few wavelengths thickness and moves from point p_{in} to point p_{out} in Fig. 3.20. It does not consume energy but it is useful to test the material with n_2. Thus thin films are investigated with this method of evanescent light waves [3.23–3.28].

Further, it should be noted that a very small displacement d occurs during total reflection. This displacement d can be calculated from [3.29]:

$$d_{\mathrm{ref}} = \frac{\lambda \sin\varphi \cos^3\varphi}{\pi(\cos^2\varphi + \sin^2\varphi - n_{\mathrm{rel}}^2)\sqrt{\sin^2\varphi - n_{\mathrm{rel}}^2}} \tag{3.95}$$

which is in the range of a few percent of the wavelength. Nevertheless, this displacement can be increased by orders of magnitude from multiple reflections in thin samples.

3.6 Relation Between Reflection, Absorption and Refraction

In the previous section absorption of light in materials was neglected in order to get analytical formulas. The influence of the absorption can be discussed in the case of perpendicular incidence from air to a material with refractive index n and absorption coefficient a (see Sect. 3.4). The simple formula (3.88) has to be extended by the absorption coefficient a to:

$$R_{0°,\mathrm{abs}} = \frac{P_{\mathrm{refl}}}{P_{\mathrm{inc}}} = \frac{I_{\mathrm{refl}}}{I_{\mathrm{inc}}} = \frac{(1 - n_{\mathrm{rel}})^2 + (a\lambda/4\pi)^2}{(1 + n_{\mathrm{rel}})^2 + (a\lambda/4\pi)^2}. \tag{3.96}$$

If the absorption is very small this equation reduces to (3.88). If the absorption coefficient a is much larger than $4\pi/\lambda$, which means a penetration depth of less than the wavelength, the second term becomes dominant and the reflectivity becomes $R_{0°}$ $(a > 4\pi/\lambda) \approx 100\%$ as, e.g. for metals.

In metals the refractive index can be below 1 as a function of the angle of incidence. Thus the phase speed can be higher than the vacuum light speed. But because of the high absorption the light wave has a penetration depth of only a few nm. For more details see [M5]. Therefore the reflection of light at metal surfaces is high over a very large spectral range as shown in Fig. 3.21.

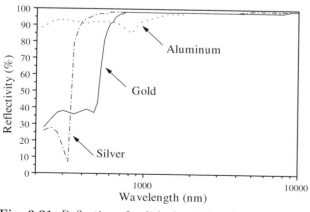

Fig. 3.21. Reflection of polished metal surfaces as a function of the wavelength

These metals are therefore used as mirrors if very wide spectral reflectivity is needed. Usually the metal is deposited at glass surfaces in thin films with good optical quality. These films are often coated with thin layers of, e.g. silicon oxide to protect the metal from oxidation.

Another possibility for mirrors is the coating of surfaces with thin layers of dielectric material. With well-designed layer thicknesses, applying the interference effects described in Sect. 2.9.6, it is possible to reach very high reflectivities, above 99.999% for reasonably large wavelength ranges of a few tens of nanometers. These mirrors are used in lasers and high-powers photonic applications. Some newer examples are given in [3.30–3.34].

3.7 Birefringence

Anisotropic materials which usually have an inner structure such as, e.g. crystals, liquid crystals, organic molecules in structured environments or thin films, can show different light speeds for different light propagation directions relative to the material. This effect is used in nonlinear optics, e.g. for phase matching in frequency conversion.

In the simplest case of optically uniaxial crystals the incident light is split into two beams, the ordinary (o) and the extraordinary (e), which have a perpendicular polarization to each other. The ordinary beam shows known refraction (see Fig. 3.22).

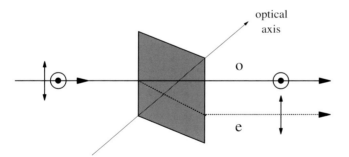

Fig. 3.22. Birefringence at an optically uniaxial crystal with ordinary (o) and extraordinary (e) beams and their polarization

This effect is called birefringence. In this case the vector of the speed of light of the extraordinary beam in the material describes a rotational ellipsoid and the ordinary beam shows equal light speed in any direction. The ellipsoid of the extraordinary beam can be narrower or wider compared to the sphere of the ordinary beam. This corresponds to an ellipsoid of the refractive index for the extraordinary light which is wider or narrower compared to the isotropic refractive index of the ordinary beam (see Fig. 3.23).

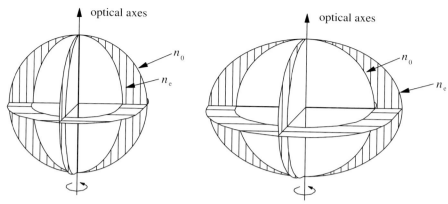

Fig. 3.23. Ellipsoid of the refractive index of the extraordinary beam in an optically uniaxial crystal in comparison to the spherical index of an ordinary beam. Left: optically negative uniaxial crystal, right: optically positive uniaxial crystal

Thus the refractive index ellipsoid for the extraordinary beam has a gradient at the surface of the crystal positioned as in Fig. 3.22, so the extraordinary beam is refracted by this gradient and so a refraction angle occurs although the beam was perpendicular to the geometrical crystal surface as shown in the figure.

The refractive indices of the two beams differ by a few percent as shown in Table 3.3 for some materials. In the direction of the optical axis of the crystal no birefringence can be obtained and perpendicular to it the birefringence is a maximum.

Table 3.3. Refractive indices for optically uniaxial crystals for the ordinary beam o and for the extraordinary beam e perpendicular to the optical axis of the crystal for light wavelength of 589 nm [M31]

Material	n_o	n_e	$1 - n_o/n_e$	Character
Ice	1.309	1.313	0.3%	+
K_2SO_4	1.455	1.515	4.0%	+
Quartz	1.544	1.553	0.6%	+
Tourmaline	1.642	1.622	−1.2%	−
Calcite	1.658	1.486	−11.6%	−
Corund	1.768	1.660	−6.5%	−

The angle of refraction can be determined for the ordinary beam as usual. The refraction of the extraordinary beam can be calculated from:

$$\frac{\sin \varphi_1}{\sin \varphi_2} = n^{\mathrm{e}}(\varphi_2 - \theta) = \frac{n_0 \sqrt{1 + \tan^2(\theta \mp \varphi_2)}}{\sqrt{1 + \left(\dfrac{n_0}{n_{\mathrm{e}}}\right)^2 \tan^2(\theta \pm \varphi_2)}} \qquad (3.97)$$

using the definitions of Fig. 3.24. The minus sign is valid for the situation of this figure and the plus sign is valid if the optical axis is mirror symmetric to the perpendicular at the incident surface. The extraordinary beam will experience a different refractive index compared to the ordinary one even at perpendicular incidence. The refractive index for the extraordinary beam will change from the value of the ordinary beam up or down to the value of perpendicular to the optical axis along the elliptical surface.

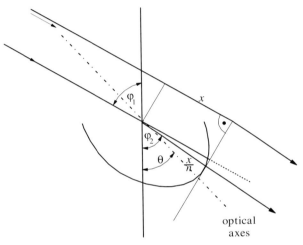

Fig. 3.24. Refraction of the extraordinary beam in an optically uniaxial crystal

For the analysis of this process the speed of extraordinary light beams will vary also as an ellipse but with inverted dimension $c_{\mathrm{vacuum}}/n_{\mathrm{e}}(\boldsymbol{r})$. The beam propagation per unit time can be used for the construction of the new wave fronts of the extraordinary beam in the material as shown in Fig. 3.24. The polarization of the extraordinary and ordinary beams will be perpendicular to each other.

Thus the intensities of the two beams will depend on the polarization of the incident light, the angle of incidence and the direction of the optical axis of the material. Fresnel formulas analogous to the set given for isotropic materials can be used but the two beams with their different polarization have to be evaluated separately with their different refractive indices.

If the material shows lower symmetry than optically uniaxial crystals two extraordinary beams and no ordinary beam can be observed. Two different rotational ellipsoids occur for the two beams similar to the single ellipsoid in uniaxial crystals. Materials with two extraordinary beams are *optical biaxial*.

In this case three refractive indices n_u, n_v and n_w will be defined as the half axes of the three-dimensional ellipsoids (see Fig. 3.25).

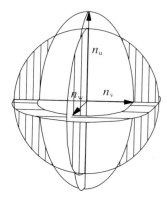

Fig. 3.25. Refractive indices of optically biaxial material

The two optical axes are defined as the directions of equal light speed for the two extraordinary beams as shown in Fig. 3.26, which gives the planar cut through Fig. 3.25 in the paper plane. The two optical axes are determined from the tangents T to the ellipse and the circle, as given. The optical axes are different to the directions S_1 and S_2, and, therefore, the propagation in the direction of the optical axis as well as in the direction of S_i will produce new beams which show a cone symmetry in three dimensions.

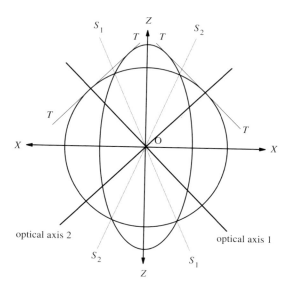

Fig. 3.26. Optical axis of optically biaxial material

Optical birefringent crystals are frequently used for changing the polarization of beams for adapting to the experimental setup. They are also used in nonlinear optics for frequency conversion of light. The different refractive indices allow for the phase matching of the different frequencies (see Sect. 4.4.2). Furthermore in nonlinear optics the material symmetry can be broken by strong laser fields and thus induced optical birefringence is applied in photonics, too. Some other applications are given in [3.35–3.38]. In laser crystals the induced birefringence is one of the limiting factors for realizing good beam quality, as described in Sect. 6.4.2.

3.8 Optical Activity (Polarization Rotation)

In some materials with helical symmetry the light polarization is rotated with propagation. Traditionally this property of matter is called *optical activity*. The rotation angle β_{oa} follows from:

rotation angle $\beta_{oa} = \kappa_{oa}(\lambda)d$ (3.98)

with the coefficient κ_{oa} describing the optical activity of the material as the angle per unit length d as a function of the wavelength. For solutions the coefficient is proportional to the concentration c_{conc} of the optically active matter. It is positive if the polarization is rotated anticlockwise with propagation and the material called right rotating. In Table 3.4 some materials with their optical activity are listed.

Table 3.4. Optical activity κ_{oa} of some materials at 589 nm

Material	κ_{oa} (grd/mm)
Quartz (Crystalline, Uniaxial)	21.7
NaBrO$_3$ (Isotropic)	2.8
Menthol (Liquid)	−0.5
Sugar (10 g l^{-1} solution in water)	0.67

The optical activity can be observed for uniaxial crystals with light propagation in the direction of the optical axis only. Otherwise the birefringence would overlay. The wavelength dependence of the optical activity is shown for crystalline quartz in Fig. 3.27 as an example.

This rotation dispersion can be used for selecting the wavelength of linear polarized light with a polarizer. As crystalline quartz is used for rotating the polarization in photonic applications such as, e.g. a 45° or 90° rotator, the wavelength has to be recognized carefully in order to choose the length of the crystal.

Optical activity can be expressed as circular birefringence. If the linearly polarized beam is represented by equal right and left circular polarized beams

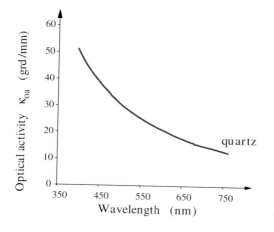

Fig. 3.27. Optical activity κ_{oa} of crystalline quartz as a function of wavelength

the positive polarization rotation is equivalent to the delay of the left circular polarized light:

$$\kappa_{oa} = \frac{\pi}{\lambda}\left(n_{left} - n_{right}\right) \tag{3.99}$$

Thus in this case the refractive index of the left circular wave n_{left} is slightly larger than n_{right} but the absolute difference is of the order of 10^{-5} as can be calculated from Table 3.4 or Fig. 3.27.

Some materials can show optical activity if magnetic fields are applied (the Faraday effect). If light propagates parallel to the magnetic field the polarization is anti-clockwise rotated. The rotation angle is dependent on Verdet's constant as given in Sect. 2.6.2. If the light is backreflected for a second pass through a Faraday rotator the total rotation of the polarization is doubled, whereas in other rotators the rotation is compensated. Therefore, Faraday rotators are used in double-pass amplifier schemes (Sect. 6.11.3) and in optical isolators between two polarizers. Some applications are given in [3.39–3.43].

3.9 Diffraction

If part of a light beam is completely absorbed or reflected, e.g. by apertures, edges or small objects, further propagation of the transmitted light will be modified. The uncertainty principle for photons, or in other words the wave character of the light, results in a change of the wave front curvature. The light will be diffracted in setups with Fresnel numbers (see Sect. 2.3.1) not large compared to one. In cases of large Fresnel numbers the geometrical optics approximation is useful.

The diffraction limits the resolution of optical devices such as magnifiers or microscopes. Usually complicated theoretical calculations are necessary for a detailed description. The resulting light beams are usually not diffraction limited behind the aperture, but if apertures with Gaussian transmission profiles are used the new beam will be diffraction limited with a new divergence.

Because apertures are widely used in photonic applications some basic configurations will be described and the diffraction results will be given. The diffraction is in general described by the Kirchhoff integral which can be simplified to expressions useful for the calculation given below. Therefore only the spatial change of the amplitude of the electric field is described, based on the SVA approximation.

Often periodical structures, such as, e.g. gratings, are used for diffracting light beams. In these cases besides the diffraction interference of the different new beams also occurs. Thus diffraction takes place as a consequence of the small dimensions of the structures and produces a wide distribution of light. The interference effect shows a narrow structuring of the signal. Both effects have to be carefully differentiated.

3.9.1 General Description: Fresnel's Diffraction Integral

It is assumed that the electric field vector $\boldsymbol{E}_{\mathrm{ap}}(\boldsymbol{r})$ is zero at the aperture and undisturbed elsewhere. Thus the detailed interaction of the field with the surface of the aperture edges is neglected. Following Huygen's principle [e.g. M5, 3.44] spherical waves are assumed at any place of the wave front. The electrical field behind the aperture can now be calculated by the superposition of all spherical waves.

If the dimension of the aperture is large compared to the wavelength and the Fresnel number is not large compared to 1 the resulting electric field amplitude $\boldsymbol{E}_{0,\mathrm{diff}}(x, y, z)$ of the diffracted light at a distance z from the aperture (see Fig. 3.28) can be calculated in the xy plane as a function of the field amplitude at the position of the aperture $\boldsymbol{E}_{0,\mathrm{ap}}(x_{\mathrm{ap}}, y_{\mathrm{ap}}, z = 0)$ by:

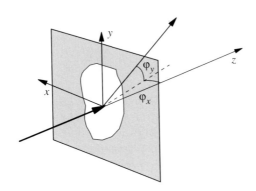

Fig. 3.28. Definitions of the directions of the electrical field amplitudes in the x and y directions, the distance in the z direction and the angles for diffraction at an aperture

$$E_{0,\text{diff}}(x, y, z) = \frac{C_{\text{diff}}}{\lambda} \iint\limits_{\text{aperture}} E_{0,\text{ap}}(x_{\text{ap}}, y_{\text{ap}})$$

$$\cdot \exp\left\{ \frac{-ik}{2z} \left[(x_{\text{ap}} - x)^2 + (y_{\text{ap}} - y)^2 \right] \right\} \, dx_{\text{ap}} \, dy_{\text{ap}}.$$

$$(3.100)$$

This is the *Fresnel integral*. The intensity of the light field follows from this integral for the amplitude of the electric field with wavelength λ and wave vector $\boldsymbol{k}$ by the square. The oscillating electric field results from multiplying this integral by $\exp(-i(2\pi\nu t + kz))$. The amplitude factor C_{diff} has to be calculated from the energy balance of the total powers at the aperture and at the screen. If the field distribution at the screen is Gaussian the diffracted field will reproduce the Gaussian shape as can be shown with this equation. Thus the Gaussian beam is a solution of this diffraction integral. For some examples see [3.45–3.48].

3.9.2 Far Field Pattern: Fraunhofer Diffraction Integral

For very large distances from the aperture the Fresnel number is much smaller than 1 ($F \ll 1$). In this case the quadratic terms of the dimension of the aperture x_{ap} and y_{ap} can be neglected in the exponent of (3.100). Thus the Fresnel integral can be simplified. By using the substitution:

$$\xi_x = k \tan \varphi_{\text{diff},x} = k\frac{x}{z} \quad \text{and} \quad \xi_y = k \tan \varphi_{\text{diff},y} = k\frac{y}{z} \qquad (3.101)$$

it follows from (3.100) that

$$E_{0,\text{diff}}(\xi_x, \xi_y, z) = i \frac{e^{\{iz(\xi_x^2 + \xi_y^2)/2k\}}}{\lambda z} \iint\limits_{\text{aperture}} E_{0,\text{ap}}(x_{\text{ap}}, y_{\text{ap}})$$

$$\cdot \exp\left\{ -ik(\xi_x + \xi_y) \right\} \, dx_{\text{ap}} \, dy_{\text{ap}}. \qquad (3.102)$$

This *Fraunhofer integral* describing the far-field of the diffraction with the field pattern $E_{0,\text{ap}}(x_{\text{ap}}, y_{\text{ap}}, z = 0)$ is the Fourier transformation of this distribution to the angle distribution of the diffracted pattern [3.49].

3.9.3 Diffraction in First Order Systems: Collins Integral

For so-called first-order systems such as lenses, mirrors and free space propagation the diffraction integral can be written based on the ray matrix elements a, b, c, d as described in Sect. 2.5. Assuming, further, constant light polarization the expression for one-dimensional systems follows as:

$$E_{0,\text{diff}}(\xi_x, \xi_y, z) = -i\frac{e^{ikz}}{b\lambda} \iint\limits_{\text{aperture}} E_{0,\text{ap}}(x_{\text{ap}}, y_{\text{ap}})$$

$$\cdot \exp\left\{i\frac{\pi}{b\lambda}\left(ax_{\text{ap}}^2 + d\xi_x^2 - 2x_{\text{ap}}\xi_x + ay_{\text{ap}}^2 + d\xi_y^2 - 2y_{\text{ap}}\xi_y\right)\right\} dx_{\text{ap}} \, dy_{\text{ap}}.$$

$$(3.103)$$

This is the Collins integral in the Fresnel approximation [3.50].

3.9.4 Diffraction at a One-Dimensional Slit

In the simplest case an indefinite light wave is diffracted by a one-dimensional slit perpendicular to the light propagation direction. In the direction parallel to the slit no diffraction will occur (see Fig. 3.29).

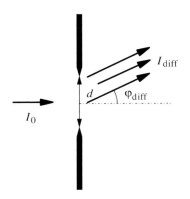

Fig. 3.29. Diffraction of light at a one-dimensional slit of the width d

Perpendicular to this direction the far-field intensity distribution I_{diff} as a function of the angle φ_{diff} is given by:

$$I_{\text{diff}}(\varphi_{\text{diff}}) = I_0 \frac{d^2}{2\lambda L} \frac{\sin^2\left(\frac{\pi d}{\lambda}\sin\varphi_{\text{diff}}\right)}{\left(\frac{\pi d}{\lambda}\sin\varphi_{\text{diff}}\right)} \qquad (3.104)$$

with the slit width d and the light wavelength λ. The graph of this formula shows characteristic maxima and minima (see Fig. 3.30).

The main maximum occurs at $\varphi_{\text{diff}} = 0°$. The mth minimum can be obtained at an angles of

$$\varphi_{\text{diff,minimum}} = \arcsin\left(\frac{m\lambda}{d}\right) \qquad (3.105)$$

with the integer value m and the maxima between these values.

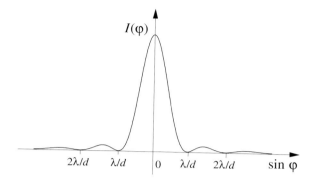

Fig. 3.30. Diffracted intensity as a function of the observation angle φ_{diff} behind a one-dimensional slit

At these angles the intensity is zero. At the angle $\arcsin(\lambda/2d)$ the intensity is $4I_{\text{diff,max}}/\pi^2 = 0.406 I_{\text{diff,max}}$. The full far-field angle for full width half maximum intensity follows from:

$$\Delta\varphi_{\text{diff,FWHM}} = 0.8855\frac{\lambda}{d} \tag{3.106}$$

and for comparison with a light beam with a Gaussian profile in the direction of the slit the full far-field angle at $1/e^2$ intensity is given by:

$$\Delta\varphi_{\text{diff},1/e^2} = 1.400\frac{\lambda}{\pi d} \tag{3.107}$$

From these formulas and Fig. 3.30 it obviously follows that the diffracted light has no Gaussian intensity distribution. The divergence of the main peak of this beam would be 0.7 times smaller than the divergence of a diffraction-limited beam.

The power content in the main maximum is for ratios of d/λ smaller 200 larger than 90.27% between the minima, 87.76% between the $1/e^2$ intensity values and 72.19% inside FWHM. The first side maxima contain than more as 4.71% each and the second maxima more than 1.65%. All other maxima have a power content of less than 1% each.

The transition between the near-field (Fresnel) and far-field (Fraunhofer) pattern is illustrated for the example of the one-dimensional slit in Fig. 3.31.

As can be seen from this figure the near-field "shadow" of the slit transforms to the far-field diffraction pattern with increasing distance. The resulting Fresnel number is 8 for 1 m distance and 0.8 for 10 m which is almost equal to the far-field pattern. At a distance of 0.1 m the Fresnel number is 80 but modulations in this near-field pattern can still be observed.

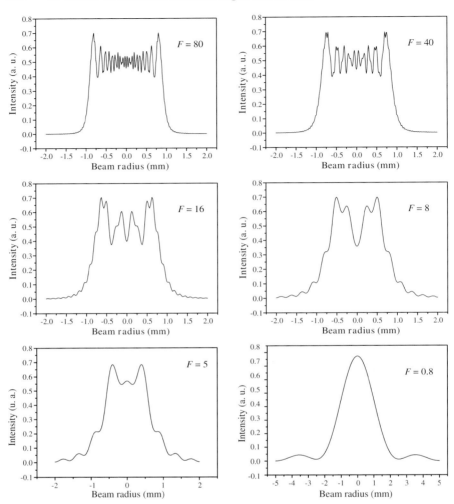

Fig. 3.31. Intensity distribution of a planar wave behind a slit of 2 mm width at a distances of 0.1 m, 0.2 m, 0.5 m, 1 m and 10 m calculated for a wavelength of 500 nm

3.9.5 Diffraction at a Two-Dimensional Slit

Diffraction in two orthogonal dimensions can be analyzed as the superposition of the two one-dimensional results of diffraction. If the right angular aperture has dimensions d_x and d_y (see Fig. 3.32) then the light beam with an equal intensity over the cross-section d_x times d_y passes the aperture perpendicular

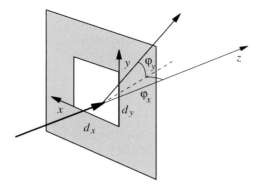

Fig. 3.32. Two-dimensional diffraction at a right angular aperture

towards the z direction. The diffracted intensity I_{diff} can be calculated from:

$$I_{\mathrm{diff}} = I_0 \frac{1}{2\lambda^2 L^2} d_x^2 \frac{\sin^2\left(\dfrac{\pi d_x}{\lambda}\sin\varphi_x\right)}{\left(\dfrac{\pi d_x}{\lambda}\sin\varphi_x\right)^2} d_y^2 \frac{\sin^2\left(\dfrac{\pi d_y}{\lambda}\sin\varphi_y\right)}{\left(\dfrac{\pi d_y}{\lambda}\sin\varphi_y\right)^2} \tag{3.108}$$

with an analogous definition as in the previous chapter. The diffraction intensity is calculated in Fig. 3.33.

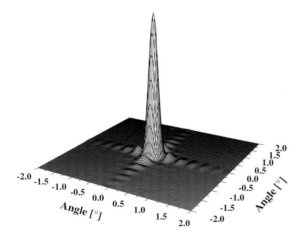

Fig. 3.33. Far-field diffraction intensity as a function of the observation direction behind a two-dimensional aperture of dimension $d_x = d_y = 200\,\mu\mathrm{m}$ and a wavelength of 532 nm

The diffracted intensity shows the same symmetry as the aperture. In the linear x or y dimension the result is similar to Fig. 3.30. Thus the far-field divergence angle can be calculated as given for the slit in the previous chapter for the x and y components of the beam separately.

For a quadratic aperture of length d the power content in the main peak is 81.49% inside the area of the zero value of the intensity, 77.03% inside the $1/e^2$ value and 52.11% inside the FWHM of the intensity.

3.9.6 Diffraction at a Circular Aperture

For the circular aperture with diameter D the diffraction pattern is given by:

$$I_{\text{diff}} = I_0 \frac{\pi^2 D^4}{128\lambda^2 L^2} \frac{J_1^2\left(\dfrac{\pi D}{\lambda}\sin\varphi_r\right)}{\left(\dfrac{\pi D}{\lambda}\sin\varphi_r\right)^2} \tag{3.109}$$

as the diffracted intensity I_{diff} observed in the direction of the angle φ_r to the z direction (see Fig. 3.34). I_0 denotes the intensity of the incident beam at the position at the aperture. J_1 stands for the first-order Bessel function and it is assumed that the incident light has a planar wave front and a constant intensity across the beam. The angles φ_r for which the first minima and maxima of the Bessel function occur can be calculated from:

$$\varphi_{r,p\text{-min/max}} = \arcsin\left(C_{\text{Bess},p\text{-min/max}}\frac{\lambda}{D}\right). \tag{3.110}$$

The values of $C_{\text{Bess},p\text{-min/max}}$ are given in Table 3.5:

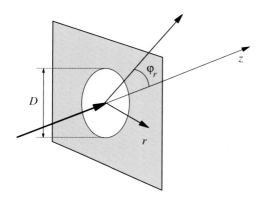

Fig. 3.34. Diffraction at a circular aperture with diameter D

Table 3.5. Values of C for the calculation of the diffraction angles for which the first-order Bessel function has minima and maxima

	$C_{\text{Bess}1,1}$	$C_{\text{Bess}1,2}$	$C_{\text{Bess}1,3}$	$C_{\text{Bess}1,4}$	$C_{\text{Bess}1,5}$	$C_{\text{Bess}1,6}$	$C_{\text{Bess}1,7}$
minimum	1.220	2.233	3.238	4.241	5.243	6.244	7.245
maximum	1.635	2.579	3.699	4.710	5.717	6.722	7.725
$I_{\max,p}/I_{\max,0}$	1.750%	0.416%	0.160%	0.078%	0.044%	0.027%	0.018%

The diffraction intensity is given in Fig. 3.35.

The shape of this diffracted light is again not Gaussian. The full far-field angle for full width half maximum intensity in the circular geometry follows from:

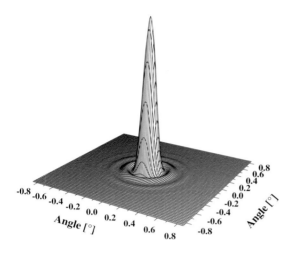

Fig. 3.35. Far-field diffraction intensity as a function of observation direction behind a circular aperture with diameter $D = 200\,\mu\text{m}$ for a light wavelength of 532 nm

$$\Delta\varphi_{\text{diff,FWHM}} = 1.029\frac{\lambda}{D} \tag{3.111}$$

and the full far-field angle at $1/e^2$ intensity of the diffracted beam is given by:

$$\text{circular aperture} \quad \Delta\varphi_{\text{diff},1/e^2} = 5.170\frac{\lambda}{\pi D} \tag{3.112}$$

The power content of the diffracted beam is 83.8% in the main peak, 76.7% in the $1/e^2$ intensity area and 47.5% inside the FWHM area [3.51]. For comparison a Gaussian beam is shown in Fig. 3.36.

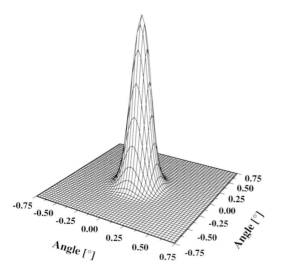

Fig. 3.36. Far-field intensity as a function of observation direction of a Gaussian beam with diameter $D = 200\,\mu\text{m}$ for a light wavelength of 532 nm

This beam with a waist diameter $2w_0 = D_{\text{Gauss}}$ shows a far-field full divergence angle of

$$\text{Gauss beam} \quad \Delta\varphi_{\text{Gauss},1/e^2} = 4\frac{\lambda}{\pi D_{\text{Gauss}}} \tag{3.113}$$

and the $1/e^2$ power content is 86.5%.

3.9.7 Diffraction at Small Objects (Babinet's Theorem)

Diffraction at small objects has to be recognized if the characteristic dimension D_{object} (wire width; sphere diameter) is not large compared to the product of the light wavelength λ multiplied by the observation distance $L_{\text{observation}}$:

$$\text{diffraction at objects} \quad D_{\text{object}} \leq \lambda L_{\text{observation}}. \tag{3.114}$$

Otherwise shadows are obtained which are given by geometrical optics (see Fig. 3.37).

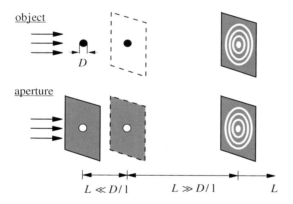

Fig. 3.37. Diffraction and shadow behind small objects or apertures

It can be shown that a small object and an aperture of the same shape and size cause the same diffraction pattern in the far-field, which is known as Babinet's theorem.

3.9.8 Spot Size of Foci and Resolution of Optical Images

Diffraction can limit the size of focused light beams or the resolution of optical images. If a diffraction-limited Gaussian beam with a plane wave front and a beam diameter of $2w_{\text{lens}} = D_{\text{lens}}$ at a lens of focal length f_{lens} is focused the resulting spot has a diameter of $2w_{\text{focus}} = d_{\text{focus}}$:

$$d_{\text{focus}} = \frac{4}{\pi}\frac{\lambda f_{\text{lens}}}{D_{\text{lens}}} \simeq 1.27\frac{\lambda f_{\text{lens}}}{D_{\text{lens}}} \tag{3.115}$$

If the beam size at the lens is limited by a conventional aperture of diameter $D_{\text{lens,ap}}$ the spot can be calculated from the value of the first minimum in Table 3.5. By focusing, the angle φ_r from the previous sections will be transformed to a spot diameter $d_{\text{focus,ap}}$:

$$d_{\text{focus,ap}} \simeq 2.44 \frac{\lambda f_{\text{lens}}}{D_{\text{lens,ap}}} \tag{3.116}$$

which is nearly twice the value of the diffraction-limited system. If the aperture has a Gaussian transversal transmission profile the beam would be diffraction limited and thus the above mentioned smaller focus could be observed.

This minimal spot size limits the resolution of optical imaging. Assuming that two images have to be at least separated by the angle difference equivalent to their diffraction spot size the limit of optical resolution, OR, can be calculated from the given angles. If small angles are assumed the sine and the angle are equal and thus the resolution can be calculated from:

optical resolution $\text{OR} = \dfrac{1}{\varphi_r}$. $\tag{3.117}$

Several approaches, for example near-field techniques, have been tried to get much higher resolution [see, for example, 3.52–3.61]. With laser techniques, resolutions of tens of nanometers are approached using visible light (see also Sect. 1.5).

3.9.9 Modulation Transfer Function (MTF)

In imaging diffraction is the principal limit of resolution. In addition all kinds of technical problems may decrease the resolution further. Thus the modulation M:

modulation $M(\nu_{\text{sp}}) = \dfrac{I_{\max} - I_{\min}}{I_{\max} + I_{\min}}$ $\tag{3.118}$

is determined via the measured high and low intensities $I_{\text{max/min}}$ as a function of the distance d resulting in a spatial frequency $\nu_{\text{sp}} = 1/d$. This modulation before and after imaging leads to the values M_{orginal} and M_{image}. The quotient of these modulations is the modulation transfer function (MTF):

MTF $\text{MTF}(\nu_{\text{sp}}) = \dfrac{M_{\text{image}}(\nu_{\text{sp}})}{M_{\text{original}}(\nu_{\text{sp}})}$ $\tag{3.119}$

which has to be measured for radial and tangential spatial frequencies ν_{sp}, separately.

The upper limit of the spatial frequencies $\nu_{\max}$ is given by the diffraction limit of the imaging system as given by (3.116) and defined as:

$$\nu_{\max} = \frac{2.44}{d_{\text{focus,ap}}} = \frac{D_{\text{lens,ap}}}{\lambda f_{\text{lens}}} = (\lambda FN)^{-1} \tag{3.120}$$

which is typically measured in lines per mm. The quotient of f_{lens} and $D_{lens,ap}$ of an imaging system, e.g. a photo lens, is called the F-number FN:

$$\textbf{F-number} \quad \text{FN} = \frac{f_{lens}}{D_{lens,ap}}. \tag{3.121}$$

The theoretical limit of the MTF for an imaging system is given by:

$$\text{MTF}(\nu_{sp}) = \frac{2}{\pi}\left[\arccos\left(\frac{\nu_{sp}}{\nu_{max}}\right) - \left(\frac{\nu_{sp}}{\nu_{max}}\right)\sqrt{1 - \left(\frac{\nu_{sp}}{\nu_{max}}\right)^2}\right]. \tag{3.122}$$

This theoretical MTF decreases up to the maximum spatial frequency ν_{max} by less than 10%, only. For optical imaging systems, values of less than 20% are possible at the highest spatial frequencies and at the border of the imaging frame. For examples see [3.62–3.64].

3.9.10 Diffraction at a Double-Slit

Diffraction at two slits of width d and distance Λ leads to a doubly modulated intensity profile:

$$I(\varphi_{diff}) \propto \frac{\sin^2\left(\dfrac{\pi d}{\lambda}\sin\varphi_{diff}\right)}{\left(\dfrac{\pi d}{\lambda}\sin\varphi_{diff}\right)^2}\cos^2\left(\frac{\pi}{\lambda}\Lambda\sin\varphi_{diff}\right) \tag{3.123}$$

which is shown in Fig. 3.38.

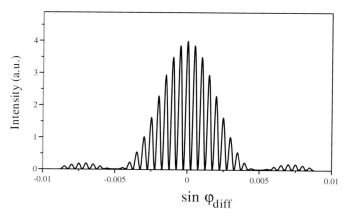

Fig. 3.38. Far-field intensity profile of a light beam with a wavelength of 500 nm behind a double slit of widths $d = 0.1$ mm at a separation $\Lambda = 1$ mm

The two periods are λ/d and λ/Λ. The maximum intensity $I_{\text{diff,max}}$ follows from the cross-sectional constant intensity I_0 of the perpendicular incident beam with planar wave front over the slits from:

$$I_{\text{diff,max}} = I_0 \frac{4d^2}{\lambda L_{\text{screen}}} \tag{3.124}$$

with the distance L_{screen} between the aperture and the screen.

The diffraction pattern can be understood as an overlay of the interference pattern of two beams resulting in the short period λ/Λ which is modulated by the diffraction resulting in the decreasing intensity in the subsidiary maxima. A modern example using an atom is given in [3.65].

3.9.11 Diffraction at One-Dimensional Slit Gratings

Diffraction at induced (slit-) gratings is repeatedly applied in photonics. Thus the diffraction pattern behind a fixed slit grating may serve as a model for these processes. The p slits are arranged at equal distances Λ and have width d (see Fig. 3.39).

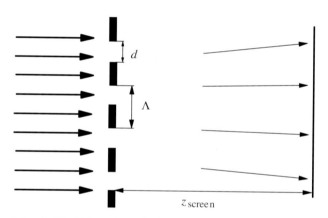

Fig. 3.39. Diffraction of planar wave at a one-dimensional slit grating with $p = 4$ slits

The diffraction intensity follows from:

$$I(\varphi_{\text{diff}}) \propto \frac{\sin^2\left(\dfrac{\pi d}{\lambda}\sin\varphi_{\text{diff}}\right)}{\left(\dfrac{\pi d}{\lambda}\sin\varphi_{\text{diff}}\right)^2} \frac{\sin^2\left(\dfrac{p\pi\Lambda}{\lambda}\sin\varphi_{\text{diff}}\right)}{\sin^2\left(\dfrac{\pi\Lambda}{\lambda}\sin\varphi_{\text{diff}}\right)}. \tag{3.125}$$

The intensity of the diffracted light I_{diff} results from the incident intensity I_0 by:

$$I_{\text{diff,max}} = I_0 \frac{p^2 d^2}{\lambda L_{\text{screen}}}. \tag{3.126}$$

The diffraction pattern is shown in Fig. 3.40. It shows periodic structure
with different periods. First the main maxima j of the pattern occur in the
directions:

$$\text{main maxima} \quad \varphi_{\text{diff,m\,max}j} = \arcsin\left(\frac{j\lambda}{\Lambda}\right) \tag{3.127}$$

and their peak intensity increases quadratically with the number of slits p^2.
The width of these maxima become narrower with $1/p$. This interference effect is overlaid by diffraction. Thus the intensity distribution over the different
main maxima is given by the diffraction curve of a single slit. Therefore the
intensity relation of the different maxima cannot be increased by changing
the number of slits but by decreasing d or Λ!

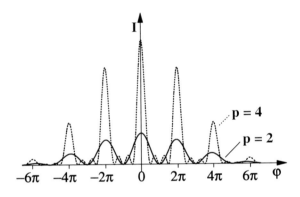

Fig. 3.40. Diffraction pattern behind a slit grating
of p slits with width d and
separation Λ

3.9.12 Diffraction at a Chain of Small Objects

If the light is diffracted at a chain of small objects or an equidistant series of
small holes the diffraction pattern will be different from the pattern of slits
because of the different symmetry. If the plane wave with wave vector in the
z direction is incident on a chain arranged in the x direction with the object
distance Λ_{ch} (see Fig. 3.41) the diffraction pattern in the xz plane will be
given by the above formula for slits (3.125).

The main intensity maxima are observable in this plane in the directions
given by (3.125), but in the y direction the diffraction intensity is not constant. Because of the cylindrical symmetry along the x axis the constant
diffraction angle φ_{diff} as, e.g. for the intensity maxima of (3.125) leads to
hyperbolas in the xy plane of the screen (see Fig. 3.42).

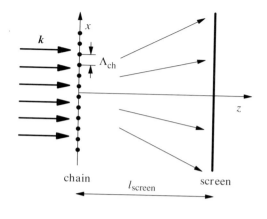

Fig. 3.41. Diffraction at a chain of small objects or holes along the x direction

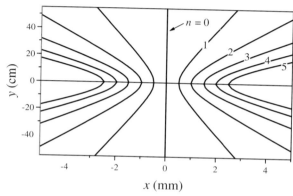

Fig. 3.42. Intensity maxima from the diffraction of light with a wavelength of 500 nm at a chain of objects at a distance of $\Lambda_{ch} = 0.1$ mm at the screen position in a distance of 10 cm of Fig. 3.41

The mth intensity maximum at the screen has, for the definitions of Fig. 3.41 the positions given by:

$$y^2 = \left[\tan\left(\arccos\frac{m\lambda}{x_{ch}}\right)\right]^2 x^2 - y^2. \tag{3.128}$$

In case of the long distances l_{screen} the hyperbolas sometimes appear nearly as straight lines in the observation field.

3.9.13 Diffraction at Two-Dimensional Gratings

If the linear chains of the last paragraph are arranged at regular distances in the y direction, too, as shown in Fig. 3.43, the diffraction at this two-dimensional grating will show a superposition of the x and y diffraction patterns [e.g. 3.66].

Thus two sets of hyperbolas occur and the intensity maxima of this pattern are given by the intersection of these two curves. They can be calculated

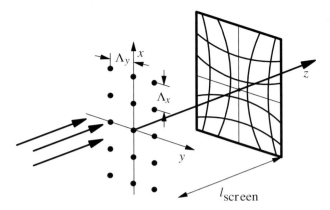

Fig. 3.43. Diffraction at a two-dimensional grating with the constants Λ_x and Λ_y

from (3.127) for the two different angles $\varphi_{\mathrm{diff},x}$ and $\varphi_{\mathrm{diff},y}$ as functions of the two grating constants Λ_x and Λ_y. The result is illustrated in Fig. 3.44. The complete intensity distribution has to be calculated by including the diffraction at the single-grating particles or holes.

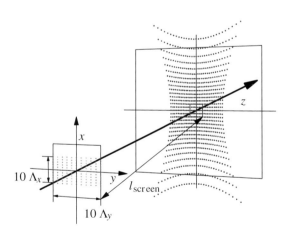

Fig. 3.44. Schematic diffraction intensity pattern of a two-dimensional grating as a superposition of two hyperbolas

Again in many practical cases the hyperbolas will be well approximated by straight lines and the complete intensity distribution has to be calculated by including the diffraction at the single grating particles or holes.

3.9.14 Diffraction at Three-Dimensional Gratings

If the diffraction grating of the previous paragraph is extended to three dimensions one additional condition has to be fulfilled. The grating in the z direction with grating constant Λ_{gr} leads to a circular distribution of the diffracted intensity maxima at the screen. The radius of the mth circle is given by the direction:

$$\varphi_{\mathrm{diff},z,m_z} = \arcsin\left(\frac{m_z\lambda}{\Lambda_{\mathrm{gr}}}\right). \tag{3.129}$$

The three angular conditions of the x, y, z grating can simultaneously be fulfilled for certain wavelengths and counters j_x, j_y, and j_z, only. Again the complete intensity distribution has to be calculated by considering diffraction at the single grating particles or holes.

3.9.15 Bragg Reflection

Three-dimensional gratings can reflect light if the grating constant in the z direction, wavelength and angle of incidence are well tuned to each other. For this purpose the interference of all scattered light waves in the propagation direction has to be destructive and the interference for the reflected light constructive. With the definitions of Fig. 3.45 the path length difference can be calculated.

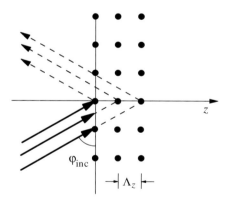

Fig. 3.45. Reflection of light at a three-dimensional grating by diffraction

Constructive interference for the reflected light will take place if the angle between the incident light beam and the grating planes with the separation Λ_{gr} fulfills the Bragg condition:

Bragg condition $\sin\varphi_{\mathrm{inc}} = \dfrac{m\lambda}{2\Lambda_{\mathrm{gr}}}.$ $\tag{3.130}$

The analogous angle of the reflected light is equal to φ_{inc}. This reflection is applied frequently at sound wave gratings or other induced gratings and thus the Bragg condition plays an important role in photonics.

3.9.16 Amplitude and Phase Gratings

Periodical structures can change the amplitude of the electric field vector of the light beam or its phase. Amplitude changes, such as from slit gratings with a square function of the transmission between 0 and 1, were discussed in the previous section, but any periodic modulation of the absorption can cause diffraction as an *amplitude grating* (see Fig. 3.46 and Sect. 7.8.3).

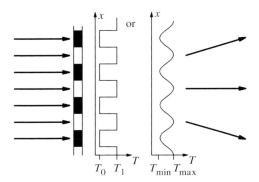

Fig. 3.46. Diffraction at absorption (amplitude) gratings with different modulation of the electric field amplitude by different transmission functions perpendicular to the propagation direction which are square or sine functions

Typically sin-modulations of the transmission are realized in photonic applications because they are produced by the interference pattern of two light beams. The diffraction maxima and minima occur in the same direction as for the slit gratings (see Sect. 3.9.11). The analysis of the amplitude distribution can be based on the superposition of the diffracted and nondiffracted shares of the light.

In the same manner emission gratings (see Sect. 6.10.4) can be described as amplitude gratings but in this case the absorption coefficient will be negative (see Fig. 3.47).

For this type of grating two interference structures will occur: one for the absorbed light and the other for the emitted light at a different wavelength. Thus the diffraction pattern will be wider for the emitted light with the longer wavelength.

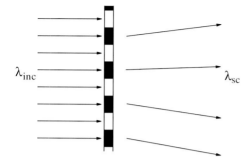

Fig. 3.47. Diffraction at emission (amplitude) grating with a modulation of the electric field amplitude of the emitted light perpendicular to the direction of propagation

Even more important for practical applications are the *phase gratings* [e.g. 3.67, 3.68] with a periodic modulation of the optical path length through the material. This different path length can be caused by modulations of (the real part of the) refractive index or by different geometrical path lengths of the transparent material (see Fig. 3.48).

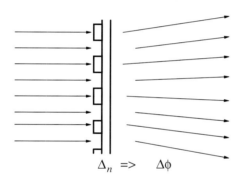

Fig. 3.48. Diffraction at refractive index or geometrical path length (phase) gratings perpendicular to the direction of propagation

These phase gratings operate almost without losses. Thus they are useful for high-power and high-efficiency applications. Efficiencies of more than 90% in diffractive optics [3.69] are possible by etching structures with many steps in the modulation function.

The main maxima and minima for an arbitrary angle of incidence θ_{inc} can be calculated from:

$$\sin(\Phi_{\text{diff},m} + \theta_{inc}) = \begin{cases} \dfrac{m\lambda_{\text{light}}}{\Lambda} + \sin\theta_{inc} & \text{maxima} \\ \dfrac{(2m+1)\lambda_{\text{light}}}{2\Lambda} + \sin\theta_{inc} & \text{minima} \end{cases} \tag{3.131}$$

with the wavelength λ_{inc} of the light and Λ as the grating constant. The diffraction order is counted by the number m. The determination of the amplitudes of the electric field or the intensity pattern needs a complete analysis of the diffraction process.

3.9.17 Diffraction at Optically Thin and Thick Gratings

If the thickness of the grating D_{gr} is sufficiently small the optical path length difference of the diffracted light from the entrance surface and the end surface of the grating is small compared to the light wavelength period λ/n in the material with refractive index n. This path length difference is a function of the angle of diffraction φ_{diff}. Thus *optically thin gratings* have a thickness $D_{thin-gr}$ of:

$$\text{thin gratings} \quad D_{thin-gr} \ll \frac{\lambda}{2\pi n}\frac{1}{1-\cos\varphi_{diff}} \tag{3.132}$$

or

$$\text{thin gratings} \quad D_{thin-gr} \ll \frac{1}{2\pi}\frac{n\Lambda^2}{\lambda}. \tag{3.133}$$

Thus the geometrical thickness of the thin gratings is roughly estimated to be not much bigger than the light wavelength.

The diffraction angles for minima and maxima intensity are the same as for slit gratings. The mth maximum occurs in the direction:

$$\sin\Theta_{diff,max-j} = \frac{m\lambda}{2\Lambda_{gr}} \tag{3.134}$$

if the distances between grating minima and maxima are Λ_{gr} (see Fig. 3.49).

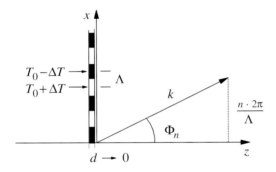

Fig. 3.49. Transmission grating of optical thickness D and grating constant Λ_{gr}

If the grating transmission is modulated between $T_0 - \Delta T$ and $T_0 + \Delta T$ the diffraction intensity at the maxima can be calculated for perpendicular incidence from:

$$I_{max,m} = \frac{1}{2}c_0\varepsilon_0 n\left|\frac{E_{inc}}{\Lambda}\int_0^\Lambda T(x)\exp\left(i\frac{m2\pi x}{\Lambda}\right)dx\right|^2 \tag{3.135}$$

For harmonic modulation of the transmission (as will occur at induced gratings, see Chap. 5):

$$T(x) = T_0 + \Delta T \cos\left(\frac{2\pi x}{\Lambda_{\mathrm{gr}}}\right) \tag{3.136}$$

only three orders of diffraction $+1$, 0, -1 occur because the integrals resulting from (3.135) are orthogonal (see Fig. 3.50).

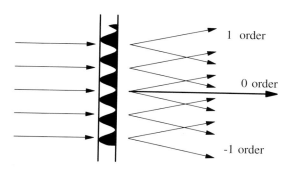

Fig. 3.50. Transmission grating with harmonic modulation shows only three orders of diffraction -1, 0, $+1$ and with total modulation between 0 and 1 a maximum diffraction intensity of 6.25% of the incident intensity in the -1 and $+1$ direction

The resulting intensities are:

0-order intensity $I_{\mathrm{diff,max}-0} = T_0^2 I_{\mathrm{inc}}$ $\tag{3.137}$

and

$+1$, -1-order intensity $I_{\mathrm{diff,max}\pm1} = \dfrac{T_0^2}{4} I_{\mathrm{inc}}.$ $\tag{3.138}$

The maximum diffraction efficiency is realized if the modulation is a maximum, which is

maximum diffraction $T_0 = \Delta T = \dfrac{1}{2}$ $\tag{3.139}$

which results in 6.25% in the two first-order diffraction directions for thin transmission gratings.

If in addition a phase modulation with the same grating constant is present the diffraction intensities can be calculated using the complex representation of the refractive index:

$$\tilde{n}_{\mathrm{total}}(x) = n_{\mathrm{phase}}(x) + i n_{\mathrm{amplitude}}(x) \tag{3.140}$$

with the modulation:

$$n_{p/a} = n_{p/a,0} + \Delta n_{p/a} \cos\left(\frac{2\pi x}{\Lambda_{\mathrm{gr}}}\right). \tag{3.141}$$

The complex transmission is:

$$\tilde{T}(x) = \exp\left[i\frac{2\pi \Delta\tilde{n}_{\mathrm{total}}D}{\lambda} \cos\left(\frac{2\pi x}{\Lambda_{\mathrm{gr}}}\right) - \frac{a}{2}D\right] \tag{3.142}$$

with the absorption coefficient a, and D as the thickness of the grating. This formula includes the complex phase shift $\tilde{\phi}$ produced by the contribution of the phase to the resulting grating:

$$\text{phase shift} \quad \tilde{\phi} = \frac{2\pi \Delta \tilde{n} D}{\lambda}. \tag{3.143}$$

Using this formula the diffracted intensity in the first-order of diffraction from a transmission and phase grating can be calculated as:

$$
\begin{aligned}
I_{\text{max},\pm 1} &= \frac{c_0 \varepsilon_0 n}{2} \left\{ \frac{E_{\text{inc},0}}{\Lambda} \, e^{-aD/2} \int_0^\Lambda e^{i(\tilde{\phi}\,\cos(2\pi x/\Lambda) \pm (2\pi x/\Lambda))} \, dx \right\} \cdot \text{c.c.} \\
&= \frac{c_0 \varepsilon_0 n}{2} E_{\text{inc},0} E_{\text{inc},0}^* \, e^{-aD/2} |J_{\pm 1}(\tilde{\phi})|^2 \\
&= I_{\text{inc}} |J_{\pm 1}(\tilde{\phi})|^2 \, e^{-aD}
\end{aligned} \tag{3.144}
$$

with the Bessel function J_m.

The maximum diffracted intensity to first-order is increased by the phase grating share. It is a maximum for a pure phase grating and reaches a value of 34% of the incident intensity. This is 5.4 times more than the pure transmission grating allows.

The diffracted intensity of a mixed transmission phase grating is shown in Fig. 3.51. In this figure the ratio of the real and imaginary part of the refractive index is considered by κ as:

$$\kappa = \frac{\text{Im}[\tilde{\phi}]}{\text{Re}[\tilde{\phi}]} = \frac{\lambda a}{4\pi \Delta n_{\text{phase}}}. \tag{3.145}$$

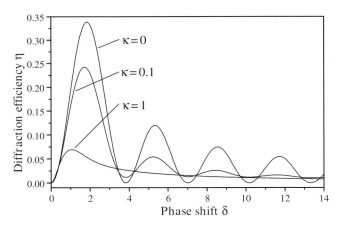

Fig. 3.51. Diffraction intensity in the first diffraction order as a function of the phase shift $\delta = \text{Re}[\tilde{\phi}]$ from the phase grating share with different modulations from the transmission and phase grating. The maximum diffraction efficiency is 33.9%, 24.3% and 6.9% for the given κ-values of 0, 0.1 and 1 indicating the transition from pure phase grating to additional absorption grating

Optically thick gratings have a thickness $D_{\text{thick-gr}}$ of:

$$\text{thick grating} \quad D_{\text{thick-gr}} \geq \frac{\lambda}{2\pi n} \frac{1}{1 - \cos \varphi_{\text{diff}}} \tag{3.146}$$

or

$$\text{thick grating} \quad D_{\text{thick-gr}} \geq \frac{1}{2\pi} \frac{n\Lambda^2}{\lambda} \tag{3.147}$$

which usually means a thickness much larger than the light wavelength λ. If the grating is thought of as thin slices of thin gratings it can finally be analyzed as a three-dimensional grating as it was described in Sect. 3.9.14. In this case the diffraction is efficient for certain angles of incidence only, because the Bragg condition has to be fulfilled. The angle of incidence φ_{inc} has to be equal to half the diffraction angle φ_{diff} and therefore:

$$\text{thick grating diffraction} \quad \varphi_{\text{inc}} = \frac{\varphi_{\text{diff}}}{2} = \arcsin\left(\frac{\lambda}{2\Lambda_{\text{gr}}}\right). \tag{3.148}$$

This formula can be illustrated as reflection at the internal "surfaces" of the thick grating (see Fig. 3.52).

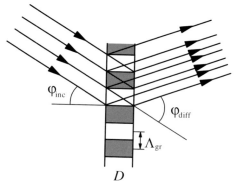

Fig. 3.52. Diffraction at a thick grating illustrated as "reflection" plus interference effect

The diffracted intensity is even more complicated to calculate. For detailed analysis see Sect. 5.9.1 and [3.70]. The maximum diffraction efficiencies defined as the quotient of the diffracted intensity and the incident intensity, to first order, are shown in Table 3.6.

Table 3.6. Maximum diffraction efficiencies with different gratings

	Amplitude grating	Phase grating
Thin grating	6.25%	33.9%
Thick grating	3.7%	100%

From this table it is obvious that phase gratings are much more efficient than amplitude gratings. In addition amplitude or transmission gratings exterminate part of the light which can cause heating of the grating if high powers are applied.

3.10 Light Scattering Processes

Light scattering as a linear interaction with matter leads to a decrease of the transmitted intensity I_{trans} similar to absorption [M3, M1]. It can be calculated from the incident intensity I_{inc}, the density of scattering particles N_{scatt} as the number per volume and the interaction length $z_{interaction}$ by:

$$I_{trans} = I_{inc} \, e^{-\sigma_{scatt} N_{scatt} z_{interaction}} \tag{3.149}$$

with the scattering cross-section σ_{scatt} measured in cm^2. This cross-section is not a function of the incident intensity in the linear case.

If the propagation direction of the photon is changed and not its energy we observe *elastic scattering* as Rayleigh and Mie scattering. If the photon energy and frequency are changed as in Brillouin and Raman scattering we have *inelastic scattering*. The photon energy loss can be as small as 10^{-7} but can also reach about 1% in other processes, too. *Coherent scattering* occurs if the scattered and incident light have a fixed phase relation and *incoherent scattering*, otherwise. Table 3.7 may serve as an overview of the different light scattering processes important in photonic applications.

Table 3.7. Light scattering processes with relative change of photon energy $\Delta E/E$

Type of scattering	Scattering at	$\Delta E/E$
Elastic:		
Rayleigh scattering	particles $\ll$ wavelength	0
Mie scattering	particles $\gg$ wavelength	0
Inelastic:		
Brillouin scattering	sound waves	10^{-7}–10^{-4}
Raman scattering	vibrations	10^{-3}–10^{-2}

All these scattering processes can disturb photonic applications, but usually the share of the scattered light is very small, typically 10^{-7}–10^{-6}. Thus, even in inelastic scattering a very small portion of the incident energy is transferred to matter.

In contrast in nonlinear optics many applications are based on stimulating these scattering processes with high intensities as in stimulated Brillouin scattering for optical phase conjugation or stimulated Raman scattering for frequency conversion. Some examples are given in [3.71–3.76] and applications are described in [3.77–3.81]. More details are given in Chaps. 4 and 5.

3.10.1 Rayleigh and Rayleigh Wing Scattering

Rayleigh scattering is observed from particles with dimensions smaller than the wavelength of the light without absorbing the light [3.81–3.97]. The oscillating electric light field (in x direction in Fig. 3.53) induces dipoles in the matter which "re-emit" the light.

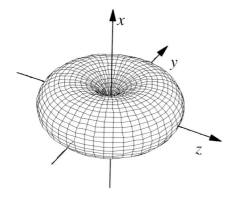

Fig. 3.53. Rayleigh scattering angle characteristics. The electric field vector of the linearly polarized light points in the x direction

The emission of this dipole takes place mostly in the plane perpendicular to the dipole, which is the direction of the electric field and is the x axis in this example. The coordinates are given in Fig. 3.54.

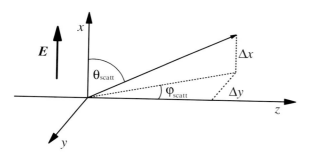

Fig. 3.54. Definition of the angles for Rayleigh scattering as given in Fig. 3.53

The differential cross-section in the space angle Ω_scatt of the scattered light towards the direction φ_scatt and θ_scatt (see Fig. 3.53) is given by:

$$\frac{\mathrm{d}\sigma_{R\,\text{scatt}}}{\mathrm{d}\Omega_\text{scatt}} = \left[\frac{\pi(n^2-1)}{N_\text{scatt}\lambda^2}\right]^2 (\cos^2\varphi_\text{scatt}\cos^2\theta_\text{scatt} + \sin^2\varphi_\text{scatt}) \qquad (3.150)$$

with the refractive index n of the matter.

Thus shorter wavelengths are scattered more, in proportion to $1/\lambda^4$. The total scattering over the whole space angle Ω is given by:

Rayleigh cross-section $\quad \sigma_{R\,\text{scatt,total}} = \dfrac{8}{3}\left[\dfrac{\pi(n^2-1)}{N_\text{scatt}\lambda^2}\right]^2 \qquad (3.151)$

As an example, a typical value is the total Rayleigh scattering cross-section of air with a density of $2.5 \cdot 10^{19}$ molecules cm^{-3} for light at 500 nm:

$$\sigma_{R\,\text{scatt,total,air}} = 6.91 \cdot 10^{-28}\,\text{cm}^2 \qquad (3.152)$$

which causes a scattering loss of $1.73 \cdot 10^{-8}\,\text{cm}^{-1}$.

In liquids and solids Rayleigh scattering takes place at stationary entropy fluctuations, which are determined for more than 94% by density fluctuations in the matter [3.97]. The decay time τ_{Rayleigh} of these fluctuations can be calculated from:

$$\tau_{\text{Rayleigh}} = \frac{\rho_{\text{matter}} c_{\text{p}}}{8 \Lambda_{\text{thermal}} k_{\text{inc}}^2 \, \sin^2(\theta_{\text{scatt}}/2)} \tag{3.153}$$

with the matter density ρ_{matter}, the specific heat c_{p}, the thermal conductivity Λ_{thermal}, the wave vector value k_{inc}, which is almost identical to k_{scatt} and the scattering angle θ_{scatt} between the incident and the scattered light. This lifetime is in the range of 10^{-8} s for liquids for backward scattering.

The broadening of the light linewidth $\Delta\nu_{\text{FWHM,scatt}}$ can be calculated from this value by:

$$\Delta\nu_{\text{FWHM,scatt}} = \frac{1}{2\pi\tau_{\text{scatt}}}. \tag{3.154}$$

It is in the range from 10^{-7} nm in gases to 10^{-5} nm in liquids at 500 nm. It is much smaller in the forward direction. The degree of polarization can be decreased to almost nonpolarized light if multiple scattering occurs.

Rayleigh wing scattering occurs as a result of the orientation fluctuations of molecules which have lifetimes of the order of 10^{-12} s. This orientation relaxation time $\tau_{\text{orientation}}$ can be estimated as given by Debye (1929) [M1]:

$$\textbf{orientation relaxation} \quad \tau_{\text{orientation}} = \frac{4\pi d_{\text{molecul}}^3 \eta_{\text{matter}}}{3 k_{\text{Boltz}} T} \tag{3.155}$$

with the average diameter of the molecules d_{molecul}, the matter viscosity η_{matter}, Boltzmann's constant k_{Boltz} and temperature T.

Thus the bandwidth of Rayleigh wing scattering is about 10^4 times larger than from Rayleigh (center) scattering. At 500 nm the line width can reach a few 0.1 nm. The total scattering cross-section of Rayleigh wing scattering is slightly larger than from Rayleigh scattering, but considering the larger bandwidth of the Rayleigh wing scattering the amplitude per bandwidth is about 10^4 times smaller than for center scattering.

3.10.2 Mie Scattering

Mie scattering takes place at particles of sizes of the order of the light wavelength [3.98–3.107]. This can be aerosols or other fine particles in the air or colloids in solution. It shows no change in wavelength. For larger particles forward scattering is more probable than backward-directed scattering. The scattering losses can be much larger than the Rayleigh scattering losses if enough particles are present. They can be in the range 10^{-8}–10^{-4} cm^{-1}.

The scattering intensity and angular distribution is a complicated function of the particle sizes, distributions and complex refractive indices. The scattering in air is proportional $(1/\lambda)^{\text{c-vis}}$ with the parameter c-vis = 0.7–1.6 indicating poor-to-excellent visibility conditions. Details are given in [3.98].

3.10.3 Brillouin Scattering

Spontaneous Brillouin scattering is the reflection of light by hyper-sound waves in matter. It plays an important role in the ignition process for stimulated Brillouin scattering (SBS) which will be described in Sect. 4.5.9. More details and references will be given there.

Each of the spontaneous sound waves, which are excited as thermal noise with different propagation directions, frequencies, phases and amplitudes, represent optically a sinusoidal modulation of the refractive index. The resulting refractive index grating of the sound wave can be described by a spatial period, which is identical with the wavelength of the sound wave Λ_{sound} and thus by the wave vector $\boldsymbol{k}_{\text{sound}}$ pointing in the propagation direction of the sound wave as shown in Fig. 3.55 and with the value $|\boldsymbol{k}_{\text{sound}}| = 2\pi/\Lambda_{\text{sound}}$.

The incident light beam with frequency ν_{inc}, wavelength λ_{inc} and wave vector $\boldsymbol{k}_{\text{inc}}$ will be scattered to the beam with analogous parameters with index Bscatt at angle Θ_{B}. The wavelength of the sound wave Λ_{sound} has to fulfill the Bragg condition for efficient superposition of the reflected light waves (see Fig. 3.55). Therefore it has to be:

hyper-sound wavelength $\Lambda_{\text{sound}} = \dfrac{\lambda_{\text{inc}}}{2\sin\dfrac{\Theta_{\text{B}}}{2}}.$ (3.156)

It is half the wavelength of the incident light for back scattering ($\Theta_{\text{B}} = 180°$). The frequency of the sound wave Ω_{sound} can be calculated from the speed of the sound wave v_{sound} by:

hyper-sound frequency $\Omega_{\text{sound}}(\Theta_{\text{B}}) = \dfrac{v_{\text{sound}}}{\Lambda_{\text{sound}}}$

$$= \frac{2v_{\text{sound}}}{\lambda_{\text{inc}}} \sin\frac{\Theta_{\text{B}}}{2}.$$ (3.157)

The frequency of the scattered light, ν_{Bscatt}, will be shifted by the Doppler effect from the moving sound wave. As a consequence of the Bragg condition the scattered light in different directions is reflected by sound waves of different frequencies. The scattered light will show the frequency:

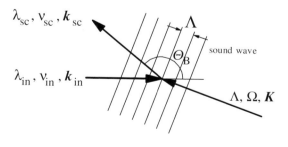

$\lambda_{\text{sc}}, \nu_{\text{sc}}, \boldsymbol{k}_{\text{sc}}$

$\lambda_{\text{in}}, \nu_{\text{in}}, \boldsymbol{k}_{\text{in}}$

Θ_{B}

sound wave

$\Lambda, \Omega, \boldsymbol{K}$

Fig. 3.55. Spontaneous Brillouin scattering of incident light (inc) at refractive index grating of a hyper-sound wave with wave length Λ_{sound}

$$\nu_{\text{Bscatt}} = \nu_{\text{inc}} \left(1 \mp 2 \frac{v_{\text{sound}}}{c} \sin \frac{\Theta_\text{B}}{2} \right) = \nu_{\text{inc}} \mp \Omega_{\text{sound}}(\Theta_\text{B}) \qquad (3.158)$$

where the decrease (minus sign) applies for sound waves moving in the same direction as the incident light and vice versa. The frequency shift is of the order of $100\,\text{MHz}$ (10^{-6} of the light frequency) for gases and $10\text{–}100\,\text{GHz}$ ($< 10^{-4}$) for liquids and solids. It is a maximum for back scattering and zero for forward scattering.

In the particle picture the Brillouin reflection of the incident photon will generate ($-$ sign) or destroy ($+$ sign) a phonon in the matter. The total momentum of the three particles have to be conserved during the scattering (see Fig. 3.56).

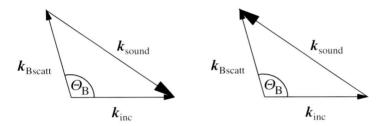

Fig. 3.56. Photon and phonon momentum in Brillouin scattering with phonon generation (left) and phonon depletion (right)

Thus for phonon generation or depletion:

$$\boldsymbol{k}_{\text{Bscatt}} = \boldsymbol{k}_{\text{inc}} - \boldsymbol{k}_{\text{sound}} \quad \text{or} \quad \boldsymbol{k}_{\text{Bscatt}} = \boldsymbol{k}_{\text{inc}} + \boldsymbol{k}_{\text{sound}}. \qquad (3.159)$$

It turns out that momentum conservation is in Brillouin scattering identical with the Bragg condition of (3.156) if the frequency shift of the scattered light is neglected.

The spectral broadening of the scattered light is a function of the observation angle as given by (3.158). In addition the lifetime of the sound wave τ_{sound} which is by definition twice as long as the lifetime of the related phonon, causes additional broadening by:

$$\Delta\Omega_{\text{sound}} = \frac{1}{2\pi\tau_{\text{sound}}}. \qquad (3.160)$$

The phonon lifetimes are of the order of magnitude of $10\,\text{ns}$ for gases and in the range of $1\,\text{ns}$ and below for liquids for Brillouin scattering of light with a wavelength of $1\,\mu\text{m}$ resulting in several $10\,\text{MHz}$ band width and a broadening of $10^{-5}\,\text{nm}$ (see also Sect. 4.5.9). This lifetime increases quadratically with λ_{inc}: $\tau_{\text{sound}} \propto \lambda_{\text{inc}}^2$.

The intensity of the scattered light is a function of the number of reflecting sound wave periods or grating planes. Thus it will usually be largest in the backward direction. The scattered intensity share is of the order of 10^{-4}–10^{-11} as a function of the geometrical conditions and the material. A waveguide geometry scatters more light in the backward direction as a result of the possible large interaction length.

3.10.4 Raman Scattering

Raman scattering [3.108–3.132] occurs as a result of the interaction of light with vibrational transitions of matter. The scattered light shows new spectral lines shifted to lower frequencies (Stokes lines) or to higher frequencies but with much less intensity (anti-Stokes lines) as shown in Fig. 3.57.

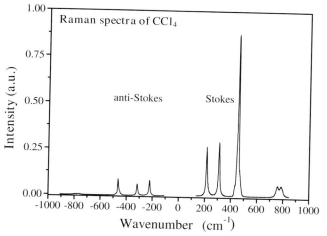

Fig. 3.57. Frequency spectrum of Raman scattering in CCl_4 measured with a conventional spectrometer based on argon laser excitation

Thus it is used in Raman spectroscopy for characterization of the structure of matter, for analysis and in nonlinear optics, e.g. for coherent Raman spectroscopy and for frequency conversion of laser radiation as stimulated Raman scattering (see Chaps. 4 and 6). It can disturb the transmission of light.

The frequency of the Raman scattered light ν_{Rscatt} follows from the frequency of the incident light ν_{inc} and the frequencies $\nu_{vib,m}$ of the (molecular) vibrations m by:

Raman frequency $\nu_{Rscatt,p} = \nu_{inc} \mp (p_{vib} \nu_{vib,m})$ (3.161)

with $m = 1, \ldots$ limited by the number of normal vibrations of the system and the number $p_{vib} = 1, \ldots$ limited by the number of vibrations up to the ionization limit of the system.

The Raman effect requires a sufficient interaction of the electric field which oscillates 1000 times faster with the vibration of the material via the nonresonant interaction of the electrons in the matter. Thus, only a few vibrations of the matter are usually Raman active and only a few of these frequencies show detectable intensities. In Table 3.8 some typical Raman active vibrations of some materials are given. Traditionally the vibrational frequencies were measured as wave numbers in cm^{-1}, which is 1/wavelength, and thus these values are given in the table, too.

Table 3.8. Raman active vibrations of some gases

Substance	ν_{vib} (THz)	ν_{vib} (cm^{-1})	$d\sigma/d\nu$ (cm^2/Ster)
N_2	69.90	2 330	$2.3 \cdot 10^{-30}$
CH_4	87.42	2 914	$1.9 \cdot 10^{-30}$
HF	118.86	3 962	$3.0 \cdot 10^{-30}$
H_2	124.65	4 155	$5.1 \cdot 10^{-30}$

With Raman scattering an excitation (Stokes scattering) or depletion of an excited (anti-Stokes scattering) vibration of the matter takes place (see Fig. 3.58). The interaction of the electric light field with the vibration of the

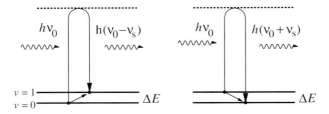

Fig. 3.58. Interaction of incident light with matter vibrations generating Stokes (left) and anti-Stokes (right) lines in the Raman process. The thick horizontal lines represent the vibrational energy levels of the material

matter is effected via the induced polarization in the substance. Thus the oscillating dipole moment $\mu_R(\nu_{vib}, t)$ can be additionally modulated by ν_{inc}:

$$\mu_R = \varepsilon_0 \alpha_0 E_0 \sin(2\pi\nu_{inc}t) + \varepsilon_0 \alpha_1 E_0 \sin(2\pi\nu_{inc}t) \sin(2\pi\nu_{vib}t)$$
$$= \varepsilon_0 \alpha_0 E_0 \sin(2\pi\nu_{inc}t)$$
$$+ \frac{1}{2}\varepsilon_0 \alpha_1 E_0 [\cos\{2\pi(\nu_{inc} - \nu_{vib})t\} - \cos\{2\pi(\nu_{inc} + \nu_{vib})t\}] \quad (3.162)$$

with the coefficient α_i representing the polarizability of the material via the vibration and the light field. The resulting difference or sum of the frequencies are the Stokes and anti-Stokes lines of the scattering process.

Because of the nonharmonic vibrational potentials the energy levels of the vibrations are not equidistant and therefore the Raman spectra are usually much more complicated although the basic processes are as simple as described.

The intensity of the scattered light I_{ram} is proportional to the incident intensity I_{inc} and the population $N_{\mathrm{vib},i}$ of the vibrational energy levels:

$$I_{\mathrm{Ram}}(\nu_{\mathrm{inc}} \mp \nu_{\mathrm{vib},i}) = \sigma_{\mathrm{ram}}(\nu_{\mathrm{inc}} \mp \nu_{\mathrm{vib},i}) I_{\mathrm{inc}}(\nu_{\mathrm{inc}}) N_{\mathrm{vib},i}. \tag{3.163}$$

Examples for measurements of the Raman cross-section are given in [3.125, 3.126, 3.108–3.113].

It is obvious that anti-Stokes signals are smaller the higher the vibrational frequency. The thermally induced occupation of the excited vibrational states can be calculated using Boltzmann's equation for the population density $N_{\mathrm{vib},i}$:

$$N_{\mathrm{vib},i} = N_0 \, \mathrm{e}^{\left(-\frac{h\nu_{\mathrm{vib},i}}{kT}\right)} \quad \text{with} \quad \sum_i N_{\mathrm{vib},i} = N_{\mathrm{total}} \tag{3.164}$$

with Boltzmann's constant $k = 1.380658 \cdot 10^{-23}\,\mathrm{J\,K^{-1}}$ and temperature T. These populations are not changed by the incident light in linear interactions with matter. Thus typical molecular vibrations with $1000\,\mathrm{cm^{-1}}$ energy show a population density of smaller than 10^{-6} at room temperature.

The main role is played by the polarizability α_i for the efficiency of the scattering process. This value is a function of the detailed structure of the material and the incident light. It can be estimated via quantum chemical calculations or measured. Some special techniques of Raman measurements are given in [3.127–3.132].

3.11 Optical Materials

Almost all kinds of materials in all kinds of preparation, from low or high temperature, under high and low pressure, with and without external fields and as mixtures, compounds or layers are used in photonics. The selection of the following incomplete list may serve as a reminder:

- glasses
- inorganic and organic crystals
- dielectric and metallic coatings
- polymers
- semiconductors
- special organic molecules in solution or in polymers
- liquid crystals

- thin films
- inorganic molecules
- gases

The materials have to be characterized, firstly, with respect to their linear optical behavior by the following properties:

$$
\begin{array}{rcl}
\text{spatially homogeneous} & \Leftrightarrow & \text{spatially inhomogeneous} \\
\text{isotropic} & \Leftrightarrow & \text{anisotropic} \\
\text{absorbing} & \Leftrightarrow & \text{transparent} \\
\text{isolating} & \Leftrightarrow & \text{conductive} \\
\text{dielectric} & \Leftrightarrow & \text{non dielectric}
\end{array}
$$

Besides this rough yes/no categorization sometimes detailed numbers for each property are necessary. For example, all materials are absorbing at all wavelengths but the absorption coefficient can be as small as $10^{-6}\,\text{cm}^{-1}$ and this can often be neglected.

Secondly, for nonlinear applications the nonlinear properties of the materials also have to be characterized. The question is then: Which property is changed by how much, by which kind of light, and of what intensity?

4. Nonlinear Interactions of Light and Matter Without Absorption

Nonlinear effects in optics offer the possibility of generating or manipulating light in almost any manner. The laser itself, producing light not available in nature, is the most obvious example. Therefore nonlinear interactions are the basis of photonics.

Because of the extremely small photon–photon interaction cross-section the direct influence of one light beam on another is not practical with today's light sources. Therefore the nonlinearity is achieved via the nonlinear interaction of light with matter. Examples are given in Chap. 1. In comparison to linear optics both the real and the imaginary part of the refractive index, or in other words both the conventional refractive index n (as described in this chapter) and the absorption coefficient a (as described in the next chapter), become functions of the light intensities I or their electric fields $\boldsymbol{E}(\boldsymbol{r}, \lambda, t, \varphi)$:

nonresonant interaction $n = f\{I\} = f\{\boldsymbol{E}(\boldsymbol{r}, \lambda, t, \varphi)\}$ (4.1)

and

resonant interaction $a = f\{I\} = f\{\boldsymbol{E}(\boldsymbol{r}, \lambda, t, \varphi)\}$ (4.2)

and thus they become functions of space, wavelength, time and polarization, dependent on the incident light. Then the transmission of the sample T becomes a complicated function of the incident intensity depending on all light and material parameters in the nonlinear case, in contrast to the linear case where it is constant while varying the light intensity in the linear regime (see Fig. 4.1).

Therefore in nonlinear optics the light has to be characterized very carefully to avoid unwanted side effects in applications and to exclude measurement errors, e.g. in nonlinear spectroscopy. The superposition of light in matter will produce new physical effects in the nonlinear regime. All the properties of newly generated light can be completely different from the properties of the incident beams.

Nonlinearity is not desired in all cases in photonic applications or in spectroscopy, but in many cases the high laser light powers and intensities generate nonlinearity as a side effect. Therefore possible nonlinear interaction have to be considered in any photonic setup.

An increasing number of photonic applications based on these nonlinear effects are routinely in use, but for other desired and in principle possible new

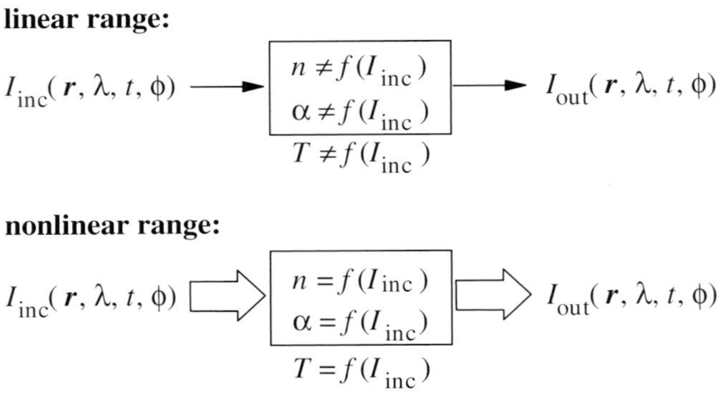

linear range:

$$I_{inc}(r, \lambda, t, \phi) \longrightarrow \boxed{\begin{array}{c} n \neq f(I_{inc}) \\ \alpha \neq f(I_{inc}) \\ \hline T \neq f(I_{inc}) \end{array}} \longrightarrow I_{out}(r, \lambda, t, \phi)$$

nonlinear range:

$$I_{inc}(r, \lambda, t, \phi) \Longrightarrow \boxed{\begin{array}{c} n = f(I_{inc}) \\ \alpha = f(I_{inc}) \\ \hline T = f(I_{inc}) \end{array}} \Longrightarrow I_{out}(r, \lambda, t, \phi)$$

Fig. 4.1. Schematics of linear and nonlinear interactions of incident light with matter

commercial devices, light-induced nonlinear effects are too small. For known matter the time constants, their long-term stability or other properties are not sufficient and thus often the demands on light intensity or power are too high. Thus the field of nonlinear interactions is still in rapid progress and new applications with strong economic and social implications can be expected.

In this and the next chapter some concepts of the nonlinear interaction of light with matter will be described, together with their experimental background and the basics of the theoretical description. For further reading for this chapter see [M1, M2, M6, M7, M9–M11, M13, M14, M20, M22, M23, M26, M28–M31, M34–M37, M40–M43 and M45–M52]. All these nonlinear effects first start from the linear interaction. In most effects the "back-interaction" of the so-modified matter with light is also linear. Thus linear optical interactions are the basis for this chapter and thus references should be taken to the previous chapter.

4.1 General Classification

There are three useful approaches to the description of the nonlinear interaction of light with matter or more precisely for the *nonlinear modification of matter* by light in the first step and the subsequent linear "back-interaction" of the changed matter to the light. One uses Maxwell's equations as the fundamental concept with the nonlinear polarization P_{nl}; the second is based on a quantum mechanical density matrix formalism; and the third neglects the coherence terms in the density matrices and results in rate equations for the population densities N_i:

- nonlinear polarization P_{nl} $(E(r, \lambda, t, \varphi))$
- density matrix formalism: ρ_{ij} $(E(r, \lambda, t, \varphi))$
- rate equations: N_i $(I(r, \lambda, t, \varphi))$

The first approach is especially useful for nonresonant (elastic) interactions in which the light is not absorbed by matter. The density matrix formalism is useful for resonant, coherent interactions. In this case the discrete structure of the energy levels of the matter and their phase-dependent occupation during the light wave period may be important for an exact description. Unfortunately, this formalism does not allow the description of systems with many energy levels. In this case the description with rate equations may be a useful tool for the analysis of nonlinear resonant interactions.

Because of the complexity of these nonlinear light–matter interactions and the difficulty of the analytical description, a classification of these interaction may be useful as given as a first approach in Table 4.1. A more detailed structure is given in Sect. 5.1.

Table 4.1. Types of nonlinear optical interactions of light with matter

Matter	Light	Useful description
Nonresonant (transparent)	Incoherent	Maxwell's equations
Nonresonant (transparent)	Coherent	Maxwell's equations
Resonant (absorbing)	Incoherent	Rate equations
Resonant (absorbing)	Coherent	Density matrix formalism

The incoherent nonresonant interaction occurs, e.g. in the self-focusing of light and coherent nonresonant interactions are, e.g. used for frequency conversion in crystals or in optical phase conjugation. Resonant interaction, which means with absorption or emission of light in the material, is achieved incoherently in nonlinear spectroscopy and applications such as, e.g. passive Q switching. Resonant coherent interaction, in which the matter oscillates in phase with the light, takes place in very fast or high intensity or coherent experiments, as, e.g. self induced transparency. Besides these extreme cases all kinds of mixed interactions are possible.

The following description of nonresonant interactions is based on Maxwell's equation. The rate equation formalism will be introduced in Sect. 5.3.6 and the density matrix formalism is described in Sect. 5.4.2. Finally some mixed cases are discussed in Sect. 5.9.

4.2 Nonresonant Interactions

Nonresonant interactions are most useful for, e.g. wavelength conversion of laser light, wave mixing and optical phase conjugation. Because of the negligible absorption almost no energy will be stored in the material. Thus they can be applied for high average powers with high efficiencies.

The nonresonant nonlinear interaction may be understood as the reaction of electric dipoles built by electrons and the positively charged atomic cores

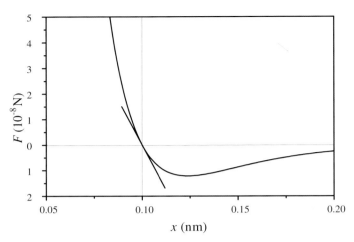

Fig. 4.2. Schematic of back driving force F as a function of the elongation x of electric matter dipoles as is typical for an organic bond with equilibrium distance $x_0 = 0.1\,\mathrm{nm}$ under the influence of strong electric fields reaching the nonlinear range. The linear response is sketched by the straight line. The reduced mass of $6\,\mathrm{g\,mol^{-1}}$ of two C atoms would lead to a resonance frequency of $2\,000\,\mathrm{cm^{-1}}$

in the matter under the influence of high electric light fields. For small fields the elongation x will be small and thus the back driving force F_b will change linearly with x in this first-order approach, as in classical mechanics. With strong electric fields the elongation will be increased and the force will become a nonlinear function of the elongation (see Fig. 4.2).

From Fig. 4.2 it is obvious that for strong electric fields the average distance of the dipole will be increased as a result of the strong repulsion forces. The polarization will become nonlinear. The potential curve as the integral over the force for this example is shown in Fig. 4.3.

With conventional light sources the electric fields are in the range of $1\,\mathrm{V\,cm^{-1}}$ and the resulting elongation is smaller than $10^{-16}\,\mathrm{m}$ which is small compared to atomic or molecular diameters of $10^{-10}\text{--}10^{-7}\,\mathrm{m}$. With laser radiation electric field values of more than $10^4\,\mathrm{V\,cm^{-1}}$ can be achieved and thus nonlinear effects are possible.

These nonlinear effects will be a function of the electric field of the incident light with increasing exponent starting with 1 for linear interactions, to 2 as the next approximation and so on. For better understanding the resulting effects can be classified for this exponent of the nonlinearity as a function of the incident field as follows.

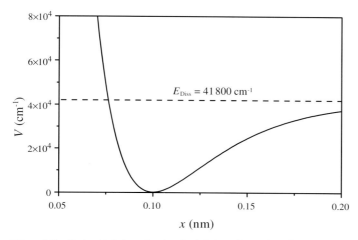

Fig. 4.3. Potential for the back driving force F of Fig. 4.2 as a function of the elongation x of electric matter dipoles with equilibrium distance $x_0 = 0.1\,\text{nm}$ illustrating as deviation from the parabolic shape the nonlinear range. The dissoziation energy is $500\,\text{kJ}\,\text{mol}^{-1}$

Second-Order Effects

- second harmonic generation (SHG)
- optical parametric amplification (OPA, OPO)
- Pockel's effect
- electro-optical beam deflection
- optical rectification

Third-Order Effects

- third harmonic generation (THG)
- Kerr effect – induced birefringence
- self-focusing
- self-diffraction
- self-phase modulation
- solitons
- four-wave mixing (FWM)
- stimulated Brillouin scattering (SBS)
- optical phase conjugation (PC)

Higher-Order Effects

- high harmonic generation

All these effects are based on the nonlinear modulation of the refractive index of the material by the incident light as a result of the strong forces from the interaction of the electric light field vector with the electrons of the matter. These forces drive the electrons out of the harmonic potential and generate anharmonic effects. Mixed cases and noninteger exponents are possible.

4.3 Nonlinear Polarization of the Medium

Based on Maxwell's equations as given in Sect. 3.1 the reaction of the matter under the influence of the electric field of the light $E(r, t)$ can be described by the polarization $P(r, t)$ as given in (3.8). The linear interaction was described there as the proportional increase of the polarization as a function of the electric field amplitude:

linear polarization $P_1(r, t) = \varepsilon_0 \chi(r) E_1(r, t)$ (4.3)

with the condition of not too high electric fields. It should be emphasized again that the polarization and the electric field are not inevitablly pointing in the same direction because of the tensor character of the susceptibility χ.

With an increase of the electric field usually up to values of several $10^4 \, \text{V cm}^{-1}$ or corresponding intensities of MW cm^{-2} (see Table 5.2) nonlinear effects occur and the formula has to be modified e.g. to:

$$P(r, t) = \varepsilon_0 \chi^{(1)} E(r, t) + \varepsilon_0 \chi^{(2)} E^2(r, t) + \varepsilon_0 \chi^{(3)} E^3(r, t) + \cdots$$
$$= P_{\text{lin}}(r, t) + P_{\text{nl}}(r, t) \qquad (4.4)$$

with the nonlinear share of the polarization:

nonlinear polarization
$$P_{\text{nl}}(r, t) = \varepsilon_0 \chi^{(2)} E^2(r, t) + \varepsilon_0 \chi^{(3)} E^3(r, t) + \cdots \qquad (4.5)$$

which describes the progressive nonlinearity of the interaction with increasing power of the electric field. Using this nonlinear polarization the equation for the electric field of the light waves can be written as:

$$\Delta E - \mu_0 \varepsilon_0 \frac{\partial^2 E}{\partial t^2} - \text{grad div } E = \mu_0 \frac{\partial^2 P_{\text{nl}}}{\partial t^2} \qquad (4.6)$$

which simplifies for plane waves (div $E = 0$) to:

$$\Delta E - \mu_0 \varepsilon_0 \frac{\partial^2 E}{\partial t^2} = \mu_0 \frac{\partial^2 P_{\text{nl}}}{\partial t^2} \qquad (4.7)$$

with the ansatz:

$$E_{\text{gen}}(z, t) = E_{\text{gen},0}(z) \cos(2\pi \nu_{\text{gen}} t - k_{\text{gen}} z) \qquad (4.8)$$

for the newly generated planar light wave. Using the assumption of slowly varying amplitudes (SVA) with

SVA approximation $\dfrac{\partial E_{\text{gen},0}}{\partial z} \ll k_{\text{gen}} E_{\text{gen},0}$ (4.9)

the second derivatives $\partial^2 E/\partial z^2$ can be neglected. The differential equation for the amplitude of the new light wave produced by the nonlinear polarization can be written as:

$$\frac{\partial E_{\text{gen},0}}{\partial z} = \mathrm{i}\frac{\mu_0}{2k_{\text{gen}}}\frac{\partial^2 P_{\text{nl}}(E)}{\partial t^2}\,\mathrm{e}^{-\mathrm{i}(2\pi\nu_{\text{gen}}t - k_{\text{gen}}z)}. \tag{4.10}$$

It should be noted that the nonlinear susceptibilities $\chi^{(m)}$ and derived coefficients are in general functions of the material and the frequencies of the applied light waves:

$$\chi^{(m)} = f(\nu_p, \boldsymbol{k}_p, \text{matter}). \tag{4.11}$$

The nonlinear susceptibilities $\chi^{(m)}$ are in general complex tensors and thus the polarization P_{nl} will be complex, too. Analog to the complex electric field the real value follows from $P_{\text{nl,real}} = 1/2(P_{\text{nl}} + P_{\text{nl}}^*)$.

Sometimes the interaction can be even more difficult. Then the description with susceptibilities $\chi^{(m)}$ as functions of the intensity, as will be shown in Sect. 5.4 as rational polynomicals, may in some cases be the best description. Therefore the theoretical analysis of the interaction can be extremely difficult and rough approximations are necessary in many cases. But in the case of nonresonant interactions the susceptibilities $\chi^{(i)}$ will be real tensors and sometimes just numbers.

On the other hand some of these susceptibilities can be zero for a given matter–light interaction. For example, centrosymmetric matter does not show a quadratic term in the nonlinear behavior.

If nonresonant interactions are considered and the susceptibility is replaced by the refractive index $n = \sqrt{1+\chi}$ the nonlinear wave equation can be simplified under the assumptions of isotropic matter, a transverse field and steady-state conditions of the response to:

$$\nabla^2 \boldsymbol{E} - \frac{1}{c^2}\frac{\partial^2}{\partial t^2}\left[(n_0 + \Delta n_{\text{nl}})^2 \boldsymbol{E}\right] = 0 \tag{4.12}$$

with the nonlinearly changed refractive index Δn_{nl} instantaneously following the incident electric field.

If the field $\boldsymbol{E}$ is further assumed to be a monochromatic cylindrically symmetric beam propagating along the z axis then

$$\boldsymbol{E} = \boldsymbol{E}_0(r, z, \xi)\,\mathrm{e}^{\mathrm{i}(2\pi\nu t - kz)} \tag{4.13}$$

with the time variable $\xi = t - zn_0/c_0$ considering the propagation of the beam with the velocity c_0/n_0 in the matter. With an amplitude varying slowly compared to the wavelength the nonlinear wave equation can further be simplified to:

$$\left(\frac{\partial^2}{\partial r^2} + \mathrm{i}2k\frac{\partial}{\partial z}\right)\boldsymbol{E}_0(r, z, \xi) + 2k^2\left(\frac{\Delta n_{\text{nl}}}{n_0}\right)\boldsymbol{E}_0(r, z, \xi) = 0. \tag{4.14}$$

In this simple form the nonlinear wave equation may be analyzed for changes in the amplitude and the phase of the beam in the nonlinear medium.

For complicated functions:

$$\Delta n_{nl} = f(\boldsymbol{E}_0(r, z, \xi)) \tag{4.15}$$

this equation may only be solvable numerically.

Some early work on nonlinear interactions may be found in [4.1–4.12] in addition to the books cited in Sect. 4.0.

4.4 Second-Order Effects

The second-order nonlinear susceptibility $\chi^{(2)}$ of used crystals reaches values of up to about $10^{-10}\,\mathrm{V\,cm}^{-1}$ and thus demands intensities of up to 0.1–1 GW cm^{-2}. These values are close to the damage threshold of optical surfaces of these commonly applied materials and thus new materials are developed as described below. New organic materials or slightly resonant matter may have some applications in the future.

In second-order nonlinear optical effects two different light waves with their electric field vectors $\boldsymbol{E}_1$ and $\boldsymbol{E}_2$ can superimpose and generate the nonlinear polarization $\boldsymbol{P}^{(2)}$:

$$\boldsymbol{P}^{(2)} = \varepsilon_0 \chi^{(2)} \boldsymbol{E}_1 \boldsymbol{E}_2. \tag{4.16}$$

In Cartesian coordinates x, y, z described by the indices m, p, q the spatial components P_x, P_y and P_z of the vector $\boldsymbol{P}^{(2)}$ can be calculated from:

$$P_m^{(2)} = \varepsilon_0 \sum_{p,q} \chi_{mqp}^{(2)} E_p E_q \tag{4.17}$$

with the components χ_{mpq} of the third-order tensor with 27 components. These components are not independent. In any case $\chi_{mpq} = \chi_{mqp}$ is valid as a consequence of the interchangebility of the two electric fields, resulting in 18 independent components.

For nonabsorbing materials these components are real and the equation $\chi_{mpq} = \chi_{pmq}$ also has to be fulfilled. In this case only 10 elements of the χ tensor are different as will be shown below.

For a certain material only a few components are relevant in a particular second-order nonlinear process.

4.4.1 Generation of the Second Harmonic

In the simplest possible case of two equal monochromatic light waves with the same polarization, frequency ν_{inc} and direction $\boldsymbol{k}_{inc}$, which can be two shares of the same wave, the second-order nonlinear polarization is determined by the product of the two electric fields. It shows terms with frequencies $\nu = 0$ and $2\nu_{inc}$.

$$P^{(2)} = \varepsilon_0 \chi^{(2)} \boldsymbol{E}_1 \boldsymbol{E}_2$$
$$= \varepsilon_0 \chi^{(2)} \boldsymbol{E}^2$$
$$= \varepsilon_0 \chi^{(2)} \{E_0(\boldsymbol{k}, \varphi) \cos(2\pi\nu_{\text{inc}}t)\}^2$$
$$= \frac{1}{2}\varepsilon_0 \chi^{(2)} E_0^2(\boldsymbol{k}, \varphi) + \frac{1}{2}\varepsilon_0 \chi^{(2)} E_0^2(\boldsymbol{k}, \varphi) \cos(4\pi\nu_{\text{inc}}t)$$
$$= P^{(2)}(0) + P^{(2)}(2\nu_{\text{inc}}). \tag{4.18}$$

The second term describes the material polarization which oscillates with twice the frequency of the incident wave. This polarization will emit a new light wave with twice the frequency of the incident wave. This wave is then called the *second harmonic (SHG)*. The first term describes the generation of a "rectified" field, resulting in a charge separation in the material.

Thus second-order nonlinearity can be used in photonic applications for the generation of light with twice the photon energy of the incident light with frequency ν_{inc} (see Fig. 4.4).

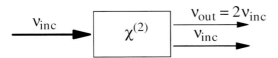

Fig. 4.4. Schematic of the generation of frequency-doubled light via second-order nonlinear susceptibility, e.g. in a suitable crystal

The energy efficiency of the second harmonic genration (SHG) is smaller than 1 and some nonconverted light will also be observed behind the nonlinear medium.

The calculation of the nonlinear polarization has to include all real second-order products of the electric field vectors:

$$\begin{pmatrix} P_x^{(2)}(2\nu_{\text{inc}}) \\ P_y^{(2)}(2\nu_{\text{inc}}) \\ P_z^{(2)}(2\nu_{\text{inc}}) \end{pmatrix} = \varepsilon_0 \begin{pmatrix} d_{11} & d_{12} & d_{13} & d_{14} & d_{15} & d_{16} \\ d_{21} & d_{22} & d_{23} & d_{24} & d_{25} & d_{26} \\ d_{31} & d_{32} & d_{33} & d_{34} & d_{35} & d_{36} \end{pmatrix} \cdot \begin{pmatrix} E_x^2 \\ E_y^2 \\ E_z^2 \\ 2E_yE_z \\ 2E_xE_z \\ 2E_xE_y \end{pmatrix} \tag{4.19}$$

leading to 18 d coefficients of the nonlinear material instead of the 27 χ_{mpq} values. For nonabsorbing materials the following relations are valid:

$$d_{14} = d_{25} = d_{36}$$
$$d_{12} = d_{26} \quad d_{13} = d_{35} \quad d_{15} = d_{31} \quad d_{16} = d_{21} \tag{4.20}$$
$$d_{23} = d_{34} \quad d_{24} = d_{32}$$

finally resulting in 10 different components. The assignment of the d_{rs} to χ_{mpq} can easily be obtained by comparing (4.19) and (4.17). It should be mentioned again that these matter parameters d_{rs} are functions of the applied light frequencies.

Because of the symmetry of commonly used crystals this number is reduced further as Table 4.2 shows:

Table 4.2. SHG crystals with symmetry group, nonlinear d_{ij} coefficients for an incident wavelength λ_{inc} of 1000 nm, transparency wavelength range $\Delta\lambda$ and damage threshold I_{dam}

Crystal	Sym. group	d_{rs} $(10^{-12}\,\mathrm{m\,V^{-1}})$	$\Delta\lambda$ (nm)	I_{dam} $(\mathrm{MW\,cm^{-2}})$
KD*P (KD$_2$PO$_4$)	$\bar{4}$ 2 m	$d_{14} = d_{25} \approx d_{36} = 0.55$	200–2 150	200
KDP (KH$_2$PO$_4$)	$\bar{4}$ 2 m	$d_{14} = d_{25} \approx d_{36} = 0.6$	200–1 700	200
ADP (NH$_4$PO$_4$)	$\bar{4}$ 2 m	$d_{14} = d_{25} \approx d_{36} = 0.76$	190–1 500	200
Banana (Ba$_2$NaNb$_5$O$_{15}$)	2 m m	$d_{15} = d_{31} = 9.2$ $d_{24} = d_{32} = 9.2$ $d_{33} = 12$	460–1 064	
KTP (KTiOPO$_4$)	2 m m	$d_{15} = d_{31} = 14$ $d_{24} = d_{32} = 14$ $d_{33} = 30$	350–450	1 000
Lithiumniobate (LiNbO$_3$)	3 m	$d_{15} = d_{24} = d_{31} = d_{32} = -6$ $d_{16} = d_{21} = -d_{22} = -3.2$ $d_{33} = 47$	420–5 200	50
Beta-Barium-borate (BBO) (β-BaB$_2$O$_4$)	3	$d_{11} = -d_{12} = -d_{26} = 2.4$ $d_{15} = d_{24} = d_{31} = d_{32} \ll 2$ $d_{16} = d_{21} = -d_{22} \ll 2$	190–3 500	5 000
Lithium-Triborate (LBO) (LiB$_3$O$_5$)	m m ^{2}	$d_{31} = 1.1$ $d_{32} = -1.0$ $d_{33} = 0.05$	160–2 600	2 000

For efficient generation of the second harmonic (SHG) phase matching conditions have to be fulfilled (see the next section).

Some theoretical aspects of second harmonic generation are given in [4.13, 4.14] in addition to the books mentioned in Chap. 4. Second-order susceptibilities for some further materials and examples of second harmonic generation can be found in [4.15–4.58]. Additional examples are given in the next chapter showing high efficiencies at low intensities and in Sect. 6.15.1.

4.4.2 Phase Matching

The generation of new frequencies via nonlinear polarization in matter is more efficient the better the incident light waves and the newly generated waves are in suitable phase over the interaction length. This can be achieved in crystals by choosing a suitable orientation of the crystal with respect to the light beam and is called phase matching.

Phase Matching for Second Harmonic Generation

Some details will be discussed here for the example of the generation of second harmonic light but the mechanisms are valid analogously for other frequency conversion setups, too. The increasing amplitude of the SHG wave can be calculated from (4.10). The nonlinear polarization for this wave is:

$$P_{nl}^{(2)}(2\nu_{inc}) = \frac{1}{2}\varepsilon_0\chi^{(2)}\,E_{inc,0}^2(z)\,e^{i2(2\pi\nu_{inc}t - k_{inc}z)}. \tag{4.21}$$

With this nonlinear polarization (4.10) results for the SHG wave in:

$$\frac{\partial E_{SHG}(z)}{\partial z} = -i\frac{k_{SHG}\chi^{(2)}}{4n_{SHG}^2}E_{inc}^2(z)\,e^{-i\Delta kz} \tag{4.22}$$

which cannot easily be solved. Assuming an undepleted incident wave, it describes an oscillation of the amplitude of the electric field of the generated second harmonic light with z as a function of $\Delta k = |k_{SHG} - k_{inc}|$. This oscillation of the intensity of the SHG light can be calculated as:

$$I_{SHG}(z) = I_{inc}^2\frac{8\pi^2 d^2}{\varepsilon_0 c_0 \lambda_{inc}^2 n_{inc}^2 n_{SHG}}\left[\frac{\sin(\Delta kz/2)}{\Delta k/2}\right]^2 \tag{4.23}$$

with the different refractive indices n_{inc} and n_{SHG} for the wavelengths λ_m of the incident wave and the SHG wave in the material and d as the relevant matrix element of the d matrix of the material.

This oscillation results from the phase mismatch of the incident and the SHG wave which are in phase at the entrance surface and at distances z_{ip}

$$z_{ip} = \frac{\pi}{\Delta k}m = \frac{\lambda_{inc}}{4(n_{SHG} - n_{inc})}m \tag{4.24}$$

and out of phase after a path length z_{oop} of:

$$z_{oop} = \frac{2\pi}{\Delta k}m = \frac{\lambda_{inc}}{2(n_{SHG} - n_{inc})}m \tag{4.25}$$

and so on with the integer m. For energy conservation the amplitude of the incident wave will oscillate, too.

This problem can be avoided by choosing

phase matching $\Delta k = |k_{SHG} - k_{inc}| \overset{!}{=} 0$ \hfill (4.26)

applying a suitable crystal orientation. In this case (4.22) can be solved analytically and the intensity of the SHG I_{SHG} follows from the pump or incident light intensity I_{inc} as:

$$I_{SHG}(z) = I_{inc}(z = 0)\tanh^2\left(z\sqrt{\frac{8\pi^2 d^2 I_{inc}}{\varepsilon_0 c_0 n_{inc}^2 n_{SHG}\lambda_{inc}^2}}\right)$$

$$= I_{inc}(z = 0)\tanh^2\left(0.88\frac{z}{z_{HM}}\right) \tag{4.27}$$

again with d as the relevant matrix element of the d matrix of the material and a characteristic length z_{HM} at which the intensity of the SHG reaches 50% of the original incident intensity $I_{inc}(z = 0)$:

$$z_{HM} = 0.31 \sqrt{\frac{\varepsilon_0 c_0 n_{inc}^2 n_{SHG} \lambda_{inc}^2}{\pi^2 d^2 I_{inc}}} \tag{4.28}$$

or for 93% conversion efficiency:

$$z_{93\%} = \sqrt{\frac{\varepsilon_0 c_0 n_{inc}^2 n_{SHG} \lambda_{inc}^2}{2\pi^2 d^2 I_{inc}}} \tag{4.29}$$

The incident intensity I_{inc} will decrease with z in non-absorbing materials by:

$$I_{inc}(z) = I_{inc}(z = 0) - I_{SHG}(z) \tag{4.30}$$

and thus the intensities of the incident and the new generated SHG wave can be plotted as in Fig. 4.5.

Phase matching can be achieved with anisotropic materials like crystals and is based on the birefringence in these materials. Therefore the orientation and the temperature of the crystal has to be chosen for equal refractive indices for the incident and the newly generated waves. Often the polarization of the incident and the newly generated waves are perpendicular as a result of the nonlinear polarization in the crystal and thus different refractive index ellipsoids are available for phase matching.

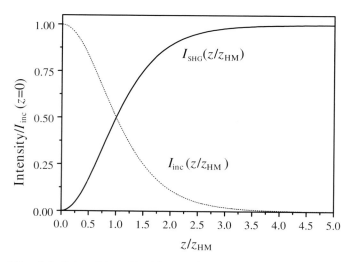

Fig. 4.5. Intensities of incident and new generated light waves in second harmonic generation SHG as a function of the interaction length in material with phase matching

For example in the case of second harmonic generation (SHG) it is possible to match the refractive index for the incident fundamental wave as an ordinary beam with the second harmonic wave as an extraordinary beam even in a uniaxial crystal (see Fig. 4.6).

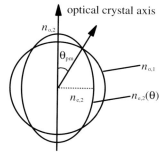

Fig. 4.6. Phase matching angle for SHG in a uniaxial crystal with refractive index sphere for an ordinary incident beam $n_{o,1}$ and refractive index ellipse for an extraordinary SHG beam $n_{e,2}$. The refractive index sphere for an extraordinary SHG beam $n_{e,2}$ is not shown. It has the radius of the extraordinary index at the optical axis

If an intersection of the two refractive index surfaces for the two waves exists then phase matching can be achieved. In this example the phase matching angle θ_{pm} follows with the definitions of Fig. 4.6 to:

$$\text{phase matching angle} \quad \sin^2 \theta_{\mathrm{pm}} = \frac{\dfrac{1}{n_{o,1}^2} - \dfrac{1}{n_{o,2}^2}}{\dfrac{1}{n_{e,2}^2} - \dfrac{1}{n_{o,2}^2}} \tag{4.31}$$

and reaches values of 40–60° in crystals such as, e.g. KDP or KTP. The two refractive indices for KDP are shown in Fig. 4.7.

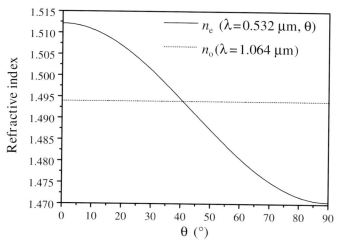

Fig. 4.7. Refractive indices for the ordinary beam of the fundamental and of the extraordinary beam of the second harmonic wave for KDP as a function of angle

Dispersion of Crystals: Sellmeier Coefficients

The dispersion of relevant crystals is available from the phenomenological Sellmeier formulas as, e.g. (see also Sect. 3.2):

$$\text{dispersion} \quad n^2(\lambda_{\text{Sell}}) = A_{\text{Sell}} + \frac{B_{\text{Sell}}}{\lambda_{\text{Sell}}^2 - C_{\text{Sell}}} + \frac{D_{\text{Sell}}\lambda_{\text{Sell}}^2}{\lambda_{\text{Sell}}^2 - E_{\text{Sell}}} \quad (4.32)$$

with the unit of measure for the wavelength λ in this equation:

$$[\lambda_{\text{Sell}}] \overset{!}{=} \mu m \quad (4.33)$$

and the coefficients as given in Table 4.3 for some commonly used crystals.

Table 4.3. Sellmeier coefficients for some commonly used crystals as used in (4.32) with (4.33). The value n_e is given for the direction perpendicular to the optical axis

		KDP	KD*P	ADP	CDA	CD*A
A_{Sell}	n_o	2.2576	2.2409	2.3041	2.4204	2.4082
	n_e	2.1295	2.1260	2.1643	2.3503	2.3458
B_{Sell}	n_o	0.0101	0.0097	0.0111	0.0163	0.0156
	n_e	0.0097	0.0086	0.0097	0.0156	0.0151
C_{Sell}	n_o	0.0142	0.0156	0.0133	0.0180	0.0191
	n_e	0.0014	0.0120	0.0129	0.0168	0.0168
D_{Sell}	n_o	1.7623	2.2470	15.1086	1.4033	2.2122
	n_e	0.7580	0.7844	5.8057	0.6853	0.6518
E_{Sell}	n_o	57.8984	126.9205	400.0000	57.8239	126.8709
	n_e	27.0535	123.4032	400.0000	127.2700	127.3309

The refractive index n_e is the value of the index ellipsoid perpendicular to the optical axis and thus the extreme value. As an example the refractive indices are given for some typical crystals for the wavelengths of the Nd lasers and their harmonics in Table 4.4.

Table 4.4. Refractive indices for the ordinary and extrordinary beams for some typical crystals for the wavelengths of Nd lasers and their harmonics. The value n_e is given for the direction perpendicular to the optical axis

		KDP	KD*P	ADP	CDA	CD*A
1064 nm	n_o	1.4942	1.4931	1.5071	1.5515	1.5499
	n_e	1.4603	1.4583	1.4685	1.5356	1.5341
532 nm	n_o	1.5129	1.5074	1.5280	1.5732	1.5692
	n_e	1.4709	1.4683	1.4819	1.5516	1.5496
355 nm	n_o	1.5317	1.5257	1.5487	1.6026	1.5974
	n_e	1.463	1.4833	1.4994	1.5788	1.5759

The refractive indices for the two beams for KDP as a function of the wavelength are shown in Fig. 4.8.

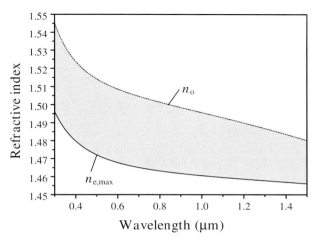

Fig. 4.8. Refractive indices for the ordinary and extraordinary beams for KDP as a function of the wavelength. By choosing the appropriate phase matching angle the refractive index for the extraordinary beam can be varied between $n_{\mathrm{e,max}}$ and n_{o} to get the same value but for different wavelengths

But not all materials allow phase matching for the required combination of wavelengths. Therefore new crystals are being developed and the user should check with the suppliers of nonlinear materials for frequency conversion for current information. A few examples are given in [4.59–4.61].

Walk-Off Angle

Even in case of perfect phase matching the directions of the beam propagation of the new and old frequencies and the propagation direction of their energy (Poynting vector) can be different because the electrical displacement $\boldsymbol{D}$ and the electric field $\boldsymbol{E}$ are not parallel in case of extraordinary beams. In this type of critical phase matching the interaction length is limited by the "walk-off" between the beam and energy propagation directions [e.g. 4.62–4.64]. In this example the walk-off angle φ_{wo} results from:

$$\textbf{walk-off angle}\quad \tan\varphi_{\mathrm{wo}} = \frac{(n_{\mathrm{e,1}}^2 - n_{\mathrm{e,2}}^2)\tan\theta_{\mathrm{pm}}}{n_{\mathrm{e,2}}^2 + n_{\mathrm{e,1}}^2\tan\theta_{\mathrm{pm}}} \qquad (4.34)$$

with a value of 1.4° for KDP and a wavelength of 1064 nm. By contrast, in *noncritical phase matching* with $\theta_{\mathrm{pm}} = 90°$ no walk-off occurs.

For high efficiency of frequency conversion the intensity should be as high as possible (but safely below damage threshold of the material), but stronger focusing reduces the interaction length. Secondly it decreases the phase matching because of the higher divergence of the incident light. The acceptance angle $\Delta\theta_{\mathrm{pm}}$ depends on the dimensions of the ellipsoids and is given for the above example by:

$$\Delta\theta_{pm} = \frac{0.442\lambda_{inc}n_{o,1}}{n_o^2(n_{o,2} - n_{e,2})L_{crystal} \sin 2\theta_{pm}} \tag{4.35}$$

as the full angle around θ_{pm} for half intensity of SHG generation. It results in values of, e.g. a few $0.1°$ for KTP and $25°$ for KDP. In noncritical phase matching this equation reduces to:

$$\Delta\theta_{pm} = n_{o,1}\sqrt{\frac{\lambda_{inc}}{(n_{o,2} - n_{e,2})L_{crystal}}}. \tag{4.36}$$

From both equations the critical wavelength range for the incident beam can be estimated by the dispersion of the crystal $\partial n_i/\partial\lambda_{inc}$. Typical values are 11 nm·cm for KDP and 0.6 nm·cm for KTP.

The temperature has to be constant in the range 25 K·cm for KTP and 4 K·cm in the case of LBO. On the other hand the phase matching can also be tuned by temperature variation for certain crystals, but especially in high-power applications this cannot easily be achieved, because of residual absorption in the material.

As with wide angles, a wide spectrum of the incident wave will not be phase matched because of dispersion. Therefore correct material selection as a function of the parameters of the application is most important for high efficiencies.

Focusing and Crystal Length

Optimal focusing has to be chosen as a function of the material and its phase matching acceptance angle, its damage threshold and its temperature sensitivity, which changes the refractive index ellipsoids. The optimal crystal length $L_{crystal}$ should be nearly three times the Rayleigh length in the crystal [4.65] which is:

$$\textbf{optimal crystal length} \quad L_{crystal} = 2.9\frac{\pi w_0^2}{\lambda_{inc}} \tag{4.37}$$

for a Gaussian beam. The optimal length can be shorter for short pulses with durations of fs or ps to avoid pulse lengthening from dispersion in the crystal [e.g. 4.66].

Using the crystal inside an optical resonator the effective intensity of the pump light can be increased using high-reflectivity mirrors for the pump wavelength. This scheme is especially useful for cw laser light if intracavity frequency doubling in the laser resonator itself is not possible (see the references of Sect. 6.15.1).

Type I and Type II Phase Matching

Because of the large number of material structures there are many different types of phase matching principles. In the given example the incident wave was used as the ordinary beam for the quadratic nonlinear effect and the second harmonic as the extraordinary beam in the material. This type of phase matching is called type I:

type I phase matching $k_{\mathrm{SHG}}(\mathrm{e}) = k_{\mathrm{inc}}(\mathrm{o}) + k_{\mathrm{inc}}(\mathrm{o}).$ (4.38)

If the incident wave is used as a mixture of an ordinary and an extraordinary beam producing an extraordinary SHG the phase matching is of type II:

type II phase matching $k_{\mathrm{SHG}}(\mathrm{e}) = k_{\mathrm{inc}}(\mathrm{o}) + k_{\mathrm{inc}}(\mathrm{e})$ (4.39)

resulting, in suitable materials, in about twice the acceptance angle compared to type I.

In high-power applications other nonlinear effects and thermal problems may disturb the phase matching and finally limit the conversion efficiency. In particular, the thermally induced birefringence will disturb the process and destroy the good beam quality of the beams. Nevertheless, conversion efficiencies of 80% have been reported (see Sect. 6.15.1).

Quasi-Phase Matching (qpm)

The phase mismatch between the original light wave and the newly generated waves as a result of the different refractive indices for the different wavelengths can also be compensated by grating structures of the orientation of the nonlinearity of the material [4.69–4.98]. With these structures the phase error can be periodically reset and thus high efficiencies are reached.

The grating is typically produced by periodic poling of ferroelectrical crystals as shown in Fig. 4.9.

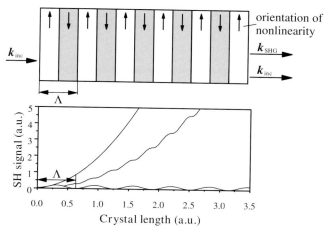

Fig. 4.9. Schematic of quasi-phase matching using periodic poling of a ferroelectric crystal

The grating period Λ_{qpm} depends on the mismatch of the refractive indices of the incident and the second harmonic waves:

qpm period $$\Lambda_{\mathrm{qpm}} = m\frac{2\pi}{|\mathbf{k}_{\mathrm{inc}} - \mathbf{k}_{\mathrm{SHG}}|} = m\frac{2\pi}{|n_{\mathrm{inc}} - n_{\mathrm{SHG}}|} \qquad (4.40)$$

with m as the order of the periodic poling. A first-order grating $(m = 1)$ means that the phase mismatch between the two waves after half the period is π. The optimum case is obtained if the sign of the nonlinearity is reversed every $\pi/|\mathbf{k}_{\mathrm{inc}} - \mathbf{k}_{\mathrm{SHG}}|$ second. Typical poling periods are in the range of tens of µm for conversion of light with wavelengths in the range around are µm.

The advantage of this quasi-phase matching compared to conventional phase matching is the higher efficiency from crystals such as $LiNbO_3$, $LiTaO_3$ and KTP. They can be applied in optimal directions showing maximum nonlinearity, which cannot be achieved with conventional phase matching. With a 53 mm long KTP crystal conversion efficiencies above 40% were achieved with an input power of 6.4 W from a cw Nd:YAG laser [4.91].

A disadvantage of this method is the limited size of crystals of about 0.5 mm in the direction of the poling field as a consequence of the poling process. Further photorefractive damage may occur but this might be overcome by codoping the crystals [4.90].

4.4.3 Frequency Mixing of Two Monochromatic Fields

Via the second-order nonlinear polarization of the material two light waves with different frequencies ν_1 and ν_2 can be mixed and new frequencies generated [4.99–4.126]. In addition to the frequencies $2\nu_1$ and $2\nu_2$ and $\nu = 0$ two new waves occur with frequencies:

sum frequency $$\nu_{\mathrm{sum}} = \nu_1 + \nu_2 \qquad (4.41)$$

and

difference frequency $$\nu_{\mathrm{diff}} = |\nu_1 - \nu_2|. \qquad (4.42)$$

These frequencies follow from the electric fields E_1 and E_2, e.g. of the two collinear waves with parallel polarization combining to give the total field E:

$$E = E_1 + E_2 = E_{0,1}\cos(2\pi\nu_1 t - k_1 z) + E_{0,2}\cos(2\pi\nu_2 t - k_2 z). \qquad (4.43)$$

Under the assumption of parallel beams the frequency terms of E^2 can be calculated and as a result the above new frequencies are obtained.

The nonlinear polarization results from:

$$\begin{pmatrix} p_x^{(2)} \\ p_y^{(2)} \\ p_z^{(2)} \end{pmatrix} = 2\varepsilon_0[d] \cdot \begin{pmatrix} E_x(\nu_1)E_x(\nu_2) \\ E_y(\nu_1)E_y(\nu_2) \\ E_z(\nu_1)E_z(\nu_2) \\ E_y(\nu_1)E_z(\nu_2) + E_y(\nu_2)E_z(\nu_1) \\ E_x(\nu_1)E_z(\nu_2) + E_x(\nu_2)E_z(\nu_1) \\ E_x(\nu_1)E_y(\nu_2) + E_x(\nu_2)E_y(\nu_1) \end{pmatrix} \qquad (4.44)$$

with the matrix $[d]$ as used in (4.19):

$$[d] = \begin{pmatrix} d_{11} & d_{12} & d_{13} & d_{14} & d_{15} & d_{16} \\ d_{21} & d_{22} & d_{23} & d_{24} & d_{25} & d_{26} \\ d_{31} & d_{32} & d_{33} & d_{34} & d_{35} & d_{36} \end{pmatrix} \qquad (4.45)$$

with the internal relations given in (4.20). These new frequencies can be achieved with parametric oscillators or amplified in parametric amplifiers.

4.4.4 Parametric Amplifiers and Oscillators

If two beams with frequencies ν_1 and ν_2 are superimposed in a suitable nonlinear material (NLM) with a total intensity reaching the nonlinear regime the additional sum frequency $\nu_{sum} = \nu_1 + \nu_2$ or difference frequency $\nu_{diff} = |\nu_1 - \nu_2|$ (or both) occur with intensities I_{sum} and I_{diff} (see Fig. 4.10) [4.127–4.133].

sum frequency generation

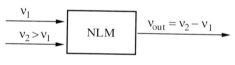

difference frequency generation

Fig. 4.10. Sum and difference frequency generation in a nonlinear material (NLM) with suitable phase matching for the new light

Depending on the phase matching conditions in the nonlinear material, we can select which of these beams with the new or old frequency will be strong after the nonlinear interaction.

In particular, if one of the two incident beams, e.g. a signal beam with frequency ν_{signal} is originally weak it can be amplified at the expense of the other strong incident pump light beam intensity with frequency ν_{pump}. Additionally, a new light frequency ν_I will occur in a so-called *idler beam* for photon energy conservation (see Fig. 4.11).

If the signal beam frequency is smaller than the pump beam frequency the idler frequency will appear as difference frequency with:

$$\nu_{idler} = \nu_{pump} - \nu_{signal}. \qquad (4.46)$$

This arrangement is called an *optical parametric amplifier (OPA)* and is increasingly being used in photonic applications to generate wavelengths not

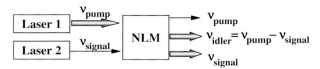

Fig. 4.11. Optical parametric amplifier (OPA): amplification of signal beam at the expense of pump beam and additional generation of idler beam in a nonlinear material (NLM) with suitable phase matching

available from the lasers directly. This process can be applied, e.g. in lithium niobate ($LiNbO_3$) crystals with laser pump light in the visible range and signal and idler frequencies in the infrared spectral region up to wavelengths of $7\,\mu m$. Conversion efficiencies can reach values larger than 50% and thus this method has become quite popular in the generation of new frequencies especially in fs lasers. With these ultra-short pulses very high intensities can be achieved without large thermal loads in the nonlinear material and the damage threshold is increased for these short pulses, too.

If only one beam with one frequency ν_{pump} is used as incident light in suitable nonlinear materials these photons can be divided into two photons with the same total energy as the pump photon:

$$\nu_{signal} + \nu_{idler} = \nu_{pump} \quad \text{with} \quad \nu_{signal} > \nu_{idler} \tag{4.47}$$

with the given convention.

In addition the phase matching condition giving momentum conservation has to be fulfilled. Again we have type I and II behavior as given in the previous chapter, and for a negative uniaxial crystal:

$$\text{type I} \quad \boldsymbol{k}^{e}_{pump} = \boldsymbol{k}^{o}_{signal} + \boldsymbol{k}^{o}_{idler} \tag{4.48}$$

and

$$\text{type II} \quad \left.\begin{array}{l} \boldsymbol{k}^{e}_{pump} = \boldsymbol{k}^{o}_{signal} + \boldsymbol{k}^{e}_{idler} \\ \boldsymbol{k}^{e}_{pump} = \boldsymbol{k}^{e}_{signal} + \boldsymbol{k}^{o}_{idler} \end{array}\right\} \tag{4.49}$$

This arrangement, shown in Fig. 4.12, is called an *optical parametric oscillator (OPO)*.

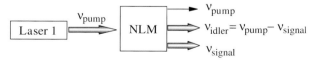

Fig. 4.12. Optical parametric oscillator (OPO): generation of signal and idler beams at the expense of pump beam in a nonlinear material (NLM) with suitable phase matching

Furthermore this nonlinear OPO scheme is increasingly being used in photonics because of the simplicity of the generation of new coherent light beams with new frequencies which are otherwise difficult to generate. Several crystals are useful for OPA and OPO applications as given in Table 4.5.

Table 4.5. Tuning ranges for the signal and idler wavelengths λ as a function of the pump wavelength for different useful OPA and OPO crystals [4.133]

λ_{pump} (nm)	laser	crystal	$\lambda_{signal}, \lambda_{idler}$ (nm)
1064	Nd:YAG	LiNbO$_3$	1400–4400
694	Ruby	LiIO$_3$	770–4000
532	Nd:YAG-SHG	KDP	957–1117
355	Nd:YAG-THG	KDP	480–580
			960–1160
266	Nd:YAG-FHG	ADP	420–730

Again phase matching can be reached by angle or temperature tuning of the crystal. As an example the possible wavelengths for the new beam are given as a function of the angle of BBO pumped with 800 nm in Fig. 4.13.

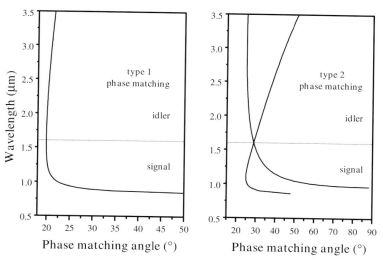

Fig. 4.13. Tuning curve of OPO pumped at 800 nm in a BBO crystal

Because the OPO process starts from noise the resulting new beams are usually not as spectrally narrow and not of as good beam quality as can be reached with the OPA scheme. But in both cases the power P or pulse energy E is higher the higher the frequency of the light. If no absorption occurs energy conservation is fulfilled and thus the Manley–Rowe conditions [4.134] with the light power P for continuous radiation and pulse energy E for the light pulses are valid:

$$\frac{(P \text{ or } E)_{\text{pump}}}{\nu_{\text{pump}}} = \frac{(P \text{ or } E)_{\text{signal}}}{\nu_{\text{signal}}} = \frac{(P \text{ or } E)_{\text{idler}}}{\nu_{\text{idler}}} \tag{4.50}$$

and

$$(P \text{ or } E)_{\text{pump,inc}} = (P \text{ or } E)_{\text{signal}} + (P \text{ or } E)_{\text{idler}}$$
$$+ (P \text{ or } E)_{\text{pump}-\text{residual}}. \tag{4.51}$$

The efficiency of these frequency conversion techniques depends on the beam quality of the pump beam, its spectrum, the degree of polarization and the pulse duration, the focusing and the nonlinear material. Values of more than 50% are reported Examples can be found in [4.135–4.182] for conventional OPA and OPO configurations and in [4.183–4.196] with poled material for low-intensity and high-efficiency applications. Splitting into three photons was reported in [4.197]. Further examples can be found in Sect. 6.15.2.

4.4.5 Pockels' Effect

Besides the opto-optical second-order nonlinear effects, some electro-optical second-order effects are important in photonics [see e.g. 4.198, 4.199]. In the nonlinear material the electric field of the light wave is superimposed on the externally applied electric field and will be influenced by the resulting nonlinear polarization.

The Pockels effect rotates the polarization of the incident light as a function of the externally applied electrical field. This can be acquired longitudinally as in Fig. 4.14 or transversally with respect to the wave vector of the light beam.

The electric light field of the incident monochromatic planar wave has components E_x and E_y as shown in Fig. 4.14 which are given by:

$$E_{\text{inc},x} = E_{\text{inc},x,0} \, e^{i(2\pi\nu_{\text{inc}} - k_{\text{inc}}z)} \tag{4.52}$$

and

$$E_{\text{inc},y} = E_{\text{inc},y,0} \, e^{i(2\pi\nu_{\text{inc}} - k_{\text{inc}}z)}. \tag{4.53}$$

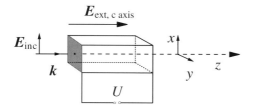

Fig. 4.14. Superposition of the electrical light field with the external field via the nonlinear polarization in the x and y directions in the Pockels effect

If a uniaxial crystal is used and its optical axis has the same direction as the external electric field $\boldsymbol{E}_{\text{ext}}$ as in Fig. 4.14 the resulting nonlinear polarization can be calculated from a formula similar to (4.19). For this geometry only the two terms with $E_y \cdot E_z$ and $E_x \cdot E_z$ will be nonzero. Thus the quadratic nonlinear polarization with the frequency ν_{inc} has the components:

$$P_x(\omega) = 2\varepsilon_0 d_{14}^{(0)} E_{\text{inc},y,0} E_{\text{ext}} \cos(2\pi\nu_{\text{inc}}t - k_{\text{inc}}z) \tag{4.54}$$

and

$$P_y(\omega) = 2\varepsilon_0 d_{25}^{(0)} E_{\text{inc},x,0} E_{\text{ext}} \cos(2\pi\nu_{\text{inc}}t - k_{\text{inc}}z) \tag{4.55}$$

with the condition $d_{14} = d_{25}$ for, e.g. KDP. These coefficients have different values because of the dispersion for different light frequencies. If this nonlinear polarization is applied to (4.10) the amplitudes of the electric field of the light wave $E_{\text{light},x}(z)$ and $E_{\text{light},y}(z)$ can be calculated to:

$$\frac{\partial E_{\text{light},x}(z)}{\partial z} = \frac{-ik_{\text{inc}} d_{14}^{(0)} E_{\text{inc},y,0} E_{\text{ext}}}{n^2} \tag{4.56}$$

and

$$\frac{\partial E_{\text{light},y}(z)}{\partial z} = \frac{-ik_{\text{inc}} d_{14}^{(0)} E_{\text{inc},x,0} E_{\text{ext}}}{n^2} \tag{4.57}$$

with the ordinary refractive index n of the crystal for the incident wavelength. These equations can be solved in the case of $E_{\text{inc},y,0} = 0$ to give the result:

$$E_{\text{light},x}(z) = E_{\text{inc},x,0} \cos\phi_{\text{Pockels}} \tag{4.58}$$

and

$$E_{\text{light},y}(z) = -i\, E_{\text{inc},x,0} \sin\phi_{\text{Pockels}} \tag{4.59}$$

where the $-i$ indicates a phase shift by $90°$ of the fast oscillating light wave. The angle ϕ_{Pockels} results from:

$$\phi_{\text{Pockels}}(z) = \frac{k_{\text{inc}} d_{14}^{(0)}}{n^2} E_{\text{ext}} z. \tag{4.60}$$

Thus in the Pockels effect the incident linearly polarized light is converted to circular polarization, to linear polarization at $90°$, to circular polarization, to linear polarization $180°$ and so on (see Fig. 4.15).

Thus for a certain external electric field the crystal works as a quarter-wave plate, producing circular polarized light, and for twice this field as a half-wave plate, producing linear but $90°$ rotated light. The necessary voltages U_i are:

quarter-wave voltage $$U_{\lambda/4} = \frac{n\lambda_{\text{inc}}}{8d_{14}^{(0)}} \tag{4.61}$$

and

half-wave voltage $$U_{\lambda/2} = \frac{n\lambda_{\text{inc}}}{4d_{14}^{(0)}} \tag{4.62}$$

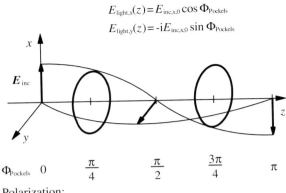

$$E_{\mathrm{light},x}(z) = E_{\mathrm{inc},x,0}\cos\Phi_{\mathrm{Pockels}}$$
$$E_{\mathrm{light},y}(z) = -\mathrm{i}E_{\mathrm{inc},x,0}\sin\Phi_{\mathrm{Pockels}}$$

Φ_{Pockels}	0	$\dfrac{\pi}{4}$	$\dfrac{\pi}{2}$	$\dfrac{3\pi}{4}$	π

Polarization:
    linear   circular   linear 90°   circular   linear 180°

Fig. 4.15. Light polarization in the Pockels effect as a function of the optical path in an optically uniaxial crystal

with λ_{inc} as the wavelength of the incident light outside the crystal.

The electro-optical Pockels effect can also be applied with uniaxial crystals which are not arranged along the optical axis as in Fig. 4.14 or with a transversal external electric field. Crystals with less symmetry such as, e.g. a two-axial material, can also be used. In any case the electro-optical effect is based on the deformation of the refractive index ellipsoid in the matter by the electric field. The theoretical description of this second-order nonlinear effect can be given in these more complicated cases similar to the example given above.

The nonlinear coefficients $d_{ij}^{(0)}$ for electro-optical applications are given for several commonly used crystals in Table 4.6.

Table 4.6. Coefficients for electro-optical applications of some widely applied nonlinear crystals from [4.198]. Refractive indices are given for the light wavelength and permittivities for temporal constant fields

crystal	λ_{inc}	n_{o}	n_{eo}	$\varepsilon_1 = \varepsilon_2$	$\varepsilon_{\mathrm{eo}}$	d_{ij} $[10^{-12}$ cm/V]	$U_{\lambda\mathrm{inc}/4}$
KDP	550 nm	1.51	1.47	42	50	$d_{14} = d_{25} = -43$ $d_{36} = -51$	2400 V
ADP	630 nm	1.52	1.48	56	15	$d_{14} = d_{25} = -120$ $d_{36} = -40$	1000 V
LiNbO$_3$	1064 nm	2.232	2.156	85	29.5	$d_{15} = d_{24} = -830$ $d_{16} = -d_{21} = d_{22} = -170$ $d_{31} = d_{32} = -242$ $d_{33} = -780$	1500 V

The change of the refractive index of the matter is in the first approximation a linear function of the external electric field, as the above equations show for the given example. Thus, this type of second-order nonlinear electro-optical effect is sometimes called linear although the interaction with the light finally shows a quadratic dependence on the total electric field. This linear change of the refractive index Δn_m can be described by:

$$\Delta n_m \simeq \frac{\overline{n}^3}{2} \sum_p r_{mp} E_p \tag{4.63}$$

with the average refractive index $\overline{n}$ and the electro-optical coefficients r_{mp} resulting from:

$$r_{mp} = -\frac{d_{pm}^{(0)}}{\overline{n}^4} \tag{4.64}$$

which are functions of the applied frequencies, too.

The Pockels effect can be applied for light modulation and optical switching if the Pockels crystal is combined with conventional polarizers as shown in Fig. 4.16.

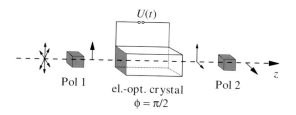

Fig. 4.16. Using the Pockels effect in combination with polarizers as an electro-optical modulator or switch

The incident unpolarized light will be vertically linearly polarized behind the polarizer Pol 1. If no voltage is applied the light polarization will stay vertical and cannot pass the crossed polarizer Pol 2. If the half-wave voltage $U_{\lambda/2}$ is used the light polarization will be changed to horizontal and the beam can pass Pol 2 undisturbed. Any voltage between 0 and $U_{\lambda/2}$ will let part of the light intensity transmit. If the incident light beam is well enough linearly polarized Pol 1 is of course not necessary.

This scheme can be simplified for Q switching of laser resonators at lower voltages as given in Fig. 4.17.

In this case a 100% reflecting mirror is placed at the position of the second polarizer Pol 2 of Fig. 4.16. If no voltage is applied the whole system will reflect linearly polarized light completely, but if the quarter-wave voltage $U_{\lambda/4}$ is used the light beam will be circularly polarized behind the nonlinear crystal. After reflection and a second pass through the crystal the light will be horizontally linearly polarized. Then it cannot pass the polarizer Pol 1. Thus the setup works as a mirror with electrically tunable reflectivity between 0

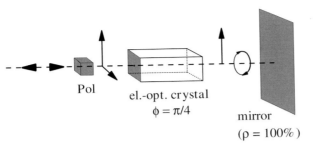

Fig. 4.17. Pockels effect for Q switching in optical resonators with quarter-wave operation

and 100%. This arrangement can be used in laser resonators for modulation of the output. In particular, it is applied for the generation of pulses with ns duration in solid-state lasers via Q switching.

The reaction time of the useful crystals is faster than 10^{-10} s and therefore the switching time is limited by the electric transient times for the necessary voltages. Transient times of less than 1 ns are possible. Because of the isolating properties of the crystals the necessary electrical energy is quite low. It is mostly determined by the required time constant which demands sufficiently low impedance of the electric circuit for recharging of the crystal capacity. This is typically a few pF, e.g. 5 pF for KD*P which has a quarter-wave voltage of 3.4 kV for 1.06 μm radiation.

4.4.6 Electro-Optical Beam Deflection

Another possibility of switching or modulating light based on the electro-optical effect [4.198] uses a refractive index step at the border of two differently oriented nonlinear crystals (see Fig. 4.18).

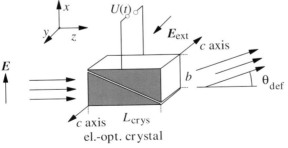

Fig. 4.18. Beam deflection via the electro-optical effect at the border of two antiparallel oriented crystals

The optical axes of the two crystals, e.g. KDP, are antiparallel and perpendicular to the incident light beam. If a voltage is applied along the optical axes of the two crystals the change of the refractive indices will add at the boundary surface. As a result the incident beam will experience different optical paths as a function of its height x in the vertical direction. The refractive index of crystal 1 will be approximately $n_1 = n_0 - d_{36}E_{\text{ext}}/2n_0$ and of crystal 2 will be $n_2 = n_0 + d_{36}E_{\text{ext}}/2n_0$. Thus the total refractive index difference will be:

$$\Delta n = \frac{1}{n_0}d_{36}E_{\text{ext}}. \tag{4.65}$$

The path length difference for the upper and lower beams in Fig. 4.18 as a consequence of the refractive index difference lead to a deflection of the light beam with angle:

$$\text{deflection angle} \quad \theta_{\text{def}} = \frac{L_{\text{crys}}}{b_{\text{crys}}}\Delta n = -\frac{L_{\text{crys}}}{b_{\text{crys}}}\frac{d_{36}}{n_0}E_{\text{ext}} \tag{4.66}$$

with the definitions of Fig. 4.18.

If this deflection is used for scanning diffraction-limited laser beams the resolution A is given by the ratio of the deflection angle and the divergence angle of the beam θ_{beam}:

$$A = \frac{\theta_{\text{def}}}{\theta_{\text{beam}}} = -\frac{\pi L_{\text{crys}}d_{36}}{2\lambda_{\text{beam}}n_0}E_{\text{ext}}. \tag{4.67}$$

In practical applications angles below $1°$ can be obtained with crystals of several $10\,\text{mm}$ length. Examples can be found in Sect. 6.10.3.

4.4.7 Optical Rectification

As mentioned in Sect. 4.4.1 the application of high electric light fields can produce a second-order nonlinear effect with frequency 0, which is a nonoscillating electric field [4.200–4.209]. The physical background for this effect is the displacement of the charges in the temporal average as a consequence of the anharmonic potential (see Fig. 4.2).

The calculation of the second-order nonlinear polarization components $P_i^{(2)}$ for this rectification with $\nu = 0$ follows by analogy to (4.19) from:

$$\begin{pmatrix} P_x^{(2)}(\nu = 0) \\ P_y^{(2)}(\nu = 0) \\ P_z^{(2)}(\nu = 0) \end{pmatrix} = \varepsilon_0 \begin{pmatrix} d_{11} & d_{12} & d_{13} & d_{14} & d_{15} & d_{16} \\ d_{21} & d_{22} & d_{23} & d_{24} & d_{25} & d_{26} \\ d_{31} & d_{32} & d_{33} & d_{34} & d_{35} & d_{36} \end{pmatrix} \cdot \begin{pmatrix} E_x E_x{}^* \\ E_y E_y{}^* \\ E_z E_z{}^* \\ 2E_y E_z{}^* \\ 2E_x E_z{}^* \\ 2E_x E_y{}^* \end{pmatrix} \tag{4.68}$$

with the same relations between the components of the d matrix as described in (4.20), but the d values are the same as for the electro-optical effects described above as given in Table 4.6. They are different from the values in Table 4.2 because of the dispersion and the different frequencies of both processes.

The experimental setup is sketched in Fig. 4.19.

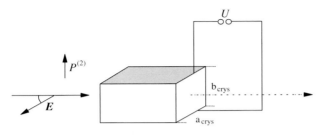

ADP-crystal

Fig. 4.19. Experimental setup for rectification of light via second-order nonlinear polarization in a suitable nonlinear crystal. The cw voltage can be detected perpendicular to the electric field vector of the light

The cw polarization $P_x^{(2)}(0)$ in the x direction can be calculated from (4.68) as:

$$P_x^{(2)}(\nu = 0) = 2\varepsilon_0 d_{14}^{(0)} E_y E_z{}^* = \varepsilon_0 d_{14} E_{\text{inc}}^2 \tag{4.69}$$

which becomes maximal for $E_y = E_z = \frac{1}{\sqrt{2}} E_{\text{inc}}$ as assumed in this equation. This nonlinear polarization generates a charge separation Q_{crys} in the crystal of:

$$Q_{\text{crys}} = a_{\text{crys}} L_{\text{crys}} P_x^{(2)}(0) = C_{\text{crys}} U. \tag{4.70}$$

This charge Q_{crys} leads to an externally observable voltage U depending on the capacity of the crystal C_{crys}:

$$C_{\text{crys}} = \varepsilon_0 \varepsilon_{\text{crys}} \frac{L_{\text{crys}} a_{\text{crys}}}{b_{\text{crys}}}. \tag{4.71}$$

If finally the light beam power P_{beam} as a function of the electric field amplitude E_{inc} is introduced:

$$P_{\text{beam}} = \frac{1}{2} \varepsilon_0 c_0 n_{\text{crys}} a_{\text{crys}} b_{\text{crys}} E_{\text{inc}}^2 \tag{4.72}$$

the voltage at the crystal can be expressed as a function of this power by:

$$U = \frac{2 d_{14}^{(0)}}{\varepsilon_0 \varepsilon_{\text{crys}} c_0 n_{\text{crys}} a_{\text{crys}}} P_{\text{beam}}. \tag{4.73}$$

In practical cases this voltage is very small. In [4.207] with a 1 MW pulsed ruby laser light source at 694 nm in a $1 \times 1 \times 2 \, \text{cm}^3$ ADP crystal a voltage of 24 mV was observed during the pulse duration of 25 ns. In another experiment in LiTaO_3 a laser pulse with a duration of 1 ps and an energy of 1 μJ produced a peak current of 0.3 A [4.204].

4.5 Third-Order Effects

Third-order nonlinear polarization $\boldsymbol{P}^{(3)}$ is a function of the third power of the incident electric light field $\boldsymbol{E}_{\text{inc}}$:

$$\boldsymbol{P}^{(3)} = \varepsilon_0 \chi^{(3)} \boldsymbol{E}_{\text{inc},1} \boldsymbol{E}_{\text{inc},2} \boldsymbol{E}_{\text{inc},3} \tag{4.74}$$

with the third-order nonlinear susceptibility $\chi^{(3)}$ which is a four-dimensional tensor which can in general be complex.

The three light fields $\boldsymbol{E}_{\text{inc},i}$ can be components of the same light beam but can also be three different light beams which overlap in the nonlinear material. If a fourth beam is used to detect the changes in the third-order nonlinear material *four wave mixing* (*FWM*) takes place (see Fig. 4.20).

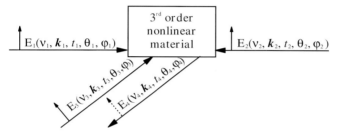

Fig. 4.20. Schematic of four-wave mixing as third order nonlinear process in suitable matter

By choosing different frequencies ν_m, propagation directions $\boldsymbol{k}_m$, timing t_m, polarization ϕ_i and phases φ_m more than hundred prominent schemes of FWM can be applied. In one of the simplest cases all light waves have the same frequency and this process is called *degenerate four-wave mixing* (*DFWM*). For more details see Sect. 5.9.2.

The three components of $\boldsymbol{P}^{(3)}$ in the x, y and z direction follow from:

$$P_m^{(3)} = \frac{\varepsilon_0}{4} \left[\sum_{p,q,r} \left\{ \chi_{mpqr}^{(3,1)} E_p E_q E_r + \chi_{mpqr}^{(3,2)} E_p^* E_q E_r \right. \right.$$

$$\left. \left. + \chi_{mpqr}^{(3,3)} E_p E_q^* E_r + \chi_{mpqr}^{(3,4)} E_p E_q E_r^* \right\} \right]$$

$$\text{with} \quad m, p, q, r = x, y, z. \tag{4.75}$$

For nonabsorbing materials the susceptibility tensor $\chi^{(3)}$ has 81 real components and the complex products of the electric field components disappear. Again as in second-order nonlinearity for symmetry reasons of the allowed permutations of the p, q, r of the electric field vectors the number of distinguishable tensor components of $\chi^{(3)}$ is reduced. Only 30 different values remain for this reason. The components of the third-order nonlinear polarization can be written as:

$$
\begin{pmatrix} P_x^{(3)} \\ P_y^{(3)} \\ P_z^{(3)} \end{pmatrix} = \varepsilon_0 \cdot \begin{pmatrix} e_{11} & e_{12} & e_{13} & e_{14} & e_{15} & e_{16} & e_{17} & e_{18} & e_{19} & e_{1\,10} \\ e_{21} & e_{22} & e_{23} & e_{24} & e_{25} & e_{26} & e_{27} & e_{28} & e_{29} & e_{2\,10} \\ e_{31} & e_{32} & e_{33} & e_{34} & e_{35} & e_{36} & e_{37} & e_{38} & e_{39} & e_{3\,10} \end{pmatrix} \cdot \begin{pmatrix} E_x^3 \\ E_y^3 \\ E_z^3 \\ 3E_x E_y^2 \\ 3E_x E_z^2 \\ 3E_y E_x^2 \\ 3E_y E_z^2 \\ 3E_z E_x^2 \\ 3E_z E_y^2 \\ 6E_x E_y E_z \end{pmatrix} \tag{4.76}
$$

with the internal relations between the e matrix elements:

$$
\begin{aligned}
& e_{12} = e_{24} \quad e_{13} = e_{35} \quad e_{14} = e_{26} \quad e_{15} = e_{38} \quad e_{16} = e_{21} \\
& e_{17} = e_{25} = e_{3\,10} \quad e_{19} = e_{2\,10} = e_{34} \quad e_{1\,10} = e_{28} = e_{36} \\
& \quad e_{18} = e_{31} \quad e_{23} = e_{37} \quad e_{27} = e_{39} \quad e_{29} = e_{32}
\end{aligned} \tag{4.77}
$$

considering the further symmetry rules $\chi_{mpqr} = \chi_{pmqr}$. Finally 15 components are relevant for describing the third-order nonlinear processes which are functions of the applied light frequency.

If the nonabsorbing crystals are of cubic symmetry only two different values are distinguishable:

$$
P_m^{(3)} = \varepsilon_0 E_m \{ e_{11} E_m^2 + 3e_{14}(E_p^2 + E_q^2) \} \quad \text{with} \quad m \neq p \neq q = x, y, z \tag{4.78}
$$

and if the material is isotropic the additional relation $e_{11} = 3e_{14}$ is valid and the third-order nonlinear polarization reduces to:

$$
P_m^{(3)} = \varepsilon_0 e_{11} E_m (\boldsymbol{EE}) \quad \text{with} \quad m = x, y, z. \tag{4.79}
$$

Isotropic matter and materials of cubic symmetry will not show any second-order nonlinear effect. In particular isotropic matter such as gases, liquids or solutions and amorphous solids can be used for third-order nonlinear effects with high efficiencies in photonic applications. Optical phase conjugation may serve as an example. On the other hand third-order nonlinear measurements may be used to characterize these materials and their internal structure. The material parameters in the third-order nonlinearity can be determined using the z-scan method as described in Sect. 7.5 and the references therin.

4.5.1 Generation of the Third Harmonic

Analogous to the generation of second harmonics (see Sect. 4.4.1) the third-order nonlinearity of (4.76) can be applied for the generation of light with three times the frequency of the incident light wave [4.210–4.225]. This process is called third harmonic generation (THG):

$$
\text{THG} \quad I_{\text{out}}(3\nu_{\text{inc}}) \xleftarrow{\chi_{\text{THG}}^{(3)}} I_{\text{inc}}(\nu_{\text{inc}}). \tag{4.80}
$$

If this process is based on the third-order susceptibility $\chi^{(3)}$ as given above the efficiency for the known materials is quite low (typically $< 10^{-4}$). Thus this method is mostly applied for the determination of third-order nonlinearity itself and the characterization of the material.

In photonics the most commonly used generation of third harmonic light is therefore based on the much more efficient two-step procedure of second harmonic generation (SHG) in a first step and frequency mixing of this second harmonic with the residual incident wave as the second step. Both steps are nonlinear to second order:

$$\text{THG II} \quad I_{\text{THG}}(3\nu_{\text{inc}}) \xleftarrow{\chi^{(2)}_{\text{mix}}} I_{\text{SHG}}(2\nu_{\text{inc}}), \quad I_{\text{res}}(\nu_{\text{inc}}) \xleftarrow{\chi^{(2)}_{\text{SHG}}} I_{\text{inc}}(\nu_{\text{inc}}). \quad (4.81)$$

The total efficiency for this type of third harmonic generation can be as high as about 80% (see Sect. 6.15.1). For optimal total efficiency, the efficiency of second harmonic generation has to be 2/3 for enough residual intensity of the fundamental wave $I_{\text{res}}(\nu_{\text{inc}})$ to achieve the optimal photon numbers for different wavelengths.

4.5.2 Kerr Effect

Based on third-order nonlinear polarization a strong applied electric field can induce optical birefringence in materials which are optically isotropic without the field [4.226–4.238]. Thus the refractive index tensor of the material becomes a function of the light intensity. Mostly this Kerr effect is achieved on the microscopic scale by the orientation of electric dipoles in the electric field. This electric field $E_{\text{ext}} = U/d$ can be applied as an oscillating field of a strong incident light wave or as an external field as in the Pockels effect but with transversal orientation to the light propagation direction as shown in Fig. 4.21. Further, polarized incident light with an electric field vector in the x direction with $E_{\text{inc},x}(\nu_{\text{inc}})$ propagating in z direction through a third-order nonlinear material is assumed as in Fig. 4.21.

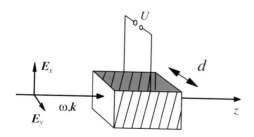

Fig. 4.21. Optical Kerr effect in an isotropic third-order nonlinear material inducing optical birefringence

The total polarization $P^{(\text{tot})}$ is the sum over the linearly induced polarization and the third-order nonlinear polarization which both have a component in the x direction:

$$P_x^{(\text{tot})}(\nu_{\text{inc}}) = P_{\text{lin},x}^{(1)} + P_{\text{nl},x}^{(3)}$$

$$= \varepsilon_0 \chi^{(1)} E_{\text{inc},x} + \varepsilon_0 \chi^{(3)} E_{\text{inc},x} \left(\frac{3}{4} E_{\text{inc},x}^2 + E_{\text{ext}}^2 \right)$$

$$= \varepsilon_0 E_{\text{inc},x} \left[\chi^{(1)} + \chi^{(3)} \left(\frac{3}{4} E_{\text{inc},x}^2 + E_{\text{ext}}^2 \right) \right] \qquad (4.82)$$

$$P_y^{(\text{tot})}(\nu_{\text{inc}}) = P_z^{(\text{tot})}(\nu_{\text{inc}}) = 0$$

with

$$P_{\text{nl},x}^{(3)}(\nu_{\text{inc}}) = \varepsilon_0 e_{11} E_{\text{inc},x} \left(\frac{3}{4} E_{\text{inc},x}^2 + E_{\text{ext}}^2 \right). \qquad (4.83)$$

The refractive index $n = \sqrt{1 + \chi}$ in the x direction is changed by this nonlinear polarization:

$$n_{\text{nl},x} = \sqrt{n_0 + e_{11} \left(\frac{3}{4} E_{\text{inc},x}^2 + E_{\text{ext}}^2 \right)} \qquad (4.84)$$

and is constant in the y and z directions. The small change Δn_x in the x direction results in:

$$\Delta n_x = \frac{e_{11}}{2n_0} \left(\frac{3}{4} E_{\text{inc},x}^2 + E_{\text{ext}}^2 \right). \qquad (4.85)$$

Assuming low light intensities the change in the refractive index as a function of the external cw electric field E_{ext}^2 results in:

$$\Delta n_x = \frac{e_{11}}{2n_0} E_{\text{ext}}^2 = K_{\text{Kerr}} \lambda_{\text{inc}} E_{\text{ext}}^2 \qquad (4.86)$$

with the Kerr constant

Kerr constant $K_{\text{Kerr}} = \dfrac{e_{11}}{2n_0 \lambda_{\text{inc}}}$ $\qquad (4.87)$

describing the Kerr effect for slowly changing fields. Thus, with the external electric field the birefringence of the Kerr material can be changed and used in a setup of a Kerr cell between two crossed polarizers as in Fig. 4.22. This is an electrically controlled light gate, a Kerr shutter. The necessary electric fields can be determined from the Kerr constants as given in Table 4.7. Thus in carbon disulfide (CS_2) an external field of the order of $50\,\text{kV}\,\text{mm}^{-1}$ has to be applied for a switching effect.

If only fast oscillating fields are applied, different parameters for the Kerr effect are observed. In this case the reorientation and diffusion processes to reach the steady-state under the applied electric field may not be completed. This fast Kerr effect is typically described with the constant γ or γ_{I} as defined in relation to e_{11} in the next section.

Using this type of induced optical birefringence of the Kerr effect very fast optical switches can be made [4.233–4.235]. The setup is sketched in Fig. 4.22.

Table 4.7. Parameters of some useful third-order nonlinear materials. The γ-values are valid for light wavelengths of 1 μm and linear polarization

Material	n_0 (1 μm)	K_{Kerr} (10^{-14} m V^{-2})	γ (10^{-22} m^2 V^{-2})	γ_I (10^{-14} cm^{2}W^{-1})	P_{cr} (kW)
CS$_2$	1.60	3.6	65	3.0	33
nitrobenzene	1.55	300	170 (ns)	8.3 (ns)	12 (ns)
			7 (ps)	0.34 (ps)	300 (ps)
benzene	1.49	0.67	14	0.7	150
glass (BK7)	1.51		0.68	0.034	3100
quartz (fused)	1.45		0.54	0.028	3900
water	1.33	137	2.2	0.13	960
air (1 bar)	1.00027		0.045	0.003	47000

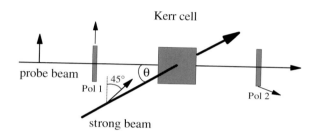

Fig. 4.22. Fast optical shutter based on the optical Kerr effect

The Kerr active material, e.g. CS$_2$, is placed between two crossed polarizers Pol 1 and Pol 2. Thus the incident probe light beam polarized in the direction of polarizer Pol 1 cannot pass the setup. In addition a short and strong light pulse is applied to switch on the transparency of this device. Therefore the direction of its polarization has to be 45° to the incident polarization and the angle θ should be as small as possible.

If the intensity of the switch beam is suitably chosen the incident probe light will be circularly polarized behind the nonlinear cell. Thus 50% of the incident probe beam intensity can pass the second polarizer at a maximum.

The reaction times of the common materials are of the order of 10^{-12} s or even shorter and thus the opening time of the shutter is mostly given by the pulse duration of the switch pulse of the strong beam. For very fast switching the dispersion in the Kerr cell may have to be considered.

The necessary intensity for this Kerr shutter can be estimated using (4.85). The phase shift has to be 90° for generating circular polarization. For a cell length or interaction length L_{cell} the necessary intensity follows from:

$$I_{\text{switch}} = \frac{n_0 c_0 \varepsilon_0}{3 K_{\text{Kerr}} L_{\text{cell}}}.$$
(4.88)

Using CS$_2$ as the Kerr medium an intensity of about $10\,\text{GW}\,\text{cm}^{-2}$ would be necessary for switching.

4.5.3 Self-Focusing

If the intensity of a transmitted light beam is sufficiently high almost every material, gases, liquids or solids, will show a nonlinear interaction. Thus the refractive index will be changed as given in (4.85) if the nonlinear range of the electric field or intensity is reached. This refractive index change will modify the light propagation not only with respect to the polarization as discussed in the previous chapter but in its geometrical properties too. In particular, if light beams with a transverse intensity profile, as, e.g. Gaussian beams, are applied this refractive index change will be different over the cross-section of the beams.

As a consequence for high-intensity beams with long interaction lengths in the matter self-focusing can occur [4.239–4.264] and for short interaction lengths self-diffraction or self-defocussing may be obtained as described in the next chapter.

Assuming a Gaussian transversal beam profile of the incident light the intensity across the beam radius r perpendicular to the beam propagation direction is given as a function of the total power P_{tot} by:

$$I_{\text{inc},r}(r,t) = \frac{2}{\pi w_{\text{beam}}^2}\, e^{-2r^2/w_{\text{beam}}^2} P_{\text{tot}}(t). \tag{4.89}$$

The refractive index $n(r,I)$ in the material will be modified across the beam diameter based on (4.85) to:

$$n(r,I) = n_0 + \Delta n_{\text{nl},r} = n_0 + \frac{1}{2} e_{11} c_0 \varepsilon_0 E_{\text{inc},r}^2$$

$$= n_0 + \gamma(\nu_{\text{inc}}) E_{\text{inc},r}^2$$

$$= n_0 + \gamma_I(\nu_{\text{inc}}) I_{\text{inc},r} \tag{4.90}$$

with the coefficients γ and γ_I as given in Table 4.7 valid for linearly polarized light. In the case of unpolarized or circularly polarized light the effective coefficients have to be reduced to $2/3$ of the given values:

$$\gamma(\nu_{\text{inc}}) = \frac{3}{8}\frac{e_{11}}{n_0} \quad \gamma(\nu_{\text{inc}})_{\text{circ}} = \frac{1}{4}\frac{e_{11}}{n_0}. \tag{4.91}$$

This Gaussian refractive index profile is shown in Fig. 4.23 resulting from a Gaussian shaped incident beam. This refractive index profile acts as a lens analogous to the quadratic refractive index profile in Table 2.6 and will focus the beam as shown in Fig. 4.24.

This focusing increases the intensity and thus the refractive index will be changed even more. The positive feedback for more and more focusing is limited by the increased divergence of the Gaussian beam with smaller diameter. After a certain length z_f an equilibrium between focusing and defocusing is reached and the beam will propagate as in a waveguide with constant diameter d_{sf}.

The theoretical description of this self-focusing can be based on the nonlinear wave (4.14) with the nonlinear refractive index of (4.90). Unfortunately,

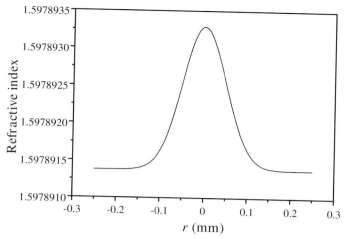

Fig. 4.23. Refractive index modification of CS_2 ($n = 1.598$) by third-order non-linear interaction of a Gaussian beam of wavelength $1.06\,\mu m$ with a beam radius $w_{beam} = 0.1\,mm$ and a power of $10\,kW$ leading to self-focusing. For the calculation $\gamma_I = 3 \cdot 10^{-18}\,m^2\,W^{-1}$ was used

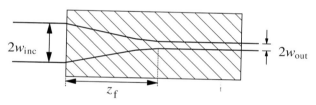

Fig. 4.24. Self-focusing of a beam in a third-order nonlinear interaction with matter. The refractive index profile will focus the beam and thus increase its intensity. Finally the divergence of the beam and the focusing compensate each other and wave guiding with a constant beam diameter takes place

the beam propagation cannot be solved analytically as a consequence of the nonlinear refractive index equation depending on the intensity and vice versa. Numerical solutions are given, in the references e.g. [4.244, 4.257, 4.259].

As a hint for the order of magnitude of the self-focusing effect the critical power P_{cr} was estimated under different assumptions, e.g. of aberration-free focusing, to give

$$\text{critical power for self-focusing} \quad P_{cr} = \frac{\varepsilon_0 c_0 \lambda_{inc}^2}{4\pi\gamma}. \tag{4.92}$$

At this critical power an incident plane wave front will stay planar or in other words the incident beam will be waveguided with unchanged diameter. Thus this power gives *self-trapping* of the beam. The compensation of smaller divergence with larger beam diameter balances out the intensity effect and thus the power is the relevant quantity. At higher powers the incident beam will be focused with a focal length z_f which was estimated in the same way as P_{cr} to give:

$$\text{self-focusing focus length} \quad z_f = \frac{\pi}{\lambda_{inc}\left(\sqrt{\dfrac{P}{P_{cr}} - 1} + 1\right)}. \tag{4.93}$$

But the Gaussian index profile leads to aberrations during focusing and thus the aberrationless approximations are only first approaches. Thus a more precise solution was numerically produced [4.259]:

$$z_f = \frac{0.734\pi w_{inc}^2}{\lambda_{inc}\sqrt{\left[\left(\dfrac{P}{P_{cr}}\right)^{1/2} - 0.852\right]^2 - 0.0219}}. \tag{4.94}$$

In practical cases the focus length z_f is in the range of several 10 cm. The spot size is theoretical almost zero but practically minimum diameters of about 5 µm were obtained.

The high intensities are even more easily achievable with short pulses. In this case the self-focusing will be a function of time and the effect of *temporarily moving foci* also has to be discussed [4.262]. It seems that the mechanisms are different for ns, ps and fs pulses.

Applying very much higher intensities for potentially tighter focusing did not work because the build-up of filaments was observed [4.249, 4.259]. As a consequence of even very small modulations in the transversal profile the beam splits into separate individual beams.

Self-focusing is used in white light generation (see Sect. 7.7.5) to achieve high intensities and large interaction lengths at the same time, which is not possible by simple focusing.

Besides reorientation, induced dipole moments and other nonlinear effects on the molecular or atomic scale causing self-focusing, additional thermal effects can also produce refractive index changes. This *thermal self-focusing* or defocusing (see Sect. 4.5.6) is based on $dn/dT \neq 0$. The thermal effect is produced by absorption of light and thus this is at least a partially resonant process. Nevertheless, the theoretical description of the self-focussing over long interaction lengths is possible with the formulas given above and the thermally induced refractive index change. The time constant of thermally induced focusing effects are in the range of ms to seconds instead of sub-ps to ns.

4.5.4 Spatial Solitons

The nonlinear refractive index change in a Kerr material can establish a waveguiding effect which compensates the self-diffraction of a propagating beam. As a result the light beam can propagate through the matter with constant beam profile and diameter. But this effect needs a certain beam profile different from the Gaussian shape. Such a beam is called a *spatial soliton* [4.265–4.288].

The nonlinear wave equation (4.14) for the electric field amplitude in the slowly varying approximation (SVA) with a nonlinear refractive index proportional to the intensity I_{inc} as given by (4.90) has a solution for the intensity of a spatial soliton I_{sol}:

$$I_{sol}(r) = \frac{1}{2}c_0\varepsilon_0 n_0 E^2_{0,sol}(r_{sol}) = I_{0,sol}(r_{sol})\sec h^2\left(\frac{r}{r_{sol}}\right) \tag{4.95}$$

with the characteristic beam radius r_{sol} where the intensity is 42% of the maximum intensity in the middle of the beam at $r = 0$. The intensity distribution is not Gaussian and for the spatial soliton the intensity $I_{0,sol}$ is a function of the beam diameter r_{sol} given by:

$$\text{spatial soliton intensity}\quad I_{0,sol} = \frac{1}{k^2\gamma_I}\frac{1}{r^2_{sol}} \tag{4.96}$$

with the Kerr coefficient γ as in previous section above and *the value of the wave vector* $\boldsymbol{k}$. The cross-section of a spatial soliton with radius $r_{sol} = 3\,\mu m$ which is about the fundamental mode diameter in an optical fiber and a wavelength of $1\,\mu m$ in CS_2 resulting in an peak soliton intensity of $41.6\,GW\,cm^{-2}$ is shown in Fig. 4.25.

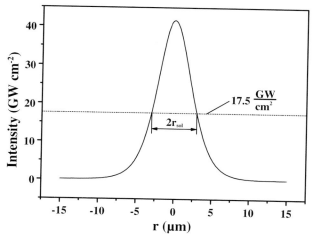

Fig. 4.25. Cross-section of a spatial soliton with sechprofile, a wavelength of $1064\,nm$ and a diameter of $6\,\mu m$ in CS_2 with $\gamma = 3\cdot10^{-18}\,m^2\,W^{-1}$

The phase velocity c_{sol} of the spatial soliton is given by:

$$\text{spatial soliton velocity}\quad c_{\mathrm{sol}} = \frac{c}{1 + \dfrac{\lambda^2}{8\pi^2 r_{\mathrm{sol}}^2}} \tag{4.97}$$

which is smaller than c (about 0.15% in the example) and approaches the velocity of the linearly diffracted light for $r_{\mathrm{sol}} \gg \lambda$.

4.5.5 Self-Diffraction

If the refractive index change is produced in a thin slice of matter by a tightly focused laser beam a lens-like index profile will occur. At this induced refractive index profile the laser beam will be diffracted as shown in Fig. 4.26.

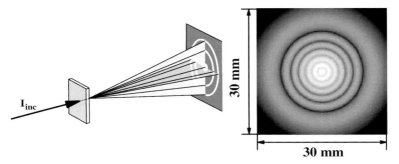

Fig. 4.26. Self-diffraction of a laser beam in a thin slice of matter resulting in a diffraction pattern behind the sample. The number of diffraction rings can be counted and thus the refractive index change determined. The schematic is on left and the calculated pattern at a distance of 0.5 m is on right. For details see text.

The resulting nonlinear change of the refractive index Δn_{nl} can be described as in the previous chapter. It can be assumed to be proportional to the incident intensity. If the incident beam has a transverse Gaussian beam profile the resulting index change will show a transverse Gauss function, too.

The resulting far-field diffraction pattern can be calculated using Fraunhofer's integral (see Sect. 3.9.3). It turns out that the number of diffraction rings N_{rings} can be used to determine the maximum change of the refractive index in the matter $\Delta n_{\mathrm{nl,max}}$ at the center of the Gaussian beam for normal incidence as:

$$\text{maximum index change}\quad \Delta n_{\mathrm{nl,max}} = \frac{\lambda_{\mathrm{beam}}}{L_{\mathrm{mat}}} N_{\mathrm{rings}} \tag{4.98}$$

with the wavelength of the light λ_{beam} and the thickness of the sample L_{mat}. N counts the number of 2π phase shifts for the transmitted light from the center to the wings. It is of course a function of the intensity of the incident light.

The calculated example in Fig. 4.26 represents the resulting self-diffraction pattern from a 100 μm thick film that is irradiated at normal incidence at a wavelength of 500 nm and observed at a distance of 50 cm behind the sample. The waist radius of the laser-induced phase shift profile amounts to 135 μm. Then $N_{rings} = 6$ bright rings in Fig. 4.26 result in $\Delta n_{nl,max} = 0.03$.

Liquid crystals are especially favorable as a nonlinear material for these experiments became of their large nonlinear refractive index change [4.289, 4.290]. This huge optical nonlinearity in, e.g. nematic liquid crystals relies on reorientation of the molecules in the optical field. This process requires a certain electric field strength. Therefore the laser intensity has to exceed a threshold value in order to obtain self-diffraction.

The observed nonlinearity of liquid crystals is strong enough to obtain a diffraction pattern as shown in Fig. 4.26 at the right side with cw laser radiation in the 10 W range [4.289].

4.5.6 Self-Focusing in Weakly Absorbing Samples

The refractive index change can be produced by very weakly absorbing samples via thermally induced refractive index changes, too. The absorption coefficient a (see Sect. 3.4) produces under steady-state conditions a thermal lens with a focal length f_{def} [4.291]:

$$f_{def} = \kappa_{mat} \left[\frac{dn_0}{dT} P_{inc}(1 - e^{-aL_{mat}}) \right]^{-1} \tag{4.99}$$

with thermal conductivity κ_{mat}, incident light beam power P_{inc} and absorption length L_{mat}. Thus again for outweighing diffraction and defocusing of a parallel input beam the incident power has a critical value $P_{cr,def}$ of:

$$P_{cr,def} = \lambda_{inc}\kappa_{mat} \left[\frac{dn_0}{dT}(1 - e^{-aL_{mat}}) \right]^{-1} . \tag{4.100}$$

This type of self-focusing can be applied for the determination of the absorption coefficients in thin films or for the measurements of very small absorption in the range $a < 10^{-6}\,\text{cm}^{-1}$. In moving media the effect can be applied for the observation of the currents in gases or liquids.

4.5.7 Self-Phase Modulation

If a light pulse of sufficiently high intensity transmits through a material the refractive index becomes a function of the temporally changing incident electric light field. The refractive index will be a function of time following the pulse shape of the intensity as a function of time $I_{inc}(t)$. This temporarily changed refractive index will change the light wavelength in the matter and thus its phase: *self-phase modulation* takes place [4.292–4.298]. As a consequence the frequency of the transmitted light will be tuned during the pulse; it has a *chirp*.

The pulse duration Δt_{FWHM} can be long compared to the reaction time of the matter, which is frequently in the order of a few ps. This may or may not lead to steady-state conditions. In non-stationary self-phase modulation the reaction of the matter will be delayed.

For a simple description a Gaussian pulse shape with the duration $\Delta t_{\mathrm{FWHM}} = 2\Delta\tau(\ln 2)^{1/2}$ is assumed:

$$I(t) = I_0 e^{-(t-t_{\mathrm{max}}/\Delta\tau)^2}. \tag{4.101}$$

During this pulse the refractive index n_{mat} will change under steady-state conditions instantaneously with the intensity (see Fig. 4.27).

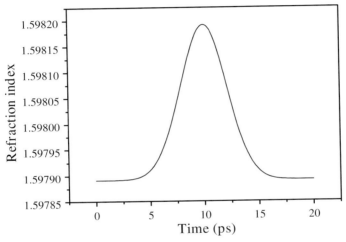

Fig. 4.27. Nonlinear refractive index change as a function of time for self-phase modulation in CS$_2$ applying a Gaussian incident pulse with a pulse FWHM duration of 5 ps, the peak at 10 ps, a wavelength of 1064 nm and a maximum power of 10 GW cm^{-2} calculated with the formulas below. In general the change of the refractive index can be delayed if the reaction time of the material is not short compared to the pulse duration

Further almost monochromatic light is presupposed for simplicity which can be achieved experimentally with $1/\Delta\tau \ll \nu_{\mathrm{inc}}$. If the beam is propagating in the z direction through a material with nonlinear refractive index $n_{\mathrm{nl}}(I)$ and length L_{mat} the electric field behind the material is given by:

$$\boldsymbol{E}(t,z) = \boldsymbol{E}_0\, e^{-\frac{1}{2}(t-t_0/\Delta\tau)^2 + i\varphi} \tag{4.102}$$

with phase factor:

$$\varphi(t, L_{\mathrm{mat}}) = 2\pi\nu_{\mathrm{inc}}t - kL_{\mathrm{mat}} = 2\pi\nu_{\mathrm{inc}}\left(t - \frac{n_{\mathrm{nl}}(I)L_{\mathrm{mat}}}{c_0}\right). \tag{4.103}$$

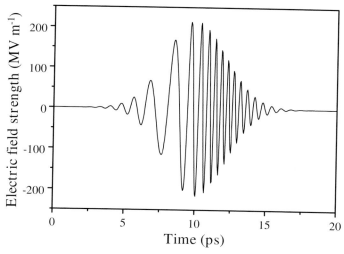

Fig. 4.28. Temporal modulation of the electric field light wave as a consequence of self-phase modulation in a third-order nonlinear material from a short pulse of high intensity (schematic)

This phase shift leads to a temporal compression and expansion of the light wave as schematically depicted in Fig. 4.28.

The refractive index will be changed by the light pulse as given above in this simple approximation, presupposing reaction times of the nonlinear material are short compared to the pulse duration.

$$n_{\mathrm{nl}}(I) = n_0 + \Delta n_{\mathrm{nl}} = n_0 + \gamma_{\mathrm{I}}(\nu_{\mathrm{inc}}) I_{\mathrm{inc},r}$$
$$= n_0 + \gamma_{\mathrm{I}}(\nu_{\mathrm{inc}}) I_0 \, e^{-(t-t_{\mathrm{max}}/\Delta\tau)^2} \tag{4.104}$$

with the nonlinear coefficient given in Table 4.7. The light frequency behind the sample can be calculated in the approximation of this simple model by:

$$\nu_m = \frac{\mathrm{d}\varphi}{\mathrm{d}t} = \nu_{\mathrm{inc}} \left(1 - \frac{L_{\mathrm{mat}}}{c_0} \frac{\mathrm{d}n_{\mathrm{nl}}}{\mathrm{d}t} \right) = \nu_{\mathrm{inc}} - \Delta\nu_{\mathrm{spm}} \tag{4.105}$$

leading to a time-dependent frequency shift $\Delta\nu_{\mathrm{spm}}$ caused by the steady-state self-phase modulation:

chirp frequency

$$\Delta\nu_{\mathrm{spm}} = \nu_{\mathrm{inc}} \gamma_{\mathrm{I}} I_0 \frac{L_{\mathrm{mat}}}{c_0} \frac{2(t-t_{\mathrm{max}})}{\Delta\tau^2} \, e^{-(t-t_{\mathrm{max}}/\Delta\tau)^2} \tag{4.106}$$

which is negative at the leading edge of the pulse and positive after its peak (see Fig. 4.29).

The maximum shift is reached at a time:

$$t_{\mathrm{spm,max/min}} = t_{\mathrm{max}} \pm \frac{\Delta t_{\mathrm{FWHM}}}{2\sqrt{2\ln 2}} \tag{4.107}$$

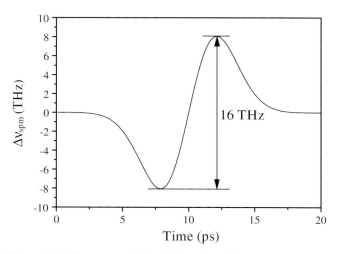

Fig. 4.29. Frequency shift of the pulse of Fig. 4.27 after self-phase modulation in CS_2 as third-order nonlinear material with a cell length of 10 cm. The shift between the minimum and the maximum is 16 THz for a peak intensity of 10 GW cm^{-2}

and reaches a total value between the frequency maximum and minimum of:

$$\Delta\nu_{\mathrm{spm,tot}} = \Delta\nu_{\mathrm{spm,max}} - \Delta\nu_{\mathrm{spm,min}}$$

$$= \nu_{\mathrm{inc}}\gamma_{\mathrm{I}}I_{\mathrm{inc}}\frac{L_{\mathrm{mat}}}{c_0}\frac{4\sqrt{2\ln 2}}{\sqrt{e}\,\Delta t_{\mathrm{FWHM}}} \qquad (4.108)$$

Thus during the main part of the pulse intensity the frequency is shifted from "red" to "blue". The pulse may be spectrally significantly broadened by this self-phase modulation but the different frequencies are well ordered in time and the pulse duration is not changed much. Thus this process is applied for the generation of very short pulses via spectral broadening followed by temporal compression (see Sect. 6.14.2).

4.5.8 Generation of Temporal Solitons: Soliton Pulses

Any linear dispersion will lead to an increase of the pulse length because pulses cannot be perfectly monochromatic. Thus over long distances, as in fiber communications, this will limit the possible modulation frequency.

The third-order nonlinearity of the materials can be used to compensate this effect and transmit specially designed pulses, *longitudinal or temporal solitons*, without any change in the temporal pulse shape over long distances, analogous to transversal or spatial solitons [4.299–4.322]. For this purpose conventional linear dispersion has to be compensated by an equal but contrary nonlinear dispersion generated by the pulse in the matter. In addition to soliton transmission other pulse deformation effects such as broadening, shortening and even splitting into multiple pulses can occur.

Nonlinear induced dispersion by self-phase modulation can delay the shorter wavelength part of the pulse only. Thus for the soliton effect an anomalous linear dispersion with $dn/d\lambda > 0$ is necessary.

Starting with the wave equation for nonlinear interaction (4.7) and assuming:

- slowly varying amplitude of the pulse
- weak dispersion
- small nonlinear effect $(n = n_0 + \gamma_I I)$

the differential equation for the temporal envelope of the pulse propagating in the z direction can be derived as [4.316]:

$$\frac{\partial E}{\partial z} + \frac{1}{c_g}\frac{\partial E}{\partial t} - i\frac{\pi}{c_0}\left\{ 2\left|\frac{dn}{d\nu}\right| + \left|\frac{d^2n}{d\nu^2}\right|\right\}\frac{\partial^2 E}{\partial t^2} - i\pi\varepsilon_0 n_0 \nu_0 \gamma |E|^2 E = 0 \qquad (4.109)$$

with group velocity c_g (see 3.2), linear dispersion $dn/d\nu$, average frequency ν_0 and nonlinear coefficient γ. This equation can be transformed to an expression similar to the nonlinear Schroedinger equation and then solved by the following function for the envelope of the electric field E_{sol}:

$$E_{sol}(z,t) = E_{sol,0} \operatorname{sech}\left(\frac{t - (z/c_g)}{\Delta t_{sol}}\right) e^{iz/4z_0} \qquad (4.110)$$

with the temporal pulse width Δt_{sol} which is measured at 65% of the maximum field amplitude. The intensity of the soliton pulse I_{sol} is given by:

$$I_{sol}(t) = I_{sol,0} \operatorname{sech}\left(\frac{t - (z/c_g)}{\Delta t_{sol}}\right) e^{iz/2z_0} \qquad (4.111)$$

which shows a half-width of $\Delta t_{FWHM,sol} = 1.76 \, \Delta t_{sol}$. The intensity maximum $I_{sol,0}$ or the pulse energy cannot be chosen freely, similar to the case of spatial solitons. For the fundamental soliton it has to fulfill the following condition:

$$\text{fundamental soliton} \quad I_{sol,0} = \frac{1}{\Delta t_{sol}^2}\frac{2}{\nu_0\gamma}\left|\frac{dn}{d\nu}\right| \qquad (4.112)$$

with the important consequence of quadratically increasing soliton pulse energy for shorter pulse duration. An example is shown in Fig. 4.30.

This fundamental soliton propagates without any changes in shape and energy through the matter, which is assumed to be without absorption. Pulse energies of an integer multiple of this value will show periodic solutions as given above with the soliton period z_{per}. Pulse energies between these discrete values may show complicated temporal evolution.

The characteristic soliton period z_{per} is given by:

$$\textbf{soliton period} \quad z_{per} = \pi z_0 = \frac{\pi c_0^2}{4}\left|\frac{\partial n}{\partial \nu}\right|^{-1}\Delta t_{sol}^2 \qquad (4.113)$$

which describes periodically the return of the original pulse shape for pulses different from the fundamental soliton by containing more energy. In between,

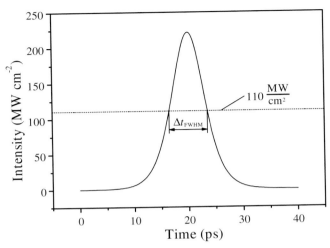

Fig. 4.30. Temporal soliton as used in fiber communication with a wavelength of 1550 nm, and a FWHM pulse duration of 7.05 ps in a quartz fiber with $\gamma_\mathrm{I} = 2.8 \cdot 10^{-20}\,\mathrm{m^2\,W^{-1}}$

the pulse may be split into different pulses [4.315, 4.316] and may then recover its original shape.

As an example a diode laser pulse of 1.55 μm wavelength and a pulse width of $\Delta t_\mathrm{FWHM} = 10$ ps propagates through a silica fiber with a refractive index of 1.45 and $\gamma = 2.8 \cdot 10^{-20}\,\mathrm{m^2\,W^{-1}}$. The soliton period would be about 6 km for a 10 ps pulse and 60 m for a pulse duration of 1 ps and the fundamental soliton needs a peak intensity of 220 MW cm^{-2}. For fibers with a diameter of a few μm peak powers in the ten-watt range are necessary for the fundamental soliton.

Solitons are solitary light wave pulses which can superimpose in nonlinear matter without disturbing each other. Thus they can cross or overtake without changing their pulse shape. Only phase changes occur in the interaction.

There are some analogies to the Π-pulses in self-induced transparency (see Sect. 5.4.8) but this effect is based on coherent bleaching of the absorption of the material, usually described by a two-level scheme. Therefore the physics of the two processes is completely different.

Solitons in optical fibers can be used for transmitting optical signals over long distances with very high bit rates based on very short pulses. They can also be applied for the generation of short pulses with a duration of a few fs in soliton lasers (see Sect. 6.10.3).

4.5.9 Stimulated Brillouin Scattering (SBS)

Stimulated Brillouin scattering (SBS) occurs as amplified spontaneous Brillouin scattering (see Sect. 3.10.3) at sufficiently high intensities of the incident light wave, which increases the sound wave amplitude by electrostriction. It occurs also as spontaneous scattering, in the nonresonant spectral range. But SBS is an inelastic optical scattering process with very small energy loss, as it excites an acoustic phonon of the hyper-sound wave.

SBS is applied in phase conjugating mirrors (PCM) (see Sect. 4.5.14), e.g. for the improvement of the beam quality of solid-state lasers (see Sects. 6.6.12 and 6.11.3). These nonlinear SBS mirrors are very easy to make because of the self-pumping of the scattering. Some examples are given in [4.323–4.341]. Although SBS is a $\chi^{(3)}$ process the phase conjugation of these SBS mirrors can be made by simply focusing the light in a cell with a suitable gas or liquid or in a solid SBS-material.

The third-order nonlinear process can be imagined in four steps which take place simultaneously:

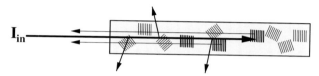

(i) a small share of the spontaneous scattered light will exactly be back-reflected towards the incident beam (scattered light is slightly frequency shifted);

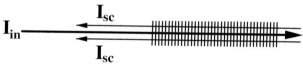

(ii) incident and reflected light waves interfere and generate a moving intensity grating;

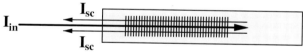

(iii) an intensity grating amplifies the suitable sound wave for back-scattering via electrostriction;

(iv) the amplified sound wave increases the share of the exactly back-scattered light which leads to more interference and the process continues at (i) with positive feedback up to saturation (reflectivity up to 100%)

Because of the perfect back-scattering in SBS the scattering angle is 180° and thus the wavelength of the sound wave Λ_{sound} results from the wavelength of the incident light wave λ_{inc}:

$$\text{sound wavelength} \quad \Lambda_{\text{sound}} = \frac{1}{2}\lambda_{\text{inc}}. \tag{4.114}$$

The sound wave frequency Ω_{sound} follows from the sound velocity v_{sound}:

$$\text{sound frequency} \quad \Omega_{\text{sound}} = 2\frac{v_{\text{sound}}}{\lambda_{\text{inc}}} = 2\frac{v_{\text{sound}} n \nu_{\text{inc}}}{c_0} \tag{4.115}$$

and is in the range of several 100 MHz for gases and up to several 10 GHz for liquids and solids. Thus the energy loss of the light in the range 10^{-4}–10^{-6} can often be neglected but it will be important if interference of the original pump beam and the reflected light is to be used in applications.

The potential linewidth of the SBS $\Delta\Omega_{\text{sound}}$ measured at $1/e$ of the peak value is a function of the lifetime of the sound wave τ_{sound}:

$$\text{SBS linewidth} \quad \Delta\Omega_{\text{sound}} = \frac{1}{2\pi\tau_{\text{sound}}} \tag{4.116}$$

which is twice the lifetime of the phonons τ_{SBS}:

$$\text{phonon lifetime} \quad \tau_{\text{SBS}} = \frac{1}{2}\tau_{\text{sound}} = \frac{K_{\text{sound}}^2 \eta}{\rho_0} \propto \frac{1}{\lambda_{\text{inc}}^2} \tag{4.117}$$

with the material viscosity η. The reciprocal quadratic wavelength dependency is a rough approximation quite well fulfilled in the near UV to IR range.

The theoretical description [4.350, 4.342–4.354] is based on the wave equation of nonlinear optics (4.6)

$$\Delta \boldsymbol{E} - \frac{1}{c_0^2}\frac{\partial^2 \boldsymbol{E}}{\partial t^2} - \text{grad div}\, \boldsymbol{E} = \frac{1}{\varepsilon_0 c_0^2}\frac{\partial^2 \boldsymbol{P}_{\text{nl}}}{\partial t^2}. \tag{4.118}$$

The nonlinear polarization P_{nl} represents the modulation of the material density ρ by electrostriction from the total electric field E_{total} of the interference pattern of the incident and the scattered light by:

$$\boldsymbol{P}_{\text{nl}} = \varepsilon_0 \boldsymbol{E}_{\text{total}}\left(\frac{\partial \chi^{(3)}}{\partial \rho_{\text{mat}}}\right)_T \overline{\rho}_{\text{mat}} = \varepsilon_0 \boldsymbol{E}_{\text{total}}\frac{\gamma^{\text{e}}}{\rho_{\text{mat},0}}\overline{\rho}_{\text{mat}} \tag{4.119}$$

with the average density modulation $\overline{\rho}_{\text{mat}}$, the average density $\rho_{\text{mat},0}$ and the coefficient of electrostriction γ^{e}. The index T indicates the derivative with constant temperature to distinguish SBS from thermally enhanced scattering STBS, which is described in the next chapter. Therefore it is assumed that the variation of χ with temperature can be neglected compared to the density modulation: $T(\partial\chi/\partial T)_\rho \ll \rho_{\text{mat}}(\partial\chi/\partial\rho_{\text{mat}})_T$. The modulation of the density $\overline{\rho}_{\text{mat}}$ can be calculated for small changes from the Navier–Stokes equation including the equation of continuity:

$$-\frac{\partial^2 \overline{\rho}_{\text{mat}}}{\partial t^2} + \left\{\frac{C_{\text{v}}}{C_{\text{p}}}v_{\text{sound}}^2 + \frac{\eta}{\rho_{\text{mat},0}}\frac{\partial}{\partial t}\right\}\Delta\overline{\rho}_{\text{mat}} = \frac{\varepsilon_0\gamma^{\text{e}}}{2}\Delta|\boldsymbol{E}_{\text{total}}|^2 \tag{4.120}$$

with the specific heats C_{v} and C_{p} for constant volume and pressure, the speed of the sound wave v_{sound} and the material viscosity $\eta = (4/3)\eta_{\text{s}} + n_{\text{b}}$ with the shear and bulk viscosities representing the damping of the sound wave. The right part of this expression is the electrostrictive force density of the electric field.

The total electric field E_{total} results from the interference of the incident and reflected light beams which are assumed monochromatic with frequencies ν_{inc} and ν_{scatt} and the wave vector values k_i propagating in the z direction:

$$E_{\text{total}} = E_{\text{inc}}(\boldsymbol{r}, t)\, \mathrm{e}^{\mathrm{i}(2\pi\nu_{\text{inc}}t - k_{\text{inc}}z)} + E_{\text{scatt}}(\boldsymbol{r}, t)\, \mathrm{e}^{\mathrm{i}(2\pi\nu_{\text{scatt}}t + k_{\text{scatt}}z)}. \tag{4.121}$$

The polarization of the light is not changed in the SBS process. The resulting sound wave will show half the wavelength of the incident light and a wave vector value of $K_{\text{SBS}} = k_{\text{inc}} + k_{\text{scatt}} \approx 2k_{\text{inc}}$. With the hyper-sound frequency Ω_{SBS} the following ansatz for the complex sound wave modulation solves the resulting Navier–Stokes equation:

$$\rho_{\text{mat,SBS}} = \rho_{\text{mat,SBS,max}}(\boldsymbol{r}, t)\, \mathrm{e}^{\mathrm{i}(2\pi\Omega_{\text{SBS}}t - K_{\text{SBS}}z)}. \tag{4.122}$$

to give the following form:

$$\begin{aligned}
&\frac{\partial^2 \rho_{\text{mat}}}{\partial t^2} - \left(\mathrm{i}4\pi\Omega_{\text{SBS}} - \frac{1}{\tau_{\text{sound}}} \right) \frac{\partial \rho_{\text{mat}}}{\partial t} \\
&- \left(4\pi^2 \Omega_{\text{SBS}}^2 + \mathrm{i}\frac{2\pi\Omega_{\text{SBS}}}{\tau_{\text{sound}}} - \frac{C_v v_{\text{sound}}^2}{C_{\text{p}}} K_{\text{SBS}}^2 \right) \rho_{\text{mat}} \\
&- \mathrm{i}2\frac{C_v v_{\text{sound}}^2}{C_{\text{p}}} K_{\text{SBS}} \frac{\partial \rho_{\text{mat}}}{\partial z} \\
&= \frac{\varepsilon_0 \gamma^{\text{e}} K_{\text{SBS}}^2}{2} E_{\text{inc}} E_{\text{scatt}}^*.
\end{aligned} \tag{4.123}$$

For solving this complicated system of partial differential equations the SVA approximation (4.9) is applied with:

$$\text{SVA approximation} \quad \frac{\partial^2}{\partial t^2} \ll \nu_{\text{inc}} \frac{\partial}{\partial t} \quad \text{and} \quad \frac{\partial^2}{\partial z^2} \ll k_{\text{inc}} \frac{\partial}{\partial z} \tag{4.124}$$

and, further, strong damping of the sound wave is assumed:

$$\text{strong damping assumption} \quad \frac{\rho_{\text{mat}}}{v_{\text{sound}}\tau_{\text{sound}}} \gg 2\frac{\partial \rho_{\text{mat}}}{\partial z} \tag{4.125}$$

For simplification the sound wave density will be replaced by a normalized sound wave amplitude S:

$$\text{sound wave amplitude} \quad S(\boldsymbol{r}, t) = \frac{2\pi\nu_{\text{inc}}\gamma^{\text{e}}}{2c_0 n \rho_{\text{mat},0}} \rho_{\text{mat}}(\boldsymbol{r}, t). \tag{4.126}$$

With these approximations the system of partial differential equations can be written as:

$$\begin{aligned}
&\frac{\partial E_{\text{inc}}(\boldsymbol{r}, t)}{\partial z} + \frac{n}{c_0}\frac{\partial E_{\text{inc}}(\boldsymbol{r}, t)}{\partial t} + \left(\frac{\partial}{\partial x} + \frac{\partial}{\partial y} \right) E_{\text{inc}}(\boldsymbol{r}, t) \\
&= \mathrm{i}\frac{S(\boldsymbol{r}, t)}{2} E_{\text{scatt}}(\boldsymbol{r}, t)
\end{aligned} \tag{4.127}$$

$$\frac{\partial E_{\text{scatt}}(\boldsymbol{r},t)}{\partial z} + \frac{n}{c_0}\frac{\partial E_{\text{scatt}}(\boldsymbol{r},t)}{\partial t} + \left(\frac{\partial}{\partial x} + \frac{\partial}{\partial y}\right) E_{\text{scatt}}(\boldsymbol{r},t)$$

$$= \mathrm{i}\frac{S^*(\boldsymbol{r},t)}{2} E_{\text{inc}}(\boldsymbol{r},t) \tag{4.128}$$

$$\frac{\partial S(\boldsymbol{r},t)}{\partial t} = \mathrm{i}\frac{g_{\text{SBS}}}{2\tau_{\text{sound}}} E_{\text{inc}}(\boldsymbol{r},t) E_{\text{scatt}}^*(\boldsymbol{r},t) - \frac{1}{2\tau_{\text{sound}}}\{S(\boldsymbol{r},t) - S_0\} \tag{4.129}$$

with the stationary Brillouin gain g_{SBS}:

stationary Brillouin gain $\quad g_{\text{SBS}} = \dfrac{(2\pi\Omega_{\text{sound}}\gamma^e)^2\tau_{\text{sound}}}{c_0^3 n\rho_{\text{mat},0}v_{\text{sound}}}$ \qquad (4.130)

which has the same Lorentzian spectral profile as the spontaneous scattering Brillouin line and the spontaneous sound amplitude S_0 relevant for the SBS process to start the self pumped scattering from noise. This spontaneous scattering amplitude can be estimated from thermodynamic calculations [4.359, 4.354–4.358] to give:

$$S_0 = (\mathrm{e}^{\{1-h\Omega_{\text{sound}}/k_{\text{Boltz}}T\}} + 1)g_{\text{SBS}}h\Omega_{\text{sound}}\frac{1}{\tau_{\text{sound}}}\frac{L_{\text{interaction}}}{4A_{\text{interaction}}}. \tag{4.131}$$

Useful parameters for several common SBS materials are given in Table 4.8a and b. More data can be found in [4.360–4.396].

The sound wave lifetime scales to a good approximation with the square of the wavelength of the incident pump light λ_{inc}^2 over the UV to near IR spectral range. For gases the gain coefficient g_{SBS} is proportional to $\rho_{\text{mat},0}^2$ and the lifetime is proportional to $\rho_{\text{mat},0}$.

The differential equations are still difficult to solve. Further approximations for the transversal modes are proposed [e.g. 4.344] but usually plane waves are assumed and thus the transversal wave front profile is neglected.

The phase between the electric fields and the sound wave can be important in long interaction lengths in the meter range or for long interaction times of μs or ms. Otherwise phase locking can be assumed.

For practical solution including focusing the scheme of Fig. 4.31 was used for the modeling. The variation of the intensity along the interaction as, e.g. from focusing the beam can be considered by the change of the beam cross-

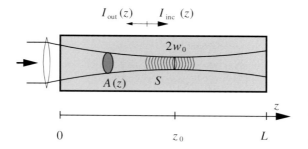

Fig. 4.31. Modeling SBS considering the intensity variation along the interaction length by changed cross-section of area $A(z)$

Table 4.8a. SBS material parameters of some useful SBS gases for several pump wavelengths

Gas	λ (nm)	p (bar)	n	ρ (g cm^{-3})	g_{SBS} (cm GW^{-1})	τ_{SBS} (ns)	$\Delta\Omega_{\mathrm{sound}}$ (MHz)	$\Delta\nu_{\mathrm{B}}$ (MHz)	S_0 (cm^{-1})
Xe	694.3	40	1.03	0.29	44	14.6	10.9	429	
N$_2$	694.3	100	1.03	0.11	14.9	6.3	25.2	1127	
CO$_2$	694.3	50	1.035	0.14	45.7	9.95	16	641	$2 \cdot 10^{-5}$
CH$_4$	694.3	100	1.05	0.075	69	7.4	21.5	1345	
SF$_6$	1064	20	1.023	0.17	14	17.3	9.2	240	$8 \cdot 10^{-5}$
C$_2$F$_6$	694.3	30		0.50	50	4.2	37.9		

Table 4.8b. SBS material parameters of some useful SBS liquids and solids for several pump wavelengths liquid

Liquid	λ (nm)	n	ρ (g cm^{-3})	g_{SBS} (cm GW^{-1})	τ_{SBS} (ns)	$\Delta\Omega_{\mathrm{sound}}$ (MHz)	$\Delta\nu_{\mathrm{B}}$ (GHz)	S_0 (cm^{-1})
CCl$_4$	1064	1.46	1.59	3.8	0.6	265.3	2.76	
GeCl$_4$	1064	1.46	1.87	12	2.3	69.2	2.1	
Methanol	532	1.33	0.79	13.7	0.4	334	5.4	
Ethanol	694.3	1.36	0.79	12	0.45	353	4.55	
Cyclohexane	694.3	1.42	0.78	6.8	0.21	774	5.55	
C$_2$Cl$_3$F$_3$	1064	1.36	1.58	6.2	0.84	189	1.74	$2 \cdot 10^{-5}$
CS$_2$	1064	1.595	1.26	130	6.4	24.8	3.76	$2 \cdot 10^{-5}$
Aceton	1064	1.355	0.79	20	2	79.6	2.67	
Solid								
SiO$_2$ (bulk)	532	1.46	2.202	2.9	1 ± 0.4	163.0	32.6	
SiO$_2$ (fiber)	532	1.46	2.202	2.5	2.3	69.2	2.1	
d-LAP	532	1.584	1.6	29.85	1.9	82.3	19.6	

section area $A(z)$ along the z axis [4.347]. Then the following equations are obtained:

$$\frac{\partial I_{\mathrm{inc}}(z,t)}{\partial z} - \frac{n}{c_0}\frac{\partial I_{\mathrm{inc}}(z,t)}{\partial t} = -S(z,t)\sqrt{I_{\mathrm{inc}}I_{\mathrm{scatt}}}$$
$$- \frac{I_{\mathrm{inc}}}{A(z)}\frac{\partial A(z)}{\partial z} \tag{4.132}$$

$$\frac{\partial I_{\mathrm{scatt}}(z,t)}{\partial z} - \frac{n}{c_0}\frac{\partial I_{\mathrm{scatt}}(z,t)}{\partial t} = -S(z,t)\sqrt{I_{\mathrm{inc}}I_{\mathrm{scatt}}}$$
$$- \frac{I_{\mathrm{scatt}}}{A(z)}\frac{\partial A(z)}{\partial z} \tag{4.133}$$

$$\frac{\partial S(z,t)}{\partial t} = \frac{1}{2\tau_{\mathrm{sound}}}\{g_{\mathrm{SBS}}\sqrt{I_{\mathrm{inc}}I_{\mathrm{scatt}}} - [S(z,t) - S_0]\} \tag{4.134}$$

with the assumption of a coherence length longer than the interaction length.

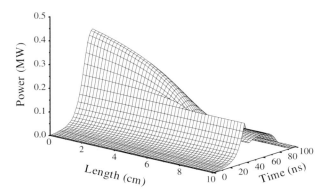

Fig. 4.32. Spatial and temporal distribution of the intensity I_{inct} of the depleted incident beam via SBS in Freon 113 for focusing a diffraction-limited Nd laser pulse with 25 ns, 11.5 mJ and 4 mm diameter with flat curvature at the lens with 120 mm focal length positioned 60 mm in front of the SBS-cell

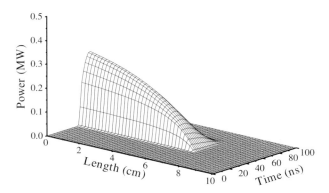

Fig. 4.33. Spatial and temporal distribution of the intensity I_{scatt} of the scattered beam via SBS in Freon 113 for focusing a diffraction-limited Nd laser pulse with 25 ns, 11.5 mJ and 4 mm diameter with flat curvature at the lens with 120 mm focal length

These equations can be solved numerically and the results are discussed in [4.347]. As an example the spatial and temporal distributions are shown in Fig. 4.32 for the depleted incident beam, in Fig. 4.33 for the generated reflected beam and in Fig. 4.34 for the sound wave amplitude for an incident intensity as given in Fig. 4.35.

For these calculations the focusing of a Nd laser beam with a pulse duration of 25 ns (FWHM) and a pulse energy of 11.5 mJ was assumed. This beam with a waist at the lens position and a diameter of 5 mm was then focused with a focal length of 150 mm into a cell filled with Freon 113. The lens was

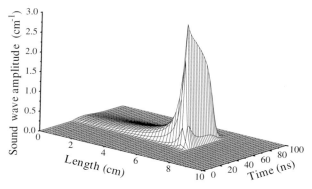

Fig. 4.34. Spatial and temporal distribution of the normalized sound wave amplitude S via SBS in Freon 113 for focusing a diffraction-limited Nd laser pulse with 25 ns, 11.5 mJ and 4 mm diameter with flat curvature at the lens with 120 mm focal length

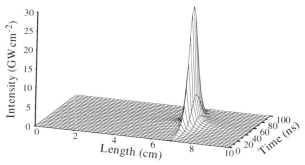

Fig. 4.35. Spatial and temporal distribution of the original intensity I_{inct} with the Gaussian temporal shape and the focusing profile for the Gaussian beam if it is not to be depleted

positioned 60 mm in front of the cell and thus the position of the focus was at 70 mm in the graph.

The parameters of Freon 113 ($C_2Cl_3F_3$) were used as given in Table 4.8. The decay time of the sound wave can easily be seen in Fig. 4.34.

Although in this modeled experiment the Rayleigh length is only 0.9 mm, the sound wave and thus the reflection is distributed over more than 10 mm. Thus the interaction length and the coherence demands have to be carefully determined in such SBS experiments.

It was shown that the reflectivity is almost not a function of the focusing, as long as the coherence length is longer than the interaction length (see also [4.378–4.381]). Different materials demand different minimal light powers for SBS characterized by an artificial "threshold" defined e.g. for 2% reflectivity.

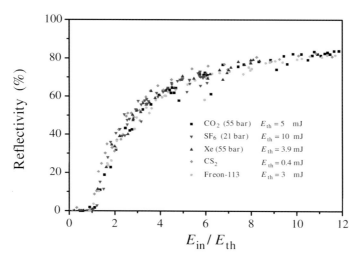

Fig. 4.36. Reflectivity curves of several materials as a function of the normalized light pulse energy

But reflectivity curves as a function of the normalized incident light power or energy in the case of pulsed light both as measured or as calculated are almost the same (see Fig. 4.36).

In the case of stationary SBS the differential equations can be approximated for nondepleted incident light. The intensity of the reflected light I_{scatt} can be written as:

$$I_{\text{scatt}}(L_{\text{interaction}}) = I_{\text{scatt,spont}}(z_0)\, e^{g_{\text{SBS}} I_{\text{inc}} L_{\text{interaction}}} \tag{4.135}$$

with the interaction length $L_{\text{interaction}}$ and assuming the incident beam and the reflected beam interfere coherently. The total stationary SBS gain G_{SBS} results from:

stationary SBS-gain $G_{\text{SBS}} = g_{\text{SBS}} I_{\text{inc}} L_{\text{interaction}}.$ $\tag{4.136}$

The stationary "SBS-threshold" power P_{th} can be estimated from this formula by considering the spontaneous reflectivity useful for starting the SBS. It is in the range of $R_{\text{spontaneous}} = 10^{-11}$ and thus the total SBS gain G_{SBS} has to be bigger than approximately 20 for 2% reflectivity:

$$\text{stationary SBS-threshold}\quad P_{\text{th}} \approx 20\,\frac{A_{\text{interaction}}}{g_{\text{SBS}} L_{\text{interaction}}}. \tag{4.137}$$

Typical cross-sections $A_{\text{interaction}}$ in SBS with focused beams are of the order of $10^{-5}\,\text{cm}^2$ and the interaction lengths are a few mm. Thus gases and solids show thresholds of several $100\,\text{kW}$ to MW and liquids can have values as low as $10\,\text{kW}$ (see [4.397–4.401] and Sect. 4.5.14 for low threshold SBS).

In nonstationary SBS with pulses shorter than the lifetime of the sound wave the threshold increases with the ratio of the phonon lifetime divided by the pulse width. Thus SBS with ps or fs laser pulses shows very small reflectivities [e.g. 4.382]. In addition in this case the coherence of the short-pulse light may be insufficient for SBS and thus the reflectivity will be reduced even more.

The $A_{\text{interaction}}/L_{\text{interaction}}$ ratio can be decreased many orders of magnitudes by using waveguide structures as SBS reflectors. In multimode quartz fibers with lengths of several meters power "thresholds" of a few 100 W were reported and with liquids in capillaries a few W were obtained [4.402–4.405]. Using such waveguides with an internal taper a very large dynamic range of 260:1 could be achieved in combination with reflectivities above 95% and a low "threshold" of 15 μJ for 30 ns pulses at 1.06 μm as shown in Fig. 4.37.

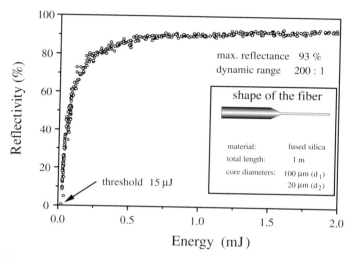

Fig. 4.37. Reflectivity of a solid-state SBS mirror based on a tapered waveguide structure as described in [4.402]

For application of SBS mirrors in optical phase conjugation the "SBS-threshold" is an important parameter. In the case of stationary scattering with pulse durations much longer than the phonon lifetime the light power is a significant value and in the case of nonstationary scattering so is the pulse energy. The temporal shape of the reflected and transmitted light for the intermediate case of comparable times as in Fig. 4.35 and Fig. 4.34 is given in Fig. 4.38.

SBS reflection in general does not change the polarization and works best for linearly polarized light. Different schemes are proposed for dealing with different polarization (see Sect. 4.5.14).

For applications several other properties of SBS materials can be important as for example:

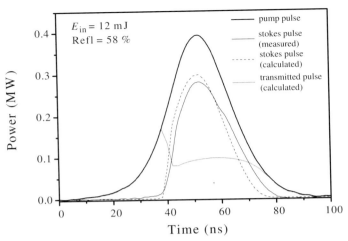

Fig. 4.38. Evolution of the intensities of the incident, scattered and transmitted light pulses for Freon 113 with the parameters as in Fig. 4.35 and Fig. 4.34

- the frequency shift $\Delta\gamma_B$ of the reflected light may disturb in interference methods and has to be considered for applying SBS mirrors in laser resonators;
- absorption can cause heating and disturb high-power applications;
- competing processes such as, e.g. stimulated Raman scattering, self-focusing, self-phase modulation, may disturb the reflectivity and fidelity in optical phase conjugation;
- damage and optical breakdown prevent good operation;
- toxicity may restrict wide application for safety reasons.

The coherence length of the pump laser light limits the maximum interaction length and thus the reflectivity. This becomes important if waveguide SBS mirrors of several meters length are to be applied. The SBS reflectivity is also decreased if the laser light bandwidth is larger than the Brillouin linewidth ($\Delta\nu_{pump} > \Delta\Omega_{sound}$) [4.383–4.396].

Some applications of SBS are described in [4.406–4.414] in addition to its use as a phase conjugating mirror (Sect. 4.5.14), in four-wave mixing (Sect. 5.9.2), or in lasers as described in Sects. 6.6.12 and 6.11.3.

4.5.10 Stimulated Thermal Brillouin Scattering (STBS)

If a small absorption in the SBS matter is present at the wavelength of the incident light the scattering can be enhanced by a temperature grating in addition to the usual sound wave or density grating [4.415]. The analysis was given by [4.416] and has additional terms in the equations given above.

The wave equation can be written as:

$$\Delta \boldsymbol{E} - \frac{1}{c_0^2}\frac{\partial^2 \boldsymbol{E}}{\partial t^2} - \frac{a}{c_0}\frac{\partial \boldsymbol{E}}{\partial t} - \mathrm{grad\,div}\,\boldsymbol{E} = \frac{1}{\varepsilon_0 c_0^2}\frac{\partial^2 \boldsymbol{P}_{\mathrm{nl}}}{\partial t^2} \tag{4.138}$$

with the absorption coefficient a. The Navier–Stokes equation including the equation of continuity then reads:

$$-\frac{\partial^2 \bar{\rho}_{\mathrm{mat}}}{\partial t^2} + \left\{ \frac{C_{\mathrm{v}}}{C_{\mathrm{p}}}v_{\mathrm{sound}}^2 + \frac{\eta}{\rho_{\mathrm{mat},0}}\frac{\partial}{\partial t}\right\}\Delta\bar{\rho}_{\mathrm{mat}} + \frac{C_{\mathrm{v}}}{C_{\mathrm{p}}}v_{\mathrm{sound}}^2\beta_{\mathrm{T}}\rho_{\mathrm{mat},0}\Delta T$$

$$= \frac{\varepsilon_0\gamma^{\mathrm{e}}}{2}\Delta|\boldsymbol{E}_{\mathrm{total}}|^2 - \frac{\varepsilon_0}{2}\left(\frac{\partial\chi^{(3)}}{\partial T}\right)_\rho \nabla(E^2\nabla T) \tag{4.139}$$

with temperature T and coefficient of thermal expansion β_{T} and all other values as given in the previous Section. With a wave ansatz for the electric field (4.121) and the density (4.122) as given above and for the temperature as:

$$T = T_{\max}(\boldsymbol{r}, t)\,\mathrm{e}^{\mathrm{i}(2\pi\Omega_{\mathrm{SBS}}t - K_{\mathrm{SBS}}z)} \tag{4.140}$$

the equation for the density is:

$$\frac{\partial^2 \rho_{\mathrm{mat}}}{\rho t^2} - \left(\mathrm{i}4\pi\Omega_{\mathrm{SBS}} - \frac{1}{\tau_{\mathrm{sound}}}\right)\frac{\partial\rho_{\mathrm{mat}}}{\partial t}$$

$$- \left(4\pi^2\Omega_{\mathrm{SBS}}^2 + \mathrm{i}\frac{2\pi\Omega_{\mathrm{SBS}}}{\tau_{\mathrm{sound}}} - \frac{C_{\mathrm{v}}v_{\mathrm{sound}}^2}{C_{\mathrm{p}}}K_{\mathrm{SBS}}^2\right)\rho_{\mathrm{mat}}$$

$$\mathrm{i}2\frac{C_{\mathrm{v}}v_{\mathrm{sound}}^2}{C_{\mathrm{p}}}K_{\mathrm{SBS}}\frac{\partial\rho_{\mathrm{mat}}}{\partial z} + \frac{C_{\mathrm{v}}v_{\mathrm{sound}}^2}{C_{\mathrm{p}}}K_{\mathrm{SBS}}^2\rho_{\mathrm{mat},0}\beta_{\mathrm{T}}$$

$$= \frac{\varepsilon_0\gamma^{\mathrm{e}}K_{\mathrm{SBS}}^2}{2}E_{\mathrm{inc}}E_{\mathrm{scatt}}^* \tag{4.141}$$

and in addition the equation for the temperature modulation T is given by:

$$\frac{\partial T}{\partial t} - \left(\mathrm{i}2\pi\Omega_{\mathrm{SBS}} - \frac{\Lambda_{\mathrm{T}}K_{\mathrm{SBS}}^2}{\rho_{\mathrm{mat},0}C_{\mathrm{v}}}\right)T$$

$$- \frac{C_{\mathrm{p}}/C_{\mathrm{v}} - 1}{\beta_{\mathrm{T}}\rho_{\mathrm{mat},0}}\left(\frac{\partial\rho_{\mathrm{mat}}}{\partial t} - \mathrm{i}2\pi\Omega_{\mathrm{SBS}}\rho_{\mathrm{mat}}\right)$$

$$= \frac{\varepsilon_0}{2C_{\mathrm{v}}\rho_{\mathrm{mat},0}}E_{\mathrm{inc}}E_{\mathrm{scatt}}^*\left[nc_0a - \mathrm{i}\pi\Omega_{\mathrm{SBS}}T_0\left(\frac{\partial\chi^{(3)}}{\partial T}\right)_{\rho_{\mathrm{mat}}}\right] \tag{4.142}$$

with the thermal conductivity Λ_{T}. Finally the equations for the electric fields of the incident and the scattered light beams are in the plane wave approximation:

$$\frac{\partial E_{\mathrm{inc}}(\boldsymbol{r}, t)}{\partial z} + \frac{n}{c_0}\frac{\partial E_{\mathrm{inc}}(\boldsymbol{r}, t)}{\partial t} + \frac{a}{2}E_{\mathrm{inc}}(\boldsymbol{r}, t)$$

$$= \mathrm{i}\frac{2\pi^2\varepsilon_0\nu_{\mathrm{inc}}}{c_0 n}\left[\frac{\gamma^{\mathrm{e}}}{\rho_{\mathrm{mat},0}}\rho_{\mathrm{mat}} + \left(\frac{\partial\chi^{(3)}}{\partial T}\right)_{\rho_{\mathrm{mat}}}T\right]E_{\mathrm{scatt}} \tag{4.143}$$

and

$$\frac{\partial E_{\text{scatt}}(\boldsymbol{r}, t)}{\partial z} + \frac{n}{c_0} \frac{\partial E_{\text{scatt}}(\boldsymbol{r}, t)}{\partial t} + \frac{a}{2} E_{\text{scatt}}(\boldsymbol{r}, t)$$

$$= \mathrm{i} \frac{2\pi^2 \varepsilon_0 \nu_{\text{scatt}}}{c_0 n} \left[\frac{\gamma^{\text{e}}}{\rho_{\text{mat},0}} \rho_{\text{mat}}^* \left(\frac{\partial \chi^{(3)}}{\partial T} \right)_{\rho_{\text{mat}}} T^* \right] E_{\text{inc}} \tag{4.144}$$

This system of four differential equations (4.141)–(4.144) may be solved numerically. It may be noticed that the sum of the incident and the scattered light is no longer constant along the z axis because of absorption. A simple approximation can be reached by using the phenomenological equations:

$$\frac{\partial I_{\text{inc}}(z, t)}{\partial z} - \frac{n}{c_0} \frac{\partial I_{\text{inc}}(z, t)}{\partial t} - a I_{\text{inc}}(z, t)$$

$$= -S(z, t) \sqrt{I_{\text{inc}} I_{\text{scatt}}} - \frac{I_{\text{inc}}}{A(z)} \frac{\partial A(z)}{\partial z} \tag{4.145}$$

and

$$\frac{\partial I_{\text{scatt}}(z, t)}{\partial z} - \frac{n}{c_0} \frac{\partial I_{\text{scatt}}(z, t)}{\partial t} - a I_{\text{scatt}}(z, t)$$

$$= -S(z, t) \sqrt{I_{\text{inc}} I_{\text{scatt}}} - \frac{I_{\text{scatt}}}{A(z)} \frac{\partial A(z)}{\partial z} \tag{4.146}$$

for the intensities of the incident I_{inc} and scattered light I_{scatt} and the equation for the sound wave amplitude of the previous chapter. The SBS gain factor g_{SBS} contains two parts from electrostriction $g_{\text{SBS}}^{\text{e}}$ as above and from absorption $g_{\text{SBS}}^{\text{a}}$:

$$g_{\text{SBS}} = g_{\text{SBS}}^{\text{e}} + g_{\text{SBS}}^{\text{a}} \tag{4.147}$$

with the additional component:

$$g_{\text{SBS}}^{\text{a}} = a \frac{\pi \nu_{\text{inc}} \tau_{\text{SBS}} \gamma^{\text{e}} \beta_{\text{T}}}{c_0 n \rho_{\text{mat},0} C_{\text{p}}} \tag{4.148}$$

with the parameters as given above. This part shows an asymmetric wavelength dependence [4.415].

4.5.11 Stimulated Rayleigh (SRLS) and Thermal Rayleigh (STRS) Scattering

Stimulated Rayleigh scattering takes place in the direct spectral neighborhood of the incident light frequency and thus it is almost elastic. The stimulation results from the change in matter density via electrostriction or thermal effects. Thus the theoretical description can be worked out analogous to SBS [4.415].

The stationary gain for stimulated Rayleigh scattering g_{SRLS} follows from [4.416]:

$$g_{SLRS} = \{g^e_{SRLS,0} - g^a_{SRLS}\} \frac{4\pi\nu_{inc}\tau_{RL}}{1 + (4\pi\nu_{inc}\tau_{RL})^2} \tag{4.149}$$

with the lifetime of the Rayleigh scattering τ_{RL} as the reciprocal spectral width:

$$\tau_{RL} = \frac{1}{2\pi\Delta\nu_{RL}} = \frac{\rho_{mat,0}C_p\lambda^2_{inc}}{16\pi^2\Lambda_T(\sin^2(\theta_{scatt}/2))} \tag{4.150}$$

which is in the range of 10 ns and with the two peak gains:

$$g^e_{SRLS,0} = \frac{4\pi^2\nu^2_{inc}\tau_{RL}\gamma^e\gamma^{RL}}{c^3 n v_{sound}\rho_{mat,0}} \tag{4.151}$$

and

$$g^a_{SRLS,0} = \frac{4\pi^2\nu^2_{inc}\tau_{RL}\gamma^e\gamma^a}{c^3 n v_{sound}\rho_{mat,0}} \tag{4.152}$$

with the coefficients:

$$\gamma^{RL} = \frac{(C_p/C_v - 1)c_0\gamma^e}{8\pi n v_{sound}\nu_{inc}\tau_{RL}} \tag{4.153}$$

and

$$\gamma^a = a\frac{c_0^2 v_{sound}\beta_T}{2\pi C_p\nu_{inc}}, \tag{4.154}$$

and γ^e as given above.

Thus even for small absorption $a \approx 10^{-3}\,\mathrm{cm}^{-1}$ the absorptive term can dominate the gain factor. Backward scattering will show shorter lifetimes and smaller gain than forward scattering. Because of the long lifetimes the Rayleigh scattering will be mostly nonstationary. Detailed analysis analogous to the description of the SBS may be necessary for evaluation.

Because of the small gain coefficients stimulated Rayleigh scattering is almost not observable in nonabsorbing material. For strongly absorbing matter with $a \approx 1\,\mathrm{cm}^{-1}$ the gain can reach values of $1\,\mathrm{cm\,MW}^{-1}$ as, e.g. in CCl_4 or other liquids. Using light-induced absorption gratings this type of scattering may find applications in photonics (see also [4.417–4.424]).

4.5.12 Stimulated Rayleigh Wing (SRWS) Scattering

The Rayleigh wing scattering produced by orientation fluctuations of particles with nonisotropic induced dipole moment such as, e.g. anisotropic molecules, can be easily stimulated with high light fields and thus SRWS can be observed [4.415, 4.425]. The anisotropic electric field vector of the light beam changes the angular orientation of these particles and thus a local variation of the refractive index will be induced.

The driving force for the SRWS results from the interaction of the electric field vector $\boldsymbol{E}_{\mathrm{inc}}$ with the vector of the polarizability μ_{mol} of the molecules resulting in a torque M as shown in Fig. 4.39.

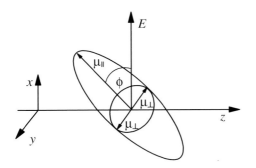

Fig. 4.39. Interaction of molecular polarizabilities μ_i with linear polarized light resulting in anisotropic orientation of matter for stimulated Rayleigh wing scattering (SRWS)

This torque can be calculated using the molecular polarizabilities:

$$M = -[\mu_{\mathrm{mol},II} - \mu_{\mathrm{mol},\perp}]|E_{\mathrm{inc}}|^2 \cos\phi \sin\phi. \tag{4.155}$$

For a simple theoretical description of SRWS, light linearly polarized in x direction and propagating in the z direction is assumed. The angular distribution function $f_{\mathrm{mol}}(\phi)$ for the molecular orientation has to fulfill the differential equation:

$$\frac{\partial f_{\mathrm{mol}}}{\partial t} - \frac{1}{2\tau_{\mathrm{orientation}} \sin\phi} \frac{\partial}{\partial\phi}\left\{\sin\phi\left[\frac{\partial f_{\mathrm{mol}}}{\partial\phi} - \frac{M}{k_{\mathrm{Boltz}}T}f_{\mathrm{mol}}\right]\right\} = 0 \tag{4.156}$$

with the relaxation time $\tau_{\mathrm{orientation}}$ of the orientation of the molecules. A solution of this equation was given for small deviations from equilibrium [4.424]. For known f the nonlinear polarization of the matter as a function of the electric field can be written as:

$$P_{\mathrm{nl}} = 2\pi\varepsilon_0 N_{\mathrm{mol}} E_{\mathrm{inc}} \int_{-1}^{1} f_{\mathrm{mol}}(\phi, E_{\mathrm{inc}}) \cos^2\phi \, \mathrm{d}(\cos\phi) \tag{4.157}$$

with the density of the molecules N_{mol}.

The total electric field consists of the incident wave, the Stokes wave with the index S and the anti-Stokes wave with the index aS:

$$\boldsymbol{E}_{\mathrm{total}} = \boldsymbol{E}_{\mathrm{inc},0}\, \mathrm{e}^{\mathrm{i}(2\pi\nu_{\mathrm{inc}}t - \boldsymbol{k}_{\mathrm{inc}}\cdot\boldsymbol{r})} + \boldsymbol{E}_{\mathrm{S},0}\, \mathrm{e}^{\mathrm{i}(2\pi\nu_{\mathrm{S}}t - \boldsymbol{k}_{\mathrm{S}}\cdot\boldsymbol{r})}$$
$$+ \boldsymbol{E}_{\mathrm{aS},0}\, \mathrm{e}^{\mathrm{i}(2\pi\nu_{\mathrm{aS}}t - \boldsymbol{k}_{\mathrm{aS}}\cdot\boldsymbol{r})} \tag{4.158}$$

and under the assumption of a nondepleted strong incident field two coupled equations for the electric fields of the Stokes E_{S} and the anti-Stokes E_{aS} wave can be derived:

$$2\mathrm{i}k_{\mathrm{S}}^z \frac{\mathrm{d}E_{\mathrm{S}}}{\mathrm{d}z} - (k_{\mathrm{S}}^x)^2 E_{\mathrm{S}} = -\frac{\Delta n_{\mathrm{SRWS}}^2 \pi^2 \nu_{\mathrm{S}}^2}{c_0^2} \frac{|E_{\mathrm{inc}}|^2 E_{\mathrm{S}} + E_{\mathrm{inc}}^2 E_{\mathrm{aS}}^*}{1 + \mathrm{i}2\pi\Delta\nu_{\mathrm{RW}}\tau_{\mathrm{RW}}} \tag{4.159}$$

and

$$2\mathrm{i}k_{\mathrm{aS}}^z \frac{\mathrm{d}E_{\mathrm{aS}}^*}{\mathrm{d}z} - (k_{\mathrm{aS}}^x)^2 E_{\mathrm{aS}}^* = \frac{\Delta n_{\mathrm{SRWS}}^2 \pi^2 \nu_{\mathrm{aS}}^2}{c_0^2} \frac{|E_{\mathrm{inc}}|^2 E_{\mathrm{aS}}^* + E_{\mathrm{inc}}^{*2} E_{\mathrm{S}}}{1 + \mathrm{i}2\pi \Delta \nu_{\mathrm{RW}} \tau_{\mathrm{RW}}} \tag{4.160}$$

with the change of permittivity Δn_{SRWS}:

$$\Delta n_{\mathrm{SRWS}}^2 = \frac{\varepsilon_0 N_{\mathrm{mol}}(\mu_{||} - \mu_\perp)^2}{90 k_{\mathrm{Boltz}} T} \tag{4.161}$$

and the relations:

$$\delta \nu_{\mathrm{RW}} = \nu_{\mathrm{inc}} - \nu_{\mathrm{S}} = \nu_{\mathrm{aS}} - \nu_{\mathrm{inc}}$$

$$k_{\mathrm{S}}^x = -k_{\mathrm{aS}}^x \ll k_{\mathrm{inc}} \tag{4.162}$$

$$\frac{c_0^2 (k_i^z)^2}{4\pi^2 \nu_i^2} - n_0^2 = \frac{1}{2} \Delta n_{\mathrm{SRWS}}^2 |E_{\mathrm{inc}}|^2 \quad \text{with} \quad i = \mathrm{S, aS.}$$

If large scattering angles between the Stokes and the anti-Stokes waves are investigated the coupling between them can be neglected in the above equations. Thus the gain factors for the case of parallel polarization of incident and scattered waves was given in [4.424] as:

$$g_{\mathrm{SRWS,S}} = \frac{4\pi^2 \nu_{\mathrm{S}} \Delta n_{\mathrm{SRWS}}^2}{c_0^2 n_0^2} \frac{2\pi \Delta \nu_{\mathrm{RW}} \tau_{\mathrm{RW}}}{1 + (2\pi \Delta \nu_{\mathrm{RW}} \tau_{\mathrm{RW}})^2} \approx -g_{\mathrm{SWRS,aS}}. \tag{4.163}$$

Some typical values are given in Table 4.9.

Table 4.9. Permittivity Δn_{SRWS}, gain factor g_{SRWS}, frequency shift $\Delta \nu_{\mathrm{RW}}$ and relaxation time τ_{RW} of some liquids

matter	Δn_{SRWS} $(10^{12}\,\mathrm{cm}^3\,\mathrm{erg}^{-1})$	g_{SRWS} $(\mathrm{cm\,MW}^{-1})$	$\Delta \nu_{\mathrm{RW}}$ (cm^{-1})	τ_{RW} $(10^{12}\,\mathrm{s})$
CS_2	41.4	30	2.65	2
Nitrobenzene	32.2	25.6	0.111	48
Bromobenzene	17.5	13.7	0.354	15
Chlorobenzene	11.9	9.7	0.624	8.5
Toluene	7.7	6.6	1.294	4.1
Benzene	6.7	5.7	1.396	3.8

As can be seen the gain factors are comparable to SBS gain factors and high enough for experiments with pulsed lasers. The orientation relaxation times are in the ps range. From newer experiments it is known that the processes determining the orientation relaxation are much more complicated and no single exponential decay is obtained if high temporal resolution is achieved.

In the near-forward direction Stokes and anti-Stokes photons will be generated by the scattering of two incident photons. The scattering is a maximum for an angle $\theta_{\text{SRWS,max}}$ of:

$$\theta_{\text{SRWS,max}} = \pm\frac{1}{\sqrt{2}} \frac{\Delta n_{\text{SRWS}}|E_{\text{inc}}|}{n_0} \tag{4.164}$$

which is in the range of a few mrad.

4.5.13 Stimulated Raman Techniques

Although stimulated Raman scattering (SRS) is usually applied in the nonab-sorbing spectral range of matter the energy change of the scattered photons compared to the incident photons can be as large as 10%. As in spontaneous Raman scattering the SRS process can be understood as scattering coupled with a matter transition between two vibrational energy states with energy difference $E_{\text{vib}} = h\nu_{\text{vib}}$ via a virtual energy state of more than ten times this energy. Different measuring techniques are used. [4.426–4.432] and examples can be found in [4.433–4.525].

Stimulated Raman Scattering (SRS)

Stimulated Raman scattering (SRS) can take place with the excitation of the vibration or of an optical phonon (Stokes SRS results in smaller photon energy) or by its depletion (anti-Stokes SRS):

> energy condition
> $$\nu_{\text{scatt,SRS}} = \nu_{\text{inc}} \mp m\nu_{\text{vib}} \quad \text{with } m = 1, 2, 3, \ldots \tag{4.165}$$

whereas subsequent scattering or nonlinear coupling of molecular vibration allows multiple frequency shifts with $m > 1$. The vibrational overtones may show slightly shifted frequencies ν_{vib} due to the anharmonicity of the vibra-tional potential. Again, as in spontaneous Raman scattering the selection rules demand polarizability for the considered vibration in contrast to the necessity of a dipole moment of this vibration to detect it in IR spectroscopy. Rotational transition and translation energies may overlay the spectra in both cases and produce complicated structures or broadening. This allows detailed analysis of the matter structure if high-resolution techniques are applied.

The intensity of the scattered light in SRS can be amplified by many orders of magnitude from spontaneous Raman scattering up to several 10% of the intensity of the incident light. Under certain circumstances strength of the anti-Stokes SRS can be comparable to the Stokes SRS. It occurs in small cones, of a few degrees in the forward and backward direction.

Therefore SRS is useful in cases of difficult spontaneous Raman scatter-ing measurements for improving the signal-to-noise ratio and shortening the measuring time by many orders of magnitude (see e.g. [4.433–4.469]).

The theoretical description can be worked out analogous to the theory of stimulated Brillouin scattering except the frequency shift is much larger in SRS and does not depend on the scattering angle [4.470–4.477].

The amplification in stimulated Raman scattering depends on the scattering cross-section $d\sigma_{SRS}/d\Omega_{scatt}$, the incident intensity I_{inc} and the interaction length $L_{interact}$. In SRS experiments the forward and backward direction will be observed with a long interaction length compared to the beam diameter and thus small angles. Therefore the amplification can be described by the Raman gain coefficient g_{SRS}. Below incident intensities causing saturation, which starts roughly at 10^{-1} of the saturation limit [4.464], the scattered intensity $I_{scatt,SRS}$ is exponentially proportional to these values:

$$\text{below saturation} \quad I_{scatt,SRS} = I_{scatt,spont} e^{g_{SRS} I_{inc} L_{interact}} \tag{4.166}$$

with the spontaneously scattered Raman intensity $I_{scatt,spont}$, which is about 10^{-6}–10^{-10} of the saturation. The stimulated new Raman photon has the same direction, frequency and polarization as the stimulating one.

Typical SRS gain coefficients are in the order of $g_{SRS} \approx 10 \, \text{cm} \, \text{GW}^{-1}$ (see Table 4.10). Thus, incident intensities of several $100 \, \text{MW} \, \text{cm}^{-2}$ are necessary for strong effects with interaction lengths in the cm range.

Table 4.10. SRS parameters of several materials wave number of the vibration, spectral width, scattering cross-section and Raman gain coefficient

material	ν_{vib} (cm^{-1})	$\Delta\nu_{vib}$ (cm^{-1})	$N_0 d\sigma/d\Omega$ $(\text{cm}^{-1}\text{ster}^{-1})$	g_{SRS} $(\text{cm} \, \text{GW}^{-1})$
Liquid O_2	1552	0.117	0.48 ± 0.14	14.5 ± 4
N_2 (liquid)	2326.5	0.067	0.29 ± 0.09	17 ± 5
Benzene	992	2.15	3.06	2.8
CS_2	655.6	0.50	7.55	24
Nitrobenzene	1345	6.6	6.4	2.1
Toluene	1003	1.94	1.1	1.2
$LiNbO_3$	256	23	381	8.9

Because of the nonlinear interaction the linewidth of the SRS is smaller than the linewidth of spontaneous Raman scattering. If the linewidth of the incident beam is small enough Doppler broadening from thermal motions can be obtained.

Anti-Stokes SRS becomes more probable with increasing incident intensity as a result of the higher population of the excited vibrational energy level from the Stokes scattering. Both intensities can reach the same order of magnitude as the incident light. Besides energy conservation for stimulated anti-Stokes Raman scattering:

$$\text{energy condition} \quad 2\nu_{inc} = \nu_{SRS,Stokes} + \nu_{SRS,anti-Stokes} \tag{4.167}$$

the conservation of momentum has to be considered. Thus the wave vectors have to fulfill the condition:

$$\text{angle condition} \quad \mathbf{k}_{inc} + \mathbf{k}_{inc} = \mathbf{k}_{SRS,Stokes} + \mathbf{k}_{SRS,anti-Stokes} \tag{4.168}$$

with the wave vectors:

$$k_m = \frac{2\pi n(\nu_m)\nu_m}{c_0} e_m \tag{4.169}$$

pointing in the propagation direction e_m of the mth beam. Thus the angle between the incident light beam and the anti-Stokes light beam can be expressed as:

anti-Stokes angle

$$\theta_{\text{SRS,anti–Stokes}} = \arccos\left[\frac{4k_{\text{inc}}^2 - k_{\text{SRS,anti–Stokes}}^2 - k_{\text{SRS,Stokes}}^2}{4k_{\text{inc}}k_{\text{SRS,anti–Stokes}}}\right] \tag{4.170}$$

which is different from zero as a consequence of the different refractive indices for the different wavelengths.

This condition can be illustrated for the SRS in nitrobenzene with a vibrational wave number of $1345\,\text{cm}^{-1}$. Scattering of ruby laser light with a wavelength of $694.3\,\text{nm}$ leads to Stokes and anti-Stokes wavelengths of $765.8\,\text{nm}$ and $635.0\,\text{nm}$. The refractive indices are 1.540 for the incident beam, 1.536 for the Stokes and 1.545 for the anti-Stokes beam. The resulting anti-Stokes angle is $1.8°$. The SRS signals occur in cones around the incident light beam with the given half-angles from the equations for the wave vectors given above.

For mathematical modeling of SRS analogous to SBS a coupled system of partial differential equation describes the change of the electric light fields and the population of the matter as functions of time and space.

Based on (4.5) for the electric field amplitude E in combination with the nonlinear polarization P_{nl} but including some absorption with the coefficient a and assuming plane waves, the wave equation is:

$$\Delta E - \mu_0\varepsilon_0 \frac{\partial^2 E}{\partial t^2} - a\sqrt{\mu_0\varepsilon_0}\frac{\partial E}{\partial t} = \mu_0\frac{\partial^2 P_{\text{nl}}}{\partial t^2} \tag{4.171}$$

With the ansatz for the electric field E of the light beam with planar wave front propagating in the z direction:

$$E(z,t) = \sum_m E_{j,0}\, e^{i(2\pi\nu_m t - k_m z)} \tag{4.172}$$

and for the nonlinear polarization P_{nl}:

$$P_{nl}(z,t) = \sum_m P_{nl_{n,0}}\, e^{i(2\pi\nu_m t)} \tag{4.173}$$

it follows that for the spectral components j using the SVA approximation and neglecting the second time derivatives of E:

$$\frac{\partial E_m}{\partial z} - \mu_0\varepsilon_0\frac{\partial^2 E_m}{\partial t^2} - \frac{a}{2}E_m = i\frac{\mu_0\pi\nu_m}{c_0}\frac{\partial^2 P_{nl_m}}{\partial t^2}\, e^{ik_m z}. \tag{4.174}$$

The interaction of the light with matter is obtained from the forced expectation value of the displacement q_{vib} in the vibration and the relative population difference $\Delta N = (N_{\text{ground}} - N_{\text{exc}})/N_{\text{ground}}$ in the vibrational two-level system. The displacement q_{vib} results from a damped wave equation:

$$\frac{\partial^2 q_{\text{vib}}}{\partial t^2} + \frac{1}{\tau_{\text{vib}}} \frac{\partial q_{\text{vib}}}{\partial t} + 4\pi^2 \nu_{\text{vib}}^2 q_{\text{vib}} = \frac{1}{2m_{\text{vib}}} \frac{\partial \mu_{\text{vib}}}{\partial q_{\text{vib}}} \Delta N E^2 \tag{4.175}$$

with the lifetime of the vibration τ_{vib}, which is inversely equal to the linewidth of the vibration $\Delta\nu_{\text{vib}}$, and its reduced mass m_{vib}. $\partial\mu_{\text{vib}}/\partial q_{\text{vib}}$ gives the polarizability of this vibration. ΔN can be calculated from a rate equation as they will be given in the next chapter but in SRS the population of the excited vibrational state can be neglected and $\Delta N = 1$ can be assumed.

The nonlinear polarization follows from the solution of this equation by:

$$P_{\text{nl}} = N_{\text{total}} \frac{\partial \mu_{\text{vib}}}{\partial q_{\text{vib}}} q_{\text{vib}} E \tag{4.176}$$

with the total density N_{total} of vibrating particles per unit volume.

Assuming forward (fw) and backward (bw) scattering along the z axis the electric field resulting from Stokes scattering of the SRS can be written as:

$$\begin{aligned}
\boldsymbol{E} = {} & \boldsymbol{E}_{\text{inc}}\, \text{e}^{\text{i}(2\pi\nu_{\text{inc}}t - k_{\text{inc}}z)} \\
& + \boldsymbol{E}_{\text{Stokes,fw}}\, \text{e}^{\text{i}(2\pi\nu_{\text{Stokes}}t - k_{\text{Stokes,fw}}z)} \\
& + \boldsymbol{E}_{\text{Stokes,bw}}\, \text{e}^{\text{i}(2\pi\nu_{\text{Stokes}}t - k_{\text{Stokes,bw}}z)}
\end{aligned} \tag{4.177}$$

with all $\boldsymbol{E}$ vectors pointing in the direction of the incident field and the resulting displacement follows from:

$$q_{\text{vib}} = \frac{1}{2}\left\{ q_{\text{fw}}\, \text{e}^{\text{i}(2\pi\nu_{\text{vib}}t - k_{\text{vib,fw}}z)} + q_{\text{bw}}\, \text{e}^{\text{i}(2\pi\nu_{\text{vib}}t - k_{\text{vib,bw}}z)} + \text{c.c.} \right\}. \tag{4.178}$$

Using this ansatz the differential equations of the electric fields of the incident light E_{inc} the forward scattered light E_{fw} and the backward scattered light E_{bw} of the Stokes wave can be written as:

$$\begin{aligned}
\frac{n_{\text{inc}}}{c_0} \frac{\partial E_{\text{inc}}}{\partial t} + \frac{\partial E_{\text{inc}}}{\partial z} = {} & \frac{\text{i}2\pi^2\nu_{\text{inc}}}{n_{\text{inc}}c_0} \Delta N \frac{\partial \mu_{\text{vib}}}{\partial q_{\text{vib}}} (q_{\text{fw}}E_{\text{fw}} + q_{\text{bw}}E_{\text{bw}}) \\
& - \frac{a}{2} E_{\text{inc}}
\end{aligned} \tag{4.179}$$

$$\frac{n_{\text{Stokes}}}{c_0} \frac{\partial E_{\text{fw}}}{\partial t} + \frac{\partial E_{\text{fw}}}{\partial z} = \frac{\text{i}2\pi^2\nu_{\text{Stokes}}}{n_{\text{Stokes}}c_0} \Delta N \frac{\partial \mu_{\text{vib}}}{\partial q_{\text{vib}}} q_{\text{fw}}^* E_{\text{inc}} - \frac{a}{2} E_{\text{fw}} \tag{4.180}$$

$$\frac{n_{\text{Stokes}}}{c_0} \frac{\partial E_{\text{bw}}}{\partial t} + \frac{\partial E_{\text{bw}}}{\partial z} = \frac{\text{i}2\pi^2\nu_{\text{Stokes}}}{n_{\text{Stokes}}c_0} \Delta N \frac{\partial \mu_{\text{vib}}}{\partial q_{\text{vib}}} q_{\text{bw}}^* E_{\text{inc}} - \frac{a}{2} E_{\text{bw}} \tag{4.181}$$

with the refractive indices n_{inc} and n_{Stokes} for the incident pump and the Stokes waves. The vibrational displacements for the Stokes Raman frequency for the two propagation directions are:

$$\frac{\partial q_{\text{fw}}}{\partial t} + \frac{1}{2\tau_{\text{vib}}} q_{\text{fw}} = \frac{\text{i}}{8\pi\nu_{\text{vib}}m} \frac{\partial \mu_{\text{vib}}}{\partial q} E_{\text{inc}} E_{\text{fw}}^* \tag{4.182}$$

$$\frac{\partial q_{\mathrm{bw}}}{\partial t} + \frac{1}{2\tau_{\mathrm{vib}}} q_{\mathrm{bw}} = \frac{i}{8\pi\nu_{\mathrm{vib}} m} \frac{\partial \mu_{\mathrm{vib}}}{\partial q} E_{\mathrm{inc}} E_{\mathrm{bw}}^* \tag{4.183}$$

If heavily damped vibrations are found, as in most experiments, the temporal derivatives of the displacements can be neglected:

$$\frac{\partial q_{\mathrm{fw}}}{\partial t} \ll \frac{q_{\mathrm{fw}}}{2\tau_{\mathrm{vib}}} \quad \text{and} \quad \frac{\partial q_{\mathrm{bw}}}{\partial t} \ll \frac{q_{\mathrm{bw}}}{2\tau_{\mathrm{vib}}}. \tag{4.184}$$

Introducing the intensities I_m of the different light beams the equations become:

SRS equations $\qquad \dfrac{n_{\mathrm{Stokes}}}{c_0} \dfrac{\partial I_{\mathrm{fw}}}{\partial t} + \dfrac{\partial I_{\mathrm{fw}}}{\partial z} = g_{\mathrm{Stokes}} I_{\mathrm{fw}} I_{\mathrm{inc}} - a I_{\mathrm{fw}}$

$$\frac{n_{\mathrm{Stokes}}}{c_0} \frac{\partial I_{\mathrm{bw}}}{\partial t} + \frac{\partial I_{\mathrm{bw}}}{\partial z} = g_{\mathrm{Stokes}} I_{\mathrm{bw}} I_{\mathrm{inc}} - a I_{\mathrm{bw}}. \tag{4.185}$$

The gain coefficient g_{SRS} follows with the relation $g_{\mathrm{Stokes}} = g_{\mathrm{inc}} \nu_{\mathrm{Stokes}} / \nu_{\mathrm{inc}}$ from:

$$\begin{aligned}
g_{\mathrm{SRS}} = g_{\mathrm{Stokes}} &= \frac{2c_0 N}{h\pi\nu_{\mathrm{vib}}^3 n^2 \Delta\overline{\nu}_{\mathrm{SRS}}} \left(\frac{\mathrm{d}\sigma_{\mathrm{SRS}}}{\mathrm{d}\Omega} \right)_{\|} \\
&= \frac{2c_0 N}{h\pi\nu_{\mathrm{vib}}^3 n^2 \Delta\overline{\nu}_{\mathrm{SRS}}} \frac{1}{1+\beta} \frac{\mathrm{d}\sigma_{\mathrm{SRS}}}{\mathrm{d}\Omega} \\
&= \frac{8\pi^2 \nu_{\mathrm{Stokes}} \Delta N \tau_{\mathrm{vib}}}{n^2 c_0^2 m \nu_{\mathrm{SRS}}} \left(\frac{\partial \mu_{\mathrm{vib}}}{\partial q_{\mathrm{vib}}} \right)^2 \frac{1}{1+\beta}
\end{aligned} \tag{4.186}$$

with the assumption of equal refractive indices $n \approx n_{\mathrm{Stokes}} \approx n_{\mathrm{inc}}$.

Thus the gain depends on two parameters: the cross-section and the vibrational lifetime or linewidth of the vibration. The cross-section σ_{SRS} can be calculated from:

$$\frac{\mathrm{d}\sigma_{\mathrm{SRS}}}{\mathrm{d}\Omega} = \frac{2\pi^2 \nu_{\mathrm{Stokes}} h}{c_0^4 m \nu_{\mathrm{SRS}}} \left(\frac{\partial \mu_{\mathrm{vib}}}{\partial q_{\mathrm{vib}}} \right)^2 \tag{4.187}$$

and the linewidth $\Delta\nu_{\mathrm{vib}}$ is related to the life time of the vibration τ_{vib} by:

$$\Delta\nu_{\mathrm{vib}} = \frac{1}{2\pi\tau_{\mathrm{vib}}} \tag{4.188}$$

These intensity equations are very similar to the SBS equations of the previous section. Therefore the numerical solutions show the same general results as the SBS calculations. The necessary SRS parameters are given in Table 4.10 for some typical materials. Strongly stimulated Raman scattering is reached in materials with large gains, meaning large dipole moments, and long lifetimes. Much stronger stimulated Raman scattering can be observed if the frequency of the incident laser is tuned closely to an absorption transition of the material. This resonance Raman scattering is used in many applications [see, for example, 4.478–4.480]. Saturation can be reached for high intensities and thus the SRS intensity can be 10^{11} times larger than spontaneous Raman scattering [4.464].

Anti-Stokes SRS can be calculated analogous to Stokes scattering with analogous formulas. A more explicit discussion is given in [4.477].

In the case of stationary Stokes scattering with almost undepleted incident light ($I_{fw} \ll I_{inc}$) and insignificant backward scattering ($I_{bw} \ll I_{fw}$) (4.185) can be solved to give:

stationary small SRS $\quad I_{fw}(z) = I_{fw}(z = 0)\, e^{(g_{SRS} I_{inc} z - az)}$ (4.189)

which is again analogous to SBS. The intensity $I_{fw}(z = 0)$ is given by the spontaneous Raman scattering at the sample entrance at $z = 0$.

A most important application of SRS is frequency conversion. A number of useful SRS materials (see, e.g. Table 3.8 and Sect. 6.15.3) can generate a wide range of new laser frequencies based on the emission wavelengths of nontunable solid-state lasers and their harmonics. Efficiencies of more than 25% were reported. Commercial solutions are available (see Sect. 6.15.3).

If the SRS cell is placed inside a suitable resonator the Stokes or anti-Stokes signal gain may be sufficient for laser action at the wavelengths $\nu_{Stokes} = \nu_{inc} - \nu_{vib}$ or $\nu_{anti\text{-}Stokes} = \nu_{inc} + \nu_{vib}$. Such Raman lasers may be important light sources in certain cases for wavelengths which are otherwise difficult to obtain (see Sect. 6.15.3).

On the other hand SRS is used for spectroscopic investigations. Material analysis is done qualitatively and quantitatively. The determination of vibrational frequencies allows the analysis of intramolecular and intermolecular forces. Using high-power laser light for exciting the samples via the electronic transition, these Raman investigations can be carried out in the excited state of the material too [see, for example, 4.481–4.484]. Thus the detailed structure of these short-lived states can be investigated by measuring the vibrational frequencies.

Another important application field may be based on surface-enhanced Raman spectroscopy (SERS) [see, for example, 4.485–4.487]. Enhancement factors for the Raman signal of up to 10^{20} were reported for molecules on cluster surfaces [4.485]. This may enable single molecule detection via Raman spectroscopy with very high selectivity.

From measurements of the anti-Stokes lines the temperature distribution of samples can be determined with this optical method, e.g. inside combustion engines [e.g. 4.489, 4.498]. The intensity of these lines can, to a good approximation be modeled via the occupation densities N_m of the vibronic states m with frequencies ν_m which are given by the Boltzmann distribution:

$$I(\nu_m) \propto e^{-h\nu_m/kT}$$
(4.190)

Inverse Raman Spectroscopy (IRS)

Inverse Raman spectroscopy (IRS) is obtained if the depletion of a weak probe light signal in the linear intensity range at the frequency of the SRS pump transition is measured while a strong laser is tuned across the Stokes frequency of the matter as schematically shown in Fig. 4.40 [M10].

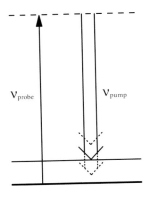

$$T_{\nu_{\text{probe}}} = f(\nu_{\text{pump}})$$

Fig. 4.40. Inverse Raman spectroscopy (IRS) with measurement of the depletion of a weak probe signal at the frequency of the SRS pump as a function of the frequency of a strong and tunable laser at the Stokes wavelength

Stimulated Raman Gain Spectroscopy (SRGS)

Another spectroscopic possibility is the measurement of the amplification via the gain coefficient g_{probe} of a weak tunable probe signal with the light frequency around the Raman Stokes signal while strong laser pumping with a suitable frequency for this Stokes signal (see Fig. 4.41) [M10]. This method is called stimulated Raman gain spectroscopy (SRGS).

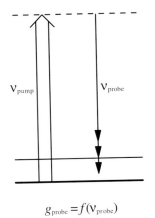

$$g_{\text{probe}} = f(\nu_{\text{probe}})$$

Fig. 4.41. Stimulated Raman gain spectroscopy (SRGS) measuring the amplification for determining the gain g_{probe} of a weak tunable probe signal around the SRS Stokes frequency under strong pumping

Coherent Anti-Stokes Raman Scattering (CARS)

The combination of simultaneous stimulated Stokes and anti-Stokes Raman scattering leads to the interaction of four photons in the matter [4.502–4.523].

In coherent anti-Stokes Raman scattering (CARS) two strong laser beams with frequencies ν_{inc} and $\nu_{\text{SRS,S}}$ are applied (see Fig. 4.42).

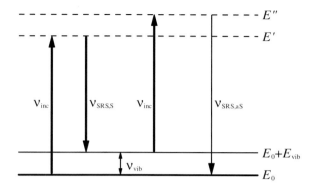

Fig. 4.42. Coherent anti-Stokes Raman scattering (CARS) pumping with two laser beams with frequencies ν_{inc} and $\nu_{\text{SRS,S}}$, obtaining the anti-Stokes Raman light with $\nu_{\text{SRS,aS}}$. For strong signals phase matching has to be achieved

The coherent scattering of these photons in CARS can be applied with very short pulses and allows highly sensitive measurements if phase matching is achieved. In addition a high spatial resolution in the μm range is possible. CARS is a four-wave mixing (FWM) process which is described in general in Sect. 5.9.2.

Phase matching is achieved if the momentum of the four attended photons are conserved and thus the wave vectors of the incident laser light $\boldsymbol{k}_{\text{inc}}$ and of the Raman Stokes light $\boldsymbol{k}_{\text{Stokes}}$ and the anti-Stokes light $\boldsymbol{k}$ have to fulfill the angle condition of Fig. 4.43.

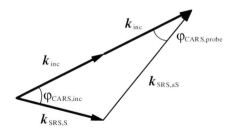

Fig. 4.43. Phase matching of the incident laser light and the generated Stokes and anti-Stokes Raman light in CARS experiments

Therefore the two incident laser beams have to be enclosed in the angle $\varphi_{\text{CARS,inc}}$ and the anti-Stokes Raman light beam can be observed at the angle $\varphi_{\text{CARS,probe}}$ to the laser with $\boldsymbol{k}_{\text{inc}}$ in the forward direction.

Thus CARS allows highly sensitive measurements of the anti-Stokes Raman signal in a spatial direction with no background light. The two strong pump lasers will populate the excited vibrational level and therefore highly efficient anti-Stokes Raman scattering will occur.

The scattering intensity $I_{\text{CARS,aS}}$ is proportional to:

$$I_{\text{CARS,aS}} \propto I_{\text{inc}}^2 I_{\text{SRS,S}} N_{\text{mat}}^2 \tag{4.191}$$

with the pump laser intensities I_{inc} of the incident and $I_{\text{SRS,S}}$ of the tuned light and particle density N_{mat}. Even continuously operating (cw) lasers can be used and then very high spectral resolution is possible.

The scattering efficiency can be increased by many orders of magnitude if the pump laser energy matches the electronic transitions of the material (*resonant CARS*) [e.g. 4.523]. In this case the virtual Raman levels of the energy schemes above will be real energy states of the matter. Absorption will take place and thus the interaction length and/or concentration are limited by the maximum optical absorption of approximately $\sigma_{\text{pump}} N L < 1$.

BOX CARS

If the CARS scattering angles are too small for safe splitting of the different signals the BOX CARS technique can be used [e.g. 4.524, 4.525]. The incident laser beam is therefore split into two beams which are applied at the angle $\varphi_{\text{BOX CARS}}$ as shown in Fig. 4.44.

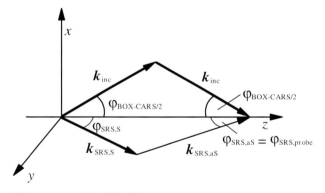

Fig. 4.44. BOX CARS angle conditions for phase matching allowing good spatial separation of the different signals

Assuming as in Fig. 4.44 the two pump laser beams with frequency ν_{inc} are applied at the angle $\varphi_{\text{BOX CARS}}$ symmetrically in the xz plane each making the angle $\varphi_{\text{BOX CARS}}/2$ each with the z axis the third laser beam with frequency $\nu_{\text{SRS,S}} = \nu_{\text{inc}} - \nu_{\text{vib}}$ can be applied in the yz plane at the angle $\varphi_{\text{SRS,S}}$ with the z axis. The resulting angle $\varphi_{\text{SRS,aS}}$ for detecting the newly generated anti-Stokes Raman light with frequency $\nu_{\text{SRS,aS}}$ can be calculated from:

$$\varphi_{\text{SRS,aS}} = \arccos \left\{ \frac{1}{1 + (\nu_{\text{vib}} \lambda_{\text{inc}}/c)} \right.$$

$$\left. \cdot \left[2 \cos \left(\frac{\varphi_{\text{BOX CARS}}}{2} \right) - \left(1 + \frac{\nu_{\text{vib}} \lambda_{\text{inc}}}{c} \right) \cos \varphi_{\text{SRS}} \right] \right\}. \quad (4.192)$$

The three beams with their different directions have to overlap in the sample. Other geometrical arrangements as in Fig. 4.44 are possible. Therefore this technique allows a wide range of different experimental setups for analytical and spectroscopic investigations [M10].

4.5.14 Optical Phase Conjugation via Stimulated Scattering

The conjugation of the phase of an optical wave is equivalent to inversion of the wave front of the light beam [4.526–4.529]. It allows the realization of phase conjugating mirrors (PCM's) which are, for example, used in lasers for improving the beam quality, as described in Sects. 6.6.12 and 6.11.3 [4.530]. Further examples can be found in [4.531–4.590]. If the incident light beam is described by:

incident light $\quad \boldsymbol{E}_{\text{inc}}(\boldsymbol{r}, t) = \text{Re}\{\boldsymbol{E}_0(\boldsymbol{r}) \, e^{i2\pi\nu t}\}$ (4.193)

with the complex amplitude

$$\boldsymbol{E}_0(\boldsymbol{r}) = \boldsymbol{A}_0(\boldsymbol{r}) \, e^{-i(kr+\varphi)} \quad (4.194)$$

the phase conjugate is given by:

phase conjugate $\quad \boldsymbol{E}_{\text{inc}}(\boldsymbol{r}, t) = \text{Re}\{\boldsymbol{E}_0^*(\boldsymbol{r}) \, e^{i2\pi\nu t}\}$ (4.195)

with the complex conjugated amplitude:

conjugated amplitude $\quad \boldsymbol{E}_0^*(\boldsymbol{r}) = \boldsymbol{A}_0(\boldsymbol{r}) \, e^{+i(kr+\varphi)}$ (4.196)

where the sign of the spatial phase is changed. The conjugate can also be written as:

$$\boldsymbol{E}_{\text{inc}}(\boldsymbol{r}, t) = \text{Re}\{\boldsymbol{E}_0(\boldsymbol{r}) \, e^{-i2\pi\nu t}\} \quad (4.197)$$

with the unchanged spatial amplitude:

$$\boldsymbol{E}_0(\boldsymbol{r}) = \boldsymbol{A}_0(\boldsymbol{r}) \, e^{-i(kr+\varphi)} \quad (4.198)$$

but there is a change of sign of the temporal phase. This corresponds formally to a change in time direction indicating that the phase front is moving perfectly backwards. It does not indicate a time direction change for the pulse shape or in general!

This phase conjugation can be achieved with nonlinear back-reflection of the beam via stimulated scattering, e.g. via stimulated Brillouin scattering (SBS) as described in Sect. 5.9.2 or via four-wave mixing (FWM) (see e.g. [4.531–4.541]). Such a volume reflector is called a phase conjugating mirror

Fig. 4.45. Optical phase conjugation of a frequency-doubled Nd:YAG laser beam with a pulse duration of 20 ns and a pulse energy of 20 mJ via SBS in a cell filled with acetone. The beam was photographed in smoky air. The incident light propagated from the left into the cell and is focused by the curved cell surface. The phase conjugated light is reflected towards the observer by the beam splitter (Foto: Menzel)

(PCM). Stimulated Brillouin scattering such as a self-pumped process allows very easy realization of this process as can be seen in Fig. 4.45.

The scheme of this setup is given in Fig. 4.46. The second harmonic light of a pulsed Nd:YAG laser beam with a duration of 20 ns and a pulse energy of 20 mJ was focused by the curved entrance window of the glass cell into the liquid acetone. As can be seen in the picture, this material is transparent for low light powers at this wavelength in the green spectral region. In the focus the sound wave grating of the SBS is established and reflects in this simple demonstration experiment about 50% of the incident light as a PCM. The reflected light observable behind the beam splitter shows the same properties as the incident light beam although the imperfect cell window introduces severe phase distortions.

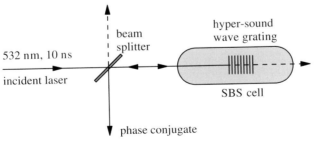

Fig. 4.46. Schematic of the setup of Fig. 4.45

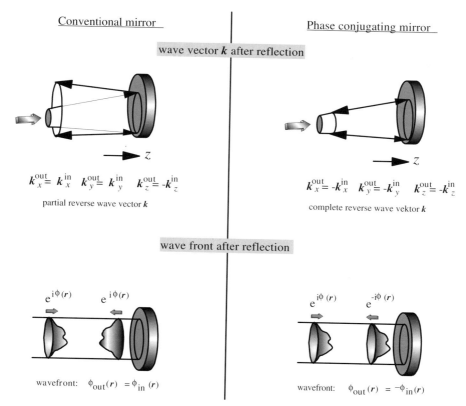

Fig. 4.47. Reflectivity properties of a phase conjugating mirror (PCM) in comparison to conventional mirrors

Such a phase conjugating mirror shows unusual properties compared to a conventional mirror (see Fig. 4.47) with important applications in photonics especially in laser technology.

Most obviously the light is perfectly back-reflected by the PCM independent of the direction of the incident light beam. Thus not only is the z component of the wave vector inverted but so are all components in the PCM. More precisely the complete wave front is inverted in the PCM with the important consequence of possible compensation of phase distortions in optical elements by applying the PCM as shown in Fig. 4.48.

A third important property of the PCM is the treatment of the polarization of the light (see Fig. 4.49).

In PCMs the linear polarization of light is unchanged and for many PCM processes linear polarized light is most efficient. In perfect phase conjugating mirrors the circular or elliptic polarization is conserved. Thus this type is called *vector phase conjugation*. It can be achieved in four-wave mixing

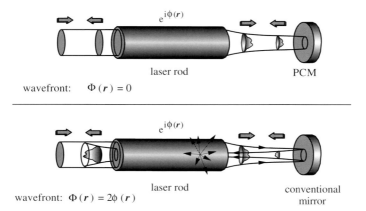

Fig. 4.48. Schematic for compensation of phase distortions via a double pass arrangement with an optical phase conjugating mirror (PCM)

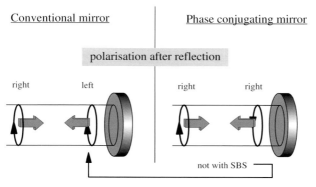

Fig. 4.49. Polarization of a light beam after reflection in a phase conjugating mirror (PCM) in comparison to a conventional mirror

(FWM) schemes, only. Phase conjugating mirrors based on stimulated scattering, e.g. SBS, will not conserve the spin direction of circular or elliptic polarization because the stimulation of the sound wave is an intensity effect. Thus PCMs based on SBS will treat light polarization in the same way as conventional mirrors. As a consequence double pass arrangements with SBS-PCM can be very easily build with a polarizer and a Faraday rotator as shown in Sect. 6.11.3.

The phase conjugating mirror based on the stimulated Brillouin scattering can be applied for almost perfect compensation of the phase distortions as they are caused by highly pumped solid state laser rods using a double pass arrangement as shown in Fig. 4.50.

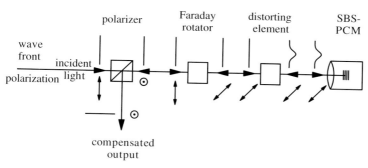

Fig. 4.50. Schematic for compensation of phase distortions of an optical "distorting" element by a double pass arrangement using a Faraday rotator for out-coupling the light using a SBS-PCM

Compensation of the phase distortions takes place if the disturber does not change during the round trip of the light via the PCM which is usually fulfilled if thermal processes are involved.

Thus PCMs based on stimulated scattering cannot compensate for distortions of polarization and amplitude in the disturbing element! For compensation of depolarization the more complicated scheme of Basov [4.531] can be used as shown in Fig. 4.51.

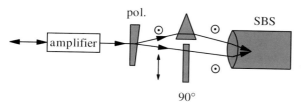

Fig. 4.51. Double pass amplifier schematic with phase conjugating SBS-mirror and compensation for depolarization in the amplifier material

Unfortunately this scheme demands quite a long distance between the amplifier and the SBS cell for the polarizer and the other elements and thus no short thermal lensing from the amplifier (see Sect. 6.4.1) can be compensated with this arrangement.

Why allows reflection via SBS optical phase conjugation? As described in Sect. 4.5.9 the sound wave grating is stimulated by the intensity inference pattern of the incident and the reflected light. Thus these intensity maxima are higher the better the overlap of the wave fronts of the incident and the reflected beams. Therefore the reflection will reach its highest values if the wave front of the reflected light is identical to the wave front of the incident light although the propagation direction is inverted. This is optical phase conjugation.

This effect can be described mathematically under the assumption of a mode mixture for both the incident and reflected light:

$$E_{\text{inc}}(\boldsymbol{r}) = \sum_j C_{\text{inc},j}(z) A_{0,\text{inc},j}(\boldsymbol{r}) \tag{4.199}$$

$$E_{\text{scatt}}(\boldsymbol{r}) = \sum_j C_{\text{scatt},j}(z) A_{0,\text{scatt},j}(\boldsymbol{r}) \tag{4.200}$$

with the electric fields E_{inc} and E_{scatt} for the incident and back-scattered light. The coefficients $C_{\text{inc/scatt},j}(z)$ determine the share of the jth mode with the shape $A_{0,\text{inc/scatt},j}$. They describe a complete orthogonal system of modes which are solutions of the wave equation:

$$\frac{\partial A_{0,\text{inc/scatt},j}(\boldsymbol{r})}{\partial z} + \left(\frac{\partial}{\partial x} + \frac{\partial}{\partial y}\right) A_{0,\text{inc/scatt},j}(\boldsymbol{r}) = 0 \tag{4.201}$$

and

$$\int_{-\infty}^{\infty} A_{0,\text{inc/scatt},j}(\boldsymbol{r}) A_{0,\text{inc/scatt},k}^*(\boldsymbol{r}) \, \mathrm{d}x \, \mathrm{d}y = \delta_{jk} \tag{4.202}$$

Substituting these electric fields in the SBS equations (4.127) and (4.128) leads under stationary conditions to:

$$\sum_j A_{0,\text{inc},j} \frac{\partial C_{\text{inc},j}}{\partial z} = -\frac{g_{\text{SBS}}}{2\tau_{\text{sound}}}$$

$$\cdot \sum_{klm} C_{\text{scatt},k}^* A_{0,\text{scatt},k}^* C_{0,\text{scatt},l} A_{0,\text{scatt},l} C_{\text{inc},m} A_{0,\text{inc},m} \tag{4.203}$$

and

$$\sum_j A_{0,\text{scatt},j} \frac{\partial C_{\text{scatt},j}}{\partial z} = -\frac{g_{\text{SBS}}}{2\tau_{\text{sound}}}$$

$$\cdot \sum_{klm} C_{\text{inc},k}^* A_{0,\text{inc},k}^* C_{\text{inc},l} A_{0,\text{inc},l} C_{\text{scatt},m} A_{0,\text{scatt},m}. \tag{4.204}$$

Because of the orthonormality of the basis system modes this equations can be simplified by multiplying each with the conjugate of one of these modes ($A_{0,\text{inc},n}^*$ and $A_{0,\text{scatt},n}^*$) and integrating over the whole space:

$$\frac{\partial C_{\text{inc},n}}{\partial z} = -\frac{g_{\text{SBS}}}{2\tau_{\text{sound}}} \sum_{klm} C_{\text{scatt},k}^* C_{\text{scatt},l} C_{\text{inc},m}$$

$$\cdot \int_{-\infty}^{\infty} A_{0,\text{scatt},k}^* A_{0,\text{scatt},l} A_{0,\text{inc},m} A_{0,\text{inc},n}^* \, \mathrm{d}x \, \mathrm{d}y \tag{4.205}$$

$$\frac{\partial C_{\text{scatt},n}}{\partial z} = -\frac{g_{\text{SBS}}}{2\tau_{\text{sound}}} \sum_{klm} C^*_{\text{inc},k} C_{\text{inc},l} C_{\text{scatt},m}$$

$$\cdot \int_{-\infty}^{\infty} A^*_{0,\text{inc},k} A_{0,\text{inc},l} A_{0,\text{scatt},m} A^*_{0,\text{scatt},n} \, \mathrm{d}x \, \mathrm{d}y \qquad (4.206)$$

As can be seen from these equations the build up of the reflected light by scattering is a function of the overlap of the wave fronts of the incident and scattered light. In cases with similar phase fronts:

$$\text{similar phase fronts} \quad A_{0,\text{scatt},l}(\boldsymbol{r}) \approx A_{0,\text{inc},m}(\boldsymbol{r}) \qquad (4.207)$$

reflectivity will be higher. If especially the incident light consists of one mode A_0 with $A_{0,\text{inc},0} = A_{0,\text{scatt},0}$, only, the increase of the reflected light in the backward $(-z)$ direction follows from:

$$\frac{\partial C_{\text{scatt},m}(z)}{\partial z} = -|C_{\text{inc},0}|^2 \sum_{l} g_{\text{SBS,nl}}(z) C_{\text{inc},l}(z) \qquad (4.208)$$

with the stationary gain coefficient $g_{\text{SBS,nl}}$:

$$g_{\text{SBS,nl}}(z) = -\frac{g_{\text{SBS}}}{2\tau_{\text{sound}}}$$

$$\cdot \int_{-\infty}^{\infty} |A_{0,\text{scatt},0}(\boldsymbol{r})|^2 A_{0,\text{scatt},n}(\boldsymbol{r}) A^*_{0,\text{scatt},l}(\boldsymbol{r}) \, \mathrm{d}x \, \mathrm{d}y. \qquad (4.209)$$

The gain coefficient $g_{\text{SBS,00}}$ for phase conjugate reflection is larger than all others, e.g. by more than a factor of 2 [4.545]. Because of the highly nonlinear increase of the reflected light this mode will be dominant. More theoretical modeling can be found in [4.526–4.529, 4.542–4.547].

As mentioned in Sect. 4.5.9 the threshold of the nonlinear phase conjugating mirror based on SBS can be reduced by using waveguide structures [4.402–4.405, 4.548–4.558]. Using long fibers for high reflectivities the coherence length of the applied light has to be large enough. This can demand values in the km range [see, for example, 4.548]. With the taper concept (see Fig. 4.37 and [4.402]) it was possible to combine low threshold with a short coherence length and a large dynamic range as well as good fidelity.

The quality of phase conjugation is given by the *fidelity* F [4.351]:

$$\textbf{fidelity} \quad F = \frac{\left| \int E_{0,\text{scatt}}(\boldsymbol{r}) E_{0,\text{inc}}(\boldsymbol{r}) \, \mathrm{d}x \, \mathrm{d}y \right|^2}{\int |E_{0,\text{scatt}}(\boldsymbol{r})|^2 \, \mathrm{d}x \, \mathrm{d}y \int |E_{0,\text{inc}}(\boldsymbol{r})|^2 \, \mathrm{d}x \, \mathrm{d}y}. \qquad (4.210)$$

This correlation function F is 1 for perfect phase conjugation and 0 for random scattering. Values above 95% are possible (see Fig. 4.53). More detailed discussion can be found in [4.526, 4.527, 4.351, 4.541, 4.543, 4.545, 4.559–4.569].

The fidelity of phase conjugation of Gaussian beams is measured frequently with the "energy in the bucket" method [4.563] (see Fig. 4.52).

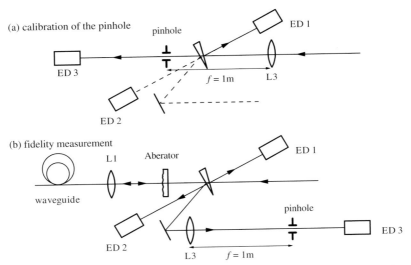

Fig. 4.52. Fidelity measurement for phase conjugation of Gaussian beams with "energy in the bucket" method (b). ED1–3 measures the energy of the incident E_1, the reflected E_2 and the phase conjugated light E_3. ED3 with pinhole has to be calibrated with the incident beam without aberrator (a) by measuring $E_{1,\text{calib}}$ and $E_{3,\text{calib}}$ to determine the factor $C_{\text{calib},F}$

The fidelity is then calculated from:

$$F = C_{\text{calib},F} \frac{E_3}{E_2} \quad \text{with} \quad C_{\text{calib},F} = \frac{E_{1,\text{calib}}}{E_{3,\text{calib}}} \tag{4.211}$$

which can reach values larger than 90% over a wide range of incident light pulse energies as shown in Fig. 4.53.

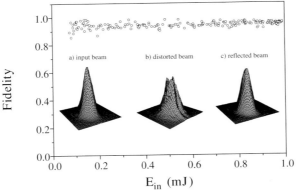

Fig. 4.53. Fidelity of phase conjugating mirror as shown in Fig. 4.37 as a function of the energy E_{in} of the incident pulse. The transversal pulse profile shows the deviations from Gaussian shape (a) after introducing phase distortions (b) and after compensation (c)

This experimentally very easily determinable fidelity value can be slightly larger than 1 if the beam diameter of the reflected light is smaller than that of the incident beam. This can occur, e.g. at low powers close to threshold in SBS resulting in very low reflectivities.

Further applications of SBS-PCM's are given in [4.570–4.590] in addition to the examples in Sect. 4.5.9 and in Sects. 6.6.12 and 6.11.3.

4.6 Higher-Order Nonlinear Effects

Nonlinear effects of higher order than 2 or 3 are present in many high-power laser experiments but are not dominant and are thus difficult to detect. Explicitly reported are the generation of higher-order frequency harmonics mainly in noble gases [4.591–4.640]. For technical applications the stepwise frequency transformation based on second and third harmonic effects is usually more efficient. This is a consequence of the much larger nonlinear coefficients $\chi^{(2)}$ and $\chi^{(3)}$ compared to, e.g. $\chi^{(4)}$, $\chi^{(5)}$ and so on for known materials. Further these materials have to be transparent over a wide spectral range because absorption will decrease the efficiency. On the other hand the resonance effect working with wavelengths close to matter absorption can increase the nonlinear effect drastically [e.g. 4.591]. Therefore a suitable compromise will enhance the harmonic output.

The usually applied atom vapors for generation of higher frequencies do not automatically give phase matching. Thus by tuned mixing of different atoms with different refractive indices at the wavelength of the fundamental and the high harmonics, phase matching can be achieved in isotropic materials, e.g. [4.592–4.598].

As an example the generation of the fifth harmonic in Ne vapor is described, which in combination with the generation of twice the second harmonic, finally results in the generation of the 20th harmonics of the original Nd:YAG laser light [M31]. The process is depicted in Fig. 4.54.

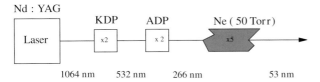

Fig. 4.54. Generation of the fifth harmonic in Ne vapor by pumping with the fourth harmonic of a Nd:YAG laser resulting in the 20th harmonic of the laser radiation

The efficiency in this experiment was less than 10^{-6} although it was performed close to the resonance of Ne atomic absorption and thus the nonlinear coefficient is distinctly enlarged.

Another example is the reported seventh harmonic generation of the radiation of a Krypton fluoride excimer laser with a wavelength of 248 nm in He vapor [4.218]. The resulting seventh harmonic shows a wavelength of 35.4 nm.

The generation of even higher harmonics for generating coherent light at wavelengths below 20 nm was obtained using very high powers in the range of TW and more with short pulses in the ps or fs range. The fourth-order nonlinear processes are discussed in [4.599–4.602], the generation of the fifth-harmonic in [4.603–4.609] and of the seventh-harmonic in [4.218, 4.610]. Much higher harmonics are observed in [4.611–4.640]. For example, in [4.594] the 221th-harmonic as discrete harmonic peak of coherent light with a wavelength of 3.6 nm was observed in He using a high-power Ti:sapphire laser pulse with a width of 26 fs, an energy of 20 mJ, a wavelength of 800 nm and a focal spot diameter of 100 µm resulting in an intensity of 6×10^{15} W cm^{-2}. Ne or He gas was used at 8 Torr. Coherent emission was observed up to the 297th harmonic of the laser light corresponding to a wavelength of 2.7 nm. Further references can be found in Sects. 6.13.5 and 6.15.1. Another technique is based on a seeded free electron laser allowing the generation of laser light with wavelengths in the nm range [4.618].

XUV generation from laser-induced plasmas should be mentioned although it is not a frequency conversion technique. The very intense light excites atoms which emit light, e.g. in the spectral range of the "water window" between 2 and 4 nm which is important for applications. These point sources are very useful for lithography and X-ray microscopy (see Sect. 1.5).

4.7 Materials for Nonresonant Nonlinear Interactions

Although many materials for applications of nonresonant nonlinear optical effects in photonics are known [4.641–4.655, M32, M33] and to some extent used in commercial devices there is still a need for better suitable materials with higher nonlinear coefficients, higher damage threshold, lower costs and higher reliability. This is especially true for wavelengths in the IR above 1.2 µm and for short wavelengths below 0.3 µm. Therefore new materials of all kinds can be expected in the next few years; and information can be obtained from scientific publications, suppliers and their catalogs.

4.7.1 Inorganic Crystals

Crystals are used for all kinds of frequency transformation technologies such as harmonic generation, frequency mixing and electro-optical effects. They may be classified into two groups:

- *grown from solution*: These crystals are hygroscopic and in contrast to the group below thermal shock sensitive, mostly fragile and comparatively soft. But they are available in large sizes of good optical quality and are mostly cheaper.
- *grown from melt*: These crystals are nonhygroscopic and thus much more useful than those above.

Known materials are, e.g. KDP, KD*P, ADP, AD*P and LiNbO$_3$. More recently developed crystals are KTP, CDA, CD*A, RDA, RDP, BBO, LBO and BANANA. the parameters of these materials are available in Sect. 4.4, 6.15 and in the references [e.g. M25, M32, 4.643, 4.644].

4.7.2 Organic Materials

Organic materials can show very high nonlinear coefficients, high damage threshold and good transparency at short wavelengths [M4, M23, M32 and references in Sect. 5.10.1]. Because of the large variety of these compounds an inestimable number of possibilities exist in principle. Molecules with a large conjugated π-electron system from a large number of multiple bonds will show a large inducable dipole moment from these delocalized electrons. This can even be enhanced by donor (N-atoms) and acceptor (O-atoms) groups.

Some of these organic molecules can be crystallized with sufficient optical quality. Known examples of crystals without an inversion center are urea [M25], DAN, MNA, MAP, COANP, PAN and MBANP.

The inversion symmetry can be broken by applying these materials at surfaces. SHG with high efficiencies was demonstrated this way. Liquid crystals are especially applied in such setups.

Amorphous organic matter will find new applications in photonics as in optical fibers, in optical switches and storage or in optical phase conjugation. Many polymer materials have been proposed and are still used.

The main problem up to now is the long-term stability of these systems. Limited photo-stability and possible chemical reactions restrict their application. New compounds will hopefully not be so restricted in the future. Using a well-designed resonance enhancement by tuning the absorption of these compounds for the desired wavelengths will allow much higher nonlinearities (see Sect. 5.10.1).

4.7.3 Liquids

Organic liquids or solutions are applied in nonresonant photonic applications for "white light" generation (see Sect. 7.7.5), optical phase conjugation (see Sect. 4.5.9 and 4.5.14) and Raman shifting (see Sect. 4.5.13 and 6.15.3) of the incident light. These are different from the nonlinear absorbers and laser materials for dye lasers in resonant applications, and in this case the transparent matter operates again by its induced dipole moments, based on the electron distribution in different electronic or vibrational states of the organic molecules in a similar way as that just described above.

Useful materials are, e.g. CS_2, CCl_4, $TiCl_4$, Freon, hexene, benzene, alcohol and almost all other solvents as described in Sect. 4.5.9. For nonlinear applications with high light powers the chemically specified purity is sometimes not sufficient. Small particles can disturb the nonlinear interaction and promote optical break down by the resulting inhomogeneous high local field. Therefore lavish cleaning with filters or "pump and freeze" procedures may be necessary before use [4.335].

4.7.4 Liquid Crystals

The geometrical orientation and order of molecular systems in liquid crystals [M23, 4.647–4.655] can be applied in photonics for changing the polarization of a transmitting light beam. This is achieved in liquid crystal displays and projectors based on electro-optically switching the orientation of the molecules by an external electric field. But the orientation of the molecules in the liquid crystal can be changed via polarized pump light, too, and thus opto-optical switching becomes possible. Furthermore, liquid crystals can be used for frequency conversion and four-wave mixing techniques (see Sect. 1.5 and 5.9).

4.7.5 Gases

Noble gases and gases of organic molecules are used in a way similar to liquids for optical phase conjugation and Raman shifting. The mechanisms are the same as described for liquids and references are given there. Gases show the advantage of easy "self-repairing" if damage threshold is exceeded and, e.g. optical breakdown occurred. Thus high-power applications are possible with nonlinear processes in gases.

Typical applied nonlinear gases are SF_6, N_2, Xenon, CH_4, C_2F_6, CO and CO_2 used at pressures of 10–100 bar. The damage threshold in gases is mostly determined by impurities. It can be improved by at least one order of magnitude by cleaning, e.g. with high electric fields [4.656].

5. Nonlinear Interactions of Light and Matter with Absorption

As described in the introduction of Chap. 4, including Sects. 4.1 and 4.2. nonlinear interactions of light with matter are of fundamental importance for photonic applications. It may be worth with reading these sections before continuing.

All matter shows some absorption in almost all spectral regions as a consequence of the Lorentzian line shape of the electronic transitions with indefinite wings. But if the absorption coefficient is smaller than about $10^{-6}\,\mathrm{cm}^{-1}$ the share of the resonant nonlinear interaction can often be neglected. This nonlinear nonresonant light-matter interaction is described in the previous chapter.

But many photonic applications are based on resonant nonlinear interactions such as, e.g. stimulated emission in lasers or passive Q-switching and mode locking. Further applications such as optical switching and storage may become important, based on nonlinear absorbing devices. In any case the resonance enhancement of very weakly absorbing materials may promote nonresonant nonlinear effects by strongly increased nonlinear coefficients.

Therefore detailed knowledge about nonlinear absorption, which is also known as nonlinear transmission and transient absorption effects, and their experimental and theoretical evaluation is essential for successful operation of nonlinear photonic devices in both resonant and nonresonant cases.

Although the resonant nonlinear interaction is always accompanied by nonresonant effects (as vice versa) under conditions of strong absorption the nonresonant part may be neglected. Therefore in this chapter the resonant interaction of light with matter will be described first and combined interactions at the end. For further reading the books [M1, M2, M4, M6, M7, M9–11, M13, M14, M22, M23, M26, M28–30, M34–37, M40–43, M45–52] can be recommended.

5.1 General Remarks

In general the nonlinear interaction can be described by the formulas given in Sect. 4.3 based on the nonlinear polarization of matter as a function of the electric field strength if all values are used as complex values. The imaginary part of the χ-tensor and the resulting imaginary part of the refractive index

n will contribute to the absorption. This method may be useful in cases of small absorption with little structure in the required spectral range.

But if nonlinear absorption effects are dominant other descriptions which consider the detailed structure of the energy levels of the matter, their transition moments and the relaxation times between them may be more useful. Again it is worth distinguishing between coherent and incoherent interactions (see Fig. 5.1).

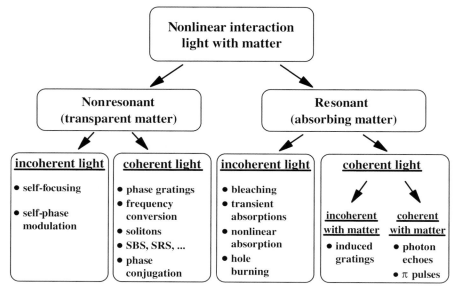

Fig. 5.1. Scheme of nonlinear interactions of light with matter which may be transparent or not at the required wavelengths

Nonlinear interactions in absorbing matter can show two levels of coherence: First, the used light fields can be coherent and thus they can produce nonlinear absorption gratings. Second, the induced dipole moment in the matter can oscillate in phase with the applied electric field. Quantum effects such as the generation of π-pulses and photon echoes can occur. Precondition for this complete coherent interaction is an internal phase coherence time of the matter, T_2, longer than the relevant experimental time, e.g. the pulse width.

5.2 Homogeneous and Inhomogeneous Broadening

In many cases matter absorption shows broad bands over a few nm up to few 100 nm. In particular mixtures of organic molecules such as, e.g. those used in dye lasers or for Q-switching and mode locking may have broad absorption bands [M4]. These bands which are easily observable with conventional UV-Vis spectrometers as the sum spectra of all participating particles in the matter. Thus for a single particle the *absorption lines may be shifted or broadened by*:

- particle–environment interactions;
- particle–particle interactions;
- combined transitions;
- Doppler shifts.

The following mechanisms may cause *additional broadening* of the observed optical absorption and emission bands of crystals or of molecular systems:

- combinations of electronic transitions;
- combinations of electronic transitions with a large number of possible simultaneous vibrational transitions;
- combinations of electronic transitions with conformational transitions of the molecules;
- participation of rotational transitions;
- molecule–solvent (intermolecular) interaction;
- molecule–molecule interactions (aggregation);
- slightly different conformations or chemical structure of the particle.

The resulting sum spectrum of the sample will show broad spectra. In contrast to conventional optical experiments under steady-state conditions with low intensities in nonlinear optics the broadening mechanisms of the absorption and emission bands can be very important.

In particular we have to be able to distinguish whether these bands are spectrally homogeneously or inhomogeneously broadened with respect to the conditions of the application or the experiment.

Homogeneously broadened absorption or emission bands change their amplitude but not their structure during excitation. *Inhomogeneously broadened* bands can change structure and amplitude under excitation.

In the case of homogeneously broadened absorption or emission bands each particle such as, e.g. the molecules shows the same absorption spectrum and therefore the sum spectrum of the sample has the same shape as the spectra of the single particles as shown in Fig. 5.2.

Inhomogeneously broadened absorption or emission bands can be caused by slightly different particle states, e.g. in slightly different environments or in different vibrational states, and then the matter is called spectrally inhomogeneously broadened. If the particles are inhomogeneously broadened as shown in Fig. 5.3 each particle shows a shifted absorption spectrum. This

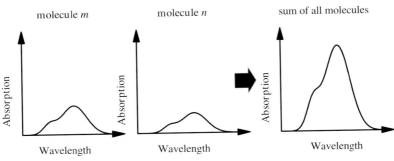

Fig. 5.2. Spectra of homogeneously broadened particle ensemble. Each particle shows the same absorption and the sum spectrum has the same shape as the single spectra

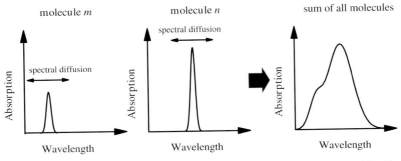

Fig. 5.3. Spectra of inhomogeneously broadened molecular ensemble: Each molecule shows a different absorption spectrum which is, e.g. shifted by environmental influences and the sum spectrum is broader than each of the single spectra

can be produced, e.g. by particle–environment interactions with slightly different arrangements of the environment around the sample particle. It can also be caused, e.g. by molecules in slightly different conformational or vibrational states. As a result the sum spectrum of the sample is broader than the spectrum of the single particle.

It is important to notice that between the different species of particles in the sample exchange processes take place. The characteristic time is called

spectral cross relaxation time T_3 (5.1)

or sometimes in molecular systems it is called the internal vibrational relaxation (IVR) time T_{IVR}. Thus in time-averaged measurements inhomogeneously broadening will often not be observable.

> *Inhomogeneous broadening* of absorption or emission bands is a function of the time scale. It occurs for characteristic experimental times shorter than the spectral cross-relaxation time, only.

The spectral cross-relaxation times are usually in the order of sub ps and some times longer. (For more details see Sect. 5.4.1 and the references therein.) In this case the inhomogeneous broadening can not be observed in conventional spectroscopy. It may be very important in nonlinear optical experiments if the characteristic time scale is short. Spectroscopic experiments, laser action, optical switching and nonlinear parameter determination may depend crucially on the inhomogeneous or homogeneous broadening of the material and the characteristic times of the investigation.

As will be shown in Chap. 7 the characteristic time scale is not always given by the pulse duration. It can be determined by the inverse of the characteristic pump rate of the nonlinear experiment (σI). This inverse pump rate can easily be in the range of sub-ps even in experiments with ns-laser pulses. Experimental techniques to investigate the homogeneous or inhomogeneous broadening of absorption bands are fractional bleaching (see Sect. 7.8.1), nonlinear polarization spectroscopy (see Sect. 7.8.2) and spectral hole burning (see Sect. 5.3.5 and 7.8.3) [5.1–5.8].

5.3 Incoherent Interaction

In the case of an incoherent resonant interaction the nonlinear behavior can be described by the change of the absorption coefficient a as a function of the incident intensity of the light beam, similar to that given in Chap. 4, (4.2):

nonlinear absorption $\quad a = f\{I\} = f\{I(r, \lambda, t, \varphi)\}$ $\qquad$ (5.2)

In simple cases such as, e.g. under stationary conditions and for optically thin samples these absorption coefficients can be given analytically as rational polynomials of the intensity. But only in the simplest cases can the intensity equation also be solved analytically.

The nonlinear incoherent absorption results from the change in population of absorbing or emitting energy states of the matter. Thus the absorption coefficient a for a given light beam may be written as:

$$a = \sum_m \pm \sigma_m(\lambda, \varphi, r) N_m(I, r, t)$$ $\qquad$ (5.3)

where σ_m is the cross-section of the mth eigenstate (or energy level) of the matter and N_m is its population density. All possible absorption (+ sign) and emission (− sign) transitions have to be summed.

In the most trivial approach the nonlinear absorption coefficient can be written as the first term of the series:

0th approach $\quad a(I) = a_0 \left\{ 1 - \dfrac{I}{I_{nl}} \right\}$ $\qquad$ (5.4)

with the

nonlinear intensity $\quad I_{nl} = \dfrac{h\nu}{2\sigma\tau} \quad$ and $\quad a_0 = \sigma N_{total} \quad$ and

$$I_{nl} = \frac{I_{nl}}{h\nu} = \frac{1}{2\sigma\tau} \tag{5.5}$$

where σ is the cross-section of the active transition, τ the recovery time of the absorption of this transition, ν its frequency and N_{total} the population density of the absorbing state without excitation. The intensity I_{nl} is measured as the photon flux density in photons $cm^{-2}\,s^{-1}$. In the case of a two-level scheme this approximation (5.4 and 5.5) is useful for intensities small compared to I_{nl} but in more complicated cases it may fail completely. Therefore the structure of the nonlinear absorption should be analyzed in detail as described in Chap. 7 to avoid fundamental errors in discussing the experimental results.

The nonlinear absorption can easily be measured if the necessary excitation intensities are available. As will be shown in detail in the following chapters as a rough estimate the intensity for nonlinear effects should possibly be as large as I_{nl}. For typical cross-sections of 10^{-16}–$10^{-20}\,cm^2$ and recovery times of ns–µs for molecular or atomic systems the resulting nonlinear intensities are in the range of 10^{22}–10^{27} photons $cm^{-2}\,s^{-1}$ or $kW\,cm^{-2}$ to $GW\,cm^{-2}$ which can easily be realized with pulsed lasers. The following effects can be obtained:

- bleaching;
- general nonlinear transmission including darkening;
- transient absorptions – excited state absorptions (ESA);
- stimulated emission – superradiance – laser action;
- spectral hole burning.

Although nonlinear absorption or transmission can be determined with high accuracy the evaluation of the experimental results for obtaining material parameters such as the transition moments or cross-sections and the decay times of all participating matter states can be very difficult. Firstly, it can be difficult to determine which states are involved in the experiment and, secondly, the population densities N_m of these states can be very difficult to work out. Sometimes the calculation with numerical models can be helpful. Simple rate equations can be sufficient (see next chapter) but for safe results some experimental strategy has to be used. Details are described in Chap. 7. A brief description of the observable effects now follows.

5.3.1 Bleaching

Optical bleaching of matter is observable in simple one-beam experiments measuring the transmission of the sample as a function of the incident intensity as shown in Fig. 5.4.

In room temperature experiments with not too short pulse durations mostly homogeneous bleaching will be observed and no spectral hole burn-

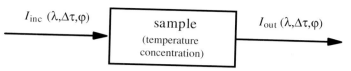

Fig. 5.4. Schematic of bleaching experiment with one beam

ing will occur. The transmission of the sample can be determined from the intensities of the transmitted beam I_{out} and the incident beam I_{inc} as:

transmission $\quad T = \dfrac{I_{out}(t_m)}{I_{inc}(t_m)}$ (5.6)

The conventional transmission as described in this expression (5.6) measuring both intensities at the same time is commonly used in nonlinear experiments as long as the pulse length is not too short. It is especially useful if steady-state conditions are realized. In experiments with ps or fs pulses the temporal pulse shape cannot be measured electronically and thus time-integrated intensities are then used (see Sect. 7.4.1). In any case the calculated transmission obtained from modeling the experiments has to use the same definition as that used in the experimental value.

A typical bleaching curve is shown in Fig. 5.5. At low intensities the transmission is constant, as expected, and then the transmission increases in this two-level model up to 1. The parameters are used as given in the figure for the two-level scheme.

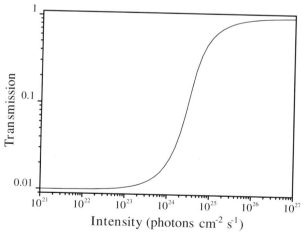

Fig. 5.5. Calculated bleaching curve of a two level system as a function of the incident intensity. The material has a cross-section of 10^{-16} cm^2, a concentration of $4.61 \cdot 10^{17}$ cm^{-3}, a absorption recovery time of 10 ns and a thickness of 1 mm. The excitation pulse length of the laser was 10 ns

The bleaching effect is used, e.g. for passive Q switching of lasers or for passive mode locking [e.g. 5.9–5.12 and references of Sects. 6.10.2 and 6.10.3]. The bleaching is often more complicated and will be discussed in Sects. 5.3.3, 5.3.5, 5.3.6 and 5.3.7 in more detail.

5.3.2 Transient Absorption: Excited State Absorption (ESA)

A large variety of nonlinear absorption effects occur in pump–probe experiments with at least two beams (see Fig. 5.6).

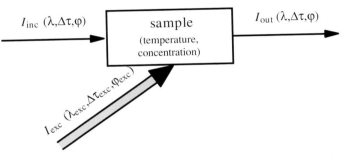

Fig. 5.6. Pump and probe technique for observing transient absorption (ESA spectroscopy)

The nonlinear effect is produced by a strong pump beam which populates excited states in the sample. This alteration of the sample has a large variety of new properties generated by the exciting light. The choice of the pump light parameters allows the appropriate population of all kinds of special material states with different new absorption characteristics, with life times from a few fs to hours, and so on. A more detailed discussion will given in Chap. 7.

Most prominently the population of excited electronic states can be achieved [5.13–5.37 and references of Sect. 7.7]. Thus the new absorption bands of these states can be measured in *excited state absorption (ESA) spectroscopy*. As an example the absorption bands of the first excited singlet and triplet states of a liquid crystal named T15 [5.21] are shown in Fig. 5.7 as a function of the time delay between the pump pulse and the probe pulse.

The longest wavelength absorption of this material has a maximum at 296 nm and almost no absorption above 340 nm. The FWHM width of this Gaussian-shaped band is 47 nm. For measuring these transient absorption bands the sample, which is transparent in the visible spectral range, was excited at 308 nm with an intensity of 6.5 MW cm^{-2} which corresponds to $5 \cdot 10^{24}$ photons cm^{-2} s^{-1}. The excitation pulse was 28 ns long (FWHM) and linearly polarized.

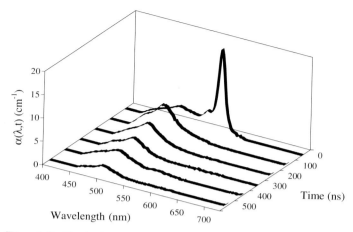

Fig. 5.7. Excited state absorption (ESA) of the singlet and triplet bands of a solution of 4'-n-pentyl-4-cyanoterphenyl (T15) in cyclohexane with a concentration of $0.09 \, \mathrm{mmol \, l^{-1}}$ in the visible range at room temperature

The free choice of the parameters of both the pump light and the probe light allows a large range of spectroscopic techniques such as, e.g. fractional bleaching or spectral hole burning (see Chap. 7).

The exciting light can be built from two or more different light beams for multiple excitation of the sample. But in any case the probe light intensity has be small enough not to disturb the sample itself. Values have to be checked in detail by investigating the nonlinear behavior of the sample (see Sect. 7.1.7). In many cases a value of less than 10^{19} photons $\mathrm{cm^{-2} \, s^{-1}}$ may be the upper limit.

5.3.3 Nonlinear Transmission

The nonlinear transmission of absorption bands especially of organic materials can be much more complicated than the described bleaching [e.g. 5.38–5.59]. In many cases an excited state absorption (ESA) occurs in the same wavelength range as the ground state absorption (GSA) with sometimes an even stronger cross-section than the GSA. The combination of bleaching and new transient absorptions can lead to quite complicated functions of the transmission as a function of the incident intensity in the nonlinear range. Besides variations in the slope of the bleaching curve the new nonlinear activated absorption can even cause darkening of the sample as shown in Fig. 5.8. But the same sample shows different behavior with a 20 nm longer excitation wavelength as shown in Fig. 5.9.

Therefore all kinds of transmission graphs as functions of the intensity with maxima, minima and plateau are possible as the examples in Sect. 5.3.6 and 5.3.7 illustrate. The evaluation of the nonlinear transmission curve allows

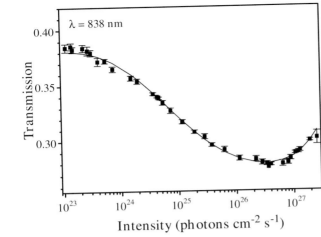

Fig. 5.8. Nonlinear transmission of the extracted antenna LH2 complex from the photosynthetic apparatus of a bacterium as a function of the excitation intensity at 838 nm after [5.50] measured with a pulse duration of 400 ps in a 1 mm cell. The solid line is the modeling of the experimental data including an excited state absorption and other processes

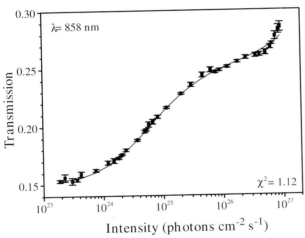

Fig. 5.9. Nonlinear transmission of the extracted antenna LH2 complex from the photosynthetic apparatus of a bacterium as a function of the excitation intensity at 838 nm after [5.50] measured with a pulse duration of 400 ps in a 1 mm cell as in Fig. 5.8. The solid line is the modeling of the experimental data including an excited state absorption and other processes

the identification of excited state absorption as will be discussed in Chap. 7 in detail. Very often at least one transient absorption is present in the wavelength range of the investigated ground state absorption band and the use of a two-level scheme for modeling the experimental results is than not sufficient. The darkening of the samples can be used for optical limiting devices [5.48–5.58].

5.3.4 Stimulated Emission: Superradiance: Laser Action

Using strong laser excitation the excited states of the matter are populated and inversion can be easily realized. In laser investigations this is expected and will be observed (see Sect. 6.2). In other photonic applications or in nonlinear spectroscopy stimulated emission will not always be in the focus of the experiment.

Nevertheless, unexpected stimulated emission such as superradiance [5.60–5.62] or laser action will change the properties of the nonlinear interaction drastically. Wide fluorescence bands will narrow to the laser line, the lifetime of the excited state can be reduced by many orders of magnitude and polarization conditions will be changed. Reabsorption of the emitted light may increase the confusion even more. On the other hand these effects can be taken advantage of, e.g. for designing special light sources or for speeding the recovery of absorption after bleaching.

The observation of stimulated emission is not always easy. The transition from fluorescence to superradiance happens in practice almost continuously (see Fig. 5.10).

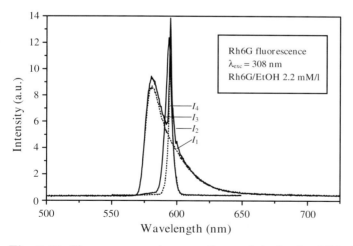

Fig. 5.10. Fluorescence and superradiance of rhodamine 6G in high concentration occurring while increasing the excitation intensity ($I_1 < I_2 < I_3 < I_4$). The spectral narrowing is accompanied by a shortening of the absorption recovery time. (The short wavelength emission is reabsorbed.)

If the cell windows act as resonator mirrors the laser spot can be somewhere in the optical setup and may be difficult to find, especially if invisible light is produced. Therefore stimulated emission require special attention in nonlinear resonant interaction measurements.

5.3.5 Spectral Hole Burning

If the optical transition of the matter is inhomogeneously broadened the change of the transmission will be different for the wavelength of the exciting beam and in the spectral neighborhood [5.63–5.116]. The change in the sum spectra will show stronger bleaching at the spectral position of excitation as shown in Fig. 5.11.

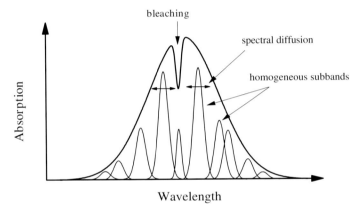

Fig. 5.11. Spectral hole burning in bleaching experiment of inhomogeneously broadened absorption band

As a consequence of the spectral diffusion processes [5.117–5.133] the other parts of the absorption band will also be bleached, but the bleaching will be weaker than at the excitation wavelength. The spectral cross-relaxation time determines the hole life time.

Finally the relation of bleaching at the excitation wavelength and in the spectral surroundings of it will depend on the relation of the spectral diffusion or spectral cross-relaxation time T_3 on one hand and the energy relaxation time T_1 on the other (see Fig. 5.12).

If the spectral cross-relaxation is much slower than the energy relaxation $T_3 \gg T_1$ maximum hole burning will be observed and thus the bleaching at the excitation wavelength will be a maximum and the rest of the absorption band will be unchanged. If the spectral cross-relaxation is much faster than the energy relaxation $T_3 \ll T_1$ almost no hole burning can be obtained, even using high intensities in the nonlinear experiments.

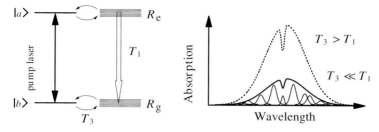

Fig. 5.12. Influence of the relation of T_1 and T_3 for the bleaching of the absorption band in spectral hole burning experiments

The homogeneous linewidth determines the minimum hole width. At low temperatures of a few Kelvin the hole width for molecular systems is of the order of a few ten GHz [5.63] or below. At room temperature the same systems show a homogeneous linewidth of several nm [5.90].

The mechanisms of spectral hole burning are at least as diverse as the reasons for inhomogeneous broadening. Finally, *chemical and photo-physical hole burning* can be distinguished with drastically different hole life times.

Spectral hole burning may find photonic applications in communications such as in spectrally coded switching and storage. In principle the limited spatial resolution of light techniques determined by the wavelength can be compensated by the additional spectral coding (see references of Sect. 1.5). Theoretically, factors of 10^4–10^6 seem to be possible. As a consequence of the short reaction times of optical systems very high processing speeds seem to be possible. But there are still a large number of physical and technological problems that have to be solved before possible applications appear.

5.3.6 Description with Rate Equations

The nonlinear interaction of a light beam with matter can be well described with rate equations if coherence effects are not important. This is usually fulfilled if the phase relaxation time T_2 (Sect. 5.4.1) is much shorter than the incident pulse duration, the inverse pump rate $1/(\sigma I)$ and the decay times τ. Mathematically all nondiagonal elements of the density matrix (see Sect. 5.4.1) are neglected and the nonlinear interaction is described with population densities N_m of the matter states and with the intensities I of the light fields. More details are given in [5.134–5.137].

Basic Equations

Between the two energy states or energy levels l and m of matter (see Fig. 5.13) only three types of transitions are recognized.

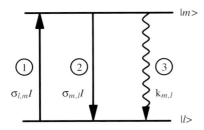

Fig. 5.13. Two levels or energy states as part of a possibly complicated level scheme of matter with two stimulated transitions, absorption (1) and stimulated emission (2), and one spontaneous transition (3) which can be radiationless or spontaneous emission or the sum effect of both

In an easily understandable way each single transition of one particle will decrease the population density of the initial state by 1 and also increase the final state by 1:

(1) *absorption* from l to m $(N_l \rightarrow N_m)$:

$$\frac{\partial N_l}{\partial t} = -\sigma_{l,m} I N_l \quad \text{and} \quad \frac{\partial N_m}{\partial t} = +\sigma_{l,m} I N_l \tag{5.7}$$

(2) *stimulated emission* from m to l $(N_l \rightarrow N_m)$:

$$\frac{\partial N_l}{\partial t} = +\sigma_{m,l} I N_m \quad \text{and} \quad \frac{\partial N_m}{\partial t} = -\sigma_{m,l} I N_m \tag{5.8}$$

(3) *spontaneous relaxation* from m to l $(N_l \rightarrow N_m)$:

$$\frac{\partial N_l}{\partial t} = +k_{m,l} N_m \quad \text{and} \quad \frac{\partial N_m}{\partial t} = -k_{m,l} N_m \tag{5.9}$$

with cross-sections $\sigma_{l,m}$ and $\sigma_{m,l}$, the intensity (photon flux density) I and decay rates $k_{m,l}$ which are the inverse of the decay times $\tau_{m,l}$:

$$k_{m,l} = \frac{1}{\tau_{m,l}}. \tag{5.10}$$

The decay rates of different channels between the same states, e.g. spontaneous emission and radiationless decay, can simply be added to the total decay rate:

$$k_{m,l}^{\text{total}} = k_{m,l}^{\text{spont.emission}} + k_{m,l}^{\text{radiationless}} + \cdots \tag{5.11}$$

meaning that the inverse of the decay times have to be summed.

Using these three basic processes and their mathematical description almost any level scheme can be modeled by combining such energy level pairs via radiationless transitions. Some examples will be given below. More complicated incoherent interactions such as, e.g. exciton formation can be adapted to this formalism [5.134].

The sum N_0 over all population densities in all states has to be constant:

$$\text{particle conservation} \quad N_{\text{total}} = \sum_l N_l = \text{const.} \tag{5.12}$$

The influence of the nonlinear interaction on the photon field can be described in the rate equation approximation under the assumption of a light beam propagating in the z direction with a photon transport equation for the intensity or, better, photon flux intensity I as a function of time t and coordinate z:

(4) *photon transport equation* $(N_l \to N_m)$:

$$\left(\frac{\partial}{\partial z} + \frac{1}{c}\frac{\partial}{\partial t}\right) I(x, y, z, t)$$
$$= (-\sigma_{l,m} N_l(x, y, z, t) + \sigma_{m,l} N_m(x, y, z, t)) I(x, y, z, t) \quad (5.13)$$

which can usually be approximated with the assumption $\sigma_{l,m} = \sigma_{m,l}$ which is the Einstein relation for a perfect two-level scheme; and by postulating a flat-top transverse intensity profile at the sample (e.g. by using an aperture in a small distance) to give:

$$\left(\frac{\partial}{\partial z} + \frac{1}{c}\frac{\partial}{\partial t}\right) I(z, t) = \sigma_{l,m}\left(-N_l(z, t) + N_m(z, t)\right) I(z, t). \quad (5.14)$$

This equation considers the stimulated emission which occurs between these two states at exactly the same wavelength as that of the incident light. If stimulated emission occurs at a different transition a second photon transport equation for this intensity at the other wavelength is required (see below).

The meaning of the equations of the processes (1)–(4) becomes very clear if the single-particle transition is related to the absorption or stimulated emission of one photon. Therefore useful units for I and N_j are:

$$[I] = \frac{\text{photons}}{\text{cm}^2\text{s}} = \frac{[P]}{[h\nu]\,\text{cm}^2} \quad \text{and} \quad [N_l] = \frac{\text{particles}}{\text{cm}^3} \quad (5.15)$$

With these modules any level scheme and the interaction with laser beams can be composed. In the case of organic matter very often at least the lowest excited singlet and triplet levels and the intersystem crossing rate have to be included. But in some cases level schemes with several radiatiative transitions at the same wavelength may be necessary for modeling experiments with high intensities [e.g. 5.44, 5.50].

The *solution of this system* of partial differential rate equations is analytically possible only in the simplest cases. At least stationary conditions are necessary. The numerical solution is also not trivial, because the coefficients can vary over more than six orders of magnitude. Runge–Kutta procedures will need tremendous computing times. Some useful strategies for solving these stiff equations are given below.

Stationary Solutions of Rate Equations

Stationary conditions with respect to the matter are fulfilled if during the nonlinear interaction all decay times of all involved states of the material are sufficiently shorter than the fastest relative changes of intensity:

$$\textbf{stationary interaction} \quad \frac{1}{I}\frac{\partial I}{\partial t} \ll \left\{\frac{1}{\tau_{l,m}}\right\}_{\min} = \{k_{l,m}\}_{\min} \qquad (5.16)$$

which in most cases means an incident laser pulse length $\Delta t_{\text{pulse}} \gg \tau_{l,m}$ for all relevant decay times $\tau_{l,m}$ of the experiment.

Therefore, e.g. in experiments with ns pulses organic matter can show stationary nonlinear interaction only as long as triplet states are not involved. Experiments with ps or fs pulses will be almost always nonstationary.

If stationary matter conditions are fulfilled the system of differential equations for the population densities can be simplified by:

$$\text{stationary approximation 1} \quad \frac{\partial N_l}{\partial t} \overset{!}{=} 0 \qquad (5.17)$$

to a simple algebraic linear system with as many equations as unknown populations densities. This can be solved analytically by using analytical computer programs such as MathCad or Mathematica.

A further important approximation is possible if the optical path length L in the matter is small compared to the pulse modulations:

$$\text{stationary length condition} \quad \frac{1}{L} \gg \frac{1}{cI}\frac{\partial I}{\partial t} \qquad (5.18)$$

or

$$c\Delta t_{\text{pulse}} \gg L \qquad (5.19)$$

with c the speed of light in the matter. This condition means that the intensity is approximately the same over the whole sample at one time. This allows the approximation:

$$\text{stationary approximation 2} \quad \frac{1}{c}\frac{\partial I}{\partial t} \overset{!}{=} 0 \qquad (5.20)$$

This approximation is not essential in the description of nonlinear experiments as long as the light beams propagate through the sample in one direction. Then a simple transformation of the time t to a new time θ_{mov} of an observer moving with the speed of light can be used:

$$\theta_{\text{mov}} = t - \frac{z}{c}. \qquad (5.21)$$

In both cases the temporal derivative in the intensity equation can be neglected. The transformation is not useful if experiments with counter-propagating beams such as, e.g. in resonators, is to be modeled.

Based on these assumptions the analytical solution of the rate equations of a two-level scheme is completely possible, and in more complicated schemes at least sometimes the system of the population density equations can be solved analytically.

Stationary Two-Level Model

The stationary two-level scheme shows nonlinear bleaching as a consequence of the increasing population of the first excited state with increasing intensity as soon as the pump rate $\sigma_{12}I_{\mathrm{inc}}$ becomes comparable to τ_{21}. The model is depicted in Fig. 5.14.

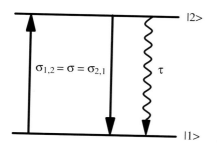

Fig. 5.14. Two-level scheme for describing bleaching and the population of the first excited state. (The scheme is also useful for more complicated situations, if one pump process and one absorption recovery time is dominant)

Under stationary conditions the rate equations for the population densities are given by:

$$0 = -N_1(z)\sigma_{12}I(z) + N_2(z)\sigma_{21}I(z) + k_{21}N_2(z) \tag{5.22}$$

$$0 = +N_1(z)\sigma_{12}I(z) - N_2(z)\sigma_{21}I(z) + k_{21}N_2(z) \tag{5.23}$$

and

$$N_1(z) + N_2(z) = N_{\mathrm{total}} \tag{5.24}$$

with the total particle density N_{total}. The intensity follows from:

$$\frac{\partial I}{\partial z} = -(N_1 - N_2)\sigma I. \tag{5.25}$$

The difference in population density $(N_1 - N_2)$ can be calculated as a function of the intensity as:

$$(N_1 - N_2) = N_{\mathrm{total}}\frac{1}{1 + 2\sigma\tau I}. \tag{5.26}$$

With this value the intensity equation can be integrated over the length L of the sample to give:

$$\ln\left(\frac{I_{\mathrm{out}}}{I_{\mathrm{inc}}}\right) = -\sigma N_{\mathrm{total}}L + 2\sigma\tau(I_{\mathrm{inc}} - I_{\mathrm{out}}) \tag{5.27}$$

which is the Lambert–Beers law for $I_{\mathrm{inc}} \rightarrow 0$ showing the transmission T_0. This equation can be written as:

$$\text{nonlinear transmission} \quad \ln T = -\sigma N_{\mathrm{total}}L + 2\sigma\tau I_{\mathrm{inc}}(1 - T)$$

$$= \ln T_0 + 2\sigma\tau I_{\mathrm{inc}}(1 - T) \tag{5.28}$$

or in calculable form:

$$I_{inc} = \frac{1}{\sigma\tau(1-T)} \ln \frac{T}{T_0}.$$ (5.29)

This solution is nonlinear for values of I_{inc} which are approximately $1/2\sigma\tau$ which was called I_{nl} above. At this intensity the population density of the excited state is $N_{total}/4$. The population of the first excited state in this model is given by:

$$\text{population } N_2 \quad N_2(I) = N_{total}\frac{\sigma\tau I}{(1+2\sigma\tau I)} = N_{total} - N_1(I)$$ (5.30)

and the maximum population density in this state is 50%.

It should be noted that the nonlinear behavior is a function of the product of the cross-section and the decay time only. Thus the shape of the bleaching curve of a stationary two-level model cannot be changed by changing the material parameters (see Fig. 5.15).

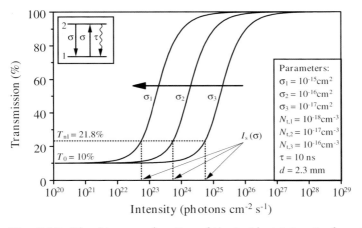

Fig. 5.15. Bleaching as a function of the incident intensity for a two-level scheme with realistic parameters for molecular systems and varying cross-sections

Only the start of the nonlinearity varies. Larger cross-sections and slower recovery times allow smaller nonlinear intensities I_{nl}. The population density of the excited state is given for the same parameters in Fig. 5.16.

The intensity I_{nl} can be found in these graphs at population densities of $N_2 = N_0/4$. At this intensity the transmissions is $T(I_{nl}) = 0.218$ in this example.

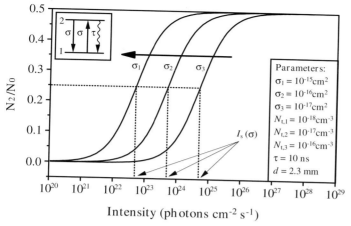

Fig. 5.16. Relative population density of the upper level 2 as a function of the incident intensity for a two-level scheme with varying cross-sections as used in Fig. 5.15

Stationary Four-Level Model

The stationary four energy level scheme with one absorption is typical of many organic molecules in the singlet system containing the electronic ground state and the excited state in the vibrational ground and excited state. Its nonlinear bleaching can be calculated similar to the two-level scheme. The four-level model is shown in Fig. 5.17.

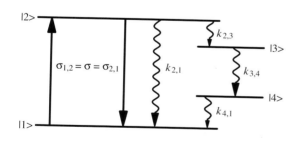

Fig. 5.17. Four-level scheme with one absorption

The population densities can be calculated under stationary conditions from:

$$\frac{\partial}{\partial t} N_1 = -\sigma_{1,2} I N_1 + \sigma_{2,1} I N_2 + k_{2,1} N_2 + k_{4,1} N_4 \overset{!}{=} 0 \tag{5.31}$$

$$\frac{\partial}{\partial t} N_2 = +\sigma_{1,2} I N_1 - \sigma_{2,1} I N_2 - k_{2,1} N_2 - k_{2,3} N_2 \overset{!}{=} 0 \tag{5.32}$$

$$\frac{\partial}{\partial t} N_3 = +k_{2,3} N_2 - k_{3,4} N_3 \overset{!}{=} 0 \tag{5.33}$$

$$\frac{\partial}{\partial t} N_4 = +k_{3,4} N_3 - k_{4,1} N_4 \overset{!}{=} 0 \tag{5.34}$$

and

$$N_{\text{total}} = N_1 + N_2 + N_3 + N_4. \tag{5.35}$$

The stationary solution of this system of equations is given by:

$$N_1 = \frac{\dfrac{\sigma_{2,1}}{\sigma_{1,2}}\left(1 + \dfrac{k_{2,1} + k_{2,3}}{\sigma_{2,1} I}\right)}{\dfrac{\sigma_{2,1}}{\sigma_{1,2}}\left(2 + \dfrac{k_{2,1} + k_{2,3}}{\sigma_{2,1} I}\right) + \dfrac{k_{2,3}}{k_{3,4}} + \dfrac{k_{2,3}}{k_{4,1}}} N_{\text{total}} \tag{5.36}$$

$$N_2 = \left[\frac{\sigma_{2,1}}{\sigma_{1,2}}\left(2 + \frac{k_{2,1} + k_{2,3}}{\sigma_{2,1} I}\right) + \frac{k_{2,3}}{k_{3,4}} + \frac{k_{2,3}}{k_{4,1}}\right]^{-1} N_{\text{total}} \tag{5.37}$$

$$N_3 = \frac{k_{2,3}}{k_{3,4}}\left[\frac{\sigma_{2,1}}{\sigma_{1,2}}\left(2 + \frac{k_{2,1} + k_{2,3}}{\sigma_{2,1} I}\right) + \frac{k_{2,3}}{k_{3,4}} + \frac{k_{2,3}}{k_{4,1}}\right]^{-1} N_{\text{total}} \tag{5.38}$$

$$N_4 = \frac{k_{2,3}}{k_{4,1}}\left[\frac{\sigma_{2,1}}{\sigma_{1,2}}\left(2 + \frac{k_{2,1} + k_{2,3}}{\sigma_{2,1} I}\right) + \frac{k_{2,3}}{k_{3,4}} + \frac{k_{2,3}}{k_{4,1}}\right]^{-1} N_{\text{total}}. \tag{5.39}$$

These solutions should not diverge if cross-sections or time constants are zero. Thus these equations may be transformed before these parameters are set zero. The intensity can be calculated from:

$$\frac{\partial}{\partial z} I = -\sigma_{1,2} I N_1 + \sigma_{2,1} I N_2. \tag{5.40}$$

The transmission follows analogously to the solution for the two-level scheme from:

$$\ln T = \ln T_0 - \frac{2 + \dfrac{k_{2,3}}{k_{3,4}} + \dfrac{k_{2,3}}{k_{4,1}}}{\dfrac{\sigma_{2,1}}{\sigma_{1,2}}}$$

$$\cdot \ln\left\{\frac{1}{1 + (\sigma_{2,1} - \sigma_{1,2})(k_{2,3} + k_{2,1})} + \frac{(\sigma_{2,1} - \sigma_{1,2})(k_{2,3} - k_{2,1})}{(\sigma_{2,1} - \sigma_{1,2})(k_{2,3} + k_{2,1}) + 1/I_0} T\right\}. \tag{5.41}$$

Thus energy levels which are only populated radiationless, do not change the characteristic shape of the bleaching curve.

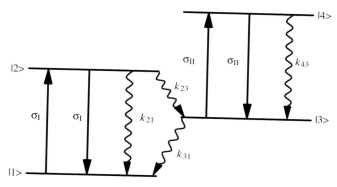

Fig. 5.18. Energy level scheme with two absorptive transitions at the same wavelength. The cross-sections of absorption and stimulated emission were assumed to be equal for each transition

Stationary Model with Two Absorptions

Energy schemes with two absorptive transitions at the same wavelength consist of at least three, but usually four energy levels as shown in Fig. 5.18.

The rate equations for the population densities are given by:

$$\frac{\partial}{\partial t} N_1 = -(N_1 - N_2)\sigma_I \mathcal{I} + k_{21} N_2 + k_{31} N_3 \overset{!}{=} 0 \tag{5.42}$$

$$\frac{\partial}{\partial t} N_2 = +(N_1 - N_2)\sigma_I \mathcal{I} - (k_{21} + k_{23}) N_2 \overset{!}{=} 0 \tag{5.43}$$

$$\frac{\partial}{\partial t} N_3 = -(N_3 - N_4)\sigma_{II} \mathcal{I} + k_{23} N_2 - k_{31} N_3 + k_{43} N_4 \overset{!}{=} 0 \tag{5.44}$$

$$\frac{\partial}{\partial t} N_4 = +(N_3 - N_4)\sigma_{II} \mathcal{I} - k_{34} N_4 \overset{!}{=} 0 \tag{5.45}$$

and the intensity can be calculated from:

$$\frac{\partial}{\partial t} \mathcal{I} = -\left[\sigma_I(N_1 - N_2) + \sigma_{II}(N_3 - N_4)\right] \mathcal{I}. \tag{5.46}$$

The stationary solution of any model with two absorptive transitions with any number of levels and radiationless transitions can be given as:

nonlinear transmission $\quad \ln T = -C_1 N_{\text{total}} L + C_2 \mathcal{I}_{\text{inc}}(1 - T)$

$$+ C_3 \ln \left(\frac{1 + C_4 \mathcal{I}_{\text{inc}}}{1 + C_4 \mathcal{I}_{\text{inc}} T} \right) \tag{5.47}$$

As can be seen from the transmission equation a further term is added in comparison to the nonlinear model with one absorption (5.28). This allows new shapes of the bleaching curve. As shown in Fig. 5.19 such a model can have a plateau in the bleaching curve as a function of the incident intensity.

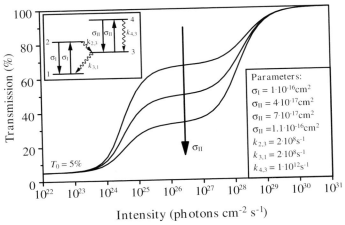

Fig. 5.19. Nonlinear transmission as a function of the incident intensity for a term scheme with two absorptions showing a plateau in bleaching ($N = 3 \cdot 10^{17}$ cm^{-3}, $d = 1$ mm)

If the cross-section of the excited state is larger than the cross-section of the ground state the nonlinear transmission can show a minimum before bleaching and thus darkening is observed first as can be seen in Fig. 5.20.

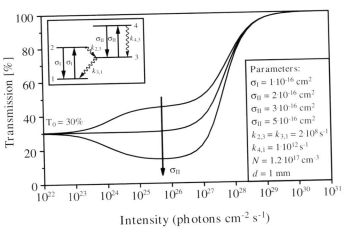

Fig. 5.20. Nonlinear transmission as a function of the incident intensity for a term scheme with two absorptions showing a minimum before bleaching

In any case the nonlinear transmission finally shows bleaching at transmission values of up to 1 as long as no further nonlinear processes are activated. In practical cases with excitation intensities below the damage threshold of the material the maximum transmission values are usually lower than 1. This indicates further excited state absorptions from higher excited states or other nonresonant interactions.

Although an analytical solution is possible for stationary models with two absorptive transitions it may be easier to use numerical calculations to describe the experiments. As shown below the numerical method is much more flexible than the analytical calculation and not fixed to stationary cases. Nevertheless, the fundamental discussion of possible structures in nonlinear transmission is better based on analytical formulas.

General Stationary Models

The population densities N_l of all m energy levels of any stationary model consisting of the above three processes will be described by a system of p equations of the form:

$$\frac{\partial}{\partial t} N_l(x,t) = I(x,t) \sum_{m(\neq l)} \{ -\sigma_{l,m} N_l(x,t) + \sigma_{m,l} N_m(x,t) \}$$

$$+ \sum_{m(\neq l)} \{ -k_{l,m} N_l(x,t) + k_{m,l} N_m(x,t) \}$$

$$\overset{!}{=} 0 \tag{5.48}$$

and

$$\sum_{l=1}^{p} N_l = N_{\text{total}} \tag{5.49}$$

for all indices l and m from 1 to p. This linear system can be solved in general to give:

$$N_l(I) = \frac{C_{l,no,0} + C_{l,no,1} I + \cdots + C_{l,no,r} I^r}{C_{l,de,0} + C_{l,de,1} I + \cdots + C_{l,de,r} I^r} N_{\text{total}} \tag{5.50}$$

with the number r of absorptive transitions of the model and numerator coefficients $C_{l,no,i}$ and denominator coefficients $C_{l,de,i}$. These coefficients can be complicated algebraic functions of the matter cross-sections and decay times. This result shows the maximum order of nonlinearity which can be reached for the absorption of any of these levels. The power of the highest possible nonlinearity is r. In known experiments usually much lower orders in the range ≤ 2 were observed. The reason for this smaller nonlinearity is the usually successive activation of the different absorptive transitions as a function of the excitation intensity.

If the model contains only a few levels, e.g. one triplet level, with non-stationary lifetimes (e.g. longer than the pulse length) and a large system of stationary levels the system can be analyzed as partly stationary. Thus it can be solved as a very small nonstationary system and a large stationary one [5.136].

But for the modeling of experiments again the direct numerical calculation may be more efficient because of more flexibility and the general approach.

The photon transport equation is given by:

$$\frac{\partial}{\partial z} I(z,t) = I(z,t) \sum_{l} \sum_{m(>l)} (-\sigma_{l,m} N_l(z,t) + \sigma_{m,l} N_m(z,t)) \tag{5.51}$$

with a total of r terms in the sum. The time derivative was neglected. A general solution is not possible. A time-saving numerical method with high accuracy is given below.

If the intensity of a (second) probe beam is sufficiently low and therefore does not disturb the population of the system, such an equation can simply be integrated over the sample length L to give the result:

$$T_{\text{probe}} = \frac{I_{\text{probe}}(L,t)}{I_{\text{probe}}(0,t)} = e^{-L \left\{ \sum_{l} \sum_{m(>l)} (-\sigma_{l,m}(\lambda_{\text{probe}})N_l + \sigma_{m,l}(\lambda_{\text{probe}})N_m) \right\}}.$$

$$\tag{5.52}$$

Numerical Solution

With today's computers the numerical solution of the partial differential system of rate equations is possible even for large models in a few minutes or seconds. But short computation times require numerical methods suitable for this special mathematical problem. The coefficients of these stiff systems of differential equations can vary from fs to seconds or longer. Thus the usually used Runge–Kutta method can lead to 10^9 or more iterations. Therefore a more useful method producing results with less than 1% error in 10–100 steps will be sketched [5.136, 5.134].

The mathematical problem consists of $p+1$ population equations:

$$\frac{\partial}{\partial t} N_l(x,t) = I(x,t) \sum_{m(\neq l)} \{-\sigma_{l,m} N_l(x,t) + \sigma_{m,l} N_m(x,t)\}$$
$$+ \sum_{m(\neq l)} \{-k_{l,m} N_l(x,t) + k_{m,l} N_m(x,t)\}$$

$$\sum_{l=1}^{p} N_l = N_{\text{total}} \tag{5.53}$$

and the intensity equation:

$$\left(\frac{\partial}{\partial z} + \frac{1}{c} \frac{\partial}{\partial t} \right) I(z,t) = I(z,t) \sum_{l} \sum_{m(>l)} (-\sigma_{l,m} N_l(z,t) + \sigma_{m,l} N_m(z,t)). \tag{5.54}$$

The starting values of the population densities are usually:

$$N_1\left(0 \leq z \leq L,\ t \leq \frac{z}{c}\right) = N_{\text{total}} \quad \text{and} \quad N_{1\neq 0}\left(0 \leq z \leq L,\ t \leq \frac{z}{c}\right) = 0 \tag{5.55}$$

the lowest energy level $l = 1$ is occupied, only. In experiments the incident light pulse is often used with Gaussian temporal profile:

$$\mathcal{I}(t) = \mathcal{I}_{\text{max}} \left\langle \frac{1.01t}{t + 0.01t_0} \right\rangle e^{-\left(\frac{4\ln 2(t - t_0)^2}{\Delta t_{\text{pulse}}^2}\right)} \tag{5.56}$$

in which the expression in brackets improves the numerical stability at $t = 0$. Otherwise the digitized experimental pulse shape $\mathcal{I}_{\text{inc}}(t)$ may be used for the calculation directly.

The population density equations can be written in matrix form as:

$$\frac{\partial}{\partial t} S(t) = P(t)S(t) + p \tag{5.57}$$

with the vector S with p elements, the $p \times p$ matrix P containing all material parameters such as cross-sections and decay rates, and the vector p as the inhomogeneity, resulting from the total sum of the population densities. The development of the population densities at time step t_{i+1} can be calculated from the values at time t_i assuming a constant step width Δt with the approximation:

$$S(t_n + 1) = S(t_n) + \frac{\Delta t}{2}\{P(t_n)S(t_n) + P(t_{n+1})S(t_{n+1}) + 2p\} \tag{5.58}$$

to give

$$S(t_n + 1) = \frac{\{E + (\Delta t/2)P(t_n)S(t_n) + \Delta t p\}}{\{E - (\Delta t/2)P(t_{n+1})\}} \tag{5.59}$$

with unity $p \times p$ matrix E. Whereas the P matrix contains the intensity and the parameters only, this expression can be straightforward to calculate reducing the caculation time by orders of magnitude. The computation time can be reduced by a further 30% if the denominator of this equation is calculated as:

$$\frac{1}{\{E - (\Delta t/2)P(t)\}} = \frac{\Omega_{n,0} + \Omega_{n,1}I(t) + \cdots + \Omega_{n,r}\mathcal{I}^r(t)}{\Omega_{d,0} + \Omega_{d,1}I(t) + \cdots + \Omega_{d,r}\mathcal{I}^r(t)} \tag{5.60}$$

and the matrices $\Omega_{n/d,i}$ are precalculated once in advance for the given parameters.

The intensity equation can be numerically solved by Runge–Kutta procedures of second or higher order.

As an example the results of modeling the nonlinear behavior of a solution of the molecule T15 (4'-n-pentyl-4-cyanoterphenyl) will be given. The parameters are taken from [5.21]. In Fig. 5.21 the necessary level scheme is depicted.

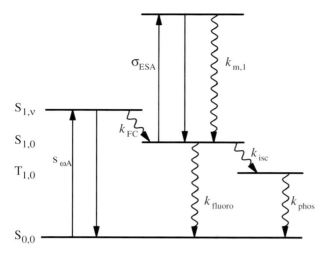

Fig. 5.21. Level scheme for modeling the nonlinear transmission of T15 [5.21]

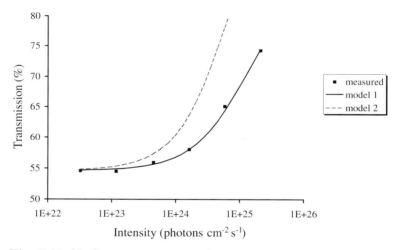

Fig. 5.22. Nonlinear transmission of T15 in cyclohexane as a function of the incident intensity at a wavelength of 308 nm measured with a pulse duration of 28 ns

The calculated nonlinear transmission of this molecular solution of T15 is given in Fig. 5.22. It shows bleaching behavior, but the slope of the experimental curve is slower than it would be in case of a two-level scheme (model 2). Therefore a two step absorption can be assumed as shown in Fig. 5.21 (model 1).

The population density in the $S_{1,0}$ state of the molecule as a function of time and intruding depth is calculated for an excitation intensity of

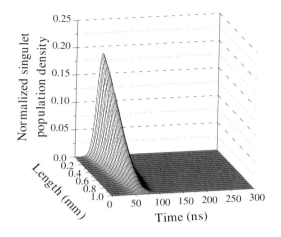

Fig. 5.23. Population density in the S_1 state of T15 exciting at 308 nm with an intensity of $3.4 \cdot 10^{24}$ photons cm^{-2} s^{-1} as a function of time and intruding depth. The concentration was $4.11 \cdot 10^{17}$ cm^{-3}, the cross-section $1.12 \cdot 10^{-16}$ cm^2 and the pulse duration 28 ns

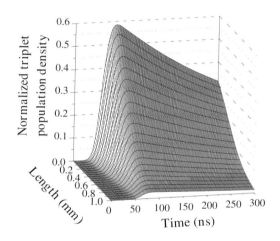

Fig. 5.24. Population density in the T_1 state of T15 exciting at 308 nm with parameters as in Fig. 5.23 as a function of time and intruding depth

$3.4 \cdot 10^{24}$ photons cm^{-2} s^{-1} as shown in Fig. 5.23. The fluorescence lifetime of this material was determined to be 1 ns [5.21].

In contrast to this figure the T_1 population density shows a much longer lifetime of 340 ns as taken from the data of Fig. 5.7 and thus an integrative character over the exciting laser pulse. The result of the calculation is given in Fig. 5.24.

These population densities can be measured via excited state absorption (ESA) spectra at different wavelengths as shown in Fig. 5.7.

Considering Spectral Hole Burning with Rate Equations

If in spectral hole burning experiments no coherent interaction takes place, e.g. the light pulses are not too short, rate equations can be used to de-

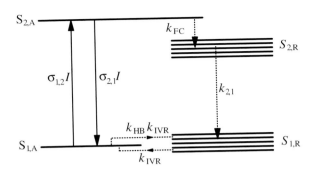

Fig. 5.25. Energy level scheme for modeling bleaching including the spectral hole burning effect

scribe the bleaching experiment in principle [5.136]. A simple four-level energy scheme as shown in Fig. 5.25 can be used.

In this scheme it is assumed that one level 1A is absorbing at the selected wavelength of the laser and a (large) number κ_{HB} of nonabsorbing levels build a reservoir R. This absorbing level is populated via spectral cross-relaxation from the reservoir with rate k_{IVR}. The probability for the decay of this level into the reservoir is then given by the rate $\kappa_{HB} \cdot k_{IVR}$. The population density without excitation is then given by:

$$N_{1A}(I = 0) = \frac{1}{1 + \kappa_{HB}} N_{\text{total}} \tag{5.61}$$

and

$$N_{1R}(I = 0) = \frac{\kappa_{HB}}{1 + \kappa_{HB}} N_{\text{total}} = \kappa_{HB} N_{1A}. \tag{5.62}$$

The population densities in the four-levels can be calculated from the following system of differential equations:

$$\frac{\partial N_{1A}}{\partial t} = -\kappa_{HB}\sigma_{1,2} I N_{1A} + \kappa_{HB}\sigma_{2,1} I N_{2A} + k_{IVR} N_{1R}$$
$$- \kappa_{HB} K_{IVR} N_{1A} \tag{5.63}$$

$$\frac{\partial N_{2A}}{\partial t} = +\kappa_{HB}\sigma_{1,2} I N_{1A} - \kappa_{HB}\sigma_{2,1} I N_{2A} - k_{FC} N_{2A} \tag{5.64}$$

$$\frac{\partial N_{2R}}{\partial t} = +k_{FC} N_{2A} - k_{T1} N_{2R} \tag{5.65}$$

$$\frac{\partial N_{1R}}{\partial t} = +k_{T1} N_{2R} + \kappa_{HB} k_{IVR} N_{1A} - k_{IVR} N_{1R} \tag{5.66}$$

and the intensity for the exciting light E_{exc} has to be calculated from:

$$\left(\frac{1}{c} \frac{\partial}{\partial t} + \frac{\partial}{\partial z} \right) I_{\text{exc}} = -\kappa_{HB}\sigma_{1,2} I_{\text{exc}} N_{1A} + \kappa_{HB}\sigma_{2,1} I_{\text{exc}} N_{2A}. \tag{5.67}$$

Its transmission is also the transmission of the probe light at the wavelength of excitation. The transmission for the probe light I_P at other wavelengths detecting the bleaching of the reservoir state follows from:

$$\left(\frac{1}{c}\frac{\partial}{\partial t} + \frac{\partial}{\partial z}\right) I_P = -\sigma_{1,2} I_P N_{1R}. \tag{5.68}$$

With this model the spectral hole burning effect results e.g. in the bleaching curves of Fig. 5.26.

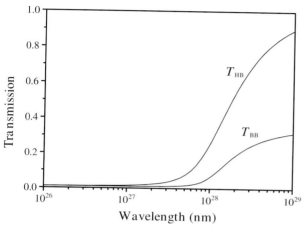

Fig. 5.26. Nonlinear transmission of an inhomogeneously broadened absorption band. The number of homogeneous subbands was assumed to be 100 and the exchange time to be 100 fs. The lifetime of the upper state is 1.5 ps and the conventional cross-section was assumed to be $5 \cdot 10^{-15}\,\mathrm{cm}^2$

As can be seen in the figure the transmission for the exciting laser is larger than the bleached transmission of the reservoir. Thus even under steady-state conditions short-lived hole burning can be observed if the ground state absorption recovery time is not much longer than the exchange rate k_{IVR}. The hole burning effect is of course easier to detect if the laser pulse duration is shorter than the hole lifetime. This demands fs pulses at room temperature but allows very slow measurements below 1 K for molecular systems.

The modeling of hole burning effects with rate equations has to be checked carefully for competing effects such as induced gratings or coherent interaction. Both can be considered explicitly as will be shown in Sects. 5.3.8 and 5.4.3.

5.3.7 Coherent Light Fields

If two or more coherent intensive light beams are applied in incoherent interactions in the absorption range of the matter the nonlinear effects will be spatially structured by the interference pattern of the light. If the phase relaxation time of the matter is too short for coherent interaction (see Sect. 5.4) just the changes in the population densities in the matter states have to be recognized for the theoretical description of this interaction.

But in addition the intensity pattern has to be determined in its non-linear interaction with the matter absorption. This can demand extensive numerical calculations of the coupled partial differential equations for the time-dependent intensity field in three dimensions (see Sect. 2.9):

$$I(\boldsymbol{r}, t) = \frac{c_0 \varepsilon_0 n}{2 h \bar{\nu}} \left\{ \sum_i \boldsymbol{E}_i(\nu_i, \boldsymbol{k}_i, \varphi_i) \right\}^2 \tag{5.69}$$

coupling with the absorptive transitions of the matter:

$$\left(\frac{\partial}{\partial z} + \frac{1}{c} \frac{\partial}{\partial t} \right) I(\boldsymbol{r}, t) = I(\boldsymbol{r}, t) \sum_l \sum_{m(>l)} (-\sigma_{l,m} N_l(\boldsymbol{r}, t) + \sigma_{m,l} N_m(\boldsymbol{r}, t)) \tag{5.70}$$

in combination with rate equations of population densities depending on time as in the previous chapter. But the dependency in space can more compli-cated:

$$\frac{\partial}{\partial t} N_l(\boldsymbol{r}, t) = I(\boldsymbol{r}, t) \sum_{m(\neq l)} \{ -\sigma_{l,m} N_l(\boldsymbol{r}, t) + \sigma_{m,l} N_m(\boldsymbol{r}, t) \}$$
$$= + \sum_{m(\neq l)} \{ -k_{l,m} N_l(\boldsymbol{r}, t) + k_{m,l} N_m(\boldsymbol{r}, t) \} \tag{5.71}$$

$$\sum_{l=1}^{n} N_1(\boldsymbol{r}, t) = N_{\text{total}}(\boldsymbol{r}, t) \tag{5.72}$$

while considering all relevant radiative matter transitions for the more or less averaged frequencies of the incident and possibly generated light fields.

This incoherent nonlinear interaction of coherent light beams with ab-sorbing matter has found several applications in photonics, especially using light patterns with a well-defined spatial grating structure.

Bleaching or nonlinear transmission of matter will lead to spatial *trans-mission gratings* of the sample (see Sect. 3.9.16 and Sect. 5.9.1). These can be used for deflection of light beams in optical switching.

If the incident light is used to excite the matter *absorption gratings* from the generated excited state absorptions (ESA) can be observed in pump and probe experiments (see Sect. 7.8.3). The diffraction of the probe beam at these often short-lived gratings can be used for highly sensitive measurements of very small absorption changes.

If the incident light pattern is applied for the pumping of laser material an *inversion grating* will be generated. This is applied, e.g. in distributed feedback dye (DFB) lasers for the spectral tuning of the laser and for pulse shortening (see Sect. 6.10.4).

If the modulation results from the standing wave in a laser resonator *spatial hole burning* (see Sect. 5.3.10) can occur and can cause instabilities in laser operation.

In general this nonlinear incoherent interaction based on intensity patterns in the sample can be used for the coupling of light beams via the nonlinear absorption (or emission) of matter in wave-mixing experiments.

5.3.8 Induced Transmission and Excited State Absorption Gratings

In the simplest case two equal spectrally degenerate light waves with parallel polarization but different propagation direction are used for illumination of the sample. In the interference region of these two intensive pump beams a spatial sine grating modulation for the intensity pattern would occur without the interaction (see Fig. 5.27).

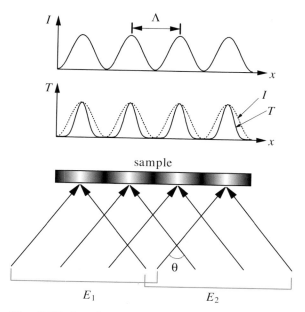

Fig. 5.27. Interference grating pattern of two intensive equal coherent light beams produces gratings of the ground state transmission, excited state absorption and possibly inversion in the sample

At positions of high excitation intensities the matter absorption at the excitation wavelength can be bleached and/or new excited state absorptions may occur at other wavelengths. Because of the nonlinear interaction the resulting transmission and absorption gratings will in general not be sinusoidal.

The distance Λ of the grating structure follows from the angle θ between the beams and the wavelength λ_{exc} of the pump light as:

$$\Lambda = \frac{\lambda_{inc}}{2 \sin\left(\dfrac{\theta}{2}\right)}. \tag{5.73}$$

In the case of sufficiently low intensities, the grating structure can be assumed to be almost sine modulated. Therefore the diffraction of a third light beam at this grating structure can be determined from the formulas of Sect. 5.9.1. The direction of the maxima and minima of the diffracted intensity can easily be calculated from the simple Bragg conditions. The intensity of the diffracted light in these directions can be very difficult to determine in special cases if, e.g. the grating structure is a complicated nonlinear function of the excitation. Numerical calculations may then be necessary. Examples for spectroscopy based on induced gratings can be found in, for example, [5.138, 5.139] and in Sect. 5.9.1 as well as in Sect. 7.8.4.

5.3.9 Induced Inversion Gratings

If laser active material is excited by the interference pattern of the pump beams an inversion grating may be generated. The laser action will be determined from this structure and the longitudinal modes may be selected. This is applied in distributed feedback (DFB) dye lasers for tuning the laser light (see Sect. 6.10.4) as schematically shown in Fig. 5.28.

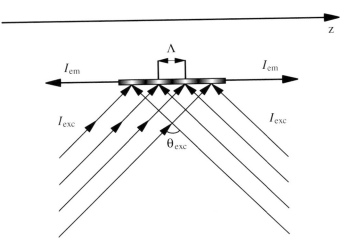

Fig. 5.28. Schematic of distributed feedback (DFB) dye laser for tuning the laser light by varying the angle between the pump beams

Because of the interference of the laser light along the z axis in the sample only a certain wavelength λ_{DFBL} can be amplified. This results from the angle θ_{exc} between the two pump beams and their wavelength λ_{exc}:

$$\lambda_{\mathrm{DFBL}} = \frac{n_{\mathrm{mat}}\lambda_{\mathrm{exc}}}{2 \sin\left(\dfrac{\theta_{\mathrm{exc}}}{2}\right)} \tag{5.74}$$

with refractive index of the material n_{mat}.

The spectral width of the laser is approximately given by:

$$\Delta\lambda_{\mathrm{laser}} = \frac{\lambda_{\mathrm{laser}}^2}{2L_{\mathrm{exc}}} \tag{5.75}$$

where L_{exc} is the length of the excited matter which determines the number of "grating lines".

In addition to this tuning effect the DFB dye laser can be operated at threshold and thus generates short pulses (Sect. 6.10.4).

5.3.10 Spatial Hole Burning

If the inversion is depopulated by the standing wave of the laser radiation in the resonator the effect is called spatial hole burning (in contrast to spectral hole burning described in Sect. 5.3.5). In this case the positions of the intensity maxima are coupled with the minima of the inversion and thus the minima of the gain g (see Fig. 5.29).

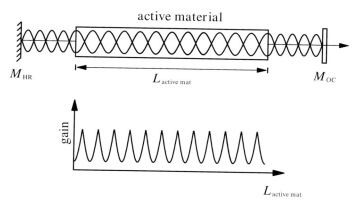

Fig. 5.29. Spatial hole burning in the active material of a laser via the standing light wave for a given longitudinal mode and the resulting inversion structure along the active material

The standing wave occurs in the laser resonator as a result of the steady-state field condition with field nodes on the mirror surfaces (see Sect. 6.7).

Thus the optical length L_{res} of the resonator defines the possible wavelengths $\lambda_{\text{laser}}(p)$ of the axial or longitudinal laser modes:

$$\text{possible axial modes}\quad \lambda_{\text{laser}}(p) = \frac{2L_{\text{res}}}{p} \tag{5.76}$$

where the optical length results, in the simple example of Fig. 5.29, from:

$$L_{\text{res}} = L_{\text{geom}} + L_{\text{active mat}}(n_{\text{active mat}} - 1) \tag{5.77}$$

and p as the axial mode order which is in practical cases of the order of 10^6.

If spatial hole burning occurs the gain of the active axial mode will drop. Because the gain minima are differently located for the different axial modes other possible active modes within the gain profile will start to oscillate. In consequence an oscillation between the axial modes can take place (*mode hopping*) and the laser output power will fluctuate. Usually the transversal modes are affected by this process too (see Sect. 6.7.5).

If the gain profile of the active matter is spectrally inhomogeneously broadened the different axial modes may oscillate completely independently and laser intensity fluctuations can take place for each axial mode separately only within the homogeneous linewidth. In the case of a laser with a spectrally inhomogeneously broadened active material usually several axial modes will be active if no further precautions are applied in the resonator (see Sect. 6.7.4).

Spatial hole burning in laser resonators can be avoided by preventing the standing waves in the laser. Several types of resonators are known to solve this problem. Examples are ring resonators with an optical valve forcing the light wave to travel in one direction only, as can be seen in Sect. 6.7.5, and unstable resonators as depicted in Sect. 6.5.2.

5.3.11 Induced Grating Spectroscopy

The nonlinear optical properties of matter especially of organic molecules can be investigated by inducing a transmission grating, an excited state absorption grating or an inversion grating in the sample with two pump pulses and using a third probe light beam for spectroscopy (see Fig. 5.30).

The probe light will be diffracted at the induced grating structure if the intensity of the two pump beams is high enough to reach the nonlinear range. Although the diffraction efficiency of the probe beam is usually smaller than 10% (see Sect. 3.9.16) this method allows a significant increase of detection sensitivity in nonlinear absorption measurements.

The experiment can be set up in a way that no pump light will be detectable in the observation direction of the scattered probe light and without nonlinear interaction also no probe light will be scattered. Thus the observation of scattered probe light has almost no background signal and can therefore reach a comparatively high sensitivity as in fluorescence measurements. For this purpose the angles of the incident pump beams and probe light beam have to be chosen appropriately.

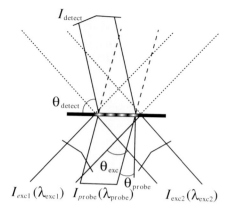

I_{detect}

θ_{detect}

θ_{exc}

θ_{probe}

$I_{exc1}(\lambda_{exc1})\quad I_{probe}(\lambda_{probe})\qquad I_{exc2}(\lambda_{exc2})$

Fig. 5.30. Induced grating spectroscopy: measuring the deflection of a probe light beam at an induced optical grating as a function of the excitation intensity

Because of this high sensitivity this type of spectroscopy (see Sect. 7.8.3) via induced nonlinear optical gratings can be used for the observation of very weak transient absorption changes in the range below 10^{-4} even in ultra-short pulse measurements with ps or fs time resolution.

5.4 Coherent Resonant Interaction

If dephasing in the matter after excitation is slower than the characteristic time constant of the interaction, such as, e.g. the pulse duration of the coherent laser light, coherent interaction can take place and new effects are observable [e.g. 5.140–5.154, M22, M28, M30, M33, M36, M37, M41, M42, M45–M52]. Dephasing times in optical experiments are typically below 1 ps. Usually a quantum theoretical description is necessary. The simplest and most useful formalism is based on the density matrix of the system and the visualizing technique based on Feynman diagrams. Unfortunately even the density matrix formalism allows analytical solutions only in the very simplest cases and numerical solutions can be obtained only for small systems with one or two optical transitions. Therefore the above methods based on Maxwell's equations (χ tensor formalism) and rate equations should be used, if no coherent interaction is present.

Up to now no important photonic applications have been based on coherent resonant interactions with a large importance. But new ideas in quantum computing or in quantum cryptography and other information technologies may change the situation in the near future.

5.4.1 Dephasing Time T_2

As will be shown in more detail below the coherent electric field of the excitation light beam induces electrical transition dipole moments μ_m in each particle, e.g. in the atom or molecule, of the material. The macroscopic non-linear polarization P_{nl} is the sum over all these microscopic dipole moments. In the case of coherent interactions these dipole moments oscillate with fixed phases. This takes place as long as the system is not distorted, e.g. by collisions (see Fig. 5.31).

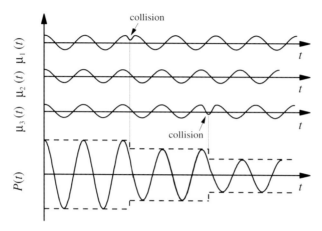

Fig. 5.31. Schematic of the decrease of nonlinear polarization P_{nl} of a system of many particles as a consequence of the phase disturbing collisions resulting in dephasing

The dephasing time T_2 is defined as the time during which the macroscopic polarization is decreased by $1/e$ after coherent excitation at time $t = 0$:

dephasing time T_2 $P_{nl}(t) = P_{nl}(0)\,e^{-t/T_2}$.   (5.78)

These dephasing times are usually smaller than 10^{-12} s for molecules at room temperature and usually in other materials shorter than µs [5.155–5.170]. Only at times shorter than this dephasing time can coherent interaction take place. Especially in molecular systems the dephasing can also result from the interaction of the electronic transition with the vibrations. Therefore, the spectral cross-relaxation and internal vibrational relaxation times (T_3, T_{IVR}) are limiting the coherent interaction too (see also Sect. 5.3.5). Some examples are given in [5.171–5.197].

5.4.2 Density Matrix Formalism

For the analysis of the coherent interaction the whole system of all partici-pating particles has to be taken into account. Therefore an averaging process over the single particles seems to be adequate. Therefore the quantum the-oretical description of the single particle is extended for averaging over the ensemble by the introduction of the density matrix ρ [see e.g. M28, M36, M37]. The elements of this matrix characterize the population of the different states of the matter including the transition states. This is based on the wavefunction of the single particles.

The potentially absorbing matter without any interaction is described in quantum theory by eigenstates numbered here by m with a certain energy E_m and a characteristic wavefunction Ψ_m which is a spatial function of all elements of the matter. In the case of a single atom or molecule the wavefunction depends on the positions of the atomic core(s) and electrons of this particle. Therefore this function can be visualized only in highly simplified cases.

Both values can be calculated for the stationary case, e.g. without interaction, from the stationary Schroedinger equation by:

stationary Schroedinger equation $H\psi_m(\boldsymbol{r}) = E_m\psi_m(\boldsymbol{r})$ (5.79)

In this differential equation H stands for the Hamilton operator which contains all the kinetic and potential energies of the system. As reminder the Hydrogen atom is described by:

$$H = -\frac{\hbar^2}{2m_{\text{sys}}}\nabla^2 + V(\boldsymbol{r})$$ (5.80)

with reduced system mass m_{sys} and the electrostatic potential V.

Whereas the energies of different parts of the system which can be analyzed independently have to be added:

$$E_{m,\text{total}} = E_{m,1} + E_{m,2} + E_{m,3} + \cdots$$ (5.81)

the wavefunctions have to be multiplied:

$$\psi_{m,\text{total}} = \psi_{m,1} \cdot \psi_{m,2} \cdot \psi_{m,3} \cdot \ \cdots$$ (5.82)

The energy of the different states m can be easily depicted in energy level schemes (Jablonski diagram) as shown earlier. For example, for molecules the energy levels for the different singlet and triplet or doublet electronic states and the additional vibrational energies for each of them are shown in Fig. 5.32.

It is often assumed that all particles of a system have the same energy levels and the energy levels are not shifted even under the influence of strong light. Only the population of these levels is assumed to be changed by the laser light.

Under these assumptions the transition of particles between these levels can be calculated with the time-dependent Schroedinger equation:

time-dependent Schroedinger Equation

$$i\hbar\frac{\partial\Psi(\boldsymbol{r},t)}{\partial t} = H\Psi(\boldsymbol{r},t)$$ (5.83)

with the new Hamilton operator containing the light field and thus the interaction with matter:

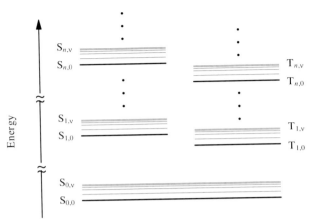

Fig. 5.32. Energy level scheme of a single molecule with an even number of electrons showing singlet (S) and triplet (T) electronic levels with additional vibrational levels. The first index counts the electronic states and the second the vibrational states

$$H = H_0 + H' \tag{5.84}$$

with

$$H' = -\mu \boldsymbol{E} = -q \boldsymbol{r}_{\text{dipole}} \boldsymbol{E} \tag{5.85}$$

where q describes the charge of the interacting dipole and E the electric field of the light wave (compare also Sect. 3.3.1).

With the assumption of only a small distortion of the system by the light beam the expectation value of the dipole moment μ for the transition between the states m and p can be calculated from:

transition dipole moment expectation value

$$\langle \mu_{p \leftarrow m} \rangle = \int_{-\infty}^{\infty} \psi_p q^* \boldsymbol{r}_{\text{dipole}} \psi_m \, dV. \tag{5.86}$$

For finally averaging over all participating particles it is helpful to expand the wavefunction of the single particle $\Psi(t)$ in a complete set of orthonormal functions, e.g. the eigen-wavefunctions of the particle in the different energy states ψ_m :

$$\Psi(\boldsymbol{r}, t) = \sum_m c_m(t) \psi_m(\boldsymbol{r}) \tag{5.87}$$

here the coefficients $c_m(t)$ describe the temporal evolution of the share of the m-th wavefunction of this particle in the total wavefunction of this particle.

The density matrix ρ of the entire system of particles can then be defined based on these coefficients $c_m(t)$ by:

density matrix $\rho_{mp}(t) = \overline{c_m^*(t) c_p(t)} \tag{5.88}$

These elements of the density matrix, which is as large as the number of considered eigenstates (or wavefunctions) of the single particle squared, are the average over all particles of the system, noted by the bar above the product. This matrix is obviously a Hermite matrix:

$$\rho_{mp} = \rho_{pm}^* \tag{5.89}$$

The density matrix elements have an easily understood meaning. The diagonal elements ρ_{mm} give the total average population density share in the m eigenstates:

$$\text{population of state } m \quad N_m(t) = \rho_{mm}(t)N_{\text{total}} \tag{5.90}$$

whereas the non-diagonal elements ρ_{mp} describe the population density share of the coherent transition between the states m and p:

$$\text{transition } m \to p \quad \rho_{mp} \tag{5.91}$$

If the nondiagonal elements of the density matrix are zero, no coherent interaction takes place and the rate equations can be used to describe the process.

The so-defined density matrix ρ can be used to determine the expectation value of any physical quantity P as the average over all particles from the expressions:

expectation value of P

$$\begin{aligned}
\langle \mathsf{P} \rangle_{\text{ensemble}} &= \text{Tr}(\rho\mathsf{P}) \\
&= \overline{\int_{-\infty}^{\infty} \Psi^*(\boldsymbol{r},t)\mathsf{P}\Psi(\boldsymbol{r},t)\,\mathrm{d}V} \\
&= \overline{\langle \Psi(\boldsymbol{r},t)\,|\mathsf{P}|\,\Psi(\boldsymbol{r},t)\rangle} \\
&= \sum_l p_l \langle \Psi_l(\boldsymbol{r},t)|\mathsf{P}|\Psi_1(\boldsymbol{r},t)\rangle \\
&= \sum_l p_l \left(\sum_{m,p} c_m^* c_p P_{m,p} \right)
\end{aligned} \tag{5.92}$$

where in the third line the bracket formalism is used to express the second line. In the fourth line p_l represents the probability of the system being in state l or the share of particles being in state l. The density matrix operator ρ can be written as:

density matrix operator

$$\rho = \overline{|\Psi(\boldsymbol{r},t)\rangle\langle\Psi(\boldsymbol{r},t)|} = \sum_l p_l |\Psi(\boldsymbol{r},t)\rangle\langle\Psi(\boldsymbol{r},t)| \tag{5.93}$$

Using this formula, e.g. the nonlinear polarization P_{nl} of the system under light irradiation can be calculated including the coherent interaction and averaging over the dephasing.

The temporal development of the density matrix is given by the Liouville equation:

Liouville equation $\quad \dfrac{\partial \rho}{\partial t} = \dfrac{\mathrm{i}}{\hbar}[\rho, \mathsf{H}] - \dfrac{1}{2}\{\rho, \Gamma\}$ $\hfill$ (5.94)

which can be derived from the Schroedinger equation:

$$\mathrm{i}\hbar \sum_m \frac{\partial c_m(t)}{\partial t} \psi_m(\boldsymbol{r}) = \sum_l c_m(t)\mathsf{H}\psi_m(\boldsymbol{r}) \hfill (5.95)$$

by multiplying by ψ_p^*, integration and considering the Hamilton operator is hermitic. This leads to

$$\mathrm{i}\hbar\frac{\partial c_m(t)}{\partial t} = \sum_p c_p(t)H_{pm} \quad \text{and} \quad \mathrm{i}\hbar\frac{\partial c_m^*(t)}{\partial t} = \sum_p c_p^*(t)H_{mp} \hfill (5.96)$$

if the relation

$$\frac{\partial \rho}{\partial t} = \overline{c_p \frac{\partial c_m^*}{\partial t}} + \overline{c_m^* \frac{\partial c_p}{\partial t}} = \sum_l p_l \left(c_p \frac{\partial c_m^*}{\partial t} + c_m^* \frac{\partial c_p}{\partial t} \right) \hfill (5.97)$$

is used.

Using the second term in (5.94), relaxation processes or other randomization processes such as the population decay of excited states, spontaneous emission or collisions can be considered. The phenomenologically introduced relaxation operator Γ fulfills the condition $\{\rho, \Gamma\} = \rho\Gamma + \Gamma\rho$.

Some further examples using the density matrix formalism will be given below and can be found in [5.198–5.200].

5.4.3 Modeling Two-Level Scheme

As an example of applying the density matrix formalism the nonlinear polarization of a two-level scheme with inhomogeneous broadening of the absorptive transition shall be described (see Fig. 5.33).

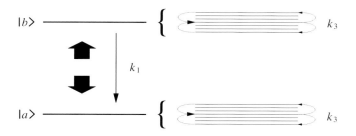

k_2: phase relaxation

Fig. 5.33. Energy level scheme with two-levels a and b separated by the energetic distance $h\nu_0$ with a spectral broadening and an internal relaxation time T_3 within these levels. The energy relaxation time for $b \rightarrow a$ is T_1 and the phase relaxation time for coherent coupling is T_2

In this model the absorption occurs at the frequency ν_0 and the inhomogeneous broadening is considered by the distribution of the absorption and stimulated emission probability with weight function $g(\nu_0)$. Therefore the density matrix becomes a function of the time and frequency $\rho(\nu, t)$. Consideration of the assumed relaxation processes of the model of Fig. 5.33 leads to the following system of Liouville differential equations for the elements of the density matrix:

$$\frac{\partial \rho_{aa}}{\partial t} = \frac{i}{\hbar}[\rho, \mathsf{H}]_{aa} - k_1(\rho_{aa} - \rho_{aa}^0) - k_3\rho_{aa}$$
$$+ k_3 \frac{g(\nu_0)}{\Delta\nu_{\mathrm{FWHM}}} \int_{-\infty}^{\infty} \rho_{aa}(\nu')\, d\nu' \qquad (5.98)$$

$$\frac{\partial \rho_{bb}}{\partial t} = \frac{i}{\hbar}[\rho, \mathsf{H}]_{bb} - k_1(\rho_{bb} - \rho_{bb}^0) - k_3\rho_{bb}$$
$$+ k_3 \frac{g(\nu_0)}{\Delta\nu_{\mathrm{FWHM}}} \int_{-\infty}^{\infty} \rho_{bb}(\nu')\, d\nu' \qquad (5.99)$$

$$\frac{\partial \rho_{ba}}{\partial t} = \frac{i}{\hbar}[\rho, \mathsf{H}]_{ba} - k_2\rho_{ba} \qquad (5.100)$$

with ρ^0 as the density matrix of the particle without interaction. In these equations the first expression describes pumping with the external light. The second term gives the decay of the population in the first two equations and of the phase in the third. The third and fourth terms in the first two equations describe the exchange between, for example, vibrational levels with the rate k_3 which allows hole burning. Thus, levels a and b, which are coupled to the light, decay into the reservoir and all reservoir levels decay into these two levels to some extent. For very large k_3 values no hole burning will be obtained. The lineshape of the whole band is described by $g(\nu_0, \nu)$ and the full-width half-maximum of this band is given by $\Delta\nu_{\mathrm{FWHM}}$. The integral over $g(\nu)$ divided by $\Delta\nu_{\mathrm{FWHM}}$ is normalized to 1.

This problem can be solved by perturbation theory assuming that the energy levels of the particle are not changed by the interaction. Therefore the particle–light interaction can be written as an additive value H' in the Hamiltonian:

$$\mathsf{H} = \mathsf{H}_0 + \mathsf{H}' \quad \text{with} \quad \mathsf{H}' = -\mu\boldsymbol{E} \qquad (5.101)$$

with the matrix elements of the dipole moment operator

$$\mu_{aa} = \mu_{bb} = 0 \quad \text{and} \quad \mu_{ab} = \mu_{ba}^* \qquad (5.102)$$

not considering the orientation of the dipole explicitly. This is equivalent to an averaging over all particle orientations and neglecting rotational relaxation of the particles. This assumption is more valid the longer the orientational relaxation time of the particles compared with the internal relaxation times. For molecular systems in solution this time is of the order of a few ps up to a few hundred ps as a function of the viscosity and the permanent dipole moments. In pump and probe spectroscopy the magic angle can be used for dealing with this point (see Sect. 7.1.5).

With respect to pump and probe experiments especially, e.g. in the case of hole burning investigations, the light is described by the sum over two monochromatic waves with frequency ν_1 for the pump beam and ν_2 for the probe beam:

$$\boldsymbol{E} = \boldsymbol{E}_{\text{pump}}(\boldsymbol{r})\,e^{(-i2\pi\nu_1 t)} + \boldsymbol{E}_{\text{probe}}(\boldsymbol{r})\,e^{(-i2\pi\nu_2 t)}. \tag{5.103}$$

With these expressions the Liouville equation for the difference of the population density ρ_D:

$$\rho_D = \rho_{aa} - \rho_{bb} \tag{5.104}$$

can be derived:

$$
\begin{aligned}
\frac{\partial \rho_D}{\partial t} = {} & -\frac{2i}{\hbar}(H'_{ab}\rho_{ba} - \rho_{ab}H'_{ba}) - k_1(\rho_D - \rho_D^{(0)}) \\
& - k_3\rho_D + k_3 g(\nu_0)\frac{1}{\Delta\nu_{\text{FWHM}}}\int_{-\infty}^{\infty}\rho_D(\nu')\,d\nu
\end{aligned}
\tag{5.105}
$$

and the nondiagonal elements follow from:

$$\frac{\partial \rho_{ba}}{\partial t} = \frac{\partial \rho_{ab}^*}{\partial t} = -\frac{i}{\hbar}H'_{ba}\rho_D - (k_2 + i\nu_0)\rho_{ba} \tag{5.106}$$

The density matrix can be expanded in this approximation as a sum of the undistorted value ρ^0 and higher-order values $\rho^{(l)}$ as a function of the different frequency components:

$$\rho_{aa} = \rho_{aa}^{(0)} + \rho_{aa}^{(2)}(0) + \rho_{aa}^{(2)}(\nu_2 - \nu_1)\,e^{-i2\pi(\nu_2 - \nu_1)t} + \cdots \tag{5.107}$$

and

$$\rho_{bb} = \rho_{bb}^{(2)}(0) + \rho_{bb}^{(2)}(\nu_2 - \nu_1)\,e^{-i2\pi(\nu_2 - \nu_1)t} + \cdots \tag{5.108}$$

for the diagonal elements, and for the nondiagonal elements:

$$
\begin{aligned}
\rho_{ml} = \rho_{lm}^* = {} & \left(\rho_{ml}^{(1)}(\nu_1) + \rho_{ml}^{(3)}(\nu_1) + \cdots\right)e^{-i2\pi\nu_1 t} \\
& + \left(\rho_{ml}^{(1)}(\nu_2) + \rho_{ml}^{(3)}(\nu_2) + \cdots\right)e^{-i2\pi\nu_2 t}.
\end{aligned}
\tag{5.109}
$$

With these values the relevant third-order nonlinear polarization $P_{\text{nl}}^{(3)}$ can be calculated, e.g. for the probe light frequency ν_2, from:

$$P_{\text{nl}}^{(3)}(\nu_2) = 2\pi N_0\mu\int_{-\infty}^{\infty}\overline{\rho_{ba}^{(3)}(\nu_0, \nu_2)}\,d\nu_0 \tag{5.110}$$

and finally the experimentally easily accessible change absorption coefficient Δa:

$$\Delta a(\nu_2) = \frac{2\pi\nu_2}{c}\,\text{Im}\left(\frac{P_{\text{nl}}^{(3)}(\nu_2)}{\varepsilon_0 E_2(\nu_2)}\right). \tag{5.111}$$

In the following calculation of the third-order term of the density matrix as a function of the probe light frequency $\rho_{ba}^{(3)}(\nu_2)$ for the stationary case, rapidly oscillating terms with $2\nu_1$ and $2\nu_2$ and so on were neglected.

As result *spectral hole burning with incoherent interaction* occurs:

$$\frac{\partial \rho_{ba}^{(3)}(\nu_2)\, e^{-i2\pi\nu_2 t}}{\partial t} = -\frac{i}{\hbar} H_{ba}'(\nu_2)\rho_D^{(2)}(0)\, e^{-i2\pi 0 t}$$
$$-(k_2 + i2\pi\nu_0)\rho_{ba}^{(3)}(\nu_2)\, e^{-i2\pi\nu_2 t} \tag{5.112}$$

with the stationary population density difference

$$\rho_D^{(2)}(0)\, e^{-i2\pi 0 t} = \rho_D^{(2)} = \text{const.} \tag{5.113}$$

which is produced by the strong excitation with frequency ν_1 and is not changed by the weak probe light with the frequency ν_2. The solution of this equation can be integrated (see (5.110)) with the assumption of a much larger bandwidth $g(\nu)$ compared to the width of the burned hole. The solution of the nonlinear polarization $P_{\text{nl}}^{(3)}$ can be used to calculate the change in absorption Δa from (5.111)). There are two terms counting incoherent and coherent interactions. The incoherent part is given by:

$$\Delta a_{\text{ic}}(\nu_2) = -\frac{N_0 \Omega_R^2 \rho_D^{(0)}(\nu_2)}{12(k_1 + k_3)} \left[\frac{k_3}{k_1} \frac{1}{\Delta\nu_{\text{FWHM}}} + \frac{2k_2}{(\nu_1 - \nu_2)^2 + (2k_2)^2} \right] \tag{5.114}$$

with N_0 as the total population density of the particles and the Rabi frequency for the pump light:

Rabi frequency $\Omega_R = 2\pi\dfrac{\mu_{ba}E_1}{h} \propto \sqrt{I_{\text{pump}}} \tag{5.115}$

which describes the oscillation of the population density in the upper state [5.201–5.204]. Its square is proportional to the incident pump intensity I_{pump}.

These formulas describe the spectral behavior of the bleaching of the inhomogeneous broadened absorption transition of the sample. The first part of (5.114) gives the homogeneous share of the bleaching which is constant over the whole band. The second part describes the possible additional bleaching of the burned hole.

The relation between the stronger bleaching of the burned hole and the bleaching of the whole band is dependent on the relation of k_3/k_1 (see (5.114)) as found with rate equations, too.

Spectral hole burning with coherent interaction has to be calculated from:

$$\frac{\partial \rho_{ba}^{(3)}(\nu_2)\, e^{-i2\pi\nu_2 t}}{\partial t} = -\frac{i2\pi}{h} H_{ba}'(\nu_1)\rho_D^{(2)}(\nu_2 - \nu_1)\, e^{-i2\pi(\nu_2-\nu_1)t}$$
$$-(k_2 + i2\pi\nu_0)\rho_{ba}^{(3)}(\nu_2)\, e^{-i2\pi\nu_2 t} \tag{5.116}$$

containing an oscillating population density difference $\rho_D^{(2)}$ beating with the difference frequency $\nu_2 - \nu_1$. This coherence effect broadens the hole further. A discussion of this effect is given in [5.205].

In this approximation the burned hole shows a Lorentzian line shape with a width given by the phase relaxation rate k_2 or by the inverse of the phase relaxation time $1/T_2$ which is identical to the homogeneous linewidth of the subbands.

It can be shown that the burned hole will be broadened if saturation of the transition is considered. The hole shape can be described by:

$$\Delta a = \Delta a_{max} \frac{k_2^2 (I/I_{nl-HB})}{4\pi^2 (\nu_2 - \nu_1)^2 + k_2^2 (1 + I/I_{nl-HB})} \tag{5.117}$$

with the intensity I_{nl-HB} given by:

$$I_{nl-HB} = \frac{\varepsilon_0 c_0 h^2 k_1 k_2}{32\pi^3 n \mu_{ba}^2} \tag{5.118}$$

where n stands for the refractive index of the sample.

The resulting change of the shape of the burned hole is depicted in Fig. 5.34.

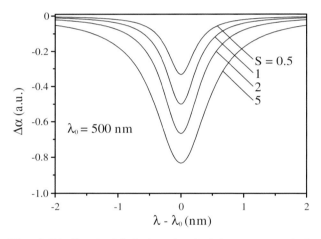

Fig. 5.34. Spectral hole burning in inhomogeneously broadened absorption of a fictive material with $T_2 = 50\,$fs. The hole width and depth increase with increasing excitation intensity I_{exc} which is given by S denoting for intensity times above nonlinear intensity of the system

With increasing pump intensity I_{exc} the hole width is broadened proportionally to $\sqrt{1 + I/I_{nl-HB}}$ and as expected the hole depth is increased. The maximum hole depth is Δa_{max}. The modeling of the results of spectral hole burning experiments allows the determination of the parameters of the sample in a coherent interaction. Further detailed mathematical description is given in [e.g. 5.206–5.211].

5.4.4 Feynman Diagrams for Nonlinear Optics

For the evaluation of the density matrix in nonlinear optical processes with several light beams a special form of Feynman diagrams may be used as introduced in [5.212] and described in detail in [5.213].

The density matrix operator (see Sect. 5.4.2) can be written as the average over the product of the ket and the bra vectors of the quantum state characterized by the wavefunction ψ:

density matrix operator $\rho = \overline{|\Psi\rangle\langle\Psi|}.$ \hfill (5.119)

Thus two lines of propagation are necessary to describe the ket and the bra side of this operator (see Fig. 5.35 for examples).

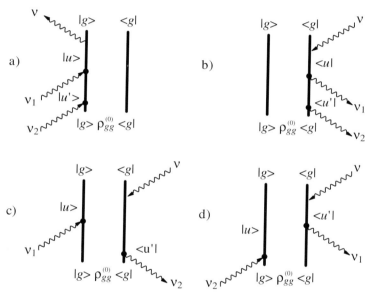

Fig. 5.35. Basic set of four Feynman diagrams describing the nonlinear optical interaction for the second-order density matrix $\rho^{(2)}$ in sum frequency generation (after [5.213])

The rules for using these diagrams to write down the expression for the temporal evolution of the density matrix $\rho^{(l)}$ and thus for the nonlinear susceptibility $\chi^{(l)}$ are:

i)   System start is $|g\rangle\rho_{gg}^{(0)}\langle g|$ (bottom of the diagram).
ii)   For propagation of the ket state factors are multiplied on the left and the bra state factors on the right.

iii)   Absorption on the left side (ketside):

$|b\rangle$ $\langle g|$

$\left(\dfrac{1}{i\hbar}\right)\langle b|H'(\nu_n)|a\rangle$

$|a\rangle$ $\langle g|$

iv)   Emission on the left side (ketside):

$|b\rangle$ $\langle g|$

$\left(\dfrac{1}{i\hbar}\right)\langle b|H'^{(+)}(\nu_n)|a\rangle$

$|a\rangle$ $\langle g|$

v)   Absorption on the right side (braside):

$|g\rangle$ $\langle b|$

$-\left(\dfrac{1}{i\hbar}\right)\langle a|H'^{(+)}(\nu_n)|b\rangle$

$|g\rangle$ $\langle a|$

vi)   Emission on the right side (braside):

$|g\rangle$ $\langle b|$

$-\left(\dfrac{1}{i\hbar}\right)\langle a|H'(\nu_n)|b\rangle$

$|g\rangle$ $\langle a|$

vii)   Propagation of the system in time along the double lines $|l\rangle\langle k|$ from the mth to the $(m+1)$th interaction is described by the propagator Π:

$$\Pi_m = \pm\left[i\left(\sum_{n=1}^{m} 2\pi(\nu_n - \nu_{lk} + ik_{lk})\right)\right]^{-1} \tag{5.120}$$

in which the frequency ν_n is taken positive if absorption at this frequency occurs at the left or emission at the right, and negative otherwise.

viii)   The final state density matrix operator of the system results from the product of the final ket and bra states (e.g. $|u'\rangle\langle u|$).

ix)   Propagation of the system via a particular set of states of the diagram is described by the product of all factors from the initial state to the final state (e.g. $|g\rangle\langle g|\cdots|u'\rangle\langle u|$). The sum over all possible products results in a density matrix which contains contributions from all states.

The absorption on the ketside appears as emission on the braside and vice versa. The Hamilton operator H' for the interaction of the photon with frequency ν_l with the matter is proportional to:

$$H'(\nu_1) \propto e^{-i2\pi\nu_1 t}$$

$$(5.121)$$

and it will annihilate a photon on the ketside and will create a photon if applied on the braside.

If a diagram can be obtained from a permutation of another diagram because identical photons are present they would result in identical expressions for the density matrix $\rho^{(u)}$. Thus the different diagrams and therefore the different interactions can be described by a degeneracy factor for this share of the density matrix. Thus the second-order process $\nu = \nu_1 + \nu_2$ leads to eight diagrams but only four of them are necessary if the possible permutation of ν_1 and ν_2 is considered (see Fig. 5.36).

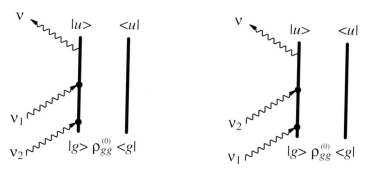

Fig. 5.36. Two of the Feynman diagrams for sum frequency generation in which the incident photons 1 and 2 can be permuted resulting in identical diagrams leading to a degeneracy factor of 2 for this share of the density matrix

Using the expression for the density matrix from the example of Fig. 5.35 the components of the second-order nonlinear susceptibility $\chi^{(2)}$ can be calculated from the components of the nonlinear polarization $P^{(2)}$ and the electric fields E:

$$\chi_{ijk}^{(2)}(\nu = \nu_1 + \nu_2) = \frac{P_i^{(2)}(\nu)}{\varepsilon_0 E_j(\nu_1) E_k(\nu_2)}$$

$$(5.122)$$

and the nonlinear polarization follows from:

$$P_i^{(2)} = \mathrm{Tr}(\rho^{(2)} P_i)$$

$$(5.123)$$

The result was given in [5.213]. The second-order nonlinear susceptibility follows as:

$$\chi_{lmv}^{(2)}(\nu = \nu_1 + \nu_2) = \frac{P_l^{(2)}(\lambda)}{\varepsilon_0 E_m(\nu_1) E_v(\nu_2)}$$

$$= -N_0 \frac{4\pi^2 e_e^3}{\varepsilon_0 h^2} \sum_{g,u,u'} \left[\frac{(r_1)_{gu}(r_m)_{uu'}(r_v)_{u'g}}{(2\pi(\nu - \nu_{ug}) + ik_{ug})(2\pi(\nu_2 - \nu_{u'g}) + ik_{u'g})} \right.$$

$$+ \frac{(r_1)_{gu}(r_v)_{uu''}(r_m)_{u'g}}{(2\pi(\nu - \nu_{ug}) + ik_{ug})(2\pi(\nu_1 - \nu_{u'g}) + ik_{u'g})}$$

$$+ \frac{(r_v)_{gu'}(r_m)_{u'u}(r_1)_{ug}}{(2\pi(\nu + \nu_{ug}) + ik_{ug})(2\pi(\nu_2 + \nu_{u'g}) + ik_{u'g})}$$

$$+ \frac{(r_m)_{gu'}(r_v)_{u'u}(r_1)_{ug}}{(2\pi(\nu + \nu_{ug}) + ik_{ug})(2\pi(\nu_1 + \nu_{u'g}) + ik_{u'g})}$$

$$- \frac{(r_m)_{ug}(r_1)_{u'u}(r_v)_{gu'}}{(2\pi(\nu - \nu_{uu'}) + ik_{uu'})}$$

$$\cdot \left(\frac{1}{2\pi(\nu_1 + \nu_{u'g}) + ik_{u'g}} + \frac{1}{2\pi(\nu_1 - \nu_{ug}) + ik_{ug}} \right)$$

$$- \frac{(r_v)_{ug}(r_1)_{u'u}(r_m)_{gu'}}{(2\pi(\nu - \nu_{uu'}) + ik_{uu'})}$$

$$\left. \cdot \left(\frac{1}{2\pi(\nu_2 - \nu_{ug}) + ik_{ug}} + \frac{1}{2\pi(\nu_1 + \nu_{u'g}) + ik_{u'g}} \right) \right] \rho_g^{(0)}. \qquad (5.124)$$

This method can also be applied for higher-order calculation. But should only be used for cases demanding the quantum description. Otherwise the above calculations based on Maxwell's equations or rate equations may be much easier to use and more appropriate to describe nonlinear optical experiments, especially if more than two-levels are involved or nonresonant interactions are investigated.

5.4.5 Damped Rabi Oscillation and Optical Nutation

In the simplest picture of an optical transition the material is described quantum mechanically by a two-level model with resonance frequency ν_0. If a near-resonant light beam with electric field amplitude E_{exc} and frequency ν_{exc} is applied the transmission of the sample will be modulated with the Rabi frequency Ω_R:

Rabi frequency $\quad \Omega_R = \sqrt{2\pi(\nu_{exc} - \nu_0)^2 + \left(\frac{\mu_{ba} E_{exc}}{2\pi h} \right)^2}. \qquad (5.125)$

If the particle relaxes from the excited level to the ground level with energy relaxation time T_1 and phase disturbing mechanisms are present resulting in the phase relaxation time T_2 a damping of these oscillations will occur. The damping time τ_R is given for the exponential decay after resonant excitation by:

$$\text{damping time} \quad \tau_{\mathrm{R}} = 2\left(\frac{1}{2T_1} + \frac{1}{T_2}\right)^{-1} \tag{5.126}$$

and nonresonant excitation results in a slightly nonexponential decay of the form [5.214]:

$$\mu_{ba} \propto \frac{e^{-t/\tau_{\mathrm{R}}}}{[1 + C_{\mathrm{R}}t^2(\nu_{\mathrm{exc}} - \nu_0)^4]^{1/4}} \tag{5.127}$$

with a small value of C_R depending on E_{exc} and thus the correction can mostly be neglected.

To obtain the Rabi oscillations in the transmission the damping time has to be longer than the inverse Rabi oscillation frequency. Thus the intensity has to be high enough. In most cases the phase relaxation is the fastest process and thus the T_2 time is the relevant value to compare. For atoms in the gas phase T_2 may be in the range of several 10 ns and intensities above some $\mathrm{W\,cm^{-2}}$ may be sufficient. For the observation of Rabi oscillations in molecular vibrational transitions intensities in the range of some $\mathrm{kW\,cm^{-2}}$ may be necessary and for molecular electronic transitions with T_2 times in the sub-ps range even $\mathrm{MW\,cm^{-2}}$ to $\mathrm{GW\,cm^{-2}}$ are necessary.

The observable effect of the periodic absorption (particles in the ground state) and amplification (particles in the excited state) of the incident light as a function of time is observable as optical nutation (see Fig. 5.37) or optical free induction decay.

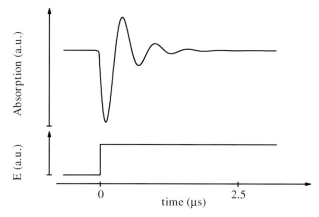

Fig. 5.37. Damped Rabi oscillations (schematic) as, e.g. observable from optical vibrational nutation in $^{13}\mathrm{CH_3F}$ with an excitation wavelength of 9.7 μm [M10]

This formalism of coherent optical interaction is developed similar to the description of coherent NMR interactions where the time constants are in the μs to ms range and thus all processes are observable much more easily than in optics [5.214–5.220].

5.4.6 Quantum Beat Spectroscopy

If the particles have two closely spaced energy levels such as, e.g. two excited states and the exciting light excites both levels at the same time with a spectrally wide short laser pulse a coherent interaction between these two states will occur. As a result the absorption and emission of this particle will show temporal oscillations which are known as *quantum beats* [5.221–5.237].

The two states (see Fig. 5.38) have energies E_1 and E_2 in addition to the ground state energy E_0.

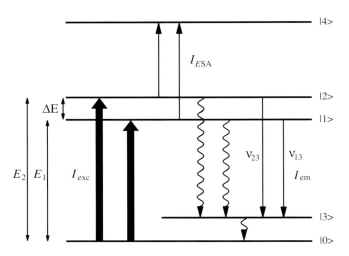

Fig. 5.38. Energy level scheme with two closely spaced states 1 and 2 resulting in quantum beats after simultaneous excitation

The energetic spacing between these levels 1 and 2 is ΔE. If the excitation pulse duration is shorter than $(\Delta E/h^{-1})$ a coherent state of the two excited levels will be prepared. Assuming both states decay with the same decay time τ_{exc} and the absorption cross-section for the excitation of the two-levels is the same the resulting wavefunction of the superposition of these upper state(s) can be written as:

$$\psi(t) = \psi_1(0)\,\mathrm{e}^{-(\mathrm{i}2\pi\nu_{13}+t/2\tau_{\mathrm{exc}})} + \psi_2(0)\,\mathrm{e}^{-(\mathrm{i}2\pi\nu_{23}+t/2\tau_{\mathrm{exc}})}$$

$$= [\psi_1(0) + \psi_2(0)]\,\mathrm{e}^{-\mathrm{i}2\pi\Delta\nu t}\,\mathrm{e}^{-t/\tau_{\mathrm{exc}}}. \tag{5.128}$$

Since the emission intensity I_{em} as well as the absorbed intensity I_{ESA} with a spectral resolution not discriminating the transitions from the two states are proportional to the expression:

$$I_{\text{em}} \propto |\langle\psi_3|\boldsymbol{E}\boldsymbol{\mu}|\psi(t)\rangle|^2 \quad \text{and} \quad I_{\text{ESA}} \propto |\langle\psi_4|\boldsymbol{E}\boldsymbol{\mu}|\psi(t)\rangle|^2 \tag{5.129}$$

this spectrally unresolved measurement of the emission or absorption intensities from these states will show an exponential decay, but in addition an oscillation with the beat frequency $\Delta\nu$:

$$I_{\text{em}}(t), I_{\text{ESA}}(t) \propto e^{-t/\tau_{\text{exc}}}[C_1 + C_2\cos(2\pi\Delta\nu t)]. \tag{5.130}$$

This modulation in the decay function (see Fig. 5.39) for the population of this doublet of states represents the coherent superposition of the quantum states.

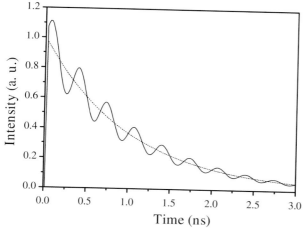

Fig. 5.39. Oscillation (quantum beat) in the decay measurement via emission or excited state absorption from a doublet of an energetically closely spaced excited level (by 3 GHz) occurring if the signals from both transitions are not spectrally resolved and the sample decays with a decay time of 1 ns

Therefore from this temporal measurement the energy difference of the two states can be determined with higher accuracy the closer the levels are. This method allows a resolution better than Doppler broadening.

5.4.7 Photon Echoes

If an ensemble of atoms or molecules is excited with two short and intensive laser pulses of a certain energy with a sufficiently small delay Δt between them a new pulse can be obtained as a photon echo after a second delay Δt if several conditions are fulfilled [5.238–5.271].

As a result of the excitation with the first short pulse the dipole moments of all particles will first oscillate in phase. But these particles may have slightly different resonance frequencies distributed over the range $\Delta\nu$ and thus the oscillation will dephase in time. But for times not longer than T_2, the initial phase information is still present in the system. With the second short pulse of a certain energy applied at the time Δt_1 the oscillation can be phase-shifted by 180°. As a result the oscillations will rephase and after the time Δt_2 all particles are in phase again except for the phase information loss resulting from T_2 collisions. Thus after that time a light pulse can be emitted as an echo of the first two pulses (see Fig. 5.40).

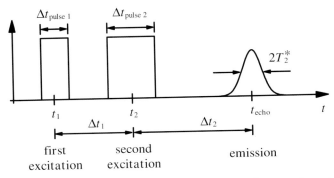

Fig. 5.40. Light pulse sequence in photon echo experiments

The phases of the Rabi oscillations may be visualized as given in Fig. 5.41. The temporal development of the phases is depicted as the angle of the polar coordinates of the dipole moment vector in the xy plane for an exciting light pulse propagating in the z direction.

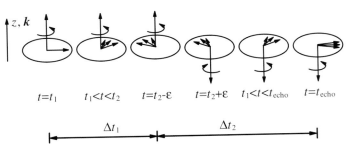

Fig. 5.41. Phases of the Rabi oscillations of particles excited at t_1 and inverted with a second pulse at the time t_2 resulting in a minimal phase difference at time t_{echo} for the pulsed excitation after Fig. 5.40

If the exciting pulse is short and intensive enough to provide a sufficiently strong electric field:

$$E_{\text{exc}} \gg \frac{h}{\mu_{ba}} |\nu - \nu_{\text{exc}}| \tag{5.131}$$

all dipoles will be excited after the pulse with the same phase. The phase angle θ_{phase} develops in time as:

$$\theta_{\text{phase}} = \int_0^{t_{\text{FWHM,pulse}}} \frac{\mu_{ba}}{2\pi h} E_{\text{exc}} \, dt. \tag{5.132}$$

For the photon echo experiment the first pulse has to produce a phase shift of:

1. excitation pulse $\theta_{\text{phase,pulse1}} = \dfrac{\pi}{2}$. $\tag{5.133}$

During the following time $t < t_2$ the phase angles become more and more different. Usually many oscillations take place. Coherent oscillation is destroyed after the fanning out time T_2^*:

$$\text{fanning out time} \quad T_2^* = \frac{1}{2\pi\Delta\nu} \tag{5.134}$$

for the distributed resonance frequencies over the range $\Delta\nu$. T_2^* is in photon echo experiments shorter than $\Delta t_{1/2}$.

Than a second short intensive laser pulse is applied after a time Δt_1. It has to produce a phase shift of:

2. excitation pulse $\theta_{\text{phase,pulse2}} = \pi$ $\tag{5.135}$

and thus the phase is inverted and the following process can be illustrated as a "back-rotation" in the phase diagram as depicted in Fig. 5.41. Finally after a second time interval Δt_2 which is obviously the same as $\Delta t_1 = \Delta t_2$ all dipoles are in phase again except for the fanning out time resulting from the spectral broadening of the particles. This results in a spreading of the phase and thus the observed emission, the photon echo, of the sample will have a pulse width of $2T_2^*$ as illustrated in Fig. 5.40.

Of course this type of coherence experiment can be carried out only for delay times $\Delta t_{1/2}$ shorter than the T_2 describing the phase disturbing collisions:

photon echo condition $\Delta t_{1/2} < T_2$ $\tag{5.136}$

whereas the measurement of the echo intensity as a function of the delays $\Delta t_{1/2}$ shows an exponential decay which is described by the T_2 time which can be determined this way.

In photon echo experiments phase shifts different from π and $\pi/2$ also are possible. In any case the intensity of the excitation has to be strong enough to produce enough transition dipole moments in competition to the relaxation processes. Thus the necessary intensities are in the range:

photon echo intensities $I_{\text{exc}} \geq \dfrac{c_0 \varepsilon_0 n h}{2\mu_{ba}^2} \left(\dfrac{1}{T_1} + 2\pi \Delta\nu \right)$ (5.137)

with the transition dipole moment μ_{ab} which can be determined from the cross-section spectrum as described in Sect. 3.3.1 and the energy relaxation time of the transition T_1.

Because of the different T_2 times for different materials such as atoms, solids and molecules, photon echo experiments need laser pulses of different lengths ranging from µs to fs.

In addition to temporal and energetic conditions an angular condition for the wave vectors has to be fulfilled:

wave vector condition $\mathbf{k}_{\text{echo}} = 2\mathbf{k}_{\text{pulse1}} - \mathbf{k}_{\text{pulse2}}$ (5.138)

which allows background-free detection of the weak scattering signals as a function of the delay of the second excitation pulse.

5.4.8 Self-Induced Transparency: Π Pulses

If optically thick samples with transmissions in the range of some 10% or below are used in coherent nonlinear experiments the shape of short pulses will be changed by matter absorption. It was observed that light pulses of a certain pulse profile (hyperbolic secant) and related pulse energies can transmit even through strongly absorbing samples without any change of their shape and energy [5.272–5.285]. Such pulses are called Π *pulses*.

In this case the front part (leading edge) of the pulse is absorbed producing an inversion in the sample. This inversion amplifies the trailing part of the pulse and thus the energy is conserved for the pulse. But the process causes a delay of the pulse which is longer the further the pulse transmits through the sample and thus the light seems to move much slower than c_0/n (see Fig. 5.42).

The description of this self-induced transparency is based on the nonlinear wave equation (4.7). Assuming the pulse propagates unchanged with velocity v_{pulse} in the z direction through the material with induced dipole moment μ_{mat} the temporal part of the wave equation can be written as [5.285]:

$$\frac{\partial E(t - z/v_{\text{pulse}})}{\partial t} = \frac{1}{\tau_{\text{setr}}} E \sqrt{1 - \left(\frac{2\pi \mu_{ba} \tau_{\text{setr}}}{h} E \right)^2}$$ (5.139)

with

$$\tau_{\text{setr}} = \sqrt{\frac{\varepsilon_0 h}{4\pi c_0 \nu N_0 \mu_{ba}^2} \left(\frac{1}{v_{\text{pulse}}} - \frac{1}{c_0} \right)}.$$ (5.140)

The solution of this equation gives for the shape of the unchanged pulse:

$$E(t - z/v_{\text{pulse}}) = \frac{h}{2\pi \mu_{ba} \tau_{\text{setr}}} \text{sec h} \left[\frac{1}{\tau_{\text{pulse}}} \left(t - \frac{z}{v_{\text{pulse}}} \right) \right]$$ (5.141)

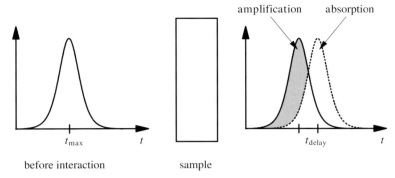

Fig. 5.42. Transmission of a Π pulse through an absorbing sample without change of pulse shape and energy via absorption of the leading edge and amplification of the trailing edge resulting in a delay t_{delay}

and the integral over this pulse shape has the value:

$$\int_{-\infty}^{\infty} 2\pi \frac{\mu_{ba}}{h} E \, \mathrm{d}t = 2\pi \tag{5.142}$$

which explains the name Π pulse.

It can be shown that pulses of any shape will be changed by the sample absorption to the sech shape of the Π pulses as long as the conditions for self-induced transparency are fulfilled. If the temporal integral of the incident pulse is larger than n times 2π the incident pulse is split into n Π pulses.

The pulse velocity is approximately given by:

$$\boldsymbol{\Pi} \textbf{ pulse velocity} \quad v_{\mathrm{pulse}} = \frac{c_0}{1 + 2c_0 a_{\mathrm{mat}} \Delta\nu \tau_{\mathrm{setr}}^2} \tag{5.143}$$

with the absorption coefficient a_{mat} and the spectral width of the transition $\Delta\nu$. Thus the pulse velocity can be reduced by more than a factor of 10^3 compared to the speed of light and thus time delays $t_{\mathrm{delay}} = L_{\mathrm{sample}}/v_{\mathrm{pulse}}$ of 100 ns or longer can be obtained in samples with lengths L_{sample} in the cm range.

For obtaining self-induced transparency with Π pulses coherent interaction has to be achieved and thus the pulse duration has to be shorter than the T_2 and the T_1 times of the material.

5.4.9 Superradiance (Superfluorescence)

Coherent coupling of light-emitting particles (without a laser resonator) results in coherent radiation which is called superradiance or superfluorescence [5.286–5.294]. The resulting intensity $I_{\mathrm{radiation}}$ is quadratically proportional to the number of particles as long as no saturation effects occur, in contrast to stochastic emitters (see Table 5.1).

Table 5.1. Properties of superradiance and stochastic emitted light

	Superradiance	Incoherent emitters
Particle density N	$I_{\text{radiation}} \propto N^2$	$I_{\text{radiation}} \propto N$
Spatial distribution	$I_{\text{radiation}} \propto \cos^2 \varphi$ (higher in the direction of most particles)	isotropic (for isotropic emitters)
Temporal distribution	short pulses	proportional excitation and population lifetime

The coherence can be established by coherent excitation of the emitters or it can be based on the sufficient high population of the excited emitting state. Based on spontaneous fluorescence the superradiance can then develop by coherent amplification. Finally a high gain in the material will be obtained. In resonators this superradiance passes to coherent laser radiation.

A mathematical description is possible with the fundamental equation of nonlinear optics (see Sect. 4.3) including the coherent interaction (see Sect. 5.4.2).

In excimer and nitrogen lasers the influence of the external resonator is sometimes so small that some of these lasers are named superradiators.

5.4.10 Amplification Without Inversion

Amplification of light in matter usually demands the inversion of the population densities of the participating energy levels. This inversion is the basic requirement for lasers as usually applied in photonics.

But if three or more energy levels are coherently coupled via the light field the absorption can be decreased in these materials showing a very high refractive index. In some cases even amplification is possible although the population densities are not inverted.

This process is based on the double-Λ-scheme as shown in Fig. 5.43.

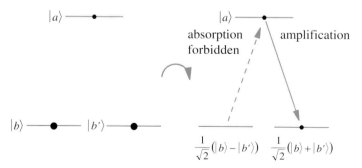

Fig. 5.43. Double-Λ-scheme with coherently coupled quantum states b allowing for amplification without inversion

The lower states $|b\rangle$ and $|b'\rangle$ can be superimposed to the orthogonal states $1/\sqrt{2}(|b\rangle - |b'\rangle)$ and $1/\sqrt{2}(|b\rangle + |b'\rangle)$. If the absorption of the first state to $|a\rangle$ is forbidden it is called a trapped state. If most of the particles are in this trapped state and the emission of $|a\rangle$ to the second combined ground state is allowed then amplification can be observed without having more particles in $|a\rangle$ than in $|b\rangle$.

A detailed theoretical discussion is given in [5.295–5.297]. More experiments are described in [5.298–5.309]. Based on this amplification, *laser without inversion* can be investigated. For photonic applications these lasers are not used today.

5.5 Two-Photon and Multiphoton Absorption

Absorption and emission of two [5.310–5.354], three [5.355–5.364] or several [5.365–5.374] photons can occur stepwise or simultaneously (see Fig. 5.44).

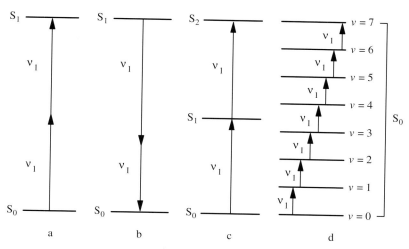

Fig. 5.44. Absorption and emission of photons: (**a**) two photons of the same frequency are absorbed simultaneously (two-photon absorption TPA), (**b**) two photons of different frequencies are emitted simultaneously, (**c**) stepwise absorption of two equal photons and (**d**) stepwise absorption of seven equal photons, e.g. between vibrational energy levels

Stepwise multiphoton absorption as in Fig. 5.44 (c) and (d) does not require coherence of the exciting light beam(s) and can be described by simple rate equations as given in Sect. 5.3.6. As long as the population densities of the involved lower states of the transitions are sufficiently high further absorption to higher states will occur. It turns out that, e.g. for many molecules the lowest energy ground state absorption band occurs in the first excited state in a similar form. Thus in laser experiments with such molecules stepwise absorption may happen.

But both absorptions via an intermediate energy state or without may occur as coherent interaction between light and the transition dipole moment. Using rate equations the two-photon transition probability from the state l to the state m can be phenomenological described by:

(1) two-photon absorption from l to m $(N_l \rightarrow N_m)$:

$$\frac{\partial N_l}{\partial t} = -\sigma_{l,m}^{(2)} I_1 I_2 N_l \quad \text{and} \quad \frac{\partial N_m}{\partial t} = +\sigma_{l,m}^{(2)} I_1 I_2 N_l \qquad (5.144)$$

(2) two photon emission from m to l $(N_l \leftarrow N_m)$:

$$\frac{\partial N_l}{\partial t} = +\sigma_{m,l}^{(2)} I_1 I_2 N_m \quad \text{and} \quad \frac{\partial N_m}{\partial t} = -\sigma_{m,l}^{(2)} I_1 I_2 N_m \qquad (5.145)$$

leading to a quadratic intensity dependence if just one light beam is applied. The cross-section $\sigma_{l,m}^{(2)}$ is then measured in cm^{-4} s if the intensity I is measured in photons $cm^{-2} s^{-1}$. Typical values for the cross-section in interactions without an intermediate state are in the range of 10^{-48}–10^{-50} cm^4 s for molecules and even 100–10^4 times smaller for atoms.

Usually only small population densities are reached in the excited state and therefore the photon transport equation can then be written as:

$$\frac{\partial I}{\partial z} = -N_{\text{total}} \sigma^{(2)} I_{\text{exc}}^2 \qquad (5.146)$$

which can be integrated to give:

two photon absorption $I(z) = I_0 \dfrac{1}{1 + \sigma^{(2)} N_{\text{total}} I_0 z}$ $\qquad (5.147)$

with the incident excitation intensity I_0 at $z = 0$ and the total number of absorbing particles N_{total}. This approximation is valid only far below saturation or bleaching. This is usually fulfilled because of the small cross-sections. Therefore intensities in the GW cm^{-2} range are required for this type of experiment. Such high intensities are close to the damage threshold of optical glasses and can produce optical breakdown in the sample (see Sect. 5.6). Thus excitation is limited. Suitable intermediate energy levels can increase the cross-section value by 10^6. In all cases resonance of the energy of the two photons with the energy levels of the sample is assumed.

The line shape is for parallel beams the same as the single photon band profile. In the case of two antiparallel beams the broadening effects from the moving particles can be avoided and the resulting two-photon spectra are Doppler-free, allowing, e.g. the resolution of the hyperfine structure of atoms [e.g. 5.333, 5.337–5.345]. This can result in strong signals although the two-photon cross-section is much smaller than the single-photon one. In this Doppler-free spectroscopy no inhomogeneous broadening takes place and thus all particles contribute to the transition at the light wavelength. In any case the polarization of the two photons should be the same for the maximum transition probability.

A theoretical description of the two-photon processes as in Fig. 5.44 (a) and (b) can be based on the combined successive transition from the ground state $|g\rangle$ to a virtual state $|v\rangle$ which is reached by the first photon and then from the virtual state $|v\rangle$ to the upper state $|u\rangle$ representing the transition from the second photon [M36, M37]. The transition state is then represented by the linear combination of the wavefunctions of all real energy states of the matter which have allowed single photon transitions from the ground state and the transition probability for this transition results from the sum of all these single transition probabilities. The second transition is treated in the same manner and thus the final transition results from the product of these two single step probabilities.

Therefore the two-photon process has different selection rules. Compared to the single-photon transitions which are allowed between states of opposite parity two-photon transitions are strong between states of the same parity:

allowed single photon transitions:

$$\text{even} \rightarrow \text{odd and odd} \rightarrow \text{even} \tag{5.148}$$

but

allowed two photon transitions:

$$\text{even} \rightarrow \text{even and odd} \rightarrow \text{odd} \tag{5.149}$$

and thus with two-photon spectroscopy states can be reached which are forbidden in conventional spectroscopy. But in many practical cases, especially for molecules, the selection rules are not so strongly fulfilled and thus even "forbidden" transitions are measurable from single-photon spectroscopy.

If the two or multiphoton transition occurs almost resonantly via a real energy level of the sample as in Fig. 5.44 (c) and 5.44 (d) the transition probability is strongly enhanced and reaches values of the usual single-photon transitions.

Two photon emission as in Fig. 5.44 (b) has to be stimulated by an incident light beam of frequency ν_1 or ν_2 and the second photon with frequency ν_2 or ν_1 respectively will be emitted in the same direction. The spontaneous two photon emission is usually too weak to be observable.

Two-photon absorption is used in several photonic applications in science and may become increasingly important, e.g. in laser medicine. Thus, e.g. samples with "blue" absorption can be excited with "red" lasers (see Sect. 7.6.4 and 6.2). As a possible measuring error in this case single-photon excited state absorption at the wavelength of the pump laser can occur with 10^3 times higher cross-section [e.g. 5.21] and simulate a very strong two photon cross-section.

This two-photon absorption can be used for exciting molecules as tumor markers or in photodynamic therapy (see Sect. 1.5). This avoids the distortions from much less intense sunlight. Using different directions of the two pump light beams a high spatial resolution can be achieved.

Conventional fluorescence can be excited via two-photon absorption, called two-photon induced fluorescence (TPF). This easily to observe second-order nonlinear process is used for the determination of the pulse length of short laser pulses [5.346] as shown in Fig. 5.45.

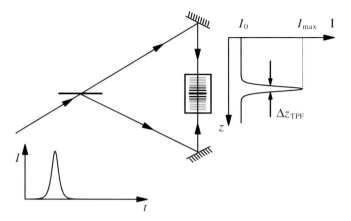

Fig. 5.45. Two-photon induced fluorescence for measuring the duration of short laser pulses from the spatial expansion Δz_{TPF}

Two equal shares of an incident beam are guided antiparallel into a cell with strongly fluorescing material (e.g. a laser dye) which can be excited by two-photon absorption with the laser wavelength. In the temporal overlap region of the two pulses the fluorescence will be much stronger over the FWHM range $\Delta z_{\mathrm{TPF\text{-}FWHM}}$ which can easily be measured using a magnifier or a microscope in the sub-mm range. The resulting pulse length in the ps or sub-ps range can be evaluated from:

$$\textbf{pulse width (Gauss)}\quad \Delta t_{\mathrm{FWHM}} = \frac{1}{c\sqrt{2}}\Delta z_{\mathrm{TPF\text{-}FWHM}} \tag{5.150}$$

with $\Delta z_{\mathrm{TPF\text{-}FWHM}}$ as the full width at half maximum of the TPF signal after background subtraction and the speed of light c in the matter. For hyperbolic secans pulse shapes the pulse width of the exciting signal is smaller for the same spatial width of the TPF signal and follows from:

$$\textbf{pulse width (sech)}\quad \Delta t_{\mathrm{FWHM}} = \frac{1}{1.5429\cdot c}\Delta z_{\mathrm{TPF\text{-}FWHM}}. \tag{5.151}$$

Laser pulse asymmetries can be detected using three-photon absorption [5.355].

Other applications of TPF are high-resolution microscopy (see Sect. 1.5 and, for example, [5.311, 5.327]) or single molecule imaging based on two-photon absorption processes (see Sect. 1.5 and, for example, [5.312, 5.316]).

The simultaneous absorption of more than two-photons becomes less probable by a factor of 10^{-6} with each additional photon and thus multi-photon absorption as in Fig. 5.44 (d) is easily observable only with resonant energy levels. Such equal steps of energy levels are provided, e.g. from the vibrational energies of molecules. Because of the multiple resonance condition this type of excitation can be very selective. It was shown that even different conformations of molecules or different isotopes in these molecules can be differentiated via a multi-photon excitation with 20–50 photons resulting in the dissociation of the molecule. This offers a new approach for very selective photochemical reactions. Laser chemistry may develop based on these multiphoton excitations in addition to femtochemistry (see Sect. 1.5).

Multiphoton excitations allow the preparation of very special quantum states of matter. Dressed atoms with electrons in very high orbits can be prepared this way. The investigation of atoms in highly excited Rydberg states which can be achieved with multiphoton excitation has given some insides in the their structure [e.g. 5.358].

5.6 Photoionization and Optical Breakdown (OBD)

If the intensity is increased to very high values above $10^{10}\,\mathrm{W\,cm^{-2}\,s^{-1}}$ all absorbing or nonabsorbing materials can be ionized and possibly damaged [5.365–5.430]. These high intensities can be easily achieved with focusing of short pulses. Besides the effects caused by strong heating the consequences of the extremely high electric fields and possible photoionization via multiphoton excitation can also be observed. In practical cases mostly a combination of these effects occurs.

If photoionization takes place during a short period many free electrons will be generated and a hot plasma will be obtained. This optical breakdown can be observed by focusing, e.g. a Q switched laser pulse of good beam quality with a few mm diameter, a pulse duration of 10 ns and a pulse energy of 100 mJ with a lens with a focal length of 100 mm in air (see Fig. 5.46).

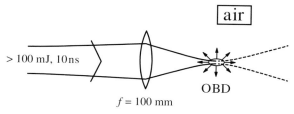

Fig. 5.46. Scheme for observing optical breakdown in air

With shorter pulse lengths the pulse energy can be correspondingly smaller and vice versa. This impressive and noisy effect works in gases with repetition rates of more than 100 Hz, in liquids with a few 10 Hz and in solids just one time per volume.

The obtained "white" light covers a large spectral range as a function of the material and may be used for illumination. In the generated plasma, optical phase conjugation (see Sect. 4.5.14) can occur and thus the laser source can be damaged by the back-scattered light, especially if oscillator amplifier laser systems are used. In this case optical isolation with a Faraday rotator and a polarizer may be applied to protect the laser source (see Sect. 6.11.3).

Both experimental and theoretical investigations of this effect are difficult to carry out because of the competing effects in this intensity range. Multiphoton excitation as discussed in the previous chapter with subsequent ionization leads for the ionization, measured e.g. as the number of generated ions $N_{\text{OBD-ions}}$, as a function of the exciting intensity I_{exc} to a high-power k_{OBD}:

$$N_{\text{OBD-ions}} \propto \Delta t_{\text{pulse,FWHM}} I_{\text{exc}}^{k_{\text{OBD}}}. \tag{5.152}$$

The pulse duration Δt_{FWHM} increases linearly the probability of this ionization.

The exponent k_{OBD} is of the size of the quotient of the ionization energy $E_{\text{ionization}}$ divided by the photon energy $h\nu_{\text{exc}}$:

$$k_{\text{OBD}} \approx \frac{E_{\text{ionization}}}{h\nu_{\text{exc}}} \tag{5.153}$$

but the observed k_{OBD} values may be smaller. In [5.388] a value of $k_{\text{OBD}} = 6.5$ was observed for the ionization of krypton atoms, which have an ionization energy of 14 eV, with a pulsed ruby laser light of the wavelength of 694 nm. The theoretical value of k_{OBD} is approximately 10. This discrepancy may be explained by the supporting field emission from the very high electric fields in the light beam and other processes.

The electric field strength E_{av} as calculated from (2.44) can exceed the field emission strength E_{FE} (see Table 5.2). With static fields this value is of the order of 10^7 V cm^{-1}.

Table 5.2. Strength of the electric field E_{av} of a light beam as a function of the intensity I

I (W cm^{-2})	I (photons cm^{-2} s^{-1}) (with $\lambda = 1\,\mu$m)	E_{av} (V cm^{-1})
10	$5.0341 \cdot 10^{19}$	86.80
10^3	$5.0341 \cdot 10^{21}$	$8.680 \cdot 10^2$
10^6	$5.0341 \cdot 10^{24}$	$2.745 \cdot 10^4$
10^9	$5.0341 \cdot 10^{27}$	$8.680 \cdot 10^5$
10^{12}	$5.0341 \cdot 10^{30}$	$2.745 \cdot 10^7$
10^{15}	$5.0341 \cdot 10^{33}$	$8.680 \cdot 10^8$

Because of the high power of I_{OBD}^k this effect appears with a threshold character. This "threshold intensity" depends on many parameters. Measured values for xenon, krypton, argon and neon gases at normal pressure are in the range of 10^{10}–$10^{11}\,\mathrm{W\,cm^{-2}\,s^{-1}}$ for red light and ns pulse duration. The OBD "threshold" shows a minimum as a function of gas pressure in the range of $100\,\mathrm{MPa}$–$1\,\mathrm{GPa}$ [5.405] because of too few particles at low pressures and stronger damping at high pressures.

Impurities can especially decrease this "threshold" by orders of magnitude. Therefore the "threshold" can be increased by purification by at least one order of magnitude as shown in [5.384]. In these experiments a transversal electric field was applied in gases to extract all particles out of the beam cross-section. OBD can be, e.g. the limiting factor for good phase conjugation based on stimulated Brillouin scattering (see Sect. 4.5.9). Liquids can be distilled in a special procedure to increase the OBD threshold [4.335].

Photoionization and the resulting plasma can disturb photonic applications as, for example, stimulated Brillouin scattering (SBS), as discussed in Sect. 4.5.9 and the references therein, but they can also be used for other applications. Thus the laser-induced plasma can be used as new broadband UV and XUV light sources (see Sect. 6.13.5 and [5.389–5.406]). Microparticles can be detected [5.409], laser pulses can be shortened [5.385] and laser spark ignitions can be realized [5.428]. Some more-detailed investigations of the breakdown mechanism are given in [5.413–5.427].

5.7 Optical Damage

Damage of optical materials by laser radiation with high powers depends on the matter and light parameters [5.431–5.467]. It is difficult to measure and to model because of the very high nonlinearity of the process and thus because of the high sensitivity towards impurities. Therefore a phenomenological description of the "damage threshold" is applied and is mostly sufficient.

Optical damage threshold is different for bulk materials and surfaces. It is also a function of almost all radiation parameters such as pulse duration, wavelength, intensity and energy, mode structure, beam size and even polarization for nonperpendicular incidence at surfaces. Most crucial are material impurities such as absorbing particles at the surface or in the bulk or inner tensions from the production process. Optical damage is investigated for optical components such as, e.g. mirrors, laser and frequency converting crystals, coatings and other nonlinear materials.

As a rough guideline the damage intensity I_{damage} is inversely proportional the square root of the pulse duration Δt [5.432]:

$$\textbf{pulse width}\quad I_{\mathrm{damage}}(\Delta t) = I_{\mathrm{damage,ref}}\sqrt{\frac{\Delta t_{\mathrm{ref}}}{\Delta t}} + I_{\mathrm{damage,cw}} \qquad (5.154)$$

with the reference damage intensity $I_{\text{damage,ref}}$ of a pulse with the duration Δt_{ref} and the damage intensity $I_{\text{damage,cw}}$ for continuos radiation. Sometimes instead of the damage intensity the damage fluence $F_{\text{damage}} = I_{\text{damage}}\Delta t_{\text{pulse}}$ is given for a certain pulse width. The damage fluence will increase with the square root of the pulse duration Δt_{pulse}.

In Table 5.3 some characteristic values of the damage intensity and fluence are given [M25].

Table 5.3. Rough estimates of damage thresholds for transparent optical components for light pulses of different pulse durations

	I_{damage} (GW cm^{-2}) $(10\,\text{ns})$	I_{damage} (GW cm^{-2}) $(1\,\text{ns})$	I_{damage} (GW cm^{-2}) $(30\,\text{ps})$	I_{damage} (GW cm^{-2}) $(100\,\text{fs})$	F_{damage} (J cm^{-2}) $(1\,\text{ns})$
Bulk damage	16	50	290	5.000	50
Surface damage (uncoated)	3	10	58	1.000	10
Dielectric mirrors	3	10	58	1.000	10
Antireflection coatings	1.6	5	3	500	5

These values may be used to estimate the useful intensity ranges for optical components in nonlinear optics. Intensities of a factor of 10–100 smaller than the damage threshold usually guarantee safe operation under clean conditions.

For continuous radiation (cw) the damage intensity is in the range of a few $10\,\text{kW cm}^{-2}$.

The damage threshold is decreased by the roughness of the optical surface drastically. For the rms roughness δ_{surface} between 1 and $30\,\text{nm}$ a simple relation was given for the decrease of the damage threshold:

$$\textbf{roughness} \quad I_{\text{damage}} \propto \frac{1}{\delta_{\text{surface}}^{m}} \tag{5.155}$$

with an exponent m between 1 and 1.5 [M25]. Thus the damage threshold can be decreased by more than two orders of magnitude. Thus in high-intensity measurements highly polished surfaces should be used, e.g. for cuvettes and all components.

Further the damage threshold a decreases with increasing spot area A_{spot} as [M25]:

$$\textbf{spot size} \quad I_{\text{damage}} \propto \frac{1}{\sqrt{A_{\text{spot}}}}. \tag{5.156}$$

For absorbing matter the damage threshold decreases with increasing absorption. For absorption coefficients $a = -1/L \ln T$ between 0.01 and 50 cm^{-1} the following formula was observed [5.432]:

$$\text{absorbing matter} \quad I_{\text{damage}} = \frac{264 \,\text{kW/cm}^2}{(a/\text{cm}^{-1})^{0.74}}. \tag{5.157}$$

Single molecules can be stable for several hundred absorption cycles but they can also be destroyed after a few absorptions, see [5.457–5.464]; this is a function of their structure. In solid materials melting or other degradation processes can occur (see the next Section and, for example, [5.465–5.467]). In any case self-focusing of the laser light (see Sects. 4.5.3 and 4.5.6) may increase the intensity in the material and thus produce optical damage [5.468–5.472].

The valuable side of the optical damage is optical material processing.

5.8 Laser Material Processing

In most cases optical material processing is based on the thermal effects of the absorbed light in the matter resulting in characteristic time constants of μs up to seconds.

In general three main types of material processing applications may be distinguished:

- *heat treatment* (10^4–10^5 W cm^{-2}): hardening of steel, surface oxidation;
- *melting* (10^5–10^7 W cm^{-2}): soldering, welding, marking, labeling, cutting, surface treatment;
- *vaporization* (10^7–10^{10} W cm^{-2}): drilling, cutting, labeling, trimming.

The first two methods can be applied with cw lasers or pulsed systems with pulse widths from ms to ns. The vaporization technique demands pulsed operation in the μs to ps range. The use of fs laser pulses resulted in drilling and micromachining to very high accuracy and may find applications, too, depending on the price of these lasers and the advantage in precision.

The theoretical description of these material processing applications has to be based on the modeling of the light energy absorption and the following heat distribution in the matter using differential transport equations.

The temperature difference distribution ΔT as a function of the radial distance r and time t inside a sample which is excited at a small spot can be calculated from:

$$\Delta T(r, t) = \Delta E \frac{1}{\rho c_{\text{P}}} (4\pi\kappa t)^{-1.5} \left\{ 1 + \left(\frac{r^2}{\kappa t} - 6 \right) \frac{s^2}{40\kappa t} \right\} \,\text{e}^{-r^2/4\kappa t} \tag{5.158}$$

where the sample was assumed to be indefinitely large compared to the excited volume dimensions. The computation of the temperature distribution was based on the superposition of spherical heat sources providing an instantaneous energy deposition ΔE. ρ denotes the density of the medium, c_p the heat capacity and κ the heat diffusion coefficient. The radius of the heat sources has to be small compared to r which is in practical cases at the order of a few $0.1\,\mathrm{mm}$.

Typical material parameters are given in Table 5.4. With these temperature and heat values the necessary laser pulse energy or power can be estimated.

Table 5.4. Material parameters relevant for material processing: density, specific heat c_p, heat conductivity k_h, melting temperature T_m, vaporization temperature T_v, melt heat Q_m, vaporization heat Q_v, absorption $1 - R$ (for $1\,\mu\mathrm{m}$ light) (from [5.475])

Material		ρ $(\mathrm{gcm^{-3}})$	c_p $(\mathrm{JgK^{-1}})$	κ $(\mathrm{Wcm^{-1}K^{-1}})$	T_m $(^\circ\mathrm{C})$	T_v $(^\circ\mathrm{C})$	Q_m $(\mathrm{Jg^{-1}})$	Q_v $(\mathrm{Jg^{-1}})$	$1-R$
Aluminium	Al	2.70	0.90	2.22	660	2450	395	10470	0.10
Chromium	Cr	7.19	0.46	0.67	1875	2220	282	5860	0.40
Copper	Cu	8.96	0.39	3.94	1083	2595	212	4770	0.06
Gold	Au	19.3	0.13	2.97	1063	2966	67	1550	0.02
Iron	Fe	7.87	0.46	0.75	1537	2735	274	6365	0.35
Lead	Pb	11.3	0.13	0.33	327	1750	26	858	
Manganese	Mn	7.43	0.48	0.5	1245	2095	267	4100	
Nickel	Ni	8.9	0.44	0.92	1453	2840	309	6450	0.28
Platinum	Pt	21.5	0.13	0.72	1769	3800	113	2615	0.27
Silicon	Si	2.33	0.76	1.49	1410	2355	337	1446	
Silver	Ag	10.5	0.23	4.29	961	2212	105	2387	0.03
Tin	Sn	7.30	0.23	0.63	232	2270	61	1945	0.45
Titanium	Ti	4.51	0.52	0.17	1668	3280	403	8790	0.40
Tungsten	W	19.3	0.14	1.67	3410	5930	193	3980	0.41
Vanadium	V	6.07	0.50	0.29	1900	3530	330	10260	0.40
Zinc	Zn	7.13	0.39	1.13	420	907	102	1760	0.50

It has to be noticed that the absorption $(1 - R)$ can change drastically if the matter is melted or if in pulsed operation a plasma is formed at the surface.

From these values it can be estimated that the melt energy for $1\,\mathrm{mm}^3$ of material is of the order of 1–$10\,\mathrm{J}$ and the vaporization energy is about 10 times larger. For a variation of the cut or drill diameter of less than 10% along the laser beam the Rayleigh length of the focused light should be three times larger than the material thickness.

Therefore the cut or welding speed for materials depends on the average laser power absorbed in the material. For a copper sheet with a thickness of $1\,\mathrm{mm}$ with a laser power of $60\,\mathrm{W}$ at $1\,\mu\mathrm{m}$ wavelength in a pulsed regime resulting in a peak power of $10\,\mathrm{kW}$ the possible cut speed is larger than $20\,\mathrm{mm/s}$. For commercial laser material processing in, e.g. car production the used laser show average output powers of 2–$6\,\mathrm{kW}$ at $1\,\mu\mathrm{m}$ (Nd laser) or even higher values at $10.6\,\mu\mathrm{m}$ (CO_2 laser).

Laser ablation is especially investigated in [5.479–5.496] and laser writing or prototyping can be found in [5.497–5.503]. The melting process, especially using short pulses, is described in [5.504–5.522]. Some biological and medical aspects are studied in [5.523–5.526] in addition to the references in Sect. 1.5. The use of very short pulses, and, thus, very high powers is described in [5.527–5.533] and the beam delivery is discussed in [5.534–5.538]. Some more applications are given in [5.539–5.553].

5.9 Combined Interactions with Diffraction and Absorption Changes

If the absorption of matter is changed the real part of the refractive index and thus the diffraction will be changed, too. Thus both the amplitude and the phase of the applied light will be changed.

change of absorption and refractive index

$\Downarrow$

change of intensity and phase of the light

In absorption measurements such as pump and probe techniques the phase change will not always be detected. But if interference structures are used and thus absorption and phase gratings are applied in the matter both changes have to be considered in the interaction of the probe light with the matter.

In this type of measurements two, three or four light beams interact more or less simultaneously with the matter. These experiments are in general described by the formalism of four-wave mixing (FWM). The theory of this FWM can be very complicated if the applied beams have different spectra, different temporal and spatial structure, different polarization and coherence. Simple cases more easily described will be reported below.

As the simplest example the generation of transient or permanent gratings by two equal light beams with different propagation directions will be outlined, first.

5.9.1 Induced Amplitude and Phase Gratings

If the production of the amplitude and phase grating in the material with two pump beams and the detection of changes in the induced matter with the probe beam are obtained separately with temporally nonoverlapping light pulses the description of the previous sections may be sufficient for the analysis. The probing was described in Sect. 3.9.16. The grating production can be calculated with the interference formulas from Sect. 2.9 and the nonlinear interaction formulas given in Chaps. 4 and 5 describing the nonlinear change of the refractive index and the nonlinear transmission.

If the diffraction at the transient grating is observed, simultaneously, using a third beam as probe resulting in a fourth beam which is need to detect the four-wave mixing process, then a description such as that given in the next section has to be used. But here it is assumed here that only two beams are applied (see Fig. 5.47).

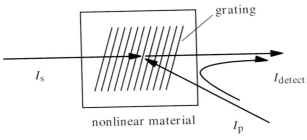

Fig. 5.47. Production and detection of induced gratings via the nonlinear interaction of two beams with absorbing matter

If, via the nonlinear interaction of the interference pattern from the two beams, an absorption or phase grating is induced in the matter some light of the pump beam p can be diffracted into the direction of the transmitted beam s and vice versa. Some examples for induced gratings and their theoretical treatment are given in [5.554–5.589]. Additional examples can be found in the next Section.

If the two beams have distinctively different intensities, an "amplification" of the weaker beam can be obtained. This process is also called two-wave mixing (TWM) or beam coupling [5.579–5.589]. It can be used for beam cleanup. Therefore, a weak beam with good and/or desired properties is amplified via two-wave mixing by a strong pump beam with bad beam quality [5.587–5.589].

The analysis is easy if the following assumptions about the two beams are justified:

- they are plane waves (e.g. in the waist region of Gaussian beams);
- coherence is sufficient for complete interference;
- they are linearly polarized perpendicular to the interaction plane;
- light frequencies are only slightly different;
- intensities are not to high (below saturation of the nonlinear effect).

The electric fields of the two beams are then given by:

$$E_p = E_p^0 \cos(2\pi\nu_p t - \boldsymbol{k}_p \cdot \boldsymbol{r} + \varphi_p) \tag{5.159}$$

and

$$E_s = E_s^0 \cos(2\pi\nu_s t - \boldsymbol{k}_s \cdot \boldsymbol{r} + \varphi_s) \tag{5.160}$$

with the resulting intensity modulation in the sample:

$$I_{ges} = I_p + I_s + 2\sqrt{I_p I_s}\,\cos\{2\pi(\nu_s - \nu_p)t - (\boldsymbol{k}_s - \boldsymbol{k}_p)\cdot\boldsymbol{r} + \Delta\varphi\} \quad (5.161)$$

and with $\Delta\varphi = \varphi_s - \varphi_p$.

If the two beams have the same frequency, i.e. they are *degenerate*, the spatial modulation is a function of the angle between the two incident light waves and their relative phase, only.

The diffraction efficiency of the pump beam with I_p towards the direction of I_{detect} is a function of the type of grating and the phase difference between I_p and I_s as given in Table 5.5.

Table 5.5. Diffraction of pump beam I_p towards the direction of I_{detect} as given in Fig. 5.47 as a function of the relative phase $\Delta\varphi$ between the beams I_p and I_s

phase grating			
$n(z)$:	$\Delta\varphi = 0$	$I_{detect} = $ min.	
	$\Delta\varphi = \dfrac{\pi}{2}$	$I_{detect} = $ max.	
amplitude grating			
$a(z)$:	$\Delta\varphi = 0$	$I_{detect} = $ max. $\Uparrow$ bleaching	$I_{detect} = $ min. $\Uparrow$ induced absorption
	$\Delta\varphi = \dfrac{\pi}{2}$	$I_{detect} = $ min.	

If the two intensities are equal the minimum intensity I_{detect} can be zero. The maximum diffraction efficiency occurs for refractive index gratings for phase shifts of $\pi/2$ between the two beams and for zero phase difference in the case of bleached absorption gratings.

For optimal phases the diffraction efficiency η_{diff}, related to the diffracted light can be calculated from:

diffraction efficiency

$$\eta_{diff} = \frac{I_{detect}}{I_p}$$
$$= \left(\frac{\pi\Delta n L_{mat}}{\lambda}\right)^2 + \left(\frac{\Delta a L_{mat}}{4}\right)^2 \quad (5.162)$$

with the sample thickness L_{mat}. The change of the refractive index Δn describes the phase grating and the change of the absorption Δa the amplitude grating.

As an example the maximum effect for an absorption grating produced by an absorption change of $\Delta a = 0.03\,\mathrm{cm}^{-1}$ with an interaction length $d = 1\,\mathrm{mm}$ would lead to a diffraction efficiency of $\eta_{diff} = 5.6\cdot10^{-7}$. If this change was produced, as a realistic case, by a pump intensity I_p of $4\cdot10^{26}$ photons $\mathrm{cm}^{-2}\,\mathrm{s}^{-1}$

and an intensity I_s of 10^{25} photons cm^{-2} s^{-1} for the second beam the additional signal intensity in the detection would be $2 \cdot 10^{-5}$ of the intensity I_s, only, and thus very difficult to detect.

The phase grating diffraction effect can easily be much stronger. If, e.g. both beams have wavelengths of 500 nm an index change of $\Delta n = 7 \cdot 10^{-5}$ results in a diffraction efficiency of $\eta_{\text{diff}} = 0.2$. In this case the scattered intensity would be four times larger than the intensity I_s and thus very easy to detect.

For small absorption and small changes of absorption and refractive index:

$$\text{assumptions} \quad a, \ \Delta a \ll 4\pi \frac{n}{\lambda} \quad \text{and} \quad \Delta n \ll n \tag{5.163}$$

the change of the refractive index accompanying the absorption change can be calculated from:

$$\Delta n = \frac{1}{2\pi^2} \int_0^\infty \frac{\Delta a(\lambda)}{1 - (\lambda'/\lambda)^2} \, d\lambda' \tag{5.164}$$

and with the approximations:

$$\chi = \left(n + i\frac{\lambda}{4\pi} \right)^2 - 1 \approx n^2 - 1 + i\frac{\lambda}{2\pi} n \tag{5.165}$$

and thus:

$$\Delta\chi \approx 2n\Delta n + i\left(\frac{\lambda}{2\pi} n \right) \Delta\alpha \tag{5.166}$$

and so:

$$\eta_{\text{diff}} = \left(\frac{\pi d}{2n\lambda} \right)^2 \cdot |\Delta\chi|^2 \tag{5.167}$$

which allows the calculation of the diffraction efficiency based on the susceptibility change of the matter.

5.9.2 Four-Wave Mixing (FWM)

In the four-wave mixing process (FWM) in general two pump beams I_{p1} and I_{p2} are used to interfere and induce phase or amplitude gratings via a nonlinear optical process together with a third beam I_{s3}. These beams are diffracted at the resulting gratings forming the fourth beam I_{d4} which is detected (see Fig. 5.48).

This scheme is used in many applications such as, e.g. holography, optical phase conjugation and coherent anti-Stokes Raman scattering (CARS) as described in Sect. 4.5.13. Some examples for degenerate four wave mixing (DFWM), where all beams have the same frequency, are given in [5.590–5.604] and for non-degenerate FWM in [5.605–5.626].

Whereas the grating in conventional holography is static and thus the writing of the grating and the read-out are temporally separated in FWM all processes take place at the same time. Thus FWM is sometimes called *real-time* or *dynamic holography*.

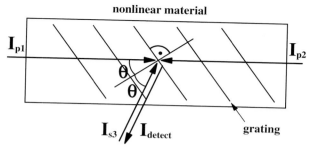

Fig. 5.48. Schematic of four-wave mixing (FWM) using two pump beams p1 and p2 to induce reflection gratings in the matter together with a third beam s3 resulting in a the fourth beam detect

This four-wave mixing (FWM) is obviously a third-order nonlinear optical process with nonlinear polarization:

$$P_{nl}^{FWM}(r, t) = \frac{\varepsilon_0}{2} \chi^{(3)} E_{p1}(r, t) E_{p2}(r, t) E_{s3}(r, t) \tag{5.168}$$

with a complex third-order susceptibility $\chi^{(3)}$ describing phase and amplitude changes of the material as a function of the orientation of the material, the polarization and the frequencies of the light beams with electric fields E_{p1}, E_{p2} and E_{s3}.

Because all beams have to be coherent during the time of interaction in four-wave mixing at least four gratings are induced in the matter. The most important beam p2 forms a reflection grating with beam s3 as depicted by the grating planes in Fig. 5.48. Beam p1 can be thought of as "reflected" at this grating towards the direction of I_{d4}. In addition three more gratings are important in FWM. A reflection grating results from the interference of the two pump beams p1 and p2 (see Fig. 5.49a). This grating will have the shortest grating constant and it allows the reflection of the two pump beams p1 and p2 in the direction of the other. Another reflection grating is formed by s3 and d4 (see Fig. 5.49b). The fourth is a transmission grating generated by the beams p1 and s3 as well as by p2 and d4 (see Fig. 5.49c). Grating planes will always occur in the direction of the half-angle plane between the beams (see Fig. 5.49).

Energy and momentum conservation have to be fulfilled for the photons in the FWM scattering process and therefore

$$\nu_{p1} + \nu_{p2} = \nu_{s3} + \nu_{detect} \tag{5.169}$$

and

$$k_{p1} + k_{p2} = k_{s3} + k_{detect} \tag{5.170}$$

are required.

If the signal beam with intensity I_{s3} is equal to one of the pump beams four-wave mixing is observed as a *two-wave mixing process (TWM)* as de-

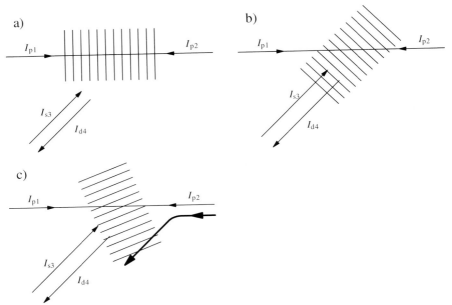

Fig. 5.49. Additional gratings occurring in four-wave mixing: (**a**) reflection grating build by beams p1 and p2; (**b**) reflection grating built by beams s3 and d4; and (**c**) transmission grating built by beams p1 and s3 as well as by p2 and d4

scribed in the previous section. The stimulated scattering processes such as stimulated Brillouin scattering (SBS) or stimulated Raman scattering (SRS) can be described as a four-wave mixing process in the TWM formalism. But these nonlinear processes can be used for real FWM, too, as, e.g. in *Brillouin enhanced four-wave mixing* (*BEFWM*) [5.627–5.633].

If the frequencies of all four waves in FWM are the same as in *degenerate four-wave mixing* (*DFWM*) the mathematical description is much easier. If the frequencies are different (as in SBS) the gratings will move in time.

In holography the phase front of the incident read-out wave is restored to produce the picture of the original object. Thus it is easy to imagine that real-time holography or four wave mixing produces the phase conjugate of the incident signal wave with I_{s3}.

Because of the importance of FWM in many fields of photonics a large number of investigations are made in the last 20 years [e.g. 5.590–5.685]. FWM is obtained with both transparent materials (Kerr-like media) as well as with absorbing matter. It is applied for small powers in the range of mW with cw-lasers up to high powers in the MW range. Thus all types of matter are used as the nonlinear material, dependent on the use. Apart from the nonlinear "threshold" of the matter their time constants are relevant parameters. Usually low necessary intensities as a result of a high nonlinearity of the matter are combined with long lifetimes in the range of µs to ms.

Theoretical descriptions of four-wave mixing and possible optical phase conjugation are given, for example, in [5.634-5.650]. In general the FWM can be described based on the fundamental equation of nonlinear optics as:

$$\Delta E - \mu_0 \varepsilon_0 \frac{\partial^2 E}{\partial t^2} - \mu_0 a_{\text{abs}} \frac{\partial E}{\partial t} = \mu_0 \frac{\partial^2 P_{\text{nl}}^{\text{FWM}}}{\partial t^2} \tag{5.171}$$

with a_{abs} accounting for the possible absorption of the matter. This equation has to be solved for the four coupled electric fields E_{p1}, E_{p1}, E_{s3} and E_{d3}.

If some of the photons are resonant with matter transitions the analysis of $\chi^{(3)}$ can be visualized with Feynman diagrams as, e.g. shown in Fig. 5.50 for the example of doubly resonant four-wave mixing.

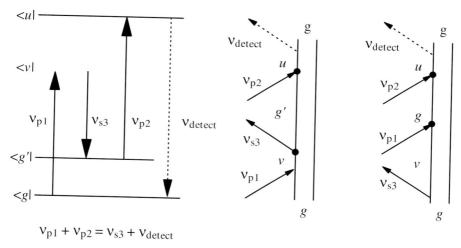

$$\nu_{\text{p1}} + \nu_{\text{p2}} = \nu_{\text{s3}} + \nu_{\text{detect}}$$

Fig. 5.50. Term schematic and Feynman diagram for a doubly resonant four-wave mixing process [5.644]

The resonant part of $\chi^{(3)}$ can then be calculated quantum mechanically from:

$$\left[X_{\text{res}}^{(3)}(\nu_{\text{detect}} = \nu_{\text{p1}} + \nu_{\text{p2}} + \nu_{\text{s3}}) \right]_{ijkl} = -\frac{(2\pi)^3 N_{\text{totale}} e^4}{\varepsilon_0 h^3}$$

$$\cdot \sum_v \frac{\langle g|r_i|u\rangle \langle u|r_j|g'\rangle}{(\nu_{\text{p1}} + \nu_{\text{p2}} - \nu_{\text{s3}} - \nu_{\text{detect}} + ik_{u \to g})(\nu_{\text{p1}} - \nu_{\text{s3}} - \nu_{jg} + ik_{g' \to g})}$$

$$\cdot \left[\frac{\langle u|r_j|v\rangle \langle v|r_l|g\rangle}{(\nu_{\text{p1}} - \nu_{vg})} + \frac{\langle g'|r_l|v\rangle \langle v|r_k|g\rangle}{-(\nu_{\text{s3}} - \nu_{vg})} \right] \rho_{gg}^0$$

$$+ \text{terms with } j \text{ and } l \text{ interchanged} \tag{5.172}$$

as given in [5.644]. Other examples of singly or triply resonant cases are given there, too. Saturation effects are not involved in this type of theory.

Therefore, sometimes especially in cases with absorption and high excitation intensities it may be easier to achieve wave mixing and determine the nonlinear parameters experimentally.

For some cases the differential equations for the field amplitudes $E_{0,l}$ can be given explicitly. If all light beams have the same frequency with a linear polarization to the interaction plane and thus *degenerate four-wave mixing* (*DFWM*) is obtained, the equations for the field amplitudes are as given in [5.648]:

$$\frac{\partial E_{p1}^0}{\partial z} = -aE_{p1}^0 - i2\kappa_{pm}E_{p1}^0 \left(|E_{p1}^0|^2 + |E_{p2}^0|^2 + |E_{s3}^0|^2 + |E_{d4}^0|^2\right)$$
$$-i2\kappa_t E_{s3}^0 \left(E_{p1}^0 E_{s3}^{0*} + E_{p2}^{0*} E_{d4}^0\right) - i2\kappa_r E_{d4}^0 \left(E_{p1}^0 E_{d4}^{0*} + E_{p2}^{0*} E_{s3}^0\right)$$
$$-i2\kappa_r' E_{p2}^0 (E_{p1}^0 E_{p2}^{0*})$$
$$-i\kappa_{2p} \left[E_{p1}^{0*} \left(E_{p1}^0 E_{p1}^0\right) + 2E_{p2}^{0*} \left(E_{p1}^0 E_{p2}^0 + E_{s3}^0 E_{d4}^0\right)\right.$$
$$\left.+2E_{s3}^{0*} \left(E_{p1}^0 E_{s3}^0\right) + 2E_{d4}^{0*} \left(E_{p1}^+ E_{d4}^0\right)\right] \tag{5.173}$$

and

$$\frac{\partial E_{p2}^0}{\partial z} = -aE_{p2}^0 - i2\kappa_{pm}E_{p2}^0 \left(|E_{p1}^0|^2 + |E_{p2}^0|^2 + |E_{s3}^0|^2 + |E_{d4}^0|^2\right)$$
$$+i2\kappa_t E_{d4}^0 \left(E_{p1}^0 E_{s3}^{0*} + E_{p2}^{0*} E_{d4}^0\right) + i2\kappa_t E_{s3}^0 \left(E_{p1}^{0*} E_{d4}^0 + E_{p2}^0 E_{s3}^0\right)$$
$$+i2\kappa_r' E_{p1}^0 \left(E_{p1}^{0*} E_{p2}^0\right)$$
$$+i\kappa_{2p} \left[2E_{p1}^{0*} \left(E_{p1}^0 E_{p2}^0 + E_{s3}^0 E_{d4}^0\right) + E_{p2}^{0*} \left(E_{p2}^0 E_{p2}^0\right)\right.$$
$$\left.+2E_{s3}^{0*} \left(E_{p2}^0 E_{s3}^0\right) + 2E_{d4}^{0*} \left(E_{p1}^0 E_{d4}^0\right)\right] \tag{5.174}$$

and

$$\frac{\partial E_{s3}^0}{\partial z} = -aE_{s3}^0 - i2\kappa_{pm}E_{s3}^0 \left(|E_{p1}^0|^2 + |E_{p2}^0|^2 + |E_{s3}^0|^2 + |E_{d4}^0|^2\right)$$
$$-i2\kappa_t E_{p1}^0 \left(E_{p1}^{0*} E_{s3}^0 + E_{p2}^0 E_{d4}^{0*}\right) - i2\kappa_r E_{p2}^0 \left(E_{p1}^0 E_{d4}^{0*} + E_{p2}^{0*} E_{s3}^0\right)$$
$$-i2\kappa_r' E_{d4}^0 \left(E_{s3}^0 E_{d4}^{0*}\right) - i\kappa_{2p} \left[2E_{p1}^{0*} \left(E_{p1}^0 E_{s3}^0\right) + 2E_{p2}^{0*} \left(E_{p2}^0 E_{s3}^0\right)\right.$$
$$\left.+E_{s3}^{0*} \left(E_{s3}^0 E_{s3}^0\right) + 2E_{d4}^{0*} \left(E_{p1}^0 E_{p2}^0 + E_{s3}^0 E_{d4}^0\right)\right] \tag{5.175}$$

and

$$\frac{\partial E_{d4}^0}{\partial z} = -aE_{d4}^0 + i2\kappa_{pm}E_{d4}^0 \left(|E_{p1}^0|^2 + |E_{p2}^0|^2 + |E_{s3}^0|^2 + |E_{d4}^0|^2\right)$$
$$+i2\kappa_t E_{p2}^0 \left(E_{p1}^0 E_{s3}^{0*} + E_{p2}^{0*} E_{d4}^0\right) + i2\kappa_r E_{p1}^0 \left(E_{p1}^{0*} E_{d4}^0 + E_{p2}^0 E_{s3}^{0*}\right)$$
$$+i2\kappa_r' E_{s3}^0 \left(E_{s3}^{0*} E_{d4}^0\right) + i\kappa_{2p} \left[2E_{p1}^{0*} \left(E_{p1}^0 E_{d4}^0\right) + 2E_{p2}^{0*} \left(E_{p2}^0 E_{d4}^0\right)\right.$$
$$\left.+2E_{s3}^{0*} \left(E_{p1}^0 E_{p2}^0 + E_{s3}^0 E_{d4}^0\right) + E_{d4}^{0*} \left(E_{s3}^0 E_{s3}^0\right)\right] . \tag{5.176}$$

In these equations the z direction was assumed to be in the direction of the bisectrix between the beams p1 and s3 or p2 and d4 as shown in Fig. 5.48. Weak absorption with the absorption coefficient a was considered:

$$a = \frac{N_{total}\sigma}{2\cos\theta} \sqrt{\frac{1}{\varepsilon}} . \tag{5.177}$$

Besides the effect of the reflection (r) and transmission (t) gratings, a self and crossed phase modulation (pm) and a two-photon excitation (2p) was considered in these equations. The nonlinear coefficients κ follow from:

$$\kappa_{\mathrm{pm}} = \frac{2\pi\nu\chi_{\mathrm{pm}}^{(3)}}{8\cos\theta}\sqrt{\frac{\mu_0}{\varepsilon}} \qquad \kappa_{\mathrm{t}} = \frac{2\pi\nu\chi_{\mathrm{t}}^{(3)}}{8\cos\theta}\sqrt{\frac{\mu_0}{\varepsilon}} \tag{5.178}$$

$$\kappa_{\mathrm{r}} = \frac{2\pi\nu\chi_{\mathrm{r}}^{(3)}}{8\cos\theta}\sqrt{\frac{\mu_0}{\varepsilon}} \qquad \kappa_{\mathrm{r}}' = \frac{2\pi\nu\chi_{\mathrm{r}}'^{(3)}}{8\cos\theta}\sqrt{\frac{\mu_0}{\varepsilon}} \tag{5.179}$$

$$\kappa_{\mathrm{2p}} = \frac{2\pi\nu\chi_{\mathrm{2p}}^{(3)}}{8\cos\theta}\sqrt{\frac{\mu_0}{\varepsilon}} \tag{5.180}$$

If all nonlinear coefficients are equal the differential equations become much simpler:

$$\frac{\partial E_{\mathrm{p1}}^0}{\partial z} = -aE_{\mathrm{p1}}^0 - \mathrm{i}\kappa\left[E_{\mathrm{p1}}^0\left(3|E_{\mathrm{p1}}^0|^2 + 6|E_{\mathrm{p2}}^0|^2 + 6|E_{\mathrm{s3}}^0|^2 + 6|E_{\mathrm{d4}}^0|^2\right)\right.$$
$$\left. + 6E_{\mathrm{p2}}^{0*}E_{\mathrm{s3}}^0 E_{\mathrm{d4}}^0\right] \tag{5.181}$$

$$\frac{\partial E_{\mathrm{p2}}^0}{\partial z} = +aE_{\mathrm{p2}}^0 + \mathrm{i}\kappa\left[E_{\mathrm{p2}}^0\left(6|E_{\mathrm{p1}}^0|^2 + 3|E_{\mathrm{p2}}^0|^2 + 6|E_{\mathrm{s3}}^0|^2 + 6|E_{\mathrm{d4}}^0|^2\right)\right.$$
$$\left. + 6E_{\mathrm{p1}}^{0*}E_{\mathrm{s3}}^0 E_{\mathrm{d4}}^0\right] \tag{5.182}$$

$$\frac{\partial E_{\mathrm{s3}}^0}{\partial z} = -aE_{\mathrm{s3}}^0 - \mathrm{i}\kappa\left[E_{\mathrm{s3}}^0\left(6|E_{\mathrm{p1}}^0|^2 + 6|E_{\mathrm{p2}}^0|^2 + 3|E_{\mathrm{s3}}^0|^2 + 6|E_{\mathrm{d4}}^0|^2\right)\right.$$
$$\left. + 6E_{\mathrm{p1}}^0 E_{\mathrm{p2}}^0 E_{\mathrm{d4}}^*\right] \tag{5.183}$$

$$\frac{\partial E_{\mathrm{d4}}^0}{\partial z} = -aE_{\mathrm{d4}}^0 - \mathrm{i}\kappa\left[E_{\mathrm{d4}}^0\left(6|E_{\mathrm{p1}}^0|^2 + 6|E_{\mathrm{p2}}^0|^2 + 6|E_{\mathrm{s3}}^0|^2 + 3|E_{\mathrm{d4}}^0|^2\right)\right.$$
$$\left. + 6E_{\mathrm{p1}}^0 E_{\mathrm{p2}}^0 E_{\mathrm{d4}}^{0*}\right] \tag{5.184}$$

which can be solved numerically. Further approximations such as neglecting absorption and the two-photon process are often useful.

For negligible absorption and two equal and undepleted pump intensities with I_{pump} at the entrance windows, the new generation of the beam d4 was calculated analytically in [5.648]. With the new variables for the beams:

$$\tilde{E}_{\mathrm{s3}}^0 = E_{\mathrm{s3}}^0\, \mathrm{e}^{\mathrm{i}6\kappa z|E_{\mathrm{pump}}|^2} \tag{5.185}$$

$$\tilde{E}_{\mathrm{d4}}^0 = E_{\mathrm{d4}}^0\, \mathrm{e}^{-\mathrm{i}6\kappa z|E_{\mathrm{pump}}|^2} \tag{5.186}$$

and for the nonlinear coefficient:

$$\tilde{\kappa} = 6\kappa|E_{\mathrm{pump}}|^2\, \mathrm{e}^{-\mathrm{i}3\kappa L3|E_{\mathrm{pump}}|^2} \tag{5.187}$$

with the interaction length L in the z direction. The new fields are then given as a function of z as:

$$\tilde{E}^0_{s3}(z) = \frac{\cos[|\tilde{\kappa}|(L-z)]}{\cos[|\tilde{\kappa}|L]} \tilde{E}^0_{s3}(z=0) \tag{5.188}$$

and

$$\tilde{E}^0_{d4}(z) = -i\frac{\tilde{\kappa}}{|\tilde{\kappa}|}\frac{\sin[|\tilde{\kappa}|(L-z)]}{\cos[|\tilde{\kappa}|L]} \tilde{E}^{0*}_{s3}(z=0). \tag{5.189}$$

The last equation shows that the generated wave d4 is proportional to the complex conjugate of the incident probe signal s3 and thus *optical phase conjugation* takes place with this scheme of DFWM. Examples of optical phase conjugation based on four-wave mixing are given in [5.646–5.676].

With these formulas the transmission T_{s3} for the beam s3 and the reflectivity R_{s3} accounting for the amount of light reflected into the direction d4 can be calculated:

$$T_{s3} = \frac{|E^0_{s3}(z=L)|^2}{|E^0_{s3}(z=0)|^2} = \frac{1}{\cos^2(|\tilde{\kappa}|L)} \tag{5.190}$$

and

$$R_{s3} = \frac{|E^0_{d4}(z=0)|^2}{|E^0_{s3}(z=0)|^2} = \tan^2(|\tilde{\kappa}|L). \tag{5.191}$$

The transmission of s3 is always larger than 1 (absorption was neglected) and the reflection can be larger than 1 if $|\tilde{\kappa}|L \geq \pi/4$ which is easily possible. Reflectivity values above 10 or 100 are observed. Even values above 10^3 are possible with suitable materials and very small signal intensities in s3. The maximum amplification and reflection will be limited by the depletion of the pump beams.

With reflectivity values larger than 1, laser oscillators can be built using the FWM phase conjugating material as one of the resonator mirrors and the active material at the same time. By including an active material different schemes have been proposed, as described in Sect. 6.6.12 and 6.7.6.

Besides these applications in lasers, spectroscopy and other measuring techniques potential applications in communication technologies and computing have been discussed. Thus OR, NOR, XOR and AND gates have been demonstrated (see Sect. 1.5). In other schemes six-wave mixing has been achieved [5.677–5.679]. Some more examples of the application of four-wave mixing are described in [5.680–5.685].

5.9.3 Optical Bistability

Optical bistability [5.686–5.706] is exhibited by optical devices which show ideally two stable transmission values T_1 and T_2 as a function of the input beam parameters, especially its intensity. These devices can be used for optical switching.

Achieving optical bistability requires an optical nonlinear element and optical feedback (see Fig. 5.51).

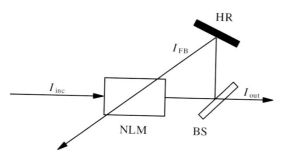

Fig. 5.51. Schematic of an optical bistable device. Part of the output light with I_{out} is split off and fed back into the nonlinear optical material (NLM). The transmission of the device shows two stable values

The transmission T_{bs} of the material is then given as a function of the output intensity $T_{bs} = T(I_{out})$. If this is, e.g. a bell-shaped function (see Fig. 5.52, left side), the output intensity I_{out} as a function of the incident intensity I_{inc} will reflect this behavior in a nonmonotonic shape (see Fig. 5.52, right side).

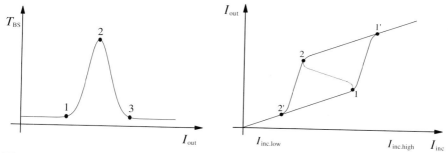

Fig. 5.52. Optical bistability produced by a bell-shaped function of matter transmission as a function of the intensity (see text for explanation)

This results in an ambiguous function of the output intensity as a function of the incident intensity which crosses an unstable region between points 1 and 2 (see Fig. 5.52, right side). Thus the output intensity will jump from point 1 to point 1' if the input intensity is increased and from point 2 to point 2' if the input intensity is decreased. At these points the transmission is changed rapidly. Thus the device should be used for incident intensities between $I_{inc,low}$ and $I_{inc,high}$. Many kinds of nonlinear effect can be applied for such bistable optical devices.

As an example a nonlinear absorber inside a Fabry–Perot interferometer can be used for this optical bistability (see Fig. 5.53).

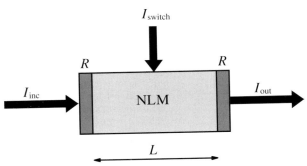

Fig. 5.53. Optical bistability using a Fabry–Perot interferometer filled with a nonlinear material. This device can also be used as an opto-optical switch applying a small switch intensity I_{sw}

Using (5.4) as a simple approach to describe the nonlinear transmission of the sample by the low signal absorption coefficient a_0 and the nonlinear intensity I_{nl} the transmission of the bistable device can be written as:

$$T(I_{out}) = \frac{I_{out}}{I_{inc}} = \left[1 + \frac{a_0 L}{(1-R)(1 + 2I_{out}/I_{nl})}\right]^{-2} \tag{5.192}$$

if both the reflectivity R of the Fabry–Perot mirrors and the transmission T of the material are close to one. In this formula L is the length of the active material. The operation of this device is better the higher the quotient $(1-T)/(1-R)$.

Besides nonlinear absorption the nonlinear Kerr effect can also be applied for optical bistability [5.686]. If the Kerr material is used as the nonlinear material in the scheme after Fig. 5.53 the nonlinear transmission of the device can be written as:

$$T(I_{out}) = \frac{(1-R)^2}{(1-R)^2 + 4R\,\sin^2[2\pi L/\lambda_0(n + 2\gamma_I I_{out}(1-R)]} \tag{5.193}$$

with the Kerr constant γ_I, refractive index n and wavelength λ_0 of the incident light. It is assumed that for low intensities the length L is tuned to resonance. Because of multiple resonance with increasing intensity this element shows periodically instabilities.

As nonlinear materials such as organic molecules or semiconductors can be used in this type of optical bistable element. The reaction time is given by the round-trip time of the Fabry–Perot interferometer and the reaction time of the matter. Therefore thermal effects which are also possible are usually too slow for the desired application. But again known materials with fast reaction times demand high intensities for switching. Thus the usefulness of these optical switches are difficult to estimate.

5.10 Materials in Resonant Nonlinear Optics

Potentially many kinds of gases, liquids, solutions and solids are useful for nonlinear optical applications based on nonlinear absorption and emission. Most important are laser materials which are described in the next chapter. Also important are nonlinear absorbers applied for Q switching and mode locking in laser oscillators. Organic and inorganic systems such as solids or liquids are used for this purpose. Optical switching, storage and new display technologies may become more important in the next few years. For these applications better knowledge about the nonlinear optical processes is demanded.

Organic matter seems to have the greatest variability to fulfill the different functions, but their sometimes limited stability has to be increased significantly. New concepts using structures in the μm and nm scale reaching molecular dimension will be developed, but also new inorganic matter or combinations such as, e.g. colloides may be obtained.

In any case new materials for these new photonic techniques are required for wide use in science and industry and knowledge about the nonlinear optical interaction of these systems has to be improved. Some methods for this purpose are described in Chap. 7.

5.10.1 Organic Molecules

The structure of organic molecules and their arrangement with more or less strong coupling to the environment can be designed in nearly indefinite variations. Thus it seems to be possible to develop almost all required new functions for photonic applications based on these compounds.

The parameters of molecular systems can vary in wide ranges:

- spectral absorption bands: UV-Vis-IR ($150\,\text{nm}-10\,\mu\text{m}$);
- cross-sections (molecules): $< 10^{-15}\,\text{cm}^2$;
- cross-sections (aggregates): $> 10^{-15}\,\text{cm}^2$;
- lifetimes: fs – years;
- saturation intensities: $< \text{kW cm}^{-2} \rightarrow \text{GW cm}^{-2}$.

Organic matter is used in photonics, e.g. as a laser material (e.g. dye lasers), as an optical switch (e.g. Q switch and mode locker) and as a waveguide. Potential applications may be advanced optical switching and optical storage, nonlinear phototherapy in medicine and new solar energy techniques.

Nature uses this potential, e.g. in photosynthesis and in the process of seeing. But in these examples a very complex layout of thousands of molecules achieves the function whereas the nonlinear behavior of single molecules such as the chlorophyll and plastochinon (see Fig. 5.54) is quite simple.

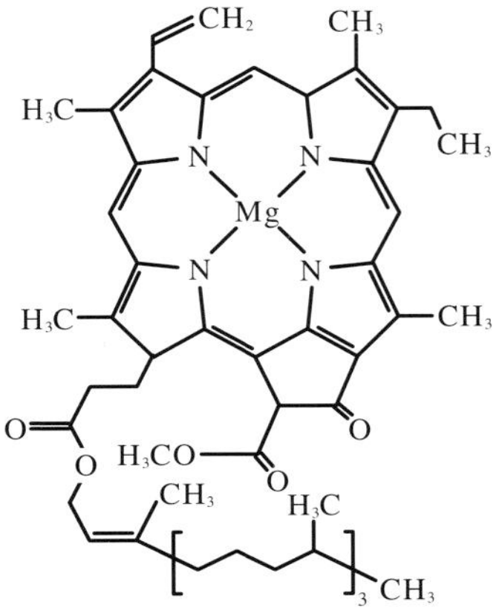

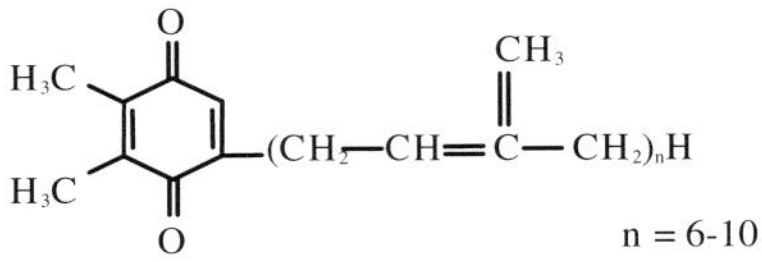

Fig. 5.54. Molecular structure of chlorophyll and plastochinon

Although many details about these complex structures are known we are still far from building similar structures for photonic applications with comparably good and stable operation. Nevertheless, these examples challenge further research activities in this field.

But as a simple rule the single molecule as a basic unit of these structures should have favorably high cross-sections σ in the applied spectral range for nonlinear optics. The higher the cross-section the lower the necessary excitation intensity I_{exc} for the required σI_{exc} products. The cross-section of the particles in the sample can be increased by aggregation and other special arrangements of single molecules by many orders of magnitude.

In stationary cases with pulse widths and modulation times much longer than the characteristic relaxation time τ of the matter the product $\sigma\tau$ can be as large as possible to allow small excitation intensities. Thus the relaxation time can be large for low intensities but on the other hand fast enough for the required application. For excitation long relaxation times, much longer than the pulse integration of the excited particles will occur. In this case the cross-section alone is responsible for the onset of nonlinearity.

Frequently used and investigated molecules for Q switching and mode locking are cryptocyanine, DTTC, phthalocyanine for ruby lasers and Kodak dyes #9860, #9740 and #14015 for neodymium lasers. For mode locking and fs generation of dye laser cresylviolet and DODCI were used. Further details are given in Sects. 6.10.2 and 6.10.3.

Structure and Optical Properties

Some general simple rules will be discussed, with examples of the simplest molecules of their kind. The linear and nonlinear optical behavior of more complex molecules may be estimated from this simple picture (see [5.707–5.712, M4, M23]). For a precise theoretical description of the wavelength and oscillator strength of the optical transitions quantum chemical calculations are needed, as they are possible, e.g. with the Gaussian package or the ZINDO-S (see Sect. 7.13.3). The theoretical description of the bandwidth and the relaxation times is still difficult for the interesting large molecules. A statistical approach is described in [5.713].

The absorption wavelength is a function of the electronic structure of the molecule. The larger the geometrical dimension of the electronic orbitals of the binding electrons the longer the wavelength of the absorption. Therefore the strongly localized σ bonds (see Fig. 5.55) show very short absorption wavelengths in the UV, typically below 200 nm. The additional much less localized π bonds formed by the π electron system from the double bonds or triple bonds shift this absorption to longer wavelengths.

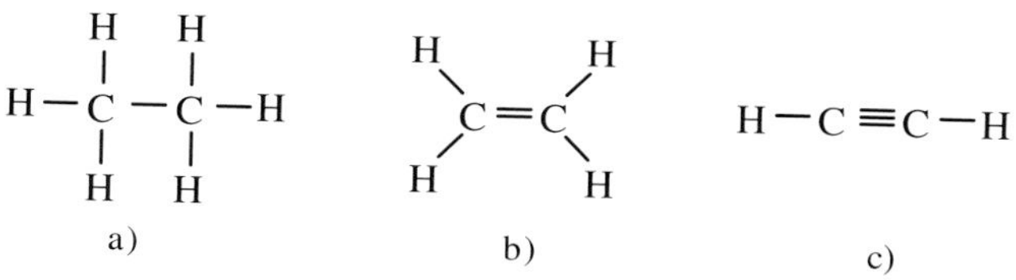

Fig. 5.55. Molecular structure of the three molecules with two C atoms: ethane (**a**), ethylene (**b**) and acetylene (**c**) which have zero, two and four π electrons

Molecules with π electron bonds are called *conjugated molecules*. They can show molecular absorption in the whole range from UV to the visible up to the near IR spectral range.

Molecules of a geometrically similar structure can also be different depending on the number of H atoms and π electrons in cyclic form, as shown in Fig. 5.56 for cyclohexane and benzene.

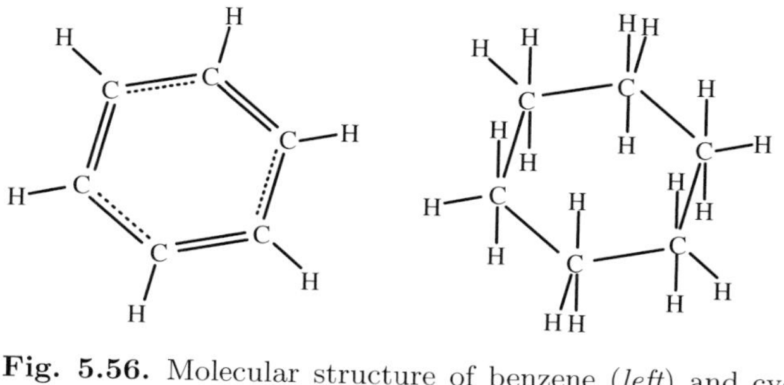

Fig. 5.56. Molecular structure of benzene (*left*) and cyclohexane (*right*) with 0 and 12 π electrons. The π electrons are not localized in benzene and form an electron cloud above and below the molecule ring which is responsible for the longest wavelength absorption

The π electron bonds are weaker than the σ bonds and thus the photostability of these compounds can be too low for the desired applications in the nonlinear optical range. Choosing suitable environment, pump intensity, polarization, pulse duration and wavelengths may be essential to overcome this problem.

With respect to their structure and their resulting optical properties more or less linear and cyclic molecules may be distinguished (see Fig. 5.57, 5.58, 5.60 and 5.62).

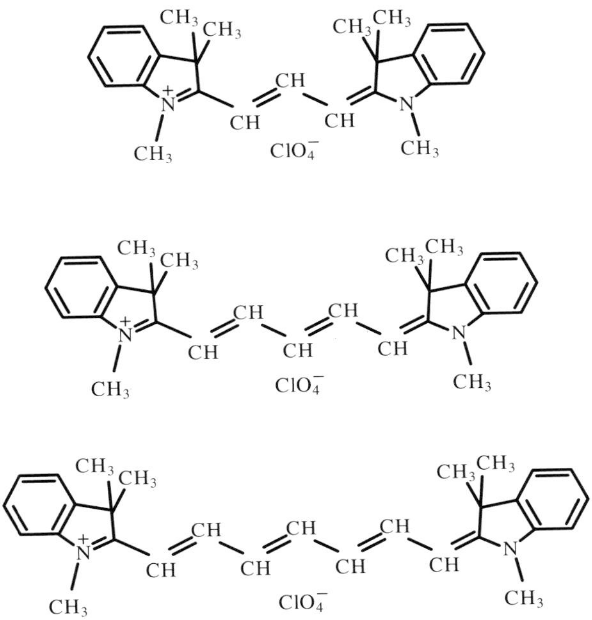

Fig. 5.57. Structure of polymethine with different chain lengths. The π electron system is located above and below the chain. The 120° bond angle results from the s-p hybridization in the C atoms. These molecules can build isomers if the bond angle folds to the other side

The polymethine dyes as shown in Fig. 5.57 and similar molecules show a π electron system which is located above and below the chain of the C atoms. The long-wavelength absorption can be roughly estimated from the length of this π electron system. Even a simple quantum mechanical box model using the length of the molecule as the box length can be applied to estimate the absorption wavelengths and the shift of the longest wavelength absorption band. For the example of the molecules given in Fig. 5.57 the longest wavelength absorption peaks are measured at 552 nm, 640 nm and 744 nm, respectivily.

These molecules can build isomers if the bond angle folds to the other side (see Fig. 5.65). This can be detected from a blue shift of the absorption spectra. The dipole moment of the longest-wavelength transition is usually directed along the long axis of the molecule and thus the light should be polarized in this direction. These molecules can be aligned by stretching polymer foils containing the molecules. The electron density alternates often from one C atom to the neighboring one. In the first excited state this electron modulation density is often exchanged between the C-atoms and thus the Stokes shift, which characterizes the spectral difference between the maxima of the absorption and emission bands, can be changed by a more or less polar solvent. This may influence the time constants, too [5.714].

The shape and broadness of the electronic absorption spectra is different for rigid molecules as, e.g. shown in Fig. 5.58 or flexible molecules as, e.g. in Fig. 5.60.

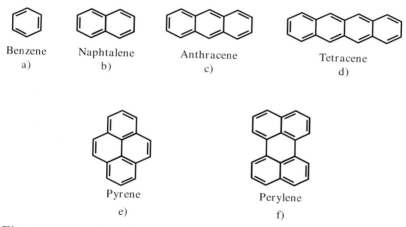

Benzene
a)

Naphtalene
b)

Anthracene
c)

Tetracene
d)

Pyrene
e)

Perylene
f)

Fig. 5.58. Structure of aromatic molecules: (**a**) benzene, (**b**) naphthalene, (**c**) anthracene, (**d**) tetracene, (**e**) pyrene and (**f**) perylene

The absorption spectra of these rigid molecules are shown in Fig. 5.59. A modulation of the longest-wavelength absorption can be observed in all cases due to the vibrational progression with energy steps in the range of $1.000\,cm^{-1}$. The longest-wavelength absorption shifts from 265 nm for benzene, via 315 nm for naphthalene and 380 nm for anthracene to 480 nm for tetracene. The cross-section is increased in this sequence, too. The absorption wavelength of pyrene is not much shifted in comparison to anthracene but the cross-section is increased drastically. In perylene both values are larger as for anthracene. These facts can be illustrated using the simple box model and considering the number of π electrons responsible for the strength of the transition.

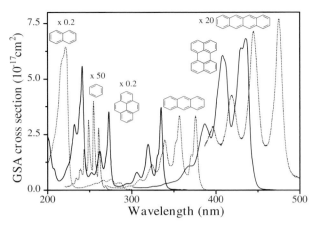

Fig. 5.59. Absorption spectra of the molecules from Fig. 5.58 dissolved in cyclohexane

Nevertheless, the detailed analysis has to done with quantum chemical calculations as discussed in Sect. 7.13. The question of allowed and forbidden transitions (as, e.g. the longest wavelength transition of naphthalene) can be answered in this way.

The structures of some simple and flexible aromatic molecules are given in Fig. 5.60. In these molecules the phenyl rings can change their twist angle along the bonds between the phenyl rings [5.715].

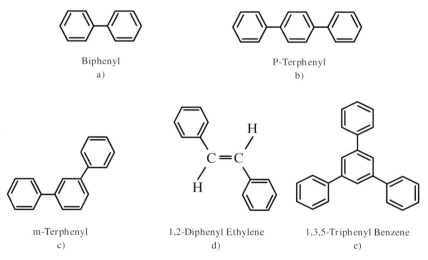

Fig. 5.60. Structure of some phenyls and derivatives: (**a**) biphenyl, (**b**) p-terphenyl, (**c**) p-quaterphenyl, (**d**) 1,1-diphenyl ethylene and (**e**) 1,3,5-triphenyl benzene

The absorption spectra of these molecules are given in Fig. 5.61.

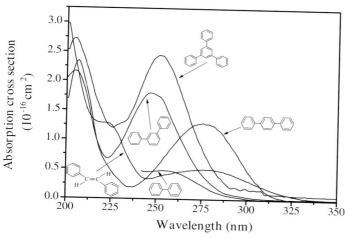

Fig. 5.61. Absorption spectra of the molecules from Fig. 5.60 dissolved in cyclohexane

The spectra of these flexible molecules are much less structured and broader than the spectra of the "related" rigid molecules. This probably results from the stronger coupling of vibrational and rotational transitions to the electronic transition and a stronger coupling between these vibrations.

As a consequence the transitions are not only broadened but the relaxation times can be shortened by orders of magnitude via competing radiationless transitions. Thus the ground state absorption recovery time of malachit green and crystal violet (see Fig. 5.62 (d) and (e)) can be as short as a few ps. The fluorescence will then be quenched almost completely. The flexibility is a function of the viscosity of the solvent and thus the recovery time can be varied over about two orders of magnitude by changing from ethanol to the highly viscous glycerol [5.716].

Because of the steric hindrance from the H atoms between the phenyls the rings are tilted by about 20–30° in the electronic ground state but it is assumed that they are more planar in the first excited state.

As a result of the weaker coupling along the bonds between the phenyl rings of the molecules the wavelength shift of the absorption bands with increasing number of phenyl rings is smaller than for the rigid molecules. The long wavelength absorption edge is about 265 nm for biphenyl, 295 nm for p-terphenyl, 315 nm for p-quaterphenyl, 265 nm for 1,1-diphenyl ethylene and 270 nm for 1,3,5-triphenyl benzene.

If electron donor atoms such as N atoms or acceptor atoms such as O atoms are integrated into the molecular structure their wavefunction and their electron density can be strongly changed. As examples the structures of acridine yellow, uranine, rhodamine 6G, malachite green and crystal violet are given in Fig. 5.62.

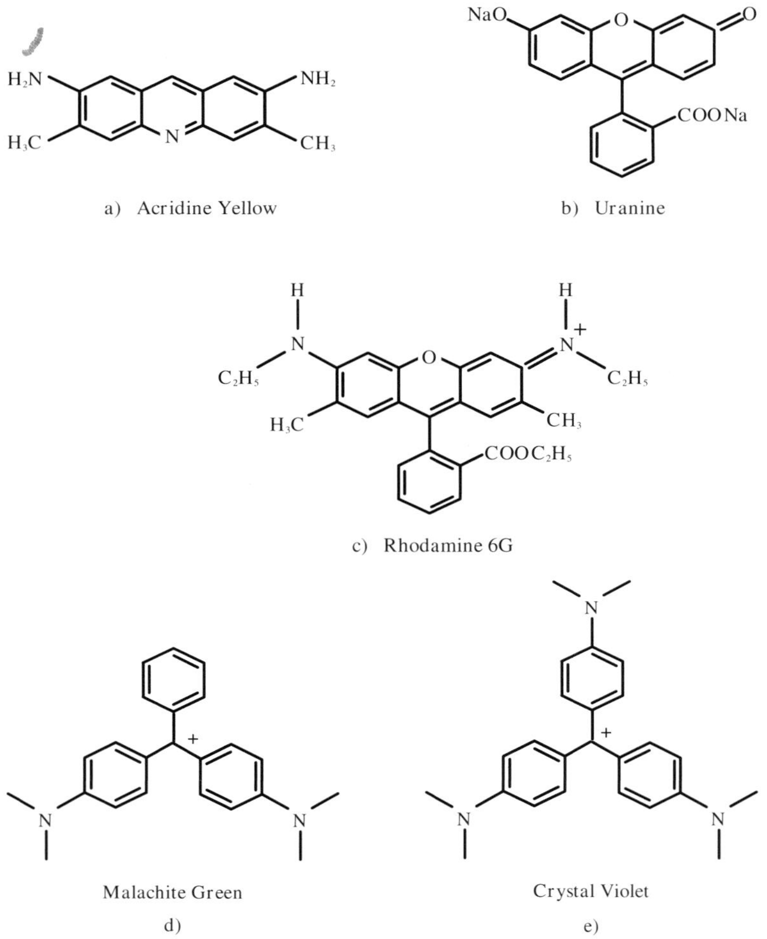

a) Acridine Yellow

b) Uranine

c) Rhodamine 6G

Malachite Green

d)

Crystal Violet

e)

Fig. 5.62. Molecular structure of acridine yellow (**a**), uranine (**b**), rhodamine 6G (**c**), malachit green (**d**) and crystal violet (**e**)

The absorption spectra of these molecules are given in Figs. 5.63 and 5.64. Compared to anthracene the cross-sections of acridine yellow, uranine and rhodamine 6G are increased by about an order of magnitude and reach values above 10^{-16} cm^2. The lowest energy absorption is shifted towards longer wavelengths by more than 100 nm.

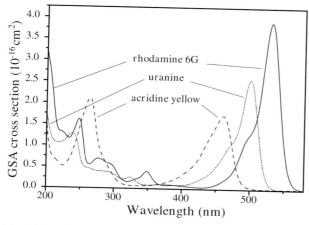

Fig. 5.63. Absorption spectra of the molecules of Fig. 5.62 dissolved in ethanol: acridine yellow, uranine and rhodamine 6G

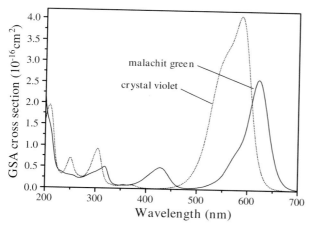

Fig. 5.64. Absorption spectra of malachit green and crystal violet molecules of Fig. 5.62 dissolved in ethanol

The π electron system can also be enlarged by including metal atoms, e.g. Fe, Ni, Mn, Cu, Co, Zn or Pt in the structure of the organic molecules. As an example, the structure of chlorophyll is given in Fig. 5.54.

Because of the high electron densities in the π electron system the absorption bands can also be influenced by polar solvents. In some cases spectral shifts of more than 100 nm can be obtained by highly polar solvents such as tetrahydrofuran in comparison to nonpolar ones such as cyclohexane.

Some of the flexible molecules can undergo changes in their structure. As a molecular engine, the photoinduced trans–cis isomerization may find applications [e.g. 5.717–5.737]. The principle is sketched in Fig. 5.65.

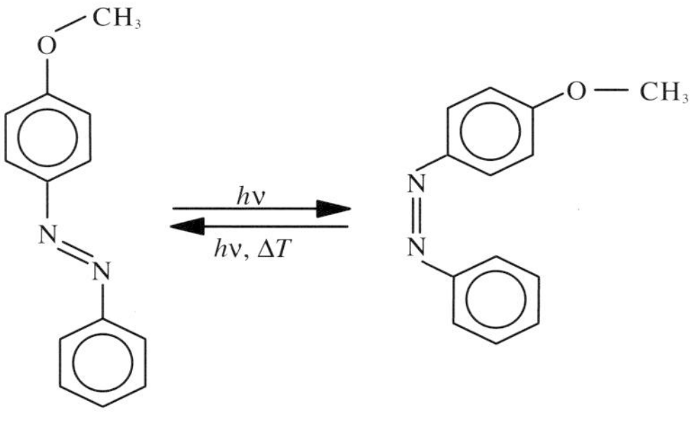

Fig. 5.65. Molecular structures of the two conformers: the trans- and the cis- isomer can be converted to each other by photoexcitation

The photoinduced structural changes can be used potentially for information storage and optical switching. Another example is twisted intramolecular charge transfer (TICT) states of molecules with self-stabilization of the twisted conformer after excitation [e.g. 5.738–5.763]. Light-induced changes of the electrical and mechanical properties of matter can be imagined by these effects, too.

Preparation of the Samples

Organic molecules can be used in photonic applications in many forms such as, e.g. dissolved in liquids, prepared in thin films, used as crystals or enclosed in polymers.

The solubility is usually higher the more similar the structure of the solvent and the agent. Thus polar molecules are solvable in polar solvents and the geometrical structure of both should be similar for high concentrations. Too high concentrations, usually above 10^{-3} mol l^{-1} (but sometimes also much less), can lead to the formation of dimers, exciplexes or other aggregates with different linear and nonlinear optical properties. Purity of the solvent is more crucial in nonlinear optics than in linear cases. Uvasol grade is usually demanded. Even these solvents may need to be cleaned further especially for particles [4.335].

Thin films and polymers [e.g. 5.764–5.779] allow the orientation of the molecules to be controlled. This can be achieved by the Langmuir–Blodgett technique, spin coating or epitaxy. Thus polarization-dependent applications can be achieved. Thin films above surfaces change the symmetry of the system, and this can be used for nonlinear effects, e.g. for second harmonic generation. Some further organic materials in different preparations are investigated in [5.780–5.839]. Some work on organic light-emitting diodes is described in [5.823–5.835].

5.10.2 Anorganic Crystals

In principle all laser materials (see Chap. 6) can be used as nonlinear optical switches or for other photonic applications, too. For Q switching the absorption wavelength has to fit the emission wavelength of the active laser material and the absorption cross-section has to be larger than the emission cross-section of the laser material (see Sect. 6.10.2). But in most of these laser materials the ground state absorption recovery time is too long for the desired application. Therefore only a few crystals have been applied in the absorbing range.

E.g. Cr^{4+}:YAG crystals are very useful for passive Q switching of Nd:YAG lasers. The nonlinear transmission of these broadband absorbing material is based on a comparatively short lifetime of 4 µs and their cross-section is $3 \cdot 10^{-18}$ cm^2 at the wavelength of 1064 nm.

Further color centers in crystals are used as nonlinear devices for Q switching solid-state lasers. For example F_2^- centers in LiF crystals are used at room temperature to generate Q switch pulses in Nd:YAG lasers.

5.10.3 Photorefractive Materials

Photorefractive materials [e.g. 5.840–5.870] can show strong nonlinear optical effects at low intensities based on local charge displacement followed by a refractive index change. Typical time constants are in the range of ms up to minutes. These materials are typically used with only very small absorption at the applied wavelength which increases the nonlinear reaction.

High nonlinearity is, e.g. obtained in electro-optic crystals such as lithiumniobate (LiNbO$_3$) and bariumtitanate (BaTiO$_3$) as well as from the semiconductor crystals GaAs, CdTe and InP.

Therefore the photorefractive material is illuminated with an excitation intensity grating leading finally to a refractive index grating. At this refractive index grating another beam can be scattered. Thus the typical application is carried out with four-wave mixing (FWM).

This process takes place in four steps as depicted in Fig. 5.66. These steps are:

(i) The excitation intensity grating $I(z)$ over the spatial coordinate z with the grating constant Λ_g as a result of light interference of two pump beams produces charge carriers in the illuminated areas proportional to the excitation intensity (Fig. 5.66, top).

(ii) These electrons with density n_e (Fig. 5.66, second top) migrate due to electrostatic forces via drift and diffusion to the less illuminated, dark areas.

(iii) The photovoltaic effect produces a space charge separation of ρ (Fig. 5.66, second bottom) with density of photoexcitations N_p.

(iv) The resulting difference of the electric field δE leads to a modulation of the refractive index Δn in the matter with the amplitude n_1 (Fig. 5.66, bottom).

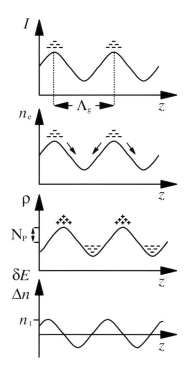

Fig. 5.66. Schematic of the refractive index generation in a photorefractive material after illumination with an intensity grating. For explanation see text

⇒ The refractive index grating is spatially shifted relative to the intensity grating by $\pi/2$.

Therefore this index grating is well suited for the diffraction of the excitation beams or further probe beams in two- or four-wave mixing schemes.

The minimum time constant τ_{photoref} for obtaining the index changes is a function of the intensity I_{exc}, the absorption coefficient a_{mat} and the desired density of photoexcitations N_{p}:

$$\textbf{buildup time} \quad \tau_{\text{photoref}} \geq \frac{N_{\text{p}}}{a_{\text{mat}} I_{\text{exc}}} \tag{5.194}$$

where it is assumed that the decay of the space charge separation is much longer than the buildup and therefore an integration occurs.

The density of photoexcitations N_{p} yields the modulation depth n_2 of the refractive index n_0 of the material:

$$\textbf{index modulation} \quad n_2 = \frac{1}{4} n_0 r_{\text{eo}} \frac{N_{\text{p}} \eta_{\text{photoexc}} e_{\text{e}}}{\varepsilon} \Lambda_{\text{g}} \tag{5.195}$$

with the conventional refractive index n_0 at the applied wavelength, r_{eo} as the electro-optical coefficient of the material, η_{photoexc} as the efficiency of charge separations from the excitation, e_{e} as the electronic charge, $\varepsilon_{\text{photoref}}$ as the relevant dielectric constant and Λ_{g} as the grating constant of the experiment.

For example using $BaTiO_3$ as photorefractive material the following values can be obtained:

$\lambda = 500\,\text{nm}$:

$r_{eo} = 16404\,\text{pm}\,\text{V}^{-1}$ $a_{mat} = 0.1\,\text{cm}^{-1}$

$\varepsilon = 3600\varepsilon_0$ $\Lambda_g = 5\,\mu\text{m}$

$\eta_{photoexc} = 10\%$ $I_{exc} = 100\,\text{W}\,\text{cm}^{-2}$

                         $f_{exc} = 2.7 \cdot 10^{20}\,\text{photons}\,\text{cm}^{-2}\,\text{s}^{-1}$

result in:

$N_p = 8.4 \cdot 10^{18}\,\text{cm}^{-3}$

$n_1 = 5\%\,\, n_0 = 0.12$

$\tau_{photoref} = 0.3\,\text{s}$

Most possible photonic applications of photorefractive materials are not yet commercialized. Good-quality material in large geometry is still not easy to make. In the laboratory the recording of volume holograms with very high storage densities has been demonstrated (see Sect. 1.5). Two-wave mixing has been applied for laser beam clean-up where the energy from a powerful laser beam with poor beam quality was transferred to another beam with very good beam quality (see Sect. 5.9.1). Further optical phase conjugation in photorefractive materials has been applied as laser resonator mirrors for improved beam quality. These adaptive mirrors can be used as self-pump devices which are easy to make. In image processing this type of optical phase conjugation can also be used. Optical switching is possible and further neural networks are demonstrated with these materials (see Sect. 1.5).

5.10.4 Semiconductors

Besides the rapidly increasing importance of semiconductors in diode lasers (see Chap. 6) they are also used as optical nonlinear devices [e.g. 5.871–5.875]. High cross-sections of up to $5 \cdot 10^{-19}\,\text{cm}^2$ as well as high carrier concentrations of up to $10^{22}\,\text{cm}^3$ are possible. Lifetimes can be as short as ns. The possibility of shifting the long-wavelength absorption band edge from the visible to near infrared wavelengths by varying the concentration of suitable mixtures make them adaptable to desired applications.

Both III–V semiconductors such as GaAs, as well as II–VI semiconductors such as CdSe are used for this purpose. They are used in all types of nanometer structures, as described in the next Section and the references given therein.

5.10.5 Nanometer Structures

Geometrical structures with dimensions in the nm range and thus less than the light wavelength can show strong absorption in the UV-Vis-NIR spectral range. The small dimension can be realized in one dimension resulting in a *quantum well* [e.g. 5.876–5.882], in two dimensionsm, which is a *quantum wire* [e.g. 5.883–5.889], and in all three dimensions representing a *quantum dot* [e.g. 5.890–5.925] (see Fig. 5.67).

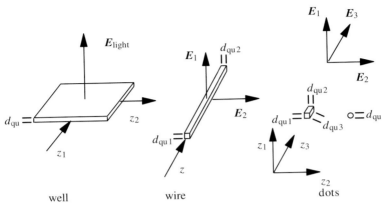

Fig. 5.67. Nanometer structures resulting in quantum wells, quantum wires and quantum dots

The applied light should have its electric field vector E in the direction of the nm dimension which results in one possibility for the well and three for the dot. The propagation direction z is of course perpendicular to this and allows only one direction for the wire.

The energy of the pth discrete levels $E_{qu,p}$ can be calculated with quantum models and they are a function of the linear dimension d_{qu} of the nm structure and the material parameters. The energy level difference and thus the frequency of the absorbed light is often roughly inversely proportional to the square of the dimension. A very rough estimate can be based on the quantum box model which results in the energy $E_{dot,p}$ of the pth level of a quantum dot:

$$E_{dot,p} = \frac{h^2 p^2}{8 m_{red} d_{qu}^2} \tag{5.196}$$

with the reduced mass m_{red} of the system. But the practically relevant dimensions have to be calculated considering the wavefunctions of the material in detail.

It can then be found that for absorption wavelengths in the visible, typical dimensions of small spheres are in the range of, e.g. 1.5–3 nm for CdTe and around 12 nm for gold.

Such nm-structures can be designed in many ways and they can have good optical stability. They can be built with different semiconductors or metals. The surface can be covered with organic and inorganic matter based on thin films and thus the optical properties can be changed over wide ranges.

Other nanometer structures such as, for example, nanotubes, were investigated [e.g. 5.926–5.930]. Quantum wells are applied, for example, in semiconductor lasers (see Sect. 6.3.2).

A new perspective may be opened by photonic band gap materials or photonic crystals [e.g. 5.931–5.936]. In these materials periodic structures in the nm-range were fabricated, and, thus, a strong modulation of the electric light field occurs as a consequence of the refractive index modulations, which have to be large enough (≥ 2.9). The periodic structure can be built in two or three dimensions. These materials can be applied potentially, for example, as high-reflectivity mirrors, waveguides or beamsplitters in conventional optics. In the nonlinear optical range they will show new properties based on the quantum confinement of the radiation in the structure. Thus, the lifetime of the excited states of test atoms or molecules may be changed by orders of magnitude. Opto-optical switches may become possible with low light intensities and new quantum optical effects may be observed.

6. Lasers

The LASER light source, whose name is based on *"Light Amplification by Stimulated Emission of Radiation"*, is the most important device in almost all photonic applications. First built in 1960 [6.1–6.5] it allows the generation of light with properties not available from natural light sources. Modern commercially available laser systems allow output powers of up to 10^{20} W for short times with good beam quality and of several kW in continuos operation, usually with less good beam quality. Very short pulses with durations smaller than $5 \cdot 10^{-15}$ s, wavelengths from a few nm in the XUV to the far IR with several $10 \, \mu m$, pulse energies of up to 10^4 J and frequency stability's and resolutions of better 10^{-13} can be generated. The laser prices range from \$ 1 to many millions of dollars and their size from less than a cubic mm to the dimensions of large buildings.

The good coherence and beam quality of laser light in combination with high powers and short pulses are the basis for many nonlinear interactions, but the laser is a highly nonlinear optical device itself, using nonlinear properties of materials as described in the previous chapters. Therefore, the fundamental laws treated in Chap. 2 for the description of light as well as the description of linear and nonlinear interactions of light with matter in Chaps. 3, 4 and 5 are the basis for the analysis of laser operation and its light properties.

Therefore, the theoretical description of laser devices represents an application of these laws and can be presented in this chapter in a compact form. The different lasers and their constructions, as well as the resulting relevant light and operation parameters, are described and the consequences for photonic applications are discussed. Finally, possible classifications are given and safety aspects are mentioned. For further reading see [M4, M11, M12, M15–17, M19, M20, M22, M25, M32, M33, M38, M39, M45–52].

6.1 Principle

Lasers are based on the stimulated emission of light in an active material which has been pre-excited by a pump mechanism. The stimulated emission can be carried out in laser oscillators which are always the primary source of laser light. In addition this light can be amplified via stimulated emission in light amplifiers as shown in Fig. 6.1 where a master oscillator is combined

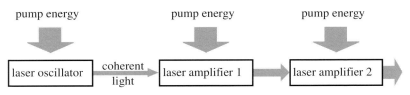

Fig. 6.1. Laser setup consisting of a laser oscillator (master oscillator) and two amplifiers (MOPA scheme)

with, e.g. two amplifiers in a MOPA (*Master Oscillator Power Amplifier*) setup.

In any case the coherent laser light has to be originally generated in a laser oscillator. This laser oscillator consists of *three basic parts* as shown in Fig. 6.2.

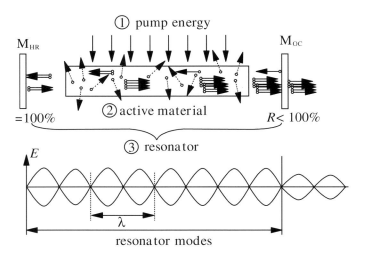

Fig. 6.2. The three basic parts of a laser oscillator: pump source ①, active material ② and resonator ③

The fundamental function of these three components is described in Table 6.1.

The laser operates in the following way:

- The pump mechanism provides enough energy in the active material and produces an *inversion* of the population density resulting in the higher population of the upper laser level compared to the lower laser level in the laser material (in fundamental contrast to a thermal population, for an exception see Sect. 5.4.10).

Table 6.1. Function and examples for the three components of lasers

Component	Function	Examples
Pump	energy/power provider	electric current
		electrical discharge
		flash or arc lamp
		other laser
		chemical reaction
Active material	possible laser	semiconductor structures (GaAs)
	light properties	atoms in gases (Ne, Ar, Kr)
		ions in crystals (Nd, Cr, Yb, Ti)
		molecules in gases (XeCl, CO_2)
		molecules in solution (dyes)
Resonator	selection of the laser	simple two-mirror resonator
	light properties	resonator with frequency selection
		resonator with internal frequency conversion
		resonator with Q switch
		resonator with mode locking
		unstable resonator
		folded resonators
		phase conjugating resonator

- *Spontaneous emission* produces photons in all directions, possibly with different polarizations and in a wide spectral range as a function of the active material.
- The resonator mirrors reflect some of these photons selectively for their propagation direction, their polarization, their wavelength and perhaps as a function of time (*laser mode selection, short pulse generation*).
- These reflected photons are cloned by the stimulated emission in the active material (*amplification*) and thus a large number of equal and coherent photons are produced forming the high-brightness laser beam.
- Part of this laser beam is coupled out of the resonator, e.g. by a partly transparent mirror at one side of the resonator (*outcoupling*).

The function of these steps will be described in detail in the following sections.

6.2 Active Materials:
Three- and Four-Level Schemes – Gain

Almost all materials (except, e.g. solid metals) can be used as the active material in lasers. Even single atom lasers were realized [see, for example, 6.6]. The efficiency and the possible laser properties are very different and therefore the number of practically used laser materials is more limited but still quite large. Therefore we can distinguish gas, liquid and solid-state lasers on one hand, and on the other, the active material can be built by molecules (CO_2, CO, N_2, excimers such as XeCl or KrF, dyes), atoms in gases (HeNe, Cu vapor), ions in gases (Ar^+, Kr^+), atoms and ions in solids (Nd, Cr, Ti, Yb, Er, Pr ...), color centers or semiconductors (GaAs, ZnSe, PbSnSe, ...). Solid-state host materials can be crystals, glasses and organic matter. Crystals can be fluoride or oxide. Typical crystals are YAG ($Y_3Al_5O_{12}$) and sapphire. The different constructions will be described in Sect. 6.13.

In any case the laser action (stimulated emission) takes place between at least two energy levels (or bands) of the matter, the upper and the lower laser level (see Fig. 6.3).

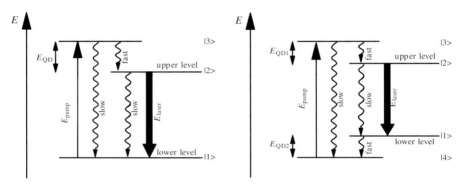

Fig. 6.3. Three (left side) and four (right side) level scheme of an active material. The laser works between levels 2 and 1 via stimulated emission as a consequence of inversion (level 2 is more highly populated than level 1). The upper laser level is populated by the pump mechanism via level 3

But for achieving inversion more than two energy levels are necessary, because the two-level scheme allows at best only equal populations of the upper and lower levels and thus transparency but no amplification.

In the *three-level scheme* (see Fig. 6.3, left side) the upper laser level (2) is populated via the higher level (3). This pump level can be a collection of several levels which can even form a pump band. If in the laser material the radiationless population channel (3) → (2) is fast compared to the possible radiationless deactivation channels (3) → (1) and (2) → (1) the upper laser level can be almost 100% populated. But the laser action populates the lower laser level (1) which will stop operation if the pump is not strong enough. Thus laser materials with a three-level scheme may have the advantage of a possibly small quantum defect (see next chapter) and therefore higher efficiency. But the strong pump demands can be difficult in cw operation. Furthermore high pumping can also favor excited state absorption from the upper levels which will decrease the laser efficiency.

In the *four-level scheme* (see Fig. 6.3, right side) the upper laser level (2) is again populated via the higher level (3) but in addition the lower laser level (1) is not identical with the ground state of the system (4). Therefore if thermal population of the lower laser level (1) can be neglected each pumped particle will produce inversion. If in addition as usual the radiationless transition (1) → (4) is fast the lower laser level will stay empty even during laser action. Therefore four-level lasers can be very easily pumped.

The amplification of the laser light in the active material can be calculated with rate equations. The gain coefficient g (negative absorption coefficient) is proportional to the inversion population density $N_2 - N_1$:

gain coefficient $g(z, t, \lambda) = \sigma(\lambda)\{N_2(z, t) - N_1(z, t)\}$ (6.1)

which is a function of the wavelength λ, the position in the propagation direction z and the time t. Its influence on the laser properties will be discussed in Sect. 6.8.

Some common and some newer laser materials and their parameters are described in [6.7–6.99]. For effective pumping energy transfer mechanisms can be used to separate the pump energy absorption and the laser operation in two different materials as, for example, in the Helium–Neon laser [6.68–6.73]. Of increasing interest are upconversion lasers which allow laser operation at shorter wavelengths as the absorption [6.74–6.99], and, thus, the generation of blue light from red diode pumping. More details are given in Sect. 6.13.

6.3 Pump Mechanism: Quantum Defect and Efficiency

The pump mechanism of the active material and its efficiency are important for the output parameters, the handling and the price of a laser system.

Almost all active materials can be *pumped by another laser* beam of suitable wavelength. The resulting opto-optical efficiency can reach high values limited by the quantum defect, radiationless transitions and excited state absorption (see Fig. 6.3).

The quantum defect energies E_{QD} in Fig. 6.3 result from:

quantum defect energy $E_{QD} = E_{pump} - E_{laser}$

$$= hc \left(\frac{1}{\lambda_{pump}} - \frac{1}{\lambda_{laser}} \right) \qquad (6.2)$$

with the wavelengths λ_i of the pump and laser light, Planck's constant h and the velocity of light c.

The quantum efficiency η_Q is the ratio between the number of emitted laser photons and the number of absorbed pump photons independent of their photon energy (see Sect. 6.3.6).

In the case of 100% quantum efficiency, i.e. each absorbed photon will generate a laser photon, the quantum defect will reduce the opto-optical efficiency to values of usual less than 90%. But in the case of Yb:YAG laser crystals emitting at 1030 nm pumped with diode lasers at 940 nm the quantum defect is as small as 9% (see Table 6.2).

Table 6.2. Quantum defects of some lasers for their strongest laser transitions

Laser material	λ_{laser} (nm)	λ_{pump} (nm)	$E_{\text{QD}}/E_{\text{pump}}$ (%)
Yb:YAG	1030	940	8.7
Nd:YAG	1064	808	24
Er:YAG	2940	532	82
Rhodamin 6G	e.g. 580	308	47
Ti:Sapphire	e.g. 800	532	34

Because of the possible choice of the pump and the emission wavelengths the quantum defect for a given material can vary drastically. As an example the absorption spectrum of Nd:YAG is given in Fig. 6.4.

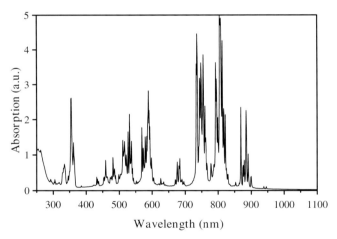

Fig. 6.4. Absorption spectrum of a Nd:YAG laser crystal

For simplicity in a laser system, direct pumping of the active material with electrical current is attempted. In *diode lasers* the resulting electro-optical efficiency can be as high as 40%. Therefore these lasers may become even more important in the very near future in high-power applications with average output powers of hundreds of watts or kilowatts. The disadvantage of diode lasers with output powers of more than ten watts is today their poor beam quality with M^2 factors of more than 10^4, which prevent these lasers being used in high-precision applications or nonlinear optics. Nevertheless, in addition to applications in surface treatment they are progressively being used for pumping of solid-state lasers. This results in a reduced thermal load by optimal adaptation of the pump wavelength to the absorption of the active matter and in high overall efficiencies of up to 20%.

6.3.1 Pumping by Other Lasers

This type of pump scheme is used to transform the wavelength or spectral width of the laser radiation or to increase the beam quality or coherence of the laser light. In the first case, as, e.g. in dye or Ti:sapphire lasers, the large spectral width of the pumped laser allows the generation of very short pulses in the ps or fs range.

Pulsed and continuously (cw) operating systems have been built and thus the pump laser can be pulsed or cw, too. A typical scheme for pumping a pulsed dye laser is a transversal geometry as shown in Fig. 6.5.

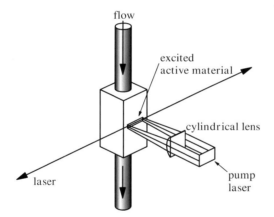

Fig. 6.5. Transversal pump laser geometry as applied in pulsed dye lasers

Because of the possible population of the triplet system in the dye, which would take these dye molecules out of the laser process, a slow flow of the dye solution is usually applied.

A much better excitation profile across the active material can be achieved using a Berthune cell for pumping as depicted in Fig. 6.6.

The totally reflecting 90° prism allows the laser material, e.g. the dye solution, to be excited from all sides in the same way. Thus power amplifier fs laser pulses pumped with ns pulses can be obtained.

Dyes in a polymer matrix are usually moved across the excitation spot to achieve average output powers in the range of a few 10 mW. This cools the active material, which was warmed up by the quantum defect energy via

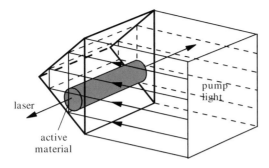

Fig. 6.6. Berthune cell for uniform transverse pumping of the active material

radiationless transitions. Typically excimer lasers (e.g. XeCl at 308 nm) or frequency doubled Nd:YAG lasers (532 nm) are used with pulse widths of 10–20 ns as pulsed pump lasers. Nitrogen lasers (337 nm) can be used both with pulse widths of a few ns (3–4 ns) or a few 100 ps (e.g. 500 ps).

If dye lasers run continuously the problems of triplet population and heating are increased and thus a strong flow is necessary. For this, dye jets are produced by injection nozzles having a very stable shape with good optical (interferometric) quality without windows as shown in Fig. 6.7.

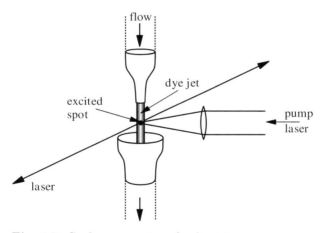

Fig. 6.7. Cw laser pumping of a dye jet

The jet has a typical thickness of 0.3 mm and a width of 5 mm. The flow speed is more than 10^2 m s^{-1}. The excitation spot has a diameter of, e.g. 50 μm. Argon (or krypton) ion lasers were first used as the pump, but diode pumped and frequency doubled solid-state lasers have been increasingly applied recently. In these cases the dye has to absorb in the green region, as e.g. rhodamine 6G does. The solvent has to be of suitable viscosity, as e.g. ethylene-glycol. A concentration of e.g. $1.4\,\mathrm{mmol\,l^{-1}}$ can be used. The excitation power is in the range of 5–10 W.

In a similar way the Titan sapphire laser can be pumped by e.g. the frequency doubled radiation of a Nd:YAG laser (see Fig. 6.8).

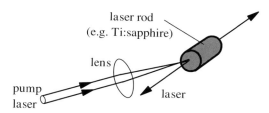

Fig. 6.8. Laser pumping of a solid-state laser (as, e.g. Ti:sapphire)

Solid-state lasers pumped by diode lasers are becoming more and more important, especially in industrial application such as welding, cutting, drilling and marking. In this case the good efficiency of the diode lasers and their high reliability is combined with the good coherence and beam quality of the solid-state lasers. Several schemes have been developed to meet the different needs in power and construction.

Solid-state rod lasers can be side-pumped and end-pumped [6.100–6.129]. A common scheme for side-pumping is shown in Fig. 6.9.

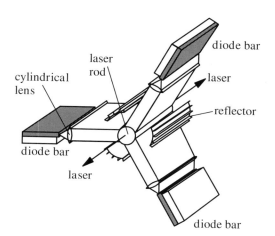

Fig. 6.9. Side-pumping of a solid-state laser rod with bars of diode lasers

The diode lasers are arranged in stripes of 10–20 diode lasers (bars) which emit about 50 W average power each at wavelengths typically above 800 nm (e.g. 808 nm for pumping of Nd:YAG and 940 nm for Yb:YAG). Because of the long rod length and the resulting high gain, as well as the large possible beam area, this type of laser is well suited for pulsed operation such as Q-switching.

Cylindrical aspherical lenses are often applied to collimate the highly divergent diode laser beam of about 90° in the axis vertical to the stripe (fast axis) before the solid-state laser rod is illuminated. Opposite to the bar on the other side of the rod a reflector usually collects the unabsorbed pump light. For uniform excitation usually three, five or seven bars are symmetrically mounted. The geometrical parameters are the important design criteria and determine to a large extent the possible quality of the laser beam.

For higher powers more than one star of diode bars can be used along the rod axis resulting in rod lengths of several cm. The efficiency of diode pumping the laser rods is higher than with lamp pumping, as a consequence of the better match of the pump laser spectrum to the absorption spectrum of the active material. Despite this reduced heat load, in high-power systems with average output powers of more than 10 W the laser rod is usually water cooled. The electro-optical efficiency can reach values of 20%.

The side-pumping of solid-state slabs [e.g. 6.123–6.127] has been applied as shown in Fig. 6.10, reaching very high average output powers of several kilowatts (see Sect. 6.13.2). In one example [6.123] the slab was e.g. 170 mm long, 36 mm high and 5 mm wide.

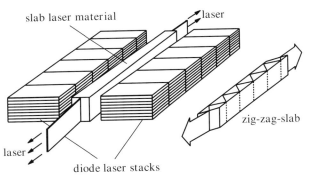

Fig. 6.10. Side-pumping of a solid-state laser slab with a stack of laser diode bars for reaching several kW average output power with good beam quality. The slab can be used as a zig-zag slab as shown on the right, decreasing thermal problems

For achieving high pump powers from diode lasers the bars were combined in arrays of 16 bars vertically resulting in a 1 cm wide and 2 cm high package. Fifteen of theses arrays were mounted in stacks along the laser axis at each side. Thus each stack contains 240 bars. Each bar consists of 20 diode lasers and has a nominal peak power of 50 W with a duty cycle of 20%. Thus the total pump average power of the 9600 laser diodes was 4.8 kW. The pulse duration could be varied from 100 μs to 1 ms. With one of the described laser heads an average output power of 1.1 kW could be obtained with a beam quality of 2.4 times the diffraction limit. Two modules allowed 2.6 kW with $M^2 = 3.2$ and three modules resulted in 3.6 kW with $M^2 = 3.5$. The maximum average output power of the three-module system in multimode operation was 5 kW.

The slab material can be used in a zig-zag geometry to overcome thermally-induced lensing and birefringence. The laser beam is totally reflected by the polished sides of the slab and in this way crosses the temperature profile in the rod which occurs between the exciting diode stacks if the slab is side-cooled. Nevertheless, carefully designed cooling has to be applied so as not to crack the strongly pumped laser material and the deformation of the end surfaces needs to be considered (see Sect. 6.4 and references there).

Solid-state lasers with output powers of less than about 100 W can be end-pumped [e.g. 6.102–6.111] as shown in Fig. 6.11.

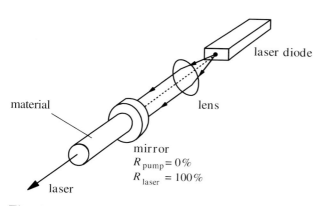

Fig. 6.11. Scheme of end-pumping a solid-state laser with laser diodes

The pump radiation excites the active material concentric to the laser beam and therefore radial symmetric inversion profiles can be achieved. If in addition the laser material is cooled at the end-faces the temperature profile occurs along the axis of the laser. In this case almost no thermally induced lensing or birefringence is obtained for the laser radiation. Thus with this simple scheme good beam quality can be achieved. The diode laser radiation can be fiber-coupled, resulting in easy maintenance (but higher prices). Because of the longitudinal temperature gradient and the cooling limitations of this pump scheme the maximum average output power is limited by the damage of the active material. For small pump spots at the active material using diode laser bars or stacks for pumping lens ducts were developed to change the beam shape of the pump light with its bad beam quality [for example, 6.109–6.112]. 40 W of output power from Nd:YVO$_4$ has been demonstrated (see Sect. 6.13.2).

In the power range below 5 W this scheme can be used in microchip lasers with a close arrangement of the diode, the solid-state laser material and if needed the interactivity SHG crystal. Output powers above 0.5 W green light have been observed from a 2 W diode (see Sect. 6.15.1).

The pumping scheme of Fig. 6.8 takes advantage of the good inversion profile available by end-pumping and is applied, e.g. in Ti:sapphire pumped by frequency doubled Nd:YAG laser radiation or from ion lasers. Fiber lasers are also mostly end-pumped (see Sect. 6.13.2.10).

Longitudinal pumping is also applied in disk lasers containing a thin slice of solid-state laser material with a thickness of, e.g. 0.2 mm and a diameter of a few mm as active matter, as shown in Fig. 6.12.

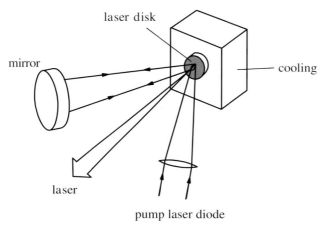

Fig. 6.12. Pumping scheme of a disk laser containing a thin slice of solid-state laser material with a thickness of, e.g. 0.2 mm and a diameter of a few mm longitudinally pumped by diode laser radiation. Because of the small absorption in the thin disk the pump beam has to be back-reflected by external mirrors

For the absorption of the diode laser pump light in the thin disk usually several passes are necessary and thus the pump beam has to be back-reflected. Four, eight or 16 passes are used in practical cases (see Sect. 6.13.2.4 and e.g. [6.128, 6.129]).

The thin disk is cooled longitudinally and thus almost no thermal lensing or birefringence occurs even at high powers. Difficulties may be caused by the mirror coatings at the disk back-side. Good reflectivity and optical quality for both the laser and the pump beams have to be combined with good thermal conductivity. Nevertheless, average output powers of several 100 W have been achieved with a Yb:YAG laser with very good beam quality and electro-optical efficiencies of more than 10% (see Sect. 6.13.2.4).

6.3.2 Electrical Pumping in Diode Lasers

Diode lasers are pumped directly by an electrical current of 10–20 mA with a voltage of about 2 V per single stripe of the laser diode as shown in Fig. 6.13. The active zone is built between a p-n junction and has a typical height of about 1 µm. In commercial diode lasers typically 10–20 stripes are arranged closely spaced resulting in a driving current of about 2 A for the whole structure (see Sect. 6.13.1).

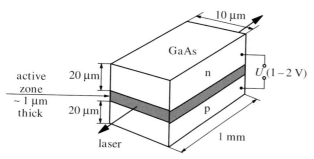

Fig. 6.13. Schematic structure of a diode laser consisting of a p-n junction of, e.g. GaAs with an active laser zone in between. Commercial diode lasers have a more complicated structure including cladding layers and waveguide channels for improved laser parameters. One diode laser consists typically of 10–20 stripes (see Sect. 6.13.1)

In the p-doped material (with less electrons than positively charged holes) the high lying conduction band is empty and the valence band is only partly occupied with electrons. In the n-doped material the valence band is complete and the conduction band is partly filled with electrons. Under the influence of the electric field across the p-n transition (produced by the external voltage U) some electrons from the upper (conduction) band of the n-doped material will move to the p-doped side. There they can recombine with the positive holes under the emission of a photon. Radiationless processes depopulate the upper laser band within about 1 ns. Nevertheless, the electro-optical efficiency of diode lasers is up to 50%.

Commercial diode lasers have a much more complicated structure. This includes, e.g. cladding layers resulting in double heterostructure lasers for improved efficiencies and waveguide channels for better laser light parameters.

The beam quality is diffraction limited in the vertical axis of Fig. 6.13 which is called the fast axis and shows a full divergence angle of about 60° (FWHM: 37°). In the horizontal axis (slow axis) the beam quality depends on the size of the gain guided structure of the electrodes. The slow axis full divergence angle is typically close to 10°. It can be almost diffraction limited for a single emitter producing an average output power of some mW. In diode lasers, as described in Sect. 6.13.1, several of these single channels are combined and the emitted radiation of these single emitters is not coherent. Thus, the beam quality in the slow axis is usually very poor for these power lasers. New concepts may increase the lateral coherence of these lasers and thus improve their beam quality.

The wavelength of the laser results from the size of the quantum confinement which is about 20 nm wide, and the doping, which is about 10^{-4}. The emission wavelength is temperature dependent. A temperature change of $+20\,\mathrm{K}$ shifts the emission wavelength by about $+6\,\mathrm{nm}$.

The voltage at the single diode $U = U_0 + IR_S$ is a function of the applied current, with $R_S \approx 200\text{–}400\,\mathrm{m\Omega}$ and $U_0 \approx 1\,\mathrm{V}$. The threshold current is in the range of a few 100 mA and the slope efficiency is about $1\,\mathrm{W/A}$.

The lifetimes of the diode lasers specified for more than 80% of the maximum output power and reach values of many tens of thousands of hours. The decrease in the output power can partially be compensated for by an adequate increase in the current.

Another concept for diode lasers is the vertical cavity surface emitting lasers or VCSELs (see Sect. 6.13.1.3). Their structure is shown schematically in Fig. 6.14.

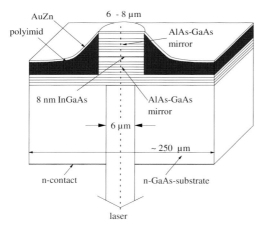

Fig. 6.14. Schematic structure of a vertical cavity surface emitting laser

In this case the laser radiation is built up in a resonator perpendicular to the p-n layer of the semiconductor structure. Because of the high gain in the active material this short amplification length is sufficient for the laser. As a consequence of the small diameter of the active zone the light is almost diffraction limited. The etalon effect of the short resonator can be used for narrow bandwidth generation.

6.3.3 Electrical Discharge Pumping

In electrically excited gas lasers, such as e.g. argon or krypton lasers, the wall-plug efficiency can be as low as 0.1% but, e.g. a copper vapor laser or excimer laser shows values as high as 1% and 2%. The argon ion laser is excited with up to 35 eV and the laser emits at 488 nm. The resulting quantum defect is 94%.

In the helium-neon (He-Ne) laser the excitation of He takes place with an electron energy of about 20 eV and the laser transition in Ne has a wavelength of 632.8 nm resulting in a quantum defect of about 92%. The helium-cadmium (He-Cd) laser is excited in the same way but the emission appears at 441.8 nm resulting in a smaller quantum defect of about 89%.

In CO_2 lasers the molecules are excited with 0.28 eV. The wavelength of the CO_2 laser is 10.6 μm which results in a quantum defect of 66%.

The three-level copper and lead lasers are pumped with about 4 eV and emit at 510.5 nm/578.2 nm and 722.9 nm, respectively. The resulting quantum defects are 50%/57% and 70%.

A further reason for the limited efficiencies of these electrically pumped gas lasers is the imperfect adaptation of the speed of the accelerated electrons in the discharge with the collision cross-section of the active particles [e.g. 6.130–6.132]. The energy distribution $F_{electron}$ of the electrons as a function of their temperature T can be given as:

$$F_{electron}(E_{electron}) = 2\sqrt{\frac{E_{electron}}{\pi(k_{Boltz}T)^3}}\ \exp\left(-\frac{E_{electron}}{k_{Boltz}T}\right).$$ (6.3)

As an example the electron speed distribution and the absorption cross-section are shown for the discharge in a nitrogen laser in Fig. 6.15. As can be seen from this figure the distribution of slow electrons with kinetic energies below 10 eV and with fast electrons with more than 35 eV is not optimally adapted to the excitation cross-section of the nitrogen molecules.

Although the velocity distributions of the electrons can be modified with buffer gases of certain pressures and the density of the active matter is chosen for optimal absorption the final excitation efficiency is sometimes smaller than 1%. In addition radiationless deactivation takes place in the gas by collisions between the particles. Better efficiencies are reached with copper (Cu) or gold (Au) vapor lasers. Values of 1% for Cu and 0.2% for Au lasers reported. This is based on a quantum defect of e.g. 40% for the copper vapor laser.

The electrical discharge can be arranged transversally or longitudinally to the laser beam (see Fig. 6.16). The transversal pump geometry is more suitable for pulsed electrical excitation and longitudinal for cw operation.

In pulsed excimer or nitrogen (N_2) lasers the discharge has to take place within about 10 ns. Therefore the discharge in the gas chamber with pressures of 0.1 bar in the nitrogen laser and a few bars in the excimer laser is spread over the whole length of the electrodes. Electric circuits with very low inductivities have to be applied. Capacities of several nF charged to voltages

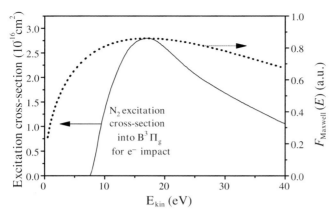

Fig. 6.15. Absorption cross-section and electron speed distribution as a function of the electron speed for the discharge in a nitrogen laser

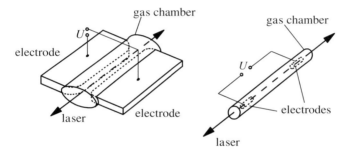

Fig. 6.16. Electrical discharge pumping of gas lasers with transversal (left side) or longitudinal (right side) geometry

of 10–30 kV are used as electrical power source. As electrical switches thyratrons are used and in the best cases trigger jitters of a few ns and delays of several 10 ns between the electric trigger and the laser pulse are obtained. Thus these lasers with ns output pulses can be synchronized electrically. For simple arrangements spark gaps can be used as high-voltage switches. Thus it is possible to built a nitrogen laser (using air as the active material) with a 500 ps pulse width at 337 nm based on a very simple construction (see Sect. 6.13.3.2).

The longitudinal discharge is typically used in cw operating He-Ne lasers or Ar and Kr ion lasers. In ion lasers with output powers of several watts the discharge tube is the most expensive part, and costs about 1/3 of the laser and losts typically for only 2 years. If the output power is reduced to less than half of the maximum the lifetime increases drastically.

6.3.4 Lamp Pumping

Flash or arc lamps are very common *for the pumping* of solid-state lasers. Typical arrangements are shown in Fig. 6.17. Also the solid-state slab arrangement of Fig. 6.10 can be pumped using lamps from both sides, instead of the diode laser stacks. Using lamps for pumping, the laser material acts as a light converter producing monochromatic and coherent light with good beam quality and polarization in possibly short pulses.

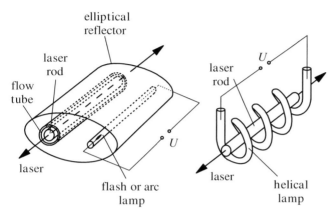

Fig. 6.17. Lamp pumping of solid-state laser rods with linear lamps (left) and helical lamps (right)

Helical lamps were used in pioneer times but may find new applications, and linear lamps are used today. For saving the flash light the laser rod and the lamp(s) are mounted inside a pump chamber which can scatter, diffuse or reflect the light. In the latter case the rod is often mounted in one focal line of an elliptically shaped reflector. If more than one lamp is applied each has its own elliptical reflector combined into a flower-like cross-section. Diffuse pump chambers will show a more equal inversion distribution and reflecting ones show a maximum in the rod center which is more useful for light extraction with Gaussian beams.

Flash lamps are typically filled with Xe or Kr and emit their light for about 100 μs to several ms. The pulse length depends on the construction of the lamp and the design of the electrical circuit. Flash lamp pumping of dye lasers is also sometimes applied to reach high output powers. Because of the possible triplet population the flash pulse has to be as short as a few μs which demands special lamps and drivers. The duration of the laser pulse is about the same as the flash.

Lamp pumping results in an electro-optical efficiency of up to 5%. Thus 95% of the flash lamp energy is converted to heat. This often limits the laser output parameters such as power and brightness. Therefore cooling of the laser material and lamp(s) is essential. Flow tubes around the rod and lamps increase the cooling efficiency. Nevertheless the laser rod will show a temperature profile with highest values in the rod center and the temperature of the cooling liquid at the surface. The refractive index of the active laser rod then shows a quadratic profile and sometimes even higher orders as a function of the rod radius. Thermally induced lensing, birefringence and depolarization occur as a consequence of the refractive index modulation (see Sect. 6.4). The resulting phase shifts cause amplitude distortions after the propagation of the light and thus the beam quality of such laser is decreased in the high-power regime. In the worst case thermally induced tension can cause damage to the laser rod.

Flash or arc lamps emit light in a wide spectral range which is absorbed in the active material via several different transitions or bands. Thus different excited states are populated and the quantum defect will be different for them. To increase the efficiency of the pumping process and to avoid the distraction of the laser material by short-wavelength radiation sometimes quantum converters, such as e.g. Ce atoms, are used to transform UV light into the visible and red spectral region which the laser rod can absorb. These materials can be introduced into the flow tubes and the pump efficiency can be increased by 20–50%. Further details are described in [6.133–6.168].

6.3.5 Chemical Pumping

Chemical lasers are pumped by the excess energy of a chemical reaction. Typical lasers are based on the reaction of fluorine and hydrogen to HF* in specially designed flow chambers (see Fig. 6.18).

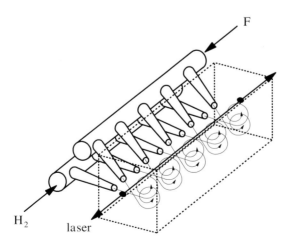

Fig. 6.18. Scheme of pumping an active material by a chemical reaction in a flow chamber

The reaction takes place as:

$$F + H_2 \rightarrow HF^* + H \tag{6.4}$$

or

$$F_2 + H \rightarrow HF^* + F \tag{6.5}$$

with a reaction heat of about $32\,\text{kcal mol}^{-1}$. This allows the excitation of the third vibrational level ($v = 3$) of HF which rapidly decays to the $v = 2$ level. The laser works between the levels with $v = 3$ and $v = 2$ with an efficiency of inversion relative to chemical energy of about 50%. The emission wavelengths of the transitions between the vibrational levels are in the range of 2.7–3.3 µm. Other possible laser molecules are HBr, HCl and DF with emissions in the range 3.5–5 µm (see [6.169–6.172] and Sect. 6.13.5).

Large volumes of the active material can be made in this way and thus large output powers are possible. Pulsed and cw operation is possible. A maximum of more than 2 MW average output power has been obtained from such a chemical laser. Because of the technological and environmental problems of these lasers as a result of the toxic halogens, military applications are mainly intended.

6.3.6 Efficiencies

Several efficiency values are used to characterize the pump process:

- quantum defect efficiency
- quantum efficiency
- opto-optical efficiency
- electro-optical efficiency
- slope efficiency
- total efficiency.

The quantum defect energy is, as described in the previous section, the difference between the photon energies of the pump E_{pump} and the laser E_{laser}. Small values result in high quantum defect efficiencies:

$$\textbf{quantum defect efficiency} \quad \eta_{\text{QD}} = \frac{E_{\text{laser}}}{E_{\text{pump}}}. \tag{6.6}$$

The quantum efficiency gives the share of emitted laser photons $N_{\text{photons,laser}}$ relative to the number of absorbed pump photons $N_{\text{photons,pump}}$ independent of the energy of both:

$$\textbf{quantum efficiency} \quad \eta_{\text{Q}} = \frac{N_{\text{photons,laser}}}{N_{\text{photons,pump}}}. \tag{6.7}$$

It accounts for, e.g. the radiationless transitions parallel to the laser transition in the active material. Thus the efficiency of the absorbed photons is the product of both η_{QD} and η_{Q}.

The opto-optical efficiency is the quotient of the laser light power $P_{\text{laserlight}}$ relative to the light power $P_{\text{pumplight}}$ of the pump source:

opto-optical efficiency $\eta_{\text{o-o}} = \dfrac{P_{\text{laserlight}}}{P_{\text{pumplight}}}.$ (6.8)

It is especially used for characterizing laser pumping, e.g. with diode lasers. If the diode laser is fiber-coupled it should be stated whether the pump light is measured at the output of the diode laser or at the output of the fiber. This difference can be as high as 50%.

The electro-optical efficiency relates the laser light output power $P_{\text{laserlight}}$ to the electric input power to the pump source $P_{\text{pump,electric}}$:

electro-optical efficiency $\eta_{\text{e-o}} = \dfrac{P_{\text{laserlight}}}{P_{\text{pump,electric}}}.$ (6.9)

In this efficiency the losses in the pump source and in the active material are considered. It is relevant to all lasers with all kinds of pump mechanism. In particular, solid-state lasers with flash lamp or diode laser pumping are characterized with this value and compared to other high-power lasers.

The slope efficiency characterizing the slope of the laser output power curve as a function of the electrical input power $P_{\text{laserlight}} = f(P_{\text{pump,electric}})$ is used especially for solid-state lasers:

slope efficiency $\eta_{\text{slope}} = \dfrac{\Delta P_{\text{laserlight}}}{\Delta P_{\text{pump,electric}}}.$ (6.10)

This curve shows, after a threshold of a certain electrical input power (e.g. of the flash lamps), an almost linear increase of the output power. The differences ΔP_i are used from this linear part of the curve.

Finally the *total efficiency* or wallplug efficiency of the pump process and the laser operation has to be calculated for evaluating different laser types for certain applications. The total power of the laser light $P_{\text{laserlight}}$ has to be compared to the sum of the total electric plug-in powers of the power supply and cooling system $P_{\text{wallplug,total}}$:

total efficiency $\eta_{\text{e-o,total}} = \dfrac{P_{\text{laserlight}}}{P_{\text{wallplug,total}}}.$ (6.11)

In addition to the mechanismus already discussed the efficiency of the pump source and the active material is decreased by radiationless transitions, long lifetimes of the lower laser level, population of long-living energy states such as e.g. triplet levels of laser dyes, unwanted chemical reactions, and phase and amplitude distortions. The principle limit of the laser efficiency was discussed in [6.174]. Further special demands such as complicated cooling or control systems may decrease the efficiency of the laser system. A very rough overview of these costs for different laser photons was given in Table 1.1. Finally the total efficiency has to be used to compare lasers with comparable brightness, wavelength and pulse duration with respect to their cost, in addition to purchase, installation and maintenance conditions.

6.4 Side-Effects from the Pumped Active Material

The active material changes the properties of the laser resonator by its refractive index and thus the optical length of the resonator which has to be taken into account while calculating the transversal and longitudinal mode structure of the laser. In the worst case its refractive index is a complicated function of space and time. Additional amplitude distortions may occur from the inhomogeneous gain in the active material. Thus the resonator properties may change during the laser process.

Especially in solid-state lasers, heating of the active material, which results from the thermal load from the pumping process, can cause serious problems [6.174–6.260]. The laser material can show thermal lensing and induced birefringence, as will be described below. Several concepts have been developed to avoid or minimize these thermally induced problems. Thus, slab, zig-zag-slab or thin disc geometry can be applied to the laser material. Resonator concepts with adaptive mirrors such as, for example, via optical phase conjugation or special polarization treatment, have been developed to compensate for these distortions. Therefore, detailed experimental investigations of the different parameters involved have been done [e.g. 6.174–6.191]. Modeling of the thermal effects in general can be found in [6.192–6.202].

6.4.1 Thermal Lensing

A typical distortion is the thermal lensing of solid-state laser rods [6.203–6.243]. The rod can be cooled at the rod surface, only. Thus a quadratically refractive index profile across the cross-section of the rod can be observed (see Table 2.6 for the ray matrix). The focal length f_{therm} of the resulting lens can be as short as a few 0.1 m while the rod is pumped with a few kW. The refractive power D_{therm} per pump power P_{pump} is often used for characterization of the active material:

$$\textbf{refractive power}\quad D_{\text{therm}} = \frac{1}{f_{\text{therm}}P_{\text{pump}}}\quad [D_{\text{therm}}] = \frac{\text{dpt}}{\text{kW}} \qquad (6.12)$$

with typical values of 1–$4\,\text{dpt}\,\text{kW}^{-1}$.

A typical example is shown in Fig. 6.19 for a flash lamp pumped Nd:YALO laser rod ($1.1\,\text{at}\%$) with a diameter of $4\,\text{mm}$ and a length of $79\,\text{mm}$ measured with a He-Ne-laser probe beam.

Unfortunately, the laser material is also cooled by the laser radiation, and thus the thermal lens can change more than 10% with and without laser operation. Further the refractive power can be different for the different polarization if the material was birefringent or birefringence was thermally induced. Thus for more precise design of the laser resonators the refractive power should be measured under laser conditions, e.g. using the stability limits of the resonator (see Sect. 6.6.15).

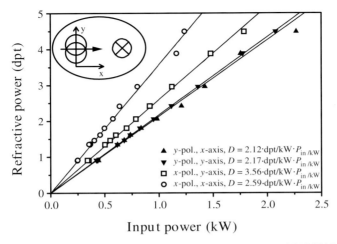

Fig. 6.19. Refractive power of a flash lamp pumped Nd:YALO laser rod (1.1 at%) with a diameter of 4 mm and a length of 79 mm measured with a He-Ne-laser probe beam. The c axis of the crystal was horizontal in the x direction in the elliptical pump chamber. The lensing was measured in the x and y direction for the two polarization directions

Thermally steady-state conditions are fulfilled if the laser is continuously operating (cw) or the repetition rate of the laser is larger than the inverse of the thermal relaxation time τ_{therm}:

thermal relaxation time $\quad \tau_{\text{term}} = \dfrac{c_{\text{heat}} \gamma_{\text{D}}}{4 K_{\text{cond}}} r_0^2$ $\qquad$ (6.13)

with the specific heat of the laser material c_{heat}, the mass density γ_{D}, the thermal conductivity K_{cond} and the radius of the rod r_0. As an example for Nd:YAG, $c_{\text{heat}} = 0.59\,\text{W s g}^{-1}\,\text{K}^{-1}$, $\gamma_{\text{D}} = 4.56\,\text{g cm}^{-3}$, $K_{\text{cond}} = 0.11$–$0.13\,\text{W cm}^{-1}\,\text{K}^{-1}$ and with a rod radius of 4 mm a thermal relaxation time of 0.83–0.98 s is obtained [6.214].

The calculation of the thermally induced lensing of laser rods can be based on the one-dimensional heat conduction differential equation for the temperature T if these steady state conditions are fulfilled:

$$\frac{\text{d}^2 T}{\text{d} r^2} + \frac{1}{r}\frac{\text{d} T}{\text{d} r} + \frac{\eta_{\text{therm}} P_{\text{pump}}}{\pi r_0^2 L_{\text{rod}} K_{\text{cond}}} = 0 \qquad (6.14)$$

with the pump power P_{pump}, the rod length L_{rod} and η_{therm} as the fraction of the pump power dissipated as heat in the rod. For this model the rod length should be more than 10 times larger than the rod radius. At the surface of the rod with radius $r = r_0$ the temperature is given by $T(r_0)$. The typical thermal time constants are of the order of magnitude of 1 s and thus this equation will hold for lasers with repetition rates larger than a few Hz or in cw operation. Other systems need more complicated analysis as, e.g. given in [6.200–6.202, 6.208]. The solution of this equation is given by:

$$T(r) = T(r_0) + \Delta T \left[1 - \left(\frac{r}{r_0} \right)^2 \right] \quad \text{with} \quad r \leq r_0 \qquad (6.15)$$

This temperature profile is quadratic, as mentioned above, with the highest value $T_{\max} = T(r_0) + \Delta T$ in the rod center using:

$$\Delta T = \frac{\eta_{\text{therm}} P_{\text{pump}}}{4\pi L_{\text{rod}} K_{\text{cond}}} \qquad (6.16)$$

From this thermal distribution across the rod a distribution of the refractive index $\Delta n(r)$ follows which is given by:

$$\Delta n(r) = [T(r) - T(r_0)] \left(\frac{dn}{dT} \right)$$

$$= \frac{\eta_{\text{therm}} P_{\text{pump}}}{4\pi L_{\text{rod}} K_{\text{cond}}} \left(\frac{dn}{dT} \right) \left[1 - \left(\frac{r}{r_0} \right)^2 \right] \qquad (6.17)$$

with the temperature-dependent refractive index change (dn/dT). This formula can be written as:

$$\Delta n(r) = n_0 - \frac{1}{2} n_{\text{therm}} r^2 \qquad (6.18)$$

with the newly defined n_{therm}:

$$n_{\text{therm}} = \frac{\eta_{\text{therm}} P_{\text{pump}}}{2\pi r_0^2 L_{\text{rod}} K_{\text{cond}}} \left(\frac{dn}{dT} \right) \qquad (6.19)$$

and can be used in the matrix ray propagation formalism with a quadratic index profile (n_2 in Tab. 2.6).

The focal length f_{rod} of thermally induced focusing in the laser rod can be given as an approximation, especially for cases with long focal lengths compared to the rod length, by:

thermal induced focal length $\quad f_{\text{rod}} = \dfrac{2\pi r_0^2 K_{\text{cond}}}{\eta_{\text{therm}} P_{\text{pump}}} \left(\dfrac{dn}{dT} \right)^{-1}$ (6.20)

which shows the influence of the parameters. The larger the radius of the laser rod the longer the focal length of the rod. The rest are material parameters.

In addition a stress-dependent component will increase the thermal effect by about 20% and the end-face curvature will add a further 5%. For more detailed analysis see [6.192–6.202, 6.236, 6.237].

The temperature-dependent change of the refractive index (dn/dT) for some solid-state laser materials is given in Table 6.3.

Table 6.3. Temperature-dependent change of the refractive index $(\mathrm{d}n/\mathrm{d}T)$ for some solid-state laser materials, their expansion coefficient α_{expan} and their thermal conductivity K_{cond}. The double numbers for Nd:YALO and Nd:YVO are for the a and the c axes of the crystal

Material	λ_{laser} (nm)	n_0	α_{expan} (K^{-1})	$\dfrac{\mathrm{d}n/\mathrm{d}T}{(\mathrm{K})}$	K_{cond} $(\mathrm{W\,cm}^{-1}\,\mathrm{K}^{-1})$
Ruby	694	1.76	$5.8 \cdot 10^{-6}$	$13 \cdot 10^{-6}$	0.42
Nd:glass	1054	1.54	$8.6 \cdot 10^{-6}$	$8.6 \cdot 10^{-6}$	0.012
Nd:YAG	1064	1.82	$7.5 \cdot 10^{-6}$	$7.3 \cdot 10^{-6}$	0.13
Nd:YALO	1080	1.90	$9\text{--}11 \cdot 10^{-6}$	$10\text{--}14 \cdot 10^{-6}$	0.11
Nd:YVO$_4$	1064	1.45	$4\text{--}11 \cdot 10^{-6}$	$8.5\text{--}3 \cdot 10^{-6}$	0.5–0.05
Nd:YLF	1047	1.45		$-3 \cdot 10^{-6}$	0.06
Nd:KGW	1067	1.94		$0.4 \cdot 10^{-6}$	0.03
Nd:Cr:GSGG	1060	1.94	$7.4 \cdot 10^{-6}$	$11 \cdot 10^{-6}$	0.06
Er:glass	1540	1.53	$12 \cdot 10^{-6}$	$6.3 \cdot 10^{-6}$	≈ 0.01
Yb:YAG	1030	1.82	$7.5 \cdot 10^{-6}$	$7.3 \cdot 10^{-6}$	0.13

In addition higher-order aberrations can occur which cannot be compensated by simple lenses or mirrors. Phase conjugating mirrors can help to solve this problem (see Sect. 6.6.12).

6.4.2 Thermally Induced Birefringence

The pump process can also cause thermally induced birefringence [6.244–6.258] in the active material and thus depolarization of the laser radiation can occur. For example, in solid-state laser rods as in Nd:YAG material thermally induced depolarization can be a serious problem for designing high-power lasers with good beam quality. In Fig. 6.20 the measuring method for determining the depolarization is shown.

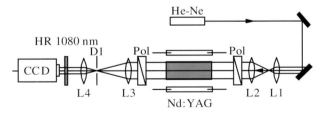

Fig. 6.20. Setup for measuring the depolarization of a flash lamp pumped Nd laser rod using a He-Ne laser probe beam and two crossed polarizer (Pol)

As an example the results for the measurement of a Nd:YAG laser rod with a diameter of 7 mm and a length of 114 mm as a function of the electrical flash lamp pump power measured between crossed polarizers, as in the scheme of Fig. 6.20, are given in Fig. 6.21.

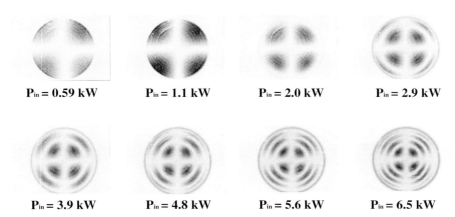

$P_{in} = 0.59 \text{ kW}$ $P_{in} = 1.1 \text{ kW}$ $P_{in} = 2.0 \text{ kW}$ $P_{in} = 2.9 \text{ kW}$

$P_{in} = 3.9 \text{ kW}$ $P_{in} = 4.8 \text{ kW}$ $P_{in} = 5.6 \text{ kW}$ $P_{in} = 6.5 \text{ kW}$

Fig. 6.21. Depolarization in a Nd:YAG laser rod with a diameter of 7 mm and a length of 114 mm as a function of the electrical flash lamp pump power measured between crossed polarizers as in the scheme of Fig. 6.20

Based on the previous Section giving the description of the thermal heating of the laser rods the depolarization can also be calculated as a function of the pump power.

The temperature difference causes mechanical strain in the laser crystal which deforms the lattice. Because of the radial symmetry of the strain the resulting refractive index change Δn will be different for the radial and the tangential components of the light polarization. The difference of the changes of the refractive indices of the radial Δn_r and the tangential Δn_ϕ polarization is given by:

$$\Delta n_\phi(r) - \Delta n_r(r) = -\frac{n_0^3 \alpha_{\text{expan}} C_{\text{bire}}}{\pi K_{\text{cond}} L_{\text{rod}}} \left(\frac{r}{r_0}\right)^2 \eta_{\text{therm}} P_{\text{pump}}$$
$$= \frac{\lambda_{\text{laser}} C_T}{2\pi L_{\text{rod}}} \left(\frac{r}{r_0}\right)^2 \eta_{\text{therm}} P_{\text{pump}} \qquad (6.21)$$

with the additional constants α_{expan} as the expansion coefficient and the dimensionless factor C_{bire} accounting for the birefringence. C_{bire} can be calculated from the photoelastic coefficients of the material but it is easier to determine it experimentally from the birefringence measurement and the determination of C_T as will be shown below. For Nd:YAG, C_{bire} is about $C_{\text{bire,Nd:YAG}} = -0.0097$ and thus $C_T = 67 \text{ kW}^{-1}$.

The difference of the refractive indices for the two polarizations of the laser light leads to a well-defined alteration of the polarization state of the light, often called depolarization. This can easily be measured, as shown in Fig. 6.21. The intensity of the light $I_{\text{out},\parallel}$ polarized parallel to the incident light behind the laser rod at position r, ϕ in polar coordinates is given by:

$$\frac{I_{\text{out},\parallel}(r,\phi)}{I_{\text{out,total}}} = 1 - \sin^2(2\phi)\sin^2\left(\frac{\delta_{\text{bire}}}{2}\right) \tag{6.22}$$

with the phase difference δ_{bire}:

$$\delta_{\text{bire}} = \frac{2\pi L_{\text{rod}}}{\lambda_{\text{laser}}}(\Delta n_\phi - \Delta n_r)$$

$$= C_{\text{T}}\eta_{\text{therm}}P_{\text{pump}}\left(\frac{r}{r_0}\right)^2. \tag{6.23}$$

This equation can be integrated over the cross-section of the rod to get the degree of polarization of the transmitted light p_{pol}:

$$p_{\text{pol}} = \frac{I_{\text{out},\parallel}^{\text{all}} - I_{\text{out},\perp}^{\text{all}}}{I_{\text{out},\parallel}^{\text{all}} + I_{\text{out},\perp}^{\text{all}}}$$

$$= \frac{1}{2} + \frac{\sin(C_{\text{T}}\eta_{\text{therm}}P_{\text{pump}})}{2C_{\text{T}}\eta_{\text{therm}}P_{\text{pump}}} \tag{6.24}$$

which can be measured as a function of the pump power.

As shown in [6.214] the agreement between these experimental and calculated results is quite satisfying.

Anisotropic materials with high natural birefringence such as Nd:YALO or Nd:YLF show only a negligible thermally induced birefringence. Therefore they emit linearly polarized light even at the highest average output powers.

The possibility of compensating can be achieved using thermally induced birefringence in highly pumped isotropic laser crystals the arrangement of two identical active materials in series with a 90° polarization rotator in between [6.244, 6.257, 6.258] as shown in Fig. 6.22.

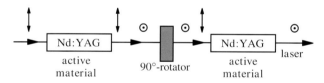

Fig. 6.22. Arrangement of two active isotropic materials, e.g. two Nd:YAG laser rods, with 90° polarization rotation in between for compensation of depolarization from birefringence

The birefringence in the first active material, which can be, e.g. a Nd:YAG rod, causes the generation in general of different elliptically polarized light across the beam. The x and y components of the polarization one interchanged by the 90° rotator (quartz plate). The depolarization is compensated during the pass through the second identical active material. Thus the depolarization loss, e.g. in Nd:YAG lasers can be reduced from the 25% level to theoretically zero and experimentally to less than 5% for pump powers of up to 16 kW [6.257].

For improved compensation a relay imaging telescope can be applied between the two laser rods. Therefore, two lenses with focal length f_{relim} are positioned in front of the rods at the distances z_{L1} and z_{L2} from the end faces. The distance between the two lenses should be $2f_{\text{relim}}$ and the condition $z_{\text{L1}} + z_{\text{L2}} = 2f_{\text{relim}} - L_{\text{rod}}/n_{\text{rod}}$ should be fulfilled [6.257]. A similar scheme can be realized with one laser rod in front of a 100% mirror and a Faraday rotator.

6.4.3 Thermal Stress Fracture Limit

The maximum output power of solid-state lasers is given by the maximum pump power and related efficiencies and thus the maximum thermally induced stress the active material can bear [e.g. 6.259, 6.260]. From the quadratic temperature profile across the diameter of the laser rods, the maximum surface stress σ_{max} as given in [6.260]:

$$\sigma_{\text{max}} = \frac{\alpha_{\text{expan}} E_{\text{young}}}{8\pi K_{\text{cond}}(1 - \nu_{\text{poisson}})} \frac{\eta_{\text{therm}} P_{\text{pump}}}{L_{\text{rod}}} \tag{6.25}$$

with Young's modulus E_{young}, Poisson's ratio ν_{poisson} and all other parameters as given above. The damage stress is typically in the range 100–200 MPa for common solid-state laser materials. The maximum power per laser rod length follows from:

$$\frac{P_{\text{pump}}}{L_{\text{rod}}} = 8\pi R_{\text{shock}} \tag{6.26}$$

with the thermal shock parameter R_{shock}:

$$R_{\text{shock}} = \frac{K_{\text{cond}}(1 - \nu_{\text{poisson}})\sigma_{\text{max}}}{\alpha_{\text{exp and}} E_{\text{yound}}} \tag{6.27}$$

for which the values in Table 6.4 were given in [M25]

Table 6.4. Shock parameter for different host materials of solid-state lasers

Host material	Glass	GSGG	YAG	Al$_2$O$_3$
R (W cm^{-1})	1	6.5	7.9	100

It should be noticed that following this consideration the rod diameter does not influence the maximum output power per rod length. An example with YAG as the host laser material involves a maximum optical pump power of about 200 W cm^{-1}.

To increase the output power of solid-state materials the extreme geometry of thin disks or fibers is applied for the active material to reach very good cooling efficiencies.

6.5 Laser Resonators

The laser resonator determines the laser light characteristics within the frames given by the active material. It consists of at least two mirrors (see Fig. 6.2) but it can contain many additional elements:

- apertures, lenses, additional mirrors and diffractive optics may be used for forming special transversal modes;
- gratings, prisms and etalons are applied for frequency selection;
- shutters, modulators, deflectors and nonlinear absorbers are used for generating short pulses;
- polarizing elements are applied for selecting certain polarizations.

Other linear and nonlinear elements are applied as well. For example, phase plates, adaptive mirrors or phase conjugating mirrors can be used for compensating phase distortions.

Because of all these options the design of a resonator with respect to a certain application producing laser light of the required properties in combination with high brightness and efficiency is a difficult task. Detailed descriptions can be found in, for example, [M16, M25, M38, M39, 6.261–6.265].

6.5.1 Stable Resonators: Resonator Modes

The laser resonator (or cavity) can be designed as a *stable resonator* producing a standing light wave from the interference of the two counterpropagating light beams with a certain transversal and longitudinal distribution of the electric field inside. These distributions are eigensolutions of Maxwell's equations for the standing light wave with the boundary conditions of the curved resonator mirror surfaces and including all further optical elements in the resonator.

The transversal structures of these eigensolutions are called *transversal resonator modes*. The transversal structure can change along the axis of the laser and a certain transversal light pattern will be observed behind the partially transparent resonator mirror, the output coupler (see Fig. 6.23). For many applications a Gaussian beam is required as the transversal mode of the laser.

The curvatures of the wave fronts of the resonator modes of the light beam at the position of the mirror surfaces is the same as the curvature of the mirrors. This condition defines the possible transversal modes of a stable resonator.

The axial structures of these eigensolutions are the *longitudinal resonator (or axial) modes* (see Fig. 6.24). The standing light wave is built by the interference of the back and forth moving light waves reflected at the mirrors. The electric field has a knot at the mirror surface and thus the longitudinal modes are selected.

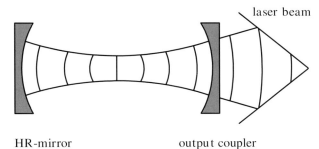

HR-mirror   output coupler

Fig. 6.23. Transversal eigenmode of a stable empty resonator consisting of the high-reflecting HR mirror ($R \simeq 100\%$) and the partially reflecting mirror, the output coupler (with, e.g. $R = 50\%$)

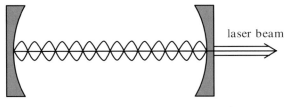

HR-mirror   output coupler

Fig. 6.24. Longitudinal eigenmode of a stable empty resonator as in Fig. 6.23

6.5.2 Unstable Resonators

In an unstable resonator the light beam diameter grows while the light is reflected back and forth in the resonator as depicted in Fig. 6.25.

In this case the near field of the out-coupled beam has a hole in the middle which has the size of the smaller mirror at its place but will be filled in the far field. With this resonator type large mode diameters in the active

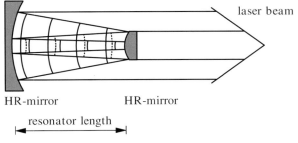

HR-mirror   HR-mirror

resonator length

Fig. 6.25. Transversal structure of a light beam in an a unstable empty resonator consisting of two high-reflecting HR mirrors ($R = 99.9\%$) of different size

material can be achieved but the beam quality is not diffraction limited. Using unstable resonators with a mirror with radially varying reflectivity, near-diffraction-limited beam quality can be achieved with large mode diameters [6.283, 6.289, 6.295–6.297, 6.301]. Diffractive optics used as structured phase plates can be placed inside the unstable resonator for improving the beam quality [6.307]. Some examples of unstable resonators are given in [6.266–6.315]. The theoretical treatment of unstable resonators is discussed in, for example, [6.266–6.281].

6.6 Transversal Modes of Laser Resonators

In general the transversal modes of a given resonator design have to be calculated as a solution of Maxwell's equations for the electric field between the two resonator mirrors including all optical elements inside this space. This problem can be solved by calculating the Kirchhoff integral including the dimensions of the resonator mirrors.

The still complicated mathematical problem is often reduced to three types of special cases:

- a fundamental mode describing a Gaussian beam as an eigensolution of a stable resonator including the optical elements in the resonator;
- higher transversal modes for an empty optical resonator with high transversal symmetry;
- simple solutions for unstable resonators.

Some resonators with an active material of very high gain and thus a very small number of round trips such as, e.g. in the excimer or nitrogen lasers, show a mixture of so many modes that the description based on geometrical optics using the geometrical dimensions of the resonator elements with their apertures can be the most efficient.

6.6.1 Fundamental Mode

The fundamental transversal mode, the TEM_{00} mode, has a Gaussian transversal profile and represents a Gaussian beam. Thus it can be derived from Gaussian beam propagation through the resonator under the condition of self-reproduction after one complete round trip. Thus the complex beam parameter of the beam $q(z)$ (see Sect. 2.4) has to be reproduced:

$$\text{eigensolution} \quad q(z_{\mathrm{oc}}) = \frac{a_{\mathrm{roundtrip}} q(z_{\mathrm{oc}}) + b_{\mathrm{roundtrip}}}{c_{\mathrm{roundtrip}} q(z_{\mathrm{oc}}) + d_{\mathrm{roundtrip}}} \tag{6.28}$$

with z_{oc} as a fixed position of observation, e.g. at the output coupler. The elements of the roundtrip matrix $M_{\mathrm{roundtrip}}$:

$$\text{roundtrip matrix} \quad M_{\mathrm{roundtrip}} = \begin{bmatrix} a_{\mathrm{roundtrip}} & b_{\mathrm{roundtrip}} \\ c_{\mathrm{roundtrip}} & d_{\mathrm{roundtrip}} \end{bmatrix} \tag{6.29}$$

have to be derived from the multiplication of all single matrices accounting for all optical elements of the resonator as described in Sect. 2.5. From the determined value of $q(z_{\mathrm{oc}})$ the beam radius $w(z)$ and wave front curvature $R(z)$ can be calculated at any position inside and outside the resonator by the propagation of the Gaussian beam.

Notice that the beam parameter outside the resonator (laser beam parameter q_{laser}) has to be calculated from $q(z_{\mathrm{OC}})$ by applying the transmission matrix M_{OC} of the output coupler which may act as a lens:

$$q_{\mathrm{laser}} = \frac{a_{\mathrm{OC}}q(z_{\mathrm{oc}}) + b_{\mathrm{OC}}}{c_{\mathrm{OC}}q(z_{\mathrm{oc}}) + d_{\mathrm{OC}}} \tag{6.30}$$

and thus the divergence angle of the laser can be determined.

In particular, numerical calculations are easily possible using personal computers. Analytical solutions can be hard because of lengthy algebraic complex expressions, but algebraic computer programs can be used for this purpose.

6.6.2 Empty Resonator

As an example the description of an empty resonator (see Fig. 6.26) will be given.

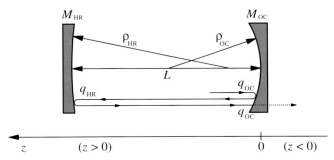

Fig. 6.26. Scheme of an empty resonator with the out-coupling mirror OC and the high-reflecting mirror HR. The curvature radii are positive for concave mirrors. The resonator length is L

This resonator with curvature ρ_{HR} of the high-reflecting mirror and ρ_{OC} of the output coupler placed at a distance L as the resonator length leads to the round trip matrix of the empty resonator:

$$M_{\mathrm{roundtrip}} = \begin{bmatrix} 1 & L \\ 0 & 1 \end{bmatrix} \cdot \begin{bmatrix} 1 & 0 \\ \dfrac{-2}{\rho_{\mathrm{HR}}} & 1 \end{bmatrix} \cdot \begin{bmatrix} 1 & L \\ 0 & 1 \end{bmatrix} \cdot \begin{bmatrix} 1 & 0 \\ \dfrac{-2}{\rho_{\mathrm{OC}}} & 1 \end{bmatrix} \tag{6.31}$$

which accounts for the first reflection at the output coupler, the path L, reflection at the high-reflecting mirror and the path L again considered as a positive value. The resulting matrix of the empty resonator is given by:

$$M_{\text{roundtrip}} = \frac{1}{\rho_{\text{OC}}\rho_{\text{HR}}}$$
$$\cdot \begin{bmatrix} -\rho_{\text{HR}}(4L - \rho_{\text{OC}}) - 2L(2L - \rho_{\text{OC}}) & 2L\rho_{\text{OC}}(\rho_{\text{HR}} - 1) \\ -2(\rho_{\text{OC}} + \rho_{\text{HR}} - 2L) & \rho_{\text{OC}}(\rho_{\text{HR}} - 2L) \end{bmatrix}. \tag{6.32}$$

An algebraic calculation of (6.28) using (6.32) which is easily done with a suitable algebraic computer program shows that the curvature R_i of the fundamental mode of this empty resonator is equal to the curvature of the mirrors ρ_i:

$$\text{wave curvature radii} \quad \begin{array}{l} R(z = z_{\text{OC}}) = \rho_{\text{OC}} \\ R(z = z_{\text{HR}}) = \rho_{\text{HR}} \end{array} \tag{6.33}$$

whereas the diameter of the beam $2w_{\text{OC}}$ at the position of the out-coupling mirror follows from:

beam diameter at OC

$$2w_{\text{OC}} = 2\sqrt{\frac{\lambda}{\pi}\rho_{\text{OC}}\sqrt{\frac{L(L - \rho_{\text{HR}})}{(\rho_{\text{OC}} + \rho_{\text{HR}} - L)(L - \rho_{\text{OC}})}}} \tag{6.34}$$

and the diameter at the high-reflecting mirror follows analogously:

beam diameter at HR

$$2w_{\text{HR}} = 2\sqrt{\frac{\lambda}{\pi}\rho_{\text{HR}}\sqrt{\frac{L(L - \rho_{\text{OC}})}{(\rho_{\text{OC}} + \rho_{\text{HR}} - L)(L - \rho_{\text{HR}})}}} \tag{6.35}$$

The position z_{waist} and diameter $2w(z_{\text{waist}})$ of the beam waist can easily be calculated from these solutions by using (2.83) and (2.84):

$$\textbf{waist position} \quad z_{\text{waist}} = \frac{L(L - \rho_{\text{HR}})}{\rho_{\text{OC}} + \rho_{\text{HR}} - 2L} \tag{6.36}$$

where the z coordinate is measured positively from the out-coupling mirror to the left in Fig. 6.26 towards the inside of the resonator and negative to the right.

waist diameter

$$2w_{\text{HR}} = 2\sqrt{\frac{\lambda}{\pi}\frac{\sqrt{L\rho_{\text{OC}}(L - \rho_{\text{HR}})(L - \rho_{\text{HR}})(\rho_{\text{OC}} + \rho_{\text{HR}} - L)}}{(\rho_{\text{OC}} + \rho_{\text{HR}} - 2L)}}. \tag{6.37}$$

As can be seen from this equation the diameter can be very large as the curvature of the mirrors is very flat. This case is usually hard to achieve because it is close to the stability limit and the misalignment sensitivity becomes very bad. Thus in this case mixtures of higher-order transversal modes are usually oscillating, resulting in bad beam quality.

Stable eigensolutions of the fundamental mode of the empty resonator are possible for:

stability condition $\rho_{OC} + \rho_{HR} \geq L$ (6.38)

and further conditions can be evaluated from the condition of a positive expression in the root (see below).

6.6.3 g Parameter and g Diagram

For a general discussion of the stability ranges the g parameters of the resonator as depicted in Fig. 6.26, can be used. These parameters are defined as:

g parameter of OC mirror $g_{OC} = 1 - \dfrac{L}{\rho_{OC}}$ (6.39)

and

g parameter of HR mirror $g_{HR} = 1 - \dfrac{L}{\rho_{HR}}$.   (6.40)

The beam waist at the output coupler follows from:

beam at OC $2w_{OC} = 2\sqrt{\dfrac{\lambda}{\pi}L}\sqrt{\dfrac{g_{HR}}{g_{OC}(1 - g_{OC}g_{HR})}}$ (6.41)

and at the high reflecting mirror:

beam at HR $2w_{HR} = 2\sqrt{\dfrac{\lambda}{\pi}L}\sqrt{\dfrac{g_{OC}}{g_{HR}(1 - g_{OC}g_{HR})}}$ (6.42)

whereas the beam waist occurs at the position:

waist position $z_{\text{waist}} = \dfrac{Lg_{HR}(L - g_{OC})}{g_{HR}(2g_{OC} - 1) - g_{OC}}$ (6.43)

with the waist diameter:

waist diameter $2w_{HR} = 2\sqrt{\dfrac{\lambda}{\pi}L}\dfrac{\sqrt{g_{OC}g_{HR}(1 - g_{OC}g_{HR})}}{g_{HR}(2g_{OC} - 1) - g_{OC}}$.   (6.44)

With these g parameters the general stability condition for the fundamental mode operation of the resonator can obviously be written as:

general stability condition $0 < g_{OC}g_{HR} < 1$.   (6.45)

This condition can be nicely visualized in the g diagram which is built by one g parameter as one coordinate (e.g. g_{OC}) and the other g parameter as the other coordinate as shown in Fig. 6.27.

At the limits of these stability ranges the Gaussian beam would have an infinite or zero diameter at the mirrors. In Fig. 6.27 some selected resonators, as described below, are marked with letters (a)–(f).

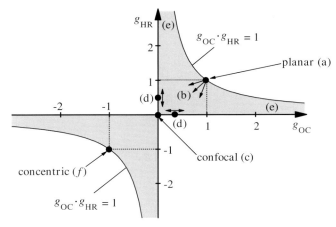

Fig. 6.27. g diagram for discussing the stability of an empty resonator. The gray area indicates the stable ranges of operation. Selected resonators are indicated (see next section)

6.6.4 Selected Stable Empty Resonators

Some of the stable empty resonators, e.g. with g_i equal 0, 1 or -1 are named for their special construction.

(a) Planar mirror resonators consist of two planar mirrors at any distance (see Fig. 6.28).

Fig. 6.28. Schematic of a planar resonator (a). The beam radius is theoretically indefinite

This resonator demands unconfined mirrors and would show an infinite beam diameter. It is at the stability limit (see Fig. 6.27) and thus it cannot be built as an empty resonator. As soon as some positive refraction occurs inside the resonator it will become stable. Therefore some times for solid-state lasers plan-plan resonators are used including the refraction of the thermal lensing of the laser rod.

If the lensing is too large the resonator will again become unstable. This effect can be used for measuring the thermal lensing of the active material under high-power pumping by measuring the output power as a function of the input power (see Sect. 6.6.15).

(b) Curved mirror resonators with concave radii larger than the resonator length as shown in Fig. 6.29 are very common.

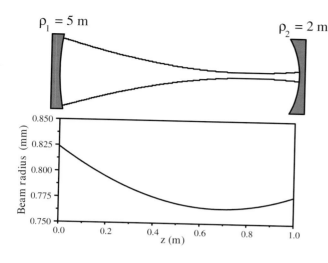

Fig. 6.29.
Schematic of a resonator with curved mirrors (b). The beam radius of the fundamental mode was calculated for 1 μm wavelength

These resonators are very stable and easy to design and align. Thus new lasers can be checked with this type of resonator, first. Usually the output coupler can be used as a planar mirror in this type of cavity and different out-coupling reflectivities can easily be obtained.

(c) Confocal resonators are symmetric with the mirror curvature radius equal the resonator length. Thus the beam waist is in the center and has a diameter of $1/\sqrt{2}$ compared to the diameter at the mirrors (see Fig. 6.30).

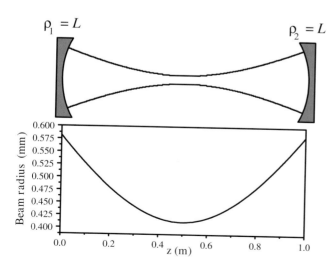

Fig. 6.30.
Schematic of a confocal resonator (c). The beam radius of the fundamental mode was calculated for 1 μm wavelength

The intensity is twice as high in the resonator center compared to the mirror position. These resonators stay stable if the resonator length is shortened as long as the resonator length is no longer than twice the curvature radius (see below).

(d) Semiconfocal resonators have the beam waist at one mirror which is planar. The curvature of the other mirror is $\rho_i = 2L$ (see Fig. 6.31).

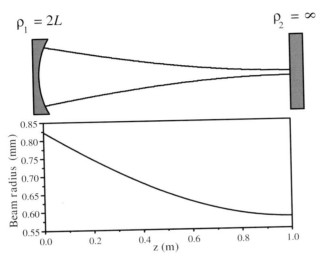

Fig. 6.31. Schematic of a semiconfocal resonator (d). The beam radius of the fundamental mode was calculated for 1 µm wavelength

The diameter of the beam is $1/\sqrt{2}$ smaller at the planar mirror compared to the curved one.

(e) Hemispherical resonators have a focus at the planar mirror (see Fig. 6.32).

The resulting high intensity at the planar mirror can be used for nonlinear effects in the resonator such as passive Q switching or passive mode locking with dyes. Care has to be taken not to damage this mirror with the resulting high intensities.

A similar transversal beam shape distribution along the resonator axis can be obtained if the two mirrors are curved and have much different radii. If, e.g. mirror 1 has a curvature radius of a few 10 mm while the radius of the other mirror 2, is in the range of m a focus will occur close to mirror 1 independent of whether the short focal length of is positive or negative (see also Fig. 6.34).

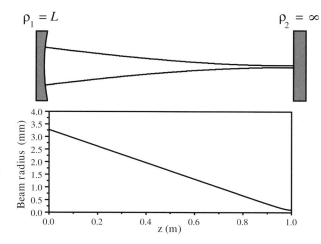

Fig. 6.32.
Schematic of a hemi-
spherical resonator
(e). The beam radius
of the fundamental
mode was calculated
for 1 µm wavelength

(f) Concentric or spherical resonators have their focus in the middle of the
cavity. The two mirrors have curvature radii equal to $L/2$ (see Fig. 6.33).

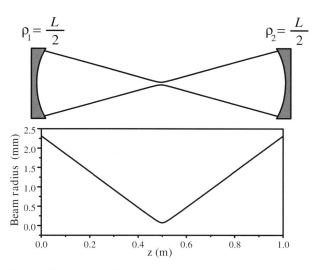

Fig. 6.33. Schematic of a concentric or spherical resonator (f). The beam radius
of the fundamental mode was calculated for 1 µm wavelength

This sharp focus can cause optical breakdown in air or damage in the
active material if this resonator is used in pulsed lasers. On the other hand
the focus in the center allows the use of an aperture for selection of the
fundamental mode in an effective way.

(g) Concave–convex resonators have a common focal point of the two mirrors outside the resonator. The curvatures radii are $\rho_1 > L$ and $\rho_2 = -(\rho_1 - L)$ as shown in Fig. 6.34.

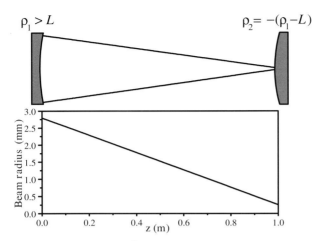

Fig. 6.34. Schematic of a concave–convex resonator (g). The beam radius of the fundamental mode was calculated for $1\,\mu m$ wavelength. The curvatures radii are $\rho_1 = 1.1\,m$ and $\rho_2 = -0.1\,m$

If the curvature radii of the two mirrors are very different the beam diameter can be very small at the mirror with the smaller radius. Thus the intensity at this mirror can cause damage similar to that in hemispherical resonators (see above), but for the application of nonlinear effects in the resonator this configuration may be well suited.

6.6.5 Higher Transversal Modes

Transversal modes of lasers with stable optical resonators higher than the fundamental can be determined analytically for empty resonators as the steady state solution for the electric field. The transversal structure of the electrical field of the light beam has to reproduce after each roundtrip. This can be described by the Kirchhoff integral equation as discussed, e.g. in [M16] in the following form:

$$E_1(x_2, y_2) = -i\frac{e^{ikL}}{2Lg_m\lambda} \cdot \int\int E_m(x_1, y_1)$$

$$\cdot \exp\left\{i\frac{\pi}{2Lg_m\lambda}\left[(x_1^2 + y_1^2 + x_2^2 + y_2^2)(2g_1g_2 - 1)\right.\right.$$

$$\left.\left. -2(x_1 + y_1 + x_2 + y_2)\right]\right\}\mathrm{d}x_m\,\mathrm{d}y_m$$

$$l, m = 1, 2, \quad l \neq m \tag{6.46}$$

with the g parameters $g_{1,2}$, the resonator length L, the wavelength λ and the wave number $k = 2\pi/\lambda$.

The solutions of this equation are the transversal modes of this empty resonator. They are called *TEM modes* for transverse electromagnetic field vectors and are numbered by indices. If the mirrors are transversally unconfined and no other apertures limit the beam diameter an infinite number of transversal modes with increasing transversal dimension and structure exist. The lowest-order mode is of course the fundamental or Gaussian or TEM_{00} mode as described above.

Eigenmodes of a resonator higher than the fundamental can have a circular or rectangular symmetry. The transversal mode structure of the intensity can be described by analytical formulas.

Examples of higher-order transversal modes are described in [6.316–6.359]. In addition to the modes described below, flat top modes are of interest for high extraction efficiency and certain applications [6.333–6.346]. Also, mode converters leading to, for example, Bessel modes, which allow a constant intensity over a certain distance and thus good efficiency in some nonlinear optical processes, are of interest [6.347–6.351]. Whispering-gallery modes occur in microlasers with microresonators [6.352–6.359]. These laser resonators may be of interest for use as diode lasers or in communication technology.

Circular Eigenmodes or Gauss–Laguerre Modes

These modes are described with Laguerre polynomials. The square of the solution for the electric field leads to the following transversal distribution for the intensity at the position of mirrors 1 and 2:

circular modes

$$I_1^{(m,p)}(r,\varphi) = \frac{I_{\max}}{F_{\max,\mathrm{circ}}} \left(\frac{2r^2}{w_l^2}\right)^p$$

$$\cdot \left[L^{(m,p)}\left(\frac{2r^2}{w_l^2}\right)\right]^2 e^{-2r^2/w_l^2} \cos(p\varphi) \tag{6.47}$$

with cylindrical coordinates r, φ, beam radius of the Gaussian beam w_l for this resonator calculated for w_{OC} from (6.34) or (6.41) and for w_{HR} from (6.35) or (6.42). The Laguerre polynomials $L^{(m,p)}(t)$ are given in mathematical textbooks. The first few are:

Laguerre polynomials:

$$L^{(0,p)}(t) = 1$$

$$L^{(1,p)}(t) = p + 1 - t$$

$$L^{(2,p)}(t) = \frac{1}{2}(p+1)(p+2) - (p+2)t + \frac{1}{2}t^2$$

$$L^{(3,p)}(t) = \frac{1}{6}(p+1)(p+2)(p+3) - \frac{1}{2}(p+2)(p+3)t + \frac{1}{2}(p+3)t^2 - \frac{1}{6}t^3$$

$$L^{(4,p)}(t) = \frac{1}{24}(p+1)(p+2)(p+3)(p+4) - \frac{1}{6}(p+2)(p+3)(p+4)t$$
$$+ \frac{1}{4}(p+3)(p+4)t^2 - \frac{1}{6}(p+4)t^3 + \frac{1}{24}t^4. \tag{6.48}$$

The maximum of these circular transversal modes $F_{\text{max,circ}}$ under the condition of equal power or energy content of all modes is given in Table 6.5.

Table 6.5. Maximum of the transversal modes $F_{\text{max,circ}}$ under the condition of equal power or energy content for all modes

$m\backslash p$	0	1	2	3	4
0	1.0	0.73	0.54	0.45	0.39
1	1.0	0.65	0.49	0.40	0.34
2	1.0	0.66	0.47	0.38	0.33
3	1.0	0.69	0.47	0.38	0.32
4	1.0	0.67	0.47	0.38	0.32

The intensity distributions of the lowest of these circular modes are shown in Fig. 6.35.

The first index in this nomenclature gives the number of maxima in the radial distribution and the shape has cylindrical symmetry for $p = 0$. The number of maxima around the circumference $\varphi = 0, \ldots, 2\pi$ is given by $2p$. Usually the modes with $p = 0$ are much easier to detect than the modes with $p > 0$.

The electric field vector is polarized antiparallel, with a phase shift of π, in neighboring peaks and their surrounding area up to the minima between them.

Rectangular or Gauss–Hermite Modes

These modes are described in Cartesian coordinates x, y using Hermite polynomials $H^{(m)}(t)$. The transversal intensity distribution is given by:

rectangular modes

$$I_1^{(m,p)}(x, y) = \frac{I_{\text{max}}}{F_{\text{max,rect}}} \left[H^{(m)} \left(\frac{2x}{w_l} \right) \right]^2$$
$$\cdot \left[H^{(p)} \left(\frac{2y}{w_l} \right) \right]^2 e^{-2(x^2+y^2)/w_l^2} \tag{6.49}$$

again with the beam radius of the Gaussian beam w_l for the resonator calculated for w_{OC} from (6.34) or (6.41) and for w_{HR} from (6.35) or (6.42). The Hermite polynomials $H^{(m/p)}(t)$ are also given in mathematical text books. The first few are:

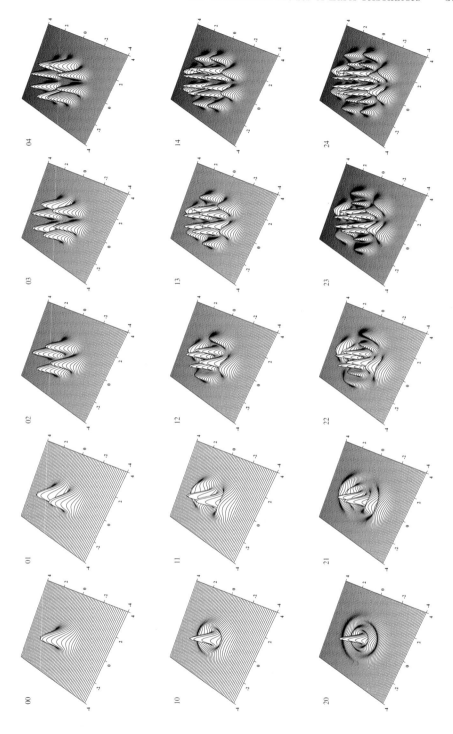

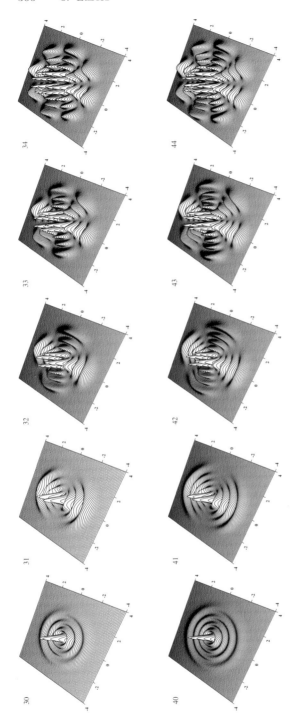

Fig. 6.35. Intensity profile of higher transversal modes of stable laser resonators with circular symmetry

Hermite polynomials:

$$H^{(0)}(t) = 1$$
$$H^{(1)}(t) = 2t$$
$$H^{(2)}(t) = 4t^2 - 2$$
$$H^{(3)}(t) = 8t^3 - 12t$$
$$H^{(4)}(t) = 16t^4 - 48t^2 + 12. \tag{6.50}$$

The maximum of these rectangular transversal modes $F_{\text{max,rect}}$ under the condition of equal power or energy content of all modes is given in Table 6.6.

Table 6.6. Maximum of the transversal modes $F_{\text{max,rect}}$ under the condition of equal power or energy content of all modes

$m\backslash p$	0	1	2	3	4
0	1	0.73	0.65	0.61	0.58
1	0.73	0.53	0.48	0.45	0.43
2	0.65	0.48	0.42	0.40	0.38
3	0.61	0.45	0.40	0.37	0.35
4	0.58	0.43	0.38	0.35	0.34

The intensity distributions of the lowest rectangular modes are shown in Fig. 6.36.

Again the mode indices account for the number of maxima in the direction of the coordinate, x and y in this case, with $m + 1$ and $p + 1$, and again the electric field vector is polarized antiparallel, with a phase shift of π, in neighboring peaks and their surrounding area up to the minima between them.

Different laser modes can occur at the same time and thus the observed mode pattern of a realistic laser may be a *superposition of several modes.* Any transversal field distribution can be written as a sum of eigenmodes because Laguerre and Hermite polynomials are each a complete set of orthogonal functions. Thus the intensity distribution $I(r, \varphi, t)$ of any circular mode can be expressed as:

$$\text{circular modes} \quad I(r, \varphi, t) = \sum_{m=0}^{\infty} \sum_{p=0}^{\infty} c_{m,p}(t) I^{(m,p)}(r, \varphi) \tag{6.51}$$

with the coefficients $c_{m,p}(t)$ accounting for the share of the m, p eigenmode. Rectangular modes can be constructed the same way using the rectangular eigenmodes:

$$\text{rectangular modes} \quad I(x/y, t) = \sum_{m=0}^{\infty} c_{x/y,m}(t) I^{(m,p)}(x/y). \tag{6.52}$$

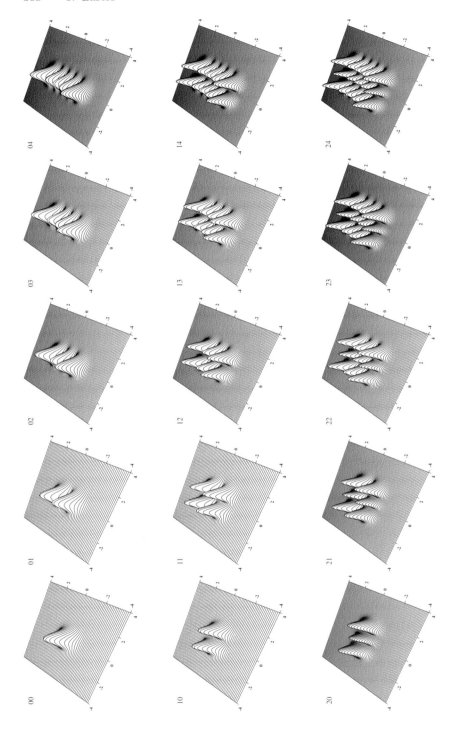

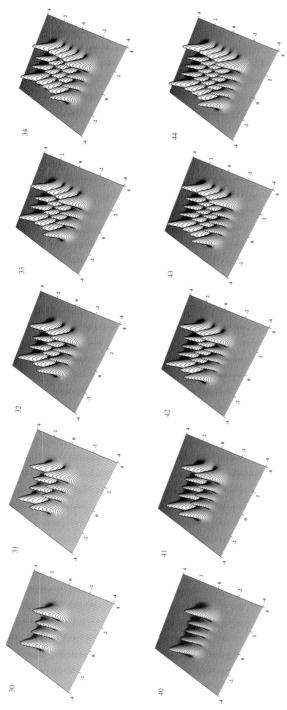

Fig. 6.36. Intensity profiles of higher transversal modes of stable laser resonators with rectangular symmetry

In some cases it may be necessary to sum over the electric field instead of over the intensity of the different modes.

Hybrid or Donut Modes

In solid-state rod lasers, hybrid or donut modes can be obtained. These transversal modes are constructed as a superposition of two circular transversal modes of the same order m, p but rotated by $90°$. The radial distribution of these modes can be calculated analogous to the circular modes from:

$$\textbf{donut modes}\quad I_1^{(m,p)}(r) = \frac{I_{\max}}{F_{\max,\text{donut}}} \left(\frac{2r^2}{w_l^2}\right)^p$$
$$\cdot \left[L^{(m,p)}\left(\frac{2r^2}{w_l^2}\right)\right]^2 \mathrm{e}^{-2r^2/w_l^2} \qquad (6.53)$$

but they show full circular symmetry in the intensity distribution. These modes are marked with an asterisk at the mode number. The maximum $F_{\max,\text{donut}}$ of the first three hybrid modes TEM_{01*}, TEM_{02*} and TEM_{03*} for the same power or energy content as the fundamental mode are 0.37, 0.34 and 0.34. The intensity distributions of these lowest three modes are shown in Fig. 6.37.

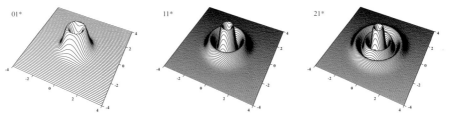

Fig. 6.37. Intensity profiles of higher transversal donut or hybrid modes of laser resonators with circular symmetry

The share of different modes can change in time and thus very complicated mode structures may be obtained with time-dependent coefficients. Mode apertures may be used for emphasizing certain modes required for the application (see Sect. 6.6.10).

6.6.6 Beam Radii of Higher Transversal Modes and Power Content

The beam radius of a higher laser mode can be defined differently. As discussed in Sect. 2.7.3 the beam radius can be given using the second intensity moment. For circular modes the beam radius follows from:

beam radius $\quad w_r^2 = \dfrac{4\pi \int r^3 I_{\text{uncal}}(r)\,\mathrm{d}r}{\int I_{\text{uncal}}(r)\,\mathrm{d}r}$ $\hspace{3cm}$ (6.54)

with the measured, not calibrated and thus relative intensity distribution $I_{\text{uncal}}(r)$ which can be measured with a CCD camera.

For rectangular modes the beam radii may be different in the x and y directions. Therefore they should be determined separately from:

$$w_x^2 = \frac{\int x^2 I_{\text{uncal}}(x)\,\mathrm{d}x}{\int I_{\text{uncal}}(x)\,\mathrm{d}x} \hspace{3cm} (6.55)$$

and

$$w_y^2 = \frac{\int y^2 I_{\text{uncal}}(y)\,\mathrm{d}y}{\int I_{\text{uncal}}(y)\,\mathrm{d}y}. \hspace{3cm} (6.56)$$

These radii are of course larger than the radius of the associated Gaussian mode w_{gauss} as follows:

circular modes $\quad w_{r,m,p} = w_{\text{gauss}}\sqrt{(2m + p + 1)}$ $\hspace{2cm}$ (6.57)

and

rectangular modes $\quad w_{x/y,m} = w_{\text{gauss}}\sqrt{(2m + 1)}$ $\hspace{1.5cm}$ (6.58)

with m and p as the indices of the Laguerre polynomials for circular modes and of the Hermite polynomials for rectangular modes as given above. The values of some circular modes are given in Table 6.7.

Table 6.7. Beam parameters for higher circular Gauss–Laguerre modes

m	p	w_r/w_{gauss}	$P_{\text{wr}}/P_{\text{total}}$	$w_{86.5\%}/w_{\text{gauss}}$	M^2	$M^2_{86.5\%}$
0	0	1	86.5%	1	1	1
0	1	1.41	90.8%	1.32	2	1.75
0	2	1.73	93.8%	1.56	3	2.44
0	3	2	95.8%	1.76	4	3.10
1	0	1.73	90.8%	1.65	3	2.71
1	1	2	92.2%	1.88	4	3.54
1	2	2.24	93.7%	2.08	5	4.32
1	3	2.45	95.0%	2.25	6	5.06
2	0	2.24	92.3%	2.12	5	4.48
2	1	2.45	93.0%	2.31	6	5.35
2	2	2.65	93.9%	2.48	7	6.17
2	3	2.83	94.9%	2.64	8	6.96
3	0	2.65	93.1%	2.51	7	6.29
3	1	2.83	93.6%	2.68	8	7.17
3	2	3	94.2%	2.83	9	8.02
3	3	3.16	94.9%	2.97	10	8.84
10	0	4.58	95.2%	4.39	21	19.3
0	10	3.32	99.7%	2.71	11	7.34

The values for the beam width w_m of rectangular modes in comparison to the associated Gaussian beam are given in Table 6.8. In this case the values along the two dimensions can be calculated, separately.

Table 6.8. Beam parameters for higher rectangular Gauss-Hermite modes

m	w_m/w_{gauss}	P_{wm}/P_{total}	$w_{86.5\%}/w_{\text{gauss}}$	M^2	$M^2_{86.5\%}$
0	1	1	1	1	1
1	1.73	99.3%	1.18	3	1.39
2	2.24	99.8%	1.50	5	2.25
3	2.65	99.7%	1.77	7	3.13
4	3	100.0%	2.01	9	4.02
5	3.32	100.0%	2.22	11	4.92
6	3.61	100.0%	2.41	13	5.83

The *power content* P_{wr} *and* P_{wm} inside the areas given by the beam radius w_r or the beam width w_m is not the same for the different higher modes. This becomes obvious from Fig. 6.38 in which these radii are shown for the lowest circular modes in relation to the radial intensity distribution.

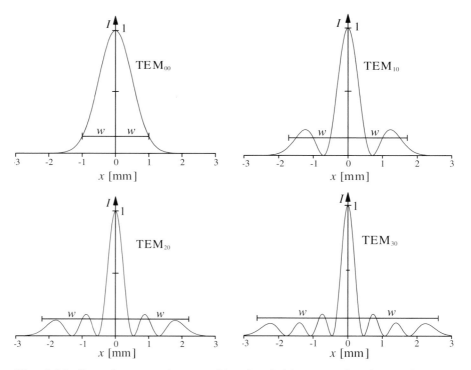

Fig. 6.38. Second-moment beam radii and radial intensity distribution for some circular modes

Therefore the power contents P_{wr} and P_{wm} relative to the total power of the beam P_{total} is also given in Tables 6.7 and 6.8. For the fundamental mode TEM_{00} the radius defined by the second moment corresponds to the intensity $I = I_{\max}/e^2$ and the area inside this radius contains 86.5% of the total energy of the beam.

Thus for comparison the beam radii or beam widths $w_{86.5\%}$ for the power content of 86.5% of the higher laser modes are given relative to the radius of the associated Gaussian mode in the Tables 6.7 and 6.8.

6.6.7 Beam Divergence of Higher Transversal Modes

Both types of these higher laser modes, circular and rectangular, are also solutions of Maxwell's equation for free space. Using the second moment radii the higher-mode beams can be transferred through optical systems using the matrix formalism.

The simplest procedure can be based on the associated Gaussian mode which has radius w_{gauss} as given relative to the radius or width of the higher mode in Tables 6.7 and 6.8 (see Sect. 6.6.9). Both beams will propagate "parallel" with a constant ratio of radii as given in Tables 6.7 and 6.8 as shown in the example of Fig. 6.39.

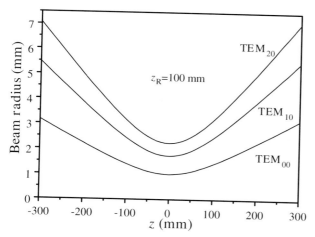

Fig. 6.39. Beam radius defined by second moment for different circular modes around the waist. The Rayleigh length is $z_R = 100\,\text{mm}$ and the wavelength is 500 nm

This results in a (far-field) *divergence angle* of the higher circular $\theta_{r,m,p}$ or rectangular $\theta_{x/y,m}$ modes which is larger than the divergence of the associated Gaussian beam by:

circular modes $\quad \theta_{r,m,p} = \theta_{\text{gauss}} \sqrt{(2m + p + 1)}$ $\hspace{2cm}$ (6.59)

and

rectangular modes $\theta_{x/y,m} = \theta_{\text{gauss}}\sqrt{(2m+1)}$ \hfill (6.60)

with the mode indices for the Laguerre polynomials m, p and for the Hermite polynomials m as given above. The resulting factors are the same as for the radii or widths and can be taken from Tables 6.7 and 6.8.

6.6.8 Beam Quality of Higher Transversal Modes

Using these values for the beam radius and the divergence angle for the higher laser modes the beam propagation factor M^2 can be given by:

circular modes $M_{m,p}^2 = 2m + p + 1$ \hfill (6.61)

and

rectangular modes $M_m^2 = 2m + 1$ \hfill (6.62)

The resulting values are also given in Tables 6.7 and 6.8, too.

It has to be noticed that this value of the beam propagation factor M^2 is based on the method of second moments and thus the power contents related to this beam propagation factor can be larger for higher modes than the 86.5% which is used for Gaussian beams. Therefore the beam propagation factor $M_{86.5\%}^2$ is also given in Tables 6.7 and 6.8, too. A further example is given in [6.360].

6.6.9 Propagating Higher Transversal Modes

The propagation of higher transversal modes through an optical system can be calculated using the following steps based on the matrix formalism for the propagation of Gaussian beams:

- determination of the beam radius $w_r(z_i)$ for circular symmetry and the beam widths in x and y directions for rectangular symmetry;
- determination of the beam divergence(s);
- determination of the beam propagation factor M^2;
- determination of the wave front radius $R(z_i)$ from these values;
- determination of the complex beam parameter $q(z_i)$ for the higher-order mode with:

$$\frac{1}{q(z_i)} = \frac{1}{R(z_i)} - \frac{i\lambda M^2}{\pi n w_r^2(z_i)};$$ \hfill (6.63)

- propagation of this beam as described in Sect. 2.4.4 using the ray matrix formalism:

$$q(z) = \frac{aq(z_i) + b}{cq(z_i) + d}$$ \hfill (6.64)

with the elements a, b, c and d of the propagation matrix;

- determination the searched beam radius $w_r(z)$ and the wave front curvature $R(z)$ of the propagated beam parameter $q(z)$.

The determination of the beam parameters can be based on the theoretical formulas given above or the values given in Tables 6.7 and 6.8 if the transversal mode structure is known. Otherwise it has to be done experimentally as described in Sects. 2.7.3 and 2.7.4.

The resulting beam radius will be at any place M times larger than the radius of the associated Gaussian beam independent of whether M^2 is calculated or measured based on the second-order moment method or on the 86.5% power content of the beam. The beam divergence will also be M times larger than the divergence of the associated Gaussian beam. The Rayleigh length will be the same for all transversal modes for the same wavelength. Other examples are given in [6.361–6.368].

6.6.10 Fundamental Mode Operation: Mode Apertures

Without any restrictions all kinds of mixtures of transversal modes can occur and thus almost any kind of mode pattern can be obtained as the laser output. Therefore methods for controlling the transversal mode structure have been developed [6.369–6.406]. For photonic applications usually low order modes are preferred in particular, the safe operation of the fundamental mode is of great interest. This TEM$_{00}$-mode gives the best possible beam quality and thus the highest brightness and best ability to focus the laser radiation.

Because the beam diameter increases with increasing mode number (see Tables 6.7 and 6.8) mode filtering can be achieved with mode apertures. In the simplest case a suitable mode aperture is applied in the resonator [6.369–6.384], e.g. near to one of the resonator mirrors as shown in Fig. 6.40.

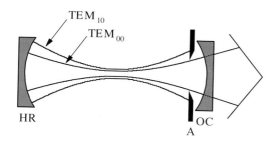

Fig. 6.40. Laser with mode aperture A which causes large losses for transversal modes higher than the fundamental mode as shown for the two lowest circular modes

The losses of the higher-order mode are larger and therefore the low-order mode will be more amplified and become dominant if the laser operates for long enough.

The *diameters of these mode apertures* have to be carefully adapted to the resonator. They have to be large enough not to cause high loss for the lower mode and have to be small enough to depress the higher ones. For pulsed lasers mode aperture radii of 1.5× the beam radius (at I_{max}/e^2) have been successfully tested. For cw lasers even larger values may be used. The best value should be determined experimentally.

But if the higher losses of the higher modes at the mode aperture are compensated by higher amplification in the active material, mode discrimination will not work satisfactorily. This can occur if the inversion at the volume of the lower and active mode is used whereas the higher and nonoperating mode may occur in other and still inverted areas of the active material. Thus mode selection cannot always be guaranteed by one mode aperture. Even oscillations between different mode patterns are possible. Therefore different concepts have been developed using two or more apertures.

The fundamental mode can almost be guaranteed using two mode apertures in the following scheme [6.369] (see Fig. 6.41).

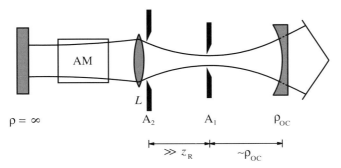

Fig. 6.41. Laser resonator with two apertures for guaranteeing fundamental mode operation

In this resonator one of the apertures A_1 is positioned in the waist position of the fundamental mode inside the cavity which is produced by a curved resonator output mirror. It should be noticed that the distance of this aperture is usually not exactly equal to the curvature ρ_{OC}. The diameter of this aperture is chosen for the desired Gaussian mode of the resonator of, e.g. 1.5 times the beam diameter in pulsed lasers. Then higher order transversal modes with approximately the same diameter at this aperture as the fundamental mode would show much higher divergence. Using a second mode aperture A_2 which is placed sufficiently away from the waist by many Rayleigh lengths z_R the beam divergence can be selected for the fundamental mode. Thus the combination of these two beams causes very high losses for all higher modes and thus the laser will operate in fundamental mode or it will not work at all [6.369].

Using this concept of the fundamental mode aperture design the potential of different inversion profiles of active materials for fundamental mode emission can be tested. Because of the nonlinear coupling of all laser modes via inversion in the active material this concept shows high efficiencies.

The apertures can be realized partly by the resonator components. For example the smaller aperture A_1 may be obtained from the inversion profile of a laser pumped active material [6.406] positioned in the waist of the beam. In other resonators the aperture A_2 may be obtained by the limited diameter of a solid-state laser rod.

Mode discrimination can also be made outside of the cavity by spatial filtering as shown in Fig. 6.42.

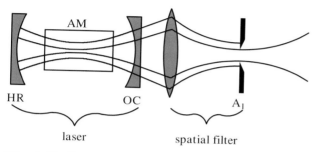

Fig. 6.42. Spatial filter for depressing higher laser modes externally

In this case one small aperture A_1 is placed in the waist of an external lens. The disadvantage of this scheme is poor efficiency. The energy of all higher modes is just wasted. Further the residual intensities from the diffraction of the higher modes may still disturb the application. Thus this scheme should be used, only for beams which already have good beam quality, or in low-power applications.

If the mode diameter is very small (below $100 \, \mu m$) optical breakdown can occur and the aperture can be damaged. In the worst case the aperture can be placed in a vacuum chamber ($<10^{-2}$ bar). Good results can be achieved using small quartz tubes with the required inner diameter as a pin-hole.

Another method for mode discrimination is based on Resonator mirrors with transversally varying reflectivity, such as, for example, Gaussian mirrors [6.385–6.393]. Further waveguides can be used for the suppression of higher-order modes [6.394–6.399]. Phase plates or more complicated diffractive optical elements can be used for mode discrimination [6.400–6.403]. Methods for smoothing the beam have been developed [6.404, 6.405].

6.6.11 Large Mode Volumes: Lenses in the Resonator

For laser wavelengths in the range of $1 \, \mu m$ the beam diameter inside empty laser resonators of 1 m length is roughly in the range of 1 mm (see Sect. 6.6.4). High-power lasers demand larger diameters for larger mode volumes in the active material [6.407–6.415]. Thus additional lenses may be applied for increasing the mode diameter. But the larger the mode diameter the more crucial is the alignment of the mirrors and the smaller is the stability range of the high-power lasers. Nevertheless, e.g. a telescope inside the cavity can be used in combination with the fundamental mode aperture design as shown in Fig. 6.43.

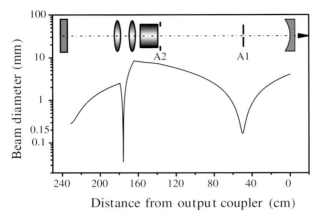

Fig. 6.43. Resonator for high-power solid-state laser with fundamental mode operation and large mode volume

With this type of resonator fundamental mode diameters of 9 mm have been achieved in stable fundamental mode operation [6.369]. The distance between the telescope lenses can be easily adapted for compensation of the thermal lensing of the laser rod. Nevertheless fluctuations of the thermal lens of the active material cause stronger fluctuations in the output power of the laser in cases of larger mode diameters.

6.6.12 Transversal Modes of Lasers with a Phase Conjugating Mirror

An ideal phase conjugating mirror (PCM) used as one of the resonator mirrors, usually the high-reflecting one, will perfectly reflect each incident beam in itself. Thus for an empty resonator with PCM all modes with a curvature equal to the curvature of the output coupler at its place are eigenmodes. In addition double or multiple roundtrip eigenmodes can occur. Therefore an indefinite number of eigenmodes exist without any further restricitions.

With apertures these modes can be discriminated. Thus the fundamental mode aperture design given in the Sect. 6.6.10 is especially useful to provide fundamental mode operation in lasers with PCM.

For the theoretical description of transversal Gaussian modes an ideal phase conjugating mirror can be described based on the matrix as given in Table 2.6 [6.416]. Real phase conjugating mirrors may decrease the beam diameter of a Gaussian mode. Their nonlinear reflectivity as a function of the intensity can result in lower reflectivity at the wings of the beam compared to the reflectivity at the center.

In particular, for lasers with phase conjugating mirrors based on stimulated Brillouin scattering (SBS) [6.417–6.448] it was suggested to calculate the roundtrip in the resonator without the PCM. The calculation of the fundamental mode can be based on the definitions of Fig. 6.44.

SBS - PCM

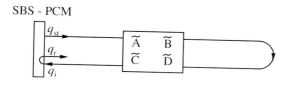

Fig. 6.44. Scheme for calculating transversal fundamental mode of lasers with phase conjugating mirror (PCM) based on stimulated Brillouin scattering (SBS)

The beam propagation matrix $\tilde{M}$ is calculated for the roundtrip through all optical elements towards the other conventional resonator mirror, the reflection there and the way back. Not included is the reflection at the phase conjugating mirror. This is considered with two assumptions about the beam parameter q (see Sect. 2.4.3) with $1/q = 1/R - i\lambda/\pi w^2$ [6.417]:

$$w_r = \beta_{\text{PCM}} w_i \quad \text{with} \quad 0 < \beta_{\text{PCM}} \leq 1 \tag{6.65}$$

and

$$R_r = -R_i \tag{6.66}$$

with the factor β_{PCM} accounting for the different nonlinear reflectivity across the beam.

The eigensolution for the fundamental transversal mode follows from:

$$q_r \overset{!}{=} q_{st} \quad \text{and} \quad q_i = \frac{\tilde{A} q_{st} + \tilde{B}}{\tilde{C} q_{st} + \tilde{D}} \tag{6.67}$$

with the matrix elements $\tilde{A}$, $\tilde{B}$, $\tilde{C}$, $\tilde{D}$ of the matrix $\tilde{M}$.

These equations have the solution for the beam radius at the PCM w_{PCM} which is equal to w_{st}, w_r and w_i:

beam radius at PCM $$w_{\text{PCM}} = \sqrt{\frac{\beta_{\text{PCM}} \lambda_{\text{laser}} \tilde{B}}{\pi}} \tag{6.68}$$

and for the curvature of the beam at the PCM moving towards the output coupler:

curvature at PCM $$R_{\text{PCM}} = -\frac{\tilde{B}}{\tilde{A}}. \tag{6.69}$$

With these eigensolutions the further beam propagation of the fundamental mode in the resonator can be done with the matrix formalism.

The phase conjugating SBS mirror can also be considered using its beam propagation matrix M_{PCM} which follows from the combination of the matrix of the ideal PCM with the matrix of a Gaussian aperture resulting in:

$$M_{\mathrm{PCM}} = \begin{pmatrix} 1 & 0 \\ -\dfrac{\mathrm{i}\lambda}{\pi a^2} & -1 \end{pmatrix} \tag{6.70}$$

where the radius of this aperture a is related to the β_{PCM} given above by:

$$a = w_{\mathrm{in}}\sqrt{\frac{\beta_{\mathrm{PCM}}^2}{1 - \beta_{\mathrm{PCM}}^2}}. \tag{6.71}$$

It turns out that β_{PCM} is almost 1 in most practical cases, but the calculation with values slightly smaller than one leads to useful results.

The optical phase conjugating mirror, e.g. based on SBS, can compensate for phase distortions in resonators as they result from the thermal lensing of solid-state laser rods if the PCM is located close to the distortion as shown in Fig. 6.45.

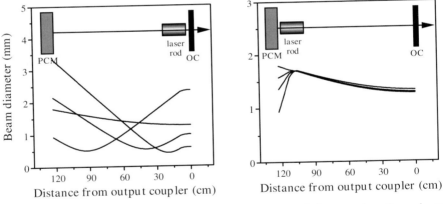

Fig. 6.45. Compensation of phase distortions with optical phase conjugating mirror (PCM) in resonators works only if the PCM is close to the disturbance (see right picture). The beam paths were calculated for an eigenmode

As shown in the right part of this figure the output beam of the laser is almost the same although thermal lensing in the laser rods shows very different values and thus the beam parameters at the PCM are very different.

For using stimulated Brillouin scattering as a self-pumped nonlinear phase conjugating mirror in the resonator a scheme as shown in Fig. 6.46 can be applied [6.418–6.421].

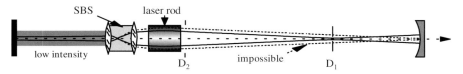

Fig. 6.46. Schematic of a laser resonator with fundamental mode discrimination and phase conjugating SBS mirror for compensation of phase distortions from the laser rod

To provide the start intensity for the nonlinear reflector a start resonator is formed by the mirror M_{start} with low reflectivity R_{start} and the output coupler M_{OC}. As soon as the reflectivity of the SBS mirror is larger than R_{start} the laser will operate mostly between the phase conjugating SBS mirror and the output coupler; mirror M_{start} becomes more and more functionless. The mode is determined by the apertures D_1 and D_2.

Using this scheme average output powers of 50 W with diffraction-limited beam quality have been obtained from a single rod flash lamp pumped Nd:YALO laser [6.418].

Besides stimulated scattering processes such as stimulated Brillouin scattering (SBS), also four-wave mixing can be applied for realizing optical phase conjugation in lasers [6.449–6.465]. These mirrors can be based on gain gratings in the active material, on absorption gratings or on third-order nonlinearity in transparent crystals. Phase distortions from the active material such as, for example, thermal lensing in solid-state lasers can also be compensated for by actively controlled adaptive mirrors [6.466–6.470].

6.6.13 Misalignment Sensitivity: Stability Ranges

Misalignment of the resonator results from tilting resonator mirrors or by changing the resonator length L. In addition the active material may change the resonator alignment by varying optical parameters. For example, solid-state laser rods can show thermally induced lensing and birefringence as a function of the pumping conditions. These effects may disturb the stable operation of the laser.

The discussion of the misalignment sensitivity and of the stability of any resonator [6.471–6.483] can be based on the equivalent g parameters g_i^* and resonator length L^* which follow from the transfer matrix M_{T} of the resonator. This transfer matrix is built by calculating a single transfer through the resonator using half of the focusing of the resonator mirrors:

transfer matrix

$$M_T = \begin{bmatrix} a_T & b_T \\ c_T & d_T \end{bmatrix}$$

$$= \begin{bmatrix} 1 & 0 \\ -1/\rho_{HR} & 1 \end{bmatrix} \cdot \begin{bmatrix} a_n & b_n \\ c_n & d_n \end{bmatrix} \cdots \begin{bmatrix} a_1 & b_1 \\ c_1 & d_1 \end{bmatrix} \cdot \begin{bmatrix} 1 & 0 \\ -1/\rho_{OC} & 1 \end{bmatrix} \quad (6.72)$$

From the elements of this transfer matrix it follows that:

g*-parameters $g^*_{OC} = a_T$ and $g^*_{HR} = d_T$ $\qquad\qquad$ (6.73)

and the equivalent optical resonator length is:

L*-length $L^* = b_T$ $\qquad\qquad\qquad\qquad\qquad\qquad\qquad\qquad$ (6.74)

Using these definitions it follows that:

$$c_T = \frac{g^*_{OC} g^*_{HR} - 1}{L^*}. \qquad\qquad\qquad\qquad\qquad\qquad (6.75)$$

The laser resonator is stable as long as:

stability condition $a_T b_T c_T d_T < 0$ $\qquad\qquad\qquad\qquad\qquad$ (6.76)

with the result of infinite beam diameters at the resonator mirrors as one of these matrix elements is zero (see Table 6.9).

Table 6.9. Beam radii at the two resonator mirrors M_{OC} and M_{HR} at the stability limits of the resonator

	w_{OC}	w_{HR}
$a_T = 0$	∞	0
$b_T = 0$	0	0
$c_T = 0$	∞	∞
$d_T = 0$	0	∞

The discussion of the whole stability range of the resonator can be based on the g^* diagram as described above for the g diagram. As an example three solid-state laser resonators with their stability ranges as a function of the thermal lensing of the active material are shown in Fig. 6.47.

Resonator (a) shows as usual two separated stability ranges, one in the upper right and the other in the lower left part of the diagram. Thus increasing the pump power leads to operation, nonoperation and operation again. In resonator (b) these two stability ranges are connected at the confocal, (0,0)-point of the diagram and result in one wide stability range. In resonator (c) the two stability ranges are connected at the concentric point with $g^*_{OC} g^*_{HR} = 1$ resulting in a wide and uninterrupted stability range. But the misalignment sensitivity of the two last resonators is much different.

Misalignment sensitivity can also be discussed based on the transfer matrix elements. Element $b_T = 0$ results in an imaging of the resonator mirror

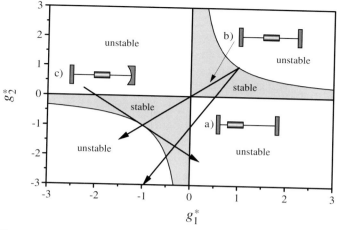

Fig. 6.47. g* diagram with the lines of operation for three resonators as a function of thermal lensing in the laser rod

M_{OC} to M_{HR} and vice versa. Thus the misalignment sensitivity is minimal for this type of resonator.

If the matrix element $c_T = 0$ the misalignment sensitivity is maximal. Figure 6.48 shows at which point of operation inside the stability ranges of the two resonators (b) and (c) this extreme occurs.

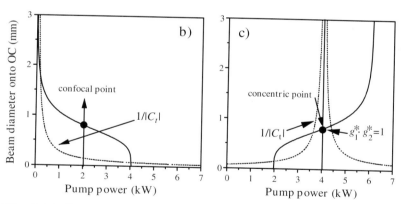

Fig. 6.48. Beam diameter at the output coupler of the resonators (b) and (c) of Fig. 6.47 as a function of the pump power P_{in} of the laser rod inducing thermal lensing. In addition the inverse matrix element $1/c_T$ is shown to describe the maximum misalignment sensitivity

Resonator (c) shows the highest misalignment sensitivity in the middle of the stability range and will therefore be difficult to operate. In resonator (b) the maximum of the misalignment sensitivity is at the left side of the stability range. This type of resonator crossing the confocal point should be used if a large stability range and low misalignment sensitivity is demanded.

The detailed discussion of the misalignment sensitivity can be based on the calculation of the resonator including the misalignment vectors for each element:

$$\text{misalignment vector} \quad \begin{pmatrix} x_{\text{element}} \\ \alpha_{\text{element}} \end{pmatrix} \tag{6.77}$$

with the misalignment shift x and the misalignment angle α of the optical element. This vector is multiplied by the resulting beam matrices [M16].

6.6.14 Dynamically Stable Resonators

Resonators designed with their point of operation at the center of the stability range are called dynamically stable resonators [6.484–6.486]. For this the beam radius is plotted as a function of the pump power, as e.g. shown in Fig. 6.49.

The beam diameter in the rod is in general a symmetric function of the pump power with its axis between the two stability ranges. As can be seen the best choice is the stability range I which has two foci at the two mirrors.

At the center-point the change of the beam radius as a function of e.g. the thermally induced changes of the refractive power of the active material is minimal and thus the fluctuation of the output power can be minimized.

The stability range of the pump power P_{pump} was calculated for dynamically stable resonators of solid-state rod lasers as [6.410, 6.485]:

$$\Delta P_{\text{pump}} = \frac{2\lambda_{\text{laser}}}{C_{\text{material}}} \left(\frac{r_{\text{rod}}}{w_{00,\text{rod}}} \right)^2 \tag{6.78}$$

with the material parameter C_{material}, the wavelength of the laser λ_{laser}, the radius of the laser rod r_{rod} and the radius of the TEM$_{00}$ mode in the rod $w_{00,\text{rod}}$. The parameter C_{material} is of the order of 10^{-5}–$10^{-6}\,\text{m kW}^{-1}$ (see Table 6.10).

Table 6.10. Material constant C_{material} defining the stability range and the TEM$_{00}$ potential for different lasers

Laser	λ_{laser} (nm)	doping (at%)	C_{material} ($\mu\text{m kW}^{-1}$)	$C_{00\text{-pot}}$ (%W μm^{-1})
Nd:YAG	1064	1.1	16	230–260
Nd:YALO	1080	0.8	36	70–150
Nd:YLF	1047	1.0	−1.3/−4.4	880
	1053	1.0	−0.89/0.72	1280

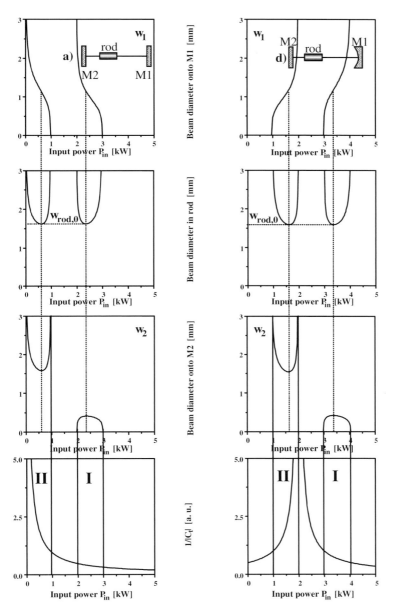

Fig. 6.49. Beam radii at the resonator mirrors M_1 and M_2 and inside the laser rod and misalignment sensitivity as a function of the pump power for two resonators. Left resonator is resonator (a) in Fig. 6.47 and resonator (d) at the right row shows the longer curvature of mirror M_1

This would demand small transversal mode diameters in the active material. On the other hand the efficiency and the maximum possible output power demands large mode volumes. Thus values of 1.5–4 have been achieved for the ratio of the rod radius divided by the mode radius.

Finally, the ratio $C_{00\text{-pot}}$ of the efficiency of the laser material η_{material} divided by the material parameter C_{material} is a measure for the TEM_{00} mode potential of the laser material:

$$C_{00\text{-pot}} = \frac{\eta_{\text{material}}}{C_{\text{material}}} \tag{6.79}$$

which is also given for flash lamp pumped Nd lasers in Table 6.10. For diode pumped lasers these values can be higher and, e.g. for Nd:YAG, values of up to $C_{00\text{-pot}} \approx 470\%\,\text{W}\,\mu\text{m}^{-1}$ were obtained. A similar value to $C_{00-\text{pot}}$ is sometimes used, namely χ_{therm}:

$$\chi_{\text{therm}} = \frac{\eta_{\text{heating}}}{\eta_{\text{excitation}}} \tag{6.80}$$

which is the quotient of the heating efficiency η_{heating} and the excitation efficiency $\eta_{\text{excitation}}$.

6.6.15 Measurement of the Thermally Induced Refractive Power

The refractive index of the active material will modify the transversal and longitudinal modes of the resonator. In high-power systems the refractive index can be a complicated function of the pump conditions and may vary in space and time (see Sect. 6.4).

As an example in rods of solid-state lasers, such as e.g. in the Nd:YAG or Nd:YALO material, thermally induced lensing will occur. The refractive power is dependent on the pump conditions, and the laser operation and may be different for the different polarizations. Thus it should be measured in the operating laser. The measurement of the stability ranges of the laser resonator allows the determination of the refractive power of the active material in an easy way [e.g. 6.487–6.490] as shown in Fig. 6.50.

Therefore the laser output power is measured as a function of the pump power for a given resonator configuration. At the stability limits the output power drops as can be seen in the figure. The thermal lens can be determined from the modeling.

The intensity cross-section pictures from the output coupler taken at the stability limits of the resonator clearly show the natural birefringence of the material leading to an astigmatic thermal lens. The two refractive powers of 0.81 and 0.84 dpt kW^{-1} in the a axis direction and 0.62 and 0.68 dpt kW^{-1} for the direction of the c axis demonstrate the accuracy of the method.

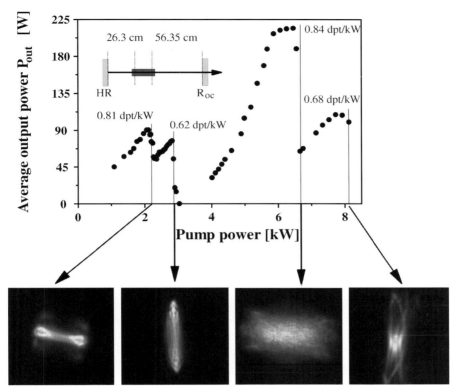

Fig. 6.50. Measurement of the output power as a function of the pump power of a solid-state laser to determine the refractive index profile as a consequence of the heating of the active material from the stability limits of the resonator. The Nd:YALO laser rod had a diameter of 8 mm and a length of 154 mm (1.1 at%). The c axis of the crystal was aligned vertical and the a axis horizontal in the pictures. The connecting line between the two flash lamps was perpendicular to the c axis. The laser light was also vertically polarized in the c direction

6.7 Longitudinal Modes

Longitudinal or axial modes of the resonator are determined by its geometry and the reflectivity of the mirrors. Which of these possible modes are activated in the operating laser depends on the properties of the active material and on possible frequency-selective losses of the resonator.

6.7.1 Mode Spacing

The eigensolution for the standing wave of the electric light field in the laser resonator shows knots at the resonator mirrors (see Fig. 6.51).

Thus the optical length of the resonator L_{opt} has to be an integer multiple p_{mode} of half the possible wavelengths λ_p of the laser:

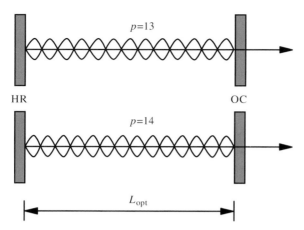

Fig. 6.51. Longitudinal modes of a laser resonator with the optical length L_{opt} which is 6.5 and 7 times as long as the light wavelength

$$\lambda_p = \frac{2}{p_{mode}} L_{opt} \tag{6.81}$$

with mode number p_{mode}. The optical length of the resonator has to be calculated from the geometrical length $L_{ith\ part,geom}$ of all path lengths from one resonator mirror to the other multiplied by the refractive index $n_{ith\ part}$ of the components, as e.g. laser rods

$$L_{opt} = \sum_{all\ parts} n_{ith\ part} L_{ithpart,geom}. \tag{6.82}$$

Thus, e.g. a Nd:YAG rod of 0.1 m length increases the optical length of the resonator by 0.082 m ($n = 1.82$).

The related mode frequencies $\nu_p = c_0/\lambda_p$ show a constant difference, the mode spacing frequency $\Delta\nu_{res}$ of the resonator:

$$\textbf{mode spacing}\quad \Delta\nu_{res} = \left| \frac{c_0}{\lambda_p} - \frac{c_0}{\lambda_{p\pm1}} \right| = \frac{c_0}{2L_{opt}} \tag{6.83}$$

and the wavelength spacing is:

$$\Delta\lambda_{res} = \frac{\lambda^2}{2L_{opt}} \tag{6.84}$$

with the vacuum speed of light c_0 and the central wavelength λ. This periodic sequence of axial modes can be generated over the possibly wide spectral range of the amplification bandwidth of the active material as shown in Fig. 6.52.

The gain of the active material always shows spectral behavior. The gain region above laser threshold defines the potential laser bandwidth and only longitudinal modes inside this laser bandwidth can be obtained.

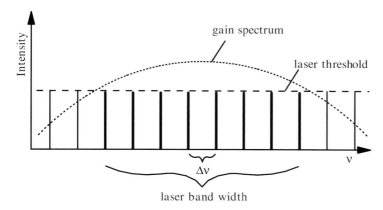

Fig. 6.52. Mode spectrum of a laser with gain spectrum and laser threshold selecting the active longitudinal laser modes

The mode spacing frequency for a resonator length of 0.5 m is 300 MHz resulting in a wavelength spacing of 0.25 pm at a central laser wavelength of 500 nm. Thus in a 0.5 m resonator of a Nd:YAG laser with a gain bandwidth of 0.5 nm about 2.000 longitudinal modes could oscillate. In practice the number of lasing axial modes is much smaller as a consequence of the nonlinear amplification and of the order of 10–100.

Because of slightly different optical path lengths inside the resonator the different transversal modes will have slightly different longitudinal mode frequencies with the mode spacing frequency $\Delta\nu_{\text{trans},m,p}$ compared to the TEM_{00}-mode:

$$\Delta\nu_{\text{trans},m,p}^{\text{circ}} = \frac{c_0}{2L_{\text{opt}}} \frac{1}{\pi} (2m + p) \arccos\left(1 - \frac{L_{\text{opt}}}{\rho_{\text{res}}}\right) \tag{6.85}$$

for circular modes and

$$\Delta\nu_{\text{trans},m,p}^{\text{rect}} = \frac{c_0}{2L_{\text{opt}}} \frac{1}{\pi} (m + p) \arccos\left(1 - \frac{L_{\text{opt}}}{\rho_{\text{res}}}\right) \tag{6.86}$$

for rectangular modes both with the transversal mode numbers m, p and the curvature of the resonator mirrors ρ_{res}.

These differences are small compared to the mode spacing $\Delta\nu_{\text{res}}$ as long as the curvature of the resonator mirrors is large compared to the resonator length. In confocal resonators the mode spacing between the transversal modes is equal to or half of the mode longitudinal mode spacing [M25].

In lasers with a phase conjugating mirror based on stimulated Brillouin scattering the longitudinal mode structure may be much more complicated as at each roundtrip all laser modes will be shifted by the Brillouin frequency shift (see Sect. 4.5.8). A rather complicated longitudinal mode pattern was observed and the temporal structure showed strong modulations (see Sect. 6.7.6).

6.7.2 Bandwidth of Single Longitudinal Modes

The empty laser resonator represents a Fabry–Perot interferometer (see Sect. 2.9.6 and [6.491]) of optical length L_{opt} formed by two mirrors M_1 and M_2 with reflectivityies R_1 and R_2 with normal incidence as depicted in Fig. 6.53.

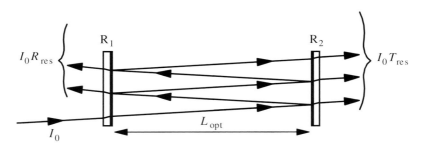

Fig. 6.53. Laser resonator as a Fabry–Perot interferometer with total transmission T_{res}

The transmitted and reflected light is a geometric series of interfering electric field contributions from the partially transmitted and reflected light traveling back and forth in the resonator with decreasing amplitude. Assuming no absorption in the two mirrors the transmittance can be written as:

$$T_{\text{res}} = \frac{(1 - R_1)(1 - R_2)}{(1 - \sqrt{R_1 R_2})^2 + 4\sqrt{R_1 R_2}\,\sin^2(2\pi L_{\text{opt}}/\lambda)} \tag{6.87}$$

and the reflectance follows from:

$$R_{\text{res}} = \frac{(\sqrt{R_1} - \sqrt{R_2})^2 + 4\sqrt{R_1 R_2}\,\sin^2(2\pi L_{\text{opt}}/\lambda)}{(1 - \sqrt{R_1 R_2})^2 + 4\sqrt{R_1 R_2}\,\sin^2(2\pi L_{\text{opt}}/\lambda)} \tag{6.88}$$

with the light wavelength λ.

The maximum transmission $T_{\text{res,max}}$ of this resonator is:

$$T_{\text{res,max}} = \frac{(1 - R_1)(1 - R_2)}{(1 - \sqrt{R_1 R_2})^2} \tag{6.89}$$

which is 1 if $R_1 = R_2$. The spectral bandwidth of the resonator follows from:

frequency bandwidth $\Delta\nu_{\text{FWHM}} = \dfrac{c}{2\pi L_{\text{opt}}}\left|\ln(\sqrt{R_1 R_2})\right|$ (6.90)

or

wavelength bandwidth $\Delta\lambda_{\text{FWHM}} = \dfrac{c^2}{2\pi \lambda^2 L_{\text{opt}}}\left|\ln(\sqrt{R_1 R_2})\right|$ (6.91)

and the intensity at the out-coupling resonator mirror inside the resonator ($I_{\text{OC,in}}$) is higher than the out-coupled intensity ($I_{\text{OC,out}}$) by:

$$I_{OC,in} = I_{OC,in} \frac{1}{(1 - R_{oc})}. \tag{6.92}$$

The finesse F and the quality Q of the empty laser resonator can be calculated from the frequency bandwidth $\Delta\nu_{FWHM}$ and the mode spacing $\Delta\nu_{res}$ by:

finesse $F_{res} = \dfrac{\Delta\nu_{res}}{\Delta\nu_{FWHM}} = \dfrac{\pi}{\left| \ln(\sqrt{R_1 R_2}) \right|}$ $\tag{6.93}$

and

quality $Q_{res} = \dfrac{\nu_{laser}}{\Delta\nu_{FWHM}} = 2\pi \Delta\nu_{res} \tau_{res}$ $\tag{6.94}$

with the life time τ_{res} of the empty resonator:

resonator life time $\tau_{res} = \dfrac{L_{opt}}{c_0 \left| \ln(\sqrt{R_1 R_2}) \right|} = \dfrac{1}{2\pi \Delta\nu_{FWHM}}$ $\tag{6.95}$

which indicates the $1/e$ decay of the light or the necessary time to reach the steady state. It is usually in the range of ns.

If the resonator contains the active material and perhaps other elements, additional losses with transmission $V < 1$ and amplification with gain $G > 1$ occur.

The resonator life time $\tau_{res,act}$ and the bandwidth $\Delta\nu_{FWHM,act}$ of the resonator with the active material will be:

$$\tau_{res,act} = \frac{L_{opt}}{c_0 \left| \ln(GV\sqrt{R_1 R_2}) \right|} = \frac{1}{2\pi \Delta\nu_{FWHM,act}} \tag{6.96}$$

and

$$\Delta\nu_{FWHM,act} = \frac{c}{2\pi L_{opt}} \left| \ln(GV\sqrt{R_1 R_2}) \right|. \tag{6.97}$$

Laser resonators with $GV > 1$ will show an increased resonator life time and a narrower spectral bandwidth. Laser threshold is reached at $GV\sqrt{R_1 R_2} = 1$ (see Sect. 6.8). In this case the resonator lifetime is infinite and the bandwidth would be zero (for more details see Sect. 6.9). It is then determined by the properties of the active material. $GV\sqrt{R_1 R_2} > 1$ can be achieved only for short times, and thus the analysis has to be made time dependent.

6.7.3 Spectral Broadening from the Active Material

The optical transitions of the laser materials have bandwidths from a few tens of pm or MHz up to more than 100 nm or 100 THz as shown in Table 6.11.

The narrow laser lines are broadened by Doppler shifts from the motion of the particles in the gas or by collisions. Molecular laser materials are broadened by the simultaneous electronic, vibrational and rotational transitions. In solids and liquids the environment of the laser active particles (atoms or

Table 6.11. Bandwidth of several laser materials as peak wavelength λ_{peak}, wavelength bandwidth $\Delta\lambda$, frequency bandwidth $\Delta\nu$ and number of modes p within this bandwidth in a 10 cm long laser resonator

Active material	Type	Mechanism	λ_{peak} (nm)	$\Delta\lambda$ (nm)	$\Delta\nu$ (GHz)	p
He-Ne	gas	Doppler	632.8	0.018	1.5	1
Ar-ion	gas	Doppler	488	0.03	4.0	4
CO_2 (10 mbar)	gas	Doppler	10 600	0.20	0.06	1
CO_2 (1 bar)	gas	collisions	10 600	14	4.0	4
CO_2 (10 bar)	gas	rotation	10 600	500	150	100
KrF	excimer	vibrations	248	0.5	2500	1700
XeCl	excimer	vibrations	308	0.7	2200	1500
Ruby	solid-state	matrix	694.3	0.5	330	220
Nd:YAG	solid-state	matrix	1064	0.45	120	80
Nd:glass	solid-state	matrix	1054	20	5400	3600
Alexandrite	solid-state	matrix	760	70	36 000	24 000
Ti:Sapphire	solid-state	matrix	790	120	58 000	38 000
Rhodamin 6G	dye	vibrations	580	60	54 000	36 000
GaAs	diode	band	800	2	100	70

molecules) may be different and produce additional broadening. The spectral broadening can be homogeneous or inhomogeneous as a function of the characteristic time constants of the experiment, as described in Sect. 5.2.

Large spectral widths of the active material allow the generation of very short pulses as a consequence of the uncertainty relation, as described in Sect. 2.1.2, down to the fs range as explained in Sect. 6.10.3.

Lasers with very large bandwidths will have very short coherence lengths and are well suited for coherence radar measurements or for optical tomography.

With specially designed resonators the spectral bandwidth of the laser radiation can be decreased to much below the bandwidth of the active material. The values are in the range of Hz (see Sect. 6.7.5). With relatively simple arrangements using etalons bandwidths of a few 100 MHz can be achieved.

6.7.4 Methods for Decreasing the Spectral Bandwidth of the Laser

The laser bandwidth can be decreased by introducing losses $V(\lambda)$ with a narrow spectral transmission width $\Delta\lambda_{filter}$ in the laser resonator [6.492–6.540]. Because of the nonlinearity of the amplification process the spectral filtering is much more effective inside the resonator compared to external filtering.

Thus all kinds of spectrally sensitive optical elements such as prisms, gratings, etalons, color filters and dielectric mirrors can be used for decreasing the bandwidth of the laser radiation. As an example a resonator using a grating for decreasing the bandwidth is shown in Fig. 6.54.

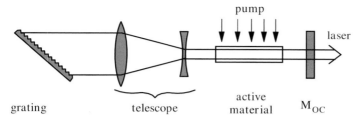

Fig. 6.54. Laser resonator with decreased bandwidth using a grating in Littrow mounting

The telescope inside the resonator is used to enlarge the illuminated area at the grating for both increasing the selectivity by using more lines and to avoid damage in the case of pulsed lasers with high peak intensities. The grating can also be realized as a gain grating such as in distributed feedback (DFB) lasers (see, for example, [6.512–6.515] and the references in Sects. 6.7.5 and 6.10.4).

An even narrower bandwidth can be reached by combining several spectral filters as shown in Fig. 6.55.

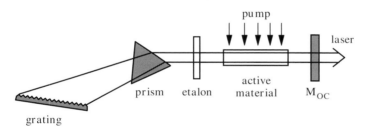

Fig. 6.55. Narrow spectral bandwidth resonator using a combination of prism, grating and etalon

The prism is used to enlarge the number of grating linewidths and in addition the grating is applied at grazing incidence. Further spectral filtering is obtained from the etalon. The wavelength of the laser can be tuned [6.516–6.540] by turning the grating. In some cases the fine tuning can be achieved by changing the air pressure inside the etalon and thus tuning the optical path length of the Fabry–Perot interferometer.

For very narrow laser linewidths etalons with a finesse above 30 000 are applied in the resonator. Often the combination of two or more etalons can be necessary to combine an effective free spectral range with a small effective bandwidth of the etalons. In these lasers with very narrow bandwidth care has to be taken over spatial hole burning in the active material as described in the next chapter.

Further care has to be taken so that the high intensities in the resonator (resonance effect) do not damage the spectral filtering devices especially in pulsed lasers.

6.7.5 Single Mode Laser

As shown in Table 6.11 the bandwidth of the active materials is usually much larger than would be necessary for the safe operation of the laser in just one single longitudinal mode. Therefore the filtering inside the resonator has to be very narrow and is usually achieved using etalons [6.541–6-656].

If the laser operates in a single longitudinal mode in a conventional two-mirror resonator the standing light wave can cause *spatial hole burning*. At the intensity maxima of the standing wave the inversion of the active material is used up for the laser process whereas the inversion in the knots would be available. Thus the gain for the neighboring longitudinal modes can be higher than for the active mode. Mode hopping can occur and the laser operation is no longer stable.

Therefore several schemes have been developed to avoid spatial hole burning in the active material when the laser is operating in a single longitudinal mode. In Fig. 6.56 a ring resonator suspending spatial hole burning is shown.

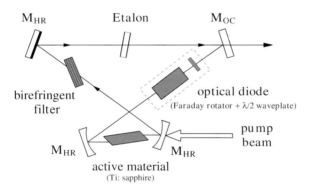

Fig. 6.56. Ring resonator for single longitudinal mode operation. The optical diode ensures the light travels in one direction, only

Spectral filtering is achieved by the etalon, which can consist of several etalons with different free spectral ranges. The optical diode guarantees that the light travels in one direction, only. Thus the laser has no standing wave and spatial hole burning cannot occur.

The very elegant and reliable concept of a ring laser emitting a longitudinal mono-mode was based on a compact laser crystal as shown in Fig. 6.57 [6.623–6.629].

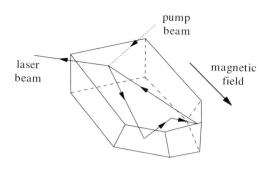

Fig. 6.57. Ring laser longitudinal mono-mode operation based on a polished laser crystal (MISER)

This laser crystal is small and thus the mode spacing is quite large. The crystal geometry can act as the mode selector. The laser is pumped by another laser beam, e.g. a diode laser, and operates very stably in the power range of a few 10 mW to 2 W. The circulation direction is determined by the magnetic field, which guarantees the alignment for one direction, only.

Another scheme for avoiding spatial hole burning is shown in Fig. 6.58.

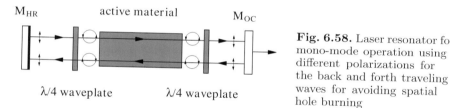

Fig. 6.58. Laser resonator for mono-mode operation using different polarizations for the back and forth traveling waves for avoiding spatial hole burning

In this resonator the back and forth travelling waves have different polarizations. Thus no intensity grating can occur and thus spatial hole burning is avoided. The elements for spectral narrowing of the laser emission are not shown in this picture.

A very simple ring resonator for avoiding spatial hole burning is shown in Fig. 6.59. This resonator is similar to the scheme of Fig. 6.56 but the propagation direction is determined by the mirror M_{AUX} in a very simple way. Again the elements for mode selection are not shown.

In all schemes the pump conditions have to be carefully controlled. The laser has to be operated not too far above threshold for achieving good mode selectivity (remember Fig. 6.52).

But stable operation in a certain single longitudinal mode with a *fixed wavelength* [6.633–6.656] usually demands further active components (piezo-

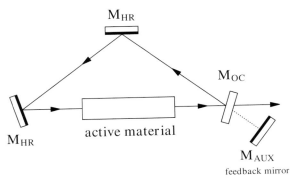

Fig. 6.59. Simple ring resonator for mono-mode operation. The propagation direction is initiated by the mirror M_{AUX}

driven devices) for the compensation of thermally induced changes of the optical lengths and other effects. Atomic transitions from spectral lamps are often used as a reference frequency normal for the laser. Stability in the range of Hz is then possible.

6.7.6 Longitudinal Modes of Resonators with an SBS Mirror

The phase conjugating mirror, like any other nonlinear element in the resonator, can cause complicated longitudinal mode structures [6.657–6.678] which may vary in time. For example, the reflectivity zone of these reflectors can move and thus the resonator length is no longer constant.

Phase conjugating mirrors based on stimulated Brillouin scattering (SBS) [6.657–6.667] shift the frequency of the light towards longer wavelengths at each reflection. This shift is equal to the Brillouin frequency of the SBS material which is in the range of 100 MHz to 50 GHz.

Thus in such lasers a whole spectrum of longitudinal modes is generated [e.g. 6.657]. For stable operation the resonator length has to be chosen carefully. Furthermore Q-switching (see Sect. 6.10.2) and modulation of the temporal pulse shape is obtained, as the intensity signal shows in Fig. 6.60.

The Brillouin shift was tuned to the roundtrip time of the resonator for resonance enhancement. The diagram of Fig. 6.60 shows the successive shift of the frequencies of the longitudinal modes at each roundtrip as measured time-dependently from the etalon picture.

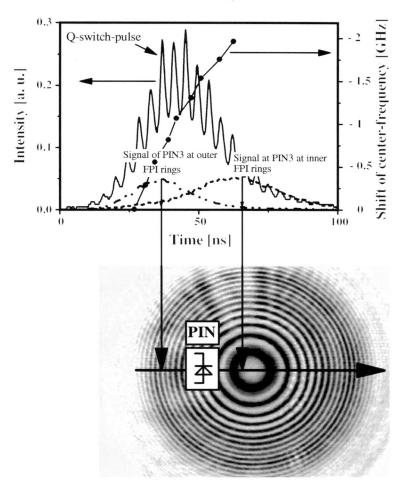

Fig. 6.60. Temporal and spectral properties of a laser with a phase conjugating mirror based on stimulated Brillouin scattering in SF_6 gas showing a Brillouin shift of about 250 MHz [6.697]

6.8 Threshold, Gain and Power of Laser Beams

Laser operation demands light amplification based on inversion in the active material. This amplification is described by the gain factor G with the gain coefficient g. It has at least to compensate the losses of the resonator – the laser has to reach its threshold. If the laser operates above threshold then temporal oscillations in the output light may occur, and spiking is observed. Short pulses with high intensities can be produced with nonlinear elements in the resonator. Thus Q switching leads to ns or ps pulses and by mode locking ps or fs pulses can be produced which will be described in Sect. 6.10.

6.8.1 Gain from the Active Material: Parameters

Light amplification is the inverse process of light absorption but with the additional effect of "cloned" photons. Thus the theoretical description is analogous to the description of nonlinear absorption as given in Chap. 5 but with gain as negative absorption [6.679–6.697].

The small-signal amplification follows directly from the inversion in the active material. The gain G_{ls} is given by:

$$\textbf{small-signal gain} \quad G_{ls} = e^{g_{ls}L_{mat}} = e^{\sigma_{laser}(N_{upper}-N_{lower})L_{mat}} \quad (6.98)$$

with the gain coefficient g_{ls}, as given in (6.1), the cross-section of the laser transition σ_{laser} the population densities of the higher N_{upper} and lower N_{lower} laser level and the length of the active material L_{mat}. This equation holds as long as the intensity is small enough not to disturb the population densities.

If the intensity increases, as required in lasers, the population densities of the different levels numbered with m, $N_m = f(I_{laser})$ become functions of the intensity I_{laser}, itself, and the inversion $\Delta N = (N_{upper} - N_{lower})$ is decreased.

$$\text{high-signal gain} \quad G_{hs} = e^{g_{hs}(I_{laser})L_{mat}}$$
$$= e^{\sigma_{laser}[N_{upper}(I_{laser})-N_{lower}(I_{laser})]L_{mat}}. \quad (6.99)$$

In this case the gain may be a complicated function of the pump process described by the pump rate W and the material parameters. It should be calculated as given in Chap. 5, whereas in many cases rate equations may be sufficient for the description.

The discussion of the general behavior is often based on the simple approximation for the gain coefficient g_{hs} similar to (5.4):

$$\textbf{gain coefficient with saturation} \quad g_{hs} = g_{ls}\frac{1}{\left(1+\dfrac{I}{I_{nl}}\right)^{h}} \quad (6.100)$$

with the nonlinear intensity I_{nl}, frequently called the saturation intensity. The exponent h is chosen to be 1.0 for homogeneously broadened lasers and 0.5 for inhomogeneously broadened lasers. This saturation intensity I_{nl} for the stimulated emission follows from the product of the cross-section σ_{laser} and the life time of the upper laser level τ_{upper} and measured in photon numbers as I_{nl} by:

$$I_{nl} = \frac{hc_0/\lambda_{laser}}{F_{lev}\sigma_{laser}\tau_{upper}} \qquad I_{nl} = \frac{\lambda_{laser}}{hc_0} \quad (6.101)$$

with Planck's constant h and the velocity of light c_0. The factor F_{lev} is equal $F_{lev,3level} = 2$ for three-level systems and $F_{lev,4level} = 1$ for four-level systems of the active material. For a more detailed discussion see Sect. 5.3.6.

Typical emission cross-sections σ_{laser} of laser materials and the life times of the upper laser level τ_{upper} for the most prominent laser wavelengths λ_{laser} of these materials are given in Table 6.12.

Table 6.12. Emission cross-sections σ_{laser}, and lifetimes of the upper laser level at room temperature τ_{upper} for the most prominent laser wavelengths λ_{laser} of some materials

Material	λ_{laser} (nm)	σ_{laser} (cm^2)	τ_{upper} (ns)
He-Ne	632.8	$3 \cdot 10^{-13}$	15 ± 5
Ar-ion	488.0	$2.5 \cdot 10^{-12}$	9
CO_2	10,600.0	$1 \cdot 10^{-16}$	10 000
KrF	248.0	$3 \cdot 10^{-16}$	5
XeCl	308.0	$3 \cdot 10^{-16}$	10
Ruby	694.3	$2.5 \cdot 10^{-20}$	3000
Nd:YAG	1064.1	$3.2 \cdot 10^{-19}$	230 000
Nd:YALO	1079.5	$2 \cdot 10^{-19}$	100 000
Nd:YLF	1047.0	$1.8 \cdot 10^{-19}$	480 000
Nd:KGW	1067.0	$3.3 \cdot 10^{-19}$	120 000
Nd:glass	1054.0	$4 \cdot 10^{-20}$	300 000
Alexandrite	e.g. 760.0	$5 \cdot 10^{-20}$	260 000
Ti:Sapphire	e.g. 800.0	$3 \cdot 10^{-19}$	3200
Rhodamin 6G	e.g. 580.0	$4 \cdot 10^{-16}$	5
GaAs	e.g. 800.0	$1 \cdot 10^{-16}$	4

The lifetime of the upper laser level is identical to the fluorescence lifetime of the laser material if no stimulated emission occurs, which will otherwise shorten it.

Using the concentration of laser atoms, ions or molecules as an order of magnitude for the inversion population density the maximum gain can be estimated. Solid-state laser can be doped by some atom% which is in the range of 10^{19} cm^{-3} and thus the gain coefficient will be in the range of 0.01–0.1 cm^{-1} and the maximum stored energy can reach J cm^{-3}. In dye lasers the gain coefficient can reach values above 1 cm^{-1}.

6.8.2 Laser Threshold

As mentioned above the laser is operating if the gain from the active material compensates all losses [6.698–6.703] in the resonator which for linear resonators results in:

laser condition $G^2 V^2 R_1 R_2 = 1$ $\hspace{2cm}$ (6.102)

with the gain in the active material G, the influence of all losses such as e.g. absorption, scattering, and diffraction in one pass through the resonator V, and the reflectivities of the two mirrors R_1 and R_2.

Because the gain of the active material is determined by the balance of the pumping on one hand and its decrease by the back and forth traveling laser light inside the resonator $I_{laser,int}$ on the other, from the laser condition of (6.102) the laser intensity for a given resonator can be calculated as follows.

By applying the simple approximation of (6.100) the gain G is given by:

gain $G(I_{\text{laser,int}}) = e^{g_{\text{hs}}L_{\text{mat}}} = \exp\left(\dfrac{g_{\text{ls}}L_{\text{mat}}}{(1 + I_{\text{laser,int}}/I_{\text{nl}})^h}\right).$ (6.103)

This value has to fulfill (6.102) and thus the laser intensity for a resonator with V, R_1 and R_2 is given by:

internal laser intensity

$$I_{\text{laser,int}} = \frac{I_{\text{nl}}}{2}\left\{\left(\frac{g_{\text{ls}}L_{\text{mat}}}{\left|\ln(V\sqrt{R_1 R_2})\right|}\right)^{1/h} - 1\right\}$$ (6.104)

considering the doubling of the intensity inside the active material from the back and forth traveling of the light in the resonator.

The *laser threshold condition* follows from this equation. This is determined by the minimal value of the small-signal gain coefficient $g_{\text{threshold}}$ that satisfies (6.102) for a given resonator:

threshold gain coefficient $g_{\text{threshold}} = \dfrac{1}{L_{\text{mat}}}\left|\ln(V\sqrt{R_1 R_2})\right|$ (6.105)

and thus the necessary inversion $\Delta N_{\text{threshold}}$ follows from:

threshold inversion $\Delta N_{\text{threshold}} = \dfrac{g_{\text{threshold}}}{\sigma_{\text{laser}}} = \dfrac{\left|\ln(V\sqrt{R_1 R_2})\right|}{L_{\text{mat}}\sigma_{\text{laser}}}$(6.106)

which determines the minimal pump rate $W_{\text{pump,threshold}}$ to reach laser operation for a given resonator. The pump rate is the density of the inverted laser particles per unit time, which will be discussed in the next chapter in more detail. At threshold the intensity of the laser is still zero. Thus in practice the threshold is determined by measuring the laser intensity as a function of the pump rate W_{pump} as sketched in Fig. 6.61.

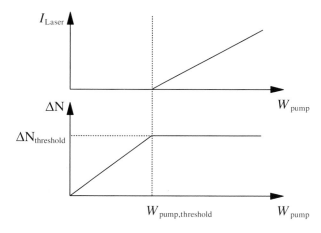

Fig. 6.61. Laser intensity and inversion as a function of the pump rate W_{pump} around the laser threshold

As an example a laser resonator with $V = 0.95$, $R_1 = 1$, $R_2 = 0.8$ and a length of the active material of $L = 0.1$ m demands a minimal gain coefficient of $g_{threshold} = 0.016$ cm^{-1}. The threshold inversion population density for Nd:YAG would be $\Delta N_{threshold} = 4.1 \cdot 10^{16}$ cm^{-3}.

The minimal pump rate for reaching threshold, which is the *threshold pump rate* $W_{pump,threshold}$ is a very useful measure for characterizing lasers experimentally. It is easy to measure and allows the characterization of how much the laser is pumped above threshold. Thus the output parameters can be related to this value. On the other hand the threshold pump power can be calculated via rate equations to optimize the parameters of the active material as will be shown in the next chapter.

6.8.3 Laser Intensity and Power

The laser with its linear and nonlinear interactions of light with matter, especially at the resonator mirrors and in the active material, can be described on different levels as discussed in Chap. 5. A density matrix formalism or wave equations may be necessary if detailed analysis of the temporal longitudinal mode structure or the photon statistics are important.

For the modeling of the output power characteristics and the temporal development of the laser pulses in the spiking or Q switched operation rate equations are usually sufficient [e.g. 6.704–6.710].

The photon transport equation for the intensity $I = I/h\nu_{Laser}$ measured in photons/cm^{2}s has to include all losses and the stimulated and spontaneous emission of the light in the active material, and is given by:

$$\frac{\partial I}{\partial z} + \frac{1}{c}\frac{\partial I}{\partial t} = [\sigma_{laser}\Delta N(I) - a]I \tag{6.107}$$

with the cross-section of the laser transition σ_{laser}, the population inversion density ΔN and possible losses in the active material described with the (absorption) coefficient a. This equation can easily be integrated, numerically. The spontaneously emitted photons are neglected in this expression because they add very low intensity in the direction of the laser light, but some starting intensity is necessary and has to be considered numerically.

To model the whole resonator, in addition the losses at the resonator mirrors and at other elements and the optical paths outside the active material also have to be taken into account during the roundtrip. The population densities as a function of the pump rate can be calculated as described in Sect. 5.3.

In addition the polarization of the light and the wavelength dependence of the intensity can be taken into account by solving this type of equation parallel for several discrete polarization directions or several discrete wavelengths (see e.g. Sect. 6.9.3.3). The coupling will be achieved via the common inversion density in the active material (see next section).

For simple three or four level laser schemes the inversion population density can be calculated using the notation of Fig. 6.62.

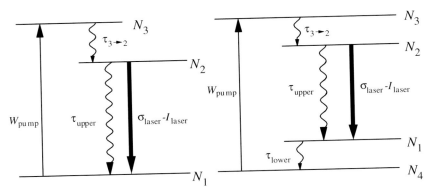

Fig. 6.62. Transitions in three- (left) and four-(right) level schemes of lasers

If the decay of level 3 in the three-level system and of levels 3 and 1 in the four-level system is fast compared to all other transitions the population density of these levels can be neglected. Under this assumption the differential equation for the inversion population densities ΔN are given by:

Three-level system:

$$
\frac{\partial \Delta N}{\partial t} = \left(W_{\text{pump}} - \frac{1}{\tau_{\text{upper}}} \right) N_0
$$
$$
- \left(W_{\text{pump}} + \frac{1}{\tau_{\text{upper}}} + 2\sigma I_{\text{laser}} \right) \Delta N \qquad (6.108)
$$

and

Four-level system:

$$
\frac{\partial \Delta N}{\partial t} = W_{\text{pump}} N_0 - \left(W_{\text{pump}} + \frac{1}{\tau_{\text{upper}}} + \sigma_{\text{laser}} I_{\text{laser}} \right) \Delta N \qquad (6.109)
$$

with the pump rate W_{pump} resulting, e.g. from the product of the pump cross-section σ_{pump} and the pump intensity E_{pump} if light is applied for pumping. N_0 describes the total population density as the sum over all population densities of all levels. These equations usually have to be solved numerically in combination with the photon transport equation, considering the losses at the mirrors and other components.

Analytical solutions are possible under steady state conditions. The cw solutions with $\partial \Delta N / \partial t = 0$ for these equations are:

Three-level system

$$
\Delta N = N_0 \frac{\tau_{\text{upper}} W_{\text{pump}} - 1}{1 + \tau_{\text{upper}} W_{\text{pump}} + 2\sigma_{\text{laser}} \tau_{\text{upper}} I_{\text{laser}}} \qquad (6.110)
$$

indicating that $\tau_{\text{upper}} W_{\text{pump}}$ has to be larger 1 for inversion in the three-level scheme. For the four-level scheme:

Four-level system

$$\Delta N = N_0 \frac{\tau_{\text{upper}} W_{\text{pump}}}{1 + \tau_{\text{upper}} W_{\text{pump}} + \sigma_{\text{laser}} \tau_{\text{upper}} I_{\text{laser}}} \qquad (6.111)$$

where all pumping leads to inversion. These steady state equations are useful for cw operation and lasers with pulse widths much larger than the lifetimes of all excited states of the laser material.

These equations have to be interpreted with $I_{\text{laser}} = 0$ for pump rates W_{pump} smaller than the threshold value $W_{\text{pump}} \leq W_{\text{pump,threshold}}$ and for $W_{\text{pump}} > W_{\text{pump,threshold}}$ as follows:

Below and at threshold

Three-level system $\Delta N(\leq \Delta N_{\text{threshold}}) = N_0 \dfrac{\tau_{\text{upper}} W_{\text{pump}} - 1}{1 + \tau_{\text{upper}} W_{\text{pump}}}$ (6.112)

and

Four-level system $\Delta N(\leq \Delta N_{\text{threshold}}) = N_0 \dfrac{\tau_{\text{upper}} W_{\text{pump}}}{1 + \tau_{\text{upper}} W_{\text{pump}}}$ (6.113)

From these equations the threshold pump rate $W_{\text{pump,threshold}}$ can be calculated for steady state operation.

Pump rate at threshold:

Three-level system $W_{\text{pump,threshold}} = \dfrac{1}{\tau_{\text{upper}}} \dfrac{N_0 + \Delta N_{\text{threshold}}}{N_0 - \Delta N_{\text{threshold}}}$ (6.114)

and

Four-level system $W_{\text{pump,threshold}} = \dfrac{1}{\tau_{\text{upper}}} \dfrac{\Delta N_{\text{threshold}}}{N_0 - \Delta N_{\text{threshold}}}$ (6.115)

for which the threshold population density $\Delta N_{\text{threshold}}$ has to be calculated from the threshold condition of (6.106).

For pump rates above the threshold value the following relations for the cw solution given above can be obtained.

Above threshold:

Three-level system   (6.116)

$$I_{\text{laser}} = \frac{1}{2\sigma_{\text{laser}} \Delta N} \left[\left(W_{\text{pump}} - \frac{1}{\tau_{\text{upper}}} \right) N_0 - \left(W_{\text{pump}} + \frac{1}{\tau_{\text{upper}}} \right) \Delta N \right]$$

and

Four-level system

$$I_{\text{laser}} = \frac{1}{\sigma_{\text{laser}} \Delta N} \left[W_{\text{pump}} N_0 - \left(W_{\text{pump}} + \frac{1}{\tau_{\text{upper}}} \right) \Delta N \right]. \qquad (6.117)$$

Under the assumed steady state conditions the population inversion density above threshold stays constant at:

above threshold

$$\Delta N_{3/4\text{-level}}(W_{\text{pump}} > W_{\text{pump,threshold}}) = \Delta N_{\text{threshold}} \tag{6.118}$$

and thus the laser intensity can be calculated as a function of the pump rate using (6.106) for determining $\Delta N_{\text{threshold}}$:

Above threshold:

Three-level system

$$I_{\text{laser}} = \frac{1}{\sigma_{\text{laser}} \tau_{\text{upper}}} \left(\frac{W_{\text{pump}} - W_{\text{pump,threshold}}}{W_{\text{pump,threshold}} - \dfrac{1}{\tau_{\text{upper}}}} \right) \tag{6.119}$$

and

Four-level system

$$I_{\text{laser}} = \frac{1}{\sigma_{\text{laser}} \tau_{\text{upper}}} \left(\frac{W_{\text{pump}} - W_{\text{pump,threshold}}}{W_{\text{pump,threshold}}} \right) \tag{6.120}$$

If the threshold pump rate and the threshold inversion are substituted in these formulas and the total loss factor $V = V_{\text{resonator}} \cdot \exp(-aL_{\text{mat}})$ is applied the following expressions are obtained:

Three-level system

$$I_{\text{laser}} = \tag{6.121}$$
$$\frac{N_0 \sigma_{\text{laser}} L_{\text{mat}} (\tau_{\text{upper}} W_{\text{pump}} - 1) - (\tau_{\text{upper}} W_{\text{pump}} + 1)|\ln(V\sqrt{R_{\text{HR}} R_{\text{OC}}})|}{2\sigma_{\text{laser}} \tau_{\text{upper}} |\ln(V\sqrt{R_{\text{HR}} R_{\text{OC}}})|}$$

and

Four-level system

$$I_{\text{laser}} = \tag{6.122}$$
$$\frac{N_0 \sigma_{\text{laser}} L_{\text{mat}} \tau_{\text{upper}} W_{\text{pump}} - (\tau_{\text{upper}} W_{\text{pump}} + 1)|\ln(V\sqrt{R_{\text{HR}} R_{\text{OC}}})|}{\sigma_{\text{laser}} \tau_{\text{upper}} |\ln(V\sqrt{R_{\text{HR}} R_{\text{OC}}})|}$$

If as a further approximation $I(L) = \text{const} = I_{\text{av}}$ is applied, assuming that the light beams in the z and $-z$ directions superimpose to an almost equal intensity distribution $2I_{\text{laser}}$ along z, from these intensities in the active material the output power $P_{\text{laser,output}}$ can be calculated. If the reflectivity of the output coupler is close to 1 the intensity in the laser material will be twice as high as the share which is propagating towards the output coupler. For lower reflectivity of the two mirrors the geometrical average may be a suitable approximation and therefore the output power is given by:

$$P_{\text{laser,output}} = I_{\text{laser}} \frac{hc_0}{\lambda_{\text{Laser}}} A_{\text{mode}} (1 - R_{\text{OC}}) \frac{V\sqrt{R_{\text{HR}} R_{\text{OC}}}}{2}. \tag{6.123}$$

Using these equations the output power can be plotted as a function of the reflectivity of the output coupler as shown in Fig. 6.63.

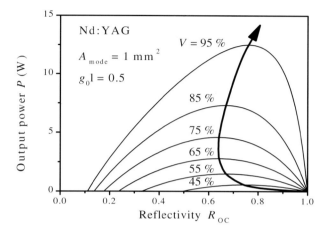

Fig. 6.63. Calculated laser output power of a Nd:YAG laser as a function of the reflectivity of the output coupler R_{OC} for different loss factors V with a pump rate resulting in $g_{ls}L_{mat} = 0.5$. The maxima of these curves (*arrow*) indicate the optimal reflectivity of the output coupler

As can be seen from this figure for each loss factor the optimal reflectivity of the output coupler leads to the maximum output power of the laser. To avoid damage it should be noticed that the internal intensity is higher by a factor of $1/(1 - R_{OC})$ than the out-coupled intensity.

For more precise modeling of the laser the calculations should be executed numerically without the used rough approximations. Nevertheless, based on this fundamental context the detailed analysis can also be carried out experimentally.

6.9 Spectral Linewidth and Position of Laser Emission

The spectral properties of the laser radiation are in general a function of the spectral resonator parameters such as the spectral reflectivity of the mirrors and the spectral curve of the spontaneous and stimulated emission of the active material. In pulsed systems this will change as a function of time. Thus both the central or peak wavelength as well as the bandwidth of the laser radiation may vary in time.

Spectral and spatial hole burning may influence the final spectra and thus homogeneous or inhomogeneous broadening of the active material may be important for understanding. Thus the longitudinal laser modes may oscillate almost independently or may be coupled via the depopulation of the inversion in the active material.

A *homogeneously broadened laser transition* will show a Lorentzian shape gain profile $g_{ls}(\nu_{laser})$ which is proportional to the spectral curve of the emission cross-section $\sigma_{laser}(\nu_{laser})$:

spectral small signal gain

$$g_{ls}(\nu) = \Delta N \sigma(\nu) = \Delta N \frac{\left(\dfrac{\Delta\nu_{FWHM}}{2}\right)^2}{\left(\dfrac{\Delta\nu_{FWHM}}{2}\right)^2 + (\nu - \nu_{max})^2} \sigma_{max}. \tag{6.124}$$

Potentially laser emission can occur over the whole part of the spectrum with low signal gain above threshold as discussed in Fig. 6.52, but because of the mode competition for inversion population density the laser will operate in the cw or long-pulse regime in a narrower range around the frequency of the gain maximum, only.

6.9.1 Minimal Spectral Bandwidth

The linewidth of a cw (continuously) laser can be very small as a result of the infinite lifetime of the radiation. The lower limit of the spectral bandwidth Δ_{min} as a result of the photon statistics was given as derived by Schalow and Townes:

minimal spectral bandwidth

$$\Delta\nu_{min} = \frac{c_0^2 h \nu_{laser} \sigma_{max} N_{upper} L_{mat} |\ln(V\sqrt{R_1 R_2})|}{4\pi P_{out} L_{res,opt}^2} \tag{6.125}$$

with the velocity of light c_0, the population density of the upper laser level N_{upper}, the length of the active material L_{mat}, the optical length of the resonator $L_{res,opt}$, the losses per transit V, the reflectivities of the two resonator mirrors $R_{1/2}$ and the output power of the laser P_{out}. Using this formula spectral bandwidths of less than $1\,\text{Hz}$ down to $10^{-4}\,\text{Hz}$ can be calculated. Realistically all values below $1\,\text{Hz}$ are very difficult to reach and the best values are in the mHz range [6.591]. Several experimental precautions have to be realized for very narrow spectral laser emission (see Sect. 6.7.5).

6.9.2 Frequency Pulling

As a consequence of the spectral bell shape of the gain profile as shown in Fig. 6.64 a slight frequency shift of the resonator modes occurs.

Only at the frequency of the gain maximum is the derivative of the gain $\partial g_{ls}/\partial\nu$ zero. At the other frequencies the gain is slightly stronger at the center-side of the resonator modes and thus the modes are pulled towards this center.

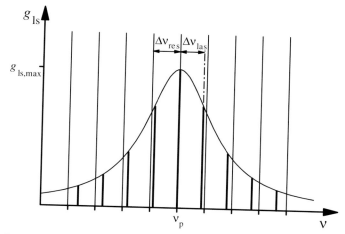

Fig. 6.64. Frequency pulling of longitudinal resonator modes towards the center of the gain curve (schematic)

In [M16] an approximate formula was given for the shift $\Delta\nu_{\text{pull}}$ of the qth resonator mode with original frequency ν_q as:

$$\text{frequency pulling}\quad \Delta\nu_{\text{pull}} = \nu_q \frac{g_{\text{ls}} L_{\text{mat}}}{2\pi} \frac{p-q}{m} \frac{1}{1+[2(p-q)/m]^2} \qquad (6.126)$$

assuming that the pth resonator mode matches the gain maximum and containing the low-signal gain coefficient g_{ls}, the length of the active material L_{mat}, the mode numbers p, q and the number of resonator modes within the gain FWHM bandwidth m.

The difference of the mode spacing with pulling compared to the original resonator modes is of the order of 10^{-4}.

6.9.3 Broad Band Laser Emission

Active materials with *inhomogeneously broadened laser transitions* can emit in a wide spectral range because of the lack of any mode competition for inversion population density. For homogeneously broadened laser material special care has to be taken to get broadband laser emission, as will be described below (see, for example, [6.711–6.718]).

Broad-Band Emission from Inhomogeneously Broadening

Inhomogeneously broadened spectra in the time scale of the laser emission (see Sect. 5.2) consist of homogeneously broadened subbands with different center wavelengths and possibly different emission cross-sections. The overall gain profile usually shows a Gaussian distribution. Thus it does not matter if the peak cross-sections $\sigma_{peak}(\nu_p)$ of the subbands show a Gaussian distribution as a function of their frequency ν_p or the associated inversion population density $\Delta N(\nu_p)$ is modulated this way. The resulting small or low signal gain g_{ls} for the frequency ν_p of the pth longitudinal resonator mode can be written as:

small–signal gain of inhomogeneously broadened active material

$$g_{ls}(\nu_p) = \Delta N_{total} \sigma_{total,max} \frac{\Delta \nu_{homogen}}{\Delta \nu_{inhomogen}}$$

$$\cdot \exp\left(-\frac{(\nu_p - \nu_{total,max})^2 4 \ln 2}{\Delta \nu_{inhomogen}^2}\right) \tag{6.127}$$

with the FWHM bandwidth $\Delta \nu_{homogen}$ of the sub bands which are assumed to be equal. The FWHM bandwidth $\Delta \nu_{inhomogen}$ describes the whole emission band containing the homogeneous subbands. ΔN_{total} is the total inversion population density of the active material independent of their distribution over the subbands and $\sigma_{total,max}$ the maximum cross-section of the whole band which is obtained at frequency $\nu_{total,max}$.

All longitudinal laser modes p with frequency ν_p which show gain above threshold can be observed in the laser:

$$V(\nu_p)\sqrt{R_{HR}(\nu_p)R_{OC}(\nu_p)}\, e^{-g_{ls}(\nu_p)L_{mat}} \geq 1 \tag{6.128}$$

where the frequency dependence of the losses V and the reflectivities of the mirrors $R_{HR/OC}$ have to be considered explicitly. For further evaluation it has to be noticed that the relaxation time $\tau(\nu_p)$ and the cross-section $\sigma(\nu_p)$ and thus the nonlinear (or saturation) intensity $I_{nl}(\nu_p)$ may be a function of the center frequency of the emission of the subbands.

In many cases exchange processes take place between the inhomogeneous subspecies as known, e.g. in molecular systems from spectral hole burning measurements. The exchange rate or internal cross-relaxation has to be much slower than the pump and stimulated emission rate in the laser to conserve the inhomogeneous behavior (see Sect. 5.3).

Broad-Band Emission from Short Pulse Generation

As a consequence of the quantum mechanical uncertainty of energy and time the spectral frequency bandwidth $\Delta \nu_{laser}$ of light pulses will be larger the shorter the pulse duration Δt_{pulse} if the light is bandwidth limited (see Sect. 2.1.2). The minimal FWHM bandwidth is given by:

$$\Delta \nu_{laser,FWHM} = \frac{1}{\pi \Delta t_{pulse}}. \tag{6.129}$$

The resulting frequency bandwidth is $3.2 \cdot 10^{11}$ Hz for a pulse width of $1\,\mathrm{ps}$ and $3.2 \cdot 10^{13}$ Hz for a $10\,\mathrm{fs}$ pulse. The resulting wavelength bandwidths are

0.27 nm and 27 nm, respectively, in the visible range at a center wavelength of 500 nm (see also Table 2.3).

For the generation of such short pulses an active laser material with a sufficiently wide homogeneously broadened gain profile is required (see Sect. 6.10.3).

Broad-Band Emission from Gain Switching

Another possibility of broad-band emission from a homogeneously broadened laser material is the fast generation of a very large inversion, the gain switching [e.g. 6.715–6.717]. Such a laser emits a short pulse without reaching the steady state conditions. Thus the gain of the active material is switched on by fast pumping, typically with ns laser pump pulses. The output pulse usually shows a pulse duration in the ns range, too, with a delay of a few 10 ns.

This nonstationary gain switching laser can be described with time-dependent rate equations. They have to contain the spectral properties of all resonator elements and of the gain.

In the simplest case using a four-level scheme as shown in Fig. 6.62 the differential equation for the inversion population density $\Delta N(t)$ as a function of time t is given by:

$$\frac{\partial \Delta N}{\partial t} = -\frac{\Delta N}{\tau_{\text{upper}}} - \sum_{\lambda_m} \Delta N \sigma_{\text{laser}}(\lambda_m) I(\lambda_m) + W(N_{\text{total}} - \Delta N) \quad (6.130)$$

with the approximations of the fast decay of levels 2 and 1. The intensity I is measured in photons $\text{cm}^{-2}\,\text{s}^{-1}$ and the wavelength λ_m was used as a discrete quantity with respect to the numerical calculation. τ_{upper} denotes the lifetime of the upper laser level and σ_{laser} is the emission cross-section of the active material. N_{total} is the total population density of laser active particles in the active material and is measured in lasing particles per cm^3.

The photon transport equation for the intensity I as a function of time and wavelength including the resonator conditions can be written as:

$$\frac{L_{\text{res}}}{c_0} \frac{\partial I}{\partial t} = I(t, \lambda_m) \left\{ \Delta N \sigma_{\text{laser}}(\lambda_m) L_{\text{mat}} - \ln\left(\frac{1}{V(\lambda_m)\sqrt{R_{\text{OC}}(\lambda_m)}}\right) \right\}$$
$$+ I_{\text{sp}}(t, \lambda_m) \quad (6.131)$$

assuming a homogeneous intensity along the active material. L_{res} is the optical length of the resonator and L_{mat} the geometrical length of the pumped active material. V describes the losses in the resonator and R_{oc} is the reflectivity of the output coupler assuming a reflectivity of the high-reflecting mirror equal to the reciprocal of the investigated spectral range. I_{sp} gives the spontaneous emission of the laser in the direction of the laser beam. Its value usually does not influence the numerical results because of the high amplification but it is necessary to start the stimulated emission at time $t = 0$. The fluorescence spectrum of the active material can be taken as its spectral distribution. In these equations the spatial hole burning is neglected and a large number of resonator modes within the gain bandwidth is presupposed.

As an example the spectrum of a gain switched Ti:sapphire laser is calculated. The emission cross-section of Ti:sapphire is shown in Fig. 6.65.

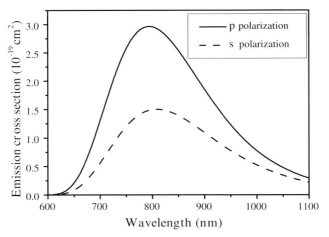

Fig. 6.65. Emission cross-section of Ti:sapphire of π and σ polarization of the light as a function of the wavelength

The reflectivity of the used commercial laser mirrors with a specified reflectivity of 0.8 as output coupler is given in Fig. 6.66.

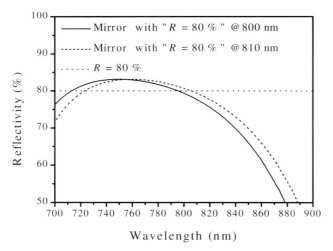

Fig. 6.66. Reflectivity of two commercial mirrors specified with a reflectivity of 0.8 at 800 nm and at 810 nm which are used as output coupler for the Ti:sapphire laser

The calculation of the emission spectra are obtained for a total population density of $3.3 \cdot 10^{19}$ cm^{-3} which is equivalent to a concentration of 0.095 wt%. $V = 0.97$ follows from the reabsorption in the laser material. With these values the minimum gain factor is $G_{\text{threshold}} = 1.15$. For nonstationary calculations a gain of 1.55 at 794 nm resulting in a starting inversion population density of $6 \cdot 10^{17}$ cm^{-3} was used in a 20 mm long crystal positioned in a 0.2 m long resonator. This laser was pumped with 1.23 mJ, 17 ns pulse with a wavelength of 532 nm and a focus radius of 250 µm. With these values the delay time between pumping and the laser pulse is 50 ns.

This laser emits a ns laser pulse of 12 ns duration and the spectrum is more than 20 nm broad. The calculated results are given in Fig. 6.67.

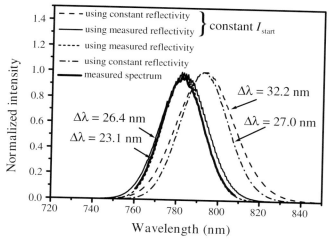

Fig. 6.67. Calculated and measured emission spectra of a Ti:sapphire laser using an artificial output mirror with a constant reflectivity of 80% or the commercial mirror I of Fig. 6.66 as output coupler while applying a spectrally constant start intensity or the spectral profile of the fluorescence

The influence of the start intensity profile as well as the reflectivity of the mirror can be seen from this figure. The FWHM bandwidth of the laser is decreased by 3 nm to 23 nm from the effect of the spectral profile of the realistic mirror compared to the flat profile mirror. Using a spectrally constant start intensity instead of the fluorescence profile the spectrum would be 9 nm wider. The theoretical results including both the spectral profile of the mirror and of the fluorescence were observed experimentally.

6.10 Intensity Modulation and Short Pulse Generation

The active material in the laser resonator shows a highly nonlinear behavior from the exponential dependence of the gain G as a function of the inversion density ΔN which is a function of the laser intensity $I_{\mathrm{laser}}(t)$ and the pump rate $W(t)$. Thus all types of temporal fluctuations are possible, pulses with durations in the µs, ns, ps and fs range can be produced with different methods applying further nonlinear devices in the resonator.

6.10.1 Spiking Operation: Intensity Fluctuations

In particular, if the pump rate is switched on, such as e.g. when flash lamp or electrical discharge pulse pumping some oscillations in the laser output can be observed [e.g. 6.719–6.722]. The modeling can usually be carried out using rate equations as given in the previous sections.

For a simplified description it is assumed that pumping with rate W is switched on at time $t = 0$ and stays constant for $t > 0$. The resulting behavior is shown in Fig. 6.68.

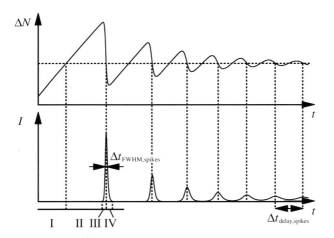

Fig. 6.68. Spiking oscillations of a laser in the laser intensity I and the population inversion density ΔN as a function of time

The following processes take place:

I pumping leads to an increase of the inversion population density (stimulated emission can be neglected);

II threshold inversion is overcome and stimulated emission takes place during the roundtrips;

III inversion population density and photon density are very high and thus the depopulation via the laser light starts up to the maximum value of the laser intensity;

IV inversion population density is strongly decreased and thus the intensity decreases to very small values with further depopulation;

V almost no laser light is present and thus the pumping increases the inversion again for the next cycle.

Based on this simple model a series of laser pulses, the *spiking*, is obtained until steady state operation is reached. If the laser is perturbed then decaying intensity oscillations will, again, occur.

The pulse widths of the spikes $\Delta t_{\text{FWHM,spike}}$ are a function of the pump rate, the gain and the resonator roundtrip time. In solid-state lasers the order of magnitude is 100 ns and in dye lasers it can be as small as sub-ps. The pulse-to-pulse delay $\Delta t_{\text{delay,spikes}}$ is one or two orders of magnitude larger than $\Delta t_{\text{FWHM,spike}}$.

As a function of the pump rate these oscillations decay towards the value related to the steady-state inversion population density. Assuming small perturbations in the inversion population density and the intensity around the steady state values the solution of the differential equation for the intensity I was given in [M38] as:

$$I_{\text{laser,spiking}} = I_{\text{laser,spiking}}(t = 0)$$

$$\cdot \exp\left(-\frac{W_{\text{pump}}}{W_{\text{pump,threshold}}} \frac{t}{\tau_{\text{upper}}} \sin(2\pi\nu_{\text{spiking}}t)\right) \qquad (6.132)$$

with the pump power W_{pump}, the threshold pump power $W_{\text{pump,threshold}}$, the fluorescence lifetime of the laser transition which is equal to the lifetime of the upper laser level τ_{upper}, and the frequency of the spiking oscillations ν_{spiking}. The initial amplitude of these oscillations $I_{\text{laser,spiking}}$ can be much higher than the laser intensity in steady state operation as can be seen from Fig. 6.68.

The frequency of the relaxation oscillations of spiking, ν_{spiking}, is determined by the fluorescence lifetime $\tau_{\text{fluore}} = \tau_{\text{upper}}$, the resonator parameters described by the resonator lifetime τ_{res} as given in (6.95) and the pump rate W_{pump}:

$$\nu_{\text{spiking}} = \frac{1}{2\pi}\sqrt{\left(\frac{W_{\text{pump}}}{W_{\text{pump,threshold}}} - 1\right)\frac{1}{\tau_{\text{upper}}\tau_{\text{res}}} - \left(\frac{W_{\text{pump}}}{2W_{\text{pump,threshold}}\tau_{\text{upper}}}\right)^2}$$

$$\simeq \frac{1}{2\pi}\sqrt{\left(\frac{W_{\text{pump}}}{W_{\text{pump,threshold}}} - 1\right)\frac{1}{\tau_{\text{upper}}\tau_{\text{res}}}} \qquad (6.133)$$

Because higher losses in the resonator result in a shorter photon lifetime in the resonator τ_{res} the spiking oscillations will occur faster. Notice that higher losses may demand higher pump rates. Stronger pumping and a shorter fluorescence lifetime will also increase the spiking frequency.

The 1/e damping time $\tau_{\text{damping,spikes}}$ of these oscillations can be given as:

$$\tau_{\text{damping,spikes}} = \frac{W_{\text{pump,threshold}}}{W_{\text{pump}}} \tau_{\text{upper}} \tag{6.134}$$

and thus stronger pumping shows faster decay.

As an example a Nd:YAG laser with fluorescence lifetime of 230 µs, an output mirror with 50% reflectivity, losses of $V = 0.95$ and a resonator length of 0.5 m which is pumped three times above threshold shows a spiking frequency of 230 kHz or a period of $1/\nu_{\text{spiking}}$ of 4.3 µs. The damping time of the oscillations is 77 µs for this laser. For comparison the resonator lifetime is 4.2 ns in this case. An example from a realistic laser showing this regular pattern is given in Fig. 6.69.

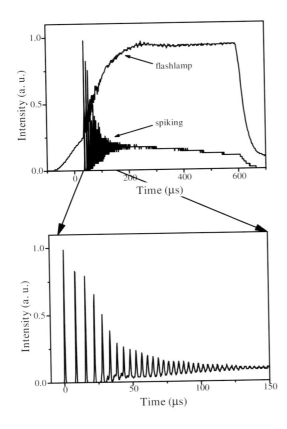

Fig. 6.69. Spiking operation of a flash lamp pumped Nd:YAG laser. In addition to the laser output the intensity of the pumping flash lamp is given

Sometimes lasers can show deviations from this regular pattern. The resulting irregular fluctuations are a consequence of the competition of longitudinal modes in the resonator. Because of the nonlinearity of the laser process this pattern can even show chaotic behavior. If a regular oscillation pattern is required the laser has to be operated in a single longitudinal mode.

In cw lasers these fluctuations can be suppressed by active stabilization of the operation by electronic controlling of the pump rate.

The fundamental limits for these fluctuations are the spontaneous emitted photons from the active material. The total noise power P_{noise} of the whole active material is given by:

$$P_{noise} = h\nu_{laser} \frac{\Delta N A_{mode} L_{mat}}{\tau_{upper}} \tag{6.135}$$

with the laser frequency ν_{laser}, the cross-section of the laser mode in the active material A_{mode}, the length of the active material L_{mat}, the inversion population density ΔN and the lifetime of the upper level.

6.10.2 Q Switching (Generation of ns Pulses)

Q switching means the quality of the laser resonator is switched from low to high values typically in short times and thus short pulses with ns pulse width and high-powers in the kW–MW range are emitted [6.723–6.809]. These intensive pulses are useful in all kinds of nonlinear optical interactions. In many cases at least Q switched lasers (or sometimes ps or fs lasers) are necessary to achieve nonlinear optical effects.

The Q switching of the laser is achieved by rapidly decreasing of artificial losses in the resonator typically with a slope in the ns range. The stored high inversion in the active material leads to a high gain for the laser radiation and thus the positive feedback of increased intensity leads to more stimulated emission which leads to more intensity. This results in a very fast rising intensity of the laser until the inversion reaches its threshold value. Then the intensity again rapidly decreases as a result of the losses from the out-coupling.

Active Q Switching and Cavity Dumping

Active Q switching can be achieved, e.g. with electro-optical devices as shown in Fig. 6.70. All electro-optical effects can be used [6.723–6.752]. Using the *Pockels effect* (see Sect. 4.4.5) the polarization of the light can be changed. In combination with polarizers the transmission of this arrangement can be varied over a wide range of more than 1:50 in ns times. Using avalanche transistor chains high voltages up to a few kV can be switched in one or a few ns.

This scheme can be extended by separating the two orthogonal polarization directions and rotating one of them by 90° before the Q switch to achieve a polarization insensitive arrangement.

Kerr cells are used for Q switching and different lasers have been synchronized by optically triggering these cell from a third laser. Today electro-optical switches allow the triggering of the Q switch laser with an accuracy better 5 ns.

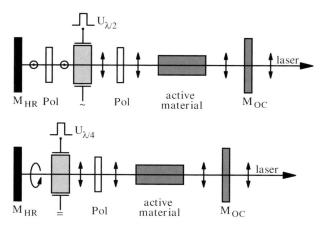

Fig. 6.70. Q switching of a laser resonator using the electro-optical Pockels effect and polarizers

For wavelengths in the IR (e.g. a Er laser) active Q switching is based on frustrated total internal reflection (*FTIR switch*) [6.724] at a very narrow slit with a thickness of some tenths of the wavelength between two glass plates. This slit width is externally decreased with a piezo-element to a width below a tenth of the wavelength. Thus the transmission is increased from a few percent to values above 50%. Another possibility is given in [6.723].

Even mechanical elements can be used for active Q switching. Rotating mirrors will be aligned only for a short time and thus allow the emission of the short pulse if they spin with about 10 000 revolutions per minute [e.g. 6.753, 6.754].

The quality of the Q-switched laser light can be improved by a prelasing configuration [e.g. 6.752]. Therefore, the losses in the cavity are tuned in the time before the Q switch pulse rises just before reaching the threshold, and, thus, almost no light will be emitted. However, during this time of typically some 10–100 μs the mode structure is formed. In the best cases a longitudinal single mode and TEM_{00} operation can be obtained. The losses can be tuned by slightly turning the polarizing elements out of the optimal position. However, very stable operation of the pump and cooling is required.

Cavity dumping works inversely to Q switching described so far. The losses of the resonator are switched on for only a short time in the ns range by out-coupling light as shown in Fig. 6.71.

Acousto-optic switches or modulators (*AOMs*) deflect the laser light wave at an index grating which is induced by hyper-sound waves in suitable crystals such as e.g. $LiNdO_3$ or quartz. The sound wave is typically induced by a piezo-driver in the MHz range. When the sound wave generator is switched on the laser beam is deflected and the quality of the resonator drops. This method is especially useful for lasers with low gain such as argon or cw dye

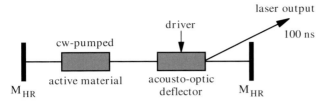

Fig. 6.71. Cavity dumping for ns pulse generation with a cw laser such as e.g. an argon laser using an acousto-optic deflector (AOM) which is switched on for a few ns. The laser light is then reflected at the sound wave grating and out-coupled

lasers. These lasers are continuously pumped and the acousto-optic switch is operated continuously, too. The laser output will consist of a periodical series of output pulses. The frequency of the acousto-optical modulator will occur as a side-band frequency in the laser light.

These pulses usually have a longer pulse duration in the range of 100 ns because the inversion is smaller in cw pumping. Repetition rates in the kHz range are possible and the peak power is about 100 times higher than in cw operation of the laser with an optimal out-coupling mirror.

Passive Q Switching

Applying the nonlinear absorption of a suitable material in the resonator leads to *passive Q switching* [6.755–6.791] as shown in Fig. 6.72.

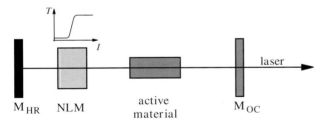

Fig. 6.72. Schematic of a laser with passive Q switching of the resonator using a nonlinear absorber material (NLM)

In this case the transmission of the absorber is progressively bleached with increasing intensity of the laser. Thus the positive feedback is amplified by the effect that more intensity causes more transmittance which results in less losses and thus more intensity. This takes place in addition to the nonlinear effect of stimulated emission in the active material.

Dyes in solution or in polymers or other solid state materials [e.g. 6.780–6.791] are used as materials for passive Q switching. The absorption cross-section of the Q switch material has to be larger at the laser wavelength than the emission cross-section of the active material. The lifetime of the upper level of the Q switch material should not be much shorter than the pulse duration.

Cyanine dyes such as cryptocyanine and phthalocyanine are used in the red region of the visible spectrum e.g. for ruby lasers. They can be solved in alcohol. Special dyes are produced for Nd lasers, such as e.g. the numbers #9740, #9860 and #14,015 by Eastman Kodak. These dyes have to be solved in 1,2-dichlorethene. Ready-to-use polymer foils for Q switching are also offered, e.g. by Kodak. On the other hand many laser materials are useful. In particular, e.g. Cr^{4+}:YAG crystals are used as a Q switch for Nd lasers. Typical low-signal transmissions of 30–70% of the Q switch are used. This additional loss also has to be compensated by the laser gain to overcame threshold. Optical phase conjugation, for example based on stimulated Brillouin scattering, can be used for Q switching, as described in Sect. 6.7.6. In this case the transparent SBS materials are applied.

For matching the laser intensity to the nonlinear transmission curve of the absorber the right position of the Q switch in the resonator has to be chosen. Sometimes telescopes in the resonator for providing suitable beam cross-section for the nonlinear absorber can be necessary.

Theoretical Description of Q Switching

The *theoretical description of Q switching* can be based on rate equations. The duration of the Q switch pulse is much shorter than the pump process. Thus the generation of the inversion need not be included in the rate equations and the spontaneous emission can be neglected. The differential equation for the inversion population density ΔN then reads as:

$$\frac{\partial \Delta N}{\partial t} = -\sigma_{\text{laser}} I \Delta N \tag{6.136}$$

and for the intensity as:

$$\frac{\partial I}{\partial t} = I \left\{ \frac{c_0 \sigma_{\text{laser}} \Delta N L_{\text{mat}}}{L_{\text{res}}} - \frac{c_0}{2L_{\text{res}}} \left| \ln(f_Q(t) V \sqrt{R_{\text{OC}}}) \right| \right\} \tag{6.137}$$

with the cross-section σ_{laser}, the geometrical length of the active material L_{mat}, the optical length of the resonator L_{res}, the resonator losses without the Q switch V, the reflectivity of the output coupler R_{OC} and a function $f_Q(t)$ describing the Q switch. This system of two equation can be solved numerically.

If $f_Q(t)$ is a simple step function with $f_Q(t > t_Q) = 1$ these equations can be solved easily. In Fig. 6.73 the temporal structure of the inversion population density ΔN and the laser intensity I_{laser} are depicted.

The quotient of these two equations results in:

$$\frac{\partial I}{\partial \Delta N} = -\frac{c_0 L_{\text{mat}}}{L_{\text{res}}} - \frac{c_0 \left| \ln(V \sqrt{R_{\text{OC}}}) \right|}{2L_{\text{res}} \sigma_{\text{laser}} I \Delta N} \tag{6.138}$$

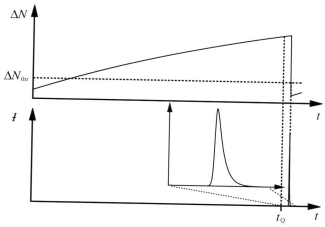

Fig. 6.73. Temporal structure of the inversion population density ΔN and the laser intensity I_{laser} during Q switching the resonator at time t_Q

which can be solved to give

$$I(\Delta N) = I_0 + \left\{ (\Delta N_{\text{start}} - \Delta N) + \Delta N_{\text{threshold}} \ln \frac{\Delta N}{\Delta N_{\text{start}}} \right\} \frac{L_{\text{mat}}}{\tau_{\text{round}}}$$

$$(6.139)$$

with the population inversion density ΔN_{start} at the beginning of the Q switch process.

From this equation the maximum intensity can be calculated under the condition that at the time of the intensity maximum the inversion population density is equal to the threshold value $\Delta N_{\text{threshold}}$ which was given in (6.106). The resulting peak intensity is equal to:

$$I_{\text{laser,max}} = \frac{L_{\text{mat}}}{\tau_{\text{round}}} \left\{ \Delta N_{\text{start}} - \Delta N_{\text{threshold}} \left(1 + \ln \frac{\Delta N_{\text{start}}}{\Delta N_{\text{threshold}}} \right) \right\}$$

$$(6.140)$$

which is higher the shorter the resonator roundtrip time τ_{round} becomes.

The maximum output power $P_{\text{laser,max}}$ of the Q switch pulse is given by:

peak power $P_{\text{laser,max}} = \dfrac{1}{2} I_{\text{laser,max}} h\nu_{\text{laser}} A_{\text{mode}} (1 - R_{\text{OC}})$ (6.141)

with the photon energy $h\nu_{\text{laser}}$ and the cross-section of the transversal mode in the active material A_{mode}.

Further the stored energy E_{pulse} of the Q switch pulse can be calculated from:

pulse energy $E_{\text{pulse}} = \dfrac{1}{2} \dfrac{A_{\text{mode}}}{\sigma_{\text{laser}}} h\nu_{\text{laser}} (1 - R_{\text{OC}}) \dfrac{\Delta N_{\text{start}} - \Delta N_{\text{end}}}{\Delta N_{\text{threshold}}}$

$$(6.142)$$

with the population inversion density N_{end} at the end of the Q switch pulse, which can be calculated from the transcendental equation:

$$\Delta N_{\text{start}} - \Delta N_{\text{end}} + \Delta N_{\text{threshold}} \ln \frac{\Delta N_{\text{end}}}{\Delta N_{\text{start}}} = 0 \qquad (6.143)$$

or estimated in a rough way, e.g. to zero.

Both the extracted pulse energy and the peak power are higher the higher the inversion population density at the beginning of the Q switching ΔN_{start} in relation to the threshold value $\Delta N_{\text{threshold}}$. The pulse width follows from:

pulse width

$$\Delta t_{\text{pulse,FWHM}} = \frac{2L_{\text{res}}(\Delta N_{\text{start}} - \Delta N_{\text{end}})}{c_0 \sigma_{\text{laser}} L_{\text{mat}} \left(\dfrac{\Delta N_{\text{start}}}{\Delta N_{\text{threshold}}} - 1 + \ln \dfrac{\Delta N_{\text{start}}}{\Delta N_{\text{threshold}}} \right)}. \qquad (6.144)$$

It is proportional to the optical length L_{res} of the resonator. The pulse built-up time can be estimated from:

built-up time $$t_{\text{buildup}} = \frac{L_{\text{res}}}{c_0 \ln(G_{\text{start}} RV)} \qquad (6.145)$$

which is again proportional to the resonator roundtrip time τ_{round} which is a function of the optical length L_{res} of the resonator:

resonator roundtrip time $$\tau_{\text{round}} = \frac{2L_{\text{res}}}{c_0} \qquad (6.146)$$

and thus typical values are of the order of tens of ns. Thus the pulse durations which can be obtained with the Q switch technique are in the range of 10 ns for solid-state lasers. The output energies are in the range of mJ for high repetition rates of kHz and go up to several J for lasers with 10 Hz repetition rate. Thus 100 MW peak power are possible. Several 100 J have been produced in large devices.

Using these formulas, as an example the results for a Nd:YAG laser with cross-section $\sigma_{\text{laser}} = 3.2 \cdot 10^{-19}$ cm^2 and a photon energy of $h\nu_{\text{laser}} = 1.87 \cdot 10^{-19}$ J at wavelength of 1064 nm will be given. For a resonator with a length of 1 m, a reflectivity of 50% for the out-coupling mirror, a loss factor $V = 0.95$ and a start gain $G_{\text{start}} = 4$ from a rod length of 80 mm and a mode diameter of 3 mm, there is a laser intensity of $2.97 \cdot 10^{26}$ photons cm^{-2} s^{-1} or 56 MW cm^{-2}, a peak power of 1 MW and a pulse energy of 48 mJ. The inversion population density ΔN_{end} at the end of the pulse is $5.4 \cdot 10^{15}$ cm^{-3}. The pulse width is 49 ns and the pulse built-up time 3.4 ns. These values can easily be achieved with a common laser.

6.10.3 Mode Locking and Generation of ps and fs Pulses

For the generation of pulses with pulse widths below 100 ps down to a few fs the locking of longitudinal modes can be applied using a periodically operating active or passive switch in the resonator. If the laser is operated close to threshold for a sufficiently long time or continuously the discrimination of nonlocked modes will be increased and thus a train of very short pulses with time intervals of the resonator roundtrip time τ_{round} between them will occur (see Fig. 6.74).

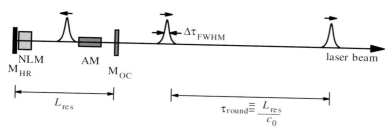

Fig. 6.74. Schematic of short pulse generation with "mode locker (ML)" in the laser resonator built by the two mirrors M_{HR} and M_{OC} and the active material AM

Different methods have been developed for mode locking in different lasers. For an introduction see [6.810–6.813, M33, M45–52]. Some general aspects are described in [6.814–6.830]. In any case the gain bandwidth of the active material has to be large enough for the generation of very short pulses $\Delta t_{FWHM} \geq 1/(2\pi\Delta\nu_{laser})$ (see Table 2.3).

Pulses with durations of 10–100 ps are produced by Nd lasers, ruby lasers, gas-ion lasers and synchronously pumped dye lasers. Shorter pulses of a few fs to 100 fs are generated, e.g. from Ti:sapphire lasers, Cr lasers and dye lasers. In any case the pulse duration can be increased by applying narrow spectral filters (e.g. etalons) in the resonator which decrease the spectral width of the light. The measurement of these short pulse durations is described in Sects. 5.5 and 7.1.5, and in the references therein.

Theoretical Description: Bandwidth-Limited Pulses

Since the frequency bandwidth for short pulses of 100 ps down to 10 fs is as large as 1.6 GHz and 16 THz more than 100 and up to many 10 000 longitudinal laser modes are active in these lasers. Therefore a statistical or continuum approach can be used, not considering each resonator mode explicitly. Thus the analysis can be carried out in the frequency or the time domain which are related by a Fourier transformation.

In general the electric field vector E of all involved longitudinal laser modes numbered with p have to be summed for the resulting intensity of the laser beam:

$$E(t) = \sum_{p=p_{min}}^{p_{max}} E_{0,p} \cos(2\pi\nu_p t + \varphi_p) \tag{6.147}$$

with the usually unknown amplitudes $E_{0,p}$ and phases φ_p of the axial laser modes with the frequencies ν_p. Many kinds of pulse shapes are possible from different amplitude distributions.

Usually bell-shaped pulses are observed and thus Gaussian or sech^2 functions were usually chosen for analysis. The Gaussian pulse can be written as a function of time t as:

$$E_{\mathrm{laser}}(t) = E_{\mathrm{laser},0}\, \mathrm{e}^{-2\,\ln 2(t/\Delta t_{\mathrm{FWHM}})^2}\, \mathrm{e}^{\mathrm{i}(2\pi\nu_p + \beta_{\mathrm{chirp}}t)t} \tag{6.148}$$

with the electric field E of the laser light, the pulse width Δt_{FWHM} and the frequency ν_p of the pth mode.

The expression $\mathrm{i}\beta_{\mathrm{chirp}}t$ causes a linear "chirp" during the pulse duration with slope β_{chirp}. This *chirp* is a linear increase of the mode frequencies during the pulse, e.g. as a result of the change of the refractive index in the active material during the pulse. This effect is similar to self-phase modulation described in Sect. 4.5.7.

The FWHM frequency bandwidth $\Delta\nu_{\mathrm{FWHM}}$ of this Gaussian pulse is given by:

$$\Delta\nu_{\mathrm{FWHM}} = \frac{1}{\pi}\sqrt{\left(\frac{2\ln 2}{\Delta t_{\mathrm{FWHM}}}\right)^2 + \beta_{\mathrm{chirp}}^2 \Delta t_{\mathrm{FWHM}}^2} \tag{6.149}$$

which leads, for chirp-free Gaussian pulses with $\beta_{\mathrm{chirp}} = 0$, to a pulse-bandwidth product of:

$$\textbf{Gauss pulse}\quad \Delta t_{\mathrm{FWHM}}\Delta\nu_{\mathrm{FWHM}} = \frac{2\ln 2}{\pi} \approx 0.44. \tag{6.150}$$

Other pulse shapes result in other values for this product. For example, sech^2 shaped pulses result in a pulse-bandwidth product of

$$\textbf{sech}^2\ \textbf{pulse}\quad \Delta t_{\mathrm{FWHM}}\Delta\nu_{\mathrm{FWHM}} \approx 0.31. \tag{6.151}$$

Laser pulses with experimentally determined pulse width Δt_{FWHM} and frequency bandwidth $\Delta\nu_{\mathrm{FWHM}}$ which fulfill this condition are called *bandwidth-limited pulses* meaning they are chirp-free and in this respect of best quality at the theoretical limit. They fulfill the time–energy uncertainty relation similar to diffraction-limited beams which fulfill the space-momentum uncertainty relation.

Because of the constructive interference of all mode-locked light waves in the short pulse the peak power or intensity is m^2 times larger than these values for a single mode pulse, whereas the peak power or intensity of the nonphase-locked pulses will be only m times larger, where m is the number of interfering modes. Thus by mode locking the peak power or intensity is m times larger than in the nonlocked case. As described above the number m of locked modes can be as high as several $10\,000$ and thus TW and PW of peak power are possible.

Another approach to model the development of these short pulses is the numerical solution of suitable time-dependent rate equations similar to the modeling of Q switching (see Sect. 6.10.2). In these equations the amplification and the nonlinear losses in the mode locker have to considered. The following processes occur during the short-pulse generation:

- In an early stage one of the intensity fluctuations shows the highest peak intensity (meaning that among the large number of the random phase mixed modes some are accidentally of similar phase); in active mode locking this highest peak is produced by the lowest losses or highest gain.
- Higher intensities will be amplified more than lower intensities and thus this pulse will grow more than the other pulses and it will be narrowed (phase locking will be rewarded).
- The passive mode locker such as e.g. the nonlinear absorption in the dye, will discriminate the lower intensities especially at the rising edge and thus this pulse will be narrowed and less absorbed than the other pulses (phase locking will be rewarded a second time).
- The active mode locker such as e.g. the AOM (Sect. 6.10.2), will show less losses at the time of this already strong pulse by the synchronization of the modulation frequency with the roundtrip time of the pulse in the resonator (phase locking is forced).
- Finally the amplification may be saturated and thus only this strongest pulse is further amplified at the expense of the rest of other pulses (phase locking is strongly forced).

Nice examples showing this process are given in [6.821]. For more details see [6.831–6.850].

Passive Mode Locking with Nonlinear Absorber

Passive mode locking is mostly used with dyes in solution or semiconductors as nonlinear absorbers which are placed directly in front of the high-reflecting mirror [6.851–6.905] as shown in Fig. 6.75.

The mechanism is described in the previous chapter. The dye has to absorb at the wavelength of the laser emission. The cross-section has to be high enough to enable bleaching with the intensities of the laser. Thus the spot size of the transversal mode has to be adapted to the necessary intensity. The recovery time of the dye absorption, which is often the fluorescence decay time of the dye, has to be at least shorter than the roundtrip time of the resonator. Recovery times of the size of the pulse duration are better. In any case only a negligible triplet or other long living state population should occur.

Therefore flow cells are often used for the dye. This laser method is typically applied in Nd or ruby lasers where the dyes have to absorb around 1060 nm or 700 nm. These dyes are often sensitive to UV radiation and thus the cells have to be shielded against daylight and the pumping flash lamps. For ruby lasers cyanine dyes are useful as in Q switch arrangements (see

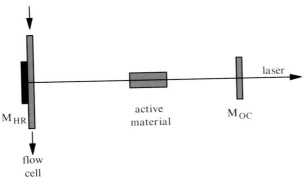

Fig. 6.75. Dye cell in front of the high-reflecting mirror as used for passive mode locking of a Nd:YAG laser

Sect. 6.10.2). For Nd lasers the Eastman Kodak dyes #9740 or #9860 dissolved in 1,2-dichlorethene can be applied. This solvent is especially crucial for impurities and should therefore be of best quality. Nevertheless, the mode locking dye has to be changed about weekly for safe operation of the lasers. Other materials as nonlinear mode lockers were also applied. For example, liquid crystals [6.851], quantum-well structures [6.855, 6.863, 6.865, 6.871, 6.883], semiconductors and semiconductor saturable-absorber mirrors (SESAMs) [6.859, 6.861, 6.870], e.g. GaAs and PbS nanocrystals [6.856, 6.864, 6.866], Cr^{4+} and Cr^{3+} ions [6.867, 6.844] and SBS mirrors [6.892] are used.

If the laser is pumped with flash lamp pulses the laser will emit a limited train of ps pulses with a bell-shaped envelope as shown in Fig. 6.76.

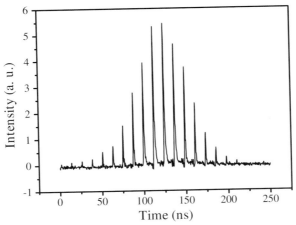

Fig. 6.76. Pulse train of ps-pulses from a Nd:YAG laser. The pulse duration is 30 ps but not resolved in this graph

The number of pulses can be determined by the optical density and thus the concentration of the dye cell and the pump rate. Less pulses usually show higher peak intensities. On the other hand only a sufficiently long generation time for the ps pulse guarantees a short pulse width and the suppression of satellite pulses.

Pulse widths of about 30 ps are common for flash lamp pumped Nd:YAG lasers whereas Nd glass lasers can show values below 5 ps. The energy content of a single pulse out of the pulse train can be as high as a few mJ resulting in a peak power of more than 30 MW. With laser amplifiers these single pulses can be increased to several 10 mJ resulting in GW powers. Several J are possible in large arrangements reaching the TW level.

Colliding Pulse Mode Locking (CPM Laser)

Even shorter pulses can be generated with the colliding pulse technique [6.905–6.912] as sketched in Fig. 6.77.

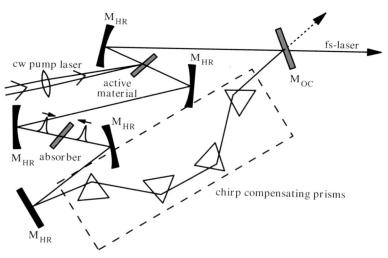

Fig. 6.77. Schematic of a colliding pulse mode locked (CPM) laser for generating fs pulses

Two counterpropagating short pulses meet in the absorber dye at the same time. By the superposition of the two intensities the dye is bleached more than in case of just one pulse. Interference between the two counterpropagating light pulse waves occurs and increases the bleaching. Thus a synchronization of these two pulses occurs in combination with pulse shortening and discrimination of all other pulses. In addition the optical path lengths are chosen for a maximum temporal distance of the two pulses in the active material for maximum amplification of both. Thus the optical path length between the absorption spot and the amplification spot has to be 1/4 of the total resonator length on one side and 3/4 on the other.

While the nonlinear absorber mostly increases the slope of the leading edge of pulse the saturated amplification in the active material decreases the negative slope of the trailing edge. Thus the combination of nonlinear absorption and amplification, both as a function of the intensity, shortens the pulse length to values in the fs range.

Because of the low intensities in this type of continuously running laser the spot sizes in both the absorber and the amplifier have to be very small with diameters of a few $10\,\mu m$. The thickness of the absorber material $L_{absorber}$ has to be not too much larger than the optical path length L_{pulse} of the short pulses. This length for the FWHM intensity part of the pulse is about:

$$\text{absorber length} \quad L_{absorber} \approx \frac{c_0}{n} \Delta t_{FWHM} \tag{6.152}$$

which is divided by the refractive index n of the absorber material to get the maximum length of the absorber. For 50 fs pulses this results in about $11\,\mu m$.

With this scheme without chirp compensation pulse durations in the range of a few 100 fs can be achieved. If in addition chirp compensation is applied, as sketched in Fig. 6.77, the pulse width can be smaller than 50 fs from this type of laser. This prism arrangement compensates the different optical path lengths for the different wavelengths from the dispersion in the active material and the absorber by simulating an anomalous dispersion.

A typical combination for a CPM laser is rhodamine 6G dye solution as active material and DODCI as nonlinear absorber. Both are used in a dye jet stream. Therefore they are dissolved in ethylene-glycol for good optical quality of the jet. The gain medium is pumped by continuously operating (cw) laser emission in the green region such as e.g. by an Ar-ion laser or a diode pumped and frequency doubled Nd laser with an output power of about 5 W.

About 60 fs pulse width can easily be achieved in stable operation. The pulse energy is in the range of a few 100 pJ which results in peak powers in the range of several kW. The average output power can be more than 30 mW at a typical repetition rate of 100 MHz. Usually these pulses are amplified with a multipath amplifier (see Sect. 6.11.3.4).

Kerr Lens Mode Locking

Today's most common technique for generating ultra-short pulses in the fs range is the passive mode locking of Ti:sapphire lasers based on the nonlinear Kerr effect in the active material [6.913–6.979] (see Fig. 6.78).

As described in Sect. 4.5.2 the refractive index of the material becomes a function of the intensity of the laser beam for sufficiently high intensities (n_2 is typically of the order of $10^{-16}\,cm^2/W$). Thus focusing of the laser mode

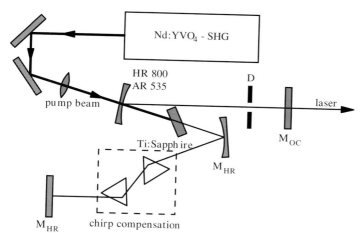

Fig. 6.78. Schematic of a fs Ti:sapphire laser with Kerr lens mode locking and chirp compensation

occurs as a function of the laser intensity in the active material because of the transversal Gaussian intensity distribution of the fundamental mode. Shorter pulses show higher intensities. If the laser resonator is designed for fundamental mode operation based on a certain Kerr lensing in the active material the associated short pulse can be selected with a simple aperture (D).

All other modes with less lensing and thus longer pulses will be discriminated by this aperture. In addition the pump light beam can be focused to a small adapted spot in the active material acting as a gain aperture for the required laser mode. Mode locking is again forced via the lensing effect.

Besides the Kerr lensing and possible self-focusing, self-phase modulation from the longitudinal Kerr effect in the active material may occur and spectral broadening may be obtained. Thus simple chirp compensation (see previous section) in the resonator using, e.g. two prisms as shown in Fig. 6.78, allows further compression of the pulses.

Recently special *chirp compensating mirrors* have been used for the compensation of linear and nonlinear dispersion [6.963–6.973].

In these dielectric mirrors the different wavelengths were reflected at different layers of the dielectric coating and thus different optical delays in the sub-μm range are achieved. Compact and reliable fs lasers can be built with these mirrors.

The chirp or dispersion compensation is the key task in reaching very short pulses in this simple arrangement. The dispersion can be discussed as a frequency-dependent propagation time $t_{\mathrm{disp}} = \sum_m \frac{1}{m!} \frac{\partial^m t_{\mathrm{sp}}}{\partial \nu^m}\bigg|_{\nu_0} (\nu - \nu_0)^m$ [6.811]. Typically, the time delay between the fastest and the slowest part of a

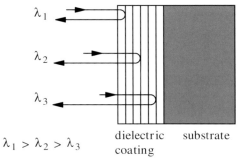

$\lambda_1 > \lambda_2 > \lambda_3$ dielectric   substrate
                        coating

Fig. 6.79. Chirp compensating mirror with dielectric coating giving different delays for different wavelengths

10-fs pulse during one round trip of a compensated resonator is about 100 fs. With prism chirp compensation the group delay dispersion (GDD) with $m = 1$ can be compensated, and, thus, 8.5 fs was reported [6.934]. However, for shorter pulses, higher-order dispersion, $m > 1$, has to be compensated, too. For fused quartz a value of about $36 \, fs^2/mm$ can be obtained at 800 nm. Therefore, double-chirped mirrors were developed which allow bandwidths of 200 THz [6.964]. In addition, good anti-reflection coating is required ($R < 10^{-4}$) over the large spectral region of the laser (690–920 nm). This demands a technical accuracy in the nm range.

The shortest pulses generated with Ti:sapphire lasers and compression showed a duration of 4.5 fs; these pulses are only 1.4 µm long (see Sect. 6.14.2). Thus experiments with such short pulses demand very high effort to keep this extremely short timing.

As pump lasers diode pumped cw Nd lasers with frequency doubling are more convenient than Ar-ion lasers. Typically 5 W of pump power are necessary. The Ti:sapphire laser shows an average output power of up to 1 W with pulse widths of 30–100 fs. The pulse repetition rate as the inverse of the resonator roundtrip time which is a function of the optical length of the resonator, is in the range of a few 10 MHz resulting in pulse energies of some 10 nJ and peak powers of more than 100 kW.

These pulse are often amplified with lower repetition rates of 10 Hz–10 kHz and then pulse energies of up to 100 mJ have been achieved. In extreme experiments pulse energies of several J were produced approaching petawatt peak powers.

A theoretical description of the Kerr lens mode locking lasers with and without chirp compensation can be found in [6.974–6.979], in addition to the textbooks given above.

Additive Pulse Mode Locking

Additive pulse mode locking (APML) [6.980–6.996] can be obtained by the feedback of the signal from self-phase modulation in an optical fiber in an arrangement of two coupled resonators as shown schematically in Fig. 6.80.

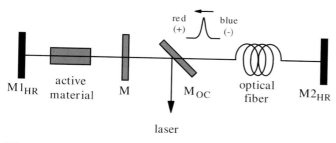

Fig. 6.80. Schematic of a laser resonator with additive pulse mode locking for generation of sub-ps pulses

Thus the main resonator between $M1_{HR}$ and M including the active material gets a feedback from the second resonator from M to $M2_{HR}$ which contains the optical fiber and the output coupler M_{OC} as a beam splitter. This feedback pulse has a red shifted trailing edge and a blue shifted leading edge from the self-phase modulation in the fiber (see Sect. 4.5.7).

The resonator lengths have to be designed for constructive interference of the leading edge of this pulse with the main pulse in the main resonator and destructive interference for the trailing edge. Thus the pulse can be shortened to values below 0.5 ps. If the lengths are tuned well no losses will occur at the beam splitter M_{OC}.

This method was applied, e.g. to color center lasers (KCl), to Ti:sapphire lasers and to Nd lasers. If these lasers are pumped with already short pulses a further shortening can be obtained to values in the range of 10 fs.

An elegant solution of this concept uses an Er-doped fiber with positive group velocity dispersion as gain medium in a ring cavity together with a standard communication fiber which shows negative group velocity dispersion at the laser wavelength of 1.55 µm. Thus the pulse is chirped inside the gain fiber and will be compressed in the standard fiber. Thus the output coupling at the end of the Er fiber shows chirped pulses in the ps range with a few 10 MHz repetition frequency. These pulses can be compressed externally to 120 fs pulse width. The Er fiber can be pumped with a diode laser at 980 nm and thus several 10 mW average power of short pulses can be achieved in a small and reliable laser. A detailed theoretical description was given in [6.987, 6.988].

In addition to the constructive and negative interference an amplitude modulation may occur in the fiber and thus solitons may be produced resulting in a further pulse shortening effect similar to the soliton laser.

Soliton Laser

Solitons as described in Sect. 4.5.7 can propagate along fibers without changing their pulse duration. A soliton generated in an optical fiber can be used

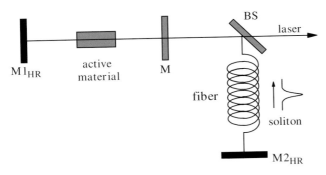

Fig. 6.81. Schematic of a soliton laser producing fs pulses

for feedback into a laser seeding this laser with short pulses as shown in Fig. 6.81.

The scheme of the soliton laser [6.997–6.1017] is similar to the arrangement of Fig. 6.80 but the feedback is tuned for the generation of a soliton in the fiber part resonator having low losses. The first-order solitons are a steady state pulse solution of the nonlinear Schroedinger equation including the effects of negative dispersion, self-phase modulation and the gain in the active material. In this case the fiber length has to be chosen carefully for stable operation. 170 fs were reported for this type of laser [6.998].

Active Mode Locking with AOM

Instead of or in addition to passive mode locking with a dye cell as shown in Fig. 6.75 an active mode locking device such as e.g. an acousto optical modulator (AOM) can be used for active mode locking [6.1018–6.1042]. This modulator adds oscillating losses into the resonator which are synchronized to the roundtrip time of the short pulses. Thus each pass will promote the short pulse and discriminate the nonsynchronized share of the radiation (see Fig. 6.82).

The modulator is driven by an electrical sine generator in resonance with the modulator crystal. Each full period twice leads to minimal loss alignment and thus the driver frequency ν_{driver} has to be half of the roundtrip frequency:

$$\textbf{driver frequency} \quad \nu_{\text{driver}} = \frac{1}{2\tau_{\text{round}}} = \frac{c_0}{4L_{\text{opt}}} \tag{6.153}$$

with the optical length of the resonator L_{opt}.

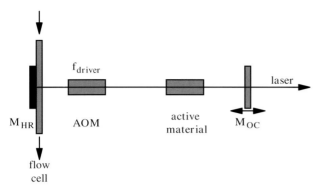

Fig. 6.82. Schematic of an active passive mode locked laser emitting ps pulses

The resonance of the modulator frequency has to be carefully adjusted with the laser roundtrip. Therefore, usually the optical length of the resonator is tuned to the frequency of the modulator with sub-mm accuracy. This can be achieved by shifting one of the resonator mirrors such as e.g. the output coupler in Fig. 6.82 or by additional variable delays using, e.g. more or less optical path through a prism arrangement. The exact alignment results in a very stable operation with ps pulses. Thus the stability of the oscillogram of the output pulses can be used as a criterion.

Active mode locking is often combined with an additional passive one. The resulting pulses are usually as short as possible by the action of the passive mode locker and very stable from the active mode locker. Pulse-to-pulse fluctuations smaller than 5% are possible even for a selected single pulse out of the train in flash lamp pumped Nd:YAG lasers with pulse lengths around 30 ps and repetition rates of 30 Hz.

Active Mode Locking by Gain Modulation

Active mode locking can be achieved by synchronously pumping the laser with ps pulses of another laser [6.1043–6.1057]. Thus the roundtrip time of the pumped laser has to be tuned very precisely to the repetition rate of the pump laser to achieve synchronization.

A typical application of this technique is the synchronous pumping of one or two dye lasers with a mode locked master oscillator such as e.g. an Ar-ion laser (see Fig. 6.83).

The resulting ps pulses of the dye lasers can be significantly shortened. For example, with about 100 ps Ar laser pulses a dye laser pulse of 10 ps can be generated. A much shorter pulse duration can be achieved by the pumping of a second dye laser with the radiation of the first dye laser. In this way pulse durations of e.g. 100 ps to 55 ps to 0.6 ps were obtained for the Ar, the first and the second dye laser [6.1050].

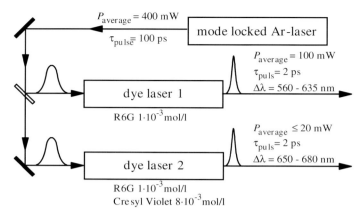

Fig. 6.83. Synchronously pumping of two dye lasers operating in the ps regime with a mode locked Ar laser

If two dye lasers are pumped synchronously the two resulting pulses are synchronized better than the pulse width and can be tuned in wavelength. In this case the resonator lengths have to be carefully controlled using a piezo-driven mirror.

6.10.4 Other Methods of Short Pulse Generation

The spatial modulation of the gain in the active material can be used for the generation of pulses in the ps range as applied in distributed feedback lasers using e.g. dyes as active material. Other methods use a small gain area for producing short pulses.

Distributed Feedback (DFB) Laser

DFB lasers allow the generation of short pulses without mode locking [6.1058–6.1068]. With two crossed pump beams a gain grating can be excited in the active material as described in Sect. 5.3.9. The two beams can be produced by a holographic grating with a sine modulation reflecting the perpendicular pump beam in the +1 and -1 order, only, as depicted in Fig. 6.84.

The angle of incidence and thus the emission wavelength of the laser can be varied by turning the mirrors M and synchronously moving the grating. The gain grating in the dye cell acts as a resonator by promoting these longitudinal laser modes with antinodes at the gain maxima. Thus a laser beam will be emitted towards the two opposite directions with a spectral resolution of:

$$\text{spectral resolution} \quad \frac{\Delta\lambda_{\text{laser}}}{\lambda_{\text{laser}}} = \frac{\lambda_{\text{laser}}}{2L_{\text{gain}}} \tag{6.154}$$

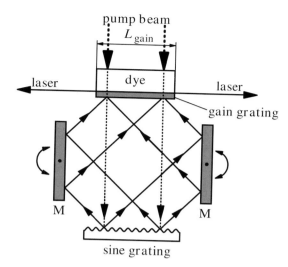

Fig. 6.84. Schematic of a distributed feedback laser pumped with ps pulses

with the optical length L_{gain} of the gain section in the material in the laser beam direction and the wavelength of the DFB laser λ_{laser}. It has to be small enough not to increase the laser pulse duration and thus L_{mat} has to be shorter than:

$$\text{gain length} \quad L_{gain} \leq \pi c_{mat} \Delta t_{FWHM,pulse} \tag{6.155}$$

with the desired pulse duration $\Delta t_{FWHM,pulse}$. E.g. a 1 ps pulse limits the maximum gain length to about 1 mm.

If this laser is pumped with short pulses it shows a spiking behavior as described in Sect. 6.10.1. If this laser is pumped just above threshold only one spike will be possible and the resulting DFB laser pulse will be much shorter than the exciting pulse. Ratios of about 50 have been reported [6.1063]. A typical laser is operated with Rhodamin 6G solution in alcohol as active material pumped by a frequency doubled Nd:YAG laser ps pulse of 25 ps. In this case the shortest observed DFB laser pulse duration is about 1 ps.

Short Resonators

If active materials with high gain are used in very short cavities a spiking operation as discussed in Sect. 5.3.9 can be observed but the pulse width of the resulting spikes will be much shorter as in solid-state lasers with resonator lengths of many 0.1 m.

Exciting dye solutions in cavities with lengths in the range of 1 mm, e.g. with the pump pulse of a Nitrogen laser at normal pressure with a duration of 500 ps can result in tunable dye laser pulses of a few 10 ps.

Even shorter wavelength tunable pulses can be produced with resonators of a few μm length filled with dye solution. These resonators act as a Fabry–

Perot system and thus by varying the resonator length the laser can be spectrally tuned. Again the bandwidth has to be large enough (see Sect. 2.9.8) to allow for the desired short pulses. Further the gain coefficient has to be high enough to provide sufficiently high gain within the short cavity gain length. Both can be achieved in dyes or semiconductors.

Such a simple short cavity dye laser can be pumped by 10–30 ps pulses, e.g. with the a pulsed Nd:YAG laser those radiation is frequency doubled or a triplet with nonlinear crystals.

Traveling Wave Excitation

In this scheme the excitation of the active material is applied by a short pulse which is moving with the same speed as the light inside the matter [6.1065, 6.1069–6.1072] as sketched in Fig. 6.85.

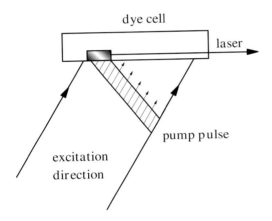

Fig. 6.85. Traveling wave excitation of a dye solution by a pump laser beam with delayed pulse across the beam for synchronized pumping of the generated laser pulse moving to the right side

Therefore the pulse has to be delayed transversely across the pump beam as shown in the figure, to be synchronized with the propagation of the laser pulse with velocity c_0/n. The laser beam is then produced by the superradiation of the excited area. Thus the transversal and longitudinal structure of the light is poor.

This scheme was used with discharge excitation in nitrogen lasers producing ns pulses. It can be applied for dye lasers in the near IR above 1 µm and it was considered for XUV generation [6.1069].

6.10.5 Chaotic Behavior

Laser emission can show temporal chaotic behavior of the emission [6.1073–6.1102]. Theoretical investigations using the laser as a model system for chaos research are based on simple model systems based on the nonlinear wave equation as given in Sect. 4.3 and the time-dependent Schroedinger equation as given in Sect. 3.3.1 considering the energy levels of the active material and the interaction as a small distortion [e.g. 6.1073]. Further, using the approximations of a plane wave, slowly varying amplitudes, homogeneous transition and unidirectional ring laser operation, the three basic equations of this theory can be derived for the relevant quantities:

field

$$\frac{\partial}{\partial t} e_{\text{th}}(t) = -\left(i2\pi\nu_{\text{res}} + \frac{1}{\tau_{\text{res}}} \right) e_{\text{th}}(t) + N_{\text{total}}^{(V)} g_{\text{th}} p_{\text{th}}(t) \tag{6.156}$$

polarization

$$\frac{\partial}{\partial t} p_{\text{th}}(t) = -\left(i2\pi\nu_{\text{mat}} + \frac{1}{\tau_{\text{trans}}} \right) p_{\text{th}}(t) + \Delta N^{(V)} g_{\text{th}} e_{\text{th}}(t) \tag{6.157}$$

inversion

$$\frac{\partial}{\partial t} \Delta N^{(V)}(t) = -\frac{1}{\tau_{\text{long}}} (\Delta N^{(V)}(t) - \Delta N_0^{(V)}) - 2g_{\text{th}} [e_{\text{th}}(t) p_{\text{th}} + \text{c.c.}] \tag{6.158}$$

with the dimensionless parameters e_{th} and p_{th} which are related to the physical values of the electric field E and polarization P by:

$$P(t) = 2\frac{N_{\text{total}}^{(V)}}{V_{\text{mat}}} \mu_{\text{mat}} p_{\text{th}}(t) \tag{6.159}$$

and

$$E(t) = -i\sqrt{\frac{2h\nu_{\text{res}}}{\varepsilon_0 V_{\text{mat}}}} e_{\text{th}}. \tag{6.160}$$

The value g_{th}, which is not a gain coefficient in this case, follows from:

$$g_{\text{th}} = \mu_{\text{mat}} \sqrt{\frac{\nu_{\text{res}}}{2h\varepsilon_0 V_{\text{mat}}}} \quad \text{with} \quad [g_{\text{th}}] = \text{s}^{-1} \tag{6.161}$$

$N_{\text{total}}^{(V)}$ describes the total number of particles in the volume V_{mat}, $\Delta N^{(V)}$ the inverted particles, $\Delta N_0^{(V)}$ the inverted particles without the laser field, ν_{res} the resonance frequency of the resonator, ν_{mat} the resonance frequency of the two-level system of the matter, τ_{res} the inverse resonator loss rate, τ_{trans} the transversal decay time which is the decay of the dipole moment, τ_{long} the longitudinal decay time which is the decay of the inversion and μ_{mat} is the dipole moment of the matter as a projection in the direction of the electric field vector. All these values are given as commonly used in these theoretical investigations.

If the corresponding relaxation rates are very fast the time derivatives of the particular differential equation can be neglected. The possible operation modes of these model lasers are determined by the relation of the resonator lifetime in comparison to the lifetimes of the inversion (in this field, called the longitudinal relaxation time) and the polarization (transversal relaxation time) of the active material.

Thus *class A, B and C lasers* are distinguished. *Class A lasers* show a much longer resonator lifetime than the other lifetimes and thus only the derivatives of the e_{th} equation have to be considered. These lasers are therefore not chaotic. *Class B lasers* have comparable resonator and longitudinal lifetimes which are both longer than the transversal relaxation time. Thus both differential equations for e_{th} and d_{th} have to be taken into account. These lasers can be forced into chaotic behavior by external influences such as e.g. external feedback. *Class C lasers* show comparable values of all three lifetimes and therefore all three differential equations are necessary for the description. Thus the system has three degrees of freedom and can therefore be chaotic by itself. For this purpose the laser has to be pumped about 20 times above threshold.

If further nonlinear elements are introduced in the laser resonator such as e.g. phase conjugating mirrors [6.1101] or crystals for frequency transformation, the laser emission can be chaotic in time and space. The theoretical description can be very complicated.

6.11 Laser Amplifier

Laser oscillators are limited in the brightness of their radiation especially if short pulses or very monochromatic light, very high output powers or pulse energies are required. Thus the amplification of laser light may be necessary. It allows the generation of peak powers in the PW range, average output powers of several 100 W to kW or pulse energies of several 10–100 J with diffraction-limited beam quality.

The laser amplifier contains an active material which is pumped as in oscillators but no resonator selects the light properties. Thus the properties of the amplified light are mostly determined by the properties of the incident light produced by the laser oscillator.

The pumping of the active material of the laser amplifier is described in the same way as given for the oscillators in the previous Sects. 6.1–6.4 and 6.8.

6.11.1 Gain and Saturation

Laser amplifiers show a gain or gain factor G_{amp} which is defined by the ratio of the intensities (powers or energies) of the light behind the amplifier I_{out} divided by the incident I_{inc}:

gain factor $$G_{\text{amp}} = \frac{I_{\text{out}}}{I_{\text{inc}}} \tag{6.162}$$

which is the same value as the transmission of the active material but shows values above 1.

The small or low signal gain $G_{\text{amp,ls}}$ which can be obtained for small incident light intensities, which almost do not change the inversion population in the active material, is given by the cross-section σ_{laser} and the population inversion density in the active material ΔN_{amp} and the length of the active material L_{amp}:

small signal gain $G_{\text{ls}} = e^{\sigma_{\text{laser}} \Delta N_{\text{amp}} L_{\text{amp}}} = e^{g_{\text{ls}} L_{\text{amp}}}$ $\qquad$ (6.163)

with the low-signal gain coefficient g_{ls}:

small-signal gain coefficient $g_{\text{ls}} = \sigma_{\text{laser}} \Delta N_{\text{amp}}$ $\qquad$ (6.164)

Losses in the active material are neglected. This is usually correct for modern laser materials. The losses can be introduced in these equations by an absorption coefficient a resulting in an expression $a L_{\text{amp}}$ in the exponent.

The low signal gain G_{ls} can be as high as 10–100. The inversion population density can be calculated using rate equations as described in Sect. 5.3.6. On the other hand the gain can be measured and thus the inversion population density calculated from the known cross-section and matter length for a given setup.

If the light intensity reaches higher values in the range of the nonlinear or saturation intensity $I_{\text{nl}} = I_{\text{nl}}/h\nu_{\text{Laser}}$ (see Sect. 5.3) the population densities will be changed and the gain will decrease as a function of the propagation coordinate z. This can be modeled using the system of rate equations as given in Sect. 5.3.6. For three- and four-level schemes the equation for the inversion population density as a function of the intensity which is a function of time and space can be written as:

Three-level amplifier system:

$$\frac{\partial \Delta N(I, t, z)}{\partial t} = \left(W_{\text{pump}}(t) - \frac{1}{\tau_{\text{upper}}} \right) N_0$$
$$- \left(W_{\text{pump}}(t) + \frac{1}{\tau_{\text{upper}}} \right) \Delta N(I, t, z)$$
$$- 2\sigma_{\text{laser}} \Delta N(I, t, z) I(t, z) \qquad (6.165)$$

and

Four-level amplifier system:

$$\frac{\partial \Delta N(I, t, z)}{\partial t} = W_{\text{pump}}(t) N_0 - \left(W_{\text{pump}}(t) + \frac{1}{\tau_{\text{upper}}} \right) \Delta N(I, t, z)$$
$$- \sigma_{\text{laser}} \Delta N(I, t, z) I(t, z) \qquad (6.166)$$

with the pump rate W_{pump}, the lifetime of the upper laser level τ_{upper} and the total population density N_0.

Using these equations and the photon transport equation:

$$\left\{ \frac{\partial I}{\partial z} + \frac{1}{c} \frac{\partial}{\partial z} \right\} I(t, z) = \sigma_{\text{laser}} \Delta N(I, t, z) I(t, z) \qquad (6.167)$$

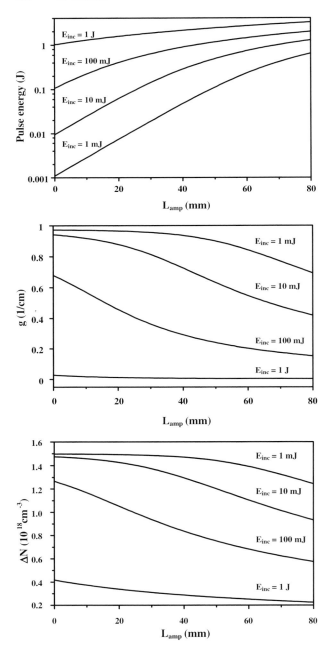

Fig. 6.86. Amplification of light from small signal value to saturation as a function of the length in the active material L_{amp}. Shown are the pulse energy E, the local gain coefficient g and the inversion population density ΔN for four different incident intensities pulse energies E_{inc} from 1 mJ to 1 J

the resulting intensity increase and depopulation of the inversion can be calculated numerically as a function of space and time. Thus the gain is a function of time and space, too. The calculated results for the amplification of a ns pulse in a Nd:YAG laser amplifier are shown in Fig. 6.86.

The Nd:YAG rod was 80 mm long and 6 mm in diameter with a Nd concentration of $1.38 \cdot 10^{20}$ atoms cm^{-3}. It was assumed to be pumped with 16 J electrical power per flash lamp pulse of 140 µs duration. The electro-optical excitation efficiency was 4%.

In the saturation regime the amplification is decreased and thus the intensity increase is slowed down, the inversion population density and the gain decreased while the light is propagating through the amplifier.

In the small-signal region of the intensity $I \ll I_{nl}$ the intensity increases exponentially with the length of the active material. It grows linearly in the saturation regime with $I \gg I_{nl}$ [6.1103–6.1112].

6.11.2 Energy or Power Content: Efficiencies

The inversion population density in the amplifier material ΔN_{amp} represents a stored energy $E_{amp/V}$ per amplifier volume V_{amp} which can be transformed to laser light during amplification:

stored energy per volume $\quad E_{amp/V} = \dfrac{hc_0}{\lambda_{laser}} \Delta N_{amp}$ $\qquad$ (6.168)

with Planck's constant h, and the laser wavelength λ_{laser}. This value can be obtained from the experimentally determined small-signal gain coefficient g_{ls} or vice versa by:

stored energy per volume $\quad E_{amp/V} = \dfrac{hc_0}{\lambda_{laser}\sigma_{laser}} g_{ls}$ $\qquad$ (6.169)

with the emission cross-section of the laser transition σ_{laser}. Values of several J per cm^3 are possible in most laser materials (see Sect. 6.13).

The total stored energy is of course:

total stored energy $\quad E_{amp} = V_{amp} E_{amp/V}$ $\qquad$ (6.170)

and the efficiency of the amplification is the share of this energy used for amplifying the incident laser light. For pulsed lasers this energy content of the stored energy in the amplifier material can be used directly for this calculation.

For continuously operating laser amplifier systems the powers or temporally averaged powers have to be used.

The efficiencies such as quantum defect efficiency, quantum efficiency, opto-optical efficiency, electro-optical efficiency and total efficiency can be calculated as Sect. 6.3.6.

The main problems in building high-power amplifiers is damage and heating which decreases the quality of the amplified light. The limited efficiency results in thermal loads of the active material (see Sect. 6.4). The wall-plug efficiencies are mostly below 10% and only a few active materials with low quantum defect such as e.g. Yb doped crystals or semiconductors, reach values of 20–50%.

6.11.3 Amplifier Schemes

Amplifier schemes are as diverse as laser oscillators with their properties such as low bandwidth, low noise, short pulses, high-powers and good beam quality. In the simplest case the amplifier is just a laser without a resonator in a single-pass supplied from a master oscillator in a *master oscillator power amplifier* (MOPA) scheme. As a semiconductor amplifier it has a size of $1\,\mu m \times 10\,\mu m \times 1\,mm$. On the other hand regenerative amplifiers for ps and/or fs pulses contain, besides the multipass amplifier, also stretchers, compressors, switches, apertures and other optical components. The largest amplifier setups such as e.g. in the National Ignition Facility (NIF, USA) and the similar system in France built for fusion experiments contains more than $10\,000$ optical elements and many of them have an aperture of almost half a meter. Thus in each of the schemes discussed below the single amplifier can be expanded to a chain of two or many amplifiers in a row.

Single Pass Amplifier

This scheme is mostly useful for pulsed lasers. The laser beam from the oscillator passes the active material once. To avoid damage in the amplifier the beam is expanded by a telescope as shown in Fig. 6.87.

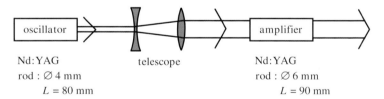

Nd:YAG
rod : $\varnothing\,4\,mm$
    $L = 80\,mm$

telescope

Nd:YAG
rod : $\varnothing\,6\,mm$
    $L = 90\,mm$

Fig. 6.87. Schematic of single-pass amplification in a Nd laser system with one amplifier

The telescope also allows the divergence of the oscillator beam to be changed. This type of amplification is often used in solid-state or dye lasers. The amplification can be as high as 10–30. In this scheme the saturation of the amplifier is often not completely reached. Therefore a second amplifier or a double-pass scheme can be applied. These single-pass amplifiers are useful for both ns and ps pulses. A typical Q switch pulse from the Nd:YAG oscillator pumped with 8 J with 25 ns pulse width and 17 mJ energy in the single and fundamental mode is, e.g. amplified to 55 mJ in the first amplifier which is also flash lamp pumped with 16 J and to 160 mJ at 1064 nm in the second amplifier which is flash lamp pumped with 16 J. A ns dye laser with two amplifiers can emit about 20 mJ with a pulse duration of 10 ns if it is pumped with about 200 mJ of a XeCl-excimer laser at 308 nm. A single 30 ps pulse from a Nd:YAG laser with an energy of 1 mJ can be amplified to about 10 mJ in a single-pass amplifier and up to 40 mJ are possible in a second amplifier. Further examples are given in [6.1113–6.1122].

If the active material is very highly pumped superradiation or laser action between the amplifier surfaces has to be avoided, e.g. by Brewster angled arrangements or antireflection coatings.

Double Pass Amplifier

If in a single-pass of the laser light through the amplifier saturation cannot be reached a double-pass arrangement [6.1123–6.1127] may be used (see Fig. 6.88)

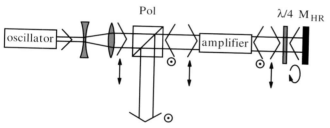

Fig. 6.88. Schematic of a double-pass amplifier setup. The polarizer in combination with the $\lambda/4$ plate achieves the out-coupling of the amplified light after the second pass

This can be necessary for longitudinal mono-mode or other special laser designs which may emit weak light powers which are below the nonlinear intensity of the amplifier material if an appropriate beam diameter is chosen.

In this scheme the incident oscillator light is linearly polarized in the plane of the paper of Fig. 6.88. Before or after the amplifier the polarization is changed to circular by a quarter-wave plate. The high-reflecting mirror produces the second pass through the amplifier. Passing the quarter-wave plate on the way back again the polarization is changed to linear again but the direction is now perpendicular to the original one which is vertical in Fig. 6.89. Thus the polarizer reflects the light out of original path. The position of the quarter-wave plate determines the polarization of the light in the active material. If necessary a Faraday rotator can be used instead.

Care has to be taken to not get to much light back into the oscillator. This amplified light can damage the oscillator components by the high intensity. Further it can disturb the mode selection in the resonator.

This double-pass scheme is again useful for ns or ps pulses. Typically it is applied for solid-state lasers. For example a 30 ps single pulse of a Nd:YAG laser with 1 mJ energy can be amplified with one double-pass amplifier to 25 mJ.

Double-pass amplifiers are especially useful in combination with phase conjugating mirrors as discussed in the last Subsection.

Multi Pass Amplifier

Several passes through the same zone of the active material of the multi pass amplifier [6.1128–6.1138] are useful for weak pulses as they are generated, e.g. in fs lasers. These pulses cannot saturate the gain in a single-pass. In particular in dye amplifiers very high gains can be achieved in short interaction lengths and thus the pulse broadening can be kept small.

The typical scheme of a multipass amplifier is given in Fig. 6.89.

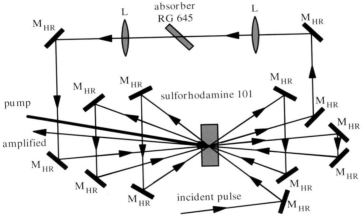

Fig. 6.89. Multipass amplifier for fs laser pulses with six transitions through the active material

In this example six passes of the laser beam are used. Even eight or twelve passes have been used. In these cases additional lenses usually have to be applied to keep the small diameter of the laser beam in the active material. For the supression of superradiation from the amplifier a nonlinear absorber can be used as shown in Fig. 6.89.

For example in such a fs laser dye amplifier with six passes and sulfor-rhodamine 101 as active material which was pumped by frequency-doubled

Nd:YAG laser light pulses with 10 ns duration and 10 mJ pulse energy, a single amplification of a factor of about 9 was reached resulting in an overall amplification of $5 \cdot 10^5$. The CPM laser pulse energy of 300 pJ at 620 nm was amplified to an output energy of 150 µJ. With a following single-pass Berthune cell dye amplifier a final output energy of 2 mJ with a pulse width of 100 fs was obtained.

Regenerative Amplifier

This type of amplifier operates as a seeded oscillator with active out-coupling after a certain number of roundtrips (see Fig. 6.90) [6.1139–6.1168]. Thus a multipass amplification with additional pulse shape regeneration is applied.

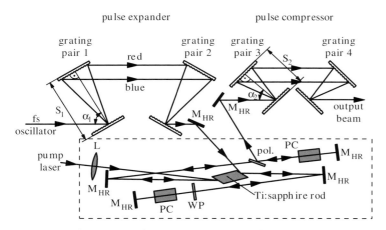

Fig. 6.90. Schematic of a regenerative amplifier

The original short laser pulse is usually temporally stretched in a grating arrangement, as shown in the figure before it is fed into the amplifier. After the required number of passes through the amplifier material it is outcoupled by switching the Pockels cell PC. Then the pulse is recompressed to its original pulse shape [6.1160–6.1168].

Very often this type of amplification is also used to reduce the repetition rate of the laser pulses from the MHz range to kHz or below. Therefore the regenerative amplifier is pumped with the lower repetition rate.

E.g. with a Ti:sapphire amplifier pumped by the frequency-doubled radiation of a Nd laser at 1 kHz an incident pulse of 8.5 nJ can be amplified to 1 mJ with a pulse duration of 100 fs or up to 2 ps. Thus the average output power is 1 W and the peak power 10 GW.

Double Pass Amplifier with Phase Conjugating Mirror

In the double-pass amplifier scheme phase conjugating mirrors can be applied to compensate phase distortions from the active material of the amplifier [6.1169–6.1206] (see also Sect. 4.5.14). Thus the beam quality can be conserved although the high-power amplifier causes strong phase distortions in the single-pass. In the simplest case self-pumped phase conjugating mirrors (PCMs) based on stimulated Brillouin scattering (SBS) can be applied. The scheme is shown in Fig. 6.91.

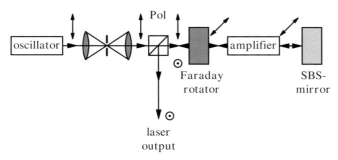

Fig. 6.91. Schematic of a double-pass amplifier with phase conjugating mirror (PCM) based on stimulated Brillouin scattering (SBS) for compensating phase distortions in the amplifier material and thus for improving the beam quality

In this double-pass scheme the constant phase distortions, e.g. from the thermally induced refractive index changes in solid-state laser rods, are compensated in the second pass through the active material because the wave front is phase conjugated in the SBS mirror. More details are given in Sect. 4.5.14.

With solid-state double-pass MOPAs with phase conjugating mirrors average output powers of more than 100 W was obtained from a single rod amplifier [6.1173] and kW were achieved with diode pumped slab amplifier chains with diffraction-limited beam quality at TRW and LLNL. Because of the high "threshold" of the SBS mirrors up to now, pulsed radiation can usually be phase conjugated only in these systems. The highest pulse energies achieved in SBS-PCM-MOPAs were above 100 J and the smallest values are less than 10 μJ. Waveguide geometry allow phase conjugating SBS mirrors for cw radiation, too. Phase conjugating mirrors based on the photo-refractive effect have very low thresholds below 1 W but they show very slow reaction times. They have not yet been used for high-power lasers.

6.11.4 Quality Problems

The laser amplifier usually shall amplify the laser light and conserve its properties, but the laser amplifier is itself a nonlinear optical device and thus it will change the properties of the light. Thus precautions are necessary to conserve low noise, beam quality, pulse duration, and polarization.

Noise

The laser amplifier produces additional noise in the laser radiation by spontaneous emission in the active material [6.1207–6.1223]. This can decrease the spectral, temporal, spatial and polarization quality of the laser light. Some improvements may be possible by using linear filters to conserve, e.g. the spectrum and polarization of the light. Further, the geometrical construction with small apertures for the laser beam can reduce the share of spontaneous emitted light in the beam.

In some cases additional nonlinear absorbers can be applied to supress the noise as, e.g. shown in Fig. 6.89. In any case superradiation from the amplifier should be avoided by choosing sufficiently short amplifier lengths or low gain coefficients. Parasitic resonators for the amplifier radiation have to be excluded by Brewster angles and/or careful antireflection coating of all relevant optical surfaces.

Beam Quality

The transversal shape of the beam can be changed by phase distortions, amplitude distortions and other diffraction effects, e.g. from the limited aperture of the amplifier. Geometrical conditions of pumping and the diameter of the incident light are important parameters for good beam quality.

Phase distortions can be compensated by phase conjugating mirrors or other devices, such as e.g. active controlled adaptive mirrors in double-pass arrangements. Diffraction-limited beam quality was obtained in this way (see Sect. 6.11.3).

Amplitude distortions can not in general be compensated and thus homogeneous pumping is a key issue for good beam quality. Thermally induced birefringence as in solid-state laser rods can be avoided using laser materials with strong natural birefringence such as e.g. Nd:YALO (see Sect. 6.4.2). Further, the birefringence can be compensated by a double amplifier scheme with 90° rotator in between as described in Sect. 6.4.2. In a double-pass amplifier the scheme can be simplified by using one amplifier and a 45° rotator in front of the mirror.

Phase distortions and birefringence can be compensated in a double-pass amplifier with a phase conjugating SBS mirror and separate polarization treatment as shown in Fig. 4.51 in Sect. 4.5.14. In this scheme the two polarization directions are interchanged after reflection in the phase conjugating SBS mirror.

Diffraction losses should be avoided by choosing a sufficiently small diameter of the incident light which should be about 1.5 times smaller than the active material. The best value has to be found experimentally. This difference in pumped-to-mode volume causes a lack in efficiency. Therefore flat-top profiles for propagation through the amplifier material have been suggested.

Finally the beam quality can be improved inside the amplifier setup or after amplification with spatial filters as described in Sect. 6.6.10 and shown in Fig. 6.41. Because of the high-powers the mode apertures usually have to be used in vacuum avoiding optical breakdown in air and sometimes it may even be necessary to cool them with a water cooler.

Some other methods for conserving the quality and polarization of the beam such as, for example, self-focusing, are discussed in [6.1224–6.1229].

Pulse Duration

As a consequence of the dispersion in the amplifier material and other associated optical components the pulse duration of ps but especially fs pulses can be lengthened in the amplifier. Thus special treatment for compensation is necessary. Chirp compensating elements or nonlinear absorbers (see Fig. 6.89) can be applied.

For high-power pulse amplification the laser pulses are often temporally stretched by more than 100 or 1000 times to decrease the peak power while containing the pulse energy (see Sects. 6.11.3, 6.14.2.1 and [6.1230–6.1243]). Thus the intensity damage threshold of the optical components becomes noncritical even for PW-pulses. A typical scheme is given as part of Fig. 6.90. Sometimes ns-pulses are compressed behind the amplifier using SBS pulse compression (see Sect. 6.14.2.2).

6.12 Laser Classification

Almost all physical and technological properties of the laser are used for classification. For practical purposes wavelength, output power and operation mode (pulsed or cw) are most prominent.

6.12.1 Classification Parameters

A list of prominent laser properties often used for classification in science and technology reads as follows:

- *Active Material*
  Lasers are often named after their active material. This is done directly as in the Nd:YAG laser, or in classes such as the excimer laser, solid-state laser and so on. All properties of the active material are used for this grouping. Details are given in Sect. 6.2.
- *Pump Mechanism*
  As described in Sect. 6.3 the active material can be pumped by other lasers, lamps or other radiation, electric discharges or chemical reactions. Pumping can be obtained continuously (cw) or pulsed. The pumping requirements are different for three- or four-level laser schemes (see Sect. 6.2).

- *Wavelength*
  The wavelength ranges can be classified as X-ray (<1 nm), XUV (1–50 nm), UV (100–300 nm), Vis (400–700 nm), NIR (800–1500 nm), IR (2–$10\,\mu$m) and far IR ($>10\,\mu$m) with the wavelengths as rough values, only.
- *Temporal Operation: Pulse Width*
  cw, quasi-cw and pulsed lasers are distinguished. Pulsed lasers can be classified for long pulses ($>1\,\mu$s) or short pulses as ns, ps and fs pulses.
- *Average Output Power*
  Average output powers may be classified in the ranges <1 mW (not dangerous), <1 W, <10 W, <50 W, <100 W, <1 kW and >1 kW.
- *Bandwidth*
  Laser bandwidths can be larger than 50 nm and smaller than 10^{-12} nm (<1 Hz). Typical values are <1 nm for molecule laser without further restrictions and 10^{-3} nm (GHz) to 10^{-6} nm (MHz) for lasers with narrow bandwidth.
- *Main Application*
  The main applications are described in Sect. 1.5 and Fig. 1.4. Material processing, spectroscopy, communication and medicine are e.g. main fields.

These properties are given for the different types of lasers in the following Section.

For theoretical investigations using the laser as a model system the possible chaotic behavior of the temporal emission is used for classification (see Sect. 6.10.5). Thus *class A, B and C lasers* are distinguished.

From the point of view of potential eye or skin damage, the lasers are divided into five safety classes, 1, 2, 3A, 3B and 4, as will be discussed in Sect. 6.16 in more detail.

6.12.2 Laser Wavelengths

From the application point of view the laser wavelength [6.1244] is often the most important parameter. In Fig. 6.92 the wavelength ranges of tunable lasers are depicted. These lasers can be grouped for semiconductor or diode lasers, solid-state lasers, color center lasers, dye lasers and excimer lasers.

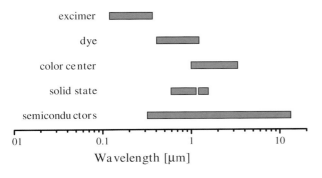

Fig. 6.92. Wavelength ranges in which tunable lasers can be achieved. The tuning range is much smaller than the total ranges shown for these lasers

The tuning ranges are much smaller than the given total ranges. They can be extended by frequency conversion as second, third, fourth harmonic generation as well as sum and difference frequency generation (see Sect. 6.15).

A general overview of typical values of laser wavelengths, tunability range per active material, the orders of magnitude of the pulse width and of the average output power, the beam quality and a rough estimate of the efficiency are given in Table 6.13.

Table 6.13. Wavelength, tunability range, pulse width range, average output power, beam quality and wall-plug efficiency of some lasers

Wavelength (μm)	Range (nm)	Laser	Pulse width	Average output	Beam quality M^2	Efficiency (%)
0.152		F_2-excimer	ns	10 W		
0.193		ArF-excimer	ns	10 W	multimode	<2
0.248	1	KrF-excimer	ns	10 W	multimode	<2
0.266		4xNd laser	μs,ns,ps	<1 W	1	0.005
0.308	1.5	XeCl-excimer	ns	10 W	multimode	<2
0.3–1.1	50	dye laser	μs,ns,ps	10 W	3	
0.3250		HeCd-laser	cw	200 mW	1	0.1
0.3371		N_2-laser	ns	100 mW	multimode	<0.1
0.351		XeF-excimer	ns	10 W	multimode	<2
0.355		3xNd laser	μs,ns,ps	5 W	1	0.04
0.4–1.0	20	dye laser	cw	W	1	0.2
0.4416		HeCd-laser	cw	10 mW	1	0.1
0.4880		Ar^+-laser	cw	10 W	1	<0.1
0.5105		Cu-vapor laser	ns	10 W	1	1
0.532		2xNd laser	cw	100 W	5	0.5
0.532		2xNd laser	μs,ns,ps	10 W	1	<0.5
0.5435		HeNe-laser	cw	1 mW	1	0.1%
0.5782		Cu-vapor laser	μs,ns	10 W	1	1%
0.6328		HeNe-laser	cw	10 mW	1	0.1
0.6471		Kr^+-laser	cw	W	1	<0.1
0.6943		ruby-laser	μs,ns	W	5	<1
0.7–0.82		alexandrite	μs,ns,ps	50 W	1	<2
0.7–1.1	300	Ti-sapphire	cw,μs–fs	50 W, 1 W	1	<1
0.72–0.84		Cr:LiCaF	cw–fs	W	1	<10
0.75–1.0		GaAs-diode	cw,ms	1 W	300	40
0.78–1.0		Cr:LiSAF	cw–fs	1 W	1	<2
1.030		Yb:YAG	cw-ps	100 W	1	10
1.04–1.08		Nd laser	cw–ps	kW	100	5
1.1–1.6		InGaAs-diode	cw–ms	mW	300	40
1.44		Nd laser	cw–μs	W	3	1
1.4–1.6		color center	μs,ns	100 mW	1	0.01
1.54	0.3	Er-fiber	cw	10 W	1	
1.55	50	Cr:YAG	μs–ps	W	1	0.5
2.06		Ho-laser	μs,ns	W	5	
2.6–3.0		HF-laser	cw–ms	100 W	10	
2.9		Er:YAG	μs,ns	10 W	1	1
5–6		CO-laser	cw	kW	1	20
9–11,10.6		CO_2-laser	cw–μs	kW	1	20

Many other laser materials and their optical transitions, and e.g. free electron lasers and XUV lasers, are not considered in this table. All laser wavelengths can be converted or shifted and thus the wavelength scale is continuously filled from UV to IR with possible laser radiation.

6.12.3 Laser Data Checklist

Laser prospects and data sheets contain all relevant parameters of the laser output and the installation and operation requirements. A description of such parameters is given in Sect. 6.12.1. The following features may be checked.

Output Data

- *temporal mode of operation*
  - cw, quasi-cw or pulsed operation
- *average output power*
  - maximum average output power
  - stability and fluctuations of output power and noise
  - variability of output power
- *pulse energy*
  - maximum pulse energy which may be a function of the repetition rate
  - stability and fluctuations of pulse energies
  - variability of pulse energies
- *pulse width and shape*
  - temporal structure such as e.g. single pulses or bursts
  - pulse width of single pulses ($>1\,\mu$s), ns, ps or fs
  - shape of the single pulse such as e.g. Gaussian, rectangular
  - substructure of single pulses such as e.g. modulations, satellites
  - background
- *repetition rate*
  - tuning range of repetition rate
  - pulse-to-pulse jitter
- *wavelength*
  - peak wavelength(s)
  - bandwidth
  - tuning range
  - background radiation
  - short and long time stability
- *wavelength conversion possibilities*
  - availability of second, third or fourth harmonics
  - available optical parametric oscillators or amplifiers
  - Raman shifter
- *beam quality and divergence*
  - beam quality should be given as M^2 (with power content)
  - far-field divergence of the laser beam

– dimensions of the cross-section of the laser beam

– position of the beam waist (inside the laser) would be helpful

Installation and Connection to Other Devices

- *trigger, jitter and delay*
  - voltage, impedance and timing of trigger signal to fire the laser pulse
  - delay between triggering and laser pulse
  - jitter and drift between trigger signal and laser pulse
  - which trigger (and pre-trigger) signals are available from the laser for triggering of other devices
- *installation requirements*
  - size of laser head, power supply and cooler
  - necessary gas supply (quality of gases)
  - electrical power and voltage
  - weight
  - vibrational isolation
  - water and/or air cooling
  - air conditioning
  - dust freeness
  - ventilation, e.g. for ozone
- *possible distortions*
  - electromagnetic fields
  - distortions at power lines
  - mechanical vibrations
  - acoustic noise

Operation and Maintenance

- *operation requirements and maintenance*
  - necessary changes of gases or liquids such as dyes
  - effort required to change components such as diodes, flash lamps, laser tubes
  - cleaning
  - realignment cycles
  - lifetime of crystals, heating requirements
  - maintenance of vacuum and other pumps
- *handling*
  - warming-up time
  - computer controlling
  - education of operator

Prices and Safety

- *prices*
  - price and lifetime of the system

   – lifetime and price of expensive and short-lived components such as flash
     lamps, diodes, laser tubes, mirrors, crystals, thyratrons
   – operating price of electrical power, cooling, air conditioning, and so on
   – price of transportation, installation and maintenance
 • *safety conditions*
   – laser safety classes
   – electrical safety

   This catalog can be used as a checklist. A clear definition of which combinations of these parameters can be obtained is important.

6.13 Common Laser Parameters

In this section some basic values for some common types of mostly commercially available lasers are collected in tables. Details, variations and new
systems should be checked from the specialized literature, the large laser con-
ferences and with the laser companies. Some perspectives may be estimated
from the given references, representing mostly scientific results.

6.13.1 Semiconductor Lasers

Semiconductor laser diodes [6.1245–6.1249] are small devices with output
powers below 3 W. They consist of quantum well structures with typically
up to 20 gain guided regions on the same chip as shown in Fig. 6.93.

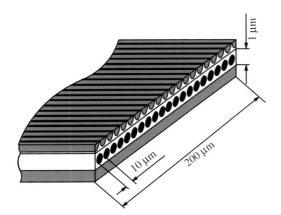

Fig. 6.93. Schematic of a single
diode laser consisting of 20 separate quantum well structures on
the same chip. The whole structure is about 1 mm deep

   These lasers are applied in data and communication technologies such as
in CD players, in fiber communication, in bar-code readers, as well as in laser
pointers. They can be coupled into fibers.

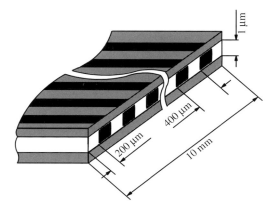

Fig. 6.94. Schematic of a diode bar emitting up to 100 W average output power. The whole structure is about 1 mm deep

Diode bars are arrangements of about 10 to 25 of such lasers on one chip as depicted in Fig. 6.94.

These bars produce more than 50 W average output power but this light shows very poor beam quality. Therefore these lasers are mostly used for diode pumping of high-power solid-state lasers. Bars cannot typically be air cooled as with single diodes. Peltier elements or water microcoolers have to be applied. Thermal expansion and other effect can lead to bending of these bars in the μm range, which is called smile.

For even higher powers these bars can be arranged in stacks of 3–20 or more bars. These stacks can be mounted together resulting in diode power block arrays containing 50 or more bars emitting many kW of diode laser light (see, e.g. Fig. 6.10). The beam quality of these arrangements decreases linearly because these lasers are not coherent with each other. Thus one of the future key questions is the coherent coupling for improving the beam quality of these lasers.

Further key issues of these lasers are lifetime, cost and reproducibility, e.g. of the center wavelength and divergence. New laser structures and new compounds may be developed in the next few years. Vertical emitting diodes (VCSELs) may serve as an example. Prices are expected to drop in the near future.

Single-Diode Lasers

Typical laser properties of today's diode lasers [6.1250–6.1274] are given in Table 6.14. These diode lasers are used as laser pointers, in CD players, for aligning purposes and for communication technologies. Laboratory setups may soon become commercial products. 1 W average output power with diffraction-limited beam quality has been reported using an external resonator [6.669]. Laser with blue emission are described in [6.1269–6.1274].

Table 6.14. Some typical properties of diode lasers

	GaAs			
Active material	GaAlAs or GaAs			
wavelength	630–1800 nm			
	780 nm for CD player, 633–675 nm pointer			
	around 1.3 and 1.55 µm for fiber communication			
Level scheme	similar to 4			
Emission cross-section	$1 \cdot 10^{-19}\,\mathrm{cm}^2$			
Lifetime upper laser level	4 ns			
Length of active material	0.2–2 mm			
Typical concentration	$10^{17}\,\mathrm{cm}^{-3}$			
Refractive index	3.5			
Operation mode	cw, quasi-cw (modulation up to 5 GHz)			
Pump mechanism	electric current at a voltage of 2–3 V			
	gain guided or index guided			
Setup	single emitter	coupled emitters	fiber coupled	external resonator
Bandwidth		4 nm		14 MHz
Average output power	1–10 mW	$\leq 2\,\mathrm{W}$	1–100 mW	$\leq 1\,\mathrm{W}$
Beam quality (M^2)	≈ 3	60 slow axis	≈ 1	≈ 1
		1 fast axis		
Wall-plug efficiency	$\leq 40\%$		$\leq 20\%$	$\leq 20\%$
Cooling system	air	air, Peltier	air, Peltier	air
Remarks		lifetime $\leq 10000\,\mathrm{h}$		

Diode Laser Bars, Arrays and Stacks

Diode laser grouping allows very high output powers from small laser devices with high wall-plug efficiency but poor beam quality [6.1275–6.1278].

Table 6.15. Some typical properties of commercial diode lasers and diode laser bars

	GaAs – bars, arrays and stacks			
Active material	GaAlAs or GaAs			
Wavelength	630–1800 nm			
	808 nm for Nd-laser pumping			
	940 nm for Yb-laser pumping			
Level scheme	similar to 4			
Emission cross-section	$1 \cdot 10^{-19} \, \mathrm{cm}^2$			
Lifetime upper laser level	4 ns			
Length of active material	0.2–2 mm			
Typical concentration	$10^{17} \, \mathrm{cm}^{-3}$			
Refractive index	3.5			
Operation mode	cw			
	or pulsed by electrical switching with duty cycles $\leq 30\%$			
Pump mechanism	electric current with voltages of 2–3 V			
	gain guided or index guided			
Setup	diode bar	stack	array,	fiber coupled
	of 25 diodes	of e.g. 10 bars	e.g. 6 stacks	
Bandwidth	≈ 4 nm			
Average output power	30–50 W	≤ 500 W	≤ 3 kW	≤ 150 W
Peak power (pulsed)	≤ 200 W	≤ 2 kW	≤ 10 kW	≤ 600 W
Beam quality (M^2)	2000	2000	12 000	≈ 700
	slow axis,	and	and	
	1 fast axis	6000	6000	
		(300^*)	(300^*)	
Wall-plug efficiency	$\leq 40\%$			
Cooling system	Peltier,	water	water	Peltier,
	water			water
Remarks	lifetime ≤ 10000 h			

* With well-designed collimators for the fast axis of the emitted beam the beam quality can be improved to this value.

Vertical Cavity Surface-Emitting Lasers (VCSEL)

The short resonator length of VCSELs [6.1279–6.1287] allows narrow bandwidths of a few GHz, high-frequency stability and fast modulation of several GHz. Wavelength uniformity of less than 2 nm can be guaranteed. Arrays of 10s of single VCSELs can be produced to increase the output power. With active resonator length tuning using a small air gap between the output coupler and the active semiconductor, wavelength tuning over about 20 nm at 960 nm has been reported. Thus these lasers seem to be well suited e.g. for wavelength division multiplexing (WDM) in communication technologies.

Table 6.16. Some typical properties of vertical cavity surface-emitting lasers (VCSEL)

	VCSEL
Active material	InGaAs
Wavelength	760–970 nm
	(950 ± 20) nm, (850 ± 10) nm, (770 ± 10) nm
Level scheme	similar to 4
Emission cross-section	$1 \cdot 10^{-19}$ cm^2
Lifetime upper laser level	4 ns
Length of active material	wavelength (e.g. 0.950 μm)
Typical concentration	10^{17} cm^{-3}
Refractive index	3.5

Operation mode	cw, quasi cw (modulation up to 5 GHz)	
Pump mechanism	electric current with voltages of 2–3 V	
Setup	single emitter	arrays (10×10)
Bandwidth	≤ 0.1 nm possible	
Average output power	≈ 1 mW	150 mW
Beam quality (M^2)	1	200
		(8*)
Wall-plug efficiency	$\leq 57\%$	
Cooling system	air, Peltier	
Remarks	Lifetime $\leq 10\,000$ h	

* With special microlens collimators the beam quality can be improved to this value representing the number of linearly coupled lasers.

6.13.2 Solid-State Lasers

Solid-state lasers cover the whole field of photonics applications and thus a wide variety of systems is offered. Overviews can be found in [M25, 6.1288–6.1293]. Microchip lasers with output powers in the mW to W range are available as well as kW systems. PW systems are built with respect to fusion experiments. New laser and host materials may become important in the near future.

Typical laser atoms (ions) are Nd, Cr, Ti, Yb, Er, Pr. Solid-state laser host materials can be crystals, glasses, ceramics and organic matter. The crystals can be fluorides or oxides. Typical crystals are YAG ($Y_3Al_5O_{12}$), YALO or YAP ($YAlO_3$), sapphire (Al_2O_3), GGG ($Gd_3Ga_5O_{12}$), GSGG ($Gd_3Sc_2Al_3O_{12}$), LiSAF ($LiSrAlF_6$), LiCaF ($LiCaAlF_6$), fosterite (Mg_2SiO_4), YLF ($LiYF_4$) and YVO (YVO_4), KGW (KGD (WO_4)). Glass materials are phosphate or silicate glasses.

Many combinations have been tried in laboratory setups (see references in Subsections). The final results are crucially dependent on the quality of the investigated material. This is important for chemical impurities and for the optical quality. Therefore new combinations may become important in future and the technology for these materials will be developed to the necessary stage if the market demands it.

Other key issues for solid-state lasers in competition with other lasers are beam quality, efficiency, lifetime, reliability and price. The beam quality for high-power applications is often required to be below $M^2 < 5$ and the wall-plug efficiency should be better 10%.

Nd:YAG Lasers

Nd laser [6.1294–6.1373] are very common because of their simple construction, reliability and large variability. Besides Nd:YAG [e.g. 6.1294–6.1339], Nd:YALO (also called Nd:YAP) [e.g. 6.38, 6.187, 6.418, 6.692, 6.768, 6.1021, 6.1340–6.1344] laser rod geometries are also applied. Nd:YAG is usually diode pumped at 808 nm and Nd:YALO should be pumped at 803 nm. Frequency doubling (SHG), tripling (THG) and double doubling (THG) is very common with these lasers, producing green and blue light (see Sect. 6.15). High powers with good beam quality are realized with slabs [6.1345–6.1352].

Similar results are obtained with Nd:YLF [6.1353–6.1361]. The better thermal conductivity, smaller thermal lensing and natural birefringence allow better beam quality of diffraction-limited pulses up to about 50 W. The last fact is also true for Nd:YALO which shows higher efficiency. The stronger thermal lens can be compensated, e.g. by phase conjugating mirrors [6.418]. Other Nd materials are investigated in [6.1362–6.1374].

Table 6.17. Some typical properties of commercial Nd:YAG lasers

		Nd:YAG laser		
Active material		$Nd^{3+}:Y_3Al_5O_{12}$		
Wavelength		1064 nm		
Level scheme		4		
Emission cross-section		$3.2 \cdot 10^{-19}\,cm^2$		
Lifetime upper laser level		230 μs		
Length of active material		5–200 mm		
Typical concentration		$10^{19}\,cm^{-3}$		
Refractive index		1.82		
Operation mode	cw	spiking	ns	ps
Pump mechanism	arc lamp diode laser		flash lamp diode laser (808 nm) ($\sigma_{808\,nm} = 4 \cdot 10^{-20}\,cm^2$)	
Pulse width	–	60 μs–10 ms	1–100 ns	25–500 ps
Bandwidth	0.001–0.1 nm	10^{11} Hz	0.1–3 GHz	10^{11} Hz
Average output power	0.1–6 kW	0.1 W–1 kW	0.5–250 W	1–30 W
Pulse energy	–	≤ 100 J	1 mJ–100 J	0.5–50 mJ
Repetition rate	–	1 Hz–50 kHz	≤ 50 kHz	1–100 Hz
Beam quality (M^2)		TEM_{00} to multimode		
Wall-plug efficiency	up to 1%	0.5–4%	0.1–2%	≤ 1%
Diode pumping	< 20%	< 20%	< 10%	< 10%
Cooling system	water	water	water	water
Remarks	multirod laser possible		amplifier systems	single pulse and amplifier

Nd:YVO Lasers

Another laser crystal with good performance is Nd:YVO [6.1375–6.1392]. It shows high efficiency in diode pumping and can thus be used to built cw lasers with probably up to 80 W average output power and very good beam quality.

This laser is typically end-pumped with diode lasers and water cooled. The available crystal size is currently limited to a few cm and thermally induced stress limits the maximum output power from these lasers. A similar promising material is Nd:KGW (Nd:KGd(WO$_4$)) [6.922, 6.1370] which shows low threshold, natural birefringence and high efficiency. Nd:YVO can be diode pumped at 809 nm.

Table 6.18. Some properties of Nd YVO lasers

	Nd YVO laser	
Active material	Nd^{3+}:YVO$_4$	
Wavelength	1,069 nm, 1,342 nm	
Level scheme	4	
Emission cross-section	a-cut: $2.5 \cdot 10^{-18}$ cm^2 at 1,069 nm, $7 \cdot 10^{-19}$ cm^2 at 1,342 nm	
Lifetime upper laser level	50–90 μs (3at%–1at%)	
Length of active material	5–70 mm	
Typical concentration	$1.7 \cdot 10^{19}$ cm^{-3}	
Refractive index	1.95 (a) 2.17 (c)	
Operation mode	cw	ns
Pump mechanism	diode laser ($\sigma_{808} = 1.4 \cdot 10^{-19}$ cm^2)	
Pulse width	60 μs–10 ms	10–100 ns
Bandwidth	up to $8 \cdot 10^{-12}$ Hz	
Average output power	≤ 40 W	≤ 5 W
Pulse energy		0.5 mJ
Repetition rate	≤ 30 kHz	
Beam quality (M^2)	TEM$_{00}$ to multimode	
Wall-plug efficiency	$\leq 4\%$	
Cooling system	Water	

Nd Glass Laser

Nd glass laser [e.g. 6.1393–6.1404] can be produced in bigger active volumes compared to Nd:YAG laser crystals and is therefore used in applications with large mode diameters as for lasers with short pulses and very high peak powers. Beam diameters of several 10 cm are used for special applications. The bad thermal conductivity of this material excludes it from applications with high average output powers in usual rod laser systems.

Table 6.19. Some typical properties of Nd glass lasers

	Nd glass laser		
Active material	Nd^{3+}:phosphate or silicate glass		
Wavelength	1054–1062 nm		
Level scheme	4		
Emission cross-section	$2.7–4 \cdot 10^{-20}\,cm^2$		
Lifetime upper laser level	290–340 µs		
Length of active material	5–500 mm		
Typical concentration	$10^{20}\,cm^{-3}$		
Refractive index	1.5–1.57		
Operation mode	spiking	ns	ps
Pump mechanism		flash lamp	
		diode laser	
Pulse width	60 µs–10 ms	1–100 ns	10–500 ps
Bandwidth		up to $8 \cdot 10^{12}\,Hz$	
Average output power		$\leq 1\,W$	
Pulse energy	$\leq 500\,J$	1 mJ–200 J	0.5–10 mJ
Repetition rate		$\leq 10\,Hz$	
Beam quality (M^2)		TEM_{00} to multimode	
Wall-plug efficiency		$\leq 1\%$	
Cooling system		water	

Yb:YAG Laser

Ytterbium:YAG as an active material [e.g. 6.1404–6.1413] has a very small quantum defect diode pumped at 940 nm and thus the thermal problems are strongly reduced. The laser crystal finds increasing applications in the high-power range, especially in material processing. The small thermal load allows very good beam quality even at the highest powers. Values close to $M^2 = 1$ are reported with average powers of 100–200 W. This crystal is used as rod or as a thin slices (see Sect. 6.3.1). Other Yb materials are investigated in [6.1414–6.1423].

Table 6.20. Some typical properties of commercial Yb:YAG lasers

	Yb:YAG laser	
Active material	$Yb^{3+}:Y_3Al_5O_{12}$	
Wavelength	1030 nm	
Level scheme	3	
Emission cross-section	$3.3 \cdot 10^{-20}$ cm^2	
Lifetime upper laser level	1160 µs	
Length of active material	10–80 mm	0.2–3 mm
Typical concentration	10^{20} cm^{-3}	$9 \cdot 10^{20}$ cm^{-3}
Refractive index	1.82	
Operation mode	cw, spiking, ns, ps, fs	
Pump mechanism	diode laser ($\sigma_{940\,nm} = 7.5 \cdot 10^{-21}$ cm^2)	
Pulse width	cw – 10 fs	
Bandwidth	monomode – nm	
Average output power	50 W	500 W
Pulse energy	0.1–8 J	0.1–1 J
Repetition rate	cw, Hz, kHz, MHz	
Beam quality (M^2)	≤ 10 (TEM$_{00}$)	
Wall-plug efficiency	$\leq 10\%$	$\leq 20\%$
Cooling system	water	water, Peltier
Remarks		

Ti:sapphire Laser

The Ti:sapphire laser [e.g. 6.1424-6.1449] is mostly used for the generation of short pulses in the ps and fs range and for tunable lasers at the red end of the visible spectrum. It is usually pumped by frequency-doubled Nd lasers around 530 nm. By doubling this radiation the blue spectral range can be covered. The green and yellow range in between can be generated by OPO or OPA systems which are commercially available, e.g. for fs and ps lasers (see Sect. 6.15.2).

The Ti:sapphire laser has become a workhorse in science and may find new applications in industry using short pulses. Thus it is very useful in measuring technologies. Another advantage may be its broad emission band enabling it to be as a broad-band laser source with good beam quality and high-powers.

Table 6.21. Some typical properties of commercial Ti:sapphire lasers

	Ti:sapphire			
Active material	$Ti:Al_2O_3$			
Wavelength	690-1100 nm			
Level scheme	4			
Emission cross-section	$3 \cdot 10^{-19} \, cm^2$ (at 800 nm)			
Lifetime upper laser level	3.2 µs			
Length of active material	3–30 mm			
Typical concentration	$10^{20} \, cm^{-3}$			
Refractive index	1.76 (birefringent)			
Operation mode	cw	ns	ps	fs
Pump mechanism	argon laser, SHG-Nd laser pump			
	longitudinal ($\sigma_{550\,nm} = 2$–$5 \cdot 10^{-20} \, cm^2$)			
Pulse width	–	2–100 ns	1–50 ps	5–100 fs
Bandwidth	≤ 2 GHz	≤ 1 nm	≤ 1 nm	60–100 nm
Average output power	≤ 50 W	1–2 W	≤ 1 W	≤ 1 W
Pulse energy	–	≤ 100 mJ	≤ 1 mJ	≤ 1 mJ
Repetition rate	–	1 Hz–40 Hz	≤ 1 kHz	≤ 1 kHz
Beam quality (M^2)	TEM_{00} to multimode		TEM_{00}	
Wall-plug efficiency	up to 8%	0.5–1%	0.1–1%	≤ 1%
Cooling system	water, Peltier cooler			
Remarks	laser pumping	flash lamp pumping	longitudinal pumped amplified	

Cr:LiCAF and Cr:LiSAF Lasers

These laser materials [6.1450–6.1473] can be pumped directly with diode lasers e.g. around 670 nm and thus very small devices can be produced. The emission is in the red and the near IR spectral region. They provide tunable or broad-band emission in the red and IR which can be frequency converted to the visible. Thus short pulses can be generated from handy lasers.

Table 6.22. Some properties of Cr:LiCAF and Cr:LiSAF lasers

	Cr:LiCAF		Cr:LiSAF
Active material	Cr^{3+}:LiCaAlF$_6$		Cr^{3+}:LiSrAlF$_6$
Wavelength	720–840 nm		780–1010 nm
Level scheme		4	
Emission cross-section	$1.3 \cdot 10^{-20}$ cm^2		$4.8 \cdot 10^{-20}$ cm^2
Lifetime upper laser level	170 µs		67 µs
Length of active material	≤ 20 mm		≤ 20 mm
Typical concentration	$7 \cdot 10^{20}$ cm^{-3}		$3 \cdot 10^{20}$ cm^{-3}
Refractive index	1.39		1.4
Operation mode	pulsed		cw, pulsed
Pump mechanism	flash lamp, laser, diode lasers (670 nm)		flash lamp, laser, diode lasers (680 nm)
Pulse width	≥ 10 ns		cw, ≥ 10 ns
Bandwidth	≤ 100 nm		≤ 200 nm
Average output power	W		1 W (cw), 100 mW (ns)
Pulse energy	≤ 100 mJ		≤ 100 mJ
Repetition rate	≤ 10 kHz		≤ 10 kHz
Beam quality (M^2)		TEM$_{00}$ to multimode	
Wall-plug efficiency	$\leq 10\%$		$\leq 2\%$
Cooling system		water, air	
Remarks		first results	

Cr:YAG [6.1474–6.1476] as active material provides laser emission in the IR around 1.5 µm. This laser can be pumped with, e.g. Nd:YAG laser radiation at 1.06 µm. The possible broad-band emission can be converted to the visible range. Besides alexandrite and ruby are other Cr lasers as Cr-fosterite [6.1477–6.1490] and as given in [6.1491, 6.1492] possible.

Alexandrite Laser

This laser crystal shows a wide spectral gain and can thus be used for tunable lasers [6.1493–6.1498]. The material can also be used in mode-locked lasers reaching 8 ps pulses.

Table 6.23. Some typical properties of alexandrite lasers

	Alexandrite laser		
Active material	Cr^{3+}:$BeAl_2O_4$		
Wavelength	700–818 nm		
Level scheme	4		
Emission cross-section	$1 \cdot 10^{-19}\,cm^2$		
Lifetime upper laser level	260 μs		
Length of active material	30–100 mm		
Typical concentration	$6 \cdot 10^{20}\,cm^{-3}$		
Refractive index	1.73–1.74 (birefringent)		
Operation mode	spiking		ns
Pump mechanism	flash lamp		
Pulse width	200 μs		20 ns
Bandwidth	≤ 100 nm, $5 \cdot 10^{-13}$ Hz		
Average output power	≤ 50 W		
Pulse energy	≤ 1 J		
Repetition rate	≤ 100 Hz		
Beam quality (M^2)	TEM_{00} (to multimode)		
Wall-plug efficiency	$\leq 2\%$		
Cooling system	water		
Remarks	twice the thermal conductivity of YAG		

Erbium (Er), Holmium (Ho), Thulium (Tm) Laser

These lasers [6.1499–6.1549] are used mainly in medical applications because of the high absorption of their radiation in water. These lasers are also eye-safe. The laser setup needs special optics for the IR wavelengths and is therefore not easy to achieve. The Q switch can be carried out with a frustrated total reflection (FTIR) shutter between two glass prisms with a narrow air gap which is modulated with a piezo-driver (see Sect. 6.10.2).

Table 6.24. Some properties of erbium and holmium lasers

	Er:YAG	CTH:YAG
Active material	$Er^{3+}:Y_3Al_5O_{12}$	$Cr^{3+}:Tm^{3+}:Ho^{3+}:Y_3Al_5O_{12}$
Wavelength	2940 nm	2080 nm
Level scheme	3	
Emission cross-section	$3 \cdot 10^{-20}\,cm^2$	$4.5 \cdot 10^{-19}\,cm^2$
Lifetime upper laser level	100 μs	3.6 ms
Length of active material	20–120 mm	50–100 mm
Typical concentration	$7 \cdot 10^{21}\,cm^{-3}$	$10^{17}\,cm^{-3}$
Refractive index	1.82	1.82
Operation mode	pulsed	pulsed
Pump mechanism	flash lamp, diode lasers (680 nm)	flash lamp
Pulse width	100–1000 μs, 20 ns	200–300 μs
Bandwidth	≤ 0.1 nm	
Average output power	≤ 50 W	≤ 40 W
Pulse energy	0.1–8 J	≤ 3.5 J
Repetition rate	1–50 Hz	20 Hz
Beam quality (M^2)	10 (TEM$_{00}$ to multimode)	
Wall-plug efficiency	0.2–3%	≤ 2%
Cooling system	water	water
Remarks	Q switch with FTIR-shutter	

Ruby Laser

The ruby laser [6.1, 6.166, 6.298, 6.620, 6.621, 6.742, 6.779, 6.850, 6.1111, 6.1791] was the first laser to be pumped with flash lamps. Despite the quite expensive ruby crystals this laser is still used because of the wavelength in the red and the high peak powers possible in Q switch operation. Rods of 10 mm diameter are available.

Table 6.25. Some typical properties of ruby lasers

		ruby laser	
Active material		$Cr^{3+}:Al_2O_3$	
Wavelength		694.3 nm	
Level scheme		3	
Emission cross-section		$2.5 \cdot 10^{-20} \, cm^2$	
Lifetime upper laser level		3 µs	
Length of active material		10–200 mm	
Typical concentration		$8 \cdot 10^{20} \, cm^{-3}$	
Refractive index		1.76 (birefringent)	
Operation mode	spiking	ns	ps
Pump mechanism		flash lamp	
Pulse width	200 µs	10–30 ns	10 ps
Bandwidth		0.53 nm (3.3 GHz)	
Average output power		1 W	
Pulse energy		≤ 100 J	
Repetition rate		≤ 5 Hz	
Beam quality (M^2)		TEM_{00} (to multimode)	
Wall-plug efficiency		$\leq 1\%$	
Cooling system		water	
Remarks		long thermal relaxation time	

Fiber Lasers

Fiber lasers [6.1550–6.1582] have good cooling conditions and thus the thermal problems are negligible. Single-mode fibers allow perfect beam quality. The small out-coupling fiber cross-section limits the maximum peak power of such lasers. Nevertheless, fiber lasers may become more important in the future. As an example the data of an Er:fiber laser are given in Table 6.26.

Table 6.26. Some properties of Er fiber lasers

	Er fiber
Active material	Er^{3+}:glass
Wavelength	550 nm, *1,550 nm*, 3,500 nm
Level scheme	3
Emission cross-section	$3 \cdot 10^{-20}\,cm^2$
Lifetime upper laser level	8 ms
Length of active material	0.3–5 m
Typical concentration	$3 \cdot 10^{20}\,cm^{-3}$
Refractive index	1.53
Operation mode	cw, pulsed
Pump mechanism	(Ti:sapphire) laser, diode lasers
Pulse width	cw, $\geq 100\,ns$
Bandwidth	$\leq 0.1\,nm$
Average output power	$\leq 1\,W$
Pulse energy	$\leq 100\,mJ$
Repetition rate	$\leq 2\,kHz$
Beam quality (M^2)	TEM_{00}
Wall-plug efficiency	depends on pump laser
Cooling system	air
Remarks	further development

Other active atoms in fibers may show similar or even better results. Thus further lasers may enter the market in the near future.

6.13.3 Gas Lasers

Gas laser are pumped via electrical discharges and inelastic collisions of the electrons with the laser molecules or atoms (see Sect. 6.3.3).

XeCl, KrF and ArF Excimer Lasers

Excimer lasers [e.g. 6.1583–6.1585] can be operated with XeCl, KrF, ArF and XeF excimer molecules. These lasers emit at least partly superradiation and thus the beam quality is very poor. Nevertheless the high average output powers in the UV make them attractive light sources, e.g. for pumping of pulsed dye lasers, for chemistry, for industrial applications in lithography and material processing as well as in medicine.

Table 6.27. Some typical properties of commercial XeCl and KrF lasers

	XeCl	KrF	ArF
Active material	XeCl-excimer	KrF-excimer	ArF-excimer
Gas mixture (example)	80 mbar of 5% HCL in He 60 mbar of Xe 2760 mbar of Ne 500 mbar of He	7.5 mbar of F_2 22.6 mbar of Kr 1100 mbar of He	1.3 mbar of F_2 40 mbar of Kr 1100 mbar of He
Wavelength	308 nm	248 nm	193 nm
Level scheme		4	
Emission cross-section	$4.5 \cdot 10^{-16}\,\mathrm{cm}^2$	$2.4 \cdot 10^{-16}\,\mathrm{cm}^2$	$2.9 \cdot 10^{-16}\,\mathrm{cm}^2$
Lifetime upper laser level	11 ns	7 ns	4.2
Length of active material		50–1500 mm	
Typical concentration		$10^{14}\,\mathrm{cm}^{-3}$	
Refractive index		$\simeq 1$	
Operation mode		ns	
Pump mechanism		transversal electrical discharge: 10–30 kV, kA, ns Electron temperature $\approx 5\,\mathrm{eV}$	
Pulse width		5–20 ns	
Bandwidth	2 nm	0.5 nm	
Average output power		1–100 W	
Pulse energy		100 mJ–10 J	
Repetition rate		10 Hz–1 kHz	
Beam quality (M^2)		multimode: beam size $\approx 5 \times 20\,\mathrm{mm}^2$, divergence $\approx 1\,\mathrm{mrad}$	
Wall-plug efficiency		$\leq 2\%$	
Cooling system		water for powers $\leq 10\,\mathrm{W}$	
Remarks		gas exhaust, gas exchange weekly to monthly	

N₂ Laser

Nitrogen lasers [e.g. 6.1586–6.1588] are useful as a low-cost UV light source for direct use or for pumping of pulsed dye lasers. Because of the long lifetime of the lower laser level the maximum average output power is limited and thus excimer lasers are often favored instead of N_2 lasers.

Table 6.28. Some typical properties of nitrogen lasers

	nitrogen laser	
Active material	N_2	
Wavelength	337.1 nm	
Level scheme	3	
Emission cross-section	$4 \cdot 10^{-13}\,\mathrm{cm}^2$	
Lifetime upper laser level	40 ns	
Length of active material	100–1000 mm	
Typical concentration	$10^{11}\,\mathrm{cm}^{-3}$	
Refractive index	$\simeq 1$	
Operation mode	ns	sub ns
Pump mechanism	transversal electrical discharge: 10–30 kV, kA, ns electron temperature 10 eV	
Gas mixture	100 mbar of N_2	1 bar of N_2
Pulse width	2–8 ns	500 ps
Bandwidth	0.1 nm	
Average output power	3 W	$\leq 1\,\mathrm{W}$
Pulse energy	$\leq 100\,\mathrm{mJ}$	$\leq 5\,\mathrm{mJ}$
Repetition rate	$\leq 200\,\mathrm{Hz}$	
Beam quality (M^2)	multimode: beam size in the range of $\leq 5 \times 20\,\mathrm{mm}^2$, divergence in the range of 1–5 mrad	
Wall-plug efficiency	$\leq 0.1\%$	
Cooling system	air	
Remarks	N_2 flow of 2 to 30 l/min at 100 mbar necessary for high-powers	relatively compact

Home Made N₂ Laser

The principle of the nitrogen laser is so simple that it is possible to built such a laser almost completely from scratch. The scheme is given in Fig. 6.95.

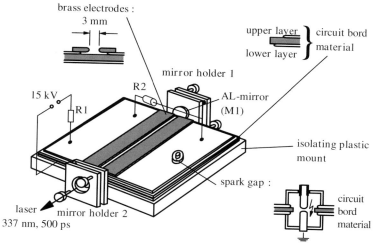

Fig. 6.95. N_2 laser with 500 ps pulse width at 337 nm as a self made construction. The foot print of this device is about 300 mm × 500 mm

Air at pressure can be used as the active material. A simple Al mirror can be used as mirror M1 and a quartz plate as mirror M2. M2 can even be leaved out and the laser will emit superradiation.

The only expensive device in this construction is the electric power supply providing more than $U_{source} \geq 10$ kV. From the resistor $R_1 \geq (U_{source}/I_{source})_{max}$ based on the maximum current I_{source} of the power supply it follows that the maximum repetition frequency of the laser with capacitance C is $f_{rep} \approx 1/2\pi(R_1 C)$. R_2 has to be somewhat smaller than R_1.

The main capacitors are made from electric circuit board material providing a capacitance of 3 pF cm^{-2}. Thus the whole capacitance is about 4 nF. Care has to be taken for avoiding sharp edges in the construction of the high-voltage elements, otherwise sparks may occur and damage the system.

Also the spark gap [6.1588] can be home-made simply from brass. Its distance has to be adjusted for breakthrough slightly below maximum voltage of the power supply.

The laser can be easily upgraded in performance and reliability by using good optics, a thyratron as switch, a sealed discharge chamber with tungsten electrodes filled with N_2 and Cu sheets of two or more mm thickness with rounded edges in the high-voltage part against unwanted discharges.

He-Ne Laser

Helium-neon lasers are used typically with red light at 633 nm for aligning or other low power applications. The good beam quality, availability of linear or circular polarization and high frequency stability of these lasers make them useful for calibration and aligment problems.

Table 6.29. Some typical properties of He-Ne lasers

	He-Ne laser
Active material	Ne
Wavelengths	543.3 nm (green), 594.1 nm (yellow), 611.8 nm,
	632.8 nm (red),
	1152.3 nm, 1523.1 nm, 2395.1 nm, 3391.3 nm
Level scheme	4
Emission cross-section	$3 \cdot 10^{-13}$ cm^2 (632 nm), $2 \cdot 10^{-14}$ cm^2 (543 nm)
Lifetime upper laser level	170 ns (633 nm)
Length of active material	100–1500 mm
Typical concentration	$3 \cdot 10^9$ cm^{-3}
Refractive index	$\simeq 1$
Operation mode	cw
Pump mechanism	longitudinal electrical discharge,
	inelastic collision He*+Ne→He+Ne*
	electron temperature ≈ 10 eV
Gas mixture	He:Ne as 5:1 (for 633 nm); and,
	e.g. as 9:1 (for 1.15 μm)
Bandwidth	1.5 GHz (633 nm), 1.75 GHz (543 nm)
Average output power	0.5–50 mW
	typical 5 mW at 632 nm
Beam quality (M^2)	TEM$_{00}$
Wall-plug efficiency	$\simeq 0.1\%$
Cooling system	air
Remarks	reliable

Mode-locked He-Ne lasers have been reported to produce ps pulses at the red line [6.1779]. The repetition rate was MHz. In cw-operation the bandwidth is limited by Doppler broadening in the active gas mixture.

He-Cd Laser

Helium-cadmium lasers are useful in low-power cw applications in the blue spectral range, (see also [6.1589]).

Table 6.30. Some typical properties of He-Cd lasers

	He-Cd laser
Active material	Cd^+ (300 C)
Wavelength	(325.0 nm, 353.6 nm) 441.6 nm
Level scheme	4
Emission cross-section	$9 \cdot 10^{-18} \, cm^2$
Lifetime upper laser level	810 ns
Length of active material	0.25 m–1.5 m
Typical concentration	$4 \cdot 10^{16} \, cm^{-3}$
Refractive index	$\simeq 1$
Operation mode	cw
Pump mechanism	longitudinal electrical discharge: 10–30 kV, kA, ns electron temperature $\approx 6 \, eV$
Gas mixture	10 mbar of He, 0.1 mbar of Cd
Bandwidth	0.1 nm
Average output power	10–200 mW
Beam quality (M^2)	TEM_{00} or multimode
Wall-plug efficiency	$\leq 0.1\%$
Cooling system	air
Remarks	

Ar and Kr Ion Lasers

Argon and krypton lasers [6.1590–6.1593] are very common in science for cw applications and quasi-cw mode-locked lasers in the ps and fs range. Thus, e.g. fs CPM dye lasers (see Sect. 6.10.3) or Kerr lens mode-locked Ti:sapphire lasers (see Sect. 6.10.3) are pumped with a 5 W Ar laser. These lasers have very high operating costs caused by the neccessary two-year exchange of the expensive laser tube, and the low efficiency.

Table 6.31. Some typical properties of commercial Ar and Kr ion lasers

	Ar and Kr ion laser	
	Ar^+	Kr^+
Active material		
Wavelength	514.5 nm, 488.0 nm	647.1 nm
Level scheme	3	
Emission cross-section	$2.5 \cdot 10^{-12}\,cm^2$	
Lifetime upper laser level	9 ns	
Length of active material	(0.5 m–)2 m	
Typical concentration	$2 \cdot 10^9\,cm^{-3}$	
Refractive index	$\simeq 1$	
Operation mode	cw, mode-locked	
Pump mechanism	longitudinal electrical low pressure discharge: 30–$150\,A\,cm^{-2}$ electron temperature $\approx 30\,eV$	
Gas pressure	0.01–1 mbar	
Pulse width	cw or 500 ps	
Bandwidth	4-12 GHz single line, multi line	
Average output power	1–3 W in single line 10 W in strong lines (e.g. 488 nm) 20 W in multiline	
Pulse energy	$\leq 100\,mJ$	$\leq 5\,mJ$
Repetition rate	MHz in mode locking regime	
Beam quality (M^2)	TEM_{00}	
Wall-plug efficiency	$\leq 0.1\%$	
Cooling system	water (up to 60 kW)	
Remarks	magnetic field within discharge tube to increase current density, automatic gas refill system, expensive gas tubes have to be replaced (2 years)	

Cu Vapor Lasers

Lasers with copper vapor as the active material [6.1594–6.1607] provide high average output powers of several 10 W in the green and yellow spectral region with high repetition rates. The beam quality can be excellent and thus these lasers find applications from spectroscopy to material processing.

Table 6.32. Some typical properties of Cu-vapor lasers

	Cu vapor laser	
Active material	Cu vapor (1480–1530°C)	
Wavelength	510.6 nm	578.2 nm
Level scheme	3	
Emission cross-section	$8.6 \cdot 10^{-14}\,\mathrm{cm}^2$	$1.25 \cdot 10^{-13}\,\mathrm{cm}^2$
Lifetime upper laser level	500 ns	610 ns
Length of active material	0.5 m–2 m	
Typical concentration	$8 \cdot 10^{13}\,\mathrm{cm}^{-3}$	
Refractive index	$\simeq 1$	
Operation mode	pulsed	
Pump mechanism	longitudinal electrical discharge tube temperature 1500°C electron temperature $\approx 5\,\mathrm{eV}$	
Gas mixture	1 mbar Cu vapor, 40 mbar buffer	
Pulse width	10–50 ns	
Bandwidth	3 GHz	
Average output power	5–70 W	
Pulse energy	1 mJ–50 mJ	
Repetition rate	1–100 kHz	
Beam quality (M^2)	TEM_{00}	
Wall-plug efficiency	1%	
Cooling system	water, air	
Remarks	maintenance each 500 h	

The same laser construction operates with gold or lead. The laser wavelengths are then 627.8 nm and 722.9 nm.

CO_2 Lasers

CO_2 lasers [e.g. 6.1608–6.1614] emit in the far IR at 10.6 μm with possibly very high average output powers and pulse energies. They are very efficient. Thus material processing in machinery especially in the car industry and in medicine are the main applications.

Table 6.33. Some typical properties of CO_2 lasers

	CO_2 laser	
Active material	CO_2	
Wavelength	10 600 nm (9400 nm)	
Level scheme	4 at low temperatures → 3 at high temperatures	
Emission cross-section	$1 \cdot 10^{-16}$ cm^2	
Lifetime upper laser level	10 μs	
Length of active material	0.3 m–2 m	
Typical concentration	$3 \cdot 10^{-17}$ cm^{-3}	
Refractive index	$\simeq 1$	
Operation mode	cw	pulsed
Pump mechanism	transversal electrical discharge, DC or AC	
	electron temperature ≈ 4 eV	
Gas mixture	rapid gas flow, CO_2:N_2 = 0.8:1	
	gas temperature 300°C	
	20 mbar	1 bar
Pulse width	cw	45 ns–15 μs
Bandwidth	$6 \cdot 10^7$ Hz	
Average output power	typical 5 kW, ≤ 100 kW	1 kW
Pulse energy		≤ 10 kJ
Repetition rate		≤ 1 kHz
Beam quality (M^2)	multimode or TEM$_{00}$	
	(long wavelength ⇒ large beam parameter product)	
Wall-plug efficiency	10–20%	$\leq 30\%$
Cooling system	water	

This laser acts between vibrational levels of the molecule. Because of the long wavelength in the far IR the focused spot size is, for a diffraction-limited beam more than 10 times larger than for visible lasers. The CO laser is operated in a similar way. Its emission wavelength is in the range of 5–6 μm.

6.13.4 Dye Lasers

Dye lasers can in principle be built with wavelengths between 300 and 1200 nm with tuning ranges of several 10–100 nm [6.1615–6.1619]. As example the tuning curves of several dyes in a commercial laser with ns pulse emission are given in Fig. 6.96.

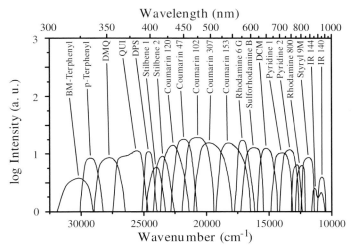

Fig. 6.96. Tuning curves of several dyes in a commercial dye laser with pulsed excitation [6.1617]

These dyes have a limited lifetime as shown for some examples in Table 6.34 and thus they are often limited to scientific applications.

Table 6.34. Life time of some laser dye solutions (after [6.1617])

Dye	Solvent	Wavelength (nm)	Excimer pumped	cw pumped
p-terphenyl	cyclohexane	340	451 Wh	
Polyphenyl 1	dioxane	380	870 Wh	
Stilbene 3	methanol	430	14 Wh	300 Wh
Coumarine 102	methanol	480	244 Wh	100 Wh
Rhodamine 6G	methanol	590	316 Wh	1000 Wh
DCM	DMSO	650	348 Wh	500 Wh
Rhodamine 700	methanol	700	80 Wh	1000 Wh
Styryl 9	DMSO	840	73 Wh	500 Wh
HITCI	DMSO	875	12 Wh	100 Wh

cw and Quasi-cw (Mode-Locked) Dye Lasers

Continuously operating dye lasers [e.g. 6.1620–6.1623] are typically pumped with ion gas lasers. They can show narrow bandwidths in cw operation. The dye laser can be mode-locked or pumped with mode-locked pulses resulting in ps or fs dye laser pulses. The dye solution is used in jets.

Table 6.35. Some typical properties of cw dye lasers

	cw-dye laser	
Active material	laser dyes	
Wavelength	410–890 nm	
Level scheme	4	
Emission cross-section	$\leq 10^{-16}\,\mathrm{cm}^2$ (maximum)	
Lifetime upper laser level	1–10 ns	
Length of active material	$\leq 0.5\,\mathrm{mm}$	
Typical concentration	$10^{17}\,\mathrm{cm}^{-3}$	
Refractive index	1.4 (solvent dependent)	
Operation mode	cw	
Pump mechanism	laser pumped: Ar-ion, SHG of Nd laser	
	dye jet, focus diameter $\approx 50\,\mu\mathrm{m}$	
Bandwidth	broad band	1 MHz ($\leq 0.5\,\mathrm{MHz}$ possible)
Average output power	1.2 W	$\leq 0.8\,\mathrm{W}$
Beam quality (M^2)	TEM$_{00}$	
Opto-optical efficiency	$\leq 13\%$	20%
Cooling system	water	
Remarks	dye jet with pump,	
	dyes need to be changed (approximately weekly),	
	frequency stabilization possible with active	
	resonator control	

Pulsed Dye Lasers

Spiking dye lasers are built with flash lamp pumping. Mode-locked fs dye lasers are usually pumped with cw lasers. Dye lasers with ns or ps pulses are pumped by ns pump lasers such as e.g. excimer, nitrogen or frequency-converted solid-state lasers [6.1624–6.1649]. Dyes in polymers may allow new lasers [6.1626–6.1641].

Table 6.36. Some typical properties of pulsed dye lasers

	pulsed dye laser			
Active material	dyes			
Wavelength	580–650 nm	300–1200 nm	400–900 nm	570–650 nm
Level scheme	4			
Emission cross-section	$\geq 10^{-16}$ cm^2 (maximum)			
Lifetime upper laser level	1–10 ns			
Length of active material	50–200 mm	5–50 mm	0.3 mm	0.3 mm
Typical concentration	10^{20} cm^{-3} (concentration of 10^{-2}–10^{-4} mol l^{-1})			
Refractive index	1.5 (solvent dependent)			
Operation mode	spiking	ns	ps	fs
Pump mechanism	flash lamp transversal dye cell	excimer or nitrogen laser dye cell	ps ion laser dye jet	cw laser dye jet
Pulse width	60 µs–10 ms	1–30 ns	1–50 ps	≥ 50 fs
Bandwidth	up to $8 \cdot 10^{-12}$ Hz possible	6 GHz 30 MHz	sub nm	nm
Average output power	several W	1 W	100 mW	50 mW
Pulse energy	several J	≤ 1 mJ	0.1 mJ	400 pJ
amplified		10 mJ	5 mJ (10 Hz)	1 mJ (10 Hz)
Repetition rate	≤ 10	1–100 Hz	≥ 50 MHz	≈ 100 MHz
Beam quality (M^2)	TEM$_{00}$ to multimode	multimode	TEM$_{00}$	TEM$_{00}$
Opto-optical efficiency	10%	$\leq 15\%$	$\leq 5\%$	$\leq 1\%$
Cooling system	dye circulation, water, air			
Remarks	high-power possible	usually amplified, dye exchange weekly		

6.13.5 Other Lasers

XUV-lasers [6.1650–6.1690] can be built by laser-induced plasma generation, e.g. with metal atoms. The resulting emission shows wavelengths of a few nm up to 40 nm. In particular, laser radiation in the transmission window of water from 2 nm to 4 nm will find applications, e.g. in microscopy of biological material. Another important field is lithography for chip production. Even before laser action is obtained point source emission can be used. For this, atoms like e.g. Al, Au and W are used. The observed laser pulses have energies in the mJ range. Incoherent light sources from table-top pump lasers have average output powers of some 10 mW in the nm region. The possible availability of well-designed solid-state lasers with high average output powers in the 100 W range, perfect beam quality and pulse energies of several J during less than 10 ns may promote these light sources in the near future.

The *free electron laser* [e.g. 6.1691–6.1696] can show a wide range of emission wavelengths from the X ray (0,1 nm), XUV (1–100 nm) and in principle to radio waves. The amplification takes place in an electron beam in series bent between the Wiggler magnets of, e.g. a synchrotron. Thus average output powers of 10 W with short pulses of 100 fs can be obtained.

Color center lasers [e.g. 6.1697–6.1704] operate in the near infrared wavelength range from 0.8 to about 4 µm with up to 100 mW average output power. The active material is some mm long and is made from crystals such as e.g. NaF, built from K, Na or Li atoms at one side and F or Cl atoms at the other. These crystals are X-ray irradiated to provide defects in the crystal structure which act as a quantum well for the charges. These F centers have quantum energy levels which provide the laser transition in a four-level scheme. They are laser pumped with wavelengths between 500 nm and 1.2 µm. Unfortunately, the available laser crystals have short lifetimes of days to months.

Far-infrared lasers [e.g. 6.1705–6.1709] can be made in the wavelength range above 30 µm using vibrational transitions of molecules in the gas phase, as e.g. HCN. These lasers can be pumped by electrical discharges or with IR lasers, e.g. CO_2 lasers. The average output power can reach a few 10 mW.

New solid-state lasers [6.1709–6.1733] with wavelengths in the visible spectral range or with better thermal properties may be developed in the future. The possibilities of diode pumping allow special constructions of microchip lasers even with frequency conversion inside the resonator (see also Sect. 6.2 and references ther).

New diode lasers [6.1730–6.1739] such as vertical emitting constructions or with new compounds may become important, soon. In particulary the green and blue spectral range may be filled. Therefore the technology of II–VI compounds such as e.g. ZnSe may be more developed. Several mW around 500 nm have already been reported.

Chemical lasers [e.g. 6.1740–6.1749] have already been mentioned in Sect. 6.3.5. They can produce very high average output powers for a short time and are therefore usually specialized for military applications.

6.14 Modification of Pulse Structure

Methods for generation of short pulses directly in lasers are described in Sect. 6.10. Some further effort may be necessary to select single short pulses from a pulse train in the ps range or to compress Q switch pulses in the ns range or to shorten fs pulses, externally. Measurement methods for determining the pulse widths of laser pulses are discussed in Sect. 7.1.5.2 and the references therein.

6.14.1 Single Pulse Selection

Mode-locked lasers mostly produce trains of ps or fs pulses (see e.g. Fig. 6.76). For some applications the repetition frequency of these pulses, typically of some 10 MHz can be too high and sometimes even single pulses are needed.

Single pulse selection out of a train of ps pulses as they are generated, e.g. from a ps solid-state laser, can be obtained by a fast gate using a Pockels cell and a polarizer as shown in Fig. 6.97.

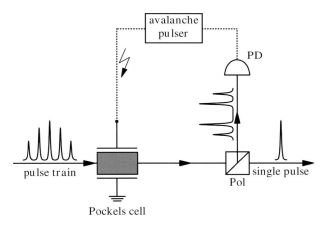

Fig. 6.97. Single pulse selection with Pockels cell and polarizer (Pol). The fast trigger is made with an avalanche transistor trigger producing ns pulses with slopes of 1 kV/ns

Another possibility is the cavity damping of a quasi-cw operated ps laser similar to that depicted for a cw-laser in Fig. 6.71.

Further, the repetition rate can be drastically decreased by a regenerative amplifier from MHz to kHz or some Hz. Amplifiers for short pulses in the ps or fs range can be pumped with ns pulses and if the lifetime of the upper laser level is in the ns range these amplifiers will amplify with their own repetition rate, only. This principle can be applied, e.g. in dye or Ti:sapphire amplifiers.

6.14.2 Pulse Compression and Optical Gates

The duration of short pulses can be decreased externally using gates or nonlinear effects for ns pulses and compressors for ns, ps or fs pulses.

Pulse Compression of fs Pulses

Compression of pulses [6.1750–6.1780] down to fs widths can be applied if these pulses show a frequency chirp. This chirp can be generated by self-phase modulation in an external device, e.g. a fiber, see Sect. 4.5.7.

Thus the combination of an optical fiber with a grating compressor allows the shortening of pulses down to some fs. The principle is depicted in Fig. 6.98.

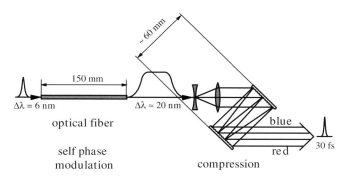

Fig. 6.98. Pulse compression using self-phase modulation in a single-mode polarization-conserving optical fiber and a grating compressor

The original laser pulse had a duration of about 90 fs and a pulse energy of 0.6 nJ at 619 nm. The fiber was 3.3 μm in diameter and thus the pulse peak power was about $5 \, \text{GW cm}^{-2}$. Shortest pulses generated with this scheme were 4.5 fs long [6.1764, 6.1771].

Pulse compression of ns Pulses

Pulses of about 10 ns pulse duration can be compressed by a factor of about 10 using stimulated Brillouin or Raman scattering with good energy conservation [6.1781–6.1798]. The scheme is shown in Fig. 6.99.

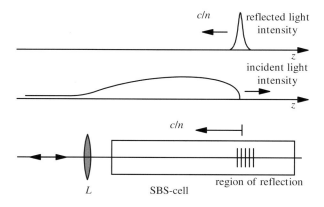

Fig. 6.99. Compression of ns pulses using stimulated Brillouin scattering in a long-focusing geometry

In this scheme the zone of large SBS reflectivity is moving towards the entrance window of the cell. Thus the incident beam is reflected at different positions and finally a compression similar to snow-shoving takes place. If the nonlinear conditions are suitable chosen, which is not simple, perfect compression occurs.

Pulse Shortening by Nonlinear Effects

Each nonlinear effect such as harmonic generation or nonlinear absorption will change the pulse duration of short pulses [e.g. 6.1799–6.1802]. If the nonlinear effect is not saturated the exponent of the nonlinear effect temporally shortens Gaussian-shaped pulses by the square root of the exponent.

In addition nonlinear absorption causes losses of the pulse energy and is therefore rarely used. Nevertheless, it keeps the original wavelength of the light and thus is useful in the low-power section of the laser system.

Pulse Shortening with Gates

Slices of pulses can be obtained using optical shutters such as Pockels cells or Kerr cells (see Sect. 4.5.2 and 6.14.1) or via other effects [6.1803–6.1808]. Electro-optic shutters are usually limited to widths larger than 1 ns. If optically driven Kerr cells are used, very fast shutters can be made and thus pulse durations of ps or even fs are possible. This method can be used to observe the dynamics of processes such as e.g. fluorescence, or to take photographs of the short pulses.

Optical Gating with Up-Conversion

Optical gating can also be achieved by up-conversion of the original light via a nonlinear frequency transformation [e.g. 6.1809–6.1811]. The scheme is given in Fig. 6.100.

The sum frequency generation occurs during the presence of the short pump pulse, only. Thus the decay of the probe signal can be obtained by delaying the pump pulse. This method can be used for the investigation of fast fluorescence decay times, e.g. of organic molecules.

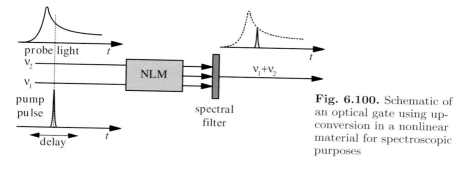

Fig. 6.100. Schematic of an optical gate using up-conversion in a nonlinear material for spectroscopic purposes

6.15 Frequency Transformation

Laser radiation can be transformed into the harmonic frequencies by doubling and mixing processes and to other frequencies by parametric devices. Further, Raman media can be used to shift the laser frequency.

Physical details are described in Chap. 4 and technical details should be extracted from catalogues. In all cases the beam quality plays a key role for the efficiency of the transformation. High values of above 50% are possible. Typical values are above 10% but in special cases, such as reaching the far UV, efficiencies below 10^{-3} are possible.

Further crucial points are the long-term stability of the nonlinear materials especially for radiation below 300 nm and temperature control of the crystals can be demanding. Moreover the design of these devices and the spot sizes have to optimized for high efficiency on one hand and no damage on the other. For material parameters see [6.1812–6.1832] and the references of Sect. 4.4.1.

6.15.1 Harmonic Generation (SHG, THG, FHG, XHG)

The generation of the second (SHG) [6.1833–6.1934], third (THG) [6.1935–6.1942] and fourth (FHG) [6.1943–6.1954] harmonics producing laser light with $\lambda_{laser}/2$, $\lambda_{laser}/3$ and $\lambda_{laser}/4$ is quite common for pulsed solid-state and dye lasers. SHG is also applied for high-power cw lasers. Further harmonics (see [6.1955] and references in Sect. 4.6) show poor efficiency.

Materials and schemes for harmonic generation are discussed in Sects. 4.4.1–4.4.3 and 4.5.1. As an example the frequency transformation of the light from a Q switched Nd:YAG laser with a transverse fundamental mode and longitudinal single mode is shown in Fig. 6.101–6.103. Figure 6.101 shows a typical parameter set for SHG.

An efficiency of to 50% was observed for this laser for stable operation. The crystal was 7 mm long and 7 mm in wide.

This configuration can be used for further third-harmonic generation (THG) but with different optimization. The scheme is shown in Fig. 6.102.

As mentioned in Sect. 4.5.1, stepwise transformation with SHG and then mixing of the fundamental with the SHG light is much more efficient than

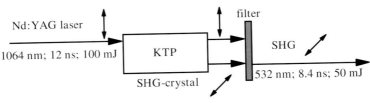

Fig. 6.101. Frequency doubling (SHG generation) of Q switched Nd:YAG laser light. The fundamental wave can be blocked with, e.g. a dielectric mirror as filter

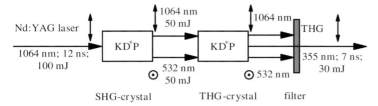

Fig. 6.102. Frequency tripling (THG generation) of Q switched Nd:YAG laser light by SHG and mixing of the second harmonics with the fundamental

direct tripling. The second crystal was KD*P and the overall efficiency for the third harmonics was 30%. Care has to be taken for the polarization demands of the crystal. In the case shown the polarization of the fundamental and the SHG are perpendicular. For example BBO crystals which are, e.g. used for the tripling of Ti:sapphire laser radiation, demand parallel polarization of these lights. In this case a complicated arrangement is necessary for the rotation of tunable short light pulses from a fs laser.

Fourth-harmonic generation (FHG) is just twice the second harmonic generation as shown in Fig. 6.103.

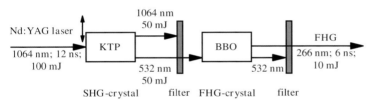

Fig. 6.103. Fourth-harmonic generation (FHG) as a series of two SHGs of Q switched Nd:YAG laser light

The overall efficiency is 10% in this case. The efficiencies of the two SHGs are different because the crystals for the different wavelength ranges have different coefficients. The lifetime of the FHG crystal is crucial for average output powers in the range of 1 W and above.

Higher harmonics can be produced, e.g. in atom vapors (see Sect. 4.6). The efficiency of these conversions is very small.

6.15.2 OPOs and OPAs

Optical parametric oscillators (OPO) and amplifiers (OPA) were described in Sect. 4.4.4 and in references [6.1998–6.2008]. Commercial devices are available for use with pulsed lasers from ns to fs.

The achievement of good frequency stability can be difficult with these devices. Further, narrow bandwidth and good beam quality are difficult to obtain. Therefore different schemes have been developed to combine the OPO with other sources such as e.g. with a dye laser as a seeder for the required radiation.

The whole spectrum from UV to IR is covered continuously by OPOs or OPAs in combination with SHG, THG and FHG. For a commercial Ti:sapphire laser the OPA is specified with following pulse energies in the fs range covering the range from 300 nm to 3 μm (see Table 6.37).

Table 6.37. Pulse energies of a commercial OPA in the fs range pumped by a 1 kHz Ti:sapphire laser of 80 fs pulse duration and 750 μJ pulse energy

Method	Wavelength (nm)	Pulse energy (μJ)	Pulse width (fs)	Stability (%)
Idler	2080	≥ 25	≤ 100	≤ 3
Signal	1300	≥ 55	≤ 100	≤ 3
SHG-idler	900	≥ 7	≤ 100	≤ 5
SHG-signal	650	≥ 10	≤ 100	≤ 5
FHG-idler	450	≥ 2	≤ 170	≤ 7.5
FHG-signal	330	≥ 2	≤ 170	≤ 7.5

This OPA is pumped with a fs laser pulse of 80 fs duration with a pulse energy of 0.75 mJ at a repetition rate of 1 kHz. Similar results are reached with the same device in the ps range. Pumping with a 1 mJ pulse of 1 ps duration at a repetition rate of 1 kHz again results in the values given in Table 6.38.

Table 6.38. Pulse energies of a commercial OPA in the ps range pumped by a 1 kHz Ti:sapphire laser of 1 ps pulse duration and 1 mJ pulse energy

Method	Wavelength (nm)	Pulse energy (μJ)	Pulse width (ps)	Stability (%)
Idler	2080	≥ 25	≤ 1.25	≤ 3
Signal	1300	≥ 60	≤ 1.25	≤ 3
SHG-idler	900	≥ 7	≤ 1.25	≤ 5
SHG-signal	650	≥ 10	≤ 1.25	≤ 5
FHG-idler	450	≥ 3	≤ 1.25	≤ 7
FHG-signal	330	≥ 3	≤ 1.25	≤ 7.5

Similar good results are obtained with ns pulses as reported, Wallenstein. Thus from a single Nd:YVO$_4$ laser with 40 W average output power, a pulse width of 7 ps, a repetition rate of 80 MHz and a beam quality of $M^2 < 1.2$ three light beams with wavelengths of 446 nm, 532 nm and 639 nm could be generated, simultaneously. The total white light power of all three beams together as usable in laser television application was 19 W.

6.15.3 Raman Shifter

Raman scattering in gases, liquids or solids can shift the laser spectrum [6.2009–6.2026] by the Raman frequency of the material (see Sects. 3.10.4 and 4.5.12) which is of the order of magnitude of $1000\,\mathrm{cm}^{-1}$ or $3\cdot10^{13}\,\mathrm{Hz}$, resulting in a few nm shift in the visible. The beam quality is usually decreased by these Raman shifters.

For high efficiencies gas cells with high pressures of about 50 bar have been applied. The light has to be strongly focused for sufficiently high intensities in the material.

Solid-state materials, such as e.g. $Ba(NO_3)_2$, allow for larger shifts with still good efficiency. Thus with intracavity Raman conversion with this material in a Q switched Nd:YAG laser the original wavelength of 1,064 nm was shifted in the region between 1160 nm and 1198 nm with an efficiency of about 25%. Frequency doubling of this radiation leads to a wavelength range from 580 to 599 nm with an output energy of 0.6 mJ of the 5 ns pulse [6.2009]. Raman lasers are reported in [6.2027–6.2037]

6.16 Laser Safety

There are laws about the correct use of laser radiation to avoid any damage [e.g. 6.2038–6.2041]. They are slightly different in different countries and should be seriously recognized.

In addition some simple rules while using laser radiation can help to avoid any eye or skin injury or damage. First, all laser radiation should stay in restricted areas. Usually all optical beams should be on the optical table at a certain height, the beam height. All unused laser reflexes have to be dumped with beam catchers. Special care has to be taken for beams leaving the plane of the optical axes as produced, e.g. by polarizers. It has to be noticed that reflection from one glass surface contains about 4 W radiation of a 100 W laser! Therefore, it helps if the operator does not wear rings or watches. Take special care of visitors in the laser lab if high-power systems are working. All beams which are not needed for direct access in the experiments should be covered. This also increases the signal-to-noise ratio.

But the main rule is:

Never look into a laser beam!

As obvious as this may appear the violation of this simple rule is still one of the main reasons for eye injury.

In many cases a pair of glasses with suitable filters can be used to avoid injury. Most dangerous are IR and UV laser radiation. IR laser radiation is not visible and UV radiation is mostly underestimated by its weak fluorescence appearance.

UV radiation can dull the eye lens after sufficient exposure. The damage will accumulate over time. Light with wavelengths between 400 and 1400 nm will reach the retina focused to a diameter of about 10 μm resulting in a 100 000 times increased intensity.

The rules about laser safety are similar in Europe and the USA and can be found in [6.2038, 6.2039]. The problem is difficult to describe in simple rules because the different wavelengths, pulse durations, powers and pulse energies, as well as mode structure, have different influences and thus many kinds of combinations have to be considered. Lasers can be classified by the possible damage to the eyes or the skin and for fire danger.

Fire danger is possible for lasers with average output powers of 500 mW or more. Thus no papers should be placed at the beam height in the lab.

Skin damage can occur above average output powers of $10\,\mathrm{mW\,cm^{-2}}$ or above pulse energies of $10\,\mathrm{mJ\,cm^{-2}}$.

Eye damage can occur even from laser pointers with output powers of 1 mW and if the eye does not blink, with even much lower powers. As rough rules the values in Table 6.39 for the maximum permissible exposure (*MPE*) of the eye may be used for choosing the optical density of protection goggles at the laser wavelengths (without any guaranty).

Table 6.39. Maximum permissible exposure (MPE) power or pulse energy of the eye as function of the pulse length and the wavelength of the laser radiation (without guaranty)

pulse length	200–620 nm	620–1050 nm	1050–1400 nm	1400 nm–1000 μm
$\geq 0.5\,\mathrm{s}$	$1\,\mathrm{\mu W\,cm^{-2}}$	$10\,\mathrm{\mu W\,cm^{-2}}$	$1\,\mathrm{mW\,cm^{-2}}$	$100\,\mathrm{mW\,cm^{-2}}$
$\geq 1\,\mathrm{ns}$	$0.5\,\mathrm{\mu J\,cm^{-2}}$	$0.5\,\mathrm{\mu J\,cm^{-2}}$	$5\,\mathrm{\mu J\,cm^{-2}}$	$1\,\mathrm{mJ\,cm^{-2}}$
$< 1\,\mathrm{ns}$	$0.5\,\mathrm{kW\,cm^{-2}}$	$0.5\,\mathrm{kW\,cm^{-2}}$	$5\,\mathrm{kW\,cm^{-2}}$	$1\,\mathrm{MW\,cm^{-2}}$

For comparison sunlight [6.2041] has a power density of about $0.12\,\mathrm{W\,cm^{-2}}$ in central Europe and would definitely damage the eye if someone looked directly into the sun. The spot diameter of the sun is about 160 μm at the retina.

The necessary optical density $\mathrm{OD_{goggles}}$ or transmission T_{goggles} of goggles at the laser wavelength λ_{laser} can be calculated from the MPE values of Table 6.39 and the maximum laser power $P_{\mathrm{laser,maximum}}$ or pulse energy $E_{\mathrm{laser,maximum}}$ as a function of the wavelength λ_{laser}, pulse duration Δt_{FWHM} and cross section of the beam A_{beam} by:

$$T_{\mathrm{toggles}}(\lambda_{\mathrm{laser}}) = \frac{\mathrm{MPE}(\lambda_{\mathrm{laser}}, \Delta t_{\mathrm{FWHM}}) \cdot A_{\mathrm{beam}}}{P_{\mathrm{laser,maximum}} \text{ or } E_{\mathrm{laser,maxmimum}}} \tag{6.171}$$

and

$$\mathrm{OD_{goggles}}(\lambda_{\mathrm{laser}}) = -\lg_{10}\{T_{\mathrm{goggles}}(\lambda_{\mathrm{laser}})\}. \tag{6.172}$$

The lasers are categorized into five safety classes, 1, 2, 3A, 3B and 4, starting from nondangerous lasers in class 1 to most dangerous lasers for eye, skin and fire danger in class 4.

7. Nonlinear Optical Spectroscopy

The nonlinear optical effects described in Chaps. 4 and 5 have to be characterized for a given material up to a certain level before they can be applied in photonics. The already wide range of applications in science, technology and medicine contains analytic aspects, questions of the structure of matter, reaction mechanisms on all time scales and the production of new states, phases or even of new matter. Thus two main questions are asked in nonlinear optics:

- Which nonlinear optical effect is, for a given material suitable for a new laser analytic method (analytic tasks)?
- Which nonlinear optical effect is most suitable for a desired photonic application (material and light modification tasks)?

The second question can be related to the problems:

- Which material is most suitable for a given nonlinear optical application or which kind of light is most suitable for a given material?

And thus nonlinear optical spectroscopy deals finally with the questions:

- Which nonlinear optical properties does a given material have?
- What are the reasons for this nonlinear optical behavior and how can materials with more useful nonlinear optical properties, such as e.g. higher nonlinear coefficients at certain wavelengths and smaller losses, be designed?

Therefore both aspects of better and new nonlinear optical methods, as well as better and new nonlinear optical materials, have to be investigated in nonlinear optical spectroscopy.

With the wide variety of nonlinear optical effects and their applications described in the previous chapters, there is also a wide variety of laser spectroscopic methods. A good overview is given e.g. in [M10]. For further reading see also [M4, M9, M22, M24, M30, M40, M45–52].

Because all kinds of nonlinear effects can appear in nonlinear optical applications and in nonlinear spectroscopy a systematic investigation may be necessary. Unwanted side-effects such as excited state absorptions, induced gratings, population of long-lived levels, photo-chemistry, damage, self-focusing, wave mixing or scattering can be detected and avoided. The right strategy is one key element in this field, as described in this chapter.

7.1 General Procedure

The material parameters characterizing the nonlinear optical behavior can be determined with these methods of nonlinear spectroscopy. Therefore cross sections, nonlinear refractive indices and decay times are investigated as a function of the light and sample parameters.

The nonlinearity can be observed as a function of the time or/and as a function of the applied pump intensities. Temporal measurements can reach resolutions down to fs with low spectral resolution. From the modeling of the nonlinear measurements as a function of the pump intensities the decay times can be determined in favorable cases, also with sub-ps resolution. In this case the spectral resolution is as good as it can be with respect to the uncertainty condition between time and energy. Thus both methods should be combined for a complete set of data.

In nonlinear optical spectroscopy each light parameter can be of crucial importance for the results of investigations. Nonlinear optical measurements almost always produce new and interesting results. But for a detailed analysis of these experimental results a well-defined procedure with careful characterization of all relevant experimental parameters is necessary.

Whereas, e.g., in conventional absorption spectroscopy the polarization, pulse width and intensity of the light can be changed in front of or behind the sample and the sample transmission will always be the same, in nonlinear spectroscopy the sample transmission will usually be different if these parameters are varied.

7.1.1 Steps of Analysis

The nonlinear optical behavior of matter can be investigated step by step starting from linear spectroscopy and changing to nonlinear measurements while increasing the light intensity. Thus parameters such as cross-sections and time constants, as well as the nonlinear refractive index, can be determined and finally modeling of the nonlinear optical behavior becomes possible. A useful sequence of experiments is given in Table 7.1.

The determination of the absorption coefficients a_i, the population densities N_i, and the nonlinear refractive index n_2 as a function of the light parameters allow the modeling of the nonlinear behavior of the material as a function of the excitation intensity, the wavelengths λ_i and the polarization. Finally the absorption and emission cross-sections σ_i, decay times τ_i and nonlinear refractive indices n_2 can be determined.

As described in the following sections all parameters of the sample and the light beams have to be traced carefully. Thus the intensities of the pump and the probe light have to be determined as a function of its spectral, temporal, geometrical, polarization and coherence properties with respect to each other and to the sample geometry. To exclude errors the setups should be checked for linearity, and dynamics with linear filters such as wire nets or optical

Table 7.1. Sequence of tasks in nonlinear optical spectroscopy to characterize the nonlinear behavior of matter with their relevant parameters. (?) indicates that these tasks are not always necessary. Not all measurement methods are possible for all samples

Task	Apparatus	Parameter
1 ground state absorption spectrum (GSA)	UV-Vis absorption spectrometer	$\sigma_{GSA}(\lambda)$ N_{total}
2 Fluorescence spectrum	fluorescence spectrometer	$\sigma_{Fluo}(\lambda)$
3 Phosphorescence spectrum (?)	fluorescence spectrometer	$\sigma_{Phos}(\lambda)$
4 Fluorescence decay time (?)	decay apparatus	τ_{Fluo}
5 Phosphorescence decay time (?)	decay apparatus	τ_{Phos}
6 Quantum yield	fluorescence spectrometer	$\tau_{rad}/\tau_{radless}$
7 Nonlinear absorption measurements in GSA-bands	single beam laser spectrometer	$\sigma_{ESA}(\lambda_{laser})$ $\tau_{recovery}$
8 z-scan (?)	z-scan laser spectrometer	n_2
9 Nonlinear emission measurements (?)	laser emission spectrometer	τ_{rad} $\sigma_{two-photon}$
10 First guess of population densities	modeling	$\approx N_{S1}, N_{T1}$
11 Pump and probe experiments	pump and probe laser apparatus	$a(I_{laser}, \lambda_{laser}, \lambda_{probe}, \dots)$
12 Fractional bleaching (FB) and nonlinear polarization spectroscopy (NLP) (?)	pump and probe spectrometer (with polarizer)	inhomogeneous broadening
13 Determination of excited state absorptions (ESA)	pump and probe spectrometer	$a_{ESA}(\lambda)$
14 Fluorescence intensity scaling	nonlinear fluorescence spectrometer	N_{S1}
15 Determination of population densities	modeling	N_{S1}, N_{T1}
16 Determination of singlet-singlet absorption cross-section	modeling	$\sigma_{S1}(\lambda)$
17 Determination of triplet-triplet absorption cross-section	modeling	$\sigma_{T1}(\lambda)$
18 Measurement 1-14 with variation of the host material	see 1-14	influence of host interaction
19 Measurement 1-15 with temperature or pressure variation	see 1-15	influence of internal geometry
20 Specialized measurements as Raman, IR, high spectral resolution, spectral hole burning, coherent measurements, SHG,...	specialized laser spectrometer	vibrational coupling, $T_2, n_2,$ $\gamma...$

filters, which have to be proven for not becoming nonlinear at the applied intensity. Noise and background light should be measured for each detector while blocking the beam at the position of the sample. All these problems are increased from the high dynamics of the nonlinear measurements.

7.1.2 Choice of Excitation Light Intensities

For evaluation of the measurements the excitation intensities have optimal values as a function of the task. These values can be determined from the nonlinear measurements. While varying the excitation intensity they have to be tested so that the smallest applied intensities are in the linear range and the result of this transmission measurement is identical with the results of conventional measurements. Both absorption and emission measurements with varying excitation intensity should be used because in bleaching measurements nonlinear behavior can be hidden as a result of active excited state absorptions.

If the excitation intensity is increased in absorption measurements the ground state of the matter will be depopulated and excited transient states will be occupied. A stepwise population of excited states can take place as shown with the energy level scheme of Fig. 7.1.

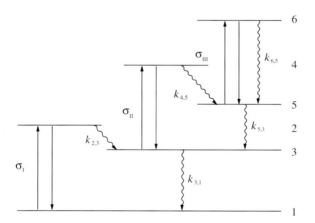

Fig. 7.1. Energy level scheme for a system with three successive absorption transitions

If these excited states absorb at the exciting light wavelength the bleaching effect will be decreased (see Sect. 5.3.3 and Figs. 5.8, 5.9). The excited states 3 and 5 of Fig. 7.1 are activated by increasing the excitation intensity. If e.g. the three absorption transitions of the given system are assumed to have the same cross-section no bleaching will occur initially. The resulting population of these levels as a function of the intensity is shown Fig. 7.2.

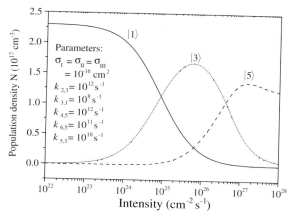

Fig. 7.2. Population densities of the energy levels of Fig. 7.1 as a function of the excitation intensity $\mathit{I} = I/h\nu_{\mathrm{Laser}}$ at the time of the maximum of the incident pulse. The excitation pulse had a FWHM duration of 10 ns and the cell length was 1 mm

The observable bleaching curve for a material with the level scheme of Fig. 7.1 is given in Fig. 7.3.

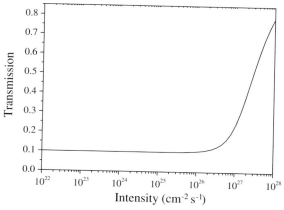

Fig. 7.3. Nonlinear transmission curve of the model of Fig. 7.1 as a function of the excitation intensity measured in photons/cm^{2}s

The transmission is almost not changed up to intensities of 10^{26} photons cm^{-2} s^{-1} although higher energy levels are already strongly populated. Thus choosing an excitation intensity above $3 \cdot 10^{23}$ photons cm^{-2} s^{-1} would include these higher states in the nonlinear measurements in this example. The detection of signals which belong to the first excited state such as e.g.

fluorescence or excited state absorptions from this state as a function of the excitation intensity can clarify these processes as will be shown in Sects. 7.6, 7.7.8 and 7.9.3.

As a rule of thumb the excitation intensity should first be varied around the value of the nonlinear intensity $I_{nl} = I_{nl}/h\nu_{Laser}$ as given in Sect. 5.3:

$$\text{rule of the thumb}\quad I_{exc} \approx \frac{1}{2\sigma_{mat}(\lambda_{exc})\tau_{mat,recovery}} \tag{7.1}$$

The cross-section σ_{mat} of the material can sometimes be difficult to determine if e.g. aggregates occur in the matter. The absorption recovery time $\tau_{mat,recovery}$ may also be unknown, but both may be estimable from similar materials for this very first approach or they may have to be measured as described below.

At very high intensities ($> 10^{26}$ photons cm^{-2} s^{-1}) all kinds of additional non-resonant nonlinear effects and scattering may occur. Thus the detailed (visual) inspection of the propagated light beams can be essential (safety rules have to be recognized!).

7.1.3 Choice of Probe Light Intensities

The probe light intensity has to be small enough not to change the sample itself. On the other hand more probe light increases the signal-to-noise ratio for the detectors and thus the tendency to increase it as much as possible has to be limited at a certain extent.

A possible rule of the thumb is to choose the probe light intensity I_{probe} at least ten times smaller than the expected nonlinear intensity I_{nl}:

$$\text{rule of the thumb}\quad I_{probe} < 0.1\frac{1}{2\sigma_{mat}(\lambda_{probe})\tau_{mat,recovery}} \tag{7.2}$$

with the parameters as in (7.1).

In any case the probe light intensity should be decreased in the measurement by a factor of 2–10 for checking its influence, and the result of the measurement e.g. the transmission should be unchanged.

7.1.4 Pump and Probe Light Overlap

The probe light has to be well inside the spatial and temporal excitation in the sample given by the excitation light volume and pulse duration, otherwise the measured transmission change considers unchanged parts of the sample leading to measuring errors.

Spatial Overlap

The spatial overlap can be achieved with *collinear or longitudinal excitation* as depicted in Fig. 7.4.

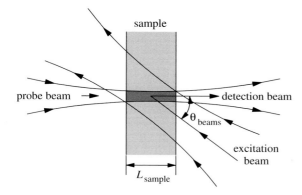

Fig. 7.4. Collinear excitation of the sample in pump and probe measurements

This type of excitation allows higher intensities but demands short samples and therefore high optical densities. The possible size of the overlap region is dependent on the angle between the two beam θ_{beams}, the spot diameters and the divergence of the beams.

In ns measurements sample lengths are in the range of a few mm, typically 1–10 mm and only in special cases a few 10 μm. In ps and fs measurements usually the lengths are below 2 mm to avoid dispersion effects. Typical angles θ_{beams} are 30° or a few degrees depending on the necessary focusing. Typical spot diameters are 1 mm if enough excitation peak power is available and down to 10 μm for low peak power lasers. Larger angles decrease the problem of scattered excitation light in the detection system because more suitable apertures can be applied. The geometry of the interaction zone should be carefully designed for the type of material, the available laser radiation and the necessary accuracy.

On the other hand the sample can be transversally excited as shown in Fig. 7.5.

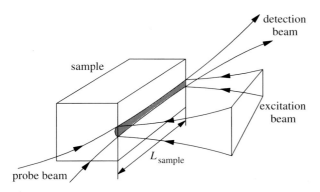

Fig. 7.5. Transversal excitation of the sample in pump and probe measurements

The excitation light is in this case usually focused with a cylinder lens along the probe light beam in the sample. Typical sample lengths are 10 mm and the excitation focus is about 1 mm to 50 μm high. Thus the probe light has to be focused well inside the excited sample volume demanding a sufficiently long Rayleigh length (see Sect. 2.4.3). This type of excitation is useful for excitation pulses longer than the sample length, and thus the pulse duration should be larger than the sample divided by the light velocity. Otherwise traveling wave excitations can be observed as described in Sect. 6.10.4.

Temporal Overlap

If the probe light pulse is not delayed, as for observing decay processes, it should be temporarily well inside the excitation pulse. Thus the probe light pulse length should be no longer than the excitation pulse length. This has to be checked for probe light pulses generated separately in white light or fluorescence materials or in other light sources such as flash lamps or other lasers.

The temporal overlap region can be determined starting with negative delays i.e. applying the probe light pulse before the excitation pulse. For times longer than 0.1 ns electronic delay generators can be applied. Below 1 ns optical delay lines are usually used.

Care has to be taken for perfect spatial overlap at all temporal delays, e.g. while changing the length of the optical delay line.

7.1.5 Light Beam Parameters

All properties of the applied light beams in nonlinear spectroscopic experiments have to be characterized carefully. For simplicity usually the assumptions of Sect. 2.1.3 can be used to reduce the four dimensions of the light intensity as a function of space, wavelength, time and polarization to the given (eight) parameters if Gaussian-shaped beams and pulses.

Coherence properties of the light beams have to be checked separately. Thus the question of which type of coherent interaction in the sense of Fig. 5.1 occur, have to be answered.

The following hints may be used for cross-checking the necessary details in the characterization of the light beam parameters.

Polarization and Magic Angle

If samples are nonlinearly excited, the polarization i.e. the geometrical distribution of the induced dipole moments of the matter, will no longer be isotropic [e.g. 7.1–7.17]. Even if the material is isotropic and the light is not polarized the resulting distribution will have a disk shape with a $\cos^2$ function because there is almost no electrical field strength in the propagation direction of the light.

Thus the interaction of the exciting and probe light with their different polarization directions of the electric field vector with respect to the sample may cause very complicated structures (see for example Figs. 2.34 and 2.35).

Even with two linearly polarized beams the situation is still complicated and in addition the induced dipole moments in the sample may change their orientation, e.g. as they relax to the isotropic distribution. This process can occur in sub-ps to hours.

Thus by choosing linearly polarized light for both pump and probe beam, the orientation relaxation of the excitation can be investigated. In all other measurements the orientation effects from the polarization of the light are usually disturbing.

To avoid these disturbing effects from light polarization in pump and probe measurements the application of the "magic angle" was proposed [7.9] as shown in Fig. 7.6.

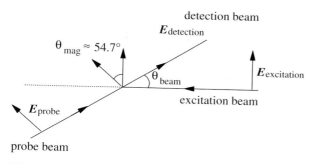

Fig. 7.6. Magic angle configuration in pump and probe measurements

For this concept both beams have to be linearly polarized and the angle θ_{mag} between the two electric field vectors $\boldsymbol{E}$ of the excitation and the probe beam has to be:

magic angle $\theta_{mag} = \arctan(\sqrt{2}) \approx 54.7°$.   (7.3)

This angle can easily be achieved e.g. if the excitation light is perpendicularly polarized relative to the plane of the propagation directions of both beams as in Fig. 7.6. It is then obviously independent of the angle between the two beams θ_{beam}. Thus also in transversal excitation configurations the magic angle can be applied.

Using the magic angle setup the orientation relaxation of all dipole moments of the matter does not influence the transmission measurement of the probe beam. This is true even if the transition dipole moment of the excitation and the probing are not parallel. This can also be applied for emission measurements.

But the initial transmission change measured with probe light polarization in the magic angle direction is slightly smaller than that compared to the value from parallel polarization of probe and excitation. But in the parallel case the measured decay of the signal is a mixture of energy and orientation relaxations. Thus the evaluation of the data needs very complicated models or substantial measuring errors can occur.

More details about considering light polarization in conventional absorption and fluorescence measurements are given in [7.1–7.17, M24]. From the formulas given there the difficulties in considering polarization effects in nonlinear spectroscopy can be estimated.

Pulse Width, Delay and Jitter

Measurements with different excitation pulse durations allow the determination of decay times in different time domains. Furthermore e.g. from the measurement of nonlinear bleaching as a function of the pulse width, sometimes the absolute cross-section of the material can be obtained. This is possible if the decay time of the matter is longer than the pulse duration of the excitation. In this case the excited particles are a function of the pulse energy density and the cross-section of the matter, only.

Thus the determination of the pulse width is essential for the evaluation of nonlinear measurements. For pulses longer than 1 ns the pulse duration can be measured directly with electronic devices.

For pulses shorter <1 ns streak cameras can be used for measurements with resolutions in the ps range. Shorter pulses can be measured using optical delays and a nonlinear optical process for detection of the superimposed shares of the delayed and nondelayed pulse [7.18–7.36]. The nonlinear process can be e.g. a two-photon emission or absorption, SHG generation or nonlinear bleaching as described in Chaps. 4 and 5. Unfortunately, these techniques do not allow the determination of the pulse shape and thus the evaluation of the nonlinear signal is slightly uncertain (see Fig. 5.45 and (5.150) and (5.151)). For bandwidth-limited pulses the duration can be determined from the spectral measurement and for all others the shortest limit of the duration can be calculated from this.

The delay of the probe pulse allows the observation of the decay mechanisms in the sample after excitation, in combination with the other probe light parameters, e.g. the orientation, spectral and spatial relaxation in the matter.

Delay lines are usually used for delays smaller than 100 ns using retroreflector arrangements as triple mirrors or prisms with two passes. The resulting delay in air is given in Table 7.2.

Delay lines longer than 50 ns can be difficult to achieve because of the necessary good beam quality of the probe light. Delay lines for fs pulses need high accuracy and no dispersive elements should be used in delay lines for pulses with duration below 100 ps.

Table 7.2. Time delay from a delay line in air passed back and forth

Length	10 m	1 m	0.1 m	10 mm	1 mm	0.1 mm	10 μm	1 μm
Delay	67 ns	6.7 ns	670 ps	67 ps	6.7 ps	670 fs	67 fs	6.7 fs

The electronic triggering of lasers usually results in jitters around 1 ns or longer and have to be measured if synchronization is important. In the worst case, measurements with too high jitters can be suppressed in the evaluation of the data if the jitter is measured, simultaneously.

Spectral Width

The spectral width of the laser light [e.g. 7.37, 7.38] has to be set in relation to the absorption and emission bandwidth of the material. Spectral hole burning can occur if the spectral bandwidth of the exciting laser is smaller than the bandwidth of the matter. Thus in such measurements spectral bleaching of the whole band has to be checked (see Sect. 7.4 and especially Sect. 7.4.5 and 7.7.9).

Focus Size and Rayleigh Length

As described in Chap. 2 the intensity increases quadratically with decreasing diameter of the excited volume in the sample but the Rayleigh length and wave front curvature radius decreases as given for Gaussian beams in Tables 2.4 and 2.5. The latter may be important for experiments with induced gratings. Thus, usually an optimum spot size is selected for a certain measurement.

As described above the beam diameters also have to be chosen for optimal overlap of the pump and probe beam in the material.

Coherence Lengths

It turns out that even in pump and probe experiments with quite incoherent light, such as e.g. from broad-band lasers, the induced absorption or phase gratings in the sample may disturb the absorption measurement. Thus the coherence properties of the laser beams and their interaction with matter should be checked carefully. As shown in Fig. 5.49 the pump light can be exactly reflected at induced grating planes towards the direction of the detection light.

7.1.6 Sample Parameters

All sample parameters including the origin and the history should be noted. Variation of temperature and pressure as well as the variation of the host material may allow new insides about the nonlinear behavior.

Preparation, Host, Solvent

Optical densities typically between 0.5 and 3 are convenient for transmission measurements of ground and transient states. Thus the thickness of the sample may be adapted. If the matter can be diluted in solution or in a glass or polymer matrix or in a host crystal the concentration may be adapted for good sensitivity of the measurement. In pump and probe measurements the sample thickness may be limited for sufficient spatial overlap.

The purity of solvents and other host materials is a key issue of nonlinear spectroscopic measurements. Two-photon or transient absorptions of impurities in the host material cannot be checked with linear spectroscopic methods. Besides chemical impurities, which are claimed on the label, small particles from the purification process can disturb nonlinear measurements. These particles are usually not specified.

Thus extensive fluorescence investigations of the solvent or host material, if possible with strong laser radiation, may be used to check for chemical impurities. Particles can be detected by scattering experiments in the visible while inspecting the material e.g. via a microscope.

Cleaning for chemicals can be done by distillation or other chemical procedures. Particles can be removed with microfilters or by a cold distillation using liquid nitrogen temperatures [e.g. 4.335].

All properties of the host material should be noted. The use of different types with e.g. different viscosities, dipole moments and geometrical structures allow the systematic study intra- and interparticle interaction of the sample.

Concentration, Aggregation

For high optical densities the concentration of the sample is often required to be as high as possible, but e.g. in the case of organic molecules the formation of aggregates [e.g. 7.39, 7.40] may appear and thus the sample will show different properties. Therefore in molecular systems concentrations above $10^{-3}\,\mathrm{mol\,l^{-1}}$ may be checked for dimers. Sometimes even concentration above $10^{-6}\,\mathrm{mol\,l^{-1}}$ were observed to be crucial for dimerization.

The aggregation in the sample can usually be observed via the ground state absorption and the fluorescence spectrum as well as the fluorescence decay. Dimerization and other aggregation can lead to a red shift and change in the structure of the long-wavelength absorption and emission bands and/or in different decay times. Thus different dilutions should be compared. In crystals the band or line structure may change for different concentrations of the sample.

In any case the precise origin of all compounds should be given as well as the concentration of the sample.

Temperature

Temperature variation of the sample (see Sect. 7.12.1) changes the probability of energy activated processes such as e.g. conformation or orientation relaxations. Thus varying the temperature usually allows Arrhenius plots of the observed spectroscopically detected decay rates k_m of the sample as a function of the temperature T in a semilogarithmic plot and thus the determination of the energetic barriers E_{barrier}:

$$k_m = k_{m,0} \, e^{-E_{\text{barrier}}/k_{\text{Boltz}} T} \tag{7.4}$$

in which k_{Boltz} is Boltzmann's constant $k_{\text{Boltz}} = 1.381 \cdot 10^{-23} \, \text{J K}^{-1}$.

In addition temperature changes also result in a change of the geometrical distances of the particles which may also strongly influence the nonlinear optical properties of the matter. Thus sometimes the comparison of results from low temperature with high pressure measurements may be helpful.

Pressure

In highly pressurized samples (see Sect. 7.12.2) the distances between different sample particles are changed. Length changes of about 10% are possible with pressures in the kbar range. Thus all kinds of energy transfer mechanisms are changed in this way (see Sect. 3.3.4). The resulting time constants can vary by orders of magnitude. Thus pressure variation can help to understand the mechanisms.

7.1.7 Possible Measuring Errors

Besides commonly known possible measuring errors caused by the *limited linearity range* of the detection system and *calibration problems* additional difficulties result from the application of very high intensities and strong nonlinearities of the samples.

Thus beam attenuation can be difficult because optical filters may show *nonlinear bleaching*. All other optical layers such as the mirror coatings may show new properties under illumination with high intensities. Thus these devices should be used at positions with large light cross-sections to decrease the intensities by some orders of magnitude. In crucial cases these elements have to be explicitly checked for their potentially nonlinear optical behavior.

In the laser setups long optical paths are often applied and thus *small wedges*, as are obtained by some filters or other "plane" optical elements, can disturb the measurement if they are changed during the experiment. Filters can be ordered with high and sufficient planarity.

On the other hand *back reflexes* from planar surfaces can crucially influence lasers by additional feedback. Thus small angles can be used for aligning these optical components in the beam.

The nonlinear refractive index change, e.g. in filters, may cause *focusing or defocusing* of the light beams. If the detector or interaction area is too small this effect may cause changes of the obtained intensities. Again large beam areas should be applied at these elements.

Another possible problem is *stimulated emission* in highly excited samples. This emission may not even be noticed because of its nonvisible wavelength. Superradiation can also be blocked by black mechanical holders or other elements and thus the reflexes are not visible. Therefore highly excited samples should not be used with planar surfaces perpendicular to the excitation path to avoid additional resonator effects. In any case the samples should be checked for stimulated emission. It can change the lifetimes and population densities of the sample by orders of magnitude.

In experiments with more than one beam, e.g. pump and probe measurements, perfect *spatial and temporal overlap* of all light pulses has to be insured.

The importance of light *polarization is often underestimated* in nonlinear optical experiments.

Because of the long coherence length of the applied laser light interference effects may occur. Thus *transmission or refractive index gratings* may be induced in the experiments. Scattering or wave mixing can occur at these gratings and the transmission measurements may be disturbed.

Spatial, spectral and temporal background light may occur and the total light power can in the worst case be dominated by this unwanted and possibly undetected light. Ten times more background power or energy means e.g. a 1% background in a $300\,\mu m$ area for a $10\,\mu m$ beam spot diameter or during a $10\,ns$ period including the $10\,ps$ pulse or over a wavelength range of $10\,nm$ for a $0.01\,nm$ spectrally broad measuring light.

In addition to the investigated nonlinear optical effect *unwanted side effects* may occur as described in Chaps. 4 and 5. for example, optical breakdown, self-focusing, self-phase modulation, damage, photo-chemistry or all kinds of nonlinear scattering may occur and have to be excluded separately.

7.2 Conventional Absorption Measurements

Absorption spectra measured with conventional light sources emitting intensities in the linear range of the matter are the basis of any nonlinear optical experiment [e.g. 7.41–7.53]. Based on these measurements the applied laser wavelengths for the nonlinear experiments are determined and the required intensities are estimated.

7.2.1 Determination of the Cross-Section

Conventional spectra allow the determination of the ground state absorption (GSA) cross-section σ_{GSA} from the ground state transmission T_{GSA} as a function of the light wavelength λ as described in Sect. 3.4:

$$\sigma_{GSA}(\lambda) = -\frac{\ln[T_{GSA}(\lambda)]}{N_{part} L_{sample}} \tag{7.5}$$

if the density of the absorbing particles N_{part} is known. L_{sample} is the length of the sample [e.g. 7.54–7.57].

In the case of inhomogeneous broadening (see Sect. 5.2) or aggregation (see above) the density of the absorbing particles is not identical to the total density of particles in the sample. It can be smaller by orders of magnitude. This can be checked with nonlinear measurements as described below.

If the cross-section is determined from several independent sample preparations the further determination of the sample concentration and/or thickness can be based on transmission measurements, which is very convenient.

Again, the conventional spectra of molecular samples should be measured for different concentrations to exclude aggregation.

After nonlinear optical experiments the (conventional) ground state spectra should be measured again, to check for photo-reactions or other degregation of the sample as a consequence of the applied high light intensities. In these measurements it should be remembered that the measuring area in conventional spectrometers is in the range of several mm^2 whereas nonlinear measurements usually have areas of less than one mm^2 and thus the disturbed volume may be much smaller than that obtained conventionally.

7.2.2 Reference Beam Method

For all linear and nonlinear absorption measurements the two-beam or reference beam method as shown in Fig. 7.7 improves the quality considerably.

With this method all variations from the light source such as fluctuations or spectral dependencies of the detectors can be excluded. Thus the accuracy of the absorption measurement is a function of the precision of the measuring devices only, and therefore can be enhanced by a factor of 10 to 1000.

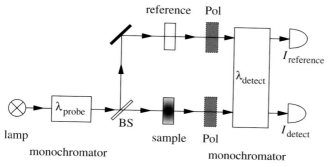

Fig. 7.7. Absorption measurement using sample and reference path for improved quality

In conventional measurements a reference sample can be applied in the reference path. Thus absorption effects from e.g. the solvent or other host materials can be suppressed. But care has to be taken for the limited linearity range of the detectors. For highly absorbing reference samples the measurement may become faulty. Therefore the additional measurement of the sample and/or the reference alone is helpful.

The second monochromator for λ_{detect} is necessary to filter for possible emission light from the sample, which usually has other wavelengths than the absorption. This light can also be suppressed by small apertures in case of good collimated probe light beams. The first monochromator can be saved resulting in a higher illumination of the sample with the whole spectrum of the lamp. In particular, UV light can cause photochemical reactions in the sample.

The polarizer Pol in both beams of Fig. 7.7 allow the absorption measurement of anisotropic effects in the sample e.g. in stretched polymer foils or crystals.

As described in Sect. 3.4 it should be checked which value is shown by commercial devices to characterize the absorption. Only the transmission or transmittance is unambiguous. Absorption or extinction can mean different values.

7.2.3 Cross-Section of Anisotropic Particles

For a detailed analysis it should be noticed that in absorption measurements, the light has no polarization component in its propagation direction. Thus these measurements do not consider absorption dipole moments of the particles aligned in the propagation direction of the light. If the particles show an anisotropic absorption (transition) dipole moment which is not independent of the orientation of the particle, as e.g. in long stretched molecules, the measured cross-section has to be corrected for the single molecule (see Fig. 7.8).

If the light propagates in the z direction and the light is linearly polarized in the y direction the absorption probability for the single molecule follows from:

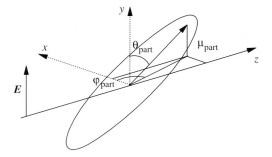

Fig. 7.8. Electric field vector and direction of the transition dipole moment of a single particle in the sample during absorption measurement

$$\sigma_{\text{measurement}} = \int_0^\pi \int_0^{2\pi} \sigma_{\text{single}} \frac{N(\varphi_{\text{part}}, \theta_{\text{part}})}{N_{\text{all}}} \cos^2(\theta_{\text{part}}) \, d\varphi_{\text{part}} \, d\theta_{\text{part}}$$

$$(7.6)$$

with the density $N(\varphi_{\text{part}}, \theta_{\text{part}})$ of particles with the transition dipole moment in the direction φ_{part}, θ_{part} and σ_{single} as cross-section of the single particle. For an isotropic distribution of linear transition dipole moments the integration results in:

$$\text{isotropic distribution} \quad \sigma_{\text{measurement,lin-pol}} = \frac{1}{3}\sigma_{\text{single}} \qquad (7.7)$$

if linearly polarized light is used. If nonpolarized light is applied the measured cross-section is twice as large ($2/3 \cdot \sigma_{\text{single}}$).

7.2.4 Further Evaluation of Absorption Spectra

Below wavelengths of 1000 nm (UV-Vis spectra) absorption is usually caused by electronic transitions in the sample. In addition vibrational, rotational or host transitions may occur and cause broadening. This often leads to band structures in the spectra.

In the case of short spectral cross-relaxation times homogeneous or inhomogeneous broadening cannot be distinguished from the conventional spectra except when very low temperatures are applied. Nonlinear measurements are necessary to answer this question (see Sect. 7.8.1 and 7.8.2). Nevertheless band shape analysis of the conventional spectra may indicate inhomogeneous broadening.

Further, from the structure of the spectra some rough estimates about the nonlinear properties of the samples may be possible as described in Sect. 5.10. Thus broad unstructured spectra may indicate flexible structures, e.g. flexible molecules, which often results in fast relaxation's.

Estimation of Excited State Absorptions (ESA)

In many cases several absorption bands can be detected indicating different optical transitions from the electronic ground state to different electronic excited states. The energetic difference between these ground state absorption (GSA) transitions can give a first hint of the energetic position of excited state absorptions (ESA). The wavelengths λ_{ESA} of the mth ESA band can thus be estimated from the wavelengths of the GSA transitions λ_{GSA} to the first $\lambda_{\text{GSA},1}$ and the mth band $\lambda_{\text{GSA},m}$:

$$\lambda_{\text{ESA},m} \approx \left[\frac{1}{\lambda_{\text{GSA},m}} - \frac{1}{\lambda_{\text{GSA},1}} \right]^{-1} \qquad (7.8)$$

As an example in Fig. 7.9 the ground state spectrum of cryptocyanine in ethanol is shown.

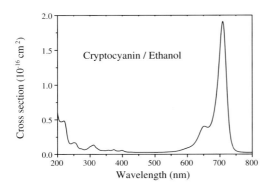

Fig. 7.9. Ground state absorption spectrum of cryptocyanine dissolved in ethanol

In molecular systems the selection rules allow transitions with an even–odd and odd–even parity change only, and thus strong GSA bands should not appear as ESA bands. But in large electronic systems these rules are usually not strongly valid and thus these ESA-bands can usually be detected.

For the given example of cryptocyanine the difference of the ground state absorptions around 700 nm, 310 nm and 270 nm would lead to ESA bands at 550 nm and 430 nm. These bands were experimentally detected around these wavelengths [5.35, 5.44].

Band Shape Analysis

Substructures such as e.g. shoulders in UV-Vis absorption spectra may indicate different electronic transitions which may have different polarizations. In molecular systems the shoulders are often vibrational substructures. Both can be important for photonic applications.

Thus a *band shape analysis* of the GSA can help to understand the experimental results of nonlinear measurements [e.g. 7.58–7.66]. The absorption spectrum is often a superposition of many transitions which may be statistically arranged. In this case the resulting spectrum can often be described by a single (or the sum) of Gaussian subbands and thus the measured spectrum is fitted using the least square method by:

Gaussian bands analysis

$$\sigma_{\text{fit}}(\nu) = \sum_{m=1}^{n} \sigma_{\text{max},m} \exp\left[-4\ln 2\left(\frac{\nu_{\text{peak},m} - \nu}{\Delta\nu_{\text{FWHM},m}}\right)^2\right] \tag{7.9}$$

with the parameters $\sigma_{\text{max},m}$ as the maximum cross-section of the mth band, $\nu_{\text{peak},m}$ as its position and $\Delta\nu_{\text{FWHM},m}$ as its bandwidth. The integral over a single Gaussian band is given by:

$$\int_{-\infty}^{\infty} \sigma_{\text{max}} \exp\left[-4\ln 2\left(\frac{\nu_{\text{peak}} - \nu}{\Delta\nu_{\text{FWHM}}}\right)^2\right] \, d\nu = \sqrt{\frac{\pi}{4\ln 2}}\sigma_{\text{max}}\Delta\nu_{\text{FWHM}}.$$

$$\tag{7.10}$$

Single transitions should show a Lorentzian shape and thus such spectra can be modeled with:

Lorentzian bands analysis

$$\sigma_{\text{fit,L}}(\nu) = \sum_{m=1}^{n} \sigma_{\text{max},m} \frac{\Delta\nu_{\text{FWHM},m}^2/4}{(\nu_{\text{peak},m} - \nu)^2 + \Delta\nu_{\text{FWHM},m}^2/4} \tag{7.11}$$

with the same parameters as given above. The integral over a single Lorentzian band is:

$$\int_{-\infty}^{\infty} \sigma_{\text{max}} \frac{\Delta\nu_{\text{FWHM}}^2/4}{(\nu_{\text{peak}} - \nu)^2 + \Delta\nu_{\text{FWHM}}^2/4} \, d\nu = \pi\sigma_{\text{max}}\Delta\nu_{\text{FWHM}}. \tag{7.12}$$

For mixed cases Voigt profiles [7.61] were sometimes applied for the description of transition bands. These describe the superposition of a large number of single Lorentzian bands weighted with a Gaussian probability distribution. The profile can be given as the integral:

Voigt profile analysis

$$\sigma_{\text{fit,V}}(\nu) = \sigma_{\text{max}}\sqrt{\ln 2}\frac{\Delta\nu_{\text{L}}}{\Delta\nu_{\text{G}}}$$

$$\cdot \int_{-\infty}^{\infty} \frac{e^{-\hat{\nu}^2}}{\left(\sqrt{\ln 2}\dfrac{\Delta\nu_{\text{L}}}{\Delta\nu_{\text{G}}}\right)^2 + \left(2\ln 2\dfrac{\nu - \nu_{\text{peak}}}{\Delta\nu_{\text{G}}} - \hat{\nu}\right)^2} \, d\hat{\nu} \tag{7.13}$$

with the selectable values of the full width half maximum band (FWHM) widths of a Gaussian shaped subfunction $\Delta\nu_{\text{G}}$ and a Lorentzian shaped sub function $\Delta\nu_{\text{L}}$. The integral over this band is:

$$\int_{-\infty}^{\infty} \sigma_{\text{fit,V}} \, d\nu = \sqrt{\pi^3 \ln 2}\sigma_{\text{max}}\frac{\Delta\nu_{\text{L}}}{\Delta\nu_{\text{G}}}. \tag{7.14}$$

The Voigt profile reduces to a Gaussian or Lorentzian profile if the opposite half-width is zero. The center part of the profile is dominated by the Gaussian function and the wings by the Lorentzian profile.

As an example Fig. 7.10 shows the band shape analysis using Gaussian-shaped curves of a GSA spectrum similar to that of crystal violet or malachite green in alcohol with a shoulder indicating two subbands.

If these two subbands belong to different electronic transitions they can be addressed by choosing a suitable excitation wavelength, but as shown in this example longer wavelengths above 650 nm would activate the short-wavelength transition peaking at 570 nm much more than the long-wavelength transition with its maximum absorption at 600 nm.

Some examples of very narrow linewidths measured with high resolution are given in [7.67–7.69].

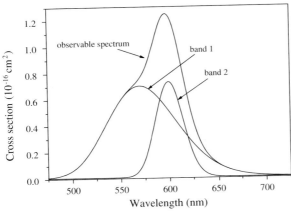

Fig. 7.10. Band shape analysis of a GSA spectrum with a shoulder indicating two subbands positioned at 570 and 600 nm with bandwidths of 84 and 34 nm and peak cross-sections of $7.0 \cdot 10^{-17} \, \mathrm{cm}^2$ and $7.3 \cdot 10^{-17} \, \mathrm{cm}^2$

7.2.5 Using Polarized Light

In conventional absorption measurements different polarization conditions will show different results for rigid or otherwise organized, nonisotropic samples, only. These can be solids such as crystals, polymer hosts or cooled samples. In gases or liquids the reorientation is usually much faster than the measuring time.

If the sample is not isotropic the different components of the absorption cross-section can be determined with two measurements of perpendicular light polarization.

Molecules can be orientated in thin films, liquid crystals or in polymers. These polymers can be stretched and thus linear molecules were aligned along the stretch direction.

7.3 Conventional Emission Measurements

Conventional emission spectra [7.70–7.73], in combination with conventional absorption spectra, allow the determination of some properties of the first excited electronic states. Thus the emission cross-sections, the emission quantum yield and the emission lifetimes can be obtained.

7.3.1 Geometry

Emission spectra are usually measured with optical excitation of the sample in a small excitation wavelength range around λ_{exc} and the observation of the emission intensity spectrum as a function of the wavelength $I_{\mathrm{emission}}(\lambda_{\mathrm{detect}})$ as shown in Fig. 7.11.

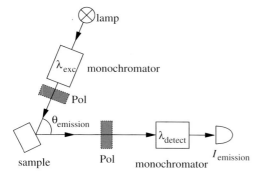

Fig. 7.11. Schematic of emission measurement with excitation–emission angle θ_{emission}

The excitation is provided by a lamp and the monochromator for selecting the excitation wavelength λ_{exc}. The emission of the sample is collected by an optical system under the angle θ_{emission} and detected behind the second monochromator selecting the detection wavelength λ_{detect}. The angle is usually chosen as 30° but sometimes 90° is applied, too. Polarizer in the excitation and detection beam can be applied. The detection path, including all optics, the monochromator and the detector with the measuring system, has to be calibrated for flat spectral sensitivity.

Measuring errors from reabsorption in the sample can be avoided by using optically thin samples in the range of the emission spectrum (OD < 0.1).

If polarizers are applied in the excitation and detection path, anisotropy effects can be obtained. If a rigid sample is used it may be possible to determine the angle between the absorption and the emission transition moments of the particle. For a detailed analysis see [7.70–7.72, M24].

7.3.2 Emission Spectra

Emission or luminescence spectra can be obtained from the singlet system of the sample resulting in fluorescence spectrum whereas the triplet system shows phosphorescence. The latter electronic transition from the excited triplet state to the singlet ground state is spin-forbidden. Thus it shows comparatively long lifetimes and often low intensities.

Fluorescence Spectrum

The fluorescence spectrum [e.g. 7.74–7.84] often shows a mirror symmetry of the longest wavelength absorption spectrum (see Fig. 7.12).

This symmetry is theoretically best if the cross-section divided by the light frequency is plotted on one side and the fluorescence intensity divided by the third power of light frequency on the other [7.70]:

$$\text{mirror plot absorption} \quad \frac{\sigma_{\text{absorption}}(\nu_{\text{light}})}{\nu_{\text{light}}} = f_{\text{absorption}}(\nu_{\text{light}}) \qquad (7.15)$$

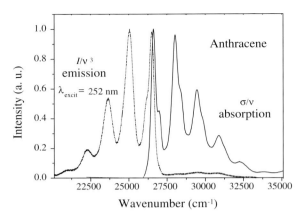

Fig. 7.12. Emission and absorption spectra of anthracene dissolved in cyclohexane showing typical mirror symmetry (spectra corrected as described in text)

and

mirror plot emission $\dfrac{I_{\text{emission}}(\nu_{\text{light}})}{\nu_{\text{light}}^3} = f_{\text{emission}}(\nu_{\text{light}})$ (7.16)

where the emission intensity is measured with a small frequency bandwidth $\Delta\nu_{\text{measure}} \ll \Delta\nu_{\text{band}}$.

The energetic difference of the applied absorption light as e.g. at the lowest energy absorption peak and the detected emission light as e.g. at the highest energy emission peak gives the quantum defect energy.

In molecular systems this energetic difference is called the Stokes shift. The Stokes shift is caused by contributions of vibrations and the environment of the molecule as e.g. solvent reorganization. The particle environment may change as a consequence of different electronic charge distributions in the ground and the excited state of the particle. Thus the pure electronic transition can be estimated as the absorption transition energy minus half the Stokes shift energy.

If this symmetry between absorption and emission bands is strongly disturbed a large difference between the electronic configurations of the ground and the excited state can be supposed as e.g. conformational changes of molecules.

Phosphorescence Spectrum: Triplet Quenching

Emission from the lowest triplet level back to the singlet ground state is spin forbidden and therefore usually shows weak intensity.

For better observation the parallel radiationless intersystem crossing transition – the triplet quenching – leading to a fast depopulation of the triplet levels should be suppressed.

Thus the extraction of oxygen from liquid samples by bubbling the solution e.g. with nitrogen can decrease this radiationless channel. Solvents with heavy atoms quench also triplet states of the sample. In both cases the resulting spin momentum from switching one electron from $+h/4\pi$ to $-h/4\pi$ or vice versa is overtaken by the environment.

From the triplet emission spectrum the energetic position of the lowest triplet level can be determined and its possible influence in photonic applications estimated. In many cases the triplet populations are unwanted, because of the long storage time of these levels. Therefore e.g. in dye lasers the active material is moved to take the molecules in the triplet state out of the beam.

7.3.3 Excitation Spectrum: Kasha's Rule

The excitation spectrum is recorded if the emission intensity at a certain wavelength, e.g. at the maximum, is plotted as a function of the wavelength of the exciting light. It is usually identical to the absorption spectrum because the particles usually decay rapidly from any higher excited state to the first excited state [7.85–7.90], which is known as Kasha's rule [7.90]. This is of course not observed in laser measurements with short pulses in the sub-ns or fs range. Other exceptions are e.g. azulene molecules which emit from the S_2 state [7.89].

If in conventional measurements the excitation spectrum shows deviations from the absorption spectrum an inhomogeneity of the spectral behavior may be indicated. Then a systematic investigation of the excitation spectra as a function of the applied emission wavelength can allow the clarification of the sample structure. Inhomogeneous broadening or impurities can cause this effect. Measurements with different polarization of excitation and emission light as well as temperature variations can be useful.

7.3.4 Emission Decay Times, Quantum Yield, Cross-Section

Fluorescence and phosphorescence decay is a consequence of the limited lifetime of the excited states. These are given by radiative and radiationless transitions [e.g. 7.91]. The decay times are important for modeling the nonlinear behavior of the material.

Fluorescence Decay Time

The fluorescence decay time can vary from sub-ps to hundreds of µs. Only the longer lifetimes can be measured with conventional spectrometers while laser measurements allow determination in the whole range as described in Sect. 7.9.2.

The fluorescence decay time is identical with the lifetime of the emitting state and can thus also be determined via time-resolved excited state observations from pump and probe measurements.

The fluorescence decay time $\tau_{\text{fluorescence}}$ is given by the natural lifetime τ_{nat} of the transition in combination with the resulting decay time $\tau_{\text{radiationless}}$ from all radiationless processes (see also Sect. 3.3.4):

fluorescence decay time $$\frac{1}{\tau_{\text{fluorescence}}} = \frac{1}{\tau_{\text{nat}}} + \frac{1}{\tau_{\text{radiationless}}}. \qquad (7.17)$$

The natural lifetime is shortened by these parallel radiationless transitions leading to the observable fluorescence decay time. Examples are given in [7.92–7.117] and in the references of Sect. 7.7.9.

Natural Lifetime

The natural fluorescence lifetime τ_{nat} of the excited state is directly connected with the natural spectral width $\Delta\nu_{\text{nat}}$ of the transition as described in Sect. 3.3.3:

natural lifetime $$\tau_{\text{nat}} = \frac{1}{2\pi\Delta\nu_{\text{mat}}}. \qquad (7.18)$$

Because of the broadening of the electronic transitions the lifetimes can often not be determined from spectral measurements and thus time-resolved laser measurements are required (see Sect. 7.9.2).

Quantum Yield

The quantum yield Φ_{yield} is the ratio of the spontaneous emitted photons $N_{\text{photons,spontaneous}}$ divided by the total number of absorbed photons $N_{\text{photons,absorbed}}$:

$$\Phi_{\text{yield}} = \frac{N_{\text{photons,spontaneous}}}{N_{\text{photons,absorbed}}} \qquad (7.19)$$

which can be calculated from the decay times as:

quantum yield $$\Phi_{\text{yield}} = \frac{\tau_{\text{radiationless}}}{\tau_{\text{radiationless}} + \tau_{\text{nat}}} = \frac{\tau_{\text{fluorescence}}}{\tau_{\text{nat}}}. \qquad (7.20)$$

Thus the measurement of the quantum yield allows the differentiation of radiationless and radiative transitions from the excited state. (Under laser conditions, i.e. with stimulated emission, this value is equal to the quantum efficiency as given in (6.7).) The natural lifetime is usually determined from the fluorescence decay time and the quantum yield.

The absolute measurement of the quantum yield is difficult because of the geometry of the emission and reabsorption. Thus it is usually measured in conventional fluorescence spectrometers in direct comparison to samples with known quantum yields [7.118–7.120, 5.707] as e.g. given in Table 7.3.

Further examples are given in [7.121–7.127]. Care has to be taken in these relative measurements for the absorption of the exciting light and the reabsorption of the fluorescence light. Thus low optical densities should be applied which are identical for the test and known material at the excitation wavelength.

Table 7.3. Quantum yields Φ_{yield}, excitation and emission wavelengths λ_{exc} and $\lambda_{fluorescence}$ and the fluorescence lifetime $\tau_{fluorescence}$ of some materials

Material	λ_{exc} (nm)	$\lambda_{fluorescence}$ (nm)	$\tau_{fluorescence}$ (ns)	Φ_{yield}
p-terphenyl in cyclohexane	275	340	0.95	0.93
Anthracene in cyclohexane	340	400	4.9	0.27
Perylene in cyclohexane	410	470	6.4	0.94
Acridine yellow in ethanol	480	500	5.1	0.86
Acridine red in ethanol	550	590	3.8	0.33

Phosphorescence Decay Time

If the phosphorescence is measurable its decay time can usually be measured with conventional spectrometers, which have choppers for the time-resolved investigations, or the decay can be observed after excitation with a short pulse by a fast detector and an oscilloscope. The decay times have values in the range from hundreds of ns up to ms.

Determination of the Emission Cross Section

The emission cross-section can be determined from the emission spectra using the relation between the life of the emitting state and its bandwidth. Therefore it has to be assumed that only one electronic transition is responsible for the band. A band shape analysis may be necessary to determine the single transition band. The emission cross-section spectrum $\sigma_{emission}$ as a function of the wavelength λ follows from:

$$\textbf{cross-section} \quad \sigma_{emission}(\lambda) = \frac{\lambda^4}{8\pi c_0 n^2} \frac{\Phi_{yield}}{\tau_{fluorescence}}$$
$$\cdot \frac{I_{emission}(\lambda)}{\int_{band} I_{emission}(\lambda')\, d\lambda'} \qquad (7.21)$$

with the velocity of light c_0, refractive index n, quantum yield Φ_{yield} of the transition, fluorescence lifetime $\tau_{fluorescence}$ and the emission intensity spectrum $I_{emission}(\lambda)$.

7.3.5 Calibration of Spectral Sensitivity of Detection

The spectral curves of the emission and detection systems of commercial spectrometers are calibrated with stored transmission and sensitivity curves including all optical elements and detectors.

Noncalibrated emission measurement systems can be calibrated based on known emission curves. In the simplest case a material with known fluorescence spectrum can be used in a reference measurement for determining the sensitivity spectrum. This calibration spectrum is then used for the correction of further emission measurements.

Another possibility is the application of blackbody radiation from a source of a known temperature. The intensity spectrum I_{bb} of the blackbody radiation as a function of the wavelength λ and the temperature of the blackbody T follows from:

$$\textbf{blackbody radiation}\quad I_{bb}(\lambda, T) = I_{bb,0}\frac{2hc_0^2}{\lambda^5}\frac{1}{e^{hc_0/\lambda kT} - 1} \qquad (7.22)$$

with Planck's constant h, light velocity c_0 and maximum intensity $I_{bb,0}$ which is a function of the observation geometry.

As an example the emission spectra for two typical temperatures of 5600 K (daylight) and 3500 K (lamplight) are given in Fig. 7.13.

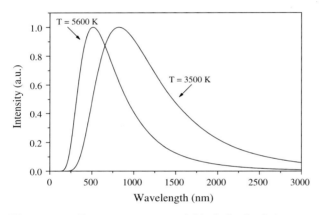

Fig. 7.13. Emission spectra of black body light source for two temperatures of 5600 K (daylight) and 3500 K (lamplight). The maximum emission signal was $2.5 \cdot 10^6$ J m^{-4} for 5600 K and $2.4 \cdot 10^5$ J m^{-4} for 3500 K

Calibrated tungsten band lamps are available as such blackbody light source with calibrated temperature curves as a function of the applied current.

7.4 Nonlinear Transmission Measurements (Bleaching Curves)

Using high light intensities in absorption measurements leads to nonlinear changes of the sample transmission (see e.g. [7.128–7.142] and references of Sect. 5.3.3). This simple one-beam method offers much important information about the nonlinear behavior of the material and allows the determination of several material parameters (see also Sect. 5.3.1 and Sect. 5.3.3).

7.4.1 Experimental Method

As described in Sect. 5.3.1 the sample is excited by the incident light beam with variable intensity and the transmitted light is measured. A typical experimental setup is shown in Fig. 7.14.

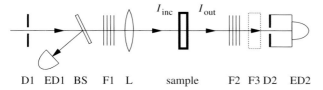

Fig. 7.14. Setup for measurements of the nonlinear transmission

For achieving the necessary high light intensities short laser pulses of ns to fs durations are usually used. With the beam splitter BS, part of the laser light is out-coupled for monitoring the incident intensity during the measurement. This beam splitter should be arranged under a small angle so as not to change the light polarization too much (see Fresnel's formulas in Sect. 3.5).

The beam is focused with lens L typically with focal lengths of 50 to 200 mm for sufficiently small changes of the beam diameter in the sample region. Spot diameters above 100 μm are usually easier to handle than much smaller values for problems of self-diffraction and damage.

The diaphragm D1 suppresses possible spatial background radiation and can be used for beam clean up. Diaphragm D2 is necessary to suppress background radiation and to decrease the share of emission light from the sample in detection. In measurements in the UV the emission light intensities can be especially pronounced by the higher spectral sensitivity of the detector at longer wavelengths. In the worst case the emission light has to be suppressed by additional spectral filters F3 or a monochromator.

The filters at position F1 are used to change the light intensity at the sample without changing any other parameter. All filters taken out at position F1 can be placed in position F2 and thus the dynamic range of the detector ED2 can be small. It has only to cover the nonlinearity of the sample.

The higher the sample absorption the larger the sensitivity of the measurement. But transmissions smaller than 0.1% are difficult to calibrate with sufficient accuracy with conventional absorption spectrometers. The transmission at low intensities has to be checked to be the same as in the conventional measurement.

The transmission can be calculated from the intensities of the transmitted beam I_{out} and the incident beam I_{inc} at a certain time, e.g. the time of pulse maximum t_m:

transmission $T(I_{\text{inc}}(t_{\text{m}})) = \dfrac{I_{\text{out}}(t_{\text{m}})}{I_{\text{inc}}(t_{\text{m}})}.$ (7.23)

This transmission determination is commonly used in nonlinear experiments with not too short pulse duration typically in the ns range or longer. It is especially useful if steady state conditions are.

In the case of peak detectors the peak intensities are used for determining the transmission:

peak transmission $T_{\text{peak}} = \dfrac{I_{\text{out}}(t_{\text{peak}})}{I_{\text{inc}}(t_{\text{peak}})}.$ (7.24)

In experiments with ps or fs pulses the temporal pulse shape cannot be measured electronically and thus temporally integrated intensities, the pulse energies, are used:

energy transmission $T_{\text{energy}} = \dfrac{\int_{\text{pulse}} I_{\text{out}}(t)\,\mathrm{d}t}{\int_{\text{pulse}} I_{\text{inc}}(t)\,\mathrm{d}t}.$ (7.25)

In any case the calculated transmission from the experiment has to be the same as that used in modeling the experiments. The difference of the transmission values can be seen from Fig. 7.15.

For the measurement the exciting light beam has to be characterized in all parameters and they have to be constant during the measurement. The sample should be carefully inspected for optical breakdown, bubbles or other damage (see Sect. 7.1.7).

As discussed in the following sections some of the pump light parameters such as the pulse duration or the spectral width can be varied from measurement to measurement for further evaluation of the sample.

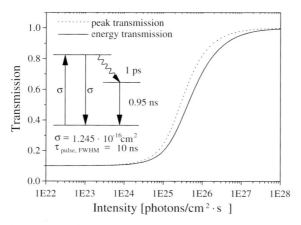

Fig. 7.15. Calculated bleaching curve as a function of the excitation intensity shown as peak or time integrated transmission of para-terphenyl in cyclohexane using a three-level model. The concentration was $1.85 \cdot 10^{17}\,\text{cm}^{-3}$ in a 1 mm cell and the pulse length of the laser was 10 ns

7.4.2 Evaluation of the Nonlinear Absorption Measurement

Important parameters of the sample such as the ground state recovery time, which can be identical with the fluorescence lifetime, or the cross-sections, can be determined from the modeling of the nonlinear absorption as a function of the applied intensity. Further, from the slope and other features of this curve possible excited state absorptions and other nonlinear processes can be discovered.

Modeling

Bleaching measurements are mostly obtained as incoherent interaction of the light wave with the matter (see Sect. 5.1) and can therefore be modeled with rate equations (see Sect. 5.3.6). With very short pulses in rare cases coherent interaction may occur (see Sect. 5.4) and the density matrix formalism may be necessary (see Sect. 5.4.2 and following).

In any case the modeling concerns the balance between the excitation (or other induced processes) of the sample with pump rate σI (or similar values depending on the cross-sections σ and powers of intensity I) on one hand and the spontaneous processes described by characteristic decay times τ on the other. Several pump or decay processes can occur simultaneously or sequentially. Thus for detailed analysis further information from other measurements are required. Then the decay times and the cross-sections can be determined even for higher excited states as will be shown in the following sections.

Modeling is most simple if rate equations can be applied and in addition all involved decay times of the matter are (much) shorter than the exciting light pulse duration. In this case steady state solutions can be used.

Examples are given in Sect. 5.3 especially the figures in Sect. 5.3.3 showing nonlinear transmission curves.

Bleaching or Darkening

The illumination of samples with high intensities reaching the nonlinear range usually leads to an increase of the transmission of the sample: it bleaches. This bleaching occurs if the matter does not strongly absorb in the occupied excited states. Thus bleaching occurs if the cross-sections of the excited states which are populated by the pump intensity are smaller than in the ground state at the wavelength of the laser light.

If the absorption cross-section of the excited state is accidentally equal to the cross-section of the ground state no transmission change will be observed although the particle has changed its quantum state and excited states are populated (see Sect. 7.1.2 and the next section).

If the excited state absorption has a larger cross-section σ_{II} as the ground state absorption with the cross-section σ_I at the applied laser wavelength the material will darken under strong light illumination (see Figs. 5.8 and 5.20). As a function of the ratio σ_{II}/σ_I this can be a strong effect, but increasing the intensity will lead to a bleaching of the sample finally in any case.

Start of Nonlinearity: Ground State Recovery Time

Although the nonlinear transmission has no threshold the change of the transmission compared to the conventional or low signal transmission becomes obvious as a function of E plotted in a logarithmic scale.

Thus the "start" of the nonlinearity in the case of bleaching can be defined as the intensity $I_{\varepsilon nl} = I_{\varepsilon nl}/h\nu_{Laser}$ for which the transmission $T_{\varepsilon nl}$ is given by:

$$T_{\varepsilon nl}(I_{\varepsilon nl}) = \varepsilon_{nl}T_0 \quad \text{with} \quad \varepsilon_{nl} \leq \frac{1}{T_0} \tag{7.26}$$

with the low signal transmission T_0. ε_{nl} should be chosen as a function of T_0 and the quality of the measurement. Smaller values are more useful for avoiding higher nonlinear absorptions in this evaluation. Unfortunately, this intensity $I_{\varepsilon nl}$ is a function of T_0 and does not only depend on ε_{nl}.

In the case of a stationary two level system $\tau \ll \Delta t_{pulse}$ the cross-section σ and the decay time τ of the ground state absorption transition can be related to a selected intensity $I_{\varepsilon nl}$ in the nonlinear range by the factor $C_{\varepsilon nl,stat}$:

stationary bleaching $\sigma\tau = C_{\varepsilon nl,stat}(\varepsilon_{nl}, T_0)\dfrac{1}{I_{\varepsilon nl}}.$ (7.27)

Some values of the factor $C_{\varepsilon nl,stat}$ which are calculated for a stationary two-level system are given in Table 7.4 for different values of ε_{nl}.

Table 7.4. Values of the factor $C_{\varepsilon nl,stat}$ as a function of ε_{nl} and T_0 for a stationary two-level system

$\varepsilon_{nl}\backslash T_0$	0.1%	1%	5%	10%	30%
1.1	0.067	0.068	0.071	0.075	0.101
1.5	0.279	0.283	0.303	0.331	0.532
2	0.468	0.477	0.526	0.601	1.327
5	1.038	1.106	1.498	2.462	–
10	1.459	1.676	3.577	–	–

These values can be used for a first rough estimate of the sample parameters reading a pair $T_{\varepsilon nl}$ and $I_{\varepsilon nl}$ directly from the experimental graph. The earlier given nonlinear intensity I_{nl} follows from $I_{\varepsilon nl}$ for a factor $C_{\varepsilon nl} = 0.5$.

As an example a sample with a small signal transmission of 1% shows a bleaching transmission of 2% at an excitation intensity of 10^{24} photons $cm^{-2}\,s^{-1}$. The $C_{\varepsilon nl,stat}$ value is 0.477 and the $\sigma\tau$ product results in $4.77 \cdot 10^{-25}\,cm^2s$. If this sample has a cross-section of $4.77 \cdot 10^{-16}\,cm^2$ the decay time can be determined from this experiment to be 1 ns.

In the case of strong nonstationary behavior of the sample, meaning that the characteristic decay time is much longer than the pulse duration $\tau \gg \Delta t_{\mathrm{pulse}}$, storage of the excited particles takes place and the cross-section σ of the sample can be determined from the start of the nonlinear absorption. In this case the cross-section follows from a selected light pulse energy density $E_{\varepsilon\mathrm{nl}} = I_{\varepsilon\mathrm{nl}}\Delta t_{\mathrm{FWHM}}$ in the nonlinear range:

$$\textbf{nonstationary bleaching} \quad \sigma = C_{\varepsilon\mathrm{nl,nonst}}(\varepsilon_{\mathrm{nl}}, T_0)\frac{1}{I_{\varepsilon\mathrm{nl}}\Delta t_{\mathrm{FWHM}}}. \quad (7.28)$$

Some values of the factor $C_{\varepsilon\mathrm{nl,nonst}}$ calculated for a two-level system are given in Table 7.5.

Table 7.5. Values of the factor $C_{\varepsilon\mathrm{nl,nonst}}$ as a function of $\varepsilon_{\mathrm{nl}}$ and T_0 for an integrating two-level system

$\varepsilon_{\mathrm{nl}}\backslash T_0$	0.1%	1%	5%	10%	30%
1.1	0.088	0.089	0.093	0.099	0.130
1.5	0.358	0.362	0.383	0.413	0.601
2	0.591	0.600	0.642	0.709	1.231
5	1.252	1.292	1.535	2.073	–
10	1.690	1.819	2.777	–	–

As an example a sample with a small signal transmission of 1% again shows a bleaching transmission of 2% while it is excited with a ps pulse of 1 ps duration and an intensity of 10^{27} photons cm^{-2} s^{-1}. The lifetime of the sample is known to be in the range of ns and thus strong nonstationary bleaching as obtained. The pulse energy density is 10^{15} photons cm^{-2}. The $C_{\varepsilon\mathrm{nl,nonst}}$ value is 0.6 in this case resulting in a cross-section of $6 \cdot 10^{-16}$ cm^2.

If the matter shows a level scheme with almost empty upper state of the absorption transition, as e.g. a three-level system, the C values have to be multiplied by 2 in both cases of stationary and nonstationary bleaching. The values from the table can also be used to estimate the necessary intensity for a required bleaching effect if the material parameters are known.

In the case of an intermediate relation of lifetime to pulse duration $\tau \approx \Delta t_{\mathrm{pulse}}$ the explicit modeling of the experimental results is necessary to determine $\sigma\tau$.

The σ in these formulas is the conventional ground state absorption cross-section at the applied wavelength which can be determined from conventional absorption measurement. But possible inhomogeneous broadening or aggregation of the sample should be considered.

The decay time τ is the ground state absorption recovery time which is not always identical to the fluorescence lifetime as shown in Fig. 7.16.

The energy level X is populated from the first excited singlet state and shows a long lifetime. The nonlinear absorption of this model is calculated for

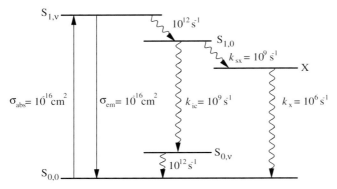

Fig. 7.16. Energy level scheme with a long-lived level X producing a long ground state absorption recovery time. The resulting ground state recovery time can be approximated by 8.1 ns if the bleaching is produced with a 10 ns laser pulse as will be shown below

a total population density of $2.3 \cdot 10^{17} \, \text{cm}^{-3}$ and 1 mm cell length as shown in Fig. 7.17.

The evaluation of this graph after the equation given above using the C values results in a recovery time of 8.1 ns which is longer than the fluorescence decay time of 1 ns. The two-level bleaching curve using this decay time for the ground state absorption recovery is also depicted in Fig. 7.17.

For further illustration the population densities of the levels $S_{0,0}$, $S_{1,0}$ and X are given in Fig. 7.18.

As described earlier, additional excited state absorptions can distort the start of the nonlinear bleaching. In the case of darkening of the sample this

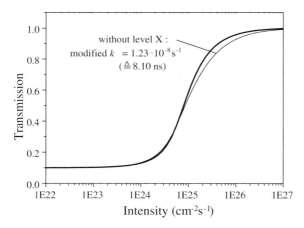

Fig. 7.17. Nonlinear transmission as a function of the incident intensity for the model of Fig. 7.16

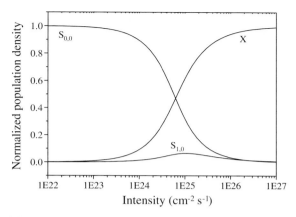

Fig. 7.18. Normalized population densities of the levels $S_{0,0}$, $S_{1,0}$ and X as a function of the incident intensity for the model of Fig. 7.16

is obvious, but in the case of bleaching the measurement of the nonlinear fluorescence or of excited state absorption may be necessary to prove. The evaluation of the slope of the nonlinear transmission may give a further hint of additional processes.

Slope, Plateaus, Minima and Maxima

For stationary two-level systems the shape of the complete nonlinear transmission graph is fixed for a given low signal transmission and only the start of the nonlinearity is varied by different $\sigma\tau$ products. For nonstationary two-level systems the slope and shape of the graph is also determined in narrow limits.

Thus slower slopes usually indicate additional excited state absorptions "switched on" by the population of higher excited states via the pumping process.

Plateaus, minima and maxima in the nonlinear transmission curve are strong indicators of additional transitions of the material (see Figs. 5.19 and 5.20). Plateaus demand at least two transitions and maxima demand three transitions at the applied laser wavelength. These transitions occur between excited states in addition to the ground state absorption. As an example a six-level scheme with three absorptions at the laser wavelength is shown in Fig. 7.19.

The resulting nonlinear transmission curve is depicted in Fig. 7.20 for a sample with a total population density of $5.76 \cdot 10^{16}$ cm^3 resulting in a linear transmission of 10% for a sample thickness of 1 mm calculated for a pulse duration of 10 ns (FWHM).

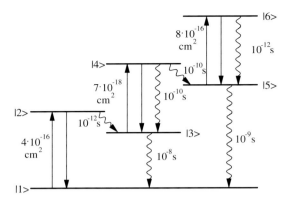

Fig. 7.19. Energy level scheme with six levels and three absorption transitions

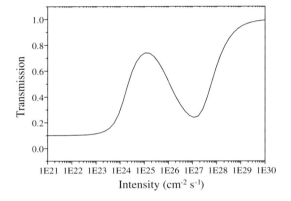

Fig. 7.20. Nonlinear transmission for the model after Fig. 7.19

As can be seen from this graph the nonlinear transmission shows a maximum and a minimum as a function of the excitation intensity. The population densities of the different states are shown in Fig. 7.21.

The different levels are sequentially populated as a function of the excitation intensity and thus as a consequence of the different cross-sections of these excited state absorptions the transmission is modulated as shown above.

7.4.3 Variation of Excitation Wavelength

By varying the applied laser wavelength across the ground state absorption spectrum of the matter in the nonlinear absorption measurements the inhomogeneous broadening of this spectrum or the detection of other nonlinear processes as such e.g. transient absorptions can be observed.

For homogeneous broadened samples with only one active absorption transition the evaluation of the nonlinear absorption curve should result in

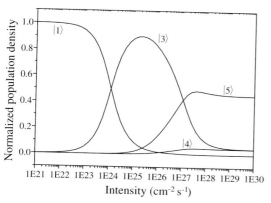

Fig. 7.21. Normalized population densities of levels 1, 3, 4 and 5 of the scheme of Fig. 7.19 as a function of the excitation intensity resulting in the nonlinear transmission of Fig. 7.20

the same spectrum of the ground state absorption cross-section as measured with conventional spectrometers.

Thus deviations of these two spectra indicate additional spectral features which can be investigated in detail with pump and probe measurements using different wavelengths.

7.4.4 Variation of Excitation Pulse Width

As described in Sect. 7.4.2 the use of excitation pulses with different duration in nonlinear transition measurements may allow the separate determination of σ and then τ from the $\sigma\tau$ product. For this purpose the pulse duration has to be varied between values comparable to τ and values much shorter than τ (see Fig. 7.22).

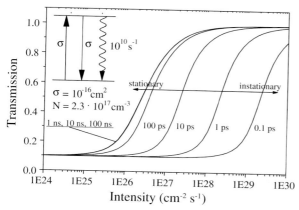

Fig. 7.22. Transition from stationary to non-stationary bleaching demonstrated with a two-level scheme with a decay time of 0.1 ns and different pulse duration's

As can be seen from this figure bleaching with pulse durations of one order of magnitude larger or smaller than τ leads to well-defined stationary or non-stationary behavior. Thus the transition from stationary to non-stationary bleaching is spread over about two orders of magnitude of pulse durations.

Measurement with an excitation pulse duration comparable to τ or longer allows the determination of the $\sigma\tau$ product. Measurement with very short pulses typically in the fs range allows the determination of the absorption cross-section σ, alone. From this value a comparison with the conventional value of σ allows e.g. the determination of inhomogeneous broadening or aggregation. From the known $\sigma\tau$ product and σ the ground state recovery time τ can be determined and compared with the fluorescence decay time. Additional levels can be detected which are radiationless in the decay path from the excited level involved [e.g. 7.131].

7.4.5 Variation of Spectral Width of Excitation Pulse

Nonlinear transmission measurements with different spectral widths of the applied laser light may allow the observation of inhomogeneously broadened ground state absorption bands [7.143]. In this case the transmission curve will show bleaching for smaller integral intensities for spectrally narrower pulses. As an example in Fig. 7.23 the bleaching curve of a two-level band scheme consisting of 10 homogeneous subbands is assumed. The spectral width of the laser covers in one case just one band and in the other broad band case five of these subbands.

As can be seen from this example the measurement shows a significant difference in the nonlinear bleaching for the two spectral widths of the laser.

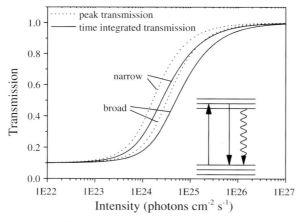

Fig. 7.23. Nonlinear transmission of an inhomogeneously broadened two-level band system with a laser beam of spectral width of one times (narrow) and five times (broad) the spectral width of the homogeneous subband of the matter

The broader laser has to saturate five transitions whereas the narrow laser saturates just one. Thus less intensity of the narrow laser results in higher bleaching.

A precondition for this result is a spectral relaxation time not much shorter than the characteristic decay time of the nonlinear absorption measurement. Otherwise the spectral cross-relaxation would simulate homogeneous broadening.

7.5 z-Scan Measurements

In z-scan measurements the sample is moved along the beam propagation direction through the waist area of an intensive Gaussian beam. Thus the matter shows a lensing effect as a consequence of the nonlinear refractive index profile transversal to the beam propagation direction similar to that discussed in Sects. 4.5.5 and 4.5.6. From the resulting self-diffraction and nonlinear absorption in the sample the real and imaginary parts of the nonlinear susceptibility of third order $\chi^{(3)}$ can be determined from this measurement. Examples of the application of this method are described in [7.144–7.166]. Other methods which allow the determination of the third-order nonlinearities are given in [7.167–7.187] in addition to the four-wave mixing measurements described in Sect. 5.9.2.

7.5.1 Experimental Method

An experimental setup for z-scan measurements to determine the real and imaginary parts of the nonlinear refractive index in third order is shown in Fig. 7.24.

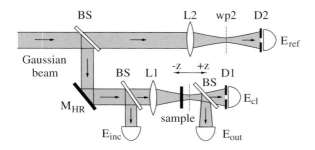

Fig. 7.24. Setup for z-scan measurements

The incident beam with transversal Gaussian profile and perfect beam quality is focused by the lens L1 and the intensity I as a function of the position z is given by:

$$I(z) = \frac{E_{\text{pulse}}}{\Delta t_{\text{FWHM}}} \frac{1}{\left[\pi w_0^2 + \dfrac{z^2 \lambda^2}{\pi w_0^2}\right]} \tag{7.29}$$

with the original light pulse energy E_{pulse}, the pulse duration Δt_{FWHM}, the beam radius w_0 at the waist and the light wavelength λ. The sample which is usually shorter than the Rayleigh length of the focusing $z_R = \pi w_0^2/\lambda$ is moved through the waist region of the beam along z. The light pulse energy is detected behind the sample through a small aperture of diameter d in the range of less than $1\,\text{mm}$ with the detector E_{cl} and in an equivalent reference channel with another detector E_{ref}. The normalized z-scan transmission $T_{\text{z-scan}}$ can then be calculated from:

$$T_{\text{z-scan}}(z) = \frac{E_{\text{cl}}(z)}{E_{\text{ref}}} \left[\frac{E_{\text{ref-withoutsample}}}{E_{\text{cl-withoutsample}}}\right] \tag{7.30}$$

where the measurement of the reference energy in an identical reference beam path decreases the measuring error for pointing instabilities and other fluctuations of the laser light substantially. Therefore both lenses L1 and L2 as well as both apertures D1 and D2 and the energy detectors should be identical.

This z-scan transmission is plotted as a function of the position z of the sample in the beam describing measurements with different intensities as shown in Fig. 7.25.

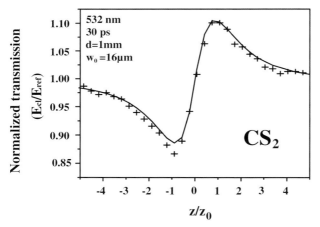

Fig. 7.25. Normalized z-scan transmission $T_{\text{z-scan}}$ as a function of the sample position around the waist at $z = 0$ for CS$_2$ in a $1\,\text{mm}$ cell. The Nd:YAG ps laser was frequency doubled. It was focused with a focal length of $150\,\text{mm}$. Aperture D1 had a diameter of $0.5\,\text{mm}$

The CS$_2$ cell is transparent at the applied wavelength of $532\,\text{nm}$. Thus without nonlinear interaction the low-signal transmission is 1. Moving the cell towards the focus of the excitation beam leads to a defocusing at the detector E_{cl} and the signal is smaller than 1. Behind the beam waist, focusing from the nonlinear refractive index distribution across the beam occurs and the normalized transmission becomes larger than 1.

The third-order nonlinear susceptibility of the sample can be described in a simple approach as:

not absorbing $\mathrm{Re}\chi^{(3)} = 2\varepsilon_0 c_0 n_0^2 \gamma_{\text{z-sc}}$ (7.31)

with

$$n(I) = n_0 + \Delta n_{\text{nl}}(I) = n_0 + \gamma_{\text{z-sc}} I \tag{7.32}$$

where n_0 is the conventional, linear refractive index, Δn_{nl} the earlier mentioned nonlinear refractive index and the coefficient $\gamma_{\text{z-sc}}$ has to be determined from the z-scan measurement.

The share of the nonlinear interaction resulting from the absorption of the sample can be described in third-order nonlinearity by:

absorbing $\mathrm{Im}\chi^{(3)} = \dfrac{1}{2\pi}\varepsilon_0 \lambda_{\text{exc}} n_0^2 \beta_{\text{z-sc}}$ (7.33)

with the absorption coefficient a:

$$a(I) = a_{\text{low signal}} + \beta_{\text{z-sc}} I \tag{7.34}$$

as a very simple approximation for the nonlinear absorption of weakly absorbing samples. The coefficient $\beta_{\text{z-sc}}$ will usually be negative resulting in bleaching of the sample. This approximation has to be checked separately as described in the following sections.

The evaluation of the measured curve results from the valley-to-peak ratio of the z-scan measuring curve. For almost nonabsorbing samples the nonlinear refractive index at the beam axis in the focus $\Delta n_{\text{nl,0}}$ follows from:

nonlinear refractive index:

$$\Delta n_{\text{nl,0}} = 0.392\frac{\lambda_{\text{exc}}}{L_{\text{sample}}}[T_{\text{z-scan}}(z_{\text{peak}}) - T_{\text{z-scan}}(z_{\text{valley}})] \tag{7.35}$$

with laser wavelength λ_{exc} length of the sample L_{sample} and the transmissions from the scan curve $T_{\text{z-scan}}$. The difference is the peak-to-valley difference of the curve. This relation holds with an accuracy of about 0.5% for small nonlinear refractive index values, negligible absorption and small detection aperture D1 [7.188 and 7.189–7.193]. The length of the sample has to be small compared to the Rayleigh length of the focused beam and the observation distance between the sample and the aperture has to be large compared to it.

Small but nonnegligible absorption can be considered by replacing the sample length by an effective sample length given by $L_{\text{sample,effective}} = (1 - \exp(-aL_{\text{sample}}))/a$. Larger diameters of the diaphragm D1 can be considered by dividing this result by $(1 - S)^{0.25}$ with $S = 1 - \exp(-2r_{\text{a}}^2/w_{\text{a}}^2)$ with r_{a} as aperture radius and w_{a} as beam radius at the aperture in the linear case. S is almost zero for very small diameters and it should not be much larger than 0.01. On the other hand the signal-to-noise ratio has to be sufficiently good. For excitation pulses short compared to the decay time of the sample the result has to be divided by 0.707.

The value of $\gamma_{\text{z-sc}}$ can be calculated from $\Delta n_{\text{nl},0}$ by:

$$\gamma_{\text{z-sc}} = \frac{\Delta n_{\text{nl},0}}{I(z=0)} \tag{7.36}$$

which results for the above example of CS_2 to $n_2 = 1.2$ esu or $1.7 \cdot 10^{-8} \, \text{m}^2 \, \text{V}^{-2}$.

The whole z-scan curve as given in Fig. 7.25 can be modeled with the fit function:

$$T_{\text{z-scan}}(z, \Delta\Phi_0) \simeq 1 - \frac{4(z/z_0)}{((z/z_0)^2 + 9)((z/z_0)^2 + 1)}\Delta\Phi_0 \tag{7.37}$$

where $\Delta\Phi_0$ is the phase shift which is described below. From this value the peak-to-valley difference follows:

$$[T_{\text{z-scan}}(z_{\text{peak}}) - T_{\text{z-scan}}(z_{\text{valley}})] \simeq 0.406|\Delta\Phi_0| \tag{7.38}$$

which can be used for further evaluation. Additional absorption has to be considered separately. The absorption effect should be subtracted before the z-scan curve is evaluated for the determination of $\gamma_{\text{z-sc}}$.

7.5.2 Theoretical Description

The theoretical description of the z-scan interaction signal was given e.g. in [7.188]. The measuring signal behind the aperture D1 of Fig. 7.24 can be calculated from a superposition of Gaussian beams which are decomposed from the signal directly behind the sample and then propagated through free space over the distance $L_{\text{sample}-\text{D1}}$ from the sample to the aperture D1. The electric field pattern at the aperture is then given by:

$$E_{\text{D1}}(r, t) = E_{\text{D1}}(r = 0, t)e^{-\frac{a L_{\text{sample}}}{2}}$$
$$\sum_{m=0}^{\infty} \frac{[i\Delta\Phi_0(t)]^m}{m!} \frac{w_{m0}}{w_m} \exp\left(-\frac{r^2}{w_m^2} - \frac{i\pi r^2}{\lambda_{\text{exc}} R_m} + i\theta_m\right) \tag{7.39}$$

with the following abbreviations:

$$w_{m0}^2 = \frac{w^2}{2m+1}, \quad d_m = \frac{\pi w_{m0}^2}{\lambda_{\text{exc}}}$$
$$w_m^2 = w_{m0}^2\left[g^2 + \frac{d^2}{d_m^2}\right], \quad R_m = d\left[1 - \frac{g}{g^2 d^2/d_m^2}\right]^{-1} \tag{7.40}$$
$$\theta_m = \tan^{-1}\left[\frac{d}{d_m g}\right], \quad g = 1 + \frac{d}{R}$$

with the phase front radius R, the beam radius w, and d as the distance between the sample and the aperture D1.

$\Delta\Phi_0$ is the phase shift on the axis at the focus of the beam which is defined as:

$$\Delta\Phi_0(t) = \frac{2\pi}{\lambda_{\mathrm{exc}}} \Delta n_{\mathrm{nl}}(t) L_{\mathrm{sample}} \qquad (7.41)$$

resulting in the phase shift across the beam in the direction r and along the z-direction of the focusing:

$$\Delta\Phi(z,r,t) = \Delta\Phi_0(t) \frac{1}{1+z^2/z_0^2}\, e^{-2r^2/w^2(z)} \qquad (7.42)$$

With this formula the transmitted power through the aperture D1 to the detector E_{cl} can be calculated:

$$\text{transmitted power} \quad P_{\mathrm{trans}}(\Delta\Phi_0(t)) = \varepsilon_0 c_0 n_0 \pi \int_0^{r_{\mathrm{a}}} |E_{\mathrm{a}}(r,t)|^2 r\,\mathrm{d}r \qquad (7.43)$$

and thus the time integrated transmission as shown in the measured curve of Fig. 7.25 can be modeled with the expression:

$$T_{\mathrm{z\text{-}scan}} = \frac{\int_{-\infty}^{\infty} P_{\mathrm{trans}}(\Delta\Phi_0(t))\,\mathrm{d}t}{S \int_{-\infty}^{\infty} P_{\mathrm{inc}}(t)\,\mathrm{d}t} \qquad (7.44)$$

with S accounting for the diameter of the aperture D1 as given above. The incident power follows from:

$$P_{\mathrm{inc}}(t) = \frac{\pi}{2} w_0^2 I_0(t) \qquad (7.45)$$

This set of formulas given a description of cubic and higher-order nonlinear behavior. Usually only a few terms of the sum are necessary for sufficient accuracy.

For small nonlinear effects of $|\Delta\Phi_0| < 1$ the maxima and minima of the transmission occur for cubic nonlinear behavior at about $0.86 z_{\mathrm{Rayleigh}}$ and for fifth-order nonlinearity at $0.6 z_{\mathrm{Rayleigh}}$. For fifth-order nonlinearity the peak and valley transmission difference follows for the given assumptions from the phase shift by $[T_{\mathrm{z\text{-}scan}}(z_{\mathrm{peak}}) - T_{\mathrm{z\text{-}scan}}(z_{\mathrm{valley}})] \simeq 0.21|\Delta\Phi_0|$ as discussed in [7.188].

The sensitivity of the z-scan method decreases slowly with aperture size as discussed above and vanishes completely for very large apertures. For sufficiently high light intensities short pulses in the ns, ps or even fs range are useful. For avoiding damage and thermal effects shorter pulses are more suitable.

7.5.3 z-Scan with Absorbing Samples

For samples with strong absorption at the applied wavelength the z-scan measurement [e.g. 7.194, 7.195] can be extended by a parallel measurement of the bleaching behavior. Therefore the energy detectors E_{inc} and E_{out} are positioned around the sample. The transmission follows from:

$$T_{\text{energy}}(I) = \frac{E_{\text{out}}(I(z))}{E_{\text{inc}}} \tag{7.46}$$

and can be plotted as a function of the intensity which is calculated from the position of the sample using (7.29). The same measurement can be carried out by removing the diaphragm D1.

This bleaching behavior can be evaluated as described previously with all variations of the parameters of the exciting light. In the z-scan measurement the nonlinear absorption will cause a superposition of an absorption around the waist of the excitation beam. In the first-order approach the peak-to-valley ratio of transmission should remain constant and can be evaluated as described above.

7.6 Nonlinear Emission Measurements

Using laser beams for excitation in emission measurements allows investigations, e.g. of weakly fluorescing samples, which are not conventionally possible. But in this case nonlinear effects may have to be considered. On the other side the detailed investigation of the spontaneous and stimulated emission as a function of the excitation intensity enables investigations of the participating excited states. This information is in some cases complementary to the data from absorption measurements. With the fluorescence intensity scaling method the population density of the emitting excited state can be determined with high accuracy as described in Sect. 7.9.3. Different nonlinear emission measurements will be described. In any case special care has to be taken for possible unwanted stimulated emission and the polarization conditions of the exciting and emitted light.

7.6.1 Excitation Intensity Variation

In the low-signal intensity range the emission signal, e.g. the fluorescence, is directly proportional to the excitation intensity, because the population density of the emitting energy level increases proportionally to the pump intensity.

At higher intensities remarkable depopulation of the ground state occurs and the population of the excited state saturates. As an example in Fig. 7.26 the emission intensity of the energy level scheme of Fig. 7.1 is shown as a function of the excitation intensity.

The emission intensity curve can be directly compared to the nonlinear absorption graph of Fig. 7.2 which describes the same excitation condition. It can be seen that the fluorescence intensity saturates at excitation intensities of 10^{25} photons $\text{cm}^{-2}\,\text{s}^{-1}$ whereas the nonlinear transmission is still constant. Thus strong excited state absorption can be supposed from this result. Further the fluorescence intensity decreases for excitation intensities larger

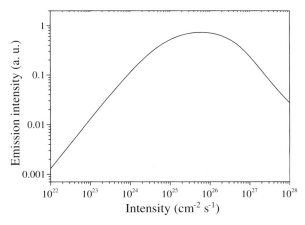

Fig. 7.26. Fluorescence intensity of the energy level scheme of the model of Fig. 7.1 as a function of the excitation intensity. Parameters as for Fig. 7.2

10^{26} photons $\mathrm{cm}^{-2}\,\mathrm{s}^{-1}$. This is another strong indicator for further excited state absorption at the pump laser wavelength.

Thus from the excitation intensity dependent measurements of the emission intensities important information about possible population densities of excited states can be determined [e.g. 7.196–7.198]. In particular, the possible incorrect interpretation from bleaching curves which do not show changes as a consequence of excited state absorptions can be avoided.

Therefore nonlinear emission measurements are an important complement of nonlinear absorption measurements. If the matter does not show sufficient emission intensity for such measurements this information has to be obtained from pump and probe measurements of the excited state absorptions.

7.6.2 Time-Resolved Measurements

The decay of the spontaneously emitted light intensity I_{emiss} of one transition should follow the exponential law for the emission intensity:

$$\text{spontaneous emission} \quad I_{\mathrm{emiss}}(t) = I_{\mathrm{emiss,max}}\mathrm{e}^{-t/\tau_{\mathrm{emiss}}} \qquad (7.47)$$

with the decay time τ_{emiss}. But time-resolved measurements of the emission decay as a function of the excitation intensity can show deviations from this simple exponential law (see e.g. [7.199–7.203] and Sect. 7.7.9). This can be caused by additional energy levels populated with high excitation intensities or other processes in the matter. Thus the superposition of relaxation chains from higher excited states to the emitting state can produce multi-exponential decays [e.g. 7.205, 7.206]. Additional relaxation processes such as orientation relaxation of the transition dipole moment or reorganization of the environment, e.g. the solvent, around the emitting particle may cause further complications.

Detailed laser measurements with a wide range of detectable decay times usually show a wide spectrum of different decay times from fs to ms indicating a large number of involved processes. Sometimes the best fit of the experimental data was reached with stretched exponential functions as:

stretched exponential decay $I_{\text{emiss}}(t) = I_{\text{emiss,max}} \, e^{-(t/\tau_{\text{emiss}})^{\beta}}$ (7.48)

with β as the stretching exponent with values between 0, or realistically 0.2, and 1. Values smaller than 1 lead to faster decay at the beginning and show a slower decay at long times. Thus a mixture of fast and slow relaxation processes is described.

The description with a stretched exponential function can be transformed to a multi-exponential decay expression which is:

multi-exponential decay $I_{\text{emiss}}(t) = \sum\limits_{m=1}^{p} I_{\text{emiss,max},m} \, e^{-t/\tau_{\text{emiss},m}}$ (7.49)

with amplitude factors $I_{\text{emiss,max},m}$ and the decay times $\tau_{\text{emiss},m}$ of the contributing emission transitions m. Two exponential decay times (with $p = 2$) can be often be fitted with high reliability. An unambiguous mathematical analysis of an experimentally observed multi-exponential decay is usually possible only with additional assumptions about the decay times.

7.6.3 Detection of Two-Photon Absorption via Fluorescence

Because of the small cross-sections for two-photon absorption it is difficult to detect it in absorption measurements. If the fluorescence is observed while the excitation light is spectrally tuned over the region of half the wavelength of the conventional absorption spectrum the simultaneous two-photon absorption can be proven much more easily [e.g. 7.210–7.211]. In this case only a few fluorescence photons have to be detected and not the difference between the transmissions of 100% and 99.999...%.

Thus the cross-section spectrum for two- or multi-photon absorption can be measured. These two-photon absorptions in the transparent spectrum of the matter can be of great importance in laser measurements if at this wavelength a possible strong excited state transition occurs as shown in Fig. 7.27.

Via this excited state absorption the laser light can be much more absorbed than via the two-photon absorption process. This type of two-photon absorption shows an emission intensity proportional to the square of the excitation intensity:

$I_{\text{emission,2-photon}} \propto I_{\text{exc}}^2$ (7.50)

This two-photon excited fluorescence can be used for the determination of the duration of short pulses as described in Sect. 5.5 in Fig. 5.45.

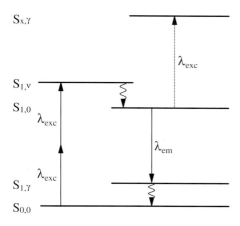

Fig. 7.27. Level scheme illustrating two-photon absorption at the wavelength λ_{exc} and conventional fluorescence with wavelength λ_{em}. In addition a possible excited state absorption at the wavelength λ_{exc} is shown

7.6.4 "Blue" Fluorescence

The appearance of fluorescence with much shorter wavelength than the wavelength of the exciting laser light is a safe indicator of two-photon absorption [e.g. 7.212–7.225]. This two-photon absorption can take place simultaneously as described in the previous section and in Sect. 5.5 or stepwise.

Stepwise absorption occurs via the population of excited states in the matter which absorb at the same wavelength as the ground state (see Fig. 7.28).

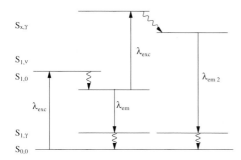

Fig. 7.28. Level scheme for stepwise absorption of two-photons leading to "blue" fluorescence with λ_{em2}

Because of the stepwise saturation of the absorption transitions in this scheme the emission intensity increases usually more slowly than the square of the excitation intensity.

This stepwise absorption can be quite efficient and thus an up-conversion of the laser light is possible (see Sect. 6.2 and references there). Thus this type of level scheme is applied in up-conversion solid-state lasers with an emission in the visible or blue spectral range.

7.7 Pump and Probe Measurements

Nonlinear absorption measurements with one or several excitation light beams and a separate probe light allow an almost indefinite variation of measuring methods for the determination of the various material properties. Some examples are given in [7.226–7.250]. The basic concepts of these measurements are described in this section. One of the main problems is the determination of the population densities of the participating excited matter states and the differentiation of superimposed absorptions. Thus the evaluation of the data can be very difficult and systematic errors are sometimes difficult to exclude. Therefore the optimal choice of the parameters of the measurements is one of the key issues.

7.7.1 Experimental Method

The sample is excited by at least one intensive light beam I_{exc} reaching the nonlinear range of the matter as shown in the experimental setup of Fig. 7.29.

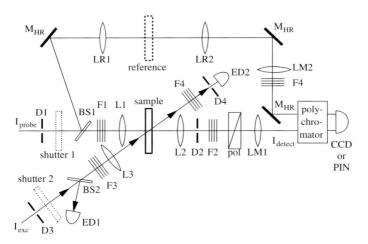

Fig. 7.29. Setup for pump and probe measurements with one excitation beam I_{exc} and detection of the probe spectrum intensity I_{detect}

Therefore the excitation light is focused to a not too small spot diameter typically of a few 100 μm for good spatial overlap as described in Sect. 7.1.4. Spatial background of the excitation light is filtered by the aperture D3. The excitation intensity can be varied by filters F3 or by changing the spot size with L3. The excitation intensity is monitored during the measurement by the energy detector ED1. For a suitable choice of the excitation parameters see the next section.

With the energy detector ED2 behind the filters F4 and the aperture D4 the transmitted excitation energy can be measured and thus the nonlinear transmission of the sample can be obtained during the measurement as described in Sect. 7.4.

The probe light beam can be generated from any light source providing enough light for a good signal-to-noise ratio in the detection system for the desired wavelengths. A more detailed description is given in Sect. 7.7.5.

The probe light beam is spatially filtered for background radiation by the diaphragm D1. The intensity can be varied for adaptation to the measuring system and for checking its influence with the filters in position F1. Lens L1 focuses the probe light beam well through the excited volume of the sample. Collinear, anticollinear and transversal geometries are possible between the pump and the probe beam (see Sect. 7.1.4).

The divergent beam is collimated with lens L2 and the diaphragm D2 spatially filters the emission light from the sample out of the detection beam path. With the filters at F2 the light can be adapted to the measuring system which has to be chosen in combination with F1. The polarizer Pol achieves the magic angle between the polarizations of the pump and the probe light (see Sect. 7.1.5).

The measuring system can be a polychromator in combination with a CCD camera as shown in the figure for measuring a transient spectrum with each excitation pulse. The measurement of the detection light can also be applied by simple (one-channel) detectors, e.g. in combination with login amplifiers for high resolution for the decay measurements of transient absorptions.

In any case the timing between the pump and the probe pulse has to be controlled with electrical (sub-ns to seconds) or optical (fs to 100 ns) delay lines to guarantee the temporal overlap of these pulses and to achieve the possible delays.

7.7.2 Measurements of Transient Spectra

With lens LM1 the light is focused for the detection system which can be e.g. a polychromator with an optical multichannel analyzer (OMA) system which is based on a CCD camera. These cameras show spatial resolutions of e.g. 1000×500 pixel, a useful dynamic range of $1:10\,000$ and a sensitivity of about one count per less than 10 photons.

With the beam splitter BS1 a reference light beam is out-coupled and via the lenses LR1, LR2 and LM2 also focused into the detection system. Both the probe light spectrum and the reference spectrum can be depicted as lines of illumination at different areas of the CCD camera, so they can simultaneously be detected and separately evaluated.

The polychromator and the CCD camera can of course be replaced by a monochromator and other light detectors such as PIN diodes, microchannel plates, photomultipliers, pyroelectric detectors or any other light detection system.

The focusing optics LM1 and LM2 have to be chosen to produce beam divergences which are adapted to the aperture of the polychromator which is of the size of e.g. 1:8. With the lenses LR1 and LR2 the probe beam divergence is compensated for the longer reference beam path. Between these two lenses a reference sample could be positioned.

The unexcited sample is usually used as reference. Thus shutters 1 and 2 are placed in the pump and probe beam and the detection light can be measured for different light conditions, e.g. in the following cycles:

(a) E_{both} pump and probe light are switched on
              (contains the transient absorption signal)
(b) E_{pump} pump light is on, probe light is off
              (contains the background signal including emission from the
              sample and noise)
(c) E_{probe} pump light is off, probe light is on
              (contains the ground state absorption of the sample)
(d) E_{noise} pump and probe light are off
              (contains the room light and noise).

If the reference beam is installed each light pulse results in the spectra of the detection pulse energy E^{detect} and the reference pulse energy E^{ref}. Thus the energy transmission of the transient spectrum is calculated from:

$$T_{\mathrm{energy}}(\lambda_{\mathrm{detect}}) = \frac{E_{\mathrm{both}}(\lambda_{\mathrm{detect}}) - E_{\mathrm{pump}}(\lambda_{\mathrm{detect}})}{E_{\mathrm{probe}}(\lambda_{\mathrm{detect}}) - E_{\mathrm{noise}}(\lambda_{\mathrm{detect}})}. \tag{7.51}$$

For comparison of fluctuations of the probe light spectrum the values of the reference spectra can be applied by calculating the transmission of the reference in the same way and dividing the sample transmission by this value.

This transient transmission spectrum is in general a function of all the parameters of the pump light and the probe light and the relation between them. Thus the measurement has to be characterized by the timing, polarization, intensities and sample conditions.

7.7.3 Coherence Effects in Pump and Probe Measurements

As described in Chap. 5 two types of coherence effects can influence the measurement. First the two light beams of the pump and probe are coherent and secondly the sample interacts coherently with the light waves.

The *coherent sample interaction* can occur if very short light pulses are applied because the sample polarization will dephase in time T_2. Thus the pulse duration has to be shorter than this T_2 time of the matter which is e.g. in the range of several 10 fs for molecular systems. The possible measuring effects are described in Sect. 5.4. The detailed analysis of these measurements demands an extensive quantum theoretical description and provides information about quantum correlation of the matter states. Some more information is given in [7.251–7.253].

Coherent light fields produce interference gratings which can result in absorption or refractive index gratings in the sample (see Sect. 5.3.8). Four-wave mixing at these gratings allows specialized pump, probe and detection beam geometry with high sensitivity as described in Sect. 7.8.3.

Unfortunately, induced gratings can occur in pump and probe measurements as unwanted side-effects. Reflection at these gratings guides the strong pump light perfectly into the direction of the detection light as shown in Fig. 7.30.

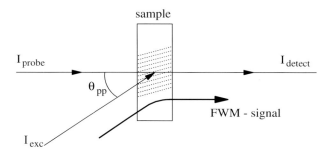

Fig. 7.30. Induced gratings and four-wave mixing in the sample from coherent pump and probe light beams. The four-wave mixing signal from the strong pump beam propagates always exactly towards the detection beam

This happens independently of the angle θ_{pp} even for transversal pumping or anticollinear schemes. Because of the much higher intensity of the pump beam this signal can dominate the detection beam.

Therefore whenever pump and probe beam have wavelength components of the same value the result of the measurement has to be checked for induced grating effects. The spectral width of the scattered FWM signal is approximately of the size of the reciprocal lifetime of the grating divided by 2π (see Sect. 7.8.3).

7.7.4 Choice of the Excitation Light

The excitation laser light beam has to be carefully designed for valuable results of the transient pump and probe measurements. The excitation intensity has to be chosen as described in Sect. 7.1.2 for a well-defined population of the selected transient matter state which is being investigated.

Care has to be taken for spatial, spectral or temporal background radiation as mentioned in Sect. 7.1.7 e.g. using diaphragms and spectral filters as far as possible.

The variation of the excitation intensity can be obtained with filters in the same way as in measurements of the nonlinear transmission (Sect. 7.4.1). They have to be proven to be linear in the intensity range used (see Sect. 7.1.7). Dielectric mirrors with different reflectivities are also useful and the combination of two polarizers can be calibrated for defined attenuation. In any case it has to be proven that attenuation does not change other parameters of the excitation light as e.g. its position or polarization.

7.7.5 Probe Light Sources and Detection

The probe light source has to provide synchronized pulses of shorter duration as the pump light pulse, with sufficient spectral width and intensity with good beam quality.

As the pump light beam the probe light has to be checked for spectral, spatial and especially for temporal background radiation. Temporal background can cause large and unnoticed errors as the probe light is detected with energy measurement devices such as e.g. CCD cameras.

Probe Light Pulse Energy

For the necessary spatial overlap the beam diameter of the probe light usually has to be smaller than 0.2 mm focused by a lens with more than 80 mm focal length for enough Rayleigh length and working distance. Thus the beam quality has to be better than about $M^2 < 20$ in the visible spectrum. This is easy to achieve with laser radiation but often causes strong losses from spatial filtering of other, especially of conventional light sources.

The necessary intensity can be estimated from the sensitivity of the applied detection system and the losses in the probe light beam path. As an example the sensitivity of a good CCD detector is about 10 photons count^{-1}. Thus for a dynamic range of $1{:}1000$ at least 10^4 photons are required. Typical minimal losses are about a factor of 10 from the polychromator, 1000 from the spectral spread, about 10 from the sample absorption and about 10 for all other optical components. Thus about 10^{10} photons are at least needed from the probe light source with this good beam quality in the short pulse. At 500 nm this corresponds to a pulse energy of about $0.01\,\mu$J. High accuracy of the measurement, polarizer, filters and a reference channel cause further demands. With laser radiation this pulse energy can usually be easily achieved. All other types of light sources have to be checked in detail.

On the other hand the probe light intensity at the sample should be small enough not to cause nonlinear effects by itself (see Sect. 7.1.3). Thus the excitation and probe light spots at the sample should be increased as much as possible for the given excitation light power to reach a good signal-to-noise ratio in detection and low probe light intensity at the sample.

Synchronized Lasers and Frequency Transformations

Lasers radiation can be converted to almost all wavelengths as described in Chap. 6, but broad-band laser radiation of several nm bandwidth is not as easy to achieve for all desired wavelengths and thus for measurements of the transient spectrum other sources are often more useful.

Nevertheless, because of the good beam quality and high pulse energy laser radiation as probe light allows very sensitive measurements at certain selected wavelengths with very small pump diameters, high signal-to-noise ratios and very small apertures [e.g. 7.254–7.255].

In particular, pump and probe measurements with the same wavelength for excitation and detection are very easy to achieve using a simple beam splitter for probe light generation and an optical delay line. Only the magic angle configuration may be difficult in this case.

Electronic synchronization of the lasers can be applied in the ns range. Thus e.g. thyratron-switched excimer or nitrogen lasers have jitters of about 2 ns and pulse durations of more than 10 ns which allows sufficient overlap. Active Q-switched solid state lasers or diode lasers can be applied in the same way. Thus time delays of almost up to infinity (ns to minutes) can be achieved electronically with this combination.

In the sub-ns to fs time domain the delay can be achieved only by optical delay lines (see Sect. 7.1.5). Thus the lasers have to be synchronized by optical coupling. As described in Sects. 6.10.3 and 6.10.4 dye lasers can be synchronized by synchronously pumping by another ps laser.

In the ps range but especially in the fs range the use of optical parametric amplifiers (OPAs) allows the synchronous generation of very short pulses. In combination with frequency transformation they emit at wavelengths from the UV to the far IR. This radiation is especially useful as probe light because it can easily be tuned over wide spectral ranges.

With special frequency conversion setups even the XUV to X-ray range with wavelengths below 10 nm can be covered. In this spectral range synchrotron radiation can also be applied as probe light.

White Light Generation with fs Duration

Very broad probe light spectra can be generated by nonlinear processes using focused fs laser pulses [7.256–7.264]. A typical scheme is shown in Fig. 7.31.

The excitation light, as e.g. from a Ti:sapphire laser, is focused with lens L1 into a sapphire plate below the damage threshold. This plate has to be short enough not to increase the pulse duration by dispersion which leads for 100 fs pulses to less than 1 mm. In this material all kinds of nonlinear optical processes take place as described in Chaps. 4 and 5 and thus frequency conversion takes place. As a result a broad spectrum ranging from about 400 to 1800 nm is produced. Choosing the SHG of the exciting laser light the lower wavelength limit can be further decreased.

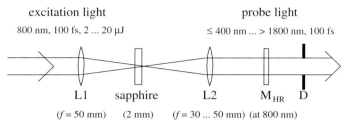

Fig. 7.31. Schematic of an experimental setup for "white light" generation using fs laser pulses

The excitation light should be filtered out of the probe light beam as much as possible. Therefore dielectric mirrors or (inversely applied) interference filters with narrow spectral bandwidth at the laser wavelength maybe used. With the lens L2 the probe light beam can be collimated and in combination with the aperture D spatially filtered for the required beam quality.

This setup usually produces light pulses not much longer than the excitation pulse. In special setups even shorter pulses can be obtained. The calibration can be applied with the pump and probe measurement using a known sample. The measured temporal slope of the transient absorption or bleaching as a function of the delay between the two pulses can be fitted with a variable pulse duration of the probe pulse and known parameters.

White Light Generation with ps Duration

In pump and probe measurements with laser pulses of more than 10 ps duration up to about 100 ps the generation of a spectrally broad probe light can be obtained in a similar way as described for fs pulses (see previous section) but the nonlinear material and its geometry are usually different [7.265–7.270]. An example is sketched in Fig. 7.32.

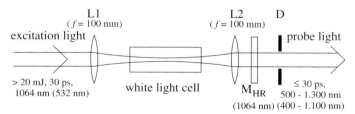

Fig. 7.32. Schematic of an experimental setup for "white light" generation using ps laser pulses

The optical elements have the same function as described for Fig. 7.31. Using ps pulses instead of fs pulses the interaction length can be longer. To avoid damaging the surfaces of the cell windows the "white light cell" should be about 100 mm long.

Liquids can be used as nonlinear matter which "repair" possible optical breakdowns by convection. Water (H_2O) and heavy water (D_2O) were used e.g. in combination with mode-locked Nd:YAG lasers. A broad and flat spectrum was observed using a mixture of CCL_4 and $CHCl_3$ in the ratio 9:1 [7.268]. The intensity of this material is smaller and thus for alignment the water cell may be used. The liquids have to be changed typically once every two weeks.

As described in [7.263] this type of "white light" generation is a mixture of several kinds of Raman scattering, self-focusing and other frequency transformation processes. Thus the exciting laser pulses should not show fluctuations above 5% for useful probe light. In any case a reference beam should be set up to reach sufficient accuracy of the measurements.

Fluorescence as Probe Light in the ns Range

For longer pulses in the ns time domain fluorescence light of selected organic materials can be used as a very useful probe light source [e.g. 7.271] as shown in Fig. 7.33.

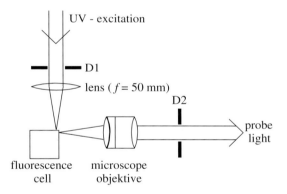

Fig. 7.33. Probe light generation for ns pump and probe experiments using the fluorescence of selected organic molecules

The fluorescence is pumped by a UV laser, e.g. an excimer or nitrogen laser in a cell of 10 mm by 10 mm. Above about 30 Hz the solution in this cell should be circulated. The excitation spot is as small as possible and the concentration of the dye solution is chosen as high as possible. Thus a "point" light source is produced. The light is 20–100 nm broad if only one dye is used. Mixtures of selected dyes allow spectral widths of up to 250 nm. The fluorescence is then collimated for high brightness and low imaging errors by a microscope lens. The resulting beam is spatially filtered with aperture D2 for the necessary beam quality of the probe light.

The dyes have to be chosen for high stability, short fluorescence lifetime and high fluorescence quantum efficiency. Useful candidates are e.g. paraterphenyl, stilbene 1, coumarin 102, coumarin 153, DCM, pyridine 1, styryl 9 and other laser dyes as shown in Fig. 6.96.

The fluorescence lifetime can be shortened if necessary if superradiation in the dye cell is obtained. Therefore the excitation light power is carefully tuned by the variable diaphragm D1. In this way the lifetime can be shortened by a factor of 10 but the spectral width of this radiation is then narrowed, too (see Fig. 5.10).

Flash Lamps

Several flash lamps [e.g. 7.272] with a pulse duration from ns to µs can be applied as probe light sources in pump and probe measurements, but it has to be checked whether the necessary beam quality of the setup and thus the necessary probe pulse energy in the desired spectral range can be achieved. In addition these lamps often show sharp emission lines which demand a high dynamic range and accuracy of the detection system. Thus flash lamps may in general be useful for investigations with a pulse duration of more than 100 ns. In pump and probe setups these lamps can be combined with gated CCD cameras or direct temporal measurements using PIN diodes or multipliers and fast oscilloscopes reaching ns resolution.

Spectral Calibration of Detection Systems

If the detector such as e.g. the CCD camera, is adapted to the polychromator the wavelength calibration has to be checked. The emission lines of atoms [e.g. 7.273–7.275] in spectral lamps are useful for this purpose. The resulting spectrum of a HgCd lamp measured with a CCD camera behind a 0.5 m polychromator with a grating of 147 lines mm^{-1} is shown as an example in Fig. 7.34.

The wavelengths of some emission lines are given in Table 7.6.

Table 7.6. Wavelengths of some high-intensity atomic and Fraunhofer (named and with color) absorption and emission lines for the calibration of detection systems

Atom	Wavelength (nm)	Atom	Wavelength (nm)	Atom	Wavelength (nm)
Hg	296.7278	Ca (K) UV	393.3666	Hg	546.0753
Hg	302.3476	Ca (H) UV	396.8468	Hg	576.959
Hg	312.5663	Hg	404.6561	Hg	578.966
Cd	325.2525	Hg	408.120	He (D$_3$) yellow	587.5618
Cd	326.1057	Fe (G) blue	430.7905	Na (D$_2$) yellow	588.9953
Hg	334.1478	Hg	435.835	Na (D$_1$) yellow	589.5923
Cd	340.3653	Cd	467.8156	Cd	643.8470
Cd	346.6201	Cd	479.9918	H (C) red	656.273
Cd	361.0510	H (F) blue-green	486.1327	O (B) red	686.72
Hg	365.4833	Cd	508.5824	O (A) IR	760.82
Hg	366.3276	Fe (E) green	527.0360	K (A') IR	766.491

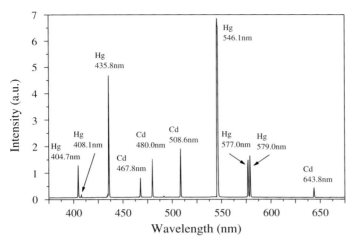

Fig. 7.34. Emission spectrum of a HgCd lamp measured with a CCD camera and a 0.5 m polychromator (grating of $147 \, \text{lines} \, \text{mm}^{-1}$)

The intensity of these lines is a function of the construction of the spectral lamp and therefore some lines may be difficult to find in a certain measured spectrum. For a fast check the line of a He-Ne laser at 632.8 nm can be used.

7.7.6 Steady-State Measurement

Steady state measurements with pulse durations longer than the longest involved decay time of the matter have the advantage of much easier evaluation of the data using much simpler mathematical models (see Sect. 5.3.6), but it has to be verified experimentally that the steady state assumptions are fulfilled under the conditions of the measurement.

Therefore the temporal shapes of the incident and transmitted pump pulses have to be compared. Asymmetrically changed transmitted pulses indicate in any case nonstationary behavior, but even symmetrical pulses are no guarantee of stationary interaction as shown in Fig. 7.35.

The calculation was applied using a nonstationary two-level model with a 10^3 times longer decay time than the pulse duration. The intensity was $1.5 \cdot 10^{24} \, \text{photons} \, \text{cm}^{-2} \, \text{s}^{-1}$. The transmitted pulse shows a symmetrical Gaussian shape but it is delayed compared to the incident pulse by about 2.2 ns. This delay does not result from the optical path but is a consequence of the nonstationary bleaching of the matter. As in stationary bleaching the duration of the transmitted pulse of 8.4 ns is shorter than the duration of the incident pulse of 10 ns.

Only a symmetrical change of the pulses around the time of the pulse maximum indicates steady state behavior of the material.

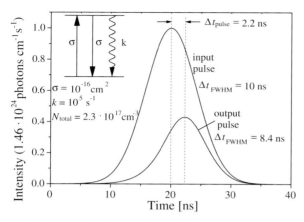

Fig. 7.35. Pulse shapes of incident and transmitted pulses in nonlinear bleaching using a nonstationary two-level model

For pulses much shorter than 1 ns the pulse shape can not usually be determined with sufficient accuracy for this purpose. Thus delayed measurements are necessary to indicate stationary or nonstationary behavior by determining the decay times directly. For pulses shorter than 10 ps stationary behavior becomes more and more unlikely.

7.7.7 Polarization Conditions

If pump and probe experiments are not particularly aimed at orientation relaxation effects [e.g. 7.276–7.284] the magic angle configuration should be applied (see Sect. 7.1.5). Although this polarization geometry is no guarantee of avoiding orientation effects in the measurement, especially if higher-order nonlinearity is involved, it solves the problem to first order.

For the detailed investigation of orientation effects such as the relative direction of the different absorption and emission dipole moments in the sample or the different intra- and interparticle orientation relaxation processes a large number of possible geometries of the pump and probe light polarizations exist. In these cases usually linearly polarized light is preferred.

In particular, in experiments with short pulses in the ps or sub-ps range the orientation relaxation can be much longer than the measurement and thus a "frozen" distribution is measured. A typical orientation relaxation time of a large molecule such as e.g. rhodamine 6G in alcohol, is in the range of a few 10 ps; in highly viscous solvents this can be increased by several orders of magnitude. Small molecules such as e.g. CS_2 can shown relaxation times in the range of 1 ps or below.

7.7.8 Excited State Absorption (ESA) Measurements

Whereas in common pump and probe measurements the transient absorption is measured as the superposition of all contributing transitions, excited state absorption (ESA) measurements are applied to extract the spectra belonging to a defined excited state in the same way as the ground state absorption spectra. Examples are given in [7.285–7.354].

Method

The determination of the spectrum of the cross-section σ_X of a transition from the excited state X to higher excited states demands the measured absorption spectrum from this state $T_X(\lambda)$ and knowledge of its population density N_X.

$$\textbf{ESA spectrum} \quad \sigma_X(\lambda) = -\frac{\ln T_X(\lambda)}{L_{mat} N_X} \tag{7.52}$$

with the length of the matter L_{mat}. These states X are often the first excited singlet S_1 or triplet T_1 state of the matter.

Thus the problem is the differentiation of the usually superimposed absorption spectra from the ground and all populated excited states and the determination of the population densities for a given excitation density.

Therefore measurements as a function of the excitation intensity and the delay time are used to differentiate the different transitions such as e.g. from the $S_{1,0}$ or $T_{1,0}$ states of organic molecules. In addition the polarization may be different for different transitions. Levels with different lifetimes may be separable by using different durations of the exciting pulses from the µs to fs range and delays (see Fig. 5.7).

As an example the level scheme of an organic molecule as shown in Fig. 7.36 is excited by pulses with pulse durations of 10 ns and 10 ps.

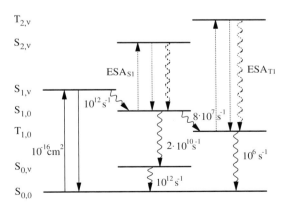

Fig. 7.36. Level scheme of an organic molecule with excited singlet and triplet state showing excited state absorptions

The resulting bleaching as a function of the excitation intensity of the pump pulses with two pulse durations is shown in Fig. 7.37.

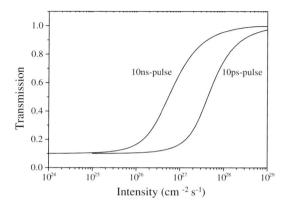

Fig. 7.37. Nonlinear transmission (bleaching) for the excitation of the level scheme of Fig. 7.36 as a function of the excitation intensity for two pulse durations of the pump pulse

The spatially averaged normalized population densities of the excited singlet $S_{1,0}$ and triplet $T_{1,0}$ levels relative to the total population density in the material of $2.30 \cdot 10^{17}$ cm^{-3} are shown for the cell length of 1 mm as a function of the excitation density for a FWHM Gaussian pulse duration of 10 ns in Fig. 7.38.

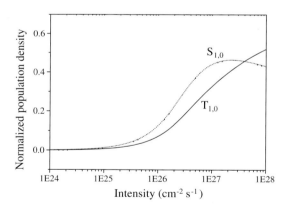

Fig. 7.38. Normalized population densities of the first excited singlet and triplet level of the level scheme of Fig. 7.36 as a function of the excitation intensity for an excitation pulse duration of 10 ns. The total concentration was $2.30 \cdot 10^{17}$ cm^{-3}

In this case the triplet population density is of the same order of magnitude as the population of the singlet level. Thus the measured ESA spectrum will show transitions in the singlet as well as in the triplet system as demonstrated in Sect. 5.3.2. Both can be measured in this time domain as will be shown below over a wide range of excitation intensities.

The superpositioned spectra can thus be separated mathematically under the assumption of the population density of the occupied states. This modeling using e.g. rate equations has to be covered by measurements of the nonlinear absorption and emission as already described.

For excitation with a much shorter pump pulse duration of 10 ps the same spatially averaged normalized population densities of the excited singlet $S_{1,0}$ and triplet $T_{1,0}$ are shown as a function of the excitation density in Fig. 7.39.

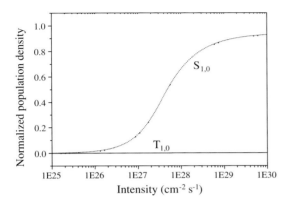

Fig. 7.39. Normalized population densities of the first excited singlet and triplet level of the level scheme of Fig. 7.36 as a function of the excitation intensity for an excitation pulse duration of 10 ps. The total concentration was $2.30 \cdot 10^{17} \, \text{cm}^{-3}$

As can be seen from this figure the population of the triplet state can be neglected using a ps excitation pulse for the given level scheme and time constants and the maximum population of the singlet level is much higher as compared to the ns excitation.

Estimate of the Population Densities

For relative small excitation intensities the population density in the first excited state may by estimated using the two-level scheme. The share of the population density in the excited state multiplied by the negative logarithm of the ground state transmission gives a value which is proportional to the measurable absorption probability from this state. For a rough estimate this value is given in Table 7.7 as $N_{\varepsilon nl}$ for stationary behavior of the system under the conditions of the measurement using a Gaussian shaped probe pulse of the same duration as the pump pulse.

These values $N_{\varepsilon nl}$ are calculated for the same bleaching values ε_{nl} as described in Table 7.4 for a two-level scheme. The population density averaged along the pump pulse in the first exited state N_{upper} as measurable with a probe pulse of the same duration as the pump pulse follows from:

$$N_{\text{upper}} = N_{\varepsilon nl} \frac{1}{\sigma L_{\text{mat}}}. \tag{7.53}$$

Table 7.7. Population density factor $N_{\varepsilon nl}$ of the first excited state of a stationary two-level scheme as a function of the bleaching parameters averaged along the excitation beam

$\varepsilon_{nl} \backslash T_0$	0.1%	1%	5%	10%	30%
1.1	0.0476	0.0474	0.0473	0.0473	0.0473
1.5	0.197	0.197	0.197	0.197	0.198
2	0.331	0.331	0.331	0.332	0.338
5	0.729	0.735	0.753	0.773	–
10	1.018	1.042	1.102	1.1513	–

As an example a sample with a small signal transmission of 1% and a length of 0.1 cm may be bleached to 2%. The resulting ε_{nl} is 2 and $N_{\varepsilon nl}$ is 0.331 and thus a sample cross-section of $10^{-16}\,\mathrm{cm}^2$ leads to a averaged population in the excited state of $0.331 \cdot 10^{17}\,\mathrm{cm}^{-3}$.

For strong nonstationary behavior with a characteristic decay time of the matter much larger than the pulse duration $\tau \gg \Delta t_{\mathrm{pulse}}$ the population densities of the first excited state of the two-level scheme are a function of the pulse energy and can be obtained from Table 7.8.

Table 7.8. Population density factor $N_{\varepsilon nl}$ of the first excited state of a non-stationary two-level scheme as a function of the bleaching parameters averaged along the excitation beam

$\varepsilon_{nl} \backslash T_0$	0.1%	1%	5%	10%	30%
1.1	0.0460	0.0458	0.0457	0.0457	0.0457
1.5	0.186	0.186	0.186	0.186	0.188
2	0.307	0.306	0.306	0.308	0.318
5	0.648	0.649	0.670	0.699	–
10	0.872	0.894	0.980	1.151	–

These values are again averaged over the sample along the longitudinal excitation and over the pump pulse duration. The absolute population densities in the excited state follow again from (7.53).

Thus from the nonlinear transmission curves in both cases of stationary and strongly nonstationary behavior the transmission in the bleached region can be read for the associated intensity. Using the values from Tables 7.7 and 7.8 a first rough estimate of the absolute population in the excited state can be executed using (7.53). Again both values have to be multiplied by 2 if no resonance emission occurs, as e.g. in a three-level scheme. As can be seen from these tables the difference between the stationary and nonstationary values is not large, as expected.

Differentiation of Singlet and Triplet Spectra

As described above the population densities of the singlet and triplet levels can be adjusted using e.g. different excitation intensities as shown in Fig. 7.38, but even more relevant are the durations of the excitation pulses. Because of the different lifetimes of the two electronic systems the singlet can be measured with short pulses and the triplet better with long ones (see Fig. 5.7). The temporal evolution of the population densities is shown in Figs. 7.40 and 7.41.

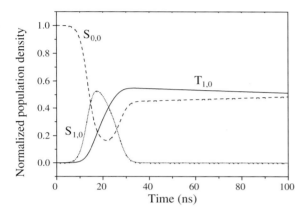

Fig. 7.40. Population densities of the first excited singlet and triplet level of the level scheme of Fig. 7.36 as a function of time for an excitation pulse duration of $10\,\text{ns}$ and an intensity of $7.2 \cdot 10^{26}\,\text{photons}\,\text{cm}^{-2}\,\text{s}^{-1}$ (compare Fig. 7.38)

Using ns excitation pulses the population of both the singlet and the triplet level can be large enough for easy measurement of both excited state absorptions. The population of the singlet level decays fast enough for a separate measurement of the spectra from the triplet level, alone after about $30\,\text{ns}$. Thus the possible superposition of the triplet absorption in the singlet spectrum can be deconvoluted by measuring the triplet decay separately and determining in this way the triplet population at short times (compare Fig. 5.7).

The measurement of the transient absorption with very short pulses can lead to negligible populations of long-lived levels such as the triplet level if the intersystem crossing rate is small compared to the direct decay rate as shown in Fig. 7.41.

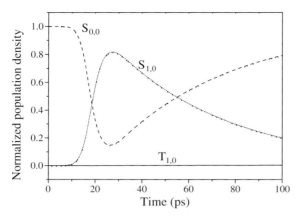

Fig. 7.41. Population densities of the first excited singlet and triplet level of the level scheme of Fig. 7.36 as a function of time for an excitation pulse duration of $10\,\mathrm{ps}$ and an intensity of $4.5 \cdot 10^{27}$ photons cm^{-2} s^{-1} (compare Fig. 7.38)

7.7.9 Decay Time Measurements

Most information is available from the measurement of the whole transient spectra as a function of the delay time between pump and probe pulse. In addition using sufficiently long negative delays a very useful baseline correction including e.g. all scattered light is possible for the transient spectra, too.

From the change of the spectrum as a function of the delay time the different superimposed components from different transitions can usually be distinguished. Therefore this type of measurement is often the precondition for all other decay measurements.

But for very high dynamic and accuracy of decay measurements, detection at one probe wavelength can be more useful [7.339–7.354]. The detection system can be designed using e.g. login amplifiers or other high-precision techniques to measure the decay of transient absorptions over many orders of magnitude with errors smaller than 3%.

If in these measurements no mono-exponential decay curves are obtained the spectra have to be measured. Then other wavelengths of the excitation and probe beam can be used to check the decay mechanism. Again care has to be taken for the polarization of the beams.

7.8 Special Pump and Probe Techniques

The optical nonlinear behavior of absorbing matter can be very complex and thus several special techniques have been developed. Thus the inhomogeneous broadening can be obtained from fractional bleaching (FB), nonlinear polarization (NLP) and hole burning (HB) measurements. Induced gratings allow very high sensitivity of the measurement. Two-photon absorption and multiphoton excitation allow the study of states with high energies.

7.8.1 Fractional Bleaching (FB) and Difference Spectra

The investigation of the spectral behavior of the bleaching of absorption bands allows the determination of the homogeneous or inhomogeneous character of the observed bands (see Sect. 5.2).

In particular, the plot of the fractional bleaching FB [7.355] as a function of the wavelength or frequency of the probe light across the absorption band easily shows inhomogeneous broadening under the conditions of the measurement. Thus this investigation is especially useful for the characterization of the ground state absorption.

The fractional bleaching FB is defined as the relative change of the absorption, i.e. the difference of absorption during excitation divided by the unexcited absorption:

fractional bleaching (FB)

$$FB(\lambda_{probe}) = \frac{a_{exc}(\lambda_{probe}) - a_{not\text{-}exc}(\lambda_{probe})}{a_{not\text{-}exc}(\lambda_{probe})} \qquad (7.54)$$

with the absorption coefficients of the sample as a_{exc} measured for the simultaneous excitation and $a_{not\text{-}exc}$ for unexciting this transition. The excitation intensity has to be high enough to achieve remarkable bleaching.

For the evaluation of the experimentally determined transmission curves of the sample with excitation resulting in T_{exc} and without excitation resulting in $T_{not\text{-}exc}$ both as a function of the wavelength of the probe light the fractional bleaching is calculated from:

fractional bleaching (FB)

$$FB(\lambda_{probe}) = 1 - \frac{\ln(T_{exc}(\lambda_{probe}))}{\ln(T_{not\text{-}exc}(\lambda_{probe}))} \qquad (7.55)$$

As can be seen from these formulas the fractional bleaching is constant as a function of the wavelength if the absorption bands of the excited particle and the unexcited particles show the same spectrum as shown in Fig. 7.42.

In the case of homogeneous broadening the decrease of the absorption coefficient to 60% results in a value of 40% for the fractional bleaching (FB). This gives the share of excited particles and thus the FB value of 40% from Fig. 7.42 means 40% of all particles were excited. If resonance emission occurs at this wavelength this number has to be divided by 2 and thus 20% of the particles would be excited for such a sample.

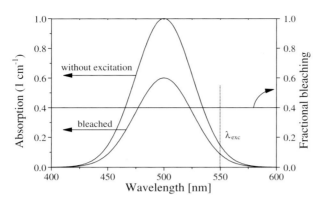

Fig. 7.42. Absorption spectra of a homogeneous transition and the result of a bleaching experiment in this band using a laser with wavelength λ_{exc} plotted as fractional bleaching (FB)

In the case when the absorption band is inhomogeneously broadened as shown as an example in Fig. 7.43 and the spectral diffusion time (see Sect. 5.2) is longer than the characteristic time of the experiment the different homogeneous absorption bands will be bleached differently.

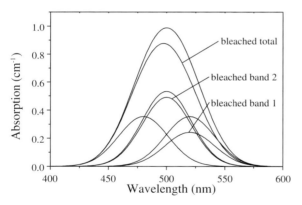

Fig. 7.43. Absorption spectra of an inhomogeneous transition consisting of three subbands positioned at 480, 500 and 520 nm with relative amplitudes of 0.355, 0.5325 and 0.355 and a width of 50 nm. As a result of a bleaching experiment these bands are decreased using a laser with wavelength $\lambda_{exc} = 550$ nm

As can be seen from this figure bleaching at the long-wavelength edge of the whole band decreases the absorption of band 3 more than bands 2 and 1. The resulting bleaching is observable at the long wavelength side. The fractional bleaching shows this result clearly as given in Fig. 7.44.

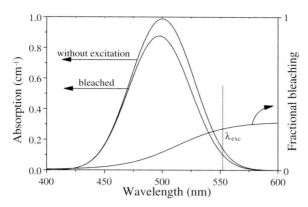

Fig. 7.44. Absorption spectra of an inhomogeneous transition as shown in Fig. 7.43 and the result of a bleaching experiment in this band using a laser with the wavelength $\lambda_{\mathrm{exc}} = 550\,\mathrm{nm}$ plotted as fractional bleaching (FB). This figure should be compared with Fig. 7.42

The fractional bleaching reaches in this case its maximum at very long wavelengths.

If inhomogeneous bleaching is detected the spectral shape of the bleaching can also be given as the difference spectrum Δa:

difference spectrum

$$\Delta a(\lambda_{\mathrm{probe}}) = a_{\mathrm{exc}}(\lambda_{\mathrm{probe}}) - a_{\mathrm{not\text{-}exc}}(\lambda_{\mathrm{probe}}) \qquad (7.56)$$

which is shown in Fig. 7.45 for the parameters of Figs. 7.43 and 7.44.

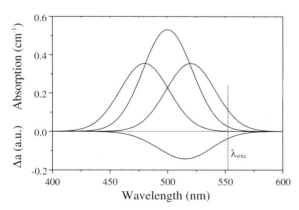

Fig. 7.45. Difference spectrum for bleaching after Fig. 7.43 and 7.44. For comparison the three subbands of the inhomogeneous absorption band are also shown

It has to be noticed that the difference spectrum is not identical with one of the subbands in the case of inhomogeneous broadening. Only for homogeneous transitions is this spectral shape identical.

Additional transient absorptions occur in the fractional bleaching spectrum as negative values and in the difference spectrum as positive values.

7.8.2 Hole Burning (HB) Measurements

Using excitation light which is spectrally narrow compared to the spectral width of the absorption band for bleaching can produce narrow spectral dips in the absorption band if the absorption band is strongly inhomogeneous (see Sect. 5.2). This hole burning [7.356–7.428] allows the observation of the different subspecies in the sample as sketched in Fig. 7.46 if the excitation pulse duration is shorter than the spectral cross-relaxation time T_3 and the energy relaxation time T_1.

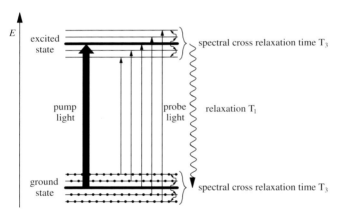

Fig. 7.46. Term scheme of a sample with many levels forming the ground and excited state which leads to inhomogeneous broadening of the absorption band

By the detailed investigation of these burned holes as a function of external parameters such as temperature or pressure the internal and external particle interactions can be studied. As a spectrally narrow nonlinear feature hole burning may find applications in communication technologies using spectral multiplexing or in analytics.

Method

Spectral hole burning is investigated with pump and probe setups as shown in Fig. 7.29 but with special attention for suppression of scattered light from excitation out of the detection system. The probe light spectrum usually covers the wavelength of the excitation. Thus anticollinear setups may be used and small apertures are placed in the detection channel.

The spectral resolution of the detection system and the bandwidth of the laser have to be narrow enough to resolve the hole burning effect. Its bandwidth is given by the inverse lifetime multiplied by 2π of the subspecies (see Sect. 2.1.2). The lifetime can be as small as 10 fs, e.g. for molecular systems at room temperature resulting in spectral widths of more than 10 nm. For low temperatures or other solid systems lifetimes of several µs or even much longer times are possible. Spectral resolutions as high as 10^{-7} nm or better are then required and more than 10 000 subspecies with different absorption bands can be present.

The general nonlinear spectroscopic features can be discussed using a simple model with an inhomogeneous absorption band consisting of nine subbands as shown in Fig. 7.47.

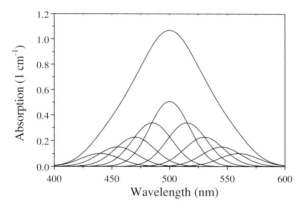

Fig. 7.47. Absorption coefficient of an inhomogeneously broadened transition consisting of nine homogeneous subbands of equal bandwidth (40 nm) equidistant around 500 nm in steps of 15 nm with amplitude ratios of 1.5

If this transition is bleached with an excitation wavelength of 530 nm the total absorption will be decreased more in the vincinity of this wavelength. The bleaching is assumed to be proportional to the absorption coefficient at the excitation wavelength. The resulting absorption is shown in Fig. 7.48.

As can be seen from the figure the bleaching leads to a "hole" in the absorption around 530 nm. The difference spectrum is shown in Fig. 7.49.

In this example the difference spectrum shows the maximal effect at a wavelength of 516 nm which is shifted from excitation towards the maximum of the whole band. The fractional bleaching has its maximum at approximately 530 nm. Both spectra are wider than the bandwidths of the subbands.

If the spectral cross-relaxation times are very long almost permanent hole burning can be occur. This can be achieved in molecular systems, e.g. if the

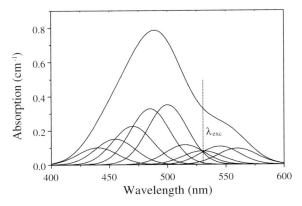

Fig. 7.48. Inhomogeneously bleached absorption coefficient of the inhomogeneous broadened transition of Fig. 7.47. The resulting absorption coefficients of the subbands are reduced proportionally to their value at 530 nm

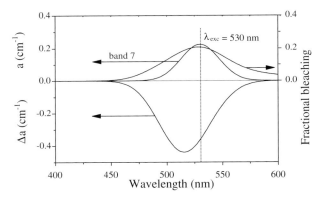

Fig. 7.49. Evaluation of the hole burning effect with excitation at 530 nm: fractional bleaching (FB), the difference of the absorption Δa, and the absorption coefficient a of subband 7 which is centered at the excitation

excited species undergoes a chemical reaction and is permanently changed. This is called *photochemical hole burning*. To avoid fast spectral cross relaxation into this absorption channel very low temperatures are applied. *Photophysical hole burning* does not involve any chemical reaction. Different molecular species are built by different influences of the arrangement of the surrounding environment and slightly different molecular conformations.

Low Temperature Hole Burning Measurements

Both photophysical and photochemical hole burning can be obtained at low temperatures. For molecular systems remarkable HB effects occur at temperatures below 10 K. Typical temperatures are 1 K or below. Burned holes of dissolved molecules showed wavelength bandwidths of the order of magnitude of 10 pm. This high spectral resolution can be achieved by using very narrow pump and probe lasers. The shape of the hole can be measured by tuning the wavelength of the probe laser.

At these temperatures the spectral cross-relaxation is in the range of seconds to minutes. Thus single-mode cw lasers with average output powers below 1 W can be used to burn the holes. The probe light can than be detected with the login technique for high dynamics. Details are described in [7.356–7.419].

Hole Burning Measurements at Room Temperature

At room temperature the spectral cross-relaxation time is in the range of fs to ps. The effect can be measured with very short pulses only. Thus it would be necessary to adapt the bandwidth and the pulse duration of the laser to the lifetime of the hole. This is difficult for new samples without knowing spectral cross-relaxation times in advance. The measured hole cannot be narrower than the spectral width of the laser pulse. Thus these experiments need very careful design [e.g. 7.420–7.428].

7.8.3 Measurement with Induced Gratings: Four-Wave Mixing

Absorption gratings can be induced via the nonlinear transmission of the sample if two coherent light beams are applied. Thus if the pump and probe beam have the same wavelength these gratings can occur and four-wave mixing can be obtained.

These induced and transient gratings can be used for a detailed analysis of the sample properties. Therefore in these experiments the process of grating production with the excitation light and the scattering of the probe light at the induced grating is usually executed, separately, as described in Sect. 5.3.11 (see especially Fig. 5.30).

If the wavelength of the probe light is in the range of the ground state absorption the measurement aims at the investigation of nonlinear bleaching of the sample as e.g. in fractional bleaching measurements. If the probe light has a different wavelength as the excitation beams new transient absorptions can be detected as e.g. in measurements of excited state absorption (ESA) spectra.

As shown in the example of Fig. 7.50 the grating of the excited state population and bleaching of the ground state absorption is obtained with two excitation beams. With a probe beam of e.g. a different wavelength the transient absorption or ESA can be measured with high sensitivity.

If a spectrally broad probe beam is applied the scattering angle will be different for the different wavelengths and thus the spectrum can be detected with a CCD camera or array, directly, behind the sample. Special care has

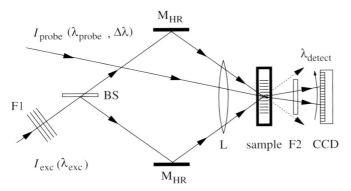

Fig. 7.50. Schematic of the experimental setup for measuring transient absorption spectra via induced grating structures of excited state population

to be taken for background light from diffuse scattering of the sample. Thus the filter F2 should block the excitation wavelength.

The detected signal is broadened by $\Delta\nu_{\text{detect,grating}}$, the limited lifetime of the grating τ_{grating}. The resulting minimal spectral width of the scattered signal follows from:

$$\Delta\nu_{\text{detect,grating}} = \frac{1}{2\pi\tau_{\text{grating}}} \tag{7.57}$$

Thus the different gratings can be differentiated by using a spectrally very narrow probe light and measuring the spectral width of the scattered signal. A rough classification is given in Table 7.9 for molecular systems.

Table 7.9. Rough classification of grating lifetimes for different decay mechanisms and the resulting spectral widths of the broadening for molecular systems

Process		lifetime	Frequency bandwidth	bandwidth at 500 nm
Dephasing	T_2	10 fs–1 ps	16 THz–160 GHz	13 nm–0.13 nm
Spectral cross-relaxation	T_3	10 fs–100 ps	16 THz–1.6 GHz	13 nm–1.3 pm
Internal conversion	T_1	1 ps–10 ns	160 GHz–16 MHz	0.13 nm–13 fm
Intersystem crossing	T_{isc}	1 ns–1 ms	160 MHz–160 Hz	130 fm–$1.3 \cdot 10^{-10}$ nm
Orientational relaxation	T_{orient}	1 ps–100 ps	160 GHz–1.6 GHz	0.13 nm–1.3 pm
Thermal	T_{thermal}	1 µs–10 s	160 Hz–16 mHz	$1.3 \cdot 10^{-7}$ nm–$1.3 \cdot 10^{-14}$ nm

These value ranges are rough estimates and can differ for special systems by many orders of magnitude. In particular the values may be changed at low temperatures or high pressures.

7.8.4 Nonlinear Polarization (NLP) Spectroscopy

In nonlinear polarization spectroscopy [7.429–7.432] the induced dipole moments from the excitation are measured between crossed polarizer Pol1 and Pol2 for the probe light beam as shown in Fig. 7.51.

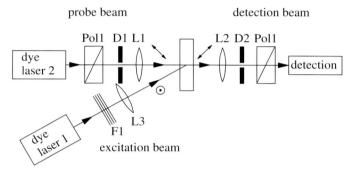

Fig. 7.51. Experimental setup for nonlinear polarization (NLP) spectroscopy. The excitation of the sample can be achieved under small angles but also transversal. The probe wavelength is usually fixed and the excitation wavelength tuned within the ground state absorption band of the sample

The linearly polarized excitation light beam has a polarization direction of $45°$ relative to the probe light polarizer. Thus no probe light will be detected in the linear case. Therefore a very good polarizer should be used with extinction ratios of better than 10^{-5} and up to 10^{-8}. In the detection beam path a monochromator can be used for suppressing scattered light.

The observed signal is a result of the third-order nonisotropic nonlinear polarization in the sample. This type of four-wave mixing process is especially useful to distinguish homogeneous and inhomogeneous broadening of absorption bands. Because of the limited spectral resolution only spectral cross-relaxation times shorter than about $100\,\mathrm{ps}$ can be obtained.

The detection signal at a certain wavelength within the absorption band is usually measured as a function of the wavelength and intensity of the excitation light which is scanned over the band. This spectrum $I_{\mathrm{detect}}(\lambda_{\mathrm{exc,NLP}})$ can be modeled and the material parameters such as the line widths and thus the relaxation times can be determined.

The theoretical description is based on four-wave mixing (FWM) in the matter based on third-order nonlinearity as given e.g. in [7.430]:

$$\frac{\partial E_{\text{probe}}}{\partial z} + \frac{a(\nu_{\text{detect}})}{2} E_{\text{probe}} = i\frac{4\pi^2\nu_{\text{detect}}}{c_0} P_{\text{nl}}^{(3)} \tag{7.58}$$

using the slowly varying amplitude approximation. The sample absorption is given by the absorption coefficient $a(\nu_{\text{detect}})$ with the frequency of the detected signal ν_{detect}. The third-order nonlinear polarization $P_{\text{nl}}^{(3)}$ results from:

$$P_{\text{nl}}^{(3)} = \frac{\varepsilon_0}{2}\chi^{(3)}(\nu_{\text{exc}}, \nu_{\text{detect}})|E_{\text{exc}}|^2 E_{\text{probe}} \tag{7.59}$$

with the electric fields E_{exc} of the excitation beam with wavelength ν_{exc} and E_{probe} of the probe beam. $\chi^{(3)}$ represents the third-order nonlinear susceptibility. The detected NLP signal intensity I_{NLP} is proportional to the square of the excitation intensity I_{exc}. It is also proportional to the square of $\chi^{(3)}$ and thus to the square of the line shape function F_{NLP} of $\chi^{(3)}$ with $\chi^{(3)} \propto F_{\text{NLP}}(\nu_{\text{exc}}, \nu_{\text{detect}})$:

$$I_{\text{NLP}}(\nu_{\text{exc}}, \nu_{\text{detect}}) \propto I_{\text{exc}}^2 I_{\text{probe}} F_{\text{NLP}}^2(\nu_{\text{exc}}, \nu_{\text{detect}}) \tag{7.60}$$

as long as no nonlinear absorption $a \neq f(I_{\text{exc}})$ occurs. The line shape function has to be calculated as given in Sect. 5.9.2. For simple cases the line shape function is given as [7.432]:

two-level model with homogeneous broadening:

$$F_{\text{NLP}}(\Delta\nu_{\text{e-d}}, \Delta\nu_{\text{r-d}}) =$$

$$\frac{i}{k_2 + i2\pi\Delta_{\text{r-d}}}\left[\frac{1}{k_1}\left(\frac{1}{k_2 + i2\pi(\Delta\nu_{\text{r-d}} - \Delta\nu_{\text{e-d}})} + \text{c.c.}\right)\right.$$

$$\left. + \frac{1}{k_1 + i2\pi\Delta\nu_{\text{e-d}}}\left(\frac{1}{k_2 + i2\pi\Delta\nu_{\text{r-d}}} + \frac{1}{k_2 - i2\pi(\Delta\nu_{\text{r-d}} - \Delta\nu_{\text{e-d}})}\right)\right] \tag{7.61}$$

two-level model with inhomogeneous broadening with distribution function g:

$$F_{\text{NLP,total}}(\Delta\nu_{\text{e-d}}, \Delta\nu_{\text{r-d}}) = \int_{-\infty}^{\infty} g(\tilde{\nu}_{\text{r}})F_{\text{NLP}}(\Delta\nu_{\text{e-d}}, \Delta\tilde{\nu}_{\text{r-d}})\,d\tilde{\nu}r. \tag{7.62}$$

With a Lorentzian line shape function g with FWHM width $1/k_{\text{L}}$ and the maximum at ν_{L} the NLP line shape function is given as:

$$F_{\text{NLP}}(\Delta\nu_{\text{e-d}}, \Delta\nu_{\text{L-e}}) =$$

$$\frac{i}{k_2 + k_{\text{L}} + i2\pi\Delta\nu_{\text{L-d}}}\left[\frac{1}{k_1}\left(\frac{1}{k_2 + k_{\text{L}} + i2\pi(\Delta\nu_{\text{L-d}} + \Delta\nu_{\text{e-d}})}\right)\right.$$

$$+ \frac{1}{(k_1 + i2\pi\Delta\nu_{\text{e-d}})(k_2 + k_{\text{L}} + i2\pi\Delta\nu_{\text{L-d}})}$$

$$+ \left(\frac{1}{k_1} + \frac{1}{k_1 + i2\pi\Delta\nu_{\text{e-d}}}\right)$$

$$\left. \cdot \left(\frac{2k_2 + 2k_{\text{L}} + i2\pi\Delta\nu_{\text{ed}}}{k_2 + k_{\text{L}} - i2\pi(\Delta\nu_{\text{L-d}} + \Delta\nu_{\text{e-d}})(2k_2 + i2\pi\Delta\nu_{\text{e-d}})}\right)\right]. \tag{7.63}$$

heterogeneous broadening from m subbands:

$$F_{\text{NLP,total}}(\Delta\nu_{\text{e-d}}, \Delta\nu_{\text{r-d},p}) = \sum_{p=1}^{m} c_p F_{\text{NLP}}(\Delta\nu_{\text{e-d}}, \Delta\nu_{\text{r-d},p}) \qquad (7.64)$$

with frequency differences:

$$\Delta\nu_{\text{e-d}} = \nu_{\text{exc}} - \nu_{\text{detect}} \quad \text{and} \quad \Delta\nu_{\text{r-d}} = \nu_{\text{resonance}} - \nu_{\text{detect}} \qquad (7.65)$$

where $\nu_{\text{resonance}}$ is the resonance frequency of the transition. In the case of heterogeneous broadening the resonance frequency of the pth transition has to be used.

As examples the normalized NLP signal as a function of the difference frequency $\Delta\nu_{\text{e-d}}$ is shown in Fig. 7.52 for a homogeneously broadened transition of a two-level scheme and an extreme inhomogeneous transition of the same conventional spectral width and peak position of a Gaussian shaped band.

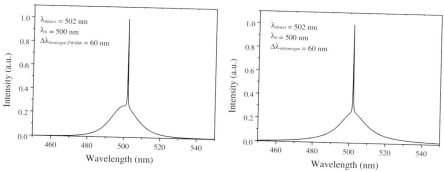

Fig. 7.52. Normalized NLP signal for a homogeneously broadened transition of a two-level scheme (left) and an extreme inhomogeneous transition (right) for the same spectral width and position of the conventionally observed Gaussian band

The conventional absorption band has a width of 60 nm and the parameters $T_1 = 1/k_1 = 1\,\text{ps}$ and $T_2 = 1/k_2 = 30\,\text{fs}$. In Fig. 7.53 the NLP signal is shown for a heterogeneous band consisting of two subband transitions. The amplitude ratio of these two bands is 1:3. The bandwidths were chosen as 8.8 nm.

As can be seen from these figures the decay times T_1 and T_2 can be determined from fitting the spectral shape of the curves. Further the inhomogeneous broadening can be detected as in hole burning or fractional bleaching measurements.

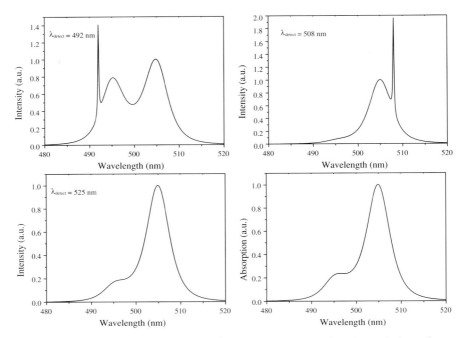

Fig. 7.53. Normalized NLP signals for a heterogeneous band consisting of two transitions as a function of the wavelength of the excitation pulse for different probe or detection pulse wavelengths as given in the figures. The lower right spectrum is the conventional absorption spectrum for comparison

7.8.5 Measurements with Multiple Excitation

Using a second strong light beam in resonance with excited state absorption as shown in Fig. 7.54 allows the bleaching of this transition between excited states $|3\rangle$ and $|5\rangle$.

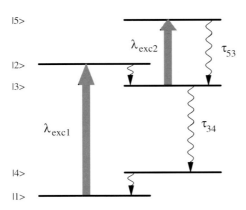

Fig. 7.54. Energy level scheme with successive excitation by absorption of two-photons with different wavelengths and thus population of a highly excited state $|5\rangle$ which can be probed for further transitions to even higher states

If the excitation intensity $I_2(\lambda_{exc2})$ is varied while the excitation intensity $I_1(\lambda_{exc1})$ is kept constant the nonlinear transmission of this transition between $|3\rangle$ and $|5\rangle$ can be evaluated and thus the decay time τ_{53} can be determined. The modeling can be carried out as described in Sect. 7.4. One example is given in [5.21]. If the population of level $|3\rangle$ is larger than about 10% the change of the population densities of levels 1–4 by the influence of the strong excited state absorption should be considered. Therefore the whole system should be taken into account using both excitation beams.

In this scheme level 5 is populated via stepwise excitation and thus the absorption from this state to even higher states can be investigated with an additional probe light beam as in pump and probe spectroscopy [5.21].

This stepwise excitation allows very specific preparation of new matter states. Thus different components in a mixture of matter with almost equal ground state absorption bands at the long-wavelength side can be distinguished with this technique e.g. for analytical purposes.

More than two excitations can be applied in the same way and much higher states can be populated and investigated in this way, but care has to be taken for reaching finally excited states above the dissociation limit of the sample.

7.8.6 Detection of Two-Photon Absorption via ESA

Excited state absorptions (ESA) can occur at half of the wavelength of the ground state absorption for particular samples [e.g. 5.22]. Thus simultaneous two-photon absorption can be achieved followed by an excited state absorption of the pump light as sketched in Fig. 7.55.

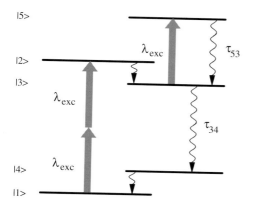

Fig. 7.55. Energy level scheme showing excited state absorption at the same wavelength as two-photon ground state absorption

In this case the cross-section for the excitation of the excited state is usually much higher than the two-photon absorption cross-section of the ground state absorption (GSA). Thus the laser will be absorbed more strongly by the

ESA than from the GSA. In this case the depopulation of the excited state $|3\rangle$ can be remarkable. This would populate the higher excited state $|5\rangle$.

This effect can disturb photonic applications designed in the nonabsorbing, transparent wavelength range of the sample based on the nonlinear refractive index n_2 as described in Chap. 4.

7.9 Determination of Population Density and Material Parameters

From the transmission T obtained in pump and probe experiments as a function of the excitation intensity I_{exc} and the wavelength λ_{probe} and delay time Δt_{probe} of the probe light the absorption coefficient a follows:

$$a(\lambda_{\text{probe}}, \Delta t_{\text{probe}}, I_{\text{exc}}) = -\frac{1}{L_{\text{sample}}} \ln[T(\lambda_{\text{probe}}, \Delta t_{\text{probe}}, I_{\text{exc}})] \qquad (7.66)$$

with the geometrical length of the sample L_{sample} in the direction of the probe light.

The cross-sections $\sigma_m(\lambda_{\text{probe}})$ of the excited state absorptions can be calculated from these experimentally determined absorption coefficients $a(\lambda_{\text{probe}}, \Delta t_{\text{probe}}, I_{\text{exc}})$ if the population densities of the involved excited states $N_m(I_{\text{exc}}, \Delta t_{\text{probe}})$ are known:

$$a(\lambda_{\text{probe}}, \Delta t_{\text{probe}}, I_{\text{exc}}) = \sum_m \sigma_m(\lambda_{\text{probe}}) N(I_{\text{exc}}, \Delta t_{\text{probe}}). \qquad (7.67)$$

Differentiation of superposing spectra from different states can usually be obtained by varying the excitation intensity and the delay time between the pump and probe light pulses. In addition the polarization and other light parameters can also be changed for this differentiation.

Thus the determination of the population densities of the involved matter states is essential for the evaluation of the pump and probe measurements and for the determination of the cross-sections of the excited states. Unfortunately, this task is difficult and therefore different methods have usually to be combined.

7.9.1 Model Calculations

Model calculations of the population densities of all populated states of the sample are in the end the best way for checking the different experimental results from different experimental conditions for the consistency of all assumptions.

In most cases rate equations are sufficient for the description discussed in Sects. 5.1–5.4, but even this simple description can be difficult because the necessary model parameters are often not known. Therefore additional measurements are necessary to determine these parameters.

But on the other hand not all involved model parameter values are crucially important for the result of the modeling with respect to the determination of the cross-sections. This can be demonstrated with a simple level scheme as used for modeling the excitation of organic molecules containing singlet and triplet levels as shown in Fig. 7.56.

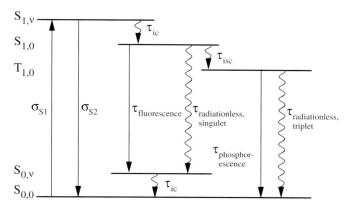

Fig. 7.56. Level scheme for the modeling of the excitation of organic molecules

The excitation occurs from the electronic and vibrational ground state $S_{0,0}$ to the electronic and vibrational excited state $S_{1,\nu}$. Fast relaxation follows with the internal conversion decay time τ_{ic} and the electronic first excited and vibrationally relaxed singlet state $S_{1,0}$ is populated. From this state the decay to the electronic ground state which is vibrational excited occurs via fluorescence and radiationless transitions. Again internal conversion follows with τ_{ic} and the electronic and vibrational ground state $S_{0,0}$ is reached. A second transition from the $S_{1,0}$ state leads to the first excited singlet state $T_{1,0}$ via intersystem crossing with decay time τ_{isc}. This state decays to the common ground state via phosphorescence and radiationless transitions.

The decay times can be compared to each other, to the pulse duration and to the inverse pump rate $[\sigma_{S1} I_{exc}]^{-1}$. Long decay times lead to integration and short decay times to rapid emptying of the levels. Thus the level scheme can be further simplified for the evaluation of a particular experiment without losing accuracy in the evaluation.

Mostly, differentiation of the emission decay time and the radiationless parallel decay time is not necessary: the upper level decays by the sum effect of these transitions the fluorescence life time.

Using e.g. very short pulses the population of the triplet system can be neglected but the internal conversion time and resonance emission may be important. Using ns pulses the internal conversion may be fast compared to the pulse duration and the pump rate. Thus the population of the $S_{1,\nu}$ level may be negligible and the resonance emission transition, too. In this case the value of τ_{ic} can be varied between 10 fs and 10 ps without changing the population of the levels $S_{1,0}$ and $T_{1,0}$.

All these assumptions can be checked by carrying out model calculations for the desired sample and varying the unknown parameters in realistic ranges.

7.9.2 Determination of Time Constants for Modeling

As far as the decay times of the material can be determined experimentally the modeling of the nonlinear behavior of the sample becomes more valid.

Fluorescence Lifetime

As described above in Sect. 7.3.4 the fluorescence decay time is built by the natural lifetime and the decay time via radiationless transitions. For nonlinear bleaching behavior and the resulting population densities of the excited states the differentiation of the two times is mostly not important. Thus the fluorescence decay should be measured for the same conditions used in the nonlinear experiment. If the decay time is too short and the fluorescence too weak the decay of the population density of the fluorescing level may be detectable via the decay of the associated exited state absorption from this level.

Very fast decaying fluorescence in the ps or fs range can be detected by optical delay techniques using fast Kerr shutters or up-conversion techniques (see Sect. 6.14.2). In any case, especially in laser measurements of fluorescence, the possible stimulated emission should be avoided. It can shorten the fluorescence lifetime by orders of magnitude.

Triplet Life Time

Phosphorescence lifetime can usually be measured with conventional spectrometers as described above in Sect. 7.3.4. If no radiation is detectable the triplet lifetime has to be measured via the triplet–triplet excited state absorption. Because of the long lifetimes of the triplet states these measurements are usually not difficult.

Care has to be taken about the sample conditions. The triplet population can be disturbed much more than the singlet population by environmental effects. In solutions the share of oxygen can change the triplet lifetime by orders of magnitude. Thus the sample should be exactly the same for all measurements or other precautions taken.

The sample could be "washed" with slow running nitrogen gas bubbles for an hour and thus the oxygen can be reduced. More sophisticated is the pump and freeze technique for taking the oxygen out. In this case the vacuum sealed sample is cooled down to the temperature of liquid nitrogen and then evacuated. During the slow warming up cycle the sample is evacuated and thus the oxygen is removed. This process can be repeated several times if necessary. As measuring criteria the triplet lifetime is measured from cycle to cycle. It stays constant if all oxygen is removed.

Ground State Absorption Recovery Time

As mentioned in Sect. 7.4.2 the ground state recovery time can be different from the fluorescence lifetime. Thus it has to be determined separately to validate the model assumptions. The necessary steps and methods are described in Sect. 7.4.2.

7.9.3 Fluorescence Intensity Scaling for Determining Population

If the sample shows fluorescence the population density of the associated excited state e.g. the $S_{1,0}$ state can be determined with high accuracy using the "fluorescence intensity scaling" method [7.271].

Because the fluorescence intensity $I_{fluoresc}$ is always proportional to the population density of the emitting state $N_{fluores}$ this method works even with very complicated nonlinear behavior of the sample. The population density of the fluorescing state follows from:

$$N_{fluores}(I_{exc}, t) = C_{fscaling} I_{fluores}(I_{exc}, t). \tag{7.68}$$

For this method the following conditions have to be fulfilled:

- The fluorescence has to be obtained exactly from the same sample volume which is transmitted by the probe beam and during the same time interval which is obtained in the pump and probe measurement.
- The reabsorption of the fluorescence by transitions from the ground state and all excited states should be negligibly small.

The experimental arrangement for the spatial overlap of the sample volumes from which the fluorescence is observed and which is transmitted by the probe light is shown in Fig. 7.57.

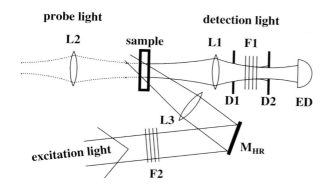

Fig. 7.57. Experimental setup for "fluorescence intensity scaling"

The fluorescence light is collimated in the same way as the probe light with lens L1. The apertures D1 and D2 filter the fluorescence light from the excited volume spatially before the detection with ED. With the filters F2 the excitation intensity is varied over many orders of magnitude from the low-signal value to the values used in the pump and probe measurements. For decreasing the linearity demands of the detector ED the filters taken out from position F2 can be placed in the detection path at position F1 if they show a flat spectral profile. For this purpose the filter transmissions have to be not much different and must be known at the two wavelengths of the excitation and fluorescence light.

For very low excitation intensities in the linear range the population density of the fluorescing state can be calculated with very high accuracy if the fluorescence lifetime τ_{fluores} is known. A simple two-level scheme is sufficient and the population density N_{fluores} at the time t_{m} is given by:

$$\text{low intensities} \quad N_{\text{fluores}} = \sigma_{\text{exc}} \tau_{\text{fluores}} N_{\text{total}} \langle I_{\text{exc,max}} \rangle f(t_{\text{m}}) \tag{7.69}$$

with the total population density of the sample N_{total}, the spatially averaged intensity maximum $\langle I_{\text{exc,max}} \rangle = \langle I_{\text{exc,max}} \rangle / h\nu_{\text{Laser}}$, along the excitation beam and a temporal function $f(t_{\text{m}})$ which has to be determined differently for stationary and nonstationary excitation from:

$$\text{steady state excitation} \quad f(t_{\text{m}}) = \frac{1}{I_{\text{exc,max}}} I(t_{\text{m}}) \tag{7.70}$$

and

nonsteady state excitation

$$f(t_{\text{m}}) = \frac{1}{I_{\text{exc,max}} \tau_{\text{fluores}}} \int_0^{t_{\text{m}}} I(t)\, e^{-t_{\text{m}} - t/\tau_{\text{fluores}}}\, dt \tag{7.71}$$

accounting for the storage effect of the fluorescing level.

The spatial averaging results in:

$$\langle I_{\text{exc,max}} \rangle = I_{\text{exc,max}} \frac{(1 - T_{\text{sample}})}{\sigma_{\text{exc}} N_{\text{total}} L_{\text{sample}}} \tag{7.72}$$

with the sample transmission T_{sample} for the excitation light and the length of the sample L_{sample}. With these equations the population density of the fluorescing state can be calculated for low excitation intensities from:

$$N_{\text{fluores}} = \frac{\tau_{\text{fluores}} (1 - T_{\text{sample}})}{L_{\text{sample}}} f(t_{\text{m}}) I_{\text{exc,max}}. \tag{7.73}$$

Thus the factor C_{fscaling} can be determined for low excitation intensities by dividing the calculated population density by the measured fluorescence intensity for given excitation intensities. With this scaling factor the population density can be directly calculated from the observed emission intensity and the graphs as e.g. given in Fig. 7.26 can be directly scaled in the population density of the $S_{1,0}$ state in this example, resulting in Fig. 7.58.

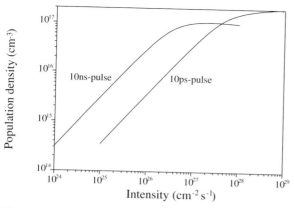

Fig. 7.58. Population density of the $S_{1,0}$ level of the model of Fig. 7.36 as it results from "fluorescence intensity scaling"

The result of "fluorescence intensity scaling" can be compared with the direct calculation of the nonlinear population density of this state. Deviations of the results from the two methods indicate incompleteness of the rate equations model. Thus this method is one of the few useful proofs of the modeling procedures. A detailed example is given in [7.271].

7.10 Practical Hints for Determination of Experimental Parameters

Because of the nonlinear behavior errors in the experimental parameters may have a more than linear influence on the evaluation of the data in laser spectroscopy. The influence of possible errors should be checked by systematic variations during the measurement and in the model calculations.

7.10.1 Excitation Light Intensities

The correct and absolute determination of the light intensities at the sample can be a difficult task. Finally the errors of the spatial and temporal distribution measurements as well as the absolute determination of the pulse energy can add up to 5–20% total error. This error influences the modeling of the measurement and the determined material parameter such as the decay times and cross-section carry at least part of this error.

Usually in modeling, the spatial and temporal profile of the light beam cannot be considered as measured functions. Often the intensity is just determined as an average value I_{average} across the transversal beam and during the pulse as discussed in Sect. 2.1.3. In this case an area A_{beam} and pulse duration Δt_{pulse} is defined for determining this average intensity by:

$$\text{intensity}\quad I_{\text{average}} = \frac{E_{\text{pulse}}}{A_{\text{beam}}\Delta t_{\text{pulse}}} \tag{7.74}$$

This approximation of rectangular intensity distributions in time and space offers the possibility of choosing the values A_{beam} and Δt_{pulse} for a given experimental distribution to describe a certain nonlinearity with smallest error. For the temporal profile the values are given in Sect. 2.7.2, especially in Table 2.11 and for the spatial distribution in Sect. 2.7.3, especially in Table 2.12.

The pulse energy can be measured with calibrated energy detectors, which are typically blackbody thermal sensors based e.g. on the pyroelectric effect. These detectors are heated by 10^{-5} K or even much less with a single pulse of a few mJ energy. The maximum repetition rate of these measuring devices is often limited to less than 100 Hz. Higher repetition rates can be measured as average power with power meters. The energy of the single pulses then follows from the duty cycle of the radiation. In this case the pulse-to-pulse fluctuations should be measured with photodetectors and oscilloscopes. If the input resistor R_{osci} of the oscilloscope is enlarged, e.g. to several $k\Omega$, the $R_{\text{osci}}C_{\text{osci}}$ time can be chosen much larger than the excitation pulse duration but still smaller than the inverse repetition rate and thus the scope shows the pulse energy as the peak value of the signal:

$$\text{energy display}\quad R_{\text{osci}} \geq F_{\text{osci}}\frac{\Delta t_{\text{pulse}}}{C_{\text{osci}}}\quad \text{with}\quad F_{\text{osci}} = 10\text{--}10^3 \tag{7.75}$$

where C_{osci} is the input capacity of the oscilloscope which is typically in the range of a few ten pF.

The absolute calibration of energy measurement devices is very difficult. One possibility is the investigation of the yield of a known chemical reaction. In practice calibration with other standards is used.

In energy measurements polarization effects at the beam splitters and possible nonlinearity of the applied filters have to be checked.

The transversal profile of the beam at the sample position can be measured with CCD cameras. If the light is scattered at a target which shows fluorescence at the sample position the nonlinear emission can "enlarge" the measured cross-section. As a first approach the transversal beam area can be determined for 86.5% energy content.

Pulse durations of about 1 ns or longer can be measured directly with fast photodiodes and oscilloscopes. The pulse duration can be chosen for 71.6% energy content of the whole pulse. Shorter pulse durations have to be measured with nonlinear optical methods as described e.g. in Sects. 5.5 and 7.1.5 [7.433].

7.10.2 Delay Time

Delays of the probe pulse in relation to the pump pulse in the ns range and above can be measured electronically. With electronic delay generators even accuracies of some ps are possible.

Delay times in the fs and ps range are usually achieved with optical delay lines. The $t = 0$ point with perfect temporal overlap of pump and probe pulse varies slightly for different alignments of the setup. Thus zero delay has to be checked for each measurement. With mechanical measurements usually an accuracy of a few ps can be achieved.

A quick way of finding the zero-point results from the measurement of the nonlinear bleaching of a known sample with the comparable long ground state recovery time as shown in Fig. 7.59.

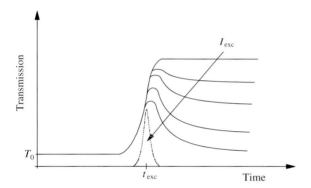

Fig. 7.59. Bleaching of a sample with long ground state recovery time by short pulses for finding the zero-point of the delay line. By increasing the excitation intensity the sample shows stimulated emission and thus the absorption recovery time is shortened

The zero-point can be found by successive decrease of the step width around the point of bleaching. By increasing the excitation intensity the recovery time of the sample can be shortened via stimulated emission and thus the accuracy of the measurement can be increased.

7.11 Examples for Spectroscopic Setups

Spectroscopic setups for nonlinear optical investigations can be designed for particular tasks or for a wide range of different methods. Because of the complexity of the nonlinear behavior of absorbing samples a combination of several methods is helpful. Measurements in different time domains have different advantages and disadvantages.

7.11.1 ns Regime

Measurements in the ns range are easy to carry out because the timing is not crucial. A change of the optical paths by a few cm does not usually disturb the temporal overlap of the pump and probe pulses and thus alignment is easy [7.434]. On the other hand the pulse intensities are high enough to achieve many nonlinear effects. In addition the data can be detected in real time using fast detectors and oscilloscopes. The number of available photons is larger such as in experiments with shorter pulses and thus the signal-to-noise ratio is good.

Further the light can be prepared in wide ranges as needed for a certain experiment with all kinds of spectral and polarization properties. Several light beams can be synchronized easily.

As an example a universal apparatus for measurements with ns pulses is shown in Fig. 7.60.

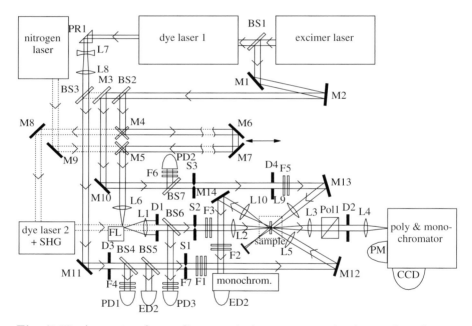

Fig. 7.60. Apparatus for nonlinear optical spectroscopy in the ns time domain allowing several methods of nonlinear absorption and emission measurement. The reference beam in the detection path is not shown

The basic light sources are an excimer laser and an electronically synchronized nitrogen laser. The first laser beam produces the excitation light which can be converted using a dye laser 1 with frequency transformation possibilities. Thus two strong excitation beams can be applied to excite the sample with two-photons of different energy. In general for the generation of the excitation light solid-state lasers can also be used, which allow the generation of any required frequency, in combination with optical parametric oscillators or amplifiers. Up to now the spectral bandwidth of these devices and the output power is not always sufficient for an universal apparatus.

The probe light from the nitrogen laser is shorter than the excitation pulse and thus temporal overlap is easy to achieve. This light can be used for generating broad spectral light using a fluorescence cell. For special purposes the nitrogen laser light can be used to pump a second dye laser with frequency conversion possibilities. For a reduced spectral range and lower pulse power short pulse lamps can also be applied as probe light sources.

The detection of the probe light is achieved by the combination of a mono/polychromator with a fast photodetector PM for temporal measurements and a CCD camera with an optical multichannel analyzer (OMA) for spectral measurements.

The whole apparatus is computer controlled and thus the different measuring methods such as nonlinear absorption and emission measurements as well as pump and probe experiments can be carried out rapidly and with high accuracy. The temporal resolution is ns for the direct measurement but down to sub-ps from the modeling of the intensity dependencies.

On the other hand many fast reactions and decays can be detected only indirectly by the intensity dependence of the nonlinear absorption. Although modeling allows the determination of these fast components sometimes even with fs resolution direct observation using shorter pulses should follow.

7.11.2 ps and fs Regime

As described above in Sect. 6.10.3 modern mode locked solid-state lasers can be operated with pulse durations of a few ps or some 10–100 fs. Regenerative amplifiers with suitable spectral filters even allow the almost continuous variation of the pulse duration in the fs and ps range. Thus the setup for nonlinear spectroscopy in the ps and fs range can be based on the same laser system as shown e.g. in Fig. 7.61.

In this example the main laser oscillator is a Kerr lens mode-locked Ti:sapphire laser with output pulses of 80 fs or 2 ps and an average output power of about 1 W. This laser is pumped by a frequency-doubled Nd:YVO laser with 5 W average output power in the green which is pumped by diode lasers.

This light is amplified in a regenerative amplifier with a repetition rate of up to 1 kHz. The output pulse has an energy of about 1 mJ. This light can be frequency double or tripled and thus used for the excitation of the sample. A share of this light is fed into an optical parametric amplifier which allows the generation of light with wavelengths between 300 nm and 3 μm which can also be used for excitation and as probe light, too. Another share of the fundamental or the harmonics is used for white light generation in a sapphire

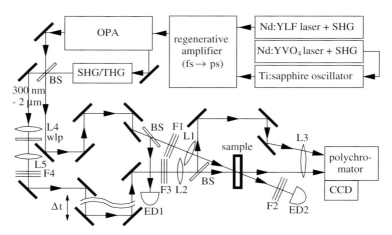

Fig. 7.61. Apparatus for nonlinear optical spectroscopy in the fs and ps time domain such as e.g. pump and probe measurements with high temporal resolution

plate. This light is used as probe light which is delayed by Δt via the delay line. The detection of the light is carried out by a polychromator and a CCD camera. Bleaching can be measured via the energy detectors ED1 and ED2 while varying the filters F1.

Using these short pulses the decay times of the sample can be measured directly by delaying the probe light beam. From the nonlinear bleaching the cross-section of the sample may be determined directly without considering the decay times. Thus the fast measurements allow new approaches in the investigation of the nonlinear behavior of the samples and thus new photonic applications using fs technology may be developed in the near future. Examples are given in [M45–M52, 7.435–7.439].

7.12 Special Sample Conditions

The nonlinear properties of the samples can be changed by changing the sample conditions. Low and high temperatures as well as low and high pressures are useful changes for characterizing the physical background causing nonlinear properties of the matter.

7.12.1 Low Temperatures

Temperatures down to 77 K can easily be achieved with liquid nitrogen cryostats. Lower temperatures demand other devices such as e.g. helium cryostats [e.g. 7.440–7.452].

Thus the temperature range between 77 K and room temperature is easy to achieve. If solutions are applied they may crack and thus disturb the optical measurements by strong scattering. Sometimes shock cooling or very slow cooling rates can solve this problem. Further solvent mixtures, such as e.g. 2, 2-dimethylbutane and n-pentane (8:3), show good flexibility in low temperature measurements even below 77 K. Other useful solvents for low temperature measurements are given in Table 7.10.

Table 7.10. Solvents for low-temperature measurements. T_{glass} is a temperature characterizing the transition from liquid to glass of the material

Solvent	Mixture	T_{glass} (K)
Isopropanole (/water)	1000:1	130
Diethylether/ethanol	1:2	130
Isopentane/triethylamine (high viscosity)	1:9	130
Ethanol/water	1:2	125
Isopentane/methylcyclohexane	1:3	98
Ethanol/methanol	4:1	77
Diethylether/propanol	5:2	77
Diethylether/isopentane/ethanol	5:5:2	77
Propionitril/butyronitril	1:1	13

The isolation vacuum of <10 mbar for the cooled samples can be sealed with conventional O-rings in the temperature range above 77 K.

Temperatures of about 8 K and above are possible with closed-cycle He cryostats which are still handy and not expensive in operation. The problem of good optical quality of the samples becomes more serious at lower temperatures. Geometrically thinner samples may help. For sample sealing against the isolation vacuum, indium seals can be used.

For temperatures below 4 K open He cryostats are available which operate down to 1 K. Lower temperatures need special techniques such as e.g. laser cooling [e.g. 7.453–7.482].

A simple technique for achieving low temperatures in the range of a few K is the application of the jet technique [e.g. 7.483–7.530] as shown in Fig. 7.62.

The sample which can be e.g. organic material is evaporated in an oven at several 100°C and mixed with a buffer gas. The gas mixture is expanded via a triggered nozzle (e.g. a car injection nozzle) into a vacuum chamber. In the hypersonic area of the gas flow right after the nozzle the transversal speed is strongly reduced and thus very low temperatures of a few K are obtained for the laser beam.

In any case the calibration of the sample temperature has to be done carefully. The lower the temperature the more important are the heating effects from the illuminating light. Thus the temperature in cryostat experiments should be detected inside the sample volume. Radiation shields may be applied with small apertures for the excitation and probe light. In high-power experiments the absorbed radiation has to be considered as a sample heater, too [7.446].

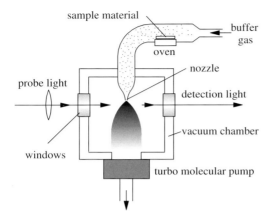

Fig. 7.62. Measuring gas samples in a hypersonic jet for achieving low temperatures in the range of a few K for the laser spectroscopic beam

Very low temperature of μK and below are achieved using laser cooling techniques for removing energy from the sample [e.g. 7.453–7.482]. This method is established for atoms and Bose–Einstein condensation is possible in this way. The cooling of molecules to such low values is in progress.

7.12.2 High Pressures

High pressures increase the density of the sample [e.g. 7.531–7.549] and thus the refractive index is increased. For example, for molecules the excited state is mostly more polar than the ground state and thus pressurizing the sample often results in a red shift of the absorption and emission spectra. This effect can be of the size of $100\,\mathrm{cm}^{-1}\,\mathrm{kbar}^{-1}$. Larger values are possible as described e.g. in [7.539].

Therefore nonlinear optical spectroscopy with high pressures allows the investigation of quantum matter states and their geometrical relations. In particular, for molecular systems the environmental influence on the molecular orbitals and conformational changes can be studied.

For a well-observable effect pressures of some kbars are necessary. At such pressures the volume of water is e.g. decreased to 80% of its normal value. These high pressures can be achieved e.g. with a chamber [7.540] as sketched in Fig. 7.63.

The sample of size about 1 cm is positioned in a very stable steel container of about 10 cm in diameter and surrounded by a liquid, e.g. water, which is pressurized to the required value. If cuvettes are used they can be sealed with flexible thin rubber which fits inside the cell if the pressure is increased.

The threads and seals of the steel container have to be designed for the applied pressures. The pressure tubes have e.g. about 6 mm outside diameter but about 1 mm the inside diameter.

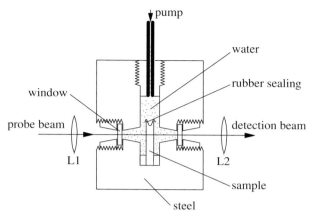

Fig. 7.63. Sample holder for achieved high pressures of several kbar

Care has to be taken for the glass windows. No crystalline material should be used because this can show birefringence if the container is pressurized. Typical useful window diameters are 5 mm with about 10 mm thickness.

7.13 Quantum Chemical Calculations

Optical transitions as measured in nonlinear optical spectroscopy can in principle be calculated with quantum chemical computer programs [7.550–7.579]. Unfortunately, the demands of large systems such as molecules with about 50 atoms are not solvable with sufficient accuracy, but the theoretical calculations may be a helpful complement to the experimental results because they usually allow the classification of the observed transitions.

7.13.1 Orbitals and Energy States of Molecules

Free electrons have no discrete quantum states but a continuous spectrum of the kinetic energy. Thus all discrete energy levels are a property of the whole quantum system e.g. the molecule or the atom (see Fig. 7.64 left side).

Transitions of the sample particles, e.g. the molecules or atoms, between these energy levels as observed in spectroscopy are coupled with transitions of electrons from one orbital to another as shown in Fig. 7.64 at the right side. Thus the energetic difference of the molecular energy levels of the system and the orbitals is equal.

But the transition of two electrons 1 and 2 between the same orbitals from the highest occupied molecular orbital (HOMO) to the lowest unoccupied molecular orbital (LUMO) as shown in the figure causes a further excitation of the molecule to the level S_x as shown in the left part of Fig. 7.64, whereas

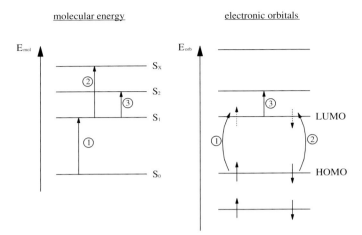

Fig. 7.64. Molecular energy levels (left) and electronic orbitals (right) for comparison

transition 3 of one electron from the LUMO to the next excited orbital leads to a transition of the molecule to the next energy level S_2 in the example.

Thus transitions of the electrons between the orbitals have to be distinguished from the transitions of the whole particle, as the molecule or atom, between the energy states or levels. Only the transitions between the energy states of the particle are obtained in optical spectroscopy. Their oscillator strength is given by the overlap of all participating orbitals as described in Sect. 3.3.1.

7.13.2 Scheme of Common Approximations

The complete theoretical description of the nonlinear interaction would be possible if the time-dependent Schroedinger equation including all components of the interaction such as the matter with all atom cores and electrons and all photons could be solved. Unfortunately this is not possible for realistic systems and thus a large number of approximations is necessary, as shown in Fig. 7.65 for a molecular system.

First it is commonly presupposed that the light does not change the energy eigenvalues of the matter but only causes transitions of the particles between these eigenstates. Thus the time-dependent problem can be divided into the problem of the calculation of these eigenstates and functions on one hand and the interaction as a small perturbation leading to the transition on the other.

For calculating the energy eigenvalues of the matter the environmental interaction effects are separated from the calculation of the isolated particle. For molecules the Born–Oppenheimer approximation allows the separate

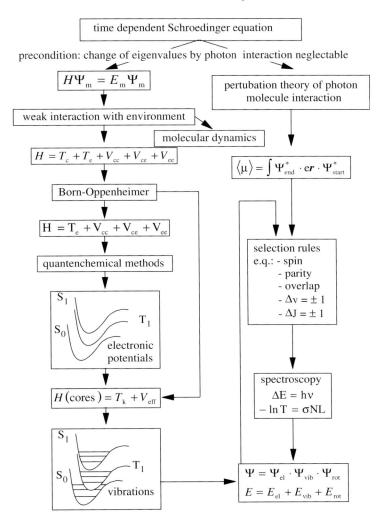

Fig. 7.65. Scheme of common approximations in quantum chemical calculations

calculation of the electronic potentials assuming fixed cores and then the vibrational states for the core energies. Thus the wavefunction results from the product of all these interactions.

With these wavefunctions the transition dipole moment can be calculated as discussed in Sect. 3.3.1. From symmetry evaluation of the wavefunctions some simple selection rules for allowed transitions can be derived.

The calculation of the electronic wavefunction demands further strong approximations if large molecules are to be calculated. Thus semi-empirical methods are developed for this purpose.

The two types of results which are important for nonlinear spectroscopy, the transition energies and the transition dipole moments, have errors as a result of these approximations. In particular the values for higher excited states such as the LUMO show errors. Therefore the results from nonlinear spectroscopy may help to improve these methods in the future.

7.13.3 Ab Initio and Semi-Empirical Calculations

Ab initio calculations consider all cores and electrons of the molecules explicitly. Thus as a result the molecular orbitals show large diversity and the user of these programs has to work out which of these solutions are of practical relevance. In addition large molecules with more than about 20 atoms (e.g. more than naphthalene) need very long computation times. Thus *ab initio* methods are useful for specialists in quantum chemical computing.

Semi-empirical calculations are based on the atomic cores in combination with the core electrons as new "cores". Thus only the outer electrons are considered for chemical bonds and optical properties. Therefore the system uses parameterized "core atoms" and thus only a much smaller number of wavefunctions has to be calculated. Therefore these semi-empirical methods are fast and allow the calculation of very large molecules. The accuracy of the spectra can be as good as about <20% for the transition energy and <50% for the transition dipole moments. Common programs for spectroscopic purposes are CNDO-S/CI, MINDO-S, AMPAQ or ZINDO/S. Unfortunately, only a few of these programs also calculate the transitions between excited states.

On mainframe computers the computation times for average size molecules such as e.g. para-terphenyl, is in the range of minutes. Therefore these programs can be a very useful tool in analyzing the nonlinear optical behavior of organic molecules.

As an example the results of a CNDOS/CI calculation of excited state absorptions for pentaphene as described in [7.297] are given in Table 7.11.

Table 7.11. Quantum chemically calculated transitions for pentaphene in comparison with the experimental data [7.297]

Transition	Wavelength (theoretical)	Oscillator strength	Wavelength (experimental)	Cross-section
$S_n \leftarrow S_1$				
1	1,720 nm	0.167	not measured	
2	596 nm	0.776	578 nm	$1.4 \cdot 10^{-16}\,\text{cm}^2$
$T_n \leftarrow T_1$				
1	424 nm	0.442	493 nm	$1.7 \cdot 10^{-16}\,\text{cm}^2$
2	371 nm	0.169	456 nm	$9.3 \cdot 10^{-17}\,\text{cm}^2$
3	365 nm	0.300	440 nm	$1.4 \cdot 10^{-16}\,\text{cm}^2$
4	352 nm	0.165		

The agreement between the calculated data and the experimental results is quite good for singlet-singlet excited state absorption, but even for the triplet excited state spectra the tendency is described. Thus the quantum chemical calculations can be very helpful in identifing the spectral ranges for pump and probe measurements.

Bibliography

A large number of references allows direct access to the detailed scientific research results in the field. The selected articles are cited with all authors, the full title and the number of pages, and are arranged in descending year order per topic. Considering this information and the title of the journal may help to select the most useful articles from the list for the reader's purpose. In addition, the related section is cited as {Sect. ...} and thus the references of a section can be read almost separately. In these references also additional effects and their applications are described. The descriptions in this book allow a general understanding of these specialized articles. It may be worth searching for a special reference in the chapters describing the basics as well as in the applications part of the book because the references are cited usually only once. These references represent mostly current research topics. The pioneering work, if not explicitly given, can be traced back from these articles. Many of the measured material parameters have slightly different values. In the sense used in this book the most probable or averaged values are given without a detailed discussion. For details the references with their cited literature shall be used.

For further general reading some selected textbooks are given (cited as monographs [M...]). The titles and publications years may be used for guidance.

Questions, comments and corrections are welcome and can be sent to the author via the e-mail address: photonics_menzel@springer.de.

Further Reading

[M1] F. T. Arecci, E.O. Schulz-Dubois (ed.): Laser Handbook Vol.1, 2 (North-Holland Publishing, Amsterdam,New York, Oxford, 1972

[M2] S. M. Barnett, P. Radmore: Methods in Theoretical Quantum Optics (Oxford University Press, New York, Oxford, 1997)

[M3] M. Bass (ed.): Handbook of Optics, Vol. I,II (McGraw-Hill, New York, 1995)

[M4] J.B. Birks (ed.): Organic Molecular Photonics (Wiley, London 1973)

[M5] M. Born, E. Wolf: Principles of Optics (Cambridge University Press, Cambridge, 1999)

[M6] R. W. Boyd: Nonlinear Optics (Academic Press, Boston, 1992)

[M7] P.N. Butcher: The Elements of Nonlinear Optics (Cambridge University Press, Cambridge, 1990)

[M8]   E. Collett: Polarized Light – Fudamentals and Applications (Marcel Dekker Inc, New York, Basel, Hong Kong, 1993)

[M9]   J. C. Dainty: Current Trends in Optics (Academic Press, San Diego, 1994)

[M10] W. Demtröder: Laser Spectroscopy (Springer, Berlin, Heidelberg, New York, 1996)

[M11] J.-C. Diels, W. Rudolph: Ultrashort Laser Pulse Phenomena (Academic Press, San Diego, 1996)

[M12] F. J. Duarte (ed.): Tunable Laser Applications (Marcel Dekker, New York, Basel, Hong Kong, 1995)

[M13] H. M. Gibbs: Nonlinear Photonics (Springer, Berlin, Heidelberg, New York, 1990)

[M14] M. C. Gupta (ed.): Handbook of Photonics (CRC Press, Boca Raton, New York, 1997)

[M15] A. R. Henderson: A Guide to Laser Safety (Chapman & Hall, London, 1997)

[M16] N. Hodgson, H. Weber: Optical Resonators (Springer, London, 1997)

[M17] P. Horowitz, W. Will: The Art of Electronics (Cambridge University Press, Cambridge, 1994)

[M18] S. Huard: Polarization of Light (John Wiley & Sons, Chichester, 1996)

[M19] K. Iga: Fundamentals of Laser Optics (Plenum Press, New York, London, 1994)

[M20] M. Inguscio, R. Wallenstein: Solid State Lasers: New Developments and Applications (Plenum Publishing Corporation, New York, 1993)

[M21] J. D. Jackson: Classical Electrodynamics (John Wiley & Sons, Chichester, 1975)

[M22] W. Kaiser: Ultrashort Laser Pulses (Springer, Berlin, Heidelberg, 1993)

[M23] I.-C. Khoo, F. Simoni: Novel Optical Materials & Applications (John Wiley & Sons, Chichester, 1996)

[M24] D.S. Kliger, J.W. Lewis, C.E. Randall: Polarized Light in Optics and Spektroscopy (Academic Press, Boston, 1990)

[M25] W. Koechner: Solid-State Laser Engineering (Springer, Berlin, Heidelberg, 1999)

[M26] D. R. Lide, Handbook of Chemistry and Physics (CRC Press, Boca Raton, New York, London, Tokyo, 1995)

[M27] L. Mandel, E. Wolf: Optical Coherence and Quantum Optics (Cambridge University Press, Cambridge, 1995)

[M28] P. Meystre, M. Sargent: Elements of Quantum Optics (Springer, Berlin, Heidelberg, 1990)

[M29] D. L. Mills: Nonlinear Optics (Springer, Berlin, Heidelberg, New York, 1998)

[M30] S. Mukamel: Principles of Nonlinear Optical Spectroscopy (Oxford University Press, Oxford, 1995)

[M31] H. Niedrig (ed.): Bergmann Schaefer, Optics of Waves and Particles (Walter de Gruyter, Berlin, New York, 1999)

[M32] D. N. Nikogosyan: Properties of Optical and Laser-Related Materials – A Handbook (John Wiley & Sons, Chichester, 1997)

[M33] C. Rulliere: Femtosecond Laser Pulses (Springer, Berlin, Heidelberg, 1998)

[M34] B. E. A. Saleh, M. C. Teich: Fundamentals of Photonics (John Wiley & Sons, New York, 1991)

[M35] E. G. Sauter: Nonlinear Optics (John Wiley & Sons, New York, 1996)

[M36] M. O. Scully, M. S. Zubairy: Quantum Optics (Cambridge University Press, 1997)

[M37] Y. R. Shen: Principles of Nonlinear Optics (John Wiley & Sons, Chichester, 1984)

[M38] A. E. Siegmann: Lasers (University Science Books, Sausalito, California, 1986)

[M39] W. T. Silfvast: Laser Fundamentals (Cambridge University Press, Cambridge, 1996)

[M40] O. Svelto: Ultrafast Processes in Spectroscopy (Plenum Press, New York, 1996)

[M41] D. F. Walls, G. J. Milburn: Quantum Optics (Springer, Berlin, Heidelberg, 1995)

[M42] A. Yariv: Quantum Electronics (John Wiley & Sons, Chichester, 1989)

[M43] M. Young: Optics and Lasers (Springer, Berlin, Heidelberg, New York, 1993)

[M44] D. Zwillinger (ed.): CRC Standard Mathematical Tables and Formulae (CRC Press, Boca Raton, New York, London, Tokyo, 1996)

[M45] Picosecond Phenomena (Springer Ser. Chem. Phys.) I, ed. by K.V. Shank, E.P. Ippen, S.L. Shapiro. Vol 4 (1978)

[M46] Picosecond Phenomena (Springer Ser. Chem. Phys.) II, ed. by R.M. Hochstrasser, W. Kaiser, C.V. Shank. Vol 14 (1980)

[M47] Picosecond Phenomena (Springer Ser. Chem. Phys.) III, ed. by E.B. Eisenthal, R.M. Hochstrasser, W. Kaiser, A. Lauberau. Vol. 38 (1982)

[M48] Ultrashort Phenomena (Springer Ser. Chem. Phys.) IV, ed. by D.H. Auston, K.B. Eisenthal. Vol. 38 (1984)

[M49] Ultrashort Phenomena (Springer Ser. Chem. Phys.) V, ed. by G.R. Fleming, A.E. Siegman. Vol. 46 (1986)

[M50] Ultrashort Phenomena (Springer Ser. Chem. Phys.) VI ed. by T. Yajima, K. Yoshihara, C.B. Harris, S. Shionoya. Vol. 48 (1988)

[M51] Ultrashort Phenomena (Springer Ser. Chem. Phys.) VII, ed. by E. Ippen, C.B. Harris, A.H. Zewail. Vol. 53 (1990)

[M52] Ultrashort Phenomena (Springer Ser. Chem. Phys.) VIII, ed. by A. Migus, J.-L. Martin, G.A. Mourou, A.H. Zewail. Vol. 55

[M53] Harnessing Light (National Academy Press, Washington, D.C, 1998)

References

1. Topics in Photonics

[1.1] {Sect. 1.0} T.H. Maiman: Stimulated Optical Radiation in Ruby, Nature 187, p.493-494 (1960)

[1.2] {Sect. 1.3} A. Zeilinger: Experiment and the foundations of quantum physics, Rev. Mod. Phys. 71, p.288-297 (1999)

[1.3] {Sect. 1.3.3} E.A. Cornell, C.E. Wieman: The Bose-Einstein Condensate, Scientific American, March, p.40-45 (1998)

[1.4] {Sect. 1.3.3} A. Griffin, D.W. Snoke, S. Stringari (ed.): Bose-Einstein-condensation, Cambridge University Press, Cambridge, 1995)

[1.5] {Sect. 1.3.3} J. Mayers: Bose-Einstein condensation and spatial correlations in He-4, Phys Rev Lett 84, p.314-317 (2000)

[1.6] {Sect. 1.3.3} E.W. Hagley, L. Deng, M. Kozuma, M. Trippenbach, Y.B. Band, M. Edwards, M. Doery, P.S. Julienne, K. Helmerson, S.L. Rolston et al.: Measurement of the coherence of a Bose-Einstein condensate, Phys Rev Lett 83, p.3112-3115 (1999)

[1.7] {Sect. 1.3.3} I. Bloch, T. W. Hänsch, T. Esslinger: Atom Laser with a cw Output Coupler, Phys. Rev. Lett. 82, p.3008-3011 (1999)

[1.8]   {Sect. 1.3.3} S. Inouye, A.P. Chikkatur, D.M. StamperKurn, J. Stenger, D.E. Pritchard, W. Ketterle: Superradiant Rayleigh scattering from a Bose-Einstein condensate, Science 285, p.571-574 (1999)

[1.9]   {Sect. 1.3.3} C.W. Gardiner, M.D. Lee, R.J. Ballagh, M.J. Davis, P. Zoller: Quantum kinetic theory of condensate growth: Comparison of experiment and theory, Phys Rev Lett 81, p.5266-5269 (1998)

[1.10] {Sect. 1.3.3} H. Gauck, M. Hartl, D. Schneble, H. Schnitzler, T. Pfau, J. Mlynek: Quasi-2D gas of laser cooled atoms in a planar matter waveguide, Phys Rev Lett 81, p.5298-5301 (1998)

[1.11] {Sect. 1.3.3} R. Graham: Decoherence of Bose-Einstein condensates in traps at finite temperature, Phys Rev Lett 81, p.5262-5265 (1998)

[1.12] {Sect. 1.3.3} C.K. Law, H. Pu, N.P. Bigelow: Quantum spins mixing in spinor Bose-Einstein condensates, Phys Rev Lett 81, p.5257-5261 (1998)

[1.13] {Sect. 1.3.3} U. Ernst, A. Marte, F. Schreck, J. Schuster. G. Rempe: Bose-Einstein condensation in a pure Ilffe-Pritchard field configuration, Europhys. Lett. 41, p.1-6 (1998)

[1.14] {Sect. 1.3.3} B. Saubamea, T.W. Hijmans, S. Kulin, E. Rasel, E. Peik, M. Leduc, C. Cohentannoudji: Direct measurement of the spatial correlation function of ultracold atoms, Phys Rev Lett 79, p.3146-3149 (1997)

[1.15] {Sect. 1.3.3} C.C. Bradley, C.A. Sackett, J.J. Tollett, R.G. Hulet: Evidence of Bose-Einstein condensation in an atomic gas with attractive interactions, Phys Rev Lett 75, p.1687-1690 (1995)

[1.16] {Sect. 1.3.3} M.H. Anderson, J.R. Ensher, M.R. Matthews, C.E. Wieman: Observation of Bose-Einstein-Condensation in a Dilute Atomic Vapor, Science269, p.198-201 (1995)

[1.17] {Sect. 1.3.3} K.B. Davis, M.O. Mewes, M.R. Andrew, N.J. Vandruten, D.S. Durfee, D.M. Kurn, W. Ketterle: Bose-Einstein condensation in a gas of sodium atoms, Phys Rev Lett 75, p.3969-3973 (1995)

[1.18] {Sect. 1.3.3} K. Helmerson, D. Hutchinson, K. Burnett, W.D. Phillips: Atom Lasers, Phys. WorldAugustp.31-35 (1999)

[1.19] {Sect. 1.3.3} M. Trippenbach, Y.B. Band, M. Edwards, M. Doery, P.S. Julienne, E.W. Hagley, L. Deng, M. Kozuma, K. Helmerson, S.L. Rolston et al.: Coherence properties of an atom laser, J Phys B At Mol Opt Phys 33, p.47-54 (2000)

[1.20] {Sect. 1.3.3} I. Bloch, T.W. Hansch, T. Esslinger: Atom laser with a cw output coupler, Phys Rev Lett 82, p.3008-3011 (1999)

[1.21] {Sect. 1.3.3} H.P. Breuer, D. Faller, B. Kappler, F. Petruccione: Non-Markovian dynamics in pulsed- and continuous-wave atom lasers, Phys Rev A 60, p.3188-3196 (1999)

[1.22] {Sect. 1.3.3} K.G. Manohar, B.N. Jagatap: Atom laser, Curr Sci 76, p.1420-1423 (1999)

[1.23] {Sect. 1.3.3} J. Schneider, A. Schenzle: Output from an atom laser: theory vs. experiment, Appl Phys B Lasers Opt 69, p.353-356 (1999)

[1.24] {Sect. 1.3.3} B. Kneer, T. Wong, K. Vogel, W.P. Schleich, D.F. Walls: Generic model of an atom laser, Phys Rev A 58, p.4841-4853 (1998)

[1.25] {Sect. 1.3.3} M. Holland, K. Burnett, C. Gardiner, J.I. Cirac, P. Zoller: Theory of an atom laser, Phys Rev A 54, p.R1757-R1760 (1996)

[1.26] {Sect. 1.3.3} M. Wilkens, R.J.C. Spreeuw, T. Pfau, U. Janicke, M. Mlynek: Towards a laser-like source of atoms, Prog Cryst Growth Charact 33, p.385-393 (1996)

[1.27] {Sect. 1.3.3} T. Pfau, U.Janicke,M. Wilkens: Laser-like scheme for atomic-matter waves, Europhys. Lett.32, p.469-474 (1995)

[1.28] {Sect. 1.3.3} W.L. Power: Atom optics: matter and waves in harmony, Phil Trans Roy Soc London A 358, p.127-135 (2000)

[1.29] {Sect. 1.3.3} M.O. Mewes, M.R. Andrews, D.M. Kum, D.S. Durfee, C.G. Townsend, W. Ketterle : Output coupler for Bose Einstein Condensation, Phys. Rev. Lett.78, p.582-585 (1997)

[1.30] {Sect. 1.3.3} C.S. Adams, M. Sigel, J. Mlynek: Atom optics, Phys. Reports 240, p.143 (1994)

[1.31] {Sect. 1.3.4} C.H. Bennett, G. Brassard, A.K. Ebert: Quantum Cryptography, Scientific AmericanOctoberp.50-59 (1992); D. Bouwmeester, A. Ekert, A. Zeilinger, The Physics at Quantum Information (Springer, Berlin, Heidelberg, 2000)

[1.32] {Sect. 1.3.4} O. Benson, C. Santori, M. Pelton, Y. Yamamoto: Regulated and entangled photons from a single quantum dot, Phys Rev Lett 84, p.2513-2516 (2000)

[1.33] {Sect. 1.3.4} M. Vasilyev, S.K. Choi, P. Kumar, G.M. DAriano: Tomographic measurement of joint photon statistics of the twin-beam quantum state, Phys Rev Lett 84, p.2354-2357 (2000)

[1.34] {Sect. 1.3.4} D. Bouwmeester, J.W. Pan, M. Daniell, H. Weinfurter, A. Zeilinger: Observation of three-photon Greenberger-Horne-Zeilinger entanglement, Phys Rev Lett 82, p.1345-1349 (1999)

[1.35] {Sect. 1.3.4} J. Brendel, N. Gisin, W. Tittel, H. Zbinden: Pulsed energy-time enangled twin-photon source for quantum communication, Phys Rev Lett 82, p.2594-2597 (1999)

[1.36] {Sect. 1.3.4} A. Kent, N. Linden, S. Massar: Optimal entanglement enhancement for mixed states, Phys Rev Lett 83, p.2656-2659 (1999)

[1.37] {Sect. 1.3.4} J.M. Merolla, Y. Mazurenko, J.P. Goedgebuer, H. Porte, W.T. Rhodes: Phase-modulation transmission system for quantum cryptography, Optics Letters 24, p.104-106 (1999)

[1.38] {Sect. 1.3.4} J.M. Merolla, Y. Mazurenko, J.P. Goedgebuer, W.T. Rhodes: Single-photon interference in sidebands of phase-modulated light for quantum cryptography, Phys Rev Lett 82, p.1656-1659 (1999)

[1.39] {Sect. 1.3.4} L. Quiroga, N.F. Johnson: Entangled Bell and Greenberger-Horne-Zeilinger states of excitons in coupled quantum dots, Phys Rev Lett 83, p.2270-2273 (1999)

[1.40] {Sect. 1.3.4} A.G. White, D.F.V. James, P.H. Eberhard, P.G. Kwiat: Non-maximally entangled states: Production, characterization, and utilization, Phys Rev Lett 83, p.3103-3107 (1999)

[1.41] {Sect. 1.3.4} D. Bouwmeester, J.W. Pan, M. Daniell, H. Weinfurter, A. Zeilinger: Observation of three-photon Greenberger-Horne-Zeilinger entanglement, Phys Rev Lett 82, p.1345-1349 (1999)

[1.42] {Sect. 1.3.4} J. Brendel, N. Gisin, W. Tittel, H. Zbinden: Pulsed energy-time enangled twin-photon source for quantum communication, Phys Rev Lett 82, p.2594-2597 (1999)

[1.43] {Sect. 1.3.4} W. Tittel, J. Brendel, N. Gisin, H. Zbinden: Long-distance Bell-type tests using energy-time entangled photons, Phys Rev A 59, p.4150-4163 (1999)

[1.44] {Sect. 1.3.4} F. Demartini: Amplification of quantum entanglement, Phys Rev Lett 81, p.2842-2845 (1998)

[1.45] {Sect. 1.3.4} T.C. Ralph, P.K. Lam: Teleportation with bright squeezed light, Phys Rev Lett 81, p.5668-5671 (1998)

[1.46] {Sect. 1.3.4} E.S. Polzik, J.L. Sorenson, J. Hald: Subthreshold tunable OPO: a source of nonclassical light for atomic physics experiments, Appl Phys. B 66, p.759-764 (1998)

[1.47] {Sect. 1.3.4} Q.A. Turchette, C.S. Wood, B.E. King, C.J. Myatt, D. Leibfried, W.M. Itano, C. Monroe, D.J. Wineland: Deterministic entanglement of two trapped ions, Phys Rev Lett 81, p.3631-3634 (1998)

[1.48] {Sect. 1.3.4} J.I. Cirac, P. Zoller, H.J. Kimble, H. Mabuchi: Quantum state transfer and entanglement distribution among distant nodes in a quantum network, Phys Rev Lett 78, p.3221-3224 (1997)

[1.49] {Sect. 1.3.4} H.B. Fei, B.M. Jost, S. Popescu, B.E.A. Saleh, M.C. Teich: Entanglement-induced two-photon transparency, Phys Rev Lett 78, p.1679-1682 (1997)

[1.50] {Sect. 1.3.4} V. Blanchet, C. Nicole, M.-A. Bouchene, B. Girard: Temporal Coherent Control in Two-Photon Transitions: From Optical Interferences to Quantum Interferences, Phys. Rev. Lett. 78, p.2716-2719 (1997)

[1.51] {Sect. 1.3.4} G. Digiuseppe, F. Demartini, D. Boschi: Experimental test of the violation of local realism in quantum mechanics without Bell inequalities, Phys Rev A 56, p.176-181 (1997)

[1.52] {Sect. 1.3.4} S.F. Huelga, C. Macchiavello, T. Pellizzari, A.K. Ekert, M.B. Plenio, J.I. Cirac: Improvement of frequency standards with quantum entanglement, Phys Rev Lett 79, p.3865-3868 (1997)

[1.53] {Sect. 1.3.4} J. Brendel, E. Mohler, W. Martienssen: Time-resolved dual-beam two-photon interferences with high visibility, Phys Rev Lett 66p.1142-1145 (1991)

[1.54] {Sect. 1.3.4} Z.Y. Ou, L. Mandel: Observation of spatial quantum beating with separated photodetectors, Phys Rev Lett 61p.54-57 (1988)

[1.55] {Sect. 1.3.4} N.J. Cerf, N. Gisin, S. Massar: Classical teleportation of a quantum bit, Phys Rev Lett 84, p.2521-2524 (2000)

[1.56] {Sect. 1.3.4} W.T. Buttler, R.J. Hughes, S.K. Lamereaux, G.L. Morgan, J.E. Nordholt, C.G. Peterson: Daylight quantum key distribution over 1.6 km, Phys. Rev. Lett.84, p. 5652-5655 (2000)

[1.57] {Sect. 1.3.4} Th. Jennewein, Ch. Simon, G. Weihs, H. Weinfurter, A. Zeilinger: Quantum cryptography with entangled photons, Phys. Rev. Lett.84, p.4729-4732 (2000)

[1.58] {Sect. 1.3.4} H. BechmannPasquinucci, N. Gisin: Incoherent and coherent eavesdropping in the six-state protocol of quantum cryptography, Phys Rev A 59, p.4238-4248 (1999)

[1.59] {Sect. 1.3.4} G. Bonfrate, V. Pruneri, P.G. Kazansky, P. Tapster, J.G. Rarity: Parametric fluorescence in periodically poled silica fibers, Appl Phys Lett 75, p.2356-2358 (1999)

[1.60] {Sect. 1.3.4} N. Gisin, S. Wolf: Quantum cryptography on noisy channels: Quantum versus classical key- agreement protocols, Phys Rev Lett 83, p.4200-4203 (1999)

[1.61] {Sect. 1.3.4} N. Lutkenhaus: Estimates for practical quantum cryptography, Phys Rev A 59, p.3301-3319 (1999)

[1.62] {Sect. 1.3.4} A.V. Sergienko, M. Atature, Z. Walton, G. Jaeger, B.E.A. Saleh, M.C. Teich: Quantum cryptography using femtosecond-pulsed parametric down-conversion, Phys Rev A 60, p.R2622-R2625 (1999)

[1.63] {Sect. 1.3.4} P.D. Townsend: Experimental investigation of the performance limits for first telecommunications-window quantum cryptography systems, IEEE Photonic Technol Lett 10, p.1048-1050 (1998)

[1.64] {Sect. 1.3.4} E. Biham, T. Mor: Bounds on information and the security of quantum cryptography, Phys Rev Lett 79, p.4034-4037 (1997)

[1.65] {Sect. 1.3.4} M. Koashi, N. Imoto: Quantum cryptography based on split transmission of one- bit information in two steps, Phys Rev Lett 79, p.2383-2386 (1997)

[1.66] {Sect. 1.3.4} A. Muller, T. Herzog, B. Huttner, W. Tittel, H. Zbinden, N. Gisin: "Plug and play" systems for quantum cryptography, Appl Phys Lett 70, p.793-795 (1997)

[1.67] {Sect. 1.3.4} B.C. Jacobs, J.D. Franson: Quantum cryptography in free space, Optics Letters 21, p.1854-1856 (1996)

[1.68] {Sect. 1.3.4} M. Koashi, N. Imoto: Quantum cryptography based on two mixed states, Phys Rev Lett 77, p.2137-2140 (1996)

[1.69] {Sect. 1.3.4} A. Peres: Quantum cryptography with orthogonal states?, Phys Rev Lett 77, p.3264 (1996)

[1.70] {Sect. 1.3.5} S. Braunstein (ed.): Quantum Computing. (Wiley-VCH, Weinheim, New York, 1999)

[1.71] {Sect. 1.3.5} H.-K. Lo, T. Spiller, S. Popescu (ed.): Introduction to Quantum Computatioj and Information. (World Scientific Pub. Co, Singapore, 1998)

[1.72] {Sect. 1.3.5} L.M.K. Vandersypen, M. Steffen, M.H. Sherwood, C.S. Yannoni, G. Breyta, I.L. Chuang: Implementation of a three-quantum-bit search algorithm, Appl Phys Lett 76, p.646-648 (2000)

[1.73] {Sect. 1.3.5} A. Imamoglu, D.D. Awschalom, G. Burkard, D.P. DiVincenzo, D. Loss, M. Sherwin, A. Small: Quantum information processing using quantum dot spins and cavity QED, Phys Rev Lett 83, p.4204-4207 (1999)

[1.74] {Sect. 1.3.5} D. Bacon, D.A. Lidar, K.B. Whaley: Robustness of decoherence-free subspaces for quantum computation, Phys Rev A 60, p.1944-1955 (1999)

[1.75] {Sect. 1.3.5} J.I. Cirac, A.K. Ekert, S.F. Huelga, C. Macchiavello: Distributed quantum computation over noisy channels, Phys Rev A 59, p.4249-4254 (1999)

[1.76] {Sect. 1.3.5} J. Eisert, M. Wilkens, M. Lewenstein: Quantum games and quantum strategies, Phys Rev Lett 83, p.3077-3080 (1999)

[1.77] {Sect. 1.3.5} S. Lloyd, S.L. Braunstein: Quantum computation over continuous variables, Phys Rev Lett 82, p.1784-1787 (1999)

[1.78] {Sect. 1.3.5} M.S. Sherwin, A. Imamoglu, T. Montroy: Quantum computation with quantum dots and terahertz cavity quantum electrodynamics, Phys Rev A 60, p.3508-3514 (1999)

[1.79] {Sect. 1.3.5} L.M.K. Vandersypen, C.S. Yannoni, M.H. Sherwood, I.L. Chuang: Realization of logically labeled effective pure states for bulk quantum computation, Phys Rev Lett 83, p.3085-3088 (1999)

[1.80] {Sect. 1.3.5} E. Farhi, J. Goldstone, S. Gutmann, M. Sipser: Limit on the speed of quantum computation in determining parity, Phys Rev Lett 81, p.5442-5444 (1998)

[1.81] {Sect. 1.3.5} E. Knill, R. Laflamme: Power of one bit of quantum information, Phys Rev Lett 81, p.5672-5675 (1998)

[1.82] {Sect. 1.3.5} P. Zanardi, F. Rossi: Quantum information in semiconductors: Noiseless encoding in a quantum-dot array, Phys Rev Lett 81, p.4752-4755 (1998)

[1.83] {Sect. 1.3.5} C. Miquel, J.P. Paz, W.H. Zurek: Quantum computation with phase drift errors, Phys Rev Lett 78, p.3971-3974 (1997)

[1.84] {Sect. 1.3.5} L.M. Duan, G.C. Guo: Preserving coherence in quantum computation by pairing quantum bits, Phys Rev Lett 79, p.1953-1956 (1997)

[1.85] {Sect. 1.3.6} W. Vogel: Nonclassical states: An observable criterion, Phys Rev Lett 84, p.1849-1852 (2000)

[1.86] {Sect. 1.3.6} A.B. Matsko, V.V. Kozlov, M.O. Scully: Backaction cancellation in quantum nondemolition measurement of optical solitons, Phys Rev Lett 82, p.3244-3247 (1999)

[1.87] {Sect. 1.3.6} V. Savalli, G.Z.K. Horvath, P.D. Featonby, L. Cognet, N. Westbrook, C.I. Westbrook, A. Aspect: Optical detection of cold atoms without spontaneous emission, Optics Letters 24, p.1552-1554 (1999)

[1.88] {Sect. 1.3.6} R.L. Dematos, W. Vogel: Quantum nondemolition measurement of the motional energy of a trapped atom, Phys Rev Lett 76, p.4520-4523 (1996)

[1.89] {Sect. 1.3.6} F.X. Kartner, H.A. Haus: Quantum-Nondemolition Measurements and the 'Collapse of the Wave Function', Phys Rev A 47, p.4585-4590 (1993)

[1.90] {Sect. 1.3.7} W. Tittel, J. Brendel, H. Zbinden, N. Gisin: Violation of bell inequalities by photons more than 10 km apart, Phys Rev Lett 81, p.3563-3566 (1998)

[1.91] {Sect. 1.3.7} A. Aspect, J. Dalibard, G. Roger: Experimental Test of Bell's Inequalities Using Time-Varying Analyzers, Phys. Rev. Lett. 49, p.1804-1807 (1982)

[1.92] {Sect. 1.3.7} A. Aspect, P. Grangier, G. Roger: Experimental Realization of Einstein-Podolsky-Rosen-Bohm Gedankenexperiment: A. New Violation of Bell's Inequalities, Phys. Rev. Lett. 49, p.91-94 (1982)

[1.93] {Sect. 1.3.7} J.F. Clauser, A. Shimony: Bell's theorem: experimental tests and implications, Rep. Prog. Phys. 41, p.1881-1927 (1978)

[1.94] {Sect. 1.3.7} J. Bell: On the Einstein-Podolsky-Rosen Paradox, Physics1, p.195-200 (1964)

[1.95] {Sect. 1.3.8} N. Bloembergen: From nanosecond to femtosecond science, Rev. Mod. Phys. 71, p.283-287 (1999)

[1.96] {Sect. 1.3.8} H. Frauenfelder, P.G. Wolynes, R.H. Austin: Biological Physics, Rev. Mod. Phys. 71, p.419-430 (1999)

[1.97] {Sect. 1.3.8} W.E. Lamb, W.P. Schleich, M.O. Scully, C.H. Townes: Laser physics: Quantum controversy in action, Rev. Mod. Phys. 71, p.263-273 (1999)

[1.98] {Sect. 1.3.8} L. Mandel: Quantum Effects in one-photon and two-photon interference, Rev. Mod. Phys. 71, p.274-282 (1999)

[1.99] {Sect. 1.3.8} R.E. Slusher: Laser technology, Rev. Mod. Phys. 71, p.471-479 (1999)

[1.100] {Sect. 1.4} B.Ya. Zel'dovich, V.I. Popovicher, V.V. Ragul'skii, F.S. Faizullow: Connection between the wavefronts of the reflected and the exciting light in stimulated Mandel'shtam-Brillouin scattering, Sov. Phys. JETP 15, p.109-112 (1972)

[1.101] {Sect. 1.4} infinity – A Revolutionary Nd:YAG Laser System, Technical digest; Coherent

[1.102] {Sect. 1.5} Harnessing Light (National Academy Press, Washington, D.C, 1998)

[1.103] {Sect. 1.5.1} A. Melloni, M. Chinello, M. Martinelli: All-optical switching in phase-shifted fiber Bragg grating, IEEE Photonic Technol Lett 12, p.42-44 (2000)

[1.104] {Sect. 1.5.1} D. Cotter, R.J. Manning, K.J. Blow, A.D. Ellis, A.E. Kelly, D. Nesset, I.D. Phillips, A.J. Poustie, D.C. Rogers: Nonlinear optics for high-speed digital information processing, Science 286, p.1523-1528 (1999)

[1.105] {Sect. 1.5.1} R.W. Eason, A. Miller (ed.): Nonlinear Optics in Signal Processing (Chapman & Hall, London, 1993)

[1.106] {Sect. 1.5.2} A. Ghatak, K. Thyagarajan: Introduction to Fiber Optics (Cambridge University Press, Cambridge, 1998)

[1.107] {Sect. 1.5.2} J. S. Sanghera, I. D. Aggarwal: Infrared Fibre Optics (CRC Press, Boca Raton, Boston, London, New York, Washington, D. C, 1998)

[1.108] {Sect. 1.5.2} J.W. Lou, J.K. Andersen, J.C. Stocker, M.N. Islam, D.A. Nolan: Polarization insensitive demultiplexing of 100-Gb/s words using a twisted fiber nonlinear optical loop mirror, IEEE Photonic Technol Lett 11, p.1602-1604 (1999)

[1.109] {Sect. 1.5.2} D.S. Govan, W. Forysiak, N.J. Doran: Long-distance 40-Gbit/s soliton transmission over standard fiber by use of dispersion management, Optics Letters 23, p.1523-1525 (1998)

[1.110] {Sect. 1.5.2} M.A. Neifeld: Information, resolution, and space-bandwidth product, Optics Letters 23, p.1477-1479 (1998)

[1.111] {Sect. 1.5.2} C.C. Chang, A.M. Weiner: Fiber transmission for sub-500-fs pulses using a dispersion-compensating fiber, IEEE J QE-33, p.1455-1464 (1997)

[1.112] {Sect. 1.5.2} T. Ono, Y. Yano: Key technologies for terabit/second WDM systems with high spectral efficiency of over 1 bit/s/Hz, IEEE J QE-34, p.2080-2088 (1998)

[1.113] {Sect. 1.5.2} E.A. Desouza, M.C. Nuss, W.H. Knox, D.A.B. Miller: Wavelength division multiplexing with femtosecond pulses, Optics Letters 20, p.1166-1168 (1995)

[1.114] {Sect. 1.5.3} A. Adibi, K. Buse, D. Psaltis: Multiplexing holograms in LiNbO3 : Fe : Mn crystals, Optics Letters 24, p.652-654 (1999)

[1.115] {Sect. 1.5.3} L. Dhar, A. Hale, H.E. Katz, M.L. Schilling, M.G. Schnoes, F.C. Schilling: Recording media that exhibit high dynamic range for digital holographic data storage, Optics Letters 24, p.487-489 (1999)

[1.116] {Sect. 1.5.3} O. Matoba, B. Javidi: Encrypted optical storage with wavelength-key and random phase codes, Appl Opt 38, p.6785-6790 (1999)

[1.117] {Sect. 1.5.3} H.H. Suh: Color-image generation by use of binary-phase holograms, Optics Letters 24, p.661-663 (1999)

[1.118] {Sect. 1.5.3} C.A. Volkert, M. Wuttig: Modeling of laser pulsed heating and quenching in optical data storage media, J Appl Phys 86, p.1808-1816 (1999)

[1.119] {Sect. 1.5.3} G. Xu, Q.G. Yang, J.H. Si, X.C. Liu, P.X. Ye, Z. Li, Y.Q. Shen: Application of all-optical poling in reversible optical storage in azopolymer films, Opt Commun 159, p.88-92 (1999)

[1.120] {Sect. 1.5.3} L. Dhar, K. Curtis, M. Tackitt, M. Schilling, S. Campbell, W. Wilson, A. Hill, C. Boyd, N. Levinos, A. Harris: Holographic storage of multiple high-capacity digital data pages in thick photopolymer systems, Optics Letters 23, p.1710-1712 (1998)

[1.121] {Sect. 1.5.3} A. Toriumi, S. Kawata, M. Gu: Reflection confocal microscope readout system for three-dimensional photochromic optical data storage, Optics Letters 23, p.1924-1926 (1998)

[1.122] {Sect. 1.5.3} H. Sasaki, K. Karaki: Direct pattern recognition of a motion picture by hole- burning holography of Eu3+:Y2SiO5, Appl Opt 36, p.1742-1746 (1997)

[1.123] {Sect. 1.5.3} E.N. Glezer, M. Milosavljevic, L. Huang, R.J. Finlay, T.H. Her, J.P. Callan, E. Mazur: Three-dimensional optical storage inside transparent materials, Optics Letters 21, p.2023-2025 (1996)

[1.124] {Sect. 1.5.3} D. Lande, J.F. Heanue, M.C. Bashaw, L. Hesselink: Digital wavelength-multiplexed holographic data storage system, Optics Letters 21, p.1780-1782 (1996)

[1.125] {Sect. 1.5.3} T. Tomiyama, I. Watanabe, A. Kuwano, M. Habiro, N. Takane, M. Yamada: Rewritable optical-disk fabrication with an optical recording material made of naphthalocyanine and polythiophene, Appl Opt 34, p.8201-8208 (1995)

[1.126] {Sect. 1.5.3} E.S. Maniloff, S.B. Altner, S. Bernet, F.R. Graf, A. Renn, U.P. Wild: Recording of 6000 holograms by use of spectral hole burning, Appl. Opt. 34, p.4140-4148 (1995)

[1.127] {Sect. 1.5.3} R. Ao, S. Jahn, L. Kümmerl, R. Weiner, D. Haarer: Spatial Resolution and Data Adddressing of Frequency Domain Optical Storage Materials in the Near IR Regime, Jpn. J. Appl. Phys. 31, p.693-698 (1992)

[1.128] {Sect. 1.5.3} H.A. Haus, A. Mecozzi: Long-Term Storage of a Bit Stream of Solitons, Optics Letters 17, p.1500-1502 (1992)

[1.129] {Sect. 1.5.3} A. Renn, U.P. Wild: Spectral hole burning and hologram storage, Appl. Opt. 26, p.4040-4042 (1987)

[1.130] {Sect. 1.5.3} U.P. Wild, S.E. Bucher, F.A. Burkhalter: Hole Burning, Stark Effect, and Data Storage, Appl. Opt. 24, p.1526-1530 (1985)

[1.131] {Sect. 1.5.4} D. Jaque, J. Capmany, J.G. Sole: Red, green, and blue laser light from a single Nd : YAl3 (BO3) (4) crystal based on laser oscillation at 1.3 mu m, Appl Phys Lett 75, p.325-327 (1999)

[1.132] {Sect. 1.5.4} A. Parfenov: Diffraction light modulator based on transverse electro-optic effect in short-pitch ferroelectric liquid crystals, Appl Opt 38, p.5656-5661 (1999)

[1.133] {Sect. 1.5.4} K. Takizawa, T. Fujii, H. Kikuchi, H. Fujikake, M. Kawakita, Y. Hirano, F. Sato: Spatial light modulators for high-brightness projection displays, Appl Opt 38, p.5646-5655 (1999)

[1.134] {Sect. 1.5.4} Q. Ye, L. Shah, J. Eichenholz, D. Hammons, R. Peale, M. Richardson, A. Chin, B.H.T. Chai: Investigation of diode-pumped, self-frequency doubled RGB lasers from Nd : YCOB crystals, Opt Commun 164, p.33-37 (1999)

[1.135] {Sect. 1.5.4} A. Bewsher, I. Powell, W. Boland: Design of single-element laser-beam shape projectors, Appl Opt 35, p.1654-1658 (1996)

[1.136] {Sect. 1.5.4} S. Maruo, A. Arimoto, S. Kobayashi: Multibeam scanning optics with single laser source for full-color printers, Appl Opt 36, p.7234-7238 (1997)

[1.137] {Sect. 1.5.5} U.P. Wild, A. Renn: Molecular Computing: a Review, J. Mol. Electron. 7, p.1-20 (1991)

[1.138] {Sect. 1.5.5} U.P. Wild, A. Renn, C. De Caro, S. Bernet: Spectral hole burning and molecular computing, Appl. Opt. 29, p.4329-4331 (1990)

[1.139] {Sect. 1.5.7} A.H. Zewail: Femtochemistry (World Scientific, Singapore 1994) Vols. I and II

[1.140] {Sect. 1.5.7} J. Manz, L. Wöste (eds.): Femtosecond Chemistry (VCH, Weinheim, 1995)

[1.141] {Sect. 1.5.7} D.L. Andrews: Lasers in Chemistry, 3rd edn. (Springer, Berlin, Heidelberg 1997)

[1.142] {Sect. 1.5.7} K.B. Eisenthal (ed.): Applications of Picosecond Spectroscopy to Chemistry (Reidel, Dordrecht 1984)

[1.143] {Sect. 1.5.7} K. Kalyanasundaram: Photochemistry in microheterogeneous systems (Academic Press Inc, Florida 1987)

[1.144] {Sect. 1.5.7} G.J. Kavarnos: Fundamentals of Photoinduced Electron Transfer (VCH Publ. Inc. 1993)

[1.145] {Sect. 1.5.7} I. Prigogine, S. Rice (ed.): Advances in Chemical Physics (Wiley, New York 1983)

[1.146] {Sect. 1.5.7} A. Callegari, J. Rebstein, R. Jost, T.R. Rizzo: State-to-state unimolecular reaction dynamics of HOCl near the dissociation threshold: The role of vibrations, rotations, and IVR probed by time- and eigenstate-resolved spectroscopy, J Chem Phys 111, p.7359-7368 (1999)

[1.147] {Sect. 1.5.7} J. Karczmarek, J. Wright, P. Corkum, M. Ivanov: Optical centrifuge for molecules, Phys Rev Lett 82, p.3420-3423 (1999)

[1.148] {Sect. 1.5.7} M. Oppel, G.K. Paramonov: Selective vibronic excitation and bond breaking by picosecond UV and IR laser pulses: application to a two-dimensional model of HONO2, Chem Phys Lett 313, p.332-340 (1999)

[1.149] {Sect. 1.5.7} J. Manz, K. Sundermann, R. deVivieRiedle: Quantum optimal control strategies for photoisomerization via electronically excited states, Chem Phys Lett 290, p.415-422 (1998)

[1.150] {Sect. 1.5.7} A. Assion, T. Baumert, M. Bergt, T. Brixner, B. Kiefer, V. Seyfried, M. Strehle, G. Gerber: Control of chemical reactions by feedback-optimized phase-shaped femtosecond laser pulses, Science 282, p.919-922 (1998)

[1.151] {Sect. 1.5.7} R.N. Zare: Laser control of chemical reactions, Science 279, p.1875-1879 (1998)

[1.152] {Sect. 1.5.7} L. Banares, T. Baumert, M. Bergt, B. Kiefer, G. Gerber: Femtosecond photodissociation dynamics of Fe (CO) (5) in the gas phase, Chem Phys Lett 267, p.141-148 (1997)

[1.153] {Sect. 1.5.7} R.J. Finlay, T.H. Her, C. Wu, E. Mazur: Reaction pathways in surface femtochemistry: routes to desorption and reaction in CO/O-2/Pt (111), Chem Phys Lett 274, p.499-504 (1997)

[1.154] {Sect. 1.5.7} W. Freyer, D. Leupold: A multiphotochromic tetraanthrapor-phyrazine based on the involvement of molecular singlet oxygen, J. Photochem. and Photobiol. A: Chemistry 105, p.153-158 (1997)

[1.155] {Sect. 1.5.7} A. Kasapi: Enhanced isotope discrimination using electromagnetically induced transparency, Phys Rev Lett 77, p.1035-1038 (1996)

[1.156] {Sect. 1.5.7} C. Desfrancois, H. Abdoulcarime, C.P. Schulz, J.P. Schermann: Laser separation of geometrical isomers of weakly bound molecular complexes, Science 269, p.1707-1709 (1995)

[1.157] {Sect. 1.5.7} V. Vaida, J.D. Simon: The photoreactivity of chlorine dioxide, Science 268, p.1443-1448 (1995)

[1.158] {Sect. 1.5.7} L.C. Zhu, V. Kleiman, X.N. Li, S.P. Lu, K. Trentelman, R.J. Gordon: Coherent laser control of the product distribution obtained in the photoexcitation of HI, Science 270, p.77-80 (1995)

[1.159] {Sect. 1.5.7} P. Siders, R.A. Marcus, R.J. Cave: A Model for Orientation Effects in Electron Transfer Reactions, J Chem Phys 81, p.5613-5624 (1984)

[1.160] {Sect. 1.5.7} A.H. Zewail: Laser selective chemistry – is it possible?, Phys. Today Nov. 1980, p.27-33 (1980)

[1.161] {Sect. 1.5.7} E.S. Yeung, C.B. Moore: Isotopic separation by photopredissociation, Appl. Phys. Lett. 21, p.109-110 (1972)

[1.162] {Sect. 1.5.8} D. Schuöcker: High-Power Lasers in Production Engineering (World Scientific Publishing, Singapore, 1998)

[1.163] {Sect. 1.5.8} W. M. Steen: Laser Material Processing (Springer, London, Berlin, Heidelberg, New York, 1998)

[1.164] {Sect. 1.5.9} M. She, D. Kim, C.P. Grigoropoulos: Liquid-assisted pulsed laser cleaning using near-infrared and ultraviolet radiation, J Appl Phys 86, p.6519-6524 (1999)

[1.165] {Sect. 1.5.9} G. Vereecke, E. Rohr, M.M. Heyns: Laser-assisted removal of particles on silicon wafers, J Appl Phys 85, p.3837-3843 (1999)

[1.166] {Sect. 1.5.9} A.A. Kolomenskii, H.A. Schuessler, V.G. Mikhalevich, A.A. Maznev: Interaction of laser-generated surface acoustic pulses with fine particles: Surface cleaning and adhesion studies, J Appl Phys 84, p.2404-2410 (1998)

[1.167] {Sect. 1.5.9} S. Siano, F. Margheri, R. Pini, P. Mazzinghi, R. Salimbeni: Cleaning processes of encrusted marbles by Nd:YAG lasers operating in free-running and Q-switching regimes, Appl Opt 36, p.7073-7079 (1997)

[1.168] {Sect. 1.5.10} R.P. Lucht, M.C. Allen, S. Downey: Laser applications to chemical and environmental analysis: An introduction, Appl Opt 36, p.3187 (1997)

[1.169] {Sect. 1.5.10} M. Bass (ed.): Handbook of Optics, Vol. I, chapter 44 (McGraw-Hill, New York, 1995)

[1.170] {Sect. 1.5.10} R.M. Measure: Laser Remote Sensing: Fundamentals and Applications (Wiley, Toronto 1984)

[1.171] {Sect. 1.5.10} A.I. Karapuzikov, A.N. Malov, I.V. Sherstov: Tunable TEA CO2 laser for long-range DIAL lidar, Infrared Phys Technol 41, p.77-85 (2000)

[1.172] {Sect. 1.5.10} A.I. Karapuzikov, I.V. Ptashnik, I.V. Sherstov, O.A. Romanovskii, G.G. Matvienko, Y.N. Ponomarev: Modeling of helicopter-borne tunable TEA CO2 DIAL system employment for detection of methane and ammonia leakages, Infrared Phys Technol 41, p.87-96 (2000)

[1.173] {Sect. 1.5.10} G.H. Pettengill, P.G. Ford: Winter clouds over the North Martian Polar Cap, Geophys Res Lett 27, p.609-612 (2000)

[1.174] {Sect. 1.5.10} P.E. Smith, N.M. Evensen, D. York: Under the volcano: A new dimension in Ar-Ar dating of volcanic ash, Geophys Res Lett 27, p.585-588 (2000)

[1.175] {Sect. 1.5.10} T. Eriksen, U.P. Hoppe, E.V. Thrane, T.A. Blix: Rocketborne Rayleigh lidar for in situ measurements of neutral atmospheric density, Appl Opt 38, p.2605-2613 (1999)

[1.176] {Sect. 1.5.10} F.J. Lubken, F. Dingler, H. vonLucke, J. Anders, W.J. Riedel, H. Wolf: MASERATI: a rocketborne tunable diode laser absorption spectrometer, Appl Opt 38, p.5338-5349 (1999)

[1.177] {Sect. 1.5.10} V. Sherlock, A. Hauchecorne, J. Lenoble: Methodology for the independent calibration of Raman backscatter water-vapor lidar systems, Appl Opt 38, p.5816-5837 (1999)

[1.178] {Sect. 1.5.10} J.H. Churnside, V.V. Tatarskii, J.J. Wilson: Oceanographic lidar attenuation coefficients and signal fluctuations measured from a ship in the Southern California Bight, Appl Opt 37, p.3105-3112 (1998)

[1.179] {Sect. 1.5.10} G.P. Gobbi: Parameterization of stratospheric aerosol physical properties on the basis of Nd:YAG lidar observations, Appl Opt 37, p.4712-4720 (1998)

[1.180] {Sect. 1.5.10} J. Kasparian, J.P. Wolf: A new transient SRS analysis method of aerosols and application to a nonlinear femtosecond lidar, Opt Commun 152, p.355-360 (1998)

[1.181] {Sect. 1.5.10} C.L. Korb, B.M. Gentry, S.X. Li, C. Flesia: Theory of the double-edge technique for Doppler lidar wind measurement, Appl Opt 37, p.3097-3104 (1998)

[1.182] {Sect. 1.5.10} A. Kouzoubov, M.J. Brennan, J.C. Thomas: Treatment of polarization in laser remote sensing of ocean water, Appl Opt 37, p.3873-3885 (1998)

[1.183] {Sect. 1.5.10} G.C. Papen, D. Treyer: Comparison of an Fe Boltzmann temperature Lidar with a Na narrow-band lidar, Appl Opt 37, p.8477-8481 (1998)

[1.184] {Sect. 1.5.10} H.R. Lange, G. Grillon, J.-F. Ripoche, M.A. Franco, B. Lamouroux, B.S. Prade, A. Mysyrowicz: Anomalous long-range propagation of femtosecond laser pulses through air: moving focus or pulse self-guiding?, Opt. Lett. 23, p.120-122 (1998)

[1.185] {Sect. 1.5.10} P. Askebjer, S.W. Barwick, L. Bergstrom, A. Bouchta, S. Carius, E. Dalberg, K. Engel, B. Erlandsson, A. Goobar, L. Gray, et al.: Optical properties of deep ice at the South Pole: Absorption, Appl Opt 36, p.4168-4180 (1997)

[1.186] {Sect. 1.5.10} Y.Y.Y. Gu, C.S. Gardner, P.A. Castleberg, G.C. Papen, M.C. Kelley: Validation of the lidar in-space technology experiment: Stratospheric temperature and aerosol measurements, Appl Opt 36, p.5148-5157 (1997)

[1.187] {Sect. 1.5.10} M.J. Mcgill, W.R. Skinner, T.D. Irgang: Analysis techniques for the recovery of winds and backscatter coefficients from a multiple-channel incoherent Doppler lidar, Appl Opt 36, p.1253-1268 (1997)

[1.188] {Sect. 1.5.10} S.H. Melfi, K.D. Evans, J. Li, D. Whiteman, R. Ferrare, G. Schwemmer: Observation of Raman scattering by cloud droplets in the atmosphere, Appl Opt 36, p.3551-3559 (1997)

[1.189] {Sect. 1.5.10} J.R. Quagliano, P.O. Stoutland, R.R. Petrin, R.K. Sander, R.J. Romero, M.C. Whitehead, C.R. Quick, J.J. Tiee, L.J. Jolin: Quantitative chemical identification of four gases in remote infrared (9-11 mu m) differential absorption lidar experiments, Appl Opt 36, p.1915-1927 (1997)

[1.190] {Sect. 1.5.10} J.D. Spinhirne, S. Chudamani, J.F. Cavanaugh, J.L. Bufton: Aerosol and cloud backscatter at 1.06, 1.54, and 0.53 mu m by airborne hard-target-calibrated Nd:YAG/methane Raman lidar, Appl Opt 36, p.3475-3490 (1997)

[1.191] {Sect. 1.5.10} P.S. Argall, F. Jacka: High-pulse-repetition-frequency lidar system using a single telescope for transmission and reception, Appl Opt 35, p.2619-2629 (1996)

[1.192] {Sect. 1.5.10} J. Roths, T. Zenker, U. Parchatka, F.G. Wienhold, G.W. Harris: Four-laser airborne infrared spectrometer for atmospheric trace gas measurements, Appl Opt 35, p.7075-7084 (1996)

[1.193] {Sect. 1.5.10} R. Targ, B.C. Steakley, J.G. Hawley, L.L. Ames, P. Forney, D. Swanson, R. Stone, R.G. Otto, V. Zarifis, P. Brockman, et al.: Coherent lidar airborne wind sensor. 2. Flight-test results at 2 and 10 mu m, Appl Opt 35, p.7117-7127 (1996)

[1.194] {Sect. 1.5.10} J. Zeyn, W. Lahmann, C. Weitkamp: Remote daytime measurements of tropospheric temperature profiles with a rotational Raman lidar, Optics Letters 21, p.1301-1303 (1996)

[1.195] {Sect. 1.5.10} V. Vaida, J.D. Simon: The photoreactivity of chlorine dioxide, Science 268, p.1443-1448 (1995)

[1.196] {Sect. 1.5.10} W. Steinbrecht, K.W. Rothe, H. Walther: Lidar setup for daytime and nighttime probing of stratospheric ozone and measurements in polar and equitorial regimes, Appl. Opt. 28, p.3616-3624 (1989)

[1.197] {Sect. 1.5.10} H. Edner, S. Svanberg, L. Uneus, W. Wendt: Gas-correlation Lidar, Opt. Lett. 9, p.493-495 (1984)

[1.198] {Sect. 1.5.10} J. Werner, K.W. Rothe, H. Walther: Monitoring of the Stratospheric Ozone Layer by Laser Radar, Appl. Phys. B 32p.113-118 (1983)

[1.199] {Sect. 1.5.11} J.H. Schon, C. Kloc, E. Bucher, B. Batiogg: Efficient organic photovoltaic diodes based on doped pentacene, Nature 403, p.408-410 (2000)

[1.200] {Sect. 1.5.11} T. Tesfamichael, E. Wackelgard: Angular solar absorptance of absorbers used in solar thermal collectors, Appl Opt 38, p.4189-4197 (1999)

[1.201] {Sect. 1.5.11} S. Hamma, P.I. RocaiCabarrocas: Determination of the mobility gap of microcrystalline silicon and of the band discontinuities at the amorphous microcrystalline silicon interface using in situ Kelvin probe technique, Appl Phys Lett 74, p.3218-3220 (1999)

[1.202] {Sect. 1.5.11} K.L. Narayanan, M. Yamaguchi: Boron ion-implanted C-60 heterojunction photovoltaic devices, Appl Phys Lett 75, p.2106-2107 (1999)

[1.203] {Sect. 1.5.11} M.K. Nazeeruddin, S.M. Zakeeruddin, R. HumphryBaker, M. Jirousek, P. Liska, N. Vlachopoulos, V. Shklover, C.H. Fischer, M. Gratzel: Acid-base equilibria of (2,2 '-bipyridyl-4,4 '-dicarboxylic acid)ruthenium (II) complexes and the effect of protonation on charge- transfer sensitization of nanocrystalline titania, Inorg Chem 38, p.6298-6305 (1999)

[1.204] {Sect. 1.5.11} A. Shah, P. Torres, R. Tscharner, N. Wyrsch, H. Keppner: Photovoltaic technology: The case for thin-film solar cells, Science 285, p.692-698 (1999)

[1.205] {Sect. 1.5.11} J.T. Warren, D.H. Johnston, C. Turro: Ground state and photophysical properties of Ru (Phen) (2)quo (+): a strong excited state electron donor, Inorg Chem Commun 2, p.354-357 (1999)

[1.206] {Sect. 1.5.11} J.T. Warren, W. Chen, D.H. Johnston, C. Turro: Ground-state properties and excited-state reactivity of 8-quinolate complexes of ruthenium (II), Inorg Chem 38, p.6187-6192 (1999)

[1.207] {Sect. 1.5.11} J.H. Zhao, A.H. Wang, M.A. Green, F. Ferrazza: 19.8% efficient "honeycomb" textured multicrystalline and 24.4% monocrystalline silicon solar cells, Appl Phys Lett 73, p.1991-1993 (1998)

[1.208] {Sect. 1.5.11} B.T. Boiko, G.S. Khripunov, V.B. Yurchenko, H.E. Ruda: Photovoltaic properties in CdS/CdTe thin-film heterosystems with graded-gap interfaces, Solar Energ Mater Solar Cells 45, p.303-308 (1997)

[1.209] {Sect. 1.5.11} K. Kalyanasundaram, M. Gratzel: Photovoltaic performance of injection solar cells and other applications of nanocrystalline oxide layers, Proc Indian Acad Sci Chem Sci 109, p.447-469 (1997)

[1.210] {Sect. 1.5.11} J.A. Quintana, P.G. Boj, J. Crespo, M. Pardo, M.A. Satorre: Line-focusing holographic mirrors for solar ultraviolet energy concentration, Appl Opt 36, p.3689-3693 (1997)

[1.211] {Sect. 1.5.11} I. Shibata, T. Nishide: Solar control coatings containing a sputter deposited SiWOx film, Solar Energ Mater Solar Cells 45, p.27-33 (1997)

[1.212] {Sect. 1.5.11} R. Memming: Photoelectrochemical Solar Energy Conversion, Topics Curr. Chem. 143, p.81-112 (1988)

[1.213] {Sect. 1.5.13} S.L. Marcus: In Lasers in Medicine, ed. by G. Petttit, R.W. Wayant (Wiley, New York 1995)

[1.214] {Sect. 1.5.13} R. Pratesi, C.A. Sacci (eds.): Lasers in Photomedicine and Photobiology, (Springer Ser. Opt. Sci, Vol.31 (Springer, Berlin, Heidelberg 1980)

[1.215] {Sect. 1.5.13} R. Steiner (ed.): Laser Lithotripsy (Springer, Berlin, Heidelberg 1988)

[1.216] {Sect. 1.5.13} A. M. Verga Scheggi, S. Martellucci, A. N. Chester, R. Pratesi (eds.): Biomedical Optical Instrumentation and Laser-Assisted Biotechnology (Kluwer Academic Publishers, Dordrecht, Boston, London, 1996)

[1.217] {Sect. 1.5.13} J. A. S. Carruth, A. L. McKenzie: Medical Lasers (Adam Hilger Ltd, Bristol, Boston, 1986)

[1.218] {Sect. 1.5.13} M. L. Wolbarsht: Laser Applications in Medicine and Biology (Plenum Publishing Corporation, New York, 1991)

[1.219] {Sect. 1.5.13} B.A. Hooper, Y. Domankevitz, C.P. Lin, R.R. Anderson: Precise, controlled laser delivery with evanescent optical waves, Appl Opt 38, p.5511-5517 (1999)

[1.220] {Sect. 1.5.13} S.R. Goldstein, P.G. McQueen, R.F. Bonner: Thermal modeling of laser capture microdissection, Appl Opt 37, p.7378-7391 (1998)

[1.221] {Sect. 1.5.13} M. Frenz, H. Pratisto, F. Konz, E.D. Jansen, A.J. Welch, H.P. Weber: Comparison of the effects of absorption coefficient and pulse duration of 2.12-mu m and 2.79-mu m radiation on laser ablation of tissue, IEEE J QE-32, p.2025-2036 (1996)

[1.222] {Sect. 1.5.13} S. Karrer, R.M.Szeimies, C. Abels, M. Landthaler: The use of photodynamic therapy for skin cancer, Onkologie 21, p.20-27 (1998)

[1.223] {Sect. 1.5.13} F. H. Blum: Photodynamic Action and Diseases Caused by Light (Hafner Publ, New ork 1964)

[1.224] {Sect. 1.5.13} J.G. Moser (ed.): Photodynamic Tumor Therapy- 2nd and 3rd Generation Photosensitizers (harwood academic publishers 1998)

[1.225] {Sect. 1.5.13} R. Cubeddu, A. Pifferi, P. Taroni, A. Torricelli, G. Valentini: Noninvasive absorption and scattering spectroscopy of bulk diffusive media: An application to the optical characterization of human breast, Appl Phys Lett 74, p.874-876 (1999)

[1.226] {Sect. 1.5.13} S. Gorti, H. Tone, G. Imokawa: Triangulation method for determining capillary blood flow and physical characteristics of the skin, Appl Opt 38, p.4914-4929 (1999)

[1.227] {Sect. 1.5.13} M. Rajadhyaksha, R.R. Anderson, R.H. Webb: Video-rate confocal scanning laser microscope for imaging human tissues in vivo, Appl Opt 38, p.2105-2115 (1999)

[1.228] {Sect. 1.5.13} G. Zacharakis, A. Zolindaki, V. Sakkalis, G. Filippidis, E. Koumantakis, T.G. Papazoglou: Nonparametric characterization of human breast tissue by the Laguerre expansion of the kernels technique applied on propagating femtosecond laser pulses through biopsy samples, Appl Phys Lett 74, p.771-772 (1999)

[1.229] {Sect. 1.5.13} K. Dowling, M.J. Dayel, M.J. Lever, P.M.W. French, J.D. Hares, A.K.L. DymokeBradshaw: Fluorescence lifetime imaging with picosecond resolution for biomedical applications, Optics Letters 23, p.810-812 (1998)

[1.230] {Sect. 1.5.13} S.L. Jacques, S.J. Kirkpatrick: Acoustically modulated speckle imaging of biological tissues, Optics Letters 23, p.879-881 (1998)

[1.231] {Sect. 1.5.13} H.Q. Shangguan, L.W. Casperson: Estimation of scattered light on the surface of unclad optical fiber tips: a new approach, Opt Commun 152, p.307-312 (1998)

[1.232] {Sect. 1.5.13} Y.C. Guo, P.P. Ho, H. Savage, D. Harris, P. Sacks, S. Schantz, F. Liu, N. Zhadin, R.R. Alfano: Second-harmonic tomography of tissues, Optics Letters 22, p.1323-1325 (1997)

[1.233] {Sect. 1.5.13} A. Joblin: Tumor contrast in time-domain, near-infrared laser breast imaging, Appl Opt 36, p.9050-9057 (1997)

[1.234] {Sect. 1.5.13} K. Konig, P.T.C. So, W.W. Mantulin, E. Gratton: Cellular response to near-infrared femtosecond laser pulses in two-photon microscopes, Optics Letters 22, p.135-136 (1997)

[1.235] {Sect. 1.5.13} Y. Guo, P.P. Ho, A. Tirksliunas, F. Liu, R.R. Alfano: Optical harmonic generation from animal tissues by the use of picosecond and femtosecond laser pulses, Appl Opt 35, p.6810-6813 (1996)

[1.236] {Sect. 1.5.13} A.P. Shepherd, P.A. Öbers (eds.): Laser Doppler Blood Flowmetry. (Klüwer, Boston 1990)

[1.237] {Sect. 1.5.13} U. Morgner, W. Drexler, F.X. Kartner, X.D. Li, C. Pitris, E.P. Ippen, J.G. Fujimoto: Spectroscopic optical coherence tomography, Optics Letters 25, p.111-113 (2000)

[1.238] {Sect. 1.5.13} Y.H. Zhao, Z.P. Chen, C. Saxer, S.H. Xiang, J.F. deBoer, J.S. Nelson: Phase-resolved optical coherence tomography and optical Doppler tomography for imaging blood flow in human skin with fast scanning speed and high velocity sensitivity, Optics Letters 25, p.114-116 (2000)

[1.239] {Sect. 1.5.13} B.E. Bouma, G.J. Tearney: Power-efficient nonreciprocal interferometer and linear-scanning fiber-optic catheter for optical coherence tomography, Optics Letters 24, p.531-533 (1999)

[1.240] {Sect. 1.5.13} W. Drexler, U. Morgner, F.X. Kartner, C. Pitris, S.A. Boppart, X.D. Li, E.P. Ippen, J.G. Fujimoto: In vivo ultrahigh-resolution optical coherence tomography, Optics Letters 24, p.1221-1223 (1999)

[1.241] {Sect. 1.5.13} A.G. Podoleanu, D.A. Jackson: Noise analysis of a combined optical coherence tomograph and a confocal scanning ophthalmoscope, Appl Opt 38, p.2116-2127 (1999)

[1.242] {Sect. 1.5.13} XA. Wax, S. Bali, J.E. Thomas: Optical phase-space distributions for low-coherence light, Optics Letters 24, p.1188-1190 (1999)

[1.243] {Sect. 1.5.13} X.J. Wang, T.E. Milner, J.F. deBoer, Y. Zhang, D.H. Pashley, J.S. Nelson: Characterization of dentin and enamel by use of optical coherence tomography, Appl Opt 38, p.2092-2096 (1999)

[1.244] {Sect. 1.5.13} S.R. Chinn, E.A. Swanson, J.G. Fujimoto: Optical coherence tomography using a frequency-tunable optical source, Optics Letters 22, p.340-342 (1997)

[1.245] {Sect. 1.5.13} B.E. Bouma, G.J. Tearney, I.P. Bilinsky, B. Golubovic, J.G. Fujimoto: Self-phase-modulated Kerr-lens mode-locked Cr:forsterite laser source for optical coherence tomography, Optics Letters 21, p.1839-1841 (1996)

[1.246] {Sect. 1.5.13} G. J. Müller, B. Chance: Medical Optical Tomography: Functional Imaging and Monitoring (SPIE Optical Engineering Press, London, 1993)

[1.247] {Sect. 1.5.13} G. Müller (ed.): Optical Tomography (SPIE Bellingham 1994)

[1.248] {Sect. 1.5.14} M.A. El-Sayed, I. Tanaka, Y. Molin: Ultrafast Processes in Chemistry and Biology (lBlackwell, Oxford 1995)

[1.249] {Sect. 1.5.14} T. Kobayashi: Primary Processes in Photobiology (Springer, Berlin, Heidelberg, 1987)

[1.250] {Sect. 1.5.14} E. Kohen, R. Santus, J. Hirschberg: Photobiology (Academic Press, San Diego, 1995)

[1.251] {Sect. 1.5.14} C.B. Moore (ed.): Chemical and Biochemical Applications of Lasers, Vols. 1-5 (Academic, New York 1974-1984)

[1.252] {Sect. 1.5.14} XL. Moreaux, O. Sandre, M. BlanchardDesce, J. Mertz: Membrane imaging by simultaneous second-harmonic generation and two- photon microscopy, Optics Letters 25, p.320-322 (2000)

[1.253] {Sect. 1.5.14} D. Kelly, K.M. Grace, X. Song, B.I. Swanson, D. Frayer, S.B. Mendes, N. Peyghambarian: Integrated optical biosensor for detection of multivalent proteins, Optics Letters 24, p.1723-1725 (1999)

[1.254] {Sect. 1.5.14} S. Shikano, K. Horio, Y. Ohtsuka, Y. Eto: Separation of a single cell by red-laser manipulation, Appl Phys Lett 75, p.2671-2673 (1999)

[1.255] {Sect. 1.5.14} Y.C. Guo, P.P. Ho, F. Liu, Q.Z. Wang, R.R. Alfano: Noninvasive two-photon-excitation imaging of tryptophan distribution in highly scattering biological tissues, Opt Commun 154, p.383-389 (1998)

[1.256] {Sect. 1.5.14} M.S.Z. Kellermayer, S.B. Smith, H.L. Granzier, C. Bustamante: Folding-unfolding transitions in single titin molecules characterized with laser tweezers, Science 276, p.1112-1116 (1997)

[1.257] {Sect. 1.5.14} D. Leupold, I.E. Kochevar: Multiphoton Photochemistry in Biological Systems: Introduction, Photochem. and Photobiol. 66, p.562-565 (1997)

[1.258] {Sect. 1.5.14} S. Maiti, J.B. Shear, R.M. Williams, W.R. Zipfel, W.W. Webb: Measuring serotonin distribution in live cells with three-photon excitation, Science 275, p.530-532 (1997)

[1.259] {Sect. 1.5.14} M. Sauer, K.H. Drexhage, C. Zander, J. Wolfrum: Diode laser based detection of single molecules in solutions, Chem Phys Lett 254, p.223-228 (1996)

[1.260] {Sect. 1.5.14} G.J. Tearney, B.E. Bouma, S.A. Boppart, B. Golubovic, E.A. Swanson, J.G. Fujimoto: Rapid acquisition of in vivo biological images by use of optical coherence tomography, Optics Letters 21, p.1408-1410 (1996)

[1.261] {Sect. 1.5.14} L.H. Wang, D. Liu, N. He, S.L. Thomsen: Biological laser action, Appl Opt 35, p.1775-1779 (1996)

[1.262] {Sect. 1.5.14} W.A. Carrington, R.M. Lynch, E.D.W. Moore, G. Isenberg, K.E. Fogarty, F.S. Fredric: Superresolution three-dimensional images of fluorescence in cells with minimal light exposure, Science 268, p.1483-1487 (1995)

[1.263] {Sect. 1.5.15} C. Kung, M.D. Barnes, N. Lermer, W.B. Whitten, J.M. Ramsey: Single-molecule analysis of ultradilute solutions with guided streams of 1-mu m water droplets, Appl Opt 38, p.1481-1487 (1999)

[1.264] {Sect. 1.5.15} M. Sauer, K.H. Drexhage, U. Lieberwirth, R. Muller, S. Nord, C. Zander: Dynamics of the electron transfer reaction between an oxazine dye and DNA oligonucleotides monitored on the single-molecule level, Chem Phys Lett 284, p.153-163 (1998)

[1.265] {Sect. 1.5.15} D.S. Ko, M. Sauer, S. Nord, R. Muller, J. Wolfrum: Determination of the diffusion coefficient of dye in solution at single molecule level, Chem Phys Lett 269, p.54-58 (1997)

[1.266] {Sect. 1.5.15} S.M. Nie, S.R. Emery: Probing single molecules and single nanoparticles by surface-enhanced Raman scattering, Science 275, p.1102-1106 (1997)

[1.267] {Sect. 1.5.15} D.A. Vandenbout, W.T. Yip, D.H. Hu, D.K. Fu, T.M. Swager, P.F. Barbara: Discrete intensity jumps and intramolecular electronic energy transfer in the spectroscopy of single conjugated polymer molecules, Science 277, p.1074-1077 (1997)

[1.268] {Sect. 1.5.15} X.H. Xu, E.S. Yeung: Direct measurement of single-molecule diffusion and photodecomposition in free solution, Science 275, p.1106-1109 (1997)

[1.269] {Sect. 1.5.15} R.M. Dickson, D.J. Norris, Y.L. Tzeng, W.E. Moerner: Three-dimensional imaging of single molecules solvated in pores of poly (acrylamide) gels, Science 274, p.966-969 (1996)

[1.270] {Sect. 1.5.15} T. Plakhotnik, D. Walser, M. Pirotta, A. Renn, U.P. Wild: Nonlinear spectroscopy on a single quantum system: Two- photon absorption of a single molecule, Science 271, p.1703-1705 (1996)

[1.271] {Sect. 1.5.15} P. Astone, M. Bassan, P. Bonifazi, P. Carelli, E. Coccia, V. Fafone, S. DAntonio, S. Frasca, A. Marini, E. Mauceli et al.: Cosmic rays observed by the resonant gravitational wave detector NAUTILUS, Phys Rev Lett 84, p.14-17 (2000)

[1.272] {Sect. 1.5.15} B. Allen, J.K. Blackburn, P.R. Brady, J.D.E. Creighton, T. Creighton, S. Droz, A.D. Gillespie, S.A. Hughes, S. Kawamura, T.T. Lyons et al.: Observational limit on gravitational waves from binary neutron stars in the Galaxy, Phys Rev Lett 83, p.1498-1501 (1999)

[1.273] {Sect. 1.5.15} G. Heinzel, A. Rudiger, R. Schilling, K. Strain, W. Winkler, J. Mizuno, K. Danzmann: Automatic beam alignment in the Garching 30-m prototype of a laser- interferometric gravitational wave detector (Vol 160, pg 321, 1999), Opt Commun 164, p.161 (1999)

[1.274] {Sect. 1.5.15} C.J. Walsh, A.J. Leistner, B.F. Oreb: Power spectral density analysis of optical substrates for gravitational-wave interferometry, Appl Opt 38, p.4790-4801 (1999)

[1.275] {Sect. 1.5.15} F. Benabid, M. Notcutt, L. Ju, D.G. Blair: Rayleigh scattering in sapphire test mass for laser interferometric gravitational-wave detectors: II: Rayleigh scattering induced noise in a laser interferometric-wave detector, Opt Commun 170, p.9-14 (1999)

[1.276] {Sect. 1.5.15} T. Uchiyama, T. Tomaru, M.E. Tobar, D. Tatsumi, S. Miyoki, M. Ohashi, K. Kuroda, T. Suzuki, N. Sato, T. Haruyama et al.: Mechanical quality factor of a cryogenic sapphire test mass for gravitational wave detectors, Phys Lett A 261, p.5-11 (1999)

[1.277] {Sect. 1.5.15} P. Fritschel, N. Mavalvala, D. Shoemaker, D. Sigg, M. Zucker, G. Gonzalez: Alignment of an interferometric gravitational wave detector, Appl Opt 37, p.6734-6747 (1998)

[1.278] {Sect. 1.5.15} A.R. Agachev, A.B. Balakin, G.N. Buinov, S.L. Buchinskaya, R.A. Daishev, G.V. Kisunko, V.A. Komissaruk, S.V. Mavrin, Z.G. Murza-khanov, R.A. Rafikov et al.: Pentagonal two-loop ring interferometer, Tech Phys 43, p.591-595 (1998)

[1.279] {Sect. 1.5.15} A.Y. Ageev, I.A. Bilenko, V.B. Braginsky: Excess noise in the steel suspension wires for the laser gravitational wave detector, Phys Lett A 246, p.479-484 (1998)

[1.280] {Sect. 1.5.15} P. Fritschel, N. Mavalvala, D. Shoemaker, D. Sigg, M. Zucker, G. Gonzalez: Alignment of an interferometric gravitational wave detector, Appl Opt 37, p.6734-6747 (1998)

[1.281] {Sect. 1.5.15} M.V. Plissi, K.A. Strain, C.I. Torrie, N.A. Robertson, S. Killbourn, S. Rowan, S.M. Twyford, H. Ward, K.D. Skeldon, J. Hough: Aspects of the suspension system for GEO 600, Rev Sci Instr 69, p.3055-3061 (1998)

[1.282] {Sect. 1.5.15} T. Uchiyama, D. Tatsumi, T. Tomaru, M.E. Tobar, K. Kuroda, T. Suzuki, N. Sato, A. Yamamoto, T. Haruyama, T. Shintomi: Cryogenic cooling of a sapphire mirror-suspension for interferometric gravitational wave detectors, Phys Lett A 242, p.211-214 (1998)

[1.283] {Sect. 1.5.15} S.V. Dhurandhar, P. Hello, B.S. Sathyaprakash, J.Y. Vinet: Stability of giant Fabry-Perot cavities of interferometric gravitational-wave detectors, Appl Opt 36, p.5325-5334 (1997)

[1.284] {Sect. 1.5.15} A. Wicht, K. Danzmann, M. Fleischhauer, M. Scully, G. Muller, R.H. Rinkleff: White-light cavities, atomic phase coherence, and gravitational wave detectors, Opt Commun 134, p.431-439 (1997)

[1.285] {Sect. 1.5.15} S. Braccini, C. Bradaschia, R. Delfabbro, A. Divirgilio, I. Ferrante, F. Fidecaro, R. Flaminio, A. Gennai, A. Giazotto, P. Lapenna, et al.: Mechanical filters for the gravitational waves detector VIRGO: Performance of a two-stage suspension, Rev Sci Instr 68, p.3904-3906 (1997)

[1.286] {Sect. 1.5.15} H. Heitmann, C. Drezen: Measurement of position and orientation of optical elements in interferometric gravity wave detectors, Rev Sci Instr 68, p.3197-3205 (1997)

[1.287] {Sect. 1.5.15} P.W. Mcnamara, H. Ward, J. Hough, D. Robertson: Laser frequency stabilization for spaceborne gravitational wave detectors, Class Quantum Gravity 14, p.1543-1547 (1997)

[1.288] {Sect. 1.5.15} J. Mizuno, A. Rudiger, R. Schilling, W. Winkler, K. Danzmann: Frequency response of Michelson- and Sagnac-based interferometers, Opt Commun 138, p.383-393 (1997)

[1.289] {Sect. 1.5.15} M. Musha, K. Nakagawa, K. Ueda: Wideband and high frequency stabilization of an injection-locked Nd:YAG laser to a high-finesse Fabry-Perot cavity, Optics Letters 22, p.1177-1179 (1997)

[1.290] {Sect. 1.5.15} M. Musha, S. Telada, K. Nakagawa, M. Ohashi, K. Ueda: Measurement of frequency noise spectra of frequency- stabilized LD-pumped Nd:YAG laser by using a cavity with separately suspended mirrors, Opt Commun 140, p.323-330 (1997)

[1.291] {Sect. 1.5.15} N. Nakagawa, B.A. Auld, E. Gustafson, M.M. Fejer: Estimation of thermal noise in the mirrors of laser interferometric gravitational wave detectors: Two point correlation function, Rev Sci Instr 68, p.3553-3556 (1997)

[1.292] {Sect. 1.5.15} A. Wicht, K. Danzmann, M. Fleischhauer, M. Scully, G. Muller, R.H. Rinkleff: White-light cavities, atomic phase coherence, and gravitational wave detectors, Opt Commun 134, p.431-439 (1997)

[1.293] {Sect. 1.5.15} J. Giaime, P. Saha, D. Shoemaker, L. Sievers: A passive vibration isolation stack for LIGO: Design, modeling, and testing, Rev Sci Instr 67, p.208-214 (1996)

[1.294] {Sect. 1.5.15} G. Heinzel, J. Mizuno, R. Schilling, W. Winkler, A. Rudiger, K. Danzmann: An experimental demonstration of resonant sideband extraction for laser-interferometric gravitational wave detectors, Phys Lett A 217, p.305-314 (1996)

[1.295] {Sect. 1.5.15} L. Ju, M. Notcutt, D. Blair, F. Bondu, C.N. Zhao: Sapphire beamsplitters and test masses for advanced laser interferometer gravitational wave detectors, Phys Lett A 218, p.197-206 (1996)

[1.296] {Sect. 1.5.15} D.E. Mcclelland: An overview of recycling in laser interferometric gravitational wave detectors, Aust J Phys 48, p.953-970 (1996)

[1.297] {Sect. 1.5.15} D. Nicholson, C.A. Dickson, W.J. Watkins, B.F. Schutz, J. Shuttleworth, G.S. Jones, D.I. Robertson, N.L. Mackenzie, K.A. Strain, B.J. Meers, et al.: Results of the first coincident observations by two laser-interferometric gravitational wave detectors, Phys Lett A 218, p.175-180 (1996)

[1.298] {Sect. 1.5.15} P.J. Veitch, J. Munch, M.W. Hamilton, D. Ottaway, A. Greentree, A. Tikhomirov: High power lasers and novel optics for laser interferometric gravitational wave detectors, Aust J Phys 48, p.999-1006 (1996)

[1.299] {Sect. 1.5.15} Y. Wang, A. Stebbins, E.L. Turner: Gravitational lensing of gravitational waves from merging neutron star binaries, Phys Rev Lett 77, p.2875-2878 (1996)

[1.300] {Sect. 1.5.15} G.W. Collins, P.M. Celliers, L.B. DaSilva, D.M. Gold, R. Cauble: Laser-shock-driven laboratory measurements of the equation of state of hydrogen isotopes in the megabar regime, High Pressure Res 16, p.281-290 (2000)

[1.301] {Sect. 1.5.15} A. Mohacsi, M. Szakall, Z. Bozoki, G. Szabo, Z. Bor: High stability external cavity diode laser system for photoacoustic gas detection, Laser Phys 10, p.378-381 (2000)

[1.302] {Sect. 1.5.15} E. Beaurepaire, L. Moreaux, F. Amblard, J. Mertz: Combined scanning optical coherence and two-photon-excited fluorescence microscopy, Optics Letters 24, p.969-971 (1999)

[1.303] {Sect. 1.5.15} A. Garnache, A.A. Kachanov, F. Stoeckel, R. Planel: High-sensitivity intracavity laser absorption spectroscopy with vertical-external-cavity surface-emitting semiconductor lasers, Optics Letters 24, p.826-828 (1999)

[1.304] {Sect. 1.5.15} J. Han: Fabry-Perot cavity chemical sensors by silicon micromachining techniques, Appl Phys Lett 74, p.445-447 (1999)

[1.305] {Sect. 1.5.15} E. Lacot, R. Day, F. Stoeckel: Laser optical feedback tomography, Optics Letters 24, p.744-746 (1999)

[1.306] {Sect. 1.5.15} J. Nolte, M. Paul: ICP-OES analysis of coins using laser ablation, At Spectrosc 20, p.212-216 (1999)

[1.307] {Sect. 1.5.15} H. Okayama, L.Z. Wang: Measurement of the spatial coherence of light influenced by turbulence, Appl Opt 38, p.2342-2345 (1999)

[1.308] {Sect. 1.5.15} K.A. Peterson, D.B. Oh: High-sensitivity detection of CH radicals in flames by use of a diode- laser-based near-ultraviolet light source, Optics Letters 24, p.667-669 (1999)

[1.309] {Sect. 1.5.15} F.M. Xu, H.E. Pudavar, P.N. Prasad, D. Dickensheets: Confocal enhanced optical coherence tomography for nondestructive evaluation of paints and coatings, Optics Letters 24, p.1808-1810 (1999)

[1.310] {Sect. 1.5.15} G. Zikratov, F.Y. Yueh, J.P. Singh, O.P. Norton, R.A. Kumar, R.L. Cook: Spontaneous anti-Stokes Raman probe for gas temperature measurements in industrial furnaces, Appl Opt 38, p.1467-1475 (1999)

[1.311] {Sect. 1.5.15} C.-T. Hsieh, C.-K. Lee: Cylindrical-type nanometer-resolution laser diffractive optical encoder, Appl. Opt. 38, p.4743-4750 (1999)

[1.312] {Sect. 1.5.15} V. Lecoeuche, D.J. Webb, C.N. Pannell, D.A. Jackson: Brillouin based distributed fibre sensor incorporating a mode-locked Brillouin fibre ring laser, Opt Commun 152, p.263-268 (1998)

[1.313] {Sect. 1.5.15} K.J. Schulz, W.R. Simpson: Frequency-matched cavity ringdown spectroscopy, Chem Phys Lett 297, p.523-529 (1998)

[1.314] {Sect. 1.5.15} F. Kühnemann, K. Schneider, A. Hecker, A.A.E. Martis, W. Urban, S.Schiller, J. Mlynek: Photoacoustic trace-gas detection using a cw single-frequency parametric oscillator, Appl. Phys. B 66, p.741-745 (1998)

[1.315] {Sect. 1.5.15} Y.M. Chang, L. Xu, H.W.K. Tom: Observation of coherent surface optical phonon oscillations by time-resolved surface second-harmonic generation, Phys Rev Lett 78, p.4649-4652 (1997)

[1.316] {Sect. 1.5.15} J.C. Cotteverte, J. Poirson, A. LeFloch, F. Bretenaker, A. Chauvin: Laser magnetometer measurement of the natural remanent magnetization of rocks, Appl Phys Lett 70, p.3075-3077 (1997)

[1.317] {Sect. 1.5.15} J. Larsson, Z. Chang, E. Judd, P.J. Schuck, R.W. Falcone, P.A. Heimann, H.A. Padmore, H.C. Kapteyn, P.H. Bucksbaum, M.M. Murnane, et al.: Ultrafast x-ray diffraction using a streak-camera detector in averaging mode, Optics Letters 22, p.1012-1014 (1997)

[1.318] {Sect. 1.5.15} B.W. Lee, H.J. Jeong, B.Y. Kim: High-sensitivity modelocked fiber laser gyroscope, Optics Letters 22, p.129-131 (1997)

[1.319] {Sect. 1.5.15} R.M. Mihalcea, D.S. Baer, R.K. Hanson: Diode laser sensor for measurements of CO, CO2, and CH4 in combustion flows, Appl Opt 36, p.8745-8752 (1997)

[1.320] {Sect. 1.5.15} R.B. Rogers, W.V. Meyer, J.X. Zhu, P.M. Chaikin, W.B. Russel, M. Li, W.B. Turner: Compact laser light-scattering instrument for microgravity research, Appl Opt 36, p.7493-7500 (1997)

[1.321] {Sect. 1.5.15} E.W. Rothe, P. Andresen: Application of tunable excimer lasers to combustion diagnostics: A review, Appl Opt 36, p.3971-4033 (1997)

[1.322] {Sect. 1.5.15} A. Brockhinke, K. Kohsehoinghaus, P. Andresen: Double-pulse one-dimensional Raman and Rayleigh measurements for the detection of temporal and spatial structures in a turbulent H-2-air diffusion flame, Optics Letters 21, p.2029-2031 (1996)

[1.323] {Sect. 1.5.15} S.L. Min, A. Gomez: High-resolution size measurement of single spherical particles with a fast Fourier transform of the angular scattering intensity, Appl Opt 35, p.4919-4926 (1996)

[1.324] {Sect. 1.5.15} P. Repond, M.W. Sigrist: Photoacoustic spectroscopy on trace gases with continuously tunable CO2 laser, Appl Opt 35, p.4065-4085 (1996)

[1.325] {Sect. 1.5.15} P.A. Roos, M. Stephens, C.E. Wieman: Laser vibrometer based on optical-feedback-induced frequency modulation of a single-mode laser diode, Appl Opt 35, p.6754-6761 (1996)

[1.326] {Sect. 1.5.15} T. Dresel, G. Häusler, H. Venzke: Three-dimensional sensing of rough surfaces by coherence radar, Appl. Opt. 31, p.919-925 (1992)

[1.327] {Sect. 1.5.15} T.J. Kane, W.J. Kozlovsky, R.L. Byer, C.E. Byvik: Coherent laser radar at 1.06 μm using Nd:YAG lasers, Opt. Lett. 12, p.239-241 (1987)

[1.328] {Sect. 1.5.16} A. Guttman, T. Lengyel, M. Szoke, M. SasvariSzekely: Ultra-thin-layer agarose gel electrophoresis – II. Separation of DNA fragments on composite agarose-linear polymer matrices, J Chromatogr A 871, p.289-298 (2000)

[1.329] {Sect. 1.5.16} M. Neumann, D.P. Herten, A. Dietrich, J. Wolfrum, M. Sauer: Capillary array scanner for time-resolved detection and identification of fluorescently labelled DNA fragments, J Chromatogr A 871, p.299-310 (2000)

[1.330] {Sect. 1.5.16} H.H. Zhou, A.W. Miller, Z. Sosic, B. Buchholz, A.E. Barron, L. Kotler, B.L. Karger: DNA sequencing up to 1300 bases in two hours by capillary electrophoresis with mixed replaceable linear polyacrylamide solutions, Anal Chem 72, p.1045-1052 (2000)

[1.331] {Sect. 1.5.16} S.O. Kelley, J.K. Barton: Electron transfer between bases in double helical DNA, Science 283, p.375-381 (1999)

[1.332] {Sect. 1.5.16} G.V. Shivashankar, A. Libchaber: Biomolecular recognition using submicron laser lithography, Appl Phys Lett 73, p.417-419 (1998)

[1.333] {Sect. 1.5.16} R. Muller, D.P. Herten, U. Lieberwirth, M. Neumann, M. Sauer, A. Schulz, S. Siebert, K.H. Drexhage, J. Wolfrum: Efficient DNA sequencing with a pulsed semiconductor laser and a new fluorescent dye set, Chem Phys Lett 279, p.282-288 (1997)

[1.334] {Sect. 1.5.16} A. Anders: Selective Laser Excitation of Bases in Nucleic Acids, Appl. Phys. 20, p.257-259 (1979)

[1.335] {Sect. 1.5.16} A. Anders: Models of DNA-Dye-Complexes: Energy Transfer and Molecular Structures as Evaluated by Laser Excitation, Appl. Phys. 18, p.333-338 (1979)

[1.336] {Sect. 1.5.17} H. Daido, S. Sebban, N. Sakaya, Y. Tohyama, T. Norimatsu, K. Mima, Y. Kato, S. Wang, Y. Gu, G. Huang et al.: Experimental characterization of short-wavelength Ni-like soft-x-ray lasing toward the water window, J Opt Soc Am B Opt Physics 16, p.2295-2299 (1999)

[1.337] {Sect. 1.5.17} C. Spielmann, N.H. Burnett, S. Sartania, R. Koppitsch, M. Schnurer, C. Kan, M. Lenzner, P. Wobrauschek, F. Krausz: Generation of coherent X-rays in the water window using 5-femtosecond laser pulses, Science 278, p.661-664 (1997)

[1.338] {Sect. 1.5.17} P. Gibbon: Harmonic generation by femtosecond laser-solid interaction: A coherent "water-window" light source?, Phys Rev Lett 76, p.50-53 (1996)

[1.339] {Sect. 1.5.17} B. Lengeler, C.G. Schroer, M. Richwin, J. Tummler, M. Drakopoulos, A. Snigirev, I. Snigireva: A microscope for hard x rays based on parabolic compound refractive lenses, Appl Phys Lett 74, p.3924-3926 (1999)

[1.340] {Sect. 1.5.17} Y. Aglitskiy, T. Lehecka, S. Obenschain, S. Bodner, C. Pawley, K. Gerber, J. Sethian, C.M. Brown, J. Seely, U. Feldman et al.: High-resolution monochromatic x-ray imaging system based on spherically bent crystals, Appl Opt 37, p.5253-5261 (1998)

[1.341] {Sect. 1.5.17} J.A. Koch, O.L. Landen, T.W. Barbee, P. Celliers, L.B. DaSilva, S.G. Glendinning, B.A. Hammel, D.H. Kalantar, C. Brown, J.

Seely et al.: High-energy x-ray microscopy techniques for laser-fusion plasma research at the National Ignition Facility, Appl Opt 37, p.1784-1795 (1998)

[1.342] {Sect. 1.5.17} C.C. Gaither, E.J. Schmahl, C.J. Crannell, B.R. Dennis, F.L. Lang, L.E. Orwig, C.N. Hartman, G.J. Hurford: Quantitative characterization of the x-ray imaging capability of rotating modulation collimators with laser light, Appl Opt 35, p.6714-6726 (1996)

[1.343] {Sect. 1.5.17} I.P. Christov, M.M. Murnane, H.C. Kapteyn: Generation of single-cycle attosecond pulses in the vacuum ultraviolet, Opt Commun 148, p.75-78 (1998)

[1.344] {Sect. 1.5.17} G. Schriever, K. Bergmann, R. Lebert: Narrowband laser produced extreme ultraviolet sources adapted to silicon/molybdenum multilayer optics, J Appl Phys 83, p.4566-4571 (1998)

[1.345] {Sect. 1.5.17} I.V. Tomov, P. Chen, P.M. Rentzepis: Pulse broadening in femtosecond x-ray diffraction, J Appl Phys 83, p.5546-5548 (1998)

[1.346] {Sect. 1.5.17} Z.H. Chang, A. Rundquist, H.W. Wang, M.M. Murnane, H.C. Kapteyn: Generation of coherent soft X rays at 2.7 nm using high harmonics, Phys Rev Lett 79, p.2967-2970 (1997)

[1.347] {Sect. 1.5.17} R.W. Schoenlein, W.P. Leemans, A.H. Chin, P. Volfbeyn, T.E. Glover, P. Balling, M. Zolotorev, K.J. Kim, S. Chattopadhyay, C.V. Shank: Femtosecond x-ray pulses at 0.4 angstrom generated by 90 degrees Thomson scattering: A tool for probing the structural dynamics of materials, Science 274, p.236-238 (1996)

[1.348] {Sect. 1.5.18} H.H. Solak, D. He, W. Li, S. SinghGasson, F. Cerrina, B.H. Sohn, X.M. Yang, P. Nealey: Exposure of 38 nm period grating patterns with extreme ultraviolet interferometric lithography, Appl Phys Lett 75, p.2328-2330 (1999)

[1.349] {Sect. 1.5.18} eds. Updated Roadmap identifies technical, stategic challenges, Solid State Technologyp.43-53 (1995)

[1.350] {Sect. 1.5.20} V.I. Bespalov: Large-size monosectorial crystal elements for powerful laser systems, J Nonlinear Opt Physics Mat 6, p.467-472 (1997)

[1.351] {Sect. 1.5.20} T.R. Boehly, D.L. Brown, R.S. Craxton, R.L. Keck, J.P. Knauer, J.H. Kelly, T.J. Kessler, S.A. Kumpan, S.J. Loucks, S.A. Letzring, et al.: Initial performance results of the OMEGA laser system, Opt Commun 133, p.495-506 (1997)

[1.352] {Sect. 1.5.20} M.J. Guardalben: Conoscopic alignment methods for birefringent optical elements in fusion lasers, Appl Opt 36, p.9107-9109 (1997)

[1.353] {Sect. 1.5.20} B.M. Vanwonterghem, J.R. Murray, J.H. Campbell, D.R. Speck, C.E. Barker, I.C. Smith, D.F. Browning, W.C. Behrendt: Performance of a prototype for a large-aperture multipass Nd: glass laser for inertial confinement fusion, Appl Opt 36, p.4932-4953 (1997)

[1.354] {Sect. 1.5.21} Y. Cheng, Z.Z. Xu: Vacuum laser acceleration by an ultrashort, high-intensity laser pulse with a sharp rising edge, Appl Phys Lett 74, p.2116-2118 (1999)

[1.355] {Sect. 1.5.21} G. Malka, E. Lefebvre, J.L. Miquel: Experimental observation of electrons accelerated in vacuum to relativistic energies by a high-intensity laser, Phys Rev Lett 78, p.3314-3317 (1997)

[1.356] {Sect. 1.5.21} G. Malka, J. Fuchs, F. Amiranoff, S.D. Baton, R. Gaillard, J.L. Miquel, H. Pepin, C. Rousseaux, G. Bonnaud, M. Busquet, et al.: Suprathermal electron generation and channel formation by an ultrarelativistic laser pulse in an underdense preformed plasma, Phys Rev Lett 79, p.2053-2056 (1997)

[1.357] {Sect. 1.5.21} B. Rau, T. Tajima, H. Hojo: Coherent electron acceleration by subcycle laser pulses, Phys Rev Lett 78, p.3310-3313 (1997)

2. Properties and Description of Light

[2.1]   {Sect. 2.1.1} D.L. Burke, R.C. Field, G. Hortonsmith, J.E. Spencer, D. Walz, S.C. Berridge, W.M. Bugg, K. Shmakov, A.W. Weidemann, C. Bula, et al.: Positron production in multiphoton light-by-light scattering, Phys Rev Lett 79, p.1626-1629 (1997)

[2.2]   {Sect. 2.2.1} W.K. Kuo, Y.T. Huang, S.L. Huang: Three-dimensional electric-field vector measurement with an electro- optic sensing technique, Optics Letters 24, p.1546-1548 (1999)

[2.3]   {Sect. 2.4.0} P. Varga, P. Torok: The Gaussian wave solution of Maxwell's equations and the validity of scalar wave approximation, Opt Commun 152, p.108-118 (1998)

[2.4]   {Sect. 2.4.0} J. Durnin, J.J. Miceli, Jr, J.H. Eberly: Diffraction-Free Beams, Phys. Rev. Lett. 58, p.1499-1501 (1987)

[2.5]   {Sect. 2.4.0} J. Durnin: Exact solutions for nondiffracting beams. I. The scalar theory, J.Opt. Soc. Am. A 4, p.651-654 (1987)

[2.6]   {Sect. 2.4.0} R. Pratesi, L. Ronchi: Generalized Gaussian beams in free space, J. Opt. Soc. Am. 67, p.1274-1276 (1977)

[2.7]   {Sect. 2.4.3} V. Laude, S. Olivier, C. Dirson, J.P. Huignard: Hartmann wave-front scanner, Optics Letters 24, p.1796-1798 (1999)

[2.8]   {Sect. 2.4.3} S. Linden, J. Kuhl, H. Giessen: Amplitude and phase characterization of weak blue ultrashort pulses by downconversion, Optics Letters 24, p.569-571 (1999)

[2.9]   {Sect. 2.4.3} A. Baltuska, M.S. Pshenichnikov, D.A. Wiersma: Amplitude and phase characterization of 4.5-fs pulses by frequency- resolved optical gating, Optics Letters 23, p.1474-1476 (1998)

[2.10] {Sect. 2.4.3} J.C. Chanteloup, F. Druon, M. Nantel, A. Maksimchuk, G. Mourou: Single-shot wave-front measurements of high-intensity ultrashort laser pulses with a three-wave interferometer, Optics Letters 23, p.621-623 (1998)

[2.11] {Sect. 2.4.3} G.Y. Yoon, T. Jitsuno, M. Nakatsuka, S. Nakai: Shack Hart-mann wave-front measurement with a large F- number plastic microlens array, Appl Opt 35, p.188-192 (1996)

[2.12] {Sect. 2.4.3} J. M. Geary: Introduction to Wavefront Sensors (SPIE Optical Engineering Press, London, 1995)

[2.13] {Sect. 2.4.4} S. Gangopadhyay, S. Sarkar: ABCD matrix for reflection and refraction of Gaussian light beams at surfaces of hyperboloid of revolution and efficiency computation for laser diode to single-mode fiber coupling by way of a hyperbolic lens on the fiber tip, Appl Opt 36, p.8582-8586 (1997)

[2.14] {Sect. 2.4.4} P.A. Bélanger: Beam propagation and the ABCD ray matrices, Opt. Lett. 16, p.196-198 (1991)

[2.15] {Sect. 2.4.4} A. Yariv: Operator algebra for propagation problems involving phase conjugation and nonreciprocal elements, Appl. Opt. 26, p.4538-4540 (1987)

[2.16] {Sect. 2.4.4} K. Halbach: Matrix Representation of Gaussian Optics, Am. J. Phys. 32, p.90-108 (1964)

[2.17] {Sect. 2.4.4} A. Gerrard, J.M. Burch: Introduction to Matrix Methods. in Optics (Wiley London 1975)

[2.18] {Sect. 2.5.4} S. Ameerbeg, A.J. Langley, I.N. Ross, W. Shaikh, P.F. Taday: An achromatic lens for focusing femtosecond pulses: Direct measurement of femtosecond pulse front distortion using a second-order autocorrelation technique, Opt Commun 122, p.99-104 (1996)

[2.19] {Sect. 2.5.4} M. Gu, E. Yap: Axial imaging behaviour of a single lens illu-minated by an ultrashort pulsed beam, Opt Commun 124, p.202-207 (1996)

[2.20] {Sect. 2.5.4} M. Kempe, U. Stamm, B. Wilhelmi, W. Rudolph: Spatial and temporal transformation of femtosecond laser pulses by lenses and lens systems, J. Opt. Soc. Am. B 9, p.1158-1165 (1992)

[2.21] {Sect. 2.6.0} E. Collett: Polarized Light – Fudamentals and Applications (Marcel Dekker Inc, New York, Basel, Hong Kong, 1993)

[2.22] {Sect. 2.6.0} R.C. Jones: A new calculus for the treatment of optical systems. VIII Electromagnetic theory, J. Opt. Soc. Am. 38, p.126-131 (1956)

[2.23] {Sect. 2.6.0} R.C. Jones: A New Calculus for the Treatment of Optical Systems, J. Opt. Soc. Am. 32, p.486-493 (1942)

[2.24] {Sect. 2.6.0} S. Huard: Polarization of Light (Wiley, VCH, Chichester, 1997)

[2.25] {Sect. 2.6.0} J. Junghans, M. Keller, H. Weber: Laser Resonators with Polarizing Elements – Eigenstates and Eigenvalues of Polarization, Appl. Opt. 13, p.2793-2798 (1974)

[2.26] {Sect. 2.6.0} A.H. Carrieri: Neural network pattern recognition by means of differential absorption Mueller matrix spectroscopy, Appl Opt 38, p.3759-3766 (1999)

[2.27] {Sect. 2.6.0} E. Compain, S. Poirier, B. Drevillon: General and self-consistent method for the calibration of polarization modulators, polarimeters, and Mueller-matrix ellipsometers, Appl Opt 38, p.3490-3502 (1999)

[2.28] {Sect. 2.6.0} G. Yao, L.V. Wang: Two-dimensional depth-resolved Mueller matrix characterization of biological tissue by optical coherence tomography, Optics Letters 24, p.537-539 (1999)

[2.29] {Sect. 2.6.0} C. Ye: Photopolarimetric measurement of single, intact pulp fibers by Mueller matrix imaging polarimetry, Appl Opt 38, p.1975-1985 (1999)

[2.30] {Sect. 2.6.0} B.D. Cameron, M.J. Rakovic, M. Mehrubeoglu, G.W. Kattawar, S. Rastegar, L.V. Wang, G.L. Cote: Measurement and calculation of the two-dimensional backscattering Mueller matrix of a turbid medium (Vol 23, pg 485, 1998), Optics Letters 23, p.1630 (1998)

[2.31] {Sect. 2.6.0} B.D. Cameron, M.J. Rakovic, M. Mehrubeoglu, G.W. Kattawar, S. Rastegar, L.V. Wang, G.L. Cote: Measurement and calculation of the two-dimensional backscattering Mueller matrix of a turbid medium, Optics Letters 23, p.485-487 (1998)

[2.32] {Sect. 2.6.0} A.H. Carrieri, J.R. Bottiger, D.J. Owens, E.S. Roese: Differential absorption Mueller matrix spectroscopy and the infrared detection of crystalline organics, Appl Opt 37, p.6550-6557 (1998)

[2.33] {Sect. 2.6.2} H. Kogelnik, L.E. Nelson, J.P. Gordon, R.M. Jopson: Jones matrix for second-order polarization made dispersion, Optics Letters 25, p.19-21 (2000)

[2.34] {Sect. 2.6.2} XD. Penninckx, V. Morenas: Jones matrix of polarization mode dispersion, Optics Letters 24, p.875-877 (1999)

[2.35] {Sect. 2.7.1} G. Grönninger, A. Penzkofer: Determination of energy and duration of picosecond light pulses by bleaching of dyes, Opt. Quant. Electr. 16, p.225-233 (1984)

[2.36] {Sect. 2.7.1} A. Penzkofer, W. Falkenstein: Direct Determination of the Intensity of Picosecond Light Pulses by Two-Photon Absorption, Opt. Comm. 17, p.1-5 (1976)

[2.37] {Sect. 2.7.1} T.R. Gentile, J.M. Houston, G. Eppeldauer, A.L. Migdall, C.L. Cromer: Calibration of a pyroelectric detector at 10.6 mu m with the National Institute of Standards and Technology high- accuracy cryogenic radiometer, Appl Opt 36, p.3614-3621 (1997)

[2.38] {Sect. 2.7.1} D.N. Fittinghoff, J.L. Bowie, J.N. Sweetser, R.T. Jennings, M.A. Krumbugel, K.W. Delong, R. Trebino, I.A. Walmsley: Measurement of the intensity and phase of ultraweak, ultrashort laser pulses, Optics Letters 21, p.884-886 (1996)

[2.39]   {Sect. 2.7.3} M.A. Bolshtyansky, N.V. Tabiryan, B.Y. Zeldovich: BRIEF-ING: Beam reconstruction by iteration of an electromagnetic field with an induced nonlinearity gauge, Optics Letters 22, p.22-24 (1997)

[2.40]   {Sect. 2.7.3} A. Cutolo, R. Ferreri, T. Isernia, R. Pierri, L. Zeni: Measurements of the waist and the power distribution across the transverse modes of a laser beam, Opt. Quantum Electron. 24, p.963-971 (1992)

[2.41]   {Sect. 2.7.3} R. Borghi, M. Santarsiero: Modal decomposition of partially coherent flat-topped beams produced by multimode lasers, Optics Letters 23, p.313-315 (1998)

[2.42]   {Sect. 2.7.3} T.Y. Cherezova, S.S. Chesnokov, L.N. Kaptsov, A.V. Kudryashov: Super-Gaussian laser intensity output formation by means of adaptive optics, Opt Commun 155, p.99-106 (1998)

[2.43]   {Sect. 2.7.3} J.J. Kasinski, R.L. Burnham: Near-diffraction-limited laser beam shaping with diamond- turned aspheric optics, Optics Letters 22, p.1062-1064 (1997)

[2.44]   {Sect. 2.7.3} N. Lisi, P. Dilazzaro, F. Flora: Time-resolved divergence measurement of an excimer laser beam by the knife-edge technique, Opt Commun 136, p.247-252 (1997)

[2.45]   {Sect. 2.7.3} W. Plass, R. Maestle, K. Wittig, A. Voss, A. Giesen: High-resolution knife-edge laser beam profiling, Opt Commun 134, p.21-24 (1997)

[2.46]   {Sect. 2.7.4} D. Dragoman: Can the Wigner transform of a two-dimensional rotationally symmetric beam be fully recovered from the Wigner transform of its one- dimensional approximation?, Optics Letters 25, p.281-283 (2000)

[2.47]   {Sect. 2.7.4} B. Eppich, C. Gao, H. Weber: Determination of the ten second order intensity moments, Opt. Laser Technol.30p.337-340 (1998)

[2.48]   {Sect. 2.7.4} H. Weber: Propagation of higher-order intensity moments in quadratic-index media, Opt. Quant. Electr. 24, p.1027-1049 (1992)

[2.49]   {Sect. 2.7.4} H.O. Bartelt, K.-H. Brenner, A.W. Lohmann: The Wigner distribution function and its optical production, Opt. Commun. 32, p.32-38 (1980)

[2.50]   {Sect. 2.7.4} M.J. Bastiaans: Wigner distribution function and its application to first-order optics, J. Opt. Soc. Am. 69, p.1710-1716 (1979)

[2.51]   {Sect. 2.7.5} S. Bollanti, P. Dilazzaro, D. Murra: How many times is a laser beam diffraction-limited?, Opt Commun 134, p.503-513 (1997)

[2.52]   {Sect. 2.7.5} G. Nemes, A.E. Siegman: Measurement of all ten second-order moments of an astigmatic beam by the use of rotating simple astigmatic (anamorphic) optics, J.Opt. Soc. Am. A 11, p.2257-2264 (1994)

[2.53]   {Sect. 2.7.5} A. Caprara, G.C. Reali: Time varying M2 in Q-switched lasers, Opt. Quant. Electr. 24, p.1001-1009 (1992)

[2.54]   {Sect. 2.7.5} N. Hodgson, T. Haase, R. Kostka, H. Weber: Determination of laser beam parameters with the phase space beam analyser, Opt. Quantum Electron. 24, p.927-949 (1992)

[2.55]   {Sect. 2.7.5} N. Reng, B. Eppich: Definition and measurements of high-power laser beam parameters, Opt. Quant. Electr. 24, p.973-992 (1992)

[2.56]   {Sect. 2.7.5} Anonymus: ISO Standards Handbook 2: Units of Measurement, 2d ed. (International Organization for Standardization, 1982)

[2.57]   {Sect. 2.7.5} ISO, Norm-Manuscript ISO/DIS 11146 "Optics and optical instruments – Lasers and laser related equipment – Test methods for laser beam parameters: Beam widths, divergence angle and beam propagation factor, 1995

[2.58] {Sect. 2.7.5} D. Wright, P. Greve, J. Fleischer, L. Austin: Laser beam width, divergence and beam propagation factor – an international standardization approach, Opt. Quant. Electr. 24, p.993-1000 (1992)

[2.59] {Sect. 2.7.5} L. LeDeroff, P. Salieres, B. Carre: Beam-quality measurement of a focused high-order harmonic beam, Optics Letters 23, p.1544-1546 (1998)

[2.60] {Sect. 2.7.5} H.L. Offerhaus, C.B. Edwards, W.J. Witteman: Single shot beam quality (M-2) measurement using a spatial Fourier transform of the near field, Opt Commun 151, p.65-68 (1998)

[2.61] {Sect. 2.7.5} T.F. Johnston, J.M. Fleischer: Calibration standard for laser beam profilers: Method for absolute accuracy measurement with a Fresnel diffraction test pattern, Appl Opt 35, p.1719-1734 (1996)

[2.62] {Sect. 2.7.8} P.F. Cohadon, A. Heidmann, M. Pinard: Cooling of a mirror by radiation pressure, Phys Rev Lett 83, p.3174-3177 (1999)

[2.63] {Sect. 2.7.8} V. Chickarmane, S.V. Dhurandhar, R. Barillet, P. Hello, J.Y. Vinet: Radiation pressure and stability of interferometric gravitational- wave detectors, Appl Opt 37, p.3236-3245 (1998)

[2.64] {Sect. 2.7.8} S. Nemoto, H. Togo: Axial force acting on a dielectric sphere in a focused laser beam, Appl Opt 37, p.6386-6394 (1998)

[2.65] {Sect. 2.7.8} Y.N. Ohshima, H. Sakagami, K. Okumoto, A. Tokoyoda, T. Igarashi, K.B. Shintaku, S. Toride, H. Sekino, K. Kabuto, I. Nishio: Direct measurement of infinitesimal depletion force in a colloid-polymer mixture by laser radiation pressure, Phys Rev Lett 78, p.3963-3966 (1997)

[2.66] {Sect. 2.7.8} Y. Harada, T. Asakura: Radiation forces on a dielectric sphere in the Rayleigh scattering regime, Opt Commun 124, p.529-541 (1996)

[2.67] {Sect. 2.7.8} K. Sasaki, M. Tsukima, H. Masuhara: Three-dimensional potential analysis of radiation pressure exerted on a single microparticle, Appl Phys Lett 71, p.37-39 (1997)

[2.68] {Sect. 2.7.8} M. Trunk, J.F. Lubben, J. Popp, B. Schrader, W. Kiefer: Investigation of a phase transition in a single optically levitated microdroplet by Raman-Mie scattering, Appl Opt 36, p.3305-3309 (1997)

[2.69] {Sect. 2.7.8} A. Ashkin, J.M. Dziedzic: Feedback stabilization of optically levitated particles, Appl. Phys. Lett. 30, p.202-204 (1977)

[2.70] {Sect. 2.7.8} A. Ashkin, J.M. Dziedzic: Optical levitation in high vacuum, Appl. Phys. Lett. 28, p.333-335 (1976)

[2.71] {Sect. 2.7.8} A. Ashkin, J.M. Dziedzic: Optical Levitation by Radiation Pressure, Appl. Phys. Lett. 19, p.283-285 (1971)

[2.72] {Sect. 2.7.8} P. Zemanek, A. Jonas, L. Sramek, M. Liska: Optical trapping of nanoparticles and microparticles by a Gaussian standing wave, Optics Letters 24, p.1448-1450 (1999)

[2.73] {Sect. 2.7.8} K.M. O'Hara, S.R. Granade, M.E. Gehm, t.A. Savard, S. Bali, C. Freed, J.E. Thomas: Ultrastable $CO2$ Laser Trapping of Lithium Fermions, Phys. Rev. Lett. 82, p.4204-4207 (1999)

[2.74] {Sect. 2.7.8} S. Chang, S.S. Lee: Optical torque exerted on a sphere in the evanescent field of a circularly-polarized Gaussian laser beam, Opt Commun 151, p.286-296 (1998)

[2.75] {Sect. 2.7.8} R.C. Gauthier, M. Ashman: Simulated dynamic behavior of single and multiple spheres in the trap region of focused laser beams, Appl Opt 37, p.6421-6431 (1998)

[2.76] {Sect. 2.7.8} T. Takekoshi, B.M. Patterson, R.J. Knize: Observation of optically trapped cold cesium molecules, Phys Rev Lett 81, p.5105-5108 (1998)

[2.77] {Sect. 2.7.8} J.P. Yin, Y.F. Zhu: Dark-hollow-beam gravito-optical atom trap above an apex of a hollow optical fibre, Opt Commun 152, p.421-428 (1998)

[2.78] {Sect. 2.7.8} P. Zemanek, A. Jonas, L. Sramek, M. Liska: Optical trapping of Rayleigh particles using a Gaussian standing wave, Opt Commun 151, p.273-285 (1998)

[2.79] {Sect. 2.7.8} T. Kuga, Y. Torii, N. Shiokawa, T. Hirano: Novel optical trap of atoms with a doughnut beam, Phys Rev Lett 78, p.4713-4716 (1997)

[2.80] {Sect. 2.7.8} T. Vanderveldt, J.F. Roch, P. Grelu, P. Grangier: Nonlinear absorption and dispersion of cold Rb 87 atoms, Opt Commun 137, p.420-426 (1997)

[2.81] {Sect. 2.7.8} W.L. Power, R.C. Thompson: Laguerre-Gaussian laser beams and ion traps, Opt Commun 132, p.371-378 (1996)

[2.82] {Sect. 2.7.8} A. Ashkin: Trapping of Atoms by Resonance Radiation Pressure, Phys. Rev. Lett. 40, p.729-732 (1978)

[2.83] {Sect. 2.7.8} M.E.J. Friese, A.G. Truscott, H. RubinszteinDunlop, N.R. Heckenberg: Three-dimensional imaging with optical tweezers, Appl Opt 38, p.6597-6603 (1999)

[2.84] {Sect. 2.7.8} M.S.Z. Kellermayer, S.B. Smith, H.L. Granzier, C. Bustamante: Folding-unfolding transitions in single titin molecules characterized with laser tweezers, Science 276, p.1112-1116 (1997)

[2.85] {Sect. 2.7.8} S. Kawata, T. Tani: Optically driven Mie particles in an evanescent field along a channeled waveguide, Optics Letters 21, p.1768-1770 (1996)

[2.86] {Sect. 2.7.8} Y. Liu, G.J. Sonek, M.W. Berns, K. Konig, B.J. Tromberg: Two-photon fluorescence excitation in continuous-wave infrared optical tweezers, Optics Letters 20, p.2246-2248 (1995)

[2.87] {Sect. 2.8.0} L. Mandel: Fluctuations of light beams. Progress in Optics 2, 181 (North Holland, Amsterdam 1963)

[2.88] {Sect. 2.8.0} G. Chirico, M. Gardella: Photon cross-correlation spectroscopy to 10-ns resolution, Appl Opt 38, p.2059-2067 (1999)

[2.89] {Sect. 2.8.0} V.P. Kozich, A.I. Vodtchits, D.A. Ivanov, V.A. Orlovich: Changing the statistical properties of noisy laser radiation in a saturable absorber, Opt Commun 169, p.97-102 (1999)

[2.90] {Sect. 2.8.0} Y.J. Qu, S. Singh, C.D. Cantrell: Measurements of higher order photon bunching of light beams, Phys Rev Lett 76, p.1236-1239 (1996)

[2.91] {Sect. 2.8.0} J.M. Raimond, P. Goy, M. Gross, C. Fabre, S. Haroche: Collective absorption of blackbody radiation by Rydberg atoms in a cavity – An Experiment on Bose statistics and Brownian motion, Phys. Rev. Lett. 49, p.117-120 (1982)

[2.92] {Sect. 2.8.4} S. Kasapi, S. Lathi, Y. Yamamoto: Sub-shot-noise frequency-modulation spectroscopy by use of amplitude- squeezed light from semiconductor losers, J Opt Soc Am B Opt Physics 17, p.275-279 (2000)

[2.93] {Sect. 2.8.4} J.R. Krenn, A. Dereux, J.C. Weeber, E. Bourillot, Y. Lacroute, J.P. Goudonnet, G. Schider, W. Gotschy, A. Leitner, F.R. Aussenegg et al.: Squeezing the optical near-field zone by plasmon coupling of metallic nanoparticles, Phys Rev Lett 82, p.2590-2593 (1999)

[2.94] {Sect. 2.8.4} D. Levandovsky, M. Vasilyev, P. Kumar: Amplitude squeezing of light by means of a phase-sensitive fiber parametric amplifier, Optics Letters 24, p.984-986 (1999)

[2.95] {Sect. 2.8.4} Y.Q. Li, D. Guzun, M. Xiao: Sub-shot-noise-limited optical heterodyne detection using an amplitude-squeezed local oscillator, Phys Rev Lett 82, p.5225-5228 (1999)

[2.96] {Sect. 2.8.4} Y.Q. Li, D. Guzun, M. Xiao: Quantum-noise measurements in high-efficiency single-pass second-harmonic generation with femtosecond pulses, Optics Letters 24, p.987-989 (1999)

[2.97]   {Sect. 2.8.4} X.M. Hu, J.S. Peng: Dynamic quantum noise reduction in a Lambda quantum-beat laser, Opt Commun 154, p.152-159 (1998)

[2.98]   {Sect. 2.8.4} Z.H. Lu, S. Bali, J.E. Thomas: Observation of squeezing in the phase-dependent fluorescence spectra of two-level atoms, Phys Rev Lett 81, p.3635-3638 (1998)

[2.99]   {Sect. 2.8.4} S. Rebic, A.S. Parkins, D.F. Walls: Transfer of photon statistics in a Raman laser, Opt Commun 156, p.426-434 (1998)

[2.100] {Sect. 2.8.4} G.M. Schucan, A.M. Fox, J.F. Ryan: Femtosecond quadrature-squeezed light generation in CdSe at 1.55 mu m, Optics Letters 23, p.712-714 (1998)

[2.101] {Sect. 2.8.4} M.S. Shahriar, P.R. Hemmer: Generation of squeezed states and twin beams via non-degenerate four- wave mixing in a Lambda system, Opt Commun 158, p.273-286 (1998)

[2.102] {Sect. 2.8.4} K.C. Peng, Q. Pan, H. Wang, Y. Zhang, H. Su, C.D. Xie: Generation of two-mode quadrature-phase squeezing and intensity-difference squeezing from a cw-NOPO, Appl. Phys. B 66, p.755-758 (1998)

[2.103] {Sect. 2.8.4} S. Kakimoto, K. Shigihara, Y. Nagai: Laser diodes in photon number squeezed state, IEEE J QE-33, p.824-830 (1997)

[2.104] {Sect. 2.8.4} S. Kasapi, S. Lathi, Y. Yamamoto: Amplitude-squeezed, frequency-modulated, tunable, diode- laser-based source for sub-shot-noise FM spectroscopy, Optics Letters 22, p.478-480 (1997)

[2.105] {Sect. 2.8.4} Y.Q. Li, P. Lynam, M. Xiao, P.J. Edwards: Sub-shot-noise laser Doppler anemometry with amplitude- squeezed light, Phys Rev Lett 78, p.3105-3108 (1997)

[2.106] {Sect. 2.8.4} J. Maeda, T. Numata, S. Kogoshi: Amplitude squeezing from singly resonant frequency- doubling laser, IEEE J QE-33, p.1057-1067 (1997)

[2.107] {Sect. 2.8.4} F. Marin, A. Bramati, V. Jost, E. Giacobino: Demonstration of high sensitivity spectroscopy with squeezed semiconductor lasers, Opt Commun 140, p.146-157 (1997)

[2.108] {Sect. 2.8.4} D.K. Serkland, P. Kumar, M.A. Arbore, M.M. Fejer: Amplitude squeezing by means of quasi-phase-matched second-harmonic generation in a lithium niobate waveguide, Optics Letters 22, p.1497-1499 (1997)

[2.109] {Sect. 2.8.4} E. Giacobino, F. Marin, A. Bramati, V. Jost: Quantum noise reduction in lasers, J Nonlinear Opt Physics Mat 5, p.863-877 (1996)

[2.110] {Sect. 2.8.4} K. Schneider, R. Bruckmeier, H. Hansen, S. Schiller, J. Mlynek: Bright squeezed-light generation by a continuous-wave semimonolithic parametric amplifier, Optics Letters 21, p.1396-1398 (1996)

[2.111] {Sect. 2.8.4} J. Kitching, A. Yariv, Y. Shevy: Room temperature generation of amplitude squeezed light from a semiconductor laser with weak optical feedback, Phys Rev Lett 74, p.3372-3375 (1995)

[2.112] {Sect. 2.8.4} J. Kitching, D. Provenzano, A. Yariv: Generation of amplitude-squeezed light from a room- temperature Fabry-Perot semiconductor laser, Optics Letters 20, p.2526-2528 (1995)

[2.113] {Sect. 2.8.4} F. Marin, A. Bramati, E. Giacobino, T.C. Zhang, J.P. Poizat, J.F. Roch, P. Grangier: Squeezing and intermode correlations in laser diodes, Phys Rev Lett 75, p.4606-4609 (1995)

[2.114] {Sect. 2.8.4} K. Bergman, C.R. Doerr, H.A. Haus, M. Shirasaki: Sub-Shot-Noise Measurement with Fiber-Squeezed Optical Pulses, Optics Letters 18, p.643-645 (1993)

[2.115] {Sect. 2.8.4} C.R. Doerr, M. Shirasaki, H.A. Haus: Dispersion of Pulsed Squeezing for Reduction of Sensor Nonlinearity, Optics Letters 17, p.1617-1619 (1992)

[2.116] {Sect. 2.8.4} D.F. Walls: Squeezed states of light, Nature 306, p.141-146 (1983)

[2.117] {Sect. 2.9.0} H. Luck, K.O. Muller, P. Aufmuth, K. Danzmann: Correction of wavefront distortions by means of thermally adaptive optics, Opt Commun 175, p.275-287 (2000)

[2.118] {Sect. 2.9.0} H.P. Ho, K.M. Leung, K.S. Chan, E.Y.B. Pun: Highly stable differential phase optical interferometer using rotating Ronchi gratings, Appl Opt 37, p.3494-3497 (1998)

[2.119] {Sect. 2.9.0} J.Y. Lee, D.C. Su: High resolution central fringe identification, Opt Commun 156, p.1-4 (1998)

[2.120] {Sect. 2.9.0} A. Araya, N. Mio, K. Tsubono, K. Suehiro, S. Telada, M. Ohashi, M.K. Fujimoto: Optical mode cleaner with suspended mirrors, Appl Opt 36, p.1446-1453 (1997)

[2.121] {Sect. 2.9.0} H. Welling, B. Wellegehausen: High Resolution Michelson Interferometer for Spectral Investigations of Lasers, Appl. Opt. 11, p.1986-1990 (1972)

[2.122] {Sect. 2.9.0} D.A. Shaddock, M.B. Gray, D.E. McClelland: Experimental demonstration of resonant sideband extraction in a Sagnac interferometer, Appl Opt 37, p.7995-8001 (1998)

[2.123] {Sect. 2.9.1} L. Gallmann, D.H. Sutter, N. Matuschek, G. Steinmeyer, U. Keller, C. Iaconis, I.A. Walmsley: Characterization of sub-6-fs optical pulses with spectral phase interferometry for direct electric-field reconstruction, Optics Letters 24, p.1314-1316 (1999)

[2.124] {Sect. 2.9.1} S. Leute, T. Lottermoser, D. Frohlich: Nonlinear spatially resolved phase spectroscopy, Optics Letters 24, p.1520-1522 (1999)

[2.125] {Sect. 2.9.1} A.M. Rollins, J.A. Izatt: Optimal interferometer designs for optical coherence tomography, Optics Letters 24, p.1484-1486 (1999)

[2.126] {Sect. 2.9.1} P.T. Wilson, Y. Jiang, O.A. Aktsipetrov, E.D. Mishina, M.C. Downer: Frequency-domain interferometric second-harmonic spectroscopy, Optics Letters 24, p.496-498 (1999)

[2.127] {Sect. 2.9.1} D. Braun, P. Fromherz: Fluorescence interferometry of neuronal cell adhesion on microstructured silicon, Phys Rev Lett 81, p.5241-5244 (1998)

[2.128] {Sect. 2.9.1} W.D. Zhou, L.L. Cai: Optical readout for optical storage with phase jump, Appl Opt 38, p.5058-5065 (1999)

[2.129] {Sect. 2.9.1} D.J. Ulness, M.J. Stimson, A.C. Albrecht: High-contrast interferometry based on anti-Stokes stimulated Raman scattering with broadband and narrow-band quasi-continuous-wave laser light, Optics Letters 22, p.433-435 (1997)

[2.130] {Sect. 2.9.1} J.L.A. Chilla, J.J. Rocca, O.E. Martinez, M.C. Marconi: Soft-x-ray interferometer for single-shot laser linewidth measurements, Optics Letters 21, p.955-957 (1996)

[2.131] {Sect. 2.9.2} L. Mandel, E. Wolf: Coherence properties of optical fields, Rev. Mod. Phys. 37, p.271 (1965)

[2.132] {Sect. 2.9.2} R.F. Wuerker, J. Munch, L.O. Heflinger: Coherence length measured directly by holography, Appl. Opt. 28, p.1015-1017 (1989)

[2.133] {Sect. 2.9.2} E. Fischer, E. Dalhoff, H. Tiziani: Overcoming coherence length limitation in two wavelength interferometry – An experimental verification, Opt Commun 123, p.465-472 (1996)

[2.134] {Sect. 2.9.2} C.C. Cheng, M.G. Raymer: Long-range saturation of spatial decoherence in wave-field transport in random multiple-scattering media, Phys Rev Lett 82, p.4807-4810 (1999)

[2.135] {Sect. 2.9.7} G. Mueller, Q.Z. Shu, R. Adhikari, D.B. Tanner, D. Reitze, D. Sigg, N. Mavalvala, J. Camp: Determination and optimization of mode matching into optical cavities by heterodyne detection, Optics Letters 25, p.266-268 (2000)

[2.136] {Sect. 2.9.7} J.Y. Lee, D.C. Su: Common-path heterodyne interferometric detection scheme for measuring wavelength shift, Opt Commun 162, p.7-10 (1999)

[2.137] {Sect. 2.9.7} C.M. Wu, J. Lawall, R.D. Deslattes: Heterodyne interferometer with subatomic periodic nonlinearity, Appl Opt 38, p.4089-4094 (1999)

[2.138] {Sect. 2.9.7} S. Yoon, Y. Lee, K. Cho: Intermode beat heterodyne sensor scheme for mapping optical properties of optical media, Opt Commun 161, p.182-186 (1999)

[2.139] {Sect. 2.9.7} C. Chou, C.Y. Han, W.C. Kuo, Y.C. Huang, C.M. Feng, J.C. Shyu: Noninvasive glucose monitoring in vivo with an optical heterodyne polarimeter, Appl Opt 37, p.3553-3557 (1998)

[2.140] {Sect. 2.9.7} H. Ludvigsen, M. Tossavainen, M. Kaivola: Laser linewidth measurements using self-homodyne detection with short delay, Opt Commun 155, p.180-186 (1998)

[2.141] {Sect. 2.9.7} G.Y. Lyu, S.S. Lee, D.H. Lee, C.S. Park, M.H. Kang, K. Cho: Simultaneous measurement of multichannel laser linewidths and spacing by use of stimulated Brillouin scattering in optical fiber, Optics Letters 23, p.873-875 (1998)

[2.142] {Sect. 2.9.7} S.A. Shen, T. Liu, J.H. Guo: Optical phase-shift detection of surface plasmon resonance, Appl Opt 37, p.1747-1751 (1998)

[2.143] {Sect. 2.9.7} M.J. Snadden, R.B.M. Clarke, E. Riis: FM spectroscopy in fluorescence in laser-cooled rubidium, Opt Commun 152, p.283-288 (1998)

[2.144] {Sect. 2.9.7} J.T. Hoffges, H.W. Baldauf, T. Eichler, S.R. Helmfrid, H. Walther: Heterodyne measurement of the fluorescent radiation of a single trapped ion, Opt Commun 133, p.170-174 (1997)

[2.145] {Sect. 2.9.7} S. Matsuo, T. Tahara: Phase-stabilized optical heterodyne detection of impulsive stimulated Raman scattering, Chem Phys Lett 264, p.636-642 (1997)

[2.146] {Sect. 2.9.7} M. Pitter, E. Jakeman, M. Harris: Heterodyne detection of enhanced backscatter, Optics Letters 22, p.393-395 (1997)

[2.147] {Sect. 2.9.7} K.X. Sun, M.M. Fejer, E.K. Gustafson, R.L. Byer: Balanced heterodyne signal extraction in a postmodulated Sagnac interferometer at low frequency, Optics Letters 22, p.1485-1487 (1997)

[2.148] {Sect. 2.9.7} R. Onodera, Y. Ishii: Effect of beat frequency on the measured phase of laser- diode heterodyne interferometry, Appl Opt 35, p.4355-4360 (1996)

[2.149] {Sect. 2.9.7} R. Onodera, Y. Ishii: Two-wavelength laser-diode heterodyne interferometry with one phasemeter, Optics Letters 20, p.2502-2504 (1995)

3. Linear Interactions Between Light and Matter

[3.1]   {Sect. 3.2} M.H. Chiu, J.Y. Lee, D.C. Su: Complex refractive-index measurement based on Fresnel's equations and the uses of heterodyne interferometry, Appl Opt 38, p.4047-4052 (1999)

[3.2]   {Sect. 3.2} S.M. Mian, A.Y. Hamad, J.P. Wicksted: Refractive index measurements using a CCD, Appl Opt 35, p.6825-6826 (1996)

[3.3]   {Sect. 3.2} Y.P. Zhang, R. Kachru: Photon-echo novelty filter for measuring a sudden change in index of refraction, Appl Opt 35, p.6762-6766 (1996)

[3.4]   {Sect. 3.2} Y. Wang, Y. Abe, Y. Matsuura, M. Miyagi, H. Uyama: Refractive indices and extinction coefficients of polymers for the mid-infrared region, Appl Opt 37, p.7091-7095 (1998)

[3.5]   {Sect. 3.2} M.J. Weber (ed.): CRC Handbook of Laser Science and Technology, Vol. IV-Optical Materials (CRC Press, Inc, Boca Raton, 1986)

[3.6]   {Sect. 3.3.1} R.C. Hilborn: Einstein coefficients, cross sections, f values, dipole moments, and all that, Am. J. Phys.50, p.982-986 (1982)

[3.7]   {Sect. 3.3.1} M.C.E. Huber, R.J. Sandeman: The measurement of oscillator strengths, Rep. Progr. Phys. 49, p.397-490 (1986)

[3.8]   {Sect. 3.3.3} K. Yasui: Single-bubble sonoluminescence from hydrogen, J Chem Phys 111, p.5384-5389 (1999)

[3.9]   {Sect. 3.3.3} J. Holzfuss, M. Ruggeberg, A. Billo: Shock wave emissions of a sonoluminescing bubble, Phys Rev Lett 81, p.5434-5437 (1998)

[3.10] {Sect. 3.3.3} J.R. Willison: Sonoluminescence: Proton-tunneling radiation, Phys Rev Lett 81, p.5430-5433 (1998)

[3.11] {Sect. 3.3.4} T. Renger, V. May: Multiple exciton effects in molecular aggregates: Application to a photosynthetic antenna complex, Phys Rev Lett 78, p.3406-3409 (1997)

[3.12] {Sect. 3.3.4} S. Savikhin, D.R. Buck, W.S. Struve: Oscillating anisotropies in a bacteriochlorophyll protein: Evidence for quantum beating between exciton levels, Chem Phys 223, p.303-312 (1997)

[3.13] {Sect. 3.3.4} M. Joffre, D. Hulin, A. Migus, A. Antonietti, C. Benoit à la Guillaume, N. Peyghambarian, M. Lindberg, S.W. Koch: Coherent effects in pump-probe spectroscopy of excitons, Opt. Lett. 13, p.276-278 (1988)

[3.14] {Sect. 3.3.4} E. Morikawa, K. Shikichi, R. Katoh, M. Kotani: Transient photoabsorption by singlet excitons in p-terphenyl single crystals, Chem. Phys. Lett. 131, p.209-212 (1986)

[3.15] {Sect. 3.3.4} W.T. Simpson, D.L. Peterson: Coupling Strength for Resonance Force Transfer of Electronic Energy in Van der Waals Solids, J. Chem. Phys. 26, p.588-593 (1957)

[3.16] {Sect. 3.3.4} J. R. Lakowicz: Principles of Fluorescence Spectroscopy (Plenum Press, New York, London, 1983)

[3.17] {Sect. 3.3.4} Th. Förster: Zwischenmolekulare Energiewanderung und Fluoreszenz, Ann. Phys. 6, p.55-75 (1948)

[3.18] {Sect. 3.4.3} H. Bach, N. Neuroth (eds.): The Properties of Optical Glass (Springer, Berlin, Heidelberg, New York, 1998)

[3.19] {Sect. 3.4.3} D. N. Nikogosyan: Properties of Optical and Laser-Related Materials – A Handbook (John Wiley & Sons, Chichester, 1997)

[3.20] {Sect. 3.4.3} H. Hosono, M. Mizuguchi, L. Skuja, T. Ogawa: Fluorine-doped SiO2 glasses for F-2 excimer laser optics: fluorine content and color-center formation, Optics Letters 24, p.1549-1551 (1999)

[3.21] {Sect. 3.4.3} V. Liberman, M. Rothschild, J.H.C. Sedlacek, R.S. Uttaro, A. Grenville, A.K. Bates, C. VanPeski: Excimer-laser-induced degradation of fused silica and calcium fluoride for 193-nm lithographic applications, Optics Letters 24, p.58-60 (1999)

[3.22] {Sect. 3.5.0} M.H. Chiu, J.Y. Lee, D.C. Su: Complex refractive-index measurement based on Fresnel's equations and the uses of heterodyne interferometry, Appl Opt 38, p.4047-4052 (1999)

[3.23] {Sect. 3.5.3} B.A. Hooper, Y. Domankevitz, C.P. Lin, R.R. Anderson: Precise, controlled laser delivery with evanescent optical waves, Appl Opt 38, p.5511-5517 (1999)

[3.24] {Sect. 3.5.3} A.C.R. Pipino: Ultrasensitive surface spectroscopy with a miniature optical resonator, Phys Rev Lett 83, p.3093-3096 (1999)

[3.25] {Sect. 3.5.3} S. Chang, S.S. Lee: Optical torque exerted on a sphere in the evanescent field of a circularly-polarized Gaussian laser beam, Opt Commun 151, p.286-296 (1998)

[3.26] {Sect. 3.5.3} H. Gauck, M. Hartl, D. Schneble, H. Schnitzler, T. Pfau, J. Mlynek: Quasi-2D gas of laser cooled atoms in a planar matter waveguide, Phys Rev Lett 81, p.5298-5301 (1998)

[3.27] {Sect. 3.5.3} V.G. Bordo, C. Henkel, A. Lindinger, H.G. Rubahn: Evanescent wave fluorescence spectra of Na atoms, Opt Commun 137, p.249-253 (1997)

[3.28] {Sect. 3.5.3} X.H. Xu, E.S. Yeung: Direct measurement of single-molecule diffusion and photodecomposition in free solution, Science 275, p.1106-1109 (1997)

[3.29] {Sect. 3.5.3} R.H. Renard: Total Reflection: A New Evaluation of the Goos-Hänchen Shift, J. Opt. Soc. Am. 54, p.1190-1197 (1964)

[3.30] {Sect. 3.6} S. Walheim, E. Schaffer, J. Mlynek, U. Steiner: Nanophase-separated polymer films as high-performance antireflection coatings, Science 283, p.520-522 (1999)

[3.31] {Sect. 3.6} F. Loewenthal, R. Tommasini, J.E. Balmer: Single-shot measurement of laser-induced damage thresholds of thin film coatings, Opt Commun 152, p.168-174 (1998)

[3.32] {Sect. 3.6} Y.A. Uspenskii, V.E. Levashov, A.V. Vinogradov, A.I. Fedorenko, V.V. Kondratenko, Y.P. Pershin, E.N. Zubarev, V.Y. Fedotov: High-reflectivity multilayer mirrors for a vacuum-ultraviolet interval of 35-50 nm, Optics Letters 23, p.771-773 (1998)

[3.33] {Sect. 3.6} S.M. Xiong, Y.D. Zhang: Optical coatings for deuterium fluoride chemical laser systems, Appl Opt 36, p.4958-4961 (1997)

[3.34] {Sect. 3.6} G. Emiliani, A. Piegari, S. De Silvestri, P. Laporta, V. Magni: Optical coatings with variable reflectance for laser mirrors, Appl. Opt. 28, p.2832-2837 (1989)

[3.35] {Sect. 3.7} T.D. Goodman, M. Mansuripur: Subtle effects of the substrate in optical disk data storage systems, Appl Opt 35, p.6747-6753 (1996)

[3.36] {Sect. 3.7} Z.X. Shao: Precise and versatile formula for birefringent filters, Appl Opt 35, p.4147-4151 (1996)

[3.37] {Sect. 3.7} J.F. deBoer, T.E. Milner, M.J.C. Vangemert, J.S. Nelson: Two-dimensional birefringence imaging in biological tissue by polarization-sensitive optical coherence tomography, Optics Letters 22, p.934-936 (1997)

[3.38] {Sect. 3.7} F.S. Pavone, G. Bianchini, F.S. Cataliotti, T.W. Hansch, M. Inguscio: Birefringence in electromagnetically induced transparency, Optics Letters 22, p.736-738 (1997)

[3.39] {Sect. 3.8} E. Khazanov, N. Andreev, A. Babin, A. Kiselev, O. Palashov, D.H. Reitze: Suppression of self-induced depolarization of high-power laser radiation in glass-based Faraday isolators, J Opt Soc Am B Opt Physics 17, p.99-102 (2000)

[3.40] {Sect. 3.8} P. Denatale, L. Gianfrani, S. Viciani, M. Inguscio: Spectroscopic observation of the Faraday effect in the far infrared, Optics Letters 22, p.1896-1898 (1997)

[3.41] {Sect. 3.8} Y. Horovitz, S. Eliezer, A. Ludmirsky, Z. Henis, E. Moshe, R. Shpitalnik, B. Arad: Measurements of inverse Faraday effect and absorption of circularly polarized laser light in plasmas, Phys Rev Lett 78, p.1707-1710 (1997)

[3.42] {Sect. 3.8} T. Verbiest, M. Kauranen, A. Persoons: Light-polarization-induced optical activity, Phys Rev Lett 82, p.3601-3604 (1999)

[3.43] {Sect. 3.8} E. Westin, S. Wabnitz, R. Frey, C. Flytzanis: Polarization flip-flop operation and dissipative structure generation with nonlinear gyrotropic resonators, Opt Commun 158, p.97-100 (1998)

[3.44] {Sect. 3.9.1} P. Baues: Huygens' Principle in Inhomogeneous, Isotropic Media and a General Integral Equation Applicable to Optical Resonators, Opto-Electr. 1, p.37-44 (1969)

[3.45] {Sect. 3.9.1} J.E. Harvey, C.L. Vernold, A. Krywonos, P.L. Thompson: Diffracted radiance: a fundamental quantity in nonparaxial scalar diffraction theory, Appl Opt 38, p.6469-6481 (1999)

[3.46] {Sect. 3.9.1} Y. Takaki, H. Ohzu: Fast numerical reconstruction technique for high-resolution hybrid holographic microscopy, Appl Opt 38, p.2204-2211 (1999)

[3.47] {Sect. 3.9.1} J.X. Pu, H.H. Zhang, S. Nemoto: Spectral shifts and spectral switches of partially coherent light passing through an aperture, Opt Commun 162, p.57-63 (1999)

[3.48] {Sect. 3.9.1} W.P. Huang, C.L. Xu: A Wide-Angle Vector Beam Propagation Method, IEEE Photonics Technol. Lett. 4, p.1118-1120 (1992)

[3.49] {Sect. 3.9.2} C.J.R. Sheppard, P. Torok: Dependence of focal shift on Fresnel number and angular aperture, Optics Letters 23, p.1803-1804 (1998)

[3.50] {Sect. 3.9.3} S.A. Collins: Lens-System Diffraction Integral Written in Terms of Matrix Opitcs, J. Opt. Soc. Am. 60, p.1168-1177 (1970)

[3.51] {Sect. 3.9.6} Z.P. Jiang, R. Jacquemin, W. Eberhardt: Time dependence of Fresnel diffraction of ultrashort laser pulses by a circular aperture, Appl Opt 36, p.4358-4361 (1997)

[3.52] {Sect. 3.9.8} C.T. Hsieh, C.K. Lee: Cylindrical-type nanometer-resolution laser diffractive optical encoder, Appl Opt 38, p.4743-4750 (1999)

[3.53] {Sect. 3.9.8} G. Andersen, J. Munch, P. Veitch: Compact, holographic correction of aberrated telescopes, Appl Opt 36, p.1427-1432 (1997)

[3.54] {Sect. 3.9.8} I. Leiserson, S.G. Lipson, V. Sarafis: Superresolution in far-field imaging, Optics Letters 25, p.209-211 (2000)

[3.55] {Sect. 3.9.8} F. Dorchies, J.R. Marques, B. Cros, G. Matthieussent, C. Courtois, T. Velikoroussov, P. Audebert, J.P. Geindre, S. Rebibo, G. Hamoniaux et al.: Monomode guiding of 10 (16) W/cm (2) laser pulses over 100 Rayleigh lengths in hollow capillary dielectric tubes, Phys Rev Lett 82, p.4655-4658 (1999)

[3.56] {Sect. 3.9.8} M.K. Lewis, P. Wolanin, A. Gafni, D.G. Steel: Near-field scanning optical microscopy of single molecules by femtosecond two-photon excitation, Optics Letters 23, p.1111-1113 (1998)

[3.57] {Sect. 3.9.8} J. Tominaga, T. Nakano, N. Atoda: An approach for recording and readout beyond the diffraction limit with an Sb thin film, Appl Phys Lett 73, p.2078-2080 (1998)

[3.58] {Sect. 3.9.8} A. vonPfeil, B. Messerschmidt, V. Blumel, T. Possner: Making fast cylindrical gradient-index lenses diffraction limited by using a wave-front-correction element, Appl Opt 37, p.5211-5215 (1998)

[3.59] {Sect. 3.9.8} W.H. Yeh, L.F. Li, M. Mansuripur: Vector diffraction and polarization effects in an optical disk system, Appl Opt 37, p.6983-6988 (1998)

[3.60] {Sect. 3.9.8} A. Yoshida, T. Asakura: Propagation and focusing of Gaussian laser beams beyond conventional diffraction limit, Opt Commun 123, p.694-704 (1996)

[3.61] {Sect. 3.9.8} M. A. Paesler, P. J. Moyer: Near-Field Optics (John Wiley & Sons, Chichester, 1996)

[3.62] {Sect. 3.9.9} B.T. Teipen, D.L. MacFarlane: Modulation transfer function measurements of microjetted microlenses, Appl Opt 38, p.2040-2046 (1999)

[3.63] {Sect. 3.9.9} O. Hadar, A. Dogariu, G.D. Boreman: Angular dependence of sampling modulation transfer function, Appl Opt 36, p.7210-7216 (1997)

[3.64] {Sect. 3.9.9} S. Makki, Z. Wang, J.R. Leger: Laser beam relaying with phase-conjugate diffractive optical elements, Appl Opt 36, p.4749-4755 (1997)

[3.65] {Sect. 3.9.10} M.W. Noel, C.R. Stroud: Young's double-slit interferometry within an atom, Phys Rev Lett 75, p.1252-1255 (1995)

[3.66] {Sect. 3.9.13} K.X. He, M. Curley, A. Williams, J.C. Wang: Visible light diffraction by a monolayer periodic array of UV laser dye Bis-MSB doped polystyrene spheres, Opt Commun 139, p.39-42 (1997)

[3.67] {Sect. 3.9.16} M.A. Muriel, A. Carballar, J. Azana: Field distributions inside fiber gratings, IEEE J QE-35, p.548-558 (1999)

[3.68] {Sect. 3.9.16} N.C.R. Holme, L. Nikolova, P.S. Ramanujam, S. Hvilsted: An analysis of the anisotropic and topographic gratings in a side-chain liquid crystalline azobenzene polyester, Appl Phys Lett 70, p.1518-1520 (1997)

[3.69] {Sect. 3.9.16} G.I. Greisukh, S.T. Bobrov, S.A. Stepanov: Optics of Diffractive and Gradient-Index Elements and Systems (SPIE Optical Engineering Press, Bellingham, 1997); J. Turunen, F.Wyrowski: Diffractive Optics for Industrial and Commercial Applications (Akademie Verlag, Berlin, 1997)

[3.70] {Sect. 3.9.16} H.J. Eichler, P. Günter, D.W. Pohl: Laser-Induced Dynamic Gratings, Springer Ser. Opt. Sci, Vol. 50 (Springer, Berlin, Heidelberg, New York, Tokyo 1986)

[3.71] {Sect. 3.10.0} F. Rachet, M. Chrysos, C. GuillotNoel, Y. LeDuff: Unique case of highly polarized collision-induced light scattering: The very far spectral wing by the helium pair, Phys Rev Lett 84, p.2120-2123 (2000)

[3.72] {Sect. 3.10.0} R.L. Murry, J.T. Fourkas, W.X. Li, T. Keyes: Mechanisms of light scattering in supercooled liquids, Phys Rev Lett 83, p.3550-3553 (1999)

[3.73] {Sect. 3.10.0} G.N. Constantinides, D. Gintides, S.E. Kattis, K. Kiriaki, C.A. Paraskeva, A.C. Payatakes, D. Polyzos, S.V. Tsinopoulos, S.N. Yannopoulos: Computation of light scattering by axisymmetric nonspherical particles and comparison with experimental results, Appl Opt 37, p.7310-7319 (1998)

[3.74] {Sect. 3.10.0} D.D. Meyerhofer: High-intensity-laser-electron scattering, IEEE J QE-33, p.1935-1941 (1997)

[3.75] {Sect. 3.10.0} F.V. Hartemann, A.K. Kerman: Classical theory of nonlinear compton scattering, Phys Rev Lett 76, p.624-627 (1996)

[3.76] {Sect. 3.10.0} A. Kienle, M.S. Patterson, L. Ott, R. Steiner: Determination of the scattering coefficient and the anisotropy factor from laser Doppler spectra of liquids including blood, Appl Opt 35, p.3404-3412 (1996)

[3.77] {Sect. 3.10.0} J.D. McKinney, M.A. Webster, K.J. Webb, A.M. Weiner: Characterization and imaging in optically scattering media by use of laser speckle and a variable-coherence source, Optics Letters 25, p.4-6 (2000)

[3.78] {Sect. 3.10.0} L.L. Gurdev, T.N. Dreischuh, D.V. Stoyanov: Pulse backscattering tomography based on lidar principle, Opt Commun 151, p.339-352 (1998)

[3.79] {Sect. 3.10.0} R. Weber, G. Schweiger: Photon correlation spectroscopy on flowing polydisperse fluid-particle systems: theory, Appl Opt 37, p.4039-4050 (1998)

[3.80] {Sect. 3.10.0} G.L. Fischer, R.W. Boyd, T.R. Moore, J.E. Sipe: Nonlinear-optical Christiansen filter as an optical power limiter, Optics Letters 21, p.1643-1645 (1996)

[3.81] {Sect. 3.10.0} D.B. Brayton: Small Particle Signal Characteristics of a Dual-Scatter Laser Velocimeter, Appl. Opt. 13, p.2346-2351 (1974)

[3.82] {Sect. 3.10.1} H. Naus, W. Ubachs: Experimental verification of Rayleigh scattering cross sections, Optics Letters 25, p.347-349 (2000)

[3.83] {Sect. 3.10.1} F. Benabid, M. Notcutt, L. Ju, D.G. Blair: Rayleigh scattering in sapphire test mass for laser interferometric gravitational-wave detectors: II: Rayleigh scattering induced noise in a laser interferometric-wave detector, Opt Commun 170, p.9-14 (1999)

[3.84] {Sect. 3.10.1} J.I. Dadap, J. Shan, K.B. Eisenthal, T.F. Heinz: Second-harmonic Rayleigh scattering from a sphere of centrosymmetric material, Phys Rev Lett 83, p.4045-4048 (1999)

[3.85] {Sect. 3.10.1} C.C. Hsu, T.H. Huang, S. Liu, F.F. Yeh, B.Y. Jin, J.A. Sattigeri, C.W. Shiau, T.Y. Luh: Conformation of substituted poly-norbornene polymers studied by hyper- Rayleigh scattering at 1064 nm, Chem Phys Lett 311, p.355-361 (1999)

[3.86] {Sect. 3.10.1} R.H.C. Janssen, D.N. Theodorou, S. Raptis, M.G. Papadopoulos: Molecular simulation of static hyper-Rayleigh scattering: A calculation of the depolarization ratio and the local fields for liquid nitrobenzene, J Chem Phys 111, p.9711-9719 (1999)

[3.87] {Sect. 3.10.1} P. Kaatz, D.P. Shelton: Two-photon fluorescence cross-section measurements calibrated with hyper-Rayleigh scattering, J Opt Soc Am B Opt Physics 16, p.998-1006 (1999)

[3.88] {Sect. 3.10.1} J.N. Woodford, C.H. Wang, A.E. Asato, R.S.H. Liu: Hyper-Rayleigh scattering of azulenic donor-acceptor molecules at 1064 and 1907 nm, J Chem Phys 111, p.4621-4628 (1999)

[3.89] {Sect. 3.10.1} S.N. Yaliraki, R.J. Silbey: Hyper-Rayleigh scattering of centrosymmetric molecules in solution, J Chem Phys 111, p.1561-1568 (1999)

[3.90] {Sect. 3.10.1} S. Inouye, A.P. Chikkatur, D.M. StamperKurn, J. Stenger, D.E. Pritchard, W. Ketterle: Superradiant Rayleigh scattering from a Bose-Einstein condensate, Science 285, p.571-574 (1999)

[3.91] {Sect. 3.10.1} M. Froggatt, J. Moore: High-spatial-resolution distributed strain measurement in optical fiber with Rayleigh scatter, Appl Opt 37, p.1735-1740 (1998)

[3.92] {Sect. 3.10.1} B.W.J. McNeil, G.R.M. Robb: Collective Rayleigh scattering from dielectric particles: a classical theory of the collective atomic recoil laser, Opt Commun 148, p.54-58 (1998)

[3.93] {Sect. 3.10.1} C. Desmet, V. Gusev, W. Lauriks, C. Glorieux, J. Thoen: All-optical excitation and detection of leaky Rayleigh waves, Optics Letters 22, p.69-71 (1997)

[3.94] {Sect. 3.10.1} S.F. Hubbard, R.G. Petschek, K.D. Singer: Spectral content and dispersion of hyper-Rayleigh scattering, Optics Letters 21, p.1774-1776 (1996)

[3.95] {Sect. 3.10.1} O.F.J. Noordman, N.F. Vanhulst: Time-resolved hyper-Rayleigh scattering: Measuring first hyperpolarizabilities beta of fluorescent molecules, Chem Phys Lett 253, p.145-150 (1996)

[3.96] {Sect. 3.10.1} S.L. Shapiro, H.P. Broida: Light Scattering from Fluctuations in Orientations of CS2 in Liquids, Phys. Rev. 154, p.129-138 (1967)

[3.97] {Sect. 3.10.1} I.P. Batra, R.H. Enns: Stimulated Thermal Rayleigh Scattering in Liquids, Phys. Rev. 185, p.396-399 (1969)

[3.98] {Sect. 3.10.2} M. Bass (ed.): Handbook of Optics, Vol. I, chapter 44 (McGraw-Hill, New York, 1995)

[3.99] {Sect. 3.10.2} M. Alexander, F.R. Hallett: Small-angle light scattering: instrumental design and application to particle sizing, Appl Opt 38, p.4158-4163 (1999)

[3.100] {Sect. 3.10.2} I. Delfino, M. Lepore, P.L. Indovina: Experimental tests of different solutions to the diffusion equation for optical characterization of scattering media by time-resolved transmittance, Appl Opt 38, p.4228-4236 (1999)

[3.101] {Sect. 3.10.2} N.M. Sijtsema, R.A.L. Tolboom, N.J. Dam, J.J. terMeulen: Two-dimensional multispecies imaging of a supersonic nozzle flow, Optics Letters 24, p.664-666 (1999)

[3.102] {Sect. 3.10.2} M. Hammer, D. Schweitzer, B. Michel, E. Thamm, A. Kolb: Single scattering by red blood cells, Appl Opt 37, p.7410-7418 (1998)

[3.103] {Sect. 3.10.2} M. Quinten, A. Leitner, J.R. Krenn, F.R. Aussenegg: Electromagnetic energy transport via linear chains of silver nanoparticles, Optics Letters 23, p.1331-1333 (1998)

[3.104] {Sect. 3.10.2} A. Doicu, T. Wriedt: Computation of the beam-shape coefficients in the generalized Lorenz-Mie theory by using the translational addition theorem for spherical vector wave functions, Appl Opt 36, p.2971-2978 (1997)

[3.105] {Sect. 3.10.2} Z.L. Jiang: Phase maps based on the Lorenz-Mie theory to optimize phase Doppler particle-sizing systems, Appl Opt 36, p.1367-1375 (1997)

[3.106] {Sect. 3.10.2} J. Kasparian, B. Kramer, J.P. Dewitz, S. Vajda, P. Rairoux, B. Vezin, V. Boutou, T. Leisner, W. Hubner, J.P. Wolf, et al.: Angular dependences of third harmonic generation from microdroplets, Phys Rev Lett 78, p.2952-2955 (1997)

[3.107] {Sect. 3.10.2} G. Mie: Beiträge zur Optik trüber Medien, speziell kolloidaler Metallösungen, Ann. Phys. 25, p.377-444 (1908)

[3.108] {Sect. 3.10.4} P. J. Hendra, J.K. Agbenyega: The Raman Spectra of Polymers (John Wiley & Sons, Chichester, 1994)

[3.109] {Sect. 3.10.4} D. Lin-Vien, N. B. Colthup, W. G. Fateley, J. G. Grasselli: The Handbook of Infrared and Raman Characteristic Frequencies of Organic Molecules (Academic Press, Boston, San Diego, New York, 1991)

[3.110] {Sect. 3.10.4} D.J. Gardiner, P.R. Grawes: Practical Raman Spectroscopy (Springer, Berlin, Heidelberg 1989)

[3.111] {Sect. 3.10.4} R.J.H.Clark, R.E. Hester (eds.): Advances in Infrared and Raman Spectroscopy, Vols. 1-10 (Heyden, London 1972-1985)

[3.112] {Sect. 3.10.4} A. Wehr: High-resolution rotational Raman Spectra of gases (in A. Weber (ed.): Raman Spectroscopy of Gases and Liquids, Topics Curr. Phys, Vol. 11 (Springer Berlin, Heidelberg 1979)

[3.113] {Sect. 3.10.4} G. Herzberg: Molecular Spectra and Molecular Structure II. Infrared and Raman Spectra (Van Nostrand Reinhold, New York, 1945)

[3.114] {Sect. 3.10.4} A.A. Sirenko, I.A. Akimov, J.R. Fox, A.M. Clark, H.C. Li, W.D. Si, X.X. Xi: Observation of the first-order Raman scattering in SrTiO3 thin films, Phys Rev Lett 82, p.4500-4503 (1999)

[3.115] {Sect. 3.10.4} N.V. Surovtsev, J. Wiedersich, V.N. Novikov, E. Rossler, E. Duval: q dependence of low-frequency Raman scattering in silica glass, Phys Rev Lett 82, p.4476-4479 (1999)

[3.116] {Sect. 3.10.4} K. Wakabayashi, K.G. Nakamura, K. Kondo, M. Yoshida: Time-resolved Raman spectroscopy of polytetrafluoroethylene under laser-driven shock compression, Appl Phys Lett 75, p.947-949 (1999)

[3.117] {Sect. 3.10.4} C. Didierjean, V. DeWaele, G. Buntinx, O. Poizat: The structure of the lowest excited singlet (S-1) state of 4,4'-bipyridine: a picosecond time-resolved Raman analysis, Chem Phys 237, p.169-181 (1998)

[3.118] {Sect. 3.10.4} F. Rabenstein, A. Leipertz: One-dimensional, time-resolved Raman measurements in a sooting flame made with 355-nm excitation, Appl Opt 37, p.4937-4943 (1998)

[3.119] {Sect. 3.10.4} X.F. Wang, R. Fedosejevs, G.D. Tsakiris: Observation of Raman scattering and hard X-rays in short pulse laser interaction with high density hydrogen gas, Opt Commun 146, p.363-370 (1998)

[3.120] {Sect. 3.10.4} H. Huang, S.Q. Li: Vibrational Raman spectrum of a degenerate Boson gas, Opt Commun 144, p.331-339 (1997)

[3.121] {Sect. 3.10.4} E. Takahashi, Y. Matsumoto, K. Kuwahara, I. Matsushima, I. Okuda, Y. Owadano: Short Stokes pulse generation by mixed Raman gas, Opt Commun 136, p.429-432 (1997)

[3.122] {Sect. 3.10.4} K. van Helvoort, R. Fantoni, W.L. Meerts, J. Reuss: Internal rotation in CH3CD3: Raman spectroscopy of torsional overtones, Chem. Phys. Lett. 128, p.494-500 (1986)

[3.123] {Sect. 3.10.4} W. Knippers, K. Van Helvoort, S. Stolte: Vibrational overtones of the homonuclear diatomics N2, O2, D2 observed by the spontaneous Raman effect, Chem. Phys. Lett 121, p.279-286 (1985)

[3.124] {Sect. 3.10.4} H. W. Schrötter, J. Bofilias: On the assignment of the second-order lines in the Raman spectrum of benzene, J. Mol. Struct. 3, p.242-244 (1969)

[3.125] {Sect. 3.10.4} M. Katsuragawa, K. Hakuta: Raman gain measurement in solid parahydrogen, Optics Letters 25, p.177-179 (2000)

[3.126] {Sect. 3.10.4} S. Hadrich, S. Hefter, B. Pfelzer, T. Doerk, P. Jauernik, J. Uhlenbusch: Determination of the absolute Raman cross section of methyl, Chem Phys Lett 256, p.83-86 (1996)

[3.127] {Sect. 3.10.4} N.D. Finkelstein, A.P. Yalin, W.R. Lempert, R.B. Miles: Dispersion filter for spectral and spatial resolution of pure rotational Raman scattering, Optics Letters 23, p.1615-1617 (1998)

[3.128] {Sect. 3.10.4} H. Yamamoto, H. Uenoyama, K. Hirai, X. Dou, Y. Ozaki: Quantitative analysis of metabolic gases by multichannel Raman spectroscopy: use of a newly designed elliptic-spherical integration type of cell holder, Appl Opt 37, p.2640-2645 (1998)

[3.129] {Sect. 3.10.4} J. Bendtsen, F. Rasmussen, S. Brodersen: Fourier-transform instrument for high-resolution Raman spectroscopy of gases, Appl Opt 36, p.5526-5534 (1997)

[3.130] {Sect. 3.10.4} N.D. Finkelstein, W.R. Lempert, R.B. Miles: Narrow-linewidth passband filter for ultraviolet rotational Raman imaging, Optics Letters 22, p.537-539 (1997)

[3.131] {Sect. 3.10.4} D.F. Marran, J.H. Frank, M.B. Long, S.H. Starner, R.W. Bilger: Intracavity technique for improved Raman/Rayleigh imaging in flames, Optics Letters 20, p.791-793 (1995)

[3.132] {Sect. 3.10.4} B. Schrader: Special techniques and applications, in Infrared and Raman Spectroscopy (VCH, Weinheim 1993)

4. Nonlinear Interactions of Light and Matter Without Absorption

[4.1] {Sect. 4.3} C.Y. Fong, Y.R. Shen: Theoretical studies on the dispersion of the nonlinear optical susceptibilities in GaAs, InAs, and InSb, Phys. Rev. B 12, p.2325-2335 (1975)

[4.2] {Sect. 4.3} C.L. Tang, C. Flytzanis: Charge-Transfer Model of the Nonlinear Susceptibilities of Polar Semiconductors, Phys. Rev. B 4, p.2520-2524 (1971)

[4.3] {Sect. 4.3} C. Flytzanis, J. Ducuing: Second-Order Optical Susceptibilities of III-V Semiconductors, Phys. Rev. 178, p.1218-1228 (1969)

[4.4]   {Sect. 4.3} B.F. Levine: Electrodynamical Bond-Charge Calculation of Non-
        linear Optical Susceptibilities, Phys. Rev. Lett. 22, p.787-790 (1969)

[4.5]   {Sect. 4.3} S.S. Jha, N. Bloembergen: Nonlinear Optical Susceptibilities in
        Group-IV and III-V Semiconductors, Phys. Rev. 171, p.891-898 (1968)

[4.6]   {Sect. 4.3} Y.R. Shen: Permutation Symmetry of Nonlinear Susceptibilities
        and Energy Relation, Phys. Rev. 167, p.818-821 (1968)

[4.7]   {Sect. 4.3} P.D. Maker, T.W. Terhune: Study of Optical Effects Due to an
        Induced Polarization Third Order in the Electric Field Strength, Phys. Rev.
        137, p.A801-A818 (1965)

[4.8]   {Sect. 4.3} G. Rosen, F.C. Whitmore: Experiment for Observing the Vacuum
        Scattering of Light by Light, Phys. Rev. 137, p.B1357-B1359 (1965)

[4.9]   {Sect. 4.3} N. Bloembergen, Y.R. Shen: Quantum-Theoretical Comparision
        of Nonlinear Susceptibilities in Parametric Media, Lasers, and Raman Lasers,
        Phys. Rev. 133, p.A37-A49 (1964)

[4.10] {Sect. 4.3} J.A. Armstrong, N. Bloembergen, J. Ducuing, P.S. Pershan:
        Interactions between Light Waves in a Nonlinear Dielectric, Phys. Rev. 127,
        p.1918-1939 (1962)

[4.11] {Sect. 4.3} D.A. Kleinman: Nonlinear Dielectric Polarization in Optical Me-
        dia, Phys. Rev. 126, p.1977-1979 (1962)

[4.12] {Sect. 4.3} P.A. Franken, A.E. Hill, C.W. Peters, G. Weinreich: Generation
        of Optical Harmonics, Phys. Rev. Lett. 7, p.118-119 (1961)

[4.13] {Sect. 4.4.1} D.W. Kim, G.Y. Xiao, G.B. Ma: Temporal properties of the
        second-harmonic generation of a short pulse, Appl Opt 36, p.6788-6793
        (1997)

[4.14] {Sect. 4.4.1} D.R. White, E.L. Dawes, J.H. Marburger: Theory of Second-
        Harmonic Generation With High-Conversion Efficiency, IEEE J. QE-6, p.793-
        796 (1970)

[4.15] {Sect. 4.4.1} H. Kouta, Y. Kuwano: Attaining 186-nm light generation in
        cooled beta-BaB2O4 crystal, Optics Letters 24, p.1230-1232 (1999)

[4.16] {Sect. 4.4.1} I. Shoji, H. Nakamura, K. Ohdaira, T. Kondo, R. Ito,
        T. Okamoto, K. Tatsuki, S. Kubota: Absolute measurement of second-
        order nonlinear-optical coefficients of beta-BaB2O4 for visible to ultraviolet
        second-harmonic wavelengths, J Opt Soc Am B Opt Physics 16, p.620-624
        (1999)

[4.17] {Sect. 4.4.1} M. Tlidi, P. Mandel: Three-dimensional optical crystals and
        localized structures in cavity second harmonic generation, Phys Rev Lett 83,
        p.4995-4998 (1999)

[4.18] {Sect. 4.4.1} S. Yu, A.M. Weiner: Phase-matching temperature shifts in blue
        generation by frequency doubling of femtosecond pulses in KNbO3, J Opt
        Soc Am B Opt Physics 16, p.1300-1304 (1999)

[4.19] {Sect. 4.4.1} G. Ghosh: Sellmeier coefficients for the birefringence and re-
        fractive indices of ZnGeP2 nonlinear crystal at different temperatures, Appl
        Opt 37, p.1205-1212 (1998)

[4.20] {Sect. 4.4.1} M. Sheik-Bahae, M. Ebrahimzadeh: Measurements of nonlinear
        refraction in the second-order chi ((2)) materials KTiOPO4, KNbO3, beta-
        BaB2O4, and LiB3O5, Opt Commun 142, p.294-298 (1997)

[4.21] {Sect. 4.4.1} D.J. Armstrong, W.J. Alford, T.D. Raymond, A.V. Smith: Ab-
        solute measurement of the effective nonlinearities of KTP and BBO crystals
        by optical parametric amplification, Appl Opt 35, p.2032-2040 (1996)

[4.22] {Sect. 4.4.1} K. Hagimoto, A. Mito: Determination of the second-order sus-
        ceptibility of ammonium dihydrogen phosphate and alpha-quartz at 633 and
        1064 nm, Appl Opt 34, p.8276-8282 (1995)

[4.23] {Sect. 4.4.1} J. Jerphagnon, S.K. Kurtz: Optical Nonlinear Susceptibilities: Accurate Relative Values for Quartz, Ammonium Dihydrogen Phosphate, and Potassium Dihydrogen Phosphate, Phys. Rev. B 1, p.17391744 (1970)

[4.24] {Sect. 4.4.1} R.C. Miller, W.A. Nordland: Absolute Signs of Second-Harmonic Generation Coefficients of Piezoelectric Crystals, Phys. Rev. B 2, p.4896-4902 (1970)

[4.25] {Sect. 4.4.1} R.C. Miller: Optical Second Harmonic Generation in Piezoelectric Crystals, Appl. Phys. Lett. 5, p.17-19 (1964)

[4.26] {Sect. 4.4.1} C. Samyn, T. Verbiest, A. Persoons: Second-order non-linear optical polymers, Macromol Rapid Commun 21, p.1-15 (2000)

[4.27] {Sect. 4.4.1} W.S. Shi, Z.H. Chen, T. Zhao, H.B. Lu, Y.L. Zhou, G.Z. Yang: Second-harmonic generation in Ce : BaTiO2 nanocrystallites grown by pulsed laser deposition, J Opt Soc Am B Opt Physics 17, p.235-238 (2000)

[4.28] {Sect. 4.4.1} B.F. Henson, B.W. Asay, R.K. Sander, S.F. Son, J.M. Robinson, P.M. Dickson: Dynamic measurement of the HMX beta-delta phase transition by second harmonic generation, Phys Rev Lett 82, p.1213-1216 (1999)

[4.29] {Sect. 4.4.1} R. Masse, J.F. Nicoud, M. BagieuBeucher, C. Bourgogne: Sodium 3-methyl-4-nitrophenolate dihydrate: a crystal engineering route towards new herringbone structures for quadratic non-linear optics, Chem Phys 245, p.365-375 (1999)

[4.30] {Sect. 4.4.1} S.N. Rashkeev, S. Limpijumnong, W.R.L. Lambrecht: Theoretical evaluation of LiGaO2 for frequency upconversion to ultraviolet, J Opt Soc Am B Opt Physics 16, p.2217-2222 (1999)

[4.31] {Sect. 4.4.1} M. Yoshimura, H. Furuya, T. Kobayashi, K. Murase, Y. Mori, T. Sasaki: Noncritically phase-matched frequency conversion in GdxY1-xCa4O (BO3) (3) crystal, Optics Letters 24, p.193-195 (1999)

[4.32] {Sect. 4.4.1} D.Y. Zhang, H.Y. Shen, W. Liu, G.F. Zhang, W.Z. Chen, G. Zhang, R.R. Zeng, C.H. Huang, W.X. Lin, J.K. Liang: Study of the nonlinear optical properties of 7.5 mol% Nb : KTP crystals, IEEE J QE-35, p.1447-1450 (1999)

[4.33] {Sect. 4.4.1} Y. Furukawa, K. Kitamura, S. Takekawa, K. Niwa, H. Hatano: Stoichiometric Mg : LiNbO3 as an effective material for nonlinear optics, Optics Letters 23, p.1892-1894 (1998)

[4.34] {Sect. 4.4.1} D. Pureur, A.C. Liu, M.J.F. Digonnet, G.S. Kino: Absolute measurement of the second-order nonlinearity profile in poled silica, Optics Letters 23, p.588-590 (1998)

[4.35] {Sect. 4.4.1} T. Verbiest, S. VanElshocht, M. Kauranen, L. Hellemans, J. Snauwaert, C. Nuckolls, T.J. Katz, A. Persoons: Strong enhancement of nonlinear optical properties through supramolecular chirality, Science 282, p.913-915 (1998)

[4.36] {Sect. 4.4.1} J. Capmany, J.G. Sole: Second harmonic generation in LaBGeO5:Nd3+, Appl Phys Lett 70, p.2517-2519 (1997)

[4.37] {Sect. 4.4.1} T. Fujiwara, M. Takahashi, A.J. Ikushima: Second-harmonic generation in germanosilicate glass poled with ArF laser irradiation, Appl Phys Lett 71, p.1032-1034 (1997)

[4.38] {Sect. 4.4.1} K. Kato: Second-harmonic and sum-frequency generation in ZnGeP2, Appl Opt 36, p.2506-2510 (1997)

[4.39] {Sect. 4.4.1} Z.D. Li, B.C. Wu, G.B. Su, G.F. Huang: Blue light emission from an organic nonlinear optical crystal of 4-aminobenzophenone pumped by a laser diode, Appl Phys Lett 70, p.562-564 (1997)

[4.40] {Sect. 4.4.1} Y.C. Wu, P.Z. Fu, J.X. Wang, Z.Y. Xu, L. Zhang, Y.F. Kong, C.T. Chen: Characterization of CsB3O5 crystal for ultraviolet generation, Optics Letters 22, p.1840-1842 (1997)

[4.41] {Sect. 4.4.1} C.T. Chen, Z.Y. Xu, D.Q. Deng, J. Zhang, G.K.L. Wong, B.C. Wu, N. Ye, D.Y. Tang: The vacuum ultraviolet phase-matching characteristics of nonlinear optical KBe2BO3F2 crystal, Appl Phys Lett 68, p.2930-2932 (1996)

[4.42] {Sect. 4.4.1} G.S.G. Quirino, M.D.I. Castillo, J.J. SanchezMondragon, S. Stepanov, V. Vysloukh: Interferometric measurements of the photoinduced refractive index profiles in photorefractive Bi12TiO20 crystal, Opt Commun 123, p.597-602 (1996)

[4.43] {Sect. 4.4.1} W.L. Zhou, Y. Mori, T. Sasaki, S. Nakai: High-efficiency intracavity continuous-wave ultraviolet generation using crystals CsLiB6O10, beta P-BaB2O4 and LiB3O5, Opt Commun 123, p.583-586 (1996)

[4.44] {Sect. 4.4.1} M. Ahlheim, M. Barzoukas, P.V. Bedworth, M. Blancharddesce, A. Fort, Z.Y. Hu, S.R. Marder, J.W. Perry, C. Runser, M. Staehelin, et al.: Chromophores with strong heterocyclic accepters: A poled polymer with a large electro-optic coefficient, Science 271, p.335-337 (1996)

[4.45] {Sect. 4.4.1} F.C. Zumsteg, J.D. Bierlein, T.E. Gier: KxRb1-xTiOPO4: A new nonlinear optical material, J. Appl. Phys. 47, p.4980-4985 (1976)

[4.46] {Sect. 4.4.1} K. Tanaka, A. Narazaki, K. Hirao: Large optical second-order nonlinearity of poled WO3-TeO2 glass, Optics Letters 25, p.251-253 (2000)

[4.47] {Sect. 4.4.1} A.V. Balakin, V.A. Bushuev, N.I. Koroteev, B.I. Mantsyzov, I.A. Ozheredov, A.P. Shkurinov, D. Boucher, P. Masselin: Enhancement of second-harmonic generation with femtosecond laser pulses near the photonic band edge for different polarizations of incident light, Optics Letters 24, p.793-795 (1999)

[4.48] {Sect. 4.4.1} S.J. Lin, I.D. Hands, D.L. Andrews, S.R. Meech: Optically induced second harmonic generation by six-wave mixing: A novel probe of solute orientational dynamics, J Phys Chem A 103, p.3830-3836 (1999)

[4.49] {Sect. 4.4.1} P. LozaAlvarez, D.T. Reid, P. Faller, M. Ebrahimzadeh, W. Sibbett: Simultaneous second-harmonic generation and femtosecond-pulse compression in aperiodically poled KTiOPO4 with a RbTiOAsO4-based optical parametric oscillator, J Opt Soc Am B Opt Physics 16, p.1553-1560 (1999)

[4.50] {Sect. 4.4.1} F. Mougel, K. Dardenne, G. Aka, A. KahnHarari, D. Vivien: Ytterbium-doped Ca4GdO (BO3) (3): An efficient infrared laser and self-frequency doubling crystal, J Opt Soc Am B Opt Physics 16, p.164-172 (1999)

[4.51] {Sect. 4.4.1} S. Pearl, H. Lotem, Y. Shimony, S. Rosenwaks: Optimization of laser intracavity second-harmonic generation by a linear dispersion element, J Opt Soc Am B Opt Physics 16, p.1705-1711 (1999)

[4.52] {Sect. 4.4.1} A. Piskarskas, V. Smilgevicius, A. Stabinis, V. Jarutis, V. Pasiskevicius, S. Wang, J. Tellefsen, F. Laurell: Noncollinear second-harmonic generation in periodically poled KTiOPO4 excited by the Bessel beam, Optics Letters 24, p.1053-1055 (1999)

[4.53] {Sect. 4.4.1} P. Wang, J.M. Dawes, P. Dekker, D.S. Knowles, J.A. Piper, B.S. Lu: Growth and evaluation of ytterbium-doped yttrium aluminum borate as a potential self-doubling laser crystal, J Opt Soc Am B Opt Physics 16, p.63-69 (1999)

[4.54] {Sect. 4.4.1} O.S. Brozek, V. Quetschke, A. Wicht, K. Danzmann: Highly efficient cw frequency doubling of 854 nm GaAlAs diode lasers in an external ring cavity, Opt Commun 146, p.141-146 (1998)

[4.55] {Sect. 4.4.1} D. Fluck, P. Gunter: Efficient second-harmonic generation by lens wave-guiding in KNbO3 crystals, Opt Commun 147, p.305-308 (1998)

[4.56] {Sect. 4.4.1} C. Iaconis, I.A. Walmsley: Fundamental-harmonic phase shift compensation in an intracavity frequency doubled Nd:YLF laser, Opt Commun 149, p.61-63 (1998)

[4.57] {Sect. 4.4.1} Y. Wang, V. Petrov, Y.J. Ding, Y. Zheng, J.B. Khurgin, W.P. Risk: Ultrafast generation of blue light by efficient second-harmonic generation in periodically-poled bulk and waveguide potassium titanyl phosphate, Appl Phys Lett 73, p.873-875 (1998)

[4.58] {Sect. 4.4.1} K.L. Moore, T. Donnelly: Probing nonequilibrium electron distributions in gold by use of second-harmonic generation, Optics Letters 24, p.990-992 (1999)

[4.59] {Sect. 4.4.2} B.A. Richman, S.E. Bisson, R. Trebino, E. Sidick, A. Jacobson: All-prism achromatic phase matching for tunable second-harmonic generation, Appl Opt 38, p.3316-3323 (1999)

[4.60] {Sect. 4.4.2} H. Endoh, M. Kawaharada, E. Hasegawa: Noncritical phase-matched second-harmonic generation with an organic crystal, 4-(isopropyl-carbamoyl)nitrobenzene, Appl Phys Lett 68, p.293-295 (1996)

[4.61] {Sect. 4.4.2} R.S. Adhav, R.W. Wallace: Second Harmonic Generation in 90 Phase-Matched KDP Isomorphs, IEEE J. QE-9, p.855-856 (1973)

[4.62] {Sect. 4.4.2} J.P. Feve, J.J. Zondy, B. Boulanger, R. Bonnenberger, X. Cabirol, B. Menaert, G. Marnier: Optimized blue light generation in optically contacted walk-off compensated RbTiOAsO4 and KTiOP1-yAsyO4, Opt Commun 161, p.359-369 (1999)

[4.63] {Sect. 4.4.2} R. Schiek, Y. Baek, G.I. Stegeman, W. Sohler: One-dimensional quadratic walking solitons, Optics Letters 24, p.83-85 (1999)

[4.64] {Sect. 4.4.2} R.J. Gehr, R.W. Kimmel, A.V. Smith: Simultaneous spatial and temporal walk-off compensation in frequency- doubling femtosecond pulses in beta-BaB2O4, Optics Letters 23, p.1298-1300 (1998)

[4.65] {Sect. 4.4.2} G.D. Boyd, D.A. Kleinman: Parametric Interaction of Focused Gaussian Light Beams, J. Appl. Phys. 39, p.3597-3639 (1968)

[4.66] {Sect. 4.4.2} A.M. Weiner, A.M. Kanan, D.E. Leaird: High-efficiency blue generation by frequency doubling of femtosecond pulses in a thick nonlinear crystal, Optics Letters 23, p.1441-1443 (1998)

[4.67] {Sect. 4.4.2} K. Mori, Y. Tamaki, M. Obara, K. Midorikawa: Second-harmonic generation of femtosecond high-intensity Ti: sapphire laser pulses, J Appl Phys 83, p.2915-2919 (1998)

[4.68] {Sect. 4.4.2} T.J. Zhang, M. Yonemura: Efficient type I second-harmonic generation of subpicosecond laser pulses with a series of alternating nonlinear and delay crystals, Appl Opt 37, p.1647-1650 (1998)

[4.69] {Sect. 4.4.2} J. Capmany, E. Montoya, V. Bermudez, D. Callejo, E. Dieguez, L.E. Bausa: Self-frequency doubling in Yb3+ doped periodically poled LiNbO3 : MgO bulk crystal, Appl Phys Lett 76, p.1374-1376 (2000)

[4.70] {Sect. 4.4.2} W. Shi, C.S. Fang, Z.L. Zu, Q.W. Pan, Q.T. Gu, X. Dong, H.Z. Wei, J.Z. Yu: Poling and characterization of nonlinear polymer DCNP/PEK-c thin films, Solid State Commun 113, p.483-487 (2000)

[4.71] {Sect. 4.4.2} R.G. Batchko, V.Y. Shur, M.M. Fejer, R.L. Byer: Backswitch poling in lithium niobate for high-fidelity domain patterning and efficient blue light generation, Appl Phys Lett 75, p.1673-1675 (1999)

[4.72] {Sect. 4.4.2} C.B.E. Gawith, D.P. Shepherd, J.A. Abernethy, D.C. Hanna, G.W. Ross, P.G.R. Smith: Second-harmonic generation in a direct-bonded periodically poled LiNbO3 buried waveguide, Optics Letters 24, p.481-483 (1999)

[4.73] {Sect. 4.4.2} X.H. Gu, M. Makarov, Y.J. Ding, J.B. Khurgin, W.P. Risk: Backward second-harmonic and third-harmonic generation in a periodically poled potassium titanyl phosphate waveguide, Optics Letters 24, p.127-129 (1999)

[4.74] {Sect. 4.4.2} I. Juwiler, A. Arie, A. Skliar, G. Rosenman: Efficient quasi-phase-matched frequency doubling with phase compensation by a wedged crystal in a standing-wave external cavity, Optics Letters 24, p.1236-1238 (1999)

[4.75] {Sect. 4.4.2} X. Liu, L.J. Qian, F. Wise: Effect of local phase-mismatch on frequency doubling of high-power femtosecond laser pulses under quasi-phase-matched conditions, Opt Commun 164, p.69-75 (1999)

[4.76] {Sect. 4.4.2} M. Pierrou, F. Laurell, H. Karlsson, T. Kellner, C. Czeranowsky, G. Huber: Generation of 740 mW of blue light by intracavity frequency doubling with a first-order quasi-phase-matched KTiOPO4 crystal, Optics Letters 24, p.205-207 (1999)

[4.77] {Sect. 4.4.2} R. Schiek, L. Friedrich, H. Fang, G.I. Stegeman, K.R. Parameswaran, M.H. Chou, M.M. Fejer: Nonlinear directional coupler in periodically poled lithium niobate, Optics Letters 24, p.1617-1619 (1999)

[4.78] {Sect. 4.4.2} S. Wang, V. Pasiskevicius, J. Hellstrom, F. Laurell, H. Karlsson: First-order type II quasi-phase-matched UV generation in periodically poled KTP, Optics Letters 24, p.978-980 (1999)

[4.79] {Sect. 4.4.2} F. Laurell: Periodically poled materials for miniature light sources, Opt. Mat. 11, p.235-244 (1999)

[4.80] {Sect. 4.4.2} Y.J.J. Ding, J.U. Kang, J.B. Khurgin: Theory of backward second-harmonic and third-harmonic generation using laser pulses in quasi-phase-matched second-order nonlinear medium, IEEE J QE-34, p.966-974 (1998)

[4.81] {Sect. 4.4.2} H. Komine, W.H. Long, J.W. Tully, E.A. Stappaerts: Quasi-phase-matched second-harmonic generation by use of a total-internal-reflection phase shift in gallium arsenide and zinc selenide plates, Optics Letters 23, p.661-663 (1998)

[4.82] {Sect. 4.4.2} K. Mizuuchi, K. Yamamato: Waveguide second-harmonic generation device with broadened flat quasi- phase-matching response by use of a grating structure with located phase shifts, Optics Letters 23, p.1880-1882 (1998)

[4.83] {Sect. 4.4.2} S. Wang, V. Pasiskevicius, F. Laurell, H. Karlsson: Ultraviolet generation by first-order frequency doubling in periodically poled KTiOPO4, Optics Letters 23, p.1883-1885 (1998)

[4.84] {Sect. 4.4.2} J. Amin, V. Pruneri, J. Webjorn, P.S. Russell, D.C. Hanna, J.S. Wilkinson: Blue light generation in a periodically poled Ti:LiNbO3 channel waveguide, Opt Commun 135, p.41-44 (1997)

[4.85] {Sect. 4.4.2} A. Arie, G. Rosenman, V. Mahal, A. Skliar, M. Oron, M. Katz, D. Eger: Green and ultraviolet quasi-phase-matched second harmonic generation in bulk periodically-poled KTiOPO4, Opt Commun 142, p.265-268 (1997)

[4.86] {Sect. 4.4.2} G.D. Miller, R.G. Batchko, W.M. Tulloch, D.R. Weise, M.M. Fejer, R.L. Byer: 42%-efficient single-pass cw second-harmonic generation in periodically poled lithium niobate, Optics Letters 22, p.1834-1836 (1997)

[4.87] {Sect. 4.4.2} K. Mizuuchi, K. Yamamoto, M. Kato: Generation of ultraviolet light by frequency doubling of a red laser diode in a first-order periodically poled bulk LiTaO3, Appl Phys Lett 70, p.1201-1203 (1997)

[4.88] {Sect. 4.4.2} J.H. Si, G. Xu, X.C. Liu, Q.G. Yang, P.X. Ye, Z. Li, H. Ma, Y.Q. Shen, L. Qiu, J.X. Zhang, et al.: All-optical poling of a polyimide film with azobenzene chromophore, Opt Commun 142, p.71-74 (1997)

[4.89]   {Sect. 4.4.2} S. Sonoda, I. Tsuruma, M. Hatori: Second harmonic generation in electric poled X-cut MgO- doped LiNbO3 waveguides, Appl Phys Lett 70, p.3078-3080 (1997)

[4.90]   {Sect. 4.4.2} A. Harada, Y. Nihei, Y. Okazaki, and H. Hyuga: Intracavity frequency doubling of a diode-pumped 946-nm Nd:YAG laser with bulk periodically poled MgO-LiNbO3, Opt. Lett. 22, p.805-807 (1997)

[4.91]   {Sect. 4.4.2} G.D. Miller, R.G. Batchko, W.M. Tulloch, D.R. Weise, M.M. Fejer, and R.L. Byer: 42%-efficient single-pass cw second-harmonic generation in periodically poled lithium niobate, Opt. Lett. 22, p.1834-1836 (1997)

[4.92]   {Sect. 4.4.2} Y. Kitaoka, K. Mizuuchi, K. Yamamoto, M. Kato, T. Sasaki: Intracavity second-harmonic generation with a periodically domain-inverted LiTaO3 device, Optics Letters 21, p.1972-1974 (1996)

[4.93]   {Sect. 4.4.2} Y.L. Lu, Y.Q. Lu, C.C. Xue, N.B. Ming: Growth of Nd3+-doped LiNbO3 optical superlattice crystals and its potential applications in self-frequency doubling, Appl Phys Lett 68, p.1467-1469 (1996)

[4.94]   {Sect. 4.4.2} K. Mizuuchi, K. Yamamoto: Generation of 340-nm light by frequency doubling of a laser diode in bulk periodically poled LiTaO3, Optics Letters 21, p.107-109 (1996)

[4.95]   {Sect. 4.4.2} V. Pruneri, S.D. Butterworth, D.C. Hanna: Low-threshold picosecond optical parametric oscillation in quasi-phase-matched lithium niobate, Appl Phys Lett 69, p.1029-1031 (1996)

[4.96]   {Sect. 4.4.2} V. Pruneri, S.D. Butterworth, D.C. Hanna: Highly efficient green-light generation by quasi-phase- matched frequency doubling of picosecond pulses from an amplified mode-locked Nd:YLF laser, Optics Letters 21, p.390-392 (1996)

[4.97]   {Sect. 4.4.2} S. Tomaru, T. Watanabe, M. Hikita, M. Amano, Y. Shuto: Quasi-phase-matched second harmonic generation in a polymer waveguide with a periodic poled structure, Appl Phys Lett 68, p.1760-1762 (1996)

[4.98]   {Sect. 4.4.2} S. Yilmaz, S. Bauer, R. Gerhard-Multhaupt: Photothermal poling of nonlinear optical polymer films, Appl. Phys. Lett. 64, p.2770-2772 (1994)

[4.99]   {Sect. 4.4.3} D. Hofmann, G. Schreiber, C. Haase, H. Herrmann, W. Grundkotter, R. Ricken, W. Sohler: Quasi-phase-matched difference-frequency generation in periodically poled Ti : LiNbO3 channel waveguides, Optics Letters 24, p.896-898 (1999)

[4.100] {Sect. 4.4.3} J.A. McGuire, W. Beck, X. Wei, Y.R. Shen: Fourier-transform sum-frequency surface vibrational spectroscopy with femtosecond pulses, Optics Letters 24, p.1877-1879 (1999)

[4.101] {Sect. 4.4.3} C.Q. Wang, Y.T. Chow, W.A. Gambling, D.R. Yuan, D. Xu, G.H. Zhang, M.H. Jiang: A continuous-wave tunable solid-state blue laser based on intracavity sum-frequency mixing and pump-wavelength tuning, Appl Phys Lett 75, p.1821-1823 (1999)

[4.102] {Sect. 4.4.3} E.V. Alieva, L.A. Kuzik, V.A. Yakovlev: Sum frequency generation spectroscopy of thin organic films on silver using visible surface plasmon generation, Chem Phys Lett 292, p.542-546 (1998)

[4.103] {Sect. 4.4.3} G.C. Bhar, P. Kumbhakar, U. Chatterjee, A.M. Rudra, Y. Kuwano, H. Kouta: Efficient generation of 200-230-nm radiation in beta barium borate by noncollinear sum-frequency mixing, Appl Opt 37, p.7827-7831 (1998)

[4.104] {Sect. 4.4.3} R.A. Kaindl, D.C. Smith, M. Joschko, M.P. Hasselbeck, M. Woerner, T. Elsaesser: Femtosecond infrared pulses tunable from 9 to 18 mu m at an 88-MHz repetition rate, Optics Letters 23, p.861-863 (1998)

[4.105] {Sect. 4.4.3} A. Nazarkin, G. Korn: Generation of self-compressed laser pulses under the condition of two- photon resonant difference-frequency mixing in gases, Opt Commun 153, p.184-190 (1998)

[4.106] {Sect. 4.4.3} V. Petrov, C. Rempel, K.P. Stolberg, W. Schade: Widely tunable continuous-wave mid-infrared laser source based on difference-frequency generation in AgGaS2, Appl Opt 37, p.4925-4928 (1998)

[4.107] {Sect. 4.4.3} J.D. Vance, C.Y. She, H. Moosmuller: Continuous-wave, all-solid-state, single-frequency 400-mW source at 589 nm based on doubly resonant sum-frequency mixing in a monolithic lithium niobate resonator, Appl Opt 37, p.4891-4896 (1998)

[4.108] {Sect. 4.4.3} G.C. Bhar, U. Chatterjee, A.M. Rudra, P. Kumbhakar, R.K. Route, R.S. Feigelson: Generation of tunable 187.9-196-nm radiation in beta-Ba2BO4, Optics Letters 22, p.1606-1608 (1997)

[4.109] {Sect. 4.4.3} D. Fluck, P. Gunter: Efficient generation of CW blue light by sum-frequency mixing of laser diodes in KNbO3, Opt Commun 136, p.257-260 (1997)

[4.110] {Sect. 4.4.3} J.M. Fraser, D.K. Wang, A. Hache, G.R. Allan, H.M. vanDriel: Generation of high-repetition-rate femtosecond pulses from 8 to 18 mu m, Appl Opt 36, p.5044-5047 (1997)

[4.111] {Sect. 4.4.3} H.M. Kretschmann, F. Heine, G. Huber, T. Halldorsson: All-solid-state continuous-wave doubly resonant all- intracavity sum-frequency mixer, Optics Letters 22, p.1461-1463 (1997)

[4.112] {Sect. 4.4.3} N. Umemura, K. Kato: Ultraviolet generation tunable to 0.185 mu m in CsLiB6O10, Appl Opt 36, p.6794-6796 (1997)

[4.113] {Sect. 4.4.3} A. Balakrishnan, S. Sanders, S. Demars, J. Webjorn, D.W. Nam, R.J. Lang, D.G. Mehuys, R.G. Waarts, D.F. Welch: Broadly tunable laser-diode-based mid-infrared source with up to 31 mu W of power at 4.3-mu m wavelength, Optics Letters 21, p.952-954 (1996)

[4.114] {Sect. 4.4.3} Y.B. Band, M. Trippenbach, C. Radzewicz, J.S. Krasinski: Ultra-short pulse nonlinear optics: Second harmonic generation and sum frequency generation without group velocity mismatch broadening, J Nonlinear Opt Physics Mat 5, p.477-494 (1996)

[4.115] {Sect. 4.4.3} M. Berdah, J.P. Visticot, C. Dedonderlardeux, D. Solgadi, B. Soep: Generation of picosecond VUV radiation by four-wave mixing of nanosecond and picosecond laser radiations, Opt Commun 124, p.118-120 (1996)

[4.116] {Sect. 4.4.3} R. Danielius, A. Dubietis, A. Piskarskas, G. Valiulis, A. Varanavicius: Generation of compressed 600-720-nm tunable femtosecond pulses by transient frequency mixing in a beta-barium borate crystal, Optics Letters 21, p.216-218 (1996)

[4.117] {Sect. 4.4.3} O. Kittelmann, J. Ringling, G. Korn, A. Nazarkin, I.V. Hertel: Generation of broadly tunable femtosecond vacuum- ultraviolet pulses, Optics Letters 21, p.1159-1161 (1996)

[4.118] {Sect. 4.4.3} A. Shirakawa, H.W. Mao, T. Kobayashi: Highly efficient generation of blue-orange femtosecond pulses from intracavity-frequency-mixed optical parametric oscillator, Opt Commun 123, p.121-128 (1996)

[4.119] {Sect. 4.4.3} Y.K. Yap, M. Inagaki, S. Nakajima, Y. Mori, T. Sasaki: High-power fourth- and fifth-harmonic generation of a Nd: YAG laser by means of a CsLiB6O10, Optics Letters 21, p.1348-1350 (1996)

[4.120] {Sect. 4.4.3} B. Dick, R.M. Hochstrasser: Spectroscopic and line-narrowing properties of resonant sum and difference frequency generation, J. Chem. Phys. 78, p.3398-3409 (1983)

[4.121] {Sect. 4.4.3} J.R. Morris, Y.R. Shen: Theory of far-infrared generation by optical mixing, Phys. Rev. A 15, p.1143-1156 (1977)

[4.122] {Sect. 4.4.3} J.A. Armstrong, N. Bloembergen, J. Ducuing, P.S. Pershan: Interactions between Light Waves in a Nonlinear Dielectric, Phys. Rev. 127, p.1918-1939 (1962)

[4.123] {Sect. 4.4.3} M. Bass, P.A. Franken, A.E. Hill, C.W. Peters, G. Weinreich: Optical Mixing, Phys. Rev. Lett. 8, p.18 (1962)

[4.124] {Sect. 4.4.3} N. Bloembergen, P. S. Pershan: Light Waves at the Boundary of Nonlinear Media, Phys. Rev. 128, p.606-622 (1962)

[4.125] {Sect. 4.4.3} P.D. Maker, R.W. Terhune, M. Nisenoff, C.M. Savage: Effects of Dispersion and Focusing on the Production of optical Harmonics, Phys. Rev. Lett. 8, p.21-22 (1962)

[4.126] {Sect. 4.4.3} D. Mazzotti, P. Denatale, G. Giusfredi, C. Fort, J.A. Mitchell, L. Hollberg: Saturated-absorption spectroscopy with low-power difference-frequency radiation, Optics Letters 25, p.350-352 (2000)

[4.127] {Sect. 4.4.4} C.L. Tang: Tutorial on optical parametric processes and devices, J Nonlinear Opt Physics Mat 6, p.535-547 (1997)

[4.128] {Sect. 4.4.4} S.J. Brosnan, R.L. Byer: Optical Parametric Oscillator Threshold and Linewidth Studies, IEEE J. QE-15, p.415-431 (1979)

[4.129] {Sect. 4.4.4} J. H. Hunt: Optical Parametric Oscillators and Amplifiers and Their Applications (SPIE Optical Engineering Press, London, 1997)

[4.130] {Sect. 4.4.4} C. L. Tang, L. K. Cheng: Fundamentals of Optical Parametric Processes and Oscillators (Harwood Academic Publishers, Amsterdam, 1995)

[4.131] {Sect. 4.4.4} L. Carrion, J.P. GirardeauMontaut: Development of a simple model for optical parametric generation, J Opt Soc Am B Opt Physics 17, p.78-83 (2000)

[4.132] {Sect. 4.4.4} M.H. Dunn, M. Ebrahimzadeh: Parametric generation of tunable light from continuous-wave to femtosecond pulses, Science 286, p.1513-1517 (1999)

[4.133] {Sect. 4.4.4} Y. R. Shen: Principles of Nonlinear Optics, chapter 9 (John Wiley & Sons, Chichester, 1984)

[4.134] {Sect. 4.4.4} J.M. Manley, H.E. Rowe: General energy relations in nonlinear reactances, Proc. IRE 47p.2115-2116 (1959)

[4.135] {Sect. 4.4.4} S. Guha: Focusing dependence of the efficiency of a singly resonant optical parametric oscillator, Appl. Phys. B 66, p.663-675 (1998)

[4.136] {Sect. 4.4.4} P.E. Britton, H.L. Offerhaus, D.J. Richardson, P.G.R. Smith, G.W. Ross, D.C. Hanna: Parametric oscillator directly pumped by a 1.55-mu m erbium-fiber laser, Optics Letters 24, p.975-977 (1999)

[4.137] {Sect. 4.4.4} S.A. Diddams, L.S. Ma, J. Ye, J.L. Hall: Broadband optical frequency comb generation with a phase-modulated parametric oscillator, Optics Letters 24, p.1747-1749 (1999)

[4.138] {Sect. 4.4.4} A. Gatti, E. Brambilla, L.A. Lugiato, M.I. Kolobov: Quantum entangled images, Phys Rev Lett 83, p.1763-1766 (1999)

[4.139] {Sect. 4.4.4} V. Petrov, F. Rotermund, F. Noack, P. Schunemann: Femtosecond parametric generation in ZnGeP2, Optics Letters 24, p.414-416 (1999)

[4.140] {Sect. 4.4.4} F. Rotermund, V. Petrov, F. Noach, V. Pasiskevicius, J. Hellstrom, F. Laurell: Efficient femtosecond traveling-wave optical parametric amplification in periodically poled KTiOPO4, Optics Letters 24, p.1874-1876 (1999)

[4.141] {Sect. 4.4.4} F. Rotermund, V. Petrov, F. Noack, M. Wittmann, G. Korn: Laser-diode-seeded operation of a femtosecond optical parametric amplifier with MgO : LiNbO3 and generation of 5-cycle pulses near 3 mu m, J Opt Soc Am B Opt Physics 16, p.1539-1545 (1999)

[4.142] {Sect. 4.4.4} T.W. Tukker, C. Otto, J. Greve: Design, optimization, and characterization of a narrow-bandwidth optical parametric oscillator, J Opt Soc Am B Opt Physics 16, p.90-95 (1999)

[4.143] {Sect. 4.4.4} R. Urschel, U. Bader, A. Borsutzky, R. Wallenstein: Spectral properties and conversion efficiency of 355-nm-pumped pulsed optical parametric oscillators of beta-barium borate with noncollinear phase matching, J Opt Soc Am B Opt Physics 16, p.565-579 (1999)

[4.144] {Sect. 4.4.4} M. Vaupel, A. Maitre, C. Fabre: Observation of pattern formation in optical parametric oscillators, Phys Rev Lett 83, p.5278-5281 (1999)

[4.145] {Sect. 4.4.4} Y. Yashkir, H.M. vanDriel: Passively Q-switched 1.57-mu m intracavity optical parametric oscillator, Appl Opt 38, p.2554-2559 (1999)

[4.146] {Sect. 4.4.4} M. Bode, P.K. Lam, I. Freitag, A. Tunnermann, H.A. Bachor, H. Welling: Continuously-tunable doubly resonant optical parametric oscillator, Opt Commun 148, p.117-121 (1998)

[4.147] {Sect. 4.4.4} L. Carrion, J.P. GirardeauMontaut: Performance of a new picosecond KTP optical parametric generator and amplifier, Opt Commun 152, p.347-350 (1998)

[4.148] {Sect. 4.4.4} I.D. Lindsay, G.A. Turnbull, M.H. Dunn, M. Ebrahimzadeh: Doubly resonant continuous-wave optical parametric oscillator pumped by a single-mode diode laser, Optics Letters 23, p.1889-1891 (1998)

[4.149] {Sect. 4.4.4} M. Scheidt, M.E. Klein, K.J. Boller: Spiking in pump enhanced idler resonant optical parametric oscillators, Opt Commun 149, p.108-112 (1998)

[4.150] {Sect. 4.4.4} J.Y. Zhang, Z.Y. Xu, Y.F. Kong, C.W. Yu, Y.C. Wu: Highly efficient, widely tunable, 10-Hz parametric amplifier pumped by frequency-doubled femtosecond Ti:sapphire laser pulses, Appl Opt 37, p.3299-3305 (1998)

[4.151] {Sect. 4.4.4} R. Al-Tahtamouni, K. Bencheikh, R. Storz, K. Schneider, M. Lang, J. Mlynek, S. Schiller: Long-term stable operation and absolute frequency stabilization of a doubly resonant parametric oscillator, Appl. Phys. B 66, p.733-739 (1998)

[4.152] {Sect. 4.4.4} T. Ikegami, S. Slyusarev, T. Kurosu, Y. Fukuyama, S. Ohshima: Characteristics of a cw monolithic KTiOPO4 optical parametric oscillator, Appl. Phys. B 66, p.719-725 (1998)

[4.153] {Sect. 4.4.4} M.E. Klein, M. Scheidt, K.-J. Boller, R. Wallenstein: Dye laser pumped, continuous-wave KTP optical parametric oscillators, Appl Phys. B 66, p.727-732 (1998)

[4.154] {Sect. 4.4.4} D.-H. Lee, M.E. Klein, K.-J. Boller: Intensity noise of pump-enhanced continuous-wave optical parametric oscillators, Appl. Phys. B 66, p.747-753 (1998)

[4.155] {Sect. 4.4.4} J.L. Sorensen, E.S. Polzik: Internally pumped subthreshold OPO, Appl. Phys. B 66, p.711-718 (1998)

[4.156] {Sect. 4.4.4} J. Izawa, K. Midorikawa, M. Obara, K. Toyoda: Picosecond ultraviolet optical parametric generation using a type-II phase-matched lithium triborate crystal for an injection seed of VUV lasers, IEEE J QE-33, p.1997-2001 (1997)

[4.157] {Sect. 4.4.4} P. Rambaldi, M. Douard, B. Vezin, J.P. Wolf, D. Rytz: Broadly tunable KNbO3 OPOs pumped by Ti:sapphire lasers, Opt Commun 142, p.262-264 (1997)

[4.158] {Sect. 4.4.4} M. Scheidt, B. Beier, K.J. Boller, R. Wallenstein: Frequency-stable operation of a diode-pumped continuous- wave RbTiOAsO4 optical parametric oscillator, Optics Letters 22, p.1287-1289 (1997)

[4.159] {Sect. 4.4.4} K.L. Vodopyanov, V. Chazapis: Extra-wide tuning range optical parametric generator, Opt Commun 135, p.98-102 (1997)

[4.160] {Sect. 4.4.4} T. Wang, M.H. Dunn, C.F. Rae: Polychromatic optical parametric generation by simultaneous phase matching over a large spectral bandwidth, Optics Letters 22, p.763-765 (1997)

[4.161] {Sect. 4.4.4} S. Wu, G.A. Blake, Z.Y. Sun, J.W. Ling: Simple, high-performance type II beta-BaB2O4 optical parametric oscillator, Appl Opt 36, p.5898-5901 (1997)

[4.162] {Sect. 4.4.4} A.R. Geiger, H. Hemmati, W.H. Farr, N.S. Prasad: Diode pumped optical parametric oscillator, Optics Letters 21, p.201-203 (1996)

[4.163] {Sect. 4.4.4} T.H. Jeys: Multipass optical parametric amplifier, Optics Letters 21, p.1229-1231 (1996)

[4.164] {Sect. 4.4.4} S.A. Reid, Y. Tang: Generation of tunable, narrow-band mid-infrared radiation through a 532-nm-pumped KTP optical parametric amplifier, Appl Opt 35, p.1473-1477 (1996)

[4.165] {Sect. 4.4.4} M. Sueptitz, R.A. Kaindl, S. Lutgen, M. Woerner, E. Riedle: 1 kHz solid state laser system for the generation of 50 fs pulses tunable in the visible, Opt Commun 131, p.195-202 (1996)

[4.166] {Sect. 4.4.4} J.M. Boonengering, L.A.W. Gloster, W.E. Vanderveer, I.T. McKinnie, T.A. King, W. Hogervorst: Highly efficient single longitudinal mode beta-BaB2O4 optical parametric oscillator with a new cavity design, Optics Letters 20, p.2087-2089 (1995)

[4.167] {Sect. 4.4.4} J. Hebling, E.J. Mayer, J. Kuhl, R. Szipocs: Chirped mirror dispersion compensated femtosecond optical parametric oscillator, Optics Letters 20, p.919-921 (1995)

[4.168] {Sect. 4.4.4} C. Rauscher, T. Roth, R. Laenen, A. Laubereau: Tunable femtosecond-pulse generation by an optical parametric oscillator in the saturation regime, Optics Letters 20, p.2003-2005 (1995)

[4.169] {Sect. 4.4.4} M.J. Rosker, C.L. Tang: Widely tunable optical parametric oscillator using urea, J. Opt. Soc. Am. B 2, p.691-696 (1985)

[4.170] {Sect. 4.4.4} A. Seilmeier, K. Spanner, A. Laubereau, W. Kaiser: Narrow-Band Tunable Infrared Pulses with Sub-Picosecond Time Resolution, Opt. Comm. 24, p.237-242 (1978)

[4.171] {Sect. 4.4.4} A.H. Kung: Generation of tunable picosecond VUV radiation, Appl. Phys. Lett. 25, p.653-654 (1974)

[4.172] {Sect. 4.4.4} T.A. Rabson, H.J. Ruiz, P.L. Shah, F.K. Tittel: Stimulated parametric fluorscence induced by picosecond pump pulses, Appl. Phys.Lett. 21, p.129-131 (1972)

[4.173] {Sect. 4.4.4} K.H. Yang, P.L. Richards, Y.R. Shen: Generation of Far-Infrared Radiation by Picosecond Light Pulses in LiNbO3, Appl. Phys. Lett. 19, p.320-323 (1971)

[4.174] {Sect. 4.4.4} J. Falk, J.E. Murray: Single-Cavity Noncollinear Optical Parametric Oscillation, Appl. Phys. Lett. 14, p.245-247 (1969)

[4.175] {Sect. 4.4.4} L.B. Kreuzer: Single Mode Oscillation of a Pulsed Singly Resonant Optical Parametric Oscillator, Appl. Phys. Lett. 15, p.263-265 (1969)

[4.176] {Sect. 4.4.4} J.E. Bjorkholm: Some Spectral Properties of Doubly and Singly Resonant Pulsed Optical Parametric Oscillators, Appl. Phys. Lett. 13, p.399-401 (1968)

[4.177] {Sect. 4.4.4} J.E. Bjorkholm: Efficient Optical Parametric Oscillation Using Doubly and Singly Resonant Cavities, Appl. Phys. Lett. 13, p.53-56 (1968)

[4.178] {Sect. 4.4.4} R.L. Byer, S.E. Harris: Power and Bandwidth of Spontaneous Parametric Emission, Phys. Rev. 168, p.1064-1068 (1968)

[4.179] {Sect. 4.4.4} T.G. Giallorenzi, C.L. Tang: Quantum Theory of Spontaneous Parametric Scattering of Intense Light, Phys. Rev. 166, p.225-233 (1968)

[4.180] {Sect. 4.4.4} J.G. Edwards: Some Factors Affecting the Pumping Efficiency of Optically Pumped Lasers, Appl. Opt. 6, p.837-843 (1967)

[4.181] {Sect. 4.4.4} S.E. Harris, M.K. Oshman, R.L. Byer: Observation of Tunable Optical Parametric Fluorescence, Phys. Rev. Lett. 18, p.732-734 (1967)

[4.182] {Sect. 4.4.4} S.E. Harris: Proposed Backward Wave Oscillation in the Infrared, Appl. Phys. Lett. 9, p.114-116 (1966)

[4.183] {Sect. 4.4.4} J. Hellstrom, V. Pasiskevicius, H. Karlsson, F. Laurell: High-power optical parametric oscillation in large-aperture periodically poled KTiOPO4, Optics Letters 25, p.174-176 (2000)

[4.184] {Sect. 4.4.4} M. Missey, V. Dominic, P. Powers, K.L. Schepler: Aperture scaling effects with monolithic periodically poled lithium niobate optical parametric oscillators and generators, Optics Letters 25, p.248-250 (2000)

[4.185] {Sect. 4.4.4} G.M. Gibson, M. Ebrahimzadeh, M.J. Padgett, M.H. Dunn: Continuous-wave optical parametric oscillator based on periodically poled KTiOPO4 and its application to spectroscopy, Optics Letters 24, p.397-399 (1999)

[4.186] {Sect. 4.4.4} J. Hellstrom, V. Pasiskevicius, F. Laurell, H. Karlsson: Efficient nanosecond optical parametric oscillators based on periodically poled KTP emitting in the 1.8-2.5-mu m spectral region, Optics Letters 24, p.1233-1235 (1999)

[4.187] {Sect. 4.4.4} N. OBrien, M. Missey, P. Powers, V. Dominic, K.L. Schepler: Electro-optic spectral tuning in a continuous-wave, asymmetric-duty- cycle, periodically poled LiNbO3 optical parametric oscillator, Optics Letters 24, p.1750-1752 (1999)

[4.188] {Sect. 4.4.4} U. Bader, J. Bartschke, I. Klimov, A. Borsutzky, R. Wallen-stein: Optical parametric oscillator of quasi-phasematched LiNbO3 pumped by a compact high repetition rate single-frequency passively Q-switched Nd:YAG laser, Opt Commun 147, p.95-98 (1998)

[4.189] {Sect. 4.4.4} P.E. Britton, D. Taverner, K. Puech, D.J. Richardson, P.G.R. Smith, G.W. Ross, D.C. Hanna: Optical parametric oscillation in periodi-cally poled lithium niobate driven by a diode-pumped Q-switched erbium fiber laser, Optics Letters 23, p.582-584 (1998)

[4.190] {Sect. 4.4.4} A. Garashi, A. Arie, A. Skliar, G. Rosenman: Continuous-wave optical parametric oscillator based on periodically poled KTiOPO4, Optics Letters 23, p.1739-1741 (1998)

[4.191] {Sect. 4.4.4} L. Lefort, K. Puech, S.D. Butterworth, G.W. Ross, P.G.R. Smith, D.C. Hanna, D.H. Jundt: Efficient, low-threshold synchronously-pumped parametric oscillation in periodically-poled lithium niobate over the 1.3 mu m to 5.3 mu m range, Opt Commun 152, p.55-58 (1998)

[4.192] {Sect. 4.4.4} P.E. Powers, K.W. Aniolek, T.J. Kulp, B.A. Richman, S.E. Bis-son: Periodically poled lithium niobate optical parametric amplifier seeded with the narrow-band filtered output of an optical parametric generator, Optics Letters 23, p.1886-1888 (1998)

[4.193] {Sect. 4.4.4} D.J.M. Stothard, M. Ebrahimzadeh, M.H. Dunn: Low-pump-threshold continuous-wave singly resonant optical parametric oscillator, Optics Letters 23, p.1895-1897 (1998)

[4.194] {Sect. 4.4.4} M. Tsunekane, S. Kimura, M. Kimura, N. Taguchi, H. Inaba: Continuous-wave, broadband tuning from 788 to 1640 nm by a doubly resonant, MgO:LiNbO3 optical parametric oscillator, Appl Phys Lett 72, p.3414-3416 (1998)

[4.195] {Sect. 4.4.4} S.D. Butterworth, P.G.R. Smith, D.C. Hanna: Picosecond Ti:sapphire-pumped optical parametric oscillator based on periodically poled LiNbO3, Optics Letters 22, p.618-620 (1997)

[4.196] {Sect. 4.4.4} D.T. Reid, Z. Penman, M. Ebrahimzadeh, W. Sibbett, H. Karlsson, F. Laurell: Broadly tunable infrared femtosecond optical parametric oscillator based on periodically poled RbTiOAsO4, Optics Letters 22, p.1397-1399 (1997)

[4.197] {Sect. 4.4.4} S. Slyusarev, T. Ikegami, S. Ohshima: Phase-coherent optical frequency division by 3 of 532-nm laser light with a continuous-wave optical parametric oscillator, Optics Letters 24, p.1856-1858 (1999)

[4.198] {Sect. 4.4.5} A.Yariv: Optical Electronics (Holt, Rinehart, Winstin, Holt-Saunders, Japan, 1985)

[4.199] {Sect. 4.4.5} B.H. Hoerman, B.M. Nichols, M.J. Nystrom, B.W. Wessels: Dynamic response of the electro-optic effect in epitaxial KNbO3, Appl Phys Lett 75, p.2707-2709 (1999)

[4.200] {Sect. 4.4.7} C. Bosshard, I. Biaggio, StFischer, S. Follonier, P. Gunter: Cascaded contributions to degenerate four-wave mixing in an acentric organic crystal, Optics Letters 24, p.196-198 (1999)

[4.201] {Sect. 4.4.7} A.V. Bragas, S.M. Landi, O.E. Martinez: Laser field enhancement at the scanning tunneling microscope junction measured by optical rectification, Appl Phys Lett 72, p.2075-2077 (1998)

[4.202] {Sect. 4.4.7} S. Tomic, V. Milanovic, Z. Ikonic: Optimization of nonlinear optical rectification in quantum wells using the supersymmetric quantum mechanics, Opt Commun 143, p.214-218 (1997)

[4.203] {Sect. 4.4.7} A. Nahata, A.S. Weling, T.F. Heinz: A wideband coherent terahertz spectroscopy system using optical rectification and electro-optic sampling, Appl Phys Lett 69, p.2321-2323 (1996)

[4.204] {Sect. 4.4.7} D.H. Auston: Nonlinear Spectroscopy of Picosecond Pulses, Opt. Comm. 3, p.272-276 (1971)

[4.205] {Sect. 4.4.7} J.F. Holzrichter, R.M. Macfarlane, A. L. Schawlow: Magnetization Induced by Optical Pumping in Antiferromagnetic MnF2, Phys. Rev. Lett. 26, p.652-655 (1971)

[4.206] {Sect. 4.4.7} P.S. Pershan, J.P. van der Ziel, L.D. Malmstrom: Theoretical Discussion of the Inverse Faraday Effect, Raman Scattering, and Related Phenomena, Phys. Rev. 143, p.574-583 (1966)

[4.207] {Sect. 4.4.7} J.F. Ward: Absolute Measurement of an Optical-Rectification Coefficient in Ammonium Dihydrogen Phosphate, Phys. Rev. 143, p.569-574 (1966)

[4.208] {Sect. 4.4.7} J.P. van der Ziel, P.S. Pershan, L.D. Malmstrom: Optically-Induced Magnetization Resulting from the Inverse Faraday Effect, Phys. Rev. Lett. 15, p.190-193 (1965)

[4.209] {Sect. 4.4.7} M.Bass, P.A. Franken, J.F. Ward, G. Weinreich: Optical Rectification, Phys. Rev. Lett. 9, p.446-448 (1962)

[4.210] {Sect. 4.5.1} P.S. Banks, M.D. Feit, M.D. Perry: High-intensity third-harmonic generation in beta barium borate through second-order and third-order susceptibilities, Optics Letters 24, p.4-6 (1999)

[4.211] {Sect. 4.5.1} D. Yelin, Y. Silberberg, Y. Barad, J.S. Patel: Phase-matched third-harmonic generation in a nematic liquid crystal cell, Phys Rev Lett 82, p.3046-3049 (1999)

[4.212] {Sect. 4.5.1} D. Eimerl, J.M. Auerbach, C.E. Barker, D. Milam, P.W. Milonni: Multicrystal designs for efficient third-harmonic generation, Optics Letters 22, p.1208-1210 (1997)

[4.213] {Sect. 4.5.1} O. Pfister, J.S. Wells, L. Hollberg, L. Zink, D.A. Vanbaak, M.D. Levenson, W.R. Bosenberg: Continuous-wave frequency tripling and quadrupling by simultaneous three-wave mixings in periodically poled crystals: application to a two-step 1.19-10.71-mu m frequency bridge, Optics Letters 22, p.1211-1213 (1997)

[4.214] {Sect. 4.5.1} S. Backus, J. Peatross, Z. Zeek, A. Rundquist, G. Taft, M.M. Murnane, H.C. Kapteyn: 16-fs, 1-mu J ultraviolet pulses generated by third-harmonic conversion in air, Optics Letters 21, p.665-667 (1996)

[4.215] {Sect. 4.5.1} T.Y.F. Tsang: Surface-plasmon-enhanced third-harmonic generation in thin silver films, Optics Letters 21, p.245-247 (1996)

[4.216] {Sect. 4.5.1} T.J. Zhang, Y. Kato, H. Daido: Efficient third-harmonic generation of a picosecond laser pulse with time delay, IEEE J QE-32, p.127-136 (1996)

[4.217] {Sect. 4.5.1} G. Hilber, A. Lago, R. Wallenstein: Broadly tunable VUV/XUV-radiation generated by resonant third-order frequency conversion in Kr, J. Opt. Soc. Am. B 4, p.1753-1764 (1987)

[4.218] {Sect. 4.5.1} J. Bokor, P.H. Bucksbaum, R.R. Freeman: Generation of 35.5-nm coherent radiation, Opt. Lett. 8, p.217-219 (1983)

[4.219] {Sect. 4.5.1} H.B. Puell, C.R. Vidal: Optimum Conditions for Nonresonant Third Harmonic Generation, IEEE J. QE-14, p.364-373 (1978)

[4.220] {Sect. 4.5.1} C.M. Bloom, G.W. Bekkers, J.F. Young, S.E. Harris: Third harmonic generation in phase-matched alkali metal vapors, Appl. Phys. Lett. 26, p.687-689 (1975)

[4.221] {Sect. 4.5.1} C.M. Bloom, J.F. Young, S.E. Harris: Mixed metal vapor phase matching for third-harmonic generation, Appl. Phys. Lett. 27, p.390-392 (1975)

[4.222] {Sect. 4.5.1} R.B. Miles, S.E. Harris: Optical Third-Harmonic Generation in Alkali Metal Vapors, IEEE J. QE-9, p.470-484 (1973)

[4.223] {Sect. 4.5.1} A.H. Kung, J.F. Young, G.C. Bjorklund, S.E. Harris: Generation of Vacuum Ultraviolet Radiation in Phase-Matched Cd Vapor, Phys. Rev. Lett. 29, p.985-988 (1972)

[4.224] {Sect. 4.5.1} S.E. Harris, R.B. Miles: Proposed Third-Harmonic Generation in Phase-Matched Metal Vapors, Appl. Phys. Lett. 19, p.385-387 (1971)

[4.225] {Sect. 4.5.1} J.F. Young, G.C. Bjorklund, A.H. Kung, R.B. Miles, S.E. Harris: Third-Harmonic Generation in Phase-Matched Rb Vapor, Phys. Rev. Lett. 27, p.1551-1553 (1971)

[4.226] {Sect. 4.5.2} G. Lenz, J. Zimmermann, T. Katsufuji, M.E. Lines, H.Y. Hwang, S. Spalter, R.E. Slusher, S.W. Cheong, J.S. Sanghera, I.D. Aggarwal: Large Kerr effect in bulk Se-based chalcogenide glasses, Optics Letters 25, p.254-256 (2000)

[4.227] {Sect. 4.5.2} J.H. Cai, W. Yang, T.J. Zhou, G. Gu, Y.W. Du: Magneto-optical Kerr effect and optical properties of amorphous Co1-xSix (0.59 ⇐ x ⇐ 0.77) alloy films, Appl Phys Lett 74, p.85-87 (1999)

[4.228] {Sect. 4.5.2} M. Neelakandan, D. Pant, E.L. Quitevis: Reorientational and intermolecular dynamics in binary liquid mixtures of hexafluorobenzene and benzene: Femtosecond optical Kerr effect measurements, Chem Phys Lett 265, p.283-292 (1997)

[4.229] {Sect. 4.5.2} B.I. Greene, R.C. Farrow: The subpicosecond Kerr effect in CS2, Chem. Phys. Lett. 98, p.273-276 (1983)

[4.230] {Sect. 4.5.2} J.M.Dziedzic, R.H. Stolen, A. Ashkin: Optical Kerr effect in long fibers, Appl. Opt. 20, p.1403-1406 (1981)

[4.231] {Sect. 4.5.2} D. Waldeck, A.J. Cross, Jr, D.B. McDonald, G.R. Fleming: Picosecond pulse induced transient molecular birefringence and dichroism, J. Chem. Phys. 74, p.3381-3387 (1981)

[4.232] {Sect. 4.5.2} S.C. Cerda, J.M. Hickmann: Spatial instabilities in the propagation of a cylindrical beam in a Kerr medium, Opt Commun 156, p.347-349 (1998)

[4.233] {Sect. 4.5.2} G. Jonusauskas, J. Oberle, E. Abraham, C. Rulliere: "Fast" amplifying optical Kerr gate using stimulated emission of organic non-linear dyes, Opt Commun 137, p.199-206 (1997)

[4.234] {Sect. 4.5.2} J.-M. Halbout, C.L. Tang: Femtosecond interferometry for non-linear optics, Appl. Phys. Lett. 40, p.765-767 (1982)

[4.235] {Sect. 4.5.2} E.P. Ippen, C.V. Shank: Picosecond response of a high-repetition-rate CS2 optical Kerr gate, Appl. Phys. Lett. 26, p.92-93 (1975)

[4.236] {Sect. 4.5.2} F. Parvaneh, M. Farhadiroushan, V.A. Handerek, A.J. Rogers: Single-shot distributed optical-fiber temperature sensing by the frequency-derived technique, Optics Letters 22, p.343-345 (1997)

[4.237] {Sect. 4.5.2} D. McMorrow, W.T. Lotshaw, G.A. Kenney-Wallace: Femtosecond Raman-induced Kerr effect. Temporal evolution of the vibrational normal modes in hologenated methanes, Chem. Phys. Lett. 145, p.309-314 (1988)

[4.238] {Sect. 4.5.3} Y. R. Shen: Principles of Nonlinear Optics, chapter 17 (John Wiley & Sons, Chichester, 1984)

[4.239] {Sect. 4.5.3} G. Fibich, A.L. Gaeta: Critical power for self-focusing in bulk media and in hollow waveguides, Optics Letters 25, p.335-337 (2000)

[4.240] {Sect. 4.5.3} K. Takahashi, R. Kodama, K.A. Tanaka, H. Hashimoto, Y. Kato, K. Mima, F.A. Weber, T.W. Barbee, L.B. DaSilva: Laser-hole boring into overdense plasmas measured with soft x-ray laser probing, Phys Rev Lett 84, p.2405-2408 (2000)

[4.241] {Sect. 4.5.3} O. Buttner, M. Bauer, S.O. Demokritov, B. Hillebrands, M.P. Kostylev, B.A. Kalinikos, A.N. Slavin: Collisions of spin wave envelope solitons and self-focused spin wave packets in yttrium iron garnet films, Phys Rev Lett 82, p.4320-4323 (1999)

[4.242] {Sect. 4.5.3} J. Tsai, A. Chiou, T.C. Hsieh, K. Hsu: One-dimensional self-focusing in photorefractive Bi12SiO20 crystal: theoretical modeling and experimental demonstration, Opt Commun 162, p.237-240 (1999)

[4.243] {Sect. 4.5.3} M. Bauer, O. Buttner, S.O. Demokritov, B. Hillebrands, V. Grimalsky, Y. Rapoport, A.N. Slavin: Observation of spatiotemporal self-focusing of spin waves in magnetic films, Phys Rev Lett 81, p.3769-3772 (1998)

[4.244] {Sect. 4.5.3} Y.C. Chen, W.Z. Lin: Thick lens model for self-focusing in Kerr medium, Appl Phys Lett 73, p.429-431 (1998)

[4.245] {Sect. 4.5.3} B. Crosignani, E. DelRe, P. Diporto, A. Degasperis: Self-focusing and self-trapping in unbiased centrosymmetric photorefractive media, Optics Letters 23, p.912-914 (1998)

[4.246] {Sect. 4.5.3} J.K. Ranka, A.L. Gaeta: Breakdown of the slowly varying envelope approximation in the self- focusing of ultrashort pulses, Optics Letters 23, p.534-536 (1998)

[4.247] {Sect. 4.5.3} G. Tempea, T. Brabec: Theory of self-focusing in a hollow waveguide, Optics Letters 23, p.762-764 (1998)

[4.248] {Sect. 4.5.3} C.C. Widmayer, L.R. Jones, D. Milam: Measurement of the nonlinear coefficient of carbon disulfide using holographic self-focusing, J Nonlinear Opt Physics Mat 7, p.563-570 (1998)

[4.249] {Sect. 4.5.3} F. Castaldo, D. Paparo, E. Santamato: Chaotic and hexagonal spontaneous pattern formation in the cross section of a laser beam in a defocusing Kerr-like film with single feedback mirror, Opt Commun 143, p.57-61 (1997)

[4.250] {Sect. 4.5.3} E. Esarey, P. Sprangle, J. Krall, A. Ting: Self-focusing and guiding of short laser pulses in ionizing gases and plasmas, IEEE J QE-33, p.1879-1914 (1997)

[4.251] {Sect. 4.5.3} G. Fibich, G.C. Papanicolaou: Self-focusing in the presence of small time dispersion and nonparaxiality, Optics Letters 22, p.1379-1381 (1997)

[4.252] {Sect. 4.5.3} G.S. He, M. Yoshida, J.D. Bhawalkar, P.N. Prasad: Two-photon resonance-enhanced refractive-index change and self-focusing in a dye-solution-filled hollow fiber system, Appl Opt 36, p.1155-1163 (1997)

[4.253] {Sect. 4.5.3} M. Vaupel, C. Seror, R. Dykstra: Self-focusing in photorefractive two-wave mixing, Optics Letters 22, p.1470-1472 (1997)

[4.254] {Sect. 4.5.3} A. Drobnik, L. Wolf: Influence of self-focusing on the operation of a neodymium glass laser, Sov. J. Quant. Electron. 8, p.274-275 (1978)

[4.255] {Sect. 4.5.3} C.R. Giuliano, J.H. Marburger: Observations of Moving Self-Foci in Sapphire, Phys. Rev. Lett. 27, p.905-908 (1971)

[4.256] {Sect. 4.5.3} M.M.T. Loy, Y.R. Shen: Correlation between Backward Stimulated Raman Pulse and Moving Focus in Liquids, Phys. Rev. Lett. 19, p.285-287 (1971)

[4.257] {Sect. 4.5.3} E.L. Dawes, J.H. Marburger: Computer Studies in Self-Focusing, Phys. Rev. 179, p.862-868 (1969)

[4.258] {Sect. 4.5.3} R.G. Brewer, C.H. Lee: Self-trapping with picosecond light pulses, Phys. Rev. Lett. 21, p.267-270 (1968)

[4.259] {Sect. 4.5.3} J.H. Marburger, E.L. Dawes: Dynamical Formation of a Small-Scale Filament, Phys. Rev. Lett. 21, p.556-558 (1968)

[4.260] {Sect. 4.5.3} E. Garmire, R.Y. Chiao, C.H. Townes: Dynamics and Characteristics of the Self-Trapping of Intense Light Beams, Phys. Rev. Lett. 16, p.347-349 (1966)

[4.261] {Sect. 4.5.3} M. Hercher: Laser-Induced Damage in Transparent Media, J. Opt. Soc. Am. 54, p.563 (1964)

[4.262] {Sect. 4.5.3} A. Brodeur, C.Y. Chien, F.A. Ilkov, S.L. Chin, O.G. Kosareva, V.P. Kandidov: Moving focus in the propagation of ultrashort laser pulses in air, Optics Letters 22, p.304-306 (1997)

[4.263] {Sect. 4.5.3} M. Mlejnek, M. Kolesik, J.V. Moloney, E.M. Wright: Optically turbulent femtosecond light guide in air, Phys Rev Lett 83, p.2938-2941 (1999)

[4.264] {Sect. 4.5.3} M. Jain, A.J. Merriam, A. Kasapi, G.Y. Yin, S.E. Harris: Elimination of optical self-focusing by population trapping, Phys Rev Lett 75, p.4385-4388 (1995)

[4.265] {Sect. 4.5.4} N. Akhmediev, A. Ankiewicz: Solitons; Non-linear pulses and beams (Chapman & Hall, New York, 1997)

[4.266] {Sect. 4.5.4} J. R. Taylor: Optical Solitons (Cambridge University Press, Cambridge, 1992)

[4.267] {Sect. 4.5.4} B. E. A. Saleh, M. C. Teich: Fundamentals of Photonics, chapter 19 (John Wiley & Sons, New York, 1991)

[4.268] {Sect. 4.5.4} T.H. Coskun, D.N. Christodoulides, Y.R. Kim, Z.G. Chen, M. Soljacic, M. Segev: Bright spatial solitons on a partially incoherent background, Phys Rev Lett 84, p.2374-2377 (2000)

[4.269] {Sect. 4.5.4} Y.S. Kivshar, A. Nepomnyashchy, V. Tikhonenko, J. Christou, B. LutherDavies: Vortex-stripe soliton interactions, Optics Letters 25, p.123-125 (2000)

[4.270] {Sect. 4.5.4} A.V. Buryak, V.V. Steblina, R.A. Sammut: Solitons and collapse suppression due to parametric interaction in bulk Kerr media, Optics Letters 24, p.1859-1861 (1999)

[4.271] {Sect. 4.5.4} Y.S. Kivshar, T.J. Alexander, S. Saltiel: Spatial optical solitons resulting from multistep cascading, Optics Letters 24, p.759-761 (1999)

[4.272] {Sect. 4.5.4} X. Liu, L.J. Qian, F.W. Wise: Generation of optical spatiotemporal solitons, Phys Rev Lett 82, p.4631-4634 (1999)

[4.273] {Sect. 4.5.4} D. Mihalache, D. Mazilu, J. Dorring, L. Torner: Elliptical light bullets, Opt Commun 159, p.129-138 (1999)

[4.274] {Sect. 4.5.4} R. Morandotti, U. Peschel, J.S. Aitchison, H.S. Eisenberg, Y. Silberberg: Dynamics of discrete solitons in optical waveguide arrays, Phys Rev Lett 83, p.2726-2729 (1999)

[4.275] {Sect. 4.5.4} J. Scheuer, M. Orenstein: Interactions and switching of spatial soliton pairs in the vicinity of a nonlinear interface, Optics Letters 24, p.1735-1737 (1999)

[4.276] {Sect. 4.5.4} M.F. Shih, F.W. Sheu: Photorefractive polymeric optical spatial solitons, Optics Letters 24, p.1853-1855 (1999)

[4.277] {Sect. 4.5.4} L. Torner, J.P. Torres, D. Artigas, D. Mihalache, D. Mazilu: Soliton content with quadratic nonlinearities, Opt Commun 164, p.153-159 (1999)

[4.278] {Sect. 4.5.4} S. Trillo, M. Haelterman: Excitation and bistability of self-trapped signal beams in optical parametric oscillators, Optics Letters 23, p.1514-1516 (1998)

[4.279] {Sect. 4.5.4} V. Kutuzov, V.M. Petnikova, V.V. Shuvalov, V.A Vysloukh: Cross-modulation coupling of incoherent soliton modes in photorefractive crystals, Phys. Rev. E 57, p.6056-6065 (1998)

[4.280] {Sect. 4.5.4} G.S. Garciaquirino, M.D. Iturbecastillo, V.A. Vysloukh, J.J. SanchezMondragon, S.I. Stepanov, G. Lugomartinez, G.E. Torrescisneros: Observation of interaction forces between one-dimensional spatial solitons in photorefractive crystals, Optics Letters 22, p.154-156 (1997)

[4.281] {Sect. 4.5.4} V. Kutuzov, V.M. Petnikova, V.V. Shuvalov, V.A Vysloukh: Spatial solitons and shock waves in photorefractive crystals with nonlocal nonlinearity, J. Nonlin. Opt. Phys. & Mat. 6, p.421-442 (1997)

[4.282] {Sect. 4.5.4} G. Duree, M. Morin, G. Salamo, M. Segev, B. Crosignani, P. Di Porto, E. Sharp, A. Yariv: Dark Photorefractive Spatial Solitons and Photorefractive Vortex Solitons, Phys. Rev. Lett. 74, p.1978-1982 (1995)

[4.283] {Sect. 4.5.4} M.-F. Shi, M. Segev, G.C. Valley, G. Salamo, B. Crosignani, P. Di Porto: Observation of two-dimensional steady-state photorefractive screening solitons, Electron. Lett. 31, p.826-827 (1995)

[4.284] {Sect. 4.5.4} M.D.I. Castillo, P.A. M. Aguilar, J.J. Sanchez-Mondragon, S. Stepanov, V. Vysloukh: Spatial solitons in photorefractive Bi12TiO20 with drift mechanism of nonlinearity, Appl. Phys. Lett. 64, p.408-410 (1994)

[4.285] {Sect. 4.5.4} G.C. Duree, Jr, J.L. Shultz, G.J. Salamo: Observation of Self-Trapping of an Optical Beam Due to the Photorefractive Effect, Phys. Rev. Lett. 71, p.533-536 (1993)

[4.286] {Sect. 4.5.4} F.X. Kartner, H.A. Haus: Quantum-Mechanical Stability of Solitons and the Correspondence Principle, Phys Rev A 48, p.2361-2369 (1993)

[4.287] {Sect. 4.5.4} A. Berzanskis, A. Matijosius, A. Piskarskas, V. Smilgevicius, A. Stabinis: Sum-frequency mixing of optical vortices in nonlinear crystals, Opt Commun 150, p.372-380 (1998)

[4.288] {Sect. 4.5.4} Y.S. Kivshar, J. Christou, V. Tikhonenko, B. LutherDavies, L.M. Pismen: Dynamics of optical vortex solitons, Opt Commun 152, p.198-206 (1998)

[4.289] {Sect. 4.5.5} L.B. Au, L. Solymar, C. Dettmann, H.J. Eichler, R. Macdonald, J. Schwartz: Theoretical and Experimental Investigations of the Reorientation of Liquid Crystal Molecules induced by Laser Beams, Physica A 174, p.94-118 (1991)

[4.290] {Sect. 4.5.5} H.J. Eichler, R. Macdonald, C. Dettmann: Nonlinear Diffraction of CW-Laserbeams by Spatial Selfphase Modulation in Nematic Liquid Crystals, Mol. Cryst. Liq. Cryst.174, p.153-168 (1989)

[4.291] {Sect. 4.5.6} J.P. Gordon, R.C.C. Leite, R.S. Moore, S.P.S. Porto, J.R. Whinnery: Long-Transient Effects in Lasers with Inserted Liquid Samples, J. Appl. Phys. 36, p.3-8 (1965)

[4.292] {Sect. 4.5.7} F. Cattani, D. Anderson, A. Berntson, M. Lisak: Effect of self-phase modulation in chirped-pulse-amplification-like schemes, J Opt Soc Am B Opt Physics 16, p.1874-1879 (1999)

[4.293] {Sect. 4.5.7} N. Karasawa, R. Morita, L. Xu, H. Shigekawa, M. Yamashita: Theory of ultrabroadband optical pulse generation by induced phase modulation in a gas-filled hollow waveguide, J Opt Soc Am B Opt Physics 16, p.662-668 (1999)

[4.294] {Sect. 4.5.7} T.G. Ulmer, R.S.K. Tan, Z.P. Zhou, S.E. Ralph, R.P. Kenan, C.M. Verber, A.J. SpringThorpe: Two-photon absorption-induced self-phase modulation in GaAs-AlGaAs waveguides for surface-emitted second-harmonic generation, Optics Letters 24, p.756-758 (1999)

[4.295] {Sect. 4.5.7} S.F. Feldman, P.R. Staver, W.T. Lotshaw: Observation of spectral broadening caused by self-phase modulation in highly multimode optical fiber, Appl Opt 36, p.617-621 (1997)

[4.296] {Sect. 4.5.7} M.D. Perry, T. Ditmire, B.C. Stuart: Self phase modulation in chirped pulse amplification, Optics Letters 19, p.2149-2151 (1994)

[4.297] {Sect. 4.5.7} Q.D. Liu, J.T. Chen, Q.Z. Wang, P.P. Ho, R.R. Alfano: Single pulse degenerate cross phase modulation in a single mode optical fiber, Optics Letters 20, p.542-544 (1995)

[4.298] {Sect. 4.5.7} R.M. Rassoul, A. Ivanov, E. Freysz, A. Ducasse, F. Hache: Second-harmonic generation under phase-velocity and group-velocity mismatch: Influence of cascading self-phase and cross-phase modulation, Optics Letters 22, p.268-270 (1997)

[4.299] {Sect. 4.5.8} F.K. Abdullaev, B.B. Baizakov: Disintegration of a soliton in a dispersion-managed optical communication line with random parameters, Optics Letters 25, p.93-95 (2000)

[4.300] {Sect. 4.5.8} N. Akhmediev, A. Ankiewicz: Partially coherent solitons on a finite background, Phys Rev Lett 82, p.2661-2664 (1999)

[4.301] {Sect. 4.5.8} S.T. Cundiff, B.C. Collings, N.N. Akhmediev, J.M. SotoCrespo, K. Bergman, W.H. Knox: Observation of polarization-locked vector solitons in an optical fiber, Phys Rev Lett 82, p.3988-3991 (1999)

[4.302] {Sect. 4.5.8} S. Darmanyan, A. Kobyakov, F. Lederer: Quadratic solitons in nonconservative media, Optics Letters 24, p.1517-1519 (1999)

[4.303] {Sect. 4.5.8} M. Hanna, H. Porte, J.P. Goedgebuer, W.T. Rhodes: Soliton optical phase control by use of is-line filters, Optics Letters 24, p.732-734 (1999)

[4.304] {Sect. 4.5.8} P.S. Jian, W.E. Torruellas, M. Haelterman, S. Trillo, U. Peschel, F. Lederer: Solitons of singly resonant optical parametric oscillators, Optics Letters 24, p.400-402 (1999)

[4.305] {Sect. 4.5.8} D. Krylov, L. Leng, K. Bergman, J.C. Bronski, J.N. Kutz: Observation of the breakup of a prechirped N-soliton in an optical fiber, Optics Letters 24, p.1191-1193 (1999)

[4.306] {Sect. 4.5.8} D. Levandovsky, M. Vasilyev, P. Kumar: Perturbation theory of quantum solitons: continuum evolution and optimum squeezing by spectral filtering, Optics Letters 24, p.43-45 (1999)

[4.307] {Sect. 4.5.8} A.H. Liang, H. Toda, A. Hasegawa: High-speed soliton transmission in dense periodic fibers, Optics Letters 24, p.799-801 (1999)

[4.308] {Sect. 4.5.8} Q.H. Park, H.J. Shin: Parametric control of soliton light traffic by cw traffic light, Phys Rev Lett 82, p.4432-4435 (1999)

[4.309] {Sect. 4.5.8} I.S. Penketh, P. Harper, S.B. Alleston, A.M. Niculae, I. Bennion, N.J. Doran: 10-Gbit/s dispersion-managed soliton transmission over 16,500 km in standard fiber by reduction of soliton interactions, Optics Letters 24, p.802-804 (1999)

[4.310] {Sect. 4.5.8} K. Chan, W. Cao: Generation of ultrashort fundamental solitons from cw light using cross-phase modulation and Raman amplification in optical fibers, Opt Commun 158, p.159-169 (1998)

[4.311] {Sect. 4.5.8} M. Matsumoto: Instability of dispersion-managed solitons in a system with filtering, Optics Letters 23, p.1901-1903 (1998)

[4.312] {Sect. 4.5.8} P. Shum, S.F. Yu: Numerical analysis of nonlinear soliton propagation phenomena using the fuzzy mesh analysis technique, IEEE J QE-34, p.2029-2035 (1998)

[4.313] {Sect. 4.5.8} E.L. Buckland, R.W. Boyd, A.F. Evans: Observation of a Raman-induced interpulse phase migration in the propagation of an ultrahigh-bit-rate coherent soliton train, Optics Letters 22, p.454-456 (1997)

[4.314] {Sect. 4.5.8} B.C. Collings, K. Bergman, W.H. Knox: True fundamental solitons in a passively mode-locked short-cavity Cr4+:YAG laser, Optics Letters 22, p.1098-1100 (1997)

[4.315] {Sect. 4.5.8} H. Hatamihanza, P.L. Chu, B.A. Malomed, G.D. Peng: Soliton compression and splitting in double-core nonlinear optical fibers, Opt Commun 134, p.59-65 (1997)

[4.316] {Sect. 4.5.8} R.H. Stolen, L.F. Mollenauer: Observation of pulse restoration at the soliton period in optical fibers, Opt. Lett. 8, p.186-188 (1983)

[4.317] {Sect. 4.5.8} L.F. Mollenauer, R.H. Stolen, J.P. Gordon: Experimental Observation of Picosecond Pulse Narrowing and Solitons in Optical Fibers, Phys. Rev. Lett. 45, p.1095-1098 (1980)

[4.318] {Sect. 4.5.8} F.G. Omenetto, B.P. Luce, D. Yarotski, A.J. Taylor: Observation of chirped soliton dynamics at lambda=1.55 mu m in a single-mode optical fiber with frequency-resolved optical gating, Optics Letters 24, p.1392-1394 (1999)

[4.319] {Sect. 4.5.8} M. Piche, J.F. Cormier, X.N. Zhu: Bright optical soliton in the presence of fourth-order dispersion, Optics Letters 21, p.845-847 (1996)

[4.320] {Sect. 4.5.8} C. Deangelis, M. Santagiustina, S. Wabnitz: Stability of vector solitons in fiber laser and transmission systems, Opt Commun 122, p.23-27 (1995)

[4.321] {Sect. 4.5.8} A.E. Kaplan, P.L. Shkolnikov: Subfemtosecond high-intensity unipolar electromagnetic solitons and shock waves, J Nonlinear Opt Physics Mat 4, p.831-841 (1995)

[4.322] {Sect. 4.5.8} S.V. Bulanov, T.Z. Esirkepov, N.M. Naumova, F. Pegoraro, V.A. Vshivkov: Solitonlike electromagnetic waves behind a superintense laser pulse in a plasma, Phys Rev Lett 82, p.3440-3443 (1999)

[4.323] {Sect. 4.5.9.1} C. Labaune, H.A. Baldis, B.S. Bauer, E. Schifano, B.I. Cohen: Spatial and temporal coexistence of stimulated scattering processes under crossed-laser-beam irradiation, Phys Rev Lett 82, p.3613-3616 (1999)

[4.324] {Sect. 4.5.9.1} K. Otsuka, R. Kawai, Y. Asakawa, T. Fukazawa: Highly sensitive self-mixing measurement of Brillouin scattering with a laser-diode-pumped microchip LiNdP4O12 laser, Optics Letters 24, p.1862-1864 (1999)

[4.325] {Sect. 4.5.9.1} A.A. Fotiadi, R.V. Kiyan: Cooperative stimulated Brillouin and Rayleigh backscattering process in optical fiber, Optics Letters 23, p.1805-1807 (1998)

[4.326] {Sect. 4.5.9.1} S. Afsharvahid, V. Devrelis, J. Munch: Nature of intensity and phase modulations in stimulated Brillouin scattering, Phys. Rev. A 57, p.3961-3971 (1998)

[4.327] {Sect. 4.5.9.1} M.S. Jo, C.H. Nam: Transient stimulated Brillouin scattering reflectivity in CS2 and SF6 under multipulse employment, Appl. Opt. 36, p.1149-1154 (1997)

[4.328] {Sect. 4.5.9.1} P.E. Young, M.E. Foord, A.V. Maximov, W. Rozmus: Stimulated Brillouin scattering in multispecies laser- produced plasmas, Phys Rev Lett 77, p.1278-1281 (1996)

[4.329] {Sect. 4.5.9.1} T. Afsharrad, L.A. Gizzi, M. Desselberger, O. Willi: Effect of filamentation of Brillouin scattering in large underdense plasmas irradiated by incoherent laser light, Phys Rev Lett 75, p.4413-4416 (1995)

[4.330] {Sect. 4.5.9.1} R.L. Berger, B.F. Lasinski, A.B. Langdon, T.B. Kaiser, B.B. Afeyan, B.I. Cohen, C.H. Still, E.A. Williams: Influence of spatial and temporal laser beam smoothing on stimulated Brillouin scattering in filamentary laser light, Phys Rev Lett 75, p.1078-1081 (1995)

[4.331] {Sect. 4.5.9.1} H.J. Eichler, R. Menzel, R. Sander, M. Schulzke, J. Schwartz: SBS at different wavelengths between 308 and 725 nm, Opt. Commun. 121, p.49-54 (1995)

[4.332] {Sect. 4.5.9.1} H.J. Eichler, R. König, R. Menzel, R. Sander, J. Schwartz, H.J. Pätzold: Test of Organic SBS Liquids in the IR and the UV, Int. J. Nonlinear Optics 2, p.267-270 (1993)

[4.333] {Sect. 4.5.9.1} N.F. Andreev, E. Khazanov, G.A. Pasmanik: Applications of Brillouin Cells to High Repetition Rate Solid-State Lasers, IEEE J. QE-28, p.330-341 (1992)

[4.334] {Sect. 4.5.9.1} Yu.I. Bychkov, V.F. Losev, Yu.N. Panchenko: Experimental investigation of the efficiency of phase conjugation of an XeCl laser beam by stimulated Brillouin scattering, Sov. J. Quantum. Electron. 22 p.638-640 (1992)

[4.335] {Sect. 4.5.9.1} H.J. Eichler, R. Menzel, R. Sander, B. Smandek: Reflectivity Enhancement of Stimulated Brillouin Scattering (SBS) Liquids by Purification, Opt. Commun. 89, p.260-262 (1992)

[4.336] {Sect. 4.5.9.1} M.R. Osborn, M.A. O'Key: Temporal response of stimulated Brillouin scattering phase conjugation, Opt. Comm. 94, p.346-352 (1992)

[4.337] {Sect. 4.5.9.1} G.K.N. Wong, M.J. Damzen: Investigations of Optical Feedback Used to Enhance Stimulated Scattering, IEEE J. QE-26, p.139-148 (1990)

[4.338] {Sect. 4.5.9.1} V.I. Bespalov, E.L. Bubis, O.V. Kulagin, G.A. Pasmanik, A.A. Shilov: Stimulated Brillouin scattering and stimulated thermal scattering of microsecond pulses, Sov. J. Quantum Electron. 16, p.1348-1352 (1986)

[4.339] {Sect. 4.5.9.1} P. Narum, M.D. Skeldon, R.W. Boyd: Effect of Laser Mode Structure on Stimulated Brillouin Scattering, IEEE J. QE-22, p.2161-2167 (1986)

[4.340] {Sect. 4.5.9.1} J.M. Vaughan: Brillouin scattering in the nematic and isotropic phases of a liquid crystal, Phys. Lett. 58A, p.325-328 (1976)

[4.341] {Sect. 4.5.9.1} M. Maier: Quasisteady State in the Stimulated Brillouin Scattering of Liquids, Phys. Rev. 166, p.113-119 (1967)

[4.342] {Sect. 4.5.9.2} V.T. Tikhochuk, C. Labaune, H.A. Baldis: Modeling of a stimulated Brillouin scattering experiment with statistical distribution of speckles, Phys. Plasmas 3, p.3777-3785 (1996)

[4.343] {Sect. 4.5.9.2} R.G. Harrison, D. Yu, W. Lu, P.M. Ripley: Chaotic stimulated Brillouin scattering: theory and experiment, Physica D 86, p.182-188 (1995)

[4.344] {Sect. 4.5.9.2} A. Kummrow: Hermite-gaussian theory of focused beam SBS cells, Opt. Commun. 96, p.185-194 (1993)

[4.345] {Sect. 4.5.9.2} R. Menzel, H.J. Eichler: Computation of Stimulated Brillouin Scattering (SBS) with Focussed Beams, Int. J. Nonlinear Optics 2, p.255-260 (1993)

[4.346] {Sect. 4.5.9.2} R. Chu, M. Kanefsky, J. Falk: Numerical study of transient stimulated Brillouin scattering, J. Appl. Phys. 71, p.4653-4658 (1992)

[4.347] {Sect. 4.5.9.2} R. Menzel, H.J. Eichler: Temporal and Spatial Reflectivity of Focussed Beams in Stimulated Brillouin Scattering for Phaseconjugation, Phys. Rev. A 46, p.7139-7149 (1992)

[4.348] {Sect. 4.5.9.2} G.J. Crofts, M.J. Damzen: Steady-state analysis and design criteria of two-cell stimulated Brillouin scattering systems, Opt. Comm. 81, p.237-241 (1991)

[4.349] {Sect. 4.5.9.2} P.H. Hu, J.A. Goldstone, S.S. Ma: Theoretical study of phase conjugation in stimulated Brillouin scattering, J. Opt. Soc. Am. B 6, p.1813-1822 (1989)

[4.350] {Sect. 4.5.9.2} G.C. Valley: A Review of Stimulated Brillouin Scattering Excited with a Broad-Band Pump Laser, IEEE J. QE-22, p.704-711 (1986)

[4.351] {Sect. 4.5.9.2} R.H. Lehmberg: Numerical study of phase conjugation in stimulated Brillouin scattering from an optical waveguide, J. Opt. Soc. Am. 73, p.558-566 (1983)

[4.352] {Sect. 4.5.9.2} R.H. Lehmberg: Numerical study of phase conjugation in stimulated backscatter with pump depletion, Opt. Comm. 43, p.369-374 (1982)

[4.353] {Sect. 4.5.9.2} A. Yariv: Quantum Theory for Parametric Interactions of Light and Hypersound, IEEE J. QE-1, p.28-36 (1965)

[4.354] {Sect. 4.5.9.2} P.W. Rambo, S.C. Wilks, W.L. Kruer: Hybrid particle-in-cell simulations of stimulated Brillouin scattering including ion-ion collisions, Phys Rev Lett 79, p.83-86 (1997)

[4.355] {Sect. 4.5.9.3} N.-M. Nguyen-Vo, S.J. Pfeifer: A Model of Spontaneous Brillouin Scattering as the Noise Source for Stimulated Scattering, IEEE J. QE-29, p.508-514 (1993)

[4.356] {Sect. 4.5.9.3} Y. Glick, S. Sternklar: Reducing the noise in Brillouin amplification by mode-selective phase conjugation, Opt. Lett. 17, p.662-664 (1992)

[4.357] {Sect. 4.5.9.3} O.V. Kulagin, G.A. Pasmanik, A.A. Shilov: Amplification and phase conjugation of weak signals, Sov. Phys. Usp. 35, p.506-519 (1992)

[4.358] {Sect. 4.5.9.3} M. Shirasaki, H.A. Haus: Reduction of Guided-Acoustic-Wave Brillouin Scattering Noise in a Squeezer, Optics Letters 17, p.1225-1227 (1992)

[4.359] {Sect. 4.5.9.3} R.W. Boyd, K. Rzazewski: Noise initiation on stimulated Brillouin scattering, Phys. Rev. A 42, p.5514-5521 (1990)

[4.360] {Sect. 4.5.9.4} J.C. Fernandez, B.S. Bauer, K.S. Bradley, J.A. Cobble, D.S. Montgomery, R.G. Watt, B. Bezzerides, K.G. Estabrook, R. Focia, S.R. Goldman et al.: Increased saturated levels of stimulated brillouin scattering of a laser by seeding a plasma with an external light source, Phys Rev Lett 81, p.2252-2255 (1998)

[4.361] {Sect. 4.5.9.4} A. Melloni, M. Frasca, A. Garavaglia, A. Tonini, M. Martinelli: Direct measurement of electrostriction in optical fibers, Optics Letters 23, p.691-693 (1998)

[4.362] {Sect. 4.5.9.4} M.S. Jo, C.H. Nam: Transient stimulated Brillouin scattering reflectivity in CS2 and SF6 under multipulse employment, Appl Opt 36, p.1149-1154 (1997)

[4.363] {Sect. 4.5.9.4} D.C. Jones: Characterisation of liquid brillouin media at 532 nm, J Nonlinear Opt Physics Mat 6, p.69-79 (1997)

[4.364] {Sect. 4.5.9.4} H. Yoshida, V. Kmetik, H. Fujita, M. Nakatsuka, T. Yamanaka, K. Yoshida: Heavy fluorocarbon liquids for a phase-conjugated stimulated Brillouin scattering mirror, Appl Opt 36, p.3739-3744 (1997)

[4.365] {Sect. 4.5.9.4} H. Yoshida, M. Nakatsuka, H. Fujita, T. Sasaki, K. Yoshida: High-energy operation of a stimulated Brillouin scattering mirror in an L-Arginine phosphate monohydrate crystal, Appl. Opt. 36, p.7783-7787 (1997)

[4.366] {Sect. 4.5.9.4} H. Yoshida, V. Kmetik, H. Fujita, M. Nakatsuka, T. Yamanaka, K. Yoshida: Heavy fluorocarbon liquids for a phase-conjugated stimulated Brillouin scattering mirror, Appl. Opt. 36, p.3739-3744 (1997)

[4.367] {Sect. 4.5.9.4} H.J. Eichler, R. König, H.-J. Pätzold, J. Schwartz: SBS mirrors for XeCl lasers with a broad spectrum, Appl. Phys. B. 61, p.73-80 (1995)

[4.368] {Sect. 4.5.9.4} S.T. Animoto, R.W.F. Gross, L. Garman-DuVall, T.W. Good, J.D. Piranian: Stimulated-Brillouin-scattering properties of SnCl4, Opt. Lett. 16, p.1382-1384 (1991)

[4.369] {Sect. 4.5.9.4} A. Kummrow, H. Meng: Pressure dependence of stimulated Brillouin backscattering in gases, Opt. Commun. 83, p.342-348 (1991)

[4.370] {Sect. 4.5.9.4} D.C. Jones, M.S. Mangir, D.A. Rockwell, J.O. White: Stimulated Brillouin scattering gain variation and transient effects in a CH4:He binary gas mixture, J. Opt. Soc. Am. B 7, p.2090-2096 (1990)

[4.371] {Sect. 4.5.9.4} E.L. Bubis, V.V. Vargin, L.R. Konchalina, A.A. Shilov: Study of low-absorption media for SBS in the near-IR-spectral range, Opt. Spectrosc. (USSR) 65, p.757-759 (1989)

[4.372] {Sect. 4.5.9.4} P.E. Dyer, J.S. Leggatt: Phase conjugation studies of a quasi-cw CO2 laser in liquid CS2, Opt. Comm. 74, p.124-128 (1989)

[4.373] {Sect. 4.5.9.4} F.E. Hovis, J.D. Kelley: Phase conjugation by stimulated Brillouin scattering in CClF3 near the gas-liquid critical temperature, J.Opt. Soc. Am. B. 6, p.840-842 (1989)

[4.374] {Sect. 4.5.9.4} Y. Aoki, K. Tajima: Stimulated Brillouin scattering in a long single-mode fiber excited with a multimode pump laser, J. Opt. Soc. Am. B 5, p.358-363 (1988)

[4.375] {Sect. 4.5.9.4} M.J. Damzen, M.H.R. Hutchinson, W.A. Schroeder: Direct Measurement of the Acoustic Decay Times of Hypersonic Waves Generated by SBS, IEEE J. QE-23, p.328-334 (1987)

[4.376] {Sect. 4.5.9.4} V.M. Volynkin, K.V. Gratsianov, A.N. Kolesnikov, Yu.I. Kruzhilin, V.V. Lyubimov, S.A. Markosov, V.G. Pankov, A.I. Stepanov, S.V. Shklyarik: Reflection by stimulated Brillouin scattering mirrors based on tetrachlorides of group IV elements, Sov. J. Quantum Electron. 15, p.1641-1642 (1985)

[4.377] {Sect. 4.5.9.4} D. Pohl, W. Kaiser: Time-Resolved Investigations of Stimulated Brillouin Scattering in Transparent and Absorbing Media: Determination of Phonon Lifetimes, Phys. Rev. B 1, p.31-43 (1970)

[4.378] {Sect. 4.5.9.4} M.R.Osborne: Stimulated Brillouin scattering using cylindrical focusing optics, J. Opt. Soc. Am. B 7, p.2106-2112 (1990)

[4.379] {Sect. 4.5.9.4} J. Munch, R.F. Wuerker, M.J. LeFebvre: Interaction length for optical phase conjugation by stimulated Brillouin scattering: an experimental investigation, Appl. Opt. 28, p.3099-3105 (1989)

[4.380] {Sect. 4.5.9.4} L.P. Schelonka, C.M. Clayton: Effect of focal intensity on stimulated-Brillouin-scattering reflectivity and fidelity, Opt. Lett. 13, p.42-44 (1988)

[4.381] {Sect. 4.5.9.4} N.B. Baranova, B.Ya. Zel'dovich, V.V. Shkunov: Wavefront reversal in stimulated light scattering in a focused spatially ingomogeneous pump beam, Sov. J. Quantum Electron. 8, p.559-566 (1978)

[4.382] {Sect. 4.5.9.4} R.A. Mullen: Multiple-Short-Pulse Stimulated Brillouin Scattering for Trains of 200 ps Pulses at 1.06 μm, IEEE J. QE-26, p.1299-1303 (1990)

[4.383] {Sect. 4.5.9.4} G. Cook, K.D. Ridley: Investigation of the bandwidth dependent characteristics of stimulated Brillouin scattering using a modeless dye laser, Opt Commun 130, p.192-204 (1996)

[4.384] {Sect. 4.5.9.4} V.F. Losev, Yu. N. Panchenko: Characteristics of stimulated scattering of broad-band XeCl laser radiation, Quant. Electron. 25, p.448-449 (1995)

[4.385] {Sect. 4.5.9.4} P.C. Wait, T.P. Newson: Measurement of Brillouin scattering coherence length as a function of pump power to determine Brillouin linewidth, Opt. Commun. 117, p.142-146 (1995)

[4.386] {Sect. 4.5.9.4} H.J. Eichler, R. König, R. Menzel, H.J. Pätzold, J. Schwartz: Stimulated Brillouin Scattering of Broadband XeCl-Laser Radiation by Hydrocarbons Liquids, Int. J. Nonlinear Optics 2, p.247-253 (1993)

[4.387] {Sect. 4.5.9.4} D. Wang, G. Rivoire: Large spectral bandwidth stimulated Rayleigh-wing scattering in CS2, J. Chem. Phys.98, p.9279-9283 (1993)

[4.388] {Sect. 4.5.9.4} H.J. Eichler, R. König, R. Menzel, H.-J. Pätzold, J. Schwartz: SBS-Reflection of Broadband XeCl-Excimer-Laser-Radiation: Comparision of Suitable SBS-Liquids, J. Phys. D: Appl. Phys. 25, p.1162-1168 (1992)

[4.389] {Sect. 4.5.9.4} Y-S. Kuo, K. Choi, J.K. McIver: The effect of pump bandwidth, lens focal length and lens focal point location on Stimulated Brillouin Scattering threshold and reflectivity, Opt. Comm. 80, p.233-238 (1991)

[4.390] {Sect. 4.5.9.4} J.-Z. Zhang, G. Chen, R.K. Chang: Pumping of stimulated Raman scattering by stimulated Brillouin scattering within a single liquid droplet: input laser linewidth effects, J. Opt. Soc. Am. B 7, p.108-115 (1990)

[4.391] {Sect. 4.5.9.4} R.A. Mullen, R.C. Lind, G.C. Valley: Observaton of stimulated Brillouin scattering gain with a dual spectral-line pump, Opt. Comm. 63, p.123-128 (1987)

[4.392] {Sect. 4.5.9.4} M. Cronin-Golomb, S.-K. Kwong, A. Yariv: Multicolor passive (self-pumped) phase conjugation, Appl. Phys. Lett. 44, p.727-729 (1984)

[4.393] {Sect. 4.5.9.4} B.Ya. Zel'dovich, V.V. Shkunov: Influence of the group velocity mismatch on reproduction of the pump spectrum under stimulated scattering conditions, Sov. J. Quantum Electron. 8, p.1505-1506 (1978)

[4.394] {Sect. 4.5.9.4} I.G. Zubarev, S.I. Mikahilov: Influence of parametric effects on the stimulated scattering of nonmonochromatic pump radiation, Sov. J. Quantum Electron. 8, p.1338-1344 (1978)

[4.395] {Sect. 4.5.9.4} V.I. Kovalev, V.I. Popovichev, V.V. Ragul'skii, F.S. Faizullov: Gain and linewidth in stimulated Brillouin scattering in gases, Sov. J. Quantum Electron. 2, p.69-71 (1972)

[4.396] {Sect. 4.5.9.4} Y.E. D'yakov: Excitation of stimulated light scattering by broad-spectrum pumping, JETP Lett. 11p.243-246 (1970)

[4.397] {Sect. 4.5.9.7} W. Jinsong, T. Weizhong, Z. Wen: Stimulated Brillouin scattering initiated by thermally excited acoustic waves in absorption media, Opt. Commun. 123, p.574-576 (1996)

[4.398] {Sect. 4.5.9.7} K. Inoue: Brillouin threshold in an optical fiber with bedirectional pump lights, Opt. Comm. 120, p.34-38 (1995)

[4.399] {Sect. 4.5.9.7} M.T. Duignan, B.J. Feldman, W.T. Whitney: Theshold reduction for stimulated Brillouin scattering using a multipass Herriott cell, J. Opt. Soc. Am. B. 9, p.548-559 (1992)

[4.400] {Sect. 4.5.9.7} N.F. Andreev, V.I. Bespalov, M.A. Dvoretsky, G.A. Pasmanik: Phase Conjugation of Single Photons, IEEE J. QE-25, p.346-350 (1989)

[4.401] {Sect. 4.5.9.7} M. Maier, G. Renner: Transient Threshold Power of Stimulated Brillouin Raman Scattering, Phys. Lett. A 34, p.299-300 (1971)

[4.402] {Sect. 4.5.9.8} A. Heuer, R. Menzel: Phase conjugating SBS-mirror for low powers and reflectivities above 90 % in an internally tapered optical fiber, Opt. Lett. 23, p.834-836 (1998)

[4.403] {Sect. 4.5.9.8} D.C. Jones, M.S. Mangir, D.A. Rockwell: A stimulated Brillouin scattering phase-conjugate mirror having a peak-power threshold <100 W, Opt. Comm. 123, p.175-181 (1996)

[4.404] {Sect. 4.5.9.8} A.M. Scott, W.T. Whitney: Characteristics of a Brillouin ring resonator used for phase conjugation at 2.1µm, J.Opt. Soc. Am. B 12, p.1634-1641 (1995)

[4.405] {Sect. 4.5.9.8} G.K.N. Wong, M.J. Damzen: Enhancement of the phase-conjugate stimulated Brillouin scattering process using optical feedback, J. Mod. Opt. 35, p.483-490 (1988)

[4.406] {Sect. 4.5.9.9} B. Kralikova, J. Skala, P. Straka, H. Turcicova: Image restoration in a highly non-steady-state regime of stimulated Brillouin scattering in a photodissociation iodine laser, Optics Letters 22, p.766-768 (1997)

[4.407] {Sect. 4.5.9.9} V.F. Losev, Y.N. Panchenko: Spectral and spatial selection of XeCl laser radiation by an SBS mirror, Opt Commun 136, p.31-34 (1997)

[4.408] {Sect. 4.5.9.9} P.C. Wait, K. Desouza, T.P. Newson: A theoretical comparison of spontaneous Raman and Brillouin based fibre optic distributed temperature sensors, Opt Commun 144, p.17-23 (1997)

[4.409] {Sect. 4.5.9.9} H.J. Eichler, S. Heinrich, J. Schwartz: Self-starting short-pulse XeCl laser with a stimulated Brillouin scattering mirror, Optics Letters 21, p.1909-1911 (1996)

[4.410] {Sect. 4.5.9.9} D.L. Carrroll, R. Johnson, S.J. Pfeifer, R.H. Moyer: Experimental investigations of stimulated Brillouin scattering beam combination, J. Opt. Soc. Am. B 9, p.2214-2224 (1992)

[4.411] {Sect. 4.5.9.9} D.J. Gauthier, R.W. Boyd: Phase-conjugate Fizeau interferometer, Opt. Lett. 14, p.323-325 (1989)

[4.412] {Sect. 4.5.9.9} R.H. Moyer, M. Valley, M.C. Cimolino: Beam combination through stimulated Brillouin scattering, J. Opt. Soc. Am. B 5, p.2473-2489 (1988)

[4.413] {Sect. 4.5.9.9} R.P. Drake, R.G. Watt, K. Estabrook: Onset and Saturation of the Spectral Intensity of Stimulated Brillouin Scattering in Inhomogeneous Laser-Produced Plasmas, Phys. Rev. Lett. 77, p.79-82 (1996)

[4.414] {Sect. 4.5.9.9} R.G. Watt, J. Cobble, D.F. DuBois, J.C. Fenandez, H.A. Rose, R.P. Drake, B.S. Bauer: Dependence of stimulated Brillouin scattering on focusing optic F number in long scale-length plasmas, Phys. Plasmas 3, p.1091-1095 (1996)

[4.415] {Sect. 4.5.10} W. Kaiser, M. Maier: Stimulated Rayleigh Brillouin and Raman-spectroscopy, in Laser Handbook, ed. by F.T Arecci, E.O. Schulz-Dubois (North-Holland, Amsterdam 1972) p. 1077

[4.416] {Sect. 4.5.10} R.M. Herman, M.A. Gray: Theoretical Prediction of the Stimulated Thermal Rayleigh Scattering in Liquids, Phys. Rev. Lett. 19, p.824-828 (1967)

[4.417] {Sect. 4.5.11} G. Olbrechts, K. Wostyn, K. Clays, A. Persoons: High-frequency demodulation of multiphoton fluorescence in long-wavelength hyper-Rayleigh scattering, Optics Letters 24, p.403-405 (1999)

[4.418] {Sect. 4.5.11} T. Latz, F. Aupers, V.M. Baev, P.E. Toschek: Emission spectrum of a multimode dye laser with frequency-shifted feedback for the simulation of Rayleigh scattering, Opt Commun 156, p.210-218 (1998)

[4.419] {Sect. 4.5.11} M.M. Denariez-Roberge, G. Giuliani: High-power single-mode laser operation using stimulated Rayleigh scattering, Opt. Lett. 6, p.339-3341 (1981)

[4.420] {Sect. 4.5.11} Y. Carmel, J. Ivers, R.E. Kribel, J. Nation: Intense Coherent Cherenkov Radiation Due to the Interaction of a Relativistic Electron Beam with a Slow-Wave Structure, Phys. Rev. Lett. 33, p.1278-1282 (1974)

[4.421] {Sect. 4.5.11} W.H. Lowdermilk, N. Bloembergen: Stimulated Concentration Scattering in the Binary-Gas Mixtures Xe-He and SF6-He, Phys. Rev. A 5, p.1423-1443 (1972)

[4.422] {Sect. 4.5.11} R.H. Pantell, G. Soncini, H.E. Puthoff: Stimulated Photon-Electron Scattering, IEEE J. QE-4, p.905-907 (1968)

[4.423] {Sect. 4.5.11} D.H. Rank, C.W. Cho, N.D. Foltz, T.A. Wiggins: Stimulated Thermal Rayleigh Scattering, Phys. Rev. Lett. 19, p.828-830 (1967)

[4.424] {Sect. 4.5.11} N. Bloembergen, P. Lallemand: Complex intensity-dependent index of refraction, frequency broadening of stimulated Raman lines, and stimulated Rayleigh scattering, Phys. Rev. Lett. 16, p.81-84 (1966)

[4.425] {Sect. 4.5.12} R.Y. Chiao, P.L. Kelley, E. Garmire: Stimulated Four-Photon Interaction and ist Influence on Stimulated Rayleigh-Wing Scattering, Phys. Rev. Lett. 17, p.1158-1161 (1966)

[4.426] {Sect. 4.5.13.0} J. J. Laserna: Modern Techniques in Raman Spectroscopy ((John Wiley & Sons, Chichester, 1996)

[4.427] {Sect. 4.5.13.0} G. Marowsky, V.V. Smirnov (eds.): Coherent Raman Spectroscopy, Springer Proc. Phys, Vol. 63 (Springer, Berlin, Heidelberg 1992)

[4.428] {Sect. 4.5.13.0} D.A. Long: The polarizability and hyperpolarizability tensors, in Nonlinear Raman Spectroscopy and Ist Chemical Applications, ed. by W. Kiefer, D. A. Long (Reidel, Dordrecht 1982)

[4.429] {Sect. 4.5.13.0} W. Kiefer: Recent techniques in Raman-spectroscopy (Adv. Infrared and Raman Spectroscopy 3, 1 (Heyden, London 1977)

[4.430] {Sect. 4.5.13.0} J. Loader: Basic Laser Raman Spectroscopy (Heyden/Sadtler, London 1970)

[4.431] {Sect. 4.5.13.0} J. R. Downey, G. J. Janz: Digital methods in Raman spectroscopy (Adv. Infrared and Raman Spectroscopy 1, 1-34, Heyden, London 1975)

[4.432] {Sect. 4.5.13.0} C.S. Wang: The stimulated Raman process, in Quantum Electronics: A Treatise, Vol. 1, ed. by H. Rabin, C.L. Tang (Academic, New York 1975) Chap. 7

[4.433] {Sect. 4.5.13.1} E.C. Honea, A. Ogura, D.R. Peale, C. Felix, C.A. Murray, K. Raghavachari, W.O. Sprenger, M.F. Jarrold, W.L. Brown: Structures and covalescence behavior of size-selected silicon nanoclusters studied by surface-plasmon-polariton enhanced Raman spectroscopy, J Chem Phys 110, p.12161-12172 (1999)

[4.434] {Sect. 4.5.13.1} V. Krylov, I. Fischer, V. Bespalov, D. Staselko, A. Rebane: Transient stimulated Raman scattering in gas mixtures, Optics Letters 24, p.1623-1625 (1999)

[4.435] {Sect. 4.5.13.1} A. Nazarkin, G. Korn, M. Wittmann, T. Elsaesser: Generation of multiple phase-locked Stokes and anti-Stokes components in an impulsively excited Raman medium, Phys Rev Lett 83, p.2560-2563 (1999)

[4.436] {Sect. 4.5.13.1} V.E. Roman, J. Popp, M.H. Fields, W. Kiefer: Minority species detection in aerosols by stimulated anti-Stokes-Raman scattering and external seeding, Appl Opt 38, p.1418-1422 (1999)

[4.437] {Sect. 4.5.13.1} O.M. Sarkisov, D.G. Tovbin, V.V. Lozovoy, F.E. Gostev, A.A. Titov, S.A. Antipin, S.Y. Umanskiy: Femtosecond Raman-induced polarisation spectroscopy of coherent rotational wave packets: D-2, N-2 and NO2, Chem Phys Lett 303, p.458-466 (1999)

[4.438] {Sect. 4.5.13.1} A.S. Grabtchikov, D.E. Gakhovich, A.G. Shvedko, V.A. Orlovich, K.J. Witte: Observation of solitary waves with different phase behavior in stimulated Raman forward scattering, Phys Rev Lett 81, p.5808-5811 (1998)

[4.439] {Sect. 4.5.13.1} S. Klewitz, S. Sogomonian, M. Woerner, S. Herminghaus: Stimulated Raman scattering of femtosecond Bessel pulses, Opt Commun 154, p.186-190 (1998)

[4.440] {Sect. 4.5.13.1} S. Sogomonian, G. Grigorian, K. Grigorian: Parametric suppression of Raman gain in coherent Raman probe scattering, Opt Commun 152, p.351-354 (1998)

[4.441] {Sect. 4.5.13.1} F. Vaudelle, J. Gazengel, G. Rivoire: Experimental study of the laser and stimulated Raman scattering wave phases by a nonlinear imaging method, Opt Commun 149, p.84-88 (1998)

[4.442] {Sect. 4.5.13.1} L. Deng, W.R. Garrett, M.G. Payne, D.Z. Lee: Observation of broadband forward hyper-Raman emission with high intensity focused laser beams, Opt Commun 142, p.253-256 (1997)

[4.443] {Sect. 4.5.13.1} M. Ozaki, E. Ehrenfreund, R.E. Benner, T.J. Barton, K. Yoshino, Z.V. Vardeny: Dispersion of resonant Raman scattering in pi-conjugated polymers: Role of the even parity excitons, Phys Rev Lett 79, p.1762-1765 (1997)

[4.444] {Sect. 4.5.13.1} M.R. Perrone, V. Piccinno: On the benefits of astigmatic focusing configurations in stimulated Raman scattering processes, Opt Commun 133, p.534-540 (1997)

[4.445] {Sect. 4.5.13.1} M.R. Perrone, V. Piccinno, G. Denunzio, V. Nassisi: Dependence of rotational and vibrational Raman scattering on focusing geometry, IEEE J QE-33, p.938-944 (1997)

[4.446] {Sect. 4.5.13.1} F. Vaudelle, J. Gazengel, G. Rivoire: Experimental studies of the spatial coherence of forward stimulated Raman scattering in dense materials, Opt Commun 134, p.559-568 (1997)

[4.447] {Sect. 4.5.13.1} B.H. Bairamov, A. Aydinli, I.V. Bodnar, Y.V. Rud, V.K. Nogoduyko, V.V. Toporov: High power gain for stimulated Raman amplification in CuAlS2, J Appl Phys 80, p.5564-5569 (1996)

[4.448] {Sect. 4.5.13.1} M. Hofmann, H. Graener: Time resolved incoherent anti-Stokes Raman spectroscopy of dichloromethane, Chem Phys 206, p.129-137 (1996)

[4.449] {Sect. 4.5.13.1} V. Krylov, A. Rebane, O. Ollikainen, D. Erni, U. Wild: Stimulated Raman scattering in hydrogen by frequency- doubled amplified femtosecond Ti:sapphire laser pulses, Optics Letters 21, p.381-383 (1996)

[4.450] {Sect. 4.5.13.1} V. Krylov, A. Rebane, D. Erni, O. Ollikainen, U. Wild, V. Bespalov, D. Staselko: Stimulated Raman amplification of femtosecond pulses in hydrogen gas, Optics Letters 21, p.2005-2007 (1996)

[4.451] {Sect. 4.5.13.1} A. Lau, M. Pfeiffer, A. Kummrow: Subpicosecond two-dimensional Raman spectroscopy applying broadband nanosecond laser radiation, Chem Phys Lett 263, p.435-440 (1996)

[4.452] {Sect. 4.5.13.1} K.T. Tsen, E.D. Grann, S. Guha, J. Menendez: Electron-phonon interactions in solid C-60 studied by transient picosecond Raman spectroscopy, Appl Phys Lett 68, p.1051-1053 (1996)

[4.453] {Sect. 4.5.13.1} A.I. Vodchitz, V.P. Kozich, P.A. Apanasevich, V.A. Orlovich: Correlations between the intensities of pump, depleted pump and Stokes waves in superbroadband stimulated Raman scattering, Opt Commun 125, p.243-249 (1996)

[4.454] {Sect. 4.5.13.1} B.F. Henson, G.V. Hartland, V.A. Venturo, R.A. Hertz, P.M. Felker: Stimulated Raman spectroscopy in the xx region of isotopically substituted benzene dimers: evidence for symmetrically inequivalent benzene moieties, Chem. Phys. Lett. 176, p.91-98 (1991)

[4.455] {Sect. 4.5.13.1} J.W. Nibler, J.J. Yang: Nonlinear Raman spectroscopy of gases, Ann. Rev. Phys. Chem. 38, p.349-381 (1987)

[4.456] {Sect. 4.5.13.1} J. Chesnoy: Determination of the modulation regime for vibrational dephasing. Demonstration on the critical Raman broadening in nitrogen, Chem. Phys. Lett. 125, p.267-271 (1986)

[4.457] {Sect. 4.5.13.1} G.M. Gale, P. Guyot-Sionnest, W.Q. Zheng: Direct Picosecond Determination of the Character of Vibrational Line-Broadening in Liquids, Opt. Comm. 58, p.395-399 (1986)

[4.458] {Sect. 4.5.13.1} M.L. Geirnaer, G.M. Gale: Time-resolved coherent spectroscopy of binary liquid systems: Methyl iodide in carbon disulphide, Chem. Phys. 86, p.205-211 (1984)

[4.459] {Sect. 4.5.13.1} I.A. Walmsley, M.G. Raymer: Observation of Macroscopic Quantum Fluctuations in Stimulated Raman Scattering, Phys. Rev. Lett. 50, p.962-965 (1983)

[4.460] {Sect. 4.5.13.1} J. Eggleston, R.L. Byer: Steady State Stimulated Raman Scattering by a Multimode Laser, IEEE J. QE16, p.850-853 (1980)

[4.461] {Sect. 4.5.13.1} R. Frey, F. Pradere: High-efficiency narrow-linewidth Raman amplification and spectral compession, Opt. Lett. 5, p.374-376 (1980)

[4.462] {Sect. 4.5.13.1} J.P. Heritage, D.L. Allara: Surface picosecond Raman gain spectra of a molecular monolayer, Chem. Phys. Lett. 74, p.507-510 (1980)

[4.463] {Sect. 4.5.13.1} B.F. Levine, C.G. Bethea, A.R. Tretola, M. Korngor: Stimulated Raman scattering from 20-A layers of silicon on sapphire, Appl. Phys. Lett. 37, p.595-597 (1980)

[4.464] {Sect. 4.5.13.1} J.B. Grun, A.K. McQuillan, B.P. Stoicheff: Intensity and Gain Measurements on the Stimulated Raman Emission in Liquid O2 and N2, Phys. Rev. 180p.61-68 (1969)

[4.465] {Sect. 4.5.13.1} D. von der Linde, M. Maier, W. Kaiser: Quantitative Investigations of the Stimulated Raman Effect Using Subnanosecond Light Pulses, Phys. Rev. 178, p.11-17 (1969)

[4.466] {Sect. 4.5.13.1} N. Bloembergen, G. Bret, P. Lallemand, A. Pine, P. Simova: Controlled Stimulated Raman Amplification and Oscillation in Hydrogen Gas, IEEE J. QE-3, p.197-201 (1967)

[4.467] {Sect. 4.5.13.1} E.E. Hagenlocker, R.W. Minck, W.G. Rado: Effects of Phonon Lifetime on Stimulated Optical Scattering in Gases, Phys. Rev. 154, p.226-233 (1967)

[4.468] {Sect. 4.5.13.1} P. Lallemand, P. Simova, G. Bret: Pressure-Induced Line Shift and Collisional Narrowing in Hydrogen Gas Determined by Stimulated Raman Emission, Phys. Rev. Lett. 17, p.1239-1241 (1966)

[4.469] {Sect. 4.5.13.1} D. Cotter, D.C. Hanna, R. Wyatt: Infrared Stimulated Raman Generation Effects of Gain Focussing on Threshold and Tuning Behaviour, Appl. Phys. 8, p.333-340 (1975)

[4.470] {Sect. 4.5.13.1} XC. Rousseaux, G. Malka, J.L. Miquel, F. Amiranoff, S.D. Baton, P. Mounaix: Experimental validation of the linear theory of stimulated Raman scattering driven by a 500-fs laser pulse in a preformed underdense plasma (vol 74, pg 4655, 1995), Phys Rev Lett 76, p.4649 (1996)

[4.471] {Sect. 4.5.13.1} J.C. van den Heuvel, F.J.M. van Putten, R.J.L. Lerou: The Stimulated Raman Scattering Threshold for a Nondiffraction-Limited Pump Beam, IEEE J. QE-28, p.1930-1936 (1992)

[4.472] {Sect. 4.5.13.1} J.C. van den Heuvel: Numerical Modeling of Stimulated Raman Scattering in an Astigmatic Focus, IEEE J. QE-28, p.378-385 (1992)

[4.473] {Sect. 4.5.13.1} B. Dick: Response funcition theory of time-resolved CARS and CSRS of rotating molecules in liquids under general polarization conditions, Chem. Phys. 113, p.131-147 (1987)

[4.474] {Sect. 4.5.13.1} S.A. Akhmanov, Yu. E. D'yakov, L.I. Pavlov: Statistical phenomena in Raman scattering stimulated by a broad-band pump, Sov. Phys. JETP 39, p.249-258 (1974)

[4.475] {Sect. 4.5.13.1} R.R. Alfano, S.L. Shapiro: Explanation of a Transient Raman Gain Anomaly, Phys. Rev. A 2p.2376-2379 (1970)

[4.476] {Sect. 4.5.13.1} R.L. Carman, F. Shimizu, C.S. Wang, N. Bloembergen: Theory of Stokes Pulse Shapes in Transient Stimulated Raman Scattering, Phys. Rev. A 2, p.60-72 (1970)

[4.477] {Sect. 4.5.13.1} Y.R. Shen, N. Bloembergen: Theory of Stimulated Brillouin and Raman Scattering, Phys. Rev. 137, p.A1787-A1805 (1965)

[4.478] {Sect. 4.5.13.1} M.N. Shkunov, W. Gellermann, Z.V. Vardeny: Amplified resonant Raman scattering in conducting polymer thin films, Appl Phys Lett 73, p.2878-2880 (1998)

[4.479] {Sect. 4.5.13.1} A.S. Jeevarajan, L.D. Kispert, G. Chumanov, C. Zhou, T.M. Cotton: Resonance Raman study of carotenoid cation radicals, Chem Phys Lett 259, p.515-522 (1996)

[4.480] {Sect. 4.5.13.1} S. Nakashima, T. Kitagawa, J.S. Olson: Time-resolved resonance Raman study of intermediates generated after photodissociation of wild-type and mutant CO-myoglobins, Chem Phys 228, p.323-336 (1998)

[4.481] {Sect. 4.5.13.1} T.L. Gustafson, J.F. Palmer, D.M. Roberts: The structure of S1 diphenylbutadiene: UV resonance Raman and picosecond transient Raman studies, Chem. Phys. Lett. 127, p.505-511 (1986)

[4.482] {Sect. 4.5.13.1} S. Koshihara, T. Kobayashi: Time-resoved resonance Raman spectrum of chrysene in the S1 and T1 states, J. Chem. Phys. 85, p.1211-1219 (1986)

[4.483] {Sect. 4.5.13.1} R. Wilbrandt, N.-H. Jensen, F.W. Langkilde: Time-resolved resonance Raman spectrum of all-trans-diphenylbutadiene in the lowest excited singlet state, Chem. Phys. Lett. 111, p.123-127 (1984)

[4.484] {Sect. 4.5.13.1} H. Hamaguchi, Ch. Kato, M. Tasumi: Observation of transient resonance Raman spectra of the S1 state of trans-stilbene, Chem. Phys. Lett. 100, p.3-7 (1983)

[4.485] {Sect. 4.5.13.1} K. Kneipp, H. Kneipp, I. Itzkan, R.R. Dasari, M.S. Feld: Surface-enhanced non-linear Raman scattering at the single-molecule level, Chem Phys 247, p.155-162 (1999)

[4.486] {Sect. 4.5.13.1} S.M. Nie, S.R. Emery: Probing single molecules and single nanoparticles by surface-enhanced Raman scattering, Science 275, p.1102-1106 (1997)

[4.487] {Sect. 4.5.13.1} S.M. Nie, S.R. Emery: Probing single molecules and single nanoparticles by surface-enhanced Raman scattering, Science 275, p.1102-1106 (1997)

[4.488] {Sect. 4.5.13.1} V.E. Roman, J. Popp, M.H. Fields, W. Kiefer: Species identification of multicomponent microdroplets by seeding stimulated Raman scattering, J Opt Soc Am B Opt Physics 16, p.370-375 (1999)

[4.489] {Sect. 4.5.13.1} G. Zikratov, F.Y. Yueh, J.P. Singh, O.P. Norton, R.A. Kumar, R.L. Cook: Spontaneous anti-Stokes Raman probe for gas temperature measurements in industrial furnaces, Appl Opt 38, p.1467-1475 (1999)

[4.490] {Sect. 4.5.13.1} T. Dreier, B. Lange, J. Wolfrum, M. Zahn: Determination of Temperature and Concentration of Molecular Nitrogen, Oxygen and Methane with Coherent Anti-Stokes Raman Scattering, Appl. Phys. B 45, p.183-190 (1988)

[4.491] {Sect. 4.5.13.1} B.F. Levine, C.V. Shank, J.P. Heritage: Surface Vibrational Spectroscopy Using Stimulated Raman Scattering, IEEE J. QE-15, p.1418-1432 (1979)

[4.492] {Sect. 4.5.13.1} T.R. Loree, R.C. Sze, D.L. Barker, P.B. Scott: New Lines in the UV: SRS of Excimer Laser Wavelengths, IEEE J. QE-15, p.337-342 (1979)

[4.493] {Sect. 4.5.13.1} A. DeMartino, R. Frey, F. Pradere: Tunable Far Infrared Generation in Hydrogen Fluoride, Opt. Comm. 27, p.262-266 (1978)

[4.494] {Sect. 4.5.13.1} V. Wilke, W. Schmidt: Tunable UV-Radiation by Stimulated Raman Scattering in Hydrogen, Appl. Phys. 16, p.151-154 (1978)

[4.495] {Sect. 4.5.13.1} R. Frey, F. Pradere, J. Ducuing: Tunable Far-Infrared Raman Generation, Opt. Comm. 23, p.65-68 (1977)

[4.496] {Sect. 4.5.13.1} R.L. Byer: A 16-μm Source for Laser Isotope Enrichment, IEEE J. QE-12, p.732-739 (1976)

[4.497] {Sect. 4.5.13.1} D. von der Linde, A. Laubereau, W. Kaiser: Molecular Vibrations in Liquids: Direct Measurement of the Molecular Dephasing Time; Determination of the Shape of Picosecond Light Pulses, Phys. Rev. Lett. 26, p.954-957 (1971)

[4.498] {Sect. 4.5.13.1} L. Beardmore, H.G.M. Edwards, D.A. Long, T.K. Tan: Raman spectroscopic measurements of temperature in a natural gas/air flame, in Lasers in Chemistry, ed. by M.A. West (Elsevier, Amsterdam 1977)

[4.499] {Sect. 4.5.13.1} M.J. Everett, A. Lal, D. Gordon, K. Wharton, C.E. Clayton, W.B. Mori, C. Joshi: Evolution of stimulated Raman into stimulated Compton scattering of laser light via wave breaking of plasma waves, Phys Rev Lett 74, p.1355-1358 (1995)

[4.500] {Sect. 4.5.13.1} M.L. Geirnaert, G.M. Gale, C. Flytzanis: Time-Resolved Spectroscopy of Vibrational Overtones and Two-Phonon States, Phys. Rev. Lett. 52, p.815-818 (1984)

[4.501] {Sect. 4.5.13.1} D.S. Bethune, J.R. Lankard, P.P. Sorokin: Time-resolved infrared spectral photography, Opt. Lett. 4, p.103-105 (1979)

[4.502] {Sect. 4.5.13.4} L.A. Carreira, M.L. Horowitz: CARS in condensed media, in Non-Linear Raman Spectroscopy and Ist Chemical Applications, ed. by W. Kiefer, D.A. Long (reidel, Dordrecht 1982) p 367

[4.503] {Sect. 4.5.13.4} E.K. Gustafson, R.L. Byer: High-resolution CARS-spectroscopy, in Laser Spectroscopy VI, ed. by H.P. Weber, W. Lüthy, Springer Ser. Opt. Sci, Vol. 40 (Springer, Berlin, Heidelberg 1983) p. 326

[4.504] {Sect. 4.5.13.4} J. Bood, P.E. Bengtsson, M. Alden: Stray light rejection in rotational coherent anti-Stokes Raman spectroscopy by use of a sodium-seeded flame, Appl Opt 37, p.8392-8396 (1998)

[4.505] {Sect. 4.5.13.4} J.C. Kirkwood, D.J. Ulness, A.C. Albrecht, M.J. Stimson: Raman spectrograms in fifth order coherent Raman scattering: The sequential CARS process in liquid benzene, Chem Phys Lett 293, p.417-422 (1998)

[4.506] {Sect. 4.5.13.4} M.Schmitt, G. Knopp, A. Materny, W. Kiefer: The Application of Femtosecond Time-Resolved Coherent Anti-Stokes Raman Scattering for the Investigation of Ground and Excited State Molecular Dynamics of Molecules in the Gas Phase, J. Phys. Chem. A 102, p.4059-4065 (1998)

[4.507] {Sect. 4.5.13.4} E.J. Beiting: Coherent anti-Stokes Raman scattering velocity and translational temperature measurements in resistojets, Appl Opt 36, p.3565-3576 (1997)

[4.508] {Sect. 4.5.13.4} J.W. Hahn, C.W. Park, S.N. Park: Broadband coherent anti-Stokes Raman spectroscopy with a modeless dye laser, Appl Opt 36, p.6722-6728 (1997)

[4.509] {Sect. 4.5.13.4} M. Schmitt, G. Knopp, A. Materny, W. Kiefer: Femtosecond time-resolved coherent anti-Stokes Raman scattering for the simultaneous study of ultrafast ground and excited state dynamics: Iodine vapour, Chem Phys Lett 270, p.9-15 (1997)

[4.510] {Sect. 4.5.13.4} G.W. Baxter, M.J. Johnson, J.G. Haub, B.J. Orr: OPO CARS: Coherent anti-Stokes Raman spectroscopy using tunable optical parametric oscillators injection-seeded by external-cavity diode lasers, Chem Phys Lett 251, p.211-218 (1996)

[4.511] {Sect. 4.5.13.4} K. Ravichandran, Y. Bai, T.R. Fletcher: Techniques for stimulated Raman excitation and CARS detection of radicals created by photodissociation, Chem Phys Lett 261, p.261-266 (1996)

[4.512] {Sect. 4.5.13.4} P.P. Yaney, J.W. Parish: Coherent anti-Stokes Raman scattering measurements of N- 2 (X, v) at low pressures corrected for stimulated Raman scattering, Appl Opt 35, p.2659-2664 (1996)

[4.513] {Sect. 4.5.13.4} B. Dick: Response function theory of time-resolved CARS andn CSRS of rotating molecules in liquids under general polarization conditions, Chem. Phys. 113, p.131-147 (1987)

[4.514] {Sect. 4.5.13.4} T. Hattori, A. Terasaki, T. Kobayashi: Coherent Stokes Raman scattering with incoherent light for vibrational-dephasing-time measurement, Phys. Rev. A 35, p.715-724 (1987)

[4.515] {Sect. 4.5.13.4} H. Graener, A. Laubereau, J.W. Nibler: Picosecond coherent anti-Stokes Raman spectroscopy of molecules in free jet expansions, Opt. Lett. 9, p.165-167 (1984)

[4.516] {Sect. 4.5.13.4} E. Gustafson, R.L. Byer: Transit Time Linewidth Limitations in CW CARS Spectroscopy, Appl Phys B 28, p.85-86 (1982)

[4.517] {Sect. 4.5.13.4} E.K. Gustafson, R.L. Byer, J.C. Mcdaniel: High Resolution Continuous Wave Coherent Anti Stokes Raman Spectroscopy in a Supersonic Jet, Optics Letters 7, p.434-436 (1982)

[4.518] {Sect. 4.5.13.4} Ch. Jung, A. Lau, H.-J. Weigmann, W. Werncke, M. Pfeiffer: Interpretation of resonance CARS and Shpolskii spectra with calculated molecular geometries, vibrational frequences and relative intensities: Chrysene in its lowest excited singlet and triplet state, Chem. Phys. 72, p.327-336 (1982)

[4.519] {Sect. 4.5.13.4} S.A. Druet, J.P.E. Taran: CARS Spectroscopy, Prog. Quant. Electr. Vol. 7, p.1-72 (1981)

[4.520] {Sect. 4.5.13.4} F. Moya, S.A.J. Druet, J.P.E. Taran: Rotation-vibration spectroscopy of gases by CARS, in Laser Spectroscopy II, ed. by S. Haroche, J.C. Pebay-Peyroula, T.W. Hänsch, S.E. Harris, Lecture Notes Phys, Vol. 43 (Springer, Berlin, Heidelberg 1975) p. 66

[4.521] {Sect. 4.5.13.4} J.W. Nibler, G.V. Knighten: Coherent anti-Stokes Raman spectroscopy, in Raman Spectroscopy of Gases and Liquids,ed. by A. Weber, Topics Curr. Phys, Vol. 11 (Springer, Berlin, Heidelberg 1979) Chap. 7

[4.522] {Sect. 4.5.13.4} A. Zumbusch, G.R. Holtom, X.S. Xie: Three-dimensional vibrational imaging by coherent anti-Stokes Raman scattering, Phys Rev Lett 82, p.4142-4145 (1999)

[4.523] {Sect. 4.5.13.4} L. Ujj, F. Jager, A. Popp, G.H. Atkinson: Vibrational spectrum of the K-590 intermediate in the bacteriorhodopsin photocycle at room temperature: Picosecond time-resolved resonance coherent anti-Raman spectroscopy, Chem Phys 212, p.421-436 (1996)

[4.524] {Sect. 4.5.13.5} T.J. Vikers: Quantitative resonance Raman spectroscopy, Appl. Spectrosc. Rev. 26, p.341 (1991)

[4.525] {Sect. 4.5.13.5} M.D. Levenson: Feasibility of Measuring the Nonlinear Index of Refraction by Third-Order Frequency Mixing, IEEE J. QE-10, p.110-115 (1974)

[4.526] {Sect. 4.5.14} B. Y. Zel'dovich, N. Pilipettshii: Principles in Phase Conjugation (Springer, Heidelberg, New York, 1985)

[4.527] {Sect. 4.5.14} R. A. Fischer: Optical Phase Conjugation (Academic Press, San Diego, 1983)

[4.528] {Sect. 4.5.14} R.W. Hellwarth: Optical beam phase conjugation by stimulated backscattering, Opt. Eng. 21, p.257-262 (1982)

[4.529] {Sect. 4.5.14} Q. Gong, Y. Huang, J. Yang: Mechanism of optical phase conjugation by stimulated Brillouin scattering, Phys. Rev. A 39, p.1227-1234 (1989)

[4.530] {Sect. 4.5.14} D.A. Rockwell: A Review of Phase-Conjugate Solid-State Lasers, IEEE J. QE-24, p.1124-1140 (1988)

[4.531] {Sect. 4.5.14} N G. Basov, V F. Efimkov, I G. Zubarev, A.V. Kotov, S.I. Mikhailov, and M. G.Smirnov: Inversion of wavefront in SMBS of a depolarized pump, JETP Lett. 28, p.197-201 (1978)

[4.532] {Sect. 4.5.14} G. Gbur, E. Wolf: Phase conjugation with random fields and with deterministic and random scatterers, Optics Letters 24, p.10-12 (1999)

[4.533] {Sect. 4.5.14} G.G. Kochemasov, F.A. Starikov: Novel features of phase conjugation at SBS of beams passed through an ordered phase plate, Opt Commun 170, p.161-174 (1999)

[4.534] {Sect. 4.5.14} D.C. Jones, G. Cook, K.D. Ridley, A.M. Scott: High reflectivity phase conjugation in the visible spectrum using stimulated Brillouin scattering in alkanes, J Nonlinear Opt Physics Mat 7, p.331-344 (1998)

[4.535] {Sect. 4.5.14} A.A. Offenberger, D.C. Thompson, R. Fedosejevs, B. Harwood, J. Santiago, H.R. Manjunath: Experimental and Modeling Studies of a Brillouin Amplifier, IEEE J. QE-29, p.207-216 (1993)

[4.536] {Sect. 4.5.14} J.J. Maki, W.V. Davis, R.W. Boyd: Phase conjugation using the surface nonlinearity of a dense potassium vapor, Phys. Rev. A. 46, p.7155-7161 (1992)

[4.537] {Sect. 4.5.14} R. Saxena, P. Yeh: Mutually pumped phase conjugation in Kerr media and the effects of external seeding, J. Opt. Soc. Am. B 7, p.326-334 (1990)

[4.538] {Sect. 4.5.14} V.N. Blashuk, B.Ya. Zel'dovich, V.N. Krasheninnikov, N.A. Mel'nikov, N.F. Pilipetskii, V.V. Ragul'skii, V.V. Shkunov: SBS wave front reversal for the depolarized light-theory and experiment, Opt. Comm. 27, p.137-141 (1978)

[4.539] {Sect. 4.5.14} A. Yariv: Phase Conjugate Optics and Real-Time Holography, IEEE J. QE-14, p.650-660 (1978)

[4.540] {Sect. 4.5.14} B.Ya. Zel'dovich, V.V. Shkupov: Reversal of wave front of light in the case of depolarized pumping, Sov. Phys. JETP 48, p.214-219 (1978)

[4.541] {Sect. 4.5.14} G.G. Kochemasov, V.D. Nikolaev: Reproduction of the spatial amplitude and phase distributions of a pump beam in stimulated Brillouin scattering, Sov. J. Quantum Electron. 7, p.60-63 (1977)

[4.542] {Sect. 4.5.14} E. Bochove: Theory of a variable aperture phase conjugate mirror with application to an optical cavity, J. Appl. Phys. 59, p.3360-3362 (1986)

[4.543] {Sect. 4.5.14} P. Suni, J. Falk: Theory of phase conjugation by stimulated Brillouin scattering, J. Opt. Soc. Am. B 3, p.1681-1691 (1986)

[4.544] {Sect. 4.5.14} N.B. Baranova, B.Ya. Zel'dovich: Wavefront reversal of focused beams (theory of stimulated Brillouin backscattering), Sov. J. Quantum Electron. 10, p.555-560 (1980)

[4.545] {Sect. 4.5.14} R.W. Hellwarth: Theory of phase conjugation by stimulated scattering in a waveguide, J. Opt. Soc. Am. 68, p.1050-1056 (1978)

[4.546] {Sect. 4.5.14} B.Ya. Zel'dovich, V.V. Shkunov: Limits of existance of wavefront reversal in stimulated light scattering, p.15-20 (1978)

[4.547] {Sect. 4.5.14} G.G. Kochemasov, V.D. Nikolaev: Reproduction of the spatial amplitude and phase distributions of a pump beam in stimulated Brillouin scattering, Sov. J. Quantum Electron. 7, p.60-63 (1977)

[4.548] {Sect. 4.5.14} R.G. Harrison, V.I. Kovalev, W.P. Lu, D.J. Yu: SBS self-phase conjugation of CWNd : YAG laser radiation in an optical fibre, Opt Commun 163, p.208-211 (1999)

[4.549] {Sect. 4.5.14} H. Naruse, M. Tateda: Trade-off between the spatial and the frequency resolutions in measuring the power spectrum of the Brillouin backscattered light in an optical fiber, Appl Opt 38, p.6516-6521 (1999)

[4.550] {Sect. 4.5.14} E. Peral, A. Yariv: Degradation of modulation and noise characteristics of semiconductor lasers after propagation in optical fiber due to a phase shift induced by stimulated Brillouin scattering, IEEE J QE-35, p.1185-1195 (1999)

[4.551] {Sect. 4.5.14} H.J. Eichler, J. Kunde, B. Liu: Quartz fibre phase conjugators with high fidelity and reflectivity, Opt. Comm. 139, p.327-334 (1997)

[4.552] {Sect. 4.5.14} Ch. Lorattanasane, K.Kikuchi: Desing of Long-Distance Optical Transmission Systems Using Midway Optical Phase Conjugation, IEEE Phot. Techn. Lett. 7, p.1375-1377 (1995)

[4.553] {Sect. 4.5.14} S. Wabnitz: Nonlinear Enhancement and Optimization of Phase-Conjugation Efficiency in Optical Fibers, IEEE Phot. Techn. Lett. 7, p.652-654 (1995)

[4.554] {Sect. 4.5.14} M. Yu, G.P. Agrawal, C.J. McKinstrie: Effect of Residual Dispersion in the Phase-Conjugation Fiber on Dispersion Compensation in Optical Communication Systems, IEEE Phot. Techn. Lett. 7, p.932-934 (1995)

[4.555] {Sect. 4.5.14} X. Zhang, F. Ebskamp, B.F. Jorgensen: Long-Distance Transmission Over Standard Fiber by Use of Mid-Way Phase Conjugation, IEEE Phot. Techn. Lett. 7, p.819-821 (1995)

[4.556] {Sect. 4.5.14} P. Shalev, St. Jackel, R. Lallouz, A. Borenstein: Low-threshold phase conjugate mirrors based on position-insensitive tapered waveguides, Opt. Eng. 33, p.278-284 (1994)

[4.557] {Sect. 4.5.14} W. Wu, P. Yeh, S. Chi: Phase Conjugation by Four-Wave Mixing in Single-Mode Fibers, IEEE J. QE-6, p.1448-1450 (1994)

[4.558] {Sect. 4.5.14} E.P. Ippen, R.H. Stolen: Stimulated Brillouin scattering in optical fibers, Appl. Phys. Lett. 21, p.539-541 (1972)

[4.559] {Sect. 4.5.14} S. Jackel, P. Shalev, R. Lallouz: Experimental and theoretical investigation of statistical fluctuations in phase conjugate mirror reflectivity, Opt. Comm. 101, p.411-415 (1993)

[4.560] {Sect. 4.5.14} M.S. Mangir, D.A. Rockwell: 4.5-J Brilloin phase-conjugate mirror producing excellent mear-and far-field fidelity, J. Opt. Soc. Am. B 10, p.1396-1400 (1993)

[4.561] {Sect. 4.5.14} C.B. Dane, W.A. Neuman, L.A. Hackel: Pulse-shape dependence of stimulated-Brillouin-scattering phase-conjugation fidelity for high input energies, Opt. Lett. 17, p.1271-1273 (1992)

[4.562] {Sect. 4.5.14} R.W.F. Gross, S.T. Amimoto, L.Garman-Du Vall: Gain and phase-conjugation fidelity of a four-wave Brillouin mirror based on methane, Opt. Lett. 16, p.94-96 (1991)

[4.563] {Sect. 4.5.14} J.J. Ottusch, D.A. Rockwell: Stimulated Brillouin scattering phase-conjugation fidelity fluctuations, Opt. Lett. 16, p.369-371 (1991)

[4.564] {Sect. 4.5.14} I.Yu. Anikeev, D.A. Glazkov, A.A. Gordeev, I.G. Zubarev, S.I. Mikhailov: Polarization and aperture losses in systems with phase conjugation mirrors, Int. J. Optoelectron. 4, p.489-500 (1989)

[4.565] {Sect. 4.5.14} V.N. Alekseev, V.V. Golubev, D.I. Dmitriev, A.N. Zhilin, V.V. Lyubimov, A.A. Mak, V.I. Reshetnikov, V.S. Sirazetdinov, A.D. Starikov: Investigation of wavefront reversal in a phosphate glass laser amplifier with a 12-cm output aperture, Sov. J. Quantum Electron. 17, p.455-458 (1987)

[4.566] {Sect. 4.5.14} P. Suni, J. Falk: Measurements of stimulated Brillouin scattering phase-conjugate fidelity, Opt. Lett. 12, p.838-840 (1987)

[4.567] {Sect. 4.5.14} R.L. Abrams, C.R. Giuliano, J.F. Lam: On the equality of stimulated Brillouin scattering reflectivity to conjugate reflectivity of a weak probe beam, Opt. Lett. 6, p.131-132 (1981)

[4.568] {Sect. 4.5.14} B.Ya. Zel'dovich, T.V. Yakovleva: Small-scale distortions in wavefront reversal of a beam with incomplete spatial modulation (stimulated Brillouin backscattering, theory), Sov. J. Quantum Electron. 10, p.181-186 (1980)

[4.569] {Sect. 4.5.14} V. Wang, C.R. Giuliano: Correction of phase aberrations via stimulated Brillouin scattering, Opt. Lett. 2, p.4-6 (1978)

[4.570] {Sect. 4.5.14} M. Ostermeyer, A. Heuer, R. Menzel: 27 Watt Average Output Power with 1.2*DL Beam Quality from a Single Rod Nd:YAG-Laser with Phase Conjugating SBS-Mirror, IEEE J. QE-34, p.372-377 (1998)

[4.571] {Sect. 4.5.14} H.L. Offerhaus, H.P. Godfried, W.J. Witteman: Al solid-state diode pumped Nd:YAG MOPA with stimulated Brillouin phase conjugate mirror, Opt. Comm. 128, p.61-65 (1996)

[4.572] {Sect. 4.5.14} C.B. Dane, L.E. Zapata, W.A. Neumann, M.A. Norton, L.A. Hackel: Design and Operation of a 150 W Near Diffraction-Limited Laser Amplifier with SBS Wavefront Correction, IEEE J. QE-31, p.148-163 (1995)

[4.573] {Sect. 4.5.14} H.J. Eichler, A. Haase, R. Menzel: 100 Watt Average Output Power 1.2*Diffraction Limited Beam From Pulsed Neodym Single Rod Amplifier with SBS-Phaseconjugation, IEEE J. QE-31, p.1265-1269 (1995)

[4.574] {Sect. 4.5.14} I.C. Khoo, H. Li, P.G. LoPresti, Y. Liang: Observation of optical limiting and backscattering of nanosecond laser pulses in liquid-crystal fibers, Opt. Lett. 19, p.530-532 (1994)

[4.575] {Sect. 4.5.14} D.S. Sumida, C.J. Jones, R.A. Rockwell: An 8.2 J Phase Conjugating Solid-State Laser Coherently Combining Eight Parallel Amplifiers, IEEE J. QE-30, p.2617-2627 (1994)

[4.576] {Sect. 4.5.14} O.V. Kulagin, G.A. Pasmanik, A.A. Shilov: Amplification and phase conjugation of weak signals, Sov. Phys. Usp. 35, p.506-519 (1992)

[4.577] {Sect. 4.5.14} O.V. Kulagin, P.B. Potlov, A.A. Shilov: Phase conjugation of microsecond pulses by forward Brillouin scattering, Sov. J. Quantum Electron. 22, p.1012-1015 (1992)

[4.578] {Sect. 4.5.14} G.J. Crofts, M.J. Damzen: Experimental and theoretical investigation of two-cell stimulated-Brillouin-scattering systems, J. Opt. Soc. Am. B 8, p.2282-2288 (1991)

[4.579] {Sect. 4.5.14} I.D. Carr, D.C. Hanna: Performance of a Nd:YAG Oscillator/Amplifier with Phase-Conjugation via Stimulated Brillouin Scattering, Appl. Phys. B 36, p.83-92 (1985)

[4.580] {Sect. 4.5.14} D.T. Hon: Applications of wavefront reversal by stimulated Brillouin scattering, Opt. Eng. 21, p.252-256 (1982)

[4.581] {Sect. 4.5.14} M. Slatkine, I.J. Bigio, B.J. Feldman, R.A. Fisher: Efficient phase conjugation of an ultraviolet XeF laser beam by stimulated Brillouin scattering, Opt. Lett. 7, p.108-110 (1982)

[4.582] {Sect. 4.5.14} V.F. Efimkov, I.G. Zubarev, A.V. Kotov, A.B. Mironov, S.I. Mikhailov, M.G. Smirnov: Investigations of systems for obtaining short high-power pulses by wavefront reversal of the radiation in a stimulated Brillouin scattering mirror, Sov. J. Quant. Electron. 10, p.211-214 (1980)

[4.583] {Sect. 4.5.14} T. Omatsu, N. Hayashi, H. Watanabe, A. Hasegawa, M. Tateda: Tunable, visible phase conjugator with a saturable-amplifier polymer laser dye, Optics Letters 23, p.1432-1434 (1998)

[4.584] {Sect. 4.5.14} V.S. Sudarshanam, M. Croningolomb, P.R. Hemmer, M.S. Shahriar: Turbulence-aberration correction with high-speed high-gain optical phase conjugation in sodium vapor, Optics Letters 22, p.1141-1143 (1997)

[4.585] {Sect. 4.5.14} D. Udaiyan, K.S. Syed, R.P.M. Green, D.H. Kim, M.J. Damzen: Transient modelling of double-pumped phase conjugation in inverted Nd:YAG, Opt Commun 133, p.596-604 (1997)

[4.586] {Sect. 4.5.14} A. Grunnetjepsen, C.L. Thompson, W.E. Moerner: Spontaneous oscillation and self-pumped phase conjugation in a photorefractive polymer optical amplifier, Science 277, p.549-552 (1997)

[4.587] {Sect. 4.5.14} I.C. Khoo, H. Li, Y. Liang: Self-starting optical phase conjugation in dyed nematic liquid crystals with a stimulated thermal-scattering effect, Opt. Lett. 18, p.1490-1492 (1993)

[4.588] {Sect. 4.5.14} S.A. Korol'kov, A.V. Mamaev, V.V. Shkunov: Mutual phase conjugation of temporally nonoverlapping optical beams, Sov. J. Quantum Electron. 22, p.861-864 (1992)

[4.589] {Sect. 4.5.14} I.C. Winkler, M.A. Norton, Adaptive phase compensation in a Raman look-through configuration, Opt. Lett. 14, p.69-71 (1989)

[4.590] {Sect. 4.5.14} R.C. Desai, M.D. Levenson, J.A. Barker: Forced Rayleigh scattering: Thermal and acoustic effects in phase-conjugate wave-front generation, Phys. Rev. A 27, p.1968-1976 (1983)

[4.591] {Sect. 4.6} E. Constant, D. Garzella, P. Breger, E. Mevel, C. Dorrer, C. LeBlanc, F. Salin, P. Agostini: Optimizing high harmonic generation in absorbing gases: Model and experiment, Phys Rev Lett 82, p.1668-1671 (1999)

[4.598] {Sect. 4.6} C.G. Durfee, A.R. Rundquist, S. Backus, C. Herne, M.M. Murnane, H.C. Kapteyn: Phase matching of high-order harmonics in hollow waveguides, Phys Rev Lett 83, p.2187-2190 (1999)

[4.592] {Sect. 4.6} K. Midorikawa, Y. Tamaki, J. Itatani, Y. Nagata, M. Obara: Phase-matched high-order harmonic generation by guided intense femtosecond pulses, IEEE J Sel Top Quantum Electr 5, p.1475-1485 (1999)

[4.593] {Sect. 4.6} A. Rundquist, C.G. Durfee, Z.H. Chang, C. Herne, S. Backus, M.M. Murnane, H.C. Kapteyn: Phase-matched generation of coherent soft X-rays, Science 280, p.1412-1415 (1998)

[4.594] {Sect. 4.6} Z.H. Chang, A. Rundquist, H.W. Wang, M.M. Murnane, H.C. Kapteyn: Generation of coherent soft X rays at 2.7 nm using high harmonics, Phys Rev Lett 79, p.2967-2970 (1997)

[4.595] {Sect. 4.6} I.P. Christov, M.M. Murnane, H.C. Kapteyn: High-harmonic generation of attosecond pulses in the "single-cycle" regime, Phys Rev Lett 78, p.1251-1254 (1997)

[4.596] {Sect. 4.6} B.K. Dey, B.M. Deb: A theoretical study of the high-order harmonics of a 200 nm laser from H-2 and HeH+, Chem Phys Lett 276, p.157-163 (1997)

[4.597] {Sect. 4.6} S. Meyer, H. Eichmann, T. Menzel, S. Nolte, B. Wellegehausen, B.N. Chichkov, C. Momma: Phase-matched high-order difference-frequency mixing in plasmas, Phys Rev Lett 76, p.3336-3339 (1996)

[4.599] {Sect. 4.6} H. Ono, Y. Harato: Higher-order optical nonlinearity observed in host-guest liquid crystals, J Appl Phys 85, p.676-680 (1999)

[4.600] {Sect. 4.6} Y.S. Lee, M.C. Downer: Reflected fourth-harmonic radiation from a centrosymmetric crystal, Optics Letters 23, p.918-920 (1998)

[4.601] {Sect. 4.6} A.V. Balakin, D. Boucher, E. Fertein, P. Masselin, A.V. Pakulev, A.Y. Resniansky, A.P. Shkurinov, N.I. Koroteev: Experimental observation of the interference of three- and five-wave mixing processes into the signal of second harmonic generation in bacteriorhodopsin solution, Opt Commun 141, p.343-352 (1997)

[4.602] {Sect. 4.6} C.C. Tian, P.Q. Wang, T.H. Sun: Generation of tunable coherent VUV radiation by four-wave sum-mixing in Ne, Opt Commun 132, p.248-250 (1996)

[4.603] {Sect. 4.6} C. Altucci, R. Bruzzese, D. DAntuoni, C. deLisio, S. Solimeno: Harmonic generation in gases by use of Bessel-Gauss laser beams, J Opt Soc Am B Opt Physics 17, p.34-42 (2000)

[4.604] {Sect. 4.6} J.C. Kirkwood, A.C. Albrecht, D.J. Ulness: Fifth-order nonlinear Raman processes in molecular liquids using quasi-cw noisy light. I. Theory, J Chem Phys 111, p.253-271 (1999)

[4.605] {Sect. 4.6} Y. Tanimura: Fifth-order two-dimensional vibrational spectroscopy of a Morse potential system in condensed phases, Chem Phys 233, p.217-229 (1998)

[4.606] {Sect. 4.6} D. Sarkisyan, G. Torosyan, K. Pokhsrarian, K. Petrossian: Fifth harmonic generation and measurements of the 7th order correlation vapor, Opt Commun 127, p.205-209 (1996)

[4.607] {Sect. 4.6} Th. Tsang: Third- and fifth-harmonic generation at the interfaces of glass and liquids, Phys. Rev. A 54, p.5454-5457 (1996)

[4.608] {Sect. 4.6} K. Tominaga, K. Yoshihara: Fifth order optical response of liquid CS2 observed by ultrafast nonresonant six-wave mixing, Phys Rev Lett 74, p.3061-3064 (1995)

[4.609] {Sect. 4.6} J. Reintjes, R.C. Eckardt, C.Y. She, N.E. Karangelen, R.C. Elton, R.A. Andrews: Generation of Coherent Radiation at 53.2 nm by Fifth-Harmonic Conversion, Phys. Rev. Lett. 37, p.1540-1543 (1976)

[4.610] {Sect. 4.6} J. Reintjes, C.Y. She, R.C. Eckardt, N.E. Karangelen, R.A. Andrews, R.C. Elton : Seventh harmonic conversion of mode-locked laser pulses to 38.0 nm, Appl. Phys. Lett. 30, p.480-482 (1977)

[4.611] {Sect. 4.6} X.M. Tong, S.I. Chu: Theoretical study of multiple high-order harmonic generation by intense ultrashort pulsed laser fields: A new generalized pseudospectral time-dependent method, Chem Phys 217, p.119-130 (1997)

[4.612] {Sect. 4.6} M. Geissler, G. Tempea, A. Scrinzi, M. Schnurer, F. Krausz, T. Brabec: Light propagation in field-ionizing media: Extreme nonlinear optics, Phys Rev Lett 83, p.2930-2933 (1999)

[4.613] {Sect. 4.6} D.B. Milosevic, A.F. Starace: Magnetic-field-induced intensity revivals in harmonic generation, Phys Rev Lett 82, p.2653-2656 (1999)

[4.614] {Sect. 4.6} H.J. Shin, D.G. Lee, Y.H. Cha, K.H. Hong, C.H. Nam: Generation of nonadiabatic blueshift of high harmonics in an intense femtosecond laser field, Phys Rev Lett 83, p.2544-2547 (1999)

[4.615] {Sect. 4.6} G. vandeSand, J.M. Rost: Irregular orbits generate higher harmonics, Phys Rev Lett 83, p.524-527 (1999)

[4.616] {Sect. 4.6} P. Salieres, P. Antoine, A. deBohan, M. Lewenstein: Temporal and spectral tailoring of high-order harmonics, Phys Rev Lett 81, p.5544-5547 (1998)

[4.617] {Sect. 4.6} R. Zerne, C. Altucci, M. Bellini, M.B. Gaarde, T.W. Hansch, A. LHuillier, C. Lynga, C.G. Wahlstrom: Phase-locked high-order harmonic sources, Phys Rev Lett 79, p.1006-1009 (1997)

[4.618] {Sect. 4.6} V.V. Goloviznin, P.W. van Amersfort: Generation of ultrahigh harmonics with a two-stage free electron laser and a seed laser, Phys. Rev. E 55, p.6002-6010 (1997)

[4.619] {Sect. 4.6} D. Descamps, C. Lynga, J. Norin, A. LHuillier, C.G. Wahlstrom, J.F. Hergott, H. Merdji, P. Salieres, M. Bellini, T.W. Hansch: Extreme ultraviolet interferometry measurements with high-order harmonics, Optics Letters 25, p.135-137 (2000)

[4.620] {Sect. 4.6} A. Ishizawa, K. Inaba, T. Kanai, T. Ozaki, H. Kuroda: High-order harmonic generation from a solid surface plasma by using a picosecond laser, IEEE J QE-35, p.60-65 (1999)

[4.621] {Sect. 4.6} B. Sheehy, J.D.D. Martin, L.F. DiMauro, P. Agostini, K.J. Schafer, M.B. Gaarde, K.C. Kulander: High harmonic generation at long wavelengths, Phys Rev Lett 83, p.5270-5273 (1999)

[4.622] {Sect. 4.6} C. deLisio, C. Altucci, C. Beneduce, R. Bruzzese, F. DeFilippo, S. Solimeno, M. Bellini, A. Tozzi, G. Tondello, E. Pace: Analysis of efficient generation and spatial intensity profiles of high-order harmonic beams produced at high repetition rate, Opt Commun 146, p.316-324 (1998)

[4.623] {Sect. 4.6} A. Goehlich, U. Czarnetzki, H.F. Dobele: Increased efficiency of vacuum ultraviolet generation by stimulated anti-Stokes Raman scattering with Stokes seeding, Appl Opt 37, p.8453-8459 (1998)

[4.624] {Sect. 4.6} G. Sommerer, E. Mevel, J. Hollandt, D. Schulze, P.V. Nickles, G. Ulm, W. Sandner: Absolute photon number measurement of high-order harmonics in the extreme UV, Opt Commun 146, p.347-355 (1998)

[4.625] {Sect. 4.6} D.M. Chambers, S.G. Preston, M. Zepf, M. Castrocelin, M.H. Key, J.S. Wark, A.E. Dangor, A. Dyson, D. Neely, P.A. Norreys: Imaging of high harmonic radiation emitted during the interaction of a 20 TW laser with a solid target, J Appl Phys 81, p.2055-2058 (1997)

[4.626] {Sect. 4.6} P. Gibbon: High-order harmonic generation in plasmas, IEEE J QE-33, p.1915-1924 (1997)

[4.627] {Sect. 4.6} R. Hassner, W. Theobald, S. Niedermeier, H. Schillinger, R. Sauerbrey: High-order harmonics from solid targets as a probe for high-density plasmas, Optics Letters 22, p.1491-1493 (1997)

[4.628] {Sect. 4.6} B.F. Shen, W. Yu, G.H. Zeng, Z.Z. Xu: High order harmonic generation due to nonlinear Thomson scattering, Opt Commun 136, p.239-242 (1997)

[4.629] {Sect. 4.6} M.P. Bogdanov, S.A. Dimakov, A.V. Gorlanov, D.A. Goryachkin, A.M. Grigorev, V.M. Irtuganov, V.P. Kalinin, S.I. Klimentev, I.M. Kozlovskaya, I.B. Orlova, et al.: Correction of segmented mirror aberrations by phase conjugation and dynamic holography, Opt Commun 129, p.405-413 (1996)

[4.630] {Sect. 4.6} I.P. Christov, J. Zhou, J. Peatross, A. Rundquist, M.M. Murnane, H.C. Kapteyn: Nonadiabatic effects in high-harmonic generation with ultrashort pulses, Phys Rev Lett 77, p.1743-1746 (1996)

[4.631] {Sect. 4.6} T. Ditmire, E.T. Gumbrell, R.A. Smith, J.W.G. Tisch, D.D. Meyerhofer, M.H.R. Hutchinson: Spatial coherence measurement of soft x-ray radiation produced by high order harmonic generation, Phys Rev Lett 77, p.4756-4759 (1996)

[4.632] {Sect. 4.6} T.D. Donnelly, T. Ditmire, K. Neuman, M.D. Perry, R.W. Falcone: High-order harmonic generation in atom clusters, Phys Rev Lett 76, p.2472-2475 (1996)

[4.633] {Sect. 4.6} Y. Kobayashi, O. Yoshihara, Y. Nabekawa, K. Kondo, S. Watanabe: Femtosecond measurement of high-order harmonic pulse width and electron recombination time by field ionization, Optics Letters 21, p.417-419 (1996)

[4.634] {Sect. 4.6} I. Mercer, E. Mevel, R. Zerne, A. LHuillier, P. Antoine, C.G. Wahlstrom: Spatial mode control of high-order harmonics, Phys Rev Lett 77, p.1731-1734 (1996)

[4.635] {Sect. 4.6} Y. Nagata, K. Midorikawa, M. Obara, K. Toyoda: High-order harmonic generation by subpicosecond KrF excimer laser pulses, Optics Letters 21, p.15-17 (1996)

[4.636] {Sect. 4.6} P.A. Norreys, M. Zepf, S. Moustaizis, A.P. Fews, J. Zhang, P. Lee, M. Bakarezos, C.N. Danson, A. Dyson, P. Gibbon, et al.: Efficient extreme UV harmonics generated from picosecond laser pulse interactions with solid targets, Phys Rev Lett 76, p.1832-1835 (1996)

[4.637] {Sect. 4.6} J. Zhou, J. Peatross, M.M. Murnane, H.C. Kapteyn: Enhanced high-harmonic generation using 25 fs laser pulses, Phys Rev Lett 76, p.752-755 (1996)

[4.638] {Sect. 4.6} S. Varró, F. Ehlotzky: Higher harmonic generation at metal surfaces by powerful femtosecond laser pulses, Phys. Rev. A 54, p.3245-3249 (1996)

[4.639] {Sect. 4.6} S.E. Harris: Generation of Vacuum-Ultraviolet and Soft-X-Ray Radiation Using High-Order Nonlinear Optical Polarizabilities, Phys. Rev. Lett. 31, p.341-344 (1973)

[4.640] {Sect. 4.6} A.H. Kung, J.F. Young, S.E. Harris: Generation of 1182-A radiation in phase-matched mixtures of inert gases, Appl. Phys. Lett. 22, (Erratum: 28, 239 (1976))p.301-302 (1973)

[4.641] {Sect. 4.7} J. V. Moloney (ed.): Nonlinear Optical Materials (Springer, New York, Berlin, Heidelberg, 1998)

[4.642] {Sect. 4.7} G. P. Agrawal: Nonlinear Fiber Optics (Academic Press, San Diego, London, Boston, 1995)

[4.643] {Sect. 4.7} C. T. Chen: Development of New Nonlinear Optical Crystals in the Borate Series (Harwood Academic Publishers, Chur, 1993)

[4.644] {Sect. 4.7} V. G. Dmitriev, G. Gurzadyan: Handbook of Nonlinear Optical Crystals (DA Information Services, Pty, Ltd, Australia, 1997)

[4.645] {Sect. 4.7} Y. Shuto, S. Tomaru, M. Hikita, M. Amano: Optical Intensity Modulators Using Diazo-Dye-Substituted Polymer Channel Waveguides, IEEE J. QE-31, p.1451-1460 (1995)

[4.646] {Sect. 4.7} R.W. Hellwarth, A. Owyoung, N. George: Origin of the Nonlinear Refractive Index of Liquid CCl4, Phys. Rev. A 4, p.2342-2347 (1971)

[4.647] {Sect. 4.7} S. Chandrasekhar: Liquid Crystals 2nd ed. (Cambridge University Press, Cambridge, 1992)

[4.648] {Sect. 4.7} I.-C. Khoo, S.-T. Wu: Optics and Nonlinear Optics of Liquid Crystals (World Scientific, Singapore, New Jersey, London, Hong Kong, 1993)

[4.649] {Sect. 4.7} Y. Reznikov, O. Ostroverkhova, K.D. Singer, J.H. Kim, S. Kumar, O. Lavrentovich, B. Wang, J.L. West: Photoalignment of liquid crystals by liquid crystals, Phys Rev Lett 84, p.1930-1933 (2000)

[4.650] {Sect. 4.7} J.E. Stockley, G.D. Sharp, K.M. Johnson: Fabry-Perot etalon with polymer cholesteric liquid-crystal mirrors, Optics Letters 24, p.55-57 (1999)

[4.651] {Sect. 4.7} Y. Tabe, N. Shen, E. Mazur, H. Yokoyama: Simultaneous observation of molecular tilt and azimuthal angle distributions in spontaneously modulated liquid-crystalline Langmuir monolayers, Phys Rev Lett 82, p.759-762 (1999)

[4.652] {Sect. 4.7} D.V. Wick, T. Martinez, M.V. Wood, J.M. Wilkes, M.T. Gruneisen, V.A. Berenberg, M.V. Vasilev, A.P. Onokhov, L.A. Beresnev: Deformed-helix ferroelectric liquid-crystal spatial light modulator that demonstrates high diffraction efficiency and 370-line pairs mm resolution, Appl Opt 38, p.3798-3803 (1999)

[4.653] {Sect. 4.7} M. Saito, N. Matsumoto, J. Nishimura: Measurement of the complex refractive-index spectrum for birefringent and absorptive liquids, Appl Opt 37, p.5169-5175 (1998)

[4.654] {Sect. 4.7} S.D. Durbin, S.M. Arakelian, Y.R. Shen: Optical-Field-Induced Birefringence and Freedericksz Transition in a Nematic Liquid Crystal, Phys. Rev. Lett. 47, p.1411-1414 (1981)

[4.655] {Sect. 4.7} E.G. Hanson, Y.R. Shen, G.K.L. Wong: Optical-field-induced refractive indices and orientational relaxation times in a homologous series of isotropic nematic substances, Phys. Rev. 14, p.1281-1289 (1976)

[4.656] {Sect. 4.7} R.A. Mullen, J.N. Matossian: Quenching optical breakdown with an applied electric field, Opt. Lett. 15, p.601-603 (1990)

5. Nonlinear Interactions of Light and Matter with Absorption

[5.1]   {Sect. 5.2} H. Talon, L. Fleury, J. Bernard, M. Orrit: Fluorescence excitation of single molecules, J. Opt. Soc. Am. B 9, p.825-827 (1992)

[5.2]   {Sect. 5.2} L.L. Wald, E.L. Hahn, M. Lukac: Variation of the Pr3+ nuclear quadrupole resonance spectrum across the inhomogeneous optical line in Pr3+:LaF3, J. Opt. Soc. Am. B 9, p.789-793 (1992)

[5.3]   {Sect. 5.2} K.-P. Müller, D. Haarer: Spectral Diffusion of Optical Transitions in Doped Polymer Glasses below 1 K, Phys. Rev. Lett. 66, p.2344-2347 (1991)

[5.4]   {Sect. 5.2} W. Kaiser, A. Seilmeier: Redistribution of Vibrational Energy in Solution, Ber. Bunsenges. Phys. Chem.91, p.1201-1205 (1987)

[5.5]   {Sect. 5.2} A.B. Myers, M.O. Trulson, J.A. Pardoen, C. Heeremans, J. Lugtenburg, R.A. Methies: Absolute resonance Raman intensities demonstrat that the spectral broadening induced by the beta-ionone ring in retinal is homogeneous, J. Chem. Phys. 84, p.633-640 (1986)

[5.6]   {Sect. 5.2} J.R. Morgan, M.A. El-Sayed: Temperature dependence of the homogeneous linewidth of the 5D0-7F0 transition of Eu3+ in amorphous hosts at high temperatures, Chem. Phys. Lett. 84, p.213-216 (1981)

[5.7]   {Sect. 5.2} A.P. Marchetti, W.C. McColgin, J.H. Eberly: Inhomogeneous Broadening and Excited-Vibrational-State Lifetimes in Low-Temperature Organic Mixed Crystals, Phys. Rev. Lett. 35, p.387-390 (1975)

[5.8]   {Sect. 5.2} D.W. Vahey: Effects of spectral cross relaxation and collisional dephasing on the absorption of light by organic-dye solutions, Phys. Rev. A 10, p.1578-1590 (1974)

[5.9]   {Sect. 5.3.1} J.F. Giuliani: Saturable Absorption and Q Switching in a Triphenylmethene Dye, J. Appl. Phys. 43, p.1290-1291 (1972)

[5.10] {Sect. 5.3.1} E.G. Arthurs, D.J. Bradley, A.G. Roddie: Photoisomer Generation and Absorption Relaxation in the Mode-Locking Dye 3,3'-Diethyloxadicarbocynaine Iodide, Opt. Comm. 8, p.118-123 (1973)

[5.11] {Sect. 5.3.1} B.H. Soffer: Giant Pulse Laser Operation by a Passive, Reversible Bleachable Absorber, J. Appl. Phys. 35, p.2551 (1964)

[5.12] {Sect. 5.3.1} H.S. Loka, S.D. Benjamin, P.W.E. Smith: Optical Characterization of Low-Temperature-Grown GaAs for Ultrafast All-Optical Switching Devices, IEEE J. QE-34, p.1426-1436 (1998)

[5.13] {Sect. 5.3.2} R. BurlotLoison, J.L. Doualan, P. LeBoulanger, T.P.J. Han, H.G. Gallagher, R. Moncorge, G. Boulon: Excited-state absorption of Er3+-doped LiNbO3, J Appl Phys 85, p.4165-4170 (1999)

[5.14] {Sect. 5.3.2} F.Z. Henari, H. Manaa, K.P. Kretsch, W.J. Blau, H. Rost, S. Pfeiffer, A. Teuschel, H. Tillmann, H.H. Horhold: Effective stimulated emission and excited state absorption measurements in the phenylene-vinylene oligomer (1,4-bis- (Alpha-cyanostyryl)-2,5-dimethoxybenzene)), Chem Phys Lett 307, p.163-166 (1999)

[5.15] {Sect. 5.3.2} N.V. Kuleshov, A.V. Podlipensky, V.G. Shcherbitsky, A.A. Lagatsky, V.P. Mikhailov: Excited-state absorption in the range of pumping and laser efficiency of Cr4+:forsterite, Optics Letters 23, p.1028-1030 (1998)

[5.16] {Sect. 5.3.2} M.F. Hazenkamp, H.U. Gudel, S. Kuck, G. Huber, W. Rauw, D. Reinen: Excited state absorption and laser potential of Mn5+-doped Li3PO4, Chem Phys Lett 265, p.264-270 (1997)

[5.17] {Sect. 5.3.2} H. Miyasaka, T. Nobuto, A. Itaya, N. Tamai, M. Irie: Picosecond laser photolysis studies on a photochromic dithienylethene in solution and in crystalline phases, Chem Phys Lett 269, p.281-285 (1997)

[5.18] {Sect. 5.3.2} D.K. Palit, A.V. Sapre, J.P. Mittal: Picosecond studies on the electron transfer from pyrene and perylene excited singlet states to N-hexadecyl pyridinium chloride, Chem Phys Lett 269, p.286-292 (1997)

[5.19] {Sect. 5.3.2} K.V. Yumashev, N.V. Kuleshov, P.V. Prokoshin, A.M. Malyarevich, V.P. Mikhailov: Excited state absorption of Cr4+ ion in forsterite, Appl Phys Lett 70, p.2523-2525 (1997)

[5.20] {Sect. 5.3.2} R. Moncorge, H. Manaa, F. Deghoul, Y. Guyot, Y. Kalisky, S.A. Pollack, E.V. Zharikov, M. Kokta: Saturable and excited state absorption measurements in Cr4+:LuAG single crystals, Opt Commun 132, p.279-284 (1996)

[5.21] {Sect. 5.3.2} R. Sander, V. Herrmann, R. Menzel: Transient Absorption Spectra and Bleaching of 4-n-Pentyl-4-Cyanoterphenyl in Cyclohexane – Determination of Cross Sections and Recovery Times, J. Chem. Phys. 104, p.4390-4395 (1996)

[5.22] {Sect. 5.3.2} H.J. Eichler, R. Macdonald, R. Menzel, R. Sander: Excited State absorption of 5CB (4'-n-pentyl-4-cyanobiphenyl) in cyclohexane, Chem. Phys. 195, p.381-386 (1995)

[5.23] {Sect. 5.3.2} R. Menzel, H. Lueck: Conformation Dependent Excited State Absorptions of 3,3",5,5"-Tetramethyl -Para-Terphenyl, Chem. Phys. 124, p.417-424 (1988)

[5.24] {Sect. 5.3.2} F.E. Doany, E.J. Heilweil, R. Moore, R.M. Hochstrasser: Picosecond study of an intermediate in the trans to cis isomerization pathway of stiff stilbene, J. Chem. Phys. 80, p.201-206 (1984)

[5.25] {Sect. 5.3.2} R. Menzel, W. Rapp: Excited Singlet- and Triplet-Absorptions of Pentaphene, Chem. Phys. 89, p.445-455 (1984)

[5.26] {Sect. 5.3.2} V. Sundstrom, T. Gillbro: Dynamics of the isomerization of trans-stilbene in n-alcohols studied by ultraviolet picosecond absorption recovery, Chem. Phys. Lett. 109, p.538-543 (1984)

[5.27] {Sect. 5.3.2} D.W. Boldridge, G.W. Scott: Excited state spectroscopy of 1,5-naphthyridine: Identification of the lowest energy excited singlet state as 1Bg (1nPI*), J. Chem. Phys. 79, p.3639-3644 (1983)

[5.28] {Sect. 5.3.2} T. Sugawara, H. Iwamura, N. Nakashima, K. Yoshihara: Transient absorption spectra of the excited states of triptycene and 3-acetyltriptycene, Chem. Phys. Lett. 101, p.303-306 (1983)

[5.29] {Sect. 5.3.2} M. Sumitani, K. Yoshihara: Direct Observation of the Rate for Cis-Trans and Trans-Cis Photoisomerization of Stilbene with Picosecond Laser Photolysis, Bull. Chem. Soc. Japan 55, p.85-89 (1982)

[5.30] {Sect. 5.3.2} F.E. Doany, B.I. Greene, R.M. Hochstrasser: Excitation energy effects in the photophysics of trans-stilbene in solution, Chem. Phys. Lett. 75, p.206-208 (1980)

[5.31] {Sect. 5.3.2} B.I. Greene, R.M. Hochstrasser, R. Weisman: Picosecond dynamics of the photoisomerization of trans-stilbene under collision-free conditions, J. Chem. Phys. 71, p.544-545 (1979)

[5.32] {Sect. 5.3.2} K. Yoshihara, A. Namiki, M. Sumitami, N. Nakashima: Picosecond flash photolysis of cis- and trans-stilbene. Observation of an intense intramolecular charge-resonance transition, J. Chem. Phys. 71, p.2892-2895 (1979)

[5.33] {Sect. 5.3.2} O. Teschke, E.P. Ippen, G.R. Holtom: Picosecond dynamics of the singlet excited state of trans-and cis-stilbene, Chem. Phys. Lett. 52, p.233-235 (1977)

[5.34] {Sect. 5.3.2} D.S. Kliger, A.C. Albrecht: Nanosecond Excited-State Polarized Absorption Spectroscopy of Anthracene in the Visible Region, J. Chem. Phys. 50, p.4109-4111 (1969)

[5.35] {Sect. 5.3.2} A. Müller, E. Pflüger: Laser-flashspectroscopy of cryptocyanine, Chem. Phys. Lett. 2, p.155-159 (1968)

[5.36] {Sect. 5.3.2} A. Müller: Kinetische Laser-Blitzspektroskopie organischer Moleküle, Z. Naturforsch. 23, p.946-949 (1968)

[5.37] {Sect. 5.3.2} J.R. Novak, M.W. Windsor: Laser photolysis and spectroscopy: a new technique for the study of rapid reactions in the nanosecond time range, Proc. Roy. Soc. A. 308, p.95-110 (1968)

[5.38] {Sect. 5.3.3} J. Barroso, A. Costela, I. Garciamoreno, R. Sastre: Wavelength dependence of the nonlinear absorption properties of laser dyes in solid and liquid solutions, Chem Phys 238, p.257-272 (1998)

[5.39] {Sect. 5.3.3} M. Samoc, A. Samoc, B. LutherDavies, H. Reisch, U. Scherf: Saturable absorption in poly (indenofluorene): A picket-fence polymer, Optics Letters 23, p.1295-1297 (1998)

[5.40] {Sect. 5.3.3} S.H. Yim, D.R. Lee, B.K. Rhee, D. Kim: Nonlinear absorption of Cr4+:YAG studied with lasers of different pulsewidths, Appl Phys Lett 73, p.3193-3195 (1998)

[5.41] {Sect. 5.3.3} M. Wittmann, R. Rotermund, R. Weigand, A. Penzkofer: Saturable absorption and absorption recovery of indocyanine green J-aggregates in water, Appl. Phys. B 66, p.453-459 (1998)

[5.42] {Sect. 5.3.3} S. Oberländer, D. Leupold: Instantaneous fluorescence quantum yield of organic molecular systems: information content of ist intensity dependence, J. Luminesc. 59, p.125-133 (1994)

[5.43] {Sect. 5.3.3} R. Menzel, P. Witte: Recovery Time of the Bleached S1 – Sn – Absorption of Para-Terphenyl in Solution.Recovery Time of the Bleached S1 – Sn – Absorption of Para-Terphenyl in Solution, Chem. Phys. Lett. 164, p.27-32 (1989)

[5.44] {Sect. 5.3.3} R. Menzel, D. Leupold: Nonlinear Absorptions of Cryptocyanine, Chem. Phys. Lett. 65, p.120-126 (1979)

[5.45] {Sect. 5.3.3} J.L. Hall, C. Bordé: Measurement of Methane Hyperfine Structure Using Laser Saturated Absorption, Phys. Rev. Lett. 30, p.1101-1104 (1973)

[5.46] {Sect. 5.3.3} M. Hercher: An Analysis of Saturable Absorbers, Appl. Opt. 6, p.947-954 (1967)

[5.47] {Sect. 5.3.3} A. Peda'el, R. Daisy, M. Horowitz, B. Fischer: Beam coupling-induced transparency in a bacteriorhodopsin-based saturable absorber, Opt. Lett. 23, p.1173-1175 (1998)

[5.48] {Sect. 5.3.3} P. Chen, X. Wu, X. Sun, J. Lin, W. Ji, K.L. Tan: Electronic structure and optical limiting behavior of carbon nanotubes, Phys Rev Lett 82, p.2548-2551 (1999)

[5.49] {Sect. 5.3.3} B. Dupuis, C. Michaut, I. Jouanin, J. Delaire, P. Robin, P. Feneyrou, V. Dentan: Photoinduced intramolecular charge-transfer systems based on porphyrin-viologen dyads for optical limiting, Chem Phys Lett 300, p.169-176 (1999)

[5.50] {Sect. 5.3.3} D. Leupold, H. Stiel, J. Ehlert, F. Nowak, K. Teuchner, B. Voigt, M. Bandilla, B. Ücker, H. Scheer: Photophysical characterization of the B800-depleted light harvesting complex B850 of Rhodobacter sphaeroides Implication to the ultrafast energy transfer 800-580 nm, Chem. Phys. Lett. 301, p.537-545 (1999)

[5.51] {Sect. 5.3.3} G.S. He, C. Weder, P. Smith, P.N. Prasad: Optical power limiting and stabilization based on a novel polymer compound, IEEE J QE-34, p.2279-2285 (1998)

[5.52] {Sect. 5.3.3} M.P. Joshi, J. Swiatkiewicz, F.M. Xu, P.N. Prasad: Energy transfer coupling of two-photon absorption and reverse saturable absorption for enhanced optical power limiting, Optics Letters 23, p.1742-1744 (1998)

[5.53] {Sect. 5.3.3} W. Lozano, C.B. deAraujo, L.H. Acioli, Y. Messaddeq: Negative nonlinear absorption in Er3+-doped fluoroindate glass, J Appl Phys 84, p.2263-2267 (1998)

[5.54] {Sect. 5.3.3} S.R. Mishra, H.S. Rawat, M. Laghate: Nonlinear absorption and optical limiting IN metalloporphyrins, Opt Commun 147, p.328-332 (1998)

[5.55] {Sect. 5.3.3} M. Pittman, P. Plaza, M.M. Martin, Y.H. Meyer: Subpicosecond reverse saturable absorption in organic and organometallic solutions, Opt Commun 158, p.201-212 (1998)

[5.56] {Sect. 5.3.3} M. Brunel, F. Chaput, S.A. Vinogradov, B. Campagne, M. Canva, J.P. Boilot, A. Brun: Reverse saturable absorption in palladium and zinc tetraphenyltetrabenzoporphyrin doped xerogels, Chem Phys 218, p.301-307 (1997)

[5.57] {Sect. 5.3.3} G.S. He, L.X. Yuan, J.D. Bhawalkar, P.N. Prasad: Optical limiting, pulse reshaping, and stabilization with a nonlinear absorptive fiber system, Appl Opt 36, p.3387-3392 (1997)

[5.58] {Sect. 5.3.3} G.S. He, G.C. Xu, P.N. Prasad, B.A. Reinhardt, J.C. Bhatt, A.G. Dillard: Two photon absorption and optical limiting properties of novel organic compounds, Optics Letters 20, p.435-437 (1995)

[5.59] {Sect. 5.3.3} R.I. Ghauharali, M. Muller, A.H. Buist, T.S. Sosnowski, T.B. Norris, J. Squier, G.J. Brakenhoff: Optical saturation measurements of fluorophores in solution with pulsed femtosecond excitation and two- dimensional CCD camera detection, Appl Opt 36, p.4320-4328 (1997)

[5.60] {Sect. 5.3.4} V.A. Zuikov, A.A. Kalachev, V.V. Samartsev, A.M. Shegeda: Two-color optical superradiance and other coherent effects in the resonant propagation of a laser pulse in a LaF3 : Pr3+ crystal, Laser Phys 10, p.364-367 (2000)

[5.61] {Sect. 5.3.4} P. Goy, J.M. Raimond, M. Gross, S. Haroche: Observation of Cavity-Enhanced Single-Atom Spontaneous Emission, Phys. Rev. Lett. 50, p.1903-1906 (1983)

[5.62] {Sect. 5.3.4} A. Szabo: Laser-Induced Fluorescence-Line Narrowing in Ruby, Phys. Rev. Lett. 25, p.924-926 (1970)

[5.63] {Sect. 5.3.5} W.E. Moerner (ed.): Persistent Spectral Hole-Burning: Science and Applications, Topics Curr. Phys, Vol. 44 (Springer, Berlin, Heidelberg 1988)

[5.64] {Sect. 5.3.5} M. Nogami, Y. Abe, K. Hirao, D.H. Cho: Room temperature persistent spectra hole burning in Sm2+-doped silicate glasses prepared by the sol-gel process, Appl. Phys. Lett. 66, p.2952-2954 (1995)

[5.65] {Sect. 5.3.5} Y.-I. Pan, Y.-Y. Zhao, Y.Yin, L.-b. Chen, R.-s. Wang, F.-m. Li: The observation of photoproducts and multiple photon-gated spectral hole burning in a donor-acceptor and a donor1+donor2-acceptor system, Opt. Comm. 119, p.538-544 (1995)

[5.66] {Sect. 5.3.5} R.B. Altmann, I. Renge, L. Kador, D. Haarer: Dipole moment differences of nonpolar dyes in polymeric matrices: Stark effect and photochemical hole burning. I, J. Chem. Phys. 97, p.5316-5322 (1992)

[5.67] {Sect. 5.3.5} W.P. Ambrose, A.J. Sievers: Persistent infrared spectral hole burning of the fundamental stretching mode of SH- in alkali halides, J. Opt. Soc. Am. B 9, p.753-762 (1992)

[5.68] {Sect. 5.3.5} S. Arnold, J. Comunale: Room-temperature microparticle-based persistent hole-burning spectroscopy, J. Opt. Soc. Am. B 9, p.819-824 (1992)

[5.69] {Sect. 5.3.5} Th. Basché, W.P. Ambrose, W.E. Moerner: Optical spectra and kinetics of single impurity molecules in a polymer: spectral diffusion and persistent spectral hole burning, J. Opt. Soc. Am. B 9, p.829-836 (1992)

[5.70] {Sect. 5.3.5} R.L. Cone, P.C. Hansen, M.J.M. Leask: Eu3+ optically detected nuclear quadrupole resonance in stoichiometric europium vanadate, J. Opt. Soc. Am. B 9, p.779-783 (1992)

[5.71] {Sect. 5.3.5} R. Hirschmann, J. Friedrich: Hole burning of long-chain molecular aggregates: homogeneous line broadening, spectral-diffusion broadening, and pressure broadening, J. Opt. Soc. Am. B 9, p.811-815 (1992)

[5.72] {Sect. 5.3.5} H. Inoue, T. Iwamoto, A. Makishima, M. Ikemoto, K. Horie: Preperation and properties of sol-gel thin films with porphins, J. Opt. Soc. Am. B 9, p.816-818 (1992)

[5.73] {Sect. 5.3.5} L. Kümmerl, H. Wolfrum, D. Haarer: Hole Burning with Chelate Complexes of Quinizarin in Alcohol Glasses, J. Phys. Chem. 96, p.10688-10693 (1992)

[5.74] {Sect. 5.3.5} S.P. Love, C.E. Mungan, A.J. Sievers: Persistant infrared spectral hole burning of Tb3+ in the glasslike mixed crystal Ba1-x-yLaxTbyF2+x+y, J. Opt. Soc. Am. B 9, p.794-799 (1992)

[5.75] {Sect. 5.3.5} C.E. Mungan, A.J. Sievers: Persistent infrared spectral hole burning of the fundamental stretching mode of SH- in alkali halides, J. Opt. Soc. Am. B 9, p.746-752 (1992)

[5.76] {Sect. 5.3.5} D. Redman, S. Brown, S.C. Rand: Origin of persistent hole burning of N-V centers in diamond, J. Opt. Soc. Am. B 9, p.768-774 (1992)

[5.77] {Sect. 5.3.5} R.J. Reeves, R.M. Macfarlane: Persistent spectral hole burning induced by ion motion in DaF2:Pr3+:D- and SrF2:Pr3+:D- crystals, J. Opt. Soc. Am. B 9, p.763-767 (1992)

[5.78] {Sect. 5.3.5} I. Renge: Relationship between electron-phonon coupling and intermolecular interaction parameters in dye-doped organic glasses, J. Opt. Soc. Am. B 9, p.719-723 (1992)

[5.79] {Sect. 5.3.5} W. Richter, M. Lieberth, D. Haarer: Frequency dependence of spectral diffusion in hole-burning systems: resonant effects of infrared radiation, J. Opt. Soc. Am. B 9, p.715-718 (1992)

[5.80] {Sect. 5.3.5} N.E. Rigby, N.B. Manson: Spectral hole burning in emerald, J. Opt. Soc. Am. B 9, p.775-778 (1992)

[5.81] {Sect. 5.3.5} B. Sauter, Th. Basché, C. Bräuchle: Temperature-dependent spectral hole-burning study of dye-surface and mixed matrix-dye-surface systems, J. Opt. Soc. Am. B 9, p.804-810 (1992)

[5.82] {Sect. 5.3.5} L. Shu, G.J. Small: Mechanism of nonphotochemical hole burning: Cresyl Violet in polyvinyl alcohol films, J. Opt. Soc. Am. B 9, p.724-732 (1992)

[5.83] {Sect. 5.3.5} L. Shu, G.J. Small: Dispersive kinetics of nonphotochemical hole burning and spontaneous hole filling: Cresyl Violet in polyvinyl films, J. Opt. Soc. Am. B 9, p.733-737 (1992)

[5.84] {Sect. 5.3.5} L. Shu, G.J. Small: Laser-induced hole filling: Cresyl Violet in polyvinyl alcohol films, J. Opt. Soc. Am. B 9, p.738-745 (1992)

[5.85] {Sect. 5.3.5} D. Wang, L. Hu, H. He, J. Rong, J. Xie, J. Zhang: Systems of organic photon-gated photochemical hole burning, J. Opt. Soc. Am. B 9, p.800-803 (1992)

[5.86] {Sect. 5.3.5} L. Kador, S. Jahn, D. Haarer: Contributions of the electrostatic and the dispersion interaction to the solvent shift in a dye-polymer system, as investigated by hole-burning spectroscopy, Phys. Rev. B 41, p.12215-12228 (1990)

[5.87]   {Sect. 5.3.5} A. Renn, A.J. Meixner, U.P. Wild: II. Diffraction Properties of two Spectrally Adjacent Holograms, J. Chem. Phys. 91, p.2748-2755 (1990)

[5.88]   {Sect. 5.3.5} U.P. Wild, A. Renn, C. De Caro, S. Bernet: Spectral hole burning and molecular computing, Appl. Opt. 29, p.4329-4331 (1990)

[5.89]   {Sect. 5.3.5} P.C. Becker, H.L. Fragnito, J.Y Bigot, C.H. Brito Cruz, R.L. Fork, C.V. Shank: Femtosecond Photon Echos from Molecules in Solution, Phys. Rev. Lett. 63, p.505-507 (1989)

[5.90]   {Sect. 5.3.5} C.H. BritoCruz, J.P. Gordon, P.C. Becker, R.L. Fork, C.V. Shank: Dynamics of Spectral Hole Burning, IEEE J. QE-24, p.261-266 (1988)

[5.91]   {Sect. 5.3.5} M. Joffre, D. Hulin, A. Migus, A. Antonietti, C. Benoit à la Guillaume, N. Peyghambarian, M. Lindberg, S.W. Koch: Coherent effects in pump-probe spectroscopy of excitons, Opt. Lett. 13, p.276-278 (1988)

[5.92]   {Sect. 5.3.5} B. Fluegel, N. Peyghambarian, G. Olbright, M. Lindberg, S.W. Koch, M. Joffre, D. Hulin, A. Migus, A. Antonietti: Femtosecond Studies of Coherent Transients in Semiconductors, Phys. Rev. Lett. 59, p.2588-2591 (1987)

[5.93]   {Sect. 5.3.5} M. Maier: Persistant Spectral Holes in External Fields, Appl. Phys. B 41, p.73-90 (1986)

[5.94]   {Sect. 5.3.5} A.U. Jalmukhambetov, I.S. Osad'ko: Dependence of photochemical and photophysical hole burning on laser intensity, Chem. Phys. 77, p.247-255 (1983)

[5.95]   {Sect. 5.3.5} J. Friedrich, D. Haarer: Transient features of optical bleaching as studies by photochemical hole burning and fluorescence line narrowing, J.Chem. Phys. 76, p.61-68 (1982)

[5.96]   {Sect. 5.3.5} R.W. Olson, H.W.H. Lee, F.G. Patterson, M.D. Fayer: Non-photochemical hole burning and antihole production in the mixed molecular crystal pentacene in benzoic acid, J. Chem. Phys. 77, p.2283-2289 (1982)

[5.97]   {Sect. 5.3.5} H. de Vries, D.A. Wiersma: Photophysical and photochemical molecular hole burning theory, J. Chem. Phys. 72, p.1851-1863 (1980)

[5.98]   {Sect. 5.3.5} J. Friedrich, D. Haarer: Phonon selective low temperature photochemistry in alcohol glasses, Chem. Phys. Lett. 74, p.503-506 (1980)

[5.99]   {Sect. 5.3.5} R. M. Macfarlane, R. M. Shelby: Photochemical and Population Hole Burning in the Zero-Phonon Line of a Color Center F3+ in NaF, Phys. Rev. Lett. 42, p.788-791 (1979)

[5.100] {Sect. 5.3.5} J.M. Hayes, G.J. Small: Non-photochemical hole burning and impurity site relaxation processes in organic glasses, Chem. Phys. 27, p.151-157 (1978)

[5.101] {Sect. 5.3.5} C.L. Tang, H. Statz, G.A. DeMars, D.T. Wilson: Spectral Properties of a Single-Mode Ruby Laser: Evidence of Homogeneous Broadening of the Zero-Phonon Lines in Solids, Phys. Rev. 136, p.A1-A8 (1964)

[5.102] {Sect. 5.3.5} R.T. Brundage, W.M. Yen: Low-temperature homogeneous linewidths of Yb3+ in inorganic glasses, Phys. Rev. B 4, p.4436-4438 (1986)

[5.103] {Sect. 5.3.5} A.I.M. Dicker, L.W. Johnson, S. Völker, J.H. van der Waals: Homogeneous linewidth and optical dephasing of the S1-S0 transition of magnesium porphin in an n-octane crystal: A study by transient and photochemical hole-burning, Chem. Phys. Lett. 100, p.8-14 (1983)

[5.104] {Sect. 5.3.5} L.A. Rebane, A.A. Gorokhovskii, J.V. Kikas: Low-Temperature Spectroscopy of Organic Molecules in Solids by Photochemical Hole Burning, Appl. Phys. B 29, p.235-250 (1982)

[5.105] {Sect. 5.3.5} A.I.M. Dicker, J. Dobkowski, S. Völker: Optical dephasing of the S1-S0 transition of free-base porphin in an n-decane host studied by photochemical hole-burning: a case of slow exchange, Chem. Phys. Lett. 84, p.415-420 (1981)

[5.106] {Sect. 5.3.5} J.R. Morgan, M.A. El-Sayed: Temperature dependence of the homogeneous linewidth of the 5D0-7F0 transition of Eu3+ in amorphous hosts at high temperatures, Chem. Phys. Lett. 84, p.213-216 (1981)

[5.107] {Sect. 5.3.5} S. Völker, R.M. Macfarlane: Laser photochemistry and hole-burning of chlorin in crystalline n-alkanes at low temperatures, J. Chem. Phys. 73, p.4476-4482 (1980)

[5.108] {Sect. 5.3.5} J. Hegarty, W.M. Yen: Optical Homogeneous Linewidths of Pr+ in BeF2 and GeO2 Glasses, Phys. Rev. Lett. 43, p.1126-1130 (1979)

[5.109] {Sect. 5.3.5} R.M. Shelby, R.M. Macfarlane: Population hole-burning using a triplet reservoir: S1-S0 transition of zinc porphin in n-octane, Chem. Phys. Lett. 64, p.545-549 (1979)

[5.110] {Sect. 5.3.5} S. Voelker, R.M. Macfarlane: Photochemical hole-burning in vibronic bands of the S1-S0 transition of free-base porphin in an n-octane crystal, Chem. Phys. Lett. 61, p.421-425 (1979)

[5.111] {Sect. 5.3.5} S. Voelker, R.M. Macfarlane: Frequency shift and dephasing of the S1-S0 transition of free-base porphin in an n-octane crystal as a function of temperature, Chem. Phys. Lett. 53, p.8-13 (1979)

[5.112] {Sect. 5.3.5} P. Avouris, A. Campion, M.A. El-Sayed: Variations in homo-geneous fluorescence linewidth and electron-phonon coupling within an in-homogeneous spectral profile, J. Chem. Phys. 67, p.3397-3398 (1977)

[5.113] {Sect. 5.3.5} A.A. Gorokhovski, L.A. Rebane: The Termperature Broaden-ing of Purely Electronic Lines by the Hole Burning Technique, Opt. Comm. 20, p.144-146 (1977)

[5.114] {Sect. 5.3.5} A.P. Marchetti, M. Scozzafava, R.H. Young: Site selection, hole burning, and Stark effect on resorufin in poly (methyl methacrylate), Chem. Phys. Lett. 51, p.424-426 (1977)

[5.115] {Sect. 5.3.5} P.M. Selzer, D.L. Huber, D.S. Hamilton, W.M. Yen, M.J. Weber: Anomoulous Fluorescence Linewidth Behavior in Eu3+-Doped Sili-cate Glass, Phys. Rev. Lett. 36, p.813-816 (1976)

[5.116] {Sect. 5.3.5} A.P. Marchetti, W.C. McColgin, J.H. Eberly: Inhomogeneous Broadening and Excited-Vibrational-State Lifetimes in Low-Temperature Organic Mixed Crystals, Phys. Rev. Lett. 35, p.387-390 (1975)

[5.117] {Sect. 5.3.5} M. Ishikawa, Y. Maruyama: Femtosecond spectral hole-burning of crystal violet in methanol. New evidence for ground state conformers, Chem. Phys. Lett. 219, p.416-420 (1994)

[5.118] {Sect. 5.3.5} H.J. Bakker, P.C.M. Planken, L. Kuipers, A. Lagendijk: Ultrafast infrared saturation spectroscopy of chloroform, bromeform, and iodoform, J. Chem. Phys. 94, p.1730-1739 (1991)

[5.119] {Sect. 5.3.5} D. Blanchard, D.A. Gilmore, T.L. Brack, H. Lemaire, D. Hughes, G.H. Atkinson: Picosecond time-resolved absorption and fluo-rescence in the bacteriorhodopsin photocycle: vibrationally-excited species, Chem. Phys. 154, p.155-170 (1991)

[5.120] {Sect. 5.3.5} T.L. Brack, G.H. Atkinson: Vibrationally Excited Retinal in the Bacteriorhodopsin Photocycle: Picosecond Time-Resolved Anti-Stokes Resonance Raman Scattering, J. Phys. Chem. 95, p.2351-2356 (1991)

[5.121] {Sect. 5.3.5} T. Elsaesser, W. Kaiser: Vibrational and vibronic relaxation of large polyatomic molecules in liquids, Annu. Rev. Phys. Chem. 42, p.83-107 (1991)

[5.122] {Sect. 5.3.5} H. Graener, G. Seifert, A. Laubereau: New Spectroscopy of Water Using Tunable Picosecond Pulses in the Infrared, Phys. Rev. Lett. 66, p.2092-2095 (1991)

[5.123] {Sect. 5.3.5} H.-J. Hübner, M. Wörner, W. Kaiser, A. Seilmeier: Subpicosecond vibrational relaxation of skeletal modes in polyatomic molecules, Chem. Phys. Lett. 182, p.315-320 (1991)

[5.124] {Sect. 5.3.5} A. Mokhtari, A. Chebira, J. Chesnoy: Subpicosecond fluorescence dynamics of dye molecules, J. Opt. Soc. Am. B 7, p.1551-1557 (1990)

[5.125] {Sect. 5.3.5} U. Sukowski, A. Seilmeier, T. Elsaesser, S.F. Fischer: Picosecond energy transfer of vibrationally hot molecules in solution: Experimental studies and theoretical analysis, J. Chem. Phys. 93, p.4094-4101 (1990)

[5.126] {Sect. 5.3.5} G. Angel, R. Gagel, A. Laubereau: Femtosecond polarization spectroscopy of liquid dye solutions, Chem. Phys. 131, p.129-134 (1989)

[5.127] {Sect. 5.3.5} G. Angel, R. Gagel, A. Laubereau: Femtosecond relaxation dynamics in the electronic ground state of dye molecules studied by polarization-dependent amplification spectroscopy, Chem. Phys. Lett. 156, p.169-174 (1989)

[5.128] {Sect. 5.3.5} H. Graener, T.Q. Ye, A. Laubereau: Ultrafast vibrational predissociation of hydrogen bonds: Mode selective infrared photochemistry in liquids, J. Chem. Phys. 91, p.1043-1046 (1989)

[5.129] {Sect. 5.3.5} H. Graener, T.Q. Ye, A. Laubereau: Ultrafast dynamics of hydrogen bonds directly observed by time-resolved infrared spectroscopy, J. Chem. Phys. 90, p.3413-3416 (1989)

[5.130] {Sect. 5.3.5} F. Laermer, T. Elsaesser, W. Kaiser: Ultrashort vibronic and thermal relaxation of dye molecules after femtosecond ultraviolet excitation, Chem. Phys. Lett. 156, p.381-386 (1989)

[5.131] {Sect. 5.3.5} A. Mokhtari, J. Chesnoy, A. Laubereau: Femtosecond time- and frequency-resolved fluorescence spectroscopy of a dye molecule, Chem. Phys. Lett. 155, p.593-598 (1989)

[5.132] {Sect. 5.3.5} A. Mokhtari, L. Fini, J. Chesnoy: Ultrafast conformation equilibration in triphenyl methane dyes analyzed by time resolved induced photoabsorption, J. Chem. Phys. 87, p.3429-3435 (1987)

[5.133] {Sect. 5.3.5} M.J. Rosker, F.W. Wise, C.L. Tang: Femtosecond Relaxation Dynamics of Large Molecules, Phys. Rev. Lett. 57, p.321-324 (1986)

[5.134] {Sect. 5.3.6} J. Ehlert, H. Stiel, K. Teuchner: A numerical solver for rate euqations and photon transport equations in nonlinear laser spectroscopy, Comp. Phys. Commun.124p.330-339 (2000)

[5.135] {Sect. 5.3.6} Stiel, Teuschner, Leupold, Oberländer, Ehlert, Jahnke: Computer Aided Laser-Spectroscopic Characterization and Handling of Molecular Excited States, Intell. Instr. Comp. 9, p.79-88 (1991)

[5.136] {Sect. 5.3.6} R. Menzel: Modelling Excited State Absorption (ESA) Measurements Including the Photophysical Hole Burning Effect with Rate Equations, Mol. Phys. 68, p.161-180 (1989)

[5.137] {Sect. 5.3.6} C.J. Bardeen, J.S. Cao, F.L.H. Brown, K.R. Wilson: Using time-dependent rate equations to describe chirped pulse excitation in condensed phases, Chem Phys Lett 302, p.405-410 (1999)

[5.138] {Sect. 5.3.8} Y.C. Shen, P. Hess: Real-time detection of laser-induced transient gratings and surface acoustic wave pulses with a Michelson interferometer, J Appl Phys 82, p.4758-4762 (1997)

[5.139] {Sect. 5.3.8} N. Tamai, T. Asahi, H. Masuhara: Intersystem crossing of benzophenone by femtosecond transient grating spectroscopy, Chem. Phys. Lett. 198, p.413-418 (1992)

[5.140] {Sect. 5.4.0} T.W. Hänsch, H. Walther: Laser spectroscopy and quantum optics, Rev. Mod. Phys. 71, p.242-252 (1999)

[5.141] {Sect. 5.4.0} L. Mandel: Quantum Effects in one-photon and two-photon interference, Rev. Mod. Phys. 71, p.274-282 (1999)

[5.142] {Sect. 5.4.0} A. Zeilinger: Experiment and the foundations of quantum physics, Rev. Mod. Phys. 71, p.288-296 (1999)

[5.143] {Sect. 5.4.0} J. Mlynek, W. Lange, H. Harde, H. Burggraf: High-resolution coherence spectroscopy using pulse trains, Phys. Rev. A24, p.1099-1102 (1989)

[5.144] {Sect. 5.4.0} J. Mlynek, W. Lange: A simple method of observing coherent ground state transients, Opt. Comm. 30, p.337-340 (1979)

[5.145] {Sect. 5.4.0} J.C. Bergquist, S.A. Lee, J.L. Hall: Saturated Absorption with Spatially Separated Laser Fiels: Observation of Optical "Ramsey" Fringes, Phys. Rev. Lett. 38, p.159-161 (1977)

[5.146] {Sect. 5.4.0} M.M. Salour, C. Cohen-Tannoudji: Observation of Ramsey"s Interference Fringes in the Profile of Doppler-Free Two-Photon Resonances, Phys. Rev. Lett. 38, p.757-760 (1977)

[5.147] {Sect. 5.4.0} R.G. Brewer, A.Z. Genack: Optical Coherent Transients by Laser Frequency Switching, Phys. Rev. Lett. 36, p.959-962 (1976)

[5.148] {Sect. 5.4.0} M.E. Kaminsky, R.T. Hawkins, F.V. Kovalski, A.L. Schawlow: Identification of Absorption Lines by Modulated Lower-Level Population: Spectrum of Na2, Phys. Rev. Lett. 36, p.671-673 (1976)

[5.149] {Sect. 5.4.0} A. Schenzle, R.G. Brewer: Optical coherent transients: Generalized two-level solutions, Phys. Rev. A 14, p.1756-1765 (1976)

[5.150] {Sect. 5.4.0} R. Teets, R. Feinberg, T.W. Hänsch, A.L. Schawlow: Simplification of Spectra by Polarization Labeling, Phys. Rev. Lett. 37, p.683-686 (1976)

[5.151] {Sect. 5.4.0} C. Wieman, T.W. Hänsch: Doppler-Free Laser Polarization Spectroscopy, Phys. Rev. A 36, p.1170-1173 (1976)

[5.152] {Sect. 5.4.0} F. Biraben, B. Cagnac, G. Grynberg: Paschen-Back Effect on the 3S-4D Two-Photon Transition in Sodium Vapor, Phys. Lett. 48 A, p.469-470 (1974)

[5.153] {Sect. 5.4.0} R.G. Brewer, R.L. Shoemaker, S. Stenhom: Collision-Induced Optical Double Resonance, Phys. Rev. Lett. 33, p.63-66 (1974)

[5.154] {Sect. 5.4.0} W. P. Schleich, E. Mayr: Quantum Optics in Phase Space (John Wiley & Sons, Chichester, 1997)

[5.155] {Sect. 5.4.1} C.H. Grossman, J.J. Schwendiman: Ultrashort dephasing-time measurements in Nile Blue polymer films, Optics Letters 23, p.624-626 (1998)

[5.156] {Sect. 5.4.1} K. Holliday, C. Wie, M. Croci, U.P. Wild: Spectral hole-burning measurements of optical dephasing between 2-300 K in Sm2+ doped substitutionally disordered microcrystals, J. Luminesc. 53, p.227-230 (1992)

[5.157] {Sect. 5.4.1} R. van den Berg, A. Visser, S. Völker: Optical dephasing in organic glasses between 0.3 and 20 K. A hole-burning study of resorufin and free-base porphin, Chem. Phys. Lett. 144, p.105-113 (1988)

[5.158] {Sect. 5.4.1} Y.J. Yan, S. Mukamel: Electronic dephasing, vibrational relaxation, and solvent friction in molecular nonlinear optical line shapes, J. Chem. Phys. 89, p.5160-5176 (1988)

[5.159] {Sect. 5.4.1} T. Hattori, T. Kobayashi: Femtosecond dephasing in a polydiacetylene film observed by degenerate four-wave mixing with an incoherent nanosecond laser, J. Luminesc. 38, p.326-328 (1987)

[5.160] {Sect. 5.4.1} M.N. Sapozhnikov: Dephasing, vibronic relaxation and homogeneous spectra of porphyrins in amorphous matrices by selective excitation of luminescence and hole burning, Chem. Phys. Lett. 136, p.192-198 (1987)

[5.161] {Sect. 5.4.1} S. Völker: Optical linewidth and dephasing of organic amorphous and semi-crystalline solids studied by hole burning, J. Luminesc. 36, p.251-262 (1987)

[5.162] {Sect. 5.4.1} M. Fujiwara, R. Kuroda: Measurement of ultrafast dephasing time of Cresyl Fast Violet in cellulose by photon echoes with incoherent light, J. Opt. Soc. Am. B 2, p.1634-1639 (1985)

[5.163] {Sect. 5.4.1} A.M. Weiner, S. De Silvestri, E.P. Ippen: Three-pulse scattering for femtosecond dephasing studies: theory and experiment, J. Opt. Soc. Am. B. 2, p.654-662 (1985)

[5.164] {Sect. 5.4.1} T.P. Carter, B.L. Fearey, J.M. Hayes, G.J. Small: Optical dephasing of cresyl violet in a polyvinyl alcohol polymer by non-photochemical hole burning, Chem. Phys. Lett. 102, p.272-276 (1983)

[5.165] {Sect. 5.4.1} J. Brickmann, P. Russegger: Dephasing in isolated one-dimensional quantum systems, Chem. Phys. 68, p.369-375 (1982)

[5.166] {Sect. 5.4.1} A.I.M. Dicker, J. Dobkowski, S. Völker: Optical dephasing of the S1-S0 transition of free-base porphin in an n-decane host studied by photochemical hole-burning: a case of slow exchange, Chem. Phys. Lett. 84, p.415-420 (1981)

[5.167] {Sect. 5.4.1} D. von der Linde, A. Laubereau, W, Kaiser: Molecular Vibrations in Liquids: Direct Measurement of the Molecular Dephasing Time; Determination of the Shape of Picosecond Light Pulses, Phys. Rev. Lett. 26, p.954-957 (1971)

[5.168] {Sect. 5.4.1} W. Langbein, J.M. Hvam, R. Zimmermann: Time-resolved speckle analysis: A new approach to coherence and dephasing of optical excitations in solids, Phys Rev Lett 82, p.1040-1043 (1999)

[5.169] {Sect. 5.4.1} O.V. Prezhdo, P.J. Rossky: Relationship between quantum decoherence times and solvation dynamics in condensed phase chemical systems, Phys Rev Lett 81, p.5294-5297 (1998)

[5.170] {Sect. 5.4.1} G. Stock, W. Domcke: Detection of ultrafast molecular-excited-state dynamics with time- and frequency-resolved pump-probe spectroscopy, Phys. Rev. A 45, p.3032-3040 (1992)

[5.171] {Sect. 5.4.1} G. Cerullo, G. Lanzani, M. Muccini, C. Taliani, S. DeSilvestri: Real-time vibronic coupling dynamics in a prototypical conjugated oligomer, Phys Rev Lett 83, p.231-234 (1999)

[5.172] {Sect. 5.4.1} K. Furuya, E. Koto, T. Ogawa: Direct observation of IVR under white light excitation: Fluorescence spectra of p-difluorobenzene by controlled electron impact, Chem Phys Lett 253, p.87-91 (1996)

[5.173] {Sect. 5.4.1} T. Matsumoto, K. Ueda, M. Tomita: Femtosecond vibrational relaxation measurement of azulene using temporally incoherent light, Chem. Phys. Lett. 191, p.627-632 (1992)

[5.174] {Sect. 5.4.1} K.-P. Müller, D. Haarer: Spectral Diffusion of Optical Transitions in Doped Polymer Glasses below 1 K, Phys. Rev. Lett. 66, p.2344-2347 (1991)

[5.175] {Sect. 5.4.1} Y.M. Engel, R.D. Levine: Vibration-vibration resonance conditions in intramolecular classical dynamics of triatomic and larger molecules, Chem. Phys. Lett. 164, p.270-278 (1989)

[5.176] {Sect. 5.4.1} A. Amirav: Rotational and vibrational energy effect on energy-resolved emission of anthracene and 9-cyanoanthracene, Chem. Phys. 124, p.163-175 (1988)

[5.177] {Sect. 5.4.1} G.A. Bickel, D.R. Demmer, G.W. Leach, St.C. Wallace: Mode- and symmetry-specific, picosecond intramolecular vibrational redistribution in 1-methylindole, Chem. Phys. Lett. 145, p.423-428 (1988)

[5.178] {Sect. 5.4.1} R. Parson: Classical-quantum correspondence in vibrational energy relaxation of nonlinear systems, J. Chem. Phys. 89, p.262-271 (1988)

[5.179] {Sect. 5.4.1} B.J. Orr, I.W.M. Smith: Collision-Induced Vibrational Energy Transfer in Small Polyatomic Molecules, J. Phys. Chem. 91, p.6106-6119 (1987)

[5.180] {Sect. 5.4.1} A. Amirav, J. Jortner, S. Okajima, E.C. Lim: Manifestation of intramolecular vibrational energy redistribution on electronic relaxation in large molecules, Chem. Phys. Lett. 126, p.487-494 (1986)

[5.181] {Sect. 5.4.1} D.B. Moss, Ch.S. Parmenter: A Time-Resolved Fluorescence Observation of Intramolecular Vibrationally Redistribution within the Channel Three Region of S1 Benzene, J. Phys. Chem. 90, p.1011-1014 (1986)

[5.182] {Sect. 5.4.1} P.O.J. Scherer, A. Seilmeier, W. Kaiser: Ultrafast intra- and intermolecular energy transfer in solutions after selective infrared excitation, J. Chem. Phys. 83, p.3948-3957 (1985)

[5.183] {Sect. 5.4.1} A.M. Weiner, E.P. Ippen: Femtosecond excited state relaxation of dye molecules in solution, Chem. Phys. Lett. 114, p.456-460 (1985)

[5.184] {Sect. 5.4.1} Th. Kulp, R.Ruoff, G. Stewart, J.D. McDonald: Intramolecular vibrational relaxation in 1,4 dioxane, J. Chem. Phys. 80, p.5359-5364 (1984)

[5.185] {Sect. 5.4.1} G. Stewart, R. Ruoff, Th. Kulp, J.D. McDonald: Intramolecular vibrational relaxation in dimethyl ether, J. Chem. Phys. 80, p.5353-5358 (1984)

[5.186] {Sect. 5.4.1} A.J. Taylor, D.J. Erskine, C.L. Tang: Femtosecond vibrational relaxation of large organic molecules, Chem. Phys. Lett. 103, p.430-435 (1984)

[5.187] {Sect. 5.4.1} H. Graener, H.R. Telle, A. Lauberau: Applications of Picosecond and Sub-Picosecond Spectroscopy, p.393-401 (1983)

[5.188] {Sect. 5.4.1} W. Zinth, C. Kolmeder, B. Benna, A. Irgens-Defregger, S.F. Fischer, W. Kaiser: Fast and exceptionally slow vibrational energy transfer in acetylene and phenylacetylene in solution, J. Chem. Phys.78, p.3916-3921 (1983)

[5.189] {Sect. 5.4.1} D. Reiser, A. Laubereau: Vibrational Relaxation of Dye Molecules Investigated by Ultrafast Induced Dichroism, Appl. Phys. B 27, p.115-122 (1982)

[5.190] {Sect. 5.4.1} A. Zewail, W. Lambert, P. Felker, J. Perry, W. Warren: Laser Probing of Vibrational Energy Redistribution and Dephasing, J. Phys. Chem. 86, p.1184-1192 (1982)

[5.191] {Sect. 5.4.1} G. Venzl, S.F. Fischer: The effect of localized modes on radiationless electronic transitions. II. Dependence on impurity concentration, J. Chem. Phys. 74, p.1887-1892 (1981)

[5.192] {Sect. 5.4.1} W. Zinth, H.-J. Polland, A. Lauberau, W. Kaiser: New Results on Ultrafast Coherent Excitation of Molecular Viibrations in Liquids, Appl. Phys. B 26, p.77-88 (1981)

[5.193] {Sect. 5.4.1} A. Laubereau, W. Kaiser: Vibrational dynamics of liquids and solids investigated by picosecond light pulses, Rev. Mod. Phys. 50, p.607-685 (1978)

[5.194] {Sect. 5.4.1} C.V. Shank, E.P. Ippen, O. Teschke: Sub-picosecond relaxation of large organic molecules in solution, Chem. Phys. Lett. 45, p.291-294 (1977)

[5.195] {Sect. 5.4.1} A. Laubereau: Picosecond phase relaxation of the fundamental vibrational mode of liquid nitrogen, Chem. Phys. Lett. 27, p.600-602 (1974)

[5.196] {Sect. 5.4.1} D.W. Vahey: Effects of spectral cross relaxation and collisional sephasing on the absorption of light by organic-dye solutions, Phys. Rev. A 10, p.1578-1590 (1974)

[5.197] {Sect. 5.4.1} A. Laubereau, L. Kirschner, W. Kaiser: Direct observation in intermolecular transfer of vibrational energy in liquids, Opt. Comm. 9, p.182-185 (1973)

[5.198] {Sect. 5.4.2} Z.G. Yi, D.A. Micha, J. Sund: Density matrix theory and calculations of nonlinear yields of CO photodesorbed from Cu (001) by light pulses, J Chem Phys 110, p.10562-10575 (1999)

[5.199] {Sect. 5.4.2} P. Yeh: Two-Wave Mixing in Nonlinear Media, IEEE J. QE-25, p.484-519 (1989)

[5.200] {Sect. 5.4.2} P. Yeh: Exact solution of a nonlinear model of two-wave mixing in Kerr media, J. Opt. Soc. Am. B 3, p.747-750 (1986)

[5.201] {Sect. 5.4.3} A. Schulzgen, R. Binder, M.E. Donovan, T. Lindberg, K. Wundke, H.M. Gibbs, G. Khitrova, N. Peyghambarian: Direct observation of excitonic Rabi oscillations in semiconductors, Phys Rev Lett 82, p.2346-2349 (1999)

[5.202] {Sect. 5.4.3} O. Kittelmann, J. Ringling, A. Nazarkin, G. Korn, I.V. Hertel: Direct observation of coherent medium response under the condition of two-photon excitation of krypton by femtosecond UV-laser pulses, Phys Rev Lett 76, p.2682-2685 (1996)

[5.203] {Sect. 5.4.3} R.M. Williams, J.M. Papanikolas, J. Rathje, S.R. Leone: Quantum-state-resolved 2-level femtosecond rotational coherence spectroscopy: Determination of rotational constants at medium and high J in Li-2, a simple diatomic system, Chem Phys Lett 261, p.405-413 (1996)

[5.204] {Sect. 5.4.3} C. Wunderlich, E. Kobler, H. Figger, T.W. Hansch: Light-induced molecular potentials, Phys Rev Lett 78, p.2333-2336 (1997)

[5.205] {Sect. 5.4.3} Y. R. Shen: Principles of Nonlinear Optics, chapter 13 (John Wiley & Sons, Chichester, 1984)

[5.206] {Sect. 5.4.3} R.F. Loring, Y.J. Yan, S. Mukamel: Time-resolved fluorescence and hole-burning line shapes of solvated molecules: Longitudinal dielectric relaxation and vibrational dynamics, J. Chem. Phys. 87, p.5840-5857 (1987)

[5.207] {Sect. 5.4.3} M.N. Sapozhnikov: Hole burning in the spectra of molecules in amorphous solids: The hole shape and ist dependence on laser frequency, power, irradiation time and temperature, Chem. Phys. Lett. 135, p.398-406 (1987)

[5.208] {Sect. 5.4.3} B. Jackson, R. Silbey: Theoretical description of photochemical hole burning in soft glasses, Chem. Phys. Lett. 99, p.331-334 (1983)

[5.209] {Sect. 5.4.3} J. Klafter, R. Silbey: A conjecture of nonphotochemical hole burning in organic glasses, J. Chem. Phys. 75, p.3973-3976 (1981)

[5.210] {Sect. 5.4.3} A. v. Jena, H.E. Lessing: Coherent Coupling Effects in Picosecond Absorption Experiments, Appl. Phys. 19, p.131-144 (1979)

[5.211] {Sect. 5.4.3} D.H. Schirrmeister, V. May: Strong-field approach to ultrafast pump-probe spectra: Dye molecules in solution, Chem Phys 220, p.1-13 (1997)

[5.212] {Sect. 5.4.4} T.K. Yee, T.K. Gustafson: Diagrammatic analysis of the density operator for nonlinear optical calculations: Pulsed and cw responses, Phys. Rev. A 18, p.1597-1617 (1978)

[5.213] {Sect. 5.4.4} Y. R. Shen: Principles of Nonlinear Optics, chapter 2 (John Wiley & Sons, Chichester, 1984)

[5.214] {Sect. 5.4.5} H.C. Torrey: Transient Nutations in Nuclear Magnetic Resonance, Phys. Rev. 76, p.1059-1068 (1949)

[5.215] {Sect. 5.4.5} R.G. DeVoe, R.G. Brewer: Experimental Test of the Optical Bloch Equations for Solids, Phys. Rev. Lett. 50, p.1269-1272 (1983)

[5.216] {Sect. 5.4.5} R.G. Brewer, R.L. Shoemaker: Optical Free Induction Decay, Phys. Rev. A 6, p.2001-2007 (1972)

[5.217] {Sect. 5.4.5} R.G. Brewer, R.L. Shoemaker: Photo Echo and Optical Nutation in Molecules, Phys. Rev. Lett. 27, p.631-634 (1971)

[5.218] {Sect. 5.4.5} G.B. Hocker, C.L. Tang: Observation of the Optical Transient Nutation Effect, Phys. Rev. Lett. 21, p.591-594 (1968)

[5.219] {Sect. 5.4.5} C.L. Tang, H. Statz: Optical Analog of the Transient Nutation Effect, Appl. Phys. Lett. 10, p.145-147 (1967)

[5.220] {Sect. 5.4.5} I.I. Rabi: Space Quantization in a Gyrating Magnetic Field, Phys. Rev. 51, p.652-654 (1937)

[5.221] {Sect. 5.4.6} T. Aoki, G. Mohs, M. KuwataGonokami, A.A. Yamaguchi: Influence of exciton-exciton interaction on quantum beats, Phys Rev Lett 82, p.3108-3111 (1999)

[5.222] {Sect. 5.4.6} M. Joschko, M. Woerner, E. Elsaesser, E. Binder, R. Hey, H. Kostial, K. Ploog: Heavy-light hole quantum beats in the band-to-band continuum of GaAs observed in 20 femtosecond pump-probe experiments, Phys Rev Lett 78, p.737-740 (1997)

[5.223] {Sect. 5.4.6} S. Savikhin, D.R. Buck, W.S. Struve: Oscillating anisotropies in a bacteriochlorophyll protein: Evidence for quantum beating between exciton levels, Chem Phys 223, p.303-312 (1997)

[5.224] {Sect. 5.4.6} C. Leichtle, I.S. Averbukh, W.P. Schleich: Generic structure of multilevel quantum beats, Phys Rev Lett 77, p.3999-4002 (1996)

[5.225] {Sect. 5.4.6} H. Bitto: Dynamics of S1 acetone studied with single rotor vibronic level resolution, Chem. Phys. 186, p.105-118 (1994)

[5.226] {Sect. 5.4.6} H. Bitto, J.R. Huber: Molecular quantum beat spectroscopy, Opt. Commun. 80, p.184-198 (1990)

[5.227] {Sect. 5.4.6} A. Mokhtari, A. Chebira, J. Chesnoy: Subpicosecond fluorescence dynamics of dye molecules, J. Opt. Soc. Am. B 7, p.1551-1557 (1990)

[5.228] {Sect. 5.4.6} A.E.A. Mokhtari, J. Chesnoy: Terahertz Fluorescence Quantum Beats in a Dye Solution, IEEE J. QE-25, p.2528-2531 (1989)

[5.229] {Sect. 5.4.6} S. Saikan, T. Nakabayashi, Y. Kanematsu, A. Imaoka: Observation of vibronic quantum beat in dye-doped polymers using femtosecond accumulated photon echo, J. Chem. Phys. 89, p.4609-4612 (1988)

[5.230] {Sect. 5.4.6} P. Schmidt, H. Bitto, J.R. Huber: Excited state dipole moments in a polyatomic molecule determined by Stark quantum beat spectroscopy, J. Chem. Phys. 88, p.696-704 (1988)

[5.231] {Sect. 5.4.6} R. Leonhardt, W. Holzapfel, W. Zinth, W. Kaiser: Terahertz quantum beats in molecular liquids, Chem. Phys. Lett. 133, p.373-377 (1987)

[5.232] {Sect. 5.4.6} N. Ochi, H. Watanabe, S. Tsuchiya: Rotationally Resolved Laser-Induced Fluorescence and Zeeman Quantum Beat Spectroscopy of the V1B2 State of Jet-Cooled CS2, Chem. Phys. 113, p.271-285 (1987)

[5.233] {Sect. 5.4.6} M.Dubs, J.Mühlbach, H.Bitto, P.Schmidt, J.R.Huber: Hyperfine quantum beats and Zeeman spectroscopy in the polyatomic molecule propynal HCxCCHO, J. Chem. Phys. 83, p.3755-3767 (1985)

[5.234] {Sect. 5.4.6} W. Lange, J. Mlynek: Quantum Beats in Transmission by Time-Resolved Polarization Spectroscopy, Phys. Rev. Lett. 40, p.1373-1375 (1978)

[5.235] {Sect. 5.4.6} A. Laubereau, G. Wochner, W. Kaiser: Collective Beating of Molecular Vibrations in Liquids on the Picosecond Time Scale, Opt. Comm. 17, p.91-94 (1976)

[5.236] {Sect. 5.4.6} S. Haroche, J.A. Paisner, A.L. Schawlow: Hyperfine Quantum Beats Observed in Cs Vapor under Pulsed Dye Laser Excitation, Phys. Rev. Lett. 30, p.948-951 (1973)

[5.237] {Sect. 5.4.6} H.R. Schlossberg, A. Javan: Saturation Behavior of a Doppler-Broadened Transition Involving Levels with Closely Spaced Structure, Phys. Rev. 150, p.267-284 (1966)

[5.238] {Sect. 5.4.7} W.A. Hugel, M.F. Heinrich, M. Wegener, Q.T. Vu, L. Banyai, H. Haug: Photon echoes from semiconductor band-to-band continuum transitions in the regime of Coulomb quantum kinetics, Phys Rev Lett 83, p.3313-3316 (1999)

[5.239] {Sect. 5.4.7} L. Menager, I. Lorgere, J.L. LeGouet, R.K. Mohan, S. Kroll: Time-domain Fresnel-to-Fraunhofer diffraction with photon echoes, Optics Letters 24, p.927-929 (1999)

[5.240] {Sect. 5.4.7} R.K. Mohan, U. Elman, M.Z. Tian, S. Kroll: Regeneration of photon echoes with amplified photon echoes, Optics Letters 24, p.37-39 (1999)

[5.241] {Sect. 5.4.7} P. Hamm, M. Lim, R.M. Hochstrasser: Non-Markovian dynamics of the vibrations of ions in water from femtosecond infrared three-pulse photon echoes, Phys Rev Lett 81, p.5326-5329 (1998)

[5.242] {Sect. 5.4.7} B.Z. Luo, U. Elman, S. Kroll, R. Paschotta, A. Tropper: Amplification of photon echo signals by use of a fiber amplifier, Optics Letters 23, p.442-444 (1998)

[5.243] {Sect. 5.4.7} T. Wang, C. Greiner, T.W. Mossberg: Experimental observation of photon echoes and power-efficiency analysis in a cavity environment, Optics Letters 23, p.1736-1738 (1998)

[5.244] {Sect. 5.4.7} J.P. Likforman, M. Joffre, V. Thierrymieg: Measurement of photon echoes by use of femtosecond Fourier-transform spectral interferometry, Optics Letters 22, p.1104-1106 (1997)

[5.245] {Sect. 5.4.7} R.M. Macfarlane, T.L. Harris, Y. Sun, R.L. Cone, R.W. Equall: Measurement of photon echoes in Er:Y2SiO5 at 1.5 mu m with a diode laser and an amplifier, Optics Letters 22, p.871-873 (1997)

[5.246] {Sect. 5.4.7} C.W. Rella, A. Kwok, K. Rector, J.R. Hill, H.A. Schwettman, D.D. Dlott, M.D. Fayer: Vibrational echo studies of protein dynamics, Phys Rev Lett 77, p.1648-1651 (1996)

[5.247] {Sect. 5.4.7} S.B. Altner, S. Bernet, A. Renn, E.S. Maniloff, F.R. Graf, U.P. Wild: Spectral hole burning and holography VI: Photon echoes from cw spectrally programmed holograms in a Pr3+:Y2SiO5 crystal, Opt. Comm. 120, p.103-111 (1995)

[5.248] {Sect. 5.4.7} P.C. Becker, H.L. Fragnito, J.Y Bigot, C.H. Brito Cruz, R.L. Fork, C.V. Shank: Femtosecond Photon Echos from Molecules in Solution, Phys. Rev. Lett. 63, p.505-507 (1989)

[5.249] {Sect. 5.4.7} S. Saikan, T. Nakabayashi, Y. Kanematsu, N. Tato: Fourier-transform spectroscopy in dye-doped polymers using the femtosecond accumulated photon echo, Phys. Rev. B 38, p.7777-7781 (1988)

[5.250] {Sect. 5.4.7} M. Berg, C.A. Walsh, L.R. Narasimhan, M.D. Fayer: Picosecond photon echo and optical hole burning studies of chromophores in organic glasses, J. Luminesc. 38, p.9-14 (1987)

[5.251] {Sect. 5.4.7} S. Saikan, A. Fujiwara. T Kushida, Y. Kato: High-Frequency Heterodyned Detection of Picosecond Accumulated Photon Echoes, Jpn. J. Appl. Phys.26, p.L941-L943 (1987)

[5.252] {Sect. 5.4.7} S. Saikan, H. Miyamoto, Y. Tosaki, A. Fujiwara: Optical-density effect in heterodyne-detected accumulated photon echo, Phys. Rev. B 36, p.5074-5077 (1987)

[5.253] {Sect. 5.4.7} C.A. Walsh, M. Berg, L.R. Narasimhan, M.D. Fayer: A picosecond photon echo study of a chromophore in an organic glass: Temperature dependence and comparision to nonphotochemical hole burning, J. Chem. Phys. 86, p.77-87 (1987)

[5.254] {Sect. 5.4.7} L.W. Molenkamp, D.A. Wiersma: Optical dephasing in organic amorphous systems. A photon echo and hole-burning study of pentacene in polymethylmethacrylate, J. Chem. Phys. 83, p.1-9 (1985)

[5.255] {Sect. 5.4.7} S. Asaka, H. Nakatsuka, M. Fujiwara, M. Matsuoka: Accumulated photon echoes with incoherent light in Nd3+-doped silicate glass, Phys. Rev. A 29, p.2286-2289 (1984)

[5.256] {Sect. 5.4.7} R. Beach, S.R. Hartmann: Incoherent Photon Echoes, Phys. Rev. Lett. 53, p.663-666 (1984)

[5.257] {Sect. 5.4.7} H. Nakatsuka, M. Tomita, M. Fujiwara, S. Asaka: Subpicosecond Photon Echoes by Using Nanosecond Laser Pulses, Opt. Comm. 52, p.150-152 (1984)

[5.258] {Sect. 5.4.7} R.G. DeVoe, R.G. Brewer: Experimental Test of the Optical Bloch Equations for Solids, Phys. Rev. Lett. 50, p.1269-1272 (1983)

[5.259] {Sect. 5.4.7} H.W.H. Lee, F.G. Patterson, R.W. Olson, D.A. Wiersma, M.D. Fayer: Temperature-dependent dephasing of delocalized dimer states of pentacene in p-terphanyl: Picosecond photon echo experiments, Chem. Phys. Lett. 90, p.172-177 (1982)

[5.260] {Sect. 5.4.7} K. Duppen, L.W. Molenkamp, J.B.W. Morsink, D.A. Wiersma, H.P. Trommsdorff: Optical dephasing in a glass-like system: A photon echo study of pentacene in benzoic acid, Chem. Phys. Lett. 84, p.421-424 (1981)

[5.261] {Sect. 5.4.7} M. Fujita, H. Nakatsuka, H. Nakanishi, M. Matsuoka: Backward Echo in Two-Level Systems, Phys. Rev. Lett. 42, p.974-977 (1979)

[5.262] {Sect. 5.4.7} T.M. Mossberg, R. Kachru, S.R. Hartmann, A.M. Flusberg: Echoes in gaseous media. A generalized theory of rephasing phenomena, Phys. Rev. A 20, p.1976-1996 (1979)

[5.263] {Sect. 5.4.7} S.C. Rand, A. Wokaun, R.G. DeVoe, R.G. Brewer: Magic-Angle Line Narrowing in Optical Spectroscopy, Phys. Rev. Lett. 43, p.1868-1871 (1979)

[5.264] {Sect. 5.4.7} S.R. Hartmann: H-3-Photon, Spin, and Raman Echoes, IEEE J. QE-4, p.802-807 (1968)

[5.265] {Sect. 5.4.7} C.K.N. Patel, R.E. Slusher: Photon echoes in gases, Phys. Rev. Lett. 20, p.1087-1089 (1968)

[5.266] {Sect. 5.4.7} I.D. Abella, N.A. Kurnit, S.R. Hartmann: Photon Echoes, Phys. Rev. 141, p.391-406 (1966)

[5.267] {Sect. 5.4.7} N.A. Kurnit, I.D. Abella, S.R. Hartmann: Observation of a Photon Echo, Phys. Rev. Lett. 13, p.567-568 (1964)

[5.268] {Sect. 5.4.7} E.L. Hahn: Spin echoes, Phys. Rev. 80, p.580-594 (1950)

[5.269] {Sect. 5.4.7} S.R. Hartmann: Photon echoes. In Lasers and Light, Readings from Scientific American (Freeman, San Francisco 1969) S. 303

[5.270] {Sect. 5.4.7} S.M. Zakharov, E.A. Manykin: Simultaneous optical image processing by photon echoes, Int. J. Optoelectron. 9, p.333-338 (1994)

[5.271] {Sect. 5.4.7} R. Yano, N. Uesugi: Demonstration of partial erasing of picosecond temporal optical data by use of accumulated photon echoes, Optics Letters 24, p.1753-1755 (1999)

[5.272] {Sect. 5.4.8} M. Blaauboer, B.A. Malomed, G. Kurizki: Spatiotemporally localized multidimensional solitons in self-induced transparency media, Phys Rev Lett 84, p.1906-1909 (2000)

[5.273] {Sect. 5.4.8} S.E. Harris, L.V. Hau: Nonlinear optics at low light levels, Phys Rev Lett 82, p.4611-4614 (1999)

[5.274] {Sect. 5.4.8} M. Muller, V.P. Kalosha, J. Herrmann: 2 pi-pulse laser using an intracavity quantum-well absorber, Opt Commun 150, p.147-152 (1998)

[5.275] {Sect. 5.4.8} P.R. Berman, J.M. Levy, R.G. Brewer: Coherent optical transient study of molecular collisions: Theory and observations, Phys. Rev. A 11, p.1668-1688 (1975)

[5.276] {Sect. 5.4.8} M.M.T. Loy: Observation of Population Inversion by Optical Adiabatic Rapid Passage, Phys. Rev. Lett. 32, p.814-817 (1974)

[5.277] {Sect. 5.4.8} M.D. Crisp: Adiabatic-Following Approximation, Phys. Rev. A 8, p.2128-2135 (1973)

[5.278] {Sect. 5.4.8} D. Grischkowsky, E. Courtens, J.A. Armstrong: Observation of Self-Steepening of Optical Pulses with Possible Shock Formation, Phys. Rev. Lett. 31, p.422-425 (1973)

[5.279] {Sect. 5.4.8} D. Grischkowsky: Adiabatic Following and Slow Optical Pulse Propagation in Rubidium Vapor, Phys. Rev. A 7, p.2096-2102 (1973)

[5.280] {Sect. 5.4.8} R.E. Slusher, H.M. Gibbs: Self-Induced Transparenca in Atomic Rubidium, Phys. Rev. A 5, p.1634-1659 (1972)

[5.281] {Sect. 5.4.8} D. Grischkowsky: Self-Focusing of Light by Potassium Vapor, Phys. Rev. Lett. 24, p.866-869 (1970)

[5.282] {Sect. 5.4.8} S.L. McCall, E.L. Hahn: Self-Induced Transparency, Phys. Rev. 183, p.457-485 (1969)

[5.283] {Sect. 5.4.8} E.B. Treacy: Adiabatic Inversion with Light Pulses, Phys. Lett. 27A, p.421-422 (1968)

[5.284] {Sect. 5.4.8} S.L. McCall, E.L. Hahn: Self-Induced Transparency by Pulsed Coherent Light, Phys. Rev. Lett. 18, p.908-911 (1967)

[5.285] {Sect. 5.4.8} Y. R. Shen: Principles of Nonlinear Optics, chapter 21 (John Wiley & Sons, Chichester, 1984)

[5.286] {Sect. 5.4.9} S. Ozcelik, I. Ozcelik, D.L. Akins: Superradiant lasing from J-aggregated molecules adsorbed onto colloidal silver, Appl Phys Lett 73, p.1949-1951 (1998)

[5.287] {Sect. 5.4.9} F. Haake, H. King, G. Schröder, J. Haus, R. Glauber, F. Hopf: Macroscopic Quantum Fluctuations in Superfluorescence, Phys. Rev. Lett. 42, p.1740-1743 (1979)

[5.288] {Sect. 5.4.9} D. Polder, M.F.H. Schuurmans, Q.H.F. Vrehen: Superfluorescence: Quantum-mechanical derivation of Maxwell-Bloch description with fluctuating field source, Phys. Rev. A 19, p.1192-1203 (1979)

[5.289] {Sect. 5.4.9} Q.H.F. Vrehen, M.F.H. Schuurmans: Direct Measurement of the Effective Initial Tipping Angle in Superfluorescence, Phys. Rev. Lett. 42, p.224-227 (1979)

[5.290] {Sect. 5.4.9} R. Glauber, F. Haake: The Initiation of Superfluorescence, Phys. Rev. Lett. 68A, p.29-32 (1978)

[5.291] {Sect. 5.4.9} H.M. Gibbs, Q.H.F. Vrehen, H.M.J. Hikspoors: Single-Pulse Superfluorescence in Cesium, Phys. Rev. Lett. 39, p.547-549 (1977)

[5.292] {Sect. 5.4.9} J.C. MacGillivray, M.S. Field: Theory of superradiance in an extended, optically thick medium, Phys. Rev. A 14, p.1169-1189 (1976)

[5.293] {Sect. 5.4.9} R. Bonifacio, L.A. Lugiato: Cooperative radiation processes in two-level systems: Superfluorescence, Phys. Rev. A 11, p.1507-1521 (1975)

[5.294] {Sect. 5.4.9} N. Bloembergen, R.V. Pound: Radiation Damping in Magnetic Resonance Experiments, Phys. Rev. 95, p.8-12 (1954)

[5.295] {Sect. 5.4.10} E.S. Fry, X. Li, D. Nikonov, G.G. Padmabandu, M.O. Scully, A.V. Smith, F.K. Tittel, C. Wang, S.R. Wilkinson, S.Y. Zhu: Atomic Coherence Effects within the Sodium D1 Line: Lasing without Inversion via Population Trapping, Phys. Rev. Lett. 70, p.3235-3246 (1993)

[5.296] {Sect. 5.4.10} M.O. Scully: Enhancement of the Index of Refraction via Quantum Coherence, Phys. Rev. Lett. 67, p.1855-1858 (1991)

[5.297] {Sect. 5.4.10} S.E. Harris: Lasers without Inversion: Interferencee of Lifetime-Broadened Resonaces, Phys. Rev. Lett. 62, p.1033-1036 (1989)

[5.298] {Sect. 5.4.10} X.M. Hu, J.S. Peng: Squeezed cascade lasers without and with inversion, Opt Commun 154, p.203-216 (1998)

[5.299] {Sect. 5.4.10} J.T. Manassah, B. Gross: Amplification without inversion in an extended optically dense open Lambda-system, Opt Commun 148, p.404-416 (1998)

[5.300] {Sect. 5.4.10} B. Sherman, G. Kurizki, D.E. Nikonov, M.O. Scully: Universal classical mechanism of free-electron lasing without inversion, Phys Rev Lett 75, p.4602-4605 (1995)

[5.301] {Sect. 5.4.10} J. Mompart, R. Corbalan, R. Vilaseca: Lasing without inversion in the V-type three-level system under the two-photon resonance condition, Opt Commun 147, p.299-304 (1998)

[5.302] {Sect. 5.4.10} C. Fort, F.S. Cataliotti, T.W. Hansch, M. Inguscio, M. Prevedelli: Gain without inversion on the cesium D-1 line, Opt Commun 139, p.31-34 (1997)

[5.303] {Sect. 5.4.10} S.Q. Gong, S.D. Du, Z.Z. Xu: Nonlinear theory of lasing with or without inversion in a simple three-level atomic system, Opt Commun 130, p.249-254 (1996)

[5.304] {Sect. 5.4.10} J.B. Khurgin, E. Rosencher: Practical aspects of lasing without inversion in various media, IEEE J QE-32, p.1882-1896 (1996)

[5.305] {Sect. 5.4.10} D.E. Nikonov, B. Sherman, G. Kurizki, M.O. Scully: Lasing without inversion in Cherenkov free-electron lasers, Opt Commun 123, p.363-371 (1996)

[5.306] {Sect. 5.4.10} G.G. Padmabandu, G.R. Welch, I.N. Shubin, E.S. Fry, D.E. Nikonov, M.D. Lukin, M.O. Scully: Laser oscillation without population inversion in a sodium atomic beam, Phys Rev Lett 76, p.2053-2056 (1996)

[5.307] {Sect. 5.4.10} A.S. Zibrov, M.D. Lukin, D.E. Nikonov, L. Hollberg, M.O. Scully, V.L. Velichansky, H.G. Robinson: Experimental demonstration of laser oscillation without population inversion via quantum interference in Rb, Phys Rev Lett 75, p.1499-1502 (1995)

[5.308] {Sect. 5.4.10} A. Nottelmann, C. Peters, W. Lange: Inversionless Amplification of Picosecond Pulses due to Zeeman Coherence, Phys. Rev. Lett. 70, p.1783-1786 (1993)

[5.309] {Sect. 5.4.10} M.O. Scully, S.-Y. Zhu: Degenerate Quantum-Beat Laser: Lasing without Inversion and Inversion without Lasing, Phys. Rev. Lett. 62, p.2813-2816 (1989)

[5.310] {Sect. 5.5} P. Kaatz, D.P. Shelton: Two-photon fluorescence cross-section measurements calibrated with hyper-Rayleigh scattering, J Opt Soc Am B Opt Physics 16, p.998-1006 (1999)

[5.311] {Sect. 5.5} E.J. Sanchez, L. Novotny, X.S. Xie: Near-field fluorescence microscopy based on two-photon excitation with metal tips, Phys Rev Lett 82, p.4014-4017 (1999)

[5.312] {Sect. 5.5} M. Sonnleitner, G.J. Schutz, T. Schmidt: Imaging individual molecules by two-photon excitation, Chem Phys Lett 300, p.221-226 (1999)

[5.313] {Sect. 5.5} E.R. Thoen, E.M. Koontz, M. Joschko, P. Langlois, T.R. Schibli, F.X. Kartner, E.P. Ippen, L.A. Kolodziejski: Two-photon absorption in semiconductor saturable absorber mirrors, Appl Phys Lett 74, p.3927-3929 (1999)

[5.314] {Sect. 5.5} K.R. Allakhverdiev: Two-photon absorption of femtosecond laser pulses in GaS crystals, Opt Commun 149, p.64-66 (1998)

[5.315] {Sect. 5.5} C.V. Bindhu, S.S. Harilal, A. Kurian, V.P.N. Nampoori, C.P.G. Vallabhan: Two and three photon absorption in rhodamine 6G methanol solutions using pulsed thermal lens technique, J Nonlinear Opt Physics Mat 7, p.531-538 (1998)

[5.316] {Sect. 5.5} M.A. Bopp, Y. Jia, G. Haran, E.A. Morlino, R.M. Hochstrasser: Single-molecule spectroscopy with 27 fs pulses: Time-resolved experiments and direct imaging of orientational distributions, Appl Phys Lett 73, p.7-9 (1998)

[5.317] {Sect. 5.5} G.S. He, R. Signorini, P.N. Prasad: Two-photon-pumped frequency-upconverted blue losing in Coumarin dye solution, Appl Opt 37, p.5720-5726 (1998)

[5.318] {Sect. 5.5} M. Reeves, M. Musculus, P. Farrell: Confocal, two-photon laser-induced fluorescence technique for the detection of nitric oxide, Appl Opt 37, p.6627-6635 (1998)

[5.319] {Sect. 5.5} J. Swiatkiewicz, P.N. Prasad, B.A. Reinhardt: Probing two-photon excitation dynamics using ultrafast laser pulses, Opt Commun 157, p.135-138 (1998)

[5.320] {Sect. 5.5} K.L. Vodopyanov, S.B. Mirov, V.G. Voevoolin, P.G. Schunemann: Two-photon absorption in GaSe and CdGeAs2, Opt Commun 155, p.47-50 (1998)

[5.321] {Sect. 5.5} Z.P. Chen, D.L. Kaplan, K. Yang, J. Kumar, K.A. Marx, S.K. Tripathy: Two-photon-induced fluorescence from the phycoerythrin protein, Appl Opt 36, p.1655-1659 (1997)

[5.322] {Sect. 5.5} C. Dorrer, F. Nez, B. deBeauvoir, L. Julien, F. Biraben: Accurate measurement of the 2 (3)S (1)-3 (3)D (1) two-photon transition frequency in helium: New determination of the 2 (3)S (1) Lamb shift, Phys Rev Lett 78, p.3658-3661 (1997)

[5.323] {Sect. 5.5} J.E. Ehrlich, X.L. Wu, L.Y.S. Lee, Z.Y. Hu, H. Rockel, S.R. Marder, J.W. Perry: Two-photon absorption and broadband optical limiting with bis-donor stilbenes, Optics Letters 22, p.1843-1845 (1997)

[5.324] {Sect. 5.5} Y.C. Guo, Q.Z. Wang, N. Zhadin, F. Liu, S. Demos, D. Calistru, A. Tirksliunas, A. Katz, Y. Budansky, P.P. Ho, et al.: Two-photon excitation of fluorescence from chicken tissue, Appl Opt 36, p.968-970 (1997)

[5.325] {Sect. 5.5} E.J. Larson, L.A. Friesen, C.K. Johnson: An ultrafast one-photon and two-photon transient absorption study of the solvent-dependent photophysics in all-trans retinal, Chem Phys Lett 265, p.161-168 (1997)

[5.326] {Sect. 5.5} T. Munakata, T. Sakashita, M. Tsukakoshi, J. Nakamura: Fine structure of the two-photon photoemission from benzene adsorbed on Cu (111), Chem Phys Lett 271, p.377-380 (1997)

[5.327] {Sect. 5.5} G. Robertson, D. Armstrong, M.J.P. Dymott, A.I. Ferguson, G.L. Hogg: Two-photon fluorescence microscopy with a diode-pumped Cr: LiSAF laser, Appl Opt 36, p.2481-2483 (1997)

[5.328] {Sect. 5.5} T. Plakhotnik, D. Walser, A. Renn, U.P. Wild: Light induced single molecule frequency shift, Phys Rev Lett 77, p.5365-5368 (1996)

[5.329] {Sect. 5.5} P.S. Weitzman, U. Osterberg: Two-photon absorption and photoconductivity in photosensitive glasses, J Appl Phys 79, p.8648-8655 (1996)

[5.330] {Sect. 5.5} R. De Salvo, A.A. Said, D.J. Hagan, E.W. Van Stryland, M. Sheik-Bahae: Infrared to Ultraviolet Measurements of Two-Photon Absorption and n2 in Wide Bandgap Solids, IEEE J. QE-32, p.1324-1333 (1996)

[5.331] {Sect. 5.5} T. Plakhotnik, D. Walser, M. Pirotta, A. Renn, U.P. Wild: Nonlinear spectroscopy on a single quantum system: Two- photon absorption of a single molecule, Science 271, p.1703-1705 (1996)

[5.332] {Sect. 5.5} C. Xu, J. Guild, W.W. Webb, W. Denk: Determination of absolute two-photon excitation cross sections by in situ second-order autocorrelation, Optics Letters 20, p.2372-2374 (1995)

[5.333] {Sect. 5.5} K. Danzmann, K. Grützmacher, B. Wende: Doppler-free two-photon polarization spectroscopy measurement of the Stark-broadened profile of the hydrogen L alpha line in a dense plasma, Phys. Rev. Lett. 57, p.2151-2153 (1986)

[5.334] {Sect. 5.5} B.M. Pierce, R.R. Birge: The Effects of Laser Pulsewidth and Molecular Lifetime on the Experimental Determination of One-Photon and Two-Photon Excitation Spectra, IEEE J. QE-19, p.826-833 (1983)

[5.335] {Sect. 5.5} S. Chu, A.P. Mills, Jr.: Excitation of the Positronium 1 3S1-23S1 Two-Photon Transition, Phys. Rev. Lett. 48, p.1333-1337 (1982)

[5.336] {Sect. 5.5} G.I. Bekov, E.P. Vidolova-Angelova, L.N. Ivanov, V.S. Letokhov, V.I. Mishin: Double-Excited Narrow Autoionization States of Ytterbium Atom, Opt. Comm. 35, p.194-198 (1980)

[5.337] {Sect. 5.5} B.P. Stoicheff, E. Weinberger: Frequency Shifts, Line Broadenings, and Phase-Interference Effects in Rb**+Rb Collisions, Measured by Doppler-Free Two-Photon Spectroscopy, Phys. Rev. Lett. 44, p.733-736 (1980)

[5.338] {Sect. 5.5} B.P. Stoicheff, E. Weinberger: Doppler-free two-photon absorption spectrum of rubidium, Can. J. Phys. 57, p.2143-2154 (1979)

[5.339] {Sect. 5.5} K.C. Harvey, B.P. Stoicheff: Fine Structure of the n2D Series in Rubidium near the Ionization Limit, Phys. Rev. Lett. 38, p.537-540 (1977)

[5.340] {Sect. 5.5} R. Teets, J. Eckstein, T.W. Hänsch: Coherent Two-Photon Excitation by Multiple Light Pulses, Phys. Rev. Lett. 38, p.760-764 (1977)

[5.341] {Sect. 5.5} P.F. Liao, G.C. Bjorklund: Polarization Rotation Induced by Resonant Two-Photon Dispersion, Phys. Rev. Lett. 36, p.584-587 (1976)

[5.342] {Sect. 5.5} M.G. Littman, M.L. Zimmerman, T.W. Ducas, R.R. Freeman, D. Kleppner: Stucture of Sodium Rydberg States in Weak to Strong Electric Fields, Phys. Rev. Lett. 36, p.788-791 (1976)

[5.343] {Sect. 5.5} T.W. Hänsch, K.C. Harvey, G. Meisel, A.L. Schawlow: Two-Photon Spectroscopy of Na 3s-4d Without Doppler Broadening Using a CW Dye Laser, Opt. Comm. 11, p.50-53 (1974)

[5.344] {Sect. 5.5} M.D. Levenson, N. Bloembergen: Observation of Two-Photon Absorption without Doppler Broadening on the 3S-5S Transition in Sodium Vapor, Phys. Rev. Lett. 32, p.645-648 (1974)

[5.345] {Sect. 5.5} W.M. McClain: Excited State Symmetry Assignment Through Polarized Two-Photon Absorption Studies of Fluids, J. Chem. Phys. 55, p.2789-2796 (1971)

[5.346] {Sect. 5.5} W.H. Glenn: Theory of the Two-Photon Absorption-Fluorescence Method of Pulswidth Measurement, IEEE J. QE-6, p.510-515 (1970)

[5.347] {Sect. 5.5} T.R. Bader, A. Gold: Polarization Dependence of Two-Photon Absorption in Solids, Phys. Rev. 171, p.997-1003 (1968)

[5.348] {Sect. 5.5} M.W. Hamilton, D.S. Elliott: Second order interference in two photon absorption, J. Mod. Opt. 43, p.1765-1771 (1965)

[5.349] {Sect. 5.5} W. Kaiser, C.G.B. Garrett: Two-Photon Excitation in CaF2: Eu2+, Phys. Rev. Lett. 7, p.229-231 (1961)

[5.350] {Sect. 5.5} M. Göppert-Mayer: Über Elementarakte mit zwei Quantensprüngen, Ann. Phys.9, p.273-294 (1931)

[5.351] {Sect. 5.5} M. Bellini, A. Bartoli, T.W. Hansch: Two-photon Fourier spectroscopy with femtosecond light pulses, Optics Letters 22, p.540-542 (1997)

[5.352] {Sect. 5.5} V. Blanchet, C. Nicole, M.A. Bouchene, B. Girard: Temporal coherent control in two-photon transitions: From optical interferences to quantum interferences, Phys Rev Lett 78, p.2716-2719 (1997)

[5.353] {Sect. 5.5} H.-B. Fei, M. Jost, S. Popescu, B.E.A. Saleh, M.C. Teich: Entanglement-Induced Two-Photon Transparency, Phys. Rev. Lett. 78, p.1679-1682 (1997)

[5.354] {Sect. 5.5} W. Rudolph, M. Sheikbahae, A. Bernstein, L.F. Lester: Femtosecond autocorrelation measurements based on two- photon photoconductivity in ZnSe, Optics Letters 22, p.313-315 (1997)

[5.355] {Sect. 5.5} P. Langlois, E.P. Ippen: Measurement of pulse asymmetry by three-photon-absorption autocorrelation in a GaAsP photodiode, Optics Letters 24, p.1868-1870 (1999)

[5.356] {Sect. 5.5} C. Majumder, O.D. Jayakumar, R.K. Vatsa, S.K. Kulshreshtha, J.P. Mittal: Multiphoton ionisation of acetone at 355 nm: a time-of-flight mass spectrometry study, Chem Phys Lett 304, p.51-59 (1999)

[5.357] {Sect. 5.5} A. Volkmer, K. Wynne, D.J.S. Birch: Near-infrared excitation of alkane ultra-violet fluorescence, Chem Phys Lett 299, p.395-402 (1999)

[5.358] {Sect. 5.5} M.A. Baig, M. Yaseen, A. Nadeem, R. Ali, S.A. Bhatti: Three-photon excitation of strontium Rydberg levels, Opt Commun 156, p.279-284 (1998)

[5.359] {Sect. 5.5} D.J. Maas, D.I. Duncan, R.B. Vrijen, W.J. Vanderzande, L.D. Noordam: Vibrational ladder climbing in NO by (sub)picosecond frequency-chirped infrared laser pulses, Chem Phys Lett 290, p.75-80 (1998)

[5.360] {Sect. 5.5} H. Shim, M.G. Liu, H.B. Chang, G.I. Stegeman: Four-photon absorption in the single-crystal polymer bis (paratoluene) sulfonate, Optics Letters 23, p.430-432 (1998)

[5.361] {Sect. 5.5} M. Castillejo, M. Martin, R. Denalda, J. Solis: Nanosecond versus picosecond near UV multiphoton dissociation of ketene, Chem Phys Lett 268, p.465-470 (1997)

[5.362] {Sect. 5.5} J. Thogersen, J.D. Gill, H.K. Haugen: Stepwise multiphoton excitation of the 4f (2)5d configuration in Nd3+:YLF, Opt Commun 132, p.83-88 (1996)

[5.363] {Sect. 5.5} J.D. Bhawalkar, G.S. He, P.N. Prasad: Three-photon induced upconverted fluorescence from an organic compound: application to optical power limiting, Opt. Comm. 119, p.587-590 (1995)

[5.364] {Sect. 5.5} M. Hippler, M. Quack, R. Schwarz, G. Seyfang, S. Matt, T. Mark: Infrared multiphoton excitation, dissociation and ionization of C-60, Chem Phys Lett 278, p.111-120 (1997)

[5.365] {Sect. 5.5} N.P. Lockyer, J.C. Vickerman: Single photon and femtosecond multiphoton ionisation of the dipeptide valyl-valine, Int J Mass Spectrom 197, p.197-209 (2000)

[5.366] {Sect. 5.5} M.J. DeWitt, R.J. Levis: Observing the transition from a multiphoton-dominated to a field-mediated ionization process for polyatomic molecules in intense laser fields, Phys Rev Lett 81, p.5101-5104 (1998)

[5.367] {Sect. 5.5} J. Wei, B. Zhang, L. Fang, L.D. Zhang, J.Y. Cai: REMPI time-of-flight mass spectra of C2H7N isomers, Opt Commun 156, p.331-336 (1998)

[5.368] {Sect. 5.5} K.W.D. Ledingham, C. Kosmidis, S. Georgiou, S. Couris, R.P. Singhal: A comparison of the femto-, pico- and nano-second multiphoton ionization and dissociation processes of NO2 at 248 and 496 nm, Chem Phys Lett 247, p.555-563 (1995)

[5.369] {Sect. 5.5} T. Baumert, M. Grosser, R. Thalweiser, G. Gerber: Femtosecond Time-Resolved Molecular Multphoton-Ionisation: The Na2 System, Phys. Rev. Lett. 67, p.3753-3756 (1991)

[5.370] {Sect. 5.5} N.Tan-no, k. Ohkawara, H. Inaba: Coherent Transient Multiphoton Scattering in a Resonant Two-Level System, Phys. Rev. Lett. 46, p.1282-1285 (1981)

[5.371] {Sect. 5.5} P. A. Schulz, Aa. S. Sudbo, E. R. Grant, Y. R. Shen, Y. T. Lee: Multiphoton dissociation of SF6 by a molecular beam method, J. Chem. Phys. 72p.4985-4995 (1980)

[5.372] {Sect. 5.5} J. G. Black, P. Kolodner, M. J. Schulz, E. Yablonovitch, N. Bloembergen Collisionless multiphoton energy deposition and dissociation of SF6, Phys. Rev. A 19, p.704-716 (1979)

[5.373] {Sect. 5.5} P. Esherick, J.A. Armstrong, R.W. Dreyfus, J.J. Wynne: Multiphoton Ionization Spectroscopy of High-Lying, Even-Parity States in Calcium, Phys. Rev. Lett. 36, p.1296-1299 (1976)

[5.374] {Sect. 5.5} D.K. Sharma, J. Stevenson, G.J. Hoytink: The photo-ionization of mono- and di-sodium tetracene in 2-MTHF at room temperature by nanosecond ruby laser pulses, Chem. Phys. Lett. 29, p.343-348 (1974)

[5.375] {Sect. 5.6} M. Li, S. Menon, J.P. Nibarger, G.N. Gibson: Ultrafast electron dynamics in femtosecond optical breakdown of dielectrics, Phys Rev Lett 82, p.2394-2397 (1999)

[5.376] {Sect. 5.6} M. Lenzner, J. Kruger, S. Sartania, Z. Cheng, C. Spielmann, G. Mourou, W. Kautek, F. Krausz: Femtosecond optical breakdown in dielectrics, Phys Rev Lett 80, p.4076-4079 (1998)

[5.377] {Sect. 5.6} J. Noack, D.X. Hammer, G.D. Noojin, B.A. Rockwell, A. Vogel: Influence of pulse duration on mechanical effects after laser-induced breakdown in water, J Appl Phys 83, p.7488-7495 (1998)

[5.378] {Sect. 5.6} E.N. Glezer, C.B. Schaffer, N. Nishimura, E. Mazur: Minimally disruptive laser-induced breakdown in water, Optics Letters 22, p.1817-1819 (1997)

[5.379] {Sect. 5.6} V.E. Peet, R.V. Tsubin: Multiphoton ionization and optical breakdown of xenon in annular laser beams, Opt Commun 134, p.69-74 (1997)

[5.380] {Sect. 5.6} I.C.E. Turcu, M.C. Gower, P. Huntington: Measurement of KrF laser breakdown threshold in gases, Opt Commun 134, p.66-68 (1997)

[5.381] {Sect. 5.6} T. Yagi, Y.S. Huo: Laser-induced breakdown in H-2 gas at 248 nm, Appl Opt 35, p.3183-3184 (1996)

[5.382] {Sect. 5.6} A. Kummrow: Effect of optical breakdown on stimulated Brillouin scattering in focused beam cells, J. Opt. Soc. Am. B 12, p.1006-1011 (1995)

[5.383] {Sect. 5.6} R.A. Mullen: Multiple-Short-Pulse Stimulated Brillouin Scattering for Trains of 200 ps Pulses at 1.06 μm, IEEE J. QE-26, p.1299-1303 (1990)

[5.384] {Sect. 5.6} R.A. Mullen, J.N. Matossian: Quenching optical breakdown with an applied electric field, Opt. Lett. 15, p.601-603 (1990)

[5.385] {Sect. 5.6} Y.S. Huo, A.J. Alcock, O.L. Bourne: A Time-Resolved Study of Sub-Nanosecond Pulse Generation by the Combined Effects of Stimulated Brillouin Scattering and Laser-Induced Breakdown, Appl. Phys. B 38, p.125-129 (1985)

[5.386] {Sect. 5.6} S.B. Papernyi, V.F. Petrov, V.A. Serebryakov, V.R. Startsev: Competition between stimulated Brillouin scattering and optical breakdown in argon, Sov. J. Quantum Electron. 13, p.293-297 (1983)

[5.387] {Sect. 5.6} N. Bloembergen: Laser-Induced Electric Breakdown in Solids, IEEE J. QE-10, p.375-386 (1974)

[5.388] {Sect. 5.6} P.N. Voronov, G.A. Delone, N.B. Delone: Multiphoton Ionization of Atoms. II. Ionization of Krypton by Ruby-Laser Radiation, Sov. Phys. JETP 24, p.1122-1135 (1967)

[5.389] {Sect. 5.6} H. Nakano, T. Nishikawa, N. Uesugi: Strongly enhanced soft x-ray emission at 8 nm from plasma on a neodymium-doped glass surface heated by femtosecond laser pulses, Appl Phys Lett 72, p.2208-2210 (1998)

[5.390] {Sect. 5.6} M. Schnurer, C. Spielmann, P. Wobrauschek, C. Streli, N.H. Burnett, C. Kan, K. Ferencz, R. Koppitsch, Z. Cheng, T. Brabec et al.: Coherent 0.5-keV X-ray emission from helium driven by a sub-10-fs laser, Phys Rev Lett 80, p.3236-3239 (1998)

[5.391] {Sect. 5.6} Z.Z. Xu, Y.S. Wang, K. Zhai, X.X. Li, Y.Q. Liu, X.D. Yang, Z.Q. Zhang, W.Q. Zhang: Direct experimental evidence of influence of ionizations on high-order harmonic generation, Opt Commun 158, p.89-92 (1998)

[5.392] {Sect. 5.6} M. Yoshida, Y. Fujimoto, Y. Hironaka, K.G. Nakamura, K. Kondo, M. Ohtani, H. Tsunemi: Generation of picosecond hard x rays by tera watt laser focusing on a copper target, Appl Phys Lett 73, p.2393-2395 (1998)

[5.393] {Sect. 5.6} V.G. Babaev, M.S. Dzhidzhoev, V.M. Gordienko, M.A. Joukov, A.B. Savelev, V.Y. Timoshenko, A.A. Shashkov, R.V. Volkov: X-ray production and second harmonic generation by superintense femtosecond laser pulses in the solids with restricted thermal conduction, J Nonlinear Opt Physics Mat 6, p.495-505 (1997)

[5.394] {Sect. 5.6} A. Behjat, J. Lin, G.J. Tallents, A. Demir, M. Kurkcuoglu, C.L.S. Lewis, A.G. MacPhee, S.P. Mccabe, P.J. Warwick, D. Neely, et al.: The effects of multi-pulse irradiation on X-ray laser media, Opt Commun 135, p.49-54 (1997)

[5.395] {Sect. 5.6} T. Ditmire, R.A. Smith, R.S. Marjoribanks, G. Kulcsar, M.H.R. Hutchinson: X-ray yields from Xe clusters heated by short pulse high intensity lasers, Appl Phys Lett 71, p.166-168 (1997)

[5.396] {Sect. 5.6} C. Kan, N.H. Burnett, C.E. Capjack, R. Rankin: Coherent XUV generation from gases ionized by several cycle optical pulses, Phys Rev Lett 79, p.2971-2974 (1997)

[5.397] {Sect. 5.6} W.P. Leemans, R.W. Schoenlein, P. Volfbeyn, A.H. Chin, T.E. Glover, P. Balling, M. Zolotorev, K.J. Kim, S. Chattopadhyay, C.V. Shank: Interaction of relativistic electrons with ultrashort laser pulses: Generation of femtosecond X-rays and microprobing of electron beams, IEEE J QE-33, p.1925-1934 (1997)

[5.398] {Sect. 5.6} O. Meighan, A. Gray, J.P. Mosnier, W. Whitty, J.T. Costello, C.L.S. Lewis, A. Macphee, R. Allott, I.C.E. Turcu, A. Lamb: Short-pulse, extreme-ultraviolet continuum emission from a table-top laser plasma light source, Appl Phys Lett 70, p.1497-1499 (1997)

[5.399] {Sect. 5.6} J.F. Pelletier, M. Chaker, J.C. Kieffer: Soft x-ray emission produced by a sub-picosecond laser in a single- and double-pulse scheme, J Appl Phys 81, p.5980-5983 (1997)

[5.400] {Sect. 5.6} P. Celliers, L.B. DaSilva, C.B. Dane, S. Mrowka, M. Norton, J. Harder, L. Hackel, D.L. Matthews, H. Fiedorowicz, A. Bartnik, et al.: Optimization of x-ray sources for proximity lithography produced by a high average power Nd:glass laser, J Appl Phys 79, p.8258-8268 (1996)

[5.401] {Sect. 5.6} B.N. Chichkov, C. Momma, A. Tunnermann, S. Meyer, T. Menzel, B. Wellegehausen: Hard-x-ray radiation from short-pulse laser-produced plasmas, Appl Phys Lett 68, p.2804-2806 (1996)

[5.402] {Sect. 5.6} M. Fraenkel, A. Zigler, Y. Horowitz, A. Ludmirsky, S. Maman, E. Moshe, Z. Henis, S. Eliezer: Optimal x-ray source development in the spectral range 4- 14 angstrom using a Nd:YAG high power laser, J Appl Phys 80, p.5598-5603 (1996)

[5.403] {Sect. 5.6} M. Schnurer, P.V. Nickles, M.P. Kalachnikov, W. Sandner, R. Nolte, P. Ambrosi, J.L. Miquel, A. Dulieu, A. Jolas: Characteristics of hard x-ray emission from subpicosecond laser-produced plasmas, J Appl Phys 80, p.5604-5609 (1996)

[5.404] {Sect. 5.6} R.C. Spitzer, T.J. Orzechowski, D.W. Phillion, R.L. Kauffman, C. Cerjan: Conversion efficiencies from laser-produced plasmas in the extreme ultraviolet regime, J Appl Phys 79, p.2251-2258 (1996)

[5.405] {Sect. 5.6} D.H. Gill, A.A. Dougal: Breakdown Minima due to Electron-impact Ionization in Super-High-Pressure Gases Irradiated by a Focused Giant-Pulse Laser, Phys. Rev. Lett. 15, p.845-847 (1965)

[5.406] {Sect. 5.6} N.S. Kim, A. Djaoui, M.H. Key, D. Neely, S.G. Preston, M. Zepf, C.G. Smith, J.S. Wark, J. Zhang, A.A. Offenberger: Extreme ultraviolet line emission at 24.7 nm from Li-like nitrogen plasma produced by a short KrF excimer laser pulse, Appl Phys Lett 69, p.884-886 (1996)

[5.407] {Sect. 5.6} E.E.B. Campbell, K. Hansen, K. Hoffmann, G. Korn, M. Tchap-lyguine, M. Wittmann, I.V. Hertel: From above threshold ionization to statistical electron emission: The laser pulse-duration dependence of C-60 photoelectron spectra, Phys Rev Lett 84, p.2128-2131 (2000)

[5.408] {Sect. 5.6} E.D. Lancaster, K.L. McNesby, R.G. Daniel, A.W. Miziolek: Spectroscopic analysis of fire suppressants and refrigerants by laser- induced breakdown spectroscopy, Appl Opt 38, p.1476-1480 (1999)

[5.409] {Sect. 5.6} M. Saito, S. Izumida, K. Onishi, J. Akazawa: Detection efficiency of microparticles in laser breakdown water analysis, J Appl Phys 85, p.6353-6357 (1999)

[5.410] {Sect. 5.6} M. Nishiura, M. Sasao, M. Bacal: H- laser photodetachment at 1064, 532, and 355 nm in plasma, J Appl Phys 83, p.2944-2949 (1998)

[5.411] {Sect. 5.6} D.X. Hammer, R.J. Thomas, G.D. Noojin, B.A. Rockwell, P.K. Kennedy, W.P. Roach: Experimental investigation of ultrashort pulse laser-induced breakdown thresholds in aqueous media, IEEE J QE-32, p.670-678 (1996)

[5.412] {Sect. 5.6} S. Chelkowski, P.B. Corkum, A.D. Bandrauk: Femtosecond Coulomb explosion imaging of vibrational wave functions, Phys Rev Lett 82, p.3416-3419 (1999)

[5.413] {Sect. 5.6} O. Baghdassarian, B. Tabbert, G.A. Williams: Luminescence characteristics of laser-induced bubbles in water, Phys Rev Lett 83, p.2437-2440 (1999)

[5.414] {Sect. 5.6} L. Koller, M. Schumacher, J. Kohn, S. Teuber, J. Tiggesbaumker, K.H. MeiwesBroer: Plasmon-enhanced multi-ionization of small metal clusters in strong femtosecond laser fields, Phys Rev Lett 82, p.3783-3786 (1999)

[5.415] {Sect. 5.6} M. Frenz, F. Konz, H. Pratisto, H.P. Weber, A.S. Silenok, V.I. Konov: Starting mechanisms and dynamics of bubble formation induced by a Ho:Yttrium aluminum garnet laser in water, J Appl Phys 84, p.5905-5912 (1998)

[5.416] {Sect. 5.6} A.B. Fedotov, N.I. Koroteev, A.N. Naumov, D.A. Sidorovbiryu-kov, A.M. Zheltikov: Coherent four-wave mixing in a laser-preproduced plasma: Optical frequency conversion and two-dimensional mapping of atoms and ions, J Nonlinear Opt Physics Mat 6, p.387-410 (1997)

[5.417] {Sect. 5.6} Q. Feng, J.V. Moloney, A.C. Newell, E.M. Wright, K. Cook, P.K. Kennedy, D.X. Hammer, B.A. Rockwell, C.R. Thompson: Theory and simulation on the threshold of water breakdown induced by focused ultrashort laser pulses, IEEE J QE-33, p.127-137 (1997)

[5.418] {Sect. 5.6} D. Giulietti, L.A. Gizzi, A. Giulietti, A. Macchi, D. Teychenne, P. Chessa, A. Rousse, G. Cheriaux, J.P. Chambaret, G. Darpentigny: Observation of solid-density laminar plasma transparency to intense 30 femtosecond laser pulses, Phys Rev Lett 79, p.3194-3197 (1997)

[5.419] {Sect. 5.6} N. Tsuda, J. Yamada: Observation of forward breakdown mechanism in high- pressure argon plasma produced by irradiation by an excimer laser, J Appl Phys 81, p.582-586 (1997)

[5.420] {Sect. 5.6} D.E. Hinkel, E.A. Williams, C.H. Still: Laser beam deflection induced by transverse plasma flow, Phys Rev Lett 77, p.1298-1301 (1996)

[5.421] {Sect. 5.6} F.H. Loesel, M.H. Niemz, J.F. Bille, T. Juhasz: Laser-induced optical breakdown on hard and soft tissues and its dependence on the pulse duration: Experiment and model, IEEE J QE-32, p.1717-1722 (1996)

[5.422] {Sect. 5.6} J.D. Moody, B.J. Macgowan, D.E. Hinkel, W.L. Kruer, E.A. Williams, K. Estabrook, R.L. Berger, R.K. Kirkwood, D.S. Montgomery, T.D. Shepard: First optical observation of intensity dependent laser beam deflection in a flowing plasma, Phys Rev Lett 77, p.1294-1297 (1996)

[5.423] {Sect. 5.6} M. Welling, R.I. Thompson, H. Walther: Photodissociation of MgC60 (+) complexes generated and stored in a linear ion trap, Chem Phys Lett 253, p.37-42 (1996)

[5.424] {Sect. 5.6} P. Gibbon, R. Forster: Short-pulse laser-plasma interactions, Plasma.Phys. Control. Fusion 38, p.769-793 (1996)

[5.425] {Sect. 5.6} Q. Feng, J.V. Moloney, A.C. Newell, E.M. Wright: Laser-induced breakdown versus self-focusing for focused picosecond pulses in water, Optics Letters 20, p.1958-1960 (1995)

[5.426] {Sect. 5.6} P.K. Kennedy, S.A. Boppart, D.X. Hammer, B.A. Rockwell, G.D. Noojin, W.P. Roach: A first-order model for computation of laser-induced breakdown thresholds in ocular and aqueous media. 2. Comparison to experiment, IEEE J QE-31, p.2250-2257 (1995)

[5.427] {Sect. 5.6} P.K. Kennedy: A first-order model for computation of laser-induced breakdown thresholds in ocular and aqueous media. 1. Theory, IEEE J QE-31, p.2241-2249 (1995)

[5.428] {Sect. 5.6} T.X. Phuoc: Laser spark ignition: experimental determination of laser-induced breakdown thresholds of combustion gases, Opt Commun 175, p.419-423 (2000)

[5.429] {Sect. 5.6} D.X. Hammer, E.D. Jansen, M. Frenz, G.D. Noojin, R.J. Thomas, J. Noack, A. Vogel, B.A. Rockwell, A.J. Welch: Shielding properties of laser-induced breakdown in water for pulse durations from 5 ns to 125 fs, Appl Opt 36, p.5630-5640 (1997)

[5.430] {Sect. 5.6} D.X. Hammer, G.D. Noojin, R.J. Thomas, C.E. Clary, B.A. Rockwell, C.A. Toth, W.P. Roach: Intraocular laser surgical probe for membrane disruption by laser-induced breakdown, Appl Opt 36, p.1684-1693 (1997)

[5.431] {Sect. 5.7} R. M. Wood: Laser Damage in Optical Materials (SPIE Optical Engineering Press, London, 1990)

[5.432] {Sect. 5.7} E.S. Bliss: Pulse Duration Dependence of Laser Damage Mechanisms, Opto-Electr. 3, p.99-108 (1971)

[5.433] {Sect. 5.7} F. Loewenthal, R. Tommasini, J.E. Balmer: Single-shot measurement of laser-induced damage thresholds of thin film coatings, Opt Commun 152, p.168-174 (1998)

[5.434] {Sect. 5.7} A.C. Tien, S. Backus, H. Kapteyn, M. Murnane, G. Mourou: Short-pulse laser damage in transparent materials as a function of pulse duration, Phys Rev Lett 82, p.3883-3886 (1999)

[5.435] {Sect. 5.7} F. Dahmani, A.W. Schmid, J.C. Lambropoulos, S. Burns: Dependence of birefringence and residual stress near laser-induced cracks in fused silica on laser fluence and on laser-pulse number, Appl Opt 37, p.7772-7784 (1998)

[5.436] {Sect. 5.7} S. Papernov, A. Schmid, F. Dahmani: Laser damage in polymer waveguides driven purely by a nonlinear, transverse scattering process, Opt Commun 147, p.112-116 (1998)

[5.437] {Sect. 5.7} Y. Zhao, Z.C. Feng, Y. Liang, H.W. Sheng: Laser-induced coloration of WO3, Appl Phys Lett 71, p.2227-2229 (1997)

[5.438] {Sect. 5.7} J.P. Féve, B. Boulanger, G. Manier, H. Albrecht: Repetition rate dependence of gray-tracking in KTiOPO4 during second-harmonic generation at 532 nm, Appl. Phys. Lett. 70, p.277-279 (1997)

[5.439] {Sect. 5.7} B.C. Stuart, M.D. Feit, S. Herman, A.M. Rubenchik, B.W. Shore, M.D. Perry: Nanosecond-to-femtosecond laser-induced breakdown in dielectrics, Phys. Rev. B 53, p.1749-1761 (1996)

[5.440] {Sect. 5.7} V. Pruneri, P.G. Kazansky, J. Webjörn, P.St.J. Russell, D.C. Hanna: Self-organized light-induced scattering in periodically poled lithium niobate, Appl. Phys. Lett. 67, p.1957-1959 (1995)

[5.441] {Sect. 5.7} M.P. Scripsick, D.N. Lolacono, J. Rottenberg, S.H. Goellner, L.E. Halliburton, F.K. Hopkins: Defects responsible for gray tracks in flux-grown KTiOPO4, Appl. Phys. Lett. 66, p.34283430 (1995)

[5.442] {Sect. 5.7} B.C. Stuart, M.D. Feit, A.M. Rubenchik, B.W. Shore, M.D. Perry: Laser-Induced Damage in Dielectrics with Nanosecond to Subpicosecond Pulses, Phys. Rev. Lett. 74, p.2248-2251 (1995)

[5.443] {Sect. 5.7} B. Boulanger, M.M. Fejer, R. Blachman, P.F. Bordui: Study of KTiOPO4 gray-tracking at 1064, 532, and 355 nm, Appl. Phys. Lett. 65, p.2401-2403 (1994)

[5.444] {Sect. 5.7} M.P. Scripsick, G.J. Edwards, L.E. Halliburton, R.F. Belt, G.M. Loiacono: Effect of crystal growth on Ti3+ centers in KTiOPO4, J. Appl. Phys. 76, p.773-776 (1994)

[5.445] {Sect. 5.7} G.M. Loiacono, D.N. Loiacono, T. McGee, M. Babb: Laser damage formation in KTiOPO4 and KTiOAsO4 crystals: Grey tracks, J. Appl. Phys. 72, p.2705-2712 (1992)

[5.446] {Sect. 5.7} J.C. Jacco, D.R. Rockafellow, E.A. Teppo: Bulk-darkening threshold of flux-grown KTiOPO4, Opt. Lett. 16, p.1307-1309 (1991)

[5.447] {Sect. 5.7} J.K. Tyminski: Photorefractive damage in KTP used as second-harmonic generator, J. Appl. Phys. 70, p.5570-5576 (1991)

[5.448] {Sect. 5.7} G.A. Magel, M.M. Fejer, R.L. Byer: Quase-phase-matched second-harmonic generation of blue light in periodically poled LiNbO3, Appl. Phys. Lett. 56, p.108-110 (1990)

[5.449] {Sect. 5.7} K.E. Montgomery, F.P. Milanovich: High-laser-damage-threshold potassium dihydrogen phosphate crystals, J. Appl. Phys. 68, p.3979-3982 (1990)

[5.450] {Sect. 5.7} S.C. Jones, P. Braunlich, R.T. Casper, X.-A. Shen, P. Kelly: Recent progress on laser-induced modifications and intrinsic bulk damage of wide-gap optical materials, Opt. Eng. 28, p.1039-1068 (1989)

[5.451] {Sect. 5.7} D.A. Bryan, R.R. Rice, R. Gerson, H.E. Tomaschke, K.L. Sweeney, L.E. Halliburton: Magnesium-doped lithium niobate for higher optical power applications, Opt. Eng. 24, p.138-143 (1985)

[5.452] {Sect. 5.7} N. Bloembergen: Role of Cracks, Pores, and Absorbing Inclusions on Laser Induced Damage Threshold at Surfaces of Transparent Dielectrics, Appl. Opt. 12, p.661-664 (1973)

[5.453] {Sect. 5.7} N.L. Boling, G. Dubé: Laser-induced inclusion damage at surfaces of transparent dielectrics, Appl. Phys. Lett. 23, p.658-660 (1973)

[5.454] {Sect. 5.7} N.L. Boling, M.D. Crisp, G. Dubé: Laser Induced Surface Damage, Appl. Opt. 12, p.650-660 (1973)

[5.455] {Sect. 5.7} M.D. Crisp, N.L. Boling, G. Dubé: Importance of Fresnel reflections in laser surface damage transparent dielectrics, Appl. Phys. Lett. 21, p.364-366 (1972)

[5.456] {Sect. 5.7} R.W. Hopper, D.R. Uhlmann: Mechanism of Inclusion Damage in Laser Glass, J. Appl. Phys. 41, p.4023-4037 (1970)

[5.457] {Sect. 5.7} W.G. Wagner, H.A. Haus, J.H. Marburger: Large-Scale Self-Trapping of Optical Beams in the Paraxial Ray Approximation, Phys. Rev. 175, p.256-266 (1968)

[5.458] {Sect. 5.7} M. Castillejo, S. Couris, E. Koudoumas, M. Martin: Ionization and fragmentation of aromatic and single-bonded hydrocarbons with 50 fs laser pulses at 800 nm, Chem Phys Lett 308, p.373-380 (1999)

[5.459] {Sect. 5.7} S. Wennmalm, R. Rigler: On death numbers and survival times of single dye molecules, J Phys Chem B 103, p.2516-2519 (1999)

[5.460] {Sect. 5.7} M. Castillejo, S. Couris, E. Lane, M. Martin, J. Ruiz: Laser photodissociation of ketene at 230 nm, Chem Phys 232, p.353-360 (1998)

[5.461] {Sect. 5.7} S. Popov: Dye photodestruction in a solid-state dye laser with a polymeric gain medium, Appl Opt 37, p.6449-6455 (1998)

[5.462] {Sect. 5.7} T. Shibata, T. Suzuki: Photofragment ion imaging with femtosecond laser pulses, Chem Phys Lett 262, p.115-119 (1996)

[5.463] {Sect. 5.7} R.K. Talukdar, M. Hunter, R.F. Warren, J.B. Burkholder, A.R. Ravishankara: UV laser photodissociation of CF2ClBr and CF2Br2 at 298 K: Quantum yields of Cl, Br, and CF2, Chem Phys Lett 262, p.669-674 (1996)

[5.464] {Sect. 5.7} D.M. Burland, F. Carmona, J. Pacansky: The photodissociation of s-tetrazine and dimethyl-s-tetrazine, Chem. Phys. Lett. 56, p.221-226 (1978)

[5.465] {Sect. 5.7} S. Link, C. Burda, M.B. Mohamed, B. Nikoobakht, M.A. El-Sayed: Laser photothermal melting and fragmentation of gold nanorods: Energy and laser pulse-width dependence, J Phys Chem A 103, p.1165-1170 (1999)

[5.466] {Sect. 5.7} A. Saemann, K. Eidmann: X-ray emission from metallic (Al) and dielectric (glass) targets irradiated by intense ultrashort laser pulses, Appl Phys Lett 73, p.1334-1336 (1998)

[5.467] {Sect. 5.7} I.M. Hodge: Physical aging in polymer glasses, Science 267, p.1945-1947 (1995)

[5.468] {Sect. 5.7} A.J. Campillo, S.L. Shapiro, B.R. Suydam: Relationship of self-focusing to spatial instability modes, Appl. Phys. Lett. 24, p.178-180 (1974)

[5.469] {Sect. 5.7} A.J. Campillo, S.L. Shapiro, B.R. Suydam: Periodic breakup of optical beams due to self-focusing, Appl. Phys. Lett. 23, p.628-630 (1973)

[5.470] {Sect. 5.7} M.M.T. Loy, Y.R. Shen: Study of Self-Focusing and Small-Scale Filaments of Light in Nonlinear Media, IEEE J. QE-9, p.409-422 (1973)

[5.471] {Sect. 5.7} E.L. Kerr: Filamentary Tracks Formed in Transparent Optical Galss by Laser Beam Self-Focusing. II. Theoretical Analysis, Phys. Rev. A 4, p.1195-1218 (1971)

[5.472] {Sect. 5.7} E.L. Dawes, J.H. Marburger: Computer Studies in Self-Focusing, Phys. Rev. 179, p.862-868 (1969)

[5.473] {Sect. 5.8} D. Bäuerle: Laser Processing and Chemistry (Springer, Berlin, Heidelberg, New York, 1996)

[5.475] {Sect. 5.8} R.Ifflӓnder: Solid-State Lasers for Materials Processing (Springer, Heidelberg, Berlin, New York, 2001)

[5.474] {Sect. 5.8} J. C. Miller (ed.): Laser Ablation (Springer, Berlin, Heidelberg, New York, 1994)

[5.476] {Sect. 5.8} M. Haag, H. Hugel, C.E. Albright, S. Ramasamy: CO2 laser light absorption characteristics of metal powders, J Appl Phys 79, p.3835-3841 (1996)

[5.477] {Sect. 5.8} A.F.H. Kaplan: An analytical model of metal cutting with a laser beam, J Appl Phys 79, p.2198-2208 (1996)

[5.478] {Sect. 5.8} C.J. Nonhof: Material processing with Nd-lasers, Electrochem. Publ. 34p.128 (1988)

[5.479] {Sect. 5.8} C. Hahn, T. Lippert, A. Wokaun: Comparison of the ablation behavior of polymer films in the IR and UV with nanosecond and picosecond pulses, J Phys Chem B 103, p.1287-1294 (1999)

[5.480] {Sect. 5.8} T.E. Itina, W. Marine, M. Autric: Nonstationary effects in pulsed laser ablation, J Appl Phys 85, p.7905-7908 (1999)

[5.481] {Sect. 5.8} J. Muramoto, I. Sakamoto, Y. Nakata, T. Okada, M. Maeda: Influence of electric field on the behavior of Si nanoparticles generated by laser ablation, Appl Phys Lett 75, p.751-753 (1999)

[5.482] {Sect. 5.8} D. Sands, F.X. Wagner, P.H. Key: Evidence for a thermal mechanism in excimer laser ablation of thin film ZnS on Si, J Appl Phys 85, p.3855-3859 (1999)

[5.483] {Sect. 5.8} A. Cavalleri, K. SokolowskiTinten, J. Bialkowski, D. vonder-Linde: Femtosecond laser ablation of gallium arsenide investigated with time- of-flight mass spectroscopy, Appl Phys Lett 72, p.2385-2387 (1998)

[5.484] {Sect. 5.8} C. Egami, Y. Kawata, Y. Aoshima, H. Takeyama, F. Iwata, O. Sugihara, M. Tsuchimori, O. Watanabe, H. Fujimura, N. Okamoto: Visible-laser ablation on a nanometer scale using urethane-urea copolymers, Opt Commun 157, p.150-154 (1998)

[5.485] {Sect. 5.8} T.W. Hodapp, P.R. Fleming: Modeling topology formation during laser ablation, J Appl Phys 84, p.577-583 (1998)

[5.486] {Sect. 5.8} H. Schmidt, J. Ihlemann, B. WolffRottke, K. Luther, J. Troe: Ultraviolet laser ablation of polymers: spot size, pulse duration, and plume attenuation effects explained, J Appl Phys 83, p.5458-5468 (1998)

[5.487] {Sect. 5.8} J. Zhang, K. Sugioka, K. Midorikawa: Direct fabrication of microgratings in fused quartz by laser-induced plasma-assisted ablation with a KrF excimer laser, Optics Letters 23, p.1486-1488 (1998)

[5.488] {Sect. 5.8} T.G. Barton, H.J. Foth, M. Christ, K. Hormann: Interaction of holmium laser radiation and cortical bone: Ablation and thermal damage in a turbid medium, Appl Opt 36, p.32-43 (1997)

[5.489] {Sect. 5.8} D.J. Krajnovich: Near-threshold photoablation characteristics of polyimide and poly (ethylene terephthalate), J Appl Phys 82, p.427-435 (1997)

[5.490] {Sect. 5.8} X. Liu, D. Du, G. Mourou: Laser ablation and micromachining with ultrashort laser pulses, IEEE J QE-33, p.1706-1716 (1997)

[5.491] {Sect. 5.8} L.V. Zhigilei, B.J. Garrison: Velocity distributions of molecules ejected in laser ablation, Appl Phys Lett 71, p.551-553 (1997)

[5.492] {Sect. 5.8} G.B. Blanchet, C.R. Fincher: Laser ablation: Selective unzipping of addition polymers, Appl Phys Lett 68, p.929-931 (1996)

[5.493] {Sect. 5.8} C.G. Gill, T.M. Allen, J.E. Anderson, T.N. Taylor, P.B. Kelly, N.S. Nogar: Low-power resonant laser ablation of copper, Appl Opt 35, p.2069-2082 (1996)

[5.494] {Sect. 5.8} W. Kautek, J. Kruger, M. Lenzner, S. Sartania, C. Spielmann, F. Krausz: Laser ablation of dielectrics with pulse durations between 20 fs and 3 ps, Appl Phys Lett 69, p.3146-3148 (1996)

[5.495] {Sect. 5.8} C. Momma, B.N. Chichkov, S. Nolte, F. vonAlvensleben, A. Tunnermann, H. Welling, B. Wellegehausen: Short-pulse laser ablation of solid targets, Opt Commun 129, p.134-142 (1996)

[5.496] {Sect. 5.8} M.A. Shannon, B. Rubinsky, R.E. Russon: Mechanical stress power measurements during high-power laser ablation, J Appl Phys 80, p.4665-4672 (1996)

[5.497] {Sect. 5.8} J.M. FitzGerald, A. Pique, D.B. Chrisey, P.D. Rack, M. Zeleznik, R.C.Y. Auyeung, S. Lakeou: Laser direct writing of phosphor screens for high-definition displays, Appl Phys Lett 76, p.1386-1388 (2000)

[5.498] {Sect. 5.8} L.D. Wang, H.S. Kwok: Pulsed laser deposition of organic thin films, Thin Solid Films 363, p.58-60 (2000)

[5.499] {Sect. 5.8} M.C. Wanke, O. Lehmann, K. Muller, Q.Z. Wen, M. Stuke: Laser rapid prototyping of photonic band-gap microstructures, Science 275, p.1284-1286 (1997)

[5.500] {Sect. 5.8} R.L. Gordon, G.W. Forbes: Gaussian beams with optimal focal properties, Opt Commun 124, p.195-201 (1996)

[5.501] {Sect. 5.8} D.H. Lowndes, D.B. Geohegan, A.A. Puretzky, D.P. Norton, C.M. Rouleau: Synthesis of novel thin-film materials by pulsed laser deposition, Science 273, p.898-903 (1996)

[5.502] {Sect. 5.8} R.W. Mcgowan, D.M. Giltner, S.A. Lee: Light force cooling, focusing, and nanometer-scale deposition of aluminum atoms, Optics Letters 20, p.2535-2537 (1995)

[5.503] {Sect. 5.8} O. Lehmann, M. Stuke: Laser-driven movement of three-dimensional microstructures generated by laser rapid prototyping, Science 270, p.1644-1646 (1995)

[5.504] {Sect. 5.8} B. Dragnea, B. Bourguignon: Photoinduced effects in UV laser melting of Si in UHV, Phys Rev Lett 82, p.3085-3088 (1999)

[5.505] {Sect. 5.8} E.N. Sobol, M.S. Kitai, N. Jones, A.P. Sviridov, T. Milner, B.J.F. Wong: Heating and structural alterations in cartilage under laser radiation, IEEE J QE-35, p.532-539 (1999)

[5.506] {Sect. 5.8} C.W. Siders, A. Cavalleri, K. SokolowskiTinten, C. Toth, T. Guo, M. Kammler, M.H. vonHoegen, K.R. Wilson, D. vonderLinde, C.P.J. Barty: Detection of nonthermal molting by ultrafast X-ray diffraction, Science 286, p.1340-1342 (1999)

[5.507] {Sect. 5.8} S.C. Chen, C.P. Grigoropoulos, H.K. Park, P. Kerstens, A.C. Tam: Photothermal displacement measurement of transient melting and surface deformation during pulsed laser heating, Appl Phys Lett 73, p.2093-2095 (1998)

[5.508] {Sect. 5.8} M. Ii, T.P. Duffey, J. Mazumder: Spatially and temporally resolved temperature measurements of plasma generated in percussion drilling with a diode-pumped Nd: YAG laser, J Appl Phys 84, p.4122-4127 (1998)

[5.509] {Sect. 5.8} V.V. Gupta, H.J. Song, J.S. Im: Numerical analysis of excimer-laser-induced melting and solidification of thin Si films, Appl Phys Lett 71, p.99-101 (1997)

[5.510] {Sect. 5.8} XD. Lacroix, G. Jeandel: Spectroscopic characterization of laser-induced plasma created during welding with a pulsed Nd:YAG laser, J Appl Phys 81, p.6599-6606 (1997)

[5.511] {Sect. 5.8} J. Xie, A. Kar: Mathematical modeling of melting during laser materials processing, J Appl Phys 81, p.3015-3022 (1997)

[5.512] {Sect. 5.8} N. Arnold: Temperature distributions and their evolution in non- planar energy beam microprocessing: A fast algorithm, J Appl Phys 80, p.1291-1298 (1996)

[5.513] {Sect. 5.8} B.A. Mehmetli, K. Takahashi, S. Sato: Direct measurement of reflectance from aluminum alloys during CO_2 laser welding, Appl Opt 35, p.3237-3242 (1996)

[5.514] {Sect. 5.8} S. Nettesheim, R. Zenobi: Pulsed laser heating of surfaces: Nanosecond timescale temperature measurement using black body radiation, Chem Phys Lett 255, p.39-44 (1996)

[5.515] {Sect. 5.8} S. Sato, K. Takahashi, B. Mehmetli: Polarization effects of a high-power CO_2 laser beam on aluminum alloy weldability, J Appl Phys 79, p.8917-8919 (1996)

[5.516] {Sect. 5.8} P.L. Silvestrelli, A. Alavi, M. Parrinello, D. Frenkel: Ab initio molecular dynamics simulation of laser melting of silicon, Phys Rev Lett 77, p.3149-3152 (1996)

[5.517] {Sect. 5.8} K. Murakami, H.C. Gerritsen, H. van Brug, F. Bijkerk, F.W. Saris, M.J. van der Wiel: Pulsed-Laser-Irradiated Silicon Studied by Time-Resolved X-Ray Absorption (90-300 eVl, Phys. Rev. Lett. 56, p.655-658 (1986)

[5.518] {Sect. 5.8} I.W. Boyd, S.C. Moss, T.F. Bogges, A.L. Smirl: Temporally resolved imaging of silicon surfaces melted with intense picosecond 1-μm laser pulses, Appl. Phys. Lett. 46, p.366-368 (1985)

[5.519] {Sect. 5.8} M.C. Downer, R.L. Fork, C.V. Shank: Femtosecond imaging of melting and evaporation at a photoexcited silicon surface, J. Opt. Soc. Am. B 2, p.595-599 (1985)

[5.520] {Sect. 5.8} P.H. Bucksbaum, J. Bokor: Rapid Melting and Regrowth Velocities in Silicon Heated by Ultraviolet Picosecond Laser Pulses, Phys. Rev. Lett. 53, p.182-185 (1984)

[5.521] {Sect. 5.8} S. Williamson, G. Mourou, J.C.M. Li: Time-Resolved Laser-Induced Phase Transformation in Aluminium, Phys. Rev. Lett. 52, p.2364-2367 (1984)

[5.522] {Sect. 5.8} C.V. Shank, R. Yen, C. Hirlimann: Femtosecond-Time-Resolved Surface Structural Dynamics of Optically Excited Silicon, Phys. Rev. Lett. 51, p.900-902 (1983)

[5.523] {Sect. 5.8} S. Sato, H. Ashida, T. Arai, Y.W. Shi, Y. Matsuura, M. Miyagi: Vacuum-cored hollow waveguide for transmission of high-energy, nanosecond Nd : YAG laser pulses and its application to biological tissue ablation, Optics Letters 25, p.49-51 (2000)

[5.524] {Sect. 5.8} S.R. Farrar, D.C. Attril, M.R. Dickinson, T.A. King, A.S. Blinkhorn: Etch rate and spectroscopic ablation studies of Er:YAG laser-irradiated dentine, Appl Opt 36, p.5641-5646 (1997)

[5.525] {Sect. 5.8} G.H. Pettit, M.N. Ediger: Corneal-tissue absorption coefficients for 193- and 213-nm ultraviolet radiation, Appl Opt 35, p.3386-3391 (1996)

[5.526] {Sect. 5.8} P.T. Staveteig, J.T. Walsh: Dynamic 193-nm optical properties of water, Appl Opt 35, p.3392-3403 (1996)

[5.527] {Sect. 5.8} J.K. Kou, V. Zhakhovskii, S. Sakabe, K. Nishihara, S. Shimizu, S. Kawato, M. Hashida, K. Shimizu, S. Bulanov, Y. Izawa et al.: Anisotropic Coulomb explosion of C-60 irradiated with a high-intensity femtosecond laser pulse, J Chem Phys 112, p.5012-5020 (2000)

[5.528] {Sect. 5.8} K.W.D. Ledingham, I. Spencer, T. McCanny, R.P. Singhal, M.I.K. Santala, E. Clark, I. Watts, F.N. Beg, M. Zepf, K. Krushelnick et al.: Photonuclear physics when a multiterawatt laser pulse interacts with solid targets, Phys Rev Lett 84, p.899-902 (2000)

[5.529] {Sect. 5.8} A. Talebpour, A.D. Bandrauk, S.L. Chin: Fragmentation of benzene in an intense Ti : sapphire laser pulse, Laser Phys 10, p.210-215 (2000)

[5.530] {Sect. 5.8} M.K. Grimes, A.R. Rundquist, Y.S. Lee, M.C. Downer: Experimental identification of "vacuum heating" at femtosecond-laser-irradiated metal surfaces, Phys Rev Lett 82, p.4010-4013 (1999)

[5.531] {Sect. 5.8} H. Kwak, K.C. Chou, J. Guo, H.W.K. Tom: Femtosecond laser-induced disorder of the (1 x 1)-relaxed GaAs (110) surface, Phys Rev Lett 83, p.3745-3748 (1999)

[5.532] {Sect. 5.8} M.D. Perry, B.C. Stuart, P.S. Banks, M.D. Feit, V. Yanovsky, A.M. Rubenchik: Ultrashort-pulse laser machining of dielectric materials, J Appl Phys 85, p.6803-6810 (1999)

[5.533] {Sect. 5.8} A. Saemann, K. Eidmann, I.E. Golovkin, R.C. Mancini, E. Andersson, E. Forster, K. Witte: Isochoric heating of solid aluminum by ultrashort laser pulses focused on a tamped target, Phys Rev Lett 82, p.4843-4846 (1999)

[5.534] {Sect. 5.8} H. Jelinkova, J. Sulc, P. Cerny, Y.W. Shi, Y. Matsuura, M. Miyagi: High-power Nd : YAG laser picosecond pulse delivery by a polymer-coated silver hollow-glass waveguide, Optics Letters 24, p.957-959 (1999)

[5.535] {Sect. 5.8} Y. Matsuura, K. Hanamoto, S. Sato, M. Miyagi: Hollow-fiber delivery of high-power pulsed Nd : YAG laser light, Optics Letters 23, p.1858-1860 (1998)

[5.536] {Sect. 5.8} P. Dainesi, J. Ihlemann, P. Simon: Optimization of a beam delivery system for a short-pulse KrF laser used for material ablation, Appl Opt 36, p.7080-7085 (1997)

[5.537] {Sect. 5.8} H. Pratisto, M. Frenz, M. Ith, H.J. Altermatt, E.D. Jansen, H.P. Weber: Combination of fiber-guided pulsed erbium and holmium laser radiation for tissue ablation under water, Appl Opt 35, p.3328-3337 (1996)

[5.538] {Sect. 5.8} B. Richou, I. Schertz, I. Gobin, J. Richou: Delivery of 10-MW Nd:YAG laser pulses by large-core optical fibers: Dependence of the laser-intensity profile on beam propagation, Appl Opt 36, p.1610-1614 (1997)

[5.539] {Sect. 5.8} T.E. Dimmick, G. Kakarantzas, T.A. Birks, P.S. Russell: Carbon dioxide laser fabrication of fused-fiber couplers and tapers, Appl Opt 38, p.6845-6848 (1999)

[5.540] {Sect. 5.8} S. Mailis, I. Zergioti, G. Koundourakis, A. Ikiades, A. Patentalaki, P. Papakonstantinou, N.A. Vainos, C. Fotakis: Etching and printing of diffractive optical microstructures by a femtosecond excimer laser, Appl Opt 38, p.2301-2308 (1999)

[5.541] {Sect. 5.8} D. Ashkenasi, H. Varel, A. Rosenfeld, S. Henz, J. Herrmann, E.E.B. Cambell: Application of self-focusing of ps laser pulses for three-dimensional microstructuring of transparent materials, Appl Phys Lett 72, p.1442-1444 (1998)

[5.542] {Sect. 5.8} T.H. Her, R.J. Finlay, C. Wu, S. Deliwala, E. Mazur: Microstructuring of silicon with femtosecond laser pulses, Appl Phys Lett 73, p.1673-1675 (1998)

[5.543] {Sect. 5.8} T. Hessler, M. Rossi, R.E. Kunz, M.T. Gale: Analysis and optimization of fabrication of continuous-relief diffractive optical elements, Appl Opt 37, p.4069-4079 (1998)

[5.544] {Sect. 5.8} K. Baba, K. Hayashi, I. Syuaib, K. Yamaki, M. Miyagi: Write-once optical data storage media with large reflectance change with metal-island films, Appl Opt 36, p.2421-2426 (1997)

[5.545] {Sect. 5.8} G.P. Behrmann, M.T. Duignan: Excimer laser micromachining for rapid fabrication of diffractive optical elements, Appl Opt 36, p.4666-4674 (1997)

[5.546] {Sect. 5.8} X.M. Wang, J.R. Leger, R.H. Rediker: Rapid fabrication of diffractive optical elements by use of image-based excimer laser ablation, Appl Opt 36, p.4660-4665 (1997)

[5.547] {Sect. 5.8} P.A. Atanasova, V.P. Manolov: Laser cutting of wire-wound resistors: Theory and experiments, J Appl Phys 80, p.2003-2008 (1996)

[5.548] {Sect. 5.8} W. Chalupczak, C. Fiorini, F. Charra, J.M. Nunzi, P. Raimond: Efficient all-optical poling of an azo-dye copolymer using a low power laser, Opt Commun 126, p.103-107 (1996)

[5.549] {Sect. 5.8} K.M. Davis, K. Miura, N. Sugimoto, K. Hirao: Writing waveguides in glass with a femtosecond laser, Optics Letters 21, p.1729-1731 (1996)

[5.550] {Sect. 5.8} S. Lazare, J. Lopez, J.M. Turlet, M. Kufner, S. Kufner, P. Chavel: Microlenses fabricated by ultraviolet excimer laser irradiation of poly (methyl methacrylate) followed by styrene diffusion, Appl Opt 35, p.4471-4475 (1996)

[5.551] {Sect. 5.8} T. Schuster, H. Kuhn, A. Raiber, T. Abeln, F. Dausinger, H. Hugel, M. Klaser, G. Mullervogt: High-precision laser cutting of high-temperature superconductors, Appl Phys Lett 68, p.2568-2570 (1996)

[5.552] {Sect. 5.8} Y.Y. Tsui, R. Fedosejevs, C.E. Capjack: Vaporization of aluminum by 50 ps KrF laser pulses, J Appl Phys 80, p.509-512 (1996)

[5.553] {Sect. 5.8} B. Bescos, H. Buchenau, R. Hoch, H.J. Schmidtke, G. Gerber: Femtosecond laser ionization of CdTe clusters, Chem Phys Lett 285, p.64-70 (1998)

[5.554] {Sect. 5.9.1} D. Maystre: Diffraction Gratings (SPIE Optical Engineering Press, London, 1993)

[5.555] {Sect. 5.9.1} H.J. Eichler, P. Günter, D.W. Pohl: Laser-Induced Dynamic Gratings, Springer Ser. Opt. Sci, Vol. 50 (Springer, Berlin, Heidelberg, New York, Tokyo 1986)

[5.556] {Sect. 5.9.1} V.A. Zuikov, A.A. Kalachev, V.V. Samartsev, I.V. Negrashov, A.K. Rebane, I. Gallus, O. Ollikainen, U.P. Wild: Spatial and spectral properties of nonequilibrium population gratings induced in a resonant medium by femtosecond pulses, Laser Phys 10, p.368-371 (2000)

[5.557] {Sect. 5.9.1} Y. Tang, J.P. Schmidt, S.A. Reid: Nanosecond transient grating studies of jet-cooled NO2, J Chem Phys 110, p.5734-5744 (1999)

[5.558] {Sect. 5.9.1} N.C.R. Holme, L. Nikolova, P.S. Ramanujam, S. Hvilsted: An analysis of the anisotropic and topographic gratings in a side-chain liquid crystalline azobenzene polyester, Appl Phys Lett 70, p.1518-1520 (1997)

[5.559] {Sect. 5.9.1} M.J. Damzen, Y. Matsumoto, G.J. Crofts, R.P.M. Green: Bragg-selectivity of a volume gain grating, Opt Commun 123, p.182-188 (1996)

[5.560] {Sect. 5.9.1} D. Trivedi, P. Tayebati, M. Tabat: Measurement of large electro-optic coefficients in thin films of strontium barium niobate (Sr0.6Ba0.4Nb2O6), Appl Phys Lett 68, p.3227-3229 (1996)

[5.561] {Sect. 5.9.1} A. Belendez, A. Fimia, L. Carretero, F. Mateos: Self-induced phase gratings due to the inhomogeneous structure of acrylamide photopolymer systems used as holographic recording materials, Appl Phys Lett 67, p.3856-3858 (1995)

[5.562] {Sect. 5.9.1} P.R. Hemmer, D.P. Katz, J. Donoghue, M. Croningolomb, M.S. Shahriar, P. Kumar: Efficient low-intensity optical phase conjugation based on coherent population trapping in sodium, Optics Letters 20, p.982-984 (1995)

[5.563] {Sect. 5.9.1} R. Macdonald, H. Danlewski: Self induced optical gratings in nematic liquid crystals with a feedback mirror, Optics Letters 20, p.441-443 (1995)

[5.564] {Sect. 5.9.1} F.W. Deeg, M.D. Fayer: Analysis of complex molecular dynamics in an organic liquid by polarization selective subpicosecond transient grating experiments, J. Chem. Phys. 91, p.2269-2279 (1989)

[5.565] {Sect. 5.9.1} I. McMichael, P. Yeh, P. Beckwith: Nondegenerate two-wave mixing in ruby, Opt. Lett. 13, p.500-502 (1988)

[5.566] {Sect. 5.9.1} A. Marcano, O.F. Garcia-Golding, R.Rojas F.: Pump-power dependences of thermal-grating and electronic components of a polarization spectroscopy signal from dye solutions, J. Opt. Soc. Am. B 3, p.3-7 (1986)

[5.567] {Sect. 5.9.1} I.-C. Khoo, R. Normandin: The mechanism and Dynamics of Transient Thermal Grating Diffraction in Nematic Liquid Crystal Films, IEEE J. QE-21, p.329-335 (1985)

[5.568] {Sect. 5.9.1} G. Eyring, M.D. Fayer: A picosecond holographic grating approach to molecular dynamics in oriented liquid crystal films, J. Chem. Phys. 81, p.4314-4321 (1984)

[5.569] {Sect. 5.9.1} K.A. Nelson, R. Casalegno, R.J. Dwayne Miller, M.D. Fayer: Laser-induced excited state and ultrasonic wave gratings: Amplitude and phase grating contributions to diffraction, J. Chem. Phys. 77, p.1144-1152 (1982)

[5.570] {Sect. 5.9.1} J.R. Andrews, R.M. Hochstrasser: Transient grating effects in resonant four-wave mixing experiment, Chem. Phys. Lett. 76, p.213-217 (1980)

[5.571] {Sect. 5.9.1} H.J. Eichler, G. Enterlein, D. Langhans: Investigation of the Spatial Coherence of a Laser Beam by a Laser-Induced Grating Method, Appl. Phys. 23, p.299-302 (1980)

[5.572] {Sect. 5.9.1} H.J. Eichler, U. Klein, D. Langhans: Coherence Time Measurement of Picosecond Pulses by a Light-Induced Grating Method, Appl. Phys. 21, p.215-219 (1980)

[5.573] {Sect. 5.9.1} J.R. Salcedo, A.E. Siegman: Laser Induced Photoacoustic Grating Effects in Molecular Crystals, IEEE J. QE-15, p.250-258 (1979)

[5.574] {Sect. 5.9.1} A. v. Jena, H.E. Lessing: Theory of laser-induced amplitude and phase gratings including photoselection, orientational relaxation and population kinetics, Opt. Quant. Electr. 11, p.419-439 (1979)

[5.575] {Sect. 5.9.1} J.R. Salcedo, A.E. Siegman, D.D. Dlott, M.D. Fayer: Dynamics of Energy Transport in Molecular Crystals: The Picosecond Transient-Grating Method, Phys. Rev. Lett. 41, p.131-134 (1978)

[5.576] {Sect. 5.9.1} D.W. Phillion, D.J. Kuizenga, A.E. Siegman: Subnanocecond relaxation time measurements using a transient induced grating method, Appl. Phys. Lett. 27, p.85-87 (1975)

[5.577] {Sect. 5.9.1} H. Eichler, G. Salje, H. Stahl: Thermal diffusion measurements using spatially periodic temperature distributions induced by laser light, J. Appl. Phys. 44, p.5383-5388 (1973)

[5.578] {Sect. 5.9.1} H. Eichler, G. Enterlein, P. Glozbach, J. Munschau, H. Stahl: Power Requirements and Resolution of Real-Time Holograms in Saturable Absorbers and Absorbing Liquids, Appl. Opt. 11, p.372-375 (1972)

[5.579] {Sect. 5.9.1} P. Delaye, G. Roosen: Evaluation of a photorefractive two-beam coupling novelty filter, Opt Commun 165, p.133-151 (1999)

[5.580] {Sect. 5.9.1} A. Pecchia, M. Laurito, P. Apai, M.B. Danailov: Studies of two-wave mixing of very broad-spectrum laser light in BaTiO3, J Opt Soc Am B Opt Physics 16, p.917-923 (1999)

[5.581] {Sect. 5.9.1} Y. Tomita, S. Matsushima: Photorefractive beam coupling between orthogonally polarized light beams by linear dichroism in Cu-doped potassium sodium strontium barium niobate, J Opt Soc Am B Opt Physics 16, p.111-116 (1999)

[5.582] {Sect. 5.9.1} A. Brignon, I. Bongrand, B. Loiseaux, J.P. Huignard: Signal-beam amplification by two-wave mixing in a liquid-crystal light valve, Optics Letters 22, p.1855-1857 (1997)

[5.583] {Sect. 5.9.1} S. Maccormack, G.D. Bacher, J. Feinberg, S. OBrien, R.J. Lang, M.B. Klein, B.A. Wechsler: Powerful, diffraction-limited semiconductor laser using photorefractive beam coupling, Optics Letters 22, p.227-229 (1997)

[5.584] {Sect. 5.9.1} P. Yeh: Two-Wave Mixing in Nonlinear Media, IEEE J. QE-25, p.484-519 (1989)

[5.585] {Sect. 5.9.1} I. McMichael, P. Yeh, P. Beckwith: Nondegenerate two-wave mixing in ruby, Opt. Lett. 13, p.500-502 (1988)

[5.586] {Sect. 5.9.1} C.V. Heer: Small-signal gain generated by two pump waves in a nonlinear medium, Opt. Lett. 6, p.549-551 (1981)

[5.587] {Sect. 5.9.1} J.E. Heebner, R.S. Bennink, R.W. Boyd, R.A. Fisher: Conversion of unpolarized light to polarized light with greater than 50% efficiency by photorefractive two-beam coupling, Optics Letters 25, p.257-259 (2000)

[5.588] {Sect. 5.9.1} A. Brignon, J.P. Huignard, M.H. Garrett, I. Mnushkina: Spatial beam cleanup of a Nd:YAG laser operating at 1.06 mu m with two-wave mixing in Rh:BaTiO3, Appl Opt 36, p.7788-7793 (1997)

[5.589] {Sect. 5.9.1} A. Takada, M. Croningolomb: Laser beam cleanup with photorefractive two-beam coupling, Optics Letters 20, p.1459-1461 (1995)

[5.590] {Sect. 5.9.2} P.C. deSouza, G. Nader, T. Catunda, M. Muramatsu, R.J. Horowicz: Transient four-wave mixing in saturable media with a nonlinear refractive index, Opt Commun 163, p.44-48 (1999)

[5.591] {Sect. 5.9.2} F. DiTeodoro, E.F. McCormack: The effect of laser bandwidth on the signal detected in two-color, resonant four-wave mixing spectroscopy, J Chem Phys 110, p.8369-8383 (1999)

[5.592] {Sect. 5.9.2} K. Morishita, Y. Higuchi, T. Okada: Infrared laser spectroscopic imaging based on degenerate four-wave- mixing spectroscopy combined with frequency-upconversion detection, Optics Letters 24, p.688-690 (1999)

[5.593] {Sect. 5.9.2} J.A. Hudgings, K.Y. Lan: Step-tunable all-optical wavelength conversion using cavity-enhanced four-wave mixing, IEEE J QE-34, p.1349-1355 (1998)

[5.594] {Sect. 5.9.2} H.B. Liao, R.F. Xiao, H. Wang, K.S. Wong, G.K.L. Wong: Large third-order optical nonlinearity in Au:TiO2 composite films measured on a femtosecond time scale, Appl Phys Lett 72, p.1817-1819 (1998)

[5.595] {Sect. 5.9.2} K.P. Lor, K.S. Chiang: Theory of nondegenerate four-wave mixing in a birefringent optical fibre, Opt Commun 152, p.26-30 (1998)

[5.596] {Sect. 5.9.2} P. Ewart, P.G.R. Smith, R.B. Williams: Imaging of trace species distributions by degenerate four- wave mixing: diffraction effects, spatial resolution, and image referencing, Appl Opt 36, p.5959-5968 (1997)

[5.597] {Sect. 5.9.2} A. Brignon, G. Feugnet, J.P. Huignard, J.P. Pocholle: Efficient degenerate four wave mixing in a diode pumped microchip Nd:YVO4 amplifier, Optics Letters 20, p.548-550 (1995)

[5.598] {Sect. 5.9.2} A. Brignon, J.P. Huignard: Continuous wave operation of saturable gain degenerate four wave mixing in a Nd:YVO4 amplifier, Optics Letters 20, p.2096-2098 (1995)

[5.599] {Sect. 5.9.2} G.J. Crofts, R.P.M. Green, M.J. Damzen: Investigation of multipass geometries for efficient degenerate four-wave mixing in Nd:YAG, Opt. Lett. 17, p.920-922 (1992)

[5.600] {Sect. 5.9.2} W.M. Dennis, W. Blau, D.J. Bradley: Picosecond degenerate four-wave mixing in soluble polydiacetylenes, Appl. Phys. Lett. 47, p.200-202 (1985)

[5.601] {Sect. 5.9.2} D.G. Steel, J.F. Lam: Two-Photon Coherent-Transient Measurement of the Nonradiative Collisionless Dephasing Rate in SF6 via Doppler-Free Degenerate Four-Wave Mixing, Phys. Rev. Lett. 43, p.1588-1591 (1979)

[5.602] {Sect. 5.9.2} T. Yajima, H. Souma, Y. Ishida: Study of ultra-fast relaxation processes by resonant Rayleigh-type optical mixing. II. Experiment on dye solutions, Phys. Rev. A 17, p.324-334 (1978)

[5.603] {Sect. 5.9.2} A. Yariv, D.M. Pepper: Amplified reflection, phase conjugation, and oscillation in degenerate Four-wave mixing, Opt. Lett. 1, p.16-18 (1977)

[5.604] {Sect. 5.9.2} R.L. Carman, R.Y. Chiao, P.L. Kelley: Observation of Degenerate Stimulated Four-Photon Interaction and Four-Wave Parametric Amplification, Phys. Rev. Lett. 17, p.1281-1283 (1966)

[5.605] {Sect. 5.9.2} R.I. Thompson, L. Marmet, B.P. Stoicheff: Effect of counterintuitive time delays in nonlinear mixing, Optics Letters 25, p.120-122 (2000)

[5.606] {Sect. 5.9.2} Y.H. Ahn, J.S. Yahng, J.Y. Sohn, K.J. Yee, S.C. Hohng, J.C. Woo, D.S. Kim, T. Meier, S.W. Koch, Y.S. Lim et al.: From exciton resonance to frequency mixing in GaAs multiple quantum wells, Phys Rev Lett 82, p.3879-3882 (1999)

[5.607] {Sect. 5.9.2} V.P. Kalosha, J. Herrmann: Formation of optical subcycle pulses and full Maxwell-Bloch solitary waves by coherent propagation effects, Phys Rev Lett 83, p.544-547 (1999)

[5.608] {Sect. 5.9.2} H. Watanabe, T. Omatsu, T. Hirose, A. Hasegawa, M. Tateda: Highly efficient degenerate four-wave mixing with multipass geometries in a polymer laser dye saturable amplifier, Optics Letters 24, p.1620-1622 (1999)

[5.609] {Sect. 5.9.2} O.L. Antipov, A.S. Kuzhelev, D.V. Chausov: Nondegenerate four-wave-mixing measurements of a resonantly induced refractive-index grating in a Nd:YAG amplifier, Optics Letters 23, p.448-450 (1998)

[5.610] {Sect. 5.9.2} M.J. LaBuda, J.C. Wright: Vibrationally enhanced four-wave mixing in 1,8-nonadiyne, Chem Phys Lett 290, p.29-35 (1998)

[5.611] {Sect. 5.9.2} H. Palm, F. Merkt: Generation of tunable coherent extreme ultraviolet radiation beyond 19 eV by resonant four-wave mixing in argon, Appl Phys Lett 73, p.157-159 (1998)

[5.612] {Sect. 5.9.2} K.S. Chiang, K.P. Lor, Y.T. Chow: Nondegenerate four-wave mixing in a birefringent optical fiber pumped by a dye laser, Optics Letters 22, p.510-512 (1997)

[5.613] {Sect. 5.9.2} L. Deng, W.R. Garrett, M.G. Payne, D.Z. Lee: Observation of a critical concentration in laser-induced transparency and multiphoton excitation and ionization in rubidium, Optics Letters 21, p.928-930 (1996)

[5.614] {Sect. 5.9.2} W. Schmid, T. Vogtmann, M. Schwoerer: A modulation technique fo rmeasuring the optical susceptibility x5 by degenerate four-wave mixing, Opt. Comm. 121, p.55-62 (1995)

[5.615] {Sect. 5.9.2} U.P. Wild, A. Renn: Spectral hole burning and holographic image storage, Mol. Cryst. Liq. Cryst. 183, p.119-129 (1990)

[5.616] {Sect. 5.9.2} R. Beach, D. DeBeer, S.R. Hartmann: Time-delayed four-wave mixing using intense incoherent light, Phys. Rev. A 32, p.3467-3474 (1985)

[5.617] {Sect. 5.9.2} F. Vallée, S.C. Wallace, J. Lukasik: Tunable Coherent Vacuum Ultraviolet Generation in Carbon Monoxide in the 1150 A Range, Opt. Comm. 42, p.148-150 (1982)

[5.618] {Sect. 5.9.2} M.D. Duncan, P. Oesterlin, F. König, R.L. Byer: Observation of saturation broadening of the coherent anti-Stokes Raman spectrum (CARS) of Acetylene in a pulsed molecular beam, Chem. Phys. Lett. 80, p.253-256 (1981)

[5.619] {Sect. 5.9.2} Y. Prior, A.R. Bogdan, M. Dagenais, N. Bloembergen: Pressure-Induced Extra Resonances in Four-Wave Mixing, Phys. Rev. Lett. 46, p.111-114 (1981)

[5.620] {Sect. 5.9.2} J.-L. Oudar, R.W. Smith, Y.R. Shen: Polarization-sensitive coherent anti-Stokes Raman spectroscopy, Appl. Phys. Lett. 34, p.758-760 (1979)

[5.621] {Sect. 5.9.2} M.A. Henesian, L. Kulevskii, R.L. Byer: cw high resolution CARS spectroscopy of the Q (ny1) Raman line of methane, J. Chem. Phys. 65, p.5530-5531 (1976)

[5.622] {Sect. 5.9.2} J.W. Nibler, J.R. McDonald, A.B. Harvey: CARS Measurement of Vibrational Temperatures in Electric Discharges, Opt. Comm. 18, p.371-373 (1976)

[5.623] {Sect. 5.9.2} D.M. Bloom, J.R. Yardley, J.F. Young, S.E. Harris: Infrared up-conversion with resonantly two-photon pumped metal vapors, Appl. Phys. Lett. 24, p.427-428 (1974)

[5.624] {Sect. 5.9.2} S.D. Kramer, F.G. Parsons, N. Bloembergen: Interference of third-order light mixing and second-harmonic excitation generation in CuCl, Phys. Rev. B 9, p.1853-1856 (1974)

[5.625] {Sect. 5.9.2} R.R. Alfano, S.L. Shapiro: Explanation of a Transient Raman Gain Anomaly, Phys. Rev. A 2p.2376-2379 (1970)

[5.626] {Sect. 5.9.2} R.R. Alfano, S.L. Shapiro: Emission in the Region 4000 to 7000 A via Four-Photon Coupling in Glass, Phys. Rev. Lett. 24, p.584-587 (1970)

[5.627] {Sect. 5.9.2} M.W. Bowers, R.W. Boyd: Phase locking via Brillouin-enhanced four-wave-mixing phase conjugation, IEEE J QE-34, p.634-644 (1998)

[5.628] {Sect. 5.9.2} A.M. Scott, K.D. Ridley: Effect of signal frequency on four-wave mixing through stimulated Brillouin scattering, Opt. Lett. 15, p.1267-1269 (1990)

[5.629] {Sect. 5.9.2} K.D. Ridley, A.M. Scott: Comparison between theory and experiment in self-pumped Brillouin-enhanced four-wave mixing, J. Opt. Soc. Am. B 6, p.1701-1708 (1989)

[5.630] {Sect. 5.9.2} W.A. Schroeder, M.J. Damzen, M.H.R. Hutchinson: Polarization-Decoupled Brillouin-Enhanced Four-Wave Mixing, IEEE J. QE-25, p.460-469 (1989)

[5.631] {Sect. 5.9.2} A.M. Scott, K.D. Ridley: A review of Brillouin-enhanced-four-wave-mixing, IEEE J. QE-25, p.438-459 (1989)

[5.632] {Sect. 5.9.2} D.E. Watkins, K.D. Ridley, A.M. Scott: Self-pumped four-wave mixing using backward and forward Brillouin scattering, J. Opt. Soc. Am. B 6, p.1693-1700 (1989)

[5.633] {Sect. 5.9.2} A.M. Scott, P. Waggott: Low-intensity phase conjugation by self-pumped Brillouin-induced four-wave mixing, J. Mod. Opt. 35, p.473-481 (1988)

[5.634] {Sect. 5.9.2} J. Minch, S.L. Chuang: Dual-pump four-wave mixing in a double-mode distributed feedback laser, J Opt Soc Am B Opt Physics 17, p.53-62 (2000)

[5.635] {Sect. 5.9.2} C.X. Yang: Propagation and self-pumped phase conjugation of femtosecond laser pulses in BaTiO3, J Opt Soc Am B Opt Physics 16, p.871-877 (1999)

[5.636] {Sect. 5.9.2} D.H. Yu, J.H. Lee, J.S. Chang: Theory of forward degenerate four-wave mixing in two-level saturable absorbers, J Opt Soc Am B Opt Physics 16, p.1261-1268 (1999)

[5.637] {Sect. 5.9.2} M.A. Dugan, A.C. Albrecht: Radiation-matter oscillations and spectal line narrowing in field-correlated four-wave mixing. I. Theory, Phys. Rev. A 43, p.3877-3921 (1991)

[5.638] {Sect. 5.9.2} P. Yeh: Exact solution of a nonlinear model of two-wave mixing in Kerr media, J. Opt. Soc. Am. B 3, p.747-750 (1986)

[5.639] {Sect. 5.9.2} B.S. Wherrett, A.L. Smirl, Th.F. Boggess: Theory of Degenerate Four-Wave Mixing in Picosecond Excitation-Probe Experiments, IEEE J. QE-19, p.680-689 (1983)

[5.640] {Sect. 5.9.2} P. Ye, Y.R. Shen: Transient four-wave mixing and coherent transient optical phenomena, Phys. Rev. A 25, p.2183-2199 (1982)

[5.641] {Sect. 5.9.2} J.-L. Oudar, Y.R. Shen: Nonlinear spectroscopy by multiresonant four-wave mixing, Phys. Rev. A 22, p.1141-1158 (1980)

[5.642] {Sect. 5.9.2} R.W. Hellwarth: Theory of phase conjugation by stimulated scattering in a waveguide, J. Opt. Soc. Am. 68, p.1050-1056 (1978)

[5.643] {Sect. 5.9.2} T.K. Yee, T.K. Gustafson: Diagrammatic analysis of the density operator for nonlinear optical calculations: Pulsed and cw responses, Phys. Rev. A 18, p.1597-1617 (1978)

[5.644] {Sect. 5.9.2} Y. R. Shen: Principles of Nonlinear Optics, chapter 14 (John Wiley & Sons, Chichester, 1984)

[5.645] {Sect. 5.9.2} M. Lobel, P.M. Petersen, P.M. Johansen: Physical origin of laser frequency scanning induced by photorefractive phase-conjugate feedback, J Opt Soc Am B Opt Physics 16, p.219-227 (1999)

[5.646] {Sect. 5.9.2} D.M. Pepper: Nonlinear optical phase conjugation, Opt. Eng. 21, p.156-183 (1982)

[5.647] {Sect. 5.9.2} A. Yariv: Phase Conjugate Optics and Real-Time Holography, IEEE J. QE-14, p.650-660 (1978)

[5.648] {Sect. 5.9.2} M. Gower, D. Proch (ed.): Optical Phase Conjugation (Springer, Berlin, Heidelberg, New York, 1994)

[5.649] {Sect. 5.9.2} J. I. Sakai: Phase Conjugate Optics (McGraw-Hill, New York, 1992)

[5.650] {Sect. 5.9.2} B. Y. Zel'dovich, N. Pilipettshii: Principles in Phase Conjugation (Springer, Heidelberg, Berlin, New York, 1985)

[5.651] {Sect. 5.9.2} Z.D. Xu, Y.F. Liu, Y. Xiang, J. Yang, S.J. You, W.L. She: Optical phase conjugation property in azo-doped nematic liquid-crystal film, Acta Phys Sin Chinese Ed 48, p.2283-2288 (1999)

[5.652] {Sect. 5.9.2} A. Brignon, S. Senac, J.L. Ayral, J.P. Huignard: Rhodium-doped barium titanate phase-conjugate mirror for an all-solid- state, high-repetition-rate, diode-pumped Nd:YAG master-oscillator power amplifier laser, Appl Opt 37, p.3990-3995 (1998)

[5.653] {Sect. 5.9.2} G.S. He, P.N. Prasad: Phase-conjugation property of one-photon pumped backward stimulated emission from a lasing medium, IEEE J QE-34, p.473-481 (1998)

[5.654] {Sect. 5.9.2} M. Lobel: Wavelength selectivity of the complex grating structure formed in a photorefractive phase conjugator, J Appl Phys 84, p.3483-3490 (1998)

[5.655] {Sect. 5.9.2} A. Miniewicz, S. Bartkiewicz, J. Parka: Optical phase conjugation in dye-doped nematic liquid crystal, Opt Commun 149, p.89-95 (1998)

[5.656] {Sect. 5.9.2} W.L. She, W.K. Lee: Crystal-air interface enhanced self-pumped phase conjugation in photorefractive crystals, Opt Commun 146, p.249-252 (1998)

[5.657] {Sect. 5.9.2} A. Brignon, J.P. Huignard, M.H. Garrett, I. Mnushkina: Self-pumped phase conjugation in rhodium-doped BaTiO3 with 1.06-mu m nanosecond pulses, Optics Letters 22, p.215-217 (1997)

[5.658] {Sect. 5.9.2} R. Gutierrezcastrejon, K.M. Hung, T.J. Hall: Spatial evolution of the phase in resonant degenerate four- wave mixing, Opt Commun 138, p.227-234 (1997)

[5.659] {Sect. 5.9.2} R.K. Mohan, C.K. Subramanian: Transient phase conjugation in dye-doped polymer saturable absorbers, Opt Commun 144, p.322-330 (1997)

[5.660] {Sect. 5.9.2} P.P. Vasilev, I.H. White: Phase-conjugation broad area twin-contact semiconductor laser, Appl Phys Lett 71, p.40-42 (1997)

[5.661] {Sect. 5.9.2} A. Costela, I. Garciamoreno: Degenerate four-wave mixing in phenylbenzimidazole proton-transfer laser dyes, Chem Phys Lett 249, p.373-380 (1996)

[5.662] {Sect. 5.9.2} R.P.M. Green, G.J. Crofts, M.J. Damzen: Novel method for double phase conjugation in gain media, Opt Commun 124, p.488-492 (1996)

[5.663] {Sect. 5.9.2} C. Medrano, M. Zgonik, P. Bernasconi, P. Gunter: Phase conjugation in optical communication links with photorefractive Fe:KNbO3, Opt Commun 128, p.177-184 (1996)

[5.664] {Sect. 5.9.2} Y. Yang, H. Fei, Z. Wei, Q. Yang, G. Shun, L. Han: Phase conjugation in methyl orange doped polyvinyl alcohol film by DFWM based on excited state absorption, Opt. Comm. 123p.189-194 (1996)

[5.665] {Sect. 5.9.2} S. Brulisauer, D. Fluck, C. Solcia, T. Pliska, P. Gunter: Non-destructive waveguide loss-measurement method using self-pumped phase conjugation for optimum end-fire coupling, Optics Letters 20, p.1773-1775 (1995)

[5.666] {Sect. 5.9.2} G.R. Gray, D.H. Detienne, G.P. Agrawal: Mode locking in semiconductor lasers by phase-conjugate optical feedback, Optics Letters 20, p.1295-1297 (1995)

[5.667] {Sect. 5.9.2} S. Miyanaga, H. Ohtateme, K. Kawano, H. Fujiwara: Excited-state absorption and pump propagation effects on optical phase conjugation in a saturable absorber, J. Opt. Soc. Am. B 10, p.1069-1076 (1993)

[5.668] {Sect. 5.9.2} Ch. Egami, K. Nakagawa, H. Fujiwara: Efficient Optical Phase Conjugation in Methyl-Orange-Doped Polyvinyl Alcohol Film, Jpn. J. Appl. Phys. 31, p.2937-2940 (1992)

[5.669] {Sect. 5.9.2} S.S. Alimpiev, I.V. Mel'nikov, V.S. Nersisyan, S.M. Nikiforov, B.G. Sartakov: Phase conjugation of CO2 laser radiation in cryogenic liquids, Sov. J. Quantum Electron. 20, p.1507-1512 (1990)

[5.670] {Sect. 5.9.2} V.I. Bespalov, A.A. Betin, E.A. Zhukov, O.V. Mitropol'sky, N.Yu. Rusov: Phase Conjugation of CO2 Laser Radiation in a Medium with Thermal Nonlinearity, IEEE J. QE-25, p.360-367 (1989)

[5.671] {Sect. 5.9.2} Y. Tomita, R. Yahalom, A. Yariv: Phase shift and cross talk of a self-pumped phase-conjugate mirror, Opt. Comm. 73, p.413-418 (1989)

[5.672] {Sect. 5.9.2} I.M. Bel'dyugin, M.V. Zolotarev, S.E. Kireev, A.I. Odintsov: Copper vapor laser with a self-pumped wavefront-reversing mirror, Sov. J. Quantum Electron. 16, p.535-537 (1986)

[5.673] {Sect. 5.9.2} R.G. Caro, M.C. Gower: Phase conjugation of KrF laser radiation, Opt. Lett. 6, p.557-559 (1981)

[5.674] {Sect. 5.9.2} B.J. Feldman, R.A. Fisher, S.L. Shapiro: Ultraviolet phase conjugation, Opt. Lett. 6, p.84-86 (1981)

[5.675] {Sect. 5.9.2} R.W. Hellwarth: Generation of time-reversed wave fronts by nonlinear refraction, J. Opt. Soc. Am. 67, p.1-3 (1977)

[5.676] {Sect. 5.9.2} B.Ya. Zel'dovich, V.I. Popovicher, V.V. Ragul'skii, F.S. Faizullow: Connection between the wavefronts of the reflected and the exciting light in stimulated Mandel'shtam-Brillouin scattering, Sov. Phys. JETP 15, p.109-112 (1972)

[5.677] {Sect. 5.9.2} H.C. Barr, S.J. Berwick, P. Mason: Six-wave forward scattering of short-pulse laser light at relativistic intensities, Phys Rev Lett 81, p.2910-2913 (1998)

[5.678] {Sect. 5.9.2} I.D. Hands, S.J. Lin, S.R. Meech, D.L. Andrews: A quantum electrodynamical treatment of second harmonic generation through phase conjugate six-wave mixing: Polarization analysis, J Chem Phys 109, p.10580-10586 (1998)

[5.679] {Sect. 5.9.2} J.N. Sweetser, J.L. Durant, R. Trebino: Ultrafast spectroscopy of high-lying excited states via eight-wave mixing, Opt Commun 150, p.180-184 (1998)

[5.680] {Sect. 5.9.2} A.B. Myers, R.M. Hochstrasser: Comparison of Four-Wave Mixing Techniques for Studying Orientational Relaxation, IEEE J. QE-22, p.1482-1492 (1986)

[5.681] {Sect. 5.9.2} M. Golombok, G.A. Kenney-Wallace, S.C. Wallace: Pulsed Laser Studies of Molecular Interactions and Reorientation of CS2 in Organic Liquids via Phase Conjugation, J. Phys. Chem. 89, p.5160-5167 (1985)

[5.682] {Sect. 5.9.2} H.C. Praddaude, D.W. Scudder, B. Lax: Coherent four-wave scattering in plasmas – application to plasma diagnostics, Appl. Phys. Lett. 35, p.766-768 (1979)

[5.683] {Sect. 5.9.2} L.A. Rahn, L.J. Zych, P.L. Mattern: Background-Free CARS Studies of Carbon Monoxide in a Flame, Opt. Comm. 30, p.249-252 (1979)

[5.684] {Sect. 5.9.2} T. Yajima, H. Souma, Y. Ishida: Study of ultra-fast relaxation processes by resonant Rayleigh-type optical mixing. II. Experiment on dye solutions, Phys. Rev. A 17, p.324-334 (1978)

[5.685] {Sect. 5.9.2} R.T. Hodgson, P.P. Sorokin, J.J. Wynne: Tunable Coherent Vacuum-Ultraviolet Generation in Atomic Vapors, Phys. Rev. Lett. 32, p.343-346 (1974)

[5.686] {Sect. 5.9.3} H.M. Gibbs, S.L.McCall, T.N.C. Venkatesan, A.C. Gossard, A. Passner, W. Wiegmann: Optical bistability in semiconductors, Appl. Phys. Lett. 35, p.451-453 (1979)

[5.687] {Sect. 5.9.3} A. Kuditcher, M.P. Hehlen, C.M. Florea, K.W. Winick, S.C. Rand: Intrinsic bistability of luminescence and stimulated emission in Yb- and Tm-doped glass, Phys Rev Lett 84, p.1898-1901 (2000)

[5.688] {Sect. 5.9.3} S. Coen, M. Haelterman: Competition between modulational instability and switching in optical bistability, Optics Letters 24, p.80-82 (1999)

[5.689] {Sect. 5.9.3} S. Coen, M. Tlidi, P. Emplit, M. Haelterman: Convection versus dispersion in optical bistability, Phys Rev Lett 83, p.2328-2331 (1999)

[5.690] {Sect. 5.9.3} Z.Z. Zhuang, Y.J. Kim, J.S. Patel: Bistable twisted nematic liquid-crystal optical switch, Appl Phys Lett 75, p.3008-3010 (1999)

[5.691] {Sect. 5.9.3} Y. Hong, K.A. Shore: Observation of optical bistability in a GaAlAs semiconductor laser under intermodal injection locking, Optics Letters 23, p.1689-1691 (1998)

[5.692] {Sect. 5.9.3} X.H. Lu, Y.X. Bai, S.Q. Li, T.J. Chen: Optical bistability and beam reshaping in nonlinear multilayered structures, Opt Commun 156, p.219-226 (1998)

[5.693] {Sect. 5.9.3} L.G. Luo, R.F. Peng, P.L. Chu: Optical bistability in a passive erbium-doped fibre ring resonator, Opt Commun 156, p.275-278 (1998)

[5.694] {Sect. 5.9.3} Y.M. Golubev, M.I. Kolobov: Noiseless transfer of nonclassical light through bistable systems, Phys Rev Lett 79, p.399-402 (1997)

[5.695] {Sect. 5.9.3} K. Hane, M. Suzuki: Bistability of a self-standing film caused by photothermal displacement, Appl Opt 36, p.5006-5009 (1997)

[5.696] {Sect. 5.9.3} L.L. Li: Optical bistability in semiconductor lasers under intermodal light injection, IEEE J QE-32, p.248-256 (1996)

[5.697] {Sect. 5.9.3} M. Okada, K. Nishio: Bistability and optical switching in a polarization- bistable laser diode, IEEE J QE-32, p.1767-1776 (1996)

[5.698] {Sect. 5.9.3} J.H. Si, Y.G. Wang, J. Zhao, B.S. Zou, P.X. Ye, L. Qui, Y.Q. Shen, Z.G. Cai, J.Y. Zhou: Picosecond optical bistability in metallophthalocyanine- doped polymer film waveguides, Optics Letters 21, p.357-359 (1996)

[5.699] {Sect. 5.9.3} H.J. Eichler, A. Haase, K. Janiak, A. Kummrow, A. Wahi, A. Wappelt: Absorption bistability and nonlinearity in evaporated thin films, Opt. Comm. 88, p.298-304 (1992)

[5.700] {Sect. 5.9.3} R. Bonifacio, L.A. Lugiato: Dispersive Bistability in Homogeneously Broadened Systems, Nuovo Cimento B 53, p.311-333 (1979)

[5.701] {Sect. 5.9.3} J.G. Chen, D.Y. Li, Y. Li, Y. Lu, X.H. Zhou: Analytical expression for the hysteresis loop width of bistable tunable external cavity semiconductor lasers, Appl Opt 38, p.6333-6336 (1999)

[5.702] {Sect. 5.9.3} M.P. Hehlen, A. Kuditcher, S.C. Rand, S.R. Luthi: Siteselective, intrinsically bistable luminescence of Yb3+ ion pairs in CsCdBr3, Phys Rev Lett 82, p.3050-3053 (1999)

[5.703] {Sect. 5.9.3} I. Towers, R. Sammut, A.V. Buryak, B.A. Malomed: Soliton multistability as a result of double-resonance wave mixing in chi ((2)) media, Optics Letters 24, p.1738-1740 (1999)

[5.704] {Sect. 5.9.3} L.G. Luo, T.J. Tee, P.L. Chu: Bistability of erbium-doped fiber laser, Opt Commun 146, p.151-157 (1998)

[5.705] {Sect. 5.9.3} D.B. Shire, C.L. Tang, M.A. Parker, C. Lei, L. Hodge: Bistable operation of coupled in-plane and oxide-confined vertical-cavity laser 1xN routing switches, Appl Phys Lett 71, p.3039-3041 (1997)

[5.706] {Sect. 5.9.3} B.M. Jost: Photorefractive two-wave mixing bistability in Fe:KNbO3 without external feedback: Increasing gain bistability, Appl Phys Lett 69, p.1346-1348 (1996)

[5.707] {Sect. 5.10.1} I. B. Berlmann: Handbook of Flourescence Spectra of Aromatic Molecules (Academic Press, New York, London, 1971)

[5.708] {Sect. 5.10.1} B.R. Henry, W. Siebrand: Radiationless Transitions, in Organic Molecular Photophysics, ed. J.B. Birks, Vol. 1, Wiley, London 1973, p. 153

[5.709] {Sect. 5.10.1} H. S. Nalwa, S. Miyata: Nonlinear Optics of Organic Molecules and Polymeric Materials (Springer, Berlin, Heidelberg, New York, 1996)

[5.710] {Sect. 5.10.1} P. N. Prasad, D. Williams: Introduction to Nonlinear Optical Effects in Molecules and Polymers (John Wiley & Sons, Chichester, 1991)

[5.711] {Sect. 5.10.1} J. Saltiel, J.L. Charlton: Rearrangement in Ground and Excited States, ed. by P. DeMeyo (Academic, New York 1980) Vol. III, p.25

[5.712] {Sect. 5.10.1} J. Zyss: Molecular Nonlinear Optics (Academic Press, Boston, 1994)

[5.713] {Sect. 5.10.1} R. Menzel, K.-H. Naumann: Towards a Theoretical Description of UV-Vis Absorption Bands of Organic Molecules, Ber. Bunsenges. Phys. Chem. 95, p.834-837 (1991)

[5.714] {Sect. 5.10.1} W. Sibbett, J.R. Taylor, D. Welford: Substituent and Environmental Effects on the Picosecond Lifetimes of the Polymethine Cyanine Dyes, IEEE J. QE-17, p.500-509 (1981)

[5.715] {Sect. 5.10.1} G. Swiatkowski, R. Menzel, W. Rapp: Hindrance of the Rotational Relaxation in the Excited Singlet State of Biphenyl and Para-Terphenyl in Cooled Solutions by Methyl Substituents, J. Luminesc. 37, p.183-189 (1987)

[5.716] {Sect. 5.10.1} V. Sundström, T. Gillbro, H. Bergström: Picosecond Kinetics of Radiationless Relaxations of Triphenyl Methane Dyes. Evidence for a Rapid Excited-State Equilibrium Between States of Differing Geometry, Chem. Phys. 73, p.439-458 (1982)

[5.717] {Sect. 5.10.1} S. Reindl, A. Penzkofer: Higher excited-state photoisomerization and singlet to triplet intersystem-crossing in DODCI, Chem Phys 230, p.83-96 (1998)

[5.718] {Sect. 5.10.1} F. Gai, K.C. Hasson, J.C. McDonald, P.A. Anfinrud: Chemical dynamics in proteins: The photoisomerization of retinal in bacteriorhodopsin, Science 279, p.1886-1891 (1998)

[5.719] {Sect. 5.10.1} T. Nagele, R. Hoche, W. Zinth, J. Wachtveitl: Femtosecond photoisomerization of cis-azobenzene, Chem Phys Lett 272, p.489-495 (1997)

[5.720] {Sect. 5.10.1} N.C.R. Holme, P.S. Ramanujam, S. Hvilsted: 10,000 optical write, read, and erase cycles in an azobenzene sidechain liquid-crystalline polyester, Optics Letters 21, p.902-904 (1996)

[5.721] {Sect. 5.10.1} C. Desfrancois, H. Abdoulcarime, C.P. Schulz, J.P. Schermann: Laser separation of geometrical isomers of weakly bound molecular complexes, Science 269, p.1707-1709 (1995)

[5.722] {Sect. 5.10.1} J. Troe: Quantitative analysis of photoisomerization rates in trans-stilbene and 4-methyl-trans-stilbene, Chem. Phys. Lett. 114, p.241-247 (1985)

[5.723] {Sect. 5.10.1} F.E. Doany, E.J. Heilweil, R. Moore, R.M. Hochstrasser: Picosecond study of an intermediate in the trans to cis isomerization pathway of stiff stilbene, J. Chem. Phys. 80, p.201-206 (1984)

[5.724] {Sect. 5.10.1} B.I. Greene, T.W. Scott: Time-resolved multiphoton ionization in the organic condensed phase: picosecond conformational dynamics of cis-stilbene and tetraphenylethylene, Chem. Phys. Lett. 106, p.399-402 (1984)

[5.725] {Sect. 5.10.1} T.J. Majors, U. Even, J. Jortner: Dynamics of trans-cis photoisomerization of large molecules in supersonic jets, J. Chem. Phys. 81, p.2330-2338 (1984)

[5.726] {Sect. 5.10.1} V. Sundstrom, T. Gillbro: Dynamics of the isomerization of trans-stilbene in n-alcohols studied by ultraviolet picosecond absorption recovery, Chem. Phys. Lett. 109, p.538-543 (1984)

[5.727] {Sect. 5.10.1} J.A. Syage, P.M. Felker, A.H. Zewail: Picosecond dynamics and photoisomerization of stilbene in supersonic beams. II. Reaction rates and potential energy surface, J. Chem. Phys. 81, p.4706-4723 (1984)

[5.728] {Sect. 5.10.1} D.A. Cremers, T.L. Cremers: Picosecond Dynamics of Conformation Changes in Malachite Green Dye Produced by Photoinonization of Malachite Green Leucocyanide, Chem. Phys. Lett. 94, p.102-106 (1983)

[5.729] {Sect. 5.10.1} B.I. Greene, R.C. Farrow: Subpicosecond time resolved multiphoton ionization: Excited state dynamics of cis-stilbene under collision free conditions, J. Chem. Phys. 78, p.3336-3338 (1983)

[5.730] {Sect. 5.10.1} M. Sumitani, K. Yoshihara: Direct Observation of the Rate for Cis-Trans and Trans-Cis Photoisomerization of Stilbene with Picosecond Laser Photolysis, Bull. Chem. Soc. Japan 55, p.85-89 (1982)

[5.731] {Sect. 5.10.1} J.A. Syage, W.R. Lambert, P.M. Felker, A.H. Zewail, R.M. Hochstrasser: Picosecond excitation and trans-cis isomerization of stilbene in a supersonic jet: dynamics and spectra, Chem. Phys. Lett. 88, p.266-270 (1982)

[5.732] {Sect. 5.10.1} F.E. Doany, B.I. Greene, R.M. Hochstrasser: Excitation energy effects in the photophysics of trans-stilbene in solution, Chem. Phys. Lett. 75, p.206-208 (1980)

[5.733] {Sect. 5.10.1} B.I. Greene, R.M. Hochstrasser, R. Weisman: Picosecond dynamics of the photoisomerization of trans-stilbene under collision-free conditions, J. Chem. Phys. 71, p.544-545 (1979)

[5.734] {Sect. 5.10.1} K. Yoshihara, A. Namiki, M. Sumitani, N. Nakashima: Picosecond flash photolysis of cis- and trans-stilbene. Observation of an intense intramolecular charge-resonance transition, J. Chem. Phys. 71, p.2892-2895 (1979)

[5.735] {Sect. 5.10.1} J.B. Birks: Horizontal radiationless transitions, Chem. Phys. Lett. 54, p.430-434 (1978)

[5.736] {Sect. 5.10.1} M. Sumitani, N. Nakashima, K. Yoshihara, S. Nagakura: Temperature Dependence of fluorescence lifetimes of trans-stilbene, Chem. Phys. Lett. 51, p.183-185 (1977)

[5.737] {Sect. 5.10.1} O. Teschke, E.P. Ippen, G.R. Holtom: Picosecond dynamics of the singlet excited state of trans-and cis-stilbene, Chem. Phys. Lett. 52, p.233-235 (1977)

[5.738] {Sect. 5.10.1} F. Schael, H.G. Lohmannsroben: The deactivation of singlet excited all-trans-1,6- diphenylhexa-1,3,5-triene by intermolecular charge transfer processes. 1. Mechanisms of fluorescence quenching and of triplet and cation formation, Chem Phys 206, p.193-210 (1996)

[5.739] {Sect. 5.10.1} R.A. Marcus: Elektronentransferreaktionen in der Chemie – Theorie und Experiment (Nobel-Vortrag), Angew. Chem. 105, p.1161-1280 (1993)

[5.740] {Sect. 5.10.1} H. Lueck, M.W. Windsor, W. Rettig: Picosecond kinetic studies of charge separation in 9,9'-bianthryl as a function of solvent viscosity and comparisions with electron transfer in bacterial photosynthesis, J. Luminesc. 48 & 49, p.425-429 (1991)

[5.741] {Sect. 5.10.1} E. Gilabert, R. Lapouyade, C. Rullière: Dual fluorescence in trans-4-dimethylamino-4'-cyanostilbene revealed by picosecond time-resolved spectroscopy: A possible new "TICT" compound, Chem. Phys. Lett. 145, p.262-268 (1988)

[5.742] {Sect. 5.10.1} D. Huppert, V. Ittah, E. M. Kosower: New insights into the mechanism of fast intramolecular electron transfer, Chem. Phys. Lett. 144, p.15-23 (1988)

[5.743] {Sect. 5.10.1} K. Nakatani, T. Okada, N. Mataga, F.C. de Schryver, M. van der Auweraer: Picosecond time-resolved transient absorption spectral studies of omega- (1-pyrenyl)-alpha-N,N-dimethylaminoalkanes in acetonitrile, Chem. Phys. Lett. 145, p.81-84 (1988)

[5.744] {Sect. 5.10.1} K. Nakatani, T. Okada, N. Mataga, F.C. de Schryver: Photoinduced intramolecular electron transfer and exciplex formation of 1-(1-pyrenyl)-3-(N-skatolyl)propane in polar solvents, Chem. Phys. 121, p.87-92 (1988)

[5.745] {Sect. 5.10.1} M. Vogel, W. Rettig, R. Sens, K.H. Drexhage: Evidence for the formation of biradicaloid charge-transfer (BCT) states in xanthene and related dyes, Chem. Phys. Lett. 147, p.461-465 (1988)

[5.746] {Sect. 5.10.1} R. Hayashi, S. Tazuke: Pressure effects on the twisted in-tramolecular charge transfer (TICT) phenomenon, Chem. Phys. Lett. 135, p.123-127 (1987)

[5.747] {Sect. 5.10.1} T. Kakitani, N. Mataga: Comprehensive Study on the Role of Coordinated Solvent Mode Played in Electron-Transfer Reactions in Polar Solutions, J. Phys. Chem. 91, p.6277-6285 (1987)

[5.748] {Sect. 5.10.1} T. Kobayashi, M. Futakami, O. Kajimoto: The charge-transfer state of 4-dimethylamino-3,5-dimethylbenzonitrile studied in a free jet, Chem. Phys. Lett. 141p.450-454 (1987)

[5.749] {Sect. 5.10.1} E. Lippert, W. Rettig, V. Bonacic-Koutecky, F. Heisel, J.A. Miehé: Photophysics of internal twisting, Adv. Chem. Phys, p.76-139 (1987)

[5.750] {Sect. 5.10.1} N. Mataga, H. Shioyama, Y. Kanda: Dynamics of Charge Recombination Processes in the Singlet Electron-Transfer State of Pyrene-Pyromellitic Dianhydride Systems in Various Solvents. Picosecond Laser Photolysis Studies, J. Phys. Chem. 91, p.314-317 (1987)

[5.751] {Sect. 5.10.1} T. Ohno, A. Yoshimura, H. Shioyama, N. Mataga: Energy Gap Dependence of Spin-Inverted Electron Transfer within Geminate Radical Pairs Formed by the Quenching of Phosphorescent States in Polar Solvents, J. Phys. Chem. 91, p.4365-4370 (1987)

[5.752] {Sect. 5.10.1} T. Kakitani, N. Mataga: Different Energy Gap Laws for the Three Types of Electron-Transfer Reactions in Polar Solvents, J. Phys. Chem. 90, p.993-995 (1986)

[5.753] {Sect. 5.10.1} N. Mataga, Y. Kanda, T. Okada: Dynamics of Aromatic Hydrocarbon Cation-Tetracyanoethylene Anion Geminate Ion Pairs in Ace-tonitrile Solution with Implications to the Mechanism of the Strongly Exothermic Charge Separation Reaction in the Excited Singlet State, J. Phys. Chem. 90, p.3880-3882 (1986)

[5.754] {Sect. 5.10.1} T. Ohno, A. Yoshimura, N. Mataga: Bell-Shaped Energy Gap Dependence of Backward Electron-Transfer Rate of Geminate Radical Pairs Produced by Electron-Transfer Quenching of Ru (II) Complexes by Aro-matic Amines, J. Phys. Chem. 90, p.3295-3297 (1986)

[5.755] {Sect. 5.10.1} W. Rettig, A. Klock: Intramolecular fluorescence quenching in aminocoumarines. Identification of an excited state with full charge sep-aration, Can. J. Chem. 63, p.1649-1653 (1985)

[5.756] {Sect. 5.10.1} N. Mataga: Photochemical charge transfer phenomena – picosecond laser photolysis studies, Pure & Appl. Chem. 56, p.1255-1268 (1984)

[5.757] {Sect. 5.10.1} Y. Wang, M. McAuliffe, K.B. Eisenthal: Picosecond Dynamics of Twisted Internal Charge-Transfer Phenomena, J. Phys. Chem. 85, p.3736-3739 (1981)

[5.758] {Sect. 5.10.1} W. Rapp: Classical treatment on intramolecular twisting relaxations of dissolved molecules, Chem. Phys. Lett. 27, p.187-190 (1974)

[5.759] {Sect. 5.10.1} R.A. Marcus: On the Theory of Electron-Transfer Reactions. VI. Unified Treatment for Homogeneous and Electrode Reactions, J. Chem. Phys. 43, p.679-701 (1965)

[5.760] {Sect. 5.10.1} R.A. Marcus: Chemical and electrochemical electron-transfer theory, Annu. Rev. Phys. Chem. 15, p.155-196 (1964)

[5.761] {Sect. 5.10.1} R.A. Marcus: On the Theory of Oxidation-Reduction Reac-tions Involving Electron Transfer. I, J. Chem. Phys. 24, p.966-978 (1956)

[5.762] {Sect. 5.10.1} J.R. Bolton, N. Mataga, Mc. Lendon (ed.): Electron Transfer in Inorganic, Organic, and Biological Systems, Adv. in Chem. Ser. 228 (Am Chem. Soc.1991)

[5.763] {Sect. 5.10.1} M.A. Fox, M. Channon (ed.): Photoinduced Electron Transfer, Part A-D (Elsevier 1988)

[5.764] {Sect. 5.10.1} M.N. Slyadnev, T. Inoue, A. Harata, T. Ogawa: A rhodamine and a cyanine dye on the water surface as studied by laser induced fluorescence microscopy, Colloid Surface A 164, p.155-162 (2000)

[5.765] {Sect. 5.10.1} A. Imhof, M. Megens, J.J. Engelberts, D.T.N. deLang, R. Sprik, W.L. Vos: Spectroscopy of fluorescein (FITC) dyed colloidal silica spheres, J Phys Chem B 103, p.1408-1415 (1999)

[5.766] {Sect. 5.10.1} K. Kitaoka, J. Si, T. Mitsuyu, K. Hirao: Optical poling of azo-dye-doped thin films using an ultrashort pulse laser, Appl Phys Lett 75, p.157-159 (1999)

[5.767] {Sect. 5.10.1} C.S. Wang, H.S. Fei, Y.Q. Yang, Z.Q. Wei, Y. Qiu, Y.M. Chen: Photoinduced anisotropy and polarization holography in azobenzene side-chain polymer, Opt Commun 159, p.58-62 (1999)

[5.768] {Sect. 5.10.1} S. Walheim, E. Schaffer, J. Mlynek, U. Steiner: Nanophase-separated polymer films as high-performance antireflection coatings, Science 283, p.520-522 (1999)

[5.769] {Sect. 5.10.1} L.M. Blinov, G. Cipparrone, S.P. Palto: Phase grating recording on photosensitive Langmuir-Blodgett films, J Nonlinear Opt Physics Mat 7, p.369-383 (1998)

[5.770] {Sect. 5.10.1} D.J. Welker, J. Tostenrude, D.W. Garvey, B.K. Canfield, M.G. Kuzyk: Fabrication and characterization of single-mode electro-optic polymer optical fiber, Optics Letters 23, p.1826-1828 (1998)

[5.771] {Sect. 5.10.1} K.T. Weitzel, U.P. Wild, V.N. Mikhailov, V.N. Krylov: Hologram recording in DuPont photopolymer films by use of pulse exposure, Optics Letters 22, p.1899-1901 (1997)

[5.772] {Sect. 5.10.1} D. Gu, Q. Chen, X. Tang, F. Gan, S. Shen, K. Liu, H. Xu: Application of phtalocyanine thin films in optical recording, Opt. Comm. 121, p.125-129 (1995)

[5.773] {Sect. 5.10.1} J.R. Kulisch, H. Franke, R. Irmscher, Ch. Buchal: Opto-optical switching in ion-implanted poly (methyl methacrylate)-waveguides, J. Appl. Phys. 71, p.3123-3126 (1992)

[5.774] {Sect. 5.10.1} E. Gross, B. Ehrenberg: The partition and distribution of porphyrins in liposomal membranes. A spectroscopic study, Biochim. Biophys. Acta 983, p.118-122 (1989)

[5.775] {Sect. 5.10.1} V. Tsukanova, A. Harata, T. Ogawa: Orientational arrangement of long-chain fluorescein molecules within the monolayer at the air/water interface studied by the SHG technique, Langmuir 16, p.1167-1171 (2000)

[5.776] {Sect. 5.10.1} L. Xu, Z.J. Hou, L.Y. Liu, Z.L. Xu, W.C. Wang, F.M. Li, M.X. Ye: Optical nonlinearity and structural phase-transition observation of organic dye-doped polymer-silica hybrid material, Optics Letters 24, p.1364-1366 (1999)

[5.777] {Sect. 5.10.1} R.J. Kruhlak, M.G. Kuzyk: Side-illumination fluorescence spectroscopy. I. Principles, J Opt Soc Am B Opt Physics 16, p.1749-1755 (1999)

[5.778] {Sect. 5.10.1} R.J. Kruhlak, M.G. Kuzyk: Side-illumination fluorescence spectroscopy. II. Applications to squaraine-dye-doped polymer optical fibers, J Opt Soc Am B Opt Physics 16, p.1756-1767 (1999)

[5.779] {Sect. 5.10.1} Y. Takeoka, A.N. Berker, R. Du, T. Enoki, A. Grosberg, M. Kardar, T. Oya, K. Tanaka, G.Q. Wang, X.H. Yu et al.: First order phase transition and evidence for frustrations in polyampholytic gels, Phys Rev Lett 82, p.4863-4865 (1999)

[5.780] {Sect. 5.10.1} A. Hoischen, H.S. Kitzerow, K. Kurschner, P. Strohriegl: Optical storage effect due to photopolymerization of mesogenic twin molecules, J Appl Phys 87, p.2105-2109 (2000)

[5.781] {Sect. 5.10.1} C.J. Brabec, F. Padinger, N.S. Sariciftci, J.C. Hummelen: Photovoltaic properties of conjugated polymer/methanofullerene composites embedded in a polystyrene matrix, J Appl Phys 85, p.6866-6872 (1999)

[5.782] {Sect. 5.10.1} A.Y.G. Fuh, M.S. Tsai, L.J. Huang, T.C. Liu: Optically switchable gratings based on polymer-dispersed liquid crystal films doped with a guest-host dye, Appl Phys Lett 74, p.2572-2574 (1999)

[5.783] {Sect. 5.10.1} W. Holzer, M. Pichlmaier, E. Drotleff, A. Penzkofer, D.D.C. Bradley, W.J. Blau: Optical constants measurement of luminescent polymer films, Opt Commun 163, p.24-28 (1999)

[5.784] {Sect. 5.10.1} O.V. Khodykin, S.J. Zilker, D. Haarer, B.M. Kharlamov: Zinc-tetrabenzoporphyrine-doped poly (Methyl methacrylate): a new photochromic recording medium, Optics Letters 24, p.513-515 (1999)

[5.785] {Sect. 5.10.1} J.S. Kim, R.H. Friend, F. Cacialli: Improved operational stability of polyfluorene-based organic light- emitting diodes with plasma-treated indium-tin-oxide anodes, Appl Phys Lett 74, p.3084-3086 (1999)

[5.786] {Sect. 5.10.1} H. Murata, C.D. Merritt, H. Inada, Y. Shirota, Z.H. Kafafi: Molecular organic light-emitting diodes with temperature-independent quantum efficiency and improved thermal durability, Appl Phys Lett 75, p.3252-3254 (1999)

[5.787] {Sect. 5.10.1} S. Pelissier, D. Blanc, M.P. Andrews, S.I. Najafi, A.V. Tishchenko, O. Parriaux: Single-step UV recording of sinusoidal surface gratings in hybrid solgel glasses, Appl Opt 38, p.6744-6748 (1999)

[5.788] {Sect. 5.10.1} G. Rojo, G. delaTorre, J. GarciaRuiz, I. Ledoux, T. Torres, J. Zyss, F. AgulloLopez: Novel unsymmetrically substituted push-pull phthalocyanines for second-order nonlinear optics, Chem Phys 245, p.27-34 (1999)

[5.789] {Sect. 5.10.1} M.G. Schnoes, L. Dhar, M.L. Schilling, S.S. Patel, P. Wiltzius: Photopolymer-filled nanoporous glass as a dimensionally stable holographic recording medium, Optics Letters 24, p.658-660 (1999)

[5.790] {Sect. 5.10.1} A. Shukla, S. Mazumdar: Designing emissive conjugated polymers with small optical gaps: A step towards organic polymeric infrared lasers, Phys Rev Lett 83, p.3944-3947 (1999)

[5.791] {Sect. 5.10.1} W.L. Yu, Y. Cao, J.A. Pei, W. Huang, A.J. Heeger: Blue polymer light-emitting diodes from poly (9,9-dihexylfluorene-alt-co-2,5-didecyloxy-para-phenylene), Appl Phys Lett 75, p.3270-3272 (1999)

[5.792] {Sect. 5.10.1} P.K.H. Ho, D.S. Thomas, R.H. Friend, N. Tessler: All-polymer optoelectronic devices, Science 285, p.233-236 (1999)

[5.793] {Sect. 5.10.1} S. Walheim, E. Schaffer, J. Mlynek, U. Steiner: Nanophase-separated polymer films as high-performance antireflection coatings, Science 283, p.520-522 (1999)

[5.794] {Sect. 5.10.1} L.L. Hu, Z.H. Jiang: Laser action in rhodamine 6G doped titania-containing ormosils, Opt Commun 148, p.275-280 (1998)

[5.795] {Sect. 5.10.1} H. Kietzmann, R. Rochow, G. Gantefor, W. Eberhardt, K. Vietze, G. Seifert, P.W. Fowler: Electronic structure of small fullerenes: Evidence for the high stability of C-32, Phys Rev Lett 81, p.5378-5381 (1998)

[5.796] {Sect. 5.10.1} S.K. Lam, D. Lo: Delayed luminescence spectroscopy and optical phase conjugation in eosin Y-doped sol-gel silica glasses, Chem Phys Lett 297, p.329-334 (1998)

[5.797] {Sect. 5.10.1} E.I. Maltsev, D.A. Lypenko, B.I. Shapiro, M.A. Brusentseva, V.I. Berendyaev, B.V. Kotov, A.V. Vannikov: J-aggregate electroluminescence in dye doped polymer layers, Appl Phys Lett 73, p.3641-3643 (1998)

[5.798] {Sect. 5.10.1} J.H. Si, T. Mitsuyu, P.X. Ye, Z. Li, Y.Q. Shen, K. Hirao: Optical storage in an azobenzene-polyimide film with high glass transition temperature, Opt Commun 147, p.313-316 (1998)

[5.799] {Sect. 5.10.1} K. Kandasamy, P.N. Puntambekar, B.P. Singh, S.J. Shetty, T.S. Srivastava: Resonant nonlinear optical studies on porphyrin derivatives, J Nonlinear Opt Physics Mat 6, p.361-375 (1997)

[5.800] {Sect. 5.10.1} X.A. Long, A. Malinowski, D.D.C. Bradley, M. Inbasekaran, E.P. Woo: Emission processes in conjugated polymer solutions and thin films, Chem Phys Lett 272, p.6-12 (1997)

[5.801] {Sect. 5.10.1} E.S. Maniloff, D. Vacar, D.W. Mcbranch, H.L. Wang, B.R. Mattes, J. Gao, A.J. Heeger: Ultrafast holography using charge-transfer polymers, Opt Commun 141, p.243-246 (1997)

[5.802] {Sect. 5.10.1} S. Ozcelik, D.L. Akins: Extremely low excitation threshold, superradiant, molecular aggregate lasing system, Appl Phys Lett 71, p.3057-3059 (1997)

[5.803] {Sect. 5.10.1} M. Ahlheim, M. Barzoukas, P.V. Bedworth, M. Blanchard-desce, A. Fort, Z.Y. Hu, S.R. Marder, J.W. Perry, C. Runser, M. Staehelin, et al.: Chromophores with strong heterocyclic accepters: A poled polymer with a large electro-optic coefficient, Science 271, p.335-337 (1996)

[5.804] {Sect. 5.10.1} F. Hide, M.A. Diazgarcia, B.J. Schwartz, M.R. Andersson, Q.B. Pei, A.J. Heeger: Semiconducting polymers: A new class of solid-state laser materials, Science 273, p.1833-1836 (1996)

[5.805] {Sect. 5.10.1} H.S. Fei, Z.Q. Wei, Q.G. Yang, Y.L. Che, Y.Q. Shen, X.F. Fu, L. Qiu: Low power phase conjugation in push pull azobenzene compounds, Optics Letters 20, p.1518-1520 (1995)

[5.806] {Sect. 5.10.1} Y.C. Liu, H.Y. Wang, M.Z. Tian, Y.L. Lin, X.G. Kong, S.H. Huang, J.Q. Yu: Multiple-hologram storage for thin layers of Methyl Orange dyes in polyvinyl alcohol matrices, Optics Letters 20, p.1495-1497 (1995)

[5.807] {Sect. 5.10.1} Y.H. Zhang, Q.W. Song, C. Tseronis, R.R. Birge: Real-time holographic imaging with a bacteriorhodopsin film, Optics Letters 20, p.2429-2431 (1995)

[5.808] {Sect. 5.10.1} F.E. Doany, E.J. Heilweil, R. Moore, R.M. Hochstrasser: Picosecond study of an intermediate in the trans to cis isomerization pathway of stiff stilbene, J. Chem. Phys. 80, p.201-206 (1984)

[5.809] {Sect. 5.10.1} Y. Maeda, T. Okada, N. Mataga: Photoinduced Trans-Cis Isomerization and Intramolecular-Charge-Transfer Interaction. Photochemistry and Picosecond Laser Spectroscopy of 4-Substituted beta-(1-Pyrenyl)styrenes, J. Phys. Chem. 88, p.2714-2718 (1984)

[5.810] {Sect. 5.10.1} V. Sundström, T. Gillbro: Dynamics of the isomerization of trans-stilbene in n-alcohols studied by ultraviolet picosecond absorption recovery, Chem. Phys. Lett. 109, p.538-543 (1984)

[5.811] {Sect. 5.10.1} A. Amirav, J. Jortner: Dynamics of trans-cis isomerization of stilbene in supersonic jets, Chem. Phys. Lett. 95, p.295-300 (1983)

[5.812] {Sect. 5.10.1} H. Görner, D. Schult-Frohlinde: Trans-cis photoisomerization of the quaternary iodides of 4-cyano- and 4-nitro-4'-azastilbene in ethanol solution: Singlet versus triplet mechanism, Chem. Phys. Lett. 101, p.79-85 (1983)

[5.813] {Sect. 5.10.1} K.S. Schanze, T. Fleming Mattox, D.G. Whitten: Solvent Effects upon the Thermal Cis-Trans Isomerization and Charge-Transfer Absorption of 4- (Diethylamino)-4'-nitroazobenzene, J. Org. Chem. 48, p.2808-2813 (1983)

[5.814] {Sect. 5.10.1} G. Bartocci, F. Masetti, U. Mazzucato, S. Dellonte, G. Orlandi: Photophysical study of rotational isomers of mono-aza- and di-aza-stilbenes, Spectrochimica Acta 38A, p.729-735 (1982)

[5.815] {Sect. 5.10.1} M. Sumitani, K. Yoshihara: Photochemistry of the lowest excited singlet state: Acceleration of trans-cis isomerization by two consecutive picosecond pulses, J. Chem. Phys. 76, p.738-740 (1982)

[5.816] {Sect. 5.10.1} St.P. Velsko, G.R. Fleming: Solvent influence on photochemical isomerizations: Photophysics of DODCI, Chem. Phys. 65, p.59-70 (1982)

[5.817] {Sect. 5.10.1} J. Saltiel, D.W. Eaker: Lifetime and geometry of 1-phenyl-2- (2-naphthyl)ethene triplets. Evidence against the triplet mechanism for direct photoisomerization, Chem. Phys. Lett. 75, p.209-213 (1980)

[5.818] {Sect. 5.10.1} T. Kobayashi, S. Nagakura: The rates of internal conversion and photoisomerization of some carbocanine dyes as revealed from picosecond time-resolved spectroscopy, Chem. Phys. 23, p.153-158 (1977)

[5.819] {Sect. 5.10.1} S. Völker, J.H. van der Waals: Laser-induced photochemical isomerization of free base porphyrin in an n-octane crystal at 4.2 K, Mol. Phys. 32, p.1703-1718 (1976)

[5.820] {Sect. 5.10.1} M. Sumitani, S. Nagakura, K. Yoshihara: Laser photolysis study of trans-cis photoisomerization of trans-1-phenyl-2-(2-naphthyl)ethylene, Chem. Phys. Lett. 29, p.410-413 (1974)

[5.821] {Sect. 5.10.1} E.G. Arthurs, D.J. Bradley, A.G. Roddie: Picosecond measurements of 3,3'-diethyloxadicarbocyanine iodide and photoisomer fluorescence, Chem. Phys. Lett. 22, p.230-234 (1973)

[5.822] {Sect. 5.10.1} N.G. Basov, A.M. Prokhorov: Possible Methods of Obtaining Active Molecules for a Molecular Oscillator, Sov. Phys. JETP 1, p.184-185 (1955)

[5.823] {Sect. 5.10.1} C. Former, H. Wagner, R. Richert, D. Neher, K. Mullen: Orientation and dynamics of chainlike dipole arrays: Donor-acceptor-substituted oligophenylenevinylenes in a polymer matrix, Macromolecules 32, p.8551-8559 (1999)

[5.824] {Sect. 5.10.1} R. Hildebrandt, H.M. Keller, G. Marowsky, W. Brutting, T. Fehn, M. Schwoerer, J.E. Sipe: Electric-field-induced optical second-harmonic generation in poly (Phenylene vinylene) light-emitting diodes, Chem Phys 245, p.341-344 (1999)

[5.825] {Sect. 5.10.1} E.I. Maltsev, D.A. Lypenko, B.I. Shapiro, M.A. Brusentseva, G.H.W. Milburn, J. Wright, A. Hendriksen, V.I. Berendyaev, B.V. Kotov, A.V. Vannikov: Electroluminescence of polymer/J-aggregate composites, Appl Phys Lett 75, p.1896-1898 (1999)

[5.826] {Sect. 5.10.1} D.J. Pinner, R.H. Friend, N. Tessler: Transient electroluminescence of polymer light emitting diodes using electrical pulses, J Appl Phys 86, p.5116-5130 (1999)

[5.827] {Sect. 5.10.1} Y.Z. Wang, R.G. Sun, F. Meghdadi, G. Leising, A.J. Epstein: Multicolor multilayer light-emitting devices based on pyridine-containing conjugated polymers and para-sexiphenyl oligomer, Appl Phys Lett 74, p.3613-3615 (1999)

[5.828] {Sect. 5.10.1} A. Yamamori, C. Adachi, T. Koyama, Y. Taniguchi: Electroluminescence of organic light emitting diodes with a thick hole transport layer composed of a triphenylamine based polymer doped with an antimonium compound, J Appl Phys 86, p.4369-4376 (1999)

[5.829] {Sect. 5.10.1} R.H. Friend, R.W. Gymer, A.B. Holmes, J.H. Burroughes, R.N. Marks, C. Taliani, D.D.C. Bradley, D.A. Dos Santos, J.L. Brédas,

M. Lögdlund, W.R. Salaneck. Electroluminescence in conjugated polymers, Nature 397, p.121-128 (1999)

[5.830] {Sect. 5.10.1} V. Bulovic, A. Shoustikov, M.A. Baldo, E. Bose, V.G. Kozlov, M.E. Thompson, S.R. Forrest: Bright, saturated, red-to-yellow organic light-emitting devices based on polarization-induced spectral shifts, Chem Phys Lett 287, p.455-460 (1998)

[5.831] {Sect. 5.10.1} A. Kraft, A.C. Grimsdale, A.B. Holmes: Electroluminescent Conjugated Polymers – Seeing Polymers in a New Light, Angew. Chem. Int. Ed. 37, p.402-428 (1998)

[5.832] {Sect. 5.10.1} H. Sirringhaus, N. Tessler, R.H. Friend: Integrated opto-electronic devices based on conjugated polymers, Science 280, p.1741-1744 (1998)

[5.833] {Sect. 5.10.1} G.H. Gelinck, J.M. Warman, M. Remmers, D. Neher: Narrow-band emissions from conjugated-polymer films, Chem Phys Lett 265, p.320-326 (1997)

[5.834] {Sect. 5.10.1} Q.B. Pei, G. Yu, C. Zhang, Y. Yang, A.J. Heeger: Polymer light-emitting electrochemical cells, Science 269, p.1086-1088 (1995)

[5.835] {Sect. 5.10.1} U. Lemmer, R.F. Mahrt, Y. Wada, A. Greiner, H. Bässler, E.O. Göbel: Time resolved luminescence study of recombination processes in electroluminescent polymers, Appl. Phys. Lett. 62, p.2827-2829 (1993)

[5.836] {Sect. 5.10.1} T. Renger, V. May: Multiple exciton effects in molecular aggregates: Application to a photosynthetic antenna complex, Phys Rev Lett 78, p.3406-3409 (1997)

[5.837] {Sect. 5.10.1} S. Creighton, J.-K. Hwang, A. Warshel, W.W. Parson, J. Norris: Simulating the Dynamics of the Primary Charge Separation Process in Bacterial Photosynthesis, Biochem. 27, p.774-781 (1988)

[5.838] {Sect. 5.10.1} A. Ogrodnik, N. Remy-Richter, M.E. Michel-Beyerle, R. Feick: Observation of activationless recombination in reaction centers of R. sphaeroides. A new key to the primary electron-transfer mechanism, Chem. Phys. Lett. 135, p.576-581 (1987)

[5.839] {Sect. 5.10.1} A.W. Rutherford, P. Heathcote: Primary photochemistry in photosystem-I, Photosynthesis Research 6, p.295-316 (1985)

[5.840] {Sect. 5.10.3} P. Yeh, C. Gu: Photorefractive Materials, Effects, and Applications (SPIE Press, 1994)

[5.841] {Sect. 5.10.3} P. Yeh, C. Gu: Landmark Papers on Photorefractive Nonlinear Optics (World Scientific, Singapore, 1995)

[5.842] {Sect. 5.10.3} P. Bernasconi, G. Montemezzani, M. Wintermantel, I. Biaggio, P. Gunter: High-resolution, high-speed photorefractive incoherent-to-coherent optical converter, Optics Letters 24, p.199-201 (1999)

[5.843] {Sect. 5.10.3} D. Day, M. Gu: Use of two-photon excitation for erasable-rewritable three-dimensional bit optical data storage in a photorefractive polymer, Optics Letters 24, p.948-950 (1999)

[5.844] {Sect. 5.10.3} J. Imbrock, S. Wevering, K. Buse, E. Kratzig: Nonvolatile holographic storage in photorefractive lithium tantalate crystals with laser pulses, J Opt Soc Am B Opt Physics 16, p.1392-1397 (1999)

[5.845] {Sect. 5.10.3} T. Nikolajsen, P.M. Johansen: Low-temperature thermal fixing of holograms in photorefractive La3Ga5SiO14 : Pr3+ crystal, Optics Letters 24, p.1419-1421 (1999)

[5.846] {Sect. 5.10.3} X.N. Shen, J.H. Zhao, X.L. Lu, Q.Z. Jiang, J.W. Zhang, H.R. Xia, L.H. Song, S.J. Zhang, J.R. Han, H.C. Chen: Photorefractive properties of Cu-doped (K0.5Na0.5) (0.2) (Sr0.75Ba0.25) (0.9)Nb2O6 crystals with different doping levels and different dimensions, J Appl Phys 86, p.3371-3376 (1999)

[5.847] {Sect. 5.10.3} E. Soergel, W. Krieger: Profiles of light-induced charge gratings on photorefractive crystals, Phys Rev Lett 83, p.2336-2339 (1999)

[5.848] {Sect. 5.10.3} J. Wolff, S. Schloter, U. Hofmann, D. Haarer, S.J. Zilker: Speed enhancement of photorefractive polymers by means of light-induced filling of trapping states, J Opt Soc Am B Opt Physics 16, p.1080-1086 (1999)

[5.849] {Sect. 5.10.3} A. ApolinarIribe, N. Korneev, J.J. SanchezMondragon: Beam amplification resulting from non-Bragg wave mixing in photorefractive strontium barium niobate, Optics Letters 23, p.1877-1879 (1998)

[5.850] {Sect. 5.10.3} T. Nikolajsen, P.M. Johansen, E. Dubovik, T. Batirov, R. Djalalov: Photorefractive two-step recording in a piezoelectric La3Ga5SiO14 crystal doped with praseodymium, Optics Letters 23, p.1164-1166 (1998)

[5.851] {Sect. 5.10.3} B. Pesach, E. Refaeli, A.J. Agranat: Investigation of the holographic storage capacity of paraelectric K1- xLixTa1-yNbyO3:Cu,V, Optics Letters 23, p.642-644 (1998)

[5.852] {Sect. 5.10.3} X.N. Shen, T.H. Zhao, R.B. Wang, P.C. Yeh, S.J. Zhang, H.C. Chen: Photorefractive properties of Cu-doped KNSBN crystal with fluorine replacing oxygen, Optics Letters 23, p.1253-1255 (1998)

[5.853] {Sect. 5.10.3} A. Brignon, D. Geffroy, J.P. Huignard, M.H. Garrett, I. Mnushkina: Experimental investigations of the photorefractive properties of rhodium-doped BaTiO3 at 1.06 mu m, Opt Commun 137, p.311-316 (1997)

[5.854] {Sect. 5.10.3} J. Neumann, S. Odoulov: Parametric amplification of a coherent light wave in photorefractive BaTiO3 by a single pump beam, Optics Letters 22, p.1858-1860 (1997)

[5.855] {Sect. 5.10.3} P.M. Lundquist, R. Wortmann, C. Geletneky, R.J. Twieg, M. Jurich, V.Y. Lee, C.R. Moylan, D.M. Burland: Organic glasses: A new class of photorefractive materials, Science 274, p.1182-1185 (1996)

[5.856] {Sect. 5.10.3} M. Taya, M.C. Bashaw, M.M. Fejer: Photorefractive effects in periodically poled ferroelectrics, Opt. Lett. 21, p.857-859 (1996)

[5.857] {Sect. 5.10.3} A.A. Kamshilin, V.V. Prokofiev, T. Jaaskelainen: Beam Fanning and Double Phase Conjugation in a Fiber-Like Photorefractive Sample, IEEE J. QE-31, p.1642-1647 (1995)

[5.858] {Sect. 5.10.3} F. Laeri, R. Jungen, G. Angelow, U. Vietze, T. Engel, M. Würtz, D. Hilgenberg: Photorefraction in the ultraviolet: Materials and effects, Appl. Phys. B. 61, p.351-360 (1995)

[5.859] {Sect. 5.10.3} D. Psaltis, F. Mok, H.-Y. S. Li: Nonvolatile storage in photorefractive crystals, Opt. Lett. 19, p.210-212 (1994)

[5.860] {Sect. 5.10.3} J. Feinberg, D. Heiman, A.R. Tanguay,Jr, R.W. Hellwarth: Photorefractive effects and light-induced charge migration in barium titanate, J. Appl. Phys. 51, p.1297-1305 (1980)

[5.861] {Sect. 5.10.3} A.M. Glass: The Photorefractive Effect, Opt. Eng. 17, p.470-479 (1978)

[5.862] {Sect. 5.10.3} Y. Kawata, H. Ishitobi, S. Kawata: Use of two-photon absorption in a photorefractive crystal for three- dimensional optical memory, Optics Letters 23, p.756-758 (1998)

[5.863] {Sect. 5.10.3} K. Meerholz, Y. DeNardin, R. Bittner, R. Wortmann, F. Wurthner: Improved performance of photorefractive polymers based on merocyanine dyes in a polar matrix, Appl Phys Lett 73, p.4-6 (1998)

[5.864] {Sect. 5.10.3} T. Nikolajsen, P.M. Johansen, X. Yue, D. Kip, E. Kratzig: Two-step two-color recording in a photorefractive praseodymium-doped La3Ga5SiO14 crystal, Appl Phys Lett 74, p.4037-4039 (1999)

[5.865] {Sect. 5.10.3} A. Liu, M.K. Lee, L. Hesselink, S.H. Lee, K.S. Lim: Light-induced absorption of cerium-doped lead barium niobate crystals, Optics Letters 23, p.1618-1620 (1998)

[5.866] {Sect. 5.10.3} V.A. Kalinin, K. Shcherbin, L. Solymar, J. Takacs, D.J. Webb: Resonant two-wave mixing in photorefractive materials with the aid of dc and ac fields, Optics Letters 22, p.1852-1854 (1997)

[5.867] {Sect. 5.10.3} H. Ueki, Y. Kawata, S. Kawata: Three-dimensional optical bit-memory recording and reading with a photorefractive crystal: Analysis and experiment, Appl Opt 35, p.2457-2465 (1996)

[5.868] {Sect. 5.10.3} W.L. She, Z.X. Yu, H.W. Ho, H. Chan, W.K. Lee: Control of self-pumped phase conjugate reflectivity in a photorefractive crystal by another laser beam, Opt Commun 139, p.77-80 (1997)

[5.869] {Sect. 5.10.3} H. Guenther, G. Wittmann, R.M. Macfarlane, R.R. Neurgaonkar: Intensity dependence and white-light gating of two-color photorefractive gratings in LiNbO3, Optics Letters 22, p.1305-1307 (1997)

[5.870] {Sect. 5.10.3} A. Grunnetjepsen, C.L. Thompson, W.E. Moerner: Spontaneous oscillation and self-pumped phase conjugation in a photorefractive polymer optical amplifier, Science 277, p.549-552 (1997)

[5.871] {Sect. 5.10.4} J. Shah: Ultrafast Spectroscopy of Semiconductors and Semiconductor Nanostructures (Springer, Berlin, Heidelberg, New York, 1996)

[5.872] {Sect. 5.10.4} S. Kakimoto, H. Watanabe: Intervalence band absorption loss coefficients of the active layer for InP-based long wavelength laser diodes, J Appl Phys 87, p.2095-2097 (2000)

[5.873] {Sect. 5.10.4} T. Verbiest, S. VanElshocht, M. Kauranen, L. Hellemans, J. Snauwaert, C. Nuckolls, T.J. Katz, A. Persoons: Strong enhancement of nonlinear optical properties through supramolecular chirality, Science 282, p.913-915 (1998)

[5.874] {Sect. 5.10.4} D.A.B. Miller, C.T. Seaton, M.E. Prise, S.D. Smith: Band-Gap-Resonant Nonlinear Refraction in III-V Semiconductors, Phys. Rev. Lett. 47, p.197-200 (1981)

[5.875] {Sect. 5.10.4} F.J.P. Schuurmans, M. Megens, D. Vanmaekelbergh, A. Lagendijk: Light scattering near the localization transition in macroporous GaP networks, Phys Rev Lett 83, p.2183-2186 (1999)

[5.876] {Sect. 5.10.5} G.B. Serapiglia, E. Paspalakis, C. Sirtori, K.L. Vodopyanov, C.C. Phillips: Laser-induced quantum coherence in a semiconductor quantum well, Phys Rev Lett 84, p.1019-1022 (2000)

[5.877] {Sect. 5.10.5} M. Kira, F. Jahnke, S.W. Koch: Quantum theory of secondary emission in optically excited semiconductor quantum wells, Phys Rev Lett 82, p.3544-3547 (1999)

[5.878] {Sect. 5.10.5} J. Schmitt, P. Mächtle, D. Eck, H. Möhwald, C. A. Helm: Preparation and Optical Properties of Colloidal Gold Monolayers, Langmuir 15, p.3256-3266 (1999)

[5.879] {Sect. 5.10.5} D. Birkedal, J. Shah: Femtosecond spectral interferometry of resonant secondary emission from quantum wells: Resonance Rayleigh scattering in the nonergodic regime, Phys Rev Lett 81, p.2372-2375 (1998)

[5.880] {Sect. 5.10.5} D.H. Lowndes, D.B. Geohegan, A.A. Puretzky, D.P. Norton, C.M. Rouleau: Synthesis of novel thin-film materials by pulsed laser deposition, Science 273, p.898-903 (1996)

[5.881] {Sect. 5.10.5} S.V. Gaponenko, U. Woggon, A. Uhrig, W. Langbein, C. Klingshirn: Narrow-band spectral hole burning in quantum dots, J. Luminesc. 60 & 61, p.302-307 (1994)

[5.882] {Sect. 5.10.5} C. A. Foss, Jr, G. L. Hornyak, J. A. Stockert, Ch. R. Martin: Optically Transparent Nanometal Composite Membranes, Adv. Mater. 5, p.135-137 (1993)

[5.883] {Sect. 5.10.5} S. DasSarma, D.W. Wang: Many-body renormalization of semiconductor quantum wire excitons: Absorption, gain, binding, and unbinding, Phys Rev Lett 84, p.2010-2013 (2000)

[5.884] {Sect. 5.10.5} O. Mauritz, G. Goldoni, F. Rossi, E. Molinari: Local optical spectroscopy in quantum confined systems: A theoretical description, Phys Rev Lett 82, p.847-850 (1999)

[5.885] {Sect. 5.10.5} T.A. Smith, J. Hotta, K. Sasaki, H. Masuhara, Y. Itoh: Photon pressure-induced association of nanometer-sized polymer chains in solution, J Phys Chem B 103, p.1660-1663 (1999)

[5.886] {Sect. 5.10.5} F. Tassone, C. Piermarocchi: Electron-hole correlation effects in the emission of light from quantum wires, Phys Rev Lett 82, p.843-846 (1999)

[5.887] {Sect. 5.10.5} J.H. Golden, F.J. Disalvo, J.M.J. Frechet, J. Silcox, M. Thomas, J. Elman: Subnanometer-diameter wires isolated in a polymer matrix by fast polymerization, Science 273, p.782-784 (1996)

[5.888] {Sect. 5.10.5} J.P. Zhang, D.Y. Chu, S.L. Wu, S.T. Ho, W.G. Bi, C.W. Tu, R.C. Tiberio: Photonic-wire laser, Phys Rev Lett 75, p.2678-2681 (1995)

[5.889] {Sect. 5.10.5} C. A. Foss, Jr, G. L. Hornyak, J. A. Stockert, Ch. R. Martin: Optical Properties of Composite Membranes Containing Arrays of Nanoscopic Gold Cylinders, J. Phys. Chem. 96, p.7497-7499 (1992)

[5.890] {Sect. 5.10.5} M.V. Artemyev, U. Woggon: Quantum dots in photonic dots, Appl Phys Lett 76, p.1353-1355 (2000)

[5.891] {Sect. 5.10.5} T. Brunhes, P. Boucaud, S. Sauvage, A. Lemaitre, J.M. Gerard, F. Glotin, R. Prazeres, J.M. Ortega: Infrared second-order optical susceptibility in InAs/GaAs self-assembled quantum dots, Phys Rev B 61, p.5562-5570 (2000)

[5.892] {Sect. 5.10.5} M.Y. Gao, C. Lesser, S. Kirstein, H. Mohwald, A.L. Rogach, H. Weller: Electroluminescence of different colors from polycation/CdTe nanocrystal self-assembled films, J Appl Phys 87, p.2297-2302 (2000)

[5.893] {Sect. 5.10.5} T. Makimura, T. Mizuta, K. Murakami: Formation dynamics of silicon nanoparticles after laser ablation studied using plasma emission caused by second-laser decomposition, Appl Phys Lett 76, p.1401-1403 (2000)

[5.894] {Sect. 5.10.5} N. Suzuki, T. Makino, Y. Yamada, T. Yoshida, S. Onari: Structures and optical properties of silicon nanocrystallites prepared by pulsed-laser ablation in inert background gas, Appl Phys Lett 76, p.1389-1391 (2000)

[5.895] {Sect. 5.10.5} M. Ajgaonkar, Y. Zhang, H. Grebel, C.W. White: Nonlinear optical properties of a coherent array of submicron SiO_2 spheres (Opal) embedded with Si nanoparticles, Appl Phys Lett 75, p.1532-1534 (1999)

[5.896] {Sect. 5.10.5} J. Bosbach, D. Martin, F. Stietz, T. Wenzel, F. Trager: Laser-based method for fabricating monodisperse metallic nanoparticles, Appl Phys Lett 74, p.2605-2607 (1999)

[5.897] {Sect. 5.10.5} B. Damilano, N. Grandjean, F. Semond, J. Massies, M. Leroux: From visible to white light emission by GaN quantum dots on Si (111) substrate, Appl Phys Lett 75, p.962-964 (1999)

[5.898] {Sect. 5.10.5} W. Kim, V.P. Safonov, V.M. Shalaev, R.L. Armstrong: Fractals in microcavities: Giant coupled, multiplicative enhancement of optical responses, Phys Rev Lett 82, p.4811-4814 (1999)

[5.899] {Sect. 5.10.5} A. Kurita, Y. Kanematsu, M. Watanabe, K. Hirata, T. Kushida: Wavelength- and angle-selective optical memory effect by interference of multiple-scattered light, Phys Rev Lett 83, p.1582-1585 (1999)

[5.900] {Sect. 5.10.5} B. Lamprecht, J.R. Krenn, A. Leitner, F.R. Aussenegg: Resonant and off-resonant light-driven plasmons in metal nanoparticles studied by femtosecond-resolution third-harmonic generation, Phys Rev Lett 83, p.4421-4424 (1999)

[5.901] {Sect. 5.10.5} K.P. ODonnell, R.W. Martin, P.G. Middleton: Origin of luminescence from InGaN diodes, Phys Rev Lett 82, p.237-240 (1999)

[5.902] {Sect. 5.10.5} D. Orlikowski, M.B. Nardelli, J. Bernholc, C. Roland: Ad-dimers on strained carbon nanotubes: A new route for quantum dot formation?, Phys Rev Lett 83, p.4132-4135 (1999)

[5.903] {Sect. 5.10.5} L.M. Robinson, H. Rho, J.C. Kim, H.E. Jackson, L.M. Smith, S. Lee, M. Dobrowolska, J.K. Furdyna: Quantum dot exciton dynamics through a nanoaperture: Evidence for two confined states, Phys Rev Lett 83, p.2797-2800 (1999)

[5.904] {Sect. 5.10.5} P.C. Sercel, A.L. Efros, M. Rosen: Intrinsic gap states in semiconductor nanocrystals, Phys Rev Lett 83, p.2394-2397 (1999)

[5.905] {Sect. 5.10.5} W.S. Shi, Z.H. Chen, N.N. Liu, H.B. Lu, Y.L. Zhou, D.F. Cui, G.Z. Yang: Nonlinear optical properties of self-organized complex oxide Ce : BaTiO3 quantum dots grown by pulsed laser deposition, Appl Phys Lett 75, p.1547-1549 (1999)

[5.906] {Sect. 5.10.5} M.V. Wolkin, J. Jorne, P.M. Fauchet, G. Allan, C. Delerue: Electronic states and luminescence in porous silicon quantum dots: The role of oxygen, Phys Rev Lett 82, p.197-200 (1999)

[5.907] {Sect. 5.10.5} Y. Yang, V.J. Leppert, S.H. Risbud, B. Twamley, P.P. Power, H.W.H. Lee: Blue luminescence from amorphous GaN nanoparticles synthesized in situ in a polymer, Appl Phys Lett 74, p.2262-2264 (1999)

[5.908] {Sect. 5.10.5} A.E. Zhukov, A.R. Kovsh, N.A. Maleev, S.S. Mikhrin, V.M. Ustinov, A.F. Tsatsulnikov, M.V. Maximov, B.V. Volovik, D.A. Bedarev, Y.M. Shernyakov et al.: Long-wavelength lasing from multiply stacked InAs/InGaAs quantum dots on GaAs substrates, Appl Phys Lett 75, p.1926-1928 (1999)

[5.909] {Sect. 5.10.5} J. Hodak, I. Martini, G.V. Hartland: Ultrafast study of electron-phonon coupling in colloidal gold particles, Chem Phys Lett 284, p.135-141 (1998)

[5.910] {Sect. 5.10.5} H. Spocker, M. Portune, U. Woggon: Biexcitonic fingerprint in the nondegenerate four-wave-mixing signal of weakly confined cadmium sulfur quantum dots, Optics Letters 23, p.427-429 (1998)

[5.911] {Sect. 5.10.5} Z.K. Tang, G.K.L. Wong, P. Yu, M. Kawasaki, A. Ohtomo, H. Koinuma, Y. Segawa: Room-temperature ultraviolet laser emission from self-assembled ZnO microcrystallite thin films, Appl Phys Lett 72, p.3270-3272 (1998)

[5.912] {Sect. 5.10.5} Al.L. Efross, M.Rosen: Quantum size level structure of narrow-gap semiconductor nanocrystals: Effect of band coupling, Phys. Rev. B 58, p.7120-7135 (1998)

[5.913] {Sect. 5.10.5} J.M. Ballesteros, R. Serna, J. Solis, C.N. Afonso, A.K. Petfordlong, D.H. Osborne, R.F. Haglund: Pulsed laser deposition of Cu:Al2O3 nanocrystal thin films with high third-order optical susceptibility, Appl Phys Lett 71, p.2445-2447 (1997)

[5.914] {Sect. 5.10.5} B.A. Smith, J.Z. Zhang, U. Giebel, G. Schmid: Direct probe of size-dependent electronic relaxation in single-sized Au and nearly monodisperse Pt colloidal nano- particles, Chem Phys Lett 270, p.139-144 (1997)

[5.915] {Sect. 5.10.5} S. Vijayalakshmi, M.A. George, H. Grebel: Nonlinear optical properties of silicon nanoclusters, Appl Phys Lett 70, p.708-710 (1997)

[5.916] {Sect. 5.10.5} S. Vijayalakshmi, F. Shen, H. Grebel: Artificial dielectrics: Nonlinear optical properties of silicon nanoclusters at lambda=532 nm, Appl Phys Lett 71, p.3332-3334 (1997)

[5.917] {Sect. 5.10.5} J.Q. Yu, H.M. Liu, Y.Y. Wang, F.E. Fernandez, W.Y. Jia, L.D. Sun, C.M. Jin, D. Li, J.Y. Liu, S.H. Huang: Irradiation-induced luminescence enhancement effect of ZnS: Mn2+ nanoparticles in polymer films, Optics Letters 22, p.913-915 (1997)

[5.918] {Sect. 5.10.5} S.A. Empedocles, M.G. Bawendi: Quantum-confined stark effect in single CdSe nanocrystallite quantum dots, Science 278, p.2114-2117 (1997)

[5.919] {Sect. 5.10.5} G.L. Hornyak, Ch.J. Patrissi, Ch.R. Martin : Fabrication, Characterization, and Optical Properties of Gold Nanoparticle/Porous Alumina Composites: The Nonscattering Maxwell-Garnett Limit, J. Phys. Chem. B 101, p.1548-1555 (1997)

[5.920] {Sect. 5.10.5} M. Nikl, K. Nitsch, K. Polák, E. Mihókova, S. Zazubovich, G.P. Pazzi, P. Fabeni, L. Salvini, R. Aceves, M. Barbosa-Flores, R. Perez Salas, ;. Gurioli, A. Scacco: Quantum size effect in the excitone luminescence of CaPbX3-like quantum dots in CaX (X = Cl, Br) single crystal host, J. Luminesc. 72-74, p.377-379 (1997)

[5.921] {Sect. 5.10.5} C. A. Foss, Jr, G. L. Hornyak, J. A. Stockert, Ch. R. Martin: Template-Synthesized Nanoscopic Gold Particles: Optical Spectra and the Effects of Particle Size and Shape, J. Phys. Chem. 98, p.2963-2971 (1994)

[5.922] {Sect. 5.10.5} Y. Kayanuma: Quantum-size effects of interacting electrons and holes in semiconductor microcrystals with spherical shape, Phys. Rev. B 38, p.9797-9805 (1988)

[5.923] {Sect. 5.10.5} K. Tachibana, T. Someya, Y. Arakawa: Nanometer-scale InGaN self-assembled quantum dots grown by metalorganic chemical vapor deposition, Appl Phys Lett 74, p.383-385 (1999)

[5.924] {Sect. 5.10.5} X. Leyronas, J. Tworzydlo, C.W.J. Beenakker: Non-Cayley-tree model for quasiparticle decay in a quantum dot, Phys Rev Lett 82, p.4894-4897 (1999)

[5.925] {Sect. 5.10.5} M. Rohner, J.P. Reithmaier, A. Forchel, F. Schafer, H. Zull: Laser emission from photonic dots, Appl Phys Lett 71, p.488-490 (1997)

[5.926] {Sect. 5.10.5} X.C. Liu, J.H. Si, B.H. Chang, G. Xu, Q.G. Yang, Z.W. Pan, S.S. Xie, P.X. Ye, J.H. Fan, M.X. Wan: Third-order optical nonlinearity of the carbon nanotubes, Appl Phys Lett 74, p.164-166 (1999)

[5.927] {Sect. 5.10.5} A. Rubio, D. SanchezPortal, E. Artacho, P. Ordejon, J.M. Soler: Electronic states in a finite carbon nanotube: A one-dimensional quantum box, Phys Rev Lett 82, p.3520-3523 (1999)

[5.928] {Sect. 5.10.5} M.L. Terranova, S. Piccirillo, V. Sessa, S. Botti, M. Rossi: Photoluminescence from silicon nanoparticles in a diamond matrix, Appl Phys Lett 74, p.3146-3148 (1999)

[5.929] {Sect. 5.10.5} Q.Y. Wang, S.R. Challa, D.S. Sholl, J.K. Johnson: Quantum sieving in carbon nanotubes and zeolites, Phys Rev Lett 82, p.956-959 (1999)

[5.930] {Sect. 5.10.5} Y. Zhang, S. Iijima: Elastic response of carbon nanotube bundles to visible light, Phys Rev Lett 82, p.3472-3475 (1999)

[5.931] {Sect. 5.10.5} C.M. Soukoulis (ed.): Photonic Band Gap Materials (Kluwer Academic Publishers, Dordrecht, 1996)

[5.932] {Sect. 5.10.5} K. Busch, S. John: Liquid-crystal photonic-band-gap materials: The tunable electromagnetic vacuum, Phys Rev Lett 83, p.967-970 (1999)

[5.933] {Sect. 5.10.5} P. Halevi, A.A. Krokhin, J. Arriaga: Photonic crystal optics and homogenization of 2D periodic composites, Phys Rev Lett 82, p.719-722 (1999)

[5.934] {Sect. 5.10.5} M. Bayer, T. Gutbrod, J.P. Reithmaier, A. Forchel, T.L. Reinecke, P.A. Knipp, A.A. Dremin, V.D. Kulakovskii: Optical modes in photonic molecules, Phys Rev Lett 81, p.2582-2585 (1998)

[5.935] {Sect. 5.10.5} G. Feiertag, W. Ehrfeld, H. Freimuth, H. Kolle, H. Lehr, M. Schmidt, M.M. Sigalas, C.M. Soukoulis, G. Kiriakidis, T. Pedersen, et al.: Fabrication of photonic crystals by deep x-ray lithography, Appl Phys Lett 71, p.1441-1443 (1997)

[5.936] {Sect. 5.10.5} S. John, T. Quang: Resonant nonlinear dielectric response in a photonic band gap material, Phys Rev Lett 76, p.2484-2487 (1996)

6. Lasers

[6.1] {Sect. 6.0} T.H. Maiman: Stimulated Optical Radiation in Ruby, Nature 187, p.493-494 (1960)

[6.2] {Sect. 6.0} C.K.N. Patel, R.A. McFarlane, W.L. Faust: Optical Maser Action in C, N, O, S, and Br on Dissociation of Diatomic and Polyatomic Molecules, Phys. Rev. 133, p.A1244-A1248 (1964)

[6.3] {Sect. 6.0} A.L. Schawlow, C.H. Townes: Infrared and Optical Masers, Phys. Rev. 112, p.1940-1949 (1958)

[6.4] {Sect. 6.0} J.P. Gordon, H.J. Zeiger, C.H. Townes: The Maser – New Type of Microwave Amplifier, Frequency Standard, and Spectrometer, Phys. Rev. 99, p.1264-1274 (1955)

[6.5] {Sect. 6.0} K. Shimoda: Introduction to Laser Physics, 2nd edn, Springer Ser. Opt. Sci, Vol. 44 (Springer, Berlin, Heidelberg 1986)

[6.6] {Sect. 6.0} K. An, J.J. Childs, R.R. Dasari, M.S. Feld: Microlaser: A laser with one atom in an optical resonator, Phys Rev Lett 73, p.3375-3378 (1994)

[6.7] {Sect. 6.2} D. N. Nikogosyan: Properties of Optical and Laser-Related Materials – A Handbook (John Wiley & Sons, Chichester, 1997)

[6.8] {Sect. 6.2} J. Capmany, D. Jaque, J.G. Sole: Continuous wave laser radiation at 1314 and 1386 nm and infrared to red self-frequency doubling in nonlinear LaBGeO5 : Nd3+ crystal, Appl Phys Lett 75, p.2722-2724 (1999)

[6.9] {Sect. 6.2} E. Cavalli, E. Zannoni, C. Mucchino, V. Carozzo, A. Toncelli, M. Tonelli, M. Bettinelli: Optical spectroscopy of Nd3+ in KLa (MoO4) (2) crystals, J Opt Soc Am B Opt Physics 16, p.1958-1965 (1999)

[6.10] {Sect. 6.2} W.C. Choi, H.N. Lee, E.K. Kim, Y. Kim, C.Y. Park, H.S. Kim, J.Y. Lee: Violet/blue light-emitting cerium silicates, Appl Phys Lett 75, p.2389-2391 (1999)

[6.11] {Sect. 6.2} J. Dong, P.Z. Deng, J. Xu: Study of the effects of Cr ions on Yb in Cr,Yb : YAG crystal, Opt Commun 170, p.255-258 (1999)

[6.12] {Sect. 6.2} J.B. Gruber, B. Zandi, M. Ferry, L.D. Merkle: Spectra and energy levels of trivalent samarium in strontium fluorapatite, J Appl Phys 86, p.4377-4382 (1999)

[6.13] {Sect. 6.2} A. Braud, S. Girard, J.L. Doualan, R. Moncorge: Spectroscopy and fluorescence dynamics of (Tm3+, Tb3+) and (Tm3+, Eu3+) doped LiYF4 single crystals for 1.5-mu m laser operation, IEEE J QE-34, p.2246-2255 (1998)

[6.14] {Sect. 6.2} J.A. Munoz, J.O. Tocho, F. Cusso: Photoacoustic determination of the luminescent quantum efficiency of Yb3+ ions in lithium niobate, Appl Opt 37, p.7096-7099 (1998)

[6.15] {Sect. 6.2} B.M. Walsh, N.P. Barnes, B. DiBartolo: Branching ratios, cross sections, and radiative lifetimes of rare earth ions in solids: Application to Tm3+ and Ho3+ ions in LiYF4, J Appl Phys 83, p.2772-2787 (1998)

[6.16] {Sect. 6.2} J.B. Gruber, A.O. Wright, M.D. Seltzer, B. Zandi, L.D. Merkle, J.A. Hutchinson, C.A. Morrison, T.H. Allik, B.H.T. Chai: Site-selective excitation and polarized absorption and emission spectra of trivalent thulium and erbium in strontium fluorapatite, J Appl Phys 81, p.6585-6598 (1997)

[6.17] {Sect. 6.2} I.T. McKinnie, A.L. Oien, D.M. Warrington, P.N. Tonga, L.A.W. Gloster, T.A. King: Ti3+ ion concentration and Ti:sapphire laser performance, IEEE J QE-33, p.1221-1230 (1997)

[6.18] {Sect. 6.2} M. Nogami, Y. Abe: Fluorescence spectroscopy of silicate glasses codoped with Sm2+ and Al3+ ions, J Appl Phys 81, p.6351-6356 (1997)

[6.19] {Sect. 6.2} R.H. Page, K.I. Schaffers, L.D. Deloach, G.D. Wilke, F.D. Patel, J.B. Tassano, S.A. Payne, W.F. Krupke, K.T. Chen, A. Burger: Cr2+-doped zinc chalcogenides as efficient, widely tunable mid-infrared lasers, IEEE J QE-33, p.609-619 (1997)

[6.20] {Sect. 6.2} G. Tohmon, H. Sato, J. Ohya, T. Uno: Thulium:ZBLAN blue fiber laser pumped by two wavelengths, Appl Opt 36, p.3381-3386 (1997)

[6.21] {Sect. 6.2} X.H. Zhang, B.X. Jiang, Y.F. Yang, Z.G. Wang: Spectroscopic properties of anisotropic absorption in a neodymium-doped YAlO3 laser crystal, J Appl Phys 81, p.6939-6942 (1997)

[6.22] {Sect. 6.2} J.B. Gruber, C.A. Morrison, M.D. Seltzer, A.O. Wright, M.P. Nadler, T.H. Allik, J.A. Hutchinson, B.H.T. Chai: Site-selective excitation and polarized absorption spectra of Nd3+ in Sr5 (PO4)3F and Ca5 (PO4)3F, J Appl Phys 79, p.1746-1758 (1996)

[6.23] {Sect. 6.2} M.A. Khan, M.A. Gondal, M.H. Rais: Laser gain on the 4p3d F-3 – 4S3d D-3 transitions of Ca following optical excitation of the 4s4p P-3 (1) state, Opt Commun 124, p.38-44 (1996)

[6.24] {Sect. 6.2} L.D. Merkle, B. Zandi, R. Moncorge, Y. Guyot, H.R. Verdun, B. Mcintosh: Spectroscopy and laser operation of Pr, Mg:SrAl12O19, J Appl Phys 79, p.1849-1856 (1996)

[6.25] {Sect. 6.2} M. Nogami, Y. Abe: Fluorescence properties of Sm2+ ions in silicate glasses, J Appl Phys 80, p.409-414 (1996)

[6.26] {Sect. 6.2} M.B. Saisudha, K.S.R.K. Rao, H.L. Bhat, J. Ramakrishna: The fluorescence of Nd3+ in lead borate and bismuth borate glasses with large stimulated emission cross section, J Appl Phys 80, p.4845-4853 (1996)

[6.27] {Sect. 6.2} K.I. Schaffers, L.D. Deloach, S.A. Payne: Crystal growth, frequency doubling, and infrared laser performance of Yb3+:BaCaBO3F, IEEE J QE-32, p.741-748 (1996)

[6.28] {Sect. 6.2} T. Schweizer, D.W. Hewak, B.N. Samson, D.N. Payne: Spectroscopic data of the 1.8-, 2.9-, and 4.3-mu m transitions in dysprosium-doped gallium lanthanum sulfide glass, Optics Letters 21, p.1594-1596 (1996)

[6.29] {Sect. 6.2} J.M. Sutherland, P.M.W. French, J.R. Taylor, B.H.T. Chai: Visible continuous-wave laser transitions in Pr3+:YLF and femtosecond pulse generation, Optics Letters 21, p.797-799 (1996)

[6.30] {Sect. 6.2} N. Sarukura, Z.L. Liu, Y. Segawa, K. Edamatsu, Y. Suzuki, T. Itoh, V.V. Semashko, A.K. Naumov, S.L. Korableva, R. Yu, et al.: Ce3 (+):LuLiF4 as a broadband ultraviolet amplification medium, Optics Letters 20, p.294-296 (1995)

[6.31] {Sect. 6.2} G.F. Wang, T.P.J. Han, H.G. Gallagher, B. Henderson: Novel laser gain media based on Cr3+-doped mixed borates RX (3) (BO3) (4), Appl Phys Lett 67, p.3906-3908 (1995)

[6.32] {Sect. 6.2} T.S. Rose, M.S. Hopkins, R.A. Fields: Characterization and Control of Gamma and Proton Radiation Effects on the Performance of Nd:YAG and Nd:YLF Lasers, IEEE J. QE-31, p.1593-1602 (1995)

[6.33] {Sect. 6.2} N. Mermilliod, R. Romero, I. Chartier, C. Garapon, R. Moncorgé: Performance of Various Diode-Pumped Nd:Laser Materials: Influence of Inhomogeneous Broadening, IEEE J. QE-28, p.1179-1187 (1992)

[6.34] {Sect. 6.2} J. Harrison, D. Welford, P.F. Moulton: Threshold Analysis of Pulsed Lasers with Application to a Room-Temperature Co:MgF2 Laser, IEEE J. QE-25, p.1708-1711 (1989)

[6.35] {Sect. 6.2} K. Fuhrmann et al.: Effective cross section of the Nd:YAG 1.0641 μm laser transition, J. Appl. Phys. 62, p.4041-4044 (1987)

[6.36] {Sect. 6.2} N. Neuroth: Laser glass: Status and prospects, Opt. Eng. 26, p.96-101 (1987)

[6.37] {Sect. 6.2} P.F. Moulton: Spectroscopic and laser characteristics of Ti:Al2O3, J. Opt. Soc. Am. B 3, p.125-133 (1986)

[6.38] {Sect. 6.2} L. Schearer, M. Leduc: Tuning Characteristics and New Laser Lines in an Nd:YAP CW Laser, IEEE J. QE-22, p.756-758 (1986)

[6.39] {Sect. 6.2} P.F. Moulton: An Investigation of the Co:MgF2 Laser System, IEEE J. QE-21, p.1582-1595 (1985)

[6.40] {Sect. 6.2} U. Brauch, U. Dürr: KZnF3:Cr3+ – A Tunable Solid State NIR-Laser, Optics Commun. 49, p.61-64 (1984)

[6.41] {Sect. 6.2} K. Maeda, M: Aabe, H. Kuroda, N. Nakano, M. Umino, N. Wada: Concentration Dependence of Fluoroscence Lifetime onf Nd3+-doped Gd3Ga5O12 Lasers, Jap. J. Appl. Phys. 23, p.759-760 (1984)

[6.42] {Sect. 6.2} B. Struve, G. Huber: Tunable Room-Temperature cw Laser Action in Cr3+:GdScGa-Garnet, Appl. Phys. B 30, p.117-120 (1983)

[6.43] {Sect. 6.2} H.P. Christensen, H.P. Jenssen: Broad-Band Emission from Chromium Doped Germanium Garnets, IEEE J. QE-18, p.1197-1201 (1982)

[6.44] {Sect. 6.2} D. Pruss, G. Huber, A. Beimowski, V.V. Laptev, I.A. Shcherbakov, Y.V. Zharikov: Efficient Cr3+ Sensitized Nd3+:GdScGa-Garnet Laser at 1.06 μm, Appl. Phys. B 28, p.355-358 (1982)

[6.45] {Sect. 6.2} E.V. Zharikov, N.N. Il'ichev, V.V. Laptev, A.A. Malyutin, V.G. Ostroumov, P.P. Pashinin, I.A. Shcherbakov: Sensitization of neodymium ion luminescence by chromium ions in a Gd3Ga5O12 crystal, Sov. J. Quantum Electron. 12, p.338-341 (1982)

[6.46] {Sect. 6.2} E.V. Zharikov, V.V. Laptev, I.A. Shcherbakov, E.I. Sidorova, Y.P. Timofeev: Absolute Quantum Yield of Luminescence of CR 3+ Ions in Gadolinium Gallium and Gadolinium Scandium Gallium Garnet Crystals, KVANTOVAYA ELEKTRONIKA 9, p.1740-1741 (1982)

[6.47] {Sect. 6.2} J.C. Walling, H.P. Jenssen, R.C. Morris, E.W. O'Dell, O.G. Peterson: Tunable-laser performance in BeAl2O4:Cr3+, Opt. Lett. 4, p.182-183 (1979)

[6.48] {Sect. 6.2} J.G. Gualtieri, T.R. Aucoin: Laser performance of large Nd-pentaphosphate crystals, Appl. Phys. Lett. 28, p.189-192 (1976)

[6.49] {Sect. 6.2} H.P. Jenssen, R.F. Begley, R. Webb, R.C. Morris.: Spectroscopic properties and laser performance of Nd3+ in lanthanum beryllate, J. Appl. Phys. 47, p.1496-1500 (1976)

[6.50] {Sect. 6.2} R.F. Belt, J.R. Latore, R. Uhrin, J. Paxton: EPR and optical study of Fe in Nd:YAlO3 laser crystals, Appl. Phys. Lett. 25, p.218-220 (1974)

[6.51] {Sect. 6.2} L.F. Johnson, H.J. Guggenheim: Electronic- and Phonon-Terminated Laser Emission from Ho3+ in BaY2F8, IEEE J. QE-10, p.442-449 (1974)

[6.52] {Sect. 6.2} W.F. Krupke: Induced-Emission Cross Sections in Neodymium Laser Glasses, IEEE J. QE-10, p.450-457 (1974)

[6.53] {Sect. 6.2} K.B. Steinbruegge, G.D. Baldwin: Evaluation of CaLaSOAP:Nd for high-power flash-pumped Q-switched lasers, Appl. Phys. Lett. 25, p.220-222 (1974)

[6.54] {Sect. 6.2} H.P. Weber, P.F. Liao, B.C. Tofield: Emission Cross Section and Fluorescence Efficiency of Nd-Pentphosphate, IEEE J. QE-10, p.563-567 (1974)

[6.55] {Sect. 6.2} H.G. Danielmeyer, G. Huber, W.W. Krühler, J.P. Jeser: Continous Oscillation of a (Sc, Nd) Pentaphosphate Laser with 4 Milliwatts Pump Threshold, Appl. Phys. 2, p.335-338 (1973)

[6.56] {Sect. 6.2} W.W. Krühler, J.P. Jeser, H.G. Danielmeyer: Properties and Laser Oscillation of the (Nd, Y) Pentaphosphate System, Appl. Phys. 2, p.329-333 (1973)

[6.57] {Sect. 6.2} H.P. Weber, T.C. Damen, H.G. Danielmeyer, B.C. Tofield: Nd-ultraphospate laser, Appl. Phys. Lett. 22, p.534-536 (1973)

[6.58] {Sect. 6.2} M.J. Weber, M. Bass, T.E. Varitimos, D.P. Bua: Laser Action from Ho3+, Er3+, and Tm3+ in YAlO3, IEEE J. QE-9, p.1079-1086 (1973)

[6.59] {Sect. 6.2} R.V. Alves, R.A. Buchanan, K.A.Wickersheim, E.A.C. Yates: Neodymium-Activated Lanthanum Oxysulfide: A New High-Gain Laser Material, J. Appl. Phys. 42, p.3043-3048 (1971)

[6.60] {Sect. 6.2} M.J. Weber, M. Bass, K. Andringa, R.R. Monchamp, E. Comperchio: Czochralski Growths and Properties of YAlO3 Laser Crystals, Appl. Phys. Lett. 15, p.342-345 (1969)

[6.61] {Sect. 6.2} R.C. Ohlmann, K.B. Steinbruegge, R. Mazelsky: Spectroscopic and Laser Characteristics of Neodymium-doped Calcium Fluorophosphate, Appl. Opt. 7, p.905-914 (1968)

[6.62] {Sect. 6.2} L.F. Johnson, H.J. Guggenheim: Photon-Terminated Coherent Emission from V2+ Ions in MgF2, J. Appl. Phys. 38, p.4837-4839 (1967)

[6.63] {Sect. 6.2} D.C. Cronemeyer: Optical Absorption Characteristics of Pink Ruby, J. Opt. Soc. Am. 56, p.1703-1706 (1966)

[6.64] {Sect. 6.2} J.R. O'Connor: Unusual Crystal-Field Energy Levels and Efficient Laser Properties of YVO4:Nd, Appl. Phys. Lett. 9, p.407-409 (1966)

[6.65] {Sect. 6.2} D.M. Dodd, D.L. Wood, R.L. Barns: Spectrophotometric Determination of Chromium Concentration in Ruby, J. Appl. Phys. 35, p.1183-1186 (1964)

[6.66] {Sect. 6.2} K. Nassau, A.M. Broyer: Calcium Tungstate: Czochralski Growth, Perfection, and Substitution, J. Appl. Phys. 33, p.3064-3073 (1962)

[6.67] {Sect. 6.2} T.H. Maiman, R.H. Hoskins, I.J. D'Haenens, C.K. Asawa, V. Evtuhov: Stimulated Optical Emission in Fluorescent Solids. II. Spectroscopy and Stimulated Emission in Ruby, Phys. Rev. 123, p.1151-1157 (1961)

[6.68] {Sect. 6.2} A. Braud, S. Girard, J.L. Doualan, M. Thuau, R. Moncorge, A.M. Tkachuk: Energy-transfer processes in Yb : Tm-doped KY3F10, LiYF4, and BaY2F8 single crystals for laser operation at 1.5 and 2.3 mu m, Phys Rev B 61, p.5280-5292 (2000)

[6.69] {Sect. 6.2} Y. Mita, T. Ide, M. Togashi, H. Yamamoto: Energy transfer processes in Yb3+ and Tm3+ ion-doped fluoride crystals, J Appl Phys 85, p.4160-4164 (1999)

[6.70] {Sect. 6.2} M. Berggren, A. Dodabalapur, R.E. Slusher: Stimulated emission and lasing in dye-doped organic thin films with Forster transfer, Appl Phys Lett 71, p.2230-2232 (1997)

[6.71] {Sect. 6.2} C. Wyss, W. Luthy, H.P. Weber, P. Rogin, J. Hulliger: Energy transfer in Yb3+:Er3+:YLF, Opt Commun 144, p.31-35 (1997)

[6.72] {Sect. 6.2} O. Barbosagarcia, E. Jonguitudisurieta, L.A. Diaztorres, C.W. Struck: The non-radiative energy transfer in high acceptor concentration codoped Nd,Ho:YAG and Nd,Er:YAG, Opt Commun 129, p.273-283 (1996)

[6.73] {Sect. 6.2} A.J. Cox, B.K. Matise: Energy Transfer Between Coumarins in a Dye Laser, Chem. Phys. Lett. 76, p.125-128 (1980)

[6.74] {Sect. 6.2} M.P. Hehlen, A. Kuditcher, A.L. Lenef, H. Ni, Q. Shu, S.C. Rand, J. Rai, S. Rai: Nonradiative dynamics of avalanche upconversion in Tm : LiYF4, Phys Rev B 61, p.1116-1128 (2000)

[6.75] {Sect. 6.2} R. Kapoor, C.S. Friend, A. Biswas, P.N. Prasad: Highly efficient infrared-to-visible energy upconversion in Er3+: Y2O3, Optics Letters 25, p.338-340 (2000)

[6.76] {Sect. 6.2} D.S. Anker, L.D. Merkle: Ion-ion upconversion excitation of the 4f5d configuration in Pr : Y3Al5O12 – Experiments and Forster theory-based rate equation model, J Appl Phys 86, p.2933-2940 (1999)

[6.77] {Sect. 6.2} E. Pecoraro, D.F. deSousa, R. Lebullenger, A.C. Hernandes, L.A.O. Nunes: Evaluation of the energy transfer rate for the Yb3+: Pr3+ system in lead fluoroindogallate glasses, J Appl Phys 86, p.3144-3148 (1999)

[6.78] {Sect. 6.2} R.W. Mosses, J.P.R. Wells, H.G. Gallagher, T.P.J. Han, M. Yamaga, N. Kodama, T. Yosida: Czochralski growth and IR-to-visible upconversion of Ho3+- and Er3+- doped SrLaAlO4, Chem Phys Lett 286, p.291-297 (1998)

[6.79] {Sect. 6.2} D.N. Patel, R.B. Reddy, S.K. NashStevenson: Diode-pumped violet energy upconversion in BaF2:Er3+, Appl Opt 37, p.7805-7808 (1998)

[6.80] {Sect. 6.2} P.J. Deren, J. Feries, J.C. Krupa, W. Strek: Anti-stokes emission in LaCl3 doped with U3+ and Pr3+ ions, Chem Phys Lett 264, p.614-618 (1997)

[6.81] {Sect. 6.2} G.S. He, K.S. Kim, L.X. Yuan, N. Cheng, P.N. Prasad: Two-photon pumped partially cross-linked polymer laser, Appl Phys Lett 71, p.1619-1621 (1997)

[6.82] {Sect. 6.2} G.S. He, L.X. Yuan, P.N. Prasad, A. Abbotto, A. Facchetti, G.A. Pagani: Two-photon pumped frequency-upconversion lasing of a new blue-green dye material, Opt Commun 140, p.49-52 (1997)

[6.83] {Sect. 6.2} G.S. He, L.X. Yuan, Y.P. Cui, M. Li, P.N. Prasad: Studies of two-photon pumped frequency-upconverted lasing properties of a new dye material, J Appl Phys 81, p.2529-2537 (1997)

[6.84] {Sect. 6.2} G.S. He, Y.P. Cui, J.D. Bhawalkar, P.N. Prasad, D.D. Bhawalkar: Intracavity upconversion lasing within a Q-switched Nd:YAG laser, Opt Commun 133, p.175-179 (1997)

[6.85] {Sect. 6.2} P.E.A. Mobert, E. Heumann, G. Huber, B.H.T. Chai: Green Er3+:YLiF4 upconversion laser at 551 nm with Yb3+ codoping: a novel pumping scheme, Optics Letters 22, p.1412-1414 (1997)

[6.86] {Sect. 6.2} H.M. Pask, A.C. Tropper, D.C. Hanna: A Pr3+-doped ZBLAN fibre upconversion laser pumped by an Yb3+-doped silica fibre laser, Opt Commun 134, p.139-144 (1997)

[6.87] {Sect. 6.2} T. Sandrock, H. Scheife, E. Heumann, G. Huber: High-power continuous-wave upconversion fiber laser at room temperature, Optics Letters 22, p.808-810 (1997)

[6.88] {Sect. 6.2} H.M. Pask, A.C. Tropper, D.C. Hanna: APr3+-doped ZBLAN fibre upconversion laser pumped by an Yb3+-doped silica fibre laser, Opt. Comm. 134, p.139-144 (1997)

[6.89]   {Sect. 6.2} D.M. Baney, G. Rankin, K.W. Chang: Blue Pr3+-doped ZBLAN fiber upconversion laser, Optics Letters 21, p.1372-1374 (1996)

[6.90]   {Sect. 6.2} D.M. Baney, G. Rankin, K.W. Chang: Simultaneous blue and green upconversion lasing in a laser-diode-pumped Pr3+/Yb3+ doped fluoride fiber laser, Appl Phys Lett 69, p.1662-1664 (1996)

[6.91]   {Sect. 6.2} S.R. Bowman, L.B. Shaw, B.J. Feldman, J. Ganem: A 7-mu m praseodymium-based solid-state laser, IEEE J QE-32, p.646-649 (1996)

[6.92]   {Sect. 6.2} T. Chuang, H.R. Verdun: Energy transfer up-conversion and excited state absorption of laser radiation in Nd:YLF laser crystals, IEEE J QE-32, p.79-91 (1996)

[6.93]   {Sect. 6.2} G.S. He, J.D. Bhawalkar, C.F. Zhao, C.K. Park, P.N. Prasad: Upconversion dye-doped polymer fiber laser, Appl Phys Lett 68, p.3549-3551 (1996)

[6.94]   {Sect. 6.2} C. Koeppen, G. Jiang, G. Zheng, A.F. Garito: Room-temperature green upconversion fluorescence of an Er3+-doped laser liquid, Optics Letters 21, p.653-655 (1996)

[6.95]   {Sect. 6.2} G.S. He, J.D. Bhawalkar, C.F. Zhao, C.K. Park, P.N. Prasad: Two-photon-pumped cavity lasing in a dye-solution-filled hollow-fiber system, Optics Letters 20, p.2393-2395 (1995)

[6.96]   {Sect. 6.2} P. Xie, T.R. Gosnell: Room-temperature upconversion fiber laser tunable in the red, orange, green, and blue spectral regions, Optics Letters 20, p.1014-1016 (1995)

[6.97]   {Sect. 6.2} W. Kaiser, C.G.B. Garrett: Two-Photon Excitation in CaF2: Eu2+, Phys. Rev. Lett. 7, p.229-231 (1961)

[6.98]   {Sect. 6.2} X. Zhang, X.G. Liu, J.P. Jouart, G. Mary: Upconversion fluorescence of Ho3+ ions in a BaF2 crystal, Chem Phys Lett 287, p.659-662 (1998)

[6.99]   {Sect. 6.2} C.L. Pope, B.R. Reddy, S.K. NashStevenson: Efficient violet upconversion signal from a fluoride fiber doped with erbium, Optics Letters 22, p.295-297 (1997)

[6.100] {Sect. 6.3.1} N.P. Barnes, M.E. Storm, P.L. Cross, M.W. Skolaut: Efficiency of Nd Laser Materials with Laser Diode Pumping, IEEE J. QE-26, p.558-569 (1990)

[6.101] {Sect. 6.3.1} W. Streifer, D.R. Scifres, G.L. Harnagel, D.F. Welch, J. Berger, M. Sakamoto: Advances in Diode Laser Pumps, IEEE J. QE-24, p.883-894 (1988)

[6.102] {Sect. 6.3.1} Y.F. Chen, C.F. Kao, S.C. Wang: Analytical model for the design of fiber-coupled laser- diode end-pumped lasers, Opt Commun 133, p.517-524 (1997)

[6.103] {Sect. 6.3.1} W.A. Clarkson, D.C. Hanna: Efficient Nd:YAG laser end pumped by a 20-W diode-laser bar, Opt. Lett. 21, p.869-871 (1996)

[6.104] {Sect. 6.3.1} S. Yamaguchi, T. Kobayashi, Y. Saito, K. Chiba: Efficient Nd:YAG laser end pumped by a high-power multistripe laser-diode bar with multiprism array coupling, Appl. Opt. 35, p.1430-1435 (1996)

[6.105] {Sect. 6.3.1} H.R. Verdún, T. Chuang: Efficient TEM00-mode operation of a Nd:YAG laser end pumped by a three-bar high-power diode-laser array, Opt. Lett. 17, p.1000-1002 (1992)

[6.106] {Sect. 6.3.1} J. Berger, D.F. Welch, W. Streifer, D.R. Scifres, N.J. Hoffmann, J.J. Smith, D. Radecki: Fiber-bundle coupled, diode end-pumped Nd:YAG laser, Opt. Lett. 13, p.306-308 (1988)

[6.107] {Sect. 6.3.1} T.Y. Fan, R.L. Byer: Diode Laser-Pumped Solid-State Lasers, IEEE J. QE-24, p.895-912 (1988)

[6.108] {Sect. 6.3.1} D.L. Sipes: Highly efficient neodymium:yttrium aluminium garnet laser end pumped by a semiconductor laser array, Appl. Phys. Lett. 47, p.74-76 (1985)

[6.109] {Sect. 6.3.1} R.L. Fu, G.J. Wang, Z.Q. Wang, E.X. Ba, G.G. Mu, X.H. Hu: Design of efficient lens ducts, Appl Opt 37, p.4000-4003 (1998)

[6.110] {Sect. 6.3.1} R.J. Beach: Theory and optimization of lens ducts, Appl Opt 35, p.2005-2015 (1996)

[6.111] {Sect. 6.3.1} R.P. Edwin: Stripe Stacker for Use with Laser Diode Bars, Optics Letters 20, p.222-224 (1995)

[6.112] {Sect. 6.3.1} J.R. Leger, W.C. Goltsos: Geometrical Transformation of Linear Diode-Laser Arrays for Longitudinal Pumping of Solid-State Lasers, IEEE J. QE-28, p.1088-1100 (1992)

[6.113] {Sect. 6.3.1} U. Griebner, R. Grunwald, H. Schonnagel: Thermally bonded Yb : YAG planar waveguide laser, Opt Commun 164, p.185-190 (1999)

[6.114] {Sect. 6.3.1} W.J. Kessler, S.J. Davis, H.C. Miller, G.D. Hager: Optically pumped hydrogen fluoride laser, J Appl Phys 83, p.7448-7452 (1998)

[6.115] {Sect. 6.3.1} T. Kojima, K. Yasui: Efficient diode side-pumping configuration of a Nd:YAG rod laser with a diffusive cavity, Appl Opt 36, p.4981-4984 (1997)

[6.116] {Sect. 6.3.1} R.J. Koshel, I.A. Walmsley: Optimal design of optically side-pumped lasers, IEEE J QE-33, p.94-102 (1997)

[6.117] {Sect. 6.3.1} Y. Liao, K.M. Du, S. Falter, J. Zhang, M. Quade, P. Loosen, R. Poprawe: Highly efficient diode-stack, end-pumped Nd:YAG slab laser with symmetrized beam quality, Appl Opt 36, p.5872-5875 (1997)

[6.118] {Sect. 6.3.1} K. Takehisa: Scaling up of a high average power dye laser amplifier and its new pumping designs, Appl Opt 36, p.584-592 (1997)

[6.119] {Sect. 6.3.1} T. Brand: Compact 170-W continuous-wave diode-pumped Nd:YAG rod laser with a cusp-shaped reflector, Optics Letters 20, p.1776-1778 (1995)

[6.120] {Sect. 6.3.1} N. Uehara, K. Nakahara, K. Ueda: Continuous-wave TEM (00)-mode 26.5-W-output virtual-point- source diode-array-pumped Nd:YAG laser, Optics Letters 20, p.1707-1709 (1995)

[6.121] {Sect. 6.3.1} M.M. Dyer, H. Helm: Axicon amplification of a synchronously pumped subpicosecond dye laser, J. Opt. Soc. Am. B 10, p.1035-1039 (1993)

[6.122] {Sect. 6.3.1} F. Hanson, D. Haddock: Laser diode side pumping of neodymium laser rods, Appl. Opt. 27, p.80-83 (1988)

[6.123] {Sect. 6.3.1} J. Machan, R.Moyer, D. Hoffmaster, J. Zamel, D. Burchman, R. Tinti, G. Holleman, L. Marabella, H. Injeyan: Multi-Kilowatt, High Brightnesss Diode-Pumped Laser for Precision Laser Machining, Techn. Digest Adv. Solid-State Lasersp.263-265 (1998)

[6.124] {Sect. 6.3.1} A. Mandl, A. Zavriyev, D.E. Klimek, J.J. Ewing: Cr:LiSAF thin slab zigzag laser, IEEE J QE-33, p.1864-1868 (1997)

[6.125] {Sect. 6.3.1} J. Richards, A. McInnes: Versatile, efficient, diode-pumped miniature slab laser, Opt. Lett. 20, p.371-373 (1995)

[6.126] {Sect. 6.3.1} T.J. Kane, R.L. Byer, R.C. Eckardt: Reduced Thermal Focusing and Birefringence in Zig Zag Slab Geometry Crystalline Lasers, IEEE J. QE19, p.1351-1354 (1983)

[6.127] {Sect. 6.3.1} J.M. Eggleston, R.L. Byer, T. Kane, J. Unternahrer: Slab Geometry Solid State Lasers, Appl Phys B 28, p.236 (1982)

[6.128] {Sect. 6.3.1} U. Brauch, A. Giesen, M. Karszewski, C. Stewen, A. Voss: Multiwatt diode pumped Yb:YAG thin disk laser continuously tunable between 1018 and 1053 nm, Optics Letters 20, p.713-715 (1995)

[6.129] {Sect. 6.3.1} A. Giesen, H. Hügel, A. Voss, K. Wittig, U. Brauch, H. Opower: Scalable Concept for Diode-Pumped High-Power Solid-State Lasers, Appl. Phys. B 58, p.365-372 (1994)

[6.130] {Sect. 6.3.3} S. Nagai, H. Furuhashi, A. Kono, Y. Uchida, T. Goto: Measurement of temporal behavior of electron density in a discharge- pumped ArF excimer laser, IEEE J QE-34, p.942-948 (1998)

[6.131] {Sect. 6.3.3} D. C. Cartwright: Total Cross Sections for the Excitation of the Triplet States in Molecular Nitrogen, Phys. Rev. A 2, p.1331-1347 (1970)

[6.132] {Sect. 6.3.3} P. Coutance, J.P. Pique: Radial and time-resolved measurement of cuprous bromide concentration in a Cu-HBr laser, IEEE J QE-34, p.1340-1348 (1998)

[6.133] {Sect. 6.3.4} D.A. Haner, B.T. McGuckin, R.T. Menzies, C.J. Bruegge, V. Duval: Directional-hemispherical reflectance for Spectralon by integration of its bidirectional reflectance, Appl Opt 37, p.3996-3999 (1998)

[6.134] {Sect. 6.3.4} P. Mazzinghi, D. Bigazzi: Wavelength-dependent model of Kr flash lamp emission and absorption, Appl Opt 36, p.2473-2480 (1997)

[6.135] {Sect. 6.3.4} D.V. Pantelic, B.M. Panic, I.Z. Belic: Solid-state laser pumping with a planar compound parabolic concentrator, Appl Opt 36, p.7730-7740 (1997)

[6.136] {Sect. 6.3.4} P.J. Walsh, A. Kermani: Electrical characterization of cw Xenon arcs moderate currents, J. Appl. Phys. 61, p.4484-4491 (1987)

[6.137] {Sect. 6.3.4} B. Smith: An overview of flashlamps and CW arc lamps, Techn. Bulletin 3.ILC Technology. (1986)

[6.138] {Sect. 6.3.4} F. Docchio, L. Pallaro, O. Svelto: Pump cavities for compact pulsed Nd:YAG lasers: a comparative study, Appl. Opt. 24, p.3752-3755 (1985)

[6.139] {Sect. 6.3.4} F. Docchio: The rod image: a new method for the calculation of pump efficiency in reflecting close-coupled cavities, Appl. Opt. 24, p.3746-3751 (1985)

[6.140] {Sect. 6.3.4} A.N. Fletcher: Effect of Flashlamp Diameter on Luminescent Coolants for a Solid-State Laser, Appl. Phys. B 37, p.31-34 (1985)

[6.141] {Sect. 6.3.4} P. Laporta, V. Magni, O. Svelto: Comparative Study of the Optical Pumping Efficiency in Solid State Lasers, IEEE J. QE-21, p.1211-1218 (1985)

[6.142] {Sect. 6.3.4} D.M. Camm: Optimal reflectors for coupling cylindrical sources and targets of finite dimensions, Appl. Opt. 23, p.601-606 (1984)

[6.143] {Sect. 6.3.4} K. Yoshida, Y. Kato, H. Yoshida, C. Yamanaka: Prediction of flash lamp explosion by stress measurements, Rev. Sci. Instr. 55, p.1415-1420 (1984)

[6.144] {Sect. 6.3.4} R.G. Hohlfeld, W. Manning, D.A. MacLennan: Self-inductance effects in linear flashtubes: an extension to the Markiewicz and Emmett theory, Appl. Opt. 22, p.1986-1991 (1983)

[6.145] {Sect. 6.3.4} J. Richards, D. Rees, K. Fueloep, B.A. See: Operation of krypton-filled flashlamps at high repetition rates, Appl. Opt. 22, p.1325-1328 (1983)

[6.146] {Sect. 6.3.4} W. Lama, T. Hammond: Arc-acoustic interaction in rare gas flashlamps, Appl. Opt. 20, p.765-769 (1981)

[6.147] {Sect. 6.3.4} J.H. Kelly, D.C. Brown, K. Teegarden: Time resolved spectroscopy of large bore Xe flashlamps for use in large aperture amplifiers, Appl. Opt. 19, p.3817-3823 (1980)

[6.148] {Sect. 6.3.4} H.L. Witting: Acoustic resonances in cylindrical high-pressure arc discharges, J. Appl. Phys. 49, p.2680-2683 (1978)

[6.149] {Sect. 6.3.4} D.A. Huchital, G.N.Steinberg: Pumping of Nd:YAG with Electrodeless arc lamps, IEEE J. QE-12, p.1-9 (1976)

[6.150] {Sect. 6.3.4} M.R. Siegrist: Cusp shape reflectors to pump disk or slab lasers, Appl. Opt. 15, p.2167-2171 (1976)

[6.151] {Sect. 6.3.4} H.U. Leuenberger, G. Herziger: Optical Pump System for Mode-Controlled Laser Operation, Appl. Opt. 14, p.1190-1192 (1975)

[6.152] {Sect. 6.3.4} V.J. Corcoran, R.W. McMillan, S.K. Barnoske: Flashlamp-Pumped YAG:Nd+3 Laser Action at Kilohertz Rates, IEEE J. QE-10, p.618-620 (1974)

[6.153] {Sect. 6.3.4} R.H. Dishington, W.R. Hook, R.P. Hilberg: Flashlamp Discharge and Laser Efficiency, Appl. Opt. 13, p.2300-2312 (1974)

[6.154] {Sect. 6.3.4} D.D. Bhawalkar, L. Pandit: Improving the Pumping Efficiency of a Nd3+ Glass Laser Using Dyes, IEEE J. QE-9, p.43-46 (1973)

[6.155] {Sect. 6.3.4} W.R. Hook, R.H. Dishington, R.P. Hilberg: Xenon Flashlamp Triggering for Laser Applications, IEEE Trans. ED-19, p.308-314 (1972)

[6.156] {Sect. 6.3.4} W. Koechner, L. DeBenedictis, E. Matovich, G.E. Mevers: Characteristics and Performance of High-Power CW Krypton Arc Lamps for Nd:YAG Laser Pumping, IEEE J. QE-8, p.310-316 (1972)

[6.157] {Sect. 6.3.4} W. Koechner: Output Fluctuations of CW-Pumped Nd:YAG Lasers, IEEE J. QE-8, p.656-661 (1972)

[6.158] {Sect. 6.3.4} W.W. Morey: Active Filtering for Neodymium Lasers, IEEE J. QE-8, p.818-819 (1972)

[6.159] {Sect. 6.3.4} S. Yoshikawa, K. Iwamoto, K. Washio: Efficient Arc Lamps for Optical Pumping of Neodymium Lasers, Appl. Opt. 10, p.1620-1623 (1971)

[6.160] {Sect. 6.3.4} W.D. Fountain, L.M. Osterink, J.D. Foster: Comparision of Kr and Xe Flashlamps for Nd:YAG Lasers, IEEE J. QE-6, p.684-687 (1970)

[6.161] {Sect. 6.3.4} D.R. Skinner: The Effect of Laser-Rod Properties on the Energy Transfer Efficiency of Pumping Cavities Using Helical Flash Lamps, Appl. Opt. 8, p.1467-1470 (1969)

[6.162] {Sect. 6.3.4} J.G. Edwards: Some Factors Affecting the Pumping Efficiency of Optically Pumped Lasers, Appl. Opt. 6, p.837-843 (1967)

[6.163] {Sect. 6.3.4} K. Kamiryo, T. Kano, H. Matsuzawa: Optimum Design of Elliptical Pumping Chambers for Solid Lasers, Jap. J. Appl. Phys. 5, p.1217-1226 (1966)

[6.164] {Sect. 6.3.4} J.P. Markiewicz, J.L. Emmett: Design of flashlamp driving circuits, IEEE J. QE-2, p.707-711 (1966)

[6.165] {Sect. 6.3.4} T.B. Read: The cw pumping of YAG:Nd3+ by water-cooled krypton arcs, Appl. Phys. Lett. 9, p.342-344 (1966)

[6.166] {Sect. 6.3.4} D. Roess: Analysis of Room Temperature CW Ruby Lasers, IEEE J. QE-2, p.208-214 (1966)

[6.167] {Sect. 6.3.4} C. Bowness: On the effeciency of single and multiple elliptical laser cavities, Appl. Opt. 4, p.103-108 (1965)

[6.168] {Sect. 6.3.4} S.B. Schuldt, R.L. Aagard: An Analysis of Radiation Transfer By Means of Elliptical Cylinder Reflectors, Appl. Opt. 2, p.509-513 (1963)

[6.169] {Sect. 6.3.5} D. Furman, B.D. Barmashenko, S. Rosenwaks: Diode-laser-based absorption spectroscopy diagnostics of a jet-type O- 2 ((1)Delta) generator for chemical oxygen-iodine lasers, IEEE J QE-35, p.540-547 (1999)

[6.170] {Sect. 6.3.5} G.N. Tsikrikas, A.A. Serafetinides: Discharge and circuit simulation of a plasma cathode TEA HF laser operating with a He/SF6/C3H8 gas mixture, Opt Commun 134, p.145-148 (1997)

[6.171] {Sect. 6.3.5} I. Blayvas, B.D. Barmashenko, D. Furman, S. Rosenwaks, M.V. Zagidullin: Power optimization of small-scale chemical oxygen-iodine laser with jet-type singlet oxygen generator, IEEE J QE-32, p.2051-2057 (1996)

[6.172] {Sect. 6.3.5} G.D. Hager, C.A. Helms, K.A. Truesdell, D. Plummer, J. Erkkila, P. Crowell: A simplified analytic model for gain saturation and power extraction in the flowing chemical oxygen-iodine laser, IEEE J QE-32, p.1525-1536 (1996)

[6.173] {Sect. 6.3.6} S. Pau, G. Bjork, J. Jacobson, Y. Yamamoto: Fundamental thermodynamic limit of laser efficiency, IEEE J QE-32, p.567-573 (1996)

[6.174] {Sect. 6.4.0} A. Sennaroglu: Experimental determination of fractional thermal loading in an operating diode-pumped Nd : YVO4 minilaser at 1064 nm, Appl Opt 38, p.3253-3257 (1999)

[6.175] {Sect. 6.4.0} D.C. Brown: Heat, fluorescence, and stimulated-emission power densities and fractions in Nd:YAG, IEEE J QE-34, p.560-572 (1998)

[6.176] {Sect. 6.4.0} S. Chang, C.C. Hsu, T.H. Huang, S.W. Lin, C.Y. Leaung, T.T. Liu: Heterodyne interferometric measurement of the thermo-optic coefficients of potassium niobate, J Appl Phys 84, p.1825-1829 (1998)

[6.177] {Sect. 6.4.0} L.C.O. Dacal, A.M. Mansanares, E.C. daSilva: Heat source distribution, vertical structure, and coating influences on the temperature of operating 0.98 mu m laser diodes: Photothermal reflectance measurements, J Appl Phys 84, p.3491-3499 (1998)

[6.178] {Sect. 6.4.0} S.L. Huang, W.L. Wu, P.L. Huang: Measurement of temperature gradient in diode-laser-pumped high-power solid-state laser by low-coherence reflectometry, Appl Phys Lett 73, p.3342-3344 (1998)

[6.179] {Sect. 6.4.0} A. Sennaroglu, B. Pekerten: Experimental and numerical investigation of thermal effects in end- pumped Cr4+:forsterite lasers near room temperature, IEEE J QE-34, p.1996-2005 (1998)

[6.180] {Sect. 6.4.0} A. Sennaroglu: Comparative experimental investigation of thermal loading in continuous-wave Cr4+:forsterite lasers, Appl Opt 37, p.1627-1634 (1998)

[6.181] {Sect. 6.4.0} M. Tsunekane, N. Taguchi, H. Inaba: Reduction of thermal effects in a diode-end-pumped, composite Nd: YAG rod with a sapphire end, Appl Opt 37, p.3290-3294 (1998)

[6.182] {Sect. 6.4.0} R. Weber, B. Neuenschwander, M. MacDonald, M.B. Roos, H.P. Weber: Cooling schemes for longitudinally diode laser-pumped Nd:YAG rods, IEEE J QE-34, p.1046-1053 (1998)

[6.183] {Sect. 6.4.0} M. Mehendale, T.R. Nelson, F.G. Omenetto, W.A. Schroeder: Thermal effects in laser pumped Kerr-lens modelocked Ti: sapphire lasers, Opt Commun 136, p.150-159 (1997)

[6.184] {Sect. 6.4.0} J.J. Kasinski, R.L. Burnham: Near-diffraction-limited, high-energy, high-power, diode-pumped laser using thermal aberration correction with aspheric diamond-turned optics, Appl. Opt. 35, p.5949-4954 (1996)

[6.185] {Sect. 6.4.0} C. Pfistner, R. Weber, H.P. Weber, S. Merazzi, R. Gruber: Thermal Beam Distortions in End-Pumped Nd:YAG Nd:GSGG, and Nd:YLF Rods, IEEE J. QE-30, p.1605-1615 (1994)

[6.186] {Sect. 6.4.0} A.K. Cousins: Temperature and Thermal Stress Scaling in Finite-Length End-Pumped Laser Rods, IEEE J. QE-28, p.1057-1069 (1992)

[6.187] {Sect. 6.4.0} Z. Zeng, H. Shen, M. Huang, H. Xu, R. Zeng, Y. Zhou, G. Yu, C. Huang: Measurement of the refractive index and thermal refractive index coefficients of Nd:YAP crystal, Appl. Opt. 29, p.1281-1286 (1990)

[6.188] {Sect. 6.4.0} M.S. Mangir, D.A. Rockwell: Measurements of Heating and Energy Storage in Flashlamp-Pumped Nd:YAG and Nd-Doped Phosphate Laser Glasses, IEEE J. QE-22, p.574-581 (1986)

[6.189] {Sect. 6.4.0} E. Friedmann, L. Poole, A. Cherdak, W. Houghton: Absorption coefficient instrument for turbid natural waters, Appl. Opt. 19, p.1688-1693 (1980)

[6.190] {Sect. 6.4.0} R.F. Hotz: Thermal Transient Effects in Repetitively Pulsed Flashlamp-Pumped YAG:Nd,Lu Laser Material, Appl. Opt. 12, p.1834-1838 (1973)

[6.191] {Sect. 6.4.0} J.A. Curcio, C.C. Petty: The near infrared absorption spectrum of liquid water, J. Opt. Soc. Am. 41, p.302-304 (1951)

[6.192] {Sect. 6.4.0} N. Hodgson, H. Weber: Influence of Sperical Aberration of the Active Medium on the Performance of Nd:YAG Lasers, IEEE J. QE-29, p.2497-2507 (1993)

[6.193] {Sect. 6.4.0} M.E. Innocenzi, H.T. Yura, C.L. Fincher, R.A. Fields: Thermal modeling of continuous-wave end-pumped solid-state lasers, Appl. Phys. Lett. 56, p.1831-1833 (1990)

[6.194] {Sect. 6.4.0} U.O. Farrukh, A.M. Buoncristiani, E.C. Byvik: An Analysis of the Temperature Distribution in Finite Solid-State Laser Rods, J. Quantum Electron. 24, p.2253-2263 (1988)

[6.195] {Sect. 6.4.0} C.S. Hoefer, K.W. Kirby, L.G. DeShazer: Thermo-optic properties of Garnet laser crystals, J. Opt. Soc. Am. B 5, p.2327-2332 (1988)

[6.196] {Sect. 6.4.0} K. Mann, H. Weber: Surface heat transfer coefficient, heat efficiency and temperature of pulsed solid state lasers, J. Appl. Phys. 64, p.1015-1021 (1988)

[6.197] {Sect. 6.4.0} T.J. Kane, J.M. Eggleston, R.L. Byer: The Slab Geometry Laser – Part II: Thermal Effects in a Finite Slab, IEEE J. QE-21, p.1195-1210 (1985)

[6.198] {Sect. 6.4.0} J.M. Eggleston, T.J. Kane, K. Kuhn, J. Unternahrer, R.L. Byer: The Slab Geometry Laser – Part I: Theory, IEEE J. QE-20, p.289-301 (1984)

[6.199] {Sect. 6.4.0} S.B. Sutton, G.F. Albrecht: Optical distortion in end-pumped solid-state rod lasers, Appl. Opt. 32, p.5256-5269 (1983)

[6.200] {Sect. 6.4.0} K.R. Richter, W. Koechner: Electrical Analogy of Transient Heat Flow in Laser Rods, Appl. Phys. 3, p.205-212 (1974)

[6.201] {Sect. 6.4.0} W. Koechner: Transient thermal profile in optically pumped laser rods, J. Appl. Phys. 44, p.3162-3170 (1973)

[6.202] {Sect. 6.4.0} M.K. Chun, J.T. Bischoff: Thermal Transient Effects in Optically Pumped Repetitively Pulsed Lasers, IEEE J. QE-7, p.200-202 (1971)

[6.203] {Sect. 6.4.1} P.J. Hardman, W.A. Clarkson, G.J. Friel, M. Pollnau, D.C. Hanna: Energy-transfer upconversion and thermal lensing in high-power end- pumped Nd : YLF laser crystals, IEEE J QE-35, p.647-655 (1999)

[6.204] {Sect. 6.4.1} M. Tsunekane, N. Taguchi, H. Inaba: Improvement of thermal effects in a diode-end-pumped, composite Tm : YAG rod with undoped ends, Appl Opt 38, p.1788-1791 (1999)

[6.205] {Sect. 6.4.1} D.Y. Zhang, H.Y. Shen, W. Liu, G.F. Zhang, W.Z. Chen, G. Zhang, R.R. Zeng, C.H. Huang, W.X. Lin, J.K. Liang: The thermal refractive index coefficients of 7.5 mol % Nb : KTiOPO4 crystals, J Appl Phys 86, p.3516-3518 (1999)

[6.206] {Sect. 6.4.1} J.L. Blows, J.M. Dawes, T. Omatsu: Thermal lensing measurements in line-focus end-pumped neodymium yttrium aluminium garnet using holographic lateral shearing interferometry, J Appl Phys 83, p.2901-2906 (1998)

[6.207] {Sect. 6.4.1} J. Calatroni, A. Marcano, R. Escalona, P. Sandoz: Visualization and measurement of a stationary thermal lens using spectrally resolved white light interferometry, Opt Commun 138, p.1-5 (1997)

[6.208] {Sect. 6.4.1} M. Shimosegawa, T. Omatsu, A. Hasegawa, M. Tateta, I. Ogura: Transient thermal lensing measurement in a laser diode pumped NdxY1-xAl3 (BO3) (4) laser using a holographic shearing interferometer, Opt Commun 140, p.237-241 (1997)

[6.209] {Sect. 6.4.1} S.D. Jackson, J.A. Piper: Thermally induced strain and bire-fringence calculations for a Nd: YAG rod encapsulated in a solid pump light collector, Appl Opt 35, p.1409-1423 (1996)

[6.210] {Sect. 6.4.1} S.D. Jackson, J.A. Piper: Encapsulated rod for efficient thermal management in diode- side-pumped Nd:YAG lasers, Appl Opt 35, p.2562-2565 (1996)

[6.211] {Sect. 6.4.1} X.H. Lu, G.Y. Ru, Q. Lin, S.M. Wang: Analysis of the proper-ties of self compensation for thermal distortion in a Eckige-Schraube laser, Opt Commun 128, p.55-60 (1996)

[6.212] {Sect. 6.4.1} A. Mcinnes, J. Richards: Thermal effects in a coplanar-pumped folded-zigzag slab laser, IEEE J QE-32, p.1243-1252 (1996)

[6.213] {Sect. 6.4.1} B. Neuenschwander, R. Weber, H.P. Weber: Thermal lens and beam properties in multiple longitudinally diode laser pumped Nd:YAG slab lasers, IEEE J QE-32, p.365-370 (1996)

[6.214] {Sect. 6.4.1} H.J. Eichler, A. Haase, R. Menzel, A. Siemoneit: Thermal Lensing and Depolarization in a Highly Pumped Nd:YAG-Laser-Amplifier, J. Phys. D: Appl. Phys. 26, p.1884-1891 (1993)

[6.215] {Sect. 6.4.1} T.Y. Fan: Heat Generation in Nd:YAG and Yb:YAG, IEEE J. QE-29, p.1457-1459 (1993)

[6.216] {Sect. 6.4.1} N. Hodgson, C. Rahlff, H. Weber: Dependence of the refractive power of Nd:YAG rods on the intracavity intensity, Opt. Laser Technol. 25, p.179-185 (1993)

[6.217] {Sect. 6.4.1} J. Frauchiger, P. Albers, H.P. Weber: Modeling of Thermal Lensing and Higher Order Ring Mode Oscillation in End-Pumped CW Nd:YAG Lasers, IEEE J. QE-28, p.1046-1056 (1992)

[6.218] {Sect. 6.4.1} H. Vanherzeele: Continuous-wave dual rod Nd:YLF laser with dynamic lensing compensation, Appl. Opt. 28, p.4042-4044 (1989)

[6.219] {Sect. 6.4.1} B. Struve, P. Fuhrberg, W. Luhs, G. Litfin: Thermal lensing and laser operation of flashlamp-pumped Cr:GSAG, Opt. Comm. 65, p.291-296 (1988)

[6.220] {Sect. 6.4.1} J.C. Lee, S.D. Jacobs: Refractive index and dn/dT of Cr:Nd: GSGG at 1064 nm, Appl. Opt. 26, p.777-778 (1987)

[6.221] {Sect. 6.4.1} K.P. Driedger, W. Krause, H. Weber: Average refractive powers of an Alexandrit laserrod, Opt. Comm. 57, p.403406 (1986)

[6.222] {Sect. 6.4.1} J.S. Uppal, J.C. Monga, D.D. Bhawalkar: Study of thermal effects in an Nd doped phosphate glass laser rod, IEEE J. QE-22, p.2259-2265 (1986)

[6.223] {Sect. 6.4.1} L. Horowitz Y.B. Band, O. Kafri, D.F. Heller: Thermal lensing analysis of alexandrite laser rods by moire deflectometry, Appl. Opt. 23, p.2229-2231 (1984)

[6.224] {Sect. 6.4.1} J.E. Murray: Pulsed Gain and Thermal Lensing of Nd:LiF4, IEEE J. QE-19, p.488-491 (1983)

[6.225] {Sect. 6.4.1} D.C. Brown, J. A. Abate, L. Lund, J. Waldbillig: Passively switched double-pass active mirror system, Appl. Opt. 20, p.1588-1594 (1981)

[6.226] {Sect. 6.4.1} D.C. Brown, J.H. Kelly, J.A. Abate: Active-Mirror Amplifiers: Progress and Prospects, IEEE J. QE-17, p.1755-1765 (1981)

[6.227] {Sect. 6.4.1} H.P. Kortz, R. Iffländer, H. Weber: Stability and beam divergence of multimode lasers with internal variable lenses, Appl. Opt. 20, p.4124-4134 (1981)

[6.228] {Sect. 6.4.1} W.E. Martin, J.B. Trenholme, G.T. Linford, S.M. Yarema, C.A. Hurley: Solid-State Disk Amplifiers for Fusion-Laser Systems, IEEE J. QE-17, p.1744-1755 (1981)

[6.229] {Sect. 6.4.1} N.L. Boling et al.: Empirical Relationships for Predicting Nonlinear Refractive Index Changes in Optical Solids, IEEE J. QE-14, p.601-608 (1978)

[6.230] {Sect. 6.4.1} T.J. Gleason, J.S. Kruger, R.M. Curnutt: Thermally Induced Focusing in a Nd:YAG Laser Rod at Low Input Powers, Appl. Opt. 12, p.2942-2946 (1973)

[6.231] {Sect. 6.4.1} D.D. Young, K.C. Jungling, T.L. Williamson, E.R. Nichols: Holographic Interferometry Measurement of the Thermal Refractive Index Coefficient and the Thermal Expansion Coefficient of Nd:YAG and Nd:YALO, IEEE J. QE-8, p.720-721 (1972)

[6.232] {Sect. 6.4.1} F.A. Levine: TEM00 Enhancement in CW Nd-YAG by Thermal Lensing Compensation, IEEE J. QE-7, p.170-172 (1971)

[6.233] {Sect. 6.4.1} G. Slack, D. Oliver: Thermal conductivity of garnets and phonon scattering by rare-earth-ions, Phys. Rev. B 4, p.592-609 (1971)

[6.234] {Sect. 6.4.1} D.C. Burnham: Simple Measurement of Thermal Lensing Effects in Laser Rods, Appl. Opt. 9, p.1727-1728 (1970)

[6.235] {Sect. 6.4.1} J.D. Forster, L.M. Osterink: Thermal Effects in a Nd:YAG Laser, J. Appl. Phys. 41, p.3656-3663 (1970)

[6.236] {Sect. 6.4.1} W. Koechner: Absorbed Pump Power, Thermal Profile and Stresses in a cw Pumped Nd:YAG Crystal, Appl. Opt. 9, p.1429-1434 (1970)

[6.237] {Sect. 6.4.1} W. Koechner: Thermal Lensing in a Nd:YAG Laser Rod, Appl. Opt. 9, p.2548-2553 (1970)

[6.238] {Sect. 6.4.1} L.M. Osterink, J.D. Foster: Thermal effects and transverse mode control in a Nd:YAG laser, Appl. Phys. Lett. 12, p.128-131 (1968)

[6.239] {Sect. 6.4.1} Y. Liao, R.J.D. Miller, M.R. Armstrong: Pressure tuning of thermal lensing for high-power scaling, Optics Letters 24, p.1343-1345 (1999)

[6.240] {Sect. 6.4.1} A. Agnesi, E. Piccinini, G.C. Reali: Influence of thermal effects in Kerr-lens mode-locked femtosecond Cr4+: Forsterite lasers, Opt Commun 135, p.77-82 (1997)

[6.241] {Sect. 6.4.1} R. Koch: Self-adaptive optical elements for compensation of thermal lensing effects in diode end-pumped solid state lasers – Proposal and preliminary experiments, Opt Commun 140, p.158-164 (1997)

[6.242] {Sect. 6.4.1} J. Song, A.P. Liu, K. Okino, K. Ueda: Control of the thermal lensing effect with different pump light distributions, Appl Opt 36, p.8051-8055 (1997)

[6.243] {Sect. 6.4.1} S. Backus, C.G. Durfee, G. Mourou, H.C. Kapteyn, M.M. Murnane: 0.2-TW laser system at 1 kHz, Optics Letters 22, p.1256-1258 (1997)

[6.244] {Sect. 6.4.2} N. Kugler, S. Dong, Q. Lü, H. Weber: Investigation of the misalignment sensitivity of a birefringence-compensated two-rod Nd:YAG laser system, Appl. Opt. 36, p.9359-9366 (1997)

[6.245] {Sect. 6.4.2} M. Ohmi, M. Akatsuka, K. Ishikawa, K. Naito, Y. Yonezawa, Y. Nishida, M. Yamanaka, Y. Izawa, S. Nakai: High-sensitivity two-dimensional thermal- and mechanical-stress-induced birefringence measurements in a Nd:YAG rod, Appl. Opt. 33, p.6368-6372 (1994)

[6.246] {Sect. 6.4.2} S.Z. Kurtev, O.E. Denchev, S.D. Savov: Effects of thermally induced birefringence in high-output-power electro-optically Q-switched Nd:YAG lases and their compensation, Appl. Opt. 32, p.278-285 (1993)

[6.247] {Sect. 6.4.2} J. Richards: Birefringence compensation in polarization coupled lasers, Appl. Opt. 26, p.2514-2517 (1987)

[6.248] {Sect. 6.4.2} G. Giuliani, R.L. Byer, Y.K. Park: Radial Birefringent Element and Its Application to Laser Resonator Design, Optics Letters 5, p.491-493 (1980)

[6.249] {Sect. 6.4.2} G. Giuliani, P. Ristori: Polarization flip cavities: A new approach to laser resonators, Optics Commun. 35, p.109-113 (1980)

[6.250] {Sect. 6.4.2} G. Giuliani, Y.K. Park, R.L. Byer: Radial birefringent element and ist application to laser resonator design, Opt. Lett. 5, p.491-493 (1980)

[6.251] {Sect. 6.4.2} A. L. Bloom: Modes of a laser resonator containing tilted birefringent plates, J. Opt. Soc. Am. 64, p.447-452 (1974)

[6.252] {Sect. 6.4.2} J. Sherman: Thermal compensation of a cw-pumped Nd:YAG laser, Appl Opt 37, p.7789-7796 (1998)

[6.253] {Sect. 6.4.2} W.A. Clarkson, N.S. Felgate, D.C. Hanna: Simple method for reducing the depolarization loss resulting from thermally induced birefringence in solid-state lasers, Optics Letters 24, p.820-822 (1999)

[6.254] {Sect. 6.4.2} D. MonzonHernandez, A.N. Starodumov, A.R.B.Y. Goitia, V.N. Filippov, V.P. Minkovich, P. Gavrilovic: Stress distribution and birefringence measurement in double-clad fiber, Opt Commun 170, p.241-246 (1999)

[6.255] {Sect. 6.4.2} J. Zhang, M. Quade, Y. Liao, S. Falter, K.M. Du, P. Loosen: Polarization characteristics of a Nd:YAG laser side pumped by diode laser bars, Appl Opt 36, p.7725-7729 (1997)

[6.256] {Sect. 6.4.2} M.P. Murdough, C.A. Denman: Mode-volume and pump-power limitations in injection-locked TEM (00) Nd:YAG rod lasers, Appl Opt 35, p.5925-5936 (1996)

[6.257] {Sect. 6.4.2} Q. Lü, N. Kugler, H. Weber, S. Dong, N. Müller, U. Wittrock: A novel approach for compensation of birefringence in cylindrical Nd:YAG rods, Opt. Quant. Electron. 28, p.57-69 (1996)

[6.258] {Sect. 6.4.2} W.C. Scott, M. de Wit: Birefringence compensation and TEM00 Mode enhancement in a Nd:YAG Laser, Appl. Phys. Lett. 18, p.3-4 (1971)

[6.259] {Sect. 6.4.3} D.C. Brown: Nonlinear thermal and stress effects and scaling behavior of YAG slab amplifiers, IEEE J QE-34, p.2393-2402 (1998)

[6.260] {Sect. 6.4.3} W. Koechner: Rupture Stess and Modulus of Elasticity for Nd:YAG Crystals, Appl. Phys. 2, p.279-280 (1973)

[6.261] {Sect. 6.5.0} J.K. Watts: Theory of Multiplate Resonant Reflectors, Appl. Opt. 7, p.1621-1624 (1968)

[6.262] {Sect. 6.5.0} D.G. Peterson, A. Yariv: Interferometry and Laser Control with Solid Fabry-Perot Etalons, Appl. Opt. 5, p.985-991 (1966)

[6.263] {Sect. 6.5.0} G.D. Boyd, H. Kogelnik: Generalized Confocal Resonator Theory, Bell Syst. Tech. J. 41, p.1347-1369 (1962)

[6.264] {Sect. 6.5.0} A.G. Fox, T. Li: Resonant Modes in a Maser Interferometer, Bell Syst. Tech. J. 40, p.453-489 (1961)

[6.265] {Sect. 6.5.0} C. Palma: Complex dynamics of a beam in a Gaussian cavity, Opt Commun 129, p.120-133 (1996)

[6.266] {Sect. 6.5.2} O. Emile, D. Chauvat, A. LeFloch, F. Bretenaker: Temporal behavior of an unstable optical cavity, Optics Letters 24, p.22-24 (1999)

[6.267] {Sect. 6.5.2} M. Endo, M. Kawakami, K. Nanri, S. Takeda, T. Fujioka: Two-dimensional simulation of an unstable resonator with a stable core, Appl Opt 38, p.3298-3307 (1999)

[6.268] {Sect. 6.5.2} G.P. Karman, J.P. Woerdman: Fractal structure of eigenmodes of unstable cavity lasers, Optics Letters 23, p.1909-1911 (1998)

[6.269] {Sect. 6.5.2} A. Torre, C. Petrucci: Two-dimensional simulation of a high-gain, generalized self-filtering, unstable resonator, Appl Opt 36, p.2499-2505 (1997)

[6.270] {Sect. 6.5.2} S. De Silvestri, P. Laporta, V. Magni, G. Valentini, G. Cerullo: Comparative Analysis of Nd:YAG Unstable Resonators with Super-Gaussian Variable Reflectance Mirrors, Opt. Commun. 77, p.179-184 (1990)

[6.271] {Sect. 6.5.2} S. De Silvestri, P. Laporta, V. Magni, O. Svelto, B. Majocchi: Unstable laser resonators with super Gaussian mirrors, Opt. Lett. 13, p.201-203 (1988)

[6.272] {Sect. 6.5.2} P.G. Gobbi, G.C. Reali: Mode analysis of a self filtering unstable resonator with a gaussian transmission aperture, Opt. Commun. 57p.355-359 (1986)

[6.273] {Sect. 6.5.2} M.E. Smithers: Unstable resonator with aspherical mirrors, J. Opt. Soc. Am. 72, p.1183-1186 (1982)

[6.274] {Sect. 6.5.2} T.F. Ewanizky: Ray-transfer-matrix approach to unstable resonator analysis, Appl. Opt. 18, p.724-727 (1979)

[6.275] {Sect. 6.5.2} R.R. Butts, P.V. Avizonis: Asymptotic analysis of unstable laser resonators with circular mirrors, J. Opt. Soc. Am. 68, p.1072-1078 (1978)

[6.276] {Sect. 6.5.2} A.E. Siegman: A Canonical Formulation for Analyzing Multi-element Unstable Resonators, IEEE J. QE-12, p.35-39 (1976)

[6.277] {Sect. 6.5.2} A.E. Siegman: Unstable Optical Resonators, Appl. Opt. 13, p.353-367 (1974)

[6.278] {Sect. 6.5.2} P. Horwitz: Asymptotic theory of unstable resonator modes, J. Opt. Soc. Am. 63, p.1528-1543 (1973)

[6.279] {Sect. 6.5.2} A.E. Siegman, H.Y. Miller: Unstable Optical Resonator Loss Calculations Using the Prony Method, Appl. Opt. 9, p.2729-2736 (1970)

[6.280] {Sect. 6.5.2} R.L. Sanderson, W. Streifer: Unstable Laser Resonator Modes, Appl. Opt. 8, p.2129-2136 (1969)

[6.281] {Sect. 6.5.2} L. Bergstein: Modes of Stable and Unstable Optical Resonators, Appl. Opt. 7, p.495-504 (1968)

[6.282] {Sect. 6.5.2} G.S. McDonald, G.H.C. New, J.P. Woerdman: Excess noise in low Fresnel number unstable resonators, Opt Commun 164, p.285-295 (1999)

[6.283] {Sect. 6.5.2} J.F. Pinto, L. Esterowitz: Unstable Cr:LiSAF baser resonator with a variable reflectivity output coupler, Appl Opt 37, p.3272-3275 (1998)

[6.284] {Sect. 6.5.2} M.A. Vaneijkelenborg, A.M. Lindberg, M.S. Thijssen, J.P. Woerdman: Higher order transverse modes of an unstable-cavity laser, IEEE J QE-34, p.955-965 (1998)

[6.285] {Sect. 6.5.2} S.A. Biellak, G. Fanning, Y. Sun, S.S. Wong, A.E. Siegman: Reactive-ion-etched diffraction-limited unstable resonator semiconductor lasers, IEEE J QE-33, p.219-230 (1997)

[6.286] {Sect. 6.5.2} Y.J. Cheng, C.G. Fanning, A.E. Siegman: Transverse-mode astigmatism in a diode-pumped unstable resonator Nd:WO4 laser, Appl Opt 36, p.1130-1134 (1997)

[6.287] {Sect. 6.5.2} E. Galletti, E. Stucchi, D.V. Willetts, M.R. Harris: Transverse-mode selection in apertured super-Gaussian resonators: An experimental and numerical investigation for a pulsed CO2 Doppler lidar transmitter, Appl Opt 36, p.1269-1277 (1997)

[6.288] {Sect. 6.5.2} R. Massudi, M. Piche: Nearly flat-top laser beams from unstable resonators with internal spatial filtering, Opt Commun 142, p.61-65 (1997)

[6.289] {Sect. 6.5.2} S. Chandra, T.H. Allik, J.A. Hutchinson: Nonconfocal unstable resonator for solid-state dye lasers based on a gradient-reflectivity mirror, Optics Letters 20, p.2387-2389 (1995)

[6.290] {Sect. 6.5.2} N. Hodgson, G. Bostanjoglo, H. Weber: Multirod unstable resonators for high-power solid-state lasers, Appl. Opt. 32, p.5902-5917 (1993)

[6.291] {Sect. 6.5.2} N. Hodgson, G. Bostanjoglo, H. Weber: The near-concentric unstable resonator (NCUR) – an improved resonator design for high power solid state lasers, Opt. Commun. 99, p.75-81 (1993)

[6.292] {Sect. 6.5.2} V. Magni, S. De Silvestri, L.-J. Qian, O. Svelto: Rod-imaging superggaussian unstable resonator for high poser solid-state lasers, Opt. Commun. 94, p.87-91 (1992)

[6.293] {Sect. 6.5.2} N. Hodgson, H. Weber: High-power solid-state lasers with unstable resonators, Opt. Quantum Electron. 22, p.39-55 (1990)

[6.294] {Sect. 6.5.2} N. Hodgson, H. Weber: Unstable Resonators with Excited Converging Wave, IEEE J. QE-26, p.731-738 (1990)

[6.295] {Sect. 6.5.2} A. Parent, P. Lavigne: Variable reflectivity unstable resonators for coherent laser radar emitters, Appl. Opt. 28, p.901-903 (1989)

[6.296] {Sect. 6.5.2} S. De Silvestri, P. Laporta, V. Magni, O. Svelto: Solid state laser unstable resonators with tapered reflectivity mirrors – The super Gaussian approach, IEEE J. QE-24, p.1172-1177 (1988)

[6.297] {Sect. 6.5.2} K.J. Snell, N. McCarthy, M. Piché: Single Transverse Mode Oscillation from an unstable Resonator Nd:YAG Laser Using a Variable Reflectivity Mirror, Opt. Commun. 65, p.377-382 (1988)

[6.298] {Sect. 6.5.2} D.T. Harter, J.C. Walling: Low-magnification unstable resonators used with ruby and alexandrite lasers, Opt. Lett. 11, p.706-708 (1986)

[6.299] {Sect. 6.5.2} A.H. Paxton, W.P. Latham, Jr.: Unstable Resonators with 90 beam rotation, Appl. Opt. 25, p.2939-2946 (1986)

[6.300] {Sect. 6.5.2} P.G. Gobbi, S. Morosi, G. C. Reali, A. S. Zarkasi: Novel unstable resonator configuration with a self-filtering aperture: experimental characterization of the Nd:YAG loaded cavity, Appl. Opt. 24, p.26-33 (1985)

[6.301] {Sect. 6.5.2} N. McCarthy, P. Lavigne: Large-size Gaussian mode in unstable resonators using Gaussian mirrors, Opt. Lett. 10, p.553-555 (1985)

[6.302] {Sect. 6.5.2} A.H. Paxton: Unstable resonators with negative fresnel numbers, Opt. Lett. 11, p.76-78 (1985)

[6.303] {Sect. 6.5.2} P.G. Gobbi, G.C. Reali: A novel unstable resonator configuration with a self filtering aperture, Opt. Commun. 52p.195-198 (1984)

[6.304] {Sect. 6.5.2} M.E. Smithers, Th.R. Ferguson: Unstable optical resonators with linear magnification, Appl. Opt. 23, p.3718-3724 (1984)

[6.305] {Sect. 6.5.2} T. Kedmi, D. Treves: Injection-locking optimization in unstable resonators, Appl. Opt. 20p.2108-2112 (1981)

[6.306] {Sect. 6.5.2} O.L. Bourne, P.E. Dyer: A novel stable-unstable resonator for beam control of raregas hailde lasers, Opt. Commun. 31, p.193-196 (1979)

[6.307] {Sect. 6.5.2} W.H. Southwell: Mode discrimination of unstable resonators with spatial filters and by phase modification, Opt. Lett. 7, p.193-195 (1979)

[6.308] {Sect. 6.5.2} A.H. Paxton, T.C. Salvi: Unstable optical resonator with self-imaging aperture, Opt. Commun. 26p.305-307 (1978)

[6.309] {Sect. 6.5.2} R.L. Herbst, H. Komine, R.L. Byer: A 200mJ unstable resonator Nd:YAG oscillator, Opt. Commun. 21, p.36712 (1977)

[6.310] {Sect. 6.5.2} T.F. Ewanizky, J.M. Craig: Negative-branch unstable resonator Nd:YAG laser, Appl. Opt. 15, p.1465-1469 (1976)

[6.311] {Sect. 6.5.2} R.J. Freiberg, P.P. Chenausky, C.J. Buczek: Unidirectional Unstable Ring Lasers, Appl. Opt. 12, p.1140-1144 (1973)

[6.312] {Sect. 6.5.2} W. Streifer: Unstable optical resonators and waveguides, IEEE J. QE-4, p.229-230 (1968)

[6.313] {Sect. 6.5.2} S.R. Baron: Optical Resonators in the Unstable Region, Appl. Opt. 6, p.861-864 (1967)

[6.314] {Sect. 6.5.2} A.E. Siegman, E. Arrathoon: Modes in Unstable Optical Resonators and Lens Waveguides, IEEE J. QE-3, p.156-163 (1967)

[6.315] {Sect. 6.5.2} W.K. Kahn: Unstable Optical Resonators, Appl. Opt. 5, p.407-413 (1966)

[6.316] {Sect. 6.6.5} F. EncinasSanz, I. Leyva, J.M. Guerra: Time resolved pattern evolution in a large aperture laser, Phys Rev Lett 84, p.883-886 (2000)

[6.317] {Sect. 6.6.5} V.N. Belyi, N.S. Kazak, N.A. Khilo: Properties of parametric frequency conversion with Bessel light beams, Opt Commun 162, p.169-176 (1999)

[6.318] {Sect. 6.6.5} D.M. Fondevila, A.A. Hnilo: Coupled dye laser modes: experimental study of the dynamics, Opt Commun 162, p.324-332 (1999)

[6.319] {Sect. 6.6.5} S.P. Hegarty, G. Huyet, J.G. McInerney, K.D. Choquette: Pattern formation in the transverse section of a laser with a large fresnel number, Phys Rev Lett 82, p.1434-1437 (1999)

[6.320] {Sect. 6.6.5} M. Santarsiero, F. Gori, R. Borghi, G. Guattari: Evaluation of the modal structure of light beams composed of incoherent mixtures of Hermite-Gaussian modes, Appl Opt 38, p.5272-5281 (1999)

[6.321] {Sect. 6.6.5} M. Vallet, M. Brunel, F. Bretenaker, M. Alouini, A. LeFloch, G.P. Agrawal: Polarization self-modulated lasers with circular eigenstates, Appl Phys Lett 74, p.3266-3268 (1999)

[6.322] {Sect. 6.6.5} M.R. Wang, X.G. Huang: Subwavelength-resolvable focused non-Gaussian beam shaped with a binary diffractive optical element, Appl Opt 38, p.2171-2176 (1999)

[6.323] {Sect. 6.6.5} M.A. Clifford, J. Arlt, J. Courtial, K. Dholakia: High-order Laguerre-Gaussian laser modes for studies of cold atoms, Opt Commun 156, p.300-306 (1998)

[6.324] {Sect. 6.6.5} A. Cutolo, M. Dellanoce, L. Zeni: Real-time measurement of transverse-mode-mixing effects in a Q-switched Nd:YAG laser, Appl Opt 35, p.2544-2547 (1996)

[6.325] {Sect. 6.6.5} J.R. Marciante, G.P. Agrawal: Nonlinear mechanisms of filamentation in broad-area semiconductor lasers, IEEE J QE-32, p.590-596 (1996)

[6.326] {Sect. 6.6.5} K.P. Driedger, B. Lu, H. Weber: Multimode Resonators, insensitive against thermal lensing, Optica Acta 32, p.847-854 (1985)

[6.327] {Sect. 6.6.5} R.L. Phillips, L.C: Andrews: Spot size and divergence for Laguerre Gaussian beams of any order, Appl. Opt. 22, p.643-644 (1983)

[6.328] {Sect. 6.6.5} D. Ryter, M. Von Allmen: Intensity of Hot Spots in Multimode Laser Beams, IEEE J. QE-17, p.2015-2017 (1981)

[6.329] {Sect. 6.6.5} R. Ifflander, H.P. Kortz, H. Weber: Beam divergence and refractive power of directly coated solid state lasers, Opt. Comm. 29, p.223-226 (1979)

[6.330] {Sect. 6.6.5} M. Hercher: The Spherical Mirror Fabry-Perot Interferometer, Appl. Opt. 7, p.951-966 (1968)

[6.331] {Sect. 6.6.5} T. Li, H. Zucker: Modes of a Fabry-Perot Laser Resonator with Output-Coupling Apertures, J. Opt. Soc. Am. 57, p.984-986 (1967)

[6.332] {Sect. 6.6.5} V. Evtuhov, A.E. Siegman: A "Twisted-Mode" Technique for Obtaining Axially Uniform Energy Density in a Laser Cavity, Appl. Opt. 4, p.142-143 (1965)

[6.333] {Sect. 6.6.5} Q. Lin, L.G. Wang: Optical resonators producing partially coherent flat-top beams, Opt Commun 175, p.295-300 (2000)

[6.334] {Sect. 6.6.5} C. Gao, H. Laabs, H. Weber, T. Brand, N. Kugler: Symmetrization of astigmatic high power diode laser stacks, Opt. Quant. Electron. 31, p.1207-1218 (1999)

[6.335] {Sect. 6.6.5} X.G. Huang, M.R. Wang, C. Yu: High-efficiency flat-top beam shaper fabricated by a nonlithographic technique, Opt. Eng. 38, p.208-213 (1999)

[6.336] {Sect. 6.6.5} H. Laabs, C.Q. Gao, H. Weber: Twisting of three-dimensional Hermite-Gaussian beams, J. Mod. Optic46p.709-719 (1999)

[6.337] {Sect. 6.6.5} T.Y. Cherezova, S.S. Chesnokov, L.N. Kaptsov, A.V. Kudryashov: Super-Gaussian laser intensity output formation by means of adaptive optics, Opt Commun 155, p.99-106 (1998)

[6.338] {Sect. 6.6.5} F. Nikolajeff, S. Hard, B. Curtis: Diffractive microlenses replicated in fused silica for excimer laser-beam homogenizing, Appl Opt 36, p.8481-8489 (1997)

[6.339] {Sect. 6.6.5} C. Parigger, Y. Tang, D.H. Plemmons, J.W.L. Lewis: Spherical aberration effects in lens-axicon doublets: theoretical study, Appl Opt 36, p.8214-8221 (1997)

[6.340] {Sect. 6.6.5} K.S. Repasky, J.K. Brasseur, J.G. Wessel, J.L. Carlsten: Correcting an astigmatic, non-Gaussian beam, Appl Opt 36, p.1536-1539 (1997)

[6.341] {Sect. 6.6.5} D. Shafer: Gaussian to flat-top in diffraction far-field, Appl Opt 36, p.9092-9093 (1997)

[6.342] {Sect. 6.6.5} W.A. Clarkson, D.C. Hanna: Two-mirror beam-shaping technique for high-power diode bars, Optics Letters 21, p.375-377 (1996)

[6.343] {Sect. 6.6.5} X.G. Deng, Y.P. Li, D.Y. Fan, Y. Qiu: Pure-phase plates for super-Gaussian focal-plane irradiance profile generations of extremely high order, Optics Letters 21, p.1963-1965 (1996)

[6.344] {Sect. 6.6.5} T. Graf, J.E. Balmer: Laser beam quality, entropy and the limits of beam shaping, Opt Commun 131, p.77-83 (1996)

[6.345] {Sect. 6.6.5} K. Nemoto, T. Fujii, N. Goto, T. Nayuki, Y. Kanai: Transformation of a laser beam intensity profile by a deformable mirror, Optics Letters 21, p.168-170 (1996)

[6.346] {Sect. 6.6.5} S.G. Chuartzman, D. Krygier, A.A. Hnilo: Pattern formation in a large Fresnel number dye laser, Opt. Comm. 121, p.1-7 (1995)

[6.347] {Sect. 6.6.5} J. Courtial, M.J. Padgett: Performance of a cylindrical lens mode converter for producing Laguerre-Gaussian laser modes, Opt Commun 159, p.13-18 (1999)

[6.348] {Sect. 6.6.5} J.A. Davis, D.M. Cottrell, J. Campos, M.J. Yzuel, I. Moreno: Bessel function output from an optical correlator with a phase-only encoded inverse filter, Appl Opt 38, p.6709-6713 (1999)

[6.349] {Sect. 6.6.5} M. Arif, M.M. Hossain, A.A.S. Awwal, M.N. Islam: Two-element refracting system for annular Gaussian-to-Bessel beam transformation, Appl Opt 37, p.4206-4209 (1998)

[6.350] {Sect. 6.6.5} P. Paakkonen, J. Turunen: Resonators with Bessel-Gauss modes, Opt Commun 156, p.359-366 (1998)

[6.351] {Sect. 6.6.5} H. Sonajalg, M. Ratsep, P. Saari: Demonstration of the Bessel-X pulse propagating with strong lateral and longitudinal localization in a dispersive medium, Optics Letters 22, p.310-312 (1997)

[6.352] {Sect. 6.6.5} M. Cai, K. Vahala: Highly efficient optical power transfer to whispering-gallery modes by use of a symmetrical dual-coupling configuration, Optics Letters 25, p.260-262 (2000)

[6.353] {Sect. 6.6.5} J.C. Ahn, K.S. Kwak, B.H. Park, H.Y. Kang, J.Y. Kim, O. Kwon: Photonic quantum ring, Phys Rev Lett 82, p.536-539 (1999)

[6.354] {Sect. 6.6.5} T. Harayama, P. Davis, K.S. Ikeda: Nonlinear whispering gallery modes, Phys Rev Lett 82, p.3803-3806 (1999)

[6.355] {Sect. 6.6.5} O. Painter, R.K. Lee, A. Scherer, A. Yariv, J.D. OBrien, P.D. Dapkus, I. Kim: Two-dimensional photonic band-gap defect mode laser, Science 284, p.1819-1821 (1999)

[6.356] {Sect. 6.6.5} C. Gmachl, F. Capasso, E.E. Narimanov, J.U. Nockel, A.D. Stone, J. Faist, D.L. Sivco, A.Y. Cho: High-power directional emission from microlasers with chaotic resonators, Science 280, p.1556-1564 (1998)

[6.357] {Sect. 6.6.5} C. Gmachl, J. Faist, F. Capasso, C. Sirtori, D.L. Sivco, A.Y. Cho: Long-wavelength (9.5-11.5 mu m) microdisk quantum-cascade lasers, IEEE J QE-33, p.1567-1573 (1997)

[6.358] {Sect. 6.6.5} A. Scherer, J.L. Jewell, Y.H. Lee, J.P. Harbison, L.T. Florez: Fabrication of microlasers and microresonator optical switches, Appl. Phys. Lett. 55, p.2724-2726 (1989)

[6.359] {Sect. 6.6.5} K. Kogelnik, T. Li. Laser Beams and Resonators, Appl. Opt. 5, p.1550-1567 (1966)

[6.360] {Sect. 6.6.8} B.D. Lu, B. Zhang, H. Ma: Beam-propagation factor and mode-coherence coefficients of hyperbolic- cosine-Gaussian beams, Optics Letters 24, p.640-642 (1999)

[6.361] {Sect. 6.6.9} R. Simon, N. Mukunda, E.C.G. Sudarshan: Partially coherent beams and a generalized ABCD-law, Optics Commun. 65, p.322-328 (1988)

[6.362] {Sect. 6.6.9} S. Nemoto, T. Makimoto: Generalized spot size for a higher-order beam mode, J. Opt. Soc. Am. 69, p.578-580 (1979)

[6.363] {Sect. 6.6.9} J.P. Campbell, L.G. DeShazer: Near Fields of Truncated-Gaussian Apertures, J. Opt. Soc. Am. 59, p.1427-1429 (1969)

[6.364] {Sect. 6.6.9} W.B. Bridges: J-3-Gaussian Beam Distorsion Caused by Saturable Gain or Loss, IEEE J. QE-4, p.820-827 (1968)

[6.365] {Sect. 6.6.9} M.R. Fetterman, J.C. Davis, D. Goswami, W. Yang, W.S. Warren: Propagation of complex laser pulses in optically dense media, Phys Rev Lett 82, p.3984-3987 (1999)

[6.366] {Sect. 6.6.9} R. Martinezherrero, P.M. Mejias: On the fourth-order spatial characterization of laser beams: New invariant parameter through ABCD systems, Opt Commun 140, p.57-60 (1997)

[6.367] {Sect. 6.6.9} S.A. Amarande: Beam propagation factor and the kurtosis parameter of flattened Gaussian beams, Opt Commun 129, p.311-317 (1996)

[6.368] {Sect. 6.6.9} C. Pare, P.A. Belanger: Propagation law and quasi-invariance properties of the truncated second-order moment of a diffracted laser beam, Opt Commun 123, p.679-693 (1996)

[6.369] {Sect. 6.6.10} R. Menzel, M. Ostermeyer: Fundamental mode determination for guaranteeing diffrraction limited beam quality of lasers with high output powers, Opt. Comm. 149, p.321-325 (1998)

[6.370] {Sect. 6.6.10} S.K. Dixit, S.R. Daulatabad, P.K. Shukla, R. Bhatnagar: Diffraction filtered resonator for Rh6G dye laser transversely pumped by a copper vapor laser, Opt Commun 134, p.149-154 (1997)

[6.371] {Sect. 6.6.10} S. Szatmari, Z. Bakonyi, P. Simon: Active spatial filtering of laser beams, Opt Commun 134, p.199-204 (1997)

[6.372] {Sect. 6.6.10} D. Golla, M. Bode, S. Knoke, W. Schöne, A. Tünnermann: 62-W cw TEM00 Nd:YAG laser side-pumped by fiber-coupled diode lasers, Opt. Lett. 21, p.210-212 (1996)

[6.373] {Sect. 6.6.10} N. Hodgson, B. Ozygus, F. Schabert, H. Weber: Degenerated confocal resonator, Appl. Opt. 32, p.3190-3200 (1993)

[6.374] {Sect. 6.6.10} A. Parent, N. McCarthy, P. Lavigne: Effects of Hard Apertures on Mode Properties of Resonators with Gaussian Reflectivity Mirrors, IEEE J. QE-23, p.222-228 (1987)

[6.375] {Sect. 6.6.10} M. Piche, P. Lavigne, F. Martin, P.A. Belanger: Modes of resonators with internal apertures, Appl. Opt. 22, p.1999-2006 (1983)

[6.376] {Sect. 6.6.10} J. Dembowski, H. Weber: Optimal pinhole radius for fundamental mode operation, Opt. Comm. 42, p.133-137 (1982)

[6.377] {Sect. 6.6.10} L.W. Casperson, S.D. Lunnam: Gaussian Modes in High Loss Laser Resonators, Appl. Opt. 14, p.1193-1199 (1975)

[6.378] {Sect. 6.6.10} L.W. Casperson: Mode Stability of Lasers and Periodic Optical Systems, IEEE J. QE-10, p.629-634 (1974)

[6.379] {Sect. 6.6.10} H. Steffen, J.-P. Lörtscher, G. Herziger: Fundamental Mode Radiation With Solid-State Lasers, IEEE J. QE-8, p.239-245 (1972)

[6.380] {Sect. 6.6.10} W.C. Fricke: Fundamental Mode YAG:Nd Laser Analysis, Appl. Opt. 9, p.2045-2052 (1970)

[6.381] {Sect. 6.6.10} J.M. Moran: Coupling of Power from a Circular Confocal Laser With an Output Aperture, IEEE J. QE-6, p.93-96 (1970)

[6.382] {Sect. 6.6.10} D. Hanna: Astigmatic Gaussian Beams Produced by Axially Asymmetric Laser Cavities, IEEE J. QE-5, p.483-488 (1969)

[6.383] {Sect. 6.6.10} G.T. Mc.Nice, V.E. Derr: Analysis of the Cylindrical Confocal Laser Resonator Having a Single Circular Coupling Aperture, IEEE J. QE-5, p.569-575 (1969)

[6.384] {Sect. 6.6.10} J.G. Skinner, J.E. Geusic: A Diffraction Limited Oscillator, J. Opt. Soc. Am. 52, p.1437-1444 (1962)

[6.385] {Sect. 6.6.10} C. Siegel, T. Graf, J. Balmer, H.P. Weber: Experimental determination of the fundamental-mode diameter in solid- state lasers, Appl Opt 37, p.4902-4906 (1998)

[6.386] {Sect. 6.6.10} S. De Silvestri, V. Magni, O. Svelto, G. Valentini: Lasers with Super-Gaussian Mirrors, IEEE J. QE-26, p.1500-1509 (1990)

[6.387] {Sect. 6.6.10} G. Emiliani, A. Piegari, S. De Silvestri, P. Laporta, V. Magni: Optical coatings with variable reflectance for laser mirrors, Appl. Opt. 28, p.2832-2837 (1989)

[6.388] {Sect. 6.6.10} C. Zizzo, C. Arnone, C. Cali, S. Sciortino: Fabrication and characterization of tuned Gaussian mirrors for the visible and the near infrared, Opt. Lett. 13, p.342-344 (1988)

[6.389] {Sect. 6.6.10} D.M. Walsh, L.V. Knight: Transverse modes of a laser resonator with Gaussian mirrors, Appl. Opt. 25, p.2947-2954 (1986)

[6.390] {Sect. 6.6.10} P. Lavigne, N. McCarthy, J.-G. Demers: Design and characterization of complementary Gaussian reflectivity mirrors, Appl. Opt. 24, p.2581-2586 (1985)

[6.391] {Sect. 6.6.10} N. McCarthy, P. Lavigne: Optical resonators with Gaussian reflectivity mirrors: misalignment sensitivity, Appl. Opt. 23, p.3845-3850 (1984)

[6.392] {Sect. 6.6.10} N. McCarthy, P. Lavigne: Optical resonators with Gaussian reflectivity mirrors: output beam characteristics, Appl. Opt. 22, p.2704-2708 (1983)

[6.393] {Sect. 6.6.10} A. Yariv, R. Yeh: Confinement and stability in optical resonators employing mirrors with Gaussian reflectivity, Opt. Comm. 13, p.370-374 (1975)

[6.394] {Sect. 6.6.10} H. Lin: Suppression of transverse instabilities in a laser by use of a spatially filtered feedback, J Opt Soc Am B Opt Physics 17, p.239-246 (2000)

[6.395] {Sect. 6.6.10} E. DelGiudice, R. Mele, G. Preparata, S. Sanvito, F. Fontana: A further look at waveguide lasers, IEEE J QE-34, p.2403-2408 (1998)

[6.396] {Sect. 6.6.10} A.A. Anderson, R.W. Eason, L.M.B. Hickey, M. Jelinek, C. Grivas, D.S. Gill, N.A. Vainos: Ti:sapphire planar waveguide laser grown by pulsed laser deposition, Optics Letters 22, p.1556-1558 (1997)

[6.397] {Sect. 6.6.10} Q.D. Liu, L. Shi, P.P. Ho, R.R. Alfano: Nonlinear vector rotation and depolarization of femtosecond laser pulses propagating in non-birefringent single-mode optical fibers, Opt Commun 138, p.45-48 (1997)

[6.398] {Sect. 6.6.10} E.J. Zang: Theory of waveguide laser resonators with small curvature mirrors, IEEE J QE-33, p.955-958 (1997)

[6.399] {Sect. 6.6.10} J.R. Marciante, G.P. Agrawal: Controlling filamentation in broad-area semiconductor lasers and amplifiers, Appl Phys Lett 69, p.593-595 (1996)

[6.400] {Sect. 6.6.10} S. Makki, J. Leger: Solid-state laser resonators with diffractive optic thermal aberration correction, IEEE J QE-35, p.1075-1085 (1999)

[6.401] {Sect. 6.6.10} A.A. Napartovich, N.N. Elkin, V.N. Troschieva, D.V. Vysotsky, J.R. Leger: Simplified intracavity phase plates for increasing laser-mode discrimination, Appl Opt 38, p.3025-3029 (1999)

[6.402] {Sect. 6.6.10} S. Bischoff, S.W. Koch: Beam shaping in vertical-cavity surface-emitting laser cavities, Opt Commun 158, p.65-71 (1998)

[6.403] {Sect. 6.6.10} J.R. Leger, D. Chen, K. Dai: High modal discrimination in a Nd:YAG laser resonator with internal phase gratings, Optics Letters 19, p.1976-1978 (1994)

[6.404] {Sect. 6.6.10} T.R. Boehly, V.A. Smalyuk, D.D. Meyerhofer, J.P. Knauer, D.K. Bradley, R.S. Craxton, M.J. Guardalben, S. Skupsky, T.J. Kessler: Reduction of laser imprinting using polarization smoothing on a solid-state fusion laser, J Appl Phys 85, p.3444-3447 (1999)

[6.405] {Sect. 6.6.10} F. Druon, G. Cheriaux, J. Faure, J. Nees, M. Nantel, A. Maksimchuk, J.C. Chanteloup, G. Vdovin: Wave-front correction of femtosecond terawatt lasers by deformable mirrors, Optics Letters 23, p.1043-1045 (1998)

[6.406] {Sect. 6.6.10} F. Sanchez, A. Chardon: Pump size optimization in microchip lasers, Opt Commun 136, p.405-409 (1997)

[6.407] {Sect. 6.6.11} G. Cerullo, S. de Silvestri, V. Magni, O. Svelto: Output Power Limitations in cw Single Transverse Mode Nd:YAG Lasers with a Rod of Large Cross-Section, Opt. Quant. Electron. 25, p.489-500 (1993)

[6.408] {Sect. 6.6.11} D. Cerullo, S. De Silvestri, V. Magni: High efficiency, 40 W cw Nd:YLF laser with large TEM00 mode, Opt. Comm. 93, p.77-81 (1992)

[6.409] {Sect. 6.6.11} V. Magni: Multielement stable resonators containing a variable lens, J. Opt. Soc. Am A 4, p.1962-1969 (1987)

[6.410] {Sect. 6.6.11} V. Magni: Resonators for Solid-State Lasers with Large-Volume Fundamental Mode and High Alignment Stability, Appl. Opt. 25, p.107-117 (1986)

[6.411] {Sect. 6.6.11} D.C. Hanna, C.G. Sawyers, M.A. Yuratich: Telescopic Resonators for Large Volume TEM00-Mode Operation, Opt. Quant. Electron. 13, p.493-507 (1981)

[6.412] {Sect. 6.6.11} D.C. Hanna, C.G. Sawyers, M.A. Yuratich: Large volume TEM00 mode operation of Nd:YAG lasers, Opt. Comm. 37, p.359-362 (1981)

[6.413] {Sect. 6.6.11} D.C. Sawyers, M.A. Yuratich: Telescopic resonators for large-volume TEM00-mode operation, Opt. Quant. Electr. 13, p.493-507 (1981)

[6.414] {Sect. 6.6.11} L.W. Casperson: Mode Stability of Lasers and Periodic Optical Systems, IEEE J. QE-10, p.629-634 (1974)

[6.415] {Sect. 6.6.11} N. Kurauchi, W.K. Kahn: Rays and Ray Envelopes within Stable Optical Resonators Containing Focusing Media, Appl. Opt. 5, p.1023-1029 (1966)

[6.416] {Sect. 6.6.12} A. Yariv: Operator algebra for propagation problems involving phase conjugation and nonreciprocal elements, Appl. Opt. 26, p.4538-4540 (1987)

[6.417] {Sect. 6.6.12} G. Giuliani, M. Denariez-Roberge, P.A. Belanger: Transverse modes of a stimulated scattering phase-conjugate resonator, Appl. Opt. 21, p.3719-3724 (1982)

[6.418] {Sect. 6.6.12} M. Ostermeyer, R. Menzel: 50 Watt average output power with 1.2*DL beam quality from a single rod Nd:YALO laser with phase-conjugating SBS mirror, Opt. Comm. 171, p.85-91 (1999)

[6.419] {Sect. 6.6.12} A.V. Kiryanov, V. Aboites, N.N. Ilichev: Analysis of a large-mode neodymium laser passively Q switched with a saturable absorber and a stimulated-Brillouin-scattering mirror, J Opt Soc Am B Opt Physics 17, p.11-17 (2000)

[6.420] {Sect. 6.6.12} I.Yu. Anikeev, J. Munch: Improved output power performance of a phase conjugated laser oscillator, Opt. Quant. Electr. 31, p.545-553 (1999)

[6.421] {Sect. 6.6.12} M. Ostermeyer, A. Heuer, R. Menzel: 27 Watt Average Output Power with 1.2*DL Beam Quality from a Single Rod Nd:YAG-Laser with Phase Conjugating SBS-Mirror, IEEE J. QE-34, p.372-377 (1998)

[6.422] {Sect. 6.6.12} H.S. Kim, K.G. Han, N.S. Kim, Y.S. Shin, H.J. Kong: Beam Smoothing in a Passive Q-Switched Laser with an Additional Stimulated Brillouin Scattering output coupler, Jpn. J. Appl. Phys. 35, p.L1324-L1326 (1996)

[6.423] {Sect. 6.6.12} R.A. Lamb: Single-longitudianal-mode, phase-conjugate ring master oscillator power amplifier using external stimulated-Brillouin-scattering Q switching, J. Opt. Soc. Am. B. 13p.1758-1765 (1996)

[6.424] {Sect. 6.6.12} V.F. Losev, Yu. N. Panchenko: Formation of high-quality XeCl laser radiation in a cavity with an SBS mirror, Quant. Electron. 25, p.450-451 (1995)

[6.425] {Sect. 6.6.12} P.J. Soan, M.J. Damzen, V. Aboites, M.H.R. Hutchinson: Long-pulse self-starting stimulated-Brillouin-scattering resonator, Opt. Lett. 19, p.783-785 (1994)

[6.426] {Sect. 6.6.12} S. Seidel, G. Phillipps: Pulse lengthening by intracavity stimulated Brillouin scattering in a Q-switched, phase-conjugated Nd:YAG laser oscillator, Appl. Opt. 32, p.7408-7417 (1993)

[6.427] {Sect. 6.6.12} A.D. Case, P.J. Soan, M.J. Damzen, M.H.R. Hutchinson: Coaxial flash-lamp-pumped dye laser with a stimulated Brillouin scattering reflector, J. Opt. Soc. Am. B 9, p.374-379 (1992)

[6.428] {Sect. 6.6.12} H.J. Eichler, R. Menzel, D. Schumann: 10 Watt Single-Rod Nd-YAG-Laser with SBS-Q-Switching Mirror, Appl. Opt. 24, p.5038-5043 (1992)

[6.429] {Sect. 6.6.12} H. Meng, H.J. Eichler: Nd:YAG laser with a phase-conjugating mirror based on stimulated Brillouin scattering in SF6 gas, Opt. Lett. 16, p.569-571 (1991)

[6.430] {Sect. 6.6.12} G.K.N. Wong, M.J. Damzen: Investigations of Optical Feed-back Used to Enhance Stimulated Scattering, IEEE J. QE-26, p.139-148 (1990)

[6.431] {Sect. 6.6.12} M.R. Osborne, W.A. Schroeder, M.J. Damzen, M.H.R. Hutchinson: Low-Divergence Operation of a Long-Pulse Excimer Laser Using a SBS Phase-Conjugate Cavity, Appl. Phys. B 48, p.351-356 (1989)

[6.432] {Sect. 6.6.12} M.D. Skeldon, R.W. Boyd: Transverse-Mode Structure of a Phase-Conjugate Oscillator Baased on Brillouin-Enhanced Four-Wave Mixing, IEEE J. QE-25, p.588-594 (1989)

[6.433] {Sect. 6.6.12} P.P. Pashinin, E.J. Shklovsky: Solid-state lasers with stimulated-Brillouin-scattering mirrors operating in the repetitive-pulse mode, J. Opt. Soc. Am. B 5, p.1957-1961 (1988)

[6.434] {Sect. 6.6.12} I.M. Bel'dyugin, B.Ya. Zel'dovich, M.V. Zolotarev, V.V. Shkunov: Lasers with wavefront-reversing mirrors (review), Sov. J. Quantum Electron. 15, p.1583-1600 (1986)

[6.435] {Sect. 6.6.12} E.J. Bochove: Transverse-mode instability and chaos in an optical cavity with phase-conjugate mirror, Opt. Lett. 11, p.727-729 (1986)

[6.436] {Sect. 6.6.12} V.S. Arakelyan, G.E. Rylov: Laser with a wavefront-reversing mirror and Q switching by stimulated Brillouin backscattering, Sov. J. Quantum Electron. 15, p.433-434 (1985)

[6.437] {Sect. 6.6.12} S. Chandra, R.C. Fukuda, R. Utano: Sidearm stimulated scattering phase-conjugated laser resonator, Opt. Lett. 10, p.356-358 (1985)

[6.438] {Sect. 6.6.12} W. Shaomin, H. Weber: Fundamental modes of stimulated scattering phase-conjugate resonators, Opt. Acta 31, p.971-976 (1984)

[6.439] {Sect. 6.6.12} P.A. Bélanger, C. Pare, M. Piche: Modes of phase-conjugate resonators with bounded mirrors, p.567-571 (1983)

[6.440] {Sect. 6.6.12} G.C. Valley, D. Fink: Three-dimensional phase-conjugate-resonator performance, J. Opt. Soc. Am. 73, p.572-575 (1983)

[6.441] {Sect. 6.6.12} P.A. Bélanger: Phase conjugation and optical resonators, Opt. Eng. 21, p.266-270 (1982)

[6.442] {Sect. 6.6.12} N.N. Il'ichev, A.A. Malyutin, P.P. Pashinin: Laser with diffraction-limited divergence and Q switching by stimulated Brillouin scattering, Sov. J. Quant. Electron. 12, p.1161-1164 (1982)

[6.443] {Sect. 6.6.12} G.J. Linford, B.C. Johnson, J.S. Hildrum, W.E. Martin, K. Snyder, R.D. Boyd, W.L. Smith, C.L. Vercimak, D. Eimerl, J.T. Hunt: Large aperture harmonic conversion experiments at Lawrence Livermore National Laboratory, Appl. Opt. 21, p.3633-3643 (1982)

[6.444] {Sect. 6.6.12} I.G. Zubarev, A.B. Mironov, S.I. Mikahilov: Single-mode pulse-periodic oscillator-amplifier system with wavefront reversal, Sov. J. Quantum Electron. 10, p.1179-1181 (1981)

[6.445] {Sect. 6.6.12} P.A. Bélanger, A. Hardy, A.E. Siegman: Resonant modes of optical cavities with phase-conjugate mirrors, Appl. Opt. 19, p.602-609 (1980)

[6.446] {Sect. 6.6.12} J. Auyeung, D. Fekete, D.M. Pepper, A. Yariv: A Theoretical and Experimental Investigation of the Modes of Optical Resonators with Phase-Conjugate Mirrors, IEEE J. QE-15, p.1180-1188 (1979)

[6.447] {Sect. 6.6.12} S.A. Lesnik, M.S. Soskin, A.I. Khizhnyak: Laser with a stimulated-Brillouin-scattering complex-conjugate mirror, Sov. Phys. Tech. Phys. 24, p.1249-1250 (1979)

[6.448] {Sect. 6.6.12} U. Ganiel, A. Hardy, Y. Silberberg: Stability of optical laser resonator with mirrors of Gaussian reflectivity profiles, which contain an active medium, Opt. Comm. 14, p.290-293 (1975)

[6.449] {Sect. 6.6.12} G.J. Crofts, M.J. Damzen: Numerical modelling of continuous-wave holographic laser oscillators, Opt Commun 175, p.397-408 (2000)

[6.450] {Sect. 6.6.12} S. CamachoLopez, M.J. Damzen: Self-starting Nd : YAG holographic laser oscillator with a thermal grating, Optics Letters 24, p.753-755 (1999)

[6.451] {Sect. 6.6.12} S. Mailis, J. Hendricks, D.P. Shepherd, A.C. Tropper, N. Moore, R.W. Eason, G.J. Crofts, M. Trew, M.J. Damzen: High-phase-conjugate reflectivity (ι 800%) obtained by degenerate four-wave mixing in a continuous-wave diode-side-pumped Nd : YVO4 amplifier, Optics Letters 24, p.972-974 (1999)

[6.452] {Sect. 6.6.12} A. Minassian, G.J. Crofts, M.J. Damzen: A tunable self-pumped phase-conjugate laser using Ti : sapphire slab amplifiers, Opt Commun 161, p.338-344 (1999)

[6.453] {Sect. 6.6.12} O.L. Antipov, A.S. Kuzhelev, V.A. Vorobyov, A.P. Zinovev: Pulse repetitive Nd:YAG laser with distributed feedback by self-induced population grating, Opt Commun 152, p.313-318 (1998)

[6.454] {Sect. 6.6.12} D.S. Hsiung, X.W. Xia, T.T. Grove, M.S. Shahriar, P.R. Hemmer: Demonstration of a phase conjugate resonator using degenerate four-wave mixing via coherent population trapping in rubidium, Opt Commun 154, p.79-82 (1998)

[6.455] {Sect. 6.6.12} K. Iida, H. Horiuchi, O. Matoba, T. Omatsu, T. Shimura, K. Kuroda: Injection locking of a broad-area diode laser through a double phase- conjugate mirror, Opt Commun 146, p.6-10 (1998)

[6.456] {Sect. 6.6.12} E. Rosas, V. Aboites, M.J. Damzen: Transient evolution and spatial mode size analysis of adaptive laser oscillators, Opt Commun 156, p.419-425 (1998)

[6.457] {Sect. 6.6.12} P. Sillard, A. Brignon, J.P. Huignard, J.P. Pocholle: Self-pumped phase-conjugate diode-pumped Nd:YAG loop resonator, Optics Letters 23, p.1093-1095 (1998)

[6.458] {Sect. 6.6.12} A.A.R. Alrashed, B.E.A. Saleh: Modes of resonators with dispersive phase-conjugate mirrors, Appl Opt 36, p.3400-3412 (1997)

[6.459] {Sect. 6.6.12} A. Minassian, G.J. Crofts, M.J. Damzen: Self-starting Ti:sapphire holographic laser oscillator, Optics Letters 22, p.697-699 (1997)

[6.460] {Sect. 6.6.12} P. Sillard, A. Brignon, J.P. Huignard: Nd:YAG loop resonator with a Cr4+:YAG self-pumped phase-conjugate mirror, IEEE J QE-33, p.483-489 (1997)

[6.461] {Sect. 6.6.12} A.-A.R. Al-Rashed, B.E.A. Saleh: Modes of resonators with dispersive phase-conjugate mirrors, Appl. Opt. 36, p.3400-3412 (1997)

[6.462] {Sect. 6.6.12} R.P.M. Green, G.J. Crofts, W. Hubbard, D. Udaiyan, D.H. Kim, M.J. Damzen: Dynamic laser control using feedback from a gain grating, IEEE J QE-32, p.371-377 (1996)

[6.463] {Sect. 6.6.12} P. Kurz, R. Nagar, T. Mukai: Highly efficient phase conjugation using spatially nondegenerate four-wave mixing in a broad-area laser diode, Appl Phys Lett 68, p.1180-1182 (1996)

[6.464] {Sect. 6.6.12} M.J. Damzen, R.P.M. Green, K.S. Syed: Self-adaptive solid-state laser oscillator formed by dynamic gain-grating holograms, Optics Letters 20, p.1704-1706 (1995)

[6.465] {Sect. 6.6.12} O. Wittler, D. Udaiyan, G.J. Crofts, K.S. Syed, M.J. Damzen: Characterization of a distortion-corrected Nd : YAG laser with a self- conjugating loop geometry, IEEE J QE-35, p.656-664 (1999)

[6.466] {Sect. 6.6.12} J.C. Chanteloup, H. Baldis, A. Migus, G. Mourou, B. Loiseaux, J.P. Huignard: Nearly diffraction-limited laser focal spot obtained by use of an optically addressed light valve in an adaptive-optics loop, Optics Letters 23, p.475-477 (1998)

[6.467] {Sect. 6.6.12} M.K. Lee, W.D. Cowan, B.H. Welsh, V.M. Bright, M.C. Roggemann: Aberration-correction results from a segmented microelectro-mechanical deformable mirror and a refractive lenslet array, Optics Letters 23, p.645-647 (1998)

[6.468] {Sect. 6.6.12} I. Moshe, S. Jackal, R. Lallouz: Dynamic correction of thermal focusing in Nd:YAG confocal unstable resonators by use of a variable radius mirror, Appl Opt 37, p.7044-7048 (1998)

[6.469] {Sect. 6.6.12} J.J. Kasinski, R.L. Burnham: Near-diffraction-limited, high-energy, high-power, diode- pumped laser using thermal aberration correction with aspheric diamond-turned optics, Appl Opt 35, p.5949-5954 (1996)

[6.470] {Sect. 6.6.12} N. Pavel, T. Dascalu, V. Lupei: Variable reflectivity mirror unstable resonator with deformable mirror thermal compensation, Opt Commun 123, p.115-120 (1996)

[6.471] {Sect. 6.6.13} N. Kugler, S.L. Dong, Q.T. Lu, H. Weber: Investigation of the misalignment sensitivity of a birefringence-compensated two-rod Nd:YAG laser system, Appl Opt 36, p.9359-9366 (1997)

[6.472] {Sect. 6.6.13} R.M.R. Pillai, E.M. Garmire: Paraxial-misalignment insensitive external-cavity semiconductor-laser array emitting near-diffraction limited single-lobed beam, IEEE J QE-32, p.996-1008 (1996)

[6.473] {Sect. 6.6.13} N. Hodgson, H. Weber: Misalignment sensitivity of stable resonators in multimode operation, J. Mod. Opt. 39, p.1873-1882 (1992)

[6.474] {Sect. 6.6.13} K.P. Driedger, R.M. Iffländer, H. Weber: Multirod Resonators for High-Power Solid-State Lasers with Improved Beam Quality, IEEE J. QE-24, p.665-674 (1988)

[6.475] {Sect. 6.6.13} R. Hauck, N. Hodgson, H. Weber: Misalignment sensitivity of unstable resonators with spherical mirrors, J. Mod. Opt. 35, p.165-176 (1988)

[6.476] {Sect. 6.6.13} D. Metcalf, P. de Giovanni, J. Zachorowski, M. Leduc: Laser resonators containing self-focusing elements, Appl. Opt. 26, p.4508-4517 (1987)

[6.477] {Sect. 6.6.13} S. De Silvestri, P. Laporta, V. Magni: Misalignment sensitivity of solid-state laser resonators with thermal lensing, Opt. Comm. 59, p.43-48 (1986)

[6.478] {Sect. 6.6.13} N. McCarthy, P. Lavigne: Optical resonators with Gaussian reflectivity mirrors: misalignment sensitivity, Appl. Opt. 23, p.3845-3850 (1984)

[6.479] {Sect. 6.6.13} A. Le Floch, J.M. Lenormand, R. Le Naour, J.P. Taché: A critical geometry fo rlasers with internal lenslike effects, Le Journal de Phys. Lett. 43, p.L493-L498 (1982)

[6.480] {Sect. 6.6.13} H.P. Kortz, R. Iffländer, H. Weber: Stability and beam divergence of multimode lasers with internal variable lenses, Appl. Opt. 20, p.4124-4134 (1981)

[6.481] {Sect. 6.6.13} R. Hauck, H.P. Kortz, H. Weber: Misalignment sensitivity of optical resonators, Appl. Opt. 19, p.598-601 (1980)

[6.482] {Sect. 6.6.13} J.L. Remo: Diffraction losses for symmetrically tilted plane reflectors in open resonators, Appl. Opt. 19, p.774-777 (1980)

[6.483] {Sect. 6.6.13} P. Horwitz: Modes in misalignment unstable resonators, Appl. Opt. 15, p.167-178 (1976)

[6.484] {Sect. 6.6.14} K. Yasui: Efficient and stable operation of a high-brightness cw 500-W Nd:YAG rod laser, Appl. Opt. 35, p.2566-2569 (1996)

[6.485] {Sect. 6.6.14} V. Magni: Multielement stable resonators containing a variable lens, J. Opt. Soc. Am A 4, p.1962-1969 (1987)

[6.486] {Sect. 6.6.14} P.H. Sarkies: A stable YAG resonator yielding a beam of very low divergence and high output energy, Opt. Comm. 31, p.189-192 (1979)

[6.487] {Sect. 6.6.15} Y.F. Chen, H.J. Kuo: Determination of the thermal loading of diode-pumped Nd:YVO4 by use of thermally induced second-harmonic output depolarization, Optics Letters 23, p.846-848 (1998)

[6.488] {Sect. 6.6.15} B. Ozygus, Q.C. Zhang: Thermal lens determination of end-pumped solid-state lasers using primary degeneration modes, Appl Phys Lett 71, p.2590-2592 (1997)

[6.489] {Sect. 6.6.15} B. Neuenschwander, R. Weber, H.P. Weber: Determination of the Thermal Lens in Solid-State Lasers with Stable Cavities, IEEE J. QE-31, p.1082-1087 (1995)

[6.490] {Sect. 6.6.15} D.C. Burnham: Simple Measurement of Thermal Lensing Effects in Laser Rods, Appl. Opt. 9, p.1727-1728 (1970)

[6.491] {Sect. 6.7.2} G. Stephan: An airy function for the laser, J Nonlinear Opt Physics Mat 5, p.551-557 (1996)

[6.492] {Sect. 6.7.4} D. Cooper, L.L. Tankersley, J. Reintjes: Narrow-linewidth unstable resonator, Opt. Lett. 13, p.568-570 (1988)

[6.493] {Sect. 6.7.4} N. Konishi, T. Suzuki, Y. Taira, H. Kato, T. Kasuya: High Precision Wavelength Meter with Fabry-Perot Optics, Appl. Phys. 25, p.311-316 (1981)

[6.494] {Sect. 6.7.4} F.J. Duarte: Multiple-prism grating solid-state dye laser oscillator: optimized architecture, Appl Opt 38, p.6347-6349 (1999)

[6.495] {Sect. 6.7.4} R.M. Hofstra, F.A. vanGoor, W.J. Witteman: Linewidth reduction of a long-pulse, low-gain XeCl* laser with intracavity etalons, J Opt Soc Am B Opt Physics 16, p.1068-1071 (1999)

[6.496] {Sect. 6.7.4} T. Earles, L.J. Mawst, D. Botez: 1.1W continuous-wave, narrow spectral width, (¡1 angstrom) emission from broad-stripe, distributed-feedback diode lasers (Lambda=0.893 mu m), Appl Phys Lett 73, p.2072-2074 (1998)

[6.497] {Sect. 6.7.4} D. Lo, S.K. Lam, C. Ye, K.S. Lam: Narrow linewidth operation of solid state dye laser based on sol-gel silica, Opt Commun 156, p.316-320 (1998)

[6.498] {Sect. 6.7.4} V.V. Vassiliev, V.L. Velichansky, V.S. Ilchenko, M.L. Gorodetsky, L. Hollberg, A.V. Yarovitsky: Narrow-line-width diode laser with a high-Q microsphere resonator, Opt Commun 158, p.305-312 (1998)

[6.499] {Sect. 6.7.4} D. Wandt, M. Laschek, A. Tunnermann, H. Welling: Continuously tunable external-cavity diode laser with a double-grating arrangement, Optics Letters 22, p.390-392 (1997)

[6.500] {Sect. 6.7.4} B.W. Liby, D. Statman: Controlling the linewidth of a semiconductor laser with photorefractive phase conjugate feedback, IEEE J QE-32, p.835-838 (1996)

[6.501] {Sect. 6.7.4} J. Harrison, G.A. Rines, P.F. Moulton, J.R. Leger: Coherent summation of injection-locked, diode-pumped Nd:YAG ring lasers, Opt. Lett. 13, p.111-113 (1988)

[6.502] {Sect. 6.7.4} F.J. Duarte, R.W. Conrad Diffraction-limited single-longitudinal-mode multiple-prism flashlamp-pumped dye laser oscillator: linewidth analysis and injection of amplifier system, Appl. Opt. 26, p.2567-2571 (1987)

[6.503] {Sect. 6.7.4} E. Armandillo, G. Giuliani: Estimation of the Minimum Laser Linewidth Achievable with a Grazing Grating Configuration, Optics Letters 8, p.274-276 (1983)

[6.504] {Sect. 6.7.4} F.J. Duarte, J.A. Piper: Prism preexpanded grazing-incidence grating cavity for pulsed dye lasers, Appl. Opt. 20, p.2113-2116 (1981)

[6.505] {Sect. 6.7.4} W. R. Leeb: Losses Introduced by Tilting Intracavity Etalons. I790+I1516, Appl. Phys. 6, p.267-272 (1975)

[6.506] {Sect. 6.7.4} W. Wiesemann: Longitudinal Mode Selection in Lasers with Three-Mirror Reflectors, Appl. Opt. 12, p.2909-2912 (1973)

[6.507] {Sect. 6.7.4} H. Walther, J. L. Hall: Tunable Dye Laser with Narrow Spectral Output, Appl. Phys. Lett. 17, p.239-242 (1970)

[6.508] {Sect. 6.7.4} W.B. Tiffany: Repetitively Pulsed, Tunable Ruby Laser with Solid Etalon Mode Control, Appl. Opt. 7, p.67-72 (1968)

[6.509] {Sect. 6.7.4} E. Snitzer: Frequency Control of a Nd3+ Glass Laser, Appl. Opt. 5, p.121-126 (1966)

[6.510] {Sect. 6.7.4} F.J. McClung, D. Weiner: Longitudinal Mode Control in Giant Pulse Lasers, IEEE J. QE-1, p.94-99 (1965)

[6.511] {Sect. 6.7.4} B.B. McFarland, R.H. Hoskins, B.H. Soffer: Narrow Spectral Emission from a Passively Q-spoiled Neodymium-glass Laser, Nature 207, p.1180-1181 (1965)

[6.512] {Sect. 6.7.4} X. Li, A.D. Sadovnikov, W.P. Huang, T. Makino: A physics-based three-dimensional model for distributed feedback laser diodes, IEEE J QE-34, p.1545-1553 (1998)

[6.513] {Sect. 6.7.4} H. Kuwatsuka, H. Shoji, M. Matsuda, H. Ishikawa: Nondegenerate four-wave mixing in a long-cavity lambda/4-shifted DFB laser using its lasing beam as pump beams, IEEE J QE-33, p.2002-2010 (1997)

[6.514] {Sect. 6.7.4} J.F. Pinto, L. Esterowitz: Distributed-feedback, tunable Ce3+-doped colquiriite lasers, Appl Phys Lett 71, p.205-207 (1997)

[6.515] {Sect. 6.7.4} K. Wada, Y. Akage, H. Marui, H. Horinaka, N. Yamamoto, Y. Cho: Simple method for determining the gain saturation coefficient of a distributed feedback semiconductor laser, Opt Commun 130, p.57-62 (1996)

[6.516] {Sect. 6.7.4} X. Peng, L.Y. Liu, J.F. Wu, Y.G. Li, Z.J. Hou, L. Xu, W.C. Wang, F.M. Li: Wide-range amplified spontaneous emission wavelength tuning in a solid-state dye waveguide, Optics Letters 25, p.314-316 (2000)

[6.517] {Sect. 6.7.4} J. McKay, K.L. Schepler, G.C. Catella: Efficient grating-tuned mid-infrared Cr2+: CdSe laser, Optics Letters 24, p.1575-1577 (1999)

[6.518] {Sect. 6.7.4} J. Struckmeier, A. Euteneuer, B. Smarsly, M. Breede, M. Born, M. Hofmann, L. Hildebrand, J. Sacher: Electronically tunable external-cavity laser diode, Optics Letters 24, p.1573-1574 (1999)

[6.519] {Sect. 6.7.4} V.P. Gerginov, Y.V. Dancheva, M.A. Taslakov, S.S. Cartaleva: Frequency tunable monomode diode laser at 670 nm for high resolution spectroscopy, Opt Commun 149, p.162-169 (1998)

[6.520] {Sect. 6.7.4} R. Khare, S.R. Daulatabad, K.K. Sharangpani, R. Bhatnagar: An independently tunable, collinear, variable delay, two-wavelength dye laser, Opt Commun 153, p.68-72 (1998)

[6.521] {Sect. 6.7.4} R. Khare, S.R. Daultabad, R. Jain, R. Bhatnagar: Utilization of the yellow component of a copper-vapor laser for extending the tuning range of a Rhodamine 6G dye laser by use of an additional dye in a novel coupled resonator scheme, Appl Opt 37, p.4921-4924 (1998)

[6.522] {Sect. 6.7.4} Y. Nagumo, N. Taguchi, H. Inaba: Widely tunable continuous-wave Cr3+:LiSrAlF6 ring laser from 800 to 936 nm, Appl Opt 37, p.4929-4932 (1998)

[6.523] {Sect. 6.7.4} B. Golubovic, B.E. Bouma, G.J. Tearney, J.G. Fujimoto: Optical frequency-domain reflectometry using rapid wavelength tuning of a Cr4+:forsterite laser, Optics Letters 22, p.1704-1706 (1997)

[6.524] {Sect. 6.7.4} D. Kopf, A. Prasad, G. Zhang, M. Moser, U. Keller: Broadly tunable femtosecond Cr:LiSAF laser, Optics Letters 22, p.621-623 (1997)

[6.525] {Sect. 6.7.4} B. Pati, J. Borysow: Single-mode tunable Ti:sapphire laser over a wide frequency range, Appl Opt 36, p.9337-9341 (1997)

[6.526] {Sect. 6.7.4} P.S. Bhatia, J.W. Keto: Precisely tunable, narrow-band pulsed dye laser, Appl Opt 35, p.4152-4158 (1996)

[6.527] {Sect. 6.7.4} D.K. Ko, G. Lim, S.H. Kim, J.M. Lee: Dual-wavelength operation of a self-seeded dye laser oscillator, Appl Opt 35, p.1995-1998 (1996)

[6.528] {Sect. 6.7.4} P. Mandel, K. Otsuka, J.Y. Wang, D. Pieroux: Two-mode laser power spectra, Phys Rev Lett 76, p.2694-2697 (1996)

[6.529] {Sect. 6.7.4} K. Tamura, M. Nakazawa: Dispersion-tuned harmonically mode-locked fiber ring laser for self-synchronization to an external clock, Optics Letters 21, p.1984-1986 (1996)

[6.530] {Sect. 6.7.4} S. Wada, K. Akagawa, H. Tashiro: Electronically tuned Ti:sapphire laser, Optics Letters 21, p.731-733 (1996)

[6.531] {Sect. 6.7.4} D. Wandt, M. Laschek, K. Przyklenk, A. Tunnermann, H. Welling: External cavity laser diode with 40 nm continuous tuning range around 825 nm, Opt Commun 130, p.81-84 (1996)

[6.532] {Sect. 6.7.4} J. Harrison, A. Finch, J.H. Flint, P.F. Moulton: Broad-Band Rapid Tuning of a Single-Frequency Diode-Pumped Neodymium Laser, IEEE J. QE-28, p.1123-1130 (1992)

[6.533] {Sect. 6.7.4} J.J. Zayhowski, J.A. Keszenheimer: Frequency Tuning of Microchip Lasers Using Pump-Power Modulation, IEEE J. QE-28, p.1118-1122 (1992)

[6.534] {Sect. 6.7.4} P.A. Schultz, S.R. Henion: Frequency-modulated Nd:YAG laser, Opt. Lett. 16, p.578-580 (1991)

[6.535] {Sect. 6.7.4} W. Fuhrmann, W. Demtröder: A Continuously Tunable GaAs Diode Laser with an External Resonator, Appl. Phys. B 49, p.29-32 (1989)

[6.536] {Sect. 6.7.4} T.J. Kane, E.A.P. Cheng: Fast frequency tuning and phase locking of diode-pumped Nd:YAG ring lasers, Opt. Lett. 13, p.970-972 (1988)

[6.537] {Sect. 6.7.4} A. Owyoung, P. Esherick: Stress-induced tuning of a diode-laser-excited monolithic Nd:YAG laser, Opt. Lett. 12, p.999-1001 (1987)

[6.538] {Sect. 6.7.4} I.J. Hodgkinson, J.I. Vukusic: Birefringent Tuning Filters without Secondary Peaks, Opt. Commun. 24, p.133-134 (1978)

[6.539] {Sect. 6.7.4} M.M. Johnson, A.H. LaGrone: Continuously Tunable Resonant Ruby Laser Reflector, Appl. Opt. 12, p.510-518 (1973)

[6.540] {Sect. 6.7.4} F. J. Duarte (ed.): Tunable Lasers Handbook: Optics and Photonics (Academic Press, San Diego, California, 1995)

[6.541] {Sect. 6.7.5} M. Zhu, J.L. Hall: Stabilization of optical phase/frequency of a laser system: application to a commercial dye laser with external stabilizer, J. Opt. Soc. Am. B 10p.802-816 (1993)

[6.542] {Sect. 6.7.5} P.A. Ruprecht, J.R. Branderberg: Enhancing diode laser tuning with a short external cavity, Opt. Commun. 93p.82-86 (1992)

[6.543] {Sect. 6.7.5} R. Kallenbach, G. Zimmermann, D.H. McIntyre, T.W. Hänsch, R.G. DeVoe: A blue dye laser with sub-kilohertz stability, Opt. Commun. 70p.56-60 (1989)

[6.544] {Sect. 6.7.5} M. Houssin, M. Jardino, B. Gely, M. Desaintfuscien: Design performance of a few-kilohertz-linewidth dye laser stabilized by reflection in an optical resonator, Opt. Lett. 13p.823-825 (1988)

[6.545] {Sect. 6.7.5} Ch. Salomon, D. Hills, J.L. Hall: Laser stabilization at the millihertz level, J. Opt. Soc. Am. B 5p.1576-1587 (1988)

[6.546] {Sect. 6.7.5} A.J. Berry, D.C. Hanna, C.G. Swayers: High power single frequency operation of a Q-switched TEMoo mode Nd:YAG laser, Opt. Commun. 40, p.54-58 (1981)

[6.547] {Sect. 6.7.5} J.M. Green, J.P. Hohimer, F.K. Tittel: Traveling-wave operation of a tunable cw dye laser, Opt. Commun. 7 p.349-350 (1973)

[6.548] {Sect. 6.7.5} F.P. Schäfer, H. Müller: Tunable dye ring-laser, Opt. Commun. 26p.407409 (1971)

[6.549] {Sect. 6.7.5} L. BarteltBerger, U. Brauch, A. Giesen, H. Huegel, H. Opower: Power-scalable system of phase-locked single-mode diode lasers, Appl Opt 38, p.5752-5760 (1999)

[6.550] {Sect. 6.7.5} Y. Beregovski, A. Fardad, H. Luo, M. Fallahi: Single-mode operation of the external cavity DBR laser with sol-gel waveguide Bragg grating, Opt Commun 164, p.57-61 (1999)

[6.551] {Sect. 6.7.5} A.K. Goyal, P. Gavrilovic, H. Po: 1.35 W of stable single-frequency emission from an external-cavity tapered oscillator utilizing fiber Bragg grating feedback, Appl Phys Lett 73, p.575-577 (1999)

[6.552] {Sect. 6.7.5} T. Heil, I. Fischer, W. Elsasser, J. Mulet, C.R. Mirasso: Statistical properties of low-frequency fluctuations during single- mode operation in distributed-feedback lasers: experiments and modeling, Optics Letters 24, p.1275-1277 (1999)

[6.553] {Sect. 6.7.5} Y. Isyanova, D. Welford: Temporal criterion for single-frequency operation of passively Q-switched lasers, Optics Letters 24, p.1035-1037 (1999)

[6.554] {Sect. 6.7.5} S. Riyopoulos: Stable single-mode vertical-cavity surface-emitting laser with a photoresistive aperture, Optics Letters 24, p.768-770 (1999)

[6.555] {Sect. 6.7.5} I. Zawischa, K. Plamann, C. Fallnich, H. Welling, H. Zellmer, A. Tunnermann: All-solid-state neodymium-based single-frequency master-oscillator fiber power-amplifier system emitting 5.5 W of radiation at 1064 nm, Optics Letters 24, p.469-471 (1999)

[6.556] {Sect. 6.7.5} D.J. Binks, D.K. Ko, L.A.W. Gloster, T.A. King: Pulsed single mode laser oscillation in a new coupled cavity design, Opt Commun 146, p.173-176 (1998)

[6.557] {Sect. 6.7.5} Y.F. Chen, T.M. Huang, C.L. Wang, L.J. Lee, S.C. Wang: Theoretical and experimental studies of single-mode operation in diode pumped Nd:YVO4/KTP green laser: influence of KTP length, Opt Commun 152, p.319-323 (1998)

[6.558] {Sect. 6.7.5} R. Dalgliesh, A.D. May, G. Stephan: Polarization states of a single-mode (microchip) Nd3+:YAG laser -Part II: Comparison of theory and experiment, IEEE J QE-34, p.1493-1502 (1998)

[6.559] {Sect. 6.7.5} D. Hofstetter, R.L. Thornton, L.T. Romano, D.P. Bour, M. Kneissl, R.M. Donaldson: Room-temperature pulsed operation of an electrically injected InGaN/GaN multi-quantum well distributed feedback laser, Appl Phys Lett 73, p.2158-2160 (1998)

[6.560] {Sect. 6.7.5} E. Lafond, A. Hirth: Optimization of a single mode Q-switched oscillator at 1.34 mu m, Opt Commun 152, p.329-334 (1998)

[6.561] {Sect. 6.7.5} H. Ludvigsen, M. Tossavainen, M. Kaivola: Laser linewidth measurements using self-homodyne detection with short delay, Opt Commun 155, p.180-186 (1998)

[6.562] {Sect. 6.7.5} A.J. Tiffany, I.T. McKinnie, D.M. Warrington: Pulse amplification of a single-frequency Cr:forsterite laser, Appl Opt 37, p.4907-4913 (1998)

[6.563] {Sect. 6.7.5} D. Wandt, M. Laschek, F. vonAlvensleben, A. Tunnermann, H. Welling: Continuously tunable 0.5 W single-frequency diode laser source, Opt Commun 148, p.261-264 (1998)

[6.564] {Sect. 6.7.5} A.K. Goyal, P. Gavrilovic, H. Po: Stable single-frequency operation of a high-power external cavity tapered diode laser at 780 nm, Appl Phys Lett 71, p.1296-1298 (1997)

[6.565] {Sect. 6.7.5} R. Knappe, G. Bitz, K.J. Boller, R. Wallenstein: Compact single-frequency diode-pumped Cr:LiSAF lasers, Opt Commun 143, p.42-46 (1997)

[6.566] {Sect. 6.7.5} K.I. Martin, W.A. Clarkson, D.C. Hanna: High-power single-frequency operation, at 1064 nm and 1061.4 nm of a Nd:YAG ring laser end-pumped by a beam- shaped diode bar, Opt Commun 135, p.89-92 (1997)

[6.567] {Sect. 6.7.5} W. Nagengast, K. Rith: High-power single-mode emission from a broad-area semiconductor laser with a pseudoexternal cavity and a Fabry-Perot etalon, Optics Letters 22, p.1250-1252 (1997)

[6.568] {Sect. 6.7.5} B. Pati, J. Borysow: Single-mode tunable Ti:sapphire laser over a wide frequency range, Appl Opt 36, p.9337-9341 (1997)

[6.569] {Sect. 6.7.5} C. Pedersen, P.L. Hansen, P. Buchhave, T. Skettrup: Single-frequency diode-pumped Nd:YAG prism laser with use of a composite laser crystal, Appl Opt 36, p.6780-6787 (1997)

[6.570] {Sect. 6.7.5} M. Teshima, M. Koga, K. Sato: Accurate frequency control of a mode-locked laser diode by reference-light injection, Optics Letters 22, p.126-128 (1997)

[6.571] {Sect. 6.7.5} A.J. Tiffany, I.T. McKinnie, D.M. Warrington: Low-threshold, single-frequency, coupled cavity Ti: Sapphire laser, Appl Opt 36, p.4989-4992 (1997)

[6.572] {Sect. 6.7.5} G.H.M. Vantartwijk, G.P. Agrawal: Nonlinear dynamics in the generalized Lorenz-Haken model, Opt Commun 133, p.565-577 (1997)

[6.573] {Sect. 6.7.5} S.J.M. Kuppens, M.P. vanExter, J.P. Woerdman, M.I. Kolobov: Observation of the effect of spectrally inhomogeneous gain on the quantum-limited laser linewidth, Opt Commun 126, p.79-84 (1996)

[6.574] {Sect. 6.7.5} P. Kurz, T. Mukai: Frequency stabilization of a semiconductor laser by external phase-conjugate feedback, Optics Letters 21, p.1369-1371 (1996)

[6.575] {Sect. 6.7.5} K.I. Martin, W.A. Clarkson, D.C. Hanna: 3 W of single-frequency output at 532 nm by intracavity frequency doubling of a diode bar pumped Nd:YAG ring laser, Optics Letters 21, p.875-877 (1996)

[6.576] {Sect. 6.7.5} K.I. Martin, W.A. Clarkson, D.C. Hanna: Limitations imposed by spatial hole burning on the single-frequency performance of unidirectional ring lasers, Opt Commun 125, p.359-368 (1996)

[6.577] {Sect. 6.7.5} B. Pezeshki, F. Agahi, J.A. Kash: A gratingless wavelength stabilized semiconductor laser, Appl Phys Lett 69, p.2807-2809 (1996)

[6.578] {Sect. 6.7.5} M. Tsunekane, N. Taguchi, H. Inaba: High-power, efficient, low-noise, continuous-wave all- solid-state Ti:Sapphire laser, Optics Letters 21, p.1912-1914 (1996)

[6.579] {Sect. 6.7.5} V. Wulfmeyer, J. Bosenberg: Single-mode operation of an injection-seeded alexandrite ring laser for application in water-vapor and temperature differential absorption lidar, Optics Letters 21, p.1150-1152 (1996)

[6.580] {Sect. 6.7.5} S.F. Yu: A quasi-three-dimensional large-signal dynamic model of distributed feedback lasers, IEEE J QE-32, p.424-432 (1996)

[6.581] {Sect. 6.7.5} M. Hyodo, T. Carty, K. Sakai: Near shot-noise-level relative frequency stabilization of a laser-diode-pumped Nd:YVO4 microchip laser, Appl. Opt. 35, p.4749-4753 (1996)

[6.582] {Sect. 6.7.5} R.A. Lamb: Single-longitudianal-mode, phase-conjugate ring master oscillator power amplifier using external stimulated-Brillouin-scattering Q switching, J. Opt. Soc. Am. B. 13p.1758-1765 (1996)

[6.583] {Sect. 6.7.5} L. Viana, S.S. Vianna, M. Oriá, J.W.R. Tabosa: Diode laser mode selection using a long external cavity, Appl. Opt. 35, p.368-371 (1996)

[6.584] {Sect. 6.7.5} I. Freitag, R. Henking, A. Tunnermann, H. Welling: Quasi-three-level room-temperature Nd:YAG ring laser with high single-frequency output power at 946 nm, Optics Letters 20, p.2499-2501 (1995)

[6.585] {Sect. 6.7.5} I. Freitag, D. Golla, S. Knoke, W. Schone, H. Zellmer, A. Tunnermann, H. Welling: Amplitude and frequency stability of a diode pumped Nd:YAG laser operating: At a single frequency continuous wave output power of 20 W, Optics Letters 20, p.462-464 (1995)

[6.586] {Sect. 6.7.5} C. Pedersen, P.L. Hansen, T. Skettrup, P. Buchhave: Diode-pumped single-frequency Nd:YVO4 laser with a set of coupled resonators, Optics Letters 20, p.1389-1391 (1995)

[6.587] {Sect. 6.7.5} C.J. Flood, D.R. Walker, H.M. van Driel: Effect of spatial hole burning in a mode-locked diode end-pumped Nd:YAG laser, Opt. Lett. 20, p.58-60 (1995)

[6.588] {Sect. 6.7.5} S. Taccheo, S. Longhi, L. Pallaro, P. Laporta: Frequency stabilization to a molecular line of a diode-pumped Er-Yb laser at 1533-nm wavelength, Opt. Lett. 20, p.2420-2422 (1995)

[6.589] {Sect. 6.7.5} N. Uehara, K. Ueda: Ultrahigh-frequency stabilization of a diode-pumped Nd:YAG laser with a high-power-acceptance photodetector, Opt. Lett. 19, p.728-730 (1994)

[6.590] {Sect. 6.7.5} H. Nagai, M. Kume, Y. Yoshikawa, K, Itoh: Low-noise operation (-140 dB/Hz) in close-coupled Nd:YVO4 second-harmonic lasers pumped by single-mode laser diodes, Appl. Opt. 32, p.6610-6615 (1993)

[6.591] {Sect. 6.7.5} T. Day, E.K. Gustafson, R.L. Byer: Sub-Hertz Relative Frequency Stabilization of Two-Diode Laser-Pumped Nd:YAG Lasers. Locked to a Fabry-Perot Interferometer, IEEE J. QE-28, p.1106-1117 (1992)

[6.592] {Sect. 6.7.5} L.J. Bromley, D.C. Hanna: Single-frequency Q-switched operation of a diode-laser-pumped Nd:YAG ring laser using an acoustic-optic modulator, Opt. Lett. 16, p.378-380 (1991)

[6.593] {Sect. 6.7.5} E.S. Fry, Q. Hu, X. Li: Single frequency operation of an injection-seeded Nd:YAG laser in high noise and vibration environments, Appl. Opt. 30, p.1015-1017 (1991)

[6.594] {Sect. 6.7.5} F. Zhou, A.I. Ferguson: Frequency stabilization of a diode-laser-pumped microchip Nd:YAG laser at 1.3 μm, Opt. Lett. 16, p.79-81 (1991)

[6.595] {Sect. 6.7.5} T. Day, E.K. Gustafson, R.L. Byer: Active frequency stabilization of a 1.062-μm, Nd:GGG diode-laser-pumped nonplanar ring oscillator to less than 3 Hz of relative linewidth, Opt. Lett. 15, p.221-223 (1990)

[6.596] {Sect. 6.7.5} W.R. Trutna, Jr, D.K. Donald: Two-piece, piezoelectrically tuned, single-mode Nd:YAG ring laser, Opt. Lett. 15, p.369-371 (1990)

[6.597] {Sect. 6.7.5} D.Shoemaker, A. Brillet, C.N. Man, O. Crégut: Frequency-stabilized laser-diode-pumped Nd:YAG laser, Opt. Lett. 14, p.609-611 (1989)

[6.598] {Sect. 6.7.5} T.J. Kane, E.A.P. Cheng: Fast frequency tuning and phase locking of diode-pumped Nd:YAG ring lasers, Opt. Lett. 13, p.970-972 (1988)

[6.599] {Sect. 6.7.5} W.R. Trutna, Jr, D.K. Donald, M. Nazarathy: Unidirectional diode-laser-pumped Nd:YAG ring laser with a small magnetic field, Opt. Lett. 12, p.248-250 (1987)

[6.600] {Sect. 6.7.5} S. De Silvestri, P. Laporta, V. Magni: The Role of the Rod Position in Single-Mode Solid State Laser Resonators: Optimization of a CW Mode-Locked Nd:YAG Laser, Opt. Comm. 57, p.339-344 (1986)

[6.601] {Sect. 6.7.5} F.J. Duarte: Multiple-prism Littrow and grazing-incidence pulsed CO2 lasers, Appl. Opt. 24, p.1244-1245 (1985)

[6.602] {Sect. 6.7.5} L.A. Rahn: Feedback stabilization of an injection-seeded Nd:YAG laser, Appl. Opt. 24, p.940-942 (1985)

[6.603] {Sect. 6.7.5} F.D. Feiock, J.R. Oldenettel: Gain effects on laser mode formation, J. Opt. Soc. Am. A 1, p.1097-1102 (1984)

[6.604] {Sect. 6.7.5} M.G. Littman: Single-mode pulsed tunable dye laser, Appl. Opt. 23, p.4465-4468 (1984)

[6.605] {Sect. 6.7.5} O.E. Nanii, A.N. Shelaev: Magnetooptic effects in a YAG:Nd3+ ring laser with a nonplanar resonator, Sov. J. Quant. Electron. 14, p.638-642 (1984)

[6.606] {Sect. 6.7.5} D.W. Hall, R.A. Haas, W.F. Krupke, M.J. Weber: Spectral and Polarization Hole Burning in Neodymium Glass Lasers, IEEE J. QE-19, p.1704-1717 (1983)

[6.607] {Sect. 6.7.5} Y.K. Park, R.L. Byer, G. Giuliani: Stable Single Axial Mode Operation of an Unstable Resonator ND YAG Oscillator by Injection Locking, Optics Letters 5, p.96-98 (1980)

[6.608] {Sect. 6.7.5} G. Marowsky, K. Kaufmann: Influence of Spatial Hole Burning on the Output Power of a CW Dye Ring Laser, IEEE J. QE-12, p.207-209 (1976)

[6.609] {Sect. 6.7.5} I. V. Hertel, A. Stamatovic: Spatial Hole Burning and Oligo-Mode Distance Control in CW Dye Lasers, IEEE J. QE-11, p.210-212 (1975)

[6.610] {Sect. 6.7.5} A. L. Bloom: Modes of a laser resonator containing tilted birefringent plates, J. Opt. Soc. Am. 64, p.447-452 (1974)

[6.611] {Sect. 6.7.5} H.G. Danielmeyer, W.N. Leibolt: Stable Tunable Single-Frequency Nd:YAG Laser, Appl. Phys. 3, p.193-198 (1974)

[6.612] {Sect. 6.7.5} A.R. Clobes, M.J. Brienza: Single-frequency traveling-wave Nd:YAG laser, Appl. Phys. Lett.21, p.265-267 (1972)

[6.613] {Sect. 6.7.5} D.A. Draegert: Efficient Single-Longitudinal-Mode Nd:YAG Laser, IEEE J. QE-8p.235-239 (1972)

[6.614] {Sect. 6.7.5} H.G. Danielmeyer, E.H. Turner: Electro-Optic Elimination of Spatial Hole Burning in Lasers, Appl. Phys. Lett. 17, p.519-521 (1970)

[6.615] {Sect. 6.7.5} H.G. Danielmeyer, W.G. Nilsen: Spontaneous Single-Frequency Output From a Spatially Homogenous Nd:YAG Laser, Appl. Phys. Lett. 16, p.124-126 (1970)

[6.616] {Sect. 6.7.5} H.G. Danielmeyer: Low-Frequency Dynamics of Homogeneous Four-Level cw Lasers, J. Appl. Phys. 41, p.4014-4018 (1970)

[6.617] {Sect. 6.7.5} M. Hercher: Tunable Single Mode Operation of Gas Lasers Using Intracavity Tilted Etalons, Appl. Opt. 8, p.1103-1106 (1969)

[6.618] {Sect. 6.7.5} J. L. Hall: The Laser Absolute Wavelength Standard Problem, IEEE J. QE-4, p.638-641 (1968)

[6.619] {Sect. 6.7.5} R. Polloni, O. Svelto: Static and Dynamic Behavior of a Single-Mode Nd-YAG Laser, IEEE J. QE-4, p.481-485 (1968)

[6.620] {Sect. 6.7.5} D. Roess: Single-Mode Operation of a Room-Temperature CW-Ruby Laser, Appl. Phys. Lett. 8, p.109-111 (1966)

[6.621] {Sect. 6.7.5} M. Hercher: Single Mode Operation of a Q-Switched Ruby Laser, Appl. Phys. Lett. 7, p.39-41 (1965)

[6.622] {Sect. 6.7.5} S.A. Collins, G.R. White: Interferometer Laser Mode Selector, Appl. Opt. 2, p.448-449 (1963)

[6.623] {Sect. 6.7.5} C. Bollig, W.A. Clarkson, D.C. Hanna, D.S. Lovering, G.C.W. Jones: Single-frequency operation of a monolithic Nd:glass ring laser via the acousto-optic effect, Opt Commun 133, p.221-224 (1997)

[6.624] {Sect. 6.7.5} M. Musha, S. Telada, K. Nakagawa, M. Ohashi, K. Ueda: Measurement of frequency noise spectra of frequency- stabilized LD-pumped Nd:YAG laser by using a cavity with separately suspended mirrors, Opt Commun 140, p.323-330 (1997)

[6.625] {Sect. 6.7.5} B. Braun, U. Keller: Single-frequency Q-switched ring laser with an antiresonant Fabry-Perot saturable absorber, Optics Letters 20, p.1020-1022 (1995)

[6.626] {Sect. 6.7.5} A.C. Nilsson, E.K. Gustafson, R.L. Byer: Eigenpolarization Theory of Monolithic Nonplanar Ring Oscillators, IEEE J. QE-25, p.767-790 (1989)

[6.627] {Sect. 6.7.5} T.J. Kane, A.C. Nilsson, R.L. Byer: Frequency stability and offset locking of a laser-diode-pumpde Nd:YAG monolithic nonplanar ring oscillator, Opt. Lett. 12, p.175-177 (1987)

[6.628] {Sect. 6.7.5} T.J. Kane, R.J. Byer: Monolithic, unidirectional single-mode Nd:YAG ring laser, Opt. Lett. 10, p.65-67 (1985)

[6.629] {Sect. 6.7.5} A. Owyoung, G.R. Hadley, P. Esherick, R.L. Schmitt, L.A. Rahn: Gain switching of a monolithic single-frequency laser-diode-excited Nd:YAG laser, Opt. Lett. 10, p.484-486 (1985)

[6.630] {Sect. 6.7.5} K. Schneider, P. Kramper, S. Schiller, T. Mlynek: Toward an optical synthesizer: A single-frequency parametric oscillator using periodically poled LiNbO3, Optics Letters 22, p.1293-1295 (1997)

[6.631] {Sect. 6.7.5} D.F. Plusquellic, O. Votava, D.J. Nesbitt: Absolute frequency stabilization of an injection-seeded optical parametric oscillator, Appl Opt 35, p.1464-1472 (1996)

[6.632] {Sect. 6.7.5} S. Schiller, G. Breitenbach, R. Paschotta, J. Mlynek: Subharmonic-pumped continuous-wave parametric oscillator, Appl Phys Lett 68, p.3374-3376 (1996)

[6.633] {Sect. 6.7.5} P.B. Sellin, N.M. Strickland, J.L. Carlsten, R.L. Cone: Programmable frequency reference for subkilohertz laser stabilization by use of persistent spectral hole burning, Optics Letters 24, p.1038-1040 (1999)

[6.634] {Sect. 6.7.5} C. Greiner, B. Boggs, T. Wang, T.W. Mossberg: Laser frequency stabilization by means of optical self-heterodyne beat-frequency control, Optics Letters 23, p.1280-1282 (1998)

[6.635] {Sect. 6.7.5} U.K. Schreiber, C.H. Rowe, D.N. Wright, S.J. Cooper, G.E. Stedman: Precision stabilization of the optical frequency in a large ring laser gyroscope, Appl Opt 37, p.8371-8381 (1998)

[6.636] {Sect. 6.7.5} R. Storz, C. Braxmaier, K. Jack, O. Pradl, S. Schiller: Ultrahigh long-term dimensional stability of a sapphire cryogenic optical resonator, Optics Letters 23, p.1031-1033 (1998)

[6.637] {Sect. 6.7.5} H. Talvitie, M. Merimaa, E. Ikonen: Frequency stabilization of a diode laser to Doppler-free spectrum of molecular iodine at 633nm, Opt Commun 152, p.182-188 (1998)

[6.638] {Sect. 6.7.5} D.J. Binks, L.A.W. Gloster, T.A. King, I.T. McKinnie: Frequency locking of a pulsed single-longitudinal-mode laser in a coupled-cavity resonator, Appl Opt 36, p.9371-9377 (1997)

[6.639] {Sect. 6.7.5} M. Musha, K. Nakagawa, K. Ueda: Wideband and high frequency stabilization of an injection-locked Nd:YAG laser to a high-finesse Fabry-Perot cavity, Optics Letters 22, p.1177-1179 (1997)

[6.640] {Sect. 6.7.5} R. Paschotta, J. Nilsson, L. Reekie, A.C. Trooper, D.C. Hanna: Single-frequency ytterbium-doped fiber laser stabilized by spatial hole burning, Optics Letters 22, p.40-42 (1997)

[6.641] {Sect. 6.7.5} G. Ruoso, R. Storz, S. Seel, S. Schiller, J. Mlynek: Nd:YAG laser frequency stabilization to a supercavity at the 0.1 Hz level, Opt Commun 133, p.259-262 (1997)

[6.642] {Sect. 6.7.5} S. Seel, R. Storz, G. Ruoso, J. Mlynek, S. Schiller: Cryogenic optical resonators: A new tool for laser frequency stabilization at the 1 Hz level, Phys Rev Lett 78, p.4741-4744 (1997)

[6.643] {Sect. 6.7.5} F. Bondu, P. Fritschel, C.N. Man, A. Brillet: Ultrahigh-spectral-purity laser for the VIRGO experiment, Optics Letters 21, p.582-584 (1996)

[6.644] {Sect. 6.7.5} D.H. Sarkisyan, A.V. Papoyan: Frequency-stabilized high-power ruby laser Q switched by Rb-2 vapor, Appl Opt 35, p.3207-3209 (1996)

[6.645] {Sect. 6.7.5} C.T. Taylor, M. Notcutt, E.K. Wong, A.G. Mann: Measurement of the coefficient of thermal expansion of a cryogenic, all-sapphire, Fabry-Perot optical cavity, Opt Commun 131, p.311-314 (1996)

[6.646] {Sect. 6.7.5} S.T. Yang, Y. Imai, M. Oka, N. Eguchi, S. Kubota: Frequency-stabilized, 10-W continuous-wave, laser-diode end-pumped, injection-locked Nd:YAG laser, Optics Letters 21, p.1676-1678 (1996)

[6.647] {Sect. 6.7.5} K. Nakagawa, A.S. Shelkovnikov, T. Katsuda, M. Ohtsu: Absolute frequency stability of a diode-laser-pumped Nd:YAG laser stabilized to a high-finesse optical cavity, Appl. Opt. 33, p.6383-6386 (1994)

[6.648] {Sect. 6.7.5} P. Robrish: Single-mode electro-optically tuned Nd:YVO4 laser, Opt. Lett. 19, p.813-815 (1994)

[6.649] {Sect. 6.7.5} N. Uehara, K. Ueda: 193-mHz beat linewidth of frequency-stabilized laser-diode-pumped Nd:YAG ring laser, Opt. Lett. 18, p.505-507 (1993)

[6.650] {Sect. 6.7.5} T. Day, E.K. Gustafson, R.L. Byer: Sub-Hertz Relative Frequency Stabilization of Two-Diode Laser-Pumped Nd:YAG Lasers. Locked to a Fabry-Perot Interferometer, IEEE J. QE-28, p.1106-1117 (1992)

[6.651] {Sect. 6.7.5} B. Zhou, T.J. Kane, G.J. Dixon, R.L. Byer: Efficient, frequency-stable laser-diode-pumped Nd:YAG laser, Opt. Lett. 10, p.62-64 (1985)

[6.652] {Sect. 6.7.5} Y.L. Sun, R.L. Byer: Submegahertz Frequency Stabilized ND YAG Oscillator, Optics Letters 7, p.408-410 (1982)

[6.653] {Sect. 6.7.5} W.G. Schweitzer Jr, E.G. Kessler Jr, R.D. Deslattes, H.P. Layer, J.R. Whetstone: Description, Performance, and Wavelengths of Iodine Stabilized Lasers, Appl. Opt. 12p.2927-2938 (1973)

[6.654] {Sect. 6.7.5} H.G. Danielmeyer: Stabilized Efficient Single-Frequency Nd:YAG Laser, IEEE J. QE-6, p.101-104 (1970)

[6.655] {Sect. 6.7.5} C.C. Harb, M.B. Gray, H.-A. Bachor, R. Schilling, P. Rottengatter, I. Freitag, H. Welling: Suppression of the Intensity Noise in a Diode-Pumped Neodymium:YAG Nonplanar Ring Laser, IEEE J. QE-30, p.2907-2913 (1994)

[6.656] {Sect. 6.7.5} H.A. Haus, A. Mecozzi: Noise of Mode-Locked Lasers, IEEE J QE-29, p.983-996 (1993)

[6.657] {Sect. 6.7.6} M. Ostermeyer, K. Mittler, R. Menzel: Q switch and longitudinal modes of a laser oscillator with a stimulated-Brillouin-scattering mirror, Phys. Rev. A 59, p.3975-3985 (1999)

[6.658] {Sect. 6.7.6} B. Barrientos, V. Aboites, M. Damzen: Temporal dynamics of a ring dye laser with a stimulated Brillouin scattering mirror, Appl Opt 35, p.5386-5391 (1996)

[6.659] {Sect. 6.7.6} B. Barrientos, V. Aboites, M.J. Damzen: Temporal dynamics of an external-injection dye laser with a stimulated Brillouin scattering reflector, J. Opt. (Paris) 26p.97-104 (1995)

[6.660] {Sect. 6.7.6} A. Agnesi, G.C. Reali: Passive and self-Q-switching of phase-conjugation Nd:YAG laser oscillators, Opt. Comm. 89, p.41-46 (1992)

[6.661] {Sect. 6.7.6} G.E. Nekraskova, M.V. Pyatakhin: Dynamics of stimulated emission from a multimode laser considered allowing for stimulated Brillouin scattering, Sov. J. Quant. Electron. 22, p.794-797 (1992)

[6.662] {Sect. 6.7.6} W.A. Schroeder, M.J. Hutchinson: Studies of a single-frequency stimulated-Brillouin-scattering phase-conjugate Nd:YAG laser oscillator, J. Opt. Soc. Am. B 6, p.171-179 (1989)

[6.663] {Sect. 6.7.6} M.J. Damzen, M.H.R. Hutchinson, W.A. Schroeder: Single-frequency phase-conjugate laser resonator using stimulated Brillouin scattering, Opt. Lett. 12, p.45-47 (1987)

[6.664] {Sect. 6.7.6} A.T. Friberg, M. Kauranen, R.Salomaa: Dynamics of Fabry-Perot resonators with a phase-conjugate mirror, J. Opt. Soc. Am. B 3, p.1656-1672 (1986)

[6.665] {Sect. 6.7.6} H. Vanherzeele, J.L. Van Eck, A.E. Siegman: Mode-locked laser oscillation using self-pumped phase-conjugate reflection, Opt. Lett. 6, p.467-469 (1981)

[6.666] {Sect. 6.7.6} V.I. Bezrodnyi, F.I. Ibragimov, V.I. Kislenko, R.A. Petrenko, V.L. Strizhevskii, E.A. Tikhonov: Mechanism of laser Q switching by intra-cavtiy stimulated scattering, Sov. J. Quant. Elevtron. 10, p.382-383 (1980)

[6.667] {Sect. 6.7.6} D. Pohl: A new laser Q-switch-technique using stimulated Brillouin scattering, Phys. Lett. 24A, p.239-241 (1967)

[6.668] {Sect. 6.7.6} M. Lobel, P.M. Petersen, P.M. Johansen: Suppressing self-induced frequency scanning of a phase conjugate diode laser array with using counterbalance dispersion, Appl Phys Lett 72, p.1263-1265 (1998)

[6.669] {Sect. 6.7.6} M. Lobel, P.M. Petersen, P.M. Johansen: Single-mode operation of a laser-diode array with frequency-selective phase-conjugate feedback, Optics Letters 23, p.825-827 (1998)

[6.670] {Sect. 6.7.6} A. Murakami, J. Ohtsubo: Dynamics and linear stability analysis in semiconductor lasers with phase-conjugate feedback, IEEE J QE-34, p.1979-1986 (1998)

[6.671] {Sect. 6.7.6} T. Omatsu, A. Katoh, K. Okada, S. Hatano, A. Hasegawa, M. Tateda, I. Ogura: Investigation of photorefractive phase conjugate feedback on the lasing spectrum of a broad-stripe laser diode, Opt Commun 146, p.167-172 (1998)

[6.672] {Sect. 6.7.6} W.A. vanderGraaf, L. Pesquera, D. Lenstra: Stability of a diode laser with phase-conjugate feedback, Optics Letters 23, p.256-258 (1998)

[6.673] {Sect. 6.7.6} D.H. Detienne, G.R. Gray, G.P. Agrawal, D. Lenstra: Semiconductor laser dynamics for feedback from a finite-penetration-depth phase-conjugate mirror, IEEE J QE-33, p.838-844 (1997)

[6.674] {Sect. 6.7.6} A. Shiratori, M. Obara: Frequency-stable, narrow linewidth oscillation of red diode laser with phase-conjugate feedback using stimulated photorefractive backscattering, Appl Phys Lett 69, p.1515-1516 (1996)

[6.675] {Sect. 6.7.6} M. Ohtsu, I. Koshishi, Y. Teramachi: A Semiconductor Laser as a Stable Phase Conjugate Mirror for Linewidth Reduction of Another Semiconductor Laser, Jap. J. Appl. Phys. 29, p.L2060-L2062 (1990)

[6.676] {Sect. 6.7.6} G.C. Valley, G.J. Dunning: Observation of optical chaos in a phase-conjugate resonator, Opt. Lett. 9, p.513-515 (1984)

[6.677] {Sect. 6.7.6} M.M. Denariez-Roberge, G. Giuliani: High-power single-mode laser operation using stimulated Rayleigh scattering, Opt. Lett. 6, p.339-3341 (1981)

[6.678] {Sect. 6.7.6} R.C. Lind, D.G. Steel: Demonstration of the longitudinal modes and aberration-correction properties of a continuous-wave dye laser with a phase-conjugate mirror, Opt. Lett. 6, p.554-556 (1981)

[6.679] {Sect. 6.8.1} M. Azadeh, L.W. Casperson: Field solutions for bidirectional high-gain laser amplifiers and oscillators, J Appl Phys 83, p.2399-2407 (1998)

[6.680] {Sect. 6.8.1} T. Taira, W.M. Tulloch, R.L. Byer: Modeling of quasi-three-level lasers and operation of cw Yb:YAG lasers, Appl Opt 36, p.1867-1874 (1997)

[6.681] {Sect. 6.8.1} J.M. Eggleston, L.M. Frantz, H. Injeyan: Derivation of the Frantz-Nodvik Equation for Zig-Zag Optical Path, Slab Geometry Laser Amplifiers, IEEE J. QE-25, p.1855-1862 (1989)

[6.682] {Sect. 6.8.1} J. Eicher, N. Hodgson, H. Weber: Output power and efficiencies of slab laser systems, J. Appl. Phys. 66, p.4608-4613 (1989)

[6.683] {Sect. 6.8.1} N. Hodgson, H. Weber: Measurement of extraction efficiency and excitation efficiency of lasers, J. Mod. Opt. 35, p.807-813 (1988)

[6.684] {Sect. 6.8.1} J.A. Caird, M.D. Shinn, T.A. Kirchoff, L.K. Smith, R.E. Wilder: Measurements of losses and lasing efficiency in GSGG:Cr, Nd and YAG:Nd laser rods, Appl. Opt. 25, p.4294-4305 (1986)

[6.685] {Sect. 6.8.1} L.W. Casperson: Power characteristics of high magnification semiconductor lasers, Opt. Quant. Electron. 18, p.155-157 (1986)

[6.686] {Sect. 6.8.1} R.S. Galeev, S.I. Krasnov: Approximate method for calculations of unstable telescopic resonators, Sov. J. Quantum Electron. 12, p.802-804 (1982)

[6.687] {Sect. 6.8.1} G.J. Linford, R.A. Saroyan, J.B. Trenholme, M.J. Weber: Measurements and Modeling of Gain Coefficients for Neodymium Laser Glasses, IEEE J. QE-15, p.510-523 (1979)

[6.688] {Sect. 6.8.1} B.K. Sina: A new method for the estimation of pumping coefficient for a Ruby laser, IEEE J. QE-15, p.1083-1085 (1979)

[6.689] {Sect. 6.8.1} W.W. Rigrod: Homogeneously broadened CW laser with uniform distributed loss, IEEE J. QE-14, p.377-381 (1978)

[6.690] {Sect. 6.8.1} H.G. Danielmeyer: Low-Frequency Dynamics of Homogeneous Four-Level cw Lasers, J. Appl. Phys. 41, p.4014-4018 (1970)

[6.691] {Sect. 6.8.1} T. Kimura, K. Otsuka: Response of a CW Nd3+:YAG Laser to Sinusoidal Cavity Perturbations, IEEE J. QE-6, p.764-769 (1970)

[6.692] {Sect. 6.8.1} J.F. Nester: Dynamic Optical Properties of CW Nd:YAlG Lasers, IEEE J. QE-6 p.97-100 (1970)

[6.693] {Sect. 6.8.1} A.Y. Cabezas, R.P. Treat: Effect of Spectral Hole-Burning and Cross Relaxation on the Gain Saturation of Laser Amplifiers, J. Appl. Phys. 37, p.3556-3563 (1966)

[6.694] {Sect. 6.8.1} D. Findlay, R.A. Clay: The measurement of internal losses in 4-level lasers, Phys. Lett. 20, p.277-278 (1966)

[6.695] {Sect. 6.8.1} D. Roess: Analysis of Room Temperature CW Ruby Lasers, IEEE J. QE-2, p.208-214 (1966)

[6.696] {Sect. 6.8.1} W.W. Rigrod: Saturation Effects in High-Gain Lasers, J. Appl. Phys. 36, p.2487-2490 (1965)

[6.697] {Sect. 6.8.1} S.J. Cooper: Systematic errors in laser gain, saturation irradiance, and cavity loss measurements and comparison with a HCN laser, Appl Opt 38, p.3258-3265 (1999)

[6.698] {Sect. 6.8.2} K. Joosten, G. Nienhuis: Loss rates of laser cavities, Opt Commun 166, p.65-69 (1999)

[6.699] {Sect. 6.8.2} S. Ozcelik, D.L. Akins: Extremely low excitation threshold, superradiant, molecular aggregate lasing system, Appl Phys Lett 71, p.3057-3059 (1997)

[6.700] {Sect. 6.8.2} G.Z.Z. Zhang, D.W. Tokaryk: Lasing threshold reduction in grating-tuned cavities, Appl Opt 36, p.5855-5858 (1997)

[6.701] {Sect. 6.8.2} J.A. Caird, M.D. Shinn, T.A. Kirchoff, L.K. Smith, R.E. Wilder: Measurements of losses and lasing efficiency in GSGG:Cr, Nd and YAG:Nd laser rods, Appl. Opt. 25, p.4294-4305 (1986)

[6.702] {Sect. 6.8.2} D. Findlay, R.A. Clay: The measurement of internal losses in 4-level lasers, Phys. Lett. 20, p.277-278 (1966)

[6.703] {Sect. 6.8.2} A.G. Fox, T. Li: Effect of Gain Saturation on the Oscillating Modes of Optical Masers, IEEE J. QE-2, p.774-783 (1966)

[6.704] {Sect. 6.8.3} M. Stanghini, M. Basso, R. Genesio, A. Tesi, R. Meucci, M. Ciofini: A new three-equation model for the CO2 laser, IEEE J QE-32, p.1126-1131 (1996)

[6.705] {Sect. 6.8.3} P. Laporta, V. Magni, O. Svelto: Comparative Study of the Optical Pumping Efficiency in Solid State Lasers, IEEE J. QE-21, p.1211-1218 (1985)

[6.706] {Sect. 6.8.3} M. Mindak, J. Szydlak: Examples of operating characteristics and power balance in pump cavity of cw Nd:YAG laser, Appl. Opt. 13, p.407-419 (1983)

[6.707] {Sect. 6.8.3} G.M.Schindler: Optimum Output Efficiency of Homogeneously Broadened Lasers with Constant Loss, IEEE J. QE-16, p.546-549 (1980)

[6.708] {Sect. 6.8.3} G.A. Massey: Criterion for selection of cw laser host materials to increase available power in the fundamental mode, Appl. Phys. Lett. 17, p.213-215 (1970)

[6.709] {Sect. 6.8.3} T.J. Karr: Power and stability of phase-conjugate lasers, J. Opt. Soc. Am. 73, p.600-609 (1983)

[6.710] {Sect. 6.8.3} G. Lescroart, R. Muller, G. Bourdet: Experimental investigations and theoretical modeling of a Tm: YVO4 microchip laser, Opt Commun 143, p.147-155 (1997)

[6.711] {Sect. 6.9.3} M. Horowitz, Y. Barad, Y. Silberberg: Noiselike pulses with a broadband spectrum generated from an erbium-doped fiber laser, Optics Letters 22, p.799-801 (1997)

[6.712] {Sect. 6.9.3} K. Shimizu, T. Horiguchi, Y. Koyamada: Broad-band absolute frequency synthesis of pulsed coherent lightwaves by use of a phase-modulation amplified optical ring, IEEE J QE-33, p.1268-1277 (1997)

[6.713] {Sect. 6.9.3} C.A. Kapetanakos, B. Hafizi, H.M. Milchberg, P. Sprangle, R.F. Hubbard, A. Ting: Generation of high-average-power ultrabroad-band infrared pulses, IEEE J QE-35, p.565-576 (1999)

[6.714] {Sect. 6.9.3} M. Brown: Increased spectral bandwidths in nonlinear conversion processes by use of multicrystal designs, Optics Letters 23, p.1591-1593 (1998)

[6.715] {Sect. 6.9.3} D. Lorenz, R. Menzel: Broadband operation of frequency doubled Cr4+:YAG laser with high beam quality, OSA TOPS Vol. 19 Advanced Solid State Lasers, p.92-96 (1998)

[6.716] {Sect. 6.9.3} V. Valerii, Ter-Mikirtychev, T. Tsubo: Ultrabroadband LiF:F2+* color center laser using two-rism spatially-disperse resonator, Opt. Comm. 137, p.74-76 (1997)

[6.717] {Sect. 6.9.3} L.W. Casperson: Analytic modeling of gain-switched lasers. I. Laser oscillators, J. Appl. Phys. 47, p.4555-4562 (1976)

[6.718] {Sect. 6.9.3} Y.H. Cha, Y.I. Kang, C.H. Nam: Generation of a broad amplified spectrum in a femtosecond terawatt Ti : sapphire laser by a long-wavelength injection method, J Opt Soc Am B Opt Physics 16, p.1220-1223 (1999)

[6.719] {Sect. 6.10.1} R. Bohm, V.M. Baev, P.E. Toschek: Measurements of operation parameters and nonlinearity of a Nd3+-doped fibre laser by relaxation oscillations, Opt Commun 134, p.537-546 (1997)

[6.720] {Sect. 6.10.1} R. Stemme, G. Herziger, H. Weber: Power and halfwidth of first laser spike, Opt. Comm. 10, p.221-225 (1974)

[6.721] {Sect. 6.10.1} H. Statz, G.A. DeMars, D.T. Wilson, C.L. Tang: Problem of Spike Elimination in Lasers, J. Appl. Phys. 36, p.1510-1514 (1965)

[6.722] {Sect. 6.10.1} R. Dunsmuir: Theory of Relaxation Oscillations in Optical Masers, J. Electron. Control 10, p.453-458 (1961)

[6.723] {Sect. 6.10.2.1} M. Ozolinsh, K. Stock, R. Hibst, R. Steiner: Q-switching of Er:YAG (2.9 mu m) solid-state laser by PLZT electrooptic modulator, IEEE J QE-33, p.1846-1849 (1997)

[6.724] {Sect. 6.10.2.1} A. Hogele, G. Horbe, H. Lubatschowski, H. Welling, W. Ertmer: 2.70 mu m CrEr: YSGG laser with high output energy and FTIR-Q-switch, Opt Commun 125, p.90-94 (1996)

[6.725] {Sect. 6.10.2.1} T. Chuang, A.D. Hays, H.R. Verdun: Effect of dispersion on the operation of a KTP electro-optic Q switch, Appl. Opt. 33, p.8355-8360 (1994)

[6.726] {Sect. 6.10.2.1} S.Z. Kurtev, O.E. Denchev, S.D. Savov: Effects of thermally induced birefringence in high-output-power electro-optically Q-switched Nd:YAG lases and their compensation, Appl. Opt. 32, p.278-285 (1993)

[6.727] {Sect. 6.10.2.1} J. Richards: Unpolarized EO Q-switched laser, Appl. Opt. 22, p.1306-1308 (1983)

[6.728] {Sect. 6.10.2.1} M.K. Chun, E.A. Teppo: Laser resonator: an electrooptical Q-switched Porro prism device, Appl. Opt. 15, p.1942-1946 (1976)

[6.729] {Sect. 6.10.2.1} H.A. Kruegle, L. Klein: High peak power output, high PRF by cavity dumping a Nd:YAG laser, Appl. Opt. 15, p.466-471 (1976)

[6.730] {Sect. 6.10.2.1} D. Cheng: Instability of Cavity-Dumped YAG Laser Due to Time-Varying Reflections, IEEE J. QE-9, p.585-588 (1973)

[6.731] {Sect. 6.10.2.1} D. Milam: Brewster-Angle Pockels Cell Design, Appl. Opt. 12, p.602-606 (1973)

[6.732] {Sect. 6.10.2.1} C.W. Reno: High Data Rate YAG Laser Techniques, Appl. Opt. 12, p.883-885 (1973)

[6.733] {Sect. 6.10.2.1} L.L. Steinmetz, T.W. Pouliot, B.C. Johnson: Cylindrical, Ring-Electrode KD*P Electrooptic Modulator, Appl. Opt. 12, p.1468-1471 (1973)

[6.734] {Sect. 6.10.2.1} M.K. Chun, J.T. Bischoff: Multipulsing Behavior of Electrooptically Q-Switched Lasers, IEEE J. QE-8, p.715-716 (1972)

[6.735] {Sect. 6.10.2.1} D.C. Hanna, B. Luther Davis, R.C. Smith: Active Q switching technique for producing high laser power in a single longitudinal mode, Electron. Lett. 8, p.369-370 (1972)

[6.736] {Sect. 6.10.2.1} R.B. Chesler, D.A. Pinnow, W.W. Benson: Suitability of PbMoO4 for Nd:YAlG Intracavity Acoustooptic Modulation, Appl. Opt. 10, p.2562 (1971)

[6.737] {Sect. 6.10.2.1} M.G. Cohen, R.T. Daly, R.A. Kaplan: Resonant Acoustooptic Q Switching of High-Gain Lasers, IEEE J. QE-7, p.316-317 (1971)

[6.738] {Sect. 6.10.2.1} W.R. Hook, R.P. Hilberg: Lossless KD*P Pockels Cell for High-Power Q Switching, Appl. Opt. 10, p.1179-1180 (1971)

[6.739] {Sect. 6.10.2.1} W. Buchman, W. Koechner, D. Rice: Vibrating Mirror as a Repetitive Q Switch, IEEE J. QE-6, p.747-749 (1970)

[6.740] {Sect. 6.10.2.1} R.P. Hilberg, W.R. Hook: Transient Elastooptic Effects and Q-Switching Performance in Lithium Niobate and KD*P Pockels Cells, Appl. Opt. 9, p.1939-1940 (1970)

[6.741] {Sect. 6.10.2.1} D. Maydan: Acoustooptical Pulse Modulators, IEEE J. QE-6, p.15-24 (1970)

[6.742] {Sect. 6.10.2.1} R.M. Schotland: A Mode Controlled Q-Switched Tuneable Ruby Laser, Appl. Opt. 9, p.1211-1213 (1970)

[6.743] {Sect. 6.10.2.1} I.W. Mackintosh: Double Etalon Q-Switching of a Continuously Pumped Nd/YAG Laser, Appl. Opt. 8, p.1991-1998 (1969)

[6.744] {Sect. 6.10.2.1} M.B. Davies, P.H. Sarkies, J.K. Wright: Operaton of a Lithium Niobate Electrooptic Q Switch at 1.06 µ, IEEE J. QE-4, p.533-535 (1968)

[6.745] {Sect. 6.10.2.1} R.W. Dixon: Acoustic Diffraction of Light in Anisotropic Media, IEEE J. QE-3, p.85-93 (1967)

[6.746] {Sect. 6.10.2.1} M. Dore: A Low Drive-Power Light Modular Using a Readily Available Material ADP, IEEE J. QE-3, p.555-560 (1967)

[6.747] {Sect. 6.10.2.1} W.R. Hook, R.H. Dishington, R.P. Hilberg: Laser cavity dumping using time variable reflection, Appl. Phys. Lett. 9, p.125-127 (1966)

[6.748] {Sect. 6.10.2.1} I.P. Kaminow, E.H. Turner: Electrooptic Light Modulators, Appl. Opt. 5, p.1612-1627 (1966)

[6.749] {Sect. 6.10.2.1} R.A. Phillips: Temperature Variation of the Index of Refraction of ADP, KDP, and Deuterated KDP*, J. Opt. Soc. Am. 56, p.629-632 (1966)

[6.750] {Sect. 6.10.2.1} E.L. Steele, W.C. Davis, R.L. Treuthart: A Laser Output Using Frustrated Total Internal Reflection, Appl. Opt. 5, p.5-8 (1966)

[6.751] {Sect. 6.10.2.1} M. Yamazaki, T. Ogawa: Temperature Dependences of the Refractive Indices of NH4H2PO4, KH2PO4, and Partially Deuterated KH2PO4, J. Opt. Soc. Am. 56, p.1407-1408 (1966)

[6.752] {Sect. 6.10.2.1} T. Crawford, C. Lowrie, J.R. Thompson: Prelase stabilization of the polarization state and frequency of a Q-switched, diode-pumped, Nd:YAG laser, Appl Opt 35, p.5861-5869 (1996)

[6.753] {Sect. 6.10.2.1} M. Marincek, M. Lukac: Development of EM Field in Lasers with Rotating Mirror Q-Switch, IEEE J. QE-29, p.2405-2412 (1993)

[6.754] {Sect. 6.10.2.1} C. Wyss, W. Luthy, H.P. Weber: Modulation and single-spike switching of a diode-pumped Er3+: LiYF4 laser at 2.8 mu m, IEEE J QE-34, p.1041-1045 (1998)

[6.755] {Sect. 6.10.2.2} I.P. Bilinsky, J.G. Fujimoto, J.N. Walpole, L.J. Missaggia: InAs-doped silica films for saturable absorber applications, Appl Phys Lett 74, p.2411-2413 (1999)

[6.756] {Sect. 6.10.2.2} P. Peterson, A. Gavrielides, M.P. Sharma, T. Erneux: Dynamics of passively Q-switched microchip lasers, IEEE J QE-35, p.1247-1256 (1999)

[6.757] {Sect. 6.10.2.2} K.L. Vodopyanov, R. Shori, O.M. Stafsudd: Generation of Q-switched Er:YAG laser pulses using evanescent wave absorption in ethanol, Appl Phys Lett 72, p.2211-2213 (1998)

[6.758] {Sect. 6.10.2.2} A. Agnesi, S. Dell'Acqua, E Piccinini, G. Reali, G. Piccinno: Efficient Wavelength Conversion with High-Power Passively Q-Switched Diode-Pumped Neodymium Lasers, IEEE J. QE-34, p.1480-1484 (1998)

[6.759] {Sect. 6.10.2.2} R.S. Afzal, A.W. Yu, T.J. Zayhowski, T.Y. Fan: Single-mode high-peak-power passively Q-switched diode- pumped Nd:YAG laser, Optics Letters 22, p.1314-1316 (1997)

[6.760] {Sect. 6.10.2.2} B. Braun, F.X. Kartner, G. Zhang, M. Moser, U. Keller: 56-ps passively Q-switched diode-pumped microchip laser, Optics Letters 22, p.381-383 (1997)

[6.761] {Sect. 6.10.2.2} R. Fluck, B. Braun, E. Gini, H. Melchior, U. Keller: Passively Q-switched 1.34-mu m Nd:YVO4 microchip laser with semiconductor saturable-absorber mirrors, Optics Letters 22, p.991-993 (1997)

[6.762] {Sect. 6.10.2.2} R.Z. Hua, L.J. Qian, T.T. Zhi, X.M. Deng: Short pulse generation in a Nd:YAG laser by silicon, Opt Commun 143, p.47-52 (1997)

[6.763] {Sect. 6.10.2.2} T.T. Kajava, A.L. Gaeta: Intra-cavity frequency-doubling of a Nd:YAG laser passively Q-switched with GaAs, Opt Commun 137, p.93-97 (1997)

[6.764] {Sect. 6.10.2.2} B. Braun, F.X. Kartner, U. Keller, J.P. Meyn, G. Huber: Passively Q-switched 180-ps Nd:LaSc3 (BO3) (4) microchip laser, Optics Letters 21, p.405-407 (1996)

[6.765] {Sect. 6.10.2.2} T.T. Kajava, A.L. Gaeta: Q switching of a diode-pumped Nd:YAG laser with GaAs, Optics Letters 21, p.1244-1246 (1996)

[6.766] {Sect. 6.10.2.2} Y. Shimony, Z. Burshtein, A.B. Baranga, Y. Kalisky, M. Strauss: Repetitive Q-Switching of a CW Nd:YAG laser using Cr4+:YAG saturable absorbers, IEEE J QE-32, p.305-310 (1996)

[6.767] {Sect. 6.10.2.2} Y. Shimony, Z. Burshtein, Y. Kalisky: Cr4+:YAG as Passive Q-Switch and Brewster Plate in a Pulsed Nd:YAG Laser, IEEE J. QE-31, p.1738-1741 (1995)

[6.768] {Sect. 6.10.2.2} H.J. Eichler, A. Haase, R. Menzel: Cr4+:YAG as Passive Q-Switch for a Nd:YALO Oscillator with an Average Repetition Rate of 2.7 kHz, TEM00 Mode and 13 W Output, Appl. Phys. B 58, p.409-411 (1994)

[6.769] {Sect. 6.10.2.2} Y. Jingguo, J. Hongwei: Self-Q-switching Nd:YAG laser operation using stimulated thermal Rayleigh scattering, Opt. Quant. Electron. 26, p.929-932 (1994)

[6.770] {Sect. 6.10.2.2} J.A. Morris, C.R. Pollock: Passive Q switching of a diode-pumped Nd:YAG laser with a saturable absorber, Opt. Lett. 15, p.440-442 (1990)

[6.771] {Sect. 6.10.2.2} E. Reed: A flashlamp-Pumped, Q-Switched Cr:Nd:GSGG Laser, IEEE J. QE-21, p.1625-1629 (1985)

[6.772] {Sect. 6.10.2.2} V.I. Bezrodnyi, F.I. Ibragimov, V.I. Kislenko, R.A. Petrenko, V.L. Strizhevskii, E.A. Tikhonov: Mechanism of laser Q switching by intracavtiy stimulated scattering, Sov. J. Quant. Elevtron. 10, p.382-383 (1980)

[6.773] {Sect. 6.10.2.2} B. Kopainsky, W. Kaiser, K.H. Drexhage: New Ultrafast Saturable Absorbers for Nd:lasers, Opt. Comm. 32, p.451-455 (1980)

[6.774] {Sect. 6.10.2.2} W.E. Schmid: Pulse Stretching in a Q-Switched Nd:YAG Laser, IEEE J. QE-16, p.790-794 (1980)

[6.775] {Sect. 6.10.2.2} J.R. Lakowicz, G. Weber: Quenching of Fluorescence by Oxygen. A Probe for Structural Fluctuations in Macromolecules, Biochem. 12, p.4161-4170 (1973)

[6.776] {Sect. 6.10.2.2} M. Hercher: An Analysis of Saturable Absorbers, Appl. Opt. 6, p.947-954 (1967)

[6.777] {Sect. 6.10.2.2} C.H. Thomas, E.V. Price: Feedback Control of a Q-Switched Ruby Laser, IEEE J. QE-2, p.617-623 (1966)

[6.778] {Sect. 6.10.2.2} B.H. Soffer: Giant Pulse Laser Operation by a Passive, Reversible Bleachable Absorber, J. Appl. Phys. 35, p.2551 (1964)

[6.779] {Sect. 6.10.2.2} F.J. McClung, R.W. Hellwarth: Characteristics of giant optical pulsations from ruby, Proc. IEEE 51, p.46 (1963)

[6.780] {Sect. 6.10.2.2} A.G. Okhrimchuk, A.V. Shestakov: Absorption saturation mechanism for YAG : Cr4+ crystals, Phys Rev B 61, p.988-995 (2000)

[6.781] {Sect. 6.10.2.2} L.G. Luo, P.L. Chu: Passive Q-switched erbium-doped fibre laser with saturable absorber, Opt Commun 161, p.257-263 (1999)

[6.782] {Sect. 6.10.2.2} P. Petropoulos, H.L. Offerhaus, D.J. Richardson, S. Dhanjal, N.I. Zheludev: Passive Q-switching of fiber lasers using a broadband liquefying gallium mirror, Appl Phys Lett 74, p.3619-3621 (1999)

[6.783] {Sect. 6.10.2.2} P. Petropoulos, S. Dhanjal, D.J. Richardson, N.I. Zheludev: Passive Q-switching of an Er3+: Yb3+ fibre laser with a fibrised liquefying gallium mirror, Opt Commun 166, p.239-243 (1999)

[6.784] {Sect. 6.10.2.2} A.V. Podlipensky, V.G. Shcherbitsky, N.V. Kuleshov, V.P. Mikhailov, V.I. Levchenko, V.N. Yakimovich: Cr2+: ZnSe and Co2+: ZnSe saturable-absorber Q switches for 1.54-mu m Er : glass lasers, Optics Letters 24, p.960-962 (1999)

[6.785] {Sect. 6.10.2.2} K.V. Yumashev: Saturable absorber Co2+: MgAl2O4 crystal for Q switching of 1.34-mu m Nd3+: YAlO (3) and 1.54-mu m Er3+: glass lasers, Appl Opt 38, p.6343-6346 (1999)

[6.786] {Sect. 6.10.2.2} Z.G. Zhang, K. Torizuka, T. Itatani, K. Kobayashi, T. Sugaya, T. Nakagawa, H. Takahashi: Broadband semiconductor saturable-absorber mirror for a self-starting mode-locked Cr:forsterite laser, Optics Letters 23, p.1465-1467 (1998)

[6.787] {Sect. 6.10.2.2} Z. Burshtein, P. Blau, Y. Kalisky, Y. Shimony, M.R. Kokta: Excited-State Absorption Studies of Cr4+ Ions in Several Garnet Host Crystals, IEEE J. QE-34, p.292-299 (1998)

[6.788] {Sect. 6.10.2.2} J. Popp, M.H. Fields, R.K. Chang: Q switching by saturable absorption in microdroplets: elastic scattering and laser emission, Optics Letters 22, p.1296-1298 (1997)

[6.789] {Sect. 6.10.2.2} Y.K. Kuo, M. Birnbaum, F. Unlu, M.F. Huang: Ho:CaF2 solid-state saturable-absorber Q switch for the 2- mu m Tm,Cr:Y3Al5O12 laser, Appl Opt 35, p.2576-2579 (1996)

[6.790] {Sect. 6.10.2.2} Y. Shimony, Z. Burshtein, Y. Kalisky, A.B. Baranga, M. Strauss: Progress in Q-switching of Nd:YAG lasers using Cr4+:YAG saturable absorber, J Nonlinear Opt Physics Mat 5, p.495-504 (1996)

[6.791] {Sect. 6.10.2.2} B.C. Weber, A. Hirth: Presentation of a new and simple technique of Q-switching with a LiSrAlf (6):Cr3+ oscillator, Opt Commun 149, p.301-306 (1998)

[6.792] {Sect. 6.10.2.3} X.Y. Zhang, S.Z. Zhao, Q.P. Wang, B. Ozygus, H. Weber: Modeling of diode-pumped actively Q-switched lasers, IEEE J QE-35p.1912-1918 (1999)

[6.793] {Sect. 6.10.2.3} S. Georgescu, V. Lupei: Q-switch regime of 3-mu m Er:YAG lasers, IEEE J QE-34, p.1031-1040 (1998)

[6.794] {Sect. 6.10.2.3} H. Su, H.Y. Shen, W.X. Lin, R.R. Zeng, C.H. Huang, G. Zhang: Computational model of Q-switch Nd : YAlO3 dual-wavelength laser, J Appl Phys 84, p.6519-6522 (1998)

[6.795] {Sect. 6.10.2.3} E. Tanguy, C. Larat, J.P. Pocholle: Modelling of the erbium-ytterbium laser, Opt Commun 153, p.172-183 (1998)

[6.796] {Sect. 6.10.2.3} G.H. Xiao, M. Bass, M. Acharekar: Passively Q-switched solid-state lasers with intracavity optical parametric oscillators, IEEE J QE-34, p.2241-2245 (1998)

[6.797] {Sect. 6.10.2.3} G.H. Xiao, M. Bass: Additional experimental confirmation of the predictions of a model to optimize passively Q-switched lasers, IEEE J QE-34, p.1142-1143 (1998)

[6.798] {Sect. 6.10.2.3} G.H. Xiao, M. Bass: A generalized model for passively Q-switched lasers including excited state absorption in the saturable absorber, IEEE J QE-33, p.41-44 (1997)

[6.799] {Sect. 6.10.2.3} X.Y. Zhang, S.Z. Zhao, Q.P. Wang, Q.D. Zhang, L.K. Sun, S.J. Zhang: Optimization of Cr4+-doped saturable-absorber Q-switched lasers, IEEE J QE-33, p.2286-2294 (1997)

[6.800] {Sect. 6.10.2.3} B. Ozygus, K. Ziegler: Determination of losses, gain, and pumping-beam mode overlap for Q-switched end-pumped lasers, Appl Phys Lett 68, p.582-583 (1996)

[6.801] {Sect. 6.10.2.3} J.J. Degnan: Optimization of Passively Q-Switched Lasers, IEEE J. QE-31, p.1890-1901 (1995)

[6.802] {Sect. 6.10.2.3} J.J. Degnan: Theory of the Optimally Coupled Q-Switched Laser, IEEE J. QE-25, p.214-220 (1989)

[6.803] {Sect. 6.10.2.3} A.E. Siegman: An Antiresonant Ring Interferometer for Coupled Laser Cavities, Laser Output Coupling, Mode Locking, and Cavity Dumping, IEEE J. QE-9, p.247-250 (1973)

[6.804] {Sect. 6.10.2.3} G.D. Baldwin: Output Power Calculations for a Continously Pumped Q-switched YAG:Nd+3 Laser, IEEE J. QE-7, p.220-224 (1971)

[6.805] {Sect. 6.10.2.3} R.B. Kay, G.S. Waldmann: Complete Solutions to the Rate Equations Describing Q-Spoiled and PTM Laser Operation, J. Appl. Phys. 36, p.1319-1323 (1965)

[6.806] {Sect. 6.10.2.3} J.E. Midwinter: The theory of Q-switching applied to slow switching and pulse shaping for solid state lasers, Brit. J. Appl. Phys. 16, p.1125-1133 (1965)

[6.807] {Sect. 6.10.2.3} W.R. Sooy: The Natural Selection of Modes in a Passive Q-Switched Laser, Appl. Phys. Lett. 7, p.36-37 (1965)

[6.808] {Sect. 6.10.2.8} Z.T. Chen, A.B. Grudinin, J. Porta, J.D. Minelly: Enhanced Q switching in double-clad fiber lasers, Optics Letters 23, p.454-456 (1998)

[6.809] {Sect. 6.10.2.8} R.S. Conroy, T. Lake, G.T. Friel, A.T. Kemp, B.D. Sinclair: Self-Q-switched Nd:YVO4 microchip lasers, Optics Letters 23, p.457-459 (1998)

[6.810] {Sect. 6.10.3.0} M.S. Demokan: Mode-Locking in Solid State and Semiconductor-Lasers (Wiley, New York 1982)

[6.811] {Sect. 6.10.3.0} G. Steinmeyer, D.H. Sutter, L. Gallmann, N. Matuschek, U. Keller: Frontiers in ultrashort pulse generation: Pushing the limits in linear and nonlinear optics, Science 286, p.1507-1512 (1999)

[6.812] {Sect. 6.10.3.0} F. Krausz, M.E. Fermann, T. Brabec, P.F. Curley, M. Hofer, M.H. Ober, C. Spielmann, E. Wintner, A.J. Schmidt Femtosecond solid state laser, IEEE J. QE-28, p.2097-2122 (1992)

[6.813] {Sect. 6.10.3.0} S. A. Akhmanov, V. A. Vysloukh, A. S. Chirkin: Optics of Femtosecond Laser Pulses (American Institute of Physics, New York, 1992)

[6.814] {Sect. 6.10.3.0} C. Rouyer, É. Mazataud, I. Allais, A. Pierre, S. Seznec, C. Sauteret, G. Mourou, A. Migus: Generation of 50-TW femtosecond pulses in a Ti:sapphire/Nd:glass chain, Opt. Lett. 18, p.214-216 (1993)

[6.815] {Sect. 6.10.3.0} M. Piché: Beam reshaping and self-mode-locking in nonlinear laser resonators, Opt. Commun. 86, p.156-160 (1991)

[6.816] {Sect. 6.10.3.0} A. Sullivan, H. Hamster, H.C. Kapteyn, S. Gordon, W. White, H. Nathel, R.J. Blair, R. W. Falcone: Multiterawatt, 100-fs laser, Opt. Lett. 16, p.1406-1408 (1991)

[6.817] {Sect. 6.10.3.0} J.P. Gordon, R.L. Fork: Optical resonator with negative dispersion, Opt. Lett. 9, p.153-155 (1984)

[6.818] {Sect. 6.10.3.0} S.R. Rotman, C. Roxlo, D. Bebelaar, T.K. Yee, M.M. Salour: Generation, Stabilization and Amplification of Subpicosecond Pulses, Appl. Phys. B 28, p.319-326 (1982)

[6.819] {Sect. 6.10.3.0} G.R. Flemming, G.S. Beddard: CW mode-locked dye lasers for ultra fast spectroscopic studies, Opt. Laser Technol. 10, p.257-264 (1978)

[6.820] {Sect. 6.10.3.0} A.E. Siegmann, D.J. Kuizenga: Active mode-coupling phenomena in pulsed and continuous lasers, Opto-Electr. 6, p.43-66 (1974)

[6.821] {Sect. 6.10.3.0} D.J. Bradley, W. Sibbett: Streak-Camera Studies of Picosecond Pulses from a Mode-Locked Nd:Glass Laser, Opt. Commun. 9, p.17-20 (1973)

[6.822] {Sect. 6.10.3.0} G. Girard, M. Michon: Transmission of a Kodak 9740 Dye Solution Under Picosecond Pulses, IEEE J. QE-9, p.979-984 (1973)

[6.823] {Sect. 6.10.3.0} D.J. Kuizenga, D.W. Phillion, T. Lund, A.E. Siegman: Simultaneous Q-Switching and Mode-Locking in the CW Nd:YAG Laser, Opt. Commun. 9, p.221-226 (1973)

[6.824] {Sect. 6.10.3.0} D. von der Linde, K.F. Rodgers: Recovery Time of Saturable Absorbers for 1.06 µ, IEEE J. QE-9, p.960-961 (1973)

[6.825] {Sect. 6.10.3.0} D. von der Linde: Mode-Locked Lasers and Ultrashort Light Pulses, Appl. Phys. 2, p.281-296 (1973)

[6.826] {Sect. 6.10.3.0} D. von der Linde: Experimental Study of Single Picosecond Light Pulses, IEEE J. QE-8, p.328-338 (1972)

[6.827] {Sect. 6.10.3.0} J.A. Fleck: Ultrashort-Pulse Generation by Q-Switched Lasers, Phys. Rev. B 1, p.84-100 (1970)

[6.828] {Sect. 6.10.3.0} D.J. Kuizenga, A.E. Siegman: FM and AM Mode Locking of the Homogeneous Laser – Part II: Experimental Results in a Nd:YAG Laser With Internal FM Modulation, IEEE J. QE-6, p.709-715 (1970)

[6.829] {Sect. 6.10.3.0} D. von der Linde, O. Bernecker, W. Kaiser: Experimental Investigation of Single Picosecond Pulses, Opt. Comm. 2, p.149-152 (1970)

[6.830] {Sect. 6.10.3.0} G.R. Huggett: Mode-Locking of CW Lasers by Regenerative RF Feedback, Appl. Phys. Lett. 13, p.186-187 (1968)

[6.831] {Sect. 6.10.3.1} M. Nakazawa, H. Kubota, A. Sahara, K. Tamura: Time-domain ABCD matrix formalism for laser mode-locking and optical pulse transmission, IEEE J QE-34, p.1075-1081 (1998)

[6.832] {Sect. 6.10.3.1} J. Theimer, M. Hayduk, M.F. Krol, J.W. Haus: Mode-locked Cr4+:YAG laser: model and experiment, Opt Commun 142, p.55-60 (1997)

[6.833] {Sect. 6.10.3.1} S. Arahira, Y. Matsui, Y. Ogawa: Mode-locking at very high repetition rates more than terahertz in passively mode-locked distributed-Bragg- reflector laser diodes, IEEE J QE-32, p.1211-1224 (1996)

[6.834] {Sect. 6.10.3.1} R.G.M.P. Koumans, R. Vanroijen: Theory for passive mode-locking in semiconductor laser structures including the effects of self-phase modulation, dispersion, and pulse collisions, IEEE J QE-32, p.478-492 (1996)

[6.835] {Sect. 6.10.3.1} J.A. Leegwater: Theory of mode-locked semiconductor lasers, IEEE J QE-32, p.1782-1790 (1996)

[6.836] {Sect. 6.10.3.1} L. Xu, C. Spielmann, A. Poppe, T. Brabec, F. Krausz, T.W. Hansch: Route to phase control of ultrashort light pulses, Optics Letters 21, p.2008-2010 (1996)

[6.837] {Sect. 6.10.3.1} R.E. Bridges, R.W. Boyd, G.P. Agrawal: Effect of beam ellipticity on self-mode locking in lasers, Opt. Lett. 18, p.2026-2028 (1993)

[6.838] {Sect. 6.10.3.1} H.A. Haus, U. Keller, W.H. Knox: Theory of Coupled Cavity Mode Locking with a Resonant Nonlinearity, J OPT SOC AM B-OPT PHYSICS 8, p.1252-1258 (1991)

[6.839] {Sect. 6.10.3.1} J. Hermann, F. Weidner, B. Wilhelmi: Influence of the Inversion Depletion in the Active Medium on the Evolution of Ultrashort Pulses in Passively Mode-Locked Solid-State Lasers, Appl. Phys. 20, p.237-245 (1979)

[6.840] {Sect. 6.10.3.1} G.H.C. New, T.B. O'Hare: A Simple Criterion for Passive Q-Switching of Lasers, Phys. Lett. 68A, p.27-28 (1978)

[6.841] {Sect. 6.10.3.1} W. Zinth, A. Lauberau, W. Kaiser: Generation of Chirp-Free Picosecond Pulses, Opt. Comm. 22, p.161-176 (1977)

[6.842] {Sect. 6.10.3.1} D. von der Linde, K.F. Rodgers: Suppression of the Spectral Narrowing Effect in Lasers Mode-Locked by Saturable Absorbers, Opt. Comm. 8, p.91-94 (1973)

[6.843] {Sect. 6.10.3.1} G.H.C. New: Mode-Locking of Quasi-Continuous Lasers, Opt. Comm. 6, p.188-192 (1972)

[6.844] {Sect. 6.10.3.1} D. Bradley, G.H.C. New, S.J. Caughey: Subpicosecond Structure in Mode-Locked Nd:Glass Lasers, Phys. Lett. 30A, p.78-79 (1969)

[6.845] {Sect. 6.10.3.1} V.S. Letokhov: Ultrashort Fluctuation Pulsed of Light in a Laser, Soviet. Phys. JETP 28, p.1026-1027 (1969)

[6.846] {Sect. 6.10.3.1} V.S. Letokhov: Generation of Ultrafast Light Pulses in a Laser with a Nonlinear Absorber, Soviet. Phys. JETP 28, p.562-568 (1969)

[6.847] {Sect. 6.10.3.1} J.A. Fleck, Jr.: Mode-Locked Pulse Generation in Passively Switched Lasers, Appl. Phys. Lett. 12, p.178-181 (1968)

[6.848] {Sect. 6.10.3.1} J.A. Fleck, Jr.: Origin of Short-Pulse Emission by Passively Switched Lasers, J. Appl. Phys. 39, p.3318-3327 (1968)

[6.849] {Sect. 6.10.3.1} H. Weber: Generation and Measurement of Ultrashort Light Pulses, J. Appl. Phys. 39, p.6041-6044 (1968)

[6.850] {Sect. 6.10.3.1} H.W. Mocker, R.J. Collins: Mode competition and self-locking effects in a Q-switched ruby laser, Appl. Phys. Lett. 7, p.270-273 (1965)

[6.851] {Sect. 6.10.3.2} D.I. Chang, H.Y. Kim, M.Y. Jeon, H.K. Lee, D.S. Lim, K.H. Kim, I. Kim, S.T. Kim: Short pulse generation in the mode-locked fibre laser using cholesteric liquid crystal, Opt Commun 162, p.251-255 (1999)

[6.852] {Sect. 6.10.3.2} V. Couderc, F. Louradour, A. Barthelemy: 2.8 ps pulses from a mode-locked diode pumped Nd : YVO4 laser using quadratic polarization switching, Opt Commun 166, p.103-111 (1999)

[6.853] {Sect. 6.10.3.2} P. Glas, M. Naumann, A. Schirrmacher, L. Daweritz, R. Hey: Self pulsing versus self locking in a cw pumped neodymium doped double clad fiber laser, Opt Commun 161, p.345-358 (1999)

[6.854] {Sect. 6.10.3.2} M. Jiang, G. Sucha, M.E. Fermann, J. Jimenez, D. Harter, M. Dagenais, S. Fox, Y. Hu: Nonlinearly limited saturable-absorber mode locking of an erbium fiber laser, Optics Letters 24, p.1074-1076 (1999)

[6.855] {Sect. 6.10.3.2} V.P. Kalosha, M. Muller, J. Herrmann: Theory of solid-state laser mode locking by coherent semiconductor quantum-well absorbers, J Opt Soc Am B Opt Physics 16, p.323-338 (1999)

[6.856] {Sect. 6.10.3.2} M. Leitner, P. Glas, T. Sandrock, M. Wrage, G. Apostolopoulos, A. Riedel, H. Kostial, J. Herfort, K.J. Friedland, L. Daweritz: Self-starting mode locking of a Nd : glass fiber laser by use of the third-order nonlinearity of low-temperature-grown GaAs, Optics Letters 24, p.1567-1569 (1999)

[6.857] {Sect. 6.10.3.2} J.T. Ahn, H.K. Lee, K.H. Kim, M.Y. Jeon, E.H. Lee: A passively mode-locked fibre laser with a delayed optical path for increasing the repetition rate, Opt Commun 148, p.59-62 (1998)

[6.858] {Sect. 6.10.3.2} Y.M. Chang, R. Maciejko, R. Leonelli, A.S. Thorpe: Self-starting passively mode-locked tunable Cr4+:yttrium-aluminum- garnet laser with a single prism for dispersion compensation, Appl Phys Lett 73, p.2098-2100 (1998)

[6.859] {Sect. 6.10.3.2} J.M. Hopkins, G.J. Valentine, W. Sibbett, J.A. derAu, F. MorierGenoud, U. Keller, A. Valster: Efficient, low-noise, SESAM-based femtosecond Cr3+:LiSrAlF6 laser, Opt Commun 154, p.54-58 (1998)

[6.860] {Sect. 6.10.3.2} M.J. Lederer, B. LutherDavies, H.H. Tan, C. Jagadish: An antiresonant Fabry-Perot saturable absorber for passive mode-locking fabricated by metal-organic vapor phase epitaxy and ion implantation design, characterization, and mode-locking, IEEE J QE-34, p.2150-2161 (1998)

[6.861] {Sect. 6.10.3.2} X. Liu, L.J. Qian, F. Wise, Z.G. Zhang, T. Itatani, T. Sugaya, T. Nakagawa, K. Torizuka: Diode-pumped Cr:forsterite laser mode locked by a semiconductor saturable absorber, Appl Opt 37, p.7080-7084 (1998)

[6.862] {Sect. 6.10.3.2} V. Magni, M. ZavelaniRossi: Nd:YVO4 laser mode locked by cascading of second order nonlinearities, Opt Commun 152, p.45-48 (1998)

[6.863] {Sect. 6.10.3.2} J.M. Shieh, T.C. Huang, K.F. Huang, C.L. Wang, C.L. Pan: Broadly tunable self-starting passively mode-locked Ti:sapphire laser with triple-strained quantum-well saturable Bragg reflector, Opt Commun 156, p.53-57 (1998)

[6.864] {Sect. 6.10.3.2} H.S. Loka, S.D. Benjamin, P.W.E. Smith: Optical Characterization of Low-Temperature-Grown GaAs for Ultrafast All-Optical Switching Devices, IEEE J. QE-34, p.1426-1436 (1998)

[6.865] {Sect. 6.10.3.2} S. Gee, R. Coffie, P.J. Delfyett, G. Alphonse, J. Connolly: Intracavity gain and absorption dynamics of hybrid modelocked semiconductor lasers using multiple quantum well saturable absorbers, Appl Phys Lett 71, p.2569-2571 (1997)

[6.866] {Sect. 6.10.3.2} P.T. Guerreiro, S. Ten, N.F. Borreli, J. Butty, G.E. Jabbour, N. Peyghambarian: PbS quantum-dot doped grasses as saturable absorbers for mode locking of a Cr:forsterite laser, Appl Phys Lett 71, p.1595-1597 (1997)

[6.867] {Sect. 6.10.3.2} M.J. Hayduk, S.T. Johns, M.F. Krol, C.R. Pollock, R.P. Leavitt: Self-starting passively mode-locked tunable femtosecond Cr4+:YAG laser using a saturable absorber mirror, Opt Commun 137, p.55-58 (1997)

[6.868] {Sect. 6.10.3.2} S. Namiki, H.A. Haus: Noise of the stretched pulse fiber laser. 1. Theory, IEEE J QE-33, p.649-659 (1997)

[6.869] {Sect. 6.10.3.2} C.X. Yu, S. Namiki, H.A. Haus: Noise of the stretched pulse fiber laser. 2. Experiments, IEEE J QE-33, p.660-668 (1997)

[6.870] {Sect. 6.10.3.2} Z.G. Zhang, K. Torizuka, T. Itatani, K. Kobayashi, T. Sugaya, T. Nakagawa: Self-starting mode-locked femtosecond forsterite laser with a semiconductor saturable-absorber mirror, Optics Letters 22, p.1006-1008 (1997)

[6.871] {Sect. 6.10.3.2} Z.G. Zhang, K. Torizuka, T. Itatani, K. Kobayashi, T. Sugaya, T. Nakagawa: Femtosecond Cr:forsterite laser with mode locking initiated by a quantum-well saturable absorber, IEEE J QE-33, p.1975-1981 (1997)

[6.872] {Sect. 6.10.3.2} J. Aus der Au, D. Kopf, F. Morier-Genoud, M. Moser, U. Keller: 60-fs pulses from a diode-pumped Nd:glass laser, Opt. Lett. 22, p.307-309 (1997)

[6.873] {Sect. 6.10.3.2} I.D. Jung, F.X. Kärtner, N. Matuschek, D.H. Sutter, F. Morier-Genoud, Z. Shi, V. Scheuer, M. Tilsch, T. Tschudi, U. Keller: Semiconductor saturable absorber mirrors supporting sub-10-fs pulses, Appl. Phys. B 65, p.137-150 (1997)

[6.874] {Sect. 6.10.3.2} B.C. Collings, J.B. Stark, S. Tsuda, W.H. Knox, J.E. Cunningham, W.Y. Jan, R. Pathak, K. Bergman: Saturable Bragg reflector self-starting passive mode locking of a Cr4+:YAG laser pumped with a diode-pumped Nd: YVO4 laser, Optics Letters 21, p.1171-1173 (1996)

[6.875] {Sect. 6.10.3.2} R. Fluck, I.D. Jung, G. Zhang, F.X. Kartner, U. Keller: Broadband saturable absorber for 10-fs pulse generation, Optics Letters 21, p.743-745 (1996)

[6.876] {Sect. 6.10.3.2} D. Kopf, G. Zhang, R. Fluck, M. Moser, U. Keller: All-in-one dispersion-compensating saturable absorber mirror for compact femtosecond laser sources, Optics Letters 21, p.486-488 (1996)

[6.877] {Sect. 6.10.3.2} R.C. Sharp, D.E. Spock, N. Pan, J. Elliot: 190-fs passively mode-locked thulium fiber laser with a low threshold, Optics Letters 21, p.881-883 (1996)

[6.878] {Sect. 6.10.3.2} M. Wegmuller, W. Hodel, H.P. Weber: Diode pumped mode-locked Nd3+ doped fluoride fiber laser emitting at 1.05 mu m, Opt Commun 127, p.266-272 (1996)

[6.879] {Sect. 6.10.3.2} S. Tsuda, W.H. Knox, S.T. Cundiff: High efficinecy diode pumping of a saturable Bragg reflector-mode-locked Cr:LiSAF femtosecond laser, Appl. Phys. Lett. 69, p.1538-1540 (1996)

[6.880] {Sect. 6.10.3.2} C. Honninger, G. Zhang, U. Keller, A. Giesen: Femtosecond Yb:YAG laser using semiconductor saturable absorbers, Optics Letters 20, p.2402-2404 (1995)

[6.881] {Sect. 6.10.3.2} D. Kopf, K.J. Weingarten, L.R. Brovelli, M. Kamp, U. Keller: Diode-pumped 100-fs passively mode-locked Cr:LiSAF laser with an antiresonant Fabry-Perot saturable absorber, Opt. Lett. 19, p.2143-2145 (1994)

[6.882] {Sect. 6.10.3.2} J.R. Lincoln, A.I. Ferguson: All-solid-state self-mode locking of a Nd:YLF laser, Opt. Lett. 19, p.2119-2121 (1994)

[6.883] {Sect. 6.10.3.2} S. Ruan, J.M. Sutherland, P.M.W. French, J.R. Taylor, P.J. Delfyett, L.T. Florez: Pulse evolution in cw femtosecond Cr (3+):LiSrAlF6 lasers mode-locked with MQW saturable absorbers, Opt. Commun. 110, p.340-344 (1994)

[6.884] {Sect. 6.10.3.2} J.C. Chen, H.A. Haus, E.P. Ippen: Stability of Lasers Mode Locked by 2 Saturable Absorbers, IEEE J QE-29, p.1228-1232 (1993)

[6.885] {Sect. 6.10.3.2} H.A. Haus, J.D. Moores, L.E. Nelson: Effect of 3rd-Order Dispersion on Passive Mode Locking, Optics Letters 18, p.51-53 (1993)

[6.886] {Sect. 6.10.3.2} H.A. Haus: Gaussian Pulse Wings with Passive Modelocking, Opt Commun 97, p.215-218 (1993)

[6.887] {Sect. 6.10.3.2} J. Herrmann: Starting dynamic, self-starting condition and mode-locking threshold in passive, coupled-cavity or Kerr-lens mode-locked solid-state lasers, Opt. Commun. 98, p.111-116 (1993)

[6.888] {Sect. 6.10.3.2} K. Tamura, J. Jacobson, E.P. Ippen, H.A. Haus, J.G. Fujimoto: Unidirectional Ring Resonators for Self-Starting Passively Mode-Locked Lasers, Optics Letters 18, p.220-222 (1993)

[6.889] {Sect. 6.10.3.2} D. Huang, M. Ulman, L.H. Acioli, H.A. Haus, J.G. Fujimoto: Self-focusing-induced saturable loss for laser mode locking, Opt. Lett. 17, p.511-513 (1992)

[6.890] {Sect. 6.10.3.2} D.W. Hughes, M.W. Phillips, J.R.M. Barr, D.C. Hanna: A Laser-Diode-Pumped Nd:Glass Laser: Mode-Locked, High Power, and Single Frequency Performance, IEEE J. QE-28, p.1010-1017 (1992)

[6.891] {Sect. 6.10.3.2} S. Chen, J. Wang: Self-starting issues of passive self-focusing mode locking, Opt. Lett. 16, p.1689-1691 (1991)

[6.892] {Sect. 6.10.3.2} M.J. Damzen, R.A. Lamb, G.K.N. Wong: Ultrashort pulse generation by phase locking of multiple stimulated Brillouin scattering, Opt. Comm. 82, p.337-341 (1991)

[6.893] {Sect. 6.10.3.2} H.A. Haus, E.P. Ippen: Self-starting of passively mode-locked lasers, Opt. Lett. 16, p.1331-1333 (1991)

[6.894] {Sect. 6.10.3.2} U. Keller, T.K. Woodward, D.L. Sivco, A.Y. Cho: Coupled Cavity Resonant Passive Mode Locked Nd Yttrium Lithium Fluoride Laser, Optics Letters 16, p.390-392 (1991)

[6.895] {Sect. 6.10.3.2} F. Krausz, C. Spielmann, T. Brabec, E. Wintner, A.J. Schmidt: Subpicosecond pulse generation from a Nd:glass laser using a nonlinear external cavity, Opt. Lett. 15, p.737-739 (1990)

[6.896] {Sect. 6.10.3.2} S. De Silvestri, P. Laporta, V. Magni: 14-W continuous-wave mode-locked Nd:YAG laser, Opt. Lett. 11p.785-787 (1986)

[6.897] {Sect. 6.10.3.2} D. Kühlke, V. Herpers, D. von der Linde: Characteristics of a Hybridly Mode-Locked cw Dye Laser, Appl. Phys. B 38, p.233-240 (1985)

[6.898] {Sect. 6.10.3.2} P.G. May, W. Sibbett, K. Smith, J.R. Taylor, J.P. Willson: Simultaneous Autocorrelation and Synchroscan Streak Camera Measurement of Cavity Length Detuning Effects in a Synchronously Pumped CW Dye Laser, Opt. Comm. 42, p.285-290 (1982)

[6.899] {Sect. 6.10.3.2} H.A. Haus: Theory of Mode Locking with a Slow Saturable Absorber, IEEE J. QE-11, p.736-746 (1975)

[6.900] {Sect. 6.10.3.2} E.P. Ippen, C.V. Shank, A. Dienes: Passive mode-locking of the cw dye laser, Appl. Phys. Lett. 21, p.348-350 (1972)

[6.901] {Sect. 6.10.3.2} A.J. DeMaria, D.A. Stetser, H. Heynau: Self Mode-Locking of Lasers with Saturable Absorbers, Appl. Phys. Lett. 8, p.174-176 (1966)

[6.902] {Sect. 6.10.3.2} A.J.DeMaria, C.M. Ferrar, G.E. Danielson, Jr.: Mode Locking of a Nd3+-Doped Glass Laser, Appl. Phys. Lett. 8, p.22-24 (1966)

[6.903] {Sect. 6.10.3.2} J.A.D. Au, D. Kopf, F. MorierGenoud, M. Moser, U. Keller: 60-fs pulses from a diode-pumped Nd:glass laser, Optics Letters 22, p.307-309 (1997)

[6.904] {Sect. 6.10.3.2} M.J. Lederer, B. LutherDavies, H.H. Tan, C. Jagadish: GaAs based anti-resonant Fabry-Perot saturable absorber fabricated by metal organic vapor phase epitaxy and ion implantation, Appl Phys Lett 70, p.3428-3430 (1997)

[6.905] {Sect. 6.10.3.2} E. Garmire, A. Yariv: Laser Mode-Locking with Saturable Absorbers, IEEE J. QE-3, p.222-226 (1967)

[6.906] {Sect. 6.10.3.3} M. Hofmann, S. Bischoff, T. Franck, L. Prip, S.D. Brorson, J. Mork, K. Frojdh: Chirp of monolithic colliding pulse mode-locked diode lasers, Appl Phys Lett 70, p.2514-2516 (1997)

[6.907] {Sect. 6.10.3.3} S. Bischoff, M.P. Sorensen, J. Mork, S.D. Brorson, T. Franck, J.M. Nielsen, A. Mollerlarsen: Pulse-shaping mechanism in colliding-pulse mode-locked laser diodes, Appl Phys Lett 67, p.3877-3879 (1995)

[6.908] {Sect. 6.10.3.3} G.T. Harvey, M.S. Heutmaker, P.R. Smith, M.C. Nuss, U. Keller, J.A. Valdmanis: Timing Jitter and Pump Induced Amplitude Modulation in the Colliding Pulse Mode Locked (cpm) Laser, IEEE J QE-27, p.295-301 (1991)

[6.909] {Sect. 6.10.3.3} M.C. Nuss, R. Leonhardt, W. Zinth: Stable operatioon of a synchronously pumped colliding-pulse mode-locked ring dye laser, Opt. Lett. 10, p.16-18 (1985)

[6.910] {Sect. 6.10.3.3} R.L. Fork, Ch.V. Shank, R. Yen, C.A. Hirlimann: Femtosecond Optical Pulses, IEEE J. QE-19, p.500-506 (1983)

[6.911] {Sect. 6.10.3.3} C.V. Shank, C. Hirlimann: New experiments in femtosecond condensed matter spectroscopy, Helv. Phys. Acta 56, p.373-381 (1983)

[6.912] {Sect. 6.10.3.3} R.L. Fork, B.I. Greene, C.V. Shank: Generation of optical pulses shorter than 0.1 psec by colliding pulse mode locking, Appl. Phys. Lett. 38, p.671-672 (1981)

[6.913] {Sect. 6.10.3.4} I.P. Bilinsky, R.P. Prasankumar, J.G. Fujimoto: Self-starting mode locking and Kerr-lens mode locking of a Ti : Al2O3 laser by use of semiconductor-doped glass structures, J Opt Soc Am B Opt Physics 16, p.546-549 (1999)

[6.914] {Sect. 6.10.3.4} M.J. Bohn, R.J. Jones, J.C. Diels: Mutual Kerr-lens mode-locking, Opt Commun 170, p.85-92 (1999)

[6.915] {Sect. 6.10.3.4} L.J. Qian, X. Liu, F.W. Wise: Femtosecond Kerr-lens mode locking with negative nonlinear phase shifts, Optics Letters 24, p.166-168 (1999)

[6.916] {Sect. 6.10.3.4} B. Henrich, R. Beigang: Self-starting Kerr-lens mode locking of a Nd:YAG-laser, Opt Commun 135, p.300-304 (1997)

[6.917] {Sect. 6.10.3.4} X.G. Huang, F.R. Huang, W.K. Lee, M.R. Wang: Cavity design of a compact Kerr-lens mode-locking laser, Opt Commun 142, p.249-252 (1997)

[6.918] {Sect. 6.10.3.4} I.D. Jung, F.X. Kartner, N. Matuschek, D.H. Sutter, F. MorierGenoud, G. Zhang, U. Keller, V. Scheuer, M. Tilsch, T. Tschudi: Self-starting 6.5-fs pulses from a Ti:Sapphire laser, Optics Letters 22, p.1009-1011 (1997)

[6.919] {Sect. 6.10.3.4} A. Ritsataki, P.M.W. French, G.H.C. New: A numerical model of Kerr-lens mode-locking, Opt Commun 142, p.315-321 (1997)

[6.920] {Sect. 6.10.3.4} G.J. Valentine, J.M. Hopkins, P. LozaAlvarez, G.T. Kennedy, W. Sibbett, D. Burns, A. Valster: Ultralow-pump-threshold, femtosecond Cr3+:LiSrAlF6 laser pumped by a single narrow-stripe AlGaInP laser diode, Optics Letters 22, p.1639-1641 (1997)

[6.921] {Sect. 6.10.3.4} B.E. Bouma, J.G. Fujimoto: Compact Kerr-lens mode-locked resonators, Optics Letters 21, p.134-136 (1996)

[6.922] {Sect. 6.10.3.4} M. Lettenberger, K. Wolfrum: Optimized Kerr lens mode-locking of a pulsed Nd:KGW laser, Opt Commun 131, p.295-300 (1996)

[6.923] {Sect. 6.10.3.4} K. Read, F. Blonigen, N. Riccelli, M.E. Murnane, H. Kapteyn: Low-threshold operation of an ultrashort-pulse mode-locked Ti: sapphire laser, Optics Letters 21, p.489-491 (1996)

[6.924] {Sect. 6.10.3.4} J. Solis, J. Siegel, C.N. Afonso, N.P. Barry, R. Mellish, P.M.W. French: Experimental study of a self-starting Kerr-lens mode- locked titanium-doped sapphire laser, Opt Commun 123, p.547-552 (1996)

[6.925] {Sect. 6.10.3.4} I.T. Sorokina, E. Sorokin, E. Wintner, A. Cassanho, H.P. Jenssen, M.A. Noginov: Efficient continuous wave TEM (00) and femtosecond Kerr lens mode-locked Cr:LiSrGaF laser, Optics Letters 21, p.204-206 (1996)

[6.926] {Sect. 6.10.3.4} I.T. Sorokina, E. Sorokin, E. Wintner, A. Cassanho, H.P. Jenssen, R. Szipocs: Prismless passively mode-locked femtosecond Cr:LiSGaF laser, Optics Letters 21, p.1165-1167 (1996)

[6.927] {Sect. 6.10.3.4} Y.P. Tong, J.M. Sutherland, P.M.W. French, J.R. Taylor, A.V. Shestakov, B.H.T. Chai: Self-starting Kerr-lens mode-locked femtosecond Cr4+:YAG and picosecond Pr3+:YLF solid-state lasers, Optics Letters 21, p.644-646 (1996)

[6.928] {Sect. 6.10.3.4} M.J.P. Dymott, A.I. Ferguson: Self mode locked diode pumped Cr:LiSAF laser producing 34- fs pulses at 42-mW average power, Optics Letters 20, p.1157-1159 (1995)

[6.929] {Sect. 6.10.3.4} G. Cerullo, S. DeSilvestri, V. Magni: Self-Starting Kerr-Lens Mode Locking of a Ti-Sapphire Laser, Optics Letters 19, p.1040-1042 (1994)

[6.930] {Sect. 6.10.3.4} G. Cerullo, S. De Silvestri, V. Magni, L. Pallaro: Resonators for Kerr-lens mode-locked femtosecond Ti:sapphire lasers, Opt. Lett. 19, p.807-809 (1994)

[6.931] {Sect. 6.10.3.4} M.J.P. Dymott, A.I. Ferguson: Self-mode-locked diode-pumped Cr:LiSAF laser, Opt. Lett. 19, p.1988-1990 (1994)

[6.932] {Sect. 6.10.3.4} D. Kopf, K.J. Weingarten, L.R. Brovelli, M. Kamp, U. Keller: Diode-pumped 100-fs passively mode-locked Cr:LiSAF laser with an antiresonant Fabry-Perot saturable absorber, Opt. Lett. 19, p.2143-2145 (1994)

[6.933] {Sect. 6.10.3.4} P.M. Mellish, P.M.W. French, J.R. Taylor, P.J. Delfyett, L.T. Florez: All-solid-state femtosecond diode-pumped Cr:LiSAF laser, Electron. Lett. 30, p.223-224 (1994)

[6.934] {Sect. 6.10.3.4} J. Zhou, G. Taft, C.-P. Huang, M.M. Murnane, H.C. Kapteyn, I.P. Christov: Pulse evolution in a broad-bandwidth Ti:sapphire laser,, Opt. Lett. 19, p.1149-1151 (1994)

[6.935] {Sect. 6.10.3.4} J. Zhou, C.-P. Huang, C. Shi, M.M. Murnane, H.C. Kapteyn: Generation of 21-fs millijoule-energy pulses by use of Ti:sapphire, Opt. Lett. 19, p.126-128 (1994)

[6.936] {Sect. 6.10.3.4} P. Beaud, M. Richardson, E.J. Miesak, B.H.T. Chai: 8-TW 90-fs Cr:LiSAF laser, Opt. Lett. 18, p.1550-1552 (1993)

[6.937] {Sect. 6.10.3.4} T. Brabec, P.F. Curley, Ch. Spielmann, E. Wintner, A.J. Schmidt: Hard-aperture Kerr-lens mode locking, J. Opt. Soc. Am. B 10, p.1029-1034 (1993)

[6.938] {Sect. 6.10.3.4} P.F. Curley, C. Spielmann, T. Brabec, F. Krausz, E. Wintner, A.J. Schmidt: Operation of a femtosecond Ti:sapphire solitary laser in the vicinity of zero group-delay dispersion, Opt. Lett. 18, p.54-56 (1993)

[6.939] {Sect. 6.10.3.4} P.M.W. French, R. Mellish, J.R. Taylor, P.J. Delfyett, L.T. Florez: Mode-locked all-solid-state diode-pumped Cr:LiSAF laser, Opt. Lett. 18, p.1934-1936 (1993)

[6.940] {Sect. 6.10.3.4} Y.M. Liu, P.R. Prucnal: Slow Amplitude Modulation in the Pulse Train of a Self-Mode-Locked Ti:Sapphire Laser, IEEE J. QE-29, p.2663-2669 (1993)

[6.941] {Sect. 6.10.3.4} V. Magni, G. Cerullo, S. DeSilvestri: ABCD matrix analysis of propagation of gaussian beams through Kerr media, Opt. Commun. 96, p.348-355 (1993)

[6.942] {Sect. 6.10.3.4} V. Magni, G. Cerullo, S. DeSilvestri: Closed form gaussian beam analysis of resonators containing a Kerr medium for femtosecond lasers, Opt. Commun. 101, p.365-370 (1993)

[6.943] {Sect. 6.10.3.4} Y. Pang, V. Yanovsky, F. Wise, B.I. Minkov: Self-mode-locked Cr:forsterite laser, Opt. Lett. 18, p.1168-1170 (1993)

[6.944] {Sect. 6.10.3.4} A. Seas, V. Petricevic, R.R. Alfano: Self-mode-locked chromium-doped forsterite laser generates 50-fs pulses-, Opt. Lett. 18, p.891-893 (1993)

[6.945] {Sect. 6.10.3.4} A. Sennaroglu, C.R. Pollock, H. Nathel: Generation of 48-fs pulses and measurement of crystal dispersion by using a regeneratively initiated self-mode-locked chromium-doped forsterite laser, Opt. Lett. 18, p.826-828 (1993)

[6.946] {Sect. 6.10.3.4} V. Yanovsky, Y. Pang, F. Wise, B.I. Minkov: Generation of 25-fs pulses from a self-mode-locked Cr:forsterite laser with optimized group-delay dispersion, Opt. Lett. 18, p.1541-1543 (1993)

[6.947] {Sect. 6.10.3.4} T. Brabec, C. Spielmann, P.F. Curley, F. Krausz: Kerr lens mode locking, Opt. Lett. 17, p.1292-1294 (1992)

[6.948] {Sect. 6.10.3.4} T. Brabec, C.H. Spielmann, F. Krausz: Limits of pulse shortening in solitary lasers, Opt. Lett. 17, p.748-750 (1992)

[6.949] {Sect. 6.10.3.4} J.M. Jacobson, K. Naganuma, H.A. Haus, J.G. Fujimoto, A.G. Jacobson: Femtosecond Pulse Generation in a Ti-Al2O3 Laser by Using 2nd-Order and 3rd-Order Intracavity Dispersion, Optics Letters 17, p.1608-1610 (1992)

[6.950] {Sect. 6.10.3.4} K.X. Liu, C.J. Flood, D.R. Walker, H.M. van Driel: Kerr lens mode locking of a diode-pumped Nd:YAG laser, Opt. Lett. 17, p.1361-1363 (1992)

[6.951] {Sect. 6.10.3.4} Y.M. Liu, K.W. Sun, P.R. Prucnal, S.A. Lyon: Simple method to start and maintain self-mode-locking of a Ti:sapphire laser, Opt. Lett. 17, p.1219-1221 (1992)

[6.952] {Sect. 6.10.3.4} A. Seas, V. Petricevic, R.R. Alfano: Generation of sub-100-fs pulses from a cw mode-locked chromium-doped forsterite laser, Opt. Lett. 17, p.937-939 (1992)

[6.953] {Sect. 6.10.3.4} U. Keller, G.W. Thooft, W.H. Knox, J.E. Cunningham: Femtosecond Pulses from a Continuously Self Starting Passively Mode Locked Ti Sapphire Laser, Optics Letters 16, p.1022-1024 (1991)

[6.954] {Sect. 6.10.3.4} J.D. Kmetec, J.J. Macklin, J.F. Young: 0.5-TW, 125-fs Ti:sapphire laser, Opt. Lett. 16, p.1001-1003 (1991)

[6.955] {Sect. 6.10.3.4} F. Salin, J. Squier, M. Piché: Mode locking of Ti:Al2O3 lasers and self-focusing: a Gaussian approximation, Opt. Lett. 16, p.1674-1676 (1991)

[6.956] {Sect. 6.10.3.4} D.E. Spence, J.M. Evans, W.E. Sleat, W. Sibbett: Regeneratively initiated self-mode-locked Ti:sapphire laser, Opt. Lett. 16, p.1762-1764 (1991)

[6.957] {Sect. 6.10.3.4} D.E. Spence, P.N. Kean, W. Sibbett: 60-fsec pulse generation from a self-mode-locked Ti:sapphire laser, Opt. Lett. 16, p.42-44 (1991)

[6.958] {Sect. 6.10.3.4} C. Spielmann, F. Krausz, T. Brabec, E. Wintner, A.J. Schmidt: Femtosecond pulse generation from a synchronously pumped Ti:sapphire laser, Opt. Lett. 16, p.1180-1182 (1991)

[6.959] {Sect. 6.10.3.4} J. Goodberlet, J. Wang, J.G. Fujimoto, P.A. Schulz: Femtosecond passively mode-locked Ti:Al2O3 laser with a nonlinear external cavity, Opt. Lett. 14, p.1125-1127 (1989)

[6.960] {Sect. 6.10.3.4} J. Jasapara, W. Rudolph, V.L. Kalashnikov, D.O. Krimer, I.G. Polyoko, M. Lenzner: Automodulations in Kerr-lens mode-locked solid-state lasers, J Opt Soc Am B Opt Physics 17, p.319-326 (2000)

[6.961] {Sect. 6.10.3.4} G.R. Boyer, G. Kononovitch: Gain optimization of a Kerr-lens mode-locked Cr:forsterite laser in the CW regime: Theory and experiments, Opt Commun 133, p.205-210 (1997)

[6.962] {Sect. 6.10.3.4} L. Xu, G. Tempea, A. Poppe, M. Lenzner, Ch. Spielmann, R. Krausz, A. Stingl, K. Ferencz: High-power sub-10-fs Ti:sapphire oscillators, Appl. Phys. B 65, p.151-159 (1997)

[6.963] {Sect. 6.10.3.4} B. Golubovic, R.R. Austin, M.K. SteinerShepard, M.K. Reed, S.A. Diddams, D.J. Jones, A.G. VanEngen: Double Gires-Tournois interferometer negative-dispersion mirrors for use in tunable mode-locked lasers, Optics Letters 25, p.275-277 (2000)

[6.964] {Sect. 6.10.3.4} R. Paschotta, G.J. Spuhler, D.H. Sutter, N. Matuschek, U. Keller, M. Moser, R. Hovel, V. Scheuer, G. Angelow, T. Tschudi: Double-chirped semiconductor mirror for dispersion compensation in femtosecond lasers, Appl Phys Lett 75, p.2166-2168 (1999)

[6.965] {Sect. 6.10.3.4} K. Gabel, P. Russbuldt, R. Lebert, P. Loosen, R. Poprawe, H. Heyer, A. Valster: Diode pumped, chirped mirror dispersion compensated, fs-laser, Opt Commun 153, p.275-281 (1998)

[6.966] {Sect. 6.10.3.4} D.H. Sutter, I.D. Jung, F.X. Kärtner, N. Matuschek, F. Morier-Genoud, V. Scheuer, M. Tilsch, T. Tschudi, U. Keller: Self-starting 6.5-fs pulses from a Ti:sapphire laser using a semiconductor saturable absorber and double-chirped mirrors, IEEE J. QE-4, p.169-178 (1998)

[6.967] {Sect. 6.10.3.4} F.X. Kartner, N. Matuschek, T. Schibli, U. Keller, H.A. Haus, C. Heine, R. Morf, V. Scheuer, M. Tilsch, T. Tschudi: Design and fabrication of double-chirped mirrors, Opt. Lett. 22, p.831-833 (1997)

[6.968] {Sect. 6.10.3.4} A.P. Kovacs, K. Osvay, Z. Bor, R. Szipocs: Group delay measurement on laser mirrors by spectrally resolved white light interferometry, Optics Letters 20, p.788-790 (1995)

[6.969] {Sect. 6.10.3.4} A. Stingl, M. Lenzner, C. Spielmann, F. Krausz, R. Szipocs: Sub-1O-fs mirror dispersion controlled Ti:sapphire laser, Optics Letters 20, p.602-604 (1995)

[6.970] {Sect. 6.10.3.4} U. Keller: Ultrafast All-Solid-State Laser Technology, Appl. Phys. B 58, p.347-363 (1994)

[6.971] {Sect. 6.10.3.4} W.H. Knox, N.M. Pearson, K.D. Li, Ch.A. Hirlimann: Interferometric measurements of femtosecond group delay in optical components, Opt. Lett. 13, p.574-576 (1988)

[6.972] {Sect. 6.10.3.4} E. Spiller: Broadening of Short Light Pulses by Many Reflections from Multilayer Dielectric Coatings, Appl. Opt. 10, p.557-566 (1971)

[6.973] {Sect. 6.10.3.4} D. Kopf, G.J. Spuhler, K.J. Weingarten, U. Keller: Mode-locked laser cavities with a single prism for dispersion compensation, Appl Opt 35, p.912-915 (1996)

[6.974] {Sect. 6.10.3.4} A.M. Dunlop, W.J. Firth, E.M. Wright: Master equation for spatio-temporal beam propagation and Kerr lens mode-locking, Opt Commun 138, p.211-226 (1997)

[6.975] {Sect. 6.10.3.4} J. Herrmann, V.P. Kalosha, M. Muller: Higher-order phase dispersion in femtosecond Kerr-lens mode-locked solid-state lasers: Sideband generation and pulse splitting, Optics Letters 22, p.236-238 (1997)

[6.976] {Sect. 6.10.3.4} I.P. Christov, V.D. Stoev, M.M. Murnane, H.C. Kapteyn: Sub-10-fs operation of Kerr-lens mode-locked lasers, Optics Letters 21, p.1493-1495 (1996)

[6.977] {Sect. 6.10.3.4} S. Gatz, J. Herrmann, M. Muller: Kerr-lens mode locking without dispersion compensation, Optics Letters 21, p.1573-1575 (1996)

[6.978] {Sect. 6.10.3.4} I.P. Christov, V.D. Stoev, M.M. Murnane, H.C. Kapteyn: Mode locking with a compensated space time astigmatism, Optics Letters 20, p.2111-2113 (1995)

[6.979] {Sect. 6.10.3.4} H.A. Haus, J.G. Fujimoto, E.P. Ippen: Analytic Theory of Additive Pulse and Kerr Lens Mode Locking, IEEE J QE-28, p.2086-2096 (1992)

[6.980] {Sect. 6.10.3.5} D.W. Huang, G.C. Lin, C.C. Yang: Fiber-grating-based self-matched additive-pulse mode-locked fiber lasers, IEEE J QE-35, p.138-146 (1999)

[6.981] {Sect. 6.10.3.5} T.M. Jeong, E.C. Kang, C.H. Nam: Temporal and spectral characteristics of an additive-pulse mode-locked Nd : YLF laser with Michelson-type configuration, Opt Commun 166, p.95-102 (1999)

[6.982] {Sect. 6.10.3.5} G. Sucha, D.S. Chemla, S.R. Bolton: Effects of cavity topology on the nonlinear dynamics of additive-pulse mode-locked lasers, J Opt Soc Am B Opt Physics 15, p.2847-2853 (1998)

[6.983] {Sect. 6.10.3.5} P. Heinz, A. Seilmeier: Pulsed diode-pumped additive-pulse mode-locked high-peak- power Nd:YLF laser, Optics Letters 21, p.54-56 (1996)

[6.984] {Sect. 6.10.3.5} I.V. Melnikov, A.V. Shipulin: Solitary-pulse regimes of solid-state laser additively mode locked by a cascading nonlinearity, Appl Phys Lett 69, p.299-301 (1996)

[6.985] {Sect. 6.10.3.5} S. Namiki, E.P. Ippen, H.A. Haus, K. Tamura: Relaxation oscillation behavior in polarization additive pulse mode-locked fiber ring lasers, Appl Phys Lett 69, p.3969-3971 (1996)

[6.986] {Sect. 6.10.3.5} G. Lenz, K. Tamura, H.A. Haus, E.P. Ippen: All-solid-state femtosecond source at 1.55 mu m, Optics Letters 20, p.1289-1291 (1995)

[6.987] {Sect. 6.10.3.5} K. Tamura, E.P. Ippen, H.A. Haus, L.E. Nelson: 77-fs pulse generation from generation from a stretched-pulse mode-locked all-fiber ring laser, Opt. Lett. 18, p.1080-1082 (1993)

[6.988] {Sect. 6.10.3.5} H.A. Haus, J.G. Fujimoto, E.P. Ippen: Analytic Theory of Additive Pulse and Kerr Lens Mode Locking, IEEE J. QE-28, p.2086-2096 (1992)

[6.989] {Sect. 6.10.3.5} H.A. Haus, J.G. Fujimoto, E.P. Ippen: Structures for additive pulse mode locking, J. Opt. Soc. Am. B 8, p.2068-2076 (1991)

[6.990] {Sect. 6.10.3.5} J. Goodberlet, J. Jacobson, J.G. Fujimoto, P.A. Schulz, T.Y. Fan: Self-starting additive-pulse mode-locked diode-pumped Nd:YAG laser, Opt. Lett. 15, p.504-506 (1990)

[6.991] {Sect. 6.10.3.5} F. Krausz, Ch. Spielmann, T. Brabec, E. Wintner, A.J. Schmidt: Self-starting additive-pulse mode locking of a Nd:glass laser, Opt. Lett. 15, p.1082-1084 (1990)

[6.992] {Sect. 6.10.3.5} L.Y. Liu, J.M. Huxley, E.P. Ippen, H.A. Haus: Self-starting additive-pulse mode locking of a Nd:YAG laser, Opt. Lett. 15, p.553-555 (1990)

[6.993] {Sect. 6.10.3.5} G.P.A. Malcolm, P.F. Curley, A.I. Ferguson: Addidive-pulse mode locking of a diode-pumped Nd:YLF laser, Opt. Lett. 15, p.1303-1305 (1990)

[6.994] {Sect. 6.10.3.5} E.P. Ippen, H.A. Haus, L.Y. Liu: Additive pulse mode locking, J. Opt. Soc. Am. B 6, p.1736-1745 (1989)

[6.995] {Sect. 6.10.3.5} U. Morgner, L. Rolefs, F. Mitschke: Dynamic instabilities in an additive-pulse mode-locked Nd: YAG laser, Optics Letters 21, p.1265-1267 (1996)

[6.996] {Sect. 6.10.3.5} V. Cautaerts, D.J. Richardson, R. Paschotta, D.C. Hanna: Stretched pulse Yb3+:silica fiber laser, Optics Letters 22, p.316-318 (1997)

[6.997] {Sect. 6.10.3.6} W.S. Man, H.Y. Tan, M.S. Demokan, P.K.A. Wai, D.Y. Tang: Mechanism of intrinsic wavelength tuning and sideband asymmetry in a passively mode-locked soliton fiber ring laser, J Opt Soc Am B Opt Physics 17, p.28-33 (2000)

[6.998] {Sect. 6.10.3.6} M.E. Fermann, A. Galvanauskas, M.L. Stock, K.K. Wong, D. Harter, L. Goldberg: Ultrawide tunable Er soliton fiber laser amplified in Yb-doped fiber, Optics Letters 24, p.1428-1430 (1999)

[6.999] {Sect. 6.10.3.6} M.E. Grein, L.A. Jiang, Y. Chen, H.A. Haus, E.P. Ippen: Timing restoration dynamics in an actively mode-locked fiber ring laser, Optics Letters 24, p.1687-1689 (1999)

[6.1000] {Sect. 6.10.3.6} M.J. Lederer, B. LutherDavies, H.H. Tan, C. Jagadish, N.N. Akhmediev, J.M. SotoCrespo: Multipulse operation of a Ti : sapphire laser mode locked by an ion- implanted semiconductor saturable-absorber mirror, J Opt Soc Am B Opt Physics 16, p.895-904 (1999)

[6.1001] {Sect. 6.10.3.6} A.M. Dunlop, E.M. Wright, W.J. Firth: Spatial soliton laser, Opt Commun 147, p.393-401 (1998)

[6.1002] {Sect. 6.10.3.6} G. Boyer: Dispersive wave generation in a Cr4+:forsterite femtosecond soliton-like laser, Opt Commun 141, p.279-282 (1997)

[6.1003] {Sect. 6.10.3.6} S. Gray, A.B. Grudinin: Soliton fiber laser with a hybrid saturable absorber, Optics Letters 21, p.207-209 (1996)

[6.1004] {Sect. 6.10.3.6} D.J. Jones, H.A. Haus, E.P. Ippen: Subpicosecond solitons in an actively mode-locked fiber laser, Optics Letters 21, p.1818-1820 (1996)

[6.1005] {Sect. 6.10.3.6} D.O. Culverhouse, D.J. Richardson, T.A. Birks, P.S.J. Russell: All-fiber sliding-frequency Er3+/Yb3+ soliton laser, Optics Letters 20, p.2381-2383 (1995)

[6.1006] {Sect. 6.10.3.6} M.E. Fermann, K. Sugden, I. Bennion: High Power Soliton Fiber Laser Based on Pulse Width Control with Chirped Fiber Bragg Gratings, Optics Letters 20, p.172-174 (1995)

[6.1007] {Sect. 6.10.3.6} M. Hofer, M.H. Ober, R. Hofer, M.E. Fermann, G. Sucha, D. Harter, K. Sugden, I. Bennion, C.A.C. Mendonca, T.H. Chiu: High-power neodymium soliton fiber laser that uses a chirped fiber grating, Optics Letters 20, p.1701-1703 (1995)

[6.1008] {Sect. 6.10.3.6} I.D. Jung, F.X. Kartner, L.R. Brovelli, M. Kamp, U. Keller: Experimental verification of soliton mode locking using only a slow saturable absorber, Optics Letters 20, p.1892-1894 (1995)

[6.1009] {Sect. 6.10.3.6} C.R. Doerr, H.A. Haus, E.P. Ippen: Asynchronous soliton mode locking, Optics Letters 19, p.1958-1960 (1994)

[6.1010] {Sect. 6.10.3.6} F.M. Mitschke, L.F. Mollenauer: Ultrashort pulses from the soliton laser, Opt. Lett. 12, p.407-409 (1987)

[6.1011] {Sect. 6.10.3.6} F.M. Mitschke, L.F. Mollenauer: Stabilizing the soliton laser, IEEE J. QE-22, p.2242-2252 (1986)

[6.1012] {Sect. 6.10.3.6} L.F. Mollenauer, R.H. Stolen: The soliton laser, Opt. Lett. 9, p.13-15 (1984)

[6.1013] {Sect. 6.10.3.6} H.J. Polland, T. Elsaesser, A. Seilmeier, W. Kaiser, M. Kussler, N.J. Marx, B. Sens, K.H. Drexhage: Picosecond Dye Laser Emission in the Infrared between 1.4 and 1.8 µm, Appl. Phys. B 32, p.53-57 (1983)

[6.1014] {Sect. 6.10.3.6} D. Marcuse Pulse distortion in single-mode fibers, Appl. Opt. 19, p.1653-1660 (1980)

[6.1015] {Sect. 6.10.3.6} L.F. Mollenauer, R.H. Stolen, J.P. Gordon: Experimental Observation of Picosecond Pulse Narrowing and Solitons in Optical Fibers, Phys. Rev. Lett. 45, p.1095-1098 (1980)

[6.1016] {Sect. 6.10.3.6} T.F. Carruthers, I.N. Duling, M. Horowitz, C.R. Menyuk: Dispersion management in a harmonically mode-locked fiber soliton laser, Optics Letters 25, p.153-155 (2000)

[6.1017] {Sect. 6.10.3.6} D. Huhse, O. Reimann, E.H. Bottcher, D. Bimberg: Generation of 290 fs laser pulses by self-seeding and soliton compression, Appl Phys Lett 75, p.2530-2532 (1999)

[6.1018] {Sect. 6.10.3.7} A.V. Muravjov, S.H. Withers, R.C. Strijbos, S.G. Pavlov, V.N. Shastin, R.E. Peale: Actively mode-locked p-Ge laser in Faraday configuration, Appl Phys Lett 75, p.2882-2884 (1999)

[6.1019] {Sect. 6.10.3.7} S. Longhi, S. Taccheo, P. Laporta: High-repetition-rate picosecond pulse generation at 1.5 mu m by intracavity laser frequency modulation, Optics Letters 22, p.1642-1644 (1997)

[6.1020] {Sect. 6.10.3.7} T.F. Carruthers, I.N. Duling: 10-GHz, 1.3-ps erbium fiber laser employing soliton pulse shortening, Optics Letters 21, p.1927-1929 (1996)

[6.1021] {Sect. 6.10.3.7} O. Guy, V. Kubecek, A. Barthelemy: Mode-locked diode-pumped Nd:YAP laser, Opt Commun 130, p.41-43 (1996)

[6.1022] {Sect. 6.10.3.7} D. Kopf, F.X. Kartner, K.J. Weingarten, U. Keller: Pulse shortening in a Nd:glass laser by gain reshaping and soliton formation, Optics Letters 19, p.2146-2148 (1994)

[6.1023] {Sect. 6.10.3.7} J.L. Dallas: Frequency-modulation mode-locking performance for four Nd3+-doped laser crystals, Appl. Opt. 33, p.6373-6376 (1994)

[6.1024] {Sect. 6.10.3.7} U. Keller, K.D. Li, B.T. Khuri-Yakub, D.M. Bloom, K.J. Weingarten, D.C. Gerstenberger: High-frequency acousto-optic mode locker for picosecond pulse generation, Opt. Lett. 15, p.45-47 (1990)

[6.1025] {Sect. 6.10.3.7} F. Krausz, L. Turi, Cs. Kuti, A.J. Schmidt: Active mode locking of lasers by piezoelectrically induced diffraction modulation, Appl. Phys. Lett. 56, p.1415-1417 (1990)

[6.1026] {Sect. 6.10.3.7} L. Turi, Cs. Kuti, F. Krausz: Piezoelectrically Induced Diffraction Modulation of Light, IEEE J. QE-26, p.1234-1240 (1990)

[6.1027] {Sect. 6.10.3.7} P. Heinz, M. Fickenscher, A. Lauberau: Elektro-optic gain control and cavity dumping of a Nd:glass laser with active-passive mode-locking, Opt. Comm. 62, p.343-347 (1987)

[6.1028] {Sect. 6.10.3.7} E.O. Gobel, J. Kuhl, G. Veith: Synchronous Mode Locking of Semiconductor Laser Diodes by a Picosecond Optoelectronic Switch, J. Appl Phys 56, p.862-864 (1984)

[6.1029] {Sect. 6.10.3.7} C.J. Kennedy: Pulse Chirping in a Nd:YAG Laser, IEEE J. QE-10, p.528-530 (1974)

[6.1030] {Sect. 6.10.3.7} R.H. Johnson: Characteristics of Acoustooptic Cavity Dumping in a Mode-Locked Laser, IEEE J. QE-9, p.255-257 (1973)

[6.1031] {Sect. 6.10.3.7} M.F. Becker, D.J. Kuizenka, A.E. Siegman: Harmonic Mode Locking of the Nd:YAG Laser, IEEE J. QE-8p.687-693 (1972)

[6.1032] {Sect. 6.10.3.7} L.E. Hargrove, R.L.Fork, M.A. Pollack: Locking of He-Ne laser modes induced by synchronous intracavity modulation, Appl. Phys. Lett. 5, p.4-7 (1964)

[6.1033] {Sect. 6.10.3.7} S.E. Harris, O.P. McDuff: FM Laser Oscillation-Theory, Appl. Phys. Lett. 5, p.205-206 (1964)

[6.1034] {Sect. 6.10.3.7} M. Horowitz, C.R. Menyuk: Analysis of pulse dropout in harmonically mode-locked fiber lasers by use of the Lyapunov method, Optics Letters 25, p.40-42 (2000)

[6.1035] {Sect. 6.10.3.7} F.X. Kartner, D.M. Zumbuhl, N. Matuschek: Turbulence in mode-locked lasers, Phys Rev Lett 82, p.4428-4431 (1999)

[6.1036] {Sect. 6.10.3.7} R. Kiyan, O. Deparis, O. Pottiez, P. Megret, M. Blondel: Stabilization of actively mode-locked Er-doped fiber lasers in the rational-harmonic frequency-doubling mode-locking regime, Optics Letters 24, p.1029-1031 (1999)

[6.1037] {Sect. 6.10.3.7} A.A. Mani, P. Hollander, P.A. Thiry, A. Peremans: All-solid-state 12 ps actively passively mode-locked pulsed Nd : YAG laser using a nonlinear mirror, Appl Phys Lett 75, p.3066-3068 (1999)

[6.1038] {Sect. 6.10.3.7} M.Y. Jeon, H.K. Lee, K.H. Kim, E.H. Lee, W.Y. Oh, B.Y. Kim, H.W. Lee, Y.W. Koh: Harmonically mode-locked fiber laser with an acousto-optic modulator in a Sagnac loop and Faraday rotating mirror cavity, Opt Commun 149, p.312-316 (1998)

[6.1039] {Sect. 6.10.3.7} S. Pajarola, G. Guekos, H. Kawaguchi: Dual-polarization optical pulse generation using a mode-locked two- arm external cavity diode laser, Opt Commun 154, p.39-42 (1998)

[6.1040] {Sect. 6.10.3.7} K.S. Abedin, N. Onodera, M. Hyodo: Repetition-rate mul-tiplication in actively mode-locked fiber lasers by higher-order FM mode locking using a high-finesse Fabry-Perot filter, Appl Phys Lett 73, p.1311-1313 (1998)

[6.1041] {Sect. 6.10.3.7} D.T. Chen, H.R. Fetterman, A.T. Chen, W.H. Steier, L.R. Dalton, W.S. Wang, Y.Q. Shi: Demonstration of 110 GHz electro-optic polymer modulators, Appl Phys Lett 70, p.3335-3337 (1997)

[6.1042] {Sect. 6.10.3.7} T. Khayim, M. Yamauchi, D.S. Kim, T. Kobayashi: Fem-tosecond optical pulse generation from a CW laser using an electrooptic phase modulator featuring lens modulation, IEEE J QE-35, p.1412-1418 (1999)

[6.1043] {Sect. 6.10.3.8} S.N. Vainshtein, G.S. Simin, J.T. Kostamovaara: Deriv-ing of single intensive picosecond optical pulses from a high- power gain-switched laser diode by spectral filtering, J Appl Phys 84, p.4109-4113 (1998)

[6.1044] {Sect. 6.10.3.8} J.D. Simon: Ultrashort light pulses, Rev. Sci. Instrum. 60, p.3597-3624 (1989)

[6.1045] {Sect. 6.10.3.8} J.M. Catherall, G.H.C. New, P.M. Radmore: Approach to the theory of mode locking by synchronous pumping, Opt. Lett. 7, p.319-321 (1982)

[6.1046] {Sect. 6.10.3.8} C.P. Ausschnitt, R.K. Jain, J.P. Heritage: Cavity Length Detuning Characteristics of the Synchronously Mode-Locked CW Dye Laser, IEEE J. QE-15, p.912-917 (1979)

[6.1047] {Sect. 6.10.3.8} J. Juhl, H. Klingenberg, D. von der Linde: Picosecond and Subpicosecond Pulse Generation in Synchronously Pumped Mode-Locked cw Dye Lasers, Appl. Phys. 18, p.279-284 (1979)

[6.1048] {Sect. 6.10.3.8} J. Falk. Y.C. See: Internal cw parametric upconversion, Appl. Phys. Lett. 32, p.100-101 (1978)

[6.1049] {Sect. 6.10.3.8} G.W. Fehrenbach, K.J. Gruntz, R.G. Ulbrich: Subpicosec-ond light pulses from synchronously mode-locked dye laser with composite gain and absorber medium, Appl. Phys. Lett. 33, p.159-160 (1978)

[6.1050] {Sect. 6.10.3.8} J.P. Heritage, R.K. Jain: Subpicosecond pulses from a tun-able cw mode-locked dye laser, Appl. Phys. Lett. 32, p.101-103 (1978)

[6.1051] {Sect. 6.10.3.8} D.M. Kim, J. Kuhl, R. Lambrich, D. von der Linde: Char-acteristics of Picosecond Pulses Generated from Synchronously Pumped CW Dye Laser System, Opt. Comm. 27, p.123-126 (1978)

[6.1052] {Sect. 6.10.3.8} J.P. Ryan, L.S. Goldberg, D.J. Bradley: Comparision of Synchronous Pumping and Passive Mode-Locking of CW Dye Lasers for the Generation of Picosecond and Subpicosecond Light Pulses, Opt. Comm. 27, p.127-132 (1978)

[6.1053] {Sect. 6.10.3.8} N.J. Frigo, T. Daly, H. Mahr: A Study of Forced Mode Locked CW Dye Laser, IEEE J. QE-13p.101-109 (1977)

[6.1054] {Sect. 6.10.3.8} Z.A. Yasa, O. Teschke: Picosecond Pulse Generation in Synchronously Pumped Dye Lasers, Opt. Comm. 15, p.169-172 (1975)

[6.1055] {Sect. 6.10.3.8} C.K. Chan, S.O. Sari: Tunable dye laser pulse converter for production of picosecond pulses, Appl. Phys. Lett. 25, p.403-406 (1974)

[6.1056] {Sect. 6.10.3.8} D.J. Kuizenga, A.E. Siegman: FM and AM Mode Locking of the Homogeneous Laser – Part I: Theory, IEEE J. QE-6, p.694-708 (1970)

[6.1057] {Sect. 6.10.3.8} D.J. Kuizenga, A.E. Siegman: FM Laser Operation of the Nd:YAG Laser, IEEE J. QE-6, p.673-677 (1970)

[6.1058] {Sect. 6.10.4.1} P.P. Yaney, D.A.V. Kliner, P.E. Schrader, R.L. Farrow: Distributed-feedback dye laser for picosecond ultraviolet and visible spec-troscopy, Rev Sci Instr 71, p.1296-1305 (2000)

[6.1059] {Sect. 6.10.4.1} M. Maeda, Y. Oki, K. Imamura: Ultrashort pulse generation from an integrated single-chip dye laser, IEEE J QE-33, p.2146-2149 (1997)

[6.1060] {Sect. 6.10.4.1} A. Müller: Two independently tunable distributed feedback dye lasers pumped by a single picosecond Nd:YAG laser, Appl. Phys. B 63, p.443-450 (1996)

[6.1061] {Sect. 6.10.4.1} F. Raksi, W. Heuer, H. Zacharias: A High-Power Subpicosecond Distributed Feedback Dye Laser System Pumped by a Mode-Locked Nd:YAG Laser, Appl. Phys. B. 53, p.97-100 (1991)

[6.1062] {Sect. 6.10.4.1} G. Szabó, Z. Bor: 300 Femtosecond Pulses at 497 Nanometer Generated by an Excimer Laser Pumped Cascade of Distributed Feedback Dye Lasers, Appl. Phys. B. 47, p.299-302 (1988)

[6.1063] {Sect. 6.10.4.1} S. Szatmári, B. Rász: Generation of 320 fs Pulses with a Distributed Feedback Dye Laser, Appl. Phys. B 43, p.93-97 (1987)

[6.1064] {Sect. 6.10.4.1} J. Hebling, Z. Bor: Distributed Feedback Dye Laser Pumped by a Laser Having a Low Degree of Coherence, J Phys E-SCIENTIFIC INSTRUMENTS 17, p.1077-1080 (1984)

[6.1065] {Sect. 6.10.4.1} G. Szabo, Z. Bor, A. Muller, B. Nikolaus, B. Racz: Travelling Wave Pumped Ultrashort Pulse Distributed Feedback Dye Laser, Appl Phys B 34, p.145-147 (1984)

[6.1066] {Sect. 6.10.4.1} S. Szatmari, Z. Bor: Directional and Wavelength Sweep of Distributed Feedback Dye Laser Pulses, Appl Phys B 34, p.29-31 (1984)

[6.1067] {Sect. 6.10.4.1} Zs. Bor, B. Rácz, G. Szabó: Picosecond Pulse Generation by Distributed Feedback Dye Lasers, Helvetica Physica Acta 56, p.383-392 (1983)

[6.1068] {Sect. 6.10.4.1} C.V. Shank, J.E. Bjorkholm, H. Kogelnik: Tunable distributed-feedback dye lasers, Appl. Phys. Lett. 18, p.395-396 (1971)

[6.1069] {Sect. 6.10.4.3} J.C. Chanteloup, E. Salmon, C. Sauteret, A. Migus, P. Zeitoun, A. Klisnick, A. Carillon, S. Hubert, D. Ros, P. Nickles et al.: Pulse-front control of 15-TW pulses with a tilted compressor, and application to the subpicosecond traveling-wave pumping of a soft-x- ray laser, J Opt Soc Am B Opt Physics 17, p.151-157 (2000)

[6.1070] {Sect. 6.10.4.3} P.O.J. Scherer, A. Seilmeier, W. Kaiser: Ultrafast intra- and intermolecular energy transfer in solutions after selective infrared excitation, J. Chem. Phys. 83, p.3948-3957 (1985)

[6.1071] {Sect. 6.10.4.3} Zs. Bor, S. Szatmári, A. Müller: Picosecond Pulse Shortening by Travelling Wave Amplified Spontaneous Emission, Appl. Phys. B 32, p.101-104 (1983)

[6.1072] {Sect. 6.10.4.3} D.H. Auston: Transverse Mode Locking, IEEE J. QE-4, p.420-422 (1968)

[6.1073] {Sect. 6.10.5} C.O.Weiss, F.Vilaseca: Dynamics of Lasers (VCH, Weinheim, 1991)

[6.1075] {Sect. 6.10.5} W. Gadomski, B. RatajskaGadomska: Homoclinic orbits and chaos in the vibronic short-cavity standing- wave alexandrite laser, J Opt Soc Am B Opt Physics 17, p.188-197 (2000)

[6.1076] {Sect. 6.10.5} H. Cao, Y.G. Zhao, S.T. Ho, E.W. Seelig, Q.H. Wang, R.P.H. Chang: Random laser action in semiconductor powder, Phys Rev Lett 82, p.2278-2281 (1999)

[6.1077] {Sect. 6.10.5} G.J. deValcarcel, E. Roldan, F. Prati: Risken-Nummedal-Graham-Haken instability in class-B lasers, Opt Commun 163, p.5-8 (1999)

[6.1078] {Sect. 6.10.5} J.B. Geddes, K.M. Short, K. Black: Extraction of signals from chaotic laser data, Phys Rev Lett 83, p.5389-5392 (1999)

[6.1079] {Sect. 6.10.5} A. Hohl, A. Gavrielides: Bifurcation cascade in a semiconductor laser subject to optical feedback, Phys Rev Lett 82, p.1148-1151 (1999)

[6.1080] {Sect. 6.10.5} A. Imhof, W.L. Vos, R. Sprik, A. Lagendijk: Large dispersive effects near the band edges of photonic crystals, Phys Rev Lett 83, p.2942-2945 (1999)

[6.1081] {Sect. 6.10.5} H.D.I. Abarbanel, M.B. Kennel: Synchronizing high-dimensional chaotic optical ring dynamics, Phys Rev Lett 80, p.3153-3156 (1998)

[6.1082] {Sect. 6.10.5} V. Espinosa, F. Silva, G.J. deValcarcel, E. Roldan: Class-B two-photon Fabry-Perot laser, Opt Commun 155, p.292-296 (1998)

[6.1083] {Sect. 6.10.5} A. Hohl, A. Gavrielides: Experimental control of a chaotic semiconductor laser, Optics Letters 23, p.1606-1608 (1998)

[6.1084] {Sect. 6.10.5} L. Larger, J.P. Goedgebuer, J.M. Merolla: Chaotic oscillator in wavelength: A new setup for investigating differential difference equations describing nonlinear dynamics, IEEE J QE-34, p.594-601 (1998)

[6.1085] {Sect. 6.10.5} C. Szwaj, S. Bielawski, D. Derozier, T. Erneux: Faraday instability in a multimode laser, Phys Rev Lett 80, p.3968-3971 (1998)

[6.1086] {Sect. 6.10.5} A. Uchida, T. Sato, F. Kannari: Suppression of chaotic oscillations in a microchip laser by injection of a new orbit into the chaotic attractor, Optics Letters 23, p.460-462 (1998)

[6.1087] {Sect. 6.10.5} G. Vaschenko, M. Giudici, J.J. Rocca, C.S. Menoni, J.R. Tredicce, S. Balle: Temporal dynamics of semiconductor lasers with optical feedback, Phys Rev Lett 81, p.5536-5539 (1998)

[6.1088] {Sect. 6.10.5} A.G. Vladimirov: Bifurcation analysis of a bidirectional class B ring laser, Opt Commun 149, p.67-72 (1998)

[6.1089] {Sect. 6.10.5} G. Levy, A.A. Hardy: Chaotic effects in flared lasers: A numerical analysis, IEEE J QE-33, p.26-32 (1997)

[6.1090] {Sect. 6.10.5} J.T. Malos, R. Dykstra, M. Vaupel, C.O. Weiss: Vortex streets in a cavity with higher-order standing waves, Optics Letters 22, p.1056-1058 (1997)

[6.1091] {Sect. 6.10.5} S.V. Sergeyev, G.G. Krylov: Dynamics operations and chaos control for an anisotropic A-class laser with a saturable absorber, Opt Commun 139, p.270-286 (1997)

[6.1092] {Sect. 6.10.5} Q.S. Yang, P.Y. Wang, H.W. Yin, J.H. Dai, D.J. Zhang: Global stability and oscillation properties of a two-level model for a class-B laser with feedback, Opt Commun 138, p.325-329 (1997)

[6.1093] {Sect. 6.10.5} I. Fischer, G.H.M. Vantartwijk, A.M. Levine, W. Elsasser, E. Gobel, D. Lenstra: Fast pulsing and chaotic itinerancy with a drift in the coherence collapse of semiconductor lasers, Phys Rev Lett 76, p.220-223 (1996)

[6.1094] {Sect. 6.10.5} P. Khandokhin, Y. Khanin, J.C. Celet, D. Dangoisse, P. Glorieux: Low frequency relaxation oscillations in class B lasers with feedback, Opt Commun 123, p.372-384 (1996)

[6.1095] {Sect. 6.10.5} J.T. Malos, K. Staliunas, M. Vaupel, C.O. Weiss: Three-dimensional representation of two-dimensional vortex dynamics in lasers, Opt Commun 128, p.123-135 (1996)

[6.1096] {Sect. 6.10.5} D.Y. Tang, N.R. Heckenberg: Spontaneous self-organisation in chaotic laser mode-mode interaction, Opt Commun 131, p.89-94 (1996)

[6.1097] {Sect. 6.10.5} M. Sanmiguel: Phase instabilities in the laser vector complex Ginzburg- Landau equation, Phys Rev Lett 75, p.425-428 (1995)

[6.1098] {Sect. 6.10.5} C. Serrat, A. Kulminskii, R. Vilaseca, R. Corbalan: Polarization chaos in an optically pumped laser, Optics Letters 20, p.1353-1355 (1995)

[6.1099] {Sect. 6.10.5} G.C. Valley, G.J. Dunning: Observation of optical chaos in a phase-conjugate resonator, Opt. Lett. 9, p.513-515 (1984)

[6.1074] {Sect. 6.10.5} R. Hauck, F. Hollinger, H. Weber: Chaotic and Periodic Emission of high power solid state lasers, Opt. Commun. 47, p.141-145 (1983)

[6.1100] {Sect. 6.10.5} F.T. Arecchi, A. Bern'e, P. Bulamacchi: High-order fluctuations in a single-mode laser field, Phys. Rev. Lett. 16, p.32-35 (1966)

[6.1101] {Sect. 6.10.5} K. Otsuka, J.L. Chern, J.S. Lih: Experimental suppression of chaos in a modulated multimode laser, Optics Letters 22, p.292-294 (1997)

[6.1102] {Sect. 6.10.5} E.M. Wright, P. Meystre, W.J. Firth: Nonlinear Theory of Self-Oscillations in a Phase-Conjugate Resonator, Opt. Comm. 51, p.428-432 (1984)

[6.1103] {Sect. 6.11.1} D.W. Hall, M.J. Weber: Modeling Gain Saturation in Neodymium Laser Glasses, IEEE J. QE-20, p.831-834 (1984)

[6.1104] {Sect. 6.11.1} W.E. Martin, D. Milam: Gain Saturation in Nd:Doped Laser Materials, IEEE J. QE-18, p.1155-1163 (1982)

[6.1105] {Sect. 6.11.1} S.M. Yarema, D. Milam: Gain Saturation in Phosphate Laser Glasses, IEEE J. QE-18, p.1941-1946 (1982)

[6.1106] {Sect. 6.11.1} J. Bunkenberg, J. Boles, D.C. Brown, J. Eastman, J. Hoose, R. Hopkins, L. Iwan, S.D. Jacobs, J.H. Kelly, S. Kumpan, S. Letzring, D. Lonobile, L.D. Lund, G. Mourou, S. Refermat, W. Seka, J.M. Soures, K. Walsh: The Omega High-Power Phosphate-Glass System: Design and Performance, IEEE J. QE-17, p.1620-1628 (1981)

[6.1107] {Sect. 6.11.1} D.R. Speck, E.S. Bliss, J.A. Glaze, J.W. Herris, F.W. Holloway, J.T. Hunt, B.C. Johnson, D.J. Kuizenga, R.G. Ozarski, H.G. Patton, P.R. Rupert, G.J. Suski, C.D. Swift, C.E. Thompson: The Shiva Laser-Fusion Facility, IEEE J. QE-17, p.1599-1619 (1981)

[6.1108] {Sect. 6.11.1} C. Yamanaka,Y. Kato, Y. Izawa, K. Yoshida, T. Yamanaka, T. Sasaki, M. Nakatsuka, T. Mochizuki, J. Kuroda, S. Nakai: Nd-Doped Phosphate Glass Laser Systems for Laser-Fusion Research, IEEE J. QE-17, p.1639-1649 (1981)

[6.1109] {Sect. 6.11.1} P. Labudde, W. Seka, H.P. Weber: Gain increase in laser amplifiers by suppression of parasitic oscillations, Appl. Phys. Lett. 29, p.732-734 (1976)

[6.1110] {Sect. 6.11.1} A.N. Chester: Gain Thresholds for Diffuse Parasitic Laser Modes, Appl. Opt. 12, p.2139-2146 (1973)

[6.1111] {Sect. 6.11.1} J.I. Davis, W.R. Sooy: The Effects of Saturation and Regeneration in Ruby Laser Amplifiers, Appl. Opt. 3, p.715-718 (1964)

[6.1112] {Sect. 6.11.1} W.W. Rigrod: Gain Saturation and Output Power of Optical Masers, J. Appl. Phys. 34, p.2602-2609 (1963)

[6.1113] {Sect. 6.11.3.1} A. Brignon, G. Feugnet, J.P. Huignard, J.P. Pocholle: Compact Nd:YAG and Nd:YVO4 amplifiers end-pumped by a high-brightness stacked array, IEEE J QE-34, p.577-585 (1998)

[6.1114] {Sect. 6.11.3.1} A.C. Wilson, J.C. Sharpe, C.R. McKenzie, P.J. Manson, D.M. Warrington: Narrow-linewidth master-oscillator power amplifier based on a semiconductor tapered amplifier, Appl Opt 37, p.4871-4875 (1998)

[6.1115] {Sect. 6.11.3.1} A. Brignon, G. Feugnet, J.-P. Huignard, J.-P. Pocholle: Compact Nd:YAG and Nd:YVO4 Amplifiers End-Pumped by a High-Brightness Stacked Array, IEEE J. QE-34, p.577-585 (1998)

[6.1116] {Sect. 6.11.3.1} Z. Dai, R. Michalzik, P. Unger, K.J. Ebeling: Numerical simulation of broad-area high-power semiconductor laser amplifiers, IEEE J QE-33, p.2240-2254 (1997)

[6.1117] {Sect. 6.11.3.1} R. Paschotta, D.C. Hanna, P. Denatale, G. Modugno, M. Inguscio, P. Laporta: Power amplifier for 1083 nm using ytterbium doped fibre, Opt Commun 136, p.243-246 (1997)

[6.1118] {Sect. 6.11.3.1} A. Sugiyama, T. Nakayama, M. Kato, Y. Maruyama: Characteristics of a dye laser amplifier transversely pumped by copper vapor lasers with a two-dimensional calculation model, Appl Opt 36, p.5849-5854 (1997)

[6.1119] {Sect. 6.11.3.1} F. Hosoi, M. Shimura, Y. Nabekawa, K. Kondo, S. Watanabe: High-power dye laser using steady-state amplification with chirped pulses, Appl Opt 35, p.1404-1408 (1996)

[6.1120] {Sect. 6.11.3.1} G.J. Linford, E.R. Peressini, W.R. Sooy, M.L. Spaeth: Very Long Lasers, Appl. Opt. 13, p.379-390 (1974)

[6.1121] {Sect. 6.11.3.1} L.M. Frantz, J.S. Nodvik: Theory of Pulse Propagation in a Laser Amplifier, J. Appl. Phys. 34, p.2346-2349 (1963)

[6.1122] {Sect. 6.11.3.1} C. Pare: Optimum laser beam profile for maximum energy extraction from a saturable amplifier, Opt Commun 123, p.762-776 (1996)

[6.1123] {Sect. 6.11.3.2} J.M. Casperson, F.G. Moore, L.W. Casperson: Double-pass high-gain laser amplifiers, J Appl Phys 86, p.2967-2973 (1999)

[6.1124] {Sect. 6.11.3.2} Y. Hirano, N. Pavel, S. Yamamoto, Y. Koyata, T. Tajime: 100-W class diode-pumped Nd : YAG MOPA system with a double-stage relay-optics scheme, Opt Commun 170, p.275-280 (1999)

[6.1125] {Sect. 6.11.3.2} A. Brignon, G. Feugnet, J.P. Huignard, J.P. Pocholle: Compact Nd:YAG and Nd:YVO4 Amplifiers End-Pumped by a High-Brightness Stacked Array, IEEE J. QE-34, p.577-585 (1998)

[6.1126] {Sect. 6.11.3.2} A. Brignon, G. Feugnet, J.P. Huignard, J.P. Pocholle: Large-field-of-view, high-gain, compact diode-pumped Nd: YAG amplifier, Optics Letters 22, p.1421-1423 (1997)

[6.1127] {Sect. 6.11.3.2} S.D. Butterworth, W.A. Clarkson, N. Moore, G.J. Friel, D.C. Hanna: High-power quasi-cw laser pulses via high-gain diode- pumped bulk amplifiers, Opt Commun 131, p.84-88 (1996)

[6.1128] {Sect. 6.11.3.3} J.W. Hahn, Y.S. Yoo: Suppression of amplified spontaneous emission from a four-pass dye laser amplifier, Appl Opt 37, p.4867-4870 (1998)

[6.1129] {Sect. 6.11.3.3} M. Zitelli, E. Fazio, M. Bertolotti: On the design of multi-pass dye laser amplifiers, IEEE J QE-34, p.609-615 (1998)

[6.1130] {Sect. 6.11.3.3} P. Heinz, A. Seilmeier, A. Piskarskas: Picosecond Nd:YLF laser-multipass amplifier source pumped by pulsed diodes for the operation of powerful OPOs, Opt Commun 136, p.433-436 (1997)

[6.1131] {Sect. 6.11.3.3} P.F. Curley, C. LeBlanc, G. Cheriaux, G. Darpentigny, P. Rousseau, F. Salin, J.P. Chambaret, A. Antonetti: Multi-pass amplification of sub-50 fs pulses up to the 4 TW level, Opt Commun 131, p.72-76 (1996)

[6.1132] {Sect. 6.11.3.3} E.S. Lee, J.W. Hahn: Four-pass amplifier for the pulsed amplification of a narrow-bandwidth continuous-wave dye laser, Optics Letters 21, p.1836-1838 (1996)

[6.1133] {Sect. 6.11.3.3} S. Petit, O. Cregut, C. Hirlimann: A tunable femtosecond pulses amplifier, Opt Commun 124, p.49-55 (1996)

[6.1134] {Sect. 6.11.3.3} M. Lenzner, C. Spielmann, E. Wintner, F. Krausz, A.J. Schmidt: Sub-20-fs, kilohertz-repetition-rate Ti:sapphire amplifier, Optics Letters 20, p.1397-1399 (1995)

[6.1135] {Sect. 6.11.3.3} J.P. Zhou, C.P. Huang, M.M. Murnane, H.C. Kapteyn: Amplification of 26-fs, 2-TW Pulses Near the Gain Narrowing Limit in Ti-Sapphire, Optics Letters 20, p.64-66 (1995)

[6.1136] {Sect. 6.11.3.3} M. Michon, R. Auffret, R. Dumanchin: Selection and Multiple-Pass Amplification of a Single Mode-Locked Optical Pulse, J. Appl. Phys. 41, p.2739-2740 (1970)

[6.1137] {Sect. 6.11.3.3} D.T. Du, J. Squier, S. Kane, G. Korn, G. Mourou, C. Bogusch, C.T. Cotton: Terawatt Ti:sapphire laser with a spherical reflective optic pulse expander, Optics Letters 20, p.2114-2116 (1995)

[6.1138] {Sect. 6.11.3.3} W.H. Lowdermilk, J.E. Murray: The multipass amplifier: Theory and numerical analysis, J. Appl. Phys. 51, p.2436-2444 (1980)

[6.1139] {Sect. 6.11.3.4} J. Faure, J. Itatani, S. Biswal, G. Cheriaux, L.R. Bruner, G.C. Templeton, G. Mourou: A spatially dispersive regenerative amplifier for ultrabroadband pulses, Opt Commun 159, p.68-73 (1999)

[6.1140] {Sect. 6.11.3.4} H. Liu, S. Biswal, J. Paye, J. Nees, G. Mourou, C. Honninger, U. Keller: Directly diode-pumped millijoule subpicosecond Yb: glass regenerative amplifier, Optics Letters 24, p.917-919 (1999)

[6.1141] {Sect. 6.11.3.4} V. Shcheslavskiy, F. Noack, V. Petrov, N. Zhavoronkov: Femtosecond regenerative amplification in Cr : forsterite, Appl Opt 38, p.3294-3297 (1999)

[6.1142] {Sect. 6.11.3.4} J. Itatani, J. Faure, M. Nantel, G. Mourou, S. Watanabe: Suppression of the amplified spontaneous emission in chirped-pulse- amplification lasers by clean high-energy seed-pulse injection, Opt Commun 148, p.70-74 (1998)

[6.1143] {Sect. 6.11.3.4} P.J. Delfyett, A. Yusim, S. Grantham, S. Gee, K. Gabel, M. Richardson, G. Alphonse, J. Connolly: Ultrafast semiconductor laser-diode-seeded Cr:LiSAF regenerative amplifier system, Appl Opt 36, p.3375-3380 (1997)

[6.1144] {Sect. 6.11.3.4} T.R. Nelson, W.A. Schroeder, C.K. Rhodes, F.G. Omenetto, J.W. Longworth: Short-pulse amplification at 745 nm in Ti:sapphire with a continuously tunable regenerative amplifier, Appl Opt 36, p.7752-7755 (1997)

[6.1145] {Sect. 6.11.3.4} H. Takada, K. Miyazaki, K.J. Torizuka: Flashlamp-pumped Cr:LiSAF laser amplifier, IEEE J QE-33, p.2282-2285 (1997)

[6.1146] {Sect. 6.11.3.4} V.A. Venturo, A.G. Joly, D. Ray: Pulse compression with a high-energy Nd:YAG regenerative amplifier system, Appl Opt 36, p.5048-5052 (1997)

[6.1147] {Sect. 6.11.3.4} A. Rundquist, C. Durfee, Z. Chang, G. Taft, E. Zeek, S. Backus, M.M. Murnane, H.C. Kapteyn, I. Christov, V. Stoev: Ultrafast laser and amplifier sources, Appl. Phys. B 65, p.161-174 (1997)

[6.1148] {Sect. 6.11.3.4} C.P.J. Barty, T. Guo, C. LeBlanc, F. Raksi, C. Rosepetruck, J. Squier, K.R. Wilson, V.V. Yakovlev, K. Yamakawa: Generation of 18-fs, multiterawatt pulses by regenerative pulse shaping and chirped-pulse amplification, Optics Letters 21, p.668-670 (1996)

[6.1149] {Sect. 6.11.3.4} T. Joo, Y. Jia, G.R. Fleming: Ti:sapphire regenerative amplifier for ultrashort high-power multikilohertz pulses without an external stretcher, Opt. Lett. 20, p.389-391 (1995)

[6.1150] {Sect. 6.11.3.4} L. Turi, T. Juhasz: High-power longitudinally and-diode-pumped Nd:YLF regenerative amplifier, Opt. Lett. 20, p.154-156 (1995)

[6.1151] {Sect. 6.11.3.4} M.D. Selker, R.S. Afzal, J.L. Dallas, A.W. Yu: Efficient, diode-laser-pumped, diode-laser-seeded, high-peak-power Nd:YLF regenerative amplifier, Opt. Lett. 19, p.551-553 (1994)

[6.1152] {Sect. 6.11.3.4} N.P. Barnes, J.C. Barnes: Injection Seeding I: Theory, IEEE J. QE-29, p.2670-2683 (1993)

[6.1153] {Sect. 6.11.3.4} M. Gifford, K.J. Weingarten: Diode-pumped Nd:YLF regenerative amplifier, Opt. Lett. 17, p.1788-1790 (1992)

[6.1154] {Sect. 6.11.3.4} T.E. Dimmick: Semiconductor-laser-pumped, cw mode-locked Nd:phosphate glass laser oscillator and regenerative amplifier, Opt. Lett. 15, p.177-179 (1990)

[6.1155] {Sect. 6.11.3.4} M. Saeed, D. Kim, L.F. DiMauro: Optimization and characterization of a high repetition rate, high intensity Nd:YLF regenerative amplifier, Appl. Opt. 29, p.1752-1757 (1990)

[6.1156] {Sect. 6.11.3.4} P. Bado, M. Bouvier, J. Scott Coe: Nd:YLF mode-locked oscillator and regenerative amplifier, Opt. Lett. 12, p.319-321 (1987)

[6.1157] {Sect. 6.11.3.4} I.N. Duling III, T. Norris, T. Sizer II, P. Bado, G.A. Mourou: Kilohertz synchronous amplification of 85-femtosecond optical pulses, J. Opt. Soc. Am. B 2, p.616-618 (1985)

[6.1158] {Sect. 6.11.3.4} R.L. Fork, C.V. Shank, R.T. Yen: Amplification of 70-fs optical pulses to gigawatt powers, Appl. Phys. Lett. 41, p.223-225 (1982)

[6.1159] {Sect. 6.11.3.4} J.E. Murray, W.H. Lowdermilk: ND:YAG regenerative amplifier, J. Appl. Phys. 51, p.3548-3555 (1980)

[6.1160] {Sect. 6.11.3.4} J. Squier, C.P.J. Barty, F. Salin, C. LeBlanc, S. Kane: Use of mismatched grating pairs in chirped-pulse amplification systems, Appl Opt 37, p.1638-1641 (1998)

[6.1161] {Sect. 6.11.3.4} O.E. Martinez, C.M.G. Inchauspe: Compact curved-grating stretcher for laser pulse amplification, Optics Letters 22, p.811-813 (1997)

[6.1162] {Sect. 6.11.3.4} I.N. Ross, M. Trentelman, C.N. Danson: Optimization of a chirped-pulse amplification Nd:glass laser, Appl Opt 36, p.9348-9358 (1997)

[6.1163] {Sect. 6.11.3.4} M. Trentelman, I.N. Ross, C.N. Danson: Finite size compression gratings in a large aperture chirped pulse amplification laser system, Appl Opt 36, p.8567-8573 (1997)

[6.1164] {Sect. 6.11.3.4} A. Galvanauskas, M.E. Fermann, D. Harter: High-Power Amplification of Femtosecond Optical Pulses in a Diode-Pumped Fiber System, Opt.Lett. 19, p.1201-1203 (1994)

[6.1165] {Sect. 6.11.3.4} J.V. Rudd, G. Korn, S. Kane, J. Squier, G. Mourou, P. Bado: Chirped-pulse amplification of 55-fs pulses at a 1-kHz repetition rate in a Ti:Al2O3 regenerative amplifier, Opt. Lett. 18, p.2044-2046 (1993)

[6.1166] {Sect. 6.11.3.4} R.L. Fork, O.E. Martinez, J.P. Gordon: Negative dispersion using pairs of prisms, Opt. Lett. 9, p.150-152 (1984)

[6.1167] {Sect. 6.11.3.4} A. Braun, S. Kane, T. Norris: Compensation of self-phase modulation in chirped-pulse amplification laser systems, Optics Letters 22, p.615-617 (1997)

[6.1168] {Sect. 6.11.3.4} D. Strickland, G. Mourou: Compression of Amplified Chirped Optical Pulses, Opt. Commun. 56, p.219-221 (1985)

[6.1169] {Sect. 6.11.3.5} Y. Tzuk, Y. Glick, M.M. Tilleman: Compact ultra-high gain multi-pass Nd : YAG amplifier with a low passive reflection phase conjugate mirror, Opt Commun 165, p.237-244 (1999)

[6.1170] {Sect. 6.11.3.5} S. Seidel, N. Kugler: Nd:YAG 200-W average-power oscillator-amplifier system with stimulated-Brillouin-scattering phase conjugation and depolarization compensation, J. Opt. Soc. Am. B 14, p.1885-1888 (1997)

[6.1171] {Sect. 6.11.3.5} C.K. Ni, A.H. Kung: Effective suppression of amplified spontaneous emission by stimulated Brillouin scattering phase conjugation, Optics Letters 21, p.1673-1675 (1996)

[6.1172] {Sect. 6.11.3.5} H.L. Offerhaus, H.P. Godfried, W.J. Witteman: All solid-state diode pumped Nd:YAG MOPA with stimulated brillouin phase conjugate mirror, Opt Commun 128, p.61-65 (1996)

[6.1173] {Sect. 6.11.3.5} H.J. Eichler, A. Haase, R. Menzel: High beam quality of a single rod neodym amplifier by SBS-phase conjugation up to 140 Watt average output, Opt. Quant. Electron. 28, p.261-265 (1996)

[6.1174] {Sect. 6.11.3.5} H.L. Offerhaus, H.P. Godfried, W.J. Witteman: Al solid-state diode pumped Nd:YAG MOPA with stimulated Brillouin phase conjugate mirror, Opt. Comm. 128, p.61-65 (1996)

[6.1175] {Sect. 6.11.3.5} D.M. Pepper, D.A. Rockwell, H.W. Bruesselbach: Phase Conjugation: Reversing Laser Aberrations, Photonics Spectra Aug. 1996, p.95-104 (1996)

[6.1176] {Sect. 6.11.3.5} E.V. Voskoboinik, A.V. Kir'yanov, P.P. Pashinin, V.S. Sidorin, V.V. Tumorin, E.I. Shklovskii: Repetitively pulsed Nd:YAG laser with an SBS mirror, Quantum Electron. 26, p.31-33 (1996)

[6.1177] {Sect. 6.11.3.5} C.B. Dane, L.E. Zapata, W.A. Neumann, M.A. Norton, L.A. Hackel: Design and Operation of a 150 W Near Diffraction-Limited Laser Amplifier with SBS Wavefront Correction, IEEE J. QE-31, p.148-163 (1995)

[6.1178] {Sect. 6.11.3.5} H.J. Eichler, A. Haase, R. Menzel: 100 Watt Average Output Power 1.2*Diffraction Limited Beam From Pulsed Neodym Single Rod Amplifier with SBS-Phaseconjugation, IEEE J. QE-31, p.1265-1269 (1995)

[6.1179] {Sect. 6.11.3.5} E.J. Shklovsky, V.V. Tumorin: Generation of long laser pulses in the scheme of a double-pass amplifier with SBS mirror, Opt. Comm. 120, p.303-306 (1995)

[6.1180] {Sect. 6.11.3.5} H.J. Eichler, A. Haase, R. Menzel, J. Schwartz: Depolarization treatment and optimization of high power double pass neodym-rod amplifiers with SBS mirror, Pure Appl. Opt. 3, p.585-591 (1994)

[6.1181] {Sect. 6.11.3.5} H.J. Eichler, A. Haase, R. Menzel: SBS-Phase Conjugation for Thermal Lens Compensation in 100 Watt Average Power Solid-State Lasers, Int. J. Nonlinear Optics 3, p.339-345 (1994)

[6.1182] {Sect. 6.11.3.5} D.S. Sumida, C.J. Jones, R.A. Rockwell: An 8.2 J Phase Conjugating Solid-State Laser Coherently Combining Eight Parallel Amplifiers, IEEE J. QE-30, p.2617-2627 (1994)

[6.1183] {Sect. 6.11.3.5} N.F. Andreev, E. Khazanov, G.A. Pasmanik: Applications of Brillouin Cells to High Repetition Rate Solid-State Lasers, IEEE J. QE-28, p.330-341 (1992)

[6.1184] {Sect. 6.11.3.5} N.F. Andreev, S.V. Kuznetsov, O.V. Palashov, G.A. Pasmanik, E.A. Khazanov: Four-pass YAG:Nd laser amplifier with compensation for aberration and polarization distortions of the wavefront, Sov. J. Quantum Electron. 22, p.800-802 (1992)

[6.1185] {Sect. 6.11.3.5} N.F. Andreev, E.A. Khazanov, S.V. Kuznetsov, G.A. Pasmanik, E.I. Shklovsky, V.S. Sidorin: Locked Phase Conjugation for Two-Beam Coupling of Pulse Repetition Rate Solid-State Lasers, IEEE J. QE-27, p.135-141 (1991)

[6.1186] {Sect. 6.11.3.5} J.-L. Ayral, J. Montel, T. Verny, J.-P. Huignard: Phase-conjugate Nd:YAG laser with internal acousto-optic beam steering, Opt. Lett. 16, p.1225-1227 (1991)

[6.1187] {Sect. 6.11.3.5} A.F. Vasil'ev, S.B. Gladin, V.E. Yashin: Pulse-periodic Nd:YAIO3 laser with a phase-locked aperture under conditions of phase conjugation by stimulated Brillouin scattering, Sov. J. Quantum Electron. 21, p.494-497 (1991)

[6.1188] {Sect. 6.11.3.5} A.A. Babin, F.I. Fel'dshtein, G.I. Freidman: Double-pass amplifier with a stimulated Brillouin scattering mirror for a subnanosecond pulse train, Sov. J. Quantum Electron. 19, p.1303-1304 (1989)

[6.1189] {Sect. 6.11.3.5} N.G. Basov, D.A. Glazkov, V.F. Efimkov, I.G. Zubarev, S.A. Pastukhov, V.B. Sobolev: Hypersonic Phase-Conjugation Mirror for the Reflection of High-Power Nanosecond Pulses, IEEE J. QE-25, p.470-478 (1989)

[6.1190] {Sect. 6.11.3.5} N.G. Basov, V.F. Efimkov, I.G. Zubarev, V.V. Kolobrodov, S.A. Pastukhov, M.G. Smirnov, V.B. Sobolev: Pulsed neodymium amplifier with phase conjugation and direct amplification, Sov. J. Quantum Electron. 18, p.1593-1595 (1989)

[6.1191] {Sect. 6.11.3.5} P. Fairchild, K. Davis, M. Valley: Coherent beam combination in barium titanate, J. Opt. Soc. Am. B 5, p.1758-1762 (1988)

[6.1192] {Sect. 6.11.3.5} D.A. Rockwell: A Review of Phase-Conjugate Solid-State Lasers, IEEE J. QE-24, p.1124-1140 (1988)

[6.1193] {Sect. 6.11.3.5} V.N. Alekseev, V.V. Golubev, D.I. Dmitriev, A.N. Zhilin, V.V. Lyubimov, A.A. Mak, V.I. Reshetnikov, V.S. Sirazetdinov, A.D. Starikov: Investigation of wavefront reversal in a phosphate glass laser amplifier with a 12-cm output aperture, Sov. J. Quantum Electron. 17, p.455-458 (1987)

[6.1194] {Sect. 6.11.3.5} M. Sugii, O. Sugihara, M. Ando, K. Sasaki: High locking efficiency XeCl ring amplifier injection locked by backward stimulated Brillouin scattering, J. Appl. Phys. 62, p.3480-3482 (1987)

[6.1195] {Sect. 6.11.3.5} K. Kyuma, A. Yariv: Polarization recovery in phase conjugation by modal dispersion, Appl. Phys. Lett. 49, p.617-619 (1986)

[6.1196] {Sect. 6.11.3.5} D.A. Rockwell, C.R. Giuliano: Coherent coupling of laser gain media using phase conjugation, Opt. Lett. 11p.147-149 (1986)

[6.1197] {Sect. 6.11.3.5} M. Valley, G. Lombardi, R. Aprahamian: Beam combination by stimulated Brillouin scattering, J. Opt. Soc. Am. B 3, p.1492-1497 (1986)

[6.1198] {Sect. 6.11.3.5} I.D. Carr, D.C. Hanna: Performance of a Nd:YAG Oscillator/Amplifier with Phase-Conjugation via Stimulated Brillouin Scattering, Appl. Phys. B 36, p.83-92 (1985)

[6.1199] {Sect. 6.11.3.5} M.C. Gower, R.G. Caro: KrF laser with a phase-conjugate Brillouin mirror, Opt. Lett. 7, p.162-164 (1982)

[6.1200] {Sect. 6.11.3.5} M.C. Gower: KrF laser amplifier with phase-conjugate Brillouin retroreflectors, Opt. Lett. 7, p.423-425 (1982)

[6.1201] {Sect. 6.11.3.5} D.T. Hon: Applications of wavefront reversal by stimulated Brillouin scattering, Opt. Eng. 21, p.252-256 (1982)

[6.1202] {Sect. 6.11.3.5} I.G. Zubarev, A.B. Mironov, S.I. Mikahilov: Single-mode pulse-periodic oscillator-amplifier system with wavefront reversal, Sov. J. Quantum Electron. 10, p.1179-1181 (1981)

[6.1203] {Sect. 6.11.3.5} N. Basov, I. Zubarev: Powerful Laser Systems with Phase Conjugation by SMBS Mirrror, Appl. Phys. 20, p.261-264 (1979)

[6.1204] {Sect. 6.11.3.5} infinity – A Revolutionary Nd:YAG Laser System, Prospekt Fa. Coherent

[6.1205] {Sect. 6.11.3.5} G.J. Crofts, X. Banti, M.J. Damzen: Tunable phase conjugation in a Ti:sapphire amplifier, Optics Letters 20, p.1634-1636 (1995)

[6.1206] {Sect. 6.11.3.5} J.H. Kelly, S.D. Jacobs, J.C. Lambropoulos, J.C. Lee, M.J. Shoup, D.J. Smith, D.L. Smith High repetition rate Cr:Nd:GSGG active mirror amplifier, Opt. Lett. 12, p.996-998 (1987)

[6.1207] {Sect. 6.11.4.1} L.W. Casperson, J.M. Casperson: Power self-regulation in double-pass high-gain laser amplifiers, J Appl Phys 87, p.2079-2083 (2000)

[6.1208] {Sect. 6.11.4.1} E.H. Huntington, T.C. Ralph, I. Zawischa: Sources of phase noise in an injection-locked solid-state laser, J Opt Soc Am B Opt Physics 17, p.280-292 (2000)

[6.1209] {Sect. 6.11.4.1} S.R. Friberg, S. Machida: Ultrafast optical pulse noise suppression using a nonlinear spectral filter: 23 dB reduction of fiber laser 1/f noise, Appl Phys Lett 73, p.1934-1936 (1998)

[6.1210] {Sect. 6.11.4.1} D.J. Ottaway, P.J. Veitch, M.W. Hamilton, C. Hollitt, D. Mudge, J. Munch: A compact injection-locked Nd:YAG laser for gravitational wave detection, IEEE J QE-34, p.2006-2009 (1998)

[6.1211] {Sect. 6.11.4.1} U. Roth, T. Graf, E. Rochat, K. Haroud, J.E. Balmer, H.P. Weber: Saturation, gain, and noise properties of a multipass diode-laser-pumped Nd:YAG CW amplifier, IEEE J QE-34, p.1987-1991 (1998)

[6.1212] {Sect. 6.11.4.1} W.M. Tulloch, T.S. Rutherford, E.H. Huntington, R. Ewart, C.C. Harb, B. Willke, E.K. Gustafson, M.M. Fejer, R.L. Byer, S. Rowan et al.: Quantum noise in a continuous-wave laser-diode-pumped Nd : YAG linear optical amplifier, Optics Letters 23, p.1852-1854 (1998)

[6.1213] {Sect. 6.11.4.1} A. Hardy, D. Treves: Amplified Spontaneous Emission in Sperical and Disk-Shaped Laser Media, IEEE J. QE-15, p.887-895 (1979)

[6.1214] {Sect. 6.11.4.1} S. Guch, Jr.: Parasitic suppression in large aperture disk lasers employing liquid edge claddings, Appl. Opt. 15, p.1453-1457 (1976)

[6.1215] {Sect. 6.11.4.1} J.A. Glaze, S. Guch, J.B. Trenholme: Parasitic Suppression in Large Aperture Nd:Glass Disk Laser Amplifiers, Appl. Opt. 13, p.2808-2811 (1974)

[6.1216] {Sect. 6.11.4.1} G.D. Baldwin, I.T. Basil: Parasitic Noise on the Output of a CW YAG:Nd+3 Laser, IEEE J. QE-7, p.179-181 (1971)

[6.1217] {Sect. 6.11.4.1} W. Imajuku, A. Takada: Gain characteristics of coherent optical amplifiers using a Mach-Zehnder interferometer with Kerr media, IEEE J QE-35, p.1657-1665 (1999)

[6.1218] {Sect. 6.11.4.1} S.A.E. Lewis, S.V. Chernikov, J.R. Taylor: Temperature-dependent gain and noise in fiber Raman amplifiers, Optics Letters 24, p.1823-1825 (1999)

[6.1219] {Sect. 6.11.4.1} F.G. Patterson, J. Bonlie, D. Price, B. White: Suppression of parasitic lasing in large-aperture Ti : sapphire laser amplifiers, Optics Letters 24, p.963-965 (1999)

[6.1220] {Sect. 6.11.4.1} H.J. Briegel, W. Dur, J.I. Cirac, P. Zoller: Quantum repeaters: The role of imperfect local operations in quantum communication, Phys Rev Lett 81, p.5932-5935 (1998)

[6.1221] {Sect. 6.11.4.1} P. DiTrapani, A. Berzanskis, S. Minardi, S. Sapone, W. Chinaglia: Observation of optical vortices and J (0) Bessel-like beams in quantum- noise parametric amplification, Phys Rev Lett 81, p.5133-5136 (1998)

[6.1222] {Sect. 6.11.4.1} W.K. Marshall, B. Crosignani, A. Yariv: Laser phase noise to intensity noise conversion by lowest-order group- velocity dispersion in optical fiber: exact theory, Optics Letters 25, p.165-167 (2000)

[6.1223] {Sect. 6.11.4.1} G. Heinzel, K.A. Strain, J. Mizuno, K.D. Skeldon, B. Willke, W. Winkler, R. Schilling, A. Rudiger, K. Danzmann: Experimental demonstration of a suspended dual recycling interferometer for gravitational wave detection, Phys Rev Lett 81, p.5493-5496 (1998)

[6.1224] {Sect. 6.11.4.2} A. Efimov, M.D. Moores, N.M. Beach, J.L. Krause, D.H. Reitze: Adaptive control of pulse phase in a chirped-pulse amplifier, Optics Letters 23, p.1915-1917 (1998)

[6.1225] {Sect. 6.11.4.2} B. Kohler, V.V. Yakovlev, K.R. Wilson, J. Squier, K.W. Delong, R. Trebino: Phase and intensity characterization of femtosecond pulses from a chirped pulse amplifier by frequency resolved optical gating, Optics Letters 20, p.483-485 (1995)

[6.1226] {Sect. 6.11.4.2} J.T. Hunt, J.A. Glaze, W.W. Simmons, P.A. Renard: Suppression of self-focusing through low-pass spatial filtering and relay imaging, Appl. Opt. 17, p.2053-2057 (1978)

[6.1227] {Sect. 6.11.4.2} J.T. Hunt, P.A. Renard, W.W. Simmons: Improved performance of fusion lasers using the imaging properties of multiple spatial filters, Appl. Opt. 16, p.779-782 (1977)

[6.1228] {Sect. 6.11.4.2} J.F. Holzrichter, D.R. Speck: Laser focusing limitations from nonlinear beam instabilities, J. Appl. Phys. 47, p.2459-2461 (1976)

[6.1229] {Sect. 6.11.4.2} B. Willke, N. Uehara, E.K. Gustafson, R.L. Byer, P.J. King, S.U. Seel, R.L. Savage: Spatial and temporal filtering of a 10-W Nd:YAG laser with a Fabry- Perot ring-cavity premode cleaner, Optics Letters 23, p.1704-1706 (1998)

[6.1230] {Sect. 6.11.4.3} G. Cerullo, M. Nisoli, S. Stagira, S. DeSilvestri, G. Tempea, F. Krausz, K. Ferencz: Mirror-dispersion-controlled sub-10-fs optical parametric amplifier in the visible, Optics Letters 24, p.1529-1531 (1999)

[6.1231] {Sect. 6.11.4.3} C. Dorrer, B. deBeauvoir, C. LeBlanc, S. Ranc, J.P. Rousseau, P. Rousseau, J.P. Chambaret: Single-shot real-time characterization of chirped-pulse amplification systems by spectral phase interferometry for direct electric-field reconstruction, Optics Letters 24, p.1644-1646 (1999)

[6.1232] {Sect. 6.11.4.3} A. Galvanauskas, D. Harter, M.A. Arbore, M.H. Chou, M.M. Fejer: Chirped-pulse amplification circuits for fiber amplifiers, based on chirped-period quasi-phase-matching gratings, Optics Letters 23, p.1695-1697 (1998)

[6.1233] {Sect. 6.11.4.3} J. Badziak, S.A. Chizhov, A.A. Kozlov, J. Makowski, M. Paduch, K. Tomaszewski, A.B. Vankov, V.E. Yashin: Picosecond, terawatt, all-Nd:glass CPA laser system, Opt Commun 134, p.495-502 (1997)

[6.1234] {Sect. 6.11.4.3} Q. Fu, F. Seier, S.K. Gayen, R.R. Alfano: High-average-power kilohertz-repetition-rate sub-100-fs Ti: sapphire amplifier system, Optics Letters 22, p.712-714 (1997)

[6.1235] {Sect. 6.11.4.3} G. Lenz, W. Gellermann, D.J. Dougherty, K. Tamura, E.P. Ippen: Femtosecond fiber laser pulses amplified by a KCl:Tl+ color-center amplifier for continuum generation in the 1.5- mu m region, Optics Letters 21, p.137-139 (1996)

[6.1236] {Sect. 6.11.4.3} C. Lozano, P. Garciafernandez, C.R. Mirasso: Analytical study of nonlinear chirped pulses: Propagation in dispersive optical fibers, Opt Commun 123, p.752-761 (1996)

[6.1237] {Sect. 6.11.4.3} S. Backus, J. Peatross, C.P. Huang, M.M. Murnane, H.C. Kapteyn: Ti:sapphire amplifier producing millijoule-level, 21-fs pulses at 1 kHz, Optics Letters 20, p.2000-2002 (1995)

[6.1238] {Sect. 6.11.4.3} N. Blanchot, C. Rouyer, C. Sauteret, A. Migus: Amplification of sub-100-TW femtosecond pulses by shifted amplifying Nd:glass amplifiers: Theory and experiments, Optics Letters 20, p.395-397 (1995)

[6.1239] {Sect. 6.11.4.3} Ch. Spielmann, M. Lenzner, F. Krausz, R. Szipöcs: Compact, high-throughput expansion-compression scheme for chirped pulse amplification in the 10 fs range, Opt. Comm. 120, p.321-324 (1995)

[6.1240] {Sect. 6.11.4.3} W.H. Knox: Femtosecond Optical Pulse Amplification, IEEE J. QE-24, p.388-397 (1988)

[6.1241] {Sect. 6.11.4.3} F.De Martini, C.H. Townes, T.K. Gustafson, P.L. Kelley: Self-Steepening of Light Pulses, Phys. Rev. 164, p.312-323 (1967)

[6.1242] {Sect. 6.11.4.3} F. Shimizu: Frequency Broadening in Liquids by a Short Light Pulse, Phys. Rev. Lett. 19, p.1097-1100 (1967)

[6.1243] {Sect. 6.11.4.3} P.J. Delfyett, H. Shi, S. Gee, I. Nitta, J.C. Connolly, G.A. Alphonse: Joint time-frequency measurements of mode-locked semiconductor diode lasers and dynamics using fuequency-resolved optical gating, IEEE J QE-35, p.487-500 (1999)

[6.1244] {Sect. 6.12.2} M. J. Weber: Handbook of Laser Wavelengths (CRC Press, Boca Raton, Boston, London, New York, Washington, D.C, 1999)

[6.1245] {Sect. 6.13.1.0} G. P. Agrawal (ed.): Semiconductor Lasers (American Institute of Physics, Woodbury, N. Y, 1995)

[6.1246] {Sect. 6.13.1.0} J. Carrol, J. Whiteaway, D. Plumb: Distributed Feedback Semiconductor Lasers (SPIE Optical Engineering Press, London, 1998)

[6.1247] {Sect. 6.13.1.0} W. W. Chow, S. W. Koch, M. Sargent III: Semiconductor-Laser Physics (Springer, Berlin, Heidelberg, New York, 1994)

[6.1248] {Sect. 6.13.1.0} C. F. Klingshirn: Semiconductor Optics (Springer, Berlin, Heidelberg, New York, 1995)

[6.1249] {Sect. 6.13.1.0} F. K. Kneubühl: Theories on Distributed Feedback Lasers (Harwood Academic Publishers, Chur, 1993)

[6.1250] {Sect. 6.13.1.1} P. Modh, N. Eriksson, A. Larsson, T. Suhara: Semiconductor laser with curved deep-etched distributed Bragg reflectors supporting a planar Gaussian mode, Optics Letters 25, p.108-110 (2000)

[6.1251] {Sect. 6.13.1.1} M. Achtenhagen, M. McElhinney, S. Nolan, A. Hardy: High-power 980-nm pump laser modules for erbium-doped fiber amplifiers, Appl Opt 38, p.5765-5767 (1999)

[6.1252] {Sect. 6.13.1.1} P. Raisch, R. Winterhoff, W. Wagner, M. Kessler, H. Schweizer, T. Riedl, R. Wirth, A. Hangleiter, F. Scholz: Investigations on the performance of multiquantum barriers in short wavelength (630 nm) AlGaInP laser diodes, Appl Phys Lett 74, p.2158-2160 (1999)

[6.1253] {Sect. 6.13.1.1} Y. Sidorin, P. Korioja, M. Blomberg: Novel tunable laser diode arrangement with a micromachined silicon filter: feasibility, Opt Commun 164, p.121-127 (1999)

[6.1254] {Sect. 6.13.1.1} C. Gmachl, A. Tredicucci, D.L. Sivco, A.L. Hutchinson, F. Capasso, A.Y. Cho: Bidirectional semiconductor laser, Science 286, p.749-752 (1999)

[6.1255] {Sect. 6.13.1.1} B. Boggs, C. Greiner, T. Wang, H. Lin, T.W. Mossberg: Simple high-coherence rapidly tunable external-cavity diode laser, Optics Letters 23, p.1906-1908 (1998)

[6.1256] {Sect. 6.13.1.1} A.K. Goyal, P. Gavrilovic, H. Po: 1.35 W of stable single-frequency emission from an external-cavity tapered oscillator utilizing fiber Bragg grating feedback, Appl Phys Lett 73, p.575-577 (1998)

[6.1257] {Sect. 6.13.1.1} M.P. Nesnidal, T. Earles, L.J. Mawst, D. Botez, J. Buus: 0.45 W diffraction-limited beam and single-frequency operation from antiguided phase-locked laser array with distributed feedback grating, Appl Phys Lett 73, p.587-589 (1998)

[6.1258] {Sect. 6.13.1.1} C. Sirtori, C. Gmachl, F. Capasso, J. Faist, D.L. Sivco, A.L. Hutchinson, A.Y. Cho: Long-wavelength (lambda approximate to 8-11.5 mu m) semiconductor lasers with waveguides based on surface plasmons, Optics Letters 23, p.1366-1368 (1998)

[6.1259] {Sect. 6.13.1.1} D.M. Cornwell, H.J. Thomas: High-power (>0.9 W cw) diffraction-limited semiconductor laser based on a fiber Bragg grating external cavity, Appl Phys Lett 70, p.694-695 (1997)

[6.1260] {Sect. 6.13.1.1} J. Diaz, H.J. Yi, M. Razeghi, G.T. Burnham: Long-term reliability of Al-free InGaAsP/GaAs (lambda=808 nm) lasers at high-power high-temperature operation, Appl Phys Lett 71, p.3042-3044 (1997)

[6.1261] {Sect. 6.13.1.1} S. Nakamura, M. Senoh, S. Nagahama, N. Iwasa, T. Yamada, T. Matsushita, Y. Sugimoto, H. Kiyoku: Longitudinal mode spectra and ultrashort pulse generation of InGaN multiquantum well structure laser diodes, Appl Phys Lett 70, p.616-618 (1997)

[6.1262] {Sect. 6.13.1.1} J.K. Wade, L.J. Mawst, D. Botez, R.F. Nabiev, M. Jansen: 5 W continuous wave power, 0.81-mu m-emitting, Al-free active-region diode lasers, Appl Phys Lett 71, p.172-174 (1997)

[6.1263] {Sect. 6.13.1.1} A. Leitenstorfer, C. Furst, A. Laubereau, W. Kaiser, G. Trankle, G. Weimann: Femtosecond carrier dynamics in GaAs far from equilibrium, Phys Rev Lett 76, p.1545-1548 (1996)

[6.1264] {Sect. 6.13.1.1} L.J. Mawst, A. Bhattacharya, J. Lopez, D. Botez, D.Z. Garbuzov, L. Demarco, J.C. Connolly, M. Jansen, F. Fang, R.F. Nabiev: 8 W continuous wave front-facet power from broad-waveguide Al-free 980 nm diode lasers, Appl Phys Lett 69, p.1532-1534 (1996)

[6.1265] {Sect. 6.13.1.1} S.B. Ross, S.I. Kanorsky, A. Weis, T.W. Hänsch: A single mode, cw, diode laser at the cesium D1 (894.59 nm) transition, Opt. Comm. 120, p.155-157 (1995)

[6.1266] {Sect. 6.13.1.1} J.-H. Kim, R.J. Lang, A. Larson, L.P. Lee, A.A. Narayanan: High-power AlGaAs/GaAs single quantum well surface-emitting lasers with integrated 45 beam deflectors, Appl. Phys. Lett. 57, p.2048-2050 (1990)

[6.1267] {Sect. 6.13.1.1} S. Murata, I. Mito: Frequency-tunable semiconductor lasers, Opt. Quantum Electr. 22, p.1-15 (1990)

[6.1268] {Sect. 6.13.1.1} N. W. Carlson: Monolithic Diode-Laser Arrays (Springer, Berlin, Heidelberg, New York, 1994)

[6.1269] {Sect. 6.13.1.1} T. Someya, R. Werner, A. Forchel, M. Catalano, R. Cingolani, Y. Arakawa: Room temperature lasing at blue wavelengths in gallium nitride microcavities, Science 285, p.1905-1906 (1999)

[6.1270] {Sect. 6.13.1.1} J. Nishio, L. Sugiura, H. Fujimoto, Y. Kokubun, K. Itaya: Characterization of InGaN multiquantum well structures for blue semiconductor laser diodes, Appl Phys Lett 70, p.3431-3433 (1997)

[6.1271] {Sect. 6.13.1.1} R.L. Aggarwal, P.A. Maki, R.J. Molnar, Z.L. Liau, I. Melngailis: Optically pumped GaN/Al0.1Ga0.9N double-heterostructure ultraviolet laser, J Appl Phys 79, p.2148-2150 (1996)

[6.1272] {Sect. 6.13.1.1} C.C. Chu, T.B. Ng, J. Han, G.C. Hua, R.L. Gunshor, E. Ho, E.L. Warlick, L.A. Kolodziejski, A.V. Nurmikko: Reduction of structural defects in II-VI blue green laser diodes, Appl Phys Lett 69, p.602-604 (1996)

[6.1273] {Sect. 6.13.1.1} S. Nakamura, M. Senoh, S. Nagahama, N. Iwasa, T. Yamada, T. Matsushita, Y. Sugimoto, H. Kiyoku: Ridge-geometry InGaN multi-quantum-well-structure laser diodes, Appl Phys Lett 69, p.1477-1479 (1996)

[6.1274] {Sect. 6.13.1.1} N. Yokouchi, N. Yamanaka, N. Iwai, Y. Nakahira, A. Kasukawa: Tensile-strained GaInAsP-InP quantum-well lasers emitting at 1.3 mu m, IEEE J QE-32, p.2148-2155 (1996)

[6.1275] {Sect. 6.13.1.2} R. Beach, W.J. Benett, B.L. Freitas, D. Mundinger, B.J. Comaskey, R.W. Solarz, M.A. Emanuel: Modular microchannel cooled heatsinks for high average power laser diode arrays, IEEE J. QE-28, p.966-976 (1992)

[6.1276] {Sect. 6.13.1.2} B.J. Comaskey, R. Beach, G. Albrecht, W.J. Benett, B.L. Freitas, C. Petty, D. Vavlue, D. Mundinger, R.W. Solarz: High average power diode pumped slab laser, IEEE J. QE-28, p.992-996 (1992)

[6.1277] {Sect. 6.13.1.2} J.G. Endriz, M. Vakili, G.S. Browder, M. DeVito, J.M. Haden, G.L. Harnagel, W.E. Plano, M. Sakamoto, D.F. Welch, S. Willing,

D.P. Worland, H.C. Yao: High Power Diode Laser Arrays, IEEE J. QE-28, p.952-965 (1992)

[6.1278] {Sect. 6.13.1.2} K.A. Forrest, J.B. Abshire: Time Evolution of Pulsed Far-Field Patterns of GaAlAs Phase-Locked Laser-Diode Arrays, IEEE J. QE-23, p.1287-1290 (1987)

[6.1279] {Sect. 6.13.1.3} A. Bramati, J.P. Hermier, A.Z. Khoury, E. Giacobino, P. Schnitzer, R. Michalzik, K.J. Ebeling, J.P. Poizat, P. Grangier: Spatial distribution of the intensity noise of a vertical-cavity surface-emitting semiconductor laser, Optics Letters 24, p.893-895 (1999)

[6.1280] {Sect. 6.13.1.3} M.A. Holm, D. Burns, P. Cusumano, A.I. Ferguson, M.D. Dawson: High-power diode-pumped AlGaAs surface-emitting laser, Appl Opt 38, p.5781-5784 (1999)

[6.1281] {Sect. 6.13.1.3} I.L. Krestnikov, W.V. Lundin, A.V. Sakharov, V.A. Semenov, A.S. Usikov, A.F. Tsatsulnikov, Z.I. Alferov, N.N. Ledentsov, A. Hoffmann, D. Bimberg: Room-temperature photopumped InGaN/GaN/ AlGaN vertical-cavity surface- emitting laser, Appl Phys Lett 75, p.1192-1194 (1999)

[6.1282] {Sect. 6.13.1.3} M.V. Maximov, Y.M. Shernyakov, A.F. Tsatsulnikov, A.V. Lunev, A.V. Sakharov, V.M. Ustinov, A.Y. Egorov, A.E. Zhukov, A.R. Kovsh, P.S. Kopev et al.: High-power continuous-wave operation of a In-GaAs/AlGaAs quantum dot laser, J Appl Phys 83, p.5561-5563 (1998)

[6.1283] {Sect. 6.13.1.3} T. Milster, W. Jiang, E. Walker, D. Burak, P. Claisse, P. Kelly, R. Binder: A single-mode high-power vertical cavity surface emitting laser, Appl Phys Lett 72, p.3425-3427 (1998)

[6.1284] {Sect. 6.13.1.3} W.T. Hu, H. Ye, C.D. Li, Z.H. Jiang, F.Z. Zhou: All-solid-state tunable DCM dye laser pumped by a diode- pumped Nd:YAG laser, Appl Opt 36, p.579-583 (1997)

[6.1285] {Sect. 6.13.1.3} D.V. Plant, B. Robertson, H.S. Hinton, M.H. Ayliffe, G.C. Boisset, W. Hsiao, D. Kabal, N.H. Kim, Y.S. Liu, M.R. Otazo, et al.: 4x4 vertical-cavity surface-emitting laser (VCSEL) and metal-semiconductor-metal (MSM) optical backplane demonstrator system, Appl Opt 35, p.6365-6368 (1996)

[6.1286] {Sect. 6.13.1.3} N.W. Carlson, G.A. Evans, D.P. Bour, S.K. Liew: Demonstration of a grating-surface-emitting diode laser with low-threshold current density, Appl. Phys. Lett. 56, p.16-18 (1990)

[6.1287] {Sect. 6.13.1.3} M.B. Willemsen, M.U.F. Khalid, M.P. vanExter, J.P. Woerdman: Polarization switching of a vertical-cavity semiconductor laser as a Kramers hopping problem, Phys Rev Lett 82, p.4815-4818 (1999)

[6.1288] {Sect. 6.13.2.0} W.F. Krupke, L.L. Chase: Ground state depleted solid state laser principles, characteristics and scaling, Opt. Quant. Electron. 22, p.1-22 (1990)

[6.1289] {Sect. 6.13.2.0} Kitaeva et al.: The properties of Crystals with Garnet structure, Phys. stat. sol. (a) 92, p.475-488 (1985)

[6.1290] {Sect. 6.13.2.0} L. DeShazer, M. Bass, U. Ranon, T.K. Guka, E.D. Reed, T.W. Strozyk, L. Rothrock: Laser operation of neodymium in YVO4 and gadolinium gallium garnet (GGG) and of holmium in YVO4. 8th International Electr. Conf, San Francisco, CA (1974)

[6.1291] {Sect. 6.13.2.0} A. A. Kaminskii: Laser Crystals (Springer, Berlin, Heidelberg, New York, 1990)

[6.1292] {Sect. 6.13.2.0} S.E. Stokowski: Glass lasers, in Handbook of Laser Science and Technology, ed. by M.J. Weber (CRC Press, Boca Raton, FL 1982) pp.215-264

[6.1293] {Sect. 6.13.2.0} Y.N. Xu, W.Y. Ching, B.K. Brickeen: Electronic structure and bonding in garnet crystals Gd3Sc2Ga3O12, Gd3Sc2Al3O12, and Gd3Ga3O12 compared to Y3Al3O12, Phys Rev B 61, p.1817-1824 (2000)

[6.1294] {Sect. 6.13.2.1} J.L. Blows, J.M. Dawes, J.A. Piper: A simple, thermally-stabilised, diode end-pumped, planar Nd : YAG laser, Opt Commun 162, p.247-250 (1999)

[6.1295] {Sect. 6.13.2.1} Y. Hirano, Y. Koyata, S. Yamamoto, K. Kasahara, T. Tajime: 208-W TEM00 operation of a diode-pumped Nd : YAG rod laser, Optics Letters 24, p.679-681 (1999)

[6.1296] {Sect. 6.13.2.1} H.J. Moon, J. Yi, J.M. Han, B.H. Cha, J. Lee: Efficient diffusive reflector-type diode side-pumped Nd : YAG rod laser with an optical slope efficiency of 55%, Appl Opt 38, p.1772-1776 (1999)

[6.1297] {Sect. 6.13.2.1} N. Moore, W.A. Clarkson, D.C. Hanna, S. Lehmann, J. Bosenberg: Efficient operation of a diode-bar-pumped Nd : YAG laser on the low- gain 1123-nm line, Appl Opt 38, p.5761-5764 (1999)

[6.1298] {Sect. 6.13.2.1} G.J. Spuhler, R. Paschotta, U. Keller, M. Moser, M.J.P. Dymott, D. Kopf, J. Meyer, K.J. Weingarten, J.D. Kmetec, J. Alexander et al.: Diode-pumped passively mode-locked Nd : YAG laser with 10-W average power in a diffraction-limited beam, Optics Letters 24, p.528-530 (1999)

[6.1299] {Sect. 6.13.2.1} A. Agnesi, S. DellAcqua, C. Pennacchio, G. Reali, P.G. Gobbi: High-repetition-rate Q-switched diode-pumped Nd:YAG laser at 1.444 mu m, Appl Opt 37, p.3984-3986 (1998)

[6.1300] {Sect. 6.13.2.1} M. Bode, S. Spiekermann, C. Fallnich, H. Welling, I. Freitag: Ultraviolet single-frequency pulses with 110 mW average power using frequency-converted passively Q-switched miniature Nd:YAG ring lasers, Appl Phys Lett 73, p.714-716 (1998)

[6.1301] {Sect. 6.13.2.1} K.M. Du, N.L. Wu, J.D. Xu, J. Giesekus, P. Loosen, R. Poprawe: Partially end-pumped Nd:YAG slab laser with a hybrid resonator, Optics Letters 23, p.370-372 (1998)

[6.1302] {Sect. 6.13.2.1} T. Kellner, F. Heine, G. Huber, S. Kuck: Passive Q switching of a diode-pumped 946-nm Nd:YAG laser with 1.6-W average output power, Appl Opt 37, p.7076-7079 (1998)

[6.1303] {Sect. 6.13.2.1} Y. Lutz, O. Musset, J.P. Boquillon, A. Hirth: Efficient pulsed 946-nm laser emission from Nd:YAG pumped by a titanium-doped sapphire laser, Appl Opt 37, p.3286-3289 (1998)

[6.1304] {Sect. 6.13.2.1} M. Tsunekane, N. Taguchi, H. Inaba: Efficient 946-nm laser operation of a composite Nd:YAG rod with undoped ends, Appl Opt 37, p.5713-5719 (1998)

[6.1305] {Sect. 6.13.2.1} T. Graf, J.E. Balmer, R. Weber, H.P. Weber: Multi-Nd:YAG-rod variable-configuration resonator (VCR) end pumped by multiple diode-laser bars, Opt Commun 135, p.171-178 (1997)

[6.1306] {Sect. 6.13.2.1} S. Konno, S. Fujikawa, K. Yasui: 80 W cw TEM00 1064 nm beam generation by use of a laser- diode-side-pumped Nd:YAG rod laser, Appl Phys Lett 70, p.2650-2651 (1997)

[6.1307] {Sect. 6.13.2.1} H.M. Kretschmann, F. Heine, V.G. Ostroumov, G. Huber: High-power diode-pumped continuous-wave Nd3+ lasers at wavelengths near 1.44 mu m, Optics Letters 22, p.466-468 (1997)

[6.1308] {Sect. 6.13.2.1} M. Ostermeyer, R. Menzel: 34 Watt flash lamp pumped single rod ND:YAG laser with 1.2 * DL beam quality via special resonator design, Appl. Phys. B 65, p.669-671 (1997)

[6.1309] {Sect. 6.13.2.1} W.A. Clarkson, D.C. Hanna: Efficient Nd:YAG laser end pumped by a 20-W diode-laser bar, Optics Letters 21, p.869-871 (1996)

[6.1310] {Sect. 6.13.2.1} D. Golla, M. Rode, S. Knoke, W. Schone, A. Tunnermann: 62-W cw TEM (00) Nd:YAG laser side pumped by fiber coupled diode lasers, Optics Letters 21, p.210-212 (1996)

[6.1311] {Sect. 6.13.2.1} K. Yasui: Efficient and stable operation of a high-brightness w 500- W Nd: YAG rod laser, Appl Opt 35, p.2566-2569 (1996)

[6.1312] {Sect. 6.13.2.1} J.L. Dallas, R.S. Afzal, M.A. Stephen: Demonstration and characterization of a multibillion-shot, 2.5-mJ, 4-ns, Q-switched Nd:YAG laser, Appl. Opt. 35, p.1427-1429 (1996)

[6.1313] {Sect. 6.13.2.1} D. Golla, S. Knoke, W. Schone, G. Ernst, M. Bode, A. Tunnermann, H. Welling: 300-W cw diode-laser side pumped Nd:YAG rod laser, Optics Letters 20, p.1148-1150 (1995)

[6.1314] {Sect. 6.13.2.1} R.S. Afzal, M.D. Selker: Simple high-efficiency TEM00 diode-laser-pumped Q-switched laser, Opt. Lett. 20, p.465-467 (1995)

[6.1315] {Sect. 6.13.2.1} T. Brand: Compact 170-W continous-wave diode-pumped Nd:YAG rod laser with a cusp-shaped reflector, Opt. Lett. 20, p.1776-1778 (1995)

[6.1316] {Sect. 6.13.2.1} D. Golla, S. Knoke, W. Schöne, A. Tünnermann, H. Schmidt: High Power Continuous-Wave Diode-Laser-Pumped Nd:YAG Laser, Appl. Phys. B 58, p.389-392 (1994)

[6.1317] {Sect. 6.13.2.1} S.C. Tidewell, J.F. Seamans, M.S. Bowers: Highly efficient 60-W TEM00 cw diode-end-pumped Nd:YAG laser, Opt. Lett. 18, p.116-118 (1993)

[6.1318] {Sect. 6.13.2.1} S.C. Tidwell, J.F. Seamans, M.S. Bowers: Highly efficient 60-W TEM00 cw diode-end-pumped Nd:YAG laser, Opt. Lett. 18, p.116-118 (1993)

[6.1319] {Sect. 6.13.2.1} S.C. Tidwell, J.F. Seamans, M.S. Bowers, A.K. Cousins: Scaling CW Diode-End-Pumped Nd:YAG Lasers to High Average Powers, IEEE J. QE-28, p.997-1009 (1992)

[6.1320] {Sect. 6.13.2.1} H.R. Verdún, T. Chuang: Efficient TEM00-mode operation of a Nd:YAG laser end pumped by a three-bar high-power diode-laser array, Opt. Lett. 17, p.1000-1002 (1992)

[6.1321] {Sect. 6.13.2.1} D.C. Shannon, R.W. Wallace: High-power Nd:YAG laser end pumped by a cw, 10 mm x 1 μm aperture, 10-W laser-diode bar, Opt. Lett. 16, p.318-320 (1991)

[6.1322] {Sect. 6.13.2.1} S.C. Tidwell, J.F. Seamans, C.E. Hamilton, C.H. Muller, D.D. Lowenthal: Efficient, 15 W Output Power, Diode End Pumped Nd YAG Laser, Optics Letters 16, p.584-586 (1991)

[6.1323] {Sect. 6.13.2.1} N.P. Barnes, D.J. Gettemy, L. Esterowitz, R.A. Allen: Comparision of Nd 1.06. and 1.33 μm Operation in Various Hosts, IEEE J. QE-23, p.1434-1451 (1987)

[6.1324] {Sect. 6.13.2.1} W.F. Krupke, M.D. Shinn, J.E. Marion, J.A. Caird, S.E. Stokowski: Spectroscopic, optical and thermomechanical properties of neodymium- and chromium-doped Gadolinium Scandium Gallium Garnet, J. Opt. Soc. Am B 3, p.102-113 (1986)

[6.1325] {Sect. 6.13.2.1} R.L. Schmitt, L.A. Rahn: Diode-laser-pumped Nd:YAG laser injection seeding system, Appl. Opt. 25, p.629-633 (1986)

[6.1326] {Sect. 6.13.2.1} H. Shen, Y. Zhou, R. Zeng, G. Yu, Q. Ye, C. Huang, X. Huang, H. Liao.: High power 1.3414 μm Nd:YAG cw laser, Optics and laser technology 18, p.193-197 (1986)

[6.1327] {Sect. 6.13.2.1} J. Marling: 1.05-1.44 μm Tunability and Performance of the CW Nd3+:YAG Laser, IEEE J. QE-14, p.56-62 (1978)

[6.1328] {Sect. 6.13.2.1} H.P. Jenssen, R.F. Begley, R. Webb, R.C. Morris.: Spectroscopic properties and laser performance of Nd3+ in lanthanum beryllate, J. Appl. Phys. 47, p.1496-1500 (1976)

[6.1329] {Sect. 6.13.2.1} S. Singh, R.G. Smith, L.G. Van Uitert:, Phys. Rev. B 10, p.2566-2572 (1974)

[6.1330] {Sect. 6.13.2.1} C.G. Bethea: Megawatt Power at 1.318 μ in Nd3+:YAG and Simultaneous Oscillation at Both 1.06 and 1.318 μ, IEEE J. QE-9, p.254 (1973)

[6.1331] {Sect. 6.13.2.1} H.G. Danielmeyer, M. Blätte, P. Balmer: Fluorescence Quenching in Nd:YAG, Appl. Phys. 1, p.269-274 (1973)

[6.1332] {Sect. 6.13.2.1} R.W. Wallace: Oscillation of the 1.833-μ Line in Nd3+:YAG, IEEE J. QE-7, p.203-204 (1971)

[6.1333] {Sect. 6.13.2.1} M.J. Weber, T.E. Varitimos: Optical Spectra and Intensities of Nd3+ in YAlO3, J. Appl. Phys. 42, p.4996-5005 (1971)

[6.1334] {Sect. 6.13.2.1} W. Koechner: Multihundred Watt Nd:YAG Continuous Laser, Rev. Sci. Instr. 41, p.1699-1706 (1970)

[6.1335] {Sect. 6.13.2.1} H.F. Mahlein, G. Schollmeier: Periodic Multiplate Resonant Reflector for a YAG:Nd3+ Laser at 1.318 μ, IEEE J. QE-6, p.529-530 (1970)

[6.1336] {Sect. 6.13.2.1} R.W. Wallace, S.E. Harris: Oscillation and Doubling of the 0.946-μ Line in Nd3+:YAG, Appl. Phys. Lett. 15, p.111-112 (1969)

[6.1337] {Sect. 6.13.2.1} T. Kushida, J.E. Geusic: Optical Refrigeration in Nd-Doped Yttrium Aluminum Garnet, Phys. Rev. Lett. 21, p.1172-1175 (1968)

[6.1338] {Sect. 6.13.2.1} P.H. Klein, W.J. Croft: Thermal Conductivity, Diffusivity, and Expansion of Y2O3, Y3Al5O12, and LaF3 in the Range 77-300K, J. Appl. Phys. 38, p.1603-1607 (1967)

[6.1339] {Sect. 6.13.2.1} J.K. Neeland, V. Evtuhov: Measurement of the Laser Transition Cross Section for Nd+3 in Yttrium Aluminum Garnet, Phys. Rev. 156, p.244-246 (1967)

[6.1340] {Sect. 6.13.2.1} M. Ostermeyer, R. Menzel: Single rod efficient Nd:YAG and Nd:YALO-lasers with average output powers of 46 and 47 W in diffraction limited beams with M2 ¡ 1.2 and 100 W with M2 ¡ 3.7, Opt. Comm. 160, p.251-254 (1999)

[6.1341] {Sect. 6.13.2.1} P. Poirier, F. Hanson: Discretely tunable multiwavelength diode-pumped Nd:YALO laser, Appl Opt 35, p.364-367 (1996)

[6.1342] {Sect. 6.13.2.1} S.L. Xue, Q.H. Lou: Passive mode-locking of a Nd:YAP laser at 1.3414 mu m by using a convex-antiresonant ring unstable resonator, Opt Commun 123, p.543-546 (1996)

[6.1343] {Sect. 6.13.2.1} G.A. Massey: Measurement of Device Parameters for Nd:YAIO3 Lasers, IEEE J. QE-8, p.669-674 (1972)

[6.1344] {Sect. 6.13.2.1} G.A. Massey, J.M. Yarborough: High average power operation and nonlinear optical generation with the Nd:YAIO3 laser, Appl. Phys. Lett. 18, p.576-579 (1971)

[6.1345] {Sect. 6.13.2.1} K. Tei, M. Kato, Y. Niwa, S. Harayama, Y. Maruyama, T. Matoba, T. Arisawa: Diode-pumped 250-W zigzag slab Na:YAG oscillator-amplifier system, Optics Letters 23, p.514-516 (1998)

[6.1346] {Sect. 6.13.2.1} E. Armandillo, C. Norrie, A. Cosentino, P. Laporta, P. Wazen, P. Maine: Diode-pumped high-efficiency high-brightness Q-switched ND: YAG slab laser, Optics Letters 22, p.1168-1170 (1997)

[6.1347] {Sect. 6.13.2.1} M. Seguchi, K. Kuba: 1.4-kW Nd:YAG slab laser with a diffusive closed coupled pump cavity, Optics Letters 20, p.300-302 (1995)

[6.1348] {Sect. 6.13.2.1} R.J. Shine, A.J. Alfrey, R.L. Byer: 40-W cw, TEM (00)-mode, diode laser pumped, Nd:YAG miniature-slab laser, Optics Letters 20, p.459-461 (1995)

[6.1349] {Sect. 6.13.2.1} R.J. Shine, Jr, A.J. Alfrey, R.L. Byer: 40-W cw, TEM00-mode, diode-laser-pumped, Nd:YAG miniature-slab laser, Opt. Lett. 20, p.459-461 (1995)

[6.1350] {Sect. 6.13.2.1} N. Hodgson, S. Dong, Q. Lü: Performance of a 2.3-kW Nd:YAG slab laser system, Opt. Lett. 18, p.1727-1729 (1993)

[6.1351] {Sect. 6.13.2.1} B.J. Comaskey, R. Beach, G. Albrecht, W.J. Benett, B.L. Freitas, C. Petty, D. VanLue, D. Mundinger, R.W. Solarz: High Average Power Diode Pumped Slab Laser, IEEE J. QE-28, p.992-996 (1992)

[6.1352] {Sect. 6.13.2.1} G.F. Albrecht, J.M. Eggleston, J.J. Ewing: Design and Characterization of a High Average Power Slab YAG Laser, IEEE J. QE-22, p.2099-2106 (1986)

[6.1353] {Sect. 6.13.2.1} M. Armstrong, X. Zhu, S. Gracewski, R.J.D. Miller: Development of a 25 W TEM00 diode-pumped Nd : YLF laser, Opt Commun 169, p.141-148 (1999)

[6.1354] {Sect. 6.13.2.1} W.A. Clarkson, P.J. Hardman, D.C. Hanna: High-power diode-bar end-pumped Nd:YLF laser at 1.053 mu m, Optics Letters 23, p.1363-1365 (1998)

[6.1355] {Sect. 6.13.2.1} P.J. Hardman, W.A. Clarkson, D.C. Hanna: High-power diode-bar-pumped intracavity-frequency-doubled Nd:YLF ring laser, Opt Commun 156, p.49-52 (1998)

[6.1356] {Sect. 6.13.2.1} I. Will, A. Liero, D. Mertins, W. Sandner: Feedback-stabilized Nd:YLF amplifier system for generation of picosecond pulse trains of an exactly rectangular envelope, IEEE J QE-34, p.2020-2028 (1998)

[6.1357] {Sect. 6.13.2.1} Th. Graf, J.E. Balmer: High-power Nd:YLF laser end pumped by a diode-laser bar, Opt. Lett. 18, p.1317-1319 (1993)

[6.1358] {Sect. 6.13.2.1} T.M. Baer, D.F. Head, P. Gooding, G.J. Kintz, S. Hutchison: Performance of Diode-Pumped Nd:YAG and Nd:YLF Lasers in a Tightly Folded Resonator Configuration, IEEE J. QE-28, p.1131-1138 (1992)

[6.1359] {Sect. 6.13.2.1} H. Zbinden, J.E. Balmer: Q-switched Nd:YLF laser end pumped by a diode-laser bar, Opt. Lett. 15, p.1014-1016 (1990)

[6.1360] {Sect. 6.13.2.1} M.G. Knights, M.D. Thomas, E.P. Chicklis, G.A. Rines, W. Seka: Very High Gain Nd:YLF Amplifiers, IEEE J. QE-24, p.712-715 (1988)

[6.1361] {Sect. 6.13.2.1} T.M. Pollak, W.F. Wing, R.J. Grasso, E.P. Chicklis, H.P. Jenssen: CW Laser Operation of Nd:YLF, IEEE J. QE-18, p.159-163 (1982)

[6.1362] {Sect. 6.13.2.1} A. Rapaport, O. Moteau, M. Bass, L.A. Boatner, C. Deka: Optical spectroscopy and lasing properties of neodymium-doped lutetium orthophosphate, J Opt Soc Am B Opt Physics 16, p.911-916 (1999)

[6.1363] {Sect. 6.13.2.1} F.C. Cruz, B.C. Young, J.C. Bergquist: Diode-pumped Nd:FAP laser at 1.126 mu m: a possible local oscillator for a Hg+ optical frequency standard, Appl Opt 37, p.7801-7804 (1998)

[6.1364] {Sect. 6.13.2.1} P. Dekker, Y.J. Huo, J.M. Dawes, J.A. Piper, P. Wang, B.S. Lu: Continuous wave and Q-switched diode-pumped neodymium, lutetium: yttrium aluminium borate lasers, Opt Commun 151, p.406-412 (1998)

[6.1365] {Sect. 6.13.2.1} I. Moshe, S. Jackel, R. Lallouz: Working beyond the static limits of laser stability by use of adaptive and polarization-conjugation optics, Appl Opt 37, p.6415-6420 (1998)

[6.1366] {Sect. 6.13.2.1} X.Y. Zhang, S.Z. Zhao, Q.P. Wang, L.K. Sun, S.J. Zhang, G.T. Yao, Z.Y. Zhang: Laser diode pumped Cr4+:YAG passively Q-switched Nd3+:S-FAP laser, Opt Commun 155, p.55-60 (1998)

[6.1367] {Sect. 6.13.2.1} Y.M. Chen, L. Major, V. Kushawaha: Efficient laser operation of diode-pumped Nd:KGd (WO4)2 crystal at 1.067 mu m, Appl Opt 35, p.3203-3206 (1996)

[6.1368] {Sect. 6.13.2.1} N. Lei, B. Xu, Z.H. Jiang: Ti:sapphire laser pumped Nd:tellurite glass laser, Opt Commun 127, p.263-265 (1996)

[6.1369] {Sect. 6.13.2.1} Q.P. Wang, S.Z. Zhao, X.Y. Zhang, L.K. Sun, S.J. Zhang: Laser demonstration of a diode-laser-pumped Nd:Sr-5 (PO4)3F crystal, Opt Commun 128, p.73-75 (1996)

[6.1370] {Sect. 6.13.2.1} C.J. Flood, D.R. Walker, H.M. van Driel: CW diode pumping and FM mode locking of a Nd:KGW laser, Appl. Phys. B 60, p.309-312 (1995)

[6.1371] {Sect. 6.13.2.1} E. Reed: A flashlamp-Pumped, Q-Switched Cr:Nd:GSGG Laser, IEEE J. QE-21, p.1625-1629 (1985)

[6.1372] {Sect. 6.13.2.1} J.E. Murray: Pulsed Gain and Thermal Lensing of Nd:LiF4, IEEE J. QE-19, p.488-491 (1983)

[6.1373] {Sect. 6.13.2.1} A. Beimowski, G. Huber, D. Pruss, V.V. Laptev, I.A. Shcherbakov, E.V. Zharikov: Efficient Cr3+ Sensitized Nd3+:GdScGa-Garnet Laser at 1.06 µm, Appl. Phys. B 28, p.234-235 (1982)

[6.1374] {Sect. 6.13.2.1} E.J. Sharp, D.J. Horowitz, J.E. Miller: High-efficiency Nd3+:LiYF4 laser, J. Appl. Phys. 44, p.5399-5401 (1973)

[6.1375] {Sect. 6.13.2.2} C. Becher, K.T. Boller: Low-intensity-noise operation of Nd : YVO4 microchip lasers by pump- noise suppression, J Opt Soc Am B Opt Physics 16, p.286-295 (1999)

[6.1376] {Sect. 6.13.2.2} Y.F. Chen, L.J. Lee, T.M. Huang, C.L. Wan: Study of high-power diode-end-pumped Nd : YVO4 laser at 1.34 mu m: influence of Auger upconversion, Opt Commun 163, p.198-202 (1999)

[6.1377] {Sect. 6.13.2.2} Y.F. Chen: Design criteria for concentration optimization in scaling diode end- pumped lasers to sigh powers: Influence of thermal fracture, IEEE J QE-35, p.234-239 (1999)

[6.1378] {Sect. 6.13.2.2} Y.F. Chen: High-power diode-pumped Q-switched intracavity frequency-doubled Nd : YVO4 laser with a sandwich-type resonator, Optics Letters 24, p.1032-1034 (1999)

[6.1379] {Sect. 6.13.2.2} T. Graf, A.I. Ferguson, E. Bente, D. Burns, M.D. Dawson: Multi-Watt Nd : YVO4 laser, mode locked by a semiconductor saturable absorber mirror and side-pumped by a diode-laser bar, Opt Commun 159, p.84-87 (1999)

[6.1380] {Sect. 6.13.2.2} A. Sennaroglu: Efficient continuous-wave operation of a diode-pumped Nd : YVO4 laser at 1342 nm, Opt Commun 164, p.191-197 (1999)

[6.1381] {Sect. 6.13.2.2} C. Becher, K.J. Boller: Intensity noise properties of Nd:YVO4 microchip lasers pumped with an amplitude squeezed diode laser, Opt Commun 147, p.366-374 (1998)

[6.1382] {Sect. 6.13.2.2} A. Agnesi, C. Pennacchio, G.C. Reali, V. Kubecek: High-power diode-pumped picosecond Nd3+:YVO4 laser, Optics Letters 22, p.1645-1647 (1997)

[6.1383] {Sect. 6.13.2.2} R.S. Conroy, A.J. Kemp, G.J. Friel, B.D. Sinclair: Microchip Nd:vanadate lasers at 1342 and 671 nm, Optics Letters 22, p.1781-1783 (1997)

[6.1384] {Sect. 6.13.2.2} K.M. Du, Y. Liao, P. Loosen: Nd:YAG slab laser end-pumped by laser-diode stacks and its beam shaping, Opt Commun 140, p.53-56 (1997)

[6.1385] {Sect. 6.13.2.2} E. Armandillo, C. Norrie, A. Cosentino, P. Laporta, P. Wazen, P. Maine: Diode-pumped high-efficiency high-brightness Q-switched ND:YAG slab laser, Opt. Lett. 22, p.1168-1170 (1997)

[6.1386] {Sect. 6.13.2.2} D.C. Brown, R. Nelson, L. Billings: Efficient cw end-pumped, end-cooled Nd:YVO4 diode-pumped, Appl. Opt. 36, p.8611-8613 (1997)

[6.1387] {Sect. 6.13.2.2} D.G. Matthews, J.R. Boon, R.S. Conroy, B.D. Sinclair: A comparative study of diode pumped microchip laser materials: Nd-doped YVO4, YOS, SFAP and SVAP, J. Mod. Opt. 43, p.1079-1087 (1996)

[6.1388] {Sect. 6.13.2.2} G. Feugnet, C. Bussac, C. Larat, M. Schwarz, J.P. Pocholle: High Efficiency TEM (00) NdYVO4 Laser Longitudinally Pumped by a High Power Array, Optics Letters 20, p.157-159 (1995)

[6.1389] {Sect. 6.13.2.2} J.E. Bernard, A.J. Alcock: High-repetition-rate diode-pumped Nd:YVO4 slab laser, Opt. Lett. 19, p.1861 (1994)

[6.1390] {Sect. 6.13.2.2} J.E. Bernard, A.J. Alcock: High-efficiency diode-pumped Nd:YVO4 slab laser, Opt. Lett. 18, p.968-970 (1993)

[6.1391] {Sect. 6.13.2.2} R.A. Fields, M. Birnbaum, C.L. Fincher: Highly efficient Nd:YVO4 diode-laser end-pumped laser, Appl. Phys. Lett. 51, p.1885-1886 (1987)

[6.1392] {Sect. 6.13.2.2} A.W. Tucker, M. Birnbaum, C.L. Fincher, J.W. Erler: Stimulated-emission cross section at 1064 and 1342 nm in Nd:YVO4, J. Appl. Phys. 48, p.4907-4911 (1977)

[6.1393] {Sect. 6.13.2.3} J.A. derAu, F.H. Loesel, F. MorierGenoud, M. Moser, U. Keller: Femtosecond diode-pumped Nd:glass laser with more than 1 W of average output power, Optics Letters 23, p.271-273 (1998)

[6.1394] {Sect. 6.13.2.3} C. Horvath, A. Braun, H. Liu, T. Juhasz, G. Mourou: Compact directly diode-pumped femtosecond Nd:glass chirped-pulse-amplification laser system, Optics Letters 22, p.1790-1792 (1997)

[6.1395] {Sect. 6.13.2.3} S. Basu, T.J. Kane, R.L. Byer: A Proposed 1 kW Average Power Moving Slab Nd:Glass Laser, IEEE J. QE-22, p.2052-2057 (1986)

[6.1396] {Sect. 6.13.2.3} J.M. Eggleston, G.F. Albrecht, R.A. Petr, J.F. Zumdieck: A High Average Power Dual Slab Nd:Glass Zigzag Laser System, IEEE J. QE-22, p.2092-2098 (1986)

[6.1397] {Sect. 6.13.2.3} J.M. Eggleston, R.L. Byer, T.J. Kane, J. Unternahrer: Slab Geometry ND Glass Laser Performance Studies, Optics Letters 7, p.405-407 (1982)

[6.1398] {Sect. 6.13.2.3} T.J. Kane, R.L. Byer: Proposed Kilowatt Average Power ND Glass Laser, J Opt Soc Am 72, p.1755 (1982)

[6.1399] {Sect. 6.13.2.3} S.M. Yarema, D. Milam: Gain Saturation in Phosphate Laser Glasses, IEEE J. QE-18, p.1941-1946 (1982)

[6.1400] {Sect. 6.13.2.3} W.W. Simmons, J.T. Hunt, W.E. Warren: Light Propagation Through Large Laser Systems, IEEE J. QE-17, p.1727-1744 (1981)

[6.1401] {Sect. 6.13.2.3} M.J. Weber, D. Milam, W.L. Smith: Nonlinear Refractive Index of Glasses and Crystals, Opt. Eng. 17, p.463-469 (1978)

[6.1402] {Sect. 6.13.2.3} D. Duston, K. Rose: Measurement of Terminal Level Lifetime in Nd-Doped Laser Glass, IEEE J. QE-6, p.3 (1970)

[6.1403] {Sect. 6.13.2.3} M. Naftaly, A. Jha: Nd3+-doped fluoroaluminate glasses for a 1.3 mu m amplifier, J Appl Phys 87, p.2098-2104 (2000)

[6.1404] {Sect. 6.13.2.4} J. Aus der Au, G.J. Spühler, T. Südmeyer, R. Paschotta, R. Hövel, M. Moser, S. Erhard, M. Karszewski, A. Giesen, U. Keller: 16.2-

W average power from a diode-pumped femtosecond Yb:YAG thin disk laser, Opt. Lett. 25, p.859-861 (2000)

[6.1405] {Sect. 6.13.2.4} E.C. Honea, R.J. Beach, S.C. Mitchell, J.A. Skidmore, M.A. Emanuel, S.B. Sutton, S.A. Payne, P.V. Avizonis, R.S. M. Monroe, D.G. Harris: High-power dual-rod Yb:YAG laser, Opt. Lett. 25, p.805-807 (2000)

[6.1406] {Sect. 6.13.2.4} J. AusderAu, S.F. Schaer, R. Paschotta, C. Honninger, U. Keller, M. Moser: High-power diode-pumped passively mode-locked Yb : YAG lasers, Optics Letters 24, p.1281-1283 (1999)

[6.1407] {Sect. 6.13.2.4} E.C. Honea, R.J. Beach, S.C. Mitchell, P.V. Avizonis: 183-W, M-2 = 2.4 Yb : YAG Q-switched laser, Optics Letters 24, p.154-156 (1999)

[6.1408] {Sect. 6.13.2.4} C. Bibeau, R.J. Beach, S.C. Mitchell, M.A. Emanuel, J. Skidmore, C.A. Ebbers, S.B. Sutton, K.S. Jancaitis: High-average-power 1-mu m performance and frequency conversion of a diode-end-pumped Yb:YAG laser, IEEE J QE-34, p.2010-2019 (1998)

[6.1409] {Sect. 6.13.2.4} H.W. Bruesselbach, D.S. Sumida, R.A. Reeder, R.W.Byren: Low-Heat High-Power Scaling Using InGaAs-Diode-Pumped Yb:YAG Lasers, IEEE J. QE-3p.105-116 (1997)

[6.1410] {Sect. 6.13.2.4} H. Bruesselbach, D.S. Sumida: 69-W-average-power Yb:YAG laser, Optics Letters 21, p.480-482 (1996)

[6.1411] {Sect. 6.13.2.4} U.Brauch, A. Giesen, M. Karszewski, Chr. Stewen, A. Voss: Multiwatt diode-pumped Yb:YAG thinh disk laser continuously tunable between 1018 and 1053 nm, Opt. Lett. 20, p.713-715 (1995)

[6.1412] {Sect. 6.13.2.4} D.S. Sumida, T.Y. Fan: Room-temperature 50-mJ/pulse side-diode-pumped Yb:YAG laser, Opt. Lett. 20, p.2384-2386 (1995)

[6.1413] {Sect. 6.13.2.4} E. Snitzer, R. Woodcock: Yb3+-Er3+ Glass Laser, Appl. Phys. Lett. 6, p.45-46 (1965)

[6.1414] {Sect. 6.13.2.4} F. Druon, F. Auge, F. Balembois, P. Georges, A. Brun, A. Aron, F. Mougel, G. Aka, D. Vivien: Efficient, tunable, zero-line diode-pumped, continuous-wave Yb3+: Ca (4)LnO (BO3) (3) (Ln = Gd,Y) lasers at room temperature and application to miniature lasers, J Opt Soc Am B Opt Physics 17, p.18-22 (2000)

[6.1415] {Sect. 6.13.2.4} X. Feng, C.H. Qi, F.Y. Lin, H.F. Hu: Spectroscopic properties and laser performance assessment of Yb3+ in borophosphate glasses, J Amer Ceram Soc 82, p.3471-3475 (1999)

[6.1416] {Sect. 6.13.2.4} A.A. Lagatsky, N.V. Kuleshov, V.P. Mikhailov: Diode-pumped CW lasing of Yb : KYW and Yb : KGW, Opt Commun 165, p.71-75 (1999)

[6.1417] {Sect. 6.13.2.4} E. Montoya, J. Capmany, L.E. Bausa, T. Kellner, A. Diening, G. Huber: Infrared and self-frequency doubled laser action in Yb3+-doped LiNbO3 : MgO, Appl Phys Lett 74, p.3113-3115 (1999)

[6.1418] {Sect. 6.13.2.4} L.A.W. Gloster, P. Cormont, A.M. Cox, T.A. King, B.H.T. Chai: Diode-pumped Q-switched Yb:S-FAP laser, Opt Commun 146, p.177-180 (1998)

[6.1419] {Sect. 6.13.2.4} D.A. Hammons, J.M. Eichenholz, Q. Ye, B.H.T. Chai, L. Shah, R.E. Peale, M. Richardson, H. Qiu: Laser action in (Yb3+:YCOB (Yb3+:YCa (4)OiBO (3)) (3)), Opt Commun 156, p.327-330 (1998)

[6.1420] {Sect. 6.13.2.4} N.V. Kuleshov, A.A. Lagatsky, A.V. Podlipensky, V.P. Mikhailov, G. Huber: Pulsed laser operation of Yb-doped KY (WO4) (2) and KGd (WO4) (2), Optics Letters 22, p.1317-1319 (1997)

[6.1421] {Sect. 6.13.2.4} V. Petrov, U. Griebner, D. Ehrt, W. Seeber: Femtosecond self mode locking of Yb:fluoride phosphate glass laser, Optics Letters 22, p.408-410 (1997)

[6.1422] {Sect. 6.13.2.4} C.D. Marshall, L.K. Smith, R.J. Beach, M.A. Emanuel, K.I. Schaffers, J. Skidmore, S.A. Payne, B.H.T. Chai: Diode-pumped ytterbium-doped Sr-5 (PO4) (3)F laser performance, IEEE J QE-32, p.650-656 (1996)

[6.1423] {Sect. 6.13.2.4} H.B. Yin, P.Z. Deng, F.X. Gan: Defects in YAG:Yb crystals, J Appl Phys 83, p.3825-3828 (1998)

[6.1424] {Sect. 6.13.2.5} T. Beddard, W. Sibbett, D.T. Reid, J. GardunoMejia, N. Jamasbi, M. Mohebi: High-average-power, 1-MW peak-power self-mode-locked Ti : sapphire oscillator, Optics Letters 24, p.163-165 (1999)

[6.1425] {Sect. 6.13.2.5} S.H. Cho, B.E. Bouma, E.P. Ippen, J.G. Fujimoto: Low-repetition-rate high-peak-power Kerr-lens mode-locked Ti : Al2O3 laser with a multiple-pass cavity, Optics Letters 24, p.417-419 (1999)

[6.1426] {Sect. 6.13.2.5} J.R. Demers, F.C. DeLucia: Modulating and scanning the mode-lock frequency of an 800-MHz femtosecond Ti : sapphire laser, Optics Letters 24, p.250-252 (1999)

[6.1427] {Sect. 6.13.2.5} J.H. Geng, S. Wada, Y. Urata, H. Tashiro: Widely tunable, narrow-linewidth, subnanosecond pulse generation in an electronically tuned Ti : sapphire laser, Optics Letters 24, p.676-678 (1999)

[6.1428] {Sect. 6.13.2.5} Z.L. Liu, S. Izumida, S. Ono, H. Ohtake, N. Sarukura: High-repetition-rate, high-average-power, mode-locked Ti : sapphire laser with an intracavity continuous-wave amplification scheme, Appl Phys Lett 74, p.3622-3623 (1999)

[6.1429] {Sect. 6.13.2.5} M.D. Perry, D. Pennington, B.C. Stuart, G. Tietbohl, J.A. Britten, C. Brown, S. Herman, B. Golick, M. Kartz, J. Miller et al.: Petawatt laser pulses, Optics Letters 24, p.160-162 (1999)

[6.1430] {Sect. 6.13.2.5} F. Siebe, K. Siebert, R. Leonhardt, H.G. Roskos: A fully tunable dual-color CWTi : Al2O3 laser, IEEE J QE-35, p.1731-1736 (1999)

[6.1431] {Sect. 6.13.2.5} W.J. Wadsworth, D.W. Coutts, C.E. Webb: Kilohertz pulse repetition frequency slab Ti : sapphire lasers with high average power (10 W), Appl Opt 38, p.6904-6911 (1999)

[6.1432] {Sect. 6.13.2.5} H. Wang, S. Backus, Z. Chang, R. Wagner, K. Kim, X. Wang, D. Umstadter, T. Lei, M. Murnane, H. Kapteyn: Generation of 10-W average-power, 40-TW peak-power, 24-fs pulses from a Ti : sapphire amplifier system, J Opt Soc Am B Opt Physics 16, p.1790-1794 (1999)

[6.1433] {Sect. 6.13.2.5} Y. Nabekawa, Y. Kuramoto, T. Togashi, T. Sekikawa, S. Watanabe: Generation of 0.66-TW pulses at 1 kHz by a Ti:sapphire laser, Optics Letters 23, p.1384-1386 (1998)

[6.1434] {Sect. 6.13.2.5} L. Xu, G. Tempea, C. Spielmann, F. Krausz, A. Stingl, K. Ferencz, S. Takano: Continuous-wave mode-locked Ti:sapphire laser focusable to 5 x 10 (13) W/cm (2), Optics Letters 23, p.789-791 (1998)

[6.1435] {Sect. 6.13.2.5} K. Yamakawa, M. Aoyama, S. Matsuoka, T. Kase, Y. Akahane, H. Takuma: 100-TW sub-20-fs Ti:sapphire laser system operating at a 10-Hz repetition rate, Optics Letters 23, p.1468-1470 (1998)

[6.1436] {Sect. 6.13.2.5} K. Yamakawa, M. Aoyama, S. Matsuoka, H. Takuma, C.P.J. Barty, D. Fittinghoff: Generation of 16-fs, 10-TW pulses at a 10-Hz repetition rate with efficient Ti:sapphire amplifiers, Optics Letters 23, p.525-527 (1998)

[6.1437] {Sect. 6.13.2.5} M. Aoyama, K. Yamakawa: Noise characterization of an all-solid-state mirror- dispersion-controlled 10-fs Ti:sapphire laser, Opt Commun 140, p.255-258 (1997)

[6.1438] {Sect. 6.13.2.5} A. Hoffstadt: Design and performance of a high-average-power flashlamp- pumped Ti:Sapphire laser and amplifier, IEEE J QE-33, p.1850-1863 (1997)

[6.1439] {Sect. 6.13.2.5} B.C. Stuart, M.D. Perry, J. Miller, G. Tietbohl, S. Herman, J.A. Britten, C. Brown, D. Pennington, V. Yanovsky, K. Wharton: 125-TW Ti:sapphire/Nd:glass laser system, Optics Letters 22, p.242-244 (1997)

[6.1440] {Sect. 6.13.2.5} A. Hoffstädt: Design and Performance of a High-Average-Power Flashlamp-Pumped Ti:Sapphire Laser and Amplifier, IEEE J. QE-33, p.1850-1863 (1997)

[6.1441] {Sect. 6.13.2.5} Y. Nabekawa, K. Sajiki, D. Yoshitomi, K. Kondo, S. Watanabe: High-repetition-rate high-average-power 300-fs KrF/Ti: sapphire hybrid laser, Optics Letters 21, p.647-649 (1996)

[6.1442] {Sect. 6.13.2.5} A. Sullivan, J. Bonlie, D.F. Price, W.E. White: 1.1-J, 120-fs laser system based on Nd:glass-pumped Ti: sapphire, Optics Letters 21, p.603-605 (1996)

[6.1443] {Sect. 6.13.2.5} D.S. Knowles, D.J.W. Brown: Compact 24-kHz copper laser pumped Ti:sapphire laser, Optics Letters 20, p.569-571 (1995)

[6.1444] {Sect. 6.13.2.5} G. Guochang, L. Ziyao: A multi-joule Ti:sapphire laser with coaxial flashlamp excitation, Opt. Comm. 120, p.63-64 (1995)

[6.1445] {Sect. 6.13.2.5} J. Harrison, A. Finch, D.M. Rines, G.A. Rines, P.F. Moulton: Low-threshold, cw, all-solid-state Ti:Al2O3 laser, Opt. Lett. 16, p.581-583 (1991)

[6.1446] {Sect. 6.13.2.5} T.R. Steele, D.C. Gerstenberger, A. Drobshoff, R.W. Wallace: Broadly tunable high-power operation of an all-solid-state titanium-doped sapphire laser system, Opt. Lett. 16, p.399-401 (1991)

[6.1447] {Sect. 6.13.2.5} G.T. Maker, A.I. Ferguson: Ti:sapphire laser pumped by a frequency-doubled diode-pumped Nd:YLF laser, Opt. Lett. 15, p.375-377 (1990)

[6.1448] {Sect. 6.13.2.5} J.M. Eggleston, L.G. DeShazer, K.W. Kangas: Characteristics and Kinetics of Laser-Pumped Ti:Sapphire Oscillators, IEEE J. QE-24, p.1009-1015 (1988)

[6.1449] {Sect. 6.13.2.5} G.F. Albrecht, J.M. Egglestone, J.J. Ewing: Measurements of Ti3+:Al2O3 as material, Opt. Comm. 52, p.401-404 (1985)

[6.1450] {Sect. 6.13.2.6} D. ParsonsKaravassilis, R. Jones, M.J. Cole, P.M.W. French, J.R. Taylor: Diode-pumped all-solid-state ultrafast Cr : LiSGAF laser oscillator- amplifier system applied to laser ablation, Opt Commun 175, p.389-396 (2000)

[6.1451] {Sect. 6.13.2.6} A. Robertson, U. Ernst, R. Knappe, R. Wallenstein, V. Scheuer, T. Tschudi, D. Burns, M.D. Dawson, A.I. Ferguson: Prismless diode-pumped mode-locked femtosecond Cr : LiSAF laser, Opt Commun 163, p.38-43 (1999)

[6.1452] {Sect. 6.13.2.6} H. Tsuchida: Pulse timing stabilization of a mode-locked Cr : LiSAF laser, Optics Letters 24, p.1641-1643 (1999)

[6.1453] {Sect. 6.13.2.6} S. Uemura, K. Torizuka: Generation of 12-fs pulses from a diode-pumped Kerr-lens mode-locked Cr : LiSAF laser, Optics Letters 24, p.780-782 (1999)

[6.1454] {Sect. 6.13.2.6} N. Zhavoronkov, V. Petrov, F. Noack: Powerful and tunable operation of a 1-2-kHz repetition-rate gain-switched Cr : forsterite laser and its frequency doubling, Appl Opt 38, p.3285-3293 (1999)

[6.1455] {Sect. 6.13.2.6} K.M. Gabel, P. Russbuldt, R. Lebert, A. Valster: Diode pumped Cr3+: LiCAF fs-laser, Opt Commun 157, p.327-334 (1998)

[6.1456] {Sect. 6.13.2.6} A. Mandl, A. Zavriyev, D.E. Klimek: Flashlamp-pumped Cr:LiSAF thin-slab zigzag laser, IEEE J QE-34, p.1992-1995 (1998)

[6.1457] {Sect. 6.13.2.6} A. Robertson, R. Knappe, R. Wallenstein: Diode-pumped broadly tunable (809-910 nm) femtosecond Cr:LiSAF laser, Opt Commun 147, p.294-298 (1998)

[6.1458] {Sect. 6.13.2.6} M. Tsunekane, M. Ihara, N. Taguchi, H. Inaba: Analysis and design of widely tunable diode-pumped Cr:LiSAF lasers with external grating feedback, IEEE J QE-34, p.1288-1296 (1998)

[6.1459] {Sect. 6.13.2.6} N.J. Vasa, H. Parhat, T. Okada, M. Maeda, O. Uchino: Performance of an optical fiber butt-coupled Cr3+:LiSrAlF6 laser, Opt Commun 147, p.196-202 (1998)

[6.1460] {Sect. 6.13.2.6} J.M. Hopkins, G.J. Valentine, W. Sibbett, J. Aus der Au, F. Morier-Genoud, U. Keller, A. Valster: Efficient, low-noise, SESAM-based femtosecond Cr3+:LiSrAlF6 laser, Opt. Comm. 154, p.54-58 (1998)

[6.1461] {Sect. 6.13.2.6} F. Balembois, F. Druon, F. Falcoz, P. Georges, A. Brun: Performances of Cr:LiSrAlF6 and Cr:LiSrGaF6 for continuous- wave diode-pumped Q-switched operation, Optics Letters 22, p.387-389 (1997)

[6.1462] {Sect. 6.13.2.6} D. Kopf, U. Keller, M.A. Emanuel, R.J. Beach, J.A. Skidmore: 1.1-W cw Cr:LiSAF laser pumped by a 1-cm diode array, Optics Letters 22, p.99-101 (1997)

[6.1463] {Sect. 6.13.2.6} S. Uemura, K. Miyazaki: Operation of a femtosecond Cr:LiSAF solitary laser near zero group-delay dispersion, Opt Commun 133, p.201-204 (1997)

[6.1464] {Sect. 6.13.2.6} S. Uemura, K. Miyazaki: Femtosecond Cr:LiSAF laser pumped by a single diode laser, Opt Commun 138, p.330-332 (1997)

[6.1465] {Sect. 6.13.2.6} S. Uemura, K. Miyazaki: Femtosecond Cr:LiSAF laser pumped by a single diode laser, Opt. Comm. 138, p.330-332 (1997)

[6.1466] {Sect. 6.13.2.6} D. Burns, M.P. Critten, W. Sibbett: Low-threshold diode-pumped femtosecond Cr3+:LiSrAlF6 laser, Optics Letters 21, p.477-479 (1996)

[6.1467] {Sect. 6.13.2.6} P.A. Beaud, M. Richardson, E.J. Miesak: Multi-Terawatt Femtosecond Cr:LiSAF Laser, IEEE J. QE-31, p.317-325 (1995)

[6.1468] {Sect. 6.13.2.6} P.M.W. French, R. Mellish, J.R. Taylor, P.J. Delfyett, L.T. Florez: All-solid-state diode-pumped modelocked Cr:LiSAF laser, Electron. Lett. 29, p.1262-1263 (1993)

[6.1469] {Sect. 6.13.2.6} S.A. Payne, W.F. Krupke, L.K. Smith, W.L. Kway, L.D. DeLoach, J.B. Tassano: 752 nm Wing-Pumped Cr:LiSAF Laser, IEEE J. QE-28, p.1188-1196 (1992)

[6.1470] {Sect. 6.13.2.6} R. Scheps, J.F. Myers, H. Serreze, A. Rosenberg, R.C. Morris, M. Long: Diode-pumped Cr:LiSrAlF6 laser, Opt. Lett. 16, p.820-822 (1991)

[6.1471] {Sect. 6.13.2.6} M. Stalder, B.H.T. Chai, M. Bass: Flashlamp pumped Cr:LiSrAlF6 laser, Appl. Phys. Lett. 58, p.216-218 (1991)

[6.1472] {Sect. 6.13.2.6} S.A. Payne, L.L. Chase, L.K. Smith, W.L. Kway, H.W. Newkirk: Laser performance of LiSrAlF6:Cr3+, J. Appl. Phys. 66, p.1051-1056 (1989)

[6.1473] {Sect. 6.13.2.6} S.A. Payne, L.L. Chase, H.W. Newkirk, L.K. Smith, W.F. Krupke: LiCaAlF6:Cr3+: A Promising New Solid-State Laser Material, IEEE J. QE-24, p.2243-2252 (1988)

[6.1474] {Sect. 6.13.2.6} I.T. Sorokina, S. Naumov, E. Sorokin, E. Wintner, A.V. Shestakov: Directly diode-pumped tunable continuous-wave room-temperature Cr4+: YAG laser, Optics Letters 24, p.1578-1580 (1999)

[6.1475] {Sect. 6.13.2.6} Z. Zhang, T. Nakagawa, K. Torizuka, T. Sugaya, K. Kobayashi: Self-starting mode-locked Cr4+: YAG laser with a low-loss broadband semiconductor saturable-absorber mirror, Optics Letters 24, p.1768-1770 (1999)

[6.1476] {Sect. 6.13.2.6} Y. Ishida, K. Naganuma: Compact diode-pumped all-solid-state femtosecond Cr4+:YAG laser, Optics Letters 21, p.51-53 (1996)

[6.1477] {Sect. 6.13.2.6} A. Agnesi, E. Piccinini, G. Reali: Threshold optimization of all-solid-state Cr : forsterite lasers, J Opt Soc Am B Opt Physics 17, p.198-201 (2000)

[6.1478] {Sect. 6.13.2.6} A.J.S. McGonigle, D.W. Coutts, C.E. Webb: 530-mW 7-kHz cerium LiCAF laser pumped by the sum-frequency-mixed output of a copper-vapor laser, Optics Letters 24, p.232-234 (1999)

[6.1479] {Sect. 6.13.2.6} B. Chassagne, G. Jonusauskas, J. Oberle, C. Rulliere: Multipulse operation regime in a self-mode-locked Cr4+: forsterite femtosecond laser, Opt Commun 150, p.355-362 (1998)

[6.1480] {Sect. 6.13.2.6} J.M. Evans, V. Petricevic, R.R. Alfano, Q. Fu: Kilohertz Cr:forsterite regenerative amplifier, Optics Letters 23, p.1692-1694 (1998)

[6.1481] {Sect. 6.13.2.6} G. Jonusauskas, J. Oberle, C. Rulliere: 54-fs, 1-GW, 1-kHz pulse amplification in Cr : forsterite, Optics Letters 23, p.1918-1920 (1998)

[6.1482] {Sect. 6.13.2.6} J.M. Evans, V. Petricevic, A.B. Bykov, A. Delgado, R.R. Alfano: Direct diode-pumped continuous-wave near-infrared tunable laser operation of Cr4+:forsterite and Cr4+:Ca2GeO4, Optics Letters 22, p.1171-1173 (1997)

[6.1483] {Sect. 6.13.2.6} L.J. Qian, X. Liu, F. Wise: Cr:forsterite laser pumped by broad-area laser diodes, Optics Letters 22, p.1707-1709 (1997)

[6.1484] {Sect. 6.13.2.6} N. Zhavoronkov, A. Avtukh, V. Mikhailov: Chromium-doped forsterite laser with 1.1 W of continuous- wave output power at room temperature, Appl Opt 36, p.8601-8605 (1997)

[6.1485] {Sect. 6.13.2.6} A. Agnesi, S. DellAcqua, P.G. Gobbi: All-solid-state gain-switched Cr4+: Forsterite laser, Opt Commun 127, p.273-276 (1996)

[6.1486] {Sect. 6.13.2.6} B. Golubovic, B.E. Bouma, I.P. Bilinsky, J.G. Fujimoto, V.P. Mikhailov: Thin crystal, room-temperature Cr4+:forsterite laser using near-infrared pumping, Optics Letters 21, p.1993-1995 (1996)

[6.1487] {Sect. 6.13.2.6} I.T. McKinnie, L.A.W. Gloster, Z.X. Jiang, T.A. King: Chromium-doped forsterite: The influence of crystal characteristics on laser performance, Appl Opt 35, p.4159-4165 (1996)

[6.1488] {Sect. 6.13.2.6} V. Petricevic, S.K. Gayen, R.R. Alfano, K. Yamagishi, H. Anzai, Y. Yamaguchi: Laser action in chromium-doped forsterite, Appl. Phys. Lett. 52, p.1040-1042 (1988)

[6.1489] {Sect. 6.13.2.6} V. Petricevic, S.K. Gayen, R.R. Alfano: Laser action in chromium-activated forsterite for near-infrared excitation: Is Cr4+ the lasing ion?, Appl. Phys. Lett. 53, p.2590-2592 (1988)

[6.1490] {Sect. 6.13.2.6} H.R. Verdun, L.M. Thomas, D.M. Andrauskas, T. McCollum, A. Pinto: Chromium-doped forsterite laser pumped with 1.06 μm radiation, Appl. Phys. Lett. 53, p.2593-2595 (1988)

[6.1491] {Sect. 6.13.2.6} U. Hommerich, X. Wu, V.R. Davis, S.B. Trivedi, K. Grasza, R.J. Chen, S. Kutcher: Demonstration of room-temperature laser action at 2.5 mu m from Cr2+:Cd0.85Mn0.15Te, Optics Letters 22, p.1180-1182 (1997)

[6.1492] {Sect. 6.13.2.6} V. Petricevic, A.B. Bykov, J.M. Evans, R.R. Alfano: Room-temperature near-infrared tunable laser operation of Cr4+: Ca2GeO4, Optics Letters 21, p.1750-1752 (1996)

[6.1493] {Sect. 6.13.2.7} S. Imai, H. Ito: Long-pulse ultraviolet-laser sources based on tunable alexandrite lasers, IEEE J QE-34, p.573-576 (1998)

[6.1494] {Sect. 6.13.2.7} R.C. Sam, J.J. Yeh, K.R. Leslie, W.R. Rapoport: Design and Performance of a 250 Hz Alexandrite Laser, IEEE J. QE-24, p.1151-1166 (1988)

[6.1495] {Sect. 6.13.2.7} J.C. Walling, J.A. Pete, H. Samelson, D.J. Harter, R.C. Morris, D.F. Heller: Tunable Alexandrite Lasers: Development and Performance, IEEE J. QE-21, p.1568-1581 (1985)

[6.1496] {Sect. 6.13.2.7} M.L. Shand, H.P. Jenssen: Temperature Dependence of the Excited-State Absorption of Alexandrite, IEEE J. QE-19, p.480-483 (1983)

[6.1497] {Sect. 6.13.2.7} S. Guch, C.E. Jones: Alexandrite-laser performance at high temperature, Opt. Lett. 7, p.608-610 (1982)

[6.1498] {Sect. 6.13.2.7} J.C. Walling, O.G. Peterson, H.P. Jenssen, R.C. Morris, E.W. O'Dell: Tunable Alexandrit Lasers, IEEE J. QE-16, p.1302-1315 (1980)

[6.1499] {Sect. 6.13.2.8} G.L. Bourdet, G. Lescroart: Theoretical modeling and design of a Tm, Ho : YLiF4 microchip laser, Appl Opt 38, p.3275-3281 (1999)

[6.1500] {Sect. 6.13.2.8} C. Li, D.Y. Shen, J. Song, Y.H. Cao, N.S. Kim, K. Ueda: Flash-lamp pumped normal-mode and Q-switched Cr-Tm : YAG laser performance at room temperature, Opt Commun 164, p.63-67 (1999)

[6.1501] {Sect. 6.13.2.8} C. Bollig, W.A. Clarkson, R.A. Hayward, D.C. Hanna: Efficient high-power Tm:YAG laser at 2 mu m, end-pumped by a diode bar, Opt Commun 154, p.35-38 (1998)

[6.1502] {Sect. 6.13.2.8} G.L. Bourdet, G. Lescroart: Theoretical modelling of mode formation in Tm3+:YVO4 microchip lasers, Opt Commun 150, p.136-140 (1998)

[6.1503] {Sect. 6.13.2.8} G.L. Bourdet, G. Lescroart: Theoretical modelling and design of a Tm:YVO4 microchip laser, Opt Commun 149, p.404-414 (1998)

[6.1504] {Sect. 6.13.2.8} D. Bruneau, S. Delmonte, J. Pelon: Modeling of Tm,Ho : YAG and Tm,Ho : YLF 2-mu m lasers and calculation of extractable energies, Appl Opt 37, p.8406-8419 (1998)

[6.1505] {Sect. 6.13.2.8} A. Diening, P.E.A. Mobert, G. Huber: Diode-pumped continuous-wave, quasi-continuous-wave, and Q-switched laser operation of Yb3+, Tm3+: YLiF4 at 1.5 and 2.3 mu m, J Appl Phys 84, p.5900-5904 (1998)

[6.1506] {Sect. 6.13.2.8} I.F. Elder, M.J.P. Payne: YAP versus YAG as a diode-pumped host for thulium, Opt Commun 148, p.265-269 (1998)

[6.1507] {Sect. 6.13.2.8} F.F. Heine, G. Huber: Tunable single frequency thulium: YAG microchip laser with external feedback, Appl Opt 37, p.3268-3271 (1998)

[6.1508] {Sect. 6.13.2.8} F. Matsuzaka, T. Yokozawa, H. Hara: Saturation parameter and small-signal gain of a laser-diode-pumped Tm:YAG laser, Appl Opt 37, p.5710-5712 (1998)

[6.1509] {Sect. 6.13.2.8} T. Rothacher, W. Luthy, H.P. Weber: Diode pumping and laser properties of Yb:Ho:YAG, Opt Commun 155, p.68-72 (1998)

[6.1510] {Sect. 6.13.2.8} A. Sato, K. Asai, T. Itabe: Double-pass-pumped Tm:YAG laser with a simple cavity configuration, Appl Opt 37, p.6395-6400 (1998)

[6.1511] {Sect. 6.13.2.8} T.M. Taczak, D.K. Killinger: Development of a tunable, narrow-linewidth, cw 2.066-mu m Ho : YLF laser for remote sensing of atmospheric CO2 and H2O, Appl Opt 37, p.8460-8476 (1998)

[6.1512] {Sect. 6.13.2.8} C.P. Wyss, W. Luthy, H.P. Weber, V.I. Vlasov, Y.D. Zavartsev, P.A. Studenikin, A.I. Zagumennyi, I.A. Shcherbakov: A diode-pumped 1.4-w Tm3+:GdVO4 microchip laser at 1.9 mu m, IEEE J QE-34, p.2380-2382 (1998)

[6.1513] {Sect. 6.13.2.8} C.P. Wyss, W. Luthy, H.P. Weber, V.I. Vlasov, Y.D. Zavartsev, P.A. Studenikin, A.I. Zagumennyi, I.A. Shcherbakov: Performance of a Tm3+: (G)dVO (4) microchip laser at 1.9 mu m, Opt Commun 153, p.63-67 (1998)

[6.1514] {Sect. 6.13.2.8} J.R. Yu, U.N. Singh, N.P. Barnes, M. Petros: 125-mJ diode-pumped injection-seeded Ho:Tm:YLF laser, Optics Letters 23, p.780-782 (1998)

[6.1515] {Sect. 6.13.2.8} N.P. Barnes, K.E. Murray, M.G. Jani: Flash-lamp-pumped Ho:Tm:Cr:YAG and Ho:Tm:Er:YLF lasers: Modeling of a single, long pulse length comparison, Appl Opt 36, p.3363-3374 (1997)

[6.1516] {Sect. 6.13.2.8} M.G. Jani, N.P. Barnes, K.E. Murray, D.W. Hart, G.J. Quarles, V.K. Castillo: Diode-pumped Ho:Tm:LuLiF4 laser at room temperature, IEEE J QE-33, p.112-115 (1997)

[6.1517] {Sect. 6.13.2.8} Y. Takenaka, J. Nishimae, M. Tanaka, Y. Motoki: High-power CO2 laser with a Gauss-core resonator for high- speed cutting of thin metal sheets, Optics Letters 22, p.37-39 (1997)

[6.1518] {Sect. 6.13.2.8} T.Y. Fan, G. Huber, R.L. Byer, Mitzscherlich: Spectroscopy and Diode Laser-Pumped Operation of Tm, Ho:YAG, IEEE J. QE-24, p.924-933 (1988)

[6.1519] {Sect. 6.13.2.8} G. Huber, E.W. Duczynski, K. Petermann: Laser pumping of Ho-, Tm-, Er-doped garnet at room temperature, IEEE J. QE-24, p.920-923 (1988)

[6.1520] {Sect. 6.13.2.8} M. Dätwyler, W. Lüthy, H.P. Weber: New wavelengths of the YALO3:Er Laser, IEEE J. QE-23, p.158-159 (1987)

[6.1521] {Sect. 6.13.2.8} N.P. Barnes, R.E. Allen, E.P. Chicklis, L. Esterowitz, H.P. Jensen, M.G. Knights: Operation of an Er:YLF laser at 1.73 µm, IEEE J. QE-22, p.337343 (1986)

[6.1522] {Sect. 6.13.2.8} N.P. Barnes, D.J. Gettemy: Pulsed Ho:YAG Oscillator and Amplifier, IEEE J. QE-17, p.1303-1308 (1981)

[6.1523] {Sect. 6.13.2.8} W.F. Krupke, J.B. Gruber: Energy Levels of Er3+ in LaF3 and Coherent Emission at 1.61 µ, J. Chem. Phys. 41, p.1225-1232 (1964)

[6.1524] {Sect. 6.13.2.8} M. Pollnau, C. Ghisler, W. Luthy, H.P. Weber, J. Schneider, U.B. Unrau: Three-transition cascade erbium laser at 1.7, 2.7, and 1.6 mu m, Optics Letters 22, p.612-614 (1997)

[6.1525] {Sect. 6.13.2.8} S. Georgescu, V. Lupei, M. Trifan, R.J. Sherlock, T.J. Glynn: Population dynamics of the three-micron emitting level of Er3+ in YAlO3, J Appl Phys 80, p.6610-6613 (1996)

[6.1526] {Sect. 6.13.2.8} B. Majaron, T. Rupnik, M. Lukac: Temperature and gain dynamics in flashlamp-pumped Er:YAG, IEEE J QE-32, p.1636-1644 (1996)

[6.1527] {Sect. 6.13.2.8} S. Wittwer, M. Pollnau, R. Spring, W. Luthy, H.P. Weber, R.A. Mcfarlane, C. Harder, H.P. Meier: Performance of a diode-pumped BaY2F8:Er3+ (7.5at.%) laser at 2.8 mu m, Opt Commun 132, p.107-110 (1996)

[6.1528] {Sect. 6.13.2.8} C.E. Hamilton, R.J. Beach, S.B. Sutton, L.H. Furu, W.F. Krupke: 1-W average power levels and tunability from a diode-pumped 2.94-µm Er:YAG oscillator, Opt. Lett. 19, p.1627-1629 (1994)

[6.1529] {Sect. 6.13.2.8} C.E. Hamilton, R.J. Beach, S.B. Sutton, L.H. Furu, W.F. Krupke: 1-W average power levels and tunability from a diode-pumped 2.94-µm Er:YAG oscillator, Opt. Lett. 19, p.1627-1629 (1994)

[6.1530] {Sect. 6.13.2.8} Y. Morishige, S. Kishida, K. Washio, H. Toratani, M. Nakazawa: Output-stabilized high-repetition-rate 1.545-µm Q-switched Er:glass laser, Opt. Lett. 9, p.147-149 (1984)

[6.1531] {Sect. 6.13.2.8} M.J. Weber, M. Bass, G.A. deMars: Laser Action and Spectroscopic Properties of Er3+ in YAlO3, J. Appl. Phys. 42, p.301-305 (1971)

[6.1532] {Sect. 6.13.2.8} E. Snitzer, R.F. Woodcock, J. Segre: Phosphate Glass Er3+ Laser, IEEE J. QE-4, p.360 (1968)

[6.1533] {Sect. 6.13.2.8} C. Bollig, R.A. Hayward, W.A. Clarkson, D.C. Hanna: 2-W Ho:YAG laser intracavity pumped by a diode-pumped Tm:YAG laser, Optics Letters 23, p.1757-1759 (1998)

[6.1534] {Sect. 6.13.2.8} M.E. Storm: Holmium YLF Amplifier Performance and the Prospects for Multi-Joule Energies Using Diode-Laser Pumping, IEEE J. QE-29, p.440-451 (1993)

[6.1535] {Sect. 6.13.2.8} B.T. McGuckin, R.T. Menzies: Efficient CW Diode-Pumped Tm, Ho:YLF Laser with Tunability Near 2.067 µm, IEEE J. QE-28, p.1025-1028 (1992)

[6.1536] {Sect. 6.13.2.8} D.P. Devor, B.H. Soffer: 2.1-µm Laser of 20-W Output Power and 4-Percent Efficiency from Ho3+ in Sensitized YAG, IEEE J. QE-8, p.231-234 (1972)

[6.1537] {Sect. 6.13.2.8} E.P. Chicklis, C.S. Naiman, R.C. Folweiler: High-Efficiency Room-Temperature 2.06-µm Laser Using Sensitized Ho3+:YLF, Appl. Phys. Lett. 19, p.119-121 (1971)

[6.1538] {Sect. 6.13.2.8} D.W. Chen, C.L. Fincher, T.S. Rose, F.L. Vernon, R.A. Fields: Diode-pumped 1-W continuous-wave Er : YAG 3-mu m laser, Optics Letters 24, p.385-387 (1999)

[6.1539] {Sect. 6.13.2.8} I.F. Elder, J. Payne: Diode-pumped, room-temperature Tm:YAP laser, Appl Opt 36, p.8606-8610 (1997)

[6.1540] {Sect. 6.13.2.8} E.C. Honea, R.J. Beach, S.B. Sutton, J.A. Speth, S.C. Mitchell, J.A. Skidmore, M.A. Emanuel, S.A. Payne: 115-W Tm:YAG diode-pumped solid-state laser, IEEE J QE-33, p.1592-1600 (1997)

[6.1541] {Sect. 6.13.2.8} I.V. Mochalov, G.T. Petrovskii, A.V. Sandulenko, V.A. Sandulenko, M. Cervantes, V.S. Terpugov: Investigation of Cr:Tm:Er:YAG laser crystals in a resonator with various degrees of spectral selectivity, Appl Opt 36, p.4090-4093 (1997)

[6.1542] {Sect. 6.13.2.8} R. Moncorge, N. Garnier, P. Kerbrat, C. Wyon, C. Borel: Spectroscopic investigation and two-micron laser performance of Tm3+: CaYAlO4 single crystals, Opt Commun 141, p.29-34 (1997)

[6.1543] {Sect. 6.13.2.8} XA. Rameix, C. Borel, B. Chambaz, B. Ferrand, D.P. Shepherd, T.J. Warburton, D.C. Hanna, A.C. Tropper: An efficient, diode-pumped, 2 mu m Tm:YAG waveguide laser, Opt Commun 142, p.239-243 (1997)

[6.1544] {Sect. 6.13.2.8} N.P. Barnes, E.D. Filer, C.A. Morrison, C.J. Lee: Ho:Tm lasers. 1. Theoretical, IEEE J QE-32, p.92-103 (1996)

[6.1545] {Sect. 6.13.2.8} I.J. Booth, C.J. Mackechnie, B.F. Ventrudo: Operation of diode laser pumped Tm3+ ZBLAN upconversion fiber laser at 482 nm, IEEE J QE-32, p.118-123 (1996)

[6.1546] {Sect. 6.13.2.8} C.J. Lee, G.W. Han, N.P. Barnes: Ho:Tm lasers. 2. Experiments, IEEE J QE-32, p.104-111 (1996)

[6.1547] {Sect. 6.13.2.8} T. Yokozawa, H. Hara: Laser-diode end-pumped Tm3+: YAG eye-safe laser, Appl Opt 35, p.1424-1426 (1996)

[6.1548] {Sect. 6.13.2.8} P.J.M. Suni, S.W. Henderson: 1-mJ/pulse Tm:YAG laser pumped by a 3-W diode laser, Opt. Lett. 16, p.817-819 (1991)

[6.1549] {Sect. 6.13.2.8} R.C. Stoneman, L. Esterowitz: Efifcient, broadly tunable, laser-pumped Tm:YAG and Tm:YSGG cw lasers, Opt. Lett. 15, p.486-488 (1990)

[6.1550] {Sect. 6.13.2.10} J.A. AlvarezChavez, H.L. Offerhaus, J. Nilsson, P.W. Turner, W.A. Clarkson, D.J. Richardson: High-energy, high-power ytterbium-doped Q-switched fiber laser, Optics Letters 25, p.37-39 (2000)

[6.1551] {Sect. 6.13.2.10} B. Srinivasan, R.K. Jain, G. Monnom: Indirect measurement of the magnitude of ion clustering at high doping densities in Er: ZBLAN fibers, J Opt Soc Am B Opt Physics 17, p.178-181 (2000)

[6.1552] {Sect. 6.13.2.10} C.J. daSilva, M.T. deAraujo, E.A. Gouveia, A.S. GouveiaNeto: Fourfold output power enhancement and threshold reduction through thermal effects in an Er3+/Yb3+-codoped optical fiber laser excited at 1.064 mu m, Optics Letters 24, p.1287-1289 (1999)

[6.1553] {Sect. 6.13.2.10} L. Goldberg, J.P. Koplow, D.A.V. Kliner: Highly efficient 4-W Yb-doped fiber amplifier pumped by a broad-stripe laser diode, Optics Letters 24, p.673-675 (1999)

[6.1554] {Sect. 6.13.2.10} R. Hofer, M. Hofer, G.A. Reider: High energy, subpicosecond pulses from a Nd-doped double-clad fiber laser, Opt Commun 169, p.135-139 (1999)

[6.1555] {Sect. 6.13.2.10} V.A. Kozlov, J. HernandezCordero, T.F. Morse: All-fiber coherent beam combining of fiber lasers, Optics Letters 24, p.1814-1816 (1999)

[6.1556] {Sect. 6.13.2.10} R. Paschotta, R. Haring, E. Gini, H. Melchior, U. Keller, H.L. Offerhaus, D.J. Richardson: Passively Q-switched 0.1-mJ fiber laser system at 1.53 mu m, Optics Letters 24, p.388-390 (1999)

[6.1557] {Sect. 6.13.2.10} T. Sandrock, D. Fischer, P. Glas, M. Leitner, M. Wrage, A. Diening: Diode-pumped 1-W Er-doped fluoride glass M-profile fiber laser emitting at 2.8 mu m, Optics Letters 24, p.1284-1286 (1999)

[6.1558] {Sect. 6.13.2.10} A. Cucinotta, S. Selleri, L. Vincetti, M. Zoboli: Numerical and experimental analysis of erbium-doped fiber linear cavity lasers, Opt Commun 156, p.264-270 (1998)

[6.1559] {Sect. 6.13.2.10} P. Glas, M. Naumann, A. Schirrmacher, S. Unger, T. Pertsch: Short-length 10-W cw neodymium-doped M-profile fiber laser, Appl Opt 37, p.8434-8437 (1998)

[6.1560] {Sect. 6.13.2.10} P. Glas, M. Naumann, A. Schirrmacher, T. Pertsch: The multicore fiber – a novel design for a diode pumped fiber laser, Opt Commun 151, p.187-195 (1998)

[6.1561] {Sect. 6.13.2.10} S.D. Jackson, T.A. King: CW operation of a 1.064-mu m pumped Tm-Ho-doped silica fiber laser, IEEE J QE-34, p.1578-1587 (1998)

[6.1562] {Sect. 6.13.2.10} D.S. Lim, H.K. Lee, K.H. Kim, S.B. Kang, J.T. Ahn, M.Y. Jeon: Generation of multiorder Stokes and anti-Stokes lines in a Brillouin erbium fiber laser with a Sagnac loop mirror, Optics Letters 23, p.1671-1673 (1998)

[6.1563] {Sect. 6.13.2.10} R. Naftali, B. Fischer, J.R. Simpson: Large core-area erbium-doped fibre laser, Opt Commun 149, p.317-320 (1998)

[6.1564] {Sect. 6.13.2.10} Y. Nishida, M. Yamada, T. Kanamori, K. Kobayashi, J. Temmyo, S. Sudo, Y. Ohishi: Development of an efficient praseodymium-doped fiber amplifier, IEEE J QE-34, p.1332-1339 (1998)

[6.1565] {Sect. 6.13.2.10} H.L. Offerhaus, N.G. Broderick, D.J. Richardson, R. Sammut, J. Caplen, L. Dong: High-energy single-transverse-mode Q-switched fiber laser based on a multimode large-mode-area erbium-doped fiber, Optics Letters 23, p.1683-1685 (1998)

[6.1566] {Sect. 6.13.2.10} J. Porta, A.B. Grudinin, Z.J. Chen, J.D. Minelly, N.J. Traynor: Environmentally stable picosecond ytterbium fiber laser with a broad tuning range, Optics Letters 23, p.615-617 (1998)

[6.1567] {Sect. 6.13.2.10} C.T.A. Brown, J. Amin, D.P. Shepherd, A.C. Tropper, M. Hempstead, J.M. Almeida: 900-nm Nd:Ti:LiNbO3 waveguide laser, Optics Letters 22, p.1778-1780 (1997)

[6.1568] {Sect. 6.13.2.10} P. Glas, M. Naumann, A. Schirrmacher, S. Unger, T. Pertsch: A high power neodymium-doped fiber laser using a novel fiber geometry, Opt Commun 141, p.336-342 (1997)

[6.1569] {Sect. 6.13.2.10} R. Hofer, M. Hofer, G.A. Reider, M. Cernusca, M.H. Ober: Modelocking of a Nd-fiber laser at 920 nm, Opt Commun 140, p.242-244 (1997)

[6.1570] {Sect. 6.13.2.10} R. Paschotta, J. Nilsson, A.C. Tropper, D.C. Hanna: Ytterbium-doped fiber amplifiers, IEEE J QE-33, p.1049-1056 (1997)

[6.1571] {Sect. 6.13.2.10} J. Schneider, C. Carbonnier, U.B. Unrau: Characterization of a Ho3+-doped fluoride fiber laser with a 3.9-mu m emission wavelength, Appl Opt 36, p.8595-8600 (1997)

[6.1572] {Sect. 6.13.2.10} M.E. Fermann, D. Harter, J.D. Minelly, G.G. Vienne: Cladding-pumped passively mode-locked fiber laser generating femtosecond and picosecond pulses, Optics Letters 21, p.967-969 (1996)

[6.1573] {Sect. 6.13.2.10} M.E. Fermann, J.D. Minelly: Cladding-pumped passive harmonically mode-locked fiber laser, Optics Letters 21, p.970-972 (1996)

[6.1574] {Sect. 6.13.2.10} C. Ghisler, W. Luthy, H.P. Weber: Cladding-pumping of a Tm3+:Ho3+ silica fibre laser, Opt Commun 132, p.474-478 (1996)

[6.1575] {Sect. 6.13.2.10} P. Glas, M. Naumann, A. Schirrmacher: A novel design for a high brightness diode pumped fiber laser source, Opt Commun 122, p.163-168 (1996)

[6.1576] {Sect. 6.13.2.10} K. Hattori, T. Kitagawa, Y. Ohmori: Gain switching of an erbium-doped silica-based planar waveguide laser, J Appl Phys 79, p.1238-1243 (1996)

[6.1577] {Sect. 6.13.2.10} W.H. Loh, L. Dong, J.E. Caplen: Single-sided output Sn/Er/Yb distributed feedback fiber laser, Appl Phys Lett 69, p.2151-2153 (1996)

[6.1578] {Sect. 6.13.2.10} M. Pollnau, R. Spring, C. Ghisler, S. Wittwer, W. Luthy, H.P. Weber: Efficiency of erbium 3-mu m crystal and fiber lasers, IEEE J QE-32, p.657-663 (1996)

[6.1579] {Sect. 6.13.2.10} B. Desthieux, R.I. Laming, D.N. Payne: 111 kW (0.5 mJ) pulse amplification at 1.5 μm using a gated cascade of three erbium-doped fiber amplifiers, Appl. Phys. Lett. 63, p.586-588 (1993)

[6.1580] {Sect. 6.13.2.10} M.J.F. Digonnet, C.J. Gaeta: Theoretical analysis of optical fiber laser amplifiers and oscillators, Appl. Opt. 24, p.333-342 (1985)

[6.1581] {Sect. 6.13.2.10} C.A. Burrus, J. Stone: Single-crystal fiber optical devices: A Nd:YAG fiber laser, Appl. Phys. Lett.26, p.318-320 (1975)

[6.1582] {Sect. 6.13.2.10} S. Sudo: Optical Fiber Amplifiers (Artech House, Boston, London, 1997)

[6.1583] {Sect. 6.13.3.1} T. Kasamatsu, M. Tsunekane, H. Sekita, Y. Morishige, S. Kishida: 1 pm spectrally narrowed ArF excimer laser injection locked by fourth harmonic seed source of 773.6 nm Ti: sapphire laser, Appl Phys Lett 67, p.3396-3398 (1995)

[6.1584] {Sect. 6.13.3.1} S. Izawa, A. Suda, M. Obara: Experimental observation of unstable resonator mode evolution in a high-power KrF laser, J. Appl. Phys. 58, p.3987-3990 (1985)

[6.1585] {Sect. 6.13.3.1} Y. Nabekawa, Y. Kuramoto, T. Sekikawa, S. Watanabe: High-power sub-100-fs UV pulse generation from a spectrally controlled KrF laser, Optics Letters 22, p.724-726 (1997)

[6.1586] {Sect. 6.13.3.2} P. Richter, J.D. Kimel, G.C. Moulton: Pulsed uv nitrogen laser: dynamical behavior, Appl. Opt. 15, p.756-760 (1976)

[6.1587] {Sect. 6.13.3.2} S.V. Kukhlevsky, L. Kozma: Diffraction-limited transverscoherent radiation of pulsed capillary gas lasers with waveguide resonators, Opt. Comm. 122, p.35-39 (1995)

[6.1588] {Sect. 6.13.3.3} H. Golnabi: Reliable spark gap switch for laser triggering, Rev. Sci.Instrum. 63, p.5804-5805 (1992)

[6.1589] {Sect. 6.13.3.5} J. Bonnetgamard, J. Bleuse, N. Magnea, J.L. Pautrat: Optical gain and laser emission in HgCdTe heterostructures, J Appl Phys 78, p.6908-6915 (1995)

[6.1590] {Sect. 6.13.3.6} D. Gay, N. Mccarthy: Improvement of the pulse and spectrum characteristics of a mode-locked argon laser with a phase-conjugating external cavity, Opt Commun 137, p.83-88 (1997)

[6.1591] {Sect. 6.13.3.6} N.A. Robertson, S. Hoggan, J.B. Mangan, J. Hough: Intensity Stabilization of an Argon Laser Using an Electro-Optic Modulator – Performance and Limitations, Appl. Phys. B 39, p.149-153 (1986)

[6.1592] {Sect. 6.13.3.6} L.L. Steinmetz, J.H. Richardson, B.W. Wallin: A mode-locked krypton ion laser with a 50-psec pulse width in the near uv, Appl. Phys. Lett. 33, p.163-165 (1978)

[6.1593] {Sect. 6.13.3.6} R.J. Freiberg, A.S. Halsted: Properties of Low Order Transverse Modes in Argon Ion Lasers, Appl. Opt. 8, p.355-362 (1969)

[6.1594] {Sect. 6.13.3.7} C. E. Little: Metal Vapour Lasers: Physics, Engineering, and Applications (John Wiley & Sons, Chichester, 1999)

[6.1595] {Sect. 6.13.3.7} E. LeGuyadec, P. Coutance, G. Bertrand, C. Peltier: A 280-W average power Cu-Ne-HBr laser amplifier, IEEE J QE-35, p.1616-1622 (1999)

[6.1596] {Sect. 6.13.3.7} M.J. Withford, D.J.W. Brown: A 60-W high-beam-quality single-oscillator copper vapor laser, IEEE J QE-35, p.997-1003 (1999)

[6.1597] {Sect. 6.13.3.7} R.J. Carman, M.J. Withford, D.J.W. Brown, J.A. Piper: Influence of the pre-pulse plasma electron density on the performance of elemental copper vapour lasers, Opt Commun 157, p.99-104 (1998)

[6.1598] {Sect. 6.13.3.7} D. Kapitan, D.W. Coutts, C.E. Webb: Efficient generation of near diffraction-limited beam-quality output from medium-scale copper vapor laser oscillators, IEEE J QE-34, p.419-426 (1998)

[6.1599] {Sect. 6.13.3.7} O. Prakash, P.K. Shukla, S.K. Dixit, S. Chatterjee, H.S. Vora, R. Bhatnagar: Spatial coherence of the generalized diffraction-filtered resonator copper vapor laser, Appl Opt 37, p.7752-7757 (1998)

[6.1600] {Sect. 6.13.3.7} M.J. Withford, D.J.W. Brown, J.A. Piper: Repetition-rate scaling of a kinetically enhanced copper-vapor laser, Optics Letters 23, p.1538-1540 (1998)

[6.1601] {Sect. 6.13.3.7} M.J. Withford, D.J.W. Brown, R.J. Carman, J.A. Piper: Enhanced performance of elemental copper-vapor lasers by use of H-2- HCl-Ne buffer-gas mixtures, Optics Letters 23, p.706-708 (1998)

[6.1602] {Sect. 6.13.3.7} O. Prakash, P.K. Shukla, S.K. Dixit, S. Chatterjee, H.S. Vora, R. Bhatnagar: Spatial coherence of the generalized diffraction-filtered resonator copper vapor laser, Appl. Opt. 37, p.7752-7757 (1998)

[6.1603] {Sect. 6.13.3.7} D.N. Astadjov, K.D. Dimitrov, D.R. Jones, V. Kirkov, L. Little, C.E. Little, N.V. Sabotinov, N.K. Vuchkov: Influence on operating characteristics of scaling sealed- off CuBr lasers in active length, Opt Commun 135, p.289-294 (1997)

[6.1604] {Sect. 6.13.3.7} D.N. Astadjov, K.D. Dimitrov, D.R. Jones, V.K. Kirkov, C.E. Little, N.V. Sabotinov, N.K. Vuchkov: Copper bromide laser of 120-W average output power, IEEE J QE-33, p.705-709 (1997)

[6.1605] {Sect. 6.13.3.7} D.J.W. Brown, C.G. Whyte, D.R. Jones, C.E. Little: High-beam quality, high-power copper HyBrID laser injection-seeded oscillator system, Opt Commun 137, p.158-164 (1997)

[6.1606] {Sect. 6.13.3.7} R.J. Carman: Modelling of the kinetics and parametric behaviour of a copper vapour laser: Output power limitation issues, J Appl Phys 82, p.71-83 (1997)

[6.1607] {Sect. 6.13.3.7} H. Kimura, M. Chinen, T. Nayuki, H. Saitoh: Improvement of the lasing performance of copper vapor laser by adding Sc atoms as energy donors, Appl Phys Lett 71, p.312-314 (1997)

[6.1608] {Sect. 6.13.3.8} S.Y. Tochitsky, R. Narang, C. Filip, C.E. Clayton, K.A. Marsh, C. Joshi: Generation of 160-ps terawatt-power CO2 laser pulses, Optics Letters 24, p.1717-1719 (1999)

[6.1609] {Sect. 6.13.3.8} J.J. Wendland, H.J. Baker, D.R. Hall: Operation of a cw (CO2)-C-14-O-16 laser in the 12 mu m spectral region, Opt Commun 154, p.329-333 (1998)

[6.1610] {Sect. 6.13.3.8} P. Repond, M.W. Sigrist: Continuously tunable high-pressure CO2 laser for spectroscopic studies on trace gases, IEEE J QE-32, p.1549-1559 (1996)

[6.1611] {Sect. 6.13.3.8} S.W.C. Scott, J.D. Strohschein, H.J.J. Seguin, C.E. Capjack, H.W. Reshef: Optical performance of a burst-mode multikilowatt CO2 laser, Appl Opt 35, p.4740-4748 (1996)

[6.1612] {Sect. 6.13.3.8} Y. Takenaka, Y. Motoki, J. Nishimae: High-power CO2 laser using gauss-core resonator for 6-kW large-volume TEM (00) mode operation, IEEE J QE-32, p.1299-1305 (1996)

[6.1613] {Sect. 6.13.3.8} W.F. Krupke, W.R. Sooy: Properties of an Unstable Confocal Resonator CO2 Laser System, IEEE J. QE-5, p.575-586 (1969)

[6.1614] {Sect. 6.13.3.8} H.C. Miller, J. McCord, G.D. Hager, S.J. Davis, W.J. Kessler, D.B. Oakes: Optically pumped mid-infrared vibrational hydrogen chloride laser, J Appl Phys 84, p.3467-3473 (1998)

[6.1615] {Sect. 6.13.4} F. J. Duarte (ed.): High Power Dye Lasers (Springer, Berlin, Heidelberg, New York, 1991)

[6.1616] {Sect. 6.13.4} F. P. Schäfer (ed.): Dye Lasers (Springer, Berlin, Heidelberg, New York, 1990)

[6.1617] {Sect. 6.13.4} U. Brackmann: Lambdachrome Laser Dyes (Lambda Physik GmbH, Göttingen, 1997)

[6.1618] {Sect. 6.13.4} T.G. Pavlopoulos: Spectroscopy and molecular structure of efficient laser dyes: Vibronic spin-orbit interactions in heterocyclics, Appl Opt 36, p.4969-4980 (1997)

[6.1619] {Sect. 6.13.4} Y. Assor, Z. Burshtein, S. Rosenwaks: Spectroscopy and laser characteristics of copper-vapor-laser pumped Pyrromethene-556 and Pyrromethene-567 dye solutions, Appl Opt 37, p.4914-4920 (1998)

[6.1620] {Sect. 6.13.4} F. J. Duarte, J. A. Piper: Narrow linewidth, high prf copper laser-pumped dye-laser oscillators, Appl. Opt. 23, p.1391-1394 (1984)

[6.1621] {Sect. 6.13.4} M. Yamashita, D.J. Bradley, W. Sibbett, D. Welford: Intra Cavity 2nd Harmonic Generation in a Synchronously Mode Locked CW Dye Laser, J Appl Phys 51, p.3559-3562 (1980)

[6.1622] {Sect. 6.13.4} H.W. Kogelnik, E.P. Ippen, A. Dienes, C.V. Shank: Astigmatically Compensated Cavities for CW Dye Lasers, IEEE J. QE-8, p.373-379 (1972)

[6.1623] {Sect. 6.13.4} R. Gvishi, G. Ruland, P.N. Prasad: New laser medium: Dye-doped sol-gel fiber, Opt Commun 126, p.66-72 (1996)

[6.1624] {Sect. 6.13.4} M. Schütz, U. Heitmann, A. Hese: Development of a dual-wavelength dye-laser system for the UV and ist application to simultaneous multi-elememt detection, Appl. Phys. B 61, p.339-343 (1995)

[6.1625] {Sect. 6.13.4} T. W. Hänsch: Repetitively Pulsed Tunable Dye Laser for High Resolution Spectroscopy, Appl. Opt. 11, p.895-898 (1972)

[6.1626] {Sect. 6.13.4} M. Ahmad, M.D. Rahn, T.A. King: Singlet oxygen and dye-triplet-state quenching in solid-state dye lasers consisting of Pyrromethene 567-doped poly (Methyl methacrylate), Appl Opt 38, p.6337-6342 (1999)

[6.1627] {Sect. 6.13.4} E.C. Chang, S.A. Chen: Cyano-containing phenylene vinylene-based copolymer as blue luminescent and electron transport material in polymer light-emitting diodes, J Appl Phys 85, p.2057-2061 (1999)

[6.1628] {Sect. 6.13.4} S.M. Giffin, I.T. McKinnie, W.J. Wadsworth, A.D. Woolhouse, G.J. Smith, T.G. Haskell: Solid state dye lasers based on 2-hydroxyethyl methacrylate and methyl methacrylate co-polymers, Opt Commun 161, p.163-170 (1999)

[6.1629] {Sect. 6.13.4} W.J. Wadsworth, S.M. Giffin, I.T. McKinnie, J.C. Sharpe, A.D. Woolhouse, T.G. Haskell, G.J. Smith: Thermal and optical properties of polymer hosts for solid-state dye lasers, Appl Opt 38, p.2504-2509 (1999)

[6.1630] {Sect. 6.13.4} F.J. Duarte, T.S. Taylor, A. Costela, I. Garciamoreno, R. Sastre: Long-pulse narrow-linewidth dispersive solid-state dye-laser oscillator, Appl Opt 37, p.3987-3989 (1998)

[6.1631] {Sect. 6.13.4} A.J. Finlayson, N. Peters, P.V. Kolinsky, M.R.W. Venner: Flashlamp pumped polymer dye laser containing Rhodamine 6G, Appl Phys Lett 72, p.2153-2155 (1998)

[6.1632] {Sect. 6.13.4} S. Stagira, M. ZavelaniRossi, M. Nisoli, S. DeSilvestri, G. Lanzani, C. Zenz, P. Mataloni, G. Leising: Single-mode picosecond blue laser emission from a solid conjugated polymer, Appl Phys Lett 73, p.2860-2862 (1998)

[6.1633] {Sect. 6.13.4} K.C. Yee, T.Y. Tou, S.W. Ng: Hot-press molded poly (methyl methacrylate) matrix for solid-state dye lasers, Appl Opt 37, p.6381-6385 (1998)

[6.1634] {Sect. 6.13.4} O.G. Calderon, J.M. Guerra, A. Costela, I. Garciamoreno, R. Sastre: Laser emission of a flash-lamp pumped Rhodamine 6 G solid copolymer solution, Appl Phys Lett 70, p.25-27 (1997)

[6.1635] {Sect. 6.13.4} M.J. Cazeca, X.L. Jiang, J. Kumar, S.K. Tripathy: Epoxy matrix for solid-state dye laser applications, Appl Opt 36, p.4965-4968 (1997)

[6.1636] {Sect. 6.13.4} S. Chandra, T.H. Allik, J.A. Hutchinson, J. Fox, C. Swim: Tunable ultraviolet laser source based on solid-state dye laser technology and CsLiB6O10 harmonic generation, Optics Letters 22, p.209-211 (1997)

[6.1637] {Sect. 6.13.4} M. Faloss, M. Canva, P. Georges, A. Brun, F. Chaput, J.P. Boilot: Toward millions of laser pulses with pyrromethene- and perylene-doped xerogels, Appl Opt 36, p.6760-6763 (1997)

[6.1638] {Sect. 6.13.4} A. Costela, I. Garciamoreno, J.M. Figuera, F. Amatguerri, J. Barroso, R. Sastre: Solid-state dye laser based on coumarin 540A-doped polymeric matrices, Opt Commun 130, p.44-50 (1996)

[6.1639] {Sect. 6.13.4} A. Mandl, A. Zavriyev, D.E. Klimek: Energy scaling and beam quality studies of a zigzag solid- state plastic dye laser, IEEE J QE-32, p.1723-1726 (1996)

[6.1640] {Sect. 6.13.4} T. Yamamoto, K. Fujii, A. Tagaya, E. Nihei, Y. Koike, K. Sasaki: High-power optical source using dye-doped polymer optical fiber, J Nonlinear Opt Physics Mat 5, p.73-88 (1996)

[6.1641] {Sect. 6.13.4} M.D. Rahn, T.A. King: Comparison of laser performance of dye molecules in sol- gel, polycom, ormosil, and poly (methyl methacrylate) host media, Appl Opt 34, p.8260-8271 (1995)

[6.1642] {Sect. 6.13.4} S.-L. Chen, Z.-H. Zhu, K-C. Chen: A Class of Novel Laser Dyes: Triphenodioxazinesl, Opt. Comm. 74, 84-86p.84-86 (1989)

[6.1643] {Sect. 6.13.4} F.L. Arbeloa, T.L. Arbeloa, I.L. Arbeloa, I. Garciamoreno, A. Costela, R. Sastre, F. Amatguerri: Photophysical and lasing properties of pyrromethene 567 dye in liquid solution. Environment effects, Chem Phys 236, p.331-341 (1998)

[6.1644] {Sect. 6.13.4} T.G. Pavlopoulos, J.H. Boyer, G. Sathyamoorthi: Laser action from a 2,6,8-position trisubstituted 1,3,5,7-tetramethylpyrromethene-BF2 complex: part 3, Appl Opt 37, p.7797-7800 (1998)

[6.1645] {Sect. 6.13.4} M.D. Rahn, T.A. King, A.A. Gorman, I. Hamblett: Photostability enhancement of Pyrromethene 567 and Perylene Orange in oxygen-free liquid and solid dye lasers, Appl Opt 36, p.5862-5871 (1997)

[6.1646] {Sect. 6.13.4} J.D. Bhawalkar, G.S. He, C.K. Park, C.F. Zhao, G. Ruland, P.N. Prasad: Efficient, two-photon pumped green upconverted cavity lasing in a new dye, Opt Commun 124, p.33-37 (1996)

[6.1647] {Sect. 6.13.4} G.S. He, J.D. Bhawalkar, C.F. Zhao, P.N. Prasad: Properties of two-photon pumped cavity lasing in novel dye doped solid matrices, IEEE J QE-32, p.749-755 (1996)

[6.1648] {Sect. 6.13.4} G.S. He, C.F. Zhao, J.D. Bhawalkar, P.N. Prasad: Two-photon pumped cavity lasing in novel dye doped bulk matrix rods, Appl Phys Lett 67, p.3703-3705 (1995)

[6.1649] {Sect. 6.13.4} A. Mandl, D.E. Klimek: Multipulse operation of a high average power, good beam quality zig-zag dye laser, IEEE J QE-32, p.378-382 (1996)

[6.1650] {Sect. 6.13.5.1} J. Dunn, J. Nilsen, A.L. Osterheld, Y.L. Li, V.N. Shlyaptsev: Demonstration of transient gain x-ray lasers near 20 nm for nickellike yttrium, zirconium, niobium, and molybdenum, Optics Letters 24, p.101-103 (1999)

[6.1651] {Sect. 6.13.5.1} Y. Hironaka, Y. Fujimoto, K.G. Nakamura, K. Kondo: Enhancement of hard x-ray emission from a copper target by multiple shots of femtosecond laser pulses, Appl Phys Lett 74, p.1645-1647 (1999)

[6.1652] {Sect. 6.13.5.1} J.J. Rocca, C.H. Moreno, M.C. Marconi, K. Kanizay: Soft-x-ray laser interferometry of a plasma with a tabletop laser and a Lloyd's mirror, Optics Letters 24, p.420-422 (1999)

[6.1653] {Sect. 6.13.5.1} B.R. Benware, C.D. Macchietto, C.H. Moreno, J.J. Rocca: Demonstration of a high average power tabletop soft X-ray laser, Phys Rev Lett 81, p.5804-5807 (1998)

[6.1654] {Sect. 6.13.5.1} J.Y. Lin, G.J. Tallents, J. Zhang, A.G. MacPhee, C.L.S. Lewis, D. Neely, J. Nilsen, G.J. Pert, R.M.N. ORourke, R. Smith et al.: Gain saturation of the Ni-like X-ray lasers, Opt Commun 158, p.55-60 (1998)

[6.1655] {Sect. 6.13.5.1} D. Ros, H. Fiedorowicz, B. Rus, A. Bartnik, M. Szczurek, G. Jamelot, F. Albert, A. Carillon, P. Jaegle, A. Klisnick et al.: Investigation of XUV amplification with Ni-like xenon ions using laser-produced gas puff plasmas, Opt Commun 153, p.368-374 (1998)

[6.1656] {Sect. 6.13.5.1} B.R. Benware, C.H. Moreno, D.J. Burd, J.J. Rocca: Operation and output pulse characteristics of an extremely compact capillary-discharge tabletop soft-x-ray laser, Optics Letters 22, p.796-798 (1997)

[6.1657] {Sect. 6.13.5.1} P. Jaegle, S. Sebban, A. Carillon, G. Jamelot, A. Klisnick, P. Zeitoun, B. Rus, M. Nantel, F. Albert, D. Ros: Ultraviolet luminescence of CsI and CsCl excited by soft x- ray laser, J Appl Phys 81, p.2406-2409 (1997)

[6.1658] {Sect. 6.13.5.1} M.P. Kalashnikov, P.V. Nickles, M. Schnuerer, I. Will, W. Sandner: Multi-terawatt hybrid Ti:Sa-Nd:glass dual-beam laser: A novel XUV laser driver, Opt Commun 133, p.216-220 (1997)

[6.1659] {Sect. 6.13.5.1} Y.L. Li, H. Schillinger, C. Ziener, R. Sauerbrey: Reinvestigation of the Duguay soft X-ray laser: a new parameter space for high power femtosecond laser pumped systems, Opt Commun 144, p.118-124 (1997)

[6.1660] {Sect. 6.13.5.1} Y.L. Li, P.X. Lu, G. Pretzler, E.E. Fill: Lasing in neonlike sulphur and silicon, Opt Commun 133, p.196-200 (1997)

[6.1661] {Sect. 6.13.5.1} P.V. Nickles, V.N. Shlyaptsev, M. Kalachnikov, M. Schnurer, I. Will, W. Sandner: Short pulse x-ray laser 32.6 nm based on transient gain in Ne-like titanium, Phys Rev Lett 78, p.2748-2751 (1997)

[6.1662] {Sect. 6.13.5.1} J. Nilsen, J.C. Moreno, T.W. Barbee, L.B. DaSilva: Measurement of spatial gain distribution for a neonlike germanium 19.6-nm laser, Optics Letters 22, p.1320-1322 (1997)

[6.1663] {Sect. 6.13.5.1} P.J. Warwick, C.L.S. Lewis, S. Mccabe, A.G. MacPhee, A. Behjat, M. Kurkcuoglu, G.J. Tallents, D. Neely, E. Wolfrum, S.B. Healy, et al.: A study to optimise the temporal drive pulse structure for efficient XUV lasing on the J=0-1, 19.6 nm line of Ge XXIII, Opt Commun 144, p.192-197 (1997)

[6.1664] {Sect. 6.13.5.1} J. Zhang, A.G. MacPhee, J. Nilsen, J. Lin, T.W. Barbee, C. Danson, M.H. Key, C.L.S. Lewis, D. Neely, R.M.N. ORourke, et al.: Demonstration of saturation in a Ni-like Ag x-ray laser at 14 nm, Phys Rev Lett 78, p.3856-3859 (1997)

[6.1665] {Sect. 6.13.5.1} P. V. Nickles, V. N. Shlyaptsev, M. Kalachnikov, M. Schnürer, I. Will, W. Sandner: Short Pulse X-Ray Laser at 32.6 nm Based on Transient Gain in Ne-like Titanium, Phys. Rev. Lett. 78, p.2748-2751 (1997)

[6.1666] {Sect. 6.13.5.1} J. Zhang, A.G. Macphee, J. Lin, E. Wolfrum, R. Smith, C. Danson, M.H. Key, C.L.S. Lewis, D. Neely, J. Nilsen, et al.: A saturated X-ray laser beam at 7 nanometers, Science 276, p.1097-1100 (1997)

[6.1667] {Sect. 6.13.5.1} G.F. Cairns, C.L.S. Lewis, M.J. Lamb, A.G. MacPhee, D. Neely, P. Norreys, M.H. Key, S.B. Healy, P.B. Holden, G.J. Pert, et al.: Using low and high prepulses to enhance the J=0-1 transition at 19.6 nm in the Ne-like germanium XUV laser, Opt Commun 123, p.777-789 (1996)

[6.1668] {Sect. 6.13.5.1} H. Daido, S. Ninomiya, T. Imani, R. Kodama, M. Takagi, Y. Kato, K. Murai, J. Zhang, Y. You, Y. Gu: Nickellike soft-x-ray lasing at the wavelengths between 14 and 7.9 nm, Optics Letters 21, p.958-960 (1996)

[6.1669] {Sect. 6.13.5.1} H. Fiedorowicz, A. Bartnik, Y. Li, P. Lu, E. Fill: Demonstration of soft x-ray lasing with neonlike argon and nickel-like xenon ions using a laser-irradiated gas puff target, Phys Rev Lett 76, p.415-418 (1996)

[6.1670] {Sect. 6.13.5.1} G.P. Gupta, B.K. Sinha: Estimation of optimum electron temperature for maximum x- ray laser gain from 3p-3s transitions of neonlike ions in laser plasmas, J Appl Phys 79, p.619-624 (1996)

[6.1671] {Sect. 6.13.5.1} J. Nilsen, Y.L. Li, P.X. Lu, J.C. Moreno, E.E. Fill: Relative merits of using curved targets and the prepulse technique to enhance the output of the neon-like germanium X-ray laser (vol 124, pg 287, 1996), Opt Commun 130, p.415-416 and 124,287 (1996)

[6.1672] {Sect. 6.13.5.1} J. Nilsen, H. Fiedorowicz, A. Bartnik, Y.L. Li, P.X. Lu, E.E. Fill: Self-photopumped neonlike x-ray laser, Optics Letters 21, p.408-410 (1996)

[6.1673] {Sect. 6.13.5.1} J.F. Pelletier, M. Chaker, J.C. Kieffer: Picosecond soft-x-ray pulses from a high-intensity laser- plasma source, Optics Letters 21, p.1040-1042 (1996)

[6.1674] {Sect. 6.13.5.1} J. Zhang, E.E. Fill, Y. Li, D. Schlogl, J. Steingruber, M. Holden, G.J. Tallents, A. Demir, P. Zeitoun, C. Danson, et al.: High-gain x-ray lasing at 11.1 nm in sodiumlike copper driven by a 20-J, 2-ps Nd:glass laser, Optics Letters 21, p.1035-1037 (1996)

[6.1675] {Sect. 6.13.5.1} R.W. Schoenlein, W.P. Leemans, A.H. Chin, P. Volfbeyn, T.E. Glover, P. Balling, M. Zolotorev, K.J. Kim, S. Chattopadhyay, C.V. Shank: Femtosecond x-ray pulses at 0.4 angstrom generated by 90 degrees Thomson scattering: A tool for probing the structural dynamics of materials, Science 274, p.236-238 (1996)

[6.1676] {Sect. 6.13.5.1} H. Daido, Y. Kato, K. Murai, S. Ninomiya, R. Kodama, G. Yuan, Y. Oshikane, M. Takagi, H. Takabe, F. Koike: Efficient soft x-ray lasing at 6 to 8 nm with nickel-like lanthanide ions, Phys Rev Lett 75, p.1074-1077 (1995)

[6.1677] {Sect. 6.13.5.1} B.E. Lemoff, G.Y. Yin, C.L. Gordon, C.P.J. Barty, S.E. Harris: Demonstration of a 1O-Hz femtosecond pulse-driven XUV laser at 41.8 nm in Xe IX, Phys Rev Lett 74, p.1574-1577 (1995)

[6.1678] {Sect. 6.13.5.1} Y.L. Li, G. Pretzler, E.E. Fill: Observation of lasing on the two J=0-1, 3p-3s transitions at 26.1 and 30.4 nm in neonlike vanadium, Optics Letters 20, p.1026-1028 (1995)

[6.1679] {Sect. 6.13.5.1} J. Nilsen, J.C. Moreno: Lasing at 7.9 nm in nickellike neodymium, Optics Letters 20, p.1386-1388 (1995)

[6.1680] {Sect. 6.13.5.1} J. Zhang, M.H. Key, P.A. Norreys, G.J. Tallents, A. Behjat, C. Danson, A. Demir, L. Dwivedi, M. Holden, P.B. Holden, et al.: Demonstration of high gain in a recombination XUV laser at 18.2 nm driven by a 20 J, 2 ps glass laser, Phys Rev Lett 74, p.1335-1338 (1995)

[6.1681] {Sect. 6.13.5.1} U. Teubner, J. Bergmann, B. van Wonterghem, F.P. Schäfer: Angle-Dependent X-Ray Emission and Resonance Absorption in a Laser-Produced Plasma Generated by a High Intensity Ultrashort Pulse, Phys. Rev. Lett. 70, p.794-797 (1993)

[6.1682] {Sect. 6.13.5.1} E. Fill (guest ed.): X-Ray Lasers, Appl. Phys. B 50, p.145-146 (1990)

[6.1683] {Sect. 6.13.5.1} Z.z. Xu, Z.-q. Zhang, P.-z. Fan, S.-s. Chen, L.-h. Lin, P.-x. Lu, X.-p. Feng, X.-f. Wang, J.-z. Zhou, A.-d. Qian: Soft X-Ray Amplification by Li-Like A.10+ and Si11+ Ions in Recombining Plasmas, Appl. Phys. B 50, p.147-151 (1990)

[6.1684] {Sect. 6.13.5.1} C.M. Brown, J.O. Ekberg, U. Feldman, J.F. Seely, M.C. Richardson, F.J. Marshall, W.E. Behring: Transitions in lithiumlike Cu26+ and berylliumlike Cu25+ of interest for x-ray lawer research, J. Opt Soc. Am. B 4, p.533-538 (1987)

[6.1685] {Sect. 6.13.5.1} D. L. Matthews, R. R. Freeman (guest eds.): The generation of coherent XUV and soft X-ray radiation (Introduction), J. Opt. Soc. Am. B 4, p.530 (1987)

[6.1686] {Sect. 6.13.5.1} W. Theobald, C. Wulker, J. Jasny, J.S. Bakos, J. Jethwa, F.P. Schafer: High-density lithium plasma columns generated by intense subpicosecond KrF laser pulses, Opt Commun 149, p.289-295 (1998)

[6.1687] {Sect. 6.13.5.1} K. Matsubara, U. Tanaka, H. Imajo, M. Watanabe: All-solid-state light source for generation of tunable continuous- wave coherent radiation near 202 nm, J Opt Soc Am B Opt Physics 16, p.1668-1671 (1999)

[6.1688] {Sect. 6.13.5.1} M.A. Klosner, W.T. Silfvast: Intense xenon capillary discharge extreme-ultraviolet source in the 10-16-nm-wavelength region, Optics Letters 23, p.1609-1611 (1998)

[6.1689] {Sect. 6.13.5.1} M.A. Klosner, H.A. Bender, W.T. Silfvast, J.J. Rocca: Intense plasma discharge source at 13.5 nm for extreme-ultraviolet lithography, Optics Letters 22, p.34-36 (1997)

[6.1690] {Sect. 6.13.5.1} J.J. Rocca, D.P. Clark, J.L.A. Chilla, V.N. Shlyaptsev: Energy extraction and achievement of the saturation limit in a discharge-pumped table-top soft x-ray amplifier, Phys Rev Lett 77, p.1476-1479 (1996)

[6.1691] {Sect. 6.13.5.2} C.S. Ng, A. Bhattacharjee: Ginzburg-Landau model and single-mode operation of a free-electron laser oscillator, Phys Rev Lett 82, p.2665-2668 (1999)

[6.1692] {Sect. 6.13.5.2} E.L. Saldin, E.A. Schneidmiller, M.V. Yurkov: Statistical properties of radiation from VUV and X-ray free electron laser, Opt Commun 148, p.383-403 (1998)

[6.1693] {Sect. 6.13.5.2} R. Bonifacio: A rigorous calculation of coherent noise bunching, Opt Commun 138, p.99-100 (1997)

[6.1694] {Sect. 6.13.5.2} T. Mizuno, T. Otsuki, T. Ohshima, H. Saito: Single-mode operations of a circular free-electron laser, Phys Rev Lett 77, p.2686-2689 (1996)

[6.1695] {Sect. 6.13.5.2} A. Abramovich, M. Canter, A. Gover, J. Sokolowski, Y.M. Yakover, Y. Pinhasi, I. Schnitzer, J. Shiloh: High spectral coherence in long-pulse and continuous free-electron laser: Measurements and theoretical limitations, Phys Rev Lett 82, p.5257-5260 (1999)

[6.1696] {Sect. 6.13.5.2} V. Telnov: Laser cooling of electron beams for linear colliders, Phys Rev Lett 78, p.4757-4760 (1997)

[6.1697] {Sect. 6.13.5.3} A.Y. Dergachev, S.B. Mirov: Efficient room temperature LiF:F-2 (+)** color center laser tunable in 820-1210 nm range, Opt Commun 147, p.107-111 (1998)

[6.1698] {Sect. 6.13.5.3} E.J. Mozdy, M.A. Jaspan, Z.H. Zhu, Y.H. Lo, C.R. Pollock, R. Bhat, M.W. Hong: NaCl:OH- color center laser modelocked by a novel bonded saturable Bragg reflector, Opt Commun 151, p.62-64 (1998)

[6.1699] {Sect. 6.13.5.3} V.V. Termikirtychev: Diode-pumped tunable room-temperature LiF:F-2 (-) color-center laser, Appl Opt 37, p.6442-6445 (1998)

[6.1700] {Sect. 6.13.5.3} T.T. Basiev, P.G. Zverev, V.V. Fedorov, S.B. Mirov: Multiline, superbroadband and sun-color oscillation of a LIF:F-2 (-) color-center laser, Appl Opt 36, p.2515-2522 (1997)

[6.1701] {Sect. 6.13.5.3} A. Konate, J.L. Doualan, S. Girard, J. Margerie: Tunable cw laser emission of the (a) variety of (F- 2 (+)) (H) centres in NaCl:OH-, Opt Commun 133, p.234-238 (1997)

[6.1702] {Sect. 6.13.5.3} V.V. Termikirtychev, T. Tsubo: Ultrabroadband LiF:F-2 (+*) color center laser using two- prism spatially-dispersive resonator, Opt Commun 137, p.74-76 (1997)

[6.1703] {Sect. 6.13.5.3} G. Phillips, P. Hinske, W. Demtröder, K. Möllmann, R. Beigang: NaCl-Color Center Laser with Birefringent Tuning, Appl. Phys. B 47, p.127-133 (1988)

[6.1704] {Sect. 6.13.5.3} R. Beigang, G. Litfin, H. Welling: Frequency behaviour and linewidth of cw single mode color center lasers, Opt. Comm. 22.p.269-271 (1977)

[6.1705] {Sect. 6.13.5.4} A. Bertolini, G. Carelli, A. Moretti, G. Moruzzi, F. Strumia: Laser action in hydrazine: Observation and characterization of new large offset FIR laser lines, IEEE J QE-35, p.12-14 (1999)

[6.1706] {Sect. 6.13.5.4} S.J. Cooper: Output power optimization and gain and saturation irradiance measurements on a RF-pumped HCN waveguide laser, Appl Opt 37, p.4881-4890 (1998)

[6.1707] {Sect. 6.13.5.4} J.N. Hovenier, A.V. Muravjov, S.G. Pavlov, V.N. Shastin, R.C. Strijbos, W.T. Wenckebach: Active mode locking of a p-Ge hot hole laser, Appl Phys Lett 71, p.443-445 (1997)

[6.1708] {Sect. 6.13.5.4} G.M.H. Knippels, X. Yan, A.M. MacLeod, W.A. Gillespie, M. Yasumoto, D. Oepts, A.F.G. vanderMeer: Generation and complete electric-field characterization of intense ultrashort tunable far-infrared laser pulses, Phys Rev Lett 83, p.1578-1581 (1999)

[6.1709] {Sect. 6.13.5.5} T.J. Carrig, G.J. Wagner, A. Sennaroglu, J.Y. Jeong, C.R. Pollock: Mode-locked Cr2+: ZnSe laser, Optics Letters 25, p.168-170 (2000)

[6.1710] {Sect. 6.13.5.5} J.J. Adams, C. Bibeau, R.H. Page, D.M. Krol, L.H. Furu, S.A. Payne: 4.0-4.5-mu m lasing of Fe : ZnSe below 180 K, a new mid-infrared laser material, Optics Letters 24, p.1720-1722 (1999)

[6.1711] {Sect. 6.13.5.5} A.A. Kaminskii, H.J. Eichler, K. Ueda, N.V. Klassen, B.S. Redkin, L.E. Li, J. Findeisen, D. Jaque, J. GarciaSole, J. Fernandez et al.: Properties of Nd3+-doped and undoped tetragonal PbWO4, NaY (WO4) (2), CaWO4, and undoped monoclinic ZnWO4 and CdWO4 as laser-active and stimulated Raman scattering-active crystals, Appl Opt 38, p.4533-4547 (1999)

[6.1712] {Sect. 6.13.5.5} S. Kuck, E. Heumann, T. Karner, A. Maaroos: Continuous-wave room-temperature laser oscillation of Cr3+: MgO, Optics Letters 24, p.966-968 (1999)

[6.1713] {Sect. 6.13.5.5} J.H. Liu, Z.S. Shao, X.L. Meng, H.J. Zhang, L. Zhu, M.H. Jiang: High-power CWNd : GdVO4 solid-state laser end-pumped by a diode-laser- array, Opt Commun 164, p.199-202 (1999)

[6.1714] {Sect. 6.13.5.5} G.S. Maciel, L.D. Menezes, C.B. deAraujo, Y. Messaddeq: Violet and blue light amplification in Nd3+-doped fluoroindate glasses, J Appl Phys 85, p.6782-6785 (1999)

[6.1715] {Sect. 6.13.5.5} J. Qiu, M. Shojiya, Y. Kawamoto: Sensitized Ho3+ up-conversion luminescence in Nd3+-Yb3+-Ho3+ co-doped ZrF4-based glass, J Appl Phys 86, p.909-913 (1999)

[6.1716] {Sect. 6.13.5.5} T. Schweizer, B.N. Samson, J.R. Hector, W.S. Brocklesby, D.W. Hewak, D.N. Payne: Infrared emission and ion-ion interactions in thulium- and terbium- doped gallium lanthanum sulfide glass, J Opt Soc Am B Opt Physics 16, p.308-316 (1999)

[6.1717] {Sect. 6.13.5.5} S.A. vandenBerg, R.H.V. denBezemer, H.F.M. Schoo, G.W. tHooft, E.R. Eliel: From amplified spontaneous emission to laser oscillation: dynamics in a short cavity polymer laser, Optics Letters 24, p.1847-1849 (1999)

[6.1718] {Sect. 6.13.5.5} G.J. Wagner, T.J. Carrig, R.H. Page, K.I. Schaffers, J.O. Ndap, X.Y. Ma, A. Burger: Continuous-wave broadly tunable Cr2+: ZnSe laser, Optics Letters 24, p.19-21 (1999)

[6.1719] {Sect. 6.13.5.5} R. Fluck, R. Haring, R. Paschotta, E. Gini, H. Melchior, U. Keller: Eyesafe pulsed microchip laser using semiconductor saturable absorber mirrors, Appl Phys Lett 72, p.3273-3275 (1998)

[6.1720] {Sect. 6.13.5.5} J.P. Foing, E. Scheer, B. Viana, N. Britos: Diode-pumped emission of Tm3+-doped Ca2Al2SiO7 crystals, Appl Opt 37, p.4857-4861 (1998)

[6.1721] {Sect. 6.13.5.5} E. Martins, C.B. deAraujo, J.R. Delben, A.S.L. Gomes, B.J. daCosta, Y. Messaddeq: Cooperative frequency upconversion in Yb3+-Tb3+ codoped fluoroindate glass, Opt Commun 158, p.61-64 (1998)

[6.1722] {Sect. 6.13.5.5} P. Rambaldi, R. Moncorge, J.P. Wolf, C. Pedrini, J.Y. Gesland: Efficient and stable pulsed laser operation of Ce:LiLuF4 around 308 nm, Opt Commun 146, p.163-166 (1998)

[6.1723] {Sect. 6.13.5.5} N. Sarukura, Z.L. Liu, S. Izumida, M.A. Dubinskii, R.Y. Abdulsabirov, S.L. Korableva: All-solid-state tunable ultraviolet subnanosecond laser with direct pumping by the fifth harmonic of a Nd:YAG laser, Appl Opt 37, p.6446-6448 (1998)

[6.1724] {Sect. 6.13.5.5} I. Sokolska, W. RybaRomanowski, S. Golab, M. Baba, T. Lukasiewicz: Spectroscopic assessment of LiTaO3 : Tm3+ as a potential diode-pumped laser near 1.9 mu m, J Appl Phys 84, p.5348-5350 (1998)

[6.1725] {Sect. 6.13.5.5} Y.X. Zhao, S. Fleming: Analysis of the effect of numerical aperture on Pr:ZBLAN upconversion fiber lasers, Optics Letters 23, p.373-375 (1998)

[6.1726] {Sect. 6.13.5.5} N. Djeu, V.E. Hartwell, A.A. Kaminskii, A.V. Butashin: Room-temperature 3.4-mu m Dy:BaYb2F8 laser, Optics Letters 22, p.997-999 (1997)

[6.1727] {Sect. 6.13.5.5} N. Sarukura, Z.L. Liu, H. Ohtake, Y. Segawa, M.A. Dubinskii, V.V. Semashko, A.K. Naumov, S.L. Korableva, R.Y. Abdulsabirov: Ultraviolet short pulses from an all-solid-state Ce:LiCAF master-oscillator-power-amplifier system, Optics Letters 22, p.994-996 (1997)

[6.1728] {Sect. 6.13.5.5} P.W. Binun, T.L. Boyd, M.A. Pessot, D.H. Tanimoto, D.E. Hargis: Pr:YLF, intracavity-pumped, room-temperature upconversion laser, Optics Letters 21, p.1915-1917 (1996)

[6.1729] {Sect. 6.13.5.5} L.B. Shaw, S.R. Bowman, B.J. Feldman, J. Ganem: Radiative and multiphonon relaxation of the Mid-IR transitions of Pr3+ in LaCl3, IEEE J QE-32, p.2166-2172 (1996)

[6.1730] {Sect. 6.13.5.5} R. Birkhahn, M. Garter, A.J. Steckl: Red light emission by photoluminescence and electroluminescence from Pr-doped GaN on Si substrates, Appl Phys Lett 74, p.2161-2163 (1999)

[6.1731] {Sect. 6.13.5.5} A.J. Steckl, M. Garter, D.S. Lee, J. Heikenfeld, R. Birkhahn: Blue emission from Tm-doped GaN electroluminescent devices, Appl Phys Lett 75, p.2184-2186 (1999)

[6.1732] {Sect. 6.13.5.5} R. Birkhahn, A.J. Steckl: Green emission from Er-doped GaN grown by molecular beam epitaxy on Si substrates, Appl Phys Lett 73, p.2143-2145 (1998)

[6.1733] {Sect. 6.13.5.5} N.D. Kumar, J.D. Bhawalkar, P.N. Prasad, F.E. Karasz, B. Hu: Solid-state tunable cavity lasing in a poly (para-phenylene vinylene) derivative alternating block co-polymer, Appl Phys Lett 71, p.999-1001 (1997)

[6.1734] {Sect. 6.13.5.6} K. Katayama, H. Yao, F. Nakanishi, H. Doi, A. Saegusa, N. Okuda, T. Yamada, H. Matsubara, M. Irikura, T. Matsuoka et al.: Lasing characteristics of low threshold ZnSe-based blue/green laser diodes grown on conductive ZnSe substrates, Appl Phys Lett 73, p.102-104 (1998)

[6.1735] {Sect. 6.13.5.6} V.G. Kozlov, V. Bulovic, S.R. Forrest: Temperature independent performance of organic semiconductor lasers, Appl Phys Lett 71, p.2575-2577 (1997)

[6.1736] {Sect. 6.13.5.6} S. Tanaka, H. Hirayama, Y. Aoyagi, Y. Narukawa, Y. Kawakami, S. Fujita: Stimulated emission from optically pumped GaN quantum dots, Appl Phys Lett 71, p.1299-1301 (1997)

[6.1737] {Sect. 6.13.5.6} A. Waag, F. Fischer, K. Schull, T. Baron, H.J. Lugauer, T. Litz, U. Zehnder, W. Ossau, T. Gerhard, M. Keim, et al.: Laser diodes based on beryllium-chalcogenides, Appl Phys Lett 70, p.280-282 (1997)

[6.1738] {Sect. 6.13.5.6} L.D. Deloach, R.H. Page, G.D. Wilke, S.A. Payne, W.F. Krupke: Transition metal-doped zinc chalcogenides: Spectroscopy and laser demonstration of a new class of gain media, IEEE J QE-32, p.885-895 (1996)

[6.1739] {Sect. 6.13.5.6} P.A. Ramos, E. Towe: Surface-emitted blue light from [112]-oriented (In, Ga)As/GaAs quantum well edge-emitting lasers, Appl Phys Lett 69, p.3321-3323 (1996)

[6.1740] {Sect. 6.13.5.7} T.L. Rittenhouse, S.P. Phipps, C.A. Helms: Performance of a high-efficiency 5-cm gain length supersonic chemical oxygen-iodine laser, IEEE J QE-35, p.857-866 (1999)

[6.1741] {Sect. 6.13.5.7} F. Wani, M. Endo, T. Fujioka: High-pressure, high-efficiency operation of a chemical oxygen-iodine laser, Appl Phys Lett 75, p.3081-3083 (1999)

[6.1742] {Sect. 6.13.5.7} M. Endo, S. Nagatomo, S. Takeda, M.V. Zagidullin, V.D. Nikolaev, H. Fujii, F. Wani, D. Sugimoto, K. Sunako, K. Nanri et al.: High-efficiency operation of chemical oxygen-iodine laser using nitrogen as buffer gas, IEEE J QE-34, p.393-398 (1998)

[6.1743] {Sect. 6.13.5.7} D. Sugimoto, M. Endo, K. Nanri, S. Takeda, T. Fujioka: Output power stabilization of a chemical oxygen-iodine laser with an external magnetic field, IEEE J QE-34, p.1526-1532 (1998)

[6.1744] {Sect. 6.13.5.7} D. Furman, B.D. Barmashenko, S. Rosenwaks: An efficient supersonic chemical oxygen-iodine laser operating without buffer gas and with simple nozzle geometry, Appl Phys Lett 70, p.2341-2343 (1997)

[6.1745] {Sect. 6.13.5.7} Y. Kalisky, K. Waichman, S. Kamin, D. Chuchem: Plasma cathode preionized atmospheric pressure HF chemical laser, Opt Commun 137, p.59-63 (1997)

[6.1746] {Sect. 6.13.5.7} G.N. Tsikrikas, A.A. Serafetinides: Discharge and circuit simulation of a plasma cathode TEA HF laser operating with a He/SF6/C3H8 gas mixture, Opt Commun 134, p.145-148 (1997)

[6.1747] {Sect. 6.13.5.7} I. Blayvas, B.D. Barmashenko, D. Furman, S. Rosenwaks, M.V. Zagidullin: Power optimization of small-scale chemical oxygen-iodine laser with jet-type singlet oxygen generator, IEEE J QE-32, p.2051-2057 (1996)

[6.1748] {Sect. 6.13.5.7} G.D. Hager, C.A. Helms, K.A. Truesdell, D. Plummer, J. Erkkila, P. Crowell: A simplified analytic model for gain saturation and power extraction in the flowing chemical oxygen-iodine laser, IEEE J QE-32, p.1525-1536 (1996)

[6.1749] {Sect. 6.13.5.7} S.P. Phipps, C.A. Helms, R.J. Copland, W. Rudolph, K.A. Truesdell, G.D. Hager: Mode locking of a CW supersonic chemical oxygen-iodine laser, IEEE J QE-32, p.2045-2050 (1996)

[6.1750] {Sect. 6.14.2.1} G. Boyer: High-power femtosecond-pulse reshaping near the zero-dispersion wavelength of an optical fiber, Optics Letters 24, p.945-947 (1999)

[6.1751] {Sect. 6.14.2.1} P. Ceccherini, M.G. Pelizzo, P. Villoresi, S. DeSilvestri, M. Nisoli, S. Stagira: Surface damage of extreme-ultraviolet gratings exposed to high-energy 20-fs laser pulses, Appl Opt 38, p.4720-4724 (1999)

[6.1752] {Sect. 6.14.2.1} O. Duhr, E.T.J. Nibbering, G. Korn, G. Tempea, F. Krausz: Generation of intense 8-fs pulses at 400 nm, Optics Letters 24, p.34-36 (1999)

[6.1753] {Sect. 6.14.2.1} C.G. Durfee, S. Backus, H.C. Kapteyn, M.M. Murnane: Intense 8-fs pulse generation in the deep ultraviolet, Optics Letters 24, p.697-699 (1999)

[6.1754] {Sect. 6.14.2.1} L. Lefort, K. Puech, S.D. Butterworth, Y.P. Svirko, D.C. Hanna: Generation of femtosecond pulses from order-of-magnitude pulse compression in a synchronously pumped optical parametric oscillator based on periodically poled lithium niobate, Optics Letters 24, p.28-30 (1999)

[6.1755] {Sect. 6.14.2.1} X. Liu, L.J. Qian, F. Wise: High-energy pulse compression by use of negative phase shifts produced by the cascade chi ((2)):chi ((2)) nonlinearity, Optics Letters 24, p.1777-1779 (1999)

[6.1756] {Sect. 6.14.2.1} V.N. Malkin, G. Shvets, N.J. Fisch: Fast compression of laser beams to highly overcritical powers, Phys Rev Lett 82, p.4448-4451 (1999)

[6.1757] {Sect. 6.14.2.1} N.A. Papadogiannis, B. Witzel, C. Kalpouzos, D. Charalambidis: Observation of attosecond light localization in higher order harmonic generation, Phys Rev Lett 83, p.4289-4292 (1999)

[6.1758] {Sect. 6.14.2.1} T. Sekikawa, T. Ohno, T. Yamazaki, Y. Nabekawa, S. Watanabe: Pulse compression of a high-order harmonic by compensating the atomic dipole phase, Phys Rev Lett 83, p.2564-2567 (1999)

[6.1759] {Sect. 6.14.2.1} A.V. Sokolov, D.D. Yavuz, S.E. Harris: Subfemtosecond pulse generation by rotational molecular modulation, Optics Letters 24, p.557-559 (1999)

[6.1760] {Sect. 6.14.2.1} A.V. Sokolov: Subfemtosecond compression of periodic laser pulses, Optics Letters 24, p.1248-1250 (1999)

[6.1761] {Sect. 6.14.2.1} E. Zeek, K. Maginnis, S. Backus, U. Russek, M. Murnane, G. Mourou, H. Kapteyn, G. Vdovin: Pulse compression by use of deformable mirrors, Optics Letters 24, p.493-495 (1999)

[6.1762] {Sect. 6.14.2.1} B.J. Eggleton, G. Lenz, R.E. Slusher, N.M. Litchinitser: Compression of optical pulses spectrally broadened by self-phase modulation with a fiber Bragg grating in transmission, Appl Opt 37, p.7055-7061 (1998)

[6.1763] {Sect. 6.14.2.1} M.A. Arbore, A. Galvanauskas, D. Harter, M.H. Chou, M.M. Fejer: Engineerable compression of ultrashort pulses by use of second-harmonic generation in chirped-period-poled lithium niobate, Optics Letters 22, p.1341-1343 (1997)

[6.1764] {Sect. 6.14.2.1} A. Baltuska, Z.Y. Wei, M.S. Pshenichnikov, D.A. Wiersma: Optical pulse compression to 5 fs at a 1-MHz repetition rate, Optics Letters 22, p.102-104 (1997)

[6.1765] {Sect. 6.14.2.1} N.G.R. Broderick, D. Taverner, D.J. Richardson, M. Ibsen, R.I. Laming: Experimental observation of nonlinear pulse compression in nonuniform Bragg gratings, Optics Letters 22, p.1837-1839 (1997)

[6.1766] {Sect. 6.14.2.1} A. Dreischuh, I. Buchvarov, E. Eugenieva, A. Iliev, S. Dinev: Experimental demonstration of pulse shaping and shortening by spatial filtering of an induced-phase-modulated probe wave, IEEE J QE-33, p.329-335 (1997)

[6.1767] {Sect. 6.14.2.1} A. Dubietis, G. Valiulis, G. Tamosauskas, R. Danielius, A. Piskarskas: Nonlinear second-harmonic pulse compression with tilted pulses, Optics Letters 22, p.1071-1073 (1997)

[6.1768] {Sect. 6.14.2.1} J. Itatani, Y. Nabekawa, K. Kondo, S. Watanabe: Generation of 13-TW, 26-fs pulses in a Ti:Sapphire laser, Opt Commun 134, p.134-138 (1997)

[6.1769] {Sect. 6.14.2.1} M. Nisoli, S. DeSilvestri, O. Svelto, R. Szipocs, K. Ferencz, C. Spielmann, S. Sartania, F. Krausz: Compression of high-energy laser pulses below 5 fs, Optics Letters 22, p.522-524 (1997)

[6.1770] {Sect. 6.14.2.1} S. Sartania, Z. Cheng, M. Lenzner, G. Tempea, C. Spielmann, F. Krausz, K. Ferencz: Generation of O.1-TW 5-fs optical pulses at a 1-kHz repetition rate, Optics Letters 22, p.1562-1564 (1997)

[6.1771] {Sect. 6.14.2.1} A. Baltuska, Z. Wei, M.S. Pshenichnikov, D.A. Wiersma, R Szipöcs: All-solid-state cavity-dumped sub-5-fs laser, Appl. Phys. B 65, p.175-188 (1997)

[6.1772] {Sect. 6.14.2.1} J.A. Britten, M.D. Perry, B.W. Shore, R.D. Boyd: Universal grating design for pulse stretching and compression in the 800-1200-nm range, Optics Letters 21, p.540-542 (1996)

[6.1773] {Sect. 6.14.2.1} J.P. Chambaret, C. LeBlanc, G. Cheriaux, P. Curley, G. Darpentigny, P. Rousseau, G. Hamoniaux, A. Antonetti, F. Salin: Generation of 25-TW, 32-fs pulses at 10 Hz, Optics Letters 21, p.1921-1923 (1996)

[6.1774] {Sect. 6.14.2.1} M. Nisoli, S. DeSilvestri, O. Svelto: Generation of high energy 10 fs pulses by a new pulse compression technique, Appl Phys Lett 68, p.2793-2795 (1996)

[6.1775] {Sect. 6.14.2.1} R.L. Fork, C.H. Brito Cruz, P.C. Becker, C.V. Shank: Compression of optical pulses to six femtoseconds by using cubic phase compensation, Opt. Lett. 12, p.483-485 (1987)

[6.1776] {Sect. 6.14.2.1} J.G. Fujimoto, A.M.Weiner, E.P. Ippen: Generation and measurement of optical pulses as short as 16 fs, Appl. Phys. Lett. 44, p.832-834 (1984)

[6.1777] {Sect. 6.14.2.1} B. Nicolaus, D. Grischkowsky: 90-fs tunable optical pulses optained by two-stage pulse compression, Appl. Phys. Lett. 43, p.228-230 (1983)

[6.1778] {Sect. 6.14.2.1} C.V. Shank, R.L. Fork, R.Yen, R.H. Stolen, W.J. Tomlinson: Compression of femtosecond optical pulses, Appl. Phys. Lett. 40, p.761-763 (1982)

[6.1779] {Sect. 6.14.2.1} M.A. Duguay, J.W. Hansen: Compression of pulses from a mode-locked He-Ne laser, Appl. Phys. Lett. 14, p.14-16 (1969)

[6.1780] {Sect. 6.14.2.1} E.B. Treacy: Optical pulse compression with diffraction gratings, IEEE J. QE-5, p.454-458 (1969)

[6.1781] {Sect. 6.14.2.2} S.V. Kurbasov, L.L. Losev: Raman compression of picosecond microjoule laser pulses in KGd (WO4) (2) crystal, Opt Commun 168, p.227-232 (1999)

[6.1782] {Sect. 6.14.2.2} V. Kmetik, H. Fiedorowicz, A.A. Andreev, K.J. Witte, H. Daido, H. Fujita, M. Nakatsuka, T. Yamanaka: Reliable stimulated Brillouin scattering compression of Nd:YAG laser pulses with liquid fluorocarbon for long-time operation at 10 Hz, Appl Opt 37, p.7085-7090 (1998)

[6.1783] {Sect. 6.14.2.2} S. Schiemann, W. Hogervorst, W. Ubachs: Fourier-transform-limited laser pulses tunable in wavelength and in duration (400-2000 ps), IEEE J QE-34, p.407-412 (1998)

[6.1784] {Sect. 6.14.2.2} N.G.R. Broderick, D. Taverner, D.J. Richardson, M. Ibsen, R.I. Laming: Optical pulse compression in fiber Bragg gratings, Phys Rev Lett 79, p.4566-4569 (1997)

[6.1785] {Sect. 6.14.2.2} P. Klovekorn, J. Munch: Variable stimulated Brillouin scattering pulse compressor for nonlinear optical measurements, Appl Opt 36, p.5913-5917 (1997)

[6.1786] {Sect. 6.14.2.2} S. Schiemann, W. Ubachs, W. Hogervorst: Efficient temporal compression of coherent nanosecond pulses in compact SBS generator-
amplifier setup, IEEE J QE-33, p.358-366 (1997)

[6.1787] {Sect. 6.14.2.2} P. Klövekorn, J. Munch: Variable stimulated Brillouin scattering pulse compressor for nonlinear optical measurements, Appl. Opt. 36,
p.5913-5917 (1997)

[6.1788] {Sect. 6.14.2.2} A. Galvanauskas, P.A. Krug, D. Harter: Nanosecond-to-
picosecond pulse compression with fiber gratings in a compact fiber-based
chirped-pulse- amplification system, Optics Letters 21, p.1049-1051 (1996)

[6.1789] {Sect. 6.14.2.2} Yu. Nizienko, A. Mamin, P. Nielsen, B. Brown: 300 ps
ruby laser using stimulated Brillouin scattering pulse compression, Rev.
Sci. Instrum. 65, p.2460-2463 (1994)

[6.1790] {Sect. 6.14.2.2} S. Kinoshita, W. Tsurumaki, Y. Shimada, T. Yagi: Relationship between coherent acoustic wave generation and a coherence spike
in an impulsive stimulated Brillouin scattering experiment, J. Opt. Soc.
Am. B. 10, p.1017-1024 (1993)

[6.1791] {Sect. 6.14.2.2} V.V. Krushas, A.S. Piskarskas, V.I. Smil'gyavichyus, G.P.
Shlekis: High-power subnanosecond optical parameter oscillator pumped by
a laser with a stimulated Brillouin scattering compressor, Sov. J. Quantum
Electron. 17, p.1054-1055 (1987)

[6.1792] {Sect. 6.14.2.2} O.L. Bourne, A.J. Alcock: Simplified Technique for Subnanosecond Pulse Generation and Injection Mode-Locking of a XeCl Laser,
Appl. Phys. B 36, p.181-185 (1985)

[6.1793] {Sect. 6.14.2.2} M.J. Damzen, M.H.R. Hutchinson: Pulse compression in a
phase-conjugating Brillouin cavity, Opt. Lett. 9, p.282-284 (1984)

[6.1794] {Sect. 6.14.2.2} M.J. Damzen, M.H.R. Hutchinson: High-efficiency laser-
pulse compression by stimulated Brillouin scattering, Opt. Lett. 8, p.313-
315 (1983)

[6.1795] {Sect. 6.14.2.2} M.J. Damzen, M.H.R. Hutchinson: Laser Pulse Compression by Stimulated Brillouin Scattering in Tapered Waveguides, IEEE J.
QE-19, p.7-14 (1983)

[6.1796] {Sect. 6.14.2.2} V.A. Gorbunov, S.B. Papernyl, V.F. Petrov, V.R. Startsev:
Time compression on pulses in the course of stimulated Brillouin scattering
in gases, Sov. J. Quantum Electron. 13, p.900-905 (1983)

[6.1797] {Sect. 6.14.2.2} I.V. Tomov, R. Fedosefevs, D.C.D. McKen, C. Domier,
A.A. Offenberger: Phase conjugation and pulse compression of KrF-laser
radiation by stimulated raman scattering, Opt. Lett. 8, p.9-11 (1983)

[6.1798] {Sect. 6.14.2.2} D.T. Hon: Pulse compression by stimulated Brillouin scattering, Opt. Lett. 5, p.516-518 (1980)

[6.1799] {Sect. 6.14.2.3} P. Antoine, A. LHuillier, M. Lewenstein: Attosecond pulse
trains using high-order harmonics, Phys Rev Lett 77, p.1234-1237 (1996)

[6.1800] {Sect. 6.14.2.3} P. Zhou, H. Schulz, P. Kohns: Atomic spectroscopy with ultrashort laser pulses using frequency-resolved optical gating, Opt Commun
123, p.501-504 (1996)

[6.1801] {Sect. 6.14.2.3} M.A. Duguay, J.W. Hansen: An ultrafast light gate, Appl.
Phys. Lett. 15, p.192-194 (1969)

[6.1802] {Sect. 6.14.2.3} A.N. Starodumov, L.A. Zenteno, N. Arzate, P. Gavrilovic:
Nonlinear-optical modulator for high-power fiber lasers, Optics Letters 22,
p.286-288 (1997)

[6.1803] {Sect. 6.14.2.4} J. Knittel, D.P. Scherrer, F.K. Kneubuhl: A plasma shutter
for far-infrared laser radiation, IEEE J QE-32, p.2058-2063 (1996)

[6.1804] {Sect. 6.14.2.4} T. Nagamura, T. Hamada: Novel all optical light modulation based on complex refractive index changes of organic die-doped polymer film upon photoexcitation, Appl Phys Lett 69, p.1191-1193 (1996)

[6.1805] {Sect. 6.14.2.4} T. Tsang, M.A. Krumbugel, K.W. Delong, D.N. Fittinghoff, R. Trebino: Frequency-resolved optical-gating measurements of ultrashort pulses using surface third-harmonic generation, Optics Letters 21, p.1381-1383 (1996)

[6.1806] {Sect. 6.14.2.4} S.P. Nikitin, Y.L. Li, T.M. Antonsen, H.M. Milchberg: Ionization-induced pulse shortening and retardation of high intensity femtosecond laser pulses, Opt Commun 157, p.139-144 (1998)

[6.1807] {Sect. 6.14.2.4} K. Sasaki, T. Nagamura: Ultrafast air-optical switch using complex refractive index changes of thin films containing photochromic dye, Appl Phys Lett 71, p.434-436 (1997)

[6.1808] {Sect. 6.14.2.4} L. Gallmann, G. Steinmeyer, D.H. Sutter, N. Matuschek, U. Keller: Collinear type II second-harmonic-generation frequency-resolved optical gating for the characterization of sub-10-fs optical pulses, Optics Letters 25, p.269-271 (2000)

[6.1809] {Sect. 6.14.2.5} N. Kamiya, H. Ohtani, T. Sekikawa, T. Kobayashi: Subpicosecond fluorescence spectroscopy of the M intermediate in the photocycle of bacteriorhodopsin by using up-conversion fluorometry, Chem Phys Lett 305, p.15-20 (1999)

[6.1810] {Sect. 6.14.2.5} A. Mokhtari, A. Chebira, J. Chesnoy: Subpicosecond fluorescence dynamics of dye molecules, J. Opt. Soc. Am. B 7, p.1551-1557 (1990)

[6.1811] {Sect. 6.14.2.5} G.S. Beddard, T. Doust, G. Porter: Picosecond fluorescence depolarization measured by frequency conversion, Chem. Phys. 61, p.17-23 (1981)

[6.1812] {Sect. 6.15.0} A.V. Smith, D.J. Armstrong, W.J. Alford: Increased acceptance bandwidths in optical frequency conversion by use of multiple walk-off-compensating nonlinear crystals, J. Opt. Soc. Am. B 15, p.122-141 (1998)

[6.1813] {Sect. 6.15.0} D.H. Jundt: Temperature-dependent Sellmeier equation for the index of refraction, ne, in congruent lithium niobate, Opt. Lett. 22, p.1553-1555 (1997)

[6.1814] {Sect. 6.15.0} M. Taya, M.C. Bashaw, M.M. Fejer: Photorefractive effects in periodically poled ferroelectrics, Opt. Lett. 21, p.857-859 (1996)

[6.1815] {Sect. 6.15.0} K. Asaumi: Approximate effective nonlinear coefficient of second-harmonic generation in KTiOPO4, Appl. Opt. 32, p.5983-5985 (1993)

[6.1816] {Sect. 6.15.0} L.K. Cheng, L.-T. Cheng, J.D. Bierlein, F.C. Zumsteg: Properties of doped and undoped crystals of single domain KTiOAsO4, Appl. Phys. Lett. 62, p.346-348 (1993)

[6.1817] {Sect. 6.15.0} J. Yao, W. Sheng, W. Shi: Accurate calculation of the optimum phase-matching parameters in three-wave interactions with biaxial nonlinear-optical crystals, J. Opt. Soc. Am. B 9, p.891-902 (1992)

[6.1818] {Sect. 6.15.0} D.N. Nikogosyan: Beta Barium Borate BBO – A Review of its Properties and Applications, Appl. Phys. A 52, p.359-368 (1991)

[6.1819] {Sect. 6.15.0} R.C. Eckardt, H. Masuda, Y.X. Fan, R.L. Byer: Absolute and Relative Nonlinear Optical Coefficients of KDP, KD*P, BaB2O4, LiIO3, MgO:LiNbO3, and KTP Measured by Phase-Matched Second-Harmonic Generation, IEEE J. QE-26, p.922-933 (1990)

[6.1820] {Sect. 6.15.0} S. Lin, Z. Sun, B. Wu, Ch. Chen: The nonlinear optical characteristics of a LiB3O5 crystal, J. Appl. Phys. 67, p.634-638 (1990)

[6.1821] {Sect. 6.15.0} J.D. Bierlein, H. Vanherzeele, A.A. Ballman: Linear and nonlinear properties of flux-grown KTiOAsO4, Appl. Phys. Lett. 54, p.783-785 (1989)

[6.1822] {Sect. 6.15.0} J.D. Bierlein, H. Vanherzeele: Potassium titanyl phosphate: properties and new applications, J. Opt. Soc. Am. B 6, p.622-633 (1989)

[6.1823] {Sect. 6.15.0} W.L. Bosenberg, L.K. Cheng, C.L. Tang: Ultraviolet optical parametric oscillation in beta-BaB2O4, Appl. Phys. Lett. 54, p.13-15 (1989)

[6.1824] {Sect. 6.15.0} D. Eimerl, L. Davis, S. Velsko, E.K. Graham, A. Zalkin: Optical, mechanical, and thermal properties of barium borate, J. Appl. Phys. 62, p.1968-1983 (1987)

[6.1825] {Sect. 6.15.0} D. Eimerl: Quadrature Frequency Conversion, IEEE J. QE-23, p.1361-1371 (1987)

[6.1826] {Sect. 6.15.0} T.Y. Fan, C.E. Huang, B.Q. Hu, R.C. Eckardt, Y.X. Fan, R.L. Byer, R.S. Feigelson: Second harmonic generation and accurate index of refraction measurements in flux-grown KTiOPO4, Appl. Opt. 26, p.2390-2394 (1987)

[6.1827] {Sect. 6.15.0} J.Q. Yao, T.S. Fahlen: Calculations of optimum phase match parameters for the biaxial crystal KTiOPO4, J. Appl. Phys. 55, p.65-68 (1984)

[6.1828] {Sect. 6.15.0} R. Hilbig, R. Wallenstein: Narrowband tunable VUV radiation generated by nonresonant sum- and difference-frequency mixing in xenon and krypton, Appl. Opt. 21, p.913-917 (1982)

[6.1829] {Sect. 6.15.0} R.S. Craxton, S.D. Jacobs, J.E. Rizzo, R. Boni: Basic Properties of KDP Related to the Frequency Conversion of 1 μm Laser Radiation, IEEE J. QE-17, p.1782-1786 (1981)

[6.1830] {Sect. 6.15.0} D.T. Hon: Electrooptical Compensation for Self-Heating in CD*A During Second-Harmonic Generation, IEEE J. QE-12, p.148-151 (1976)

[6.1831] {Sect. 6.15.0} A.M. Glass, D. von der Linde, T.J. Negran: High-voltage bulk photovoltaic effect and the photorefractive process in LiNbO3, Appl. Phys. Lett. 25, p.233-235 (1974)

[6.1832] {Sect. 6.15.0} K. Kato: High Efficient UV Generation at 3572 A in RDA, IEEE J. QE-10, p.622-624 (1974)

[6.1833] {Sect. 6.15.1} S. Konno, T. Kojima, S. Fujikawa, K. Yasui: High-brightness 138-W green laser based on an intracavity-frequency- doubled diode-side-pumped Q-switched Nd : YAG laser, Optics Letters 25, p.105-107 (2000)

[6.1834] {Sect. 6.15.1} J.D. Bhawalkar, Y. Mao, H. Po, A.K. Goyal, P. Gavrilovic, Y. Conturie, S. Singh: High-power 390-nm laser source based on efficient frequency doubling of a tapered diode laser in an external resonant cavity, Optics Letters 24, p.823-825 (1999)

[6.1835] {Sect. 6.15.1} J.C. Diettrich, I.T. McKinnie, D.M. Warrington: Tunable high-repetition-rate visible solid-state lasers based on intracavity frequency doubling of Cr : forsterite, IEEE J QE-35, p.1718-1723 (1999)

[6.1836] {Sect. 6.15.1} S.M. Giffin, I.T. McKinnie: Tunable visible solid-state lasers based on intracavity frequency doubling of Cr : forsterite in KTP, Optics Letters 24, p.884-886 (1999)

[6.1837] {Sect. 6.15.1} A.K. Goyal, J.D. Bhawalkar, Y. Conturie, P. Gavrilovic, Y. Mao, H. Po, J. Guerra: High beam quality of ultraviolet radiation generated through resonant enhanced frequency doubling of a diode laser, J Opt Soc Am B Opt Physics 16, p.2207-2216 (1999)

[6.1838] {Sect. 6.15.1} Y. Inoue, S. Konno, T. Kojima, S. Fujikawa: High-power red beam generation by frequency-doubling of a Nd : YAG laser, IEEE J QE-35, p.1737-1740 (1999)

[6.1839] {Sect. 6.15.1} J.G. Liu, D. Kim: Optimization of intracavity doubled passively Q-switched solid-state lasers, IEEE J QE-35, p.1724-1730 (1999)

[6.1840] {Sect. 6.15.1} M. Mlejnek, E.M. Wright, J.V. Moloney, N. Bloembergen: Second harmonic generation of femtosecond pulses at the boundary of a nonlinear dielectric, Phys Rev Lett 83, p.2934-2937 (1999)

[6.1841] {Sect. 6.15.1} K. Otsuka, R. Kawai, Y. Asakawa: Intracavity second-harmonic and sum-frequency generation with a laser- diode-pumped multi-transition-oscillation LiNdP4O12 laser, Optics Letters 24, p.1611-1613 (1999)

[6.1842] {Sect. 6.15.1} I.I. Smolyaninov, C.H. Lee, C.C. Davis: Giant enhancement of surface second harmonic generation in BaTiO3 due to photorefractive surface wave excitation, Phys Rev Lett 83, p.2429-2432 (1999)

[6.1843] {Sect. 6.15.1} T. Sugita, K. Mizuuchi, Y. Kitaoka, K. Yamamoto: 31%-efficient blue second-harmonic generation in a periodically poled MgO: LiNbO3 waveguide by frequency doubling of an AlGaAs laser diode, Optics Letters 24, p.1590-1592 (1999)

[6.1844] {Sect. 6.15.1} K. Tei, M. Kato, F. Matsuoka, Y. Niwa, Y. Maruyama, T. Matoba, T. Arisawa: High-repetition rate 1-J green laser system, Appl Opt 38, p.4548-4551 (1999)

[6.1845] {Sect. 6.15.1} D. Woll, B. Beier, K.J. Boller, R. Wallenstein, M. Hagberg, S. OBrien: 1 W of blue 465-nm radiation generated by frequency doubling of the output of a high-power diode laser in critically phase-matched LiB3O5, Optics Letters 24, p.691-693 (1999)

[6.1846] {Sect. 6.15.1} A. Agnesi, G.C. Reali, P.G. Gobbi: 430-mW single-transverse-mode diode-pumped Nd:YVO4 laser at 671 nm, IEEE J QE-34, p.1297-1300 (1998)

[6.1847] {Sect. 6.15.1} Y.F. Chen, T.M. Huang, C.L. Wang, L.J. Lee: Compact and efficient 3.2-W diode-pumped Nd:YVO4/KTP green laser, Appl Opt 37, p.5727-5730 (1998)

[6.1848] {Sect. 6.15.1} J.M. Eichenholz, M. Richardson, G. Mizell: Diode pumped, frequency doubled LiSAF microlaser, Opt Commun 153, p.263-266 (1998)

[6.1849] {Sect. 6.15.1} M. Hofer, M.E. Fermann, A. Galvanauskas, D. Harter, R.S. Windeler: High-power 100-fs pulse generation by frequency doubling of an erbium- ytterbium-fiber master oscillator power amplifier, Optics Letters 23, p.1840-1842 (1998)

[6.1850] {Sect. 6.15.1} E.C. Honea, C.A. Ebbers, R.J. Beach, J.A. Speth, J.A. Skidmore, M.A. Emanuel, S.A. Payne: Analysis of an intracavity-doubled diode-pumped Q-switched Nd: YAG laser producing more than 100 W of power at 0.532 mu m, Optics Letters 23, p.1203-1205 (1998)

[6.1851] {Sect. 6.15.1} T. Kaing, M. Houssin: Ring cavity enhanced second harmonic generation of a diode laser using LBO crystal, Opt Commun 157, p.155-160 (1998)

[6.1852] {Sect. 6.15.1} S. Konno, S. Fujikawa, K. Yasui: Highly efficient 68-W green-beam generation by use of an intracavity frequency-doubled diode side-pumped Q-switched Nd:YAG rod laser, Appl Opt 37, p.6401-6404 (1998)

[6.1853] {Sect. 6.15.1} I.D. Lindsay, M. Ebrahimzadeh: Efficient continuous-wave and Q-switched operation of a 946-nm Nd:YAG laser pumped by an injection-locked broad-area diode laser, Appl Opt 37, p.3961-3970 (1998)

[6.1854] {Sect. 6.15.1} P.E.A. Mobert, E. Heumann, G. Huber, B.H.T. Chai: 540 mW of blue output power at 425 nm generated by intracavity frequency doubling an upconversion-pumped Er3+:YLiF4 laser, Appl Phys Lett 73, p.139-141 (1998)

[6.1855] {Sect. 6.15.1} V. Pasiskevicius, S.H. Wang, J.A. Tellefsen, F. Laurell, H. Karlsson: Efficient Nd:YAG laser frequency doubling with periodically poled KTP, Appl Opt 37, p.7116-7119 (1998)

[6.1856] {Sect. 6.15.1} D.Y. Shen, A.P. Liu, J. Song, K. Ueda: Efficient operation of an intracavity-doubled Nd:YVO4/KTP laser end pumped by a high-brightness laser diode, Appl Opt 37, p.7785-7788 (1998)

[6.1857] {Sect. 6.15.1} C.L. Wang, K.H. Lin, T.M. Hwang, Y.F. Chen, S.C. Wang, C.L. Pan: Mode-locked diode-pumped self-frequency-doubling neodymium yttrium aluminum borate laser, Appl Opt 37, p.3282-3285 (1998)

[6.1858] {Sect. 6.15.1} E.C. Honea, Ch.A. Ebbers, R.J. Beach, J.A. Speth, J.A. Skidmore, M.A. Emanuel, S.A. Payne: Analysis of an intracavity-doubled diode-pumped Q-switched Nd:YAG laser producing more than 100 W of power at 0.532 µm, Opt. Lett. 23, p.1203-1205 (1998)

[6.1859] {Sect. 6.15.1} A. Agnesi, E. Piccinini, G.C. Reali, C. Solcia: All-solid-state picosecond tunable source of near-infrared radiation, Optics Letters 22, p.1415-1417 (1997)

[6.1860] {Sect. 6.15.1} M.A. Arbore, M.M. Fejer, M.E. Fermann, A. Hariharan, A. Galvanauskas, D. Harter: Frequency doubling of femtosecond erbium-fiber soliton lasers in periodically poled lithium niobate, Optics Letters 22, p.13-15 (1997)

[6.1861] {Sect. 6.15.1} J. Bartschke, R. Knappe, K.J. Boller, R. Wallenstein: Investigation of efficient self-frequency-doubling Nd:YAB lasers, IEEE J QE-33, p.2295-2300 (1997)

[6.1862] {Sect. 6.15.1} B. Beier, D. Woll, M. Scheidt, K.J. Boller, R. Wallenstein: Second harmonic generation of the output of an AlGaAs diode oscillator amplifier system in critically phase matched LiB3O5 and beta-BaB2O4, Appl Phys Lett 71, p.315-317 (1997)

[6.1863] {Sect. 6.15.1} M. Bode, I. Freitag, A. Tunnermann, H. Welling: Frequency-tunable 500-mW continuous-wave all-solid-state single-frequency source in the blue spectral region, Optics Letters 22, p.1220-1222 (1997)

[6.1864] {Sect. 6.15.1} A. Brenier: Modelling of the NYAB self-doubling laser with focused Gaussian beams, Opt Commun 141, p.221-228 (1997)

[6.1865] {Sect. 6.15.1} A. Englander, R. Lavi, M. Katz, M. Oron, D. Eger, E. Lebiush, G. Rosenman, A. Skliar: Highly efficient doubling of a high-repetition-rate diode-pumped laser with bulk periodically poled KTP, Optics Letters 22, p.1598-1599 (1997)

[6.1866] {Sect. 6.15.1} S. Falter, K.M. Du, Y. Liao, M. Quade, J. Zhang, P. Loosen, R. Poprawe: Dynamics and stability of a laser system with second-order nonlinearity, Optics Letters 22, p.609-611 (1997)

[6.1867] {Sect. 6.15.1} X. Liu, L.J. Qian, F.W. Wise: Efficient generation of 50-fs red pulses by frequency doubling in LiB3O5, Opt Commun 144, p.265-268 (1997)

[6.1868] {Sect. 6.15.1} K.I. Martin, W.A. Clarkson, D.C. Hanna: Stable, high-power, single-frequency generation at 532 nm from a diode-bar-pumped Nd:YAG ring laser with an intracavity LBO frequency doubler, Appl Opt 36, p.4149-4152 (1997)

[6.1869] {Sect. 6.15.1} K.I. Martin, W.A. Clarkson, D.C. Hanna: Self-suppression of axial mode hopping by intracavity second-harmonic generation, Optics Letters 22, p.375-377 (1997)

[6.1870] {Sect. 6.15.1} I.T. McKinnie, A.M.L. Oien: Tunable red-yellow laser based on second harmonic generation of Cr:forsterite in KTP, Opt Commun 141, p.157-161 (1997)

[6.1871] {Sect. 6.15.1} G.D. Miller, R.G. Batchko, W.M. Tulloch, D.R. Weise, M.M. Fejer, R.L. Byer: 42%-efficient single-pass cw second-harmonic generation in periodically poled lithium niobate, Optics Letters 22, p.1834-1836 (1997)

[6.1872] {Sect. 6.15.1} Y. Uchiyama, M. Tsuchiya, H.F. Liu, T. Kamiya: Efficient ultraviolet-light (345-nm) generation in a bulk LiIO3 crystal by frequency doubling of a self-seeded gain- switched AlGaInP Fabry-Perot semiconductor laser, Optics Letters 22, p.78-80 (1997)

[6.1873] {Sect. 6.15.1} A. Harada, Y. Nihei, Y. Okazaki, and H. Hyuga: Intracavity frequency doubling of a diode-pumped 946-nm Nd:YAG laser with bulk beriodically poled MgO-LiNbO3, Opt. Lett. 22, p.805-807 (1997)

[6.1874] {Sect. 6.15.1} J. P. Meyn, M. M. Fejer: Tunable ultraviolet radiation by second-harmonic generation in periodically poled lithium tantalate, Opt. Lett. 22, p.1214-1216 (1997)

[6.1875] {Sect. 6.15.1} B. Braun, C. Honninger, G. Zhang, U. Keller, F. Heine, T. Kellner, G. Huber: Efficient intracavity frequency doubling of a passively mode-locked diode-pumped neodymium lanthanum scandium borate laser, Optics Letters 21, p.1567-1569 (1996)

[6.1876] {Sect. 6.15.1} A. Brenier: Numerical investigation of the CW end-pumped NYAB and LiNbO3:MgO: Nd self-doubling lasers, Opt Commun 129, p.57-61 (1996)

[6.1877] {Sect. 6.15.1} B.J. Legarrec, G.J. Raze, P.Y. Thro, M. Gilbert: High-average-power diode-array-pumped frequency-doubled YAG laser, Optics Letters 21, p.1990-1992 (1996)

[6.1878] {Sect. 6.15.1} H. Nagai, M. Kume, A. Yoshikawa, K. Itoh, C. Hamaguchi: Periodic pulse oscillation in an intracavity-doubled Nd: YVO4 laser, Appl Opt 35, p.5392-5394 (1996)

[6.1879] {Sect. 6.15.1} K. Schneider, S. Schiller, J. Mlynek, M. Bode, I. Freitag: 1.1-W single-frequency 532-nm radiation by second-harmonic generation of a miniature Nd:YAG ring laser, Optics Letters 21, p.1999-2001 (1996)

[6.1880] {Sect. 6.15.1} K.I. Martin, W.A. Clarkson, D.C. Hanna: 3 W of single-frequency output at 532 nm by intracavity frequency doubling of a diode-bar-pumped Nd:YAG ring laser, Opt. Lett. 21, p.875-877 (1996)

[6.1881] {Sect. 6.15.1} S.H. Ashworth, M. Joschko, M. Woerner, E. Riedle, T. Elsaesser: Generation of 16-fs pulses at 425 nm by extracavity frequency doubling of a mode-locked Ti:sapphire laser, Optics Letters 20, p.2120-2122 (1995)

[6.1882] {Sect. 6.15.1} C.Y. Chien, G. Korn, J.S. Coe, J. Squier, G. Mourou: Highly efficient second harmonic generation of ultraintense Nd: glass laser pulses, Optics Letters 20, p.353-355 (1995)

[6.1883] {Sect. 6.15.1} D.W. Coutts: Optimization of line-focusing geometry for efficient nonlinear frequency conversion from copper-vapor lasers, IEEE J QE-31, p.2208-2214 (1995)

[6.1884] {Sect. 6.15.1} V. Krylov, A. Rebane, A.G. Kalintsev, H. Schwoerer, U.P. Wild: 2nd Harmonic Generation of Amplified Femtosecond Ti Sapphire Laser Pulses, Optics Letters 20, p.198-200 (1995)

[6.1885] {Sect. 6.15.1} H. Hemmati, J.R. Lesh: 3.5-W Q-switched 532-nm Nd:YAG laser pumped with fiber-coupled diode lasers, Optics Lett. 19, p.1322-1324 (1994)

[6.1886] {Sect. 6.15.1} J.-P. Meyn, G. Huber: Intracavity frequency doubling of a continouos-wave, diode-laser-pumped neodymium lanthanum scandium borate laser, Opt. Lett. 19, p.1436-1438 (1994)

[6.1887] {Sect. 6.15.1} V. Magni, G. Cerullo, S. De Silvestri, O. Svelto, L.J. Qian, M. Danailov: Intracavity frequency doubling of a cw high-power TEM00 Nd:YLF laser, Opt. Lett. 18, p.2111-2113 (1993)

[6.1888] {Sect. 6.15.1} L.R. Marshall, A. Kaz, O. Aytur: Continously tunable diode-pumped UV-blue laser source, Opt. Lett. 18, p.817-819 (1993)

[6.1889] {Sect. 6.15.1} M. M. Fejer, G. A. Magel, D. H. Jundt, R. L. Byer: Quasi-phase-matched second harmonic generation, IEEE J. QE-28, p.2631-2654 (1992)

[6.1890] {Sect. 6.15.1} G.P.A. Malcolm, J. Ebrahimzadeh, A.I. Ferguson: Efficient Frequency Conversion of Mode-Locked Diode-Pumped Lasers and Tunable All-Solid-State Laser Sources, IEEE J. QE-28, p.1172-1178 (1992)

[6.1891] {Sect. 6.15.1} W.S. Pelouch, P.E. Powers, C.L. Tang: Ti:sapphire-pumped, high-repetition-rate femtosecond optical parametric oscillator, Opt. Lett. 17, p.1070-1072 (1992)

[6.1892] {Sect. 6.15.1} C.H. Brito Cruz, A.G. Prosser, P.C. Becker: Generation of tunable femtosecond pulses in the 690-750 nm wavelength region, Opt. Comm. 86, p.65-69 (1991)

[6.1893] {Sect. 6.15.1} K.M. Yoo, Q. Xing, R.R. Alfano: Imaging objects hidden in highly scattering media using femtosecond second-harmonic-generation cross-correlation time gating, Opt. Lett. 16, p.1019-1021 (1991)

[6.1894] {Sect. 6.15.1} G.E. James, E.M. Harrell II, C. Bracikowski, K. Wiesenfeld, R. Roy: Elimination of chaos in an intracavity-doubled Nd:YAG laser, Opt. Lett. 15, p.1141-1143 (1990)

[6.1895] {Sect. 6.15.1} W.S. Pelouch, T. Ukachi, E.S. Wachman, C.L. Tang: Evaluation of LiB3O5 for second-harmonic generation of femtosecond optical pulses, Appl. Phys. Lett. 57, p.111-113 (1990)

[6.1896] {Sect. 6.15.1} J.R.M. Barr, D.W. Hughes: Coupled Cavity Modelocking of a Nd:YAG Laser Using Second-Harmonic-Generation, Appl. Phys. B 49, p.323-325 (1989)

[6.1897] {Sect. 6.15.1} K. Bratengeier, H.-G. Purucker, A. Laubereau: Free induction decay of inhomogeneously broadened lines, Opt. Comm. 70, p.393-398 (1989)

[6.1898] {Sect. 6.15.1} Y. Li, L. Wang, P. Neos, G. Zhang, X.C. Liang, R.R. Alfano: Ultrafast noncollinear secnd-harmonic-generation-based 4 x 4 optical switching array, Opt. Lett. 14, p.347-349 (1989)

[6.1899] {Sect. 6.15.1} A. Mokhtari, J. Chesnoy, A. Laubereau: Femtosecond time- and frequency-resolved fluorescence spectroscopy of a dye molecule, Chem. Phys. Lett. 155, p.593-598 (1989)

[6.1900] {Sect. 6.15.1} M. Woerner, A. Seilmeier, W. Kaiser: Reshaping of infrared picosecond pulses after passage through atmospheric CO2, Opt. Lett. 14, p.636-638 (1989)

[6.1901] {Sect. 6.15.1} D. Josse, R. Hierle, I. Ledoux, J. Zyss: Highly efficient second-harmonic generation of picosecond pulses at 1.32 µm in 3-methyl-4-nitropyridine-1-oxide, Appl. Phys. Lett. 53, p.2251-2253 (1988)

[6.1902] {Sect. 6.15.1} W.J. Kozlovsky, C.D. Nabors, R.L. Byer: Efficient Second Harmonic Generaton of a Diode-Laser-Pumped CW Nd:YAG Laser Using Monolithic MgO:LiNbO3 External Resonant Cavities, IEEE J. QE-24, p.913-919 (1988)

[6.1903] {Sect. 6.15.1} F. Laermer, J. Dobler, T. Elsaesser: Generation of Femtosecond UV Pulses by Intracavity Frequency Doubling in a Modelocked Dye Laser, Opt. Comm.67, p.58-62 (1988)

812   6. Lasers

[6.1904] {Sect. 6.15.1} M. Maroncelli, G. R. Fleming: Comparision of time-resolved fluorescence Stokes shift measurements to a molecular theory of solvation dynamics, J. Chem. Phys. 89, p.875-881 (1988)

[6.1905] {Sect. 6.15.1} M. Oka, S. Kubota: Stable intracavity doubling of orthogonal linearly polarized modes in diode-pumped Nd:YAG lasers, Opt. Lett. 13, p.805-807 (1988)

[6.1906] {Sect. 6.15.1} A. Penzkofer, F. Ossig, P. Qiu: Picosecond Third-Harmonic Light Generation in Calcite, Appl. Phys. B 47, p.71-81 (1988)

[6.1907] {Sect. 6.15.1} P. Qiu, A. Penzkofer: Picosecond Third-Harmonic Light Generation in beta-BaB2O4, Appl. Phys. B 45, p.225-236 (1988)

[6.1908] {Sect. 6.15.1} A. Seilmeier, M. Wörner, H.-J. Hübner, W. Kaiser: Distortion of infrared picosecond pulses after propagation in atmospheric air, Appl. Phys. Lett. 53, p.2468-2470 (1988)

[6.1909] {Sect. 6.15.1} J. Shah: Ultrafast Luminescence Spectroscopy Using Sum Frequency Generation, IEEE J. QE-24, p.276-288 (1988)

[6.1910] {Sect. 6.15.1} K.A. Stankov, J. Jethwa: A New Mode-Locking Technique Using a Nonlinear Mirror, Opt. Comm. 66, p.41-46 (1988)

[6.1911] {Sect. 6.15.1} R.R. Alfano, Q.Z. Wang, T. Jimbo, P.P. Ho, R.N. Bhargava, B.J. Fitzpatrick: Induced spectral broadening about a second harmonic generated by an intense primary ultrashort laser pulse in ZnSe crystals, Phys. Rev. A 35, p.459-462 (1987)

[6.1912] {Sect. 6.15.1} J. Collet, T. Amand: Picosecond Cascadable Inverter Gate Using Second Harmonic Pumping, Opt. Comm. 62, p.353-356 (1987)

[6.1913] {Sect. 6.15.1} Y. Ishida, T. Yajima: Characteristics of a New-Type SHG Crystala beta-BaB2O2 in the Femtosecond Region, Opt. Comm. 62, p.197-200 (1987)

[6.1914] {Sect. 6.15.1} J.N. Moore, P.A. Hansen, R.M. Hochstrasser: A new method for picosecond time-resolved infrared spectroscopy: applications to CO photodissociation from iron porphyrins, Chem. Phys. Lett. 138, p.110-114 (1987)

[6.1915] {Sect. 6.15.1} P.E. Perkins, T.S. Fahlen: 20-W average power KTP intracavity doubled Nd:YAG laser, J. Opt. Soc. Am. B 4, p.1066-1071 (1987)

[6.1916] {Sect. 6.15.1} T. Baer: Large-amplitude fluctuations due to longitudinal mode coupling in diode-pumped intracavity-doubled Nd:YAG lasers, J. Opt. Soc. Am. B 3, p.1175-1180 (1986)

[6.1917] {Sect. 6.15.1} K. Kato: Second-Harmonic Generation to 2048 A in Beta-BaB2O4, IEEE J. QE-22, p.1013-1014 (1986)

[6.1918] {Sect. 6.15.1} J.-C. Baumert, J. Hoffnagle, P. Günter: High-efficiency intracavity frequency doubling of a styril-9 dye laser with KNbO3 crystals, Appl. Opt. 24, p.1299-1301 (1985)

[6.1919] {Sect. 6.15.1} Ch. Chuangtian, W. Bochang, J. Aidong, Y. Giuming: A new-type ultraviolet SHG crystal BaB2O4, Scientia Sinica B 28, p.235-243 (1985)

[6.1920] {Sect. 6.15.1} Y.S. Liu, D. Dentz, R. Belt: High-average-power intracavity second-harmonic generation using KTiOPO4 in an acousto-optically Q-switched Nd:YAG laser oscillator at 5 kHz, Opt. Lett. 9, p.76-78 (1984)

[6.1921] {Sect. 6.15.1} G.J. Linford, R.D. Boyd, D. Eimerl, J.S. Hildum, J.T. Hunt, B.C. Johnson, W.E. Martin, W.L. Smith, K. Snyder, C.L. Vercimak: Large Aperture Harmonic Conversion Experiments at Lawrence Livermore National Laboratory, Appl Opt 21, p.3633-3643 (1982)

[6.1922] {Sect. 6.15.1} J. Reintjes, R.C. Echardt: Efficient harmonic generation from 532 to 266 nm in ADP und KD*P, Appl. Phys. Lett. 30, p.91-93 (1977)

[6.1923] {Sect. 6.15.1} J. Falk: A Theory of the Mode-Locked, Internally Frequency-Doubled Laser, IEEE J. QE-11, p.21-31 (1975)

[6.1924] {Sect. 6.15.1} K. Kato: Second-Harmonic Generation in CDA and CD*A, IEEE J. QE-10, p.616-618 (1974)

[6.1925] {Sect. 6.15.1} R.S. Adhav, R.W. Wallace: Second Harmonic Generation in 90 Phase-Matched KDP Isomorphs, IEEE J. QE-9, p.855-856 (1973)

[6.1926] {Sect. 6.15.1} O. Bernecker: Limitations for Mode-Locking Enhancement of Internal SHG in a Laser, IEEE J. QE-9, p.897-900 (1973)

[6.1927] {Sect. 6.15.1} K. Kato: Efficient Second Harmonic Generation in CDA, Opt. Commun. 9, p.249-251 (1973)

[6.1928] {Sect. 6.15.1} D.B. Anderson, J.T. Boyd: Wideband CO_2 Laser Second Harmonic Generation Phase Matched in GaAs Thin-Film Waveguides, Appl. Phys. Lett. 19, p.266-268 (1971)

[6.1929] {Sect. 6.15.1} C.B. Hitz, L.M. Osterink: Simultaneous Intracavity Frequency Doubling and Mode Locking in a Nd:YAG Laser, Appl. Phys. Lett. 18, p.378-380 (1971)

[6.1930] {Sect. 6.15.1} R.R. Rice, G.H. Burkhart: Efficient Mode-Locked Frequency-Doubled Operation of an Nd:YAlO3 Laser, Appl. Phys. Lett. 19, p.225-227 (1971)

[6.1931] {Sect. 6.15.1} T.R. Gurski: Simultaneous Mode-Locking and Second-Harmonic Generation by the Same Nonlinear Crystal, Appl. Phys. Lett. 15, p.36682 (1969)

[6.1932] {Sect. 6.15.1} J.E. Geusic, H.J. Levinstein, S. Singh, R.C. Smith, L.G. Van Uitert: Continous 0.532-μ Solid-State Source Using Ba2NaNb5O15, Appl. Phys. Lett. 12, p.306-308 (1968)

[6.1933] {Sect. 6.15.1} P.D. Maker, R.W. Terhune, M. Nisenoff, C.M. Savage: Effects of Dispersion and Focusing on the Production of optical Harmonics, Phys. Rev. Lett. 8, p.21-22 (1962)

[6.1934] {Sect. 6.15.1} P.A. Franken, A.E. Hill, C.W. Peters, G. Weinreich: Generation of Optical Harmonics, Phys. Rev. Lett. 7, p.118-119 (1961)

[6.1935] {Sect. 6.15.1} X.D. Mu, X.H. Gu, M.V. Makarov, Y.J. Ding, J.Y. Wang, I.Q. Wei, Y.G. Liu: Third-harmonic generation by cascading second-order nonlinear processes in a cerium-doped KTiOPO4 crystal, Optics Letters 25, p.117-119 (2000)

[6.1936] {Sect. 6.15.1} F. Druon, F. Balembois, P. Georges, A. Brun: High-repetition-rate 300-ps pulsed ultraviolet source with a passively Q-switched microchip laser and a multipass amplifier, Optics Letters 24, p.499-501 (1999)

[6.1937] {Sect. 6.15.1} J.M. Eichenholz, D.A. Hammons, L. Shah, Q. Ye, R.E. Peale, M. Richardson, B.H.T. Chai: Diode-pumped self-frequency doubling in a Nd3+: YCa4O (BO3) (3) laser, Appl Phys Lett 74, p.1954-1956 (1999)

[6.1938] {Sect. 6.15.1} D. Jaque, J. Capmany, J.G. Sole: Continuous wave laser radiation at 669 nm from a self-frequency-doubled laser of YAl3 (BO3) (4): Nd3+, Appl Phys Lett 74, p.1788-1790 (1999)

[6.1939] {Sect. 6.15.1} F. Balembois, M. Gaignet, P. Georges, A. Brun, N. Stelmakh, J.M. Lourtioz: Tunable picosecond blue and ultraviolet pulses from a diode-pumped laser system seeded by a gain-switched laser diode, Appl Opt 37, p.4876-4880 (1998)

[6.1940] {Sect. 6.15.1} G. Hilber, A. Lago, R. Wallenstein: Broadly tunable VUV/XUV-radiation generated by resonant third-order frequency conversion in Kr, J. Opt. Soc. Am. B 4, p.1753-1764 (1987)

[6.1941] {Sect. 6.15.1} R.S. Craxton: High Efficiency Frequency Tripling Schemes for High-Power Nd:Glass Lasers, IEEE J. QE-17, p.1771-1782 (1981)

[6.1942] {Sect. 6.15.1} W. Seka, S.D. Jacobs, J.E. Rizzo, R. Boni, R.S. Craxton: Demonstration of High Efficiency Third Harmonic Conversion of High Power Nd-Glass Laser Radiation, Opt. Commun. 34, p.469-473 (1980)

[6.1943] {Sect. 6.15.1} T. Kojima, S. Konno, S. Fujikawa, K. Yasui, K. Yoshizawa, Y. Mori, T. Sasaki, M. Tanaka, Y. Okada: 20-W ultraviolet-beam generation by fourth-harmonic generation of an all-solid-state laser, Optics Letters 25, p.58-60 (2000)

[6.1944] {Sect. 6.15.1} V. Petrov, F. Rotermund, F. Noack, J. Ringling, O. Kittelmann, R. Komatsu: Frequency conversion of Ti : sapphire-based femtosecond laser systems to the 200-nm spectral region using nonlinear optical crystals, IEEE J Sel Top Quantum Electr 5, p.1532-1542 (1999)

[6.1945] {Sect. 6.15.1} J.P. Koplow, D.A.V. Kliner, L. Goldberg: Development of a narrow-band, tunable, frequency-quadrupled diode laser for UV absorption spectroscopy, Appl Opt 37, p.3954-3960 (1998)

[6.1946] {Sect. 6.15.1} A.H. Kung, J.I. Lee, P.J. Chen: An efficient all-solid-state ultraviolet laser source, Appl Phys Lett 72, p.1542-1544 (1998)

[6.1947] {Sect. 6.15.1} F. Rotermund, V. Petrov: Generation of the fourth harmonic of a femtosecond Ti:sapphire laser, Optics Letters 23, p.1040-1042 (1998)

[6.1948] {Sect. 6.15.1} XS. Bourzeix, B. deBeauvoir, F. Nez, F. Detomasi, L. Julien, F. Biraben: Ultra-violet light generation at 205 nm by two frequency doubling steps of a cw titanium-sapphire laser, Opt Commun 133, p.239-244 (1997)

[6.1949] {Sect. 6.15.1} D.A.V. Kliner, J.P. Koplow, L. Goldberg: Narrow-band, tunable, semiconductor-laser-based source for deep-UV absorption spectroscopy, Optics Letters 22, p.1418-1420 (1997)

[6.1950] {Sect. 6.15.1} J. Knittel, A.H. Kung: 39.5% conversion of low-power Q-switched Nd:YAG laser radiation to 266 nm by use of a resonant ring cavity, Optics Letters 22, p.366-368 (1997)

[6.1951] {Sect. 6.15.1} R. Komatsu, T. Sugawara, K. Sassa, N. Sarukura, Z. Liu, S. Izumida, Y. Segawa, S. Uda, T. Fukuda, K. Yamanouchi: Growth and ultraviolet application of Li2B4O7 crystals: Generation of the fourth and fifth harmonics of Nd: Y3Al5O12 lasers, Appl Phys Lett 70, p.3492-3494 (1997)

[6.1952] {Sect. 6.15.1} G. Veitas, A. Dubietis, G. Valiulis, D. Podenas, G. Tamosauskas: Efficient femtosecond pulse generation at 264 nm, Opt Commun 138, p.333-336 (1997)

[6.1953] {Sect. 6.15.1} L.B. Shama, H. Daido, Y. Kato, S. Nakai, T. Zhang, Y. Mori, T. Sasaki: Fourth-harmonic generation of picosecond glass laser pulses with cesium lithium borate crystals, Appl Phys Lett 69, p.3812-3814 (1996)

[6.1954] {Sect. 6.15.1} T.J. Zhang, Y. Kato, H. Daido: Fourth harmonic generation and pulse compression of a picosecond laser pulse, Opt Commun 124, p.83-89 (1996)

[6.1955] {Sect. 6.15.1} A. Dubietis, G. Tamosauskas, A. Varanavicius, G. Valiulis, R. Danielius: Highly efficient subpicosecond pulse generation at 211 nm, J Opt Soc Am B Opt Physics 17, p.48-52 (2000)

[6.1956] {Sect. 6.15.2} J. Piel, M. Beutter, E. Riedle: 20-50-fs pulses tunable across the near infrared from a blue-pumped noncollinear parametric amplifier, Optics Letters 25, p.180-182 (2000)

[6.1957] {Sect. 6.15.2} U. Bader, J.P. Meyn, J. Bartschke, T. Weber, A. Borsutzky, R. Wallenstein, R.G. Batchko, M.M. Fejer, R.L. Byer: Nanosecond periodically poled lithium niobate optical parametric generator pumped at 532 nm by a single-frequency passively Q-switched Nd : YAG laser, Optics Letters 24, p.1608-1610 (1999)

[6.1958] {Sect. 6.15.2} R.S. Conroy, C.F. Rae, M.H. Dunn, B.D. Sinclair, J.M. Ley: Compact, actively Q-switched optical parametric oscillator, Optics Letters 24, p.1614-1616 (1999)

[6.1959] {Sect. 6.15.2} T. Graf, G. McConnell, A.I. Ferguson, E. Bente, D. Burns, M.D. Dawson: Synchronously pumped optical parametric oscillation in periodically poled lithium niobate with 1-W average output power, Appl Opt 38, p.3324-3328 (1999)

[6.1960] {Sect. 6.15.2} P. LozaAlvarez, C.T.A. Brown, D.T. Reid, W. Sibbett, M. Missey: High-repetition-rate ultrashort-pulse optical parametric oscillator continuously tunable from 2.8 to 6.8 mu m, Optics Letters 24, p.1523-1525 (1999)

[6.1961] {Sect. 6.15.2} M. Sato, T. Hatanaka, S. Izumi, T. Taniuchi, H. Ito: Generation of 6.6-mu m optical parametric oscillation with periodically poled LiNbO3, Appl Opt 38, p.2560-2563 (1999)

[6.1962] {Sect. 6.15.2} U. Strossner, A. Peters, J. Mlynek, S. Schiller, J.P. Meyn, R. Wallenstein: Single-frequency continuous-wave radiation from 0.77 to 1.73 mu m generated by a green-pumped optical parametric oscillator with periodically poled LiTaO3, Optics Letters 24, p.1602-1604 (1999)

[6.1963] {Sect. 6.15.2} P.E. Britton, N.G.R. Broderick, D.J. Richardson, P.G.R. Smith, G.W. Ross, D.C. Hanna: Wavelength-tunable high-power picosecond pulses from a fiber-pumped diode-seeded high-gain parametric amplifier, Optics Letters 23, p.1588-1590 (1998)

[6.1964] {Sect. 6.15.2} G. Cerullo, M. Nisoli, S. Stagira, S. DeSilvestri: Sub-8-fs pulses from an ultrabroadband optical parametric amplifier in the visible, Optics Letters 23, p.1283-1285 (1998)

[6.1965] {Sect. 6.15.2} T.J. Edwards, G.A. Turnbull, M.H. Dunn, M. Ebrahimzadeh, F.G. Colville: High-power, continuous-wave, singly resonant, intracavity optical parametric oscillator, Appl Phys Lett 72, p.1527-1529 (1998)

[6.1966] {Sect. 6.15.2} G.M. Gibson, R.S. Conroy, A.J. Kemp, B.D. Sinclair, M.J. Padgett, M.H. Dunn: Microchip laser-pumped continuous-wave doubly resonant optical parametric oscillator, Optics Letters 23, p.517-518 (1998)

[6.1967] {Sect. 6.15.2} G. Hansson, D.D. Smith: Mid-infrared-wavelength generation in 2-mu m pumped periodically poled lithium niobate, Appl Opt 37, p.5743-5746 (1998)

[6.1968] {Sect. 6.15.2} M. Nisoli, S. Stagira, S. DeSilvestri, O. Svelto, G. Valiulis, A. Varanavicius: Parametric generation of high-energy 14.5-fs light pulses at 1.5 mu m, Optics Letters 23, p.630-632 (1998)

[6.1969] {Sect. 6.15.2} T.W. Tukker, C. Otto, J. Greve: A narrow-bandwidth optical parametric oscillator, Opt Commun 154, p.83-86 (1998)

[6.1970] {Sect. 6.15.2} K.L. Vodopyanov: Megawatt peak power 8-13 mu m CdSe optical parametric generator pumped at 2.8 mu m, Opt Commun 150, p.210-212 (1998)

[6.1971] {Sect. 6.15.2} M.S. Webb, P.F. Moulton, J.J. Kasinski, R.L. Burnham, G. Loiacono, R. Stolzenberger: High-average-power KTiOAsO4 optical parametric oscillator, Optics Letters 23, p.1161-1163 (1998)

[6.1972] {Sect. 6.15.2} T. Chuang, R. Burnham: Multiband generation of mid infrared by use of periodically poled lithium niobate, Opt. Lett. 23, p.43-45 (1998)

[6.1973] {Sect. 6.15.2} K. Drühl: Diffractive effects in singly resonant continuous-wave parametric oscillators, Appl. Phys. B 66, p.677-683 (1998)

[6.1974] {Sect. 6.15.2} S. Guha: Focusing dependence of the efficiency of a singly resonant optical parametric oscillator, Appl. Phys. B 66, p.663-675 (1998)

[6.1975] {Sect. 6.15.2} T. Kartaloglu, K.G. Köprülü, O. Aytür, M. Sundheimer, W.P. Risk: Femtosecond optical parametric oscillator based on periodically poled KTiOPO4, Opt. Lett. 23, p.61-63 (1998)

[6.1976] {Sect. 6.15.2} M. E. Klein, D.-H. Lee, J.-P. Meyn, B. Beier, K.-J. Boller, R. Wallenstein: Diode-pumped continuous-wave widely tunable optical parametric oscillator based on periodically poled lithium tantalate, Opt. Lett. 23, p.831-833 (1998)

[6.1977] {Sect. 6.15.2} S. Schiller, J. Mlynek (guest eds.): Continous-wave optical parametric oscillators, Appl. Phys. B 52, p.661-760 (1998)

[6.1978] {Sect. 6.15.2} C. Schwob, P.F. Cohadon, C. Fabre, M.A.M. Marte, H. Ritsch, A. Gatti, L. Lugiato: Transverse effects and mode couplings in OPOS, Appl. Phys. B 66, p.685-699 (1998)

[6.1979] {Sect. 6.15.2} G.A. Turnbull, M.H. Dunn, M. Ebrahimzadeh: Continuous-wave, intracavity optical parametric oscillators: an analysis of power characteristics, Appl. Phys. B 66, p.701-710 (1998)

[6.1980] {Sect. 6.15.2} T.H. Allik, S. Chandra, D.M. Rines, P.G. Schunemann, J.A. Hutchinson, R. Utano: Tunable 7-12-mu m optical parametric oscillator using a Cr, Er: YSGG laser to pump CdSe and ZnGeP2 crystals, Optics Letters 22, p.597-599 (1997)

[6.1981] {Sect. 6.15.2} F.G. Colville, M.H. Dunn, M. Ebrahimzadeh: Continuous-wave, singly resonant, intracavity parametric oscillator, Optics Letters 22, p.75-77 (1997)

[6.1982] {Sect. 6.15.2} J.C. Deak, L.K. Iwaki, D.D. Dlott: High-power picosecond mid-infrared optical parametric amplifier for infrared Raman spectroscopy, Optics Letters 22, p.1796-1798 (1997)

[6.1983] {Sect. 6.15.2} A. Galvanauskas, M.A. Arbore, M.M. Fejer, M.E. Fermann, D. Harter: Fiber-laser-based femtosecond parametric generator in bulk periodically poled LiNbO3, Optics Letters 22, p.105-107 (1997)

[6.1984] {Sect. 6.15.2} T. Kartaloglu, K.G. Koprulu, O. Aytur: Phase-matched self-doubling optical parametric oscillator, Optics Letters 22, p.280-282 (1997)

[6.1985] {Sect. 6.15.2} S.W. Lee, S.H. Kim, D.K. Ko, J.M. Han, J.M. Lee: High-efficiency and low-threshold operation of the pump reflection configuration in the noncollinear phase matching optical parametric oscillator, Opt Commun 144, p.241-244 (1997)

[6.1986] {Sect. 6.15.2} D. Wang, C. Grasser, R. Beigang, R. Wallenstein: The generation of tunable blue ps-light-pulses from a cw mode-locked LBO optical parametric oscillator, Opt Commun 138, p.87-90 (1997)

[6.1987] {Sect. 6.15.2} K.C. Burr, C.L. Tang, M.A. Arbore, M.M. Fejer: Broadly tunable mid-infrared femtosecond optical parametric oscillator using all-solid-state-pumped periodically poled lithium niobate, Opt. Lett. 22, p.1458-1460 (1997)

[6.1988] {Sect. 6.15.2} L.E. Myers, W.R. Bosenberg: Periodically Polded Lithium Niobate and Quasi-Phase-Matched Optical Parametric Oscillators, IEEE J. QE-33, p.1663-1672 (1997)

[6.1989] {Sect. 6.15.2} W.R. Bosenberg, A. Drobshoff, J.I. Alexander, L.E. Myers, R.L. Byer: 93% pump depletion, 3.5-W continuous-wave, singly resonant optical parametric oscillator, Optics Letters 21, p.1336-1338 (1996)

[6.1990] {Sect. 6.15.2} S.D. Butterworth, V. Pruneri, D.C. Hanna: Optical parametric oscillation in periodically poled lithium niobate based on continuous-wave synchronous pumping at 1.047 mu m, Optics Letters 21, p.1345-1347 (1996)

[6.1991] {Sect. 6.15.2} R. Lavi, A. Englander, R. Lallouz: Highly efficient low-threshold tunable all-solid-state intracavity optical parametric oscillator in the mid infrared, Optics Letters 21, p.800-802 (1996)

[6.1992] {Sect. 6.15.2} L.E. Myers, R.C. Eckardt, M.M. Fejer, R.L. Byer, W.R. Bosenberg, J.W. Pierce: Quasi-phase-matched optical parametric oscillators in bulk periodically poled LiNbO3, J. Opt. Soc. Am. B 12, p.2102-2116 (1995)

[6.1993] {Sect. 6.15.2} U.Simon, S. Waltman, I. Loa, F.K. Tittel, L. Hollberg: External-cavity difference-frequency source near 3.2 µm, based on combining a tunable diode laser with a diode-pumped Nd:YAG laser in AgGaS2, J. Opt. Soc. Am. B 12, p.323-327 (1995)

[6.1994] {Sect. 6.15.2} D.R. Walker, C.J. Flood, H.M. van Driel: Kilohertz all-solid-state picosecond lithium triborate optical parametric generator, Opt. Lett. 20, p.145-147 (1995)

[6.1995] {Sect. 6.15.2} M.J.T. Milton, T.D. Gardiner, G. Chourdakis, P.T. Woods: Injection seeding of an infrared optical parametric oscillator with a tunable diode laser, Opt. Lett. 19, p.281-283 (1994)

[6.1996] {Sect. 6.15.2} R. Danielius, A. Piskarskas, A. Stabinis, G.P. Banfi, P. Di Trapani, R. Righini: Traveling-wave parametric generation of widely tunable, highly coherent femtosecond light pulses, J. Opt. Soc. Am. B 10, p.2222-2232 (1993)

[6.1997] {Sect. 6.15.2} Q. Fu, G. Mak, H.M. van Driel: High-power, 62-fs infrared optical parametric oscillator synchronously pumped by a 76-MHz Ti:sapphire laser, Opt. Lett. 17, p.1006-1008 (1992)

[6.1998] {Sect. 6.15.2} H.-J. Krause, W. Daum: Efficient parametric generation of high-power coherent picosecond pulses in lithium borate tunable from 0.405 to 2.4 µm, Appl. Phys. Lett. 60, p.2180-2182 (1992)

[6.1999] {Sect. 6.15.2} G. Mak, Q. Fu, H.M. van Driel: Externally pumped high repetition rate femtosecond infrared optical parametric oscillator, Appl. Phys. Lett. 60, p.542-544 (1992)

[6.2000] {Sect. 6.15.2} K. Kato: Parametric Oscillation at 3.2 µm in KTP Pumped at 1.064 µm, IEEE J. QE-27, p.1137-1139 (1991)

[6.2001] {Sect. 6.15.2} E.S. Wachmann, W.S. Pelouch, C.L. Tang: cw femtosecond pulses tunable in the near- and midinfrared, J. Appl. Phys. 70, p.1893-1895 (1991)

[6.2002] {Sect. 6.15.2} J.T. Lin, J.L. Montgomery: Generation of Tunable MID-IR (1.8-2.4 µm) Laser from Optical Parametric Oscillation in KTP, Opt. Commun. 75, p.315-320 (1990)

[6.2003] {Sect. 6.15.2} E.S. Wachmann, D.C. Edelstein, C.L. Tang: Continuous-wave mode-locked and dispersion-compensated femtosecond optical parametric oscillator, Opt. Lett. 15, p.136-138 (1990)

[6.2004] {Sect. 6.15.2} D.C. Edelstein, E.S. Wachmann, C.L. Tang: Brodly tunable high repetition rate femtosecond optical parametric oscillator, Appl. Phys. Lett. 54, p.1728-1730 (1989)

[6.2005] {Sect. 6.15.2} T.Y. Fan, R.C. Eckardt, R.L. Byer, J. Nolting, R. Wallenstein: Visible BaB2O4 optical parametric oscillator pumped at 355 nm by a single-axial-mode pulsed source, Appl. Phys. Lett. 53, p.2014-2016 (1988)

[6.2006] {Sect. 6.15.2} M.J. Rosker, C.L. Tang: Widely tunable optical parametric oscillator using urea, J. Opt. Soc. Am. B 2, p.691-696 (1985)

[6.2007] {Sect. 6.15.2} S.J. Brosnan, R.L. Byer: Optical Parametric Oscillator Threshold and Linewidth Studies, IEEE J. QE-15, p.415-431 (1979)

[6.2008] {Sect. 6.15.2} V. Wilke, W. Schmidt: Tunable Coherent Radiation Source Covering a Spectral Range from 185 to 880 nm, Appl Phys. 18, p.177-181 (1979)

[6.2009] {Sect. 6.15.3} J. Findeisen, H.J. Eichler, P. Peuser, A.A. Kaminskii, J. Hulliger: Diode-pumped Ba (NO3)2 and NaBrO3 Raman Lasers, Appl. Phys. B. 70, p.159-162 (2000)

[6.2010] {Sect. 6.15.3} I.G. Koprinkov, A. Suda, P.Q. Wang, K. Midorikawa: High-energy conversion efficiency of transient stimulated Raman scattering in methane pumped by the fundamental of a femtosecond Ti : sapphire laser, Optics Letters 24, p.1308-1310 (1999)

[6.2011] {Sect. 6.15.3} A.J. Merriam, S.J. Sharpe, H. Xia, D.A. Manuszak, G.Y. Yin, S.E. Harris: Efficient gas-phase VUV frequency up-conversion, IEEE J Sel Top Quantum Electr 5, p.1502-1509 (1999)

[6.2012] {Sect. 6.15.3} H.M. Pask, J.A. Piper: Efficient all-solid-state yellow laser source producing 1.2-W average power, Optics Letters 24, p.1490-1492 (1999)

[6.2013] {Sect. 6.15.3} Y. Urata, S. Wada, H. Tashiro, T. Fukuda: Fiber-like lanthanum tungstate crystal for efficient stimulated Raman scattering, Appl Phys Lett 75, p.636-638 (1999)

[6.2014] {Sect. 6.15.3} H.M. Pask, J.A. Piper: Practical 580nm source based on frequency doubling of an intracavity- Raman-shifted Nd:YAG laser, Opt Commun 148, p.285-288 (1998)

[6.2015] {Sect. 6.15.3} V. Simeonov, V. Mitev, H. vandenBergh, B. Calpini: Raman frequency shifting in a CH4:H-2:Ar mixture pumped by the fourth harmonic of a Nd:YAG laser, Appl Opt 37, p.7112-7115 (1998)

[6.2016] {Sect. 6.15.3} D.V. Wick, M.T. Gruneisen, P.R. Peterson: Phase-preserving wavefront amplification at 590 nm by stimulated Raman scattering, Opt Commun 148, p.113-116 (1998)

[6.2017] {Sect. 6.15.3} L. Deschoulepnikoff, V. Mitev, V. Simeonov, B. Calpini, H. vandenBergh: Experimental investigation of high-power single-pass Raman shifters in the ultraviolet with Nd:YAG and KrF lasers, Appl Opt 36, p.5026-5043 (1997)

[6.2018] {Sect. 6.15.3} G.G.M. Stoffels, P. Schmidt, N. Dam, J.J. terMeulen: Generation of 224-nm radiation by stimulated Raman scattering of ArF excimer laser radiation in a mixture of H-2 and D-2, Appl Opt 36, p.6797-6801 (1997)

[6.2019] {Sect. 6.15.3} J.P. Watson, H.C. Miller: Raman shifting in the absence of multiple Stokes orders with a Nd:YAG laser in hydrogen: Evidence of coupling between the forward and backward Stokes processes, IEEE J QE-33, p.1288-1293 (1997)

[6.2020] {Sect. 6.15.3} M. Jain, H. Xia, G.Y. Yin, A.J. Merriam, S.E. Harris: Efficient nonlinear frequency conversion with maximal atomic coherence, Phys Rev Lett 77, p.4326-4329 (1996)

[6.2021] {Sect. 6.15.3} D.J. Brink, H.P. Burger, T.N. de Kock, J.A. Strauss, D.R. Preussler: Importance of focusing geometry with stimulated Raman scattering of Nd:YAG laser light in methane, J. Phys. D: Appl. Phys. 19, p.1421-1427 (1986)

[6.2022] {Sect. 6.15.3} K. Ludewigt, K. Birkmann, B. Wellegehausen: Anti-Stokes Raman Laser Investigations on Atomic Tl and Sn, Appl Phys B 33, p.133-139 (1984)

[6.2023] {Sect. 6.15.3} R.L. Byer, W.R. Trunta: 16-μm generation by CO2-pumped rotational Raman scattering in H2, Opt. Lett. 3, p.144-146 (1978)

[6.2024] {Sect. 6.15.3} A.Z. Grasiuk, I.G. Zubarev: High Power Tunable IR Raman Lasers, Appl. Phys. 17, p.211-232 (1978)

[6.2025] {Sect. 6.15.3} V. Wilke, W. Schmidt: Tunable UV-Radiation by Stimulated Raman Scattering in Hydrogen, Appl. Phys. 16, p.151-154 (1978)

[6.2026] {Sect. 6.15.3} R.H. Stolen, E.P. Ippen, A.R. Tynes: Raman Oscillation in Glass Optical Waveguide, Appl. Phys. Lett. 20, p.62-64 (1972)

[6.2027] {Sect. 6.15.3} J. Findeisen, H.J. Eichler, A.A. Kaminskii: Efficient picosecond PbWO4 and two-wavelength KGd (WO4) (2) Raman lasers in the IR and visible, IEEE J QE-35, p.173-178 (1999)

[6.2028] {Sect. 6.15.3} V.I. Karpov, E.M. Dianov, V.M. Paramonov, O.I. Medvedkov, M.M. Bubnov, S.L. Semyonov, S.A. Vasiliev, V.N. Protopopov, O.N. Egorova, V.F. Hopin et al.: Laser-diode-pumped phosphosilicate-fiber Raman laser with an output power of 1 W at 1.48 mu m, Optics Letters 24, p.887-889 (1999)

[6.2029] {Sect. 6.15.3} D.I. Chang, J.Y. Lee, H.J. Kong: Raman shifting of Nd:YAP laser radiation with a Brillouin resonator coupled with a Raman half-resonator, Appl Opt 36, p.1177-1179 (1997)

[6.2030] {Sect. 6.15.3} I.K. Ilev, H. Kumagai, K. Toyoda: Ultraviolet and blue discretely tunable double-pass fiber Raman laser, Appl Phys Lett 70, p.3200-3202 (1997)

[6.2031] {Sect. 6.15.3} I.K. Ilev, H. Kumagai, K. Toyoda: A powerful and widely tunable double-pass fiber Raman laser, Opt Commun 138, p.337-340 (1997)

[6.2032] {Sect. 6.15.3} A. Suda, T. Takasaki, K. Sato, K. Nagasaka, H. Tashiro: High-power generation of 16-mu m second-Stokes pulses in an ortho-deuterium Raman laser, Opt Commun 133, p.185-188 (1997)

[6.2033] {Sect. 6.15.3} I.K. Ilev, H. Kumagai, K. Toyoda: A widely tunable (0.54-1.01 mu m) double-pass fiber Raman laser, Appl Phys Lett 69, p.1846-1848 (1996)

[6.2034] {Sect. 6.15.3} J.C. White, D. Henderson: Anti-Stokes Raman laser, Phys. Rev. A 25, p.1226-1229 (1982)

[6.2035] {Sect. 6.15.3} W. Hartig, W. Schmidt: A Broadly Tunable IR Waveguide Raman Laser Pumped by a Dye Laser, Appl. Phys. 18, p.235-241 (1979)

[6.2036] {Sect. 6.15.3} P. Rabinowitz, A. Stein, R. Brickman, A. Kaldor: Efficient tunable H2 Raman laser, Appl. Phys. Lett. 35, p.739-741 (1979)

[6.2037] {Sect. 6.15.3} E.P. Ippen: Low-Power Quasi-cw Raman Oscillator, Appl. Phys. Lett. 16, p.303-305 (1970)

[6.2038] {Sect. 6.16} A. R. Henderson: A Guide to Laser Safety (Chapman & Hall, London, 1997)

[6.2039] {Sect. 6.16} A.M. Clarke: Ocular Hazards. In Handbook of Lasers with Selected Data on Optical Technology (CRC Press, Cleveland 1977)

[6.2040] {Sect. 6.16} W.T. Ham, Jr, H.A. Mueller, J.J. Ruffolo, Jr, A.M. Clarke: Sensitivity of the Retina to Radiation Damage as a Function of Wavelength, Photochemistry and Photobiology 29, p.735-743 (1979)

[6.2041] {Sect. 6.16} A.F. Bais: Absolute spectral measurements of direct solar ultraviolet irradiance with a Brewer spectrophotometer, Appl Opt 36, p.5199-5204 (1997)

7. Nonlinear Optical Spectroscopy

[7.1] {Sect. 7.1.5.1} J. Kusba, J.R. Lakowicz: Definition and properties of the emission anisotropy in the absence of cylindrical symmetry of the emission field: Application to the light quenching experiments, J Chem Phys 111, p.89-99 (1999)

[7.2] {Sect. 7.1.5.1} I.S. Osad'ko, S.L. Soldatov, A.U. Jalmukhambetov: The intensity and polarization aspects of photochemical hole burning, Chem. Phys. Lett. 118, p.97-100 (1985)

[7.3] {Sect. 7.1.5.1} F. Pellegrino, A. Dagen, R.R. Alfano: Fluorescence polarization anisotropy and kinetics of malachite green measured as a function of solvent viscosity, Chem. Phys. 67, p.111-117 (1982)

[7.4] {Sect. 7.1.5.1} D. Reiser, A. Laubereau: Picosecond Polarization Spectroscopy of Dye Molecules, Ber. Bunsenges. Phys. Chem. 86, p.1106-1114 (1982)

[7.5] {Sect. 7.1.5.1} M.D. Barkley, A.A. Kowalczyk, L. Brand: Fluorescence decay studies of anisotropic rotations of small molecules, J. Chem. Phys. 75, p.3581-3593 (1981)

[7.6] {Sect. 7.1.5.1} D.P. Millar, R. Shah, A.H. Zewail: Picosecond saturation spectroscopy of cresyl violet: Rotational diffusion by a "sticking" boundary condition in the liquid phase, Chem. Phys. Lett. 66, p.435-440 (1979)

[7.7] {Sect. 7.1.5.1} A. v. Jena, H.E. Lessing: Rotational Diffusion of Prolate and Oblate Molecules from Absorption Relaxation, Ber. Bunsenges. Phys. Chem. 83, p.181-191 (1979)

[7.8] {Sect. 7.1.5.1} H.E. Lessing, A. von Jena: Orientation of S1-Sn transition moments of oxazine dyes from continuous picosecond photometry, Chem. Phys. Lett. 59, p.249-254 (1978)

[7.9] {Sect. 7.1.5.1} H.E. Lessing, A. von Jena: Separation of rotational diffusion and level kinetics in transient absorption spectroscopy, Chem. Phys. Lett. 42, p.213-217 (1976)

[7.10] {Sect. 7.1.5.1} H.E. Lessing, A. von Jena, M. Reichert: Orientational aspect of transient absorption in solutions, Chem. Phys. Lett. 36, p.517-522 (1975)

[7.11] {Sect. 7.1.5.1} D.W. Vahey: The effects of molecular reorientation on the absorption of intense light by organic-dye solutions, Chem. Phys. 10, p.261-270 (1975)

[7.12] {Sect. 7.1.5.1} T.J. Chuang, K.B. Eisenthal: Theory of Fluorescence Depolarization by Anisotropic Rotational Diffusion, J. Chem. Phys. 57, p.5094-5097 (1972)

[7.13] {Sect. 7.1.5.1} R. Antoine, A.A. TamburelloLuca, P. Hebert, P.F. Brevet, H.H. Girault: Picosecond dynamics of Eosin B at the air/water interface by time- resolved second harmonic generation: orientational randomization and rotational relaxation, Chem Phys Lett 288, p.138-146 (1998)

[7.14] {Sect. 7.1.5.1} R.E. Dipaolo, J.O. Tocho: Polarization anisotropy applied to the determination of structural changes in the photoisomerization of DODCI, Chem Phys 206, p.375-382 (1996)

[7.15] {Sect. 7.1.5.1} J.J. Larsen, H. Sakai, C.P. Safvan, I. WendtLarsen, H. Stapelfeldt: Aligning molecules with intense nonresonant laser fields, J Chem Phys 111, p.7774-7781 (1999)

[7.16] {Sect. 7.1.5.1} D.S. Wiersma, A. Muzzi, M. Colocci, R. Righini: Time-resolved anisotropic multiple light scattering in nematic liquid crystals, Phys Rev Lett 83, p.4321-4324 (1999)

[7.17] {Sect. 7.1.5.1} Th. Kühne, P. Vöhringer: Transient Anisotropy and Fragment Rotational Excitation in the Femtosecond Photodissociation of Triiodide in Solution, J. Phys. Chem. A 102, p.4177-4185 (1998)

[7.18] {Sect. 7.1.5.2} S. Ashihara, K. Kuroda, Y. OkadaShudo, K. Jarasiunas: Autocorrelation of picosecond pulses in bacteriorhodopsin film using light self-diffraction from intensity and polarization holograms, Opt Commun 165, p.83-89 (1999)

[7.19] {Sect. 7.1.5.2} J.M. Dudley, L.P. Barry, J.D. Harvey, M.D. Thomson, B.C. Thomsen, P.G. Bollond, R. Leonhardt: Complete characterization of ultra-short pulse sources at 1550 nm, IEEE J QE-35, p.441-450 (1999)

[7.20] {Sect. 7.1.5.2} L. Gallmann, D.H. Sutter, N. Matuschek, G. Steinmeyer, U. Keller, C. Iaconis, I.A. Walmsley: Characterization of sub-6-fs optical pulses with spectral phase interferometry for direct electric-field reconstruction, Optics Letters 24, p.1314-1316 (1999)

[7.21] {Sect. 7.1.5.2} D.J. Kane: Recent progress toward real-time measurement of ultrashort laser pulses, IEEE J QE-35, p.421-431 (1999)

[7.22] {Sect. 7.1.5.2} J.W. Nicholson, F.G. Omenetto, D.J. Funk, A.J. Taylor: Evolving FROGS: phase retrieval from frequency-resolved optical gating measurements by use of genetic algorithms, Optics Letters 24, p.490-492 (1999)

[7.23] {Sect. 7.1.5.2} F.G. Omenetto, J.W. Nicholson, A.J. Taylor: Second-harmonic generation-frequency-resolved optical gating analysis of low-intensity shaped femtosecond pulses at 1.55 mu m, Optics Letters 24, p.1780-1782 (1999)

[7.24] {Sect. 7.1.5.2} P.J. Bennett, A. Malinowski, B.D. Rainford, I.R. Shatwell, Y.P. Svirko, N.I. Zheludev: Femtosecond pulse duration measurements utilizing an ultrafast nonlinearity of nickel, Opt Commun 147, p.148-152 (1998)

[7.25] {Sect. 7.1.5.2} M. Drabbels, G.M. Lankhuijzen, L.D. Noordam: Demonstration of a far-infrared streak camera, IEEE J QE-34, p.2138-2144 (1998)

[7.26] {Sect. 7.1.5.2} J.K. Ranka, A.L. Gaeta, A. Baltuska, M.S. Pshenichnikov, D.A. Wiersma: Autocorrelation measurement of 6-fs pulses based on the two-photon-induced photocurrent in a GaAsP photodiode, Optics Letters 22, p.1344-1346 (1997)

[7.27] {Sect. 7.1.5.2} K.W. Delong, D.N. Fittinghoff, R. Trebino: Practical issues in ultrashort-laser-pulse measurement using frequency-resolved optical gating, IEEE J QE-32, p.1253-1264 (1996)

[7.28] {Sect. 7.1.5.2} Y.M. Li, R. Fedosejevs: Visible single-shot autocorrelator in BaF2 for subpicosecond KrF laser pulses, Appl Opt 35, p.2583-2586 (1996)

[7.29] {Sect. 7.1.5.2} B. LutherDavies, M. Samoc, J. Swiatkiewicza, A. Samoc, M. Woodruff, R. Trebino, K.W. Delong: Diagnostics of femtosecond laser pulses using films of poly (p-phenylenevinylene), Opt Commun 131, p.301-306 (1996)

[7.30] {Sect. 7.1.5.2} A.V. Vinogradov, J. Janszky, T. Kobayashi: A single-molecule interferometer for measurement of femtosecond laser pulse duration, Opt Commun 127, p.223-229 (1996)

[7.31] {Sect. 7.1.5.2} I. Will, P. Nickles, M. Schnuerer, M. Kalashnikov, W. Sander: Compact FROG system useful for measurement of multiterawatt laser pulses, Opt Commun 132, p.101-106 (1996)

[7.32] {Sect. 7.1.5.2} D.R. Yankelevich, P. Pretre, A. Knoesen, G. Taft, M.M. Murnane, H.C. Kapteyn, R.J. Twieg: Molecular engineering of polymer films for amplitude and phase measurements of Ti:sapphire femtosecond pulses, Optics Letters 21, p.1487-1489 (1996)

[7.33] {Sect. 7.1.5.2} A. Braun, J.V. Rudd, H. Cheng, G. Mourou, D. Kopf, I.D. Jung, K.J. Weingarten, U. Keller: Characterization of short-pulse oscillators by means of a high-dynamic-range autocorrelation measurement, Optics Letters 20, p.1889-1891 (1995)

[7.34] {Sect. 7.1.5.2} G. Taft, A. Rundquist, M.M. Murnane, H.C. Kapteyn, K.W. Delong, R. Trebino, I.P. Christov: Ultrashort optical waveform measurements using frequency resolved optical gating, Optics Letters 20, p.743-745 (1995)

[7.35] {Sect. 7.1.5.2} G. Szabó, A. Müller: A sensitive single shot method to determine duration and chirp of ultrashort pulses with a streak camera, Opt. Comm. 82, p.56-62 (1991)

[7.36] {Sect. 7.1.5.2} S.A. Arakelian, R.N. Gyuzalian, S.B. Sogomonian: Comments of the Picosecond Pulse Width Measurement by the Single-Shot Second Harmonic Beam Technique, Opt. Comm. 44, p.67-72 (1982)

[7.37] {Sect. 7.1.5.3} Z. Cheng, A. Furbach, S. Sartania, M. Lenzner, C. Spielmann, F. Krausz: Amplitude and chirp characterization of high-power laser pulses in the 5-fs regime, Optics Letters 24, p.247-249 (1999)

[7.38] {Sect. 7.1.5.3} T. Udem, J. Reichert, R. Holzwarth, T.W. Hansch: Accurate measurement of large optical frequency differences with a mode-locked laser, Optics Letters 24, p.881-883 (1999)

[7.39] {Sect. 7.1.6.2} W.T. Simpson, D.L. Peterson: Coupling Strength for Resonance Force Transfer of Electronic Energy in Van der Waals Solids, J. Chem. Phys. 26, p.588-593 (1957)

[7.40] {Sect. 7.1.6.2} F. Rotermund, R. Weigand, A. Penzkofer: J-aggregation and disaggregation of indocyanine green in water, Chem Phys 220, p.385-392 (1997)

[7.41] {Sect. 7.2.0} H. -H. Perkampus: UV-VIS Spectroscopy and Its Applications (Springer, Berlin, Heidelberg, New York, 1992)

[7.42] {Sect. 7.2.0} S. Svanberg: Atomic and Molecular Spectroscopy (Springer, Berlin, Heidelberg, New York, 1997)

[7.43] {Sect. 7.2.0} Y.B. He, B.J. Orr: Ringdown and cavity-enhanced absorption spectroscopy using a continuous-wave tunable diode laser and a rapidly swept optical cavity, Chem Phys Lett 319, p.131-137 (2000)

[7.44] {Sect. 7.2.0} D.G. Lancaster, R. Weidner, D. Richter, F.K. Tittel, J. Limpert: Compact CH4 sensor based on difference frequency mixing of diode lasers in quasi-phasematched LiNbO3, Opt Commun 175, p.461-468 (2000)

[7.45] {Sect. 7.2.0} T.J. Latz, G. Weirauch, V.M. Baev, P.E. Toschek: External photoacoustic detection of a trace vapor inside a multimode laser, Appl Opt 38, p.2625-2629 (1999)

[7.46] {Sect. 7.2.0} A. Garnache, A. Campargue, A.A. Kachanov, F. Stoeckel: Intra-cavity laser absorption spectroscopy near 9400 cm (-1) with a Nd:glass laser: application to (N2O)-N-14-O-16, Chem Phys Lett 292, p.698-704 (1998)

[7.47] {Sect. 7.2.0} U. Willamowski, D. Ristau, E. Welsch: Measuring the absolute absorptance of optical laser components, Appl Opt 37, p.8362-8370 (1998)

[7.48] {Sect. 7.2.0} C. Zander, K.H. Drexhage, K.T. Han, J. Wolfrum, M. Sauer: Single-molecule counting and identification in a microcapillary, Chem Phys Lett 286, p.457-465 (1998)

[7.49] {Sect. 7.2.0} M.S. Baptista, C.D. Tran: Near-infrared thermal lens spectrometer based on an erbium-doped fiber amplifier and an acousto-optic tunable filter, and its application in the determination of nucleotides, Appl Opt 36, p.7059-7065 (1997)

[7.50] {Sect. 7.2.0} M.J. Fernee, P.F. Barker, A.E.W. Knight, H. RubinszteinDunlop: Infrared seeded parametric four-wave mixing for sensitive detection of molecules, Phys Rev Lett 79, p.2046-2049 (1997)

[7.51] {Sect. 7.2.0} L. Lehr, P. Hering: Quantitative nonlinear spectroscopy: A direct comparison of degenerate four-wave mixing with cavity ring-down spectroscopy applied to NaH, IEEE J QE-33, p.1465-1473 (1997)

[7.52] {Sect. 7.2.0} Y. Oki, K. Furukawa, M. Maeda: Extremely sensitive Na detection in pure water by laser ablation atomic fluorescence spectroscopy, Opt Commun 133, p.123-128 (1997)

[7.53] {Sect. 7.2.0} D. Romanini, A.A. Kachanov, F. Stoeckel: Diode laser cavity ring down spectroscopy, Chem Phys Lett 270, p.538-545 (1997)

[7.54] {Sect. 7.2.1} I. Derzy, V.A. Lozovsky, S. Cheskis: Absorption cross-sections and absolute concentration of singlet methylene in methane/air flames, Chem Phys Lett 313, p.121-128 (1999)

[7.55] {Sect. 7.2.1} A.C.R. Pipino: Ultrasensitive surface spectroscopy with a miniature optical resonator, Phys Rev Lett 83, p.3093-3096 (1999)

[7.56] {Sect. 7.2.1} P.H.S. Ribeiro, C. Schwob, A. Maitre, C. Fabre: Sub-shot-noise high-sensitivity spectroscopy with optical parametric oscillator twin beams, Optics Letters 22, p.1893-1895 (1997)

[7.57] {Sect. 7.2.1} C.T. Hansen, S.C. Wilks, P.E. Young: Spectral evidence for collisionless absorption in subpicosecond laser- solid interactions, Phys Rev Lett 83, p.5019-5022 (1999)

[7.58] {Sect. 7.2.4} R. Menzel, W. Kessler: Band Shape Analysis of the Absorption Bands of Four Triphenylmethane Dyes Using a Self Starting Routine, J. Mol. Liquids 39, p.279-298 (1988)

[7.59] {Sect. 7.2.4} J. Humlicek: Optimized Computation of the Voigt and Complex Probability Functions, J. Quant. Spectrosc. Radiat. Transfer 27, p.437-444 (1982)

[7.60] {Sect. 7.2.4} R. Kubo: A stochastic theory of line shape, Adv. Chem. Phys. 15, p.101-127 (1969)

[7.61] {Sect. 7.2.4} B.H. Armstrong: Spectrum Line Profiles: The Voigt Function, J. Quant. Spectrosc. Radiat. Transfer 7, p.61-88 (1967)

[7.62] {Sect. 7.2.4} D. Biswas, B. Ray, S. Dutta, P.N. Ghosh: Diode laser spectroscopic measurement of line shape of $(1 + 3\ 3)$ band transitions of acetylene, Appl. Phys. B 68, p.1125-1130 (1999)

[7.63] {Sect. 7.2.4} Y. Makdisi: Spectral line broadening of Sr under the influence of collisions with foreign gas perturbers, Opt Commun 142, p.215-219 (1997)

[7.64] {Sect. 7.2.4} R. Sander, R. Menzel, K.-H. Naumann: Solvent Induced Broadening of Fluorescent Electronic Transitions of Para-Terphenyl, Ber. Bunsenges. Phys. Chem. 96, p.188-194 (1992)

[7.65] {Sect. 7.2.4} E.T.J. Nibbering, D.A. Wiersma, K. Duppen: Femtosecond Non-Markovian Optical Dynamics in Solution, Phys. Rev. Lett. 66, p.2464-2467 (1991)

[7.66] {Sect. 7.2.4} E.T.J. Nibbering, K. Duppen, D.A. Wiersma: Optical dephasing in solution: A line shape and resonance light scattering study of azulene in isopentane and cyclohexane, J. Chem. Phys. 93, p.5477-5484 (1990)

[7.67] {Sect. 7.2.4} E.G. Myers, H.S. Margolis, J.K. Thompson, M.A. Farmer, J.D. Silver, M.R. Tarbutt: Precision measurement of the 1s2p P-3 (2)-P-3 (1) fine structure interval in heliumlike fluorine, Phys Rev Lett 82, p.4200-4203 (1999)

[7.68] {Sect. 7.2.4} B. Abel, A. Charvat, S.F. Deppe: Lifetimes of the lowest triplet state of ozone by intracavity laser absorption spectroscopy, Chem Phys Lett 277, p.347-355 (1997)

[7.69] {Sect. 7.2.4} K.S.E. Eikema, W. Ubachs, W. Vassen, W. Hogervorst: Precision measurements in helium at 58 nm: Ground state lamb shift and the 1 (1)S-2 (1)P transition isotope shift, Phys Rev Lett 76, p.1216-1219 (1996)

[7.70] {Sect. 7.3.0} J. R. Lakowicz: Principles of Fluorescence Spectroscopy (Plenum Press, New York, London, 1983)

[7.71] {Sect. 7.3.0} J. R. Lakowicz: Topics in Fluorescence Spectroscopy, Vol. 1: Techniques (Plenum Press New York, London, 1991)

[7.72] {Sect. 7.3.0} J. R. Lakowicz: Topics in Fluorescence Spectroscopy, Vol. 2: Principles (Plenum Press New York, London, 1991)

[7.73] {Sect. 7.3.0} J. R. Lakowicz: Topics in Fluorescence Spectroscopy, Vol. 3; Biomedical Applications (Plenum Press New York, London, 1992)

[7.74] {Sect. 7.3.2} J. Enderlein: New approach to fluorescence spectroscopy of individual molecules on surfaces, Phys Rev Lett 83, p.3804-3807 (1999)

[7.75] {Sect. 7.3.2} K. Palewska, Z. Ruziewicz, H. Chojnacki: Shpolskii spectra and photophysical properties of dinaphtho (1,2-a;1',2'-h)Anthracene – A Strongly non-planar, overcrowded aromatic hydrocarbon, J. Luminesc. 39, p.75-85 (1987)

[7.76] {Sect. 7.3.2} G. Swiatkowski, R. Menzel, W. Rapp: Hindrance of the Rotational Relaxation in the Excited Singlet State of Biphenyl and Para-Terphenyl in Cooled Solutions by Methyl Substituents, J. Luminesc. 37, p.183-189 (1987)

[7.77] {Sect. 7.3.2} R.A. Lampert, S.R. Meech, J. Metcalfe, D. Phillips: The Refractive Index Correction to the Radiative Rate Constant in Fluorescence Lifetime Measurements, Chem. Phys. Lett. 94, p.137-140 (1983)

[7.78] {Sect. 7.3.2} F.J. Busselle, N.D. Haig, C. Lewis: Reply to the comment on the refractive index correction in luminescence spectroscopy, Chem. Phys. Lett. 88, p.128-130 (1982)

[7.79] {Sect. 7.3.2} L.A. Bykovskaya, R.I. Personov, B.M. Kharlamov: Luminescence of solutions of 9-aminoacridine at 4.2 K: Sharp narrowing of spectral bands with laser excitation, Chem. Phys. Lett. 27, p.80-83 (1974)

[7.80] {Sect. 7.3.2} R.I. Personov, E.I Al'Shits, L.A. Bykovskaya: The effect of fine structure appearance in laser-excited fluorescence spectra of organic compounds in solid solutions, Opt. Comm. 6, p.169-173 (1972)

[7.81] {Sect. 7.3.2} J.L. Richards, S.A. Rice: Study of Impurity-Host Coupling in Shpolskii Matrices, J. Chem. Phys. 54, p.2014-2023 (1971)

[7.82] {Sect. 7.3.2} J.M.G. Levins, D.M. Benton, J. Billowes, P. Campbell, T.G. Cooper, P. Dendooven, D.E. Evans, D.H. Forest, I.S. Grant, J.A.R. Griffith et al.: First on-line laser spectroscopy of radioisotopes of a refractory element, Phys Rev Lett 82, p.2476-2479 (1999)

[7.83] {Sect. 7.3.2} A.I. Lvovsky, S.R. Hartmann, F. Moshary: Omnidirectional superfluorescence, Phys Rev Lett 82, p.4420-4423 (1999)

[7.84] {Sect. 7.3.2} M. Fukushima: Laser induced fluorescence spectroscopy of AlNC/AlCN in supersonic free expansions, Chem Phys Lett 283, p.337-344 (1998)

[7.85] {Sect. 7.3.3} K. Ohta, T.J. Kang, K. Tominaga, K. Yoshihara: Ultrafast relaxation processes from a higher excited electronic state of a dye molecule in solution: a femtosecond time-resolved fluorescence study, Chem Phys 242, p.103-114 (1999)

[7.86] {Sect. 7.3.3} T.J. Kang, K. Ohta, K. Tominaga, K. Yoshihara: Femtosecond relaxation processes from a higher excited electronic state of a dye molecule in solution, Chem Phys Lett 287, p.29-34 (1998)

[7.87] {Sect. 7.3.3} G. Berden, J. Vanrooy, W.L. Meerts, K.A. Zachariasse: Rotationally resolved electronic spectroscopy of 4-aminobenzonitrile, Chem Phys Lett 278, p.373-379 (1997)

[7.88] {Sect. 7.3.3} T.M. Woudenberg, S.K. Kulkarni, J.E. Kenny: Internal conversion rates for single vibronic levels of S2 in azulene, J. Chem. Phys. 89, p.2789-2796 (1988)

[7.89]   {Sect. 7.3.3} Z.S. Ruzevich: Fluorescence and Absorption Spectra of Azulene in Frozen Crystalline Solutions, Opt. Spektrosk. 15, p.191-193 (1962)

[7.90]   {Sect. 7.3.3} M. Kasha: Characterization of Electronic Transitions in Complex Molecules, Disc. Farady Soc. 9, p.14-19 (1950)

[7.91]   {Sect. 7.3.4.0} E.S. Medvedev, V.I. Osherov: Radiationless Transitions in Polyatomic Molecules, Springer Ser. in Chem. Phys. 57 (Springer-Verlag 1995)

[7.92]   {Sect. 7.3.4.1} N. Ito, O. Kajimoto, K. Hara: Picosecond time-resolved fluorescence depolarization of p-terphenyl at high pressures, Chem. Phys. Lett. 318, p.118-124 (2000)

[7.93]   {Sect. 7.3.4.1} S.D. Pack, M.W. Renfro, G.G. King, N.M. Laurendeau: Photon-counting technique for rapid fluorescence-decay measurement, Optics Letters 23, p.1215-1217 (1998)

[7.94]   {Sect. 7.3.4.1} A.N. Watkins, Ch.M. Ingersoll, G.A. Baker, F.V. Bright: A Parallel Multiharmonic Frequency-Domain Fluorometer for Measuring Excited-State Decay Kinetics Following One-, Two-, or Three-Photon Excitation, Anal. Chem. 70, p.3384-3396 (1998)

[7.95]   {Sect. 7.3.4.1} R. Muller, C. Zander, M. Sauer, M. Deimel, D.S. Ko, S. Siebert, J. Ardenjacob, G. Deltau, N.J. Marx, K.H. Drexhage, et al.: Time-resolved identification of single molecules in solution with a pulsed semiconductor diode laser, Chem Phys Lett 262, p.716-722 (1996)

[7.96]   {Sect. 7.3.4.1} W. Nadler, R.A. Marcus: Mean relaxation time description of quasi-dissipative behavior in finite-state quantum systems, Chem. Phys. Lett. 144, p.509-514 (1988)

[7.97]   {Sect. 7.3.4.1} W. Rettig, M. Vogel, E. Lippert: The dynamics of adiabatic photoreactions as studied by means of the time structure of synchrotron radiation, Chem. Phys. 103, p.381-390 (1986)

[7.98]   {Sect. 7.3.4.1} G. Calzaferri, Th. Hugentobler: Time-resolved fluorescence spectra derived from multiple frequency phase fluorimetry, Chem. Phys. Lett. 121, p.147-153 (1985)

[7.99]   {Sect. 7.3.4.1} K.N. Swamy, W.L. Hase: The heavy-atom effect in intramolecular vibrational energy transfer, J. Chem. Phys. 82, p.123-133 (1985)

[7.100] {Sect. 7.3.4.1} W. Wild, A. Seilmeier, N.H. Gottfried, W. Kaiser: Ultrafast investigation of vibrational hot molecules after internal conversion in solution, Chem. Phys. Lett. 119, p.259-263 (1985)

[7.101] {Sect. 7.3.4.1} J. Chesnoy, G.M. Gale: Vibrational energy relaxation in liquids, Ann. Phys. Fr. 9, p.893-949 (1984)

[7.102] {Sect. 7.3.4.1} N.H. Gottfried, A. Seilmeier, W. Kaiser: Transient internal temperature on anthracene after picosecond infrared excitation, Chem. Phys. Lett. 111, p.326-332 (1984)

[7.103] {Sect. 7.3.4.1} J.R. Lakowicz: Time-Dependent Rotational Rates of Excited Fluorophores – A Linkage Between Fluorescence Depolarization and Solvent Relaxation, Biophys. Chem. 19, p.13-23 (1984)

[7.104] {Sect. 7.3.4.1} J.R. Lakowicz, G. Laczko, H. Cherek: Analysis of fluorescence decay kinetics from variable-frequency phase shift and modulation data, Biophys. J. 46, p.463-477 (1984)

[7.105] {Sect. 7.3.4.1} V. Sundström, T. Gillbro: Effects of solvent on TMP photophysics. Transition from no barrier to barrier case, induced by solvent properties, J. Chem. Phys. 81, p.3463-3474 (1984)

[7.106] {Sect. 7.3.4.1} F. Wondrazek, A. Seilmeier, W. Kaiser: Ultrafast intramolecular redistribution and intermolecular relaxation of vibrational energy in large molecules, Chem. Phys. Lett. 104, p.121-128 (1984)

[7.107] {Sect. 7.3.4.1} W. Zinth, C. Kolmeder, B. Benna, A. Irgens-Defregger, S.F. Fischer, W. Kaiser: Fast and exceptionally slow vibrational energy transfer in acetylene and phenylacetylene in solution, J. Chem. Phys.78, p.3916-3921 (1983)

[7.108] {Sect. 7.3.4.1} V. Lopez, R.A. Marcus: Heavy mass barrier to intramolecular energy transfer, Chem. Phys. Lett. 93, p.232-234 (1982)

[7.109] {Sect. 7.3.4.1} D.P. Millar, R.J. Robbins, A.H. Zewail: Torsion and bending of nucleic acids studied by subnanosecond time-resolved fluorescence depolarization of intercalated dyes, J. Chem. Phys. 76, p.2080-2094 (1982)

[7.110] {Sect. 7.3.4.1} W. Sibbett, J.R. Taylor, D. Welford: Substituent and Environmental Effects on the Picosecond Lifetimes of the Polymethine Cyanine Dyes, IEEE J. QE-17, p.500-509 (1981)

[7.111] {Sect. 7.3.4.1} J.R. Taylor, M.C. Adams, W. Sibbett: Investigation of Viscosity Dependent Fluorescence Lifetime Using a Synchronously Operated Picosecond Streak Camera, App Phys 21, p.13-17 (1980)

[7.112] {Sect. 7.3.4.1} Th. Förster: Zwischenmolekulare Energiewanderung und Fluoreszenz, Ann. Phys. 6, p.55-75 (1948)

[7.113] {Sect. 7.3.4.1} D.V. O'Connor, D. Phillips: Time-Correlated Single-Photon Counting (Academic, New York 1989)

[7.114] {Sect. 7.3.4.1} M. Ameloot, H. Hendrikckx: Extension of the Performance of Laplace Deconvolution in the Analysis of Fluorescence Dacay Curves, Biophys. J. 44, p.27-38 (1983)

[7.115] {Sect. 7.3.4.1} D. Welford, W. Sibbett, J.R. Taylor: Dual component fluorescence lifetime of some polymethine saturable absorbing dyes, Opt. Comm. 34, p.175-180 (1980)

[7.116] {Sect. 7.3.4.1} A. Polimeno, P.L. Nordio, G. Moro: Master Equation Representation of Fokker-Planck Operators in the Energy Diffusion Regime: Strong Collision Versus Random Walk Processes, Chem. Phys. Lett. 144, p.357-361 (1988)

[7.117] {Sect. 7.3.4.1} B. Bagchi, G.R. Fleming, D.W. Oxtoby: Theory of electronic relaxation in solution in the absence of an activation barrier, J. Chem. Phys. 78, p.7375-7385 (1983)

[7.118] {Sect. 7.3.4.3} D.F. Eaton: Reference Materials for Fluorescence Measurement, J. Photochem. and Photobiol. B: Biology 2, p.523-531 (1988)

[7.119] {Sect. 7.3.4.3} M. Sonnenschein, A. Amirav, J. Jortner: Absolute fluorescence quantum yields of large molecules in supersonic expansions, J. Phys. Chem. 88, p.4214-4218 (1984)

[7.120] {Sect. 7.3.4.3} S. Hamal, F. Hirayama: Actinometric Determination of Absolute Fluorescence Quantum Yields, J. Phys. Chem. 87, p.83-89 (1983)

[7.121] {Sect. 7.3.4.3} A.I. Akimov, A.N. Solov'ev, V.I. Yuzhakov, M.A. Kirpichenok: Luminescence spectra and lasing characteristics of some new coumarins, Sov. J. Quantum Electron. 22, p.999-1001 (1992)

[7.122] {Sect. 7.3.4.3} M. Vogel, W. Rettig, R. Sens, K.H. Drexhage: Structural relaxation of rhodamine dyes with different n-substitution patterns: A study of fluorescence decay times and quantum yields, Chem. Phys. Lett. 147, p.452-460 (1988)

[7.123] {Sect. 7.3.4.3} I. Lopez Arbeloa: Solvent effects on the photophysics of the molecular forms of rhodamine B. Internal conversion mechanism, Chem. Phys. Lett. 129, p.607-614 (1986)

[7.124] {Sect. 7.3.4.3} D.C. Dong, M.A. Winnik: The Py scale of solvent polarities. Solvent effects on the vibronic fine structure of pyrene fluorescence and empirical correlations with Er and Y values, Photochem. and Photobiol. 35, p.17-21 (1982)

[7.125] {Sect. 7.3.4.3} J.R. Lakowicz, G. Weber: Quenching of Fluorescence by Oxygen. A Probe for Structural Fluctuations in Macromolecules, Biochem. 12, p.4161-4170 (1973)

[7.126] {Sect. 7.3.4.3} Th. Förster, G. Hoffmann: Die Viskositätsabhängigkeit der Fluoreszenzquantenausbeuten einiger Farbstoffsysteme, Z. Physik Chem. NF 75, p.63-76 (1971)

[7.127] {Sect. 7.3.4.3} W. Siebrand: Nonradiative processes in molecular systems, in Dynamics of Molecular Collisions, ed. W.H. Miller, Modern Theoretical Chemistry, Vol. 1, Part A (Plenum, New York 1976), p. 249-302

[7.128] {Sect. 7.4.2} F. Li, Y.L. Song, K. Yang, S.T. Liu, C.F. Li, Y.Q. Wu, X. Zuo, C.X. Yu, P.W. Zhu: Determination of nonlinear absorption mechanisms using a single pulse width laser, J Appl Phys 82, p.2004-2006 (1997)

[7.129] {Sect. 7.4.2} T. Robl, A. Seilmeier: Ground State Recovery of Electronically Excited Malachite Green via Transient Vibrational Heating, Chem. Phys. Lett. 147, p.544-550 (1988)

[7.130] {Sect. 7.4.2} M.J. Rosker, F.W. Wiese, C.L. Tang: Femtosecond Relaxation Dynamics of Large Molecules, Phys. Rev. Lett. 57, p.321-324 (1986)

[7.131] {Sect. 7.4.2} D. Leupold, M. Scholz: Determination of the energy level scheme of saturable absorbers by variation of excitation pulse duration. Demonstration with chlorophyll, Chem. Phys. Lett. 115, p.434-436 (1985)

[7.132] {Sect. 7.4.2} S. Oberländer, D. Leupold: Information contained in non-linear absorption curves with extrema, Opt. Comm. 52, p.57-62 (1984)

[7.133] {Sect. 7.4.2} R. Trebino, A.E. Siegman: Subpicosecond relaxation study of malachite green using a three-laser frequency-domain technique, J. Chem. Phys. 79, p.3621-3626 (1983)

[7.134] {Sect. 7.4.2} R.W. Eason, R.C. Greenhow, J.A.D. Matthew: Modeling of Picosecond Pump and Probe Photobleaching Experiments on Fast Saturable Absorbers, IEEE J. QE-17, p.95-102 (1981)

[7.135] {Sect. 7.4.2} A. Penzkofer: Generation of picosecond and subpicosecond light pulses with saturable absorbers, Opto-Electr. 6, p.87-98 (1974)

[7.136] {Sect. 7.4.2} G. Girard, M. Michon: Transmission of a Kodak 9740 Dye Solution Under Picosecond Pulses, IEEE J. QE-9, p.979-984 (1973)

[7.137] {Sect. 7.4.2} G. Mourou, B. Drouin, M. Bergeron, M. M. Denariez-Roberge: Kinetics of Bleaching in Polymethine Cyanine Dyes, IEEE J. QE-9, p.745-748 (1973)

[7.138] {Sect. 7.4.2} A. Zunger, K. Bar-Eli: Nonlinear Behavior of Solutions Illuminated by a Ruby Laser, J. Chem. Phys. 57, p.3558-3567 (1972)

[7.139] {Sect. 7.4.2} H. Schüller, H. Puell: Investigations of non-linear absorption of light in solutions of cryptocyanine, Opt. Comm. 3, p.352-356 (1971)

[7.140] {Sect. 7.4.2} M. Andorn, K.H. Bar-Eli: Optical Bleaching and Deviations from Beer-Lambert's Law of Solutions Illuminated by Ruby Laser. I. Crytocyanine Solutions, J. Chem. Phys. 55, p.5008-5015 (1970)

[7.141] {Sect. 7.4.2} L. Huff, L.G. DeShazer: Saturation of Optical Transitions in Organic Compounds by Laser Flux, J. Opt. Soc. Am. 60, p.157-165 (1970)

[7.142] {Sect. 7.4.2} M. Hercher: An Analysis of Saturable Absorbers, Appl. Opt. 6, p.947-954 (1967)

[7.143] {Sect. 7.4.5} D. Leupold, R. König, B. Voigt, R. Menzel: Modell des sättigbaren Absorbers Crytocyanin/Methanol, Opt. Commun. 11, p.78-82 (1974)

[7.144] {Sect. 7.5.0} G. Battaglin, P. Calvelli, E. Cattaruzza, R. Polloni, E. Borsella, T. Cesca, F. Gonella, P. Mazzoldi: Laser-irradiation effects during Z-scan measurement on metal nanocluster composite glasses, J Opt Soc Am B Opt Physics 17, p.213-218 (2000)

[7.145] {Sect. 7.5.0} A.G. Bezerra, I.E. Borissevitch, A.S.L. Gomes, C.B. deAraujo: Exploitation of the Z-scan technique as a method to optically probe pK (A) in organic materials: application to porphyrin derivatives, Optics Letters 25, p.323-325 (2000)

[7.146] {Sect. 7.5.0} A.G. Bezerra, A.S.L. Gomes, D.A. daSilva, L.H. Acioli, C.B. deAraujo, C.P. deMelo: Molecular hyperpolarizabilities of retinal derivatives, J Chem Phys 111, p.5102-5106 (1999)

[7.147] {Sect. 7.5.0} J.A. Hermann, T. Bubner, T.J. Mckay, P.J. Wilson, J. Staromlynska, A. Eriksson, M. Lindgren, S. Svensson: Optical limiting capability of thick nonlinear absorbers, J Nonlinear Opt Physics Mat 8, p.253-275 (1999)

[7.148] {Sect. 7.5.0} T. Kawazoe, H. Kawaguchi, J. Inoue, O. Haba, M. Ueda: Measurement of nonlinear refractive index by time-resolved z-scan technique, Opt Commun 160, p.125-129 (1999)

[7.149] {Sect. 7.5.0} R. QuinteroTorres, M. Thakur: Measurement of the nonlinear refractive index of polydiacetylene using Michelson interferometry and z-scan, J Appl Phys 85, p.401-403 (1999)

[7.150] {Sect. 7.5.0} W.F. Zhang, M.S. Zhang, Z. Yin, Y.Z. Gu, Z.L. Du, B.L. Yu: Large third-order optical nonlinearity in SrBi2Ta2O9 thin films by pulsed laser deposition, Appl Phys Lett 75, p.902-904 (1999)

[7.151] {Sect. 7.5.0} G. Xiao, J.H. Lim, E.V. Stryland, M. Bass, L. Weichman: Z-Scan Measurement of the Ground and Excited State Absorption Cross Sections of Cr4+ in Yttrium Aluminum Garnet, IEEE J. QE-35, p.1086-1091 (1999)

[7.152] {Sect. 7.5.0} X. Chen, B. Lavorel, T. Dreier, N. Genetier, H. Misserey, X. Michaut: Self-focusing in Terbium Gallium Garnet using Z-scan, Opt Commun 153, p.301-304 (1998)

[7.153] {Sect. 7.5.0} M. Falconieri, G. Salvetti, E. Cattaruzza, F. Gonella, G. Mattei, P. Mazzoldi, M. Piovesan, G. Battaglin, R. Polloni: Large third-order optical nonlinearity of nanocluster-doped glass formed by ion implantation of copper and nickel in silica, Appl Phys Lett 73, p.288-290 (1998)

[7.154] {Sect. 7.5.0} F.E. Hernandez, A. Marcano, Y. Alvarado, A. Biondi, H. Maillotte: Measurement of nonlinear refraction index and two-photon absorption in a novel organometallic compound, Opt Commun 152, p.77-82 (1998)

[7.155] {Sect. 7.5.0} B.M. Patterson, W.R. White, T.A. Robbins, R.J. Knize: Linear optical effects in Z-scan measurements of thin films, Appl Opt 37, p.1854-1857 (1998)

[7.156] {Sect. 7.5.0} T.H. Wei, T.H. Huang, M.S. Lin: Signs of nonlinear refraction in chloroaluminum phthalocyanine solution, Appl Phys Lett 72, p.2505-2507 (1998)

[7.157] {Sect. 7.5.0} O.V. Prhonska, J.H. Lim, D.J. Hagan, E.W. Vanstryland, M.V. Bondar, Y.L. Slominski: Nonlinear light absorption of polamethine dyes in liquid and solid media, J. Opt. Soc. Am.B15p.802-809 (1998)

[7.158] {Sect. 7.5.0} S. Bian, J. Frejlich, K.H. Ringhofer: Photorefractive saturable Kerr-type nonlinearity in photovoltaic crystals, Phys Rev Lett 78, p.4035-4038 (1997)

[7.159] {Sect. 7.5.0} S. Bian: Estimation of photovoltaic field in LiNbO3 crystal by Z- scan, Opt Commun 141, p.292-297 (1997)

[7.160] {Sect. 7.5.0} K. Kandasamy, P.N. Puntambekar, B.P. Singh, S.J. Shetty, T.S. Srivastava: Resonant nonlinear optical studies on porphyrin derivatives, J Nonlinear Opt Physics Mat 6, p.361-375 (1997)

[7.161] {Sect. 7.5.0} F. Li, Y.L. Song, K. Yang, S.T. Liu, C.F. Li: Measurements of the triplet state nonlinearity of C-60 in toluene using a Z-scan technique with a nanosecond laser, Appl Phys Lett 71, p.2073-2075 (1997)

[7.162] {Sect. 7.5.0} V. Pilla, P.R. Impinnisi, T. Catunda: Measurement of saturation intensities in ion doped solids by transient nonlinear refraction, Appl Phys Lett 70, p.817-819 (1997)

[7.163] {Sect. 7.5.0} M. Terazima, H. Shimizu, A. Osuka: The third-order nonlinear optical properties of porphyrin oligomers, J Appl Phys 81, p.2946-2951 (1997)

[7.164] {Sect. 7.5.0} F. Michelotti, F. Caiazza, G. Liakhou, S. Paoloni, M. Bertolotti: Effects of nonlinear Fabry-Perot resonator response on Z- scan measurements, Opt Commun 124, p.103-110 (1996)

[7.165] {Sect. 7.5.0} R.E. Bridges, G.L. Fischer, R.W. Boyd: Z-scan measurement technique for non-Gaussian beams and arbitrary sample thicknesses, Optics Letters 20, p.1821-1823 (1995)

[7.166] {Sect. 7.5.0} T.H. Wei, D.J. Hagan, M.J. Sence, E.W. Van Stryland, J.W. Perry, D.R. Coulter: Direct Measurement of Nonlinear Absorption and Refraction in Solutions of Phthalocyannines, Appl. Phys. B 54, p.46-51 (1992)

[7.167] {Sect. 7.5.0} P. Klovekorn, J. Munch: Investigation of transient nonlinear optical mechanisms using a variable pulselength laser, IEEE J QE-35, p.187-197 (1999)

[7.168] {Sect. 7.5.0} W.F. Sun, C.C. Byeon, C.M. Lawson, G.M. Gray, D.Y. Wang: Third-order susceptibilities of asymmetric pentaazadentate porphyrin- like metal complexes, Appl Phys Lett 74, p.3254-3256 (1999)

[7.169] {Sect. 7.5.0} M.O. Martin, L. Canioni, L. Sarger: Measurements of complex third-order optical susceptibility in a collinear pump-probe experiment, Optics Letters 23, p.1874-1876 (1998)

[7.170] {Sect. 7.5.0} J. Vanhanen, V.P. Leppanen, T. Haring, V. Kettunen, T. Jaaskelainen, S. Parkkinen, J.P.S. Parkkinen: Nonlinear refractive index change of photoactive yellow protein, Opt Commun 155, p.327-331 (1998)

[7.171] {Sect. 7.5.0} S. Dhanjal, S.V. Popov, I.R. Shatwell, Y.P. Svirko, N.I. Zheludev, V.E. Gusev: Femtosecond optical nonlinearity of metallic indium across the solid-liquid transition, Optics Letters 22, p.1879-1881 (1997)

[7.172] {Sect. 7.5.0} H.J. Huang, G. Gu, S.H. Yang, J.S. Fu, P. Yu, G.K.L. Wong, Y.W. Du: Nonlinear optical response of the higher fullerene C-90 – A comparison with C-60, Chem Phys Lett 272, p.427-432 (1997)

[7.173] {Sect. 7.5.0} I. Kang, T. Krauss, F. Wise: Sensitive measurement of nonlinear refraction and two- photon absorption by spectrally resolved two-beam coupling, Optics Letters 22, p.1077-1079 (1997)

[7.174] {Sect. 7.5.0} P. Klovekorn, J. Munch: Variable stimulated Brillouin scattering pulse compressor for nonlinear optical measurements, Appl Opt 36, p.5913-5917 (1997)

[7.175] {Sect. 7.5.0} J.Y. Wu, J. Yan, D.C. Sun, F.M. Li, L.W. Zhou, M. Sun: Third-order nonlinear optical property of a polyphenylene oligomer: Poly (2,5-dialkozyphenylene), Opt Commun 136, p.35-38 (1997)

[7.176] {Sect. 7.5.0} J. Yan, J.Y. Wu, H.Y. Zhu, X.T. Zhang, D.C. Sun, Y.M. Hu, F.M. Li, M. Sun: Excited state enhancement of the third order nonlinear optical susceptibility of nonether polyphenylquinoxaline, Optics Letters 20, p.255-257 (1995)

[7.177] {Sect. 7.5.0} N.I. Zheludev, P.J. Bennett, H. Loh, S.V. Popov, I.R. Shatwell, Y.P. Svirko, V.E. Gusev, V.F. Kamalov, E.V. Slobodchikov: Cubic optical nonlinearity of free electrons in bulk gold, Optics Letters 20, p.1368-1370 (1995)

[7.178] {Sect. 7.5.0} H. Fei, Z. Wei, Q. Yang, Y. Che: Low-power phase conjugation in push-pull azobenzene compounds, Opt. Lett. 20, p.1518-1520 (1995)

[7.179] {Sect. 7.5.0} A. Marcano O, L. Aranguren: Absolute values of the nonlinear susceptibility of dye solutions measured by polarization spectroscopy, J. Appl. Phys. 62, p.3100-3103 (1987)

[7.180] {Sect. 7.5.0} E.J. Heilweil, R.M. Hochstrasser: Nonlinear spectroscopy and picosecond transient grating study of colloidal gold, J. Chem. Phys. 82, p.4762-4770 (1985)

[7.181] {Sect. 7.5.0} J.P.Hermann, J. Ducuing: Third-order polarizabilities of long-chain molecules, J. Appl. Phys. 45, p.5100-5102 (1974)

[7.182] {Sect. 7.5.0} M.D. Levenson, N. Bloembergen: Dispersion of the nonlinear optical susceptibilities of organic liquids and solutions, J.Chem. Phys. 60, p.1323-1327 (1974)

[7.183] {Sect. 7.5.0} M.D. Levenson, N. Bloembergen: Dispersion of the nonlinear optical susceptibility tensor in centrosymmetric media, Phys. Rev. B 10, p.4447-4463 (1974)

[7.184] {Sect. 7.5.0} K.C. Rustagi, J. Ducuing: Third-order optical polarizability of conjugated organic molecules, Opt. Comm. 10, p.258-261 (1974)

[7.185] {Sect. 7.5.0} J.P. Hermann, D. Ricard: Optical nonlinearities in conjugated systems: beta-carotene, Appl. Phys. Lett. 23, p.178-180 (1973)

[7.186] {Sect. 7.5.0} A. Owyoung, R.W. Hellwarth, N. George: Intensity-Induced Changes in Optical Polarizations in Glasses, Phys. Rev. B 5, p.628-633 (1972)

[7.187] {Sect. 7.5.0} J.J. Wynne: Nonlinear Optical Spectroscopy of X (3) in LiNbO3, Phys. Rev. Lett. 29, p.650-653 (1972)

[7.188] {Sect. 7.5.2} M. Sheik-Bahae, A.A. Said, T.-H. Wei, D.J. Hagan, E.W. Van Stryland: Sensitive Measurement of Optical Nonlinearities Using a Single Beam, IEEE J. QE-26, p.760-769 (1990)

[7.189] {Sect. 7.5.2} M. Martinelli, S. Bian, J.R. Leite, R.J. Horowicz: Sensitivity-enhanced reflection Z-scan by oblique incidence of a polarized beam, Appl Phys Lett 72, p.1427-1429 (1998)

[7.190] {Sect. 7.5.2} P.B. Chapple, J. Staromlynska, J.A. Hermann, T.J. Mckay, R.G. Mcduff: Single-beam Z-scan: Measurement techniques and analysis, J Nonlinear Opt Physics Mat 6, p.251-293 (1997)

[7.191] {Sect. 7.5.2} C.R. Mendonca, L. Misoguti, S.C. Zilio: Z-scan measurements with Fourier analysis in ion-doped solids, Appl Phys Lett 71, p.2094-2096 (1997)

[7.192] {Sect. 7.5.2} P.B. Chapple, P.J. Wilson: Z-scans with near-Gaussian laser beams, J Nonlinear Opt Physics Mat 5, p.419-436 (1996)

[7.193] {Sect. 7.5.2} W. Zhao, P. Palffy-Muhoray: Z-scan measurement of X (3) using top-hat beams, Appl. Phys. Lett. 65, p.673-675 (1994)

[7.194] {Sect. 7.5.3} G. Xiao, J.H. Lim, E.V. Stryland, M. Bass, L. Weichman: Z-Scan Measurement of the Ground and Excited State Absorption Cross Sections of Cr4+ in Yttrium Aluminum Garnet, IEEE J. QE-35, p.1086-1091 (1999)

[7.195] {Sect. 7.5.3} H.S. Loka, S.D. Benjamin, P.W.E. Smith: Optical Characterization of Low-Temperature-Grown GaAs for Ultrafast All-Optical Switching Devices, IEEE J. QE-34, p.1426-1436 (1998)

[7.196] {Sect. 7.6.1} D. Leupold, I.E. Kochevar: Multiphoton Photochemistry in Biological Systems: Introduction, Photochem. and Photobiol. 66, p.562-565 (1997)

[7.197] {Sect. 7.6.1} S. Oberländer, D. Leupold: Instantaneous fluorescence quantum yield of organic molecular systems: information content of ist intensity dependence, J. Luminesc. 59, p.125-133 (1994)

[7.198] {Sect. 7.6.1} K.R. Naqvi, D.K. Sharma, G.J. Hoytink: Measurements of Sub-Nanosecond Lifetimes from Biphotonic Fluorescence Produced by Nanosecond Laser Pulses, Chem. Phys. Lett. 22, p.222-225 (1973)

[7.199] {Sect. 7.6.2} N. Kamiya, M. Ishikawa, K. Kasahara, M. Kaneko, N. Yamamoto, H. Ohtani: Picosecond fluorescence spectroscopy of the purple membrane of Halobacterium halobium in alkaline suspension, Chem Phys Lett 265, p.595-599 (1997)

[7.200] {Sect. 7.6.2} S. Reindl, A. Penzkofer: Triplet quantum yield determination by picosecond laser double-pulse fluorescence excitation, Chem Phys 213, p.429-438 (1996)

[7.201] {Sect. 7.6.2} T. Doust: Picosecond Fluorescence Decay Kinetics of Crystal Violet in Low-Viscosity Solvents, Chem. Phys. Lett. 96, p.522-525 (1983)

[7.202] {Sect. 7.6.2} E.F. Hilinski, P.M Rentzepis: Chemical Applications of Picosecond Spectroscopy, Acc. Chem. Res. 16, p.224-232 (1983)

[7.203] {Sect. 7.6.2} V. Sundström, T. Gillbro, H. Bergström: Picosecond Kinetics of Radiationless Relaxations of Triphenyl Methane Dyes. Evidence for a Rapid Excited-State Equilibrium Between States of Differing Geometry, Chem. Phys. 73, p.439-458 (1982)

[7.204] {Sect. 7.6.2} G.R. Fleming, A.E.W. Knight, J.M. Morris, R.J. Robbins, G.W. Robinson: Picosecond spectroscopic studies of spontaneous and stimulated emission in organic dye molecules, Chem. Phys. 23p.61-70 (1977)

[7.205] {Sect. 7.6.2} S.H. Lee, I.C. Chen: Non-exponential decays of the S-1 vibronic levels of acetaldehyde, Chem Phys 220, p.175-189 (1997)

[7.206] {Sect. 7.6.2} V. Sundström, T. Gillbro: A Discussion of the Problem of Determining Multiple Lifetimes from Picosecond Absorption Recovery Data as Encountered in Two Carbocyanine Dyes, Appl. Phys. B 31, p.235-247 (1983)

[7.207] {Sect. 7.6.2} J.R. Torga, J.I. Etcheverry, M.C. Marconi: Design of a fluorescence technique using double laser pulse excitation for the measurement of molecular Brownian dynamics, Opt Commun 143, p.230-234 (1997)

[7.208] {Sect. 7.6.2} B.D. Fainberg, B. Zolotov, D. Huppert: Nonlinear laser spectroscopy of nonlinear solvation, J Nonlinear Opt Physics Mat 5, p.789-807 (1996)

[7.209] {Sect. 7.6.2} J.R. Lakowicz, A. Balter: Differential-Wavelength Deconvolution of Time-Resolved Fluorescence Intensities, Biophys. Chem. 16, p.223-240 (1982)

[7.210] {Sect. 7.6.3} H. Kano, S. Kawata: Two-photon-excited fluorescence enhanced by a surface plasmon, Optics Letters 21, p.1848-1850 (1996)

[7.211] {Sect. 7.6.3} J. Mertz, C. Xu, W.W. Webb: Single-molecule detection by two-photon-excited fluorescence, Optics Letters 20, p.2532-2534 (1995)

[7.212] {Sect. 7.6.4} D. Klemp, B. Nickel: Relative quantum yield of the S2-S1 fluorescence from azulene, Chem. Phys. Lett. 130, p.493-497 (1986)

[7.213] {Sect. 7.6.4} Y. Kurabayashi, K. Kikuchi, H. Kokubun, Y. Kaizu, H. Kobayashi: S2-S0 Fluorescence of Some Metallotetraphenylporphyrins, J. Phys. Chem. 88, p.1308-1310 (1984)

[7.214] {Sect. 7.6.4} G.J. Hoytink: The "anomalous" fluorescence of 1,12-benzperylene in n-heptane, Chem. Phys. Lett. 22, p.10-12 (1983)

[7.215] {Sect. 7.6.4} A. Maciejewski, R.P. Steer: Effect of solvent on the subnanosecond decay of the second excited singlet state of tetramethylindanethione, Chem. Phys. Lett. 100, p.540-545 (1983)

[7.216] {Sect. 7.6.4} A.A. Krasheninnikov, A.V. Shablya: Determination of luminescence quantum yield from highly excited electronic states of moleculaes by the photo-acoustic effect, Opt. Spectrosc. (USSR) 52, p.159-162 (1982)

[7.217] {Sect. 7.6.4} S. Muralidharan, G. Ferraudi, L.K. Patterson: Luminescence from Upper Electronic Excited States of Phthalocyanines, Inorganica Chimica Acta 65, p.L235-L236 (1982)

[7.218] {Sect. 7.6.4} B.S. Vogt, S.G. Schulman: Anomalous Fluorescence of 9-Aminofluorene, Chem. Phys. Lett. 89, p.320-323 (1982)

[7.219] {Sect. 7.6.4} V.L. Bogdanov, V.P. Klochkov: Secondary emission of coronene molecules with excitation of higher electronic states, Opt. Spectrosc. (USSR) 50, p.479-484 (1981)

[7.220] {Sect. 7.6.4} K. Teuchner, S. Dähne: The anomalous blue fluorescence of pseudoisocyanine dyes, J. Luminesc. 23, p.413-422 (1981)

[7.221] {Sect. 7.6.4} E.N. Kaliteevskaya, T.K. Razumova: Photochemical conversions and short-wavelength luminescence of polymethine dyes. Studies of short-wavelength luminescence, Opt. Spectrosc. (USSR) 48, p.269-273 (1980)

[7.222] {Sect. 7.6.4} M. Orenstein, S. Kimel, S. Speiser: Laser excited S2-S1 and S1-S0 emission spectra and the S2-Sn absorption spectrum of azulene in solution, Chem. Phys. Lett. 58, p.582-585 (1978)

[7.223] {Sect. 7.6.4} J.R. Huber, M. Mahaney: S2-S0 fluorescence in an aromatic thioketone, xanthione, Chem. Phys. Lett. 30, p.410-412 (1975)

[7.224] {Sect. 7.6.4} J.B. Birks: Dual fluorescence of isolated aromatic molecules, Chem. Phys. Lett. 25, p.315-460 (1974)

[7.225] {Sect. 7.6.4} L. Bajema, M. Gouterman: Porphyrens XXIII. Fluorescence of the Second Excited Singlet and Quasiline Structure of Zinc Tetrabenzporphin, J. Mol. Spectr. 39, p.421-431 (1971)

[7.226] {Sect. 7.7.1} M. Assel, R. Laenen, A. Laubereau: Retrapping and solvation dynamics after femtosecond UV excitation of the solvated electron in water, J Chem Phys 111, p.6869-6874 (1999)

[7.227] {Sect. 7.7.1} V.V. Lozovoy, O.M. Sarkisov, A.S. Vetchinkin, S.Y. Umanskii: Coherent control of the molecular iodine vibrational dynamics by chirped femtosecond light pulses: theoretical simulation of the pump- probe experiment, Chem Phys 243, p.97-114 (1999)

[7.228] {Sect. 7.7.1} F. Stienkemeier, F. Meier, A. Hagele, H.O. Lutz, E. Schreiber, C.P. Schulz, I.V. Hertel: Coherence and relaxation in potassium-doped helium droplets studied by femtosecond pump-probe spectroscopy, Phys Rev Lett 83, p.2320-2323 (1999)

[7.229] {Sect. 7.7.1} S. Hashimoto: Diffuse reflectance laser photolytic studies of pyrene included in zeolites – Formation of pyrene anion radicals via excited-state electron transfer between guest molecules, Chem Phys Lett 252, p.236-242 (1996)

[7.230] {Sect. 7.7.1} J.N. Heyman, K. Unterrainer, K. Craig, J. Williams, M.S. Sherwin, K. Campman, P.F. Hopkins, A.C. Gossard, B.N. Murdin, C.J.G.M. Langerak: Far-infrared pump-probe measurements of the intersubband lifetime in an AlGaAs/GaAs coupled-quantum well, Appl Phys Lett 68, p.3019-3021 (1996)

[7.231] {Sect. 7.7.1} P. Tamarat, B. Lounis, J. Bernard, M. Orrit, S. Kummer, R. Kettner, S. Mais, T. Basche: Pump-probe experiments with a single molecule: ac-Stark effect and nonlinear optical response, Phys Rev Lett 75, p.1514-1517 (1995)

[7.232] {Sect. 7.7.1} D.S. Kliger, A.C. Albrecht: Polarized Spectroscopy of Excites States of Substituted Anthracenes on a Nanosecond Time Scale, J. Chem. Phys. 53, p.4059-4065 (1970)

[7.233] {Sect. 7.7.1} G. Porter, F.R.S. Topp, M.R. Topp: Nanosecond flash photolysis, Proc. Roy. Soc. Lond. A. 315, p.163-184 (1970)

[7.234] {Sect. 7.7.1} H. Takahashi (ed.): Transient Vibrational Spectroscopy, Springer Proc. Phys, Vol. 68 (Springer, Berlin, Heidelberg 1992)

[7.235] {Sect. 7.7.1} G. Haran, W.D. Sun, K. Wynne, R.M. Hochstrasser: Femtosecond far-infrared pump-probe spectroscopy: A new tool for studying low-frequency vibrational dynamics in molecular condensed phases, Chem Phys Lett 274, p.365-371 (1997)

[7.236] {Sect. 7.7.1} E. Budiarto, J. Margolies, S. Jeong, J. Son, J. Bokor: High-intensity terahertz pulses at 1-kHz repetition rate, IEEE J QE-32, p.1839-1846 (1996)

[7.237] {Sect. 7.7.2} N. Zhavoronkov, V. Petrov, F. Noack: Transient excited-state absorption measurements in chromium-doped forsterite, Phys Rev B 61, p.1866-1870 (2000)

[7.238] {Sect. 7.7.2} G.M. Gale, G. Gallot, F. Hache, N. Lascoux, S. Bratos, J.C. Leicknam: Femtosecond dynamics of hydrogen bonds in liquid water: A real time study, Phys Rev Lett 82, p.1068-1071 (1999)

[7.239] {Sect. 7.7.2} J.P. Likforman, M. Joffre, D. Hulin: Hyper-Raman gain due to excitons coherently driven with femtosecond pulses, Phys Rev Lett 79, p.3716-3719 (1997)

[7.240] {Sect. 7.7.2} D. Tittelbachhelmrich, R.P. Steer: Subpicosecond population decay time of the first excited singlet state of thioxanthione in fluid solution, Chem Phys Lett 262, p.369-373 (1996)

[7.241] {Sect. 7.7.2} T. Okada, N. Mataga, W. Baumann, A. Siemiarczuk: Picosecond Laser Spectroscopy of 4- (9-Anthryl)-N,N-dimethylaniline and Related Compounds, J. Phys. Chem. 91, p.4490-4495 (1987)

[7.242] {Sect. 7.7.2} T. Okada, N. Mataga, W. Baumann: Sn-S1 Absorption Spectra of 4- (N,N-Dimethylamino)benzonitrile in Various Solvents: Confirmation of the Intramolecular Ion Pair State in Polar Solvents, J. Phys. Chem. 91, p.760-762 (1987)

[7.243] {Sect. 7.7.2} E. Morikawa, K. Shikichi, R. Katoh, M. Kotani: Transient photoabsorption by singlet excitons in p-terphenyl single crystals, Chem. Phys. Lett. 131, p.209-212 (1986)

[7.244] {Sect. 7.7.2} C.V. Shank, R. Yen, J. Orenstein, G.L. Baker: Femtosecond excited-state relaxation in polyacetylene, Phys. Rev. B 28, p.6095-6096 (1983)

[7.245] {Sect. 7.7.2} S.K. Chattopadhyay, P.K. Das: Singlet-singlet absorption spectra of Diphenylpolyenes, Chem. Phys. Lett. 87, p.145-150 (1982)

[7.246] {Sect. 7.7.2} T. Okada, N. Tashita, N. Mataga: Direct observation of intermediate heteroexcimer in the photoinduced hydrogen-atom transfer reaction by picosecond laser spectroscopy, Chem. Phys. Lett. 75, p.220-223 (1980)

[7.247] {Sect. 7.7.2} C.V. Shank, E.P. Ippen, R.L. Fork, A. Migus, T. Kobayashi: Application of subpicosecond optical techniques to molecular dynamics, Phil. Trans. R. Soc. Lond. A 298, p.303-308 (1980)

[7.248] {Sect. 7.7.2} S. Tagawa, W. Schnabel: Laser flash photolysis studies on excited singlet states of benzene, toluene, p-xylene, polystyrene, and poly-alpha-methylstyrene, Chem. Phys. Lett. 75, p.120-122 (1980)

[7.249] {Sect. 7.7.2} A. Müller, J. Schulz-Hennig, H. Tashiro: Excited State Absorption of 1,3,3,1',3',3'-Hexamethylindotricarboncyanine Iodide: A Quantitative Study by Ultrafast Absorption Spectroscopy, Appl. Phys. 12, p.333-339 (1977)

[7.250] {Sect. 7.7.2} D. Magde, M.W. Windsor, D. Holten, M. Gouterman: Picosecond flash photolysis: transient absorption in Sn (IV), Pd (II), and Cu (II) porphyrins, Chem. Phys. Lett. 29, p.183-188 (1974)

[7.251] {Sect. 7.7.3} A.B. Myers, R.M. Hochstrasser Comparision of Four-Wave Mixing Techniques for Studying Orientational Relaxation, IEEE J. QE-22, p.1482-1492 (1986)

[7.252] {Sect. 7.7.3} T.F. Heinz, S.L. Palfrey, K.B. Eisenthal: Coherent Coupling Effects in pump-probe measurements with collinear, copropagating beams, Opt. Lett. 9, p.359-361 (1984)

[7.253] {Sect. 7.7.3} A. v. Jena, H.E. Lessing: Coherent Coupling Effects in Picosecond Absorption Experiments, Appl. Phys. 19, p.131-144 (1979)

[7.254] {Sect. 7.7.5} C. Brunel, B. Lounis, P. Tamarat, M. Orrit: Triggered source of single photons based on controlled single molecule fluorescence, Phys Rev Lett 83, p.2722-2725 (1999)

[7.255] {Sect. 7.7.5} A. Leitenstorfer, C. Furst, A. Laubereau: Widely tunable two color mode locked Ti:sapphire laser with pulse jitter of less than 2 fs, Optics Letters 20, p.916-918 (1995)

[7.256] {Sect. 7.7.5} N. Karasawa, R. Morita, H. Shigekawa, M. Yamashita: Generation of intense ultrabroadband optical pulses by induced phase modulation in an argon-filled single-mode hollow waveguide, Optics Letters 25, p.183-185 (2000)

[7.257] {Sect. 7.7.5} A. Brodeur, S.L. Chin: Ultrafast white-light continuum generation and self-focusing in transparent condensed media, J Opt Soc Am B Opt Physics 16, p.637-650 (1999)

[7.258] {Sect. 7.7.5} A.A. Zozulya, S.A. Diddams, A.G. VanEngen, T.S. Clement: Propagation dynamics of intense femtosecond pulses: Multiple splittings, coalescence, and continuum generation, Phys Rev Lett 82, p.1430-1433 (1999)

[7.259] {Sect. 7.7.5} J.U. Kang, R. Posey: Demonstration of supercontinuum generation in a long-cavity fiber ring laser, Optics Letters 23, p.1375-1377 (1998)

[7.260] {Sect. 7.7.5} J.P. Likforman, A. Alexandrou, M. Joffre: Intracavity white-light continuum generation in a femtosecond Ti: sapphire oscillator, Appl Phys Lett 73, p.2257-2259 (1998)

[7.261] {Sect. 7.7.5} E.T.J. Nibbering, O. Duhr, G. Korn: Generation of intense tunable 20-fs pulses near 400 nm by use of a gas-filled hollow waveguide, Optics Letters 22, p.1335-1337 (1997)

[7.262] {Sect. 7.7.5} A. Brodeur, F.A. Ilkov, S.L. Chin: Beam filamentation and the white light continuum divergence, Opt Commun 129, p.193-198 (1996)

[7.263] {Sect. 7.7.5} M. Wittmann, A. Penzkofer: Spectral superbroadening of femtosecond laser pulses, Opt Commun 126, p.308-317 (1996)

[7.264] {Sect. 7.7.5} H. Nishioka, W. Odajima, K. Ueda, H. Takuma: Ultrabroadband flat continuum generation in multichannel propagation of terrawatt Ti:sapphire laser pulses, Optics Letters 20, p.2505-2507 (1995)

[7.265] {Sect. 7.7.5} I.A. Bufetov, M.V. Grekov, K.M. Golant, E.M. Dianov, R.R. Khrapko: Ultraviolet-light generation in nitrogen-doped silica fiber, Optics Letters 22, p.1394-1396 (1997)

[7.266] {Sect. 7.7.5} I. Ilev, H. Kumagai, K. Toyoda, I. Koprinkov: Highly efficient wideband continuum generation in a single- mode optical fiber by powerful broadband laser pumping, Appl Opt 35, p.2548-2553 (1996)

[7.267] {Sect. 7.7.5} R.R. Alfano, Q.X. Li, T. Jimbo, J.T. Manassah, P.P. Ho: Induced spectral broadening of a weak picosecond pulse in glass produced by an intense picosecond pulse, Opt. Lett. 11, p.626-628 (1986)

[7.268] {Sect. 7.7.5} R. Menzel, C.W. Hoganson, M.W. Windsor: Picosecond Bleaching Behavior of the Ground-State Absorption and Excited-State Absorptions of Crystal Violet between 455 and 720 nm, Chem. Phys. Lett. 120, p.29-34 (1985)

[7.269] {Sect. 7.7.5} A. Borghese, S.S. Merola: Time-resolved spectral and spatial description of laser-induced breakdown in air as a pulsed, bright, and broadband ultraviolet-visible light source, Appl Opt 37, p.3977-3983 (1998)

[7.270] {Sect. 7.7.5} S.V. Chernikov, Y. Zhu, J.R. Taylor, V.P. Gapontsev: Supercontinuum self-Q-switched ytterbium fiber laser, Optics Letters 22, p.298-300 (1997)

[7.271] {Sect. 7.7.5} R. Menzel, W. Rapp: Excited Singlet- and Triplet-Absorptions of Pentaphene, Chem. Phys. 89, p.445-455 (1984)

[7.272] {Sect. 7.7.5} S. Kubodera, M. Kitahara, J. Kawanaka, W. Sasaki, K. Kurosawa: A vacuum ultraviolet flash lamp with extremely broadened emission spectra, Appl Phys Lett 69, p.452-454 (1996)

[7.273] {Sect. 7.7.5} T. Udem, J. Reichert, R. Holzwarth, T.W. Hansch: Absolute optical frequency measurement of the cesium D-1 line with a mode-locked laser, Phys Rev Lett 82, p.3568-3571 (1999)

[7.274] {Sect. 7.7.5} B.C. Young, F.C. Cruz, W.M. Itano, J.C. Bergquist: Visible lasers with subhertz linewidths, Phys Rev Lett 82, p.3799-3802 (1999)

[7.275] {Sect. 7.7.5} B. deBeauvoir, F. Nez, L. Julien, B. Cagnac, F. Biraben, D. Touahri, L. Hilico, O. Acef, A. Clairon, J.J. Zondy: Absolute frequency measurement of the 2S-8S/D transitions in hydrogen and deuterium: New determination of the Rydberg constant, Phys Rev Lett 78, p.440-443 (1997)

[7.276] {Sect. 7.7.7} P.A. Blanche, P.C. Lemaire, M. Dumont, M. Fischer: Photoinduced orientation of azo dye in various polymer matrices, Optics Letters 24, p.1349-1351 (1999)

[7.277] {Sect. 7.7.7} E.L. Quitevis, K.G. Casey, T.W. Sinor: Picosecond rotational reorientation of cresyl violet in polymer solution, Chem. Phys. Lett. 132, p.77-82 (1986)

[7.278] {Sect. 7.7.7} G.J. Blanchard, M.J. Wirth: A critical comparision of molecular reorientation in the ground and excited states: Cresyl violet in methanol, J. Chem. Phys. 82, p.39-44 (1985)

[7.279] {Sect. 7.7.7} L.A. Philips, S.P. Webb, J.H. Clark: High-pressure studies of rotational reorientation dynamics: The role of dielectric friction, J. Chem. Phys. 83, p.5810-5821 (1985)

[7.280] {Sect. 7.7.7} D. Reiser, A. Laubereau: Effect of electronic excitation on ultrafast rotational motion of dye molecules, Chem. Phys. Lett. 92, p.297-301 (1982)

[7.281] {Sect. 7.7.7} A. v. Jena, H.E. Lessing: Rotational Diffusion of Dyes in Solvents of Low Viscosity from Transient-Dichroism Experiments, Chem. Phys. Lett. 78, p.187-193 (1981)

[7.282] {Sect. 7.7.7} A. Penzkofer, J. Wiedmann: Orientation of transition dipole moments of Rhodamine 6G determined by excited state absorption, Opt. Comm. 35, p.81-86 (1980)

[7.283] {Sect. 7.7.7} A. Penzkofer, W. Falkenstein: Photoinduced dichroism and vibronic relaxation of rhodamine dyes, Chem. Phys. Lett. 44, p.547-552 (1976)

[7.284] {Sect. 7.7.7} H.E. Lessing, A. von Jena, M. Reichert: Orientational aspect of transient absorption in solutions, Chem. Phys. Lett. 36, p.517-522 (1975)

[7.285] {Sect. 7.7.8} D. Magde, S.T. Gaffney, B.F. Campbell: Excited Singlet Absorption in Blue Laser Dyes: Measurement by Picosecond Falsh Photolysis, IEEE J. QE-17, p.489-495 (1981)

[7.286] {Sect. 7.7.8} J.F. Shepanski, R.W. Anderson, Jr.: Chlorophyll-a excited singlet state absorption measured in the picosecond time regime, Chem. Phys. Lett. 78, p.165-173 (1981)

[7.287] {Sect. 7.7.8} F.E. Doany, B.I. Greene, R.M. Hochstrasser: Excitation energy effects in the photophysics of trans-stilbene in solution, Chem. Phys. Lett. 75, p.206-208 (1980)

[7.288] {Sect. 7.7.8} H.E. Lessing, A. von Jena: Separation of rotational diffusion and level kinetics in transient absorption spectroscopy, Chem. Phys. Lett. 42, p.213-217 (1976)

[7.289] {Sect. 7.7.8} N. Nakashima, N. Mataga: Picosecond flash photolysis and transient spectral measurements over the entire visible, near ultraviolet and near infrared regions, Chem. Phys. Lett. 35, p.487-492 (1975)

[7.290] {Sect. 7.7.8} D. Magde, M.W. Windsor: Picosecond flash photolysis and spectroscopy: 3,3'-diethyloxadicarbocyanine iodide (DODCI), Chem. Phys. Lett. 27, p.31-36 (1974)

[7.291] {Sect. 7.7.8} H. Tashiro, T. Yajima: Picosecond absorption spectroscopy of excited states of dye molecules, Chem. Phys. Lett. 25, p.582-586 (1974)

[7.292] {Sect. 7.7.8} E. Sahar, I. Wieder: Excited singlet state absorption spectrum with tunable dye lasers, Chem. Phys. Lett. 23, p.518-521 (1973)

[7.293] {Sect. 7.7.8} H. Masuhara, N. Mataga: Fluorescence spectra and excited singlet-singlet absorption spectra of s-tetracyanobenzene EDA complexes by laser excitation, Chem. Phys. Lett. 6, p.608-610 (1970)

[7.294] {Sect. 7.7.8} D.S. Kliger, A.C. Albrecht: Nanosecond Excited-State Polarized Absorption Spectroscopy of Anthracene in the Visible Region, J. Chem. Phys. 50, p.4109-4111 (1969)

[7.295] {Sect. 7.7.8} G. Porter, M.R. Topp: Nanosecond Flash Photolysis and the Absorption Spectra of Excited Singlet States, Nature 220, p.1228-1229 (1968)

[7.296] {Sect. 7.7.8} R.S. Taylor, S. Mihailov: Excited Singlet-State Absorption in Laser Dyes at the XeCl Wavelength, Appl. Phys. B 38, p.131-137 (1985)

[7.297] {Sect. 7.7.8} R. Menzel, W. Rapp: Excited Singlet- and Triplet-Absorptions of Pentaphene, Chem. Phys. 89, p.445-455 (1984)

[7.298] {Sect. 7.7.8} A. Penzkofer, W. Blau: Theoretical analysis of S1-state lifetime measurements of dyes with picosecond laser pulses, Opt. Quantum Electr. 15, p.325-347 (1983)

[7.299] {Sect. 7.7.8} Yu.I. Kiryukhin, Z.A. Sinitsyna, Kh. S. Bagdasaryan: Spectra and extinction coefficients for the Sn-S1 absorption of naphthalene and pyrene in the UV region, Opt. Spectrosc. (USSR) 46, p.517-519 (1979)

[7.300] {Sect. 7.7.8} A.V. Aristov, Yu.S. Maslyukov: Effect of the solvent on the cross section and absorption spectra of the excited states of organic luminor molecules, Opt. Spectrosc. 41, p.240-243 (1976)

[7.301] {Sect. 7.7.8} J.-P. Fouassier, D.-J. Lougnot, J. Faure: Transient absorptions in a polymethine laser dye, Chem. Phys. Lett. 35, p.189-193 (1975)

[7.302] {Sect. 7.7.8} R.M. Hochstrasser, H. Lutz, G.W. Scott: The dynamics of populating the lowest triplet state of benzophenone following singlet excitation, Chem. Phys. Lett. 24, p.162-167 (1974)

[7.303] {Sect. 7.7.8} J. Shah, R.F. Leheny: Excited-state absorption spectrum of cresyl violet perchlorate, Appl. Phys. Lett. 24, p.562-564 (1974)

[7.304] {Sect. 7.7.8} D. Lavalette, C.J. Werkhoven, D. Bebelaar, J. Langelaar, J.D.W. van Voorst: Excited singlet state polarization and absorption spectra of 1,2-benzcoronene, 1,12-benzperylene and 1,2:3,4-dibenzanthracene, Chem. Phys. Lett. 9, p.230-233 (1971)

[7.305] {Sect. 7.7.8} J.M. Larkin, W.R. Donaldson, T.H. Foster, R.S. Knox: Reverse intersystem crossing from a triplet state of rose bengal populated by sequential 532-+1064-nm laser excitation, Chem Phys 244, p.319-330 (1999)

[7.306] {Sect. 7.7.8} S. Reindl, A. Penzkofer: Higher excited-state triplet-singlet intersystem crossing of some organic dyes, Chem Phys 211, p.431-439 (1996)

[7.307] {Sect. 7.7.8} N. Kanamaru, J. Tanaka: Nanosecond Laser Photolysis of N-Methylindole in Acetonitrile, Bull. Chem. Soc. Jpn. 59, p.569-573 (1986)

[7.308] {Sect. 7.7.8} S.-ya Koshihara, T. Kobayashi: Sn-S1 and Tn-T1 absorption spectra of highly purified chrysene in solution, Chem. Phys. Lett. 124, p.331-335 (1986)

[7.309] {Sect. 7.7.8} M.R. Wasielewski: Direct measurement of the lowest excited singlet state lifetime of all-trans-beta-carotene and related carotenoids, Chem. Phys. Lett. 128, p.238-243 (1986)

[7.310] {Sect. 7.7.8} S. Mory, H.-J. Weigmann, A. Rosenfeld, M. Siegmund, R. Mitzner, J. Bendig: The S1 and T1 transient absorptions of 10-substituted acridin-9-ones measured by nanosecond laser spectroscopy, Chem. Phys. Lett. 115, p.201-204 (1985)

[7.311] {Sect. 7.7.8} R.S. Taylor, S. Mihailov: Excited Singlet-State Absorption in Laser Dyes at the XeCl Wavelength, Appl. Phys. B 38, p.131-137 (1985)

[7.312] {Sect. 7.7.8} D. Leupold, J. Ehlert, S. Oberländer, B. Wiesner: S1 absorption of chlorophyll-a in the red region, Chem. Phys. Lett. 100, p.345-350 (1983)

[7.313] {Sect. 7.7.8} J.S. Horwitz, R.A. Goldbeck, D.S. Kliger: Excited-state absorption spectroscopy and state ordering in polyenes. 1,3,5,7-octatetraene, Chem. Phys. Lett. 80, p.229-234 (1981)

[7.314] {Sect. 7.7.8} G.W. Scott, L.D. Talley: Excited state absorption spectra and intersystem crossing kinetics in diazanaphthalenes, J. Chem. Phys. 72, p.5002-5013 (1980)

[7.315] {Sect. 7.7.8} E.L. Russell, A.J. Twarowski, D.S. Kliger, E. Switkes: The excited singlet state absorption spectrum of 1,4-diphenylnaphthalene, J. Chem. Phys. 22p.167-173 (1977)

[7.316] {Sect. 7.7.8} N. Mataga, T. Okada, H. Masuhara, N. Nakashima, Y. Sakata, S. Misumi: Eletonic structure and dynamical behvior of some intramolecular exciplexes, J. Luminesc. 12/13, p.159-168 (1976)

[7.317] {Sect. 7.7.8} A. Mueller, J. Schulz-Hennig, H. Tashiro: Ultrafast absorption spectroscopy of laser dyes using a streak camera, Opt. Comm. 18, p.152-153 (1976)

[7.318] {Sect. 7.7.8} M.A. Slifkin, A.O. Al-Chalabi: S1-Sn transitions of some polycyclic aromatic hydrocarbons observed by modulation excitation spectrophotometry, Chem. Phys. Lett. 29, p.405-409 (1974)

[7.319] {Sect. 7.7.8} Ch.R. Goldschmidt, M. Ottolenghi: Excited singlet-singlet spectra of anthracene, N,N-diethylaniline and their CT complex, Chem. Phys. Lett. 4, p.570-572 (1970)

[7.320] {Sect. 7.7.8} J.R. Novak, M.W. Windsor: Laser Photolysis and Spectroscopy in the Nanosecond Time Range: Excited Singlet State Absorption in Coronene, J. Chem. Phys. 47, p.3075-3076 (1967)

[7.321] {Sect. 7.7.8} N. Tamai, T. Asahi, H. Masuhara: Intersystem crossing of benzophenone by femtosecond transient grating spectroscopy, Chem. Phys. Lett. 198, p.413-418 (1992)

[7.322] {Sect. 7.7.8} C. Kryschi, H. Kupka, H.-H. Perkampus: Triplet-triplet absorption spectra of phenanthrene and azaanalogues, Chem. Phys. 116, p.53-60 (1987)

[7.323] {Sect. 7.7.8} I. Carmichael, G.L. Hug: Triplet-Triplet Absorption Spectra of Organic Molecules in Condensed Phases, J. Phys. Chem. Ref. Data 15, p.1-250 (1986)

[7.324] {Sect. 7.7.8} K. Kikuchi, H. Fukumura, H. Kokubun: The Sm-T1 absorption spectrum of 9,10-dibromoanthracene, Chem. Phys. Lett. 123, p.226-228 (1986)

[7.325] {Sect. 7.7.8} J. Saltiel, G.R. Marchand, R. Dabestani, J.M. Pecha: The quenching of anthracene triplets by ground-state anthracene, Chem. Phys. Lett. 100, p.219-222 (1983)

[7.326] {Sect. 7.7.8} L.M. Bolotko, V.V. Gruzinskii, V.I. Danilova, T.N. Kopylova: Triplet-triplet absorption of organic compounds lasing efficiency in the ultraviolet, Opt. Spectrosc. (USSR) 52, p.379-381 (1982)

[7.327] {Sect. 7.7.8} A.P. Darmanyan: Laser photolysis study of the mechanism of rubrene quenching by molecular oxygen, Chem. Phys. Lett. 86, p.405-410 (1982)

[7.328] {Sect. 7.7.8} H. Fukumura, K, Kikuchi, H. Kokubun: Temperature effect on inverse (Tn-S1) intersystem crossing, Chem. Phys. Lett. 92, p.29-32 (1982)

[7.329] {Sect. 7.7.8} H. Görner: Triplet States of Phenylethylenes in Solution. Energies, Lifetimes, and Absorption Spectra of 1,1-Diphenyl-, Triphenyl-, and Tetraphenylethylene Triplets, J. Phys. Chem. 86, p.2028-2035 (1982)

[7.330] {Sect. 7.7.8} H. Hirano, T. Azumi: A new method to determine the quantum yield of intersystem crossing, Chem. Phys. Lett. 86, p.109-112 (1982)

[7.331] {Sect. 7.7.8} H.E. Lessing, D. Richardt, A. von Jena: Quantitative Triplet Photophysics by Picosecond Photometry, J. Mol. Struct. 84, p.281-292 (1982)

[7.332] {Sect. 7.7.8} L. J.A. Martins, T.J. Kemp: Triplet State of 2-Nitrothiophen, J. Chem. Soc, Faraday Trans. I 78, p.519-531 (1982)

[7.333] {Sect. 7.7.8} G.J. Smith: Enhanced Intersystem Crossing in the Oxygen Quenching of Aromatic Hydrocarbon Triplet States with High Energies, J. Chem. Soc, Faraday Trans. 2 78, p.769-773 (1982)

[7.334] {Sect. 7.7.8} M.A. El-Sayed: Double Resonance and the Properties of the Lowest Excited Triplet State of Organic Molecules, Annu. Rev. Phys. Chem. 26, p.235-258 (1975)

[7.335] {Sect. 7.7.8} R.W. Anderson, R.M. Hochstrasser, H. Lutz, G.W. Scott: Measurements of intersystem crossing kinetics using 3545 A picosecond pulses: nitronaphthalenes and benzophenone, Chem. Phys. Lett. 28, p.153-157 (1974)

[7.336] {Sect. 7.7.8} J.L. Laporte, Y. Rousset, P. Peretti, P. Ranson: Triplet-singlet radiationless energy transfer between benzophenone and perylene in vitreous solution, Chem. Phys. Lett. 29, p.444-446 (1974)

[7.337] {Sect. 7.7.8} A.R. Horrocks, F. Wilkinson: Triplet state formation efficiencies of aromatic hydrocarbons in solution, Proc. Roy. Soc. Lond. A. 306, p.257-273 (1968)

[7.338] {Sect. 7.7.8} B. Dick: Accessibility of the lowest quintet state of organic molecules through triplet-triplet annihilation; an indo CI study, Chem. Phys. 78, p.1-16 (1983)

[7.339] {Sect. 7.7.9} T. Freudenberg, V. Stert, W. Radloff, J. Ringling, J. Gudde, G. Korn, I.V. Hertel: Ultrafast dynamics of ammonia clusters excited by femtosecond VUV laser pulses, Chem Phys Lett 269, p.523-529 (1997)

[7.340] {Sect. 7.7.9} A. Grofcsik, M. Kubinyi, W.J. Jones: Intermolecular photoinduced proton transfer in nile blue and oxazine 720, Chem Phys Lett 250, p.261-265 (1996)

[7.341] {Sect. 7.7.9} J. Dobler, W. Zinth, W. Kaiser, D. Oesterhelt: Excited-state reaction dynamics of bacteriorhodopsin studied by femtosecond spectroscopy, Chem. Phys. Lett. 144, p.215-220 (1988)

[7.342] {Sect. 7.7.9} T. Elsaesser, W. Kaiser: Visible and infrared spectroscopy of intramolecular proton transfer using picosecond laser pulses, Chem. Phys. Lett. 128, p.231-237 (1986)

[7.343] {Sect. 7.7.9} R.W. Yip, D.K. Sharma, R. Giasson, D. Gravel: Picosecond Excited-State Absorption of Alkyl Nitrobenzenes in Solution, J. Phys. Chem. 88, p.5770-5772 (1984)

[7.344] {Sect. 7.7.9} T. Doust: Picosecond flourescence decay kinetics of crystal violet in low-viscosity solvents, Chem. Phys. Lett. 96, p.522-515 (1983)

[7.345] {Sect. 7.7.9} R. Trebino, A.E. Siegman: Subpicosecond relaxation study of malachite green using a three-laser frequency-domain technique, J. Chem. Phys. 79, p.3621-3626 (1983)

[7.346] {Sect. 7.7.9} T. Kobayashi: Picosecond time-resolved Sn-S1 absorption spectrum on the tetracyanobenzene-toluene complex, Chem. Phys. Lett. 85, p.170-174 (1982)

[7.347] {Sect. 7.7.9} B. Kopainsky, W. Kaiser: Ultrafast transient processes of monomers, dimers, and aggregates of pseudoisocyanine chloride (PIC), Chem. Phys. Lett. 88, p.337-361 (1982)

[7.348] {Sect. 7.7.9} S.K. Rentsch, D. Fassler, P. Hampe, R.V. Danielius, R.A. Gadonas: Picosecond time-resolved spectroscopic studies of a monomer-dimer system of 3.3'-diethyl thiacarcocyanine iodide in aqueous solution, Chem. Phys. Lett. 89, p.249-253 (1982)

[7.349] {Sect. 7.7.9} V. Sundström, T. Gillbro, H. Bergström: Picosecond kinetics of radiationless relaxations of triphenyl methane dyes. Evidence for a rapid excited-state equilibrium between states of differing geometry, Chem. Phys. 73, p.439-458 (1982)

[7.350] {Sect. 7.7.9} Y. Wang, E.V. Sitzmann, F. Novak, C. Dupuy, K.B. Eisenthal: Reactions of Excited Triplet Diphenylcarbene Studied with Picosecond Lasers, J. Am. Chem. Soc. 104, p.3238-3239 (1982)

[7.351] {Sect. 7.7.9} D. Huppert, S.D. Rand, P.M. Rentzepis, P.F. Barbara, W.S. Struve, Z.R. Grabowski: Picosecond kinetics of p-dimethylaminobenzonitrile, J. Chem. Phys. 75, p.5714-5719 (1981)

[7.352] {Sect. 7.7.9} S.K. Rentsch, R.V. Danielius, R.A. Gadonas, A. Piskarskas: Picosecond kinetics of transient spectra of pseudoisocyanine monomers and J-aggregates in aqueous solution, Chem. Phys. Lett. 84, p.446-449 (1981)

[7.353] {Sect. 7.7.9} M.C. Adams, D.J. Bradley, W. Sibbett, J.R. Taylor: Application of the synchroscan streak camera to real time picosecond measurements of molecular energy transfer, J. Mol. Struct. 61, p.5-10 (1980)

[7.354] {Sect. 7.7.9} T. Kobayashi, E.O. Degenkolb, R. Bersohn, P.M. Rentzepis, R. MacColl, D.S. Berns: Energy Transfer among the Chromophores in Phycocyanins Measured by Picosecond Kinetics, Biochem. 18, p.5073-5078 (1979)

[7.355] {Sect. 7.8.1} R. Menzel, C.W. Hoganson, M.W. Windsor: Picosecond Bleaching Behavior of the Ground-State Absorption and Excited-State Absorptions of Crystal Violet between 455 and 720 nm, Chem. Phys. Lett. 120, p.29-34 (1985)

[7.356] {Sect. 7.8.2.2} B.S. Ham, S.M. Shahriar, P.R. Hemmer: Electromagnetically induced transparency over spectral hole-burning temperature in a rare-earth-doped solid, J Opt Soc Am B Opt Physics 16, p.801-804 (1999)

[7.357] {Sect. 7.8.2.2} S.T. Li, G.K. Liu, W. Zhao: Converting Eu3+ between defect sites in BaFCl for persistent spectral hole burning, Optics Letters 24, p.838-840 (1999)

[7.358] {Sect. 7.8.2.2} J. Pieper, K.D. Irrgang, M. Ratsep, T. Schrotter, J. Voigt, G.J. Small, G. Renger: Effects of aggregation on trimeric light-harvesting complex II of green plants: A hole-burning study, J Phys Chem A 103, p.2422-2428 (1999)

[7.359] {Sect. 7.8.2.2} Z. Hasan, L. Biyikli, P.I. Macfarlane: Power-gated spectral holeburning in MgS:Eu2+, Eu3+: A case for high- density persistent spectral holeburning, Appl Phys Lett 72, p.3399-3401 (1998)

[7.360] {Sect. 7.8.2.2} Z. Hasan, M. Solonenko, P.I. Macfarlane, L. Biyikli, V.K. Mathur, F.A. Karwacki: Persistent high density spectral holeburning in CaS:Eu and CaS: Eu,Sm phosphors, Appl Phys Lett 72, p.2373-2375 (1998)

[7.361] {Sect. 7.8.2.2} A. Muller, W. Richter, L. Kador: Persistent spectral hole burning in the few-molecule limit: terrylene in p-terphenyl, Chem Phys Lett 285, p.92-98 (1998)

[7.362] {Sect. 7.8.2.2} H. Sasaki, K. Karaki: Optical parallel pattern recognition of multiple stored images in a persistent spectral holeburning memory, Opt Commun 153, p.9-13 (1998)

[7.363] {Sect. 7.8.2.2} Z. Hasan, L. Biyikli, P.I. Macfarlane: Power-gated spectral holeburning in MgS:Eu2+, Eu3+: A case for high-density persistent holeburning, Appl. Phys. Lett. 72, p.3399-3401 (1998)

[7.364] {Sect. 7.8.2.2} Z. Hasan, M. Solonenko, P.I. Macfarlane, L. Biyikli: Persistent high density spectral holeburning in CaS:Eu and CaS:Eu,Sm phosphors, Appl. Phys. Lett. 72, p.2373-2375 (1998)

[7.365] {Sect. 7.8.2.2} M. Nogami, Y. Abe: High-temperature persistent spectral hole burning of Eu3+- doped SiO2 glass prepared by the sol-gel process, Appl Phys Lett 71, p.3465-3467 (1997)

[7.366] {Sect. 7.8.2.2} M. Tian, F. Grelet, D. Pavolini, J.P. Galaup, J.L. LeGouet: Four-wave hole burning spectroscopy with a broadband laser source, Chem Phys Lett 274, p.518-524 (1997)

[7.367] {Sect. 7.8.2.2} J. Valenta, J. Moniatte, P. Gilliot, R. Levy, B. Honerlage, A.I. Ekimov: Hole-filling of persistent spectral holes in the excitonic absorption band of CuBr quantum dots, Appl Phys Lett 70, p.680-682 (1997)

[7.368] {Sect. 7.8.2.2} M. Nogami, Y. Abe: High-temperature persistent spectral hole burning of Eu3+-doped SiO2 glass prepared by the sol-gel process, Appl. Phys. Lett. 71, p.3465-3467 (1997)

[7.369] {Sect. 7.8.2.2} Y. Mao, P. Gavrilovic, S. Singh, A. Bruce, W.H. Grodkiewicz: Persistent spectral hole burning at liquid nitrogen temperature in Eu (3+)-doped aluminosilicate glass, Appl Phys Lett 68, p.3677-3679 (1996)

[7.370] {Sect. 7.8.2.2} M. Nogami, Y. Abe, K. Hirao, D.H. Cho: Room temperature persistent spectra hole burning in Sm2+-doped silicate glasses prepared by the sol-gel process, Appl. Phys. Lett. 66, p.2952-2954 (1995)

[7.371] {Sect. 7.8.2.2} Y.-I. Pan, Y.-Y. Zhao, Y.Yin, L.-b. Chen, R.-s. Wang, F.-m. Li: The observation of photoproducts and multiple photon-gated spectral hole burning in a donor-acceptor and a donor1+donor2-acceptor system, Opt. Comm. 119, p.538-544 (1995)

[7.372] {Sect. 7.8.2.2} R.B. Altmann, I. Renge, L. Kador, D. Haarer: Dipole moment differences of nonpolar dyes in polymeric matrices: Stark effect and photochemical hole burning. I, J. Chem. Phys. 97, p.5316-5322 (1992)

[7.373] {Sect. 7.8.2.2} W.P. Ambrose, A.J. Sievers: Persistent infrared spectral hole burning of the fundamental stretching mode of SH- in alkali halides, J. Opt. Soc. Am. B 9, p.753-762 (1992)

[7.374] {Sect. 7.8.2.2} S. Arnold, J. Comunale: Room-temperature microparticle-based persistent hole-burning spectroscopy, J. Opt. Soc. Am. B 9, p.819-824 (1992)

[7.375] {Sect. 7.8.2.2} Th. Basché, W.P. Ambrose, W.E. Moerner: Optical spectra and kinetics of single impurity molecules in a polymer: spectral diffusion and persistent spectral hole burning, J. Opt. Soc. Am. B 9, p.829-836 (1992)

[7.376] {Sect. 7.8.2.2} R.L. Cone, P.C. Hansen, M.J.M. Leask: Eu3+ optically detected nuclear quadrupole resonance in stoichiometric europium vanadate, J. Opt. Soc. Am. B 9, p.779-783 (1992)

[7.377] {Sect. 7.8.2.2} R. Hirschmann, J. Friedrich: Hole burning of long-chain molecular aggregates: homogeneous line broadening, spectral-diffusion broadening, and pressure broadening, J. Opt. Soc. Am. B 9, p.811-815 (1992)

[7.378] {Sect. 7.8.2.2} H. Inoue, T. Iwamoto, A. Makishima, M. Ikemoto, K. Horie: Preperation and properties of sol-gel thin films with porphins, J. Opt. Soc. Am. B 9, p.816-818 (1992)

[7.379] {Sect. 7.8.2.2} L. Kümmerl, H. Wolfrum, D. Haarer: Hole Burning with Chelate Complexes of Quinizarin in Alcohol Glasses, J. Phys. Chem. 96, p.10688-10693 (1992)

[7.380] {Sect. 7.8.2.2} S.P. Love, C.E. Mungan, A.J. Sievers: Persistant infrared spectral hole burning of Tb3+ in the glasslike mixed crystal Ba1-x-yLaxTbyF2+x+y, J. Opt. Soc. Am. B 9, p.794-799 (1992)

[7.381] {Sect. 7.8.2.2} C.E. Mungan, A.J. Sievers: Persistent infrared spectral hole burning of the fundamental stretching mode of SH- in alkali halides, J. Opt. Soc. Am. B 9, p.746-752 (1992)

[7.382] {Sect. 7.8.2.2} D. Redman, S. Brown, S.C. Rand: Origin of persistent hole burning of N-V centers in diamond, J. Opt. Soc. Am. B 9, p.768-774 (1992)

[7.383] {Sect. 7.8.2.2} R.J. Reeves, R.M. Macfarlane: Persistent spectral hole burning induced by ion motion in DaF2:Pr3+:D- and SrF2:Pr3+:D- crystals, J. Opt. Soc. Am. B 9, p.763-767 (1992)

[7.384] {Sect. 7.8.2.2} I. Renge: Relationship between electron-phonon coupling and intermolecular interaction parameters in dye-doped organic glasses, J. Opt. Soc. Am. B 9, p.719-723 (1992)

[7.385] {Sect. 7.8.2.2} W. Richter, M. Lieberth, D. Haarer: Frequency dependence of spectral diffusion in hole-burning systems: resonant effects of infrared radiation, J. Opt. Soc. Am. B 9, p.715-718 (1992)

[7.386] {Sect. 7.8.2.2} N.E. Rigby, N.B. Manson: Spectral hole burning in emerald, J. Opt. Soc. Am. B 9, p.775-778 (1992)

[7.387] {Sect. 7.8.2.2} B. Sauter, Th. Basché, C. Bräuchle: Temperature-dependent spectral hole-burning study of dye-surface and mixed matrix-dye-surface systems, J. Opt. Soc. Am. B 9, p.804-810 (1992)

[7.388] {Sect. 7.8.2.2} L. Shu, G.J. Small: Mechanism of nonphotochemical hole burning: Cresyl Violet in polyvinyl alcohol films, J. Opt. Soc. Am. B 9, p.724-732 (1992)

[7.389] {Sect. 7.8.2.2} L. Shu, G.J. Small: Dispersive kinetics of nonphotochemical hole burning and spontaneous hole filling: Cresyl Violet in polyvinyl films, J. Opt. Soc. Am. B 9, p.733-737 (1992)

[7.390] {Sect. 7.8.2.2} L. Shu, G.J. Small: Laser-induced hole filling: Cresyl Violet in polyvinyl alcohol films, J. Opt. Soc. Am. B 9, p.738-745 (1992)

[7.391] {Sect. 7.8.2.2} H. Talon, L. Fleury, J. Bernard, M. Orrit: Fluorescence excitation of single molecules, J. Opt. Soc. Am. B 9, p.825-827 (1992)

[7.392] {Sect. 7.8.2.2} L.L. Wald, E.L. Hahn, M. Lukac: Variation of the Pr3+ nuclear quadrupole resonance spectrum across the inhomogeneous optical line in Pr3+:LaF3, J. Opt. Soc. Am. B 9, p.789-793 (1992)

[7.393] {Sect. 7.8.2.2} D. Wang, L. Hu, H. He, J. Rong, J. Xie, J. Zhang: Systems of organic photon-gated photochemical hole burning, J. Opt. Soc. Am. B 9, p.800-803 (1992)

[7.394] {Sect. 7.8.2.2} K.-P. Müller, D. Haarer: Spectral Diffusion of Optical Transitions in Doped Polymer Glasses below 1 K, Phys. Rev. Lett. 66, p.2344-2347 (1991)

[7.395] {Sect. 7.8.2.2} L. Kador, S. Jahn, D. Haarer: Contributions of the electrostatic and the dispersion interaction to the solvent shift in a dye-polymer system, as investigated by hole-burning spectroscopy, Phys. Rev. B 41, p.12215-12226 (1990)

[7.396] {Sect. 7.8.2.2} R.F. Mahrt, H. Bässler: Vibronic hole burning in acene-doped MTHF glasses, Chem. Phys. Lett. 165, p.125-130 (1990)

[7.397] {Sect. 7.8.2.2} U.P. Wild, A. Renn: Spectral hole burning and holographic image storage, Mol. Cryst. Liq. Cryst. 183, p.119-129 (1990)

[7.398] {Sect. 7.8.2.2} J.K. Gillie, G.J. Small, J.H. Golbeck: Nonphotochemical Hole Burning of the Native Antenna Complex of Photosystem I (PSI-200), J. Phys. Chem. 93, p.1620-1627 (1989)

[7.399] {Sect. 7.8.2.2} R. Jankowiak, D. Tang, G.J. Small: Transient and Persistant Hole Burning of the Reaction Center of Photosystem II, J. Phys. Chem. 93, p.1649-1654 (1989)

[7.400] {Sect. 7.8.2.2} A.J. Meixner, A. Renn, U.P. Wild: Spectral hole-burning and holography. I. Transmission and holographic detection of spectral holes, J. Chem. Phys. 91, p.6728-6736 (1989)

[7.401] {Sect. 7.8.2.2} A. Renn, S.E. Bucher, A.J. Meixner, E.C. Meister, U.P. Wild: Spectral hole burning: electric field effect on resorufin, oxazine-4 and cresylviolet in polyvinylbutyral, J. Luminesc. 39, p.181-187 (1988)

[7.402] {Sect. 7.8.2.2} A. Elschner, H. Bässler: Site-selective fluorescence and hole-burning spectroscopy of MTHF glasses doped with tetracene or pentacene, Chem. Phys. 112, p.285-291 (1987)

[7.403] {Sect. 7.8.2.2} J.K. Gillie, B.L. Fearey, J.M. Hayes, G.J. Small: Persistent hole burning of the primary donor state of photosystem I: Strong linear electron-phonon coupling, Chem. Phys. Lett. 134, p.316-322 (1987)

[7.404] {Sect. 7.8.2.2} J.K. Gillie, J.M. Hayes, G.J. Small, J.H. Golbeck: Hole Burning Spectroscopy of a Core Antenna Complex, J. Phys. Chem. 91, p.5524-5527 (1987)

[7.405] {Sect. 7.8.2.2} R. Jankowiak, G.J. Small: Hole-Burning Spectroscopy and Relaxation Dynamics of Amorphous Solids at Low Temperatures, Science 237, p.618-625 (1987)

[7.406] {Sect. 7.8.2.2} R.F. Loring, Y.J. Yan, S. Mukamel: Hole-Burning Spectroscopy of Polar Molecules in Polar Solvents: Solvation Dynamics and Vibrational Relaxation, J. Phys. Chem. 91, p.1302-1305 (1987)

[7.407] {Sect. 7.8.2.2} R.M. Macfarlane, R.M. Shelby: Homogeneous line broadening of optical transitions of ions and molecules in glasses, J. Luminesc. 36, p.179-207 (1987)

[7.408] {Sect. 7.8.2.2} K.K. Rebane, A.A. Gorokhovskii: Hole-Burning Study of Zero-Phonon Linewidths in Organic Glasses, J. Luminesc. 36, p.237-250 (1987)

[7.409] {Sect. 7.8.2.2} S. Völker: Optical linewidth and dephasing of organic amorphous and semi-crystalline solids studied by hole burning, J. Luminesc. 36, p.251-262 (1987)

[7.410] {Sect. 7.8.2.2} A. Gorokhovskii, V. Korrovits, V. Palm, M. Trummal: Temperature broadening of a photochemical hole in the spectrum of H2-octaethylporphin in polystyrene between 0.05 and 1.5 K, Chem. Phys. Lett. 125, p.355-359 (1986)

[7.411] {Sect. 7.8.2.2} H.W.H. Lee, A.L. Huston, M. Gehrtz, W.E. Moerner: Photochemical hole-burning in a protonated phthalocynine with GaAlAs diode lasers, Chem. Phys. Lett. 114, p.491-496 (1985)

[7.412] {Sect. 7.8.2.2} M. Romagnoli, W.E. Moerner, F.M. Schellenberg, M.D. Levenson, G.C. Bjorklund: Beyond the bottleneck: submicrosecond hole burning in phthalocyanine, J. Opt. Soc. Am. B 1, p.343-348 (1984)

[7.413] {Sect. 7.8.2.2} J. Friedrich, D. Haarer: Reversible and irreversible broadening of photochemical holes in amorphous solids, Chem. Phys. Lett. 95, p.119-123 (1983)

[7.414] {Sect. 7.8.2.2} H.P.H. Thijssen, R. van den Berg, S. Völker: Thermal broadening of optical homogeneous linewidths in organic glasses adn polymers studied via photochemical hole-burning, Chem. Phys. 97, p.295-302 (1983)

[7.415] {Sect. 7.8.2.2} R.M. Shelby, D.P. Burum, R.M. Macfarlane: Nonphotochemical hole burning and antihole production in the mixed molecular crystal pentacene in benzoic acid, J. Chem. Phys. 77, p.2283-2289 (1982)

[7.416] {Sect. 7.8.2.2} J.M.J. Vankan, W.S. Veeman: Inhomogeneous triplet absorption in 1,4-dibromonaphthalene, Chem. Phys. Lett. 91, p.358-361 (1982)

[7.417] {Sect. 7.8.2.2} A.I.M. Dicker, M. Noort, H.P.H. Thijssen, S. Völker, J.H. Van der Waals: Zeeman effect of the S1-S0 transition of the two tautomeric forms of chlorin: A study by photochemical hole burning in an n-hexane host at 4.2 K, Chem. Phys. Lett. 78, p.212-218 (1981)

[7.418] {Sect. 7.8.2.2} R.M. Macfarlane, R.M. Shelby: Sub-Kilohertz Optical Linewidth of the 7F0-5D0 Transition in Y2O3:Eu3+, Opt. Comm. 39, p.169-171 (1981)

[7.419] {Sect. 7.8.2.2} S. Völker, R.M. Macfarlane, A.Z. Genack, H.P. Trommsdorf:, J. Chem. Phys. 67, p.1759-1765 (1977)

[7.420] {Sect. 7.8.2.3} H.W. Song, T. Hayakawa, M. Nogami: Room temperature spectral hole burning and electron transfer in Sm- doped aluminosilicate glasses, J Appl Phys 86, p.5619-5623 (1999)

[7.421] {Sect. 7.8.2.3} K. Fujita, K. Tanaka, K. Hirao, N. Soga: Room-temperature persistent spectral hole burning of EU3+ in sodium aluminosilicate glasses, Optics Letters 23, p.543-545 (1998)

[7.422] {Sect. 7.8.2.3} M. Benhmida, V. Netiksis, M. Robino, J.B. Grun, M. Petrauskas, B. Honerlage: Picosecond spectral hole burning in ZnCdTe layers, J Appl Phys 80, p.4632-4636 (1996)

[7.423] {Sect. 7.8.2.3} K. Hirao, S. Todoroki, N. Soga: Room temperature persistent spectral hole burning of Sm2+ in fluorohafnate glasses, J. Luminesc. 55, p.217-219 (1993)

[7.424] {Sect. 7.8.2.3} C.H. Brito Cruz, R.L. Fork, W.H. Knox, C.V. Shank: Spectral hole burning in large molecules probed with 10 fs optical pulses, Chem. Phys. Lett. 132, p.341-344 (1986)

[7.425] {Sect. 7.8.2.3} G. Mourou: Spectral Hole Burning in Dye Solutions, IEEE J. QE-11, p.1-8 (1975)

[7.426] {Sect. 7.8.2.3} D. Leupold, R. König, B. Voigt, R. Menzel: Modell des sättigbaren Absorbers Crytocyanin/Methanol, Opt. Commun. 11, p.78-82 (1974)

[7.427] {Sect. 7.8.2.3} G. Mourou, B. Drouin, M.M. Denariez-Roberge: Observation du "Hole-burning" dans une solution de cryptocyanine dans le methanol, Opt. Comm. 8, p.56-59 (1973)

[7.428] {Sect. 7.8.2.3} B.H. Soffer, B.B. McFarland: Frequency locking and dye spectral hole burning in Q-spoiled lasers, Appl. Phys. Lett. 8, p.166-169 (1966)

[7.429] {Sect. 7.8.4} B. Voigt, F.R. Nowak, W. Beenken: A new set-up for nonlinear polarization spectroscopy in the frequency domain: experimental examples and theoretical background, Meas. Sci. Technol. 10, p.N7-N11 (1999)

[7.430] {Sect. 7.8.4} W. Beenken, J. Ehlert: Subband analysis of molecular electronic transitions by nonlinear polarization spectroscopy in the frequency domain, J. Chem. Phys. 109, p.10126-10137 (1998)

[7.431] {Sect. 7.8.4} W. Beenken, V. May: Strong-field theory of nonlinear polarization spectroscopy. Fundamentals and the two-level system, J. Opt. Soc. Am B 14, p.2804-2810 (1997)

[7.432] {Sect. 7.8.4} B. Voigt, F. Nowak, J. Ehlert, W. Beenken, D. Leupold, W. Sandner: Substructures and different energy relaxation time within the first electronic transition of pinacyanol, Chem. Phys. Lett. 278, p.380-390 (1997)

[7.433] {Sect. 7.10.1} D.T. Reid, M. Padgett, C. Mcgowan, W.E. Sleat, W. Sibbett: Light-emitting diodes as measurement devices for femtosecond laser pulses, Optics Letters 22, p.233-235 (1997)

[7.434] {Sect. 7.11.1} P.R. Spyak: Beam expander, pinhole, and crosshair alignment to laser beams, Appl Opt 36, p.9111-9112 (1997)

[7.435] {Sect. 7.11.2} T. Baumert, G. Gerber: Femtosecond spectroscopy of molecules and clusters, Adv. Atom, Mol. and Opt. Phys. 35, p.163-208 (1995)

[7.436] {Sect. 7.11.2} M. Dantus, M. Rosker, A.H. Zewail: Real-time-femtosecond probing of "transition states" in chemical reactions, J. Chem. Phys. 87, p.2395-2397 (1987)

[7.437] {Sect. 7.11.2} C.V. Shank, B.I. Greene: Femtosecond Spectroscopy and Chemistry:, J. Phys. Chem. 87, p.732-734 (1983)

[7.438] {Sect. 7.11.2} C.V. Shank: Measurement of Ultrafast Phenomena in the Femtosecond Time Domain, Science 219, p.1027-1031 (1983)

[7.439] {Sect. 7.11.2} A. Bartels, T. Dekorsy, H. Kurz: Femtosecond Ti : sapphire ring laser with a 2-GHz repetition rate and its application in time-resolved spectroscopy, Optics Letters 24, p.996-998 (1999)

[7.440] {Sect. 7.12.1.1} Y.L.S. Zhang, J. Cheng: Theoretical study of transient thermal conduction and temperature distribution generated by pulsed laser, Appl. Phys. B 70p. 85-90 (2000)

[7.441] {Sect. 7.12.1.1} C. Tietz, O. Chekhlov, A. Drabenstedt, J. Schuster, J. Wrachtrup: Spectroscopy on single light-harvesting complexes at low temperature, J Phys Chem B 103, p.6328-6333 (1999)

[7.442] {Sect. 7.12.1.1} S.C. Chen, C.P. Grigoropoulos: Noncontact nanosecond-time-resolution temperature measurement in excimer laser heating of Ni-P disk substrates, Appl Phys Lett 71, p.3191-3193 (1997)

[7.443] {Sect. 7.12.1.1} K. Teuchner, M. Schulzevers, D. Leupold, D. Strehlow, W. Rudiger: The complex excited state dynamics of the early photocycle of phytochrome, Chem Phys Lett 268, p.157-162 (1997)

[7.444] {Sect. 7.12.1.1} M. Pirotta, A. Renn, M.H.V. Werts, U.P. Wild: Single molecule spectroscopy, perylene in the Shpol'skii matrix n-nonane, Chem Phys Lett 250, p.576-582 (1996)

[7.445] {Sect. 7.12.1.1} W. Ketterle, N.J. van Druten: Evaporative cooling of trapped atoms, Adv. Atom, Mol. and Opt. Phys. 37, p.181-236 (1996)

[7.446] {Sect. 7.12.1.1} H. Lueck, R. Menzel, R. Sander: Inherent sample heating and temperature calibration in excited state absorption (ESA) measurements between room temperature and 77 kelvin, Opt. Commun. 108, p.258-264 (1994)

[7.447] {Sect. 7.12.1.1} M. Pirotta, F. Güttler, H. Gygax, A. Renn, J. Sepiol, U.P. Wild: Single molecule spectroscopy. Fluorescence-lifetime measurements of pentacene in p-terphenyl, Chem. Phys. Lett. 208, p.379-384 (1993)

[7.448] {Sect. 7.12.1.1} U.P. Wild, F. Güttler, M. Pirotta, A. Renn: Single molecule spectroscopy: Stark effect of pentacene in p-terphenyl, Chem. Phys. Lett. 193, p.451-455 (1992)

[7.449] {Sect. 7.12.1.1} D. Ben-Amotz, C.B. Harris: Torsional dynamics of molecules on barrierless potentials in liquids. I. Temperature and wavelength dependent picosecond studies of triophenyl-methane dyes, J. Chem. Phys. 86, p.4856-4870 (1987)

[7.450] {Sect. 7.12.1.1} D. Ben-Amotz, C.B. Harris: Torsional dynamics of molecules on barrierless potentials in liquids. II. Test of theoretical models, J. Chem. Phys. 86, p.5433-5440 (1987)

[7.451] {Sect. 7.12.1.1} D. Ben-Amotz, R. Jeanloz, C.B. Harris: Torsional dynamics of molecules on barrierless potentials in liquids. III. Pressure dependent picosecond studies of triphenyl-methane dye solutions in a diamond anvil cell, J. Chem. Phys. 86, p.6119-6127 (1987)

[7.452] {Sect. 7.12.1.1} H. -H. Perkampus: UV-VIS Atlas of Organic Compounds (VCH, Weinheim, 1992)

[7.453] {Sect. 7.12.1.2} B.C. Edwards, J.E. Anderson, R.I. Epstein, G.L. Mills, A.J. Mord: Demonstration of a solid-state optical cooler: An approach to cryogenic refrigeration, J Appl Phys 86, p.6489-6493 (1999)

[7.454] {Sect. 7.12.1.2} T.R. Gosnell: Laser cooling of a solid by 65 K starting from room temperature, Optics Letters 24, p.1041-1043 (1999)

[7.455] {Sect. 7.12.1.2} H. Wadi, E. Pollak: Theory of laser cooling of polyatomic molecules in an electronically excited state, J Chem Phys 110, p.11890-11905 (1999)

[7.456] {Sect. 7.12.1.2} C.E. Wieman, D.E. Pritchard, D.J. Wineland: Atom cooling, trapping, and quantum manipulation, Rev. Mod. Phys. 71, p.253-262 (1999)

[7.457] {Sect. 7.12.1.2} G. Lamouche, P. Lavallard, R. Suris, R. Grousson: Low temperature laser cooling with a rare-earth doped glass, J Appl Phys 84, p.509-516 (1998)

[7.458] {Sect. 7.12.1.2} G. Lei, J.E. Anderson, M.I. Buchwald, B.C. Edwards, R.I. Epstein, M.T. Murtagh, G.H. Sigel: Spectroscopic evaluation of Yb3+-doped glasses for optical refrigeration, IEEE J QE-34, p.1839-1845 (1998)

[7.459] {Sect. 7.12.1.2} X. Luo, M.D. Eisaman, T.R. Gosnell: Laser cooling of a solid by 21 K starting from room temperature, Optics Letters 23, p.639-641 (1998)

[7.460] {Sect. 7.12.1.2} T. Esslinger, I. Bloch, T.W. Hänsch: Bose-Einstein condensation in a quadrupole-Ioffe-configuration trap, Phys. Rev. A 58, p.R2664-R2667 (1998)

[7.461] {Sect. 7.12.1.2} C.E. Mungan, M.I. Buchwald, B.C. Edwards, R.I. Epstein, T.R. Gosnell: Laser cooling of a solid by 16 K starting from room temperature, Phys Rev Lett 78, p.1030-1033 (1997)

[7.462] {Sect. 7.12.1.2} C.E. Mungan, M.I. Buchwald, B.C. Edwards, R.I. Epstein, T.R. Gosnell: Internal laser cooling of Yb3+-doped glass measured between 100 and 300 K, Appl Phys Lett 71, p.1458-1460 (1997)

[7.463] {Sect. 7.12.1.2} L.A. Rivlin, A.A. Zadernovsky: Laser cooling of semiconductors, Opt Commun 139, p.219-222 (1997)

[7.464] {Sect. 7.12.1.2} G. Morigi, J.I.Cirac, M. Lewenstein, P. Zoller: Ground-state laser cooling beyond the Lamb-Dicke limit, Europhys. Lett. 39, p.13-18 (1997)

[7.465] {Sect. 7.12.1.2} E.G. Bessonov, K.J. Kim: Radiative cooling of ion beams in storage rings by broad- band lasers, Phys Rev Lett 76, p.431-434 (1996)

[7.466] {Sect. 7.12.1.2} J.L. Clark, G. Rumbles: Laser cooling in the condensed phase by frequency up- conversion, Phys Rev Lett 76, p.2037-2040 (1996)

[7.467] {Sect. 7.12.1.2} H.J. Lee, C.S. Adams, M. Kasevich, S. Chu: Raman cooling of atoms in an optical dipole trap, Phys Rev Lett 76, p.2658-2661 (1996)

[7.468] {Sect. 7.12.1.2} M.O. Mewes, M.R. Andrews, N.J. van Druten, D.M. Kurn, D.S. Durfee, W.Ketterle: Bose-Einstein Condensation in a Tightly Confining dc Magnetic Trap, Phys. Rev. Lett. 77, p.416-419 (1996)

[7.469] {Sect. 7.12.1.2} J. Lawall, S. Kulin, B. Saubamea, N. Bigelow, M. Leduc, C. Cohentannoudji: Three-dimensional laser cooling of helium beyond the single-photon recoil limit, Phys Rev Lett 75, p.4194-4197 (1995)

[7.470] {Sect. 7.12.1.2} M.H. Anderson, J.R. Ensher, M.R. Matthews, C.E. Wieman, E.A. Cornell: Observation of Bose-Einstein Condensation in a Dilute Atomic Vapor, Science 269, p.198-201 (1995)

[7.471] {Sect. 7.12.1.2} J. Reichel, F. Bardou, M.B. Dasan, E. Peik, S. Rand, C. Salomon, C. Cohen-Tannoudji: Raman Cooling of Cesium below 3 nK: New Approach Inspired by L'evy Flight Statistics, Phys. Rev. Lett. 75, p.4575-4578 (1995)

[7.472] {Sect. 7.12.1.2} C.N. Cohen-Tannoudji, W.D. Phillips: New mechanisms for laser cooling, Phys. Today 43, p.33-40 (1990)

[7.473] {Sect. 7.12.1.2} A. Aspect, E. Arimondo, R. Kaiser, N. Vansteenkiste, C. Cohen-Tannoudji: Laser cooling below the one-photon recoil energy by velocity-selctive coherent population trapping: theoretical analysis, J. Opt. Soc. Am B 6, p.2112-2124 (1989)

[7.474] {Sect. 7.12.1.2} S. Chu, C. Wieman (guest ed.): Laser Cooling and Trapping, J. Opt. Soc. Am. B 6, p.2020 (1989)

[7.475] {Sect. 7.12.1.2} J. Dalibard, C. Cohen-Tannoudji: Laser cooling below the Doppler limit by polarization gradients: simple theoretical models, J. Opt. Soc. Am B 6, p.2023-2045 (1989)

[7.476] {Sect. 7.12.1.2} S. Stenholm: The semiclassical theory of laser cooling, Rev. Mod. Phys. 58, p.699-739 (1986)

[7.477] {Sect. 7.12.1.2} S. Chu. L. Holberg. J.E. Bjorkholm, A. Cable, A. Ashkin: Three-Dimensional Viscous Confinement and Cooling of Atoms by Resonance Radiation Pressure, Phys. Rev. Lett. 55, p.48-51 (1985)

[7.478] {Sect. 7.12.1.2} J.E. Bjorkholm, R.R. Freeman, A. Ashkin, D.B. Pearson: Experimental observation of the influence of the quantum fluctuations of resonance-radiation pressure, Opt. Lett. 5, p.111-113 (1980)

[7.479] {Sect. 7.12.1.2} A. Ashkin, J.P. Gordon: Cooling and trapping of atoms by resonance radiation pressure, Opt. Lett. 4, p.161-163 (1979)

[7.480] {Sect. 7.12.1.2} W. Neuhauser, M. Hohenstatt, P. Toschek, H. Dehmelt: Optical-Sideband Cooling of Visible Atom Cloud Confined in Parabolic Well, Phys. Rev. Lett. 41, p.233-236 (1978)

[7.481] {Sect. 7.12.1.2} W. Neuhauser, M. Hohenstatt, P. Toschek: Visiual Observation and Optical Cooling of Electrodynamically Contained Ions, Appl. Phys. 17, p.123 (1978)

[7.482] {Sect. 7.12.1.2} D.J. Wineland, R.E. Drullinger, F.L. Walls: Radiation-Pressure Cooling of Baound Resonant Absorbers, Phys. Rev. Lett. 40, p.1639-1642 (1978)

[7.483] {Sect. 7.12.1.3} M. Lorono, H.A. Cruse, P.B. Davies: Infrared laser absorption spectroscopy of the nu (7) band of jet-cooled iron pentacarbonyl, J Mol Struct 519, p.199-204 (2000)

[7.484] {Sect. 7.12.1.3} M.M. Ahern, M.A. Smith: Low temperature relaxation of OH in the X-2 Pi and A (2)Sigma states in an argon free-jet, J Chem Phys 110, p.8555-8563 (1999)

[7.485] {Sect. 7.12.1.3} M.A. Duncan, A.M. Knight, Y. Negishi, S. Nagao, Y. Nakamura, A. Kato, A. Nakajima, K. Kaya: Production of jet-cooled coronene and coronene cluster anions and their study with photoelectron spectroscopy, Chem Phys Lett 309, p.49-54 (1999)

[7.486] {Sect. 7.12.1.3} D.R. Farley, K.G. Estabrook, S.G. Glendinning, S.H. Glenzer, B.A. Remington, K. Shigemori, J.M. Stone, R.J. Wallace, G.B. Zimmerman, J.A. Harte: Radiative jet experiments of astrophysical interest using intense lasers, Phys Rev Lett 83, p.1982-1985 (1999)

[7.487] {Sect. 7.12.1.3} P. Farmanara, H.H. Ritze, V. Stert, W. Radloff: Vibrational wavepacket motion in I-2 excited with femtosecond laser pulses in the 200 nm wavelength region, Chem Phys Lett 307, p.1-7 (1999)

[7.488] {Sect. 7.12.1.3} A.L. McIntosh, Z. Wang, R.R. Lucchese, J.W. Bevan, A.C. Legon: Identification of the OC-IH isomer based on near-infrared diode laser spectroscopy, Chem Phys Lett 305, p.57-62 (1999)

[7.489] {Sect. 7.12.1.3} M. Decker, A. Schik, U.E. Meier, W. Stricker: Quantitative Raman imaging investigations of mixing phenomena in high- pressure cryogenic jets, Appl Opt 37, p.5620-5627 (1998)

[7.490] {Sect. 7.12.1.3} T. Ditmire, R.A. Smith: Short-pulse laser interferometric measurement of absolute gas densities from a cooled gas jet, Optics Letters 23, p.618-620 (1998)

[7.491] {Sect. 7.12.1.3} S. Ishiuchi, H. Shitomi, K. Takazawa, M. Fujii: Nonresonant ionization detected IR spectrum of jet-cooled phenol. Ionization mechanism and its application to overtone spectroscopy, Chem Phys Lett 283, p.243-250 (1998)

[7.492] {Sect. 7.12.1.3} G.R. Kennedy, C.L. Ning, J. Pfab: The 355 nm photodissociation of jet-cooled CH3SNO: alignment of the NO photofragment, Chem Phys Lett 292, p.161-166 (1998)

[7.493] {Sect. 7.12.1.3} A. Kumar, C.C. Hsiao, Y.Y. Lee, Y.P. Lee: Observation of saturation dip in degenerate four-wave mixing and two- color resonant four-wave mixing spectra of jet-cooled CH, Chem Phys Lett 297, p.300-306 (1998)

[7.494] {Sect. 7.12.1.3} A.M. Little, G.K. Corlett, A.M. Ellis: UV absorption of LiO in a supersonic jet, Chem Phys Lett 286, p.439-445 (1998)

[7.495] {Sect. 7.12.1.3} Z.A. Liu, R.J. Livingstone, P.B. Davies: Pulse pyrolysis infrared laser jet spectroscopy of free radicals, Chem Phys Lett 291, p.480-486 (1998)

[7.496] {Sect. 7.12.1.3} G.N. Patwari, S. Doraiswamy, S. Wategaonkar: Hole-burning spectroscopy of jet-cooled hydroquinone, Chem Phys Lett 289, p.8-12 (1998)

[7.497] {Sect. 7.12.1.3} A. Vdovin, J. Sepiol, J. Jasny, J.M. Kauffman, A. Mordzinski: Excited state proton transfer in jet-cooled 2,5-di- (2-benzoxazolyl)phenol, Chem Phys Lett 296, p.557-565 (1998)

[7.498] {Sect. 7.12.1.3} X. Yang, I. Gerasimov, P.J. Dagdigian: Electronic spectroscopy and excited state dynamics of the Al-N-2 complex, Chem Phys 239, p.207-221 (1998)

[7.499] {Sect. 7.12.1.3} A. Zehnacker, F. Lahmani, J.P. Desvergne, H. BouasLaurent: Conformation-dependent intramolecular exciplex formation in jet-cooled bichromophores, Chem Phys Lett 293, p.357-365 (1998)

[7.500] {Sect. 7.12.1.3} H.G. Kramer, M. Keil, J. Wang, R.A. Bernheim, W. Demtroder: Intercombination transitions b (3)Pi (u)¡-Chi (1)Sigma (+) (g) in Na-2, Chem Phys Lett 272, p.391-398 (1997)

[7.501] {Sect. 7.12.1.3} H.Z. Li, P. Dupre, W. Kong: Degenerate four wave mixing and laser induced fluorescence of pyrazine and pyridazine, Chem Phys Lett 273, p.272-278 (1997)

[7.502] {Sect. 7.12.1.3} K. Tanaka, Y. Tachikawa, T. Tanaka: Time-resolved infrared diode laser spectroscopy of jet-cooled FeCO and Fe (CO) (2) radicals produced by the UV photolysis of Fe (CO) (5), Chem Phys Lett 281, p.285-291 (1997)

[7.503] {Sect. 7.12.1.3} T. Troxler, B.A. Pryor, M.R. Topp: Spectroscopy and dynamics of jet-cooled 2- methoxynaphthalene, Chem Phys Lett 274, p.71-78 (1997)

[7.504] {Sect. 7.12.1.3} D.T. Anderson, S. Davis, T.S. Zwier, D.J. Nesbitt: An intense slit discharge source of jet-cooled molecular ions and radicals (T-rot¡30K), Chem Phys Lett 258, p.207-212 (1996)

[7.505] {Sect. 7.12.1.3} M. Fukushima, K. Obi: Laser-induced fluorescence spectra of jet cooled p- chlorobenzyl radical, Chem Phys Lett 248, p.269-276 (1996)

[7.506] {Sect. 7.12.1.3} Y. Nibu, D. Sakamoto, T. Satho, H. Shimada: Dispersed phosphorescence spectra in a supersonic free jet by electric discharge excitation, Chem Phys Lett 262, p.615-620 (1996)

[7.507] {Sect. 7.12.1.3} H.K. Sinha, V.J. Mackenzie, R.P. Steer: Laser-induced fluorescence excitation spectroscopy of jet-cooled tropolone carbon monoxide van der Waals complexes, Chem Phys 213, p.397-411 (1996)

[7.508] {Sect. 7.12.1.3} Y. Tang, S.A. Reid: Infrared degenerate four wave mixing spectroscopy of jet- cooled C2H2, Chem Phys Lett 248, p.476-481 (1996)

[7.509] {Sect. 7.12.1.3} A. Zehnacker, F. Lahmani, E. Breheret, J.P. Desvergne, H. BouasLaurent, A. Germain, V. Brenner, P. Millie: Laser induced fluorescence of jet-cooled non-conjugated bichromophores: Bis-phenoxymethane and bis-2,6-dimethylphenoxymethane, Chem Phys 208, p.243-257 (1996)

[7.510] {Sect. 7.12.1.3} E. Zingher, S. Kendler, Y. Haas: The photophysics of a photoreactive system in a supersonic jet. Styrene-trimethylamine, Chem Phys Lett 254, p.213-222 (1996)

[7.511] {Sect. 7.12.1.3} S.A. Wittmeyer, M.R. Topp: Spectral hole burning in free perylene and in small clusters with methane and alcyl halides, Chem. Phys. Lett. 163, p.261-268 (1989)

[7.512] {Sect. 7.12.1.3} P. Erman, O. Gustafsson, P. Lindblom: A Simple Supersonic Jet Discharge Source for Sub-Doppler Spectroscopy, Phys. Scripta 38, p.789-792 (1988)

[7.513] {Sect. 7.12.1.3} A.G. Taylor, W.G. Bouwman, A.C. Jones, C. Guo, D. Phillips: Laser-induced fluorescence of jet-cooled 7-diethylamino-4-trifluoromethyl coumarin, Chem. Phys. Lett. 145, p.71-74 (1988)

[7.514] {Sect. 7.12.1.3} S. Hirayama: A comparative study of the fluorescence lifetimes of 9-cyanoanthracene in a bulb and supersonic free jet, J. Chem. Phys. 85, p.6867-6873 (1986)

[7.515] {Sect. 7.12.1.3} J.A. Warren, E.R. Bernstein: The S2-S0 laser photoexcitation spectrum and excited state dynamics of jet-cooled acetophenone, J. Chem. Phys. 85, p.2365-2367 (1986)

[7.516] {Sect. 7.12.1.3} N.P. Ernsting; The visible spectrum of jet-cooled CCIF2NO, J. Chem. Phys. 80, p.3042-3049 (1984)

[7.517] {Sect. 7.12.1.3} P.M. Felker, A.H. Zewail: Jet spectroscopy of isoquinoline, Chem. Phys. Lett. 94.p.448-453 (1983)

[7.518] {Sect. 7.12.1.3} P.M. Felker, A.H. Zewail: Stepwise solvation of molecules as studies by picosecond-jet spectroscopy: Dynamics and spectra, Chem. Phys. Lett. 94, p.454-460 (1983)

[7.519] {Sect. 7.12.1.3} H.T. Jonkman, D.A. Wiersma: Spectroscopy and dynamics of jet-cooled 1,1'-binaphthyl, Chem. Phys. Lett. 97, p.261-264 (1983)

[7.520] {Sect. 7.12.1.3} H.Abe, N. Mikami, M. Ito: Fluorescence Excitation Spectra of Hydrogen-Bonded Phenols in a Supersonic Free Jet, J. Chem. Phys. 86, p.1768-1771 (1982)

[7.521] {Sect. 7.12.1.3} P.M. Felker, S. R. Lambert, A.H. Zewail: Picosecond excitation of jet-cooled pyrazine: Magnetic field effects on the fluorescence decay and quantum beats, Chem. Phys. Lett. 89, p.309-314 (1982)

[7.522] {Sect. 7.12.1.3} R.E. Smalley: Vibrational Randomization Measurements with Supersonic Beams, J. Phys. Chem. 86, p.3504-3512 (1982)

[7.523] {Sect. 7.12.1.3} M.D. Duncan, P. Österlin, R.L. Byer: Pulsed supersonic molecular-beam coherent anti-Stokes Raman spectroscopy of C2H2, Opt. Lett. 6, p.90-92 (1981)

[7.524] {Sect. 7.12.1.3} I. Raitt, A.M. Griffiths, P.A. Freedman: Resonance fluorescence from nitrogen dioxide cooled in a supersonic jet, Chem. Phys. Lett. 77, p.433-436 (1981)

[7.525] {Sect. 7.12.1.3} A. Amirav, U. Even, J. Jortner: Butterfly motion of the isolated pentacene molecule in its first-excited singlet state, Chem. Phys. Lett. 72, p.21-24 (1980)

[7.526] {Sect. 7.12.1.3} D. Coe, R. Robben, L. Talbot: Interferometric measurements of linewidths and spin doubling in the N2+ first negative band system in free-jet expansions, J. Opt. Soc. Am. 70, p.1238-1144 (1980)

[7.527] {Sect. 7.12.1.3} N.Mikami, A. Hiraya, I. Fujiwara, M. Ito: The fluorescence spectrum of aniline in a supersonic free jet: Double minimum potential for the inversion vibration in the excited state, Chem. Phys. Lett. 74, p.531-535 (1980)

[7.528] {Sect. 7.12.1.3} J.J. Valentini, P. Esherick, A. Owoyoung: Use of a free-expansion jet in ultra-high-resolution inverse Raman spectroscopy, Chem. Phys. Lett. 75, p.590-592 (1980)

[7.529] {Sect. 7.12.1.3} P. Huber-Wälchli, D.M. Guthals, J.W. Nibler: CARS spectra of supersonic molecular beams, Chem. Phys. Lett. 67, p.233-236 (1979)

[7.530] {Sect. 7.12.1.3} D.H. Levy, L. Wharton, R.E. Smalley: Laser spectroscopy in supersonic jets, in Chemical and Biochemical Applications of Laser, Vol. II, ed. by C.B. Moore (Academic, New York 1977)

[7.531] {Sect. 7.12.2} N. Ito, O. Kajimoto, K. Hara: Picosecond time-resolved fluorescence depolarization of p-terphenyl at high pressures, Chem. Phys. Lett. 318, p.118-124 (2000)

[7.532] {Sect. 7.12.2} Ch. Spitz, S. Dähne: Architecture of J-Aggregates Studied by Pressure-Dependent Absorption and Fluorescence Measurements, Ber. Bunsenges. Phys. Chem. 102, p.738-744 (1998)

[7.533] {Sect. 7.12.2} T.P. Russell, T.M. Allen, Y.M. Gupta: Time resolved optical spectroscopy to examine chemical decomposition of energetic materials under static high pressure and pulsed heating conditions, Chem Phys Lett 267, p.351-358 (1997)

[7.534] {Sect. 7.12.2} A. Anderson, W. Smith, J.F. Wheeldon: Infrared study of sulphur at high pressures, Chem Phys Lett 263, p.133-137 (1996)

[7.535] {Sect. 7.12.2} J. Liu, Y.K. Vohra: Fluorescence emission from high purity synthetic diamond anvil to 370 GPa, Appl Phys Lett 68, p.2049-2051 (1996)

[7.536] {Sect. 7.12.2} M. Croci, H.-J. Müschenborn, F. Güttler, A. Renn, U.P. Wild: Single molecule spectroscopy: pressure effect on pentacene in p-terphenyl, Chem. Phys. Lett. 212, p.71-77 (1993)

[7.537] {Sect. 7.12.2} R. Menzel, M.W. Windsor: Picosecond Kinetics of the Excited State Absorption of 4- (9-Anthryl)-N,N-dimethylaniline in a Pressurized Solution, Chem. Phys. Lett. 184, p.6-10 (1991)

[7.538] {Sect. 7.12.2} N. Redline, M. Windsor, R. Menzel: The Effect of Pressure on the Secondary Charge Transfer Step in Bacterial Reaction Centers of Rhodobacter Spheroides R-26, Chem. Phys. Lett. 186, p.204-209 (1991)

[7.539] {Sect. 7.12.2} H. Lueck, M.W. Windsor: Pressure Dependence of the Kinetics of Photoinduced Intramolecular Charge Separation in 9,9'-Bianthryl Monitored by Picosecond Transient Absorption: Comparision with Electron Transfer in Photosynthesis, J. Phys. Chem. 94, p.4550-4559 (1990)

[7.540] {Sect. 7.12.2} M.W. Windsor, R. Menzel: Effect of Pressure on the 12 ns Charge Recombination Step in Reduced Bacterial Reaction Centers of Rhodobacter Sphaeroides R-26, Chem. Phys. Lett. 164, p.143-150 (1989)

[7.541] {Sect. 7.12.2} R. Menzel, H. Lueck, K. Jordan, M.W. Windsor: Pressure Dependence of the Conformational Relaxation Process in the Excited State of Tetra-Methyl-Paraterphenyl in Solution, Chem. Phys. Lett. 145, p.61-66 (1988)

[7.542] {Sect. 7.12.2} K. M. Sando, Shih-I Chu: Pressure broadening and laser-induced spectral line shapes, Adv. At. Mol. Phys. 25, p.133-161 (1988)

[7.543] {Sect. 7.12.2} Th. Sesselmann, W. Richter, D. Haarer: Hole-Burning Experiments in Doped Polymers Under Uniaxial and Hydrostatic Pressure, J. Luminesc. 36, p.263-271 (1987)

[7.544] {Sect. 7.12.2} F.T. Clark, H.G. Drickamer: High-Pressure Study of Triphenylmethane Dyes in Polymeric and Aqueous Media, J. Phys. Chem. 90, p.589-592 (1986)

[7.545] {Sect. 7.12.2} H.G. Drickamer: Pressure Tuning Spectroscopy, Accounts of Chem. Research 19, p.329-344 (1986)

[7.546] {Sect. 7.12.2} F.T. Clark, H.G. Drickamer: High-pressure studies of rotational isomerism of triphenylmethane dye molecules, Chem. Phys. Lett. 115, p.173-175 (1985)

[7.547] {Sect. 7.12.2} F.T. Clark, H.G. Drickamer: The effect of pressure on the adsorption of crystal violet on oriented ZnO crystals, J. Chem. Phys. 81, p.1024-1029 (1984)

[7.548] {Sect. 7.12.2} D. Kirin, S.L. Chaplot, G.A. Mackenzie, G.S. Pawley: The pressure dependence of the low-frequency Raman spectra of crystalline biphenyl and p-terphenyl, Chem. Phys. Lett. 102, p.105-108 (1983)

[7.549] {Sect. 7.12.2} R. S. Bradley, ed.: High Pressure Physics and Chemistry (Academic Press, New York, 1963)

[7.550] {Sect. 7.13} M. Quack, W. Kutzelnigg: Molecular Spectroscopy and Molecular Dynamics: Theory and Experiment, Ber. Bunsenges. Phys. Chem. 99, p.231-245 (1995)

[7.551] {Sect. 7.13} D.C. Harris, M.D. Bertolucci: Symmetry and Spectroscopy. An Introduction to Vibrational and Electronic Spectroscopy (Oxford University Press, New York 1987)

[7.552] {Sect. 7.13} A. Longarte, J.A. Fernandez, I. Unamuno, F. Castano: Ground and first electronic excited state vibrational modes of the methyl-p-aminobenzoate molecule, Chem Phys Lett 308, p.516-522 (1999)

[7.553] {Sect. 7.13} B.A. Zon: Born-Oppenheimer approximation for molecules in a strong light field, Chem Phys Lett 262, p.744-746 (1996)

[7.554] {Sect. 7.13} G. Hohlneicher, J. Wolf: Interference between Franck-Condon and Herzberg-Teller Contributions in Naphthalene and Phenanthrene, Ber. Bunsenges. Phys. Chem. 99, p.366-370 (1995)

[7.555] {Sect. 7.13} L. Kador, S. Jahn, D. Haarer: Contributions of the electrostatic and the dispersion interaction to the solvent shift in a dye-polymer system, as investigated by hole-burning spectroscopy, Phys. Rev. B 41, p.12215-12226 (1990)

[7.556] {Sect. 7.13} M. Maroncelli, G. R. Fleming: Picosecond solvation dynamics of coumarin 153: The importance of molecular aspects of solvation, J. Chem. Phys. 86, p.6221-6239 (1987)

[7.557] {Sect. 7.13} A.C. Borin, F.R. Ornellas: The lowest triplet and singlet electronic states of the molecule SO, Chem Phys 247, p.351-364 (1999)

[7.558] {Sect. 7.13} R. Menzel, K.-H. Naumann: Towards a Theoretical Description of UV-Vis Absorption Bands of Organic Molecules, Ber. Bunsenges. Phys. Chem. 95, p.834-837 (1991)

[7.559] {Sect. 7.13} A. Smolyar, C.F. Wong: Theoretical studies of the spectroscopic properties of tryptamine, tryptophan and tyrosine, J. Mol. Struct. 488, p.51-67 (1999)

[7.560] {Sect. 7.13} M. Aoyagi, Y. Osamura, S. Iwata: An MCSCF study of the low-lying states of trans-butadiene, J. Chem. Phys. 83, p.1140-1148 (1985)

[7.561] {Sect. 7.13} R.A. Goldbeck, E. Switkes: Localized Excitation Analysis of the Singlet Excited States of Polyenes and Diphenylpolyenes, J. Phys. Chem. 89, p.2585-2591 (1985)

[7.562] {Sect. 7.13} R.J. Hemley, U. Dinur, V. Vaida, M. Karplus: Theoretical Study of the Ground and Excited Singlet States of Styrene, J. Am. Chem. Soc. 107, p.836-844 (1985)

[7.563] {Sect. 7.13} R.L. Ellis, G. Kuehnlenz, H.H. Jaffé: The Use of the CNDO Method in Spectroscopy, Theoret. chim. Acta (Berl.) 26, p.131-140 (1972)

[7.564] {Sect. 7.13} J. Lavalette, C. Tetreau, J. Langelaar: SCF MO Calculations on Excited Singlet-Singlet and Triplet-Triplet Transitions of 1,2:3,4-Dibenzanthracene, 1,12-Benzperylene and 3,4-Benzcoronene, Chem. Phys. Lett. 9, p.319-322 (1971)

[7.565] {Sect. 7.13} M. Mestechkin, L. Gutyrya, V. Poltavets: Excited states of alternant hydrocarbons in the LCAO MO approximation. II. Singlet and Triplet Absorption Spectra of Condensed Aromatic Systems, p.244-247 (1969)

[7.566] {Sect. 7.13} J.J. Bene, H.H. Jaffé: Use of the CNDO Method in Spectroscopy. I. Benzene, Pyridine, and the Diazines, J. Chem. Phys. 48, p.1807-1810 (1968)

[7.567] {Sect. 7.13} G.W. Robinson: Intensity Enhancement of Forbidden Electronic Transitions by Weak Intermolecular Interactions, J. Chem. Phys. 46, p.572-585 (1967)

[7.568] {Sect. 7.13} A. Schweig: Calculation of static electric polarizabilities of closed shell organic Pi-electron systems using a variation method, Chem. Phys. Lett. 1, p.163-166 (1967)

[7.569] {Sect. 7.13} J.A. Pople, G.A. Segal: Approximate Self-Consistent Molecular Orbital Theory. II. Calculations with Complete Neglect of Differential Overlap, J. Chem. Phys. 43, p.136-138 (1965)

[7.570] {Sect. 7.13} K.H.J. Buschow, J. Dieleman, G.J. Hoijtink: Corrrelations between the electronic spectra of alternant hydrocarbon molecules and their mono- and di-valent ions. III. Linear polyphenyls, p.1-9 (1962)

[7.571] {Sect. 7.13} R. Pariser: Theory of the Electronic Spectra and Structure of the Polyacenes and of Alternant Hydrocarbons, J. Chem Phys. 24p.250-268 (1956)

[7.572] {Sect. 7.13} F. Zerbetto, M.Z. Zgierski: Theoretical Study of the CC Stretching Vibrations in Linked Polyene Chains: Nystatin, Chem. Phys. Lett. 144, p.437-443 (1988)

[7.573] {Sect. 7.13} G.A. Voth, R.A. Marcus: Semiclassical Theory of Fermi Resonance Between Stretching and Bending Modes in Polyatomic Molecules, J Chem Phys 82, p.4064-4072 (1985)

[7.574] {Sect. 7.13} S.M. Lederman, R.A. Marcus: Densities of Vibrational States of Given Symmetry Species Linear Molecules and Rovibrational States of Nonlinear Molecules, J Chem Phys 81, p.5601-5607 (1984)

[7.575] {Sect. 7.13} G.A. Voth, A.H. Zewail, R.A. Marcus: The Highly Excited C H Stretching States of Chd3, Cht3, and Ch3D, J Chem Phys 81, p.5494-5507 (1984)

[7.576] {Sect. 7.13} A. Warshel, A. Lappicirella: Calculations of Ground- and Excited-State Potential Surfaces for Conjugated Heteroatomic Molecules, J. Am. Chem. Soc. 103, p.4664-4673 (1981)

[7.577] {Sect. 7.13} A. Warshel: The QCFF/PI+MCA Program Package Efficiency and Versatility in Molecular Mechanics, Computers & Chemistry 1, p.195-202 (1977)

[7.578] {Sect. 7.13} S. Lifson, A. Warshel: Consistent Force Field for Calculations of Conformations, Vibrational Spectra, and Enthalpies of Cycloallkane andn n-Alkane Molecules, J. Chem. Phys. 49, p.5116-5129 (1968)

[7.579] {Sect. 7.13} C.L. Tang: A Simple Molecular-Orbital Theory of the Nonlinear Optical Properties of Group III-V and II-VI Compounds, IEEE J. QE-9, p.755-762 (1973)

List of Tables

1.1 Roughly estimated costs of some lasers and their operational cost during their lifetime in relation to the photon energy and average output power 9

2.1 Characteristic values of a photon of different color 13
2.2 Energy of photons in different measuring units for comparison 14
2.3 Spectral uncertainty as a function of the time window Δt and the mid-wavelength 16
2.4 Beam radius $w(z)$, beam curvature radius $R(z)$ and local divergence $\theta_{\mathrm{loc}}(z)$ of a Gaussian beam for different distances z from waist at $z = 0$ measured in Rayleigh lengths z_{R} 31
2.5 Rayleigh length z_{R}, divergence θ, beam diameter w ($z = 0.1\,\mathrm{m}$) and curvature radius R ($z = 0.1\,\mathrm{m}$) for Gaussian beams with different wavelength λ and waist radius w_0 32
2.6 Matrices of frequently used optical elements 36
2.7 Jones vectors for some common light beam polarizations 45
2.8 Jones matrices for some common optical elements 46
2.9 Stokes vectors for some typical light polarizations 49
2.10 Mueller matrices for some common optical elements 50
2.11 Relations of power $P(t_0 - \Delta t_{\mathrm{pulse}}/2)$, power P_{FT} during $\Delta t_{\mathrm{pulse}}$ and energy $E(\Delta t_{\mathrm{pulse}})$ during $\Delta t_{\mathrm{pulse}}$ relative to the peak power P_{max} and the total energy $E_{\mathrm{pulse,tot}}$ for Gaussian pulses. The NLP-exponent describes the nonlinear process which is correctly described by P_{averge} 54
2.12 Relations of the intensity I_{FT} of a flat-top beam with the same energy as a Gaussian beam as a function of the radius of this beam w_{FT} in comparison of the intensity $I(w_{\mathrm{FT}})$ and the energy E inside w_{FT} for the Gaussian beam relative to the peak intensity I_{max} and the total energy E_{tot} of the Gaussian beam. The NLP-exponent describes the nonlinear process which is correctly described by I_{FT} 57
2.13 Coherence length and time, bandwidth $\Delta\nu$ and $\nu_0/\Delta\nu$ of light sources given for a wavelength of $500\,\mathrm{nm}$ 71

3.1 Refractive indices of some gases liquids and solids 89
3.2 Selection rules for light induced transitions in matter 97
3.3 Refractive indices for optically uniaxial crystals for the ordinary beam o and for the extraordinary beam e perpendicular to the optical axis of the crystal for light wavelength of $589\,\mathrm{nm}$ 116
3.4 Optical activity κ_{oa} of some materials at $589\,\mathrm{nm}$ 119
3.5 Values of C for the calculation of the diffraction angles for which the first-order Bessel function has minima and maxima 127
3.6 Maximum diffraction efficiencies with different gratings 142

3.7 Light scattering processes with relative change of photon energy $\Delta E/E$. 142
3.8 Raman active vibrations of some gases 148

4.1 Types of nonlinear optical interactions of light with matter 153
4.2 SHG crystals with symmetry group, nonlinear d_{ij} coefficients for an in-
     cident wavelength λ_{inc} of 1000 nm, transparency wavelength range $\Delta\lambda$
     and damage threshold I_{dam} 160
4.3 Sellmeier coefficients for some commonly used crystals 164
4.4 Refractive indices for the ordinary and extrordinary beams for some
     typical crystals for the wavelengths of Nd lasers and their harmonics . . . 164
4.5 Tuning ranges for the signal and idler wavelengths λ as a function of the
     pump wavelength for different useful OPA and OPO crystals 171
4.6 Coefficients for electro-optical applications of some widely applied non-
     linear crystals 174
4.7 Parameters of some useful third-order nonlinear materials. The γ-values
     are valid for light wavelengths of 1 μm and linear polarization 183
4.8a SBS material parameters of some useful SBS gases for several pump
     wavelengths 199
4.8b SBS material parameters of some useful SBS liquids and solids for several
     pump wavelengths liquid 199
4.9 Permittivity Δn_{SRWS}, gain factor g_{SRWS}, frequency shift $\Delta\nu_{RW}$ and re-
     laxation time τ_{RW} of some liquids 209
4.10 SRS parameters of several materials wave number of the vibration, spec-
     tral width, scattering cross-section and Raman gain coefficient 211

5.1 Properties of superradiance and stochastic emitted light 286
5.2 Strength of the electric field E_{av} of a light beam as a function of the
     intensity I 292
5.3 Rough estimates of damage thresholds for transparent optical compo-
     nents for light pulses of different pulse durations 294
5.4 Material parameters relevant for material processing: density, specific
     heat c_p, heat conductivity k_h, melting temperature T_m, vaporization
     temperature T_v, melt heat Q_m, vaporization heat Q_v, absorption $1 - R$
     (for 1 μm light) 296
5.5 Diffraction of pump beam I_p towards the direction of I_{detect} as a function
     of the relative phase $\Delta\varphi$ between the beams I_p and I_s 299

6.1 Function and examples for the three components of lasers 327
6.2 Quantum defects of some lasers for their strongest laser transitions 330
6.3 Temperature-dependent change of the refractive index $(\mathrm{d}n/\mathrm{d}T)$ for some
     important solid-state laser materials, their expansion coefficient α_{expan}
     and their thermal conductivity K_{cond} 348
6.4 Shock parameter for different host materials of solid-state lasers 351
6.5 Maximum of the transversal modes $F_{max,circ}$ under the condition of equal
     power or energy content for all modes 364
6.6 Maximum of the transversal modes $F_{max,rect}$ under the condition of equal
     power or energy content of all modes 367
6.7 Beam parameters for higher circular Gauss–Laguerre modes 371
6.8 Beam parameters for higher rectangular Gauss-Hermite modes 372
6.9 Beam radii at the two resonator mirrors M_{OC} and M_{HR} at the stability
     limits of the resonator 382
6.10 Material constant $C_{material}$ defining the stability range and the TEM_{00}
     potential for different lasers 384

6.11 Bandwidth of several laser materials as peak wavelength λ_{peak}, wavelength bandwidth $\Delta\lambda$, frequency bandwidth $\Delta\nu$ and number of modes p within this bandwidth in a 10 cm long laser resonator 392

6.12 Emission cross-sections σ_{laser}, and lifetimes of the upper laser level at room temperature τ_{upper} for the most prominent laser wavelengths λ_{laser} of some materials 399

6.13 Wavelength, tunability range, pulse width range, average output power, beam quality and wall-plug efficiency of some lasers 448

6.14 Some typical properties of diode lasers 453

6.15 Some typical properties of commercial diode lasers and diode laser bars . 454

6.16 Some typical properties of vertical cavity surface-emitting lasers (VCSEL) 455

6.17 Some typical properties of commercial Nd:YAG lasers 457

6.18 Some typical properties of Nd YVO lasers 458

6.19 Some typical properties of Nd glass lasers 459

6.20 Some typical properties of commercial Yb:YAG lasers 460

6.21 Some typical properties of commercial Ti:sapphire lasers 461

6.22 Some typical properties of Cr:LiCAF and Cr:LiSAF lasers 462

6.23 Some typical properties of alexandrite lasers 463

6.24 Some typical properties of erbium and holmium lasers 464

6.25 Some typical properties of ruby lasers 465

6.26 Some typical properties of Er fiber lasers 466

6.27 Some typical properties of commercial XeCl and KrF lasers 467

6.28 Some typical properties of nitrogen lasers 468

6.29 Some typical properties of He-Ne lasers 470

6.30 Some typical properties of He-Cd lasers 471

6.31 Some typical properties of commercial Ar and Kr ion lasers 472

6.32 Some typical properties of Cu-vapor lasers 473

6.33 Some typical properties of CO_2 lasers 474

6.34 Life time of some laser dye solutions 475

6.35 Some typical properties of cw dye lasers 476

6.36 Some typical properties of pulsed dye lasers 477

6.37 Pulse energies of a commercial OPA in the fs range pumped by a 1 kHz Ti:sapphire laser of 80 fs pulse duration and 750 µJ pulse energy 484

6.38 Pulse energies of a commercial OPA in the ps range pumped by a 1 kHz Ti:sapphire laser of 1 ps pulse duration and 1 mJ pulse energy 484

6.39 Maximum permissible exposure (MPE) power or pulse energy of the eye as function of the pulse length and the wavelength of the laser radiation (without guaranty) 486

7.1 Sequence of tasks in nonlinear optical spectroscopy to characterize the nonlinear behavior of matter with their relevant parameters 491

7.2 Time delay from a delay line in air passed back and forth 499

7.3 Quantum yields Φ_{yield}, excitation and emission wavelengths λ_{exc} and $\lambda_{fluorescence}$ and the fluorescence lifetime $\tau_{fluorescence}$ of some materials . . 513

7.4 Values of the factor $C_{\varepsilon nl,stat}$ as a function of ε_{nl} and T_0 for a stationary two-level system 518

7.5 Values of the factor $C_{\varepsilon nl,nonst}$ as a function of ε_{nl} and T_0 for an integrating two-level system 519

7.6 Wavelengths of some high-intensity atomic and Fraunhofer (named and with color) absorption and emission lines for the calibration of detection systems 542

7.7   Population density factor $N_{\varepsilon\mathrm{nl}}$ of the first excited state of a stationary two-level scheme as a function of the bleaching parameters averaged along the excitation beam 548

7.8   Population density factor $N_{\varepsilon\mathrm{nl}}$ of the first excited state of a non-stationary two-level scheme as a function of the bleaching parameters averaged along the excitation beam 548

7.9   Rough classification of grating lifetimes for different decay mechanisms and the resulting spectral widths of the broadening for molecular systems 558

7.10 Solvents for low-temperature measurements. T_{glass} is a temperature characterizing the transition from liquid to glass of the material 575

7.11 Quantum chemically calculated transitions for pentaphene in comparison with the experimental data 580

Subject Index

2×2 Jones matrices 43
4 × 4 Mueller matrices 43
4 × 4 ray matrices 26
1,1-diphenyl ethylene 314
1,2-dichlorethene 418
1,3,5-triphenyl benzene 314
221th-harmonic 228

ab initio 580 ff
absorption 101
– and emission lines 542
– coefficient a 88, 101
– cycles 295
– grades 102
– measurements 502 ff
– or emission of photons 95
– spectrum 330
acceptor groups 229
Aceton 199
acetylene 311
achromatic 91
acoustic phonon 195
Acousto-optic switches or modulators
  (AOMs) 416, 430
acridine yellow 316
active material 326, 327 ff
– bandwidths 391
active mode locking 430 ff
– by gain modulation 431 ff
active stabilization 415
adaptive mirrors 345
additive pulse mode locking 428 ff
ADP 160, 164, 171
aggregation 318, 524
Airy function 79
alcohol 230
alexandrite 392, 399
alexandrite laser 463 ff
allowed transitions 96
Aluminium 296
AMPAQ 580

amplification without inversion 286 ff
amplifier
– beam quality 445 ff
– efficiencies 439 ff
– energy or power content 439 ff
– gain 436 ff
– noise 445 ff
– quality problems 444 ff
– saturation 436 ff
amplifier schemes 440 ff
amplitude grating 136
anisotropic materials 115
anomalous dispersion 90
anthracene 313
anti-Stokes lines 147
anti-Stokes SRS 211
antireflection coatings 294
AOM driver frequency 416, 430
apochromatic 91
Argon (Ar) ion lasers 340, 392, 399,
  472 ff
atom laser 4
atom optics 4
avalanche transistor trigger 479
average intensity 55
average photon energy $\overline{E}_{\mathrm{photon}}$ 52
average power $\overline{P}_{\mathrm{pulse}}$ 52
axial resonator modes 352

Babinet's theorem 128 ff
back driving force 154
Banana 160
band shape analysis 506 ff
bandwidth 99
– of Rayleigh wing scattering 144
– of single longitudinal modes 390 ff
bandwidth-limited pulses 421 ff
bariumtitanate (BaTiO$_3$) 319, 321
BBO 483
beam cleanup 298, 321
beam coupling 298

beam delivery 297
beam diameter 28
beam diameter at HR *356*
beam diameter at OC *356*
beam divergence of higher transversal
    modes 373 ff
beam parameter product *59*
beam parameters w_i and R_i 33
beam propagation factor M^2 58, 60,
    374
beam propagation factor $M^2_{86.5\%}$ 374
beam propagation matrix 36, 380
beam propagation of higher transversal
    modes 374
beam propagation of Gaussian beams
    32
beam quality BQ 60, 219
beam quality of higher transversal
    modes 374 ff
beam radii of higher transversal modes
    370 ff
beam radius $w(z)$ 28, 29, *55*
Bell's inequalities 5
benzene 209, 211, 230, 311, 313
Berthune cell 331
Beta-Bariumborate (BBO) 160
biconvex lens *42*
bio-technologies 8
biphenyl 314
birefringence 115 ff, 577
birefringence measurement 349
bistable device 308
blackbody radiation *63, 514*
bleaching 232, 236 ff 425, 517 ff
Boltzmann energy 97
Boltzmann's constant 149
Boltzmann's equation for the popula-
    tion density 149
Born–Oppenheimer approximation
    578
Bose–Einstein condensation 4, 576
Bose–Einstein distribution 63
BOX CARS 218 ff
boxcar-integrators 54
bracket formalism 269
Bragg condition *135*
Bragg reflection 135 ff
breakdown mechanism 293
Brewster angle *111*
brightness *61*
brilliance *61*
Brillouin enhanced four-wave mixing
    (BEFWM) 302

Brillouin scattering 145
broad band laser emission 407 ff
Bromobenzene 209
buffer gases 339
built-up time *420*
bulk damage 294

C_2F_6 230
calibration of spectral sensitivity
    513 ff
Carbondioxide CO_2 199
Carbondisulfide CS_2 183, 199, 209,
    211, 230, 544
Carbontetrachloride CCl_4 148, 199,
    230
CARS 216 ff
cavity dumping 415 ff
CCl_4 148, 230
CDA 164
CdTe 319
Ce 342
centrosymmetric matter 157
CH_4 148, 230
characteristic pump rate 235
characteristic time scale 235
chemical hole burning 243
chemical lasers 342, 478
chemical pumping 342 ff
chirp 91, 189, 422
chirp compensation 426, 427
chirp frequency *191*
chlorobenzene 209
chlorophyll 309
choice of the excitation light 537 ff
chromium (Cr) 296
circuit board material 469
circular aperture
– far-field angle 127
– power content 128
circular birefringence 119
circular eigenmodes 363 ff
class A lasers 436
class B lasers 436
class C lasers 436
CNDO-S/CI 580
CO 230
CO_2 230, 392, 399
CO_2 lasers 474 ff
COANP 229
coating 114
coherence 67 ff
– conditions 68
– length l_c 67, *69*
– length of light sources 70, 499

– of lasers 70
– time 67, *69*
coherence radar measurements 392
coherent Anti-Stokes Raman scattering
  (CARS) 216 ff
coherent interaction 232
coherent light fields 259 ff
coherent resonant interaction 265 ff,
  536
coherent scattering 142
coherent state 66
colliding pulse mode locking (CPM
  laser) 425 ff
collinear or longitudinal excitation 494
Collins integral 122 ff
collision cross-section 339
colloids 145
color center formation 103
color center lasers 478
combined interactions with diffraction
  and absorption 297 ff
common laser parameters 451 ff
compensation of phase distortions 221,
  223
complex beam parameter $q(z)$ 31
complex susceptibility χ 87
Compton-backscattered photons 13
concave–convex resonators 362
concentric or spherical resonators 361
confocal resonators 359, 389
conformational changes 510
conjugated molecules 311
cooling efficiency 342
copper (Cu) 296
copper vapor laser 339, 473 ff
cost of photons 8
CPM laser 443
Cr 456
Cr lasers 421
Cr:YAG 319, 418
Cr:LiCAF and Cr:LiSAF lasers 462 ff
cresylviolet 310
critical power for self-focusing 185
cross-section σ 55, *102*, 244, 498
cross-section of stimulated Raman
  scattering σ_{SRS} 214
cross-section of anisotropic particles
  504 ff
cross-section of Rayleigh wing
  scattering 145
cross-section of the absorption band 96
cross-section of laser materials 399,
  410, 453 ff

cross-sections of excited state absorptions 564
cryptocyanine 310, 418, 506
crystal capacity 176
crystal violet 316, 507
crystals 115
crystals grown from melt 229
crystals grown from solution 229
CS_2 183, 209, 211, 230, 544
Cu vapor lasers 339, 473 ff
current density 85
curved mirror resonators 359
cutting 295
cw and quasi-cw (mode-locked) dye
  lasers 476 ff
cyclohexane 199, 311

d-LAP 199
damage stress 351
damage threshold
– pulse width *293*
– roughness *294*
– spot size *294*
damped Rabi oscillation 278 ff
DAN 229
darkening 252, 517 ff
decay mechanisms 558
decay rates 244
decay time *99*
decay time $\tau_{Rayleigh}$ 144
decay time measurements 550 ff
decreasing spectral bandwidth 392 ff
deflection angle 177
degenerate four wave mixing (DFWM)
  179, 300, 304
degree of polarization *48*
delay lines 496, 498
density 296
density matrix elements 269
density matrix formalism 152, 266 ff
density matrix operator ρ 269, 275
dephasing time T_2 265 ff
DF laser 343
DFB laser 262, 393, 432
– spectral resolution 432
diameters of mode apertures 375
Dicarbonhexafloride C_2F_6 199
dielectric material 114
dielectric mirrors 294
difference frequency *168*
difference spectra 551 ff
differential cross-section, Rayleigh
  scattering 144

differentiation of singlet and triplet
  spectra 549 ff
diffraction 120 ff
– at a chain of small objects 133 ff
– at a double-slit 130 ff
– at a one-dimensional slit 122 ff
– at a two-dimensional slit 125 ff
– at one-dimensional slit gratings
  131 ff
– at small objects 128 ff
– at three-dimensional gratings 135 ff
– at two-dimensional gratings 134 ff
diffraction efficiency η_{diff} 142, 299
diffraction in first order systems
  122 ff
diffraction integral 121
diffractive optics 137
dimers 318
diode lasers 337 ff, 452 ff
diode laser bars arrays and stacks
  454 ff
diode laser stacks 334, 452
dipole moment $\mu_R(\nu_{vib}, t)$ 148
dipole moment operator 271
dipole–dipole interaction 100
disk lasers 336
dispersion 90
– phenomenalogical description 91
dispersion compensation 427
– of crystals 164 ff
dissociation limit 563
distributed feedback (DFB) dye lasers
  262, 432 ff
distributed feedback (DFB) lasers,
  393
divergence angle θ 57, 373
divergence angle of the laser 355
DNA 8
DODCI 310, 426
donor groups 229
donut modes 370 ff
Doppler effect 146
Doppler shift 97
double pass amplifier 441 ff
double pass amplifier schemes 120
double pass amplifier with phase
  conjugating mirror 444 ff
double-chirped mirrors 428
drilling with lasers 295
DTTC 310
dye jets 332, 426
dye laser 421, 441, 475 ff, 511
dyes 417

dyes in a polymer matrix 331
dynamic holography 300
dynamically stable resonators 384 ff

Einstein's coefficients 96
Einstein–Podolski–Rosen (EPR)
  paradox 5
elastic scattering 142
electric displacement 85
electric field strength E_{av} 292
electric field vector 19
electric permittivity ε_r 19, 86
electric susceptibility χ 86
electrical charge density 85
electrical conductivity 86
electrical discharge pumping 339 ff
electrical polarization 86
electrical pumping in diode lasers
  337 ff
electro-optical beam deflection 155,
  176 ff
electro-optical coefficients r_{mp} 175
electro-optical effects 228, 415
electro-optical efficiency 344
electro-optical modulator 175
electro-optical second-order effects 172
electron density 313
electronic delay generators 571
electronic transitions 97
electrostriction 195
elementary beams 63 ff
ellipsoid of the refractive index 115
emission cross section 508, 513 ff
emission cross-sections σ_{laser} of laser
  materials 398
emission gratings 137
emission lifetimes 508
emission probability 98
emission quantum yield 508
empty resonator 355 ff
end-pumped 333
energy density 22
energy distribution of the electrons
  339
energy level scheme 268
energy of a light pulse E_{pulse} 52
energy transfer laser pumping 329
energy transmission 516
entangled states 4
environmental detection 8
environmental interaction 578
Er 456
Er:glass 348
Er:YAG 330

Erbium (Er) laser 464 ff
etalons 421
ethane 311
ethanol 199
ethylene 311
ethylene-glycol 333, 426
evanescent light waves 112
excess energy 100
excimer and nitrogen lasers 286
excimer lasers 332
exciplexes 318
excitation 538
excitation intensity variation 530 ff
excitation light intensities 569 ff
excited state absorption (ESA) 238,
   328, 517, 520, 521, 531 ff
– estimation 505 ff
– measurements 545 ff
excited state absorption (ESA)
   spectroscopy 238, 534 ff
excited state absorption gratings 261 ff
exciton formation 244
excitons 100
expectation value of P 269
extinction coefficient ε_a 102
extraordinary beam of second harmonic
   163
extraordinary beam 115
extreme inhomogeneous transition 561
eye damage 486

Fabry Perot roughness of the optical
   surfaces 81
Fabry–Perot etalon 77
Fabry–Perot interferometer 68, 77, 307
far-field 57, 124
far-field pattern 122 ff
far-field angle 123
far-infrared lasers 478
Faraday effect 120
Faraday rotator 48, 120, 222, 441
fast axis 337
fast decaying fluorescence 481, 566
fast optical switches 182
femtochemistry 8
femtosecond lasers 420, 442
Feynman diagrams for nonlinear optics
   275 ff
– degeneracy factor 277
fiber lasers 466 ff
fiber-coupling 335
fidelity F of phase conjugation 225
field emission strength E_{FE} 292

fifth harmonic 227 ff
filaments 186
finesse F 80
flash lamps 542 ff
flat-top intensity profile 56, 57
flexible aromatic molecules 314
– twist angle 314
flow tubes 342
fluctuations 415
fluorescence as probe light 541 ff
fluorescence bands 241
fluorescence decay time 511 ff
fluorescence intensity scaling 567 ff
fluorescence spectrum 509 ff
F-number (FN) 130
focusing with a lens 40 ff
Foerster mechanism 100
forbidden transitions 96
force on reflector 62
fosterite (Mg_2SiO_4) 456
four-element Stokes vectors 43
four-level amplifier system 437
four-level system 329, 402
four-wave mixing (FWM) 155, 197,
   221, 297, 300 ff, 319, 525, 557, 560
fourth (FHG) harmonics 482
fractional bleaching (FB) 235, 239,
   551 ff
Fraunhofer lines 542
Fraunhofer integral 122
free electron laser 228, 478
free spectral range 80
freon 230
freon 113 $(C_2Cl_3F_3)$ 199
frequency 12
frequency bandwidth $\Delta\nu_{FWHM}$ of
   Gaussian pulse 422
frequency conversion 158 ff, 180, 226,
   232
frequency doubled Nd:YAG lasers 332
frequency doubling 158
frequency mixing 168 ff, 228
frequency of the spiking oscillations
   413
frequency pulling 406 ff
frequency spectrum of light pulses 83 ff
frequency transformation 158, 180,
   226, 482 ff
frequency tripling 180
frequency–time uncertainty 15
Fresnel integral 121
Fresnel number F 23, 120
Fresnels formulas 105 ff

– reflectivity R 106
– transmission T 106
fringe visibility *69*
fs lasers 420, 442
FTIR switch 416
full width at half maximum power 53
fundamental limits for fluctuations
    415
fundamental mode 354 ff
fundamental mode aperture design
    376, 378
fundamental mode operation 375 ff
fundamental soliton 193

g parameter 357 ff
GaAs 319, 392, 399, 424
gain 398 ff, *400*
gain coefficient g 329, 398
gain factor G_{amp} 436
gain switching 409 ff
gas lasers 467 ff
gas-ion lasers 421
gases 230 ff
Gauss beam 27
– power content 128
– propagation 32
Gauss pulse *53*
Gauss–Hermite modes 364 ff
Gauss–Laguerre modes 363 ff
Gaussian band integral 506
Gaussian bands analysis *506*
Gaussian bandwidth *83*
Gaussian mirrors 377
generation of ns pulses 415 ff
generation of ps and fs pulses 420 ff
generation of the second harmonic
    158 ff, 482 ff
generation of the third harmonics 180,
    482 ff
generation of higher harmonics 226,
    482 ff
geometrical optics *23*
Germaniumtetrachloride GeCl$_4$ 199
GGG (Gd$_3$Ga$_5$O$_{12}$) 456
Gold 296
– laser 473
gravitational wave detection 8, 65
grazing incidence 393
ground state absorption (GSA) 505
ground state absorption recovery time
    518 ff
group velocity *90*
GSGG (Gd$_3$Sc$_2$Al$_3$O$_{12}$) 456

H$_2$ 148
half width half maximum 53
half-wave plate 47
half-wave voltage 173
Hamilton operator H 93
hardening of steel 295
harmonic generation 228, 482
harmonic modulation (sine) grating
    139
Hartmann dispersion 91
Hartmann equation 91
HBr 343
HCl 343
He-Cd laser 471 ff
He-Ne 392, 399
He-Ne laser 339, 470 ff
heat conductivity k_h 296
heat treatment 295
helium cryostats 574
helium-neon (He-Ne) laser 339, 470 ff
hemispherical resonators 360
heterodyne technique 82 ff
hexene 230
HF 148
HF* laser 342
high harmonic generation 155, 226,
    482
high pressures 576 ff
high reflectivities 114
high-resolution microscopy 291
higher transversal modes 362 ff
higher-order effects 155 ff, 226
highest occupied molecular orbital
    (HOMO) 577
hole burning 232
hole burning (HB) measurements 554
    ff
Holmium (Ho) laser 464 ff
homogeneous and inhomogeneous
    broadening 232 ff
homogeneously broadened absorption
    or emission bands 233, 561
homogeneously broadened laser
    transition 406
hybrid modes 370 ff
hyper-sound wave 195
hyper-sound wave frequency 82, 146
hyper-sound wavelength 146

idler beam 169
II–VI semiconductors 321
III–V semiconductors 321
image position 26

image size 26
impedance 20
impurities 293
incoherent interaction with absorption
    235 ff
incoherent scattering 142
index modulation *320*
indium seals 575
induced absorption or phase gratings
    499
induced birefringence 118
induced grating spectroscopy 264 ff
induced gratings 70, 232, 297 ff, 537,
    557
induced inversion gratings 262 ff
induced transmission 261 ff
inelastic scattering 142
inhomogeneous broadening 511, 524,
    561
inhomogeneously broadened bands
    233
inhomogeneously broadened laser
    transitions 407
injection nozzle 576
inorganic crystals 228 ff
InP 319
intensity I *17, 22*
intensity of the SHG 161
intensity of the soliton 193
interaction length 67 ff
interaction length SHG 165
interaction range transversal 67
interference experiments 68
interference pattern 67
interference pattern result from two
    Gaussian beams 73
interferometers 68
internal laser intensity *400*
internal vibrational relaxation (IVR)
    time, T_{IVR} (T_3) 234, 266
intracavity frequency doubling 166
inverse Raman spectroscopy (IRS)
    215 ff
inversion 326, 328
inversion population density
    329, 401
ionization energy 292
IR 103
iron 296
isomers 313

Jablonski diagram 267
Jamin interferometer 68

jet cooling 576
jitters 499
Jones matrices 46
Jones vectors 43, 43 ff

Kaliumbromide 103
Kasha's rule 511 ff
KD*P 160, 164, 176, 483
KDP 160, 163, 164, 166, 171, 177
Kerr cell 415
Kerr constant 182
Kerr effect 181 ff, 426
– induced birefringence 155
Kerr lens mode locking 426 ff
Kerr medium 183
Kerr shutter
– opto-optical 183
– electro optical 182
KGW (KGD (WO_4)) 456
knife edge method 56
Kodak dye #14015, 310
Kodak dye #9740, 310
Kodak dye #9860, 310
Kr ion lasers 340, 472 ff
Kramers–Kronig relation 88
KrF 392, 399
Krypton (Kr) ion laser 472 ff
KTP 160, 166

labeling 295
Lambert–Beer law 101
lamp pumping 341 ff
large mode volumes 377 ff
LASER 325 ff
laser ablation 297
laser action 241
laser amplifier 436 ff
laser chaotic behavior 434 ff
laser chemistry 8
laser classification 446 ff
laser cleaning 8
laser condition *399*
laser cooling 575
laser data checklist 449 ff
laser display 8
laser ignited fusion 8
laser intensity and power 401 ff
laser interferometers 8
laser light 8
laser light statistics *64*
laser material parameters 329
laser material processing 295 ff
laser medicine 8
laser molecules 343

laser pulses 293
laser resonators 352 ff
laser safety 485 ff
laser spark ignitions 293
laser threshold 391, 399 ff
laser threshold condition 400
laser wavelengths 447 ff
laser without inversion 287
laser writing 297
laser-induced plasmas 228
lasers pumping 331 ff
lasers with a phase conjugating mirror 378 ff
lasers with very high powers 8, 420 ff, 436 ff, 451 ff
lateral coherence 67, 71
lead (Pb) 296
lead vapor laser 473
lens duct 335
lenses in the resonator 377 ff
levitation force 62
LiCaF (LiCaAlF$_6$) 456
lifetime of the excited state 241
lifetime of the Rayleigh scattering τ_{RL} 207
lifetime of phonons 147
lifetime of the sound wave 147
lifetime of the upper laser level 398
lifetime of the vibration τ_{vib} 213
lifetimes of induced gratings 558
lifetimes of the diode lasers 338
light beam coherence 68
light beats 82 ff
light matter interaction 86, 231 ff
light modulation 175
light power P 52
light scattering 142
LiIO$_3$ 171
LiNbO$_3$ 171, 211
line shape function 560
linear interactions of light with matter 85
linear optics 101
linear polarization of matter 156
linewidth of the SRS 211
linewidth of the vibrational transition $\Delta\nu_{vib}$ 213
Liouville equation 270
liquid crystals 115, 230 ff, 424
liquid nitrogen cryostats 574
liquid O$_2$ 211
LiSAF (LiSrAlF$_6$) 456
Lithium-Triborate (LBO) 160

lithiumniobate (LiNbO$_3$) 160, 319
lithography 8
Littrow mounting 393
longitudinal modes 387 ff
longitudinal resonator modes 352
longitudinal single mode 416
longitudinal soliton 192
Lorentz force 87
Lorentzian band integral 507
Lorentzian bands analysis 507
Lorentzian profile 99
Loschmidt's number N_L 102
low temperature of μK 576
low temperatures 574 ff
lowest triplet level 511
lowest unoccupied molecular orbital (LUMO) 577
luminescence spectra 509

Mach–Zehnder interferometer 68
magnetic field vector 19
magnetic induction 85
magnetic permeability μ_r 19, 86
magnetic polarization 86
magnetic susceptibility 86
malachite green 316, 507
Manganese 296
Manley–Rowe conditions 172
MAP 229
marking 295
master oscillator power amplifier (MOPA) 440
material
– concentration 102
– isotropic 180
– parameters 102
– pressure 102
– temperature 102
material processing 8
material viscosity 196
materials for optics 103
materials for nonresonant nonlinear interactions 228 ff
materials in resonant nonlinear optics 309 ff
matrices of frequently used optical elements 36
maximum permissible exposure (MPE) 486 ff
Maxwell's equations 85
MBANP 229
mechanical Q-switch 416
melt heat Q_m 296
melting 295

melting process 297
melting temperature T_m 296
metal reflectivity 113
methane CH_4 148, 199, 230
methanol 199
Michelson interferometer 68
microparticles 293
Mie scattering 145
Mie scattering in air 145
MINDO-S 580
minimal bandwidth of short pulses 16
minimal spectral bandwidth 406 ff
mirror symmetry fluorescence 509
misalignment sensitivity 381 ff, 383
MISER 395
MNA 229
mode apertures 375 ff
mode hopping 264, 394
mode locking 420 ff
mode number p_mode 388
mode spacing *388*
modeling 517 ff
modulation *130*
modulation transfer function (MTF)
    130 ff
molecular computers 8
molecular energy 97
momentum conservation 147
monochromatic wave 19
MOPA (*M*aster *O*scillator *P*ower
    *A*mplifier) 326
moving foci 186
MPE (maximum permissible exposure)
    486 ff
Mueller matrices 50
multi pass amplifier 442 ff
multi-exponential decay 532
multiphoton absorption 287 ff
multiple excitation 562 ff
multiple roundtrip eigenmodes 378

N_2 148, 230
N_2 (liquid) 211
N_2 laser 468 ff
nanometer structures 321 ff
nanotubes 323
naphthalene 313
narrow linewidths 507
National Ignition Facility (NIF) 440
natural lifetime 512 ff
Nd 456
Nd:glass laser 425, 459 ff
Nd lasers 421

Nd:Cr:GSGG 348
Nd:glass 348, 392, 399
Nd:KGW 348, 399
Nd:KGW (Nd:KGd(WO_4)) 458
Nd:YAG 57, 330, 348, 392, 399, 425,
    457
Nd:YAG amplifier 439, 440
Nd:YAG laser 431, 442
– Q switch 420
Nd:YALO 348, 399, 457
Nd:YAP 457
Nd:YLF 348, 399
Nd:YVO lasers 458 ff
Nd:YVO_4 348, 484
near-field 124
near-field optics 129
net observation times 68, 70
new diode lasers 478
new solid-state lasers 478
nickel 296
nitrobenzene 209, 211
nitrogen lasers 332
nitrogen (N_2) 199
NLP-exponent 54, 57
noble gases 230
noncritical phase matching 165
nondiagonal elements of the density
    matrix 269
nonharmonic vibrational potentials
    149
nonlinear absorber 426, 442
nonlinear absorption 232, *235*
nonlinear absorption coefficient 235
nonlinear absorption measurement
    517 ff
nonlinear emission measurements
    530 ff
nonlinear intensity I_nl, I_nl 236, 398
nonlinear interaction
– nonresonant (transparent matter)
    232
– resonant (absorbing matter) 232
nonlinear interactions of light and
    matter with absorption 231 ff
nonlinear nonresonant light-matter
    interaction 231
nonlinear optical spectroscopy 489 ff
nonlinear optics 489
nonlinear polarization of matter P_nl
    152, 156 ff, 272
nonlinear polarization (NLP) spec-
    troscopy 235, 559 ff

nonlinear spectroscopy
– optimum spot size 499
– possible measuring errors 501 ff
– probe light intensity 494
– pump and probe light overlap 494 ff
– sample lengths 495
– steps of analysis 490 ff
– temporal overlap 496 ff
– transversal excitation 495
nonlinear susceptibilities $\chi^{(m)}$ 157
nonlinear transmission 239 ff, 308, 520
– maxima 521 ff
– minima 521 ff
– models with two absorption 251
– plateaus 521 ff
– slope 521 ff
– two level scheme 247
nonlinear transmission measurements
   (bleaching curves) 514 ff
nonlinear wave equation 157
nonlinearly changed refractive index
   Δn_{nl} 157
nonplanar wave 21
nonresonant interaction 151
nonresonant nonlinear interaction 153
nonstationary bleaching 519
normal dispersion 90
ns regime 572 ff
number of diffraction rings 188
number of lasing axial modes 389
number of photons 52

OBD (optical break down) 291 ff
– threshold intensity 293
observation time 67
OPA and OPO crystals 171
operational cost of light sources 9
optical activity 118
optical biaxial 117
optical bistability 306 ff
optical break down (OBD) 291 ff
optical coherence tomography (OCT)
   8
optical damage 293 ff
– threshold 293
optical density OD 102
optical diode 394
optical fibers 7
optical filter
– KG4, 104
– NG11, 104
– NG12, 104
– NG4, 104
– UG1, 104
– UG11, 104
optical free induction decay 279
optical gates 479 ff
optical gating with up-conversion,
   481 ff
optical isolators 120
optical length of resonator 388
optical limiting 241
optical materials 114 ff
– CaF_2 104
– Duran 103
– Herasil 103
– MgF_2 104
– NaCl 104
– optical glass 103
– sapphire 104
– Suprasil 103
– ZNS 104
optical measurement techniques 8
optical nutation 278 ff
optical parametric amplifiers (OPA)
   155, 169, 170, 483
optical parametric oscillator (OPO)
   170, 483
optical phase conjugation 155, 219 ff,
   229, 292, 306, 321, 418
optical rectification 155, 177 ff
optical sequencing 8
optical shutter 183
optical storage 8
optical switches 7, 175
optical tomography 392
optical trap 62
optical tweezer 62
optically thick gratings 140
optically thin gratings 138
optimal reflectivity of the output
   coupler 405
opto-optical efficiency 344
orbitals 577 ff
order of nonlinearity via absorption
   253
ordinary beams 115
organic liquids or solutions 229
organic materials 229 ff
organic molecules 115, 309 ff
orientation relaxation 144
orientation relaxation time 544
oscillator strength 96
oscilloscopes 54
output coupler 352
output power 404

Π pulse 232, 284 ff
– velocity *285*
π electron system 311
p-n junction 337
p-quaterphenyl 314
p-terphenyl 314
PAN 229
parallel decays 101
partially polarized light 48
particle acceleration 8
passive mode locking 423 ff
passive Q switching 417 ff
PbS 424
peak power *53*, *419*
peak transmission *516*
permanent hole burning 555
perpendicular incidence 107
perylene 313
phase *12*
phase coherence time 232
phase conjugate 219
phase conjugating mirror (PCM)
  219 ff, 378
phase conjugation 232
phase fluctuations 70
phase fronts of the Gaussian beam 30
phase gratings 136 ff, 137, 232
phase jump 107
phase light velocity 89
phase matching 115, 160, *161*, 169, 217
– type I *167*
– type II *167*
phase matching for second harmonic
  generation 161 ff
phase plates 377
phonon depletion 146
phonon generation 146
phonon lifetime 196
phosphorescence 509
phosphorescence decay time 513 ff
phosphorescence spectrum 510 ff
photo ionization 291 ff
– exponent k_{OBD} 292
photo-physical hole burning 243
photo-chemical hole burning 556
photo-chemical reactions 103
photodynamic therapy 8
photon 8, 11 ff
– cross section 13
– energy *12*
– momentum *12*
– spin *12*
photon absorption operator 94

photon echo 232, 281 ff
– condition *283*
– fanning out time 283
– intensities *284*
– wave vector condition *284*
photon emission operators 94
photon flux intensity I *17*
photon photon scattering 13
photon statistics 62
photon transport equation 245, 401,
  437
photonic applications 7
photonic band gap materials 323
photonic crystals 323
photonics 1
photophysical hole burning 556
photorefractive effect 319 ff
– buildup time *320*
– materials 319 ff
phthalocyanine 310, 418
planar mirror resonators 358
plano-convex lens *43*
plastochinon 309
platinum (Pt) 296
Pockels cell 479
Pockels effect 155, 172, 415
– quarter-wave voltage 173
Poisson distribution 63
polarization 12
polarization and magic angle 496 ff
polarizer 46
pollution measurements 8
polymers 103, 318
polymethine dyes 312
population densities 534, 547 ff
potential curve 154
power *22*
power content 370 ff, 372
– quadratic aperture 125
power of a flat-top profile pulse 54
power reflectivity 108
power transmission 108
Poynting vector 165
Pr 456
prism chirp compensation 428
prisms 45° 112
probe light 494, 538 ff
– pulse energy 538
– sources 538 ff
– spot 538
propagating higher transversal modes
  374 ff
propagation of a Gaussian beam 34

ps and fs regime 573 ff
pulse center t_0 53
pulse compression 479 ff
pulse compression of fs-pulses
    480 ff
pulse compression of ns pulses
    480 ff
pulse energy *53, 419*
pulse length 290, 498
pulse shortening and gating by
    nonlinear effects 481 ff
pulse width *52*, 54, *420*
pulse width of spikes 413
pulsed dye laser 331, 477 ff
pump and freeze technique 566
pump and probe measurements,
    534 ff
pump mechanism 326, 329 ff
pump rate W_{pump} 437
pump rate at threshold 403
purification 293
purity of solvents 500
pyrene 313

Q switching 175, 415 ff
– active 415 ff
– FTIR 416
– mechanical 416
– passive 417
– prelasing 416
Q switching theory 418 ff
qpm period *168*
quadratic aperture 125
quantum beat spectroscopy 280 ff
quantum chemical calculations
    577 ff
quantum computing 5
quantum converters 342
quantum cryptography 4
quantum defect 328, 329 ff
quantum defect efficiency *343*, 512
quantum defect energy *329*, 510
quantum dot 322
quantum effects 4
quantum efficiency η_Q 329
quantum mechanical box model,
    312
quantum non-demolition measurements
    5
quantum well 321 ff
quantum wire 322
quantum yield 512 ff
quantum well structures 424

quarter-wave plate 47, 441
quartz 428
quasi-phase matching (qpm) 167 ff

Rabi frequency *273, 278*
radiation momentum *61*
radiation pressure 61, *62*
radiationless conversion 100
radiationless transitions 100 ff
Raman scattering 147 ff
– anti-Stokes 148
– Stokes 148
Raman frequency *147*
Raman lasers 215, 485
Raman shifter 485 ff
Raman shifting 229
Raman spectroscopy 148
– BOX CARS 218
– CARS 216
– inverse 215
– stimulated 210
– stimulated Raman gain 216
rate equations 152, 243 ff, 422
– numerical solution 254 ff
ray matrices 27, 34 ff
ray vector *25*
ray vector (off plane) *25*
Rayleigh cross-section *144*
Rayleigh length z_R 29
Rayleigh scattering 143
Rayleigh scattering broadening 144
Rayleigh wing scattering 144
real refractive index 88
real-time holography 300
recovery time 315
rectangular modes 364 ff
red–green–blue laser (RGB) 8
reference beam method 503 ff
refraction law *91*
refractive index 87
refractive index ellipsoid 115
refractive indices for optically uniaxial
    crystals 116
refractive indices measurement 111
refractive power *345*
regenerative amplifier 443 ff
relaxation operator Γ 270
relaxation time $\tau_{orientation}$ 208
relay imaging telescope 351
resolution of optical devices 120
resolution of optical images 129
resonance condition *95*
resonance enhancement of very weakly
    absorbing materials 231

resonance enhancement SBS 396
resonance Raman scattering 214
resonant interaction *151*
resonant nonlinear interactions 231
resonator finesse *391*
resonator life time τ_{res} *391*
resonator modes 352 ff
resonator roundtrip time *420*
resonators 352 ff
– concave-convex 362
– concentric 361
– confocal 359
– dynamically stable 384 ff
– hemispherical 360
– plan plan 358
– semiconfocal 360
resonators with an SBS mirror 396 ff
rhodamin 6G 316, 330, 332, 392, 399,
    426, 544
rigid molecules 313
rotational transitions 97
rotator 47
roughness 81
roughness finesse *81*
roundtrip matrix 354
ruby 348, 392, 399
ruby laser 421, 465 ff

σ bonds 311
safety classes 1, 2, 3A, 3B and 4
    487
sample parameters 499 ff
sapphire (Al_2O_3) 456
saturation intensity 398
SBS
– focusing 201
– linewidth 196
– material parameters 199
– materials 418
– sound frequency 196
– sound wavelength 196
– spatial and temporal distribution
    201
– threshold 201
SBS mirror 380, 424
– threshold 203
SBS, SRS 232
SBS phase conjugating mirror 219 ff,
    378, 444
– taper concept 225
sech^2 pulse *422*
second harmonic generation (SHG)
    155, 158 ff, 318, 482

second intensity moment 370
second-order effects 155 ff
seeded oscillator 443
selection rules 579
self diffraction 155, 188 ff
self-focusing 155, 184, 184 ff, 232, 295
– focus length 186
– of weakly absorbing samples 189 ff
self-induced transparency 284 ff
self-phase modulation 155, 189 ff, 232,
    480
self-trapping 185
Sellmeier coefficients 164 ff
Sellmeier dispersion 91
Sellmeier equation 91
semi-empirical quantum chemical
    methods 579
semiconductors 321 ff, 424
semiconductor lasers 451 ff
semiconfocal resonators 360
SESAMs 424
seventh harmonic 228
SF_6 230
Shack Hartmann wavefront sensors 30
SHG (second harmonic generation)
    158, 482
– acceptance angle 165
– crystals 160
– optimal crystal length *166*
– optimal focusing 166
– periodic poling 167
– temperature influence 166
shock parameter R_{shock} 351
short pulse generation 412 ff
short resonators 433 ff
side-pumping 333
signal-to-noise ratio (SNR) 65
silicon (Si) 296
silver (Ag) 296
single absorption transition 88
single atom lasers 327
single mode laser 394 ff
single molecule absorption 62
single molecule imaging 291
single molecules damage 295
single molecules detection 8
single pass amplifier 440 ff
single photon counting 63
single pulse selection 479 ff
single-diode lasers 453 ff
SiO_2 199
six-wave mixing 306
skin damage 486

slab laser material 345
slope efficiency *344*
slow axis 337
small signal gain *398, 437*
– coefficient 437
Snellius law 91
solar energy converters 8
soldering 295
solid state lasers 456 ff
solid-state lasers pumped by diode
    lasers 333
solid-state slabs 334
soliton laser 429 ff
soliton period *193*
soliton pulses 192 ff
solitons 155, 192, 232
solvents for low-temperature measure-
    ments 575
sound wave amplitude 197
spark gap 469
spatial hole burning 261, 263 ff, 394
spatial overlap 534
spatial soliton 186 ff
spatial soliton velocity 188
specific heat c_p 296
spectral broadening from the active
    material 391 ff
spectral calibration 542 ff
spectral cross relaxation 266, 525
– time 234
spectral degeneration 73
spectral filtering 392
spectral hole burning 235, 239, 242 ff,
    257 ff
spectral linewidth and position of laser
    emission 405 ff
spectral resolution 80
spectral small signal gain 406
spectral uncertainty 16
spectral width Fabry Perot etalon *80*
spectral width hole burning 263
spectral width of excitation pulse
    524 ff
spectroscopic setups 571 ff
speed of photons 11
spiking 412 ff
– damping time 414
– Nd:YAG laser 414
spontaneous Brillouin scattering 145
spontaneous emission 98
spontaneously scattered Raman
    intensity 211
squeezed light 65 ff, 66

SRS (stimulated Raman scattering)
– anti-Stokes angle 212
– excited state 215
– equations *214*
– parameters 211
stability condition of lasers *357, 382*
stability limits 386
stability range of the resonator 381 ff
stable resonators 352 ff
standing wave 20
start of nonlinearity 518 ff
stationary bleaching *518*
stationary Brillouin gain *198*
stationary four-level model 249 ff
stationary interaction *246*
stationary model with two absorptions
    251 ff
stationary models 253 ff
stationary SBS 202
– gain *202*
– threshold 202
stationary Schroedinger equation 93,
    267
stationary small signal in SRS 215
stationary solutions of rate equations
    246 ff
stationary two-level model 247 ff
STBS (stimulated thermal Brillouin
    scattering) 204
– temperature modulation 205
steady state measurement 543 ff
stimulated Brillouin scattering (SBS)
    155, 194 ff, 293
stimulated emission 98, 241 ff
stimulated Raman gain spectroscopy
    (SRGS) 216 ff
stimulated Raman measuring tech-
    niques 210
stimulated Raman scattering (SRS)
    210 ff
stimulated Rayleigh scattering (SRLS)
    206 ff
stimulated Rayleigh wing (SRWS)
    scattering 207 ff
stimulated thermal Brillouin scattering
    (STBS) 204 ff
Stokes lines 147
Stokes shift 313, 510
Stokes vector 48
stored energy per volume *439*
streak cameras 54
stretched exponential decay *532*
strong damping assumption 197

sulforhexafluoride SF$_6$ 199
sum frequency *168*
sum frequency generation 481
sunlight power density 486
super-Gaussian beam *55*
superfluorescence 285 ff
superposition modes 367
superposition of light 67
superradiance, superradiation 241 ff,
    285 ff, 441
surface damage 294
surface oxidation 295
surface-enhanced Raman spectroscopy
    (SERS) 215
SVA approximation *156, 197*

T_2 232
T_3 234
TEM modes 363
TEM$_{00}$ mode 363
temporal pulse width Δt_{sol} 193
temporal solitons 192 ff
tetracene 313
thermal conductivity Λ_T 205
thermal lensing 345 ff, 386
thermal light *64*
thermal light statistics *63*
thermal noise 145
thermal Rayleigh scattering (STRS)
    206 ff
thermal relaxation time *346*
thermal self-focusing 186
thermal stress fracture limit 351 ff
thermally induced birefringence
    348 ff
thermally induced depolarization
    348
thermally induced focal length *347*
thermally induced refractive power
    386 ff
thin disc geometry 345
thin films 112, 115, 318
thin grating with phase modulation
    139
third harmonic generation (THG) 155,
    180, 482
– optimal total efficiency 181
third order NLO 179 ff
– crystals of cubic symmetry 180
– isotropic 180
third-order $\chi^{(3)}$ tensor 158
third-order effects 155 ff

third-order nonlinear polarization,
    179
third-order nonlinear susceptibility,
    527
third-order nonlinearity 560 ff
three-level amplifier system 437
three-level scheme 328, 402
threshold gain coefficient *400*
threshold inversion *400*
threshold pump rate $W_{\mathrm{pump,threshold}}$
    400, 401
Thulium (Tm) laser 464 ff
Ti 456
Ti:sapphire 330, 392, 399
Ti:sapphire amplifier 443
Ti:sapphire laser 333, 421, 426, 461 ff,
    484
TiCl$_4$ 230
tilt of lenses 38
time-dependent Schroedinger equation
    92, 267, 578
times diffraction limited 60
Tin (Sn) 296
Titanium (Ti) 296
Toluene 209, 211
total efficiency *344*
total noise power P_{noise} 415
total ray matrix 39
total reflection *111*
– displacement 113
total stored energy *439*
TPF (two photon fluorescence)
– pulse width (Gauss) *290*
– pulse width (sech) *290*
transfer matrix M_T 381
transfer rate dipole–dipole interaction
    100
transient absorption 232
transient absorption: excited state
    absorption (ESA) 238 ff
transient gratings 557
transient spectra 535 ff
transition dipole moment $\mu_{p\leftarrow m}$ 94 ff
transition dipole moment expectation
    value *268*
transition into optical thinner medium
    108 ff
transition into optically denser medium
    107 ff
transmission T *102, 237*, 515
transmission of the etalon 80
transmittance 102
transparent materials 103

transversal resonator modes 352, 354 ff
– frequency 389
traveling wave excitation 434 ff, 496
trimming 295
triplet population density 546
triplet quenching 510 ff
tunability range 448
tunable lasers 447
tungsten (W) 296
tungsten band lamps 514
twist angle 314
twisted intramolecular charge transfer (TICT) 318
TWM (two wave mixing) 298 ff
– degenerate (DTWM) 299
two photon absorption *288*, 532 ff
– selection rules 289
two-beam interference 72 ff
two-level scheme with broadening, 270 ff
two-photon detection of absorption via ESA 563 ff
two-photon induced fluorescence (TPF) 290, 532
two photon transitions 287 ff
two-wave mixing (TWM) 298 ff

uncertainty of energy and time 15 ff
uncertainty of photon number and phase 63 ff
uncertainty of position and momentum 14 ff
uniaxial crystals 115
unit wave vector 22
unstable resonators 353 ff
upconversion lasers 329
upper and the lower laser level 328
uranine 316
urea 229
UV 103
UV and XUV light sources 293

vacuum impedance 20
vacuum permeability *19*
vacuum permittivity *19*
vacuum polarization *66*
vanadium (V) 296
vaporization 295 ff
– heat Q_v 296
– temperature T_v 296
variation of excitation pulse width, 523 ff

vector phase conjugation 221
vector potential 94
velocity of sound wave v_{sound} 196
Verdet constant 48, 120
vertical cavity surface emitting lasers (VCSELs) 338, 455 ff
vibronic transitions 97
Voigt profile 507 ff
– band analysis 507
– integral 507
volume holograms 321

walk-off angle 165 ff
wallplug efficiency 344
water window 8, 228
wave monochromatic planar 19
wave front 21
wave front inversion 221
wave front radius *30*
wave function ψ 93
– superposition of the 93
wave guiding 147, 185
wave number *12*
wave vector *12*
wavelength *12*
wavelength division multiplexing (WDM) 7, 455
welding 295
white light 292 ff
white light generation 186, 229
white light generation with fs pulses 539 ff
white light generation with ps pulses 540 ff

X-ray microscopy 8
X-ray sources 8
XeCl 392, 399
XeCl KrF and ArF excimer lasers, 441, 467 ff
Xenon (Xe) 199, 230
XUV generation 228
XUV-lasers 477

YAG ($Y_3Al_5O_{12}$) 456
YALO or YAP (YAlO$_3$) 456
Yb laser 456 ff
Yb:YAG 330, 348
Yb:YAG laser 460 ff
YLF (LiYF$_4$) 456
YVO (YVO$_4$) 456

z-scan 525 ff
– theoretical description 528 ff
– with absorbing samples 529 ff
zero delay 571

zero point energy *66*
zig-zag slab 334, 345
Zinc (Zn) 296
ZINDO/S 580

Printing: Mercedes-Druck, Berlin
Binding: Buchbinderei Lüderitz & Bauer, Berlin

Light characteristics

light velocity	$c = \nu\lambda$
photon energy	$E_{ph} = h\nu$
wave vector	$\|\boldsymbol{k}\| = \dfrac{2\pi}{\lambda}$
photon momentum	$\|\boldsymbol{p}\| = \dfrac{h}{\lambda}$
intensity	$I = \dfrac{1}{2} c_0 \varepsilon_0 n \|\boldsymbol{E}\|^2$
photon flux intensity	$\mathbf{I} = \dfrac{I}{h\nu}$
Fresnel number	$F = \dfrac{D^2}{\lambda L}$
band width	$\Delta\nu = \dfrac{1}{2\pi\tau}$
beam parameter product	$w_0\Theta = \dfrac{\lambda}{\pi} M^2$
beam waist	$w = w_0 \sqrt{1 + \left(\dfrac{z\lambda_0}{w_0^2 n\pi}\right)^2}$
Rayleigh length	$z_R = \dfrac{n\pi}{\lambda_0} w_0^2$
wave front curvature	$R = z_R \left(\dfrac{z}{z_R} + \dfrac{z_R}{z}\right)$
complex beam parameter	$\dfrac{1}{q(z)} = \dfrac{1}{R(z)} - \dfrac{i\lambda_0}{\pi n w(z)^2}$
beam propagation	$q_{out} = \dfrac{a q_{in} + b}{c q_{in} + d}$

Light matter interaction

refractive index	$n^2 = \varepsilon_r = 1 + \chi; \; c_0 = \dfrac{1}{\sqrt{\varepsilon_0 \mu_0}}$
Einstein coefficients	$A_{mp} = \dfrac{8\pi h}{c_0^3} \nu^3 B_{mp}$
transition dipole moment	$\mu^2 = \dfrac{3 c_0 \varepsilon_0 h}{2\pi^2 \nu_0} \displaystyle\int \sigma \, d\nu$